ELECTRICITY

THEORY AND APPLICATIONS

Eugene Silberstein
MS, BEAP, CMHE

Publisher
The Goodheart-Willcox Company, Inc.
Tinley Park, IL
www.g-w.com

Library of Congress Control Number: 2025939671

ISBN 979-8-89737-438-0

2 3 4 5 6 7 8 9 -27- 30 29 28 27 26

Image Credits. Front cover: Courtesy Fluke Corporation (upper-left inset), WEG (upper-right inset), Kunakorn Rassadornyindee/Shutterstock.com (middle-right inset), zstock/Shutterstock.com (lower-right inset), Trzmiel/Shutterstock.com (background)

Preface

The vision for *Electricity: Theory and Applications* is to create a valuable educational resource for those who aspire to become an electrical industry professional, those who are currently employed in the industry, or those who want to learn about the exciting world of electricity. Regardless of the role you play in the electrical industry, this textbook provides valuable insight into the electrical theories and concepts that form a solid foundation onto which more advanced learning and understanding can be structured.

Throughout this comprehensive textbook you will find features that are intended to enhance the learning process while providing additional guidance and insight into the content being presented. As safety is a primary concern for everyone working on or around electric circuits, systems, and equipment, safety guidelines and important safety-related issues are provided throughout the text, as well as in the **Safety Alert** features.

Every successful individual working in the electrical field should have a good grasp of math and how to manipulate numbers. *Electricity: Theory and Applications* takes a multifaceted approach to address the math challenges that learners may face. Two chapters are dedicated to reinforcing fundamental math concepts required for electrical theory applications. The text provides a thorough overview of the necessary formulas used when evaluating circuits and explains how to perform calculations in an easy-to-understand manner. Step-by-step guidance on utilizing, manipulating, and navigating mathematical expressions are presented in **Example** features. When appropriate, this feature is immediately followed by **Now It's Your Turn** that allows the learner to perform calculations and solve problems on their own. The answers to this feature are located in Appendix D at the end of the textbook. **Math Sidebar** features provide short refreshers of foundational math concepts at the point of application. These features are presented in a manner that will not pose distractions to those who are mathematically strong, while providing guidance to those who stand to benefit from this additional content.

Each chapter in *Electricity: Theory and Applications* begins with a **Sparking Discussion** question that stimulates classroom discussion. End-of-chapter review questions are presented in three levels: Measuring Your Potential questions evaluate the learner's basic knowledge retention, Plug-In questions allow the learner to apply and analyze data, and Making the Connection questions strengthen their critical thinking skills.

This textbook took a long time to organize and write because we wanted to make certain that we got it right the first time. As the old expression goes, measure twice . . . cut once, and we did a lot of measuring on this project! I had a great time putting this textbook together, and I truly hope that you enjoy reading and learning from it as much as I have enjoyed creating it for you.

Eugene Silberstein

About the Author

Eugene Silberstein has been involved in building trades and career and technical education for over 45 years, with 30 years in front of the classroom. He has served as a field service professional, controls specialist, system designer, company owner, college professor, administrator, consultant, and industry-leading author. Over the past three decades, Mr. Silberstein has taught at a number of private and public secondary and post-secondary institutions. Mr. Silberstein has authored more than a dozen textbooks covering a wide range of building science and career and technical education topics. He can be seen discussing building science on many podcasts and webcasts and at many live industry events, conventions, and conferences.

Mr. Silberstein's undergraduate work and bachelor's degree include electrical engineering, economics, and operations management. He obtained his master of science degree in renewable/sustainable energy and environmental systems. As part of his graduate work, he created an introductory energy course in wind, solar, geothermal, biomass, and hydropower.

As a National Programs Director at the ESCO Institute, Mr. Silberstein presently works with over 1500 career and technical education programs across the country. It is in this capacity that he helps institutions start new career and technical education programs and helps others bring their existing training to a higher level.

Mr. Silberstein is an active member of many building science and electrical industry organizations including the Institute of Electrical and Electronics Engineers (IEEE) and the American Society of Heating, Refrigerating and Air-Conditioning Engineers (ASHRAE). He also had the title of CMHE (Certified Master HVACR Educators) bestowed upon him by HVAC Excellence, which classifies him as one of only 150 certified master HVACR educators in the country. He also carries ASHRAE's BEAP (Building Energy Assessment Professional) credential.

Reviewers

The author and publisher wish to thank the following industry and teaching professionals for their valuable input into the development of *Electricity: Theory and Applications.*

Bryant Abbott
Manchester Community College
Manchester, NH

Erik Bade
Madison Area Technical College
Madison, WI

Philip A. Bazicki
Hudson Valley Community College
Troy, NY

Mark Bohnet
Northwest Iowa Community College
Sheldon, IA

Robert Chaboya
Cerritos College
Norwalk, CA

Daniel Choate
Wytheville Community College
Wytheville, VA

James A. Cox
Ozarks Technical Community College
Springfield, MO

Jesse Dahl
Minnesota North College - Hibbing
Hibbing, MN

Donald Farrell
Centura College CNOR
Norfolk, VA

Lane Gunnels
Workforce Training Solutions
Carencro, LA

Jeffery S. Hamilton
Terra State Community College
Fremont, OH

William J. Hargrove
Central Carolina Technical College
Sumter, SC

Frank R. Holcomb
Tennessee College of Applied Technology
Paris, TN

Glen A. Hortizuela
College of Southern Nevada
Henderson, NV

David V. Jones
Lenoir Community College
Kinston, NC

Tom Kennedy
Milwaukee Area Technical College
Milwaukee, WI

Brian D. Khairullah
Eklektron
Yucaipa, CA

Troy Klein
Northwest Iowa Community College
Sheldon, IA

Dave Krispinsky
Rochester Institute of Technology
Rochester, NY

David Lorenz
Paul D. Camp Community College
Franklin, VA

Ron McGary
Industrial Systems Technology Training
Alliance, OH

Daniel Neff
Palm Beach State College
Lake Worth, FL

Jake Nielsen
National Technical Institute
Henderson, NV

Victoria Denise Ojeda
Central New Mexico Community College
Albuquerque, NM

Peter W. Ortner
Nashville State Community College
Nashville, TN

Michael Renzoni
Gateway Technical College
Kenosha, WI

Ronald Ropson
Northeast Wisconsin Technical College
Green Bay, WI

Barbara Salazar
El Paso Community College
El Paso, TX

Charles Sekafetz
Chemeketa Community College
Salem, OR

Steve Senty
Anoka Technical College
Anoka, MN

Tim Tewalt
Chippewa Valley Technical College
Eau Claire, WI

Duane Thormahlen
Porter County Career Center
Valparaiso, IN

John Veitch
SUNY Adirondack
Queensbury, NY

Rebecca Velasquez
Salt Lake Community College
Salt Lake City, UT

Richard Wagner
Des Moines Area Community College
Ankeny, IA

JD White
Columbus State Community College
Columbus, OH

Matthew Wilkinson
Madison Area Technical College
Madison, WI

Robert Wood
Hudson Valley Community College
Troy, NY

Acknowledgments

The author and publisher would like to thank the following individuals for their contribution and additional support in the development of *Electricity: Theory and Applications*.

Erik Bade
Madison Area Technical College
Madison, WI

Philip A. Bazicki
Hudson Valley Community College
Troy, NY

Barbara Salazar
El Paso Community College
El Paso, TX

Features of the Textbook

The instructional design of this textbook includes student-focused learning tools to help learners succeed. This visual guide highlights these features.

Chapter Opening Materials

Each chapter begins with a **Sparking Discussion** question designed to stimulate classroom discussion and help students begin to understand the larger context of the chapter topics. **Learning Outcomes** clearly identify the knowledge and skills to be gained when the chapter is completed. These outcomes are numbered and correlated to related assessment questions to facilitate reporting and remediation. **Technical Terms** list the key words to be learned in the chapter. An **Introduction** provides an overview and preview of the chapter content.

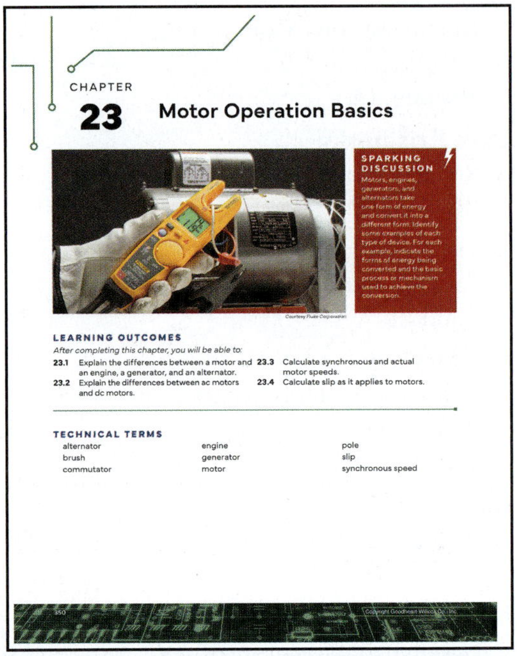
CHAPTER 23 Motor Operation Basics

Courtesy Fluke Corporation

SPARKING DISCUSSION
Motors, engines, generators, and alternators take one form of energy and convert it into a different form. Identify some examples of each type of device. For each example, indicate the forms of energy being converted and the basic process or mechanism used to achieve the conversion.

LEARNING OUTCOMES
After completing this chapter, you will be able to:
23.1 Explain the differences between a motor and an engine, a generator, and an alternator.
23.2 Explain the differences between ac motors and dc motors.
23.3 Calculate synchronous and actual motor speeds.
23.4 Calculate slip as it applies to motors.

TECHNICAL TERMS
alternator
brush
commutator
engine
generator
motor
pole
slip
synchronous speed

450

Additional Features

Additional features are used throughout the body of each chapter to further learning and knowledge. **Safety Alerts** warn students of potentially dangerous materials and practices. **Shedding Light** provides advice and guidance that is especially applicable for on-the-job situations. **Know the Code** points out specific items from typical building codes.

Examples walk students step-by-step through math calculations. Students can then practice and apply the concepts in the **Now It's Your Turn** feature. **Math Sidebars** reinforce and remind students of foundational math concepts at the point of application. **Procedures** provide clear instructions for hands-on troubleshooting and service tasks.

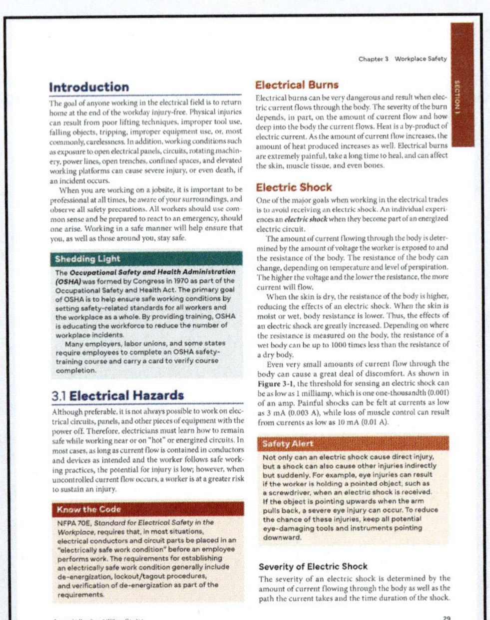
Chapter 3 Workplace Safety

SECTION 1

Introduction

The goal of anyone working in the electrical field is to return home at the end of the workday injury-free. Physical injuries can result from poor lifting techniques, improper tool use, falling objects, tripping, improper equipment use, or, most commonly, carelessness. In addition, working conditions such as exposure to open electrical panels, circuits, rotating machinery, power lines, open trenches, confined spaces, and elevated working platforms can cause severe injury, or even death, if an incident occurs.

When you are working on a jobsite, it is important to be professional at all times, be aware of your surroundings, and observe all safety precautions. All workers should use common sense and be prepared to react to an emergency, should one arise. Working in a safe manner will help ensure that you, as well as those around you, stay safe.

Shedding Light
The ***Occupational Safety and Health Administration (OSHA)*** was formed by Congress in 1970 as part of the Occupational Safety and Health Act. The primary goal of OSHA is to help ensure safe working conditions by setting safety-related standards for all workers and the workplace as a whole. By providing training, OSHA is educating the workforce to reduce the number of workplace incidents.
Many employers, labor unions, and some states require employees to complete an OSHA safety-training course and carry a card to verify course completion.

3.1 Electrical Hazards

Although preferable, it is not always possible to work on electrical circuits, panels, and other pieces of equipment with the power off. Therefore, electricians must learn how to remain safe while working near or on "hot" or energized circuits. In most cases, as long as current flow is contained in conductors and devices as intended and the worker follows safe working practices, the potential for injury is low; however, when uncontrolled current flow occurs, a worker is at a greater risk to sustain an injury.

Know the Code
NFPA 70E, *Standard for Electrical Safety in the Workplace*, requires that, in most situations, electrical conductors and circuit parts be placed in an "electrically safe work condition" before an employee performs work. The requirements for establishing an electrically safe work condition generally include de-energization, lockout/tagout procedures, and verification of de-energization as part of the requirements.

Electrical Burns

Electrical burns can be very dangerous and result when electric current flows through the body. The severity of the burn depends, in part, on the amount of current flow and how deep into the body the current flows. Heat is a by-product of electric current. As the amount of current flow increases, the amount of heat produced increases as well. Electrical burns are extremely painful, take a long time to heal, and can affect the skin, muscle tissue, and even bones.

Electric Shock

One of the major goals when working in the electrical trades is to avoid receiving an electric shock. An individual experiences an ***electric shock*** when they become part of an energized electric circuit.

The amount of current flowing through the body is determined by the amount of voltage the worker is exposed to and the resistance of the body. The resistance of the body can change, depending on temperature and level of perspiration. The higher the voltage and the lower the resistance, the more current will flow.

When the skin is dry, the resistance of the body is higher, reducing the effects of an electric shock. When the skin is moist or wet, body resistance is lower. Thus, the effects of an electric shock are greatly increased. Depending on where the resistance is measured on the body, the resistance of a wet body can be up to 1000 times less than the resistance of a dry body.

Even very small amounts of current flow through the body can cause a great deal of discomfort. As shown in **Figure 3-1**, the threshold for sensing an electric shock can be as low as 1 milliamp, which is one one-thousandth (0.001) of an amp. Painful shocks can be felt at currents as low as 3 mA (0.003 A), while loss of muscle control can result from currents as low as 10 mA (0.01 A).

Safety Alert
Not only can an electric shock cause direct injury, but a shock can also cause other injuries indirectly but suddenly. For example, eye injuries can result if the worker is holding a pointed object, such as a screwdriver, when an electric shock is received. If the object is pointing upwards when the arm pulls back, a severe eye injury can occur. To reduce the chance of these injuries, keep all potential eye-damaging tools and instruments pointing downward.

Severity of Electric Shock

The severity of an electric shock is determined by the amount of current flowing through the body as well as the path the current takes and the time duration of the shock.

Copyright Goodheart-Willcox Co., Inc.

29

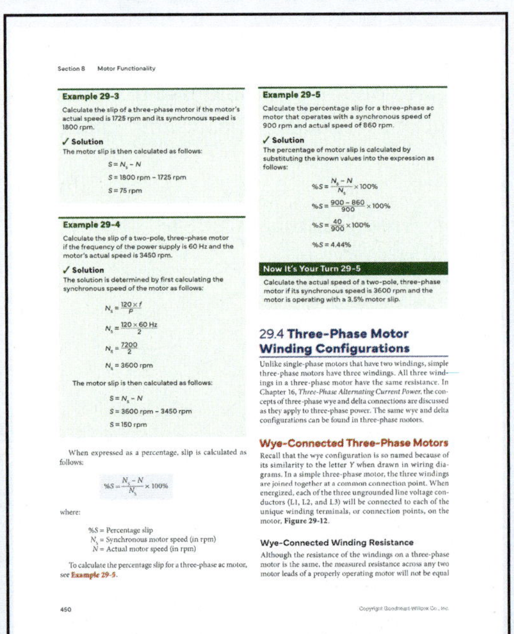
Section 8 Motor Functionality

Example 29-3
Calculate the slip of a three-phase motor if the motor's actual speed is 1725 rpm and its synchronous speed is 1800 rpm.
✓ **Solution**
The motor slip is then calculated as follows:
$$S = N_s - N$$
$$S = 1800 \text{ rpm} - 1725 \text{ rpm}$$
$$S = 75 \text{ rpm}$$

Example 29-4
Calculate the slip of a two-pole, three-phase motor if the frequency of the power supply is 60 Hz and the motor's actual speed is 3450 rpm.
✓ **Solution**
The solution is determined by first calculating the synchronous speed of the motor as follows:
$$N_s = \frac{120 \times f}{P}$$
$$N_s = \frac{120 \times 60 \text{ Hz}}{2}$$
$$N_s = \frac{7200}{2}$$
$$N_s = 3600 \text{ rpm}$$
The motor slip is then calculated as follows:
$$S = N_s - N$$
$$S = 3600 \text{ rpm} - 3450 \text{ rpm}$$
$$S = 150 \text{ rpm}$$

When expressed as a percentage, slip is calculated as follows:
$$\%S = \frac{N_s - N}{N_s} \times 100\%$$
where:
$\%S$ = Percentage slip
N_s = Synchronous motor speed (in rpm)
N = Actual motor speed (in rpm)

To calculate the percentage slip for a three-phase ac motor, see **Example 29-5**.

Example 29-5
Calculate the percentage slip for a three-phase ac motor that operates with a synchronous speed of 900 rpm and actual speed of 860 rpm.
✓ **Solution**
The percentage of motor slip is calculated by substituting the known values into the expression as follows:
$$\%S = \frac{N_s - N}{N_s} \times 100\%$$
$$\%S = \frac{900 - 860}{900} \times 100\%$$
$$\%S = \frac{40}{900} \times 100\%$$
$$\%S = 4.44\%$$

Now It's Your Turn 29-5
Calculate the actual speed of a two-pole, three-phase motor if its synchronous speed is 3600 rpm and the motor is operating with a 3.5% motor slip.

29.4 Three-Phase Motor Winding Configurations

Unlike single-phase motors that have two windings, simple three-phase motors have three windings. All three windings in a three-phase motor have the same resistance. In Chapter 16, *Three-Phase Alternating Current Power*, the concepts of three-phase wye and delta connections are discussed as they apply to three-phase power. The same wye and delta configurations can be found in three-phase motors.

Wye-Connected Three-Phase Motors

Recall that the wye configuration is so named because of its similarity to the letter *Y* when drawn in wiring diagrams. In a simple three-phase motor, the three windings are joined together at a common connection point. When energized, each of the three ungrounded line voltage conductors (L1, L2, and L3) will be connected to each of the unique winding terminals, or connection points, on the motor, **Figure 29-12**.

Wye-Connected Winding Resistance

Although the resistance of the windings on a three-phase motor is the same, the measured resistance across any two motor leads of a properly operating motor will not be equal

450

Copyright Goodheart-Willcox Co., Inc.

Illustrations

Clear and simple presentation of concepts heavily supported by illustrations prevents students from becoming lost in overly complicated, overly detailed explanations of advanced topics beyond what is required in an introductory electrical theory course.

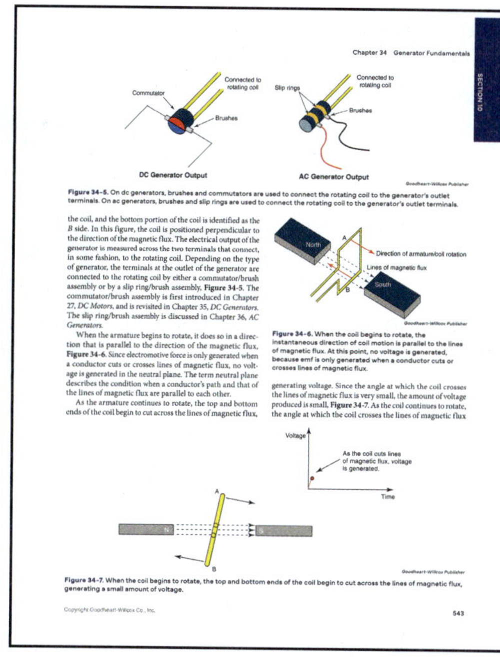
Chapter 34 Generator Fundamentals

SECTION 10

Figure 34-5. On dc generators, brushes and commutators are used to connect the rotating coil to the generator's outlet terminals. On ac generators, brushes and slip rings are used to connect the rotating coil to the generator's outlet terminals.

the coil, and the bottom portion of the coil is identified as the *B* side. In this figure, the coil is positioned perpendicular to the direction of the magnetic flux. The electrical output of the generator is measured across the two terminals that connect, in some fashion, to the rotating coil. Depending on the type of generator, the terminals at the outlet of the generator are connected to the rotating coil by either a commutator/brush assembly or by a slip ring/brush assembly, **Figure 34-5**. The commutator/brush assembly is first introduced in Chapter 27, *DC Motors*, and is revisited in Chapter 35, *DC Generators*. The slip ring/brush assembly is discussed in Chapter 36, *AC Generators*.

When the armature begins to rotate, it does so in a direction that is parallel to the direction of the magnetic flux, **Figure 34-6**. Since electromotive force is only generated when a conductor cuts or crosses lines of magnetic flux, no voltage is generated in the neutral plane. The term neutral plane describes the condition when a conductor's path and that of the lines of magnetic flux are parallel to each other.

As the armature continues to rotate, the top and bottom ends of the coil begin to cut across the lines of magnetic flux,

Figure 34-6. When the coil begins to rotate, the instantaneous direction of coil motion is parallel to the lines of magnetic flux. At this point, no voltage is generated, because emf is only generated when a conductor cuts or crosses lines of magnetic flux.

generating voltage. Since the angle at which the coil crosses the lines of magnetic flux is very small, the amount of voltage produced is small, **Figure 34-7**. As the coil continues to rotate, the angle at which the coil crosses the lines of magnetic flux

Figure 34-7. When the coil begins to rotate, the top and bottom ends of the coil begin to cut across the lines of magnetic flux, generating a small amount of voltage.

Copyright Goodheart-Willcox Co., Inc.

543

End-of-Chapter Content

End-of-chapter material provides an opportunity for review and application of concepts. A concise **Summary** provides an additional review tool and reinforces key learning outcomes. This helps students focus on important concepts presented in the text. **Measuring Your Potential** questions enable learners to demonstrate knowledge, identification, and comprehension of chapter material. **Plug-In** questions extend learning and develop abilities to use learned material in new situations and to break down material into its component parts. **Making the Connection** questions develop higher-order thinking and problem-solving abilities.

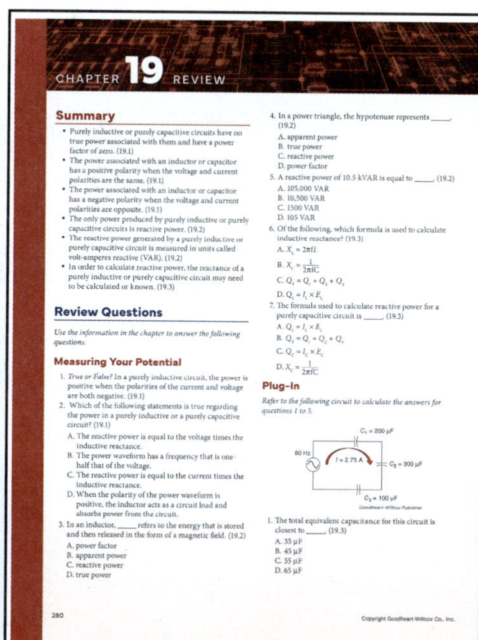
CHAPTER 19 REVIEW

Summary

- Purely inductive or purely capacitive circuits have no true power associated with them and have a power factor of zero. (19.1)
- The power associated with an inductor or capacitor has a positive polarity when the voltage and current polarities are the same. (19.1)
- The power associated with an inductor or capacitor has a negative polarity when the voltage and current polarities are opposite. (19.1)
- The only power produced by purely inductive or purely capacitive circuits is reactive power. (19.2)
- The reactive power generated by a purely inductive or purely capacitive circuit is measured in units called volt-amperes reactive (VAR). (19.2)
- In order to calculate reactive power, the reactance of a purely inductive or purely capacitive circuit may need to be calculated or known. (19.3)

Review Questions

Use the information in the chapter to answer the following questions.

Measuring Your Potential

1. *True or False?* In a purely inductive circuit, the power is positive when the polarities of the current and voltage are both negative. (19.1)
2. Which of the following statements is true regarding the power in a purely inductive or a purely capacitive circuit? (19.1)
   A. The reactive power is equal to the voltage times the inductive reactance.
   B. The power waveform has a frequency that is one-half that of the voltage.
   C. The reactive power is equal to the current times the inductive reactance.
   D. When the polarity of the power waveform is positive, the inductor acts as a circuit load and absorbs power from the circuit.
3. In an inductor, _____ refers to the energy that is stored and then released in the form of a magnetic field. (19.2)
   A. power factor
   B. apparent power
   C. reactive power
   D. true power
4. In a power triangle, the hypotenuse represents _____. (19.2)
   A. apparent power
   B. true power
   C. reactive power
   D. power factor
5. A reactive power of 10.5 kVAR is equal to _____. (19.2)
   A. 105,000 VAR
   B. 10,500 VAR
   C. 1500 VAR
   D. 105 VAR
6. Of the following, which formula is used to calculate inductive reactance? (19.3)
   A. $X_L = 2\pi fL$
   B. $X_C = \frac{1}{2\pi fC}$
   C. $Q_T = Q_1 + Q_2 + Q_3$
   D. $Q_L = I_L \times E_L$
7. The formula used to calculate reactive power for a purely capacitive circuit is _____. (19.3)
   A. $Q_L = I_L \times E_L$
   B. $Q_T = Q_1 + Q_2 + Q_3$
   C. $Q_C = I_C \times E_C$
   D. $X_C = \frac{1}{2\pi fC}$

Plug-In

Refer to the following circuit to calculate the answers for questions 1 to 5.

1. The total equivalent capacitance for this circuit is closest to _____. (19.3)
   A. 35 µF
   B. 45 µF
   C. 55 µF
   D. 65 µF

280

Copyright Goodheart-Willcox Co., Inc.

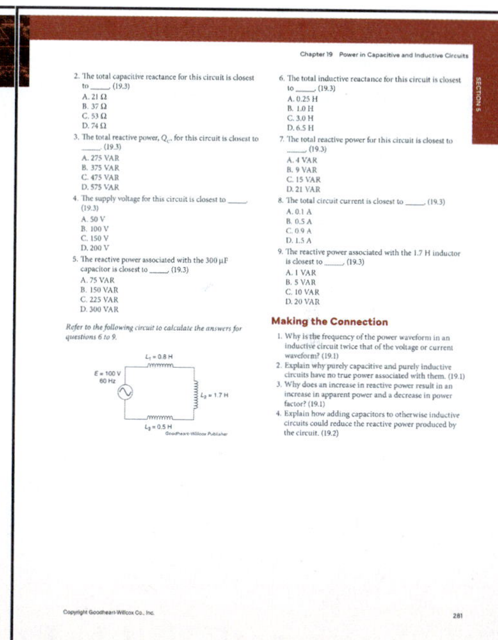
Chapter 19 Power in Capacitive and Inductive Circuits

SECTION 5

2. The total capacitive reactance for this circuit is closest to _____. (19.3)
   A. 21 Ω
   B. 37 Ω
   C. 53 Ω
   D. 74 Ω
3. The total reactive power, Q_C, for this circuit is closest to _____. (19.3)
   A. 275 VAR
   B. 375 VAR
   C. 475 VAR
   D. 575 VAR
4. The supply voltage for this circuit is closest to _____. (19.3)
   A. 50 V
   B. 100 V
   C. 150 V
   D. 200 V
5. The reactive power associated with the 300 µF capacitor is closest to _____. (19.3)
   A. 75 VAR
   B. 150 VAR
   C. 225 VAR
   D. 300 VAR

Refer to the following circuit to calculate the answers for questions 6 to 9.

6. The total inductive reactance for this circuit is closest to _____. (19.3)
   A. 0.25 H
   B. 1.0 H
   C. 3.0 H
   D. 6.5 H
7. The total reactive power for this circuit is closest to _____. (19.3)
   A. 4 VAR
   B. 9 VAR
   C. 15 VAR
   D. 21 VAR
8. The total circuit current is closest to _____. (19.3)
   A. 0.1 A
   B. 0.5 A
   C. 0.9 A
   D. 1.5 A
9. The reactive power associated with the 1.7 H inductor is closest to _____. (19.3)
   A. 1 VAR
   B. 5 VAR
   C. 10 VAR
   D. 20 VAR

Making the Connection

1. Why is the frequency of the power waveform in an inductive circuit twice that of the voltage or current waveform? (19.1)
2. Explain why purely capacitive and purely inductive circuits have no true power associated with them. (19.1)
3. Why does an increase in reactive power result in an increase in apparent power and a decrease in power factor? (19.1)
4. Explain how adding capacitors to otherwise inductive circuits could reduce the reactive power produced by the circuit. (19.2)

Copyright Goodheart-Willcox Co., Inc.

281

TOOLS FOR STUDENT AND INSTRUCTOR SUCCESS

Student Tools

Student Text

Available in print and digital, *Electricity: Theory and Applications* provides valuable insight into the electrical theories and concepts that form a solid foundation, onto which more advanced learning and understanding can be structured.

Lab Workbook

- Hands-on practice activities help students recall, apply, and analyze concepts introduced in the textbook.
- Chapter review questions provide additional practice outside of the core text review questions to solidify learning.
- Print and digital options are available.

G-W Digital Companion

- E-flash cards and vocabulary exercises allow interaction with content to create opportunities to increase achievement.

EduHub LMS-Ready Content

- Students will enjoy intuitive digital resources in your school's preferred platform.
- The digital platform encourages application and discovery.
- Electrical animations and videos enrich learning by providing a more thorough explanation of concepts from the textbook.
- Math animations provide step-by-step guidance to select the correct formula, analyze wire diagrams to find readings to insert into the formula, and then the steps to solve the problem.

Instructor Tools

LMS Integration

Integrate EduHub LMS-Ready Content into your preferred platform for a seamless user experience for both you and your students.

A Complete Teaching and Learning Package

- The Digital Textbook and Online Lab Workbook build student knowledge.
- Animations, videos, and drill and practice activities help your students remain engaged and learn effectively.
- G-W Assessment gives students easy access to assignments, tests and quizzes, instant feedback, and grades.

Online Instructor Resources

- Tap into timesaving preparation tools: answer keys, editable lesson plans, and other teaching aids.
- Take advantage of **Instructor's Presentations for PowerPoint**® with fully customizable, richly illustrated slides that help you teach and visually reinforce the key concepts from each chapter.
- Easily create, customize, edit, and print prebuilt questions and tests using G-W Assessment, our cloud-based platform. No more downloads, uploads, or imports needed! Student assessments are autograded and flow automatically into your gradebook when you integrate with your LMS.

G-W Assessment
Assess learning • Gain insights • Build skills

See www.g-w.com/electricty-theory-and-applications-2027 for a list of all available formats.

G-W EduHub®

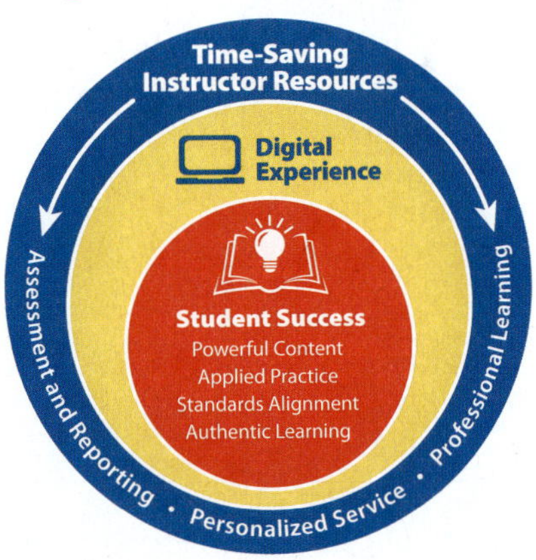

Brief Contents

Contents

SECTION

1 Getting Started

Xmentoys/Shutterstock.com

CHAPTER

1 Introduction to Electricity

taka1022/Shutterstock.com

SPARKING DISCUSSION

Considering all the possible technologies used to produce electricity, what is the best combination considering economics and sustainabilty?

LEARNING OUTCOMES

After completing this chapter, you will be able to:

1.1 Describe how earlier inventions and innovations helped to improve the predictability of electricity.

1.2 List factors that will affect the future of the electrical industry and explain the challenges related to each.

TECHNICAL TERMS

biofuel
efficiency
energy density
flow-limited
nonrenewable energy
renewable energy
sustainable energy

Introduction

Electricity is something that each of us relies on every day. Electricity provides us with light once the sun has set, cools our homes in the summer, and heats our homes in the winter. Electricity keeps our food fresh and cooks it when we are hungry. Electricity allows us to communicate with each other, even when we are far apart. Just think about medical research and space exploration: the electron microscope to observe the very small and the Hubble telescope to observe the vast expanses of space.

The bottom line is that electricity plays an active role in every aspect of our daily lives. It is sometimes difficult to imagine that, although humankind has known about electricity for thousands of years, we have only truly harnessed its power over the past two and a half centuries.

J J Osuna Caballero/Shutterstock.com

Figure 1-1. The first electric battery was invented by Alessandro Volta. A reconstruction of this first electrical battery is shown here.

1.1 History of Electricity

You are likely familiar with the key and kite experiment that was conducted by Benjamin Franklin in 1752. This was just the beginning of the new electrical frontier. Countless experiments, discoveries, and innovations helped to improve the predictability and reliability of electricity, pushing us into the digital age. Over the past 250 years, our understanding of electricity—how to harness, produce, transport, and manipulate it—has made remarkable advances. With our present-day advanced technologies, these advances are being made much more rapidly than in the decades following Franklin's discovery.

Early Inventions

Although electricity-based experiments were still being conducted after Benjamin Franklin to prove that electricity and lightning were actually the same thing, it was not until 1800 that the first energy-storing device—the battery—was invented by Italian physicist Alessandro Volta, **Figure 1-1**. The electrical unit "volt" is named after this physicist. With scientists now able to store electricity, a number of significant advances were made over the next four decades.

In 1808, Sir Humphry Davy, a British chemist, invented, as it was referred to at the time, an arc lamp. The arc lamp was a simple electric circuit that connected a piece of carbon across the terminals of a battery. When energized, the carbon material glowed and emitted light. Not quite a light bulb, but definitely an electrically created light source.

In 1820, French physicist and mathematician Andre Marie Ampere confirmed that there was a relationship between electric current and magnetism. Ampere is known as the father of electrodynamics, which is now known as electromagnetism. In a series of experiments he conducted, Ampere was able to establish that two current-carrying conductors would either attract each other or repel each other depending on the direction of current flow through each one. The term "ampere," which is used to represent the quantity of electric current flow, is named after Andre Marie Ampere.

One year after Ampere's findings, the first electric motor was invented by Michael Faraday, an English physicist and chemist. Faraday's work took Ampere's work to the next level by producing an electric current from a magnetic field. Faraday also speculated, and proved, that electricity was a force, not a fluid, as was commonly thought by many in the scientific community. In 1831, Faraday is also credited with inducing current in a coil of wire by energizing another coil of wire in close proximity to it. Today, we use this technology in transformers, which are electrical devices that "transform" or change an incoming voltage to a different outgoing voltage. In 1839, Faraday introduced the concepts of electrical conductors and insulators.

In 1826, just a few years after the first electric motor was invented, Georg Simon Ohm, a German physicist and mathematician, developed the relationship between voltage, current, and resistance. This relationship is now known as Ohm's law.

With the groundwork set, many inventions were introduced in the decades that followed. Some of these inventions include the following:

- **Electric relay (1835).** Invented by Joseph Henry, the electric relay is used to transmit electric current over long distances.
- **Fuel cell (1839).** Invented by William Grove, the fuel cell is used to produce electrical energy from hydrogen and oxygen.
- **Telegraph (1844).** The telegraph, invented by Samuel Morse, is used to send messages (Morse code) over long distances, **Figure 1-2**.

Luca Ponti/Shutterstock.com

Figure 1-2. A method of transmitting information by coded signal over distance, the telegraph machine became an important tool for the transmission of news around the country.

J J Osuna Caballero/Shutterstock.com

Figure 1-3. Although not the first electric light bulb, an incandescent light bulb, similar to the one shown here, was considered a more practical method of emitting light.

- **Generator (1876).** Invented by Charles Brush, the generator is used to produce a steady electric current.
- **Incandescent light bulb (1878).** Joseph Swan invented the first incandescent light bulb, **Figure 1-3**.
- **Tesla coil (1883).** Nikola Tesla invented the Tesla coil, which was used to increase voltage for more efficient transmission, **Figure 1-4**.
- **Electric alternator (1884).** Invented by Nikola Tesla, the electric alternator is used to produce alternating current (ac).
- **Transistor (1947).** Used in electronic circuits, the transistor was invented by Bell Laboratories.

Baloncici/Shutterstock.com

Figure 1-4. The Tesla coil was used to produce high-voltage, low-current, high-frequency alternating current. These coils were used for wireless telegraphy and in medical equipment.

- **Light emitting diode (1962).** While working for the General Electric Company, Nick Holonyak invented the light emitting diode, also known as LED, which has evolved into the most common form of lighting fixtures in use today.

Shedding Light

As you read the list of inventions, were you surprised by the actual inventor's name? For example, consider the 1878 invention of the first incandescent light bulb. Many believe that the incandescent light bulb was invented by Thomas Edison, but such is not the case. Edison did, however, improve on Joseph Swan's invention. Swan's light bulb burned for only about 12 hours before failing. By 1880, Edison improved on Swan's light bulb design and produced an incandescent light bulb that could burn for nearly 1200 hours before failing.

Formation of Power Companies

Inventions played a big part in the mainstreaming of electricity, but in order for these inventions to work, people needed electricity to power them. As a result, power companies and regulatory agencies began to form. For example, in 1879 the California Electric Light Company was the first company to sell electricity to customers. In 1882, Thomas Edison opened the first central power plant in New York City. Power generation plants began opening around the world. These generation plants used a variety of energy sources, such as waterpower, coal, and natural gas, to produce electricity. The Hoover Dam generated electricity using the power of moving water, **Figure 1-5**. The Hoover Dam started delivering usable power in 1936.

Kunal HehtaShutterstock.com

Figure 1-5. Hoover Dam's power plant generates around 4 billion kilowatt-hours of hydroelectric power annually. This is enough to provide power to 1.3 million people in Nevada, Arizona, and California.

1.2 The Future of Electricity

Newer renewable and sustainable energy sources for generating electricity, such as wind, solar, geothermal, and biomass, are now used at increasing rates to generate power all across the world, **Figure 1-6**. Not only are we expanding the ways we produce electricity, but we are also finding ways to produce it more efficiently, cleanly, and with far less of an environmental impact. The future of electricity seems to be revolving around the following key areas:

- Advances in energy storage
- Increased reliability on renewable and sustainable energy
- Increases in electrical production efficiency
- Regulatory changes
- Increased efficiency of electricity-consuming devices

These areas are interrelated, as they each have the ability to affect the others.

Energy Storage

Energy storage is a major factor in the supply chain of electricity. Without the ability to store electrical energy, any electricity not used immediately is lost. This, of course, may result in increased costs for the end users of electricity.

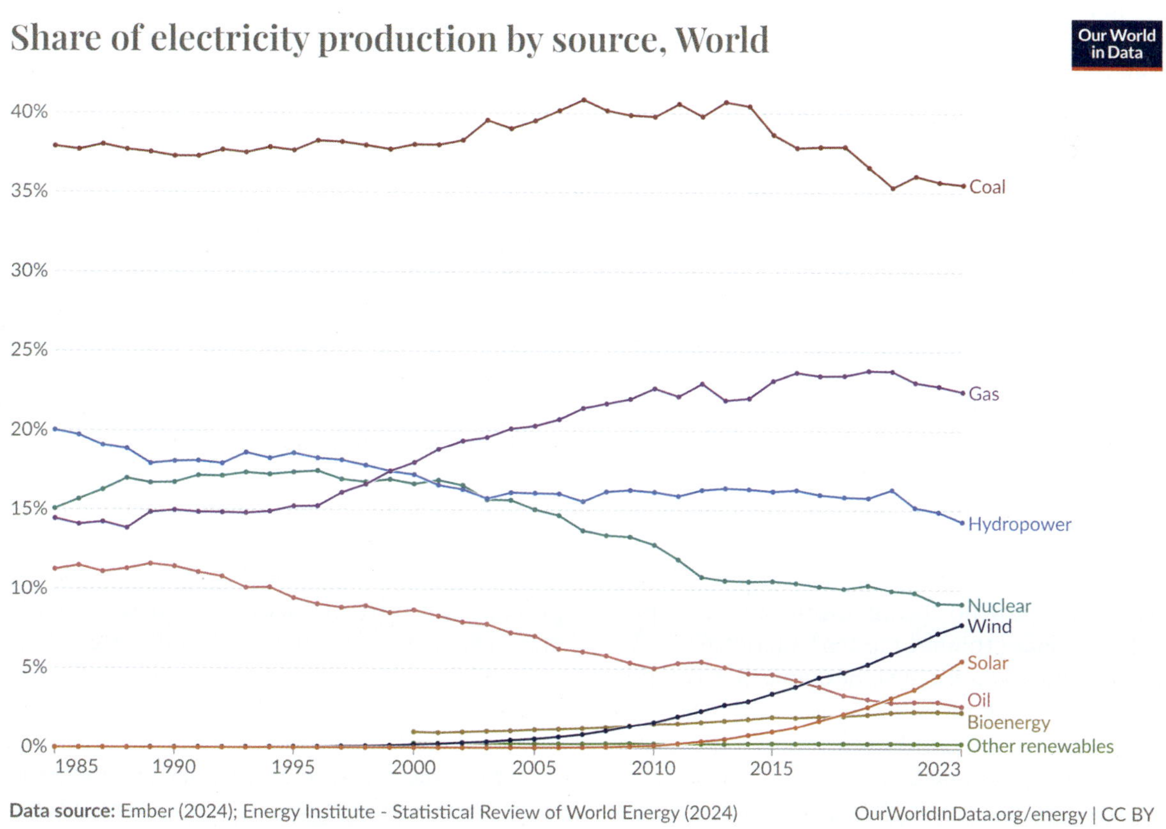

OurWorldInData.org/CC BY

Figure 1-6. The use of renewable sources, such as solar and wind, will continue to increase in electric power generation.

Shedding Light

The importance of energy production and storage can be illustrated by thinking of the fishing industry. Before refrigeration (storage), if a fishing company caught 10,000 pounds of fish but could sell only 4000 pounds, the remaining 6000 pounds would spoil. This would increase the cost of fish for customers and deplete the supply of fish moving forward. With advances in refrigeration and freezing technology, a fishing company that caught 10,000 pounds of fish could now sell all 10,000 pounds. The fishing industry became more efficient and profitable, and the cost of fish decreased for customers. The energy industry is similar: as energy storage technologies improve, the industry will become more efficient and profitable, and costs for customers will decrease.

petrmalinak/Shutterstock.com

Figure 1-7. A lithium-air battery, used in electric vehicles, is a lightweight battery that uses air as an anode.

Energy storage has historically been expensive and, for the most part, it still is today. As the costs associated with energy storage drop, the overall cost of energy will drop as well. Batteries are the most common form of electrical energy storage. Some of the challenges associated with storage batteries are the weight of batteries and the costs associated with mining and processing the materials needed in their production.

Research is continually being performed to improve the ***energy density*** of batteries. Energy density is the amount of energy, in kilowatt-hours, that can be provided by each pound of a battery's weight. By improving a battery's energy density, more energy can be stored in each pound of a battery. This will ultimately lead to smaller batteries that have the ability to store more energy. Experiments are being conducted, for example, on a lithium-air battery, **Figure 1-7**. The lithium-air battery has, pound for pound, over 14 times the energy storage capacity of a lithium-ion battery. This means that a one-pound lithium-air battery could provide the same amount of power (kilowatt-hours) as a 14-pound lithium-ion battery.

Advances are also being made regarding the materials used to manufacture batteries. Research is being conducted with more earth-abundant materials as opposed to those that require large amounts of energy to mine and process. By using more commonly found substances, the energy required to produce batteries will decrease, lowering the cost of storage for the consumer. For example, cobalt is a rare and expensive material used in most lithium-ion batteries. Other battery technologies that are presently being researched, developed, and/or perfected include lithium-tungsten batteries, nanowire gel electrolyte batteries, and zinc-magnesium oxide batteries. As sulfur is much more abundant than lithium and is easier to mine and process, the use of sulfur in batteries as opposed to lithium is also being explored.

Manufacturers are also looking at ways to reduce battery degradation. Prolonging their useful life will reduce the flow of batteries into the waste stream.

Thermal storage is another area that is experiencing large growth. A great deal of research is being done to store thermal power from the sun. One such research project is concentrating on storing solar energy in molten silicon. When needed, the energy in silicon can be converted to electricity. Some large air conditioning systems are using thermal storage to balance electrical demand. These thermal storage systems can operate in the evening, when power demand is lower, to produce ice that is then used during the day to cool a building. Energy storage technologies continue to advance and, as they do so, will pave the way for increased and less expensive energy production.

Renewable and Sustainable Energy

When we flip a switch to turn on a light, we hardly give a thought to how the electricity we are using was produced. What energy source was converted? Was the electricity wind-generated or nonrenewable? What we do know is as we move forward, there is an increased likelihood that the energy you use will come from a renewable energy source.

The United States Energy Information Administration (EIA) predicts that the percentage of US electricity produced from renewable sources will more than double in the period extending from 2024 through 2050, **Figure 1-8**. During the same period, the percentage of US electricity produced by fossil fuels and nuclear energy is expected to decrease by 16% and 7%, respectively. Of the three main renewable energy types, solar energy is expected to see the most growth, **Figure 1-9**.

When power sources such as wind, solar, nuclear, hydropower, and geothermal are discussed, the terms "renewable" and "sustainable" are often used. Quite often, these two terms are used interchangeably, but they are not the same thing.

Terms that are often used to describe sustainable generally include "long-term availability" and "natural replenishment." Generally speaking, when we refer to a ***sustainable energy*** source, we are referring to an energy option that has the ability to help meet the growing demand for energy today without compromising the ability of those in the future from relying

US Electricity Generation by Energy Source Type

Energy Source	2024	2050 (predicted)
Fossil fuels	60%	44%
Nuclear	19%	12%
Renewables	21%	44%

Source: US Energy Information Administration

Figure 1-8. Expected changes in energy use in the United States.

US Electricity Generation by Renewable Source Type

Energy Source	2024	2050 (predicted)
Solar	18%	51%
Wind	48%	31%
Hydroelectric	27%	12%
Other	7%	6%

Source: US Energy Information Administration

Figure 1-9. Expected changes in renewable energy use in the United States.

on the same resource. The term ***renewable energy*** source is often used to describe energy sources that do not cause any harm to the surrounding environment, are widely available, and replenish naturally at a rate that exceeds that of their use, **Figure 1-10**. In other words, there is no fixed quantity or amount of renewable energy sources. Solar, wind, geothermal, and hydropower are both renewable and sustainable energy sources since they have a minimal impact on the environment. Nuclear energy, however, is nonrenewable because it requires fuel to generate electricity. Many energy experts may have differing opinions regarding the threshold of sustainability, namely when exactly does an energy source become sustainable or not sufficiently sustainable.

Some renewable energy sources are naturally replenishing but ***flow-limited***. The term flow-limited refers to the fact that the amount of energy available at any given time might be limited. Take for instance wind and solar energy. Wind is a varying force and, when there is no wind, there is no energy being generated. Similarly, the sun's rays are not acting on a solar panel 24 hours a day, therefore, energy is not being produced at night.

In a technical sense, fossil fuels may be classified as a renewable energy source, as gas and oil are constantly being created. However, it takes far longer for these fuels to be created than to be used. Although they are technically renewable, from a practical standpoint, fossil fuels are considered as ***nonrenewable energy***, meaning they cannot be replaced by natural means. A line has been drawn in the sand by energy experts who state that, if a resource cannot be replenished within our lifetime, it should be classified as a nonrenewable resource. Fossil fuels are also classified as unsustainable energy sources for much the same reasons. The continued use of fossil fuels is not sustainable, as we are using these fuels faster than they are replenishing. From a practical standpoint, fossil fuels are both nonrenewable and unsustainable energy sources. Biofuels are an exception to this. ***Biofuels*** are fuels that are created from biomass, just as fossil fuels are, but there is one significant difference. Biofuels are produced using faster, modern technologies than the very slow geological processes that produce traditional fossil fuels.

Production and Distribution Efficiency

For electricity to be made available to us, many refining, processing, combusting, generating, transmitting, and distributing processes have to take place first. During each of these processes, losses occur so that by the time the energy reaches us, there is less of it than when the process started.

Shedding Light

Although overly simplified, consider the example of running with a bucket full of water. As you run, some water spills out. When you finally reach your destination, there is less water in the bucket than when you started. However, if you devised a way to keep the water in the bucket, such as using a tight-fitting lid, you would be able to minimize the amount of water lost. By limiting the loss of water during the water-transmission process, the efficiency of the process is increased. For electricity production and distribution, continuously improving the many activities throughout the process will lead to greater efficiency in the future.

Generally speaking, the ***efficiency*** of a process is the ratio of the output produced divided by the input required to produce the output. Mathematically, the efficiency formula looks like this:

$$\text{efficiency (\%)} = \frac{\text{units of output}}{\text{units of input}} \times 100$$

For example, if 10 units of input are required to produce 8 units of output, the efficiency can be calculated as:

$$\text{efficiency (\%)} = \frac{\text{units of output}}{\text{units of input}} \times 100$$

$$\text{efficiency (\%)} = \frac{8}{10} \times 100$$

$$\text{efficiency (\%)} = 0.8 \times 100$$

$$\text{efficiency (\%)} = 80\%$$

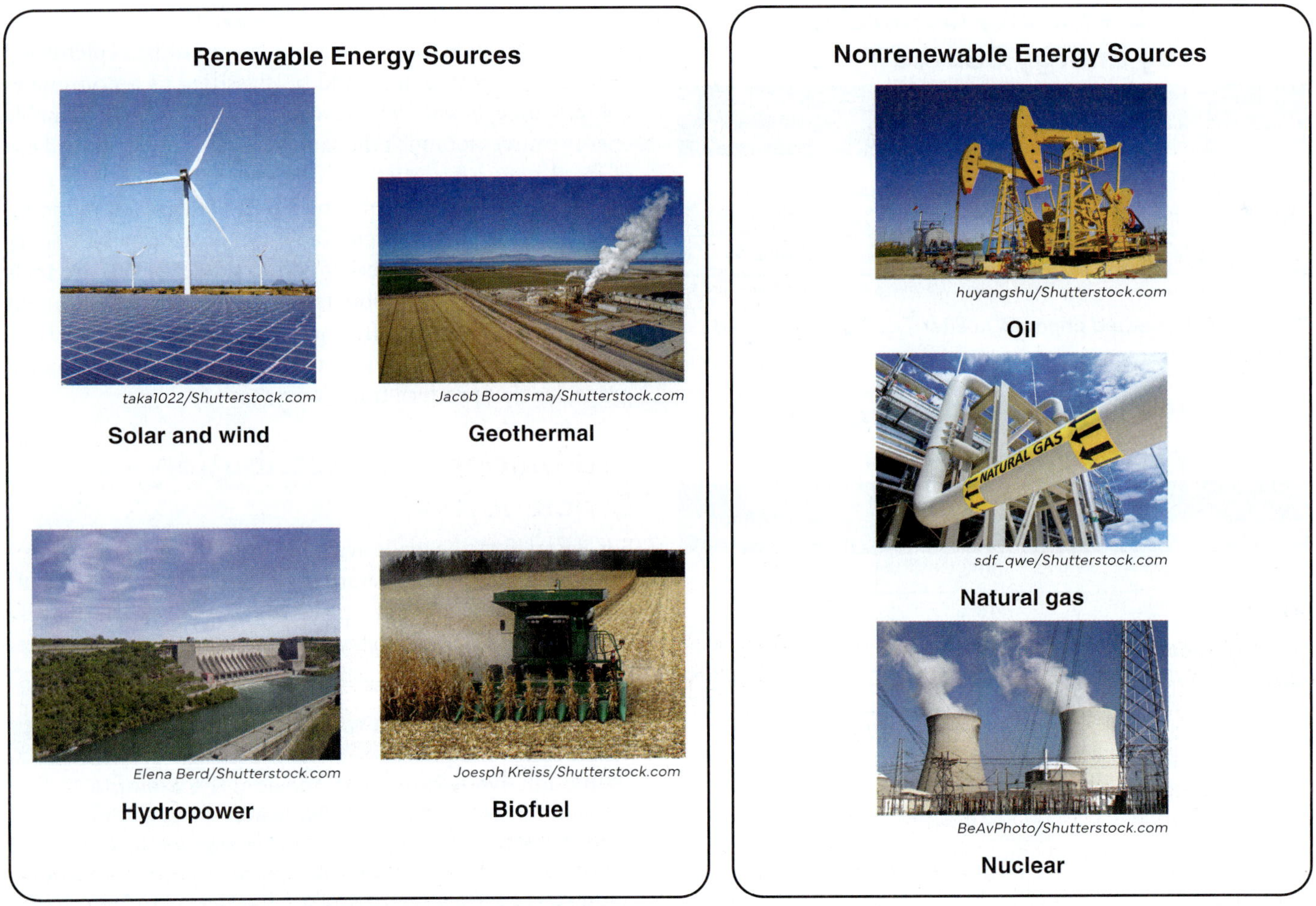

Figure 1-10. Renewable and nonrenewable energy is available in many different sources.

Similarly, if 10 units of input produce 10 units of output, the efficiency of the process would be 100%. To increase the efficiency of a process, the goal is to obtain as much output as possible for each unit of input.

Consider the process chart shown in **Figure 1-11**. In this process, which represents a simple-cycle natural gas power plant, a fuel/air mixture is ignited in a boiler so that water can be heated to create steam. The steam is then used to turn a turbine that generates electricity. The electricity is then transmitted and distributed to the consumer. A simple-cycle plant represents the most basic of power plant designs and the efficiency of plants such as these are 30% to 40%. This does not take into account any additional losses resulting from power transmission, distribution, or any inefficiencies in the appliances that ultimately are used by the consumer. Transmission and distribution losses, for example, can account for an additional 5% loss of efficiency.

There are many steps to supply energy to the consumer. Small increases in efficiency along the way are additive and have the potential to greatly increase the overall efficiency of energy production, transmission, and distribution. The following are some examples of methods used to improve efficiency:

- **Recovering waste heat.** A great deal of heat is lost as water is converted to steam in the boiler. Some power plants, such as combined-cycle gas plants, use this waste heat to turn secondary turbines, which produce additional electricity. Since this increases the energy output with no additional fuel requirement, efficiency is increased.
- **Increased turbine efficiency.** A number of advances have been made that increase the efficiency of the turbine itself. These include new-generation bearings that further reduce internal turbine friction and new blade designs, which can increase the speed of the turbine while better withstanding the stresses placed upon them.
- **Reducing transmission losses.** A great deal of energy transmission losses may be eliminated by improving the quality of the connections made in the power lines. Additional measures, such as replacing power lines with larger-diameter lines, also helps reduce losses. Another

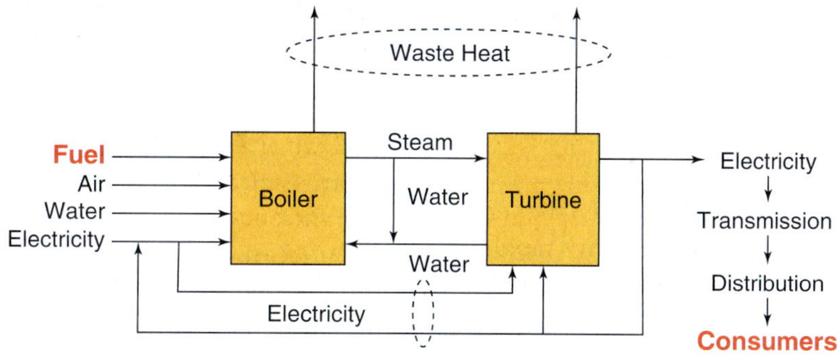

Goodheart-Willcox Publisher

Figure 1-11. Process cycle of a simple-cycle natural gas power plant.

method to reduce losses is to increase the voltage of the energy transmitted, which results in lower current flows at a given power level. New developments in transformer and power line design will also contribute to a reduction in transmission losses.

- **Upgrades to the power grid.** The existing power grid in the United States is a network of systems, many of which are outdated and prone to failure. In addition, some of the older grid technologies were not designed to manage or work in conjunction with renewable energy resources and smart grid technologies. Completely redesigning and reconfiguring the power grid all at once would be impossible but, as advances and changes are implemented, improved efficiency levels will result, helping us move closer to reaching our broader sustainability goals.

Regulatory Changes

Regulatory changes greatly affect the future of the electrical industry. Although many regulations are in place to protect the environment, there may be negative, unintended effects. These negative effects may limit the amount of energy produced, driving the cost of energy up. For example, the Environmental Protection Agency (EPA) proposed the Clean Power Plan that would limit the amount of carbon dioxide emissions from existing power plants. This rule would, in effect, cause many coal-fired power plants to close, as they produce very large amounts of carbon dioxide. The Clean Power Plan is expected to reduce carbon dioxide pollution from the energy sector by 32% by the year 2030.

In 2007, the US Department of Energy put forth energy efficiency mandates for a number of technologies including light bulbs. These regulations, which were originally scheduled to take effect in January of 2020, were discontinued in September of 2019. Some states, including California, Nevada, Washington, and Colorado, enacted their own light bulb efficiency standards when it became known that the proposed mandates would be rolled back.

Regulations have changed in the past, and they will continue to change in the future. It is the resiliency and continued dedication of the engineers, manufacturers, and electrical workers that keep the lights burning brightly, even under the most uncertain conditions.

Electricity-Consuming Devices

Minimizing losses during the processes of producing and distributing electrical energy to the end users is one of the main elements that must be considered in helping to ensure a bright future for electricity. Another often overlooked factor is the amount of electricity we consume based on the overall efficiency of the devices and appliances we use. Consider one of the most-used electrical devices, the light bulb. Advances in lighting technologies have resulted in significant savings in energy usage while producing illumination levels that are historically consistent over time. In the mid-1980s, almost 25% of the average homeowner's energy bill was attributed to lighting. This percentage dropped to 13% in 2018 as compact fluorescent (CFL) and light-emitting diode (LED) technologies became more popular. This percentage is expected to drop to as low as 8% by 2030 as the cost of these technologies drops. Older-generation incandescent light bulbs converted only 5% of their consumed power into light, while newer-generation lighting fixtures convert over 85% of their consumed power into light, while at the same time producing more light per watt consumed. For comparison, a 150-W incandescent light bulb consumes 150 W, while an LED bulb that produces the same amount of light consumes less than 20 W. In addition, consider the life expectancy of newer-generation lighting fixtures. An incandescent bulb is expected to function for approximately 1000 hours, while a CFL bulb operates for about 10,000 hours and an LED light lasts for about 30,000 hours. Although LED lights are more expensive than incandescent bulbs, the amount of energy used to produce LED lights, on an annualized basis, is much lower and, as a result, additional energy savings are realized.

In addition to the lighting sector, advances in electric motor design and motor control technologies have also paved the way for significant savings resulting from utilizing more efficient electricity-consuming devices.

CHAPTER 1 REVIEW

Summary

- The electricity-based discoveries made between the years 1750 and 1950 formed the foundation for today's rapid advances in electrical and electronics technologies. (1.1)
- Advances in energy storage and increases in electricity production efficiency will help bring energy costs down. (1.2)
- Renewable energy sources have long-term availability and are naturally replenishing. (1.2)
- Nonrenewable energy cannot be replaced in a lifetime by natural means. (1.2)
- Efficiency is the ratio of a process output to the process input. (1.2)
- Regulatory changes in the electrical industry serve to increase efficiency of the power-generation process as well as that of the end-use products. (1.2)
- Advances in lighting technologies have resulted in ways to minimize losses during the process of producing and distributing electrical energy. (1.2)

Review Questions

Use the information in the chapter to answer the following questions.

Measuring Your Potential

1. *True or False?* The incandescent light bulb was invented by Thomas Edison. (1.1)
2. Nikola Tesla is credited with having invented the electric _____. (1.1)
   A. dc motor
   B. electric alternator
   C. telegraph
   D. light bulb
3. *True or False?* The electrical term "volt" is named after Italian physicist Alessandro Volta. (1.1)
4. Scientists are experimenting with which element to potentially replace the lithium presently found in many batteries? (1.2)
   A. Cobalt
   B. Nickel
   C. Iron
   D. Sulfur
5. *True or False?* Lithium-air batteries have the potential to store about 14 times as much power as lithium-ion batteries. (1.2)
6. *True or False?* In a simple-cycle power plant, waste heat is recovered to produce additional power. (1.2)
7. *True or False?* Electric power line transmission losses are typically in the 5% range. (1.2)
8. Efficiency is most commonly expressed as a _____. (1.2)
   A. fraction
   B. percentage
   C. whole number
   D. decimal
9. A simple-cycle natural gas power plant typically operates at an efficiency that ranges from _____. (1.2)
   A. 10% to 20%
   B. 20% to 30%
   C. 30% to 40%
   D. 40% to 50%
10. The US Clean Power Plan is expected to reduce carbon pollution from energy production by approximately _____. (1.2)
    A. 10% by 2040
    B. 20% by 2050
    C. 30% by 2030
    D. 40% by 2060

Plug-In

1. Nuclear power is considered _____. (1.2)
   A. renewable and sustainable energy
   B. renewable, but flow-limited
   C. nonrenewable and unsustainable
   D. nonrenewable, but sustainable
2. Solar power is considered _____. (1.2)
   A. renewable and sustainable energy
   B. renewable, but flow-limited
   C. nonrenewable and unsustainable
   D. nonrenewable, but sustainable
3. Hydropower is considered _____. (1.2)
   A. renewable and sustainable energy
   B. renewable, but flow-limited
   C. nonrenewable and unsustainable
   D. nonrenewable, but sustainable
4. At an efficiency of 100%, an input of 10 units will produce an output of how many units? (1.2)
   A. 10
   B. 7
   C. 5
   D. 0

Making the Connection

1. The use of solar photovoltaic to produce electricity on a large scale is becoming mainstream. However, the sun does not always shine. How are utilities meeting electrical demand when this is the case? (1.2)
2. If 10 units of input are required to produce 9 units of output, how would you estimate the efficiency of the process? (1.2)

CHAPTER

2 Careers in the Electrical Field

Phovoir/Shutterstock.com

SPARKING DISCUSSION

Discuss the journey from electrical helper to apprentice to journey worker to master craftsman.

LEARNING OUTCOMES

After completing this chapter, you will be able to:

2.1 List the levels of education required for various careers in the electrical industry.

2.2 Explain the process involved when applying for a position.

2.3 Discuss methods to prepare for a job interview.

2.4 List traits that are desirable in the workplace.

TECHNICAL TERMS

apprentice
apprenticeship
diversity
internship
job shadowing
letter of recommendation
lifelong learning
mentor
nonverbal communication
portfolio
résumé
self-motivation
verbal communication

Introduction

Choosing a career is one of the most important decisions you will ever make; important, yet difficult. The field of electricity offers many opportunities. The first step to achieving your career goal begins with identifying the things you enjoy doing and do well.

No matter what career path you decide to follow, your electrical skills and knowledge can be applied in many different industries, **Figure 2-1**, including the following:

- Aviation
- Automotive
- Chemical
- Construction
- Manufacturing
- Medical
- Computers
- Heating, ventilation, air conditioning, and refrigeration (HVACR)
- Communications
- Petroleum
- Television and movie production
- Low-voltage wiring for communications infrastructure

This chapter will help you determine the education and job skills you will need to prepare for your career, as well as steps to getting a job and succeeding within the electrical field.

2.1 Preparing for Your Career

As part of the decision-making process, you will need to consider what educational level is necessary for entering each career you investigate. How much training or experience is needed? Can people enter the field with less training and acquire expertise while working on the job? Are special certificates, licenses, or credentials needed? Will your career path require a license or certification? Other programs may require a college degree, **Figure 2-2**.

Investigating Education and Training

Before deciding on a specific career, you may wish to shadow someone who holds the type of job you desire. ***Job shadowing*** is the process of observing a person in the workplace to learn more about their job and its requirements. In addition, you may seek to have a mentor's assistance. A ***mentor*** is someone with greater experience and knowledge who guides you in your career.

You might also consider assisting someone who knows how to do the tasks well, such as working in an internship or apprenticeship. An ***internship*** is an arrangement with an educational institution whereby a student is supervised while working with a more experienced jobholder. An ***apprenticeship*** is the most common path to becoming

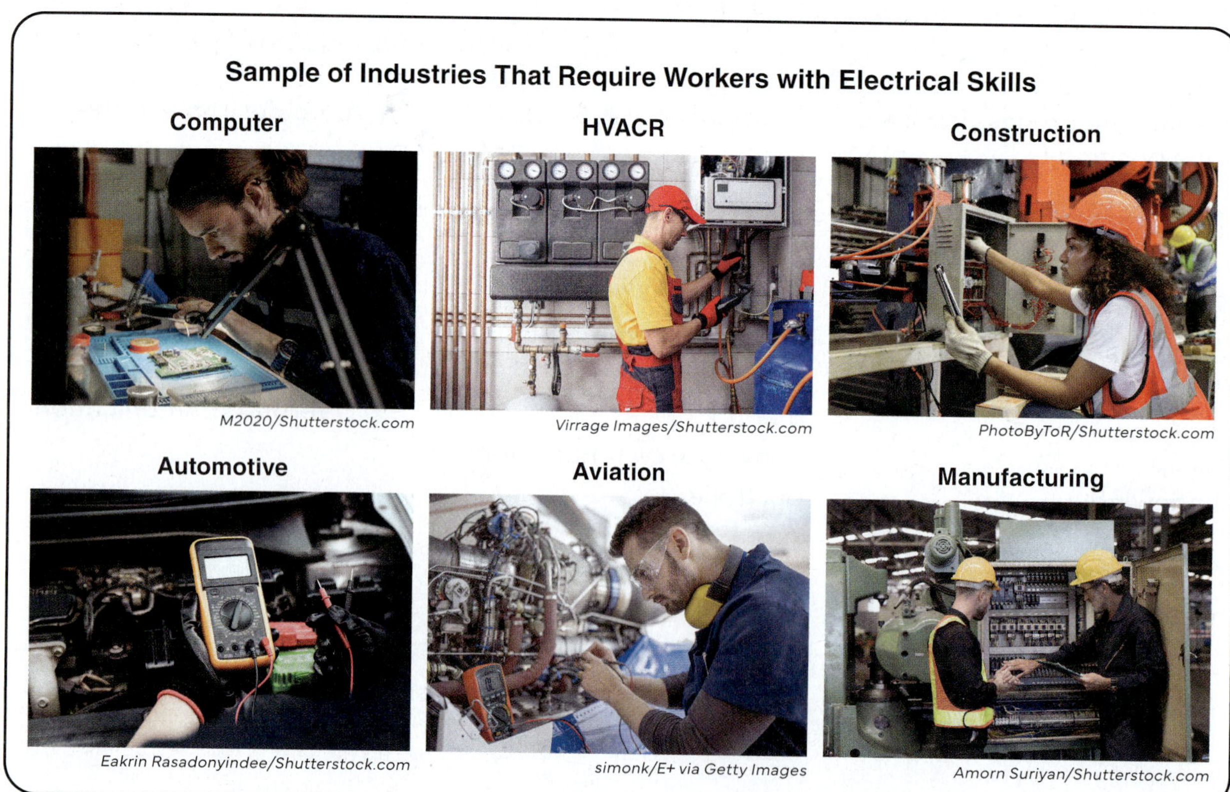

Figure 2-1. Your electrical skills and knowledge can be applied in a variety of different industries.

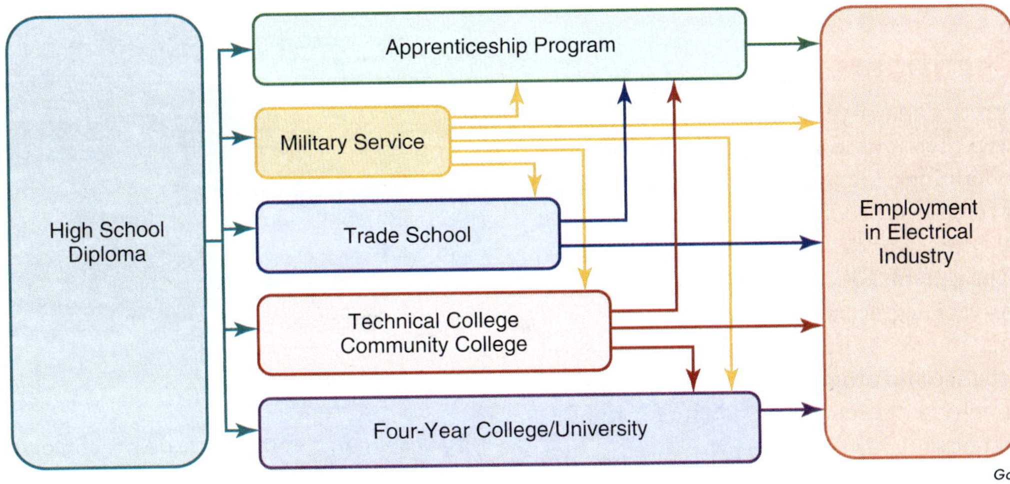

Goodheart-Willcox Publisher

Figure 2-2. There are various paths that will lead you to a career in the electrical field. As not every state requires an apprenticeship, check with your state and/or local regulations to find out what is required to be certified as an electrician.

an electrician. It involves learning a trade under the direction and guidance of an experienced electrician. You can also gain valuable experience through working at a part-time job or even volunteering at community or charitable organizations.

High School Diploma or a One-Year Certificate

It is possible to start your career in the electrical field with a high school diploma or a one-year certificate from a trade school or technical college. An employer will generally favor a candidate who has earned a one-year certificate over a candidate with only a high school diploma. These jobs will be entry-level positions:

- **Helper.** An electrician's helper is responsible for assisting an electrician on the job. Helpers work with their hands and perform tasks that require less skill, such as cutting and stripping wire, installing different cable types, measuring, and cleaning. The best helpers learn quickly by asking questions, listening, and observing. As they learn, helpers progress and perform more advanced tasks. Starting as a helper or a beginning electrician will often lead to an apprenticeship and a path to becoming a journey worker or master electrician.
- **Apprentice.** An electrical ***apprentice*** works under a licensed electrician, typically three to five years, to learn the trade, **Figure 2-3**. The US Department of Labor Office of Apprenticeship sets the minimum standards for training programs across the country. An electrical apprenticeship requires about four years of training, including 576 classroom hours and 8000 hours of on-the-job learning. Apprentices perform many of the more advanced tasks required of an electrician, such as print and schematic diagram reading, installing and replacing electrical devices, taking and evaluating electrical measurements, installing distribution panels, and performing preventive maintenance on existing equipment and systems.

ALPA PROD/Shutterstock.com

Figure 2-3. As an electrical apprentice, you will combine classroom learning with work in the field under the guidance of a licensed electrician.

Shedding Light

Besides hands-on training, an apprentice will require additional related classroom instruction. Often times, the classroom instruction is provided working in conjunction with a local technical college. If the employer is large enough, in-house classroom instruction may be provided. If the apprentice is working in a represented, union environment, the supervising committee will be composed of union and employer representatives. One of the common elements of all apprenticeships is pay increases as benchmarks are reached for hours and education.

Associate's Degree or a Two-Year Certificate

An associate's degree is a two-year liberal arts or science-based degree that typically is used as a stepping stone for further education in obtaining a four-year degree. Many colleges also offer two-year certificate programs, which generally have the same technical coursework as an associate's degree program but fewer of the general education requirements. Upon completion of a two-year degree or certificate, the learner will be prepared to enter the workforce, in either an entry-level position or an apprenticeship program. An associate's degree in electrical technology will prepare a graduate for the following electrical careers:

- **Electrical installer.** Electrical installers typically work on new construction or renovation projects where they follow project drawings and plans to install new electrical systems.
- **Electrical repair technician.** These technicians specialize in inspecting, troubleshooting, and repairing electrical systems. They use various pieces of test equipment to take electrical readings, and then interpret the findings to isolate system problems. Most of the time, electrical repair technicians work on the wiring network that provides power to the end users.
- **Electronics repair technician.** These technicians specialize in inspecting, troubleshooting, and repairing problems with electronic devices and systems, **Figure 2-4**. Unlike electrical repair technicians, electronics repair technicians typically work on end-use devices.
- **Electrical maintenance technician.** These technicians are responsible for keeping the electrical system up and running. Proper maintenance reduces unexpected system failures, which are much more expensive to remedy. Electrical maintenance technicians often perform some of the tasks performed by installers and repair technicians.
- **Renewable energy technician.** These technicians perform installation and repair work on solar panels, wind turbines, and other hardware associated with power generation from renewable sources, **Figure 2-5**.
- **Electrical line worker.** Typically employed by utility companies, electrical line workers are responsible for installing, servicing, and maintaining the overhead and underground power lines. These workers often specialize in either installation or repair, but many workers are able to perform multiple tasks, **Figure 2-6**.
- **Instrumentation technician.** These technicians typically specialize in calibrating test instruments used in the field. Instrumentation technicians are employed by test instrument manufacturers and are responsible for ensuring electrical readings obtained in the field are within an acceptable range. This type of job is often an in-house position, as the testing and calibration of the instruments is normally performed within the manufacturing warehouse environment.
- **Electrical technician.** An electrical technician, also known as an electrical technologist, works with, and assists, engineers that develop various pieces of electrical or electronic equipment. Electrical technicians often complete their four-year engineering degree.

chunyawut_stock/Shutterstock.com

Figure 2-4. An electronics repair technician may troubleshoot problems and perform routine maintenance on electronic devices.

Zstock/Shutterstock.com

Figure 2-5. As the use of renewable and sustainable energy sources increases, the need for qualified technicians increases as well.

Bachelor's Degree

Many high-level careers in the electrical industry require a four-year degree. It is common practice to attend college part-time to earn a four-year degree while working full-time in an electrical career. Working in the industry while going to college at the same time is a challenge, but being able to apply your real-world work experience to your studies makes the content easier to grasp and master. Electrical careers that require a four-year degree include the following:

- **Electrical engineer.** Electrical engineers perform a wide range of duties that include designing and improving electrical components and products at residential, commercial, and industrial levels. Electrical engineers are involved in many aspects of manufacturing, construction, research, and testing processes, **Figure 2-7**.
- **Electronics engineer.** An electronics engineer performs many of the same duties as an electrical engineer, but specializes in end-user products such as cell phones, computers, and appliances, **Figure 2-8**.
- **Electrical drafter.** Electrical drafters are responsible for creating wiring and electrical equipment plans and prints. These are the plans electrical contractors follow when wiring a construction or renovation project. Many successful electrical drafters have electrical and technical drawing backgrounds.

kozmoat98/E+ via Getty Images

Figure 2-6. An electrical line worker works for a utility, telecommunications, or energy company to maintain, repair, and install power lines.

Gorodenkoff/Shutterstock.com

Figure 2-7. To obtain a job as an electrical engineer, you will need to complete a four-year program that includes courses in advanced math and engineering principles as well as complete an engineering internship.

7postman/E+ via Getty Images

Figure 2-8. Essential skills for electronics engineers include a thorough knowledge of technical concepts such as electric circuit design and the operation of electronic devices.

2.2 Applying for a Position

Getting any job requires a great deal of work on your part. The ultimate goal is to find a position that you will find rewarding at a company that offers room for growth at a compensation level that you feel is appropriate. To find that perfect match, you need to research the company.

Applying for a job is a multistep process that first starts with finding an appealing open position. This process also includes preparing your résumé, tailoring a cover letter, obtaining references and letters of recommendation, and organizing a portfolio.

Preparing Your Résumé

A ***résumé*** is a brief outline that highlights your education, work experience, and other qualifications for work. A sample résumé is shown in **Figure 2-9**. The main purpose résumé is to serve as an introduction to a potential employer, presenting your qualifications with the goal of securing an interview. It is important that your résumé is well written, precise, accurate, and error-free. The key word in this description is brief. According to human resource experts, the average employer spends less than 10 seconds looking at an applicant's résumé and, based on that short inspection, determines whether an interview will be granted. Your résumé should include the following:

- **Full legal name.** This is the name that appears on legal documents, such as your diploma. You should avoid using nicknames.

Michael J. Garcia

134 Lincoln Street
Wilton, CA 93232

(212) 555-1234
mjgarcia22@email.com

Career Objective
To obtain an entry-level position in the electrical wiring field.

Work Experience
Builder's Supply, Holloton, CA — September 2025–present
Retail Sales Associate
- Providing assistance to customers in hardware, plumbing, and electrical departments
- Reviewing department inventory
- Maintaining neat and orderly displays

Receiver/Stocker — March 2025–September 2025
- Unloaded, stored, and stocked merchandise
- Operated forklifts and hand trucks

Tony's Grocery Store, Wilton, CA — May 2024–March 2025
Cashier
- Operated checkout scanner and bagged groceries
- Answered customer questions
- Packed out product on store shelves as needed
- Ensured that all spills and breakages were cleaned up

Education
Associates of Applied Science (A.A.S.) in Electrical Construction Technology — May 2025
Oceanside Community College
- GPA: 3.22/4.0
- Coursework included residential wiring, commercial wiring, and solar photovoltaic systems
- Participated in SkillsUSA local chapter competition (won second place)

Community Service
Habitat for Humanity, volunteer, summers of 2023, 2024, 2025
Wilton Food Bank, volunteer, 2022–present

References
Available upon request

Goodheart-Willcox Publisher

Figure 2-9. A résumé provides a brief overview of your education, work experience, and qualifications. A good résumé is neat and easy to read.

- **Full mailing address.** This may be needed as a method to contact you directly with any follow-up information.
- **Phone number.** This should be the best contact number to reach you. If using your personal cell phone number, you should use an appropriate message greeting.
- **Email address.** This should be a professional-sounding email address (for example, mjgarcia22@email.com). Never include a humorous or informal email address (for example, skater_king2000@email.com).
- **Objective.** This statement identifies the job you are seeking.
- **Education history.** Begin with your most recent schooling and include credentials you received.
- **Work experience.** Begin with your most recent job and then list in chronological order other job experiences.

Other items that you might want to include:

- **Community service/volunteer work.** Include any volunteer community service work completed as part of your graduation requirements as well as any extracurricular volunteer work you may have done through service and nonprofit organizations.
- **Awards and certifications.** This should align with the type of work you are applying for, such as an award from a SkillsUSA competition.

The following are some items you should *not* include on your résumé:

- **Personal details.** Do not include age, height, weight, eye color, marital status, etc.
- **Irrelevant work experience.** Keep the résumé relevant to the job sought.
- **More than one phone number.** Provide the easiest number to be reached at.
- **Hobbies.** These can be brought up during the interview discussion.
- **Personal references.** Provide these as a separate list, or note that they will be provided on request.
- **Tables, images, charts, or photos.** They distract and take up valuable space.
- **Desired salary.** This will be a part of the interview discussion.
- **Reason for leaving last job.** This will be a part of the interview discussion.
- **Opinions and lies.** Stick to the facts.
- **Unnecessary words.** Be as direct as possible.
- **Current boss's name and contact information.** This is the company already listed in work experience.
- **Why you want this position.** This will be a part of your cover letter and the interview discussion.

When preparing your résumé, keep the following in mind:

- For entry-level positions, your résumé should be no more than one page long.
- Include words and phrases that are commonly used in the industry and found in the job description.
- Remember, less is more. It is perfectly acceptable to use phrases instead of complete sentences.
- Avoid placing periods at the end of a statement that is not a complete sentence.
- Make certain the formatting is consistent, including indentations, use of font size, bullets, and spacing.
- Use the past tense to describe duties at previous jobs.
- Use the present tense to describe duties at your present job.

Once you have prepared your résumé, have a family member or friend read it. If possible, have someone in the electrical field review it and provide you with feedback. Many schools offer assistance with résumé preparation. Be sure to ask if such a service is available at your school.

Writing Your Cover Letter

When you submit your résumé to a potential employer, you will also need to send a cover letter or introductory email message. The cover letter serves as an introduction, as it is read before the employer reviews your résumé. An example of a cover letter is shown in **Figure 2-10**. This introductory letter has the potential of making a lasting impression, so it should be just as neat and error-free as the résumé itself. Before sending it to a potential employer, have several people read the letter and offer advice for any improvement. The cover letter should be brief, to the point, and should include the following items:

- Self-introduction
- Title of the job you are seeking
- Why you are a good candidate for the job
- Your strengths and skills, as they relate to the job you are seeking
- When you will be available to begin work; otherwise, it will be assumed you are available very soon, and this can be discussed in the interview
- Request for an interview

Modify your cover letter for each position you apply for by including information that is unique to the company and the specific position. A potential employer can easily identify "cookie cutter" cover letters and is far more likely to dismiss job seekers who submit them.

Obtaining References

Along with your résumé, you may be asked to submit a list of references, typically three. A reference is an individual who will provide important information about you to a prospective employer. A reference can be a teacher, school official, previous employer, or any other adult outside your family who knows you well.

Before submitting your list of references, it is important to obtain permission from each person you select. The list of references should include a name, title, phone numbers, and address for each person listed.

In some cases, potential employers will ask you for letters of recommendation. A ***letter of recommendation*** is written by an employer, supervisor, teacher, coach, or other adult that describes your personal characteristics and qualities that would be strengths in the workplace. These letters give an employer a better idea of your character, what skills you possess, and what impression others have of you. Since these letters will be used in the hiring process, be sure to choose people who know you well and are good writers, as they will be speaking on your behalf. Ask as many people as possible, so you can choose the best ones to submit.

Organizing a Portfolio

A ***portfolio*** is a collection of documents that highlights your talents, skills, awards, and abilities. Unlike a résumé,

Michael J. Garcia
134 Lincoln Street
Wilton, CA 93232
(212) 555-1234
mjgarcia22@email.com

April 23, 2026

Mr. James Williamson
Williamson Electrical
4392 East 134th Street
Wilton, CA 93232

Dear Mr. Williamson:

The apprentice electrician position you advertised on the Career Finder website is exactly the type of job I am seeking. After reviewing the job description and requirements, it was clear that my experience, skills, and interests are a perfect match for this opportunity.

While obtaining my associate's degree in electrical construction technology from Oceanside Community College, I gained both the theoretical knowledge and the hands-on skills required for this position. In my current position as a retail sales associate at Builder's Supply, I have developed strong customer service skills and gained a better understanding of electrical parts and systems. I am anxious to apply the skills I have learned and to continue gaining new skills.

Please find my résumé enclosed with this letter. I would greatly appreciate an opportunity to interview for this position. Please contact me at your convenience by phone or email to schedule an interview. I look forward to hearing from you.

Sincerely,

Michael J. Garcia
Michael J. Garcia

enclosure

Goodheart-Willcox Publisher

Figure 2-10. Use a cover letter or introductory message to introduce yourself and explain in your own words why you would be a good hire.

a portfolio provides a more complete visual representation of your accomplishments and skills. The following are some items to include in your portfolio:

- Certification completion documentation
- Industry-related credentials, such as OSHA-10
- Competition awards, such as from SkillsUSA
- Written assignments that received outstanding instructor feedback
- First-aid class completion certification, such as CPR
- Photos and descriptions of projects completed in school
- School transcripts
- School awards and similar certificates
- Dean's list letters
- Letter of appreciation from community service organizations such as Habitat for Humanity
- Lab reports completed as part of your electrical studies

Portfolios are not typically required, but should be prepared, especially if you had many accolades that you would like to show to a potential employer.

Submitting Your Application

Although job postings are found through a variety of different media, the internet may be one of the best resources to search

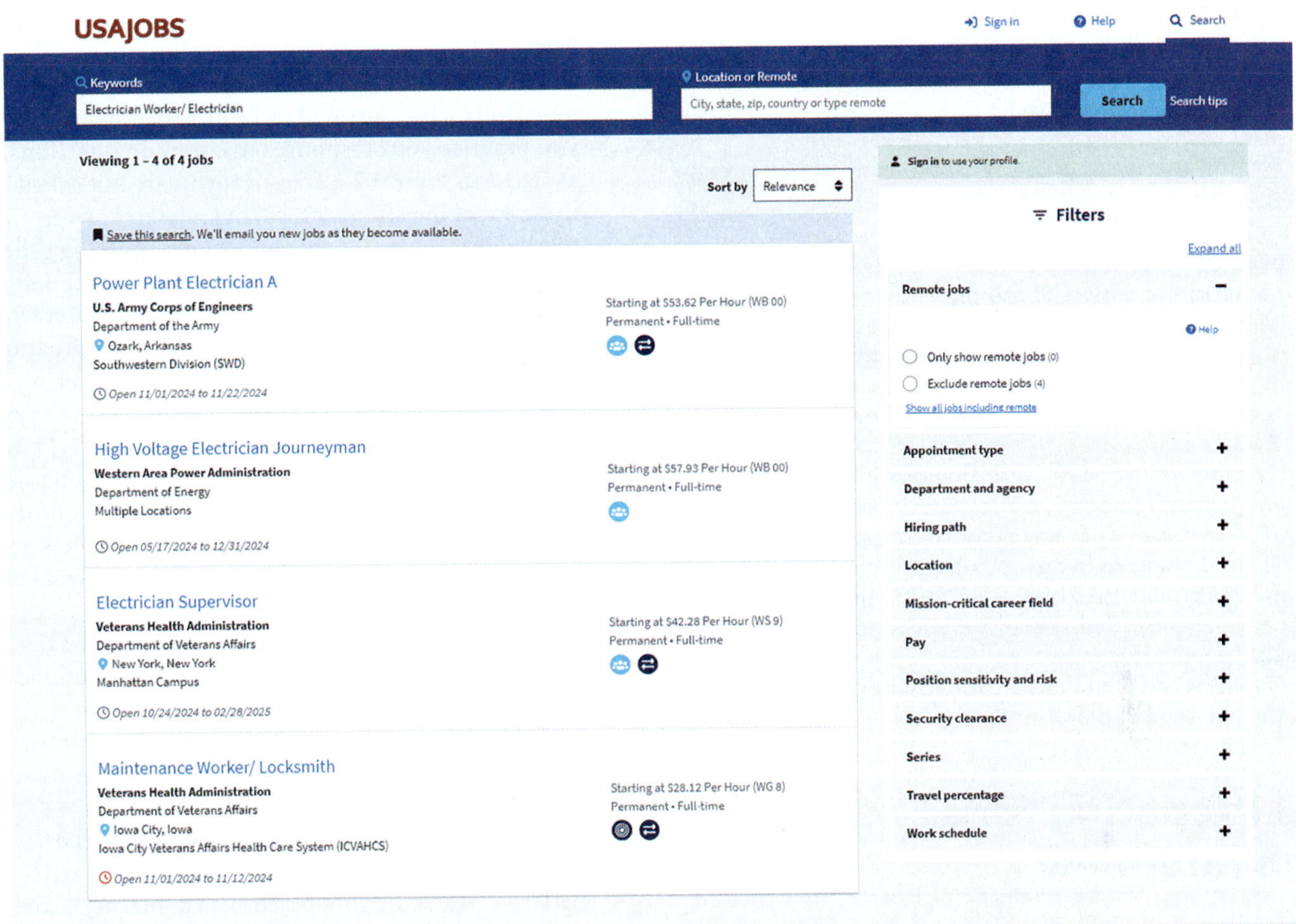

Source: USAJOBS

Figure 2-11. When researching job openings in the electrical field, online government agency sites, like the one shown here, are a valuable resource to assist you in finding and applying for jobs.

for job openings. A number of online job-search engines and government agencies are available to assist you in finding and applying for jobs, **Figure 2-11**. Another useful resource to find job opportunities is through networking and talking to friends and family.

As you read through job postings, be sure to look for those that align with the type of employment you are looking for. If you do not like to travel, for example, avoid listings that mention travel will be required.

Shedding Light

Your school's career center may be another source for identifying available job opportunities. In addition, your school or local industry may host career fairs. Career fairs provide an opportunity to speak with representatives from companies. These conversations can often lead to a more formal interview and employment opportunities.

Once you find a job offering that appeals to you, the next step is to fill out a job application. Depending on the employer and the nature of the position, this can either be completed in person or online. When you complete a job application, you should have your résumé and portfolio (if you have one) prepared as these may be required at the time you submit your application. Often, these documents are used to prescreen applicants for the skills needed on the job.

When completing a job application in person, be sure to fill out the application as accurately, completely, and as neatly as possible. Your application is your first opportunity to make a great impression on the employer. This document may be evaluated not only for your qualifications, but also for your handwriting and your ability to form grammatically correct sentences.

Many employers now require potential hires to complete an online application, either through the employer's website or through an independent job-search website. When filling out an online application, include key terms listed in the job description within your cover sheet and/or résumé. This will help your application stand out from the many other applicants.

Once you have completed filling out the application, be sure to follow all directions regarding how to submit it. In addition, make certain that you have supplied all of the requested paperwork and supporting documentation. Most employers will send a confirmation email when your online application has been received.

Shedding Light

When applying for a job, you will be one of many applying for the same position. The key is to stand out among the crowd. Remember to complete all the fields of the application even if the question does not apply to you. If a response to a question is left blank, it might appear that the question was overlooked. Instead, indicate "Not Applicable," "Does Not Apply," or "N/A."

If asked why you left your previous job, avoid saying anything negative about your previous employer. It is okay to say something along the lines of, "I am looking for a new position that will provide more of a challenge," but never say, "I got bored because all we did all day was sit around and play cards," even if it is true.

Finally, when filling out an application in person, use a pen with blue or black ink.

2.3 Preparing for a Job Interview

After a potential employer reviews your application and determines your qualifications meet the requirements of the position, you may receive a message to set up an interview. An interview gives you the opportunity to learn more about a company and to discuss your qualifications with the employer. The sole purpose of an interview is to find a perfect fit, or at least a very good fit, for both the company and the potential employee. A perfect fit is not always possible or necessary, but try to determine if you would be reasonably content and proud to do the type of work on offer. Not only is the company interviewing you, but you should be interviewing the company as well.

To help ensure a successful interview, adequate preparation is essential. The following are some ways to prepare for the interview:

- **Research the employer.** Understand the mission of the company and the specific requirements of the job. It is also beneficial to learn about the company in other ways, including its culture and any newsworthy events.
- **Prepare a list of questions you would like answered.** This is your chance to learn more about the company and your role within the company. You might want to ask how your job performance will be evaluated and how often. You might also want to know why this position became available. For example, is the company expanding and in need of more electrical installation technicians? Did the previous technician leave? Why did they leave?
- **Practice answering commonly asked interview questions.** Have a friend or family member help you practice your responses to common questions, as listed in **Figure 2-12**. Even if you are not asked these exact questions, preparing how to answer these questions will help you when answering other questions. The key is to sound confident, be honest, and focus on positives when possible. Focus on what you learned regarding difficult or negative situations. You might consider recording yourself answering these questions or watch in a mirror to observe your body language.
- **Decide what to wear.** Before the interview, whether it is in person or virtual, think carefully about the impression you want to present to the interviewer. Dress appropriately, usually one step above what is worn by your future coworkers. For instance, wearing business casual clothing (for example, pressed dress pants, button down shirt) is acceptable for individuals who will do manual labor or wear a company uniform. If the job involves greeting the public or working in an office environment, a suit may be more appropriate.
- **Know where to go for the interview.** Verify the address of the interview location and plan to arrive between 10 and 15 minutes early. An internet search will tell you how far you will have to travel and how long it will take to get there. If you find yourself at the interview location more than 15 minutes early, wait for a bit. Arriving too early can be interpreted as a deficiency in time management skills. Five or ten minutes ahead is a good time to show up.
- **Have copies of your job search documents.** Although you already submitted your résumé, you should bring a printed copy with you. If you are having a virtual interview, it is also a good idea to have your résumé and list of questions nearby. To share samples of your work, prepare an electronic portfolio or personal website.
- **Have your driver's license or other identification with you.** During the interview process, you may be asked for your driver's license. Many jobs in the electrical industry require that you drive from one job location to another. Before offering you a job, your driver's license information will likely be sent to a company's insurance carrier to determine if you are a safe driver with no disqualifying violations.

Interviewing

Interviewing is an essential part of the hiring process. This step involves either a face-to-face meeting or a virtual

Common Interview Questions

Question	Response
What can you tell me about yourself?	Briefly summarize your abilities as they relate to the job qualifications or your career goals. Do not provide a general life history.
Why do you want to work for this company?	Tell what you know about the company. Explain how your abilities match the company's needs.
Why do you think you would like this kind of work?	Relate the job requirements to your successful past experiences.
What are some of the projects you worked on in school?	Briefly summarize a project or coursework relevant to the job qualifications.
What other jobs have you had?	Focus on jobs with skills that relate to the job you are seeking.
Why did you leave your last job?	Be honest. However, avoid saying anything negative about your previous employer.
Have you ever been fired from a job? If so, why?	Answer honestly. If you have been fired, share what you learned from the experience. Avoid trying to blame others.
What are your major strengths and weaknesses?	Select a strength that relates to the job qualifications. Be honest when selecting a weakness, but give an example of how you've worked to improve on it.
Have you ever had a conflict with a coworker? How did you handle it?	Briefly describe the situation and how you handled it. Avoid placing all the blame on the other person. Explain what you learned from the experience.
What do you expect to be paid?	If possible, determine the salary range before the interview. Say that you are willing to discuss the salary or state a range you feel comfortable with.
What are your future plans?	Describe how the need to learn and grow is important to you. Confine your answer to the company with which you are interviewing.
How did you hear about this position?	Avoid one-word answers here. This might sound like a simple question, but it opens the door for you to show how excited you are about this position. If you found the listing on an internet site, be sure to mention what made the listing appeal to you. If you found out about the job from a friend who works at the company, mention the person's name and why you are interested in the position.
Why should we hire you?	This is where you get to sell yourself. Tell why you can do the work, how well you can do the work, and how you fit into the company's culture.
Give an example of a mistake you made and how you handled it.	This type of question is always asking you to explain how you learned from a situation that didn't go according to plan. Never answer this question by saying that you never made a mistake.
I see that there is a gap in your employment history. Why is that?	Be prepared to discuss any gaps. Whether the gap was to go back to school, or to care for your children, be upfront. Even if you were responding to a personal crisis, always speak about the positives that came from it, such as providing a unique learning experience.
What did you like most about your previous job?	When answering questions like these, be sure to use positive terms and phrases that the employer wants to hear. Include things such as, "all of the employees treated each other like family," "we all worked together to get the job done," or "there was mutual respect among all of the workers."
What did you like least about your previous job?	Be careful with questions like this! Answer this question in a positive way, such as, "The position that I held did not pose as much of a challenge as I originally thought it would."

Goodheart-Willcox Publisher

Figure 2-12. Prior to an interview, be prepared to answer these common questions.

online meeting between the employee and the employer. Either way, making a strong, positive impression is key. An interview can be a stressful experience, but with the right preparation, you will feel confident. **Figure 2-13** lists several virtual interview tips that will assist you when preparing for a remote interview.

Virtual Interview Tips

✔ **Test your equipment before the interview.** Check the audio and video settings on your computer to avoid technical issues during the interview. Ensure you have a strong, high-speed internet connection to avoid internet lag or a dropped connection.

✔ **Check your surrounding space.** Your space should look professional. Be sure you have a clean, uncluttered background. Use good lighting so the interviewer can clearly see you.

✔ **Do not sit too close to or too far from the screen.** Your head, shoulders, and upper chest should be visible. The proper distance will allow you to maintain good eye contact.

✔ **Limit distractions.** Go to a quiet room, silence your cell phone, and turn off the TV. Make sure others, including pets, do not interrupt your interview.

✔ **Do not read from a script.** Maintain eye contact and avoid looking down at your notes, which can be tempting during a virtual interview.

Goodheart-Willcox Publisher

Figure 2-13. In addition to dressing professionally during a virtual interview, it is also important to follow these tips to prepare your computer and surroundings prior to the interview.

Shedding Light

Making eye contact is important during an interview. However, this can be tricky when you are doing a virtual interview. A simple trick is to look directly at the camera on your computer. This gives the appearance that you are making eye contact with the interviewer. If you look directly at the person on your computer screen, you may give the impression that you are looking downward and are not fully engaged.

If you are asked to meet in person, it is recommended to turn off your cell phone or, better yet, leave it in your car. Make certain that you have a pen with you. Not having a pen with you may appear to the interviewer as an indication you might not have the best foresight or organizational skills.

The first person you will meet during the interview process may be the receptionist. Keep in mind that the interviewer might ask the receptionist how you conducted yourself upon arrival. Therefore, make sure you greet the receptionist with a smile, introduce yourself, make eye contact, and engage yourself in a professional manner when making small talk as well as waiting for your interview to begin.

When called into the office for your interview, be polite, friendly, and cheerful. Be sure to greet your interviewer appropriately and make eye contact. Remain standing until the interviewer invites you to take a seat. When seated, maintain good posture, avoid fidgeting, and keep your eyes in the direction of the interviewer. This, of course, does not mean you should be staring at the interviewer, but do not let your eyes wander around the room, as this is an indication of disinterest, nervousness, and discomfort.

Answer all questions carefully and as completely as you can. Be honest about your abilities as well as your limitations. There is nothing wrong with saying that you do not know the answer to the interviewer's questions. Lying during an interview can result in not being considered for the position. When asked a question, take a few seconds to formulate a well thought out response. Avoid using filler words such as "um," "you know," and "uh" as part of your response. It is far better to remain silent while you gather your thoughts.

Shedding Light

When an employer asks if you have questions, saying "no" can indicate a lack of interest or preparation. It is often a good idea to ask a question at the end of your response to a previous question. This gives the interview a conversational, interactive tone. For example, suppose the interviewer asked you about your ability to work overtime on holidays. After you answered that you are willing to be flexible and work on holidays when required, you might want to add a question about what are the paid holidays that the company observes. Talking about some "employee-centered" concerns often requires considerable diplomacy. You should not appear too eager to learn about what the company can do for you. Instead, you should stress how you are able to help the company fulfill its mission and reach its goals.

During the interview you may be asked to take pre-employment tests. Some employers administer tests to job candidates to measure their knowledge or skill level in industry-related or other areas. Since most employers support a drug-free workplace, you will likely be required to take a drug test if hired. You might receive a conditional offer of employment pending the results of the drug test.

At the conclusion of the interview, be sure to thank the interviewer for their time and ask when you can expect to hear from them with their hiring decision. Within 24 hours after the interview, contact the employer via email to once again thank them for their time. This is also an opportunity to ask any follow-up questions.

If you receive a job offer, respond to it quickly. Two or three days is considered reasonable in most places, along with asking for more time, if needed, to consider various factors. If you do not receive an offer after several interviews, evaluate your interview techniques and look for ways to improve them. You may want to reach out to the interviewer and ask for feedback.

Evaluating Job Offers

When considering a job offer or comparing two or more positions, you should explore the following work factors:

- **Physical location.** Where will your home base be located? How long will your commute be?
- **Work schedule.** Will the work days and work hours fit in with your lifestyle? Is occasional overtime work required?
- **Income and benefits.** Is the proposed salary fair? Will you be given the use of a company vehicle or service truck? What is the paid time off (vacation) policy? Are there medical and life insurance benefits? Is there a retirement plan? Does the employer offer tuition reimbursement for college courses or additional training programs related to your job?
- **Job obligations.** Will you be expected to join a union or other professional organization? If so, what are the costs? Will you be expected to attend meetings before or after work?
- **Advancement potential.** Is there any opportunity for advancement? Before you can advance, are there special expectations, such as a higher degree? Are training programs provided?

When comparing two or more positions, make a list of the pros and cons for each position, then weigh these against each other to determine which job opportunity is best for you. For example, the benefits offered by one company may outweigh a lower annual salary.

2.4 Succeeding in the Workplace

After securing employment, adjusting to your new duties and responsibilities will occupy your first few weeks. Your supervisor and coworkers will help you learn the routine. An introduction to company policies and procedures as well as the special safety rules that all employees must know is common for new employees.

While your coworkers will be watching what you do, they will also pay attention to how you work. How to behave in the workplace is an important lesson all employees should learn. Making an effort to do your best will help you succeed.

Workplace Ethics

Employers want employees who are punctual, dependable, and responsible. They want their employees to be capable of taking initiative and working independently. Other desirable employee qualities include organization, accuracy, and efficiency.

A punctual employee is always prompt and on time. This means not only when the workday starts, but also when returning from breaks and lunches. Being dependable means that people can rely on you to keep your word, fulfill your promises, and meet your deadlines. If you are not well, be sure to call in and let the employer know right away. If there are any reasons you cannot be at work, discuss this with your employer and work out an alternate arrangement. Many people have lost their jobs by not checking with their supervisor about time off.

Taking initiative means that you start activities on your own without being told. When you finish one task, you do not wait to hear what to do next—you do the next thing or ask for direction. Individuals who take initiative need much less supervision. They have ***self-motivation***, or an inner urge to perform well. Generally, this motivation will drive you to set goals and accomplish them. All these qualities together show that you are capable of working independently.

You are expected to be as accurate and error-free as possible in all that you do. This is why you were hired. Complete your work with precision and double-check it to ensure accuracy. Your coworkers depend on the careful completion of your tasks.

Professional Etiquette

No matter what career you choose, you will be expected to behave professionally on the job. Professional behavior makes an organization run more smoothly. It can also help you advance in your career. Keep the following tips in mind for behaving professionally:

- Show respect for your supervisor and coworkers.
- Be worthy of being respected by others.
- Arrive on time, every time.
- Practice safe work practices and encourage others to do the same.
- Avoid bringing personal problems to work.
- Be courteous, honest, patient, and friendly.
- Convey confidence without being arrogant.

Critical Thinking Skills

Employers value workers who have the ability to make good decisions and solve problems. Sometimes, you will have to make a decision on your own but, in many cases, you will be part of a team that needs to make a decision. When faced with a problem, the following six-step procedure can be useful:

1. Identify the problem or issue.
2. Brainstorm and make a list of all possible solutions.
3. List the pros and cons of each possible solution.
4. Decide which solution to implement.
5. Implement the solution.
6. Evaluate the results.

Teamwork

Although many jobs in the electrical industry involve working alone, there will be times when you will need to work as part of a team. A big advantage of a team is the ability to develop plans and complete work faster than individuals working alone. Creative ideas often develop from building on another person's idea. Try to understand the ideas of others before trying to get others to understand yours.

As a member of the workforce, you will encounter supervisors, customers, and coworkers from diverse backgrounds and cultures. ***Diversity*** is a global term that includes differences among people and can specifically refer to gender, sexuality, religion, race, nationality, and many other distinctions. Being a good team player involves accepting these differences, which includes using bias-free language in all your communications, whether verbal or written, to show respect for those with whom you come in contact. In many cases, discrimination on any of the grounds listed above is illegal and federal and state laws protect workers from it.

Communication

Communicating effectively with others is important for job success. Being a good communicator means that you can share information effectively with others. It also means that you are a good listener.

Good communication is central to the smooth operation of any business. Communication is the process of exchanging ideas, thoughts, or information. Poor communication is costly to an employer, as when time is lost because an order was entered incorrectly. Poor communication can result in lost customers, too.

The primary forms of communication are verbal and nonverbal. ***Verbal communication*** involves speaking, listening, and writing. ***Nonverbal communication*** involves body language, which includes the expression on your face and your body posture. Your body language should support what you say. If your verbal and nonverbal communications are not in agreement with each other, you will be sending out mixed messages. For example, telling someone you are interested in what they are saying as you are slouched back in your chair and yawning sends a bad sign.

Good communication skills are an absolute necessity when working with customers. The most important aspect of good customer relations are to always remain courteous and empathetic, while being a good listener. When listening, be sure to ask questions if needed and give signs, possibly nonverbal, to let them know you understand them and are interested in what they have to say. Leaning forward while a person is talking signals interest and keen listening.

Leadership Skills

Leaders are valuable employees. A good leader is able to communicate well with others, accept responsibility, and make decisions with confidence. They provide the motivation to keep everyone working together. Good leaders lead by example, are the first to respond, and always do their fair share of the work. As a result, showing leadership potential could provide opportunities for promotion.

Safety Mindset

Safety on the job is everyone's responsibility. Many workplace incidents resulting in injury occur because of careless behavior. Quite often, poor attitudes are the cause of unsafe behavior. Practicing good safety habits is essential for preventing injuries on the job. The government agency that promotes safety in the workplace is the Occupational Safety and Health Administration (OSHA). You will be required to follow the specific OSHA and/or state-mandated regulations that apply to your workplace. Safety practices and protocols are discussed in Chapter 3, *Workplace Safety*.

Positive Attitude

Your attitude often determines your success in your job. Your attitude is reflected by how you react to the events and people around you. A smile and courteous behavior can make customers and fellow employees feel good about themselves and that reflects on you. Customers prefer to do business in friendly environments.

Enthusiasm spreads easily from one person to another. Usually, enthusiasm means a person enjoys what they are doing. Whether on a job site or in an office, enthusiasm builds a team spirit for working together.

Time Management

A good employee knows how to manage time wisely. This includes the ability to prioritize assignments and complete them in a timely fashion. It also involves not wasting time. Time-wasting behaviors include visiting with coworkers, making personal phone calls, texting, sending personal emails, or doing other non-work-related activities during work hours. Time management also involves getting to work on time, every time, as well as returning from lunch or breaks in a timely manner. Showing respect for the clock also shows that you respect the company you work for.

Digital Citizenship

Being a good citizen means behaving ethically, honestly, and responsibly. The same applies for your behavior online. Digital citizenship is also applicable to the workplace. In most jobs today, computers, cell phones, and other electronic devices are essential tools. Employers often provide one or more of these devices to their employees to facilitate communications as well as performing work-related tasks such as research, completing job or work orders, or submitting time sheets. It is often a violation of company policy to use these devices for unauthorized, personal use such as playing games, shopping, social media posting, and texting.

It is also unethical to access confidential information, download copyrighted material, or harass others while using company equipment. Unapproved device use may jeopardize the integrity of the network or may divulge proprietary company data. Many organizations monitor user activity to make certain that all activity is ethical and legal. Users may also be required to sign an agreement that all devices assigned to an employee will only be used for specific work-related purposes.

Good Health and Hygiene

As an employee, you are a representative of your company. In many cases, you will be the only person from the company that your customer will meet face-to-face. Presenting yourself in a healthy, well-groomed manner tells your customers and others around you that you are professional and take your job seriously.

Good daily grooming habits include bathing or showering, shaving, using an antiperspirant or deodorant, and putting on freshly pressed clean clothes. Regularly brushing your teeth and flossing will promote healthy teeth and fresh breath. Keep your hair clean and styled in a way that will not be distracting.

Employers expect their workers to dress appropriately. Most workplaces have implemented dress codes, and some require that service personnel wear a company uniform. If the company does not have a dress code in place, use common sense concerning your dress and appearance, making certain to avoid extremes. Some employers have rules requiring that tattoos remain covered or that piercings beyond pierced ears remain unused. As each company has its own set of rules, be sure to review the company's employee handbook.

Be a Lifelong Learner

Companies look for employees who are eager to improve their skills, seek additional education and training, and keep up-to-date in their field of expertise. Continually updating your knowledge and skills is known as ***lifelong learning***. You cannot assume that the skills you have now will be all you will ever need during your career. Obtaining industry certifications or credentials is a good way to pursue lifelong learning and advance your career.

Employers sometimes provide in-house training or incentives to pursue additional training. In fact, some companies offer tuition reimbursement plans. In many cases, employees have to attend classes on their own time. People who enjoy their work will view lifelong learning as an exciting challenge.

Those who seek additional training often become more valuable to their employers and are more likely to retain their position with the organization in the event layoffs or downsizing become necessary. One way to show that you are a lifelong learner is to join industry organizations. Many states require electricians to participate in continuing education specific to the electrical construction trade. Oftentimes a lifelong learner will exceed the minimum requirements. Some will be motivated to study for and likely pass a master's electrical exam.

Join Professional Organizations

Belonging to a student or professional organization can help you reach your career goals. Your membership in an organization shows a current employer and potential employers that you are actively involved in the electrical industry and are serious about your career. Becoming involved in an industry-related or other organization can also help you develop your teamwork and leadership skills. You will learn how to work well in a group as you plan events, create projects, and accomplish goals together. In addition, taking a role as an officer or a committee chair in the organization will give you even more practice.

Organizations that may help you prepare for a career in the electrical industry include:

- **SkillsUSA.** SkillsUSA is an organization for students preparing for technical, trade, and skilled-service occupations. Its goal is to create a strong American workforce. SkillsUSA programs include the Professional Development Program, which builds employability skills such as communication and teamwork. The Work Force Ready System offers assessments for career and technical education, including areas such as technical drafting, residential wiring, and plumbing.
- **Institute of Electrical and Electronics Engineers (IEEE).** The IEEE is the world's largest technical professional organization dedicated to advancing technology for the benefit of humanity. IEEE and its members inspire a global community through its highly cited publications, conferences, technology standards, and professional and educational activities. IEEE has over 400,000 members around the world and over 120,000 student members.
- **Electric Power Research Institute (EPRI).** The EPRI is an organization that conducts research for the benefit of the world's population. They focus on electricity generation, delivery, and use in collaboration with the electricity sector to make electric power safe, reliable, affordable, and environmentally responsible. The EPRI offers a number of electrical-based courses on a wide variety of topics.
- **Independent Alliance of the Electrical Industry (IAEI).** Formerly known as the International Association of Electrical Inspectors, the IAEI is involved in the training of not only electrical inspectors but also electricians in all of the 50 states and, to some degree, internationally. Local chapters provide for much of the required continuing education needed in the electrical industry. Local chapters are made up of not only electrical inspectors but also electricians involved in every aspect of the trade.

CHAPTER 2 REVIEW

Summary

- Choosing a career is one of the most important decisions you will make. Begin by identifying the things you enjoy doing. (2.1)
- As you prepare for your career, determine how much training and education is needed to achieve your career goals. (2.1)
- Job shadowing and interning are great ways to research various careers in the electrical industry. (2.1)
- Securing a good job involves researching potential positions, preparing a résumé and cover letter, and applying for the position. (2.2)
- References and a portfolio may be needed when applying for some positions. (2.2)
- Adequate preparation is essential to perform well during an interview. (2.3)
- Traits of good employees include being punctual, reliable, organized, responsible, and dependable. (2.4)
- Once you are in the workforce, companies look for employees who will be lifelong learners. (2.4)

Review Questions

Use the information in the chapter to answer the following questions.

Measuring Your Potential

1. An arrangement with an educational institution where a student is supervised while working under a master craftsman and with journey workers is known as a(n) _____. (2.1)
   A. internship
   B. apprenticeship
   C. job shadowing
   D. mentorship
2. *True or False?* Job shadowing involves working alongside someone with greater experience and knowledge who can guide you in your career. (2.1)
3. Of the following, which is *not* part of the process of applying for a job? (2.2)
   A. Preparing a résumé
   B. Obtaining references
   C. Writing a cover letter
   D. Evaluating a job offer
4. *True or False?* Your desired salary should be included on your résumé. (2.2)
5. The purpose of a résumé is to _____. (2.2)
   A. secure an interview
   B. replace the need of filling out an online application
   C. describe your personal characteristics and qualities
   D. provide a more complete visual representation of your accomplishments and skills
6. Of the following, which is *not* a way to prepare for an interview? (2.3)
   A. Prepare a list of questions you would like addressed during the interview.
   B. Practice answering commonly asked interview questions.
   C. Show up at least 5 minutes before the interview.
   D. Review the employer's website to learn more about their mission statement.
7. *True or False?* Although you submitted your résumé when you applied for the position online, you should bring a printed copy with you. (2.3)
8. *True or False?* Body language is considered a nonverbal form of communication. (2.4)
9. The first step to problem-solving is to _____. (2.4)
   A. make a list of all possible solutions
   B. clearly identify the problem
   C. determine who caused the problem
   D. express your thoughts about a solution
10. Obtaining an industry certification to advance your career is an example of _____. (2.4)
    A. being a lifelong learner
    B. taking an online course
    C. having leadership skills
    D. being a team player

Plug-In

1. Your future goal is to become an electrical repair technician. Of the following, which is *not* a step that you should follow to prepare for your career choice? (2.1)
   A. Attend a four-year college to earn a degree in electrical engineering.
   B. Explore how much training or experience is needed.
   C. Shadow someone who holds the type of job you desire.
   D. Speak to someone already in the electrical field who can guide you on your career path.

2. Of the following, which item is *not* recommended to include in a portfolio? (2.2)
   A. CPR certification from your lifeguard training course
   B. Complete lab reports from an electrical training course
   C. Competition award from SkillsUSA
   D. Letter of recommendation from your coach
3. After the interview, you should contact the employer via email within _____ to thank them for their time. (2.3)
   A. 1 day
   B. 2 days
   C. 7 days
   D. 14 days
4. During an interview you are asked why you left your previous job. Of the following, which is an appropriate response? (2.3)
   A. I was bored.
   B. I am looking for a new position that will provide more of a challenge.
   C. The boss was always upset that I was just a few minutes late every day.
   D. I felt like my supervisor did not trust me to do anything.
5. An example of sending mixed messages is _____. (2.4)
   A. leaning toward your colleague as you nod in agreement
   B. closing your eyes and yawning while you express interest
   C. slouching in your chair while you look at your watch
   D. tapping your foot on the floor as you look around
6. One way to advance your career is taking continuing education courses. Of the following, this organization provides required continuing education as needed in the electrical industry. (2.4)
   A. Independent Alliance of the Electrical Industry
   B. Institute of Electrical and Electronics Engineers
   C. Electric Power Research Institute
   D. SkillsUSA

Making the Connection

1. How would a company's customers be affected if the technicians consistently arrived at work late, took extended lunch breaks, and practiced other types of time stealing? (2.4)
2. Would the problem-solving and decision-making process for a new electrician be the same as that for a technician with 20 years of experience? Explain your answer. (2.4)

CHAPTER

3 Workplace Safety

Somnuek saelim/Shutterstock.com

SPARKING DISCUSSION

You arrive at a jobsite and, after setting up to perform your work for the day, you realize that you do not have your safety glasses with you. What do you do? Discuss what options are available to you and why you chose the option you did.

LEARNING OUTCOMES

After completing this chapter, you will be able to:

3.1 Describe the dangers associated with working on and around electric circuits and equipment.

3.2 List common circuit protection devices and explain how they help prevent electric shock.

3.3 Identify common pieces of personal protective equipment (PPE) that an electrician might wear on the job.

3.4 Describe the different classes of fires and the types of extinguishers used to fight them.

3.5 List best practices on how to be safe when working on a jobsite.

3.6 Explain the importance of ladder safety.

3.7 Identify safe practices when working alongside multiple tradespeople on a jobsite.

3.8 List safety measures to follow when working in confined spaces.

3.9 Summarize safety precautions to implement when working in open trenches and pits.

TECHNICAL TERMS

- ampacity
- arc blast
- arc fault
- arc-fault circuit interrupter (AFCI)
- arc flash
- arc flash boundary
- asphyxiation
- confined space
- electric shock
- ground-fault circuit interrupter (GFCI)
- limited approach boundary
- lockout/tagout (LOTO)
- Occupational Safety and Health Administration (OSHA)
- personal protective equipment (PPE)
- point of egress
- restricted approach boundary
- safety data sheet (SDS)
- trade stacking

Introduction

The goal of anyone working in the electrical field is to return home at the end of the workday injury-free. Physical injuries can result from poor lifting techniques, improper tool use, falling objects, tripping, improper equipment use, or, most commonly, carelessness. In addition, working conditions such as exposure to open electrical panels, circuits, rotating machinery, power lines, open trenches, confined spaces, and elevated working platforms can cause severe injury, or even death, if an incident occurs.

When you are working on a jobsite, it is important to be professional at all times, be aware of your surroundings, and observe all safety precautions. All workers should use common sense and be prepared to react to an emergency, should one arise. Working in a safe manner will help ensure that you, as well as those around you, stay safe.

Shedding Light

The ***Occupational Safety and Health Administration (OSHA)*** was formed by Congress in 1970 as part of the Occupational Safety and Health Act. The primary goal of OSHA is to help ensure safe working conditions by setting safety-related standards for all workers and the workplace as a whole. By providing training, OSHA is educating the workforce to reduce the number of workplace incidents.

Many employers, labor unions, and some states require employees to complete an OSHA safety-training course and carry a card to verify course completion.

3.1 Electrical Hazards

Although preferable, it is not always possible to work on electrical circuits, panels, and other pieces of equipment with the power off. Therefore, electricians must learn how to remain safe while working near or on "hot" or energized circuits. In most cases, as long as current flow is contained in conductors and devices as intended and the worker follows safe working practices, the potential for injury is low; however, when uncontrolled current flow occurs, a worker is at a greater risk to sustain an injury.

Know the Code

NFPA 70E, *Standard for Electrical Safety in the Workplace*, requires that, in most situations, electrical conductors and circuit parts be placed in an "electrically safe work condition" before an employee performs work. The requirements for establishing an electrically safe work condition generally include de-energization, lockout/tagout procedures, and verification of de-energization as part of the requirements.

Electrical Burns

Electrical burns can be very dangerous and result when electric current flows through the body. The severity of the burn depends, in part, on the amount of current flow and how deep into the body the current flows. Heat is a by-product of electric current. As the amount of current flow increases, the amount of heat produced increases as well. Electrical burns are extremely painful, take a long time to heal, and can affect the skin, muscle tissue, and even bones.

Electric Shock

One of the major goals when working in the electrical trades is to avoid receiving an electric shock. An individual experiences an ***electric shock*** when they become part of an energized electric circuit.

The amount of current flowing through the body is determined by the amount of voltage the worker is exposed to and the resistance of the body. The resistance of the body can change, depending on temperature and level of perspiration. The higher the voltage and the lower the resistance, the more current will flow.

When the skin is dry, the resistance of the body is higher, reducing the effects of an electric shock. When the skin is moist or wet, body resistance is lower. Thus, the effects of an electric shock are greatly increased. Depending on where the resistance is measured on the body, the resistance of a wet body can be up to 1000 times less than the resistance of a dry body.

Even very small amounts of current flow through the body can cause a great deal of discomfort. As shown in **Figure 3-1**, the threshold for sensing an electric shock can be as low as 1 milliamp, which is one one-thousandth (0.001) of an amp. Painful shocks can be felt at currents as low as 3 mA (0.003 A), while loss of muscle control can result from currents as low as 10 mA (0.01 A).

Safety Alert

Not only can an electric shock cause direct injury, but a shock can also cause other injuries indirectly but suddenly. For example, eye injuries can result if the worker is holding a pointed object, such as a screwdriver, when an electric shock is received. If the object is pointing upwards when the arm pulls back, a severe eye injury can occur. To reduce the chance of these injuries, keep all potential eye-damaging tools and instruments pointing downward.

Severity of Electric Shock

The severity of an electric shock is determined by the amount of current flowing through the body as well as the path the current takes and the time duration of the shock.

Effects of Electric Shock

Level of Current (milliamps)	Effects on Human Body
1–3	Sensation threshold, faint tingling
3–10	Painful shock, muscle contractions
10–40	Muscle contractions, loss of muscle control
30–70	Extreme pain, respiratory paralysis, reduced breathing ability
50–150	Severe muscle contractions, loss of ability to let go from an energized circuit, possible death
100	Ventricular fibrillation (rapid, inadequate heartbeat)
200	Heart muscle clamps tight, cardiac arrest
1000	Tissues and organ burning, death likely

Goodheart-Willcox Publisher

Figure 3-1. Effects of electric current on the human body. The values in this chart are averages and will vary based on a number of factors including overall health, gender, current path, and the time duration of the shock.

Receiving a shock between two fingers on the same hand will produce a relatively mild shock when compared to one received between the left hand and right hand, **Figure 3-2**. Similarly, a shock resulting from current flow from the right hand to the right foot or from the left hand to the left foot will be less severe than one resulting from current flow from the right hand to the left foot, **Figure 3-3**. A shock that results from current flow through the chest, namely the heart, is more dangerous than a shock that flows only through one side of the body, not entering the chest. In addition, the longer the duration of an electric shock, the more severe the injury is likely to be.

Shedding Light

Electricity will flow through all available conductive paths. The lower the resistance, the greater potential for increased current flow. Damp surfaces such as concrete, damp earth, or metal grounded objects are more conducive to current flow than a dry wood floor or dry earth.

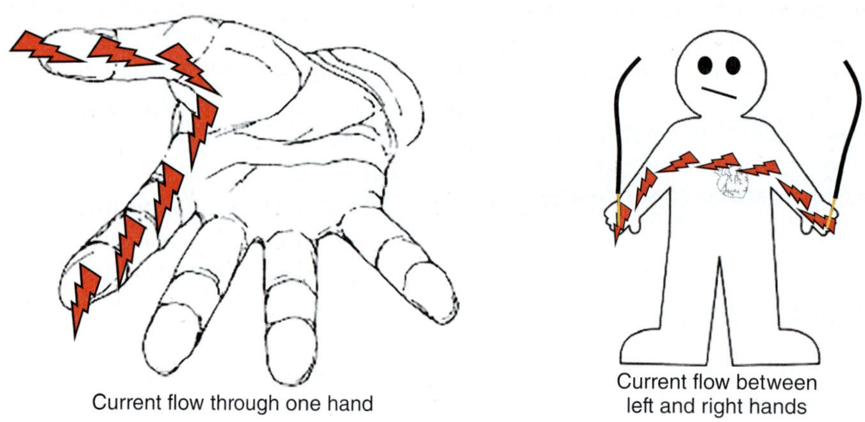

Goodheart-Willcox Publisher

Figure 3-2. Receiving an electric shock between two fingers of the same hand will produce a relatively mild shock compared to receiving an electric shock between the left and right hands.

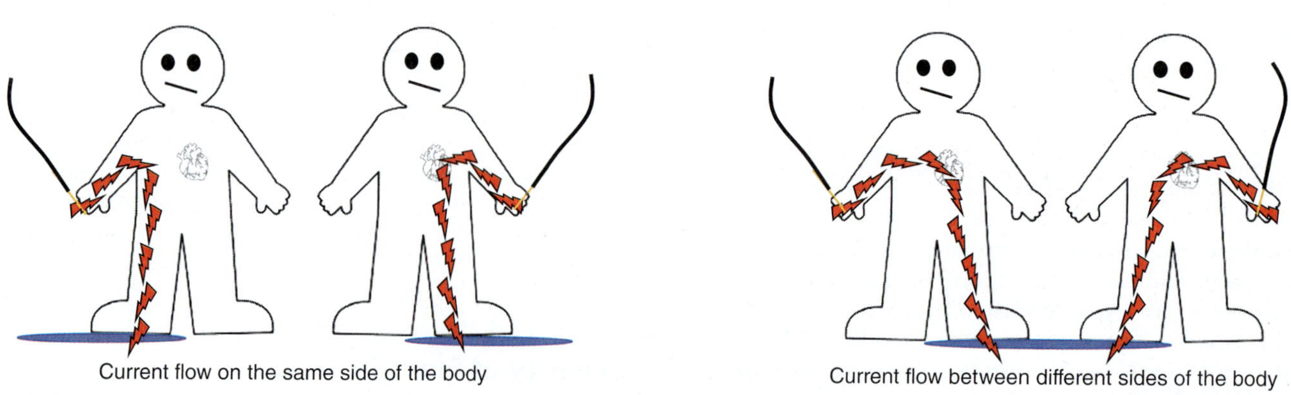

Goodheart-Willcox Publisher

Figure 3-3. Current flow between a hand and a foot on the same side of the body will be less severe than current flow between a hand and foot on different sides of the body.

Arc Faults, Flashes, and Blasts

An ***arc fault*** occurs when electric current flows in an uncontrolled path through air, from one conductor to another, or from one conductor to ground. Arc faults can be caused by loose electrical connections, improperly installed electrical components, conductor insulation deterioration, and corrosion. Arc faults are primarily a fire hazard, but can also cause personal injury. The most effective way to avoid the risk of injury resulting from an arc fault is to de-energize the equipment.

Arc Flash

An ***arc flash*** is an explosive and violent event that can cause severe injuries or death to anyone in its path. An arc flash creates extremely high temperatures that can far exceed the surface temperature of the sun. The temperatures produced during an arc flash can exceed 35,000°F, while the surface of the sun is about 10,000°F. At these temperatures, metals can easily melt and even vaporize. Although an arc flash typically lasts for a very short period of time (until the overcurrent protection device opens), a great deal of damage can be done.

Two factors that affect the severity of an arc flash injury are the following:

- The distance between the worker and the arc flash
- The duration of the arc flash

To help protect workers, the National Fire Protection Association (NFPA) through code NFPA 70E, has established specific boundaries that need to be observed when working on or near equipment, **Figure 3-4**. The boundaries differ based on the equipment and the potential dangers that are posed. These boundaries include the following:

- **Arc flash boundary.** The ***arc flash boundary*** is the boundary located the farthest from the energy source. Workers within this boundary can sustain burns resulting from the heat produced by an arc flash. No one should pass into this area without training and personal protective equipment (PPE).
- **Limited approach boundary.** The ***limited approach boundary*** is the outer distance at which a worker may be exposed to potential shock hazard. Nonqualified workers should remain outside of this boundary unless wearing PPE and escorted by a worker with specialized training. A qualified worker, on the other hand, may cross this boundary after shock and arc flash assessments are performed, with appropriate PPE as needed.
- **Restricted approach boundary.** The ***restricted approach boundary*** is the area closest to live equipment. This boundary represents the distance from a piece of equipment that exposes a worker to a

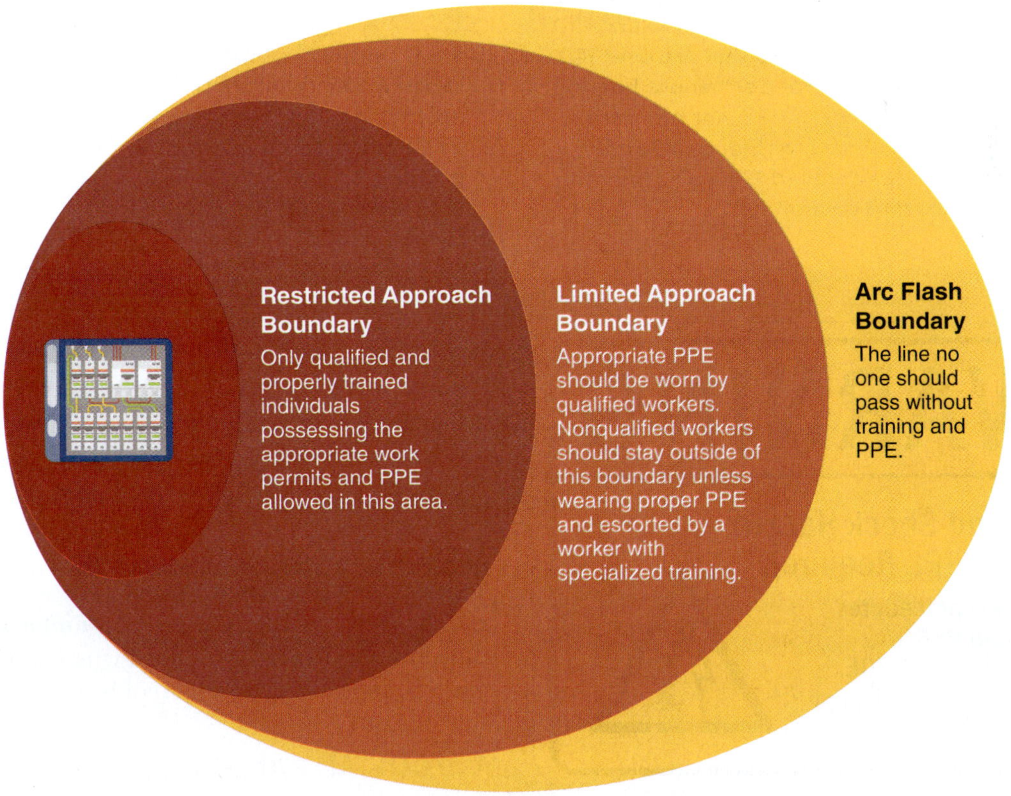

Goodheart-Willcox Publisher

Figure 3-4. For an arc flash, the electrical safety boundaries established by NFPA 70E include the arc flash boundary, limited approach boundary, and the restricted approach boundary.

high probability of electric shock or arc flash caused by inadvertent movement. Only qualified and properly trained individuals possessing the appropriate work permits and PPE can work within this boundary.

Safety Alert

Be aware of all equipment warning labels and comply with the requirements on them, **Figure 3-5**. Complying with these warnings includes wearing the appropriate pieces of PPE, as well as using insulated, voltage-rated tools, observing all distancing requirements, and receiving the proper training. Refer to *Section 110.16* of the *National Electrical Code*® *(NEC)*® for more information on arc-flash hazard warnings.

Arc Blast

An ***arc blast*** is the pressure wave created during an arc flash. An arc blast can produce pressures of over 2000 psi, which can easily throw a fully-grown person to the ground, across the room, or into other equipment. The pressure generated can propel molten or vaporized metals. This, in turn, can create additional burn and breathing dangers. Arc blasts are also very loud and can damage or rupture ear drums and impair brain function. **Figure 3-6** provides a comparison of arc flash and arc blast.

Safety Alert

An arc blast can produce noise levels that are over 140 decibels, which is the threshold for permanent hearing loss. In addition to other pieces of PPE, ear protection is required. NFPA 70E, *Standard for Electrical Safety in the Workplace*, provides detailed PPE requirements for various arc-flash hazard environments.

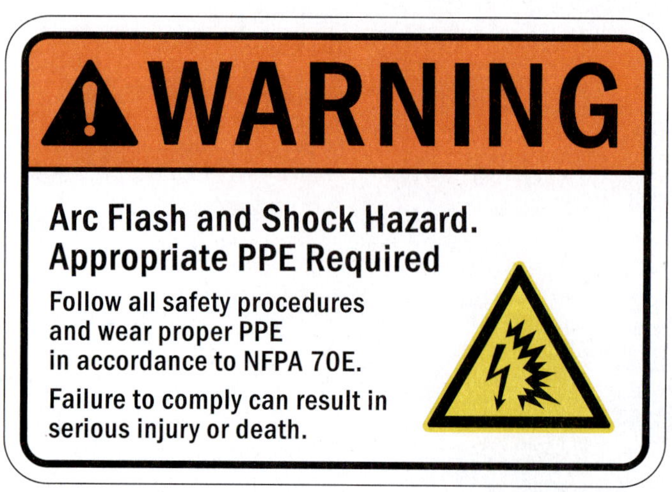

gobigoo/Shutterstock.com

Figure 3-5. To avoid a possibly serious workplace incident or injury, observe and follow any arc flash warning labels.

By-Products of Arc Flash

- Electric shocks and burns
- Creates light and heat
- Can melt/vaporize metals
- Can melt clothing onto skin
- Temperatures up to 35,000°F

By-Products of Arc Blast

- Pressure wave
- 2000+ psi pressure
- Can knock a person down
- Hearing and brain damage likely, if near
- Noise levels over 140 decibels

Goodheart-Willcox Publisher

Figure 3-6. By-products comparison of arc flashes and arc blasts.

3.2 Electrical System and Circuit Safety

Heat is generated when electric current flows through a circuit. As the amount of current increases, so does the amount of heat produced. When the current exceeds the conductor's ampacity rating, excessive heat can damage a circuit's interconnecting wires and even cause a fire.

To reduce the possibility of a fire, circuit protection devices are used to de-energize a circuit when an overcurrent condition occurs. Common circuit protection devices are fuses and circuit breakers. Other devices, such as ground-fault circuit interrupters and arc-fault circuit interrupters, are used to protect those who are working on, or are in close proximity to, the circuits.

Safety Alert

When working on live circuits, be sure to wear appropriate PPE and keep your arms and body clear of all electrical terminals and connections. Even when the power is off, always treat electrical circuits as though they are energized. Doing so will help ensure that you remain safe.

Fuses

Fuses are one-time circuit protectors that open, or "blow," when an overcurrent condition is present, **Figure 3-7**. Most fuses are constructed of a high-resistance metal strip with a low melting point that, when exposed to current flow, will heat up faster than the circuit wiring. As the circuit current increases, the heat produced increases until the metal strip melts, opening the circuit.

The size of a fuse is based on ***ampacity***, meaning the current-carrying capability of a wire or circuit. Blown fuses must be replaced as they cannot be reset.

Circuit Breakers

Similar to fuses, circuit breakers are devices that protect a circuit from an overcurrent condition, **Figure 3-8**. Unlike fuses, circuit breakers can be reset when they trip.

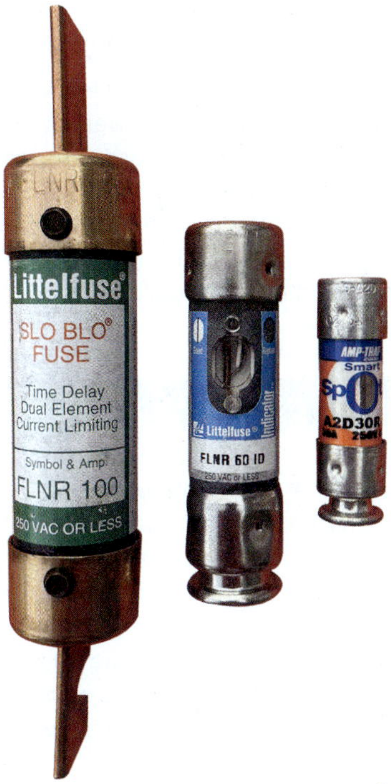

Courtesy Rob Zachariason

Figure 3-7. Typical fuses are one-time circuit protectors that blow when an overcurrent condition is present.

Courtesy Rob Zachariason

Figure 3-8. A circuit breaker that protects a circuit from an overcurrent condition can be easily reset when it trips.

The Toidi/Shutterstock.com

Figure 3-9. A GFCI receptacle provides protection to individuals from ground faults.

Most circuit breakers offer two ways to protect a circuit. One way uses a bimetal strip that bends when excessive heat is applied. A bimetal strip consists of material made of two dissimilar metals that expand and contract at different rates when exposed to heat. When the bimetal strip bends, the circuit breaker opens to de-energize the circuit. Circuit breakers also provide circuit protection with a magnetic coil that causes the circuit to open quickly when a short circuit is present.

Circuit breakers, like fuses, are sized based on the ampacity of the circuit.

Safety Alert

Never replace a fuse or circuit breaker with one that is rated at a higher ampacity than the current-carrying capability of the circuit wiring. Also, never remove (or jump out) fuses or circuit breakers from a circuit. If a fuse blows or a circuit breaker trips, be sure to identify the cause of the failure and correct the condition before restoring power to the circuit.

Ground-Fault Circuit Interrupters

A ***ground-fault circuit interrupter (GFCI)***, **Figure 3-9**, protects individuals from electric shock. Unlike a circuit breaker, which protects the electrical conductors in a circuit, a GFCI senses small current leaks to ground and de-energizes the circuit if any occur. By sensing the current flowing into a circuit from the power source as well as the current returning from the circuit to the power source, a GFCI can sense current leaks as low as 6 mA (0.006 A). Any small difference between the two indicates that there is a current leak to ground, and the GFCI de-energizes the circuit. This all happens in about 1/40 of a second.

Know the Code

Refer to *Section 210.8* in the *NEC* to learn about the applications that require GFCIs as well as guidelines and exceptions for their use. The list of the areas requiring GFCI protection in dwelling and non-dwelling units is extensive. The common characteristic for required protection areas is those that are not typically dry. If a receptacle is outside or near water, it will likely require protection by a GFCI receptacle to protect the end user or a GFCI circuit breaker to protect the entire electrical system. Some examples are bathrooms, kitchens, basements, outdoor receptacles, locker rooms, and garages.

Arc-Fault Circuit Interrupters

Arc faults can occur in a circuit due to damaged conductor insulation, loose wire connections and terminals, and damaged cords on electric devices. Arc faults should not be confused with electrical arcs, which occur all the time. For example, every time a wall switch is flipped to turn on a light, an electrical arc or spark is produced within the electrical circuit. This is normal. However, the sparking produced by an arc fault (uncontrolled flow through air) presents a fire hazard, along with the potential for personal injury.

An ***arc-fault circuit interrupter (AFCI)***, **Figure 3-10**, uses electronic circuitry to determine the difference between a safe electrical spark or arc and a dangerous arc fault. When an arc fault is detected, an AFCI provides protection by de-energizing or opening a circuit. To provide protection from both arc faults and ground faults, it is common practice to use an AFCI circuit breaker and a GFCI outlet or a combination AFCI/GFCI circuit breaker in areas where both types of protection are required.

Safety Alert

If a GFCI or AFCI has tripped, be sure to determine why it tripped and correct the situation before attempting to reset the device. More often than not, if a GFCI or an AFCI trips there is a problem that must be addressed. Rarely is the problem a result of a faulty protective device.

Courtesy Rob Zachariason

Figure 3-10. An AFCI circuit breaker provides protection from the effects of arc faults.

Know the Code

Refer to *Section 210.12* in the *NEC* to learn about the applications that require AFCIs as well as guidelines and exceptions for their use. The areas currently requiring this type of protection include most areas in a residential dwelling. Also included are dormitories, guest rooms, guest suites, and patient sleeping rooms in nursing homes and limited-care facilities.

Shedding Light

As with any device, in order for an AFCI to work properly, it must be installed correctly. When a standard circuit breaker is installed, only the ungrounded (hot) conductor passes through the breaker. Like a GFCI circuit breaker, when an AFCI circuit breaker is installed, both the ungrounded (hot) and grounded (neutral) conductors must be connected to the device. Always be sure to follow the manufacturer's installation instructions.

Equipment and Circuit Grounding

Another important way that workers and equipment users are protected from electrical injuries is equipment and circuit grounding. Grounding conductors, either bare copper wires, or conductors with green insulation or green insulation with a yellow stripe provide a low resistance, alternate current path to ground for the ungrounded or hot conductor to protect equipment and to help protect individuals from electric shock. More electric current will flow through a low-resistance path than through a higher resistance path. If an unsafe electrical condition exists, providing this low-resistance path for current can greatly reduce the severity of an electric shock. Ground wires should never be cut, disconnected, or otherwise removed from an electrical circuit.

Safety Alert

In general, all temporary receptacles installed on a jobsite must be GFCI-protected. This includes all receptacles that are not part of the permanent building or structure.

Lockout/Tagout Procedures

Whenever possible, equipment should be de-energized, the electrical power should be locked in the off position at the distribution panel, and the disconnect switch should be tagged. This process is referred to as ***lockout/tagout (LOTO)***. LOTO prevents the equipment from being energized by mistake and alerts others that work on the equipment is being performed.

SocoXbreed/Shutterstock.com

Figure 3-11. Most buildings have a LOTO station, which is a collection of uniquely keyed locks, multi-lock hasps, and tags for a number of different applications.

The disconnect switch should be locked out in an approved manner when installing, servicing, or working on equipment. Most buildings have a LOTO station, **Figure 3-11**. This station may consist of a collection of uniquely keyed locks, multi-lock hasps, and tags for a number of different applications.

OSHA provides specific details for ensuring safe working conditions, including electrical LOTO procedures. To prevent someone from turning the power on, the panel must be locked, and you should keep the key to the lock with you at all times. If multiple workers are locking out the power with a multi-lock hasp, each lock must be removed by the person that installed the lock. More elaborate LOTO stations include lockout devices for plugs, circuit breakers, and even valves.

Safety Data Sheets

A ***safety data sheet (SDS)*** is a uniformly prepared document that provides detailed information about a hazardous material. Some information provided in an SDS includes the following:

- Product name, manufacturer, and contact information
- Product hazards
- Physical/chemical properties and chemical composition of the product
- First aid, accidental release, and/or fire-fighting procedures
- Handling and storage requirements
- Exposure limits and toxicology information
- Stability and reactivity information

Whenever potentially hazardous materials and substances are used, an SDS should be readily available for reference when questions arise about the product's use or in the event of an incident.

3.3 Personal Protective Equipment

Remaining safe on the jobsite should be your number one concern. One easy method to stay safe is by wearing the required and recommended ***personal protective equipment (PPE)***. PPE is a general term used to describe the various pieces of equipment that workers wear to help prevent personal injury. Different grades of PPE are available and are selected based on the task. It is good practice to use high-quality products that meet or exceed the recommended limits to help ensure workplace safety. PPE includes, but is not limited to, the following:

- Hand protection, such as cut-resistant gloves and voltage-protective gloves, **Figure 3-12**
- Head protection, such as hard hats
- Eye protection, such as safety glasses and shields
- Ear protection, such as ear plugs and earmuffs, **Figure 3-13**
- Protective footwear, such as work boots
- Back protection, braces
- Safety harnesses
- Breathing masks and respirators
- Protective clothing, such as an arc flash suit or an arc flash kit

Safety Alert

Carefully inspect all pieces of PPE for wear or damage prior to use. Damaged or worn protective equipment may not provide adequate protection. Check the seals on breathing masks to ensure they are intact. Check that all buckles and other securing devices on braces and harnesses are fully operational. Make sure all PPE that is required to be tested, such as rubber gloves, are current and not past any expiration date.

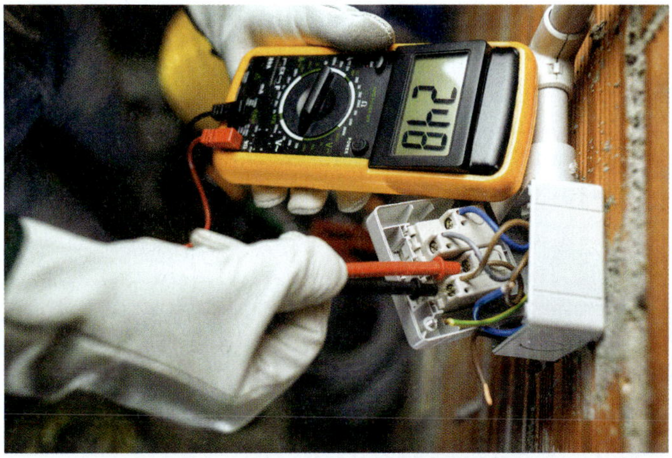

Francescomoufotografo/Shutterstock.com

Figure 3-12. Wearing protective gloves can protect hands and fingers from burns and electric shocks.

Loco/Shutterstock.com

Figure 3-13. Ear plugs and earmuffs are the most common forms of hearing protection.

3.4 Fire Safety

If not controlled quickly, fire can be a dangerous force that can cause serious personal injury and extensive property damage. In less than 30 seconds, a small controllable fire can grow and engulf an entire structure. It is best if a fire never starts, but accidents do happen. The key to preventing small fires from becoming uncontrollable is being prepared. Proper preparation involves two things: knowing the location of the nearest fire extinguisher and, just as important, knowing how to use the fire extinguisher properly.

The table in **Figure 3-14** describes the five classes of fire and the different types of fire extinguishers needed for each type of fire:

- **Class A.** Combustible common materials such as wood, paper, cardboard, cloth, or trash
- **Class B.** Flammable liquid or gas such as gasoline, propane, or oil
- **Class C.** Electrical equipment
- **Class D.** Metallic fire such as magnesium, sulfur, lithium, potassium, or titanium
- **Class K.** Kitchen grease, fat, or oil

When working on electrical current, it is important to have a basic understanding of the classification of fire extinguishers and how to use them. By choosing the incorrect type of fire extinguisher, you can actually make a fire worse.

Safety Alert

Never use water on an electrical (Class C) fire due to the high risk of electrical shock. Only use a Class C-rated fire extinguisher on electrical fires.

Proper Fire Extinguisher Use

When using a fire extinguisher, the goal is to extinguish a fire as quickly as possible. The fastest way to do this is to target the base of the fire and remember the acronym PASS (Pull, Aim, Squeeze, Sweep).

- **Pull** the pin on the extinguisher.
- **Aim** the fire extinguisher at the base of the fire.
- **Squeeze** the handle on the extinguisher.
- **Sweep** the extinguisher from left to right at the base until the fire is out.

Safety Alert

To help ensure that a fire extinguisher is ready to perform its function properly, it must be inspected on a regular basis. Fire extinguisher inspection and service records must be kept. These records include the date of service or inspection as well as the initials of the individual who performed the work.

3.5 Workplace Awareness

When working on a jobsite, always evaluate the workplace conditions and identify any potentially hazardous conditions or situations. Knowing that a potential hazard is present and being prepared to react if a mishap occurs are major steps to prevent and reduce the effects of an accident.

Safety Alert

Avoid standing in wet or damp areas whenever possible. Be sure to wear heavy shoes with an insulated sole and heel. Electricians think about the potential dangers around them and take all precautions.

Trips and falls are common workplace injuries, and can result from tripping over tools, debris, and other materials left on the floor. When carrying boxes, tools, and other items, it is not always possible to look directly downward. Keeping the work area clean and neat is one way to help keep all workers safer. Here are a few safe practices to follow:

- Clean up all liquid spills immediately to prevent slips and falls.
- Route cords and hoses so they are not in traffic paths.
- Continually remove waste and debris to minimize clutter.
- Keep materials and supplies well organized.
- Keep tools and materials off the floor.
- Never block access to fire extinguishers or electrical panels.
- Do not stack boxes or place equipment in front of exit doors.

In addition to promoting safety, workers appear more professional and are far more likely to complete jobs faster in a clean, well-organized work area.

Workplace safety is everyone's responsibility. Many workplace incidents occur because of careless behavior. Often poor

Fire Classifications

Class	Description	Requires	New Symbol	Old Symbol
A	Ordinary Combustibles (Materials such as wood, paper, and textiles)	Cooling/quenching		A
B	Flammable Liquids (Liquids such as grease, gasoline, oils, and paints)	Blanketing or smothering		B
C	Electrical Equipment (Wiring, computers, switches and any other energized electrical equipment)	A nonconducting agent		C
D	Combustible Metals (Flammable metals such as magnesium and lithium)	Blanketing or smothering		D
K	Kitchen Fires (Grease, fat, and oil fires in commercial kitchens)	Blanketing or smothering		K

Fire Extinguishers

Type	Description	Typically Approved for Use On	Not for Use On
Pressurized Water	Water under pressure	A	B C D K
Carbon Dioxide (CO_2)	Carbon dioxide (CO_2) gas under pressure	B C	A D K
Foam	Aqueous film-forming foam (AFFF) or film-forming fluoroprotein (FFFP)	A B	C D K
Dry Chemical, Multipurpose Type	Typically contains ammonium phosphate	A B C	D K
Dry Chemical, BC Type	May contain sodium bicarbonate or potassium bicarbonate	B C	A D K
Dry Powder	May contain sodium chloride, sodium carbonate, copper, or graphite	D	A B C K
Wet Chemical	May contain potassium acetate, potassium citrate, or potassium carbonate	K	B C D

Goodheart-Willcox Publisher

Figure 3-14. Before using a fire extinguisher, always check that the fire extinguisher is compatible with the type of fire.

attitudes can cause unsafe behavior, too. Common causes of incidents include the following:

- Taking chances
- Showing off
- Forgetting safety details
- Disobeying company rules
- Daydreaming
- Losing your temper
- Falling asleep

Shedding Light

Is it an accident or an incident? According to OSHA, the term "accident" has often been used to describe an unplanned, unwanted event. As a result, an accident was considered random and could not be prevented. However, since many workplace fatalities, injuries, and illnesses are preventable, OSHA uses the terminology "incident" instead.

Practicing good safety habits is essential for preventing fatalities and injuries on the job. A healthy worker is more alert and less likely to make accident-prone mistakes. Knowing how to use machines and tools properly is the responsibility of both the employer and employees. Wearing protective clothing and using safety equipment correctly helps keep workers safe. Your employer will emphasize the safety practices that employees must follow in your workplace.

3.6 Working at Heights

Ladders are used on a daily basis when installing, servicing, or maintaining electrical systems. Ladders should be made of non-conductive materials such as fiberglass. Aluminum ladders should be avoided as they conduct electricity.

Two common types of ladders are the A-frame ladder and the extension ladder. The A-frame ladder has four points of contact with the ground and is a freestanding ladder. The extension ladder, designed to lean up against a building, has two points of contact with the ground.

Before using any ladder, it should be thoroughly inspected for any damage. Under no circumstances should a damaged ladder be used. While inspecting a ladder, the following should be checked:

- Rails or rungs inspected for physical damage
- Rails or rungs inspected for grease or oil
- Foot pads and shoes examined for any damage
- Integrity noted for all hinges, pivot points, locking mechanisms, spreaders, and extension ropes

When using a ladder, always follow safety recommendations:

- Never exceed the weight limit of a ladder. This weight limit includes the weight of the worker as well as the tools or equipment being used or lifted.
- While climbing a ladder, remain centered on the ladder. Workers should not lean off to the side; doing so can cause the ladder to fall over.
- Always maintain three points of contact with the ladder. These three points can be either two feet and a hand, or two hands and a foot.
- Follow all guidelines and warnings printed on the ladder.

Safety Alert

Carrying a toolbox in your hand as you climb a ladder prevents you from adhering to the three-point rule. Since one hand is dedicated to holding the toolbox, it cannot be used to hold onto the ladder. As such, the other hand and both of your feet would need to be in constant contact with the ladder, which is not possible. If tools or other equipment must be brought to a roof or other elevated surface, hand ropes should be used to lift these items. In some instances, shoulder straps and backpack-type carriers can be used.

Proper Use of an A-Frame Ladder

In addition to general ladder safety guidelines, it is also important to follow guidelines specific to the use of A-frame ladders:

- Before using, make certain that the ladder is fully opened and that the spreaders are straight.
- All four of the ladder's feet must be resting securely on a solid surface.
- Never use an A-frame ladder by leaning the folded ladder against a wall.
- Never place an open ladder on boxes, bricks, or other objects that can shift.
- Never use the cross-bracing on the far side of stepladders for climbing. Always climb on the steps.

Proper Use of an Extension Ladder

Extension ladders are often used to access roofs or other elevated platforms. In addition to the general ladder safety guidelines, the following are precautions that apply specifically to extension ladders:

- Both feet of the extension ladder must be resting on a solid, level surface.
- The top of the extension ladder should extend at least 3′ above the roof or landing area, **Figure 3-15.**
- The bottom of the ladder should be positioned away from the building at a distance that is equal to one-quarter of the working length of the ladder. The working length is the distance from the base of the properly positioned ladder to the point where the ladder contacts the landing area.
- Once positioned, secure the ladder to the structure to prevent it from falling over.

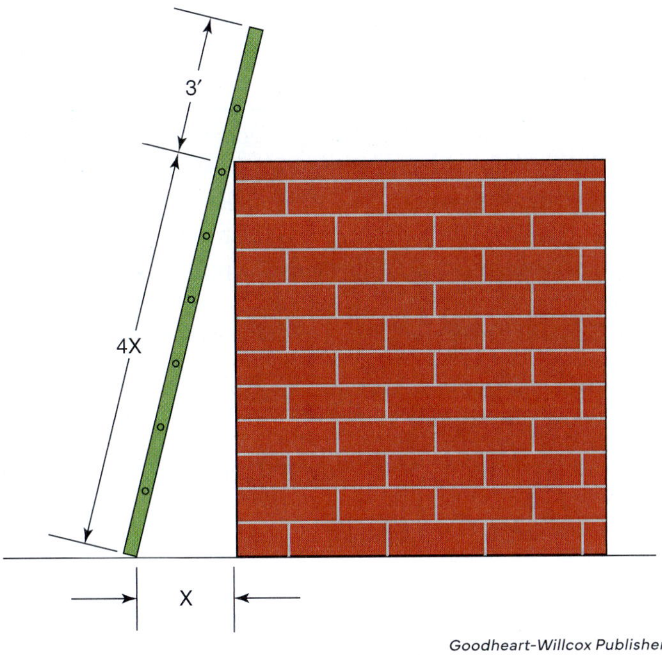

Figure 3-15. The proper positioning of an extension ladder. If 8′ of ladder (working length) is needed to reach the top of the wall, the ladder's base should be 2′ from the wall's bottom.

Scaffolding

There may be times when an electrician will need to work on scaffolding. When used properly, scaffolding provides an elevated platform on which workers can stand. Scaffolding must be assembled by individuals who have been properly trained and all manufacturer's assembly guidelines must be followed. Scaffolding must be installed on a solid, level surface and scaffolding parts and installation must be inspected by a qualified individual before use. Guidelines for workers who use scaffolding include the following:

- Workers must be properly trained on how to climb up onto and climb down from scaffolding, as well as how to conduct their behaviors when working on the scaffolding.
- Avoid mixing and matching of scaffolding parts from different manufacturers. The pieces might not fit together as intended.
- Before climbing scaffolding, ensure that the wheels, if there are any, are locked in place.
- Never exceed 25% of the rated weight limit. The weight limit of the scaffolding should provide a safety factor of at least four. For example, if the expected weight on the scaffolding is 600 lb, the scaffolding should be rated at a minimum of 2400 lb.
- The platform, or walking surface, should be completely decked. This means that the platform should be free of holes or spaces that may result in a worker falling or tripping.
- If the platform of the scaffolding is 10′ or more above ground or installation level, fall protection is required. Proper guardrails and/or personal fall protection devices may be required.

3.7 Working with Other Trades

Quite often, there will be multiple trades and multiple tradespeople working at the same jobsite at the same time. This practice is referred to as ***trade stacking***. Electricians, plumbers, HVAC contractors, carpenters, and masons may all be competing for space at the jobsite. As more workers are present, the potential for injury increases. In order to remain safe on the job, all tradespeople must work together to ensure the workspace is safe.

Safety Alert

If you notice that another tradesperson is not keeping the work area safe by cluttering walkways, blocking doorways, leaving debris behind, or continually leaving tools and materials on the floor, it is a good idea to first bring the issue to the attention of the person who seems to be directly responsible. This should be done in a matter-of-fact, but not confrontational, manner. If there is no agreement—and there must also be quick action taken—bring it to the attention of the project coordinator or manager. Never wait until an incident or injury occurs to say something because, at that time, it will have already been too late.

Workers must be aware of their surroundings, as they have no direct control over what other tradespeople are doing, and how neatly they are doing it. It is best to keep your distance from others on the job as much as reasonably possible. On larger projects, construction or project managers often conduct daily briefings with all workers at the start of the workday. During this meeting, a plan is established that minimizes the interaction of the workers in an effort to maintain the pace of the project while keeping the workers safe.

3.8 Working in Confined Spaces

While performing electrical installations or service, workers may find themselves in confined spaces. A ***confined space*** is an area with limited entry and egress points that is not suitable for human occupancy, but is large enough for a person

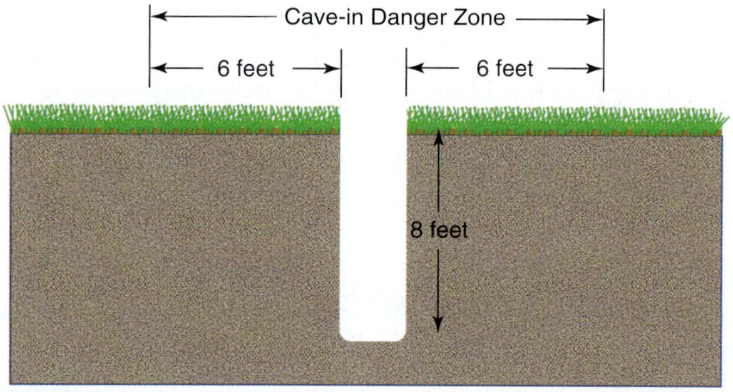

Goodheart-Willcox Publisher

Figure 3-16. If a trench is 8′ deep, the cave-in danger zone can occur as far as 6′ from the trench.

to enter. Examples of confined spaces include manholes, pits, tunnels, ductwork, and pipelines. A ***point of egress*** is simply an exit or outlet from a space or area.

When a confined space poses a serious safety or health hazard, an entry permit is required. The entry permit provides details about the space that include, but are not limited to, the following:

- The purpose for entry into the space.
- The workers who will be entering.
- The dates they will be entering.
- All safety measures (such as gas monitoring and mechanical ventilation) that are required based on the type of hazards that are likely expected.

Safety Alert

Working in confined spaces accounts for a great deal of workplace fatalities. Before entering a confined space, make certain that you are properly trained to do so, are aware of the potential dangers, and have complied with all of the required safety protocols.

3.9 Working in Open Trenches and Pits

Trenches and pits are commonly used on construction sites. Trenches are often created to provide for underground utility lines and pipes that carry water, electrical, and communication services into the structure and waste out of the structure. Pits are used to bury fuel storage or septic tanks, or serve as larger excavation areas for new structures or extension projects. Trench- and pit-related accidents often involve falling, crushing, asphyxiating, and drowning. In fact, falling into a trench is ranked as one of the top construction incidents as reported by OSHA.

Many trench and pit incidents and injuries occur as a result of a cave-in, which is when the walls of a trench collapse. A cave-in can occur as far away as 75% of the depth of the trench, **Figure 3-16.**

Severe personal injury or death can result from a cave-in, even if a worker is not completely buried in a trench. Dirt and soil are heavy and can put a great deal of pressure on the body, resulting in a loss of blood circulation. In addition, excessive pressure on the chest may cause ***asphyxiation***, which is when the body becomes deprived of oxygen.

Given the risk of cave-ins, it is good practice to keep the areas in the vicinity of the trench free from equipment, tools, and other materials that might fall into the trench in the event a cave-in does occur.

Safety Alert

Protective barriers, such as fencing, should be positioned around trenches and pits. These barriers are required for trenches deeper than 5′, but are strongly recommended for all trenches. For trenches deeper than 4′, an individual properly trained in soil analysis and trench protection should be on the jobsite. In addition, ladders must be positioned in the trench at distances no more than 25′ from each other to allow workers to climb safely out of the trenches.

CHAPTER 3 REVIEW

Summary

- Carelessness is a major cause of work-related injuries and incidents. (3.1)
- The Occupational Safety and Health Administration (OSHA) sets standards and provides education to help ensure a safe work environment. (3.1)
- Electric shocks occur when a person becomes part of an energized electric circuit. (3.1)
- The path the current takes, the amount of current, and the duration of the electric shock determine the severity of an electric shock. (3.1)
- Arc faults may be caused by loose electrical connections, corrosion, wire insulation deterioration, and improper component installation. (3.1)
- An arc flash is a massive explosion that creates a great deal of light and heat. (3.1)
- An arc blast is a pressure wave that produces up to 2000 psi of pressure. (3.1)
- Fuses and circuit breakers are typically used to protect the wiring and circuits, not those who work on or around them or the devices that are being controlled. (3.2)
- Ground-fault circuit interrupters (GFCIs) are used to protect people by detecting small current leaks to ground. (3.2)
- Arc-fault circuit interrupters (AFCIs) de-energize circuits if an arc fault is detected. (3.2)
- Work on electrical circuits and equipment with the power off whenever possible. (3.2)
- Work on energized circuits is only permitted under certain conditions per OSHA and NFPA 70E. (3.2)
- Lockout/tagout procedures ensure that equipment and circuits are not accidentally energized. (3.2)
- Personal protective equipment (PPE) is intended to help keep workers safe from a wide variety of workplace hazards. (3.3)
- Be sure to select the proper fire extinguisher for the fire being fought. (3.4)
- Remember to point, aim, squeeze, and sweep when fighting a small fire. (3.4)
- When at a jobsite, always evaluate the workplace conditions and identify any potentially hazardous conditions or situations. (3.5)
- Always follow all safety rules and use guidelines that are affixed to the ladder. (3.6)
- Always maintain three points of contact with a ladder. (3.6)
- Increased injury potential occurs when multiple tradespeople are working on the same job. (3.7)
- Proper training should be received before working in a confined space. (3.8)
- Many trench and pit accidents and injuries occur as a result of a cave-in, which is when the walls of a trench or pit collapse. (3.9)
- Protective barriers, such as fencing, should be positioned around trenches and pits. (3.9)

Review Questions

Use the information in the chapter to answer the following questions.

Measuring Your Potential

1. The primary goal of the Occupational Safety and Health Administration is to _____. (3.1)
   A. issue fines when safety rules are not met
   B. help ensure workers are constantly monitored
   C. educate the workforce about important safety issues
   D. set standards to make compliance more difficult
2. *True or False?* When electric current flows, heat is also produced. (3.1)
3. *True or False?* Arc faults can result from damaged conductor insulation. (3.1)
4. The threshold for sensing an electric shock is _____ mA. (3.1)
   A. 1–3
   B. 3–10
   C. 10–15
   D. 1000
5. You should remain outside a(n) _____ boundary, unless you are wearing proper PPE and escorted by a coworker that is properly trained. (3.1)
   A. arc flash
   B. arc blast
   C. limited approach
   D. restricted approach
6. The severity of an arc flash injury is affected most by the _____. (3.1)
   A. duration of the arc flash and the distance between the worker and the arc flash
   B. temperature of the arc flash and the distance between the worker and the arc flash
   C. pressure and temperature of the arc flash
   D. pressure of the arc flash and the distance between the worker and the arc flash
7. *True or False?* It is good field practice to treat all electrical circuits as though they are energized. (3.2)

8. *True or False?* The purpose of a multi-lock hasp is to allow more than one worker to lockout a single power supply. (3.2)
9. Of the following, which statement is *true* regarding PPE? (3.3)
   A. Select only high-quality products that meet or exceed safety standards.
   B. Supervisors should inspect all PPE for wear or damage monthly.
   C. No matter what the task, you should always use a breathing mask.
   D. You should rely on your supervisor to provide operational PPE.
10. A Class C fire is a fire that involves _____. (3.4)
    A. paper and wood
    B. electricity
    C. oil and grease
    D. gasoline and oil
11. Of the following, which may *not* result in a workplace incident? (3.5)
    A. Showing off
    B. Keeping tools organized
    C. Stacking boxes in front of a fire exit
    D. Asking a coworker to clean up an oil spill when they have a chance
12. When an A-frame ladder is properly set up for use, how many points of contact are being made with the floor? (3.6)
    A. 1
    B. 2
    C. 3
    D. 4
13. *True or False?* Daily briefings to develop a plan that minimizes the interaction of workers on a worksite is used to prevent accidents. (3.7)
14. Which of the following is *not* necessarily a characteristic of a confined space? (3.8)
    A. It is below ground.
    B. It is not intended for human occupancy.
    C. It has limited access for entry and/or egress.
    D. It is large enough for a person to enter.
15. *True or False?* If a trench is 10′ deep, the potential for a cave-in can be as far away as 10′ from the trench. (3.9)

Plug-In

1. When a worker receives an electric shock, the amount of current flowing through the body depends primarily on the _____. (3.1)
   A. resistance of the body and the voltage
   B. height and weight of the individual
   C. path of the current and the duration of the shock
   D. resistance of the body and the height of the individual
2. A _____ fire extinguisher is recommended for a Class C fire. (3.4)
   A. pressurized water
   B. foam
   C. dry chemical
   D. wet chemical
3. If the working length of an extension ladder is 20′, how far from the building should the base of the ladder be? (3.6)
   A. 5′
   B. 8′
   C. 10′
   D. 15′

Making the Connection

1. *NEC 110.26* requires the minimum working space around a panel to be at least 30″ wide and a minimum space in front of the panel to be at least 3′ and depending on the voltages, possibly more. As a safety concerned maintenance person, how would you address the storage of water bottles in the space in front of the panel? (3.2)
2. Consider a situation where you are working on a worksite along with one of your coworkers, who happens to be a very good friend. You notice that they are not wearing the required personal protective equipment. Discuss whether you would bring this topic up to your coworker. If you decide to bring it up, discuss how. If you decide not to, provide reasons to support your decision. (3.3)

CHAPTER

4 Building Blocks of Electricity and Electronics

Lane V. Erickson/Shutterstock.com

SPARKING DISCUSSION

Why are copper and aluminum the two best choices for the conductors in electrical wiring? When might platinum and gold be used?

LEARNING OUTCOMES

After completing this chapter, you will be able to:

4.1 Identify the relationship among elements, compounds, and mixtures.
4.2 Understand the law of charges and electrical charge.
4.3 Explain the relationship among current electricity, voltage, and power supply.
4.4 Discuss the differences between conductors and insulators.
4.5 Describe the differences and similarities that exist between electricity and electronics.
4.6 Describe the process of doping as it applies to semiconductors.

TECHNICAL TERMS

anion
atom
atomic mass
atomic number
bump theory
cation
compound
conductor
current electricity
diode
doping
electrical charge
electricity
electron
electronics
electron shell
element
free electron
insulator
ion
isotope
law of charges
mass
matter
mixture
molecule
N-type material
neutron
nucleus
P-type material
power supply
proton
semiconductor
static electricity
switch
valence electron
valence shell
voltage

Introduction

To be successful in the electrical industry, a solid foundation in electrical knowledge is needed. Imagine trying to build a house on a foundation that was set on sand. No matter how magnificent the house is, eventually it will crumble due to the weak foundation.

To understand electrical theories, a solid understanding of the terms, phrases, and concepts used to describe electricity is necessary. This chapter is dedicated to the very building blocks that form the solid foundation on which all future electrical education is based.

4.1 Elements, Compounds, and Mixtures

Understanding the theory of how electric current flows requires an understanding of matter. ***Matter*** is used to describe everything around us. Matter consists of any substance that occupies space and has mass. ***Mass*** is defined as an object's resistance to movement when a force acts on it. Mass should not be confused with weight, as weight takes gravity into account, while mass does not. An object's weight depends on the strength of the gravitational force acting on it.

Shedding Light

The gravitational force on the moon is one-sixth of the gravitational force on earth. This means that an object's weight on the moon will be one-sixth of its weight on earth. For example, if an object weighs 50 lb here on earth, then it would weigh only 8.3 lb on the moon. Why? Objects weigh six times more on earth due to gravitational force.

Matter exists in three common states: solid, liquid, and gas, **Figure 4-1**. Solids exert all of their force in a downward direction. Liquids exert their force in a downward and outward direction. Gases exert their force equally in all directions.

One important property of matter is its electric charge. ***Electrical charge*** refers to how a substance moves, reacts, or is acted on when exposed to a magnetic field. Charged materials are identified as either positively (+) or negatively (–) charged and will experience a force when positioned within or close to a magnetic field. Some materials do not have a charge and are said to be neutral. Materials that do not have a charge will not be acted on by a magnetic or electric field. Magnetism and magnetic fields are further discussed in Chapter 5, *Magnetism Fundamentals*.

Elements

All matter is identified as either an element or a combination of elements. An ***element*** is one of 118 pure substances known to exist. Each element has a unique set of properties and, when exposed to different conditions, each behaves differently from all the others. All of the known elements, along with their properties, are listed in the periodic table of elements, **Figure 4-2**. The periodic table is organized into 18 columns called groups and seven rows called periods. In the electrical trades, elements such as copper (Cu) and aluminum (Al) are commonly encountered. These materials are used to manufacture many of the electrical components, including wiring, in devices and systems in use today. Copper can be found in period 4 of group 11 in the periodic table, while aluminum can be found in period 3 of group 13.

Shedding Light

Water is not located on the periodic table of elements, as it consists of two elements: hydrogen and oxygen. Hydrogen (H) is located at the top of group 1, and oxygen (O) is located at the top of group 16. Table salt is not an element either, as it consists of two elements: sodium and chlorine. Sodium (Na) can be found in period 3 of group 1, while chlorine (Cl) can be found in period 3 of group 17.

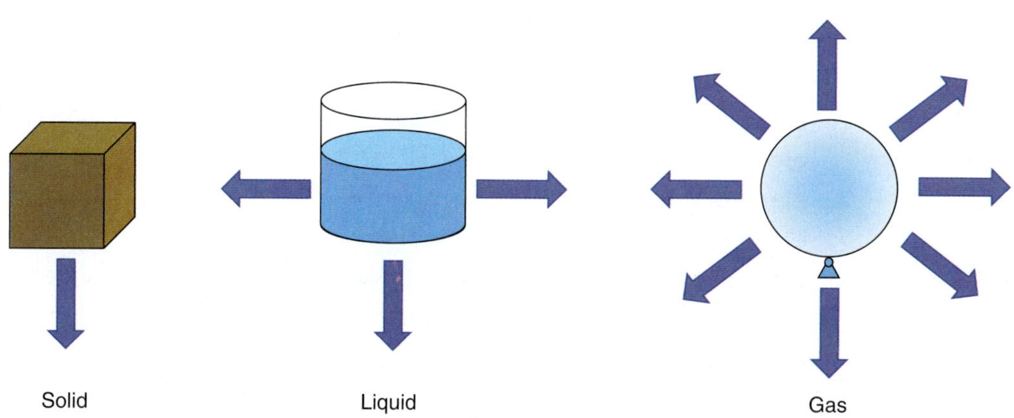

Goodheart-Willcox Publisher

Figure 4-1. Arrows indicate the direction of forces exerted by solids, liquids, and gases.

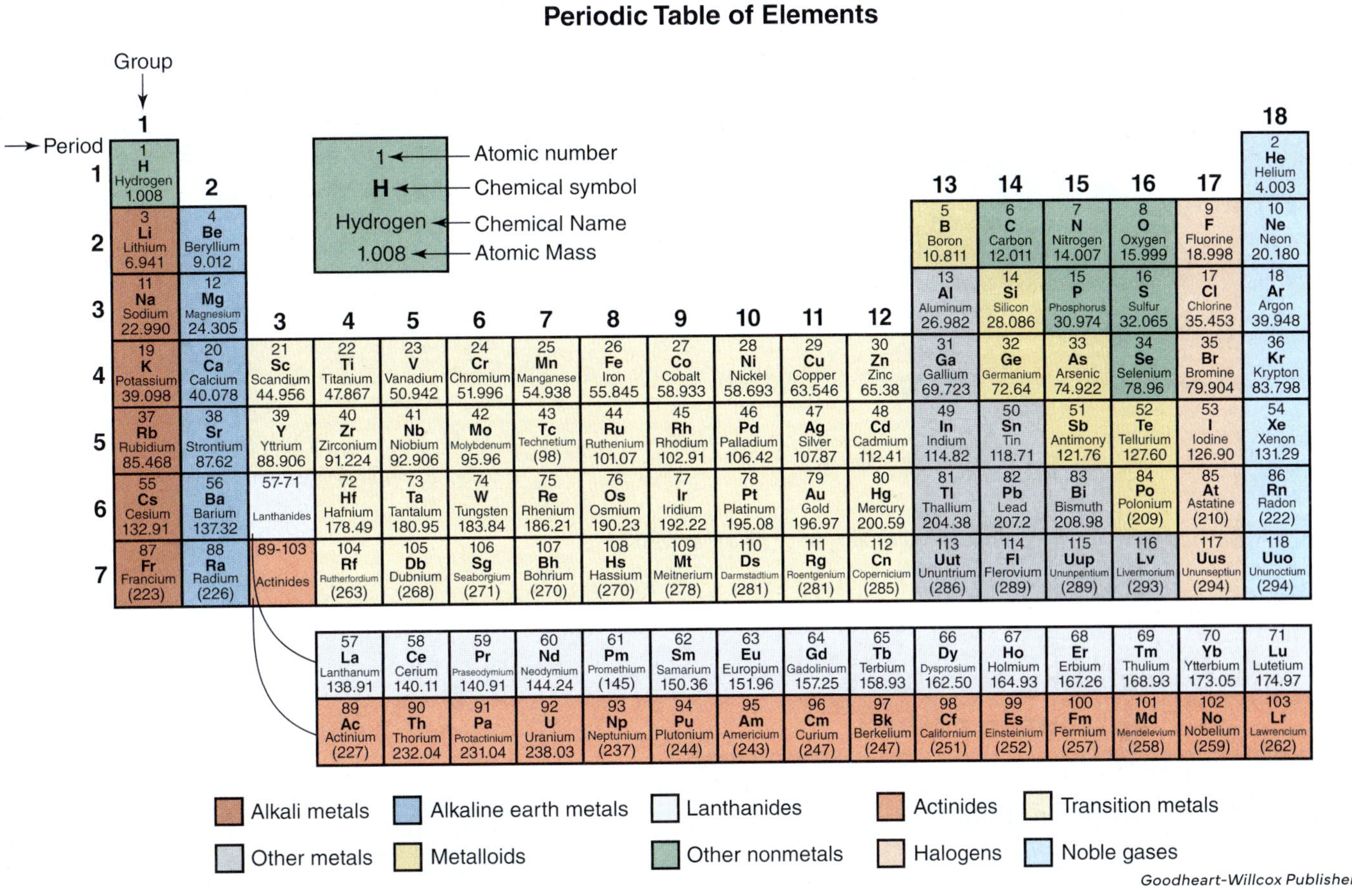

Goodheart-Willcox Publisher

Figure 4-2. Periodic table of elements.

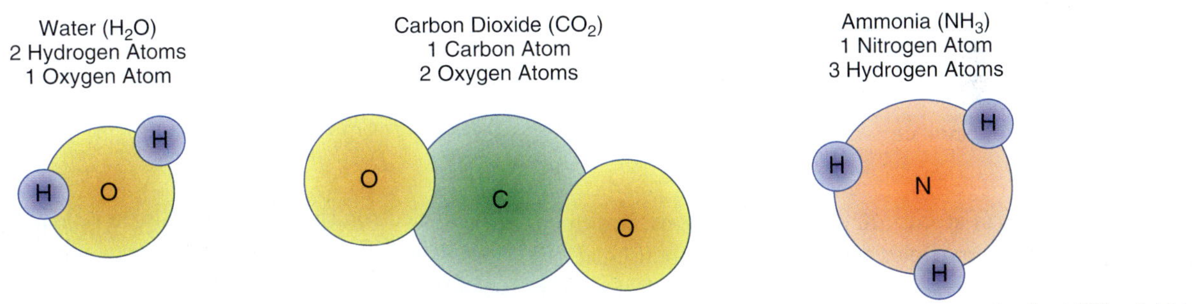

Goodheart-Willcox Publisher

Figure 4-3. Examples of molecules.

Compounds and Mixtures

Common everyday items, such as food, clothing, and household items, are made up of various combinations of elements. Elements can be combined to form compounds or mixtures. A ***compound*** is a substance that is formed when two or more elements combine as a result of a chemical reaction. Table salt, for example, is a compound created when sodium (Na) and chlorine (Cl) combine chemically to form sodium chloride, NaCl. The smallest amount of a compound that can exist, while still maintaining the properties of the compound, is called a ***molecule***, **Figure 4-3**. A ***mixture*** is a nonchemical blend of two or more elements or compounds that can be separated without a chemical reaction. Pure copper and aluminum are neither compounds nor mixtures as they are not combined with other elements.

4.2 Atoms and the Law of Charges

Matter is made up of particles called atoms. The ***atom*** is the smallest amount of an element that can exist, while still retaining all of the properties of the substance. The three

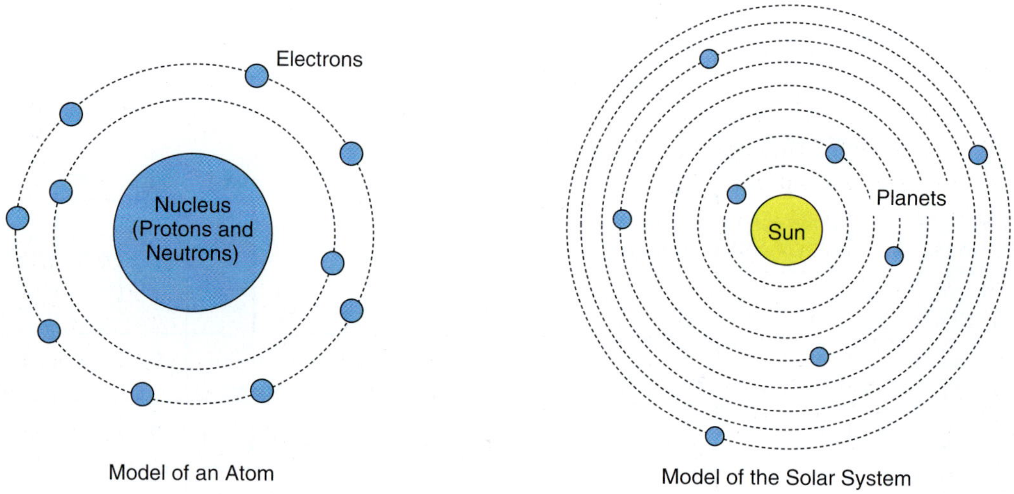

Goodheart-Willcox Publisher

Figure 4-4. The Niels Bohr atomic model resembles the solar system.

Calculating Number of Electrons in a Shell

Shell Number	Formula	Total Number of Electrons in Shell
1	$2(1)^2 = 2(1)(1)$	2
2	$2(2)^2 = 2(2)(2)$	8
3	$2(3)^2 = 2(3)(3)$	18
4	$2(4)^2 = 2(4)(4)$	32
5	$2(5)^2 = 2(5)(5)$	50
6	$2(6)^2 = 2(6)(6)$	72

Goodheart-Willcox Publisher

Figure 4-5. Electron shell capacity calculations.

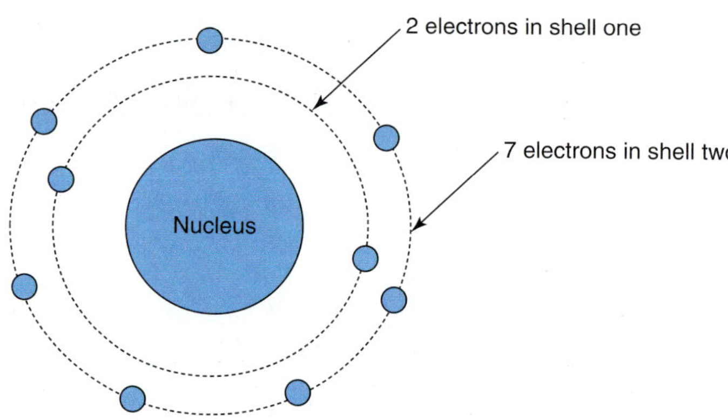

Goodheart-Willcox Publisher

Figure 4-6. An atom with nine electrons. The second shell still has room for one more electron.

main components of atoms are electrons, protons, and neutrons. These atomic particles are identified by their location in an atom as well as by their electric charge. ***Electrons*** orbit around the ***nucleus***, or center of the atom, and have a negative charge. It is the movement of electrons that is the primary concern of those working in the electrical trades. ***Protons*** are located in an atom's nucleus and have a positive charge. ***Neutrons***, also located in the atom's nucleus, do not have a charge. From a charge standpoint, neutrons have no effect on an atom's electrical characteristics. If an atom has the same number of protons and electrons, the atom is said to have no charge, and is referred to as a neutral atom. Since protons have a positive charge and neutrons have no charge, the nucleus of an atom is positively charged.

The orbits in which electrons travel around the nucleus are called ***electron shells***. In 1913, the Danish physicist Niels Bohr presented a theory that electrons orbit around the nucleus in well-defined orbits—similar to our solar system, **Figure 4-4**. Each electron shell has the ability to hold a different number of electrons. The maximum number of electrons that can be present in any given shell is determined by the following equation:

$$\text{number of electrons} = 2n^2$$

In this equation, the n represents the shell number, with shell one being closest to the nucleus. Using this equation, it can be determined that shell number one can hold a maximum of two electrons, shell number two can hold a maximum of eight electrons, and shell number three can hold a maximum of 18 electrons. **Figure 4-5** shows the calculations for the first six electron shells. In the first three electron shells, an electron cannot enter an outer shell until the inner shells are full. In the fourth shell and higher, some electrons will enter the outer shell after the inner shell has been partially filled.

Consider an atom that has nine electrons, as shown in **Figure 4-6**. The shell closest to the nucleus can hold a maximum of two electrons. The remaining seven electrons will be held in shell two, which can hold a maximum of eight electrons. If this atom were to gain an extra electron, it would reside in shell two, filling this shell. This additional electron could not enter shell three because of the vacancy in shell number two.

The outermost electron shell of an atom is called the ***valence shell***. Electrons that occupy the valence shell are called ***valence electrons***. For the atom shown in **Figure 4-6**, shell number two is the valence shell and the seven electrons in this shell are the valence electrons.

Protons are confined to the nucleus of an atom. They are not free to move from atom to atom. The electrons, depending on the conditions, can move from one atom to another, or from one electron shell to another. The number of electrons in an atom determines the electrical charge of the atom. If an atom contains the same number of protons and electrons, the atom is electrically neutral with a net charge of zero. If the number of electrons differs from the number of protons, the atom will have either a positive or a negative charge. If the atom has more electrons than protons, the atomic charge will be negative. If the atom has more protons than electrons, the atomic charge will be positive. Atoms with either positive or negative charges are called ***ions***.

Atoms and the Periodic Table of Elements

The periodic table lists an atomic number and an atomic mass for each element. The ***atomic number*** of an atom indicates the number of protons in the nucleus. In an electrically neutral atom, this is also the number of electrons. The ***atomic mass*** of an atom indicates the total number of protons and neutrons in an atom.

Consider the chemical symbol for boron, **Figure 4-7**. With an atomic number of 5, boron has five electrons and five protons. (Boron is electrically neutral.) By rounding its atomic mass of 10.811 to the nearest whole number, it can be determined that the combined total number of protons and neutrons is 11. Since there are five protons, there are six neutrons contained in the nucleus of a boron atom (11 – 5 = 6).

Atoms and the Law of Charges

One of the fundamental laws in the study of electricity is the law of charges. The ***law of charges*** states that substances with like charges repel each other, while substances with opposite charges attract each other. At the atomic level, the law of charges can be applied to protons and electrons. Two protons will repel each other, as both particles have a positive charge. Similarly, two electrons will repel each other. An electron will be attracted to a nearby proton as they have opposite charges. Since matter is made up of atoms, the law of charges applies to all matter as well.

An example of the law of charges is illustrated in **Figure 4-8**. As shown in the figure, rubbing a glass rod with a piece of silk fabric will create a negative charge in the silk fabric and a positive charge in the glass rod. These two materials will be

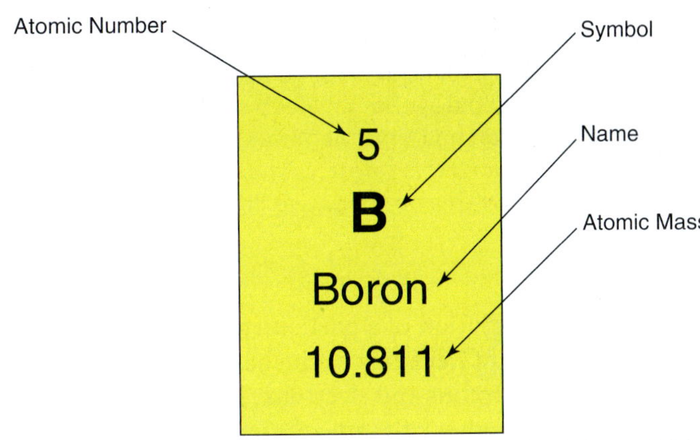

Goodheart-Willcox Publisher

Figure 4-7. Atomic breakdown of a boron atom. Boron has an atomic number of 5 and an atomic mass of 10.811.

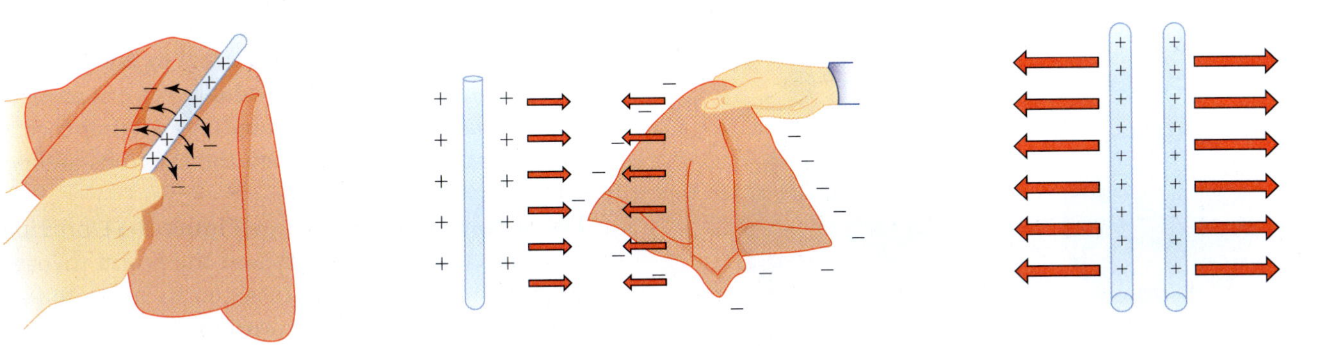

Glass rod has positive charge, silk fabric has negative charge

Ergun_Pinar/Shutterstock.com

Rod and fabric are attracted to one another due to opposite charges

Goodheart-Willcox Publisher

Two positively charged glass rod repel one another due to like charge

Goodheart-Willcox Publisher

Figure 4-8. An example of the law of charges.

attracted to each other as they have opposite electric charges. If two glass rods were both rubbed with the silk cloth, both rods would have a positive charge. If the two rods were placed near each other, they would repel each other.

Shedding Light

Under normal conditions, the number of electrons and protons in an atom is the same, resulting in a net zero charge. The atom copper (Cu) has an atomic number of 29, meaning it contains 29 electrons and 29 protons. In a copper ion, the number of electrons and the number of protons is not the same. The ion Cu^{+}, for example, has 29 protons and 28 electrons, and the ion Cu^{2+} has 29 protons and 27 electrons. The superscript indicates how many electrons have been gained or lost. The superscript + indicates that the atom has lost one electron and has a net positive charge of 1. The superscript 2+ indicates that the atom has lost two electrons and has a net positive charge of 2. The ions Cu^{+} and Cu^{2+} are classified as cations. A ***cation*** is the term used to describe ions with positive charges.

Consider a hydrogen atom, which has one electron. If a hydrogen atom gains an electron, a hydrogen ion is formed. This ion is expressed as H^{-}, where the superscript – indicates that an electron has been gained and the ion has a net negative charge of 1, **Figure 4-9**. The ion H^{-} is referred to as an anion. An ***anion*** is any ion with a net negative charge.

Isotopes

An ***isotope*** is a variation of a particular element that has a different number of neutrons as the base element, but the same number of protons and electrons. The element carbon has an atomic number of 6, meaning it has six electrons and six protons. The atomic mass of carbon is 12.01, meaning that the sum of the protons and neutrons is 12. Since there are six protons in a carbon atom, there are six neutrons (12 – 6 = 6). Consider carbon-14, which is an isotope of carbon. The 14 in carbon-14 represents the atomic mass of a carbon-14 atom. The atomic number is still 6, meaning there are six electrons and six protons in a carbon-14 atom. However, the total number of protons and neutrons is 14, not 12. Since there are six protons in carbon-14, this atom contains eight neutrons (14 – 6 = 8).

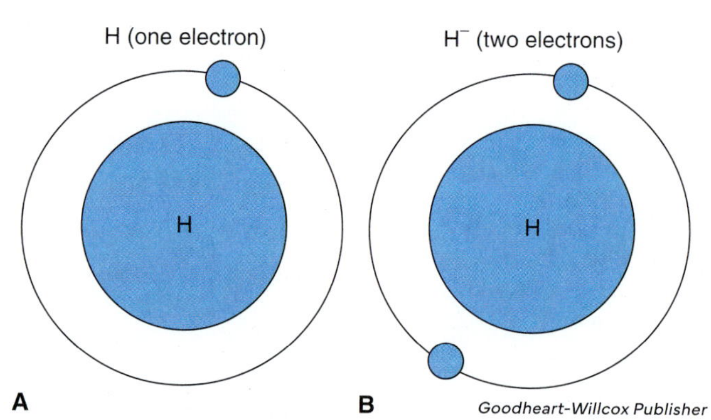

Figure 4-9. A hydrogen ion is formed when a hydrogen atom gains an electron. A—Model of a hydrogen (H) atom. B—Model of a hydrogen (H^{-}) ion.

4.3 What is Electricity?

Electricity is described as the property, condition, or set of conditions that is present when there is an accumulation of electrical charge. Two common types of electricity are current electricity and static electricity. We rely on current electricity to power our homes, cars, computers, and cell phones. We know that batteries supply electricity to run our electronic devices. Another source of electricity is outlets (receptacles) that serve as the access point to a building's electrical supply. Even the static charge created by walking across a wool carpet is a form of electricity.

Current Electricity

Current electricity refers to the continuous flow of electrons through an electric circuit. If electron flow is to be continuous, there must be a continuous driving force. This driving force is an electrical pressure called voltage. ***Voltage***, also known as electromotive force (EMF) and electrical potential, is used to describe the strength, or push of an electrical charge that facilitates the flow of electrons through an electric circuit. The source of circuit voltage is the ***power supply***. When the power supply is a battery, direct current (dc) is delivered to the circuit. When the power supply is a wall outlet, alternating current (ac) is delivered to the circuit.

The concept of current flow is made possible by what is commonly referred to as ***bump theory***. This is when electrons become free electrons. A ***free electron*** is an electron that is not bound to any particular atom. Electrons often become free when other electrons repel them and "bump" them out of an electron shell. If a free electron enters the shell of another atom, this electron can possibly "bump" another electron, causing it to leave the shell and flow to another shell in another atom, **Figure 4-10**. Individual electrons move slowly through a conductor but understanding the bump theory will explain why it appears as though the electrons are moving near the speed of light.

The simple circuit shown in **Figure 4-11** provides an example of current electricity and the bump theory. The battery has an excess of electrons on its negative (–) side. When connected, electrons flow from the negative terminal of the battery, through one of the conductors, to the light bulb. From the light bulb, electrons flow back to the positive terminal of the battery through the other conductor. The battery's positive terminal has a deficiency of electrons and attracts the electrons.

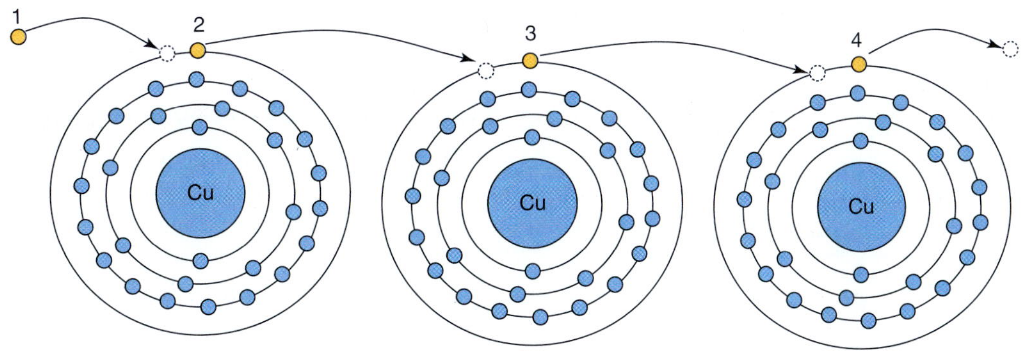

Goodheart-Willcox Publisher

Figure 4-10. Free electrons can easily be displaced. In this example of bump theory, electron one repels electron two from the first atom, which then repels electron three from the second atom. Electron three then repels electron four from the third atom.

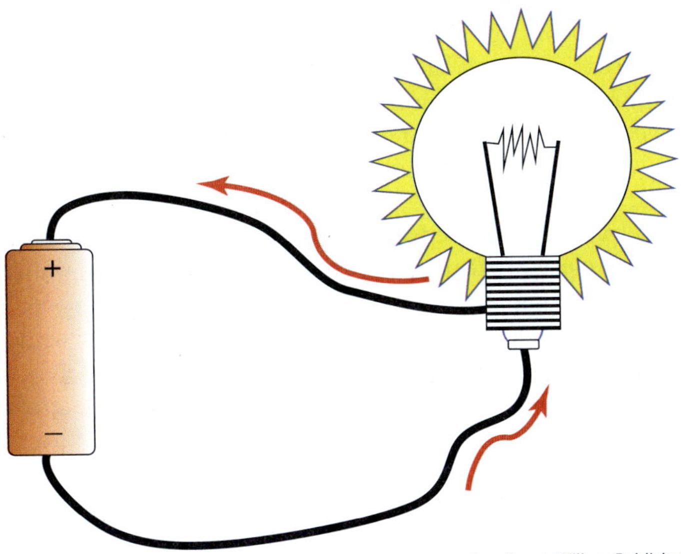

Goodheart-Willcox Publisher

Figure 4-11. A simple circuit with a battery and a light bulb. The arrows represent the direction of electron flow.

Controlling the Flow of Current Electricity

In the real world, the configuration of the circuit shown in **Figure 4-11** is not practical, as the light bulb will illuminate all the time. A more realistic approach would be to add a control device, or ***switch***, to the circuit that can start and stop the flow of electrons through the circuit at the desired time, **Figure 4-12**. When in the ON position, the circuit is complete and the light illuminates. When the switch is OFF, no current flows through the circuit and the light is off.

Switches can be manual or automatic devices and are available in a wide range of types, styles, and configurations. Throughout this text, various types of switches and control devices are discussed.

Direction of Current Flow

There are two points of view regarding the flow of current in an electric circuit. From one perspective, since current flow is defined as the flow of electrons through an electric circuit, the direction of current flow should be identified as the direction in which electrons flow, as shown in **Figure 4-12**. The arrows indicate flow from the negative (–) side of the power source to the positive (+) side of the power source. This negative-to-positive direction convention is referred to as electron flow. Electron flow is widely accepted by many of those working in the electrical industry and is the direction convention that will be used throughout this text.

The other way of looking at current flow is electrical current flows from a higher potential level to a lower one. It is often stated that electric current from a battery flows from the positive (+) terminal to the negative (–) terminal. This positive-to-negative direction convention is referred to as conventional flow. Conventional flow is commonly used in the electronics field.

Shedding Light

While electricity travels at a speed that approaches the speed of light, individual electrons in a conductor move at a much slower speed. The nearly simultaneous "bumps" of electrons along the entire length of a conductor results in the high speed of electrical current. To help visualize how electron motion differs from electrical current, imagine a tube filled with balls. When a ball is added at one end of the tube, a different ball is pushed out of the other end. It appears that the ball instantaneously emerges from the other end of the tube after being inserted into the tube. The ball being added does not travel the entire length of the tube, but its energy does.

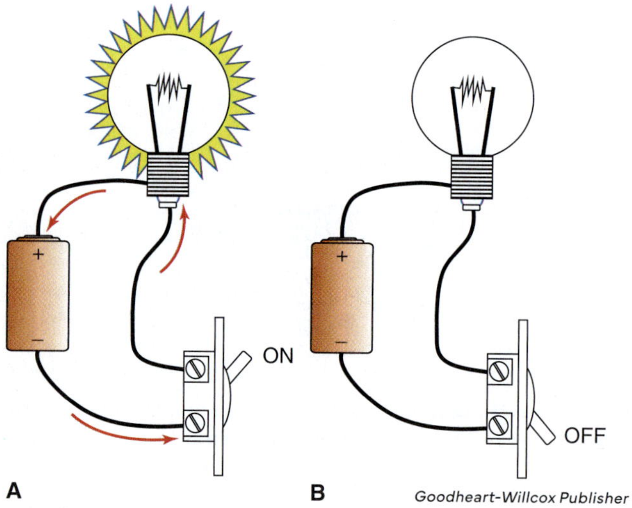

Figure 4-12. A manual switch is similar to the one used to control the lights in your home. A—A complete circuit with the switch in the ON position. B—The circuit is broken when the switch is in the OFF position.

Static Electricity

Static electricity can be an amusing phenomenon. For example, a balloon rubbed against a wool sweater will then stick to a wall. It can also be a nuisance, causing your clothing to stick to itself or other objects. Static electricity also has the potential to be quite dangerous as it can destroy electronic circuitry and can even cause massive explosions and severe personal injury.

Unlike current electricity, static electricity often builds up in materials that do not conduct electricity very well. Since the electrical charge builds up and does not flow, ***static electricity*** is generally described as an imbalance between negative and positive charges in an object. Electric charge is a fundamental property of matter. Some substances and materials are more likely to give up electrons, some are more likely to accept electrons, while others are neutral.

Shedding Light

Have you ever walked across a carpeted floor in your socks and received a stinging electric shock when you touched a metallic doorknob? As you walked across the room, your body attracted electrons, creating a negative charge, and a potential difference, or voltage, between your body and the doorknob. The generated voltage of static electricity can be high, sometimes greater than 25,000 V.

Materials that are likely to give up electrons, creating a net positive (+) charge include air, skin, aluminum, glass, wool, and nylon. Materials that are likely to accept electrons, creating a net negative (–) charge include latex, copper, gold, silver, and plastic. Electrically neutral material includes cotton, wood, paper, and steel. Neutral materials can be attracted to any charged material.

How Static Electricity is Created

Static electricity is often generated when two solid objects come in contact with each other. This process occurs when one substance or material has only a few valence electrons and the other material has vacancies in its outermost electron shells to accept these loosely held electrons. Once the objects separate, the charge imbalance remains. This was the case in the silk fabric and glass rod example discussed earlier in **Figure 4-8**.

Another example of this phenomenon is when two positively-charged metallic spheres are acted on by a negatively-charged material such as latex. When two metallic spheres, each with a neutral charge, are in contact with each other and are insulated from the earth, electrons are free to move from one sphere to another, **Figure 4-13A**. When a negatively-charged latex material is brought near sphere 1, **Figure 4-13B**, electrons in sphere 1 will be repelled by negatively-charged latex, and move into sphere 2. With electrons moving from sphere 1 to sphere 2, sphere 1 becomes positively charged, while sphere 2 becomes negatively charged. If the two spheres are moved away from each other, with the latex still in close proximity to sphere 1, each sphere maintains its charge, **Figure 4-13C**. Even when the latex is removed, **Figure 4-13D**, sphere 1 will have a positive charge, while sphere 2 will have a negative charge. Since there is a difference in charge between the two spheres, a visible and audible spark will likely result if they are brought close enough together.

Dangers of Static Electricity

Sudden high-voltage discharges of static electricity can ignite flammable substances and damage electronic circuitry. Flammable substances, such as gas and oil, can ignite if the pipe or tube that carries these substances is touched and an electric spark is created.

When working on equipment that contains solid-state (electronic) circuit boards, it is important to ground yourself before touching the boards to prevent static electricity from damaging them. Circuit boards are often stored in conductive plastic bags to protect them from static electricity. In some cases, those who work in facilities that produce solid-state boards and other electronic devices are required to wear grounding straps, which continuously dissipate any static charge from their bodies.

Shedding Light

If you ever wondered why lightning strikes are so dangerous, keep in mind that the potential difference between the clouds and the ground can range from tens of thousands of volts to millions of volts, depending on the atmospheric conditions.

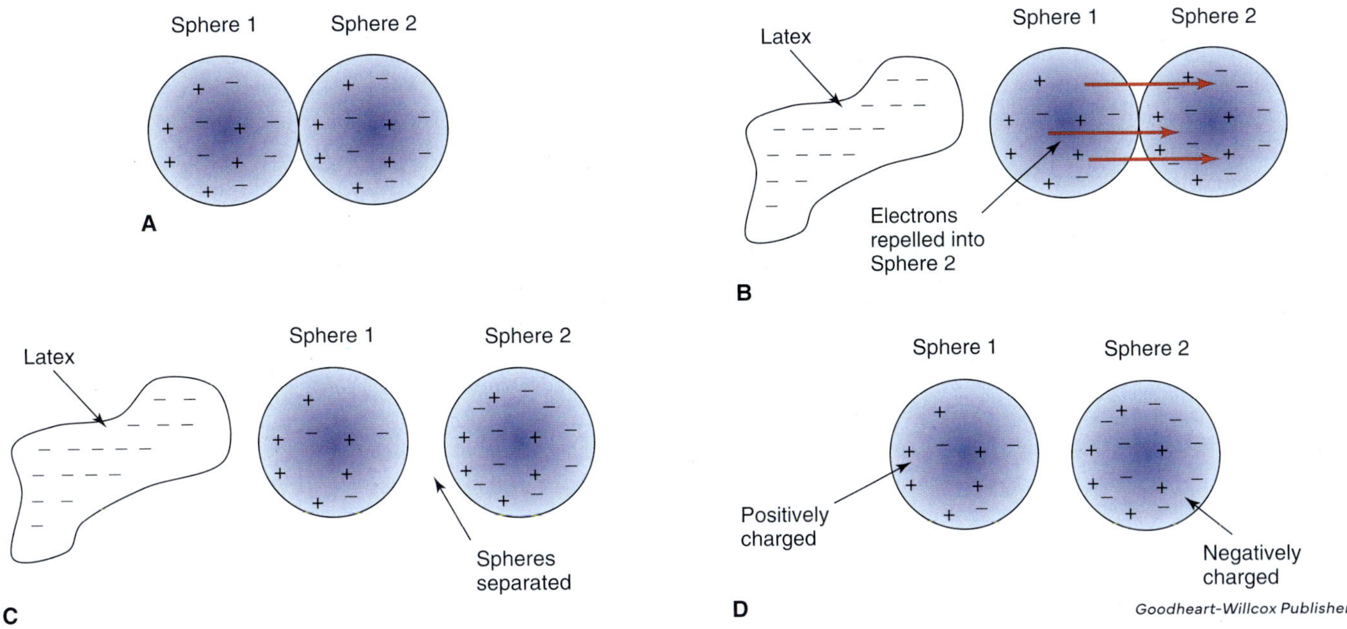

Figure 4-13. An example of how static electricity is created.

Benefits and Uses of Static Electricity

Although static electricity can damage electronic devices and result in the combustion of flammable substances, the power of static electricity can be, and is, harnessed to perform useful and beneficial tasks. Electrostatic air filters, often found in air-conditioning systems, consist of a charged material that attracts dust as the air flows through it. These filters can capture dust, pollen, and other airborne contaminants.

Electrostatic precipitators are commonly found at the inlet to the smokestacks of coal-fired power plants. These precipitators, which are similar to electrostatic air filters, trap and collect particulates, preventing them from entering the atmosphere. After these impurities and contaminants are captured, they are removed from the smoke and discarded as solid waste. This greatly reduces the amount of air pollution produced by the power plant.

4.4 Conductors and Insulators

Circuits rely on a power source to create the electrical pressure needed to push electrons toward the controlled devices and components. To ensure that the electric current flows to the right components at the right time, an interconnecting network of pathways is used. In many cases, these pathways are created with wires. The materials used to create wires are selected based on the ability of their electrons to pass freely from one atom to another. Wires used to create electrical circuits need to be good conductors of electricity. A ***conductor*** is a material that has very few valence electrons and allows electrons (and thus electricity) to flow relatively easily.

Metals are good conductors. Copper, for example, is a very good conductor of electricity as it has only one electron in its valence shell. Other metallic elements such as silver, gold, mercury, platinum, and aluminum are also good electrical conductors and have only a small number of valence electrons, **Figure 4-14**.

Metallic Elements and Conductivity

Element	Number of Valence Electron(s)	Electrical Conductivity (relative to conductivity of silver)
Silver	1	100%
Copper	1	95% as conductive as silver
Gold	1	65% as conductive as silver
Aluminum	3	56% as conductive as silver
Platinum	1	15% as conductive as silver
Mercury	2	2% as conductive as silver

Goodheart-Willcox Publisher

Figure 4-14. Metallic elements, the number of electrons in their valence shell, and their relative conductivity.

Shedding Light

Nearly all wires used in electrical wiring are made of copper or aluminum. Because aluminum is less conductive than copper, an aluminum wire will generally need to be sized larger than a copper wire to carry a load. However, the aluminum wire weighs less than a comparable copper wire and may be less expensive.

Atoms seek stability. In order for an atom to become more stable, it is desirable to have a completely full, or completely empty, outermost shell. In the case of copper, the single electron in its valence shell is weakly held in place. If a copper atom were to lose this outermost electron, the atom would have a completely empty outermost shell and would become stable. But the valence electrons in copper are loosely held, so another electron will easily enter; thus, copper is a very good conductor.

Shedding Light

Because of their atomic structure, good conductors of electricity are often good conductors of heat as well.

Know the Code

The amount of current that can be carried safely by a conductor depends on the application, temperature, material, insulating material, and the size/diameter of the conductor. Thicker conductors, for example, carry more current than thinner ones. Detailed information on conductor properties, including sizing, types, designation, insulation, and appropriate applications are all provided in the *NEC, Article 310, Conductors for General Wiring,* and *Table 8* and *Table 9* in *Chapter 9* of that text. Wire sizing is covered in more detail in Chapter 38: *Power Distribution: From the Generator to You.*

Unlike conductors, ***insulators*** are materials that do not allow electrons to flow freely through them. Materials that have full or empty outermost shells tend to hold onto their electrons tightly, making them poor conductors of electricity. Glass, air, and rubber are good insulators. Insulating material (typically plastic, a cross-linked polymer, or rubber) is commonly placed around the conducting wire, **Figure 4-15**. This prevents the conductor from coming in contact with other conductive materials and protects those handling the wires from an electric shock.

4.5 What is Electronics?

The theories that govern the behavior of electrical circuits and electronic circuits are the same. Both electrical and electronic circuits rely on the flow of electrons to operate. When electrical circuitry is mentioned, applications such as the wiring within a residential or commercial structure often come to mind, while computers and solid-state circuit boards, **Figure 4-16**, are often associated with electronic circuitry.

Electronics is an industry that concentrates on the designing, manufacturing, and servicing of equipment and circuitry that uses components such as diodes, transistors, printed circuit boards, and other similar devices. Electronics is also used to classify products that use or incorporate electronic circuitry and components into their design.

One major difference between an electrical circuit and an electronic circuit is that the electronic circuit often has the ability to make decisions. Electric circuits simply facilitate electron flow through a circuit, while electronic circuits determine how to facilitate electron flow through a circuit and then do so. Most pieces of equipment and appliances in use today rely on both electrical and electronic circuitry to perform as intended. Even appliances such as a clothes iron often uses electronic circuitry to incorporate safety features and improve its overall performance.

Another important difference between an electric circuit and an electronic circuit is the switching methods that are used to open and close current pathways. A light switch, for example, is a physical switch with mechanical components that must move from one position to another to start or stop the flow of electrons. Electronic devices, for the most part, rely on the behavior of certain materials, known as semiconductors, to either start or stop the flow of electrons. Since there are no physical moving parts in an electronic switch, these devices can start and stop electron flow many thousands of times a second. If this was attempted with the light switch on

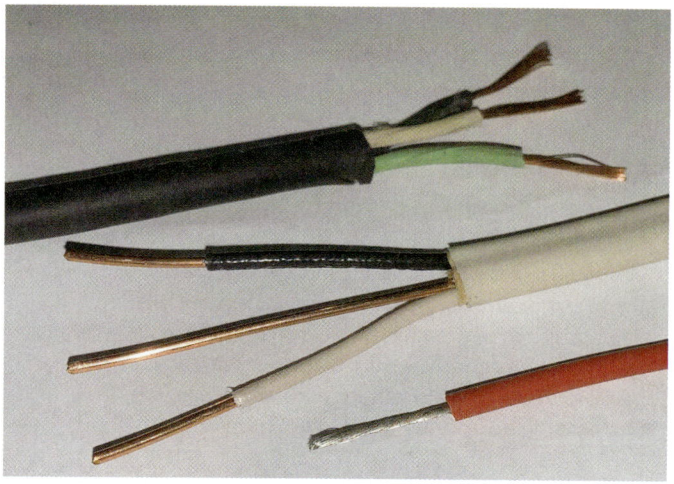

Goodheart-Willcox Publisher

Figure 4-15. Conductors are often encased in insulating material.

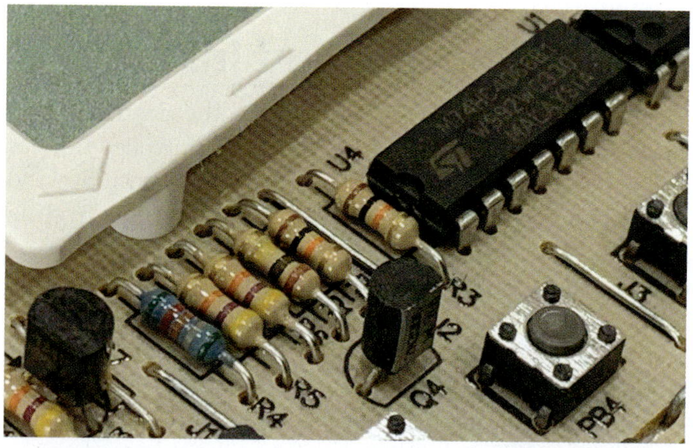

Goodheart-Willcox Publisher

Figure 4-16. A solid-state circuit board is an example of electronic circuitry.

the wall, it would likely break. Electronic circuitry has made it possible for computers and other electronic devices to operate as quickly as they do. In addition to the items discussed, some other notable differences between electricity and electronics include the following:

- For consumption by the end user, electricity is often converted to other forms of energy such as light, heat, and mechanical, while electronic circuitry uses electricity to perform computational or operational tasks.
- Electric circuits typically consume much more power than electronic circuits.
- Electronic components and devices are much smaller than similar electric devices.
- Electric devices such as lights, heaters, and motors are often controlled by electronic devices such as motion sensors, thermostats, and variable frequency drives.

Shedding Light

Although electricians do not replace individual solid-state components on circuit boards, they should be able to determine when the board is functioning as it should. If the circuit board is being powered and all of the board's input signals are correct, but the board's output is not correct, chances are there is a malfunction within the board. However, if the input signals supplied to the board are not correct, it is likely that the board is operational, and some external factor is causing the problem. Be sure to evaluate all of the external factors that affect the operation of an electronic circuit board before condemning the board!

4.6 Semiconductors

The digital age as we know it would not be possible without the advances made in the semiconductor field. ***Semiconductors*** are materials that can either allow or prevent electron flow through them depending on the external conditions acting on the material. Solid-state printed circuit boards, such as those found in cell phones, computers, televisions, and other high-tech devices, would not exist without semiconductors and the components made from them. The material commonly used in the manufacturing of semiconductor-based components is silicon. Although not as popular, another element used in semiconductors is germanium. Gallium arsenide, a compound consisting of gallium and arsenic, is also used in electronics as a semiconductor. In addition, other silicon-based compounds, such as silicon dioxide, are commonly found in semiconductor materials and components.

Shedding Light

Silicon is the second most abundant element on earth, with oxygen being the most abundant. Silicon is relatively inexpensive and easy to obtain as nearly 25% of the earth's crust is made of silicon.

In their pure form, semiconductors such as silicon are not very good conductors of electricity. A pure silicon atom has four valence electrons, and together with other silicon atoms, forms crystals that tightly bond their electrons, **Figure 4-17**. Each silicon atom shares one of its four valence electrons with each of four neighboring atoms, **Figure 4-18**. Many elements that create crystalline configurations make poor conductors for this reason. If semiconductor materials are to conduct electricity, the material must be altered. The process used to change the atomic structure of the material is called doping.

Doping

When referring to semiconductors, ***doping*** is the process of introducing impurities to the pure material to create either

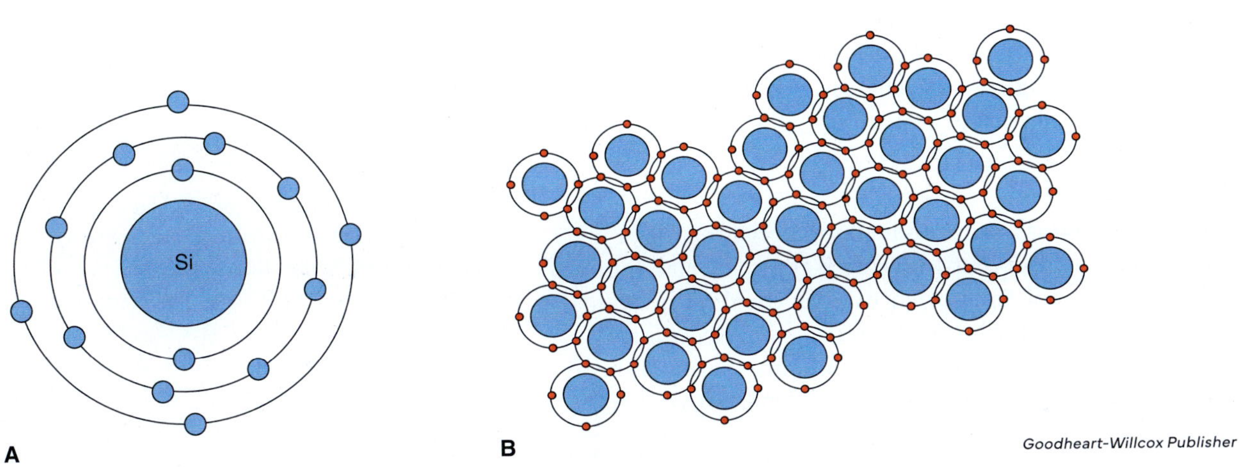

Figure 4-17. Silicon is commonly used in the manufacturing of semiconductors. A—One silicon atom showing four valence electrons. B—Crystalline structure of silicon.

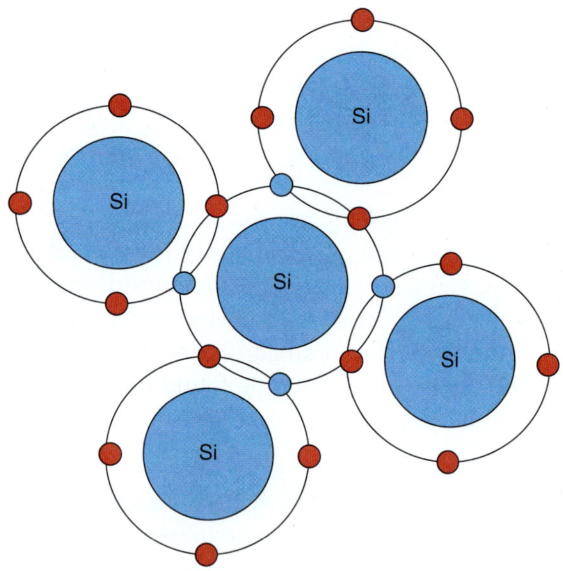

Goodheart-Willcox Publisher

Figure 4-18. A silicon atom sharing its valence electrons. The shared valence electrons in the center silicon atom have been color-coded blue, while the valence electrons in the surrounding silicon atoms are mostly color-coded red to clarify the "electron-sharing" process.

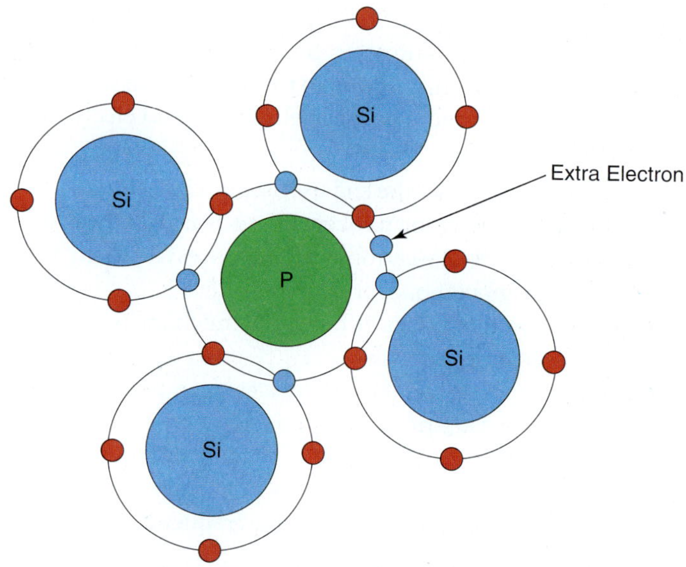

Goodheart-Willcox Publisher

Figure 4-19. The extra electron resulting from the phosphorus doping of silicon.

a positively-charged or a negatively-charged semiconductor substance. The resulting charge of the semiconductor material depends on the doping material that is used. By carefully configuring layers of semiconductor materials with opposite charges, non-mechanical switching, control, and amplification devices can be created. These devices are referred to as solid-state electronic devices.

N-Type Material

N-type material is a semiconductor material that has a net negative charge. In order to create N-type material, phosphorus (chemical symbol P) is often used as the doping agent. The valence shell for phosphorus contains five electrons.

For each phosphorus atom that is added to silicon, there will be an extra electron. This will result in creating the net negative charge, **Figure 4-19**. These free electrons are loosely held and can easily move from atom to atom.

P-Type Material

P-type material is a semiconductor material that has a net positive charge. In order to create P-type material, boron (chemical symbol B) is often used as the doping agent. The valence shell for boron contains three electrons. For each boron atom that is added to silicon, there will be a missing electron, creating a hole and a net positive charge, **Figure 4-20**.

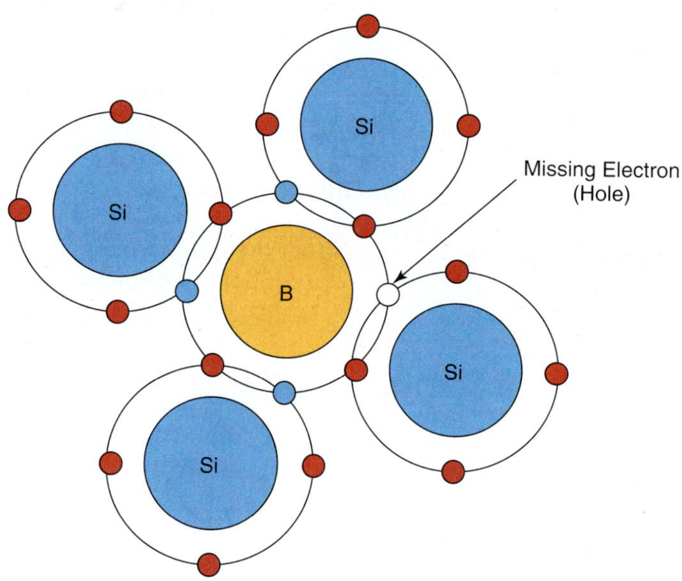

Goodheart-Willcox Publisher

Figure 4-20. The resulting hole from the boron doping of silicon.

Shedding Light

N-type and P-type materials are the building blocks for nearly all devices found in electronic circuits. ***Diodes***, which only allow electric current to flow in one direction, for example, are made up of a single layer of N-type material and a single layer of P-type material. Other electronic components, such as transistors, are made up of different combinations of these two materials.

CHAPTER 4 REVIEW

Summary

- Matter refers to everything that surrounds us and is made up of atoms. (4.1)
- Mass refers to an object's resistance to movement. (4.1)
- Matter is commonly found in the solid, liquid, or gaseous state. (4.1)
- Weight is a function of mass and gravitational force. (4.1)
- There are 118 known elements. (4.1)
- The properties of elements are presented in the periodic table. (4.1)
- Mixtures and compounds are made from various combinations of elements. (4.1)
- Mixtures are easily separated, while compounds require chemical reactions to be created or break apart. (4.1)
- A molecule is the smallest amount of a compound that retains the characteristics and properties of the compound. (4.1)
- An atom is the smallest amount of an element that retains the characteristics and properties of the element. (4.2)
- Electrons have a negative charge, protons have a positive charge, and neutrons have a neutral charge. (4.2)
- Like charges repel each other, while opposite charges attract each other. (4.2)
- Ions are charged atoms, cations are positively charged ions, and anions are negatively charged ions. (4.2)
- Isotopes are atoms that have a different number of neutrons than the base element. (4.2)
- Electricity is the property, condition, or set of conditions that is present when there is an accumulation of electrical charge. (4.3)
- Two types of electricity are current electricity and static electricity. (4.3)
- Current electricity relies on electron flow through an electric circuit or pathway. Static electricity refers to an imbalance between negative and positive charges in an object. (4.3)
- A conductor is a material that has very few electrons in its valence (outermost) shell. (4.4)
- Insulators hold onto their electrons tightly so they are not free to move from atom to atom. (4.4)
- Electronics is an industry that concentrates on designing, manufacturing, and servicing equipment and electronic circuitry. (4.5)
- An electronic circuit has the ability to make decisions about electrical flow, while an electrical circuit simply facilitates electron flow. (4.5)
- Semiconductors can either allow or prevent electron flow through them depending on the external conditions acting on the material. (4.6)
- The most common semiconductor material is silicon. (4.6)
- Impurities, such as boron and phosphorus, are added to silicon to produce N-type and P-type materials. (4.6)
- N-type and P-type materials are the building blocks for many electronic devices. (4.6)

Review Questions

Use the information in the chapter to answer the following questions.

Measuring Your Potential

1. *True or False?* The mass of an object is the same as the object's weight. (4.1)
2. A liquid exerts a force in which direction? (4.1)
   A. Downward only
   B. Downward and outward
   C. In all directions
   D. Outward only
3. *True or False?* A molecule of sodium chloride ($NaCl$) contains one atom of sodium and one atom of chlorine. (4.1)
4. An atom of carbon, which has six electrons in total, will have _____ electron(s) in its outermost shell. (4.2)
   A. one
   B. two
   C. three
   D. four
5. *True or False?* Two types of electricity are static electricity and current electricity. (4.3)
6. *True or False?* Electrons flow from the positive terminal of a battery to the negative terminal of the battery. (4.3)
7. Which of the following statements is *true* regarding good insulators? (4.4)
   A. They have many loosely held electrons.
   B. They do not allow electrons to flow freely through them.
   C. They are typically metals.
   D. They have one or two valence electrons.

8. Which of the following statements regarding electronic components is *not* correct? (4.5)
   A. Electronic circuits consume more power than electrical circuits.
   B. Electric devices are often controlled by electronic devices.
   C. Electronic devices rely on semiconductors to stop the flow of electrons.
   D. Electronic circuits have the ability to make decisions.
9. *True or False?* Materials that can either allow or prevent electron flow are known as semiconductors. (4.6)
10. The charge of a semiconductor material depends on the _____. (4.6)
    A. use of N-type material
    B. use of P-type material
    C. doping material used
    D. the power supply

Plug-In

1. Which of the following atoms would most likely make the best conductor? (4.1)
   A. An atom with a total of 10 electrons
   B. An atom with a total of 16 electrons
   C. An atom with a total of 29 electrons
   D. An atom with a total of 35 electrons
2. An element with an atomic number of 79 and an atomic mass of 197 will have _____ neutrons. (4.2)
   A. 79
   B. 197
   C. 118
   D. 276
3. Using the Periodic Table of Elements in **Figure 4-2**, find the atomic number for aluminum. Using this number, determine how many electrons are in the outer shell. (4.2)
   A. 2
   B. 3
   C. 4
   D. 5

Making the Connection

1. Why is the electricity we use to power our homes classified as a secondary energy source, but static electricity is not? (4.3)
2. List the most common elements used in the electrical and electronic world. Discuss why an electrician will have a greater concern for certain elements than an electronic technician. (4.3)

CHAPTER 5 Magnetism Fundamentals

Brookgardener/Shutterstock.com

SPARKING DISCUSSION

Magnetism is one of the fundamental forces in the universe. Can you list some of the ways that magnetism is used in our daily lives without us even knowing?

LEARNING OUTCOMES

After completing this chapter, you will be able to:

5.1 Discuss how the strength of a magnetic field is measured.

5.2 Explain the law of charges as it applies to magnets and magnetism.

5.3 Explain magnetic behavior at the atomic level, including the concepts of electron spin, magnetic dipoles, and magnetic domains.

5.4 Explain the difference between a permanent and temporary magnet.

5.5 Define magnetic properties, including permeability, reluctance, retentivity, and coercivity.

TECHNICAL TERMS

ampere-turns (At)
coercivity
diamagnetic
electromagnet
electromagnetic induction
electron spin
ferromagnetic
gauss
gilbert
left-hand rule
lodestone
magnetic dipole
magnetic domain
magnetic field
magnetic flux
magnetic flux density
magnetic induction
magnetic pole
magnetism
magnetomotive force (MMF)
maxwell
paramagnetic
permanent magnet
permeability
pinning
reluctance
retentivity
saturation
temporary magnet
tesla

Introduction

Magnetism is a key component of many practical electrical processes, applications, and components, including the generation of electricity. Many aspects of our daily lives rely on magnetism, most of the time without our knowledge. From the strip on the back of your credit card to the hard drive on your computer, magnetism is at work. Magnets are used extensively in the farming, manufacturing, and mining industries, and are even used to ensure that metallic substances do not find their way into the foods you eat.

5.1 What is Magnetism?

Magnetism is a property of matter that results in the creation of repulsive or attractive forces among substances. These attractive and repulsive forces are strongest at the ends, or ***magnetic poles***, of the magnet. The magnetic poles on a magnet are present in pairs and are identified as the north pole and the south pole. As shown in **Figure 5-1**, if a magnet is broken into pieces, no matter how small these pieces become, they will still have a north and south pole.

Magnetic Field

A ***magnetic field*** is an invisible field of force that is produced around a magnetic object or particle. Magnetic fields are also produced when electrons flow through a conductor.

Magnetic fields exert a force on magnetic substances or objects that are positioned within the magnetic field. A magnetic field is often represented by a series of lines that start at a magnet's north pole and return to the magnet's south pole, **Figure 5-2**. These lines of force repel each other and will never cross each other.

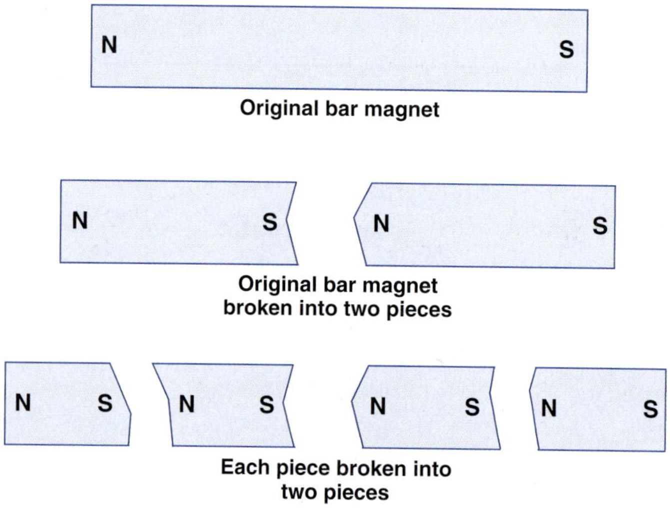

Figure 5-1. Magnetic poles on a magnet will always have a north and south pole.

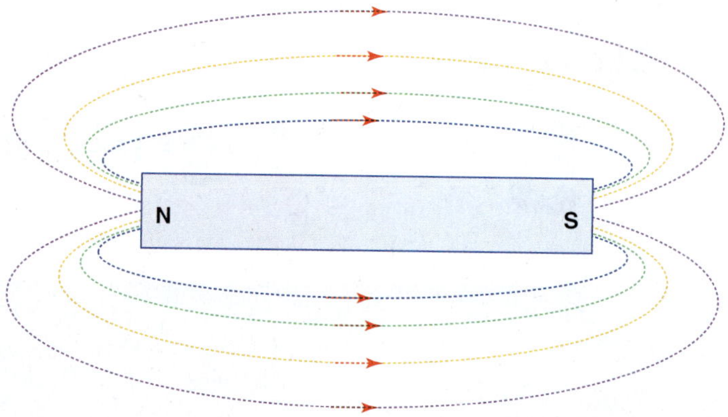

Figure 5-2. Magnetic field lines do not cross. The lines flow from a magnet's north pole to its south pole.

Magnetic Field Strength

Our discussion of magnetic field strength begins with the concept of magnetic flux. ***Magnetic flux*** is defined as the total collection of magnetic field lines (lines of force) in a magnetic field. Magnetic flux is represented by the Greek letter phi, Φ. Changes in magnetic flux produce electric current flow in a conductor. This process is called ***electromagnetic induction***.

The strength of a magnetic field refers to the amount of force it exerts on objects and particles within it. The term ***magnetic flux density*** is used to describe a magnetic field's relative strength and is directly related to the concentration of magnetic field lines that pass through a given area. Magnetic flux density is expressed in the units of lines per square inch. The flux density is highest at a magnet's poles and drops significantly as the distance from the magnet increases, **Figure 5-3**. Flux density is expressed by the following equation:

$$B = \frac{\Phi}{A}$$

where:

B = Flux density
Φ = Magnetic flux (number of lines)
A = Cross-sectional area

As the number of flux lines (Φ) per square inch (A) increases, the magnetic flux density increases.

Shedding Light

The difference between magnetic flux and magnetic flux density can be equated to the difference between a state's population and its population density. The population of Wisconsin, for example, is approximately 5,800,000 people and its population density is about 106 people per square mile. Magnetic flux relates to the total population, while magnetic flux density relates to the population per unit area. In comparison, Wisconsin's largest city, Milwaukee, is 96 square miles and has a population density of 6001 per square mile. Other areas of the state have a much lower density.

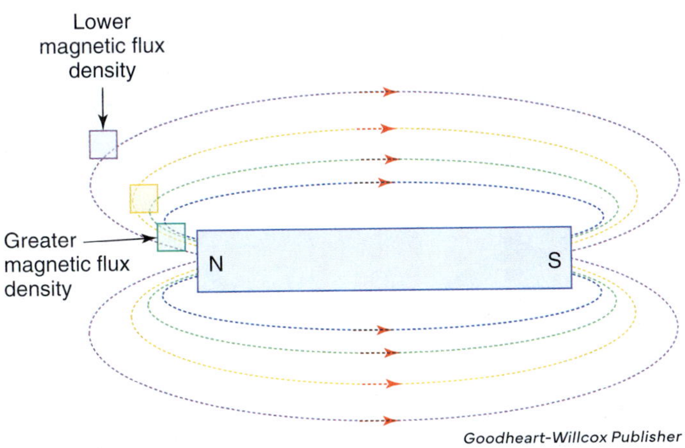

Figure 5-3. Magnetic flux density is greatest near the poles and becomes weaker as the distance from the poles increases.

Different measurement systems express the concentration of flux lines per unit area differently. Common terms used to help quantify the strength of a magnetic field are maxwell, gauss, and tesla. The term ***maxwell*** is used to describe one line of magnetic force and is a unit of magnetic flux. The term ***gauss*** refers to a magnetic force of one maxwell per square centimeter, or 6.4516 lines of force per square inch. The term ***tesla*** refers to a magnetic force of 10,000 gauss. Both gauss and tesla are units of magnetic flux density.

Shedding Light

A magnetic field's relative strength, or magnetic flux density, changes in a manner that is similar to the effects of the earth's gravitational forces. As you move further away from the earth, the gravitational force that pulls you to the earth becomes weaker. As an object moves further away from a magnet, the strength of the magnetic field acting on the object weakens.

When current flows through a conductor, a magnetic field is produced around the conductor. The strength of the generated magnetic field is directly related to the amount of current flow through the conductor. As the rate of electron flow increases, the strength of the magnetic field increases.

Shedding Light

Clamp-on meters determine the current flowing in an electrical conductor by measuring the magnetic field produced by the current. This electrical test instrument is commonly used by electricians in a number of electrical-related industries to evaluate, troubleshoot, and service electrical services, appliances, and circuits. Clamp-on ammeters and other digital multimeters are discussed in Chapter 8, *Electrical Tools and Testing Instruments.*

The magnetic field rotates around the conductor in a direction determined by the ***left-hand rule***. This rule states that if you hold a conductor in your left hand with your left thumb pointing in the direction of electron flow (from negative to positive) the direction of your curled fingers indicates the direction of the magnetic field, **Figure 5-4**.

Magnetic Field Strength in Coils

The strength of the effective magnetic field can be increased by forming the wire or conductor into a coil, **Figure 5-5**. When a coil is formed, the direction of electron flow through each turn of wire is the same. If a vertical coil of

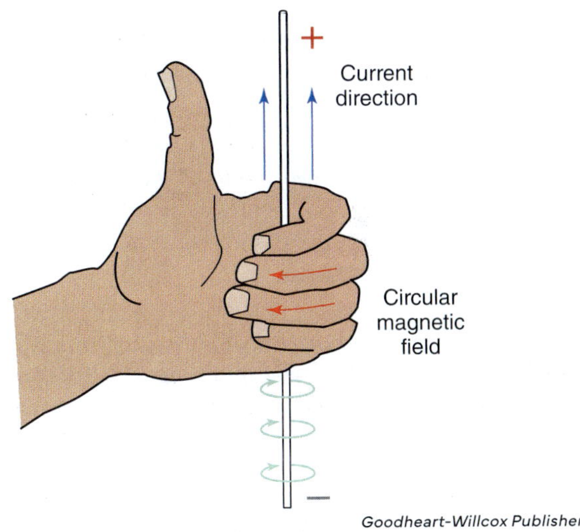

Figure 5-4. The left-hand rule demonstrates electron theory, where electrons flow from negative to positive. The circular magnetic field produces an identifiable current direction.

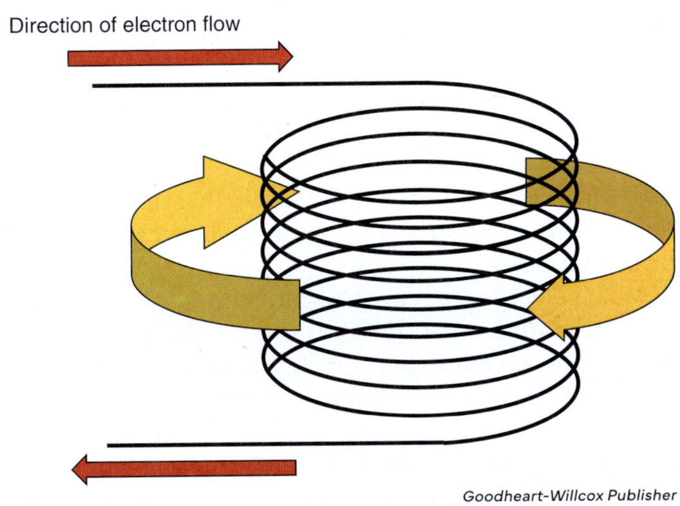

Figure 5-5. When wire is wound into a coil, the direction of electrical current in each turn of wire is consistent.

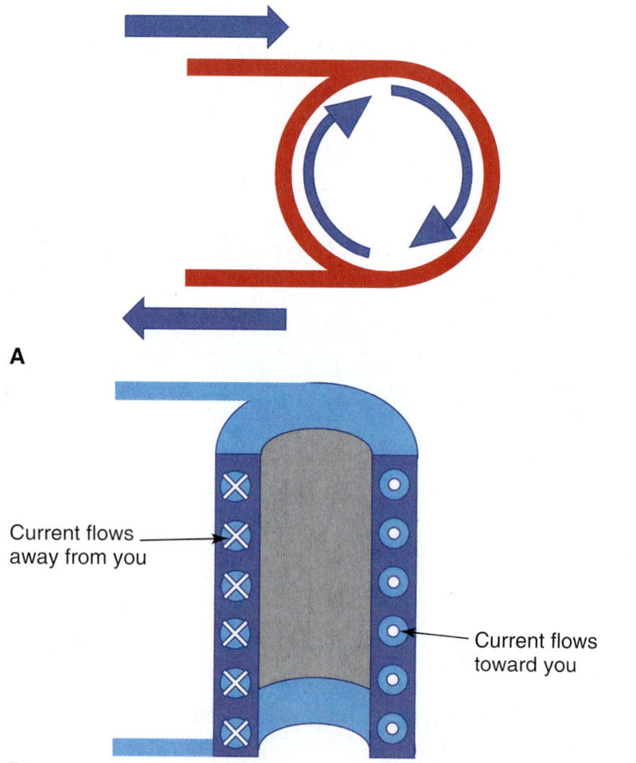

Figure 5-6. The direction of current flow is consistent in each turn of a coil. A—When viewed from above, the current flows in this coil in a clockwise direction. B—This cutaway side view shows where current is flowing "toward you" and "away from you" in the coil.

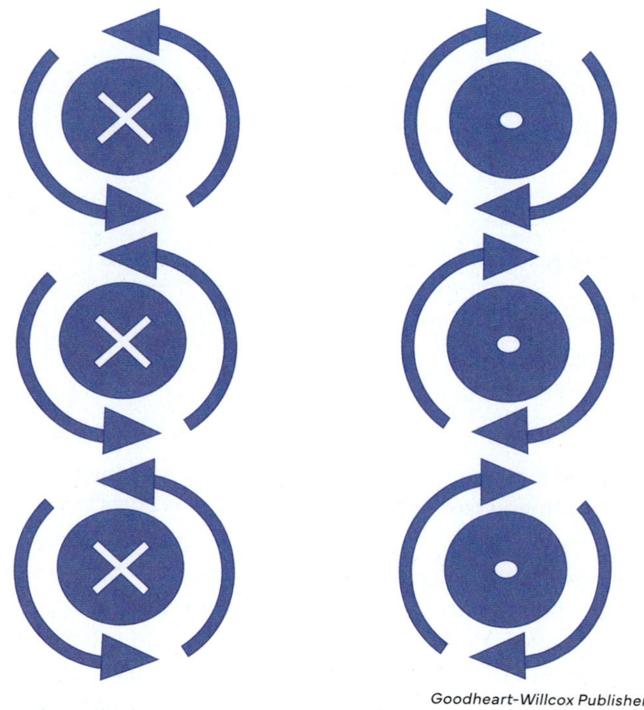

Figure 5-7. Direction of magnetic field from left-hand rule.

wire is observed from the top, the direction of electron flow will be either clockwise or counterclockwise depending on how the coil is wound, **Figure 5-6A**. If this coil were cut in half vertically, all the conductors along one side of the coil have electron flow in one direction (toward or away from the viewer), and all the conductors along the other side of the coil have electron flow in the opposite direction, **Figure 5-6B**. The left-hand rule can be used to determine the direction of the magnetic field around the conductors on either side of the coil, **Figure 5-7**. The magnetic fields in between the wires on the same side of the coil are in the opposite direction and, as such, cancel each other out, **Figure 5-8**. Each of the small magnetic fields inside the coil all point in the same direction, resulting in a much stronger, concentrated, overall magnetic field, **Figure 5-9**.

When electric current flows through a coil of wire, a magnetomotive force (MMF) is created. ***Magnetomotive force*** can be described as magnetic pressure that produces a magnetic field, just as electromotive force (EMF), or voltage, was described as electrical pressure. Magnetomotive force is expressed in ***ampere-turns (At)***. Magnetomotive force for a coil is calculated by multiplying the number of turns in the coil by the current flowing through the coil. See **Example 5-1** and **Example 5-2**.

Example 5-1

How many ampere-turns (At) are generated when a current of 10 A flows through a coil with 5 turns?

✓ Solution

Ampere-turns are calculated by multiplying the number of turns by the amount of current, in amperes, flowing through the coil. In this example, the coil has five turns and the current flowing through it is 10 A, so 50 At (5 turns × 10 A) are produced.

Example 5-2

How many ampere-turns (At) are generated when a current of 5 A flows through a coil with 10 turns?

✓ Solution

Notice that this configuration produces the same number of ampere-turns as in Example 5-1. By doubling the number of turns in the coil, the amount of current required to produce the same magnetomotive force is cut in half.

$$10 \text{ turns} \times 5 \text{ A} = 50 \text{ At}$$

Now It's Your Turn 5-2

How many ampere-turns are generated when a current of 1 A flows through a coil with 50 turns?

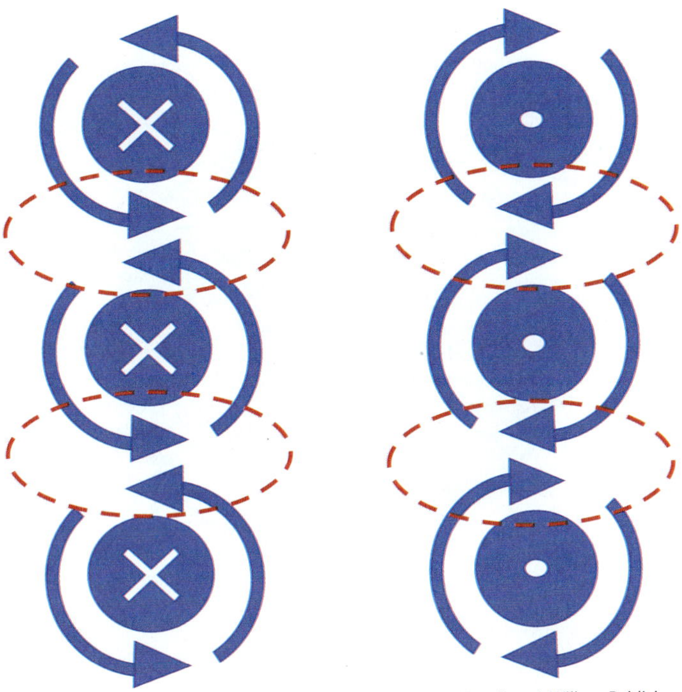

Figure 5-8 Magnetic fields between conductors cancel each other out.

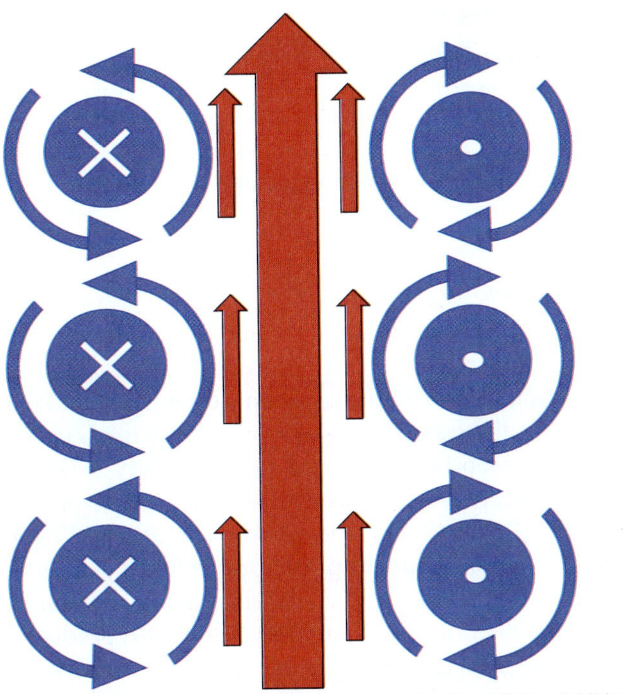

Figure 5-9. Magnetic fields are additives within a coil.

Shedding Light

Not only can electric current be used to produce a magnetic field, but magnetic fields can also be used to produce electricity. If a conductor, often in the form of a coil of wire, is rotated within a magnetic field, voltage is produced.

Although not nearly as common as ampere-turns, magnetomotive force can also be expressed in units called ***gilberts*** (Gb) where:

$$1 \text{ Gb} = 0.7958 \text{ At}$$
$$1 \text{ At} = 1.257 \text{ Gb}$$

5.2 Law of Charges in Magnetism

The law of charges states that like charges repel each other, while opposite charges attract each other. Chapter 4, *The Building Blocks of Electricity and Electronics*, explains how the law of charges applies in atomic theory: when two electrons or two protons come into contact with each other, they will repel each other. When an electron comes into contact with a proton, the two will be attracted to each other.

In magnetism, the same basic rules apply. Instead of electrons and protons, we are referring to the north pole and south pole of a magnet. When a magnet's north pole is positioned near the south pole of another magnet, there will be an attractive force from the north pole to the south pole, drawing the two magnets together. If two like poles, such as the north poles of two magnets, are placed in close proximity to each other, they will repel each other, **Figure 5-10**.

What Makes a Material Magnetic?

Magnets are not attracted to every object or substance. For example, magnets are not attracted to paper, cloth, wood, or plastic. Magnets are attracted to metallic objects that contain iron, nickel, or cobalt. These metals have the ability to align their internal magnetic fields with each other, as well as with those of external magnetic fields. As discussed later in this chapter, it is the ability of a substance, or lack of such ability, to align these internal magnetic fields that determines whether or not a particular substance of an object will be attracted to, or be able to become, a magnet.

The Earth is a Giant Magnet

The outer portion of the earth's core is made up, in part, of flowing, molten metals. Contained in these metals are charged particles and, as they move, a relatively weak magnetic field surrounding the earth is produced. Just as magnets have north and south poles, the earth also has north and south poles. In fact, the earth has four poles:

- True (geographic) north pole
- Magnetic north pole
- True (geographic) south pole
- Magnetic south pole

The earth rotates on an axis that is established by the true, or geographic, north pole and the true, or geographic, south

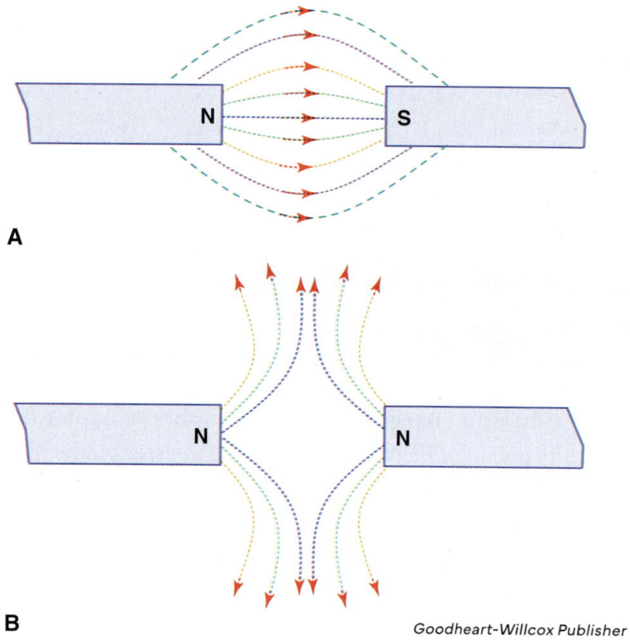

Figure 5-10. The law of charges applies to magnetic poles. A—Unlike poles are attracted to each other. B—Like poles repel each other.

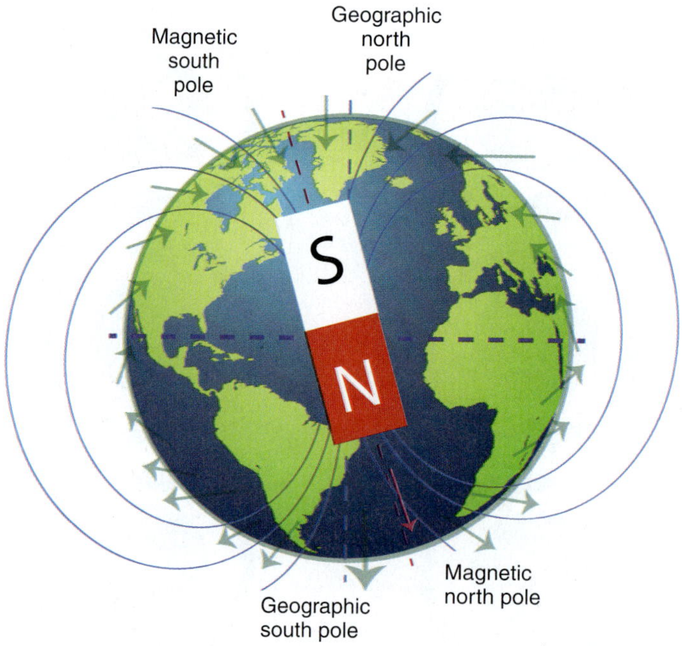

Figure 5-11. The earth's four poles.

pole. The magnetic north and south poles are the locations where the earth's magnetic field is the strongest. The earth's magnetic north pole is close to its geographic south pole, and the earth's magnetic south pole is close to its geographic north pole, **Figure 5-11**. Using a compass to find one's way would not be possible without these magnetic poles.

Shedding Light

Research has shown that magnetism plays a major role in protecting life on earth. The earth's magnetic field helps deflect solar wind, keeping our atmosphere in place. It is no secret that without our atmosphere, life as we know it would cease to exist.

The compass is a commonly used navigational tool that senses the direction of the earth's magnetic field. The compass has a free-spinning or rotating needle, which is a magnetized piece of metal. When allowed to spin, the compass needle aligns itself with the earth's magnetic field and always points toward the magnetic north pole. The opposite end of the needle, which is the end that is used to indicate direction, points in the direction we accept to be north.

Shedding Light

The earth's magnetic field results from charged particles in the flowing molten metal that is present in the earth's core. If the earth's core were to solidify, there would be no flowing molten metal and there would no longer be a magnetic field around the earth.

5.3 Magnetic Behavior at the Atomic Level

Three atomic-level interactions take place deep within objects and substances to create the magnetic effect. These interactions are electron spin, magnetic dipoles, and magnetic domains.

Electron Spin

While electrons orbit around the nucleus of an atom, they also rotate around their own axis, **Figure 5-12**. This concept, known as ***electron spin***, is similar to a planet orbiting around the sun. The earth, for example, rotates around its own axis once every 24 hours as it orbits around the sun once every 365 days.

The direction of spin determines the direction of the magnetic field. The direction of electron spin is identified by an arrow that points either upward or downward, as illustrated in **Figure 5-13**. If the majority of electrons in an atom spin in the same direction, the magnetic field that is produced will be much stronger than when the number of "up" spins and "down" spins are relatively balanced.

When two electrons with opposite spin direction pair, they cancel each other out, and the net magnetic field is zero. Most substances have roughly equal numbers of electrons spinning in each direction. Atoms that only have sets of paired electrons are called diamagnetic substances. Substances classified as ***diamagnetic*** have no magnetic field, are not attracted to magnets, and cannot be magnetized. Examples of diamagnetic substances include aluminum, phosphorus, antimony, copper, hydrogen, and stainless steel.

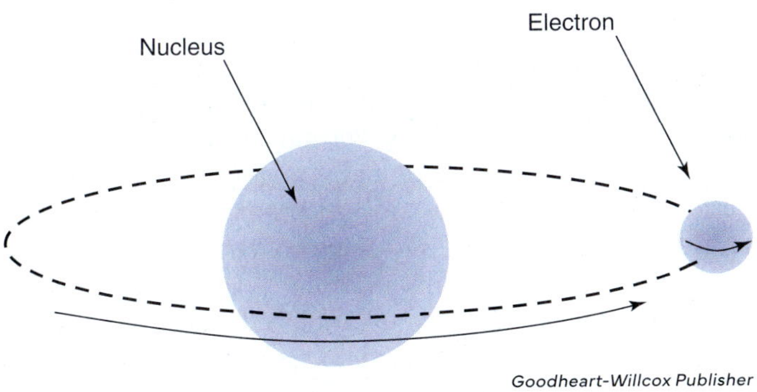

Figure 5-12. Electron spin resembles a planet's rotation on its own axis as it orbits the sun.

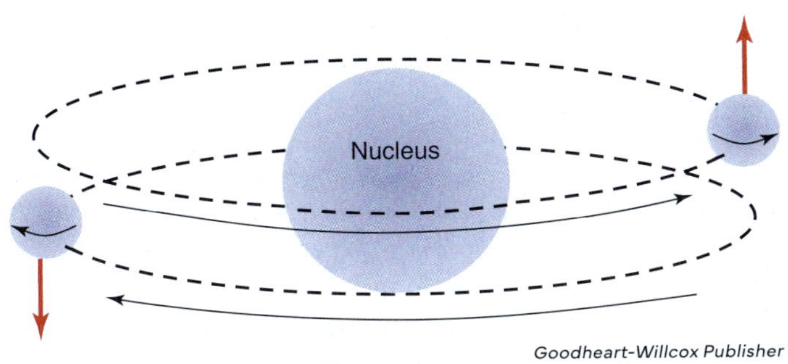

Figure 5-13. A magnetic field is canceled when electron pairs exist.

Atoms that have unpaired electrons are called paramagnetic substances. Substances classified as ***paramagnetic*** have a net magnetic field and are slightly attracted to magnets. Examples of paramagnetic substances include magnesium and lithium.

A third class of substances, called ***ferromagnetic*** substances, have a high level of attraction to magnetic fields. Examples of ferromagnetic substances include iron, cobalt, and nickel.

Magnetic Dipoles

Spinning electrons create a magnetic field, and the direction of the magnetic field is determined by the direction of the spin. When spinning, electrons behave like tiny magnets and become magnetic dipoles. A ***magnetic dipole*** is a microscopic magnet. If these magnetic dipoles line up with their north and south poles oriented in the same directions, the magnetic effect of the dipoles increase. If they do not line up, the magnetic effect decreases, and can, if extreme conditions are present, be completely negated. If many magnetic dipoles are in natural alignment with each other, they can collectively become what is known as permanent magnets.

Magnetic Domains

A ***magnetic domain*** is a physical region or space within a substance where a group of atoms has their magnetic fields pointing all in the same direction. The orientation of the magnetic field of a magnetic domain is determined by the direction of electron spin of the atoms in the domain. All substances are made up of multiple magnetic domains, each of which can be thought of as a very small, individual magnet with a north and south pole.

Different materials behave differently when exposed to an external magnetic field. In some substances, when there is no magnetic field acting on them, the domains are randomly positioned and there is no net magnetic field, as shown in **Figure 5-14A**. Substances that maintain this randomized, disorderly domain arrangement under all, or most, conditions are classified as nonmagnetic substances. These materials, which include copper, aluminum, paper, and glass, are not attracted to magnets, nor can they become magnets. However, in the presence of a magnetic field, the domains in some substances will rotate and align themselves with each other as well as with the external magnetic field. When all, or most, of the domains are aligned with each other, the object becomes magnetized, as shown in **Figure 5-14B**.

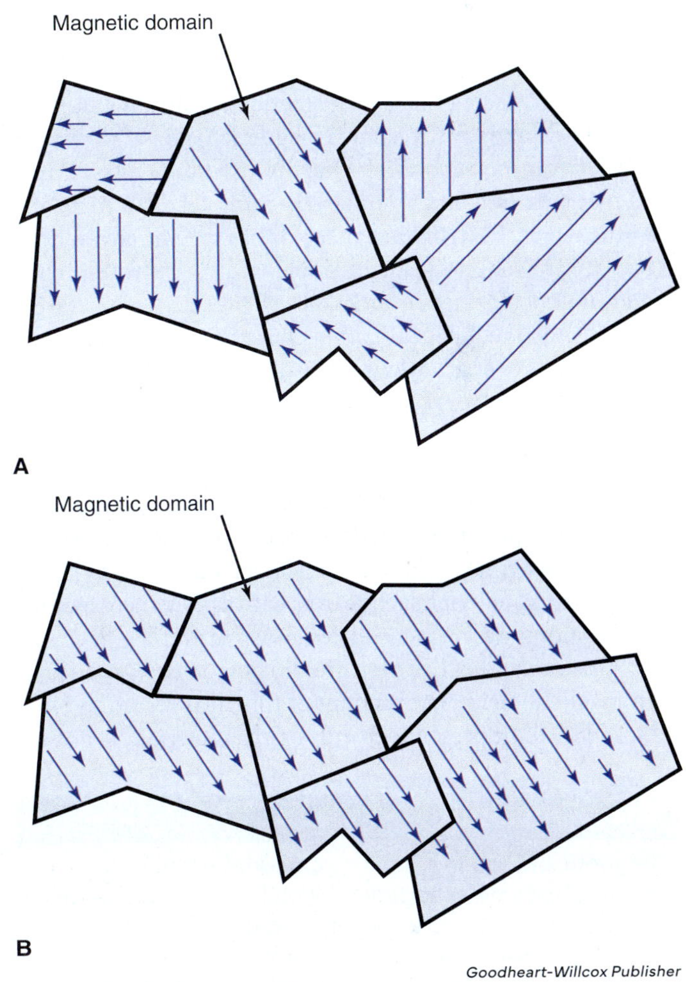

Figure 5-14. Effects of a magnetic field on domains. A—When no magnetic field is present, domains are randomly positioned. B—When a magnetic field is present, domains are aligned with each other.

Magnetic Saturation

There is a limit to the strength of a magnet. When all of the magnetic domains are in alignment, the strength of the magnetic field is at its highest point. This point is referred to as ***saturation***. Characteristics such as saturation and flux density are affected, in part, by the magnetic and physical properties of the substance itself. When an electromagnet is at the point of saturation, increasing the current will not result in a noticeable increase in the strength of the magnetic field.

5.4 Permanent and Temporary Magnets

Some materials act as permanent magnets, while others act as temporary magnets. Substances classified as ***permanent magnets*** maintain their magnetic properties for many years, while ***temporary magnets*** lose their magnetic properties shortly after the external magnetic influence on them has been removed. The most magnetic natural material is ***lodestone***, which is a rare form of the mineral magnetite. Magnetite is attracted to a common magnet, while lodestone itself acts as a magnet. Lodestone is a permanent magnet, partly due to the fact that the magnetic domains of lodestone naturally align with the earth's magnetic field.

Other common materials that are used to create permanent magnets are neodymium, samarium-cobalt, alnico, and ceramic/ferrite. Each of these magnets has specific advantages. Neodymium magnets, for example, are very strong, while samarium-cobalt magnets function well over large temperature ranges. Alnico magnets are often chosen for use in corrosive environments or applications subjected to extremely high temperatures. Ceramic/ferrite magnets are not very strong, but are the most cost effective choice for many applications.

Strong magnets and generated magnetic fields, such as those generated by current flowing in a wire, can be used to produce other magnets by a process called ***magnetic induction***. Magnetic induction can be accomplished by stroking a magnetic material, such as iron, with a permanent magnet or by positioning a magnetic material within a strong magnetic field. Once a magnet has been created, it is able to produce its own magnetic field. The amount of time this resulting magnetic field will be present depends on the material.

Shedding Light

Magnetic induction and electromagnetic induction are not the same. Magnetic induction is used to create magnets, while electromagnetic induction is used to create current flow. One easy way to distinguish between the two is to recall that electric current flow is the movement of electrons through a conductor, so electromagnetic induction pertains to the flow of electrons.

In some materials, the domains will remain lined up in the same direction long after the external magnetic field is removed. In other materials, the domains will quickly return to their random orientations. Permanent magnets have the property of keeping their domains aligned for an extended period of time (many years), while temporary magnets maintain domain alignment only briefly. Generally, permanent magnets are weaker than temporary magnets. Permanent magnets are often made from hard iron or steel and temporary magnets are typically made from soft iron.

Shedding Light

Consider the following scenarios. If a permanent magnet is brought close to a pile of paper clips, the paper clips will be attracted to the magnet. The magnetic domains within the paper clips will line up and the paper clips will also be attracted to each other. Once the permanent magnet is moved away from the pile of paper clips, the magnetic domains within the clips will return to their original, randomly pointing orientations, and the paper clips will no longer be magnetized. However, if an iron nail is stroked with the same permanent magnet, the nail will become a long-lasting magnet. The nail's magnetic domains remain aligned long after the nail is originally magnetized.

Electromagnets

One common type of temporary magnet is the electromagnet. An ***electromagnet*** is created when electric current is passed through a coil of wire. A simple electromagnet can be created by wrapping an insulated conductor or wire around an iron nail, and then connecting the ends of the wire to the terminals of a battery, **Figure 5-15**. The strength of the electromagnet can be increased by using a battery with a higher voltage or by increasing the number of wire turns around the nail.

Electromagnets are commonly used in stereo speakers, MRI machines, motors, generators, electromechanical relays, switching devices, solenoids, and computer hard drives. Electromagnets are also used to move large amounts of metallic objects, such as in a junkyard or in a large-scale production facility.

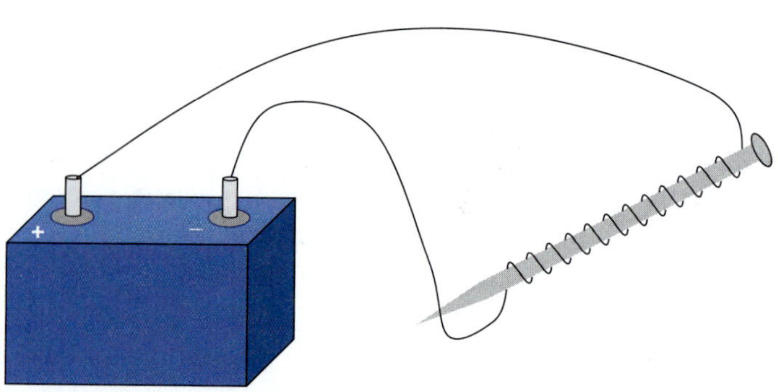

Goodheart-Willcox Publisher

Figure 5-15. A simple electromagnet.

5.5 Magnetic Properties

Magnetic materials are classified by a number of factors, including how easy it is for them to become magnetized and how long it takes for a magnetized material to lose the ability to attract other materials. Four common terms used to describe materials such as soft iron (temporary magnets) and hard steel (permanent magnets) include permeability, reluctance, retentivity, and coercivity, **Figure 5-16**.

Properties of Magnetic Materials

Property	Soft Iron (Temporary Magnets)	Hard Steel (Permanent Magnets)
Permeability	High	Low
Reluctance	Low	High
Retentivity	Low	High
Coercivity	Low	High

Goodheart-Willcox Publisher

Figure 5-16. Magnetic properties that describe soft iron and hard steel.

Permeability

Permeability is a measure of the ease with which a substance can be magnetized. Substances such as wood, brass, and plastic are not attracted to a magnet and have a permeability value of one, which is very low on the permeability scale. Substances with a high permeability value, such as soft iron, are easily magnetized, while substances with a low permeability value, such as hard steel, are difficult to magnetize.

The soft iron or soft steel used as a core in an electromagnet has a high permeability, which increases the strength of the magnetic field produced by the coil.

Shedding Light

Permeability can be looked at in a manner that is similar to the ability of water to flow through the kitchen spout when the faucet handle is in the open position. The resistance to flow is low, so water can flow easily. A similar analogy can be made regarding electric circuits. When the resistance of a circuit is low, more current can flow.

Reluctance

Reluctance is the opposition, or resistance, to the production of a magnetic field. Mathematically, reluctance is the inverse of permeability, and can be expressed as:

$$\text{reluctance} = \frac{1}{\text{permeability}}$$

$$\text{permeability} = \frac{1}{\text{reluctance}}$$

As permeability increases, the reluctance decreases. As permeability decreases, reluctance increases. The relationship between permeability and reluctance is the same as the relationship that exists between the resistance of an electric circuit and the amount of current flowing through the circuit. As circuit resistance increases, the current flow in the circuit decreases. Materials such as hard steel have high reluctance values, while materials such as soft iron have low reluctance values.

Retentivity

Retentivity describes the ability of a substance to retain a magnetic field once the initial magnetizing force is removed. The carbon content in a substance directly relates to the substance's ability to retain its magnetic field in the absence of an external magnetic field. Hard steel, for example, has a high carbon content and is able to retain the uniform, aligned position of the magnetic domains. The ability to maintain magnetic domains in alignment after the external magnetic field has been removed is referred to as ***pinning***. Soft iron has a very low carbon content, so its random domain positioning returns shortly after the external magnetic field is removed.

Coercivity

Once magnetized, it is possible to demagnetize a material. ***Coercivity*** is a material's resistance to changes in magnetization, such as its ability to resist demagnetization. A high coercivity value also indicates that the magnetic material is hard, such as in the case of permanent steel magnets, while a low coercivity level is characteristic of soft iron.

Hard magnetic materials, which are used to create permanent magnets, maintain their magnetic properties for a long period of time and are difficult to demagnetize. Soft magnetic materials, which are used to create temporary magnets, lose their magnetic properties shortly after their exposure to an external magnetic field has ended.

One method of demagnetization is positioning a magnet within a magnetic field opposite to that of the magnet. The greater the coercivity of the magnetic material, the stronger the magnetic field needed to demagnetize. Repeatedly dropping a magnetic object or striking it with a hammer can also demagnetize it, as can heating the magnet to a high temperature. All of these methods cause changes at the atomic level within the magnet.

Shedding Light

Soft iron is commonly used to manufacture motor windings and transformer cores. These electrical components experience frequent power signal polarity changes and the speed at which the magnetic field can be reduced to zero helps to improve the overall efficiency of these electrical devices. Alternating current will typically change direction at a rate of 50 Hz or 60 Hz per second. Soft iron is ideally suited to handle this rapid direction change, making it a good choice for many electrical applications.

CHAPTER 5 REVIEW

Summary

- Magnetism creates repulsive or attractive forces among substances. (5.1)
- Magnetic fields are strongest at and around the poles, or ends, of a magnet. (5.1)
- Magnetic flux measures a magnetic field's relative strength. (5.1)
- Like magnetic poles repel each other while opposite poles attract each other. (5.1)
- The left-hand rule is used to determine the direction of a magnetic field around a current-carrying conductor. (5.1)
- Common units of magnetic field strength are the tesla and the gauss. (5.1)
- Coils of wire are created to increase the effective strength of a magnetic field. (5.1)
- Units of magnetic field coil strength are called magnetomotive force, and are measured by the ampere-turn and the gilbert. (5.1)
- Similar to the law of charges, when a magnet's north pole is positioned near the south pole of another magnet, there is an attractive force. If two like poles are placed in close proximity, they will repel each other. (5.2)
- The earth's magnetic field is the result of flowing, molten metals in the earth's outer core. (5.2)
- As electrons spin, magnetic dipoles are created. (5.3)
- Substances can be classified as diamagnetic, paramagnetic, or ferromagnetic. (5.3)
- In a magnetic domain, the polarities of the magnetic dipoles are aligned. (5.3)
- Permanent magnets have the ability to remain magnetic for many years. (5.4)
- Permanent magnets are commonly made from hard steel. (5.4)
- Temporary magnets are commonly made from soft iron. (5.4)
- Lodestone is the most powerful natural magnetic material. (5.4)
- Magnets can be created by induction. (5.4)
- Four properties that describe a magnetic material are permeability, reluctance, retentivity, and coercivity. (5.5)

Review Questions

Use the information in the chapter to answer the following questions.

Measuring Your Potential

1. If electrons flow through a conductor from left to right, which of the following best describes the direction of the generated magnetic field when using the left-hand rule? (5.1)
   A. In a clockwise direction.
   B. In a counterclockwise direction.
   C. Over the top away from you.
   D. Over the top toward you.
2. The term _____ is used to describe one line of magnetic force. (5.1)
   A. tesla
   B. maxwell
   C. gauss
   D. Edison
3. *True or False?* If the north poles of two magnets are brought close to each other they will attract each other. (5.2)
4. With regards to magnetism, the terms "up" and "down" refer to _____. (5.3)
   A. electron spin
   B. magnetic flux
   C. permanent magnets
   D. permeability
5. Magnetic dipoles are aligned in regions called the magnetic _____. (5.3)
   A. domain
   B. pole
   C. flux
   D. moment
6. *True or False?* Aluminum is an example of a ferromagnetic substance. (5.3)
7. Substances classified as paramagnetic _____. (5.3)
   A. have a high level of attraction to magnetic fields
   B. do not have magnetic fields
   C. cannot be magnetized
   D. have a net magnetic field

8. The magnet that is created by wrapping a conductor around a soft iron core and applying a voltage is called a(n) _____. (5.4)
   A. neodymium magnet
   B. electromagnet
   C. permanent magnet
   D. alnico magnet
9. Induction is a process that is commonly used to _____ magnets (5.4)
   A. create
   B. test
   C. destroy
   D. classify
10. Which of the following magnetic properties are inverses of each other? (5.5)
   A. Permeability and reluctance
   B. Reluctance and coercivity
   C. Reluctance and retentivity
   D. Coercivity and retentivity

Plug-In

1. If the left-hand rule is used, the _____. (5.1)
   A. fingers will always curl in the direction of current flow
   B. fingers will always curl in the direction of the magnetic field
   C. thumb always points in the direction of the magnetic field
   D. Both A and C
2. Which of the following will result in the production of 100 ampere-turns? (5.1)
   A. 5 A flowing through a coil with 10 turns
   B. 2 A flowing through a coil with 50 turns
   C. 4 A flowing through a coil with 20 turns
   D. 3 A flowing through a coil with 25 turns
3. Electron spin can most accurately be equated to _____. (5.3)
   A. the flight of a paper airplane
   B. a ball being dropped from the roof of a building
   C. a planet orbiting around the sun
   D. the earth rotating around its axis
4. Substances with a high permeability are _____. (5.5)
   A. resistant to magnetic field production
   B. able to maintain magnetic properties
   C. easy to magnetize
   D. required to be field demagnetized
5. Substances that have a high reluctance are best described as _____. (5.5)
   A. resistant to magnetic field production
   B. able to maintain magnetic properties
   C. easy to magnetize
   D. required to be field demagnetized
6. Which type of material with a low coercivity would be used in a temporary electromagnet? (5.5)
   A. Hard copper
   B. Hard iron
   C. Soft iron
   D. Hard steel
7. If the carbon present in a metal is high, this would likely result in a _____. (5.5)
   A. permanent hard steel magnet
   B. strong, temporary hard steel magnet
   C. weak, permanent hard steel magnet
   D. strong, temporary soft steel magnet

Making the Connection

1. How does using a material with a low coercivity help to increase the operating efficiency of an electric motor? (5.5)
2. Why are there two conventions used to describe current flow? (5.1)

SECTION

2

Electrical Quantities and Relationships

nelikdulatov/Shutterstock.com

CHAPTER

6 Electrical Characteristics

Hussain Warraich/Shutterstock.com

SPARKING DISCUSSION

In this chapter, an electric circuit and its primary characteristics—voltage, resistance, and current—are compared to the flow of water through a piping system. In this analogy, water pressure equates to voltage, pipe size equates to resistance, and water flow equates to current. Discuss other systems that could be compared with an electric circuit, and identify the components of that system that equate to voltage, resistance, and current.

LEARNING OUTCOMES

After completing this chapter, you will be able to:

6.1 Describe how an electric circuit works.
6.2 Describe how electric current is measured.
6.3 Define how resistance affects the flow of electrons.
6.4 Explain how power in an electric circuit is measured.
6.5 Apply Ohm's law and Watt's law to determine power, current, voltage, and resistance in an electric circuit.

TECHNICAL TERMS

alternating current (ac)
ampere (A)
average dc voltage
coulomb
current
direct current (dc)
electromotive force (EMF)
equipment grounding conductor
ground
grounded conductor
load
neutral conductor
ohm (Ω)
Ohm's law
power
resistance
ungrounded conductor
volt (V)
voltage
watt (W)
Watt's law

Introduction

Although there are many different jobs and careers in the electrical trades, all successful workers in this great industry have one thing in common—they all have a solid understanding of electrical systems and how these systems are affected by changes to circuit configurations and component selection.

To become successful in the electrical industry, it is simply not enough to know what to do or what happens when a particular action is taken. It is important to understand installation regulations and the impact of changes on circuit performance. A successful electrician, for example, has a deeper understanding of why a light fixture illuminates when the wall switch is turned on, as opposed to just knowing that when the light switch is flipped, the light turns on. This level of understanding and comprehension begins with the basics.

This chapter introduces the electrical quantities of voltage, current, resistance, and power. The relationship among these forces creates the foundation for understanding the concepts discussed in the remainder of this text.

Electrical characteristics indicate, among other things, how much electrical pressure is being exerted on the circuit, how much current is flowing through the circuit, how much power and energy are being consumed, and how efficiently the circuit or system is operating. Voltage, current, resistance, and power are commonly measured by the electrician to evaluate and troubleshoot circuit and system problems.

As you work through this chapter, keep in mind that taking your time to carefully and systematically grasp the concepts as presented is the key to future success. By building the required strong foundation, you will reap immeasurable benefits as you progress in your career.

6.1 Voltage

In the typical electric circuit, there is a flow of electrons through a conductor. In order for electrons to flow, there must be some type of force or pressure to push the electrons through the circuit. Voltage can be equated to the water pressure behind the valve in a kitchen sink, **Figure 6-1**. When the valve is closed, no water flows out of the spout. Although water is not flowing out of the spout, there is water pressure behind the faucet, indicating that the potential for flow exists. The closed valve creates a blockage to water flow. In this situation, there is water pressure, but no water flow. By opening the valve, the blockage is removed, and water flows. When the valve is open, there is both water pressure and water flow. The same concept can easily be applied to electricity. Voltage is the pressure that pushes electrons through a circuit, just as pressure pushes water through a pipe. Voltage does not flow, electrons do.

Voltage is the difference in charge or electrical potential between two points or objects. Voltage is also the potential or ability to do work in an electric circuit. Voltage is expressed in units called ***volts (V)***. A voltage of 20 volts is written as 20 V. In algebraic formulas and expressions, voltage is represented by the letter *E*, which stands for electromotive force. ***Electromotive force (EMF)*** is the voltage that causes current to flow in an electric circuit. Electromotive force, as voltage, is the electrical pressure that pushes electrons through the wires in a circuit. The difference in potential causes current to flow, as long as a complete, uninterrupted path between the points of potential difference exists.

Shedding Light

We can compare voltage, or potential difference in charges, with velocity. You and your family are driving in a car traveling at 60 miles per hour. The car and its passengers, moving at a relatively high velocity, can be thought of as a point of greater electrical charge. If you reach out your hand and pat the person sitting next to you, little energy is exchanged between you and the passenger because you are both moving at the same velocity (comparable to having the same electrical charge). However, if a person is standing along the side of the road (comparable to a point of lesser electrical charge) and you extend your hand out of the window to pat that stationary person, a large amount of energy will be exchanged due to the differences in velocities (comparable to a difference in electrical charge). Needless to say, the person standing on the side of the road will not be very happy.

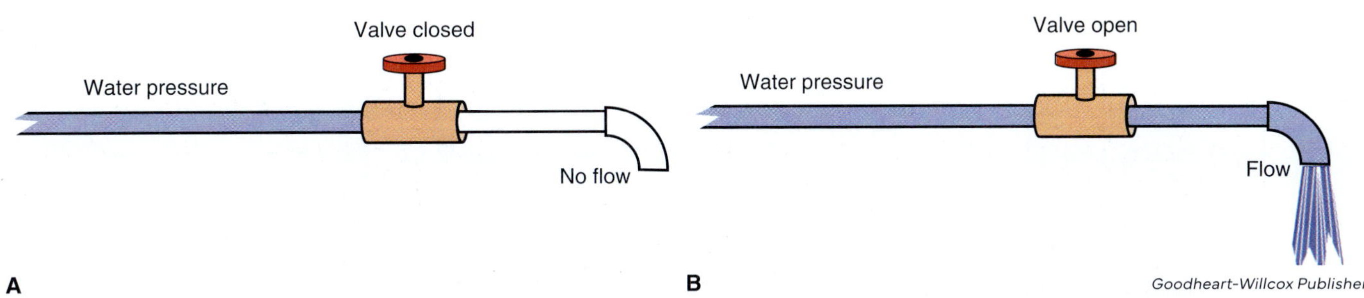

Figure 6-1. The flow of electrons in an electrical circuit can be compared to the flow of water through a piping system. A—With the valve closed, there is water pressure, but no water flow. B—With the valve open, there is water pressure and water flow.

Ground as a Voltage Reference Point

With voltage being described as the difference in charge between two points in an electric circuit, knowing the charge at one of the two points provides valuable information that electricians can use to evaluate electrical systems and circuits. Consider the situation where an electrician obtains a voltage reading of 0 V between two points, A and B, in an electric circuit, **Figure 6-2**. Armed only with this information, all that is known is that the electrical charges at both points are equal to each other, as the difference in charge between these two points is zero. Both charges can be zero, or they can be equal to any nonzero value. Electrically, these points are identical, meaning they have the same electrical charge or voltage.

Having a reference point to which obtained readings can be compared helps eliminate any confusion that might arise. Consider an electrical distribution panel that might be found in the average residence. This panel has one grounded terminal block and two ungrounded conductors, **Figure 6-3**. In an electrical system such as this one with two of the conductors intentionally ungrounded, the reference point is the earth or ***ground***, which can be thought of as having a potential of 0 V. The potential difference between the two circuit breakers connected to the red ungrounded conductor is 0 V, as is the potential difference between the two circuit breakers connected to the black ungrounded conductor, **Figure 6-4**. However, measuring the potential difference between the

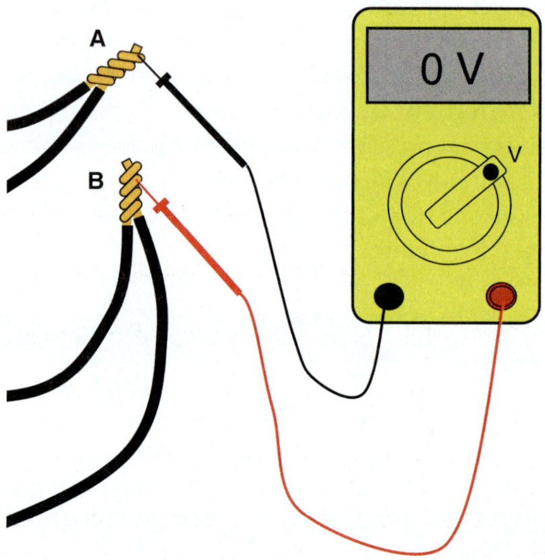

Goodheart-Willcox Publisher

Figure 6-2. A reading of 0 V is obtained between two points in a circuit. This indicates that the charge at points A and B is identical.

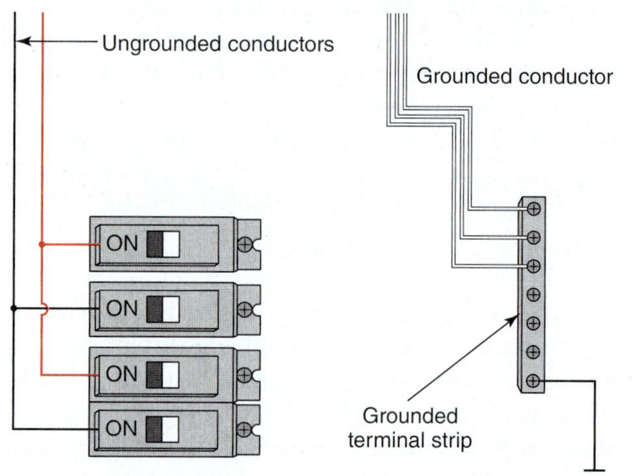

Goodheart-Willcox Publisher

Figure 6-3. A residential electrical distribution panel with circuit breakers, a grounded terminal strip, and two ungrounded conductors.

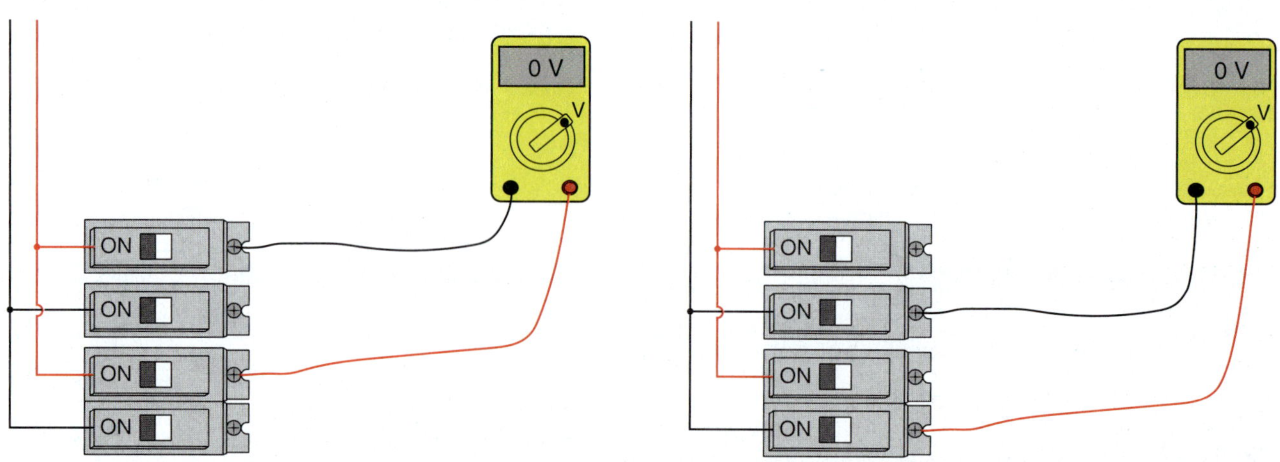

Goodheart-Willcox Publisher

Figure 6-4. When two circuit breakers are fed by the same ungrounded conductor, the potential difference between the load sides of the two circuit breakers will be 0 V.

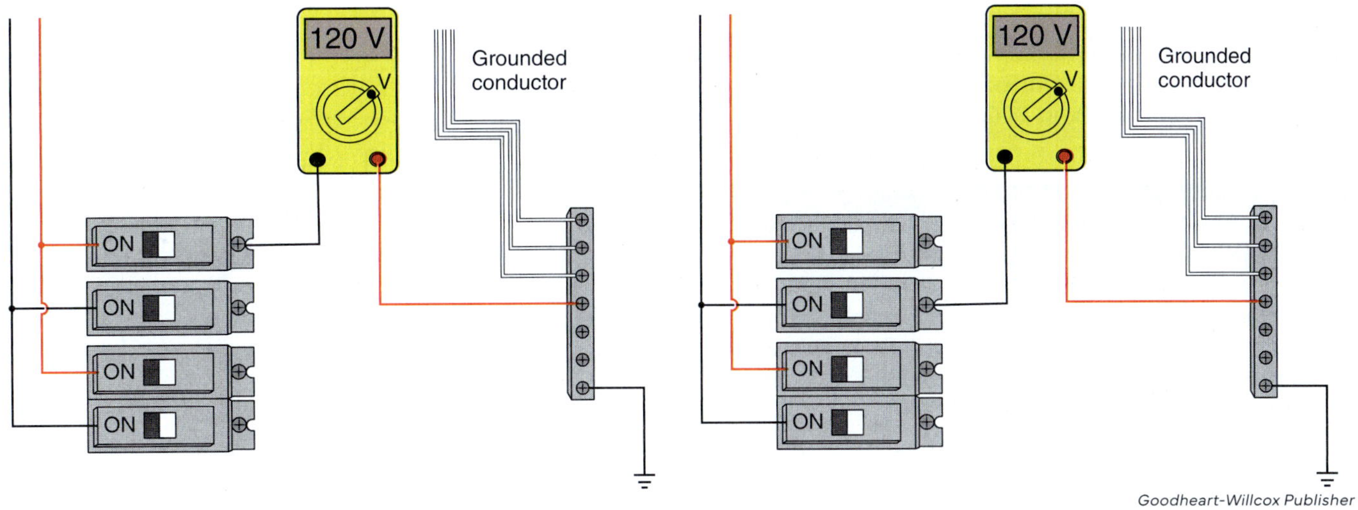

Figure 6-5. In this electrical panel, the potential difference between an ungrounded conductor, either red or black, and a grounded conductor is 120 V.

red ungrounded conductor and the grounded terminal block yields a reading of 120 V. The same holds true for a measurement between the black ungrounded terminal and the grounded terminal block, **Figure 6-5**.

Shedding Light

Most electrical distribution systems will have one of the conductors intentionally grounded. Systems without an intentionally grounded conductor might operate in some industrial or health care situations, but are not common.

A system or piece of equipment is said to be grounded when there is an intentional connection made between the equipment and ground, or to another conductor that is connected to ground. In an electrical system, an ***equipment grounding conductor*** is used to connect conductive components that are not intended to carry current with ground. Equipment grounding conductors are either bare copper wires or conductors with green insulation or green insulation with a yellow stripe. Grounding conductors are intended to carry current only when a fault condition is present, not during the normal operation of the system.

Know the Code

In most instances, electrical services and systems (power supplies) are required by the *NEC* to be grounded. *NEC Sections 250.20*, *250.21*, and *250.22* list situations in which the grounding rules are modified to allow ungrounded installations. Some exceptions involve control transformers, cranes over combustible materials, circuits in parts of health care facilities, and larger ungrounded delta systems in industrial installations.

Electrical systems also often include a grounded conductor. A ***grounded conductor*** is a current-carrying conductor that is connected to ground. The difference in electrical charge or potential between a grounded conductor and an ungrounded conductor creates the voltage that powers the electrical devices in a circuit.

Shedding Light

Grounded conductors and equipment grounding conductors (ground) are both connected to ground at the main service panel. Beyond the service panel, the two systems (grounded and equipment grounding) are separate entities and generally do not connect again in the wiring system. The exception to this occurs when a separately derived system is installed (*NEC Section 250.30(A)(1)*). Examples of separately derived systems include solar photovoltaic systems, some generators, and transformers where the primary is isolated from the secondary.

Voltage at a Standard Wall Outlet

Consider a standard electrical outlet, or receptacle, in a home or office. In most cases, the outlet has two parallel, rectangular slots and one round hole. One slot is longer than the other. These two slots are connected to conductors that carry current to and from the appliance or system that is plugged into the receptacle. The longer slot is connected to the grounded conductor. In residential applications, the grounded conductor is almost always color-coded white and correctly referred to as the neutral conductor. A ***neutral conductor*** is a conductor that carries current during normal operation and is connected to the neutral point of the system.

Know the Code

Coloring options for grounded conductors are listed in *NEC Section 200.6(A)* and *200.6 (B)*.

Shedding Light

Although an equipment grounding conductor is connected to ground, it is not considered to be a grounded conductor. Grounded conductors carry current during the normal operation of a circuit, while equipment grounding conductors should only carry current when a fault condition is present.

The shorter rectangular slot in a receptacle is connected to the ***ungrounded conductor***. An ungrounded conductor, like a grounded conductor, carries circuit current during normal circuit operation. However, an ungrounded conductor is not connected to ground and is at a potential that is higher than that of the grounded conductor. Ungrounded conductors often have black, red, or blue insulation, with black and red insulation being common in 120-V systems. An ungrounded conductor is referred to as a "hot" conductor. The equipment grounding, grounded, and ungrounded conductors are identified in **Figure 6-6**.

A standard receptacle supplies 120 V to appliances and other devices that are plugged into it. The 120 V is the potential difference between the ungrounded conductor and the grounded and grounding conductors, **Figure 6-7**. The ungrounded conductor has a potential of approximately 120 V with respect to ground, while the grounded conductor has a potential of about 0 V with respect to ground. We know from our life experiences that electric current is not shooting out of the wall, but when an appliance or other device is plugged into the outlet, electric current will flow between the ungrounded and grounded conductors as long as there is a complete conductive path for the current to take.

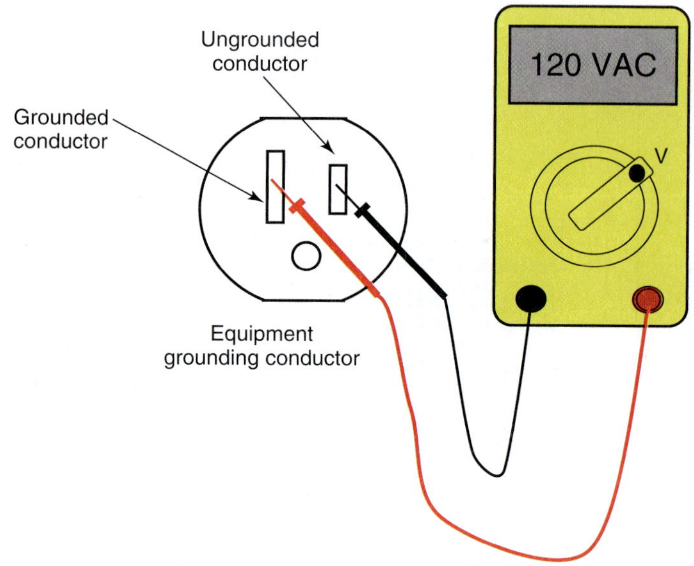

Goodheart-Willcox Publisher

Figure 6-7. The measurement between the ungrounded conductor and the grounded conductor is 120 V.

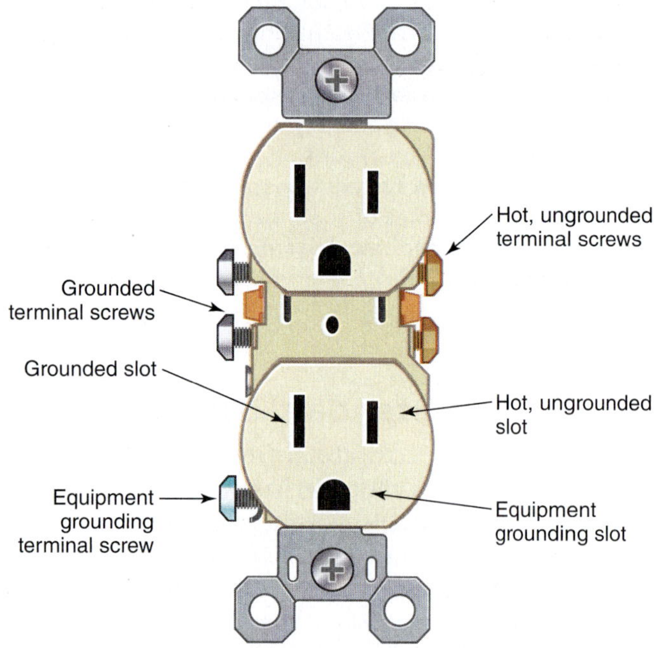

Goodheart-Willcox Publisher

Figure 6-6. Diagram of a typical residential wall outlet or receptacle showing the ungrounded conductor, the grounded conductor, and the grounding conductor.

AC and DC Voltage

There are two common types of voltage that we rely on every day, namely direct current and alternating current. The type of power most likely found in your home is called ***alternating current (ac)***. In alternating current, the polarity of the voltage continuously changes from negative to positive. This changing voltage results in current flow that alternates in direction. In a typical residential electrical system in the United States, the voltage reverses polarity 60 times per second. An ac waveform often takes on the shape of a sine wave, **Figure 6-8**. AC waveform is discussed in detail in Chapter 15, *Single-Phase Alternating Current Power*.

Unlike alternating current, ***direct current (dc)*** travels in one direction and does not change polarity, **Figure 6-9**. The voltage supplied over time by the dc power signal is always +V volts. In most cases, however, there are fluctuations in the voltage, **Figure 6-10**. The actual dc power signal is not typically represented by a straight, horizontal line. The most common source of dc voltage is batteries that are used

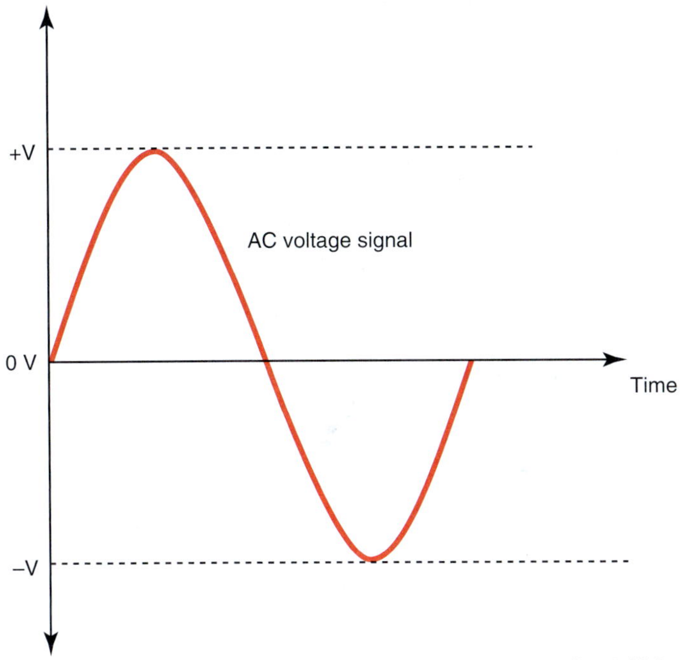

Goodheart-Willcox Publisher

Figure 6-8. Diagram of an ac sine wave; the voltage changes from positive to negative repeatedly over time.

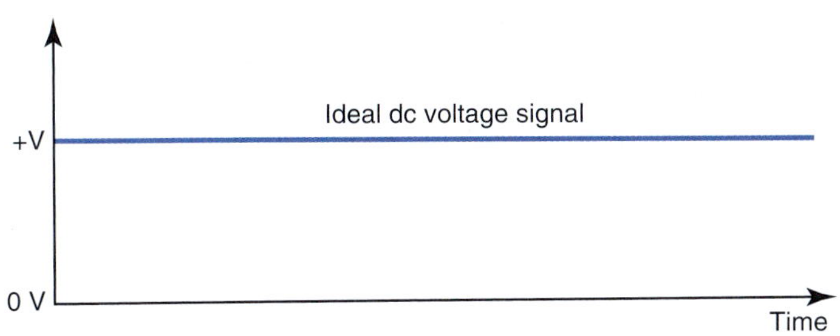

Goodheart-Willcox Publisher

Figure 6-9. Diagram of an ideal dc voltage supply.

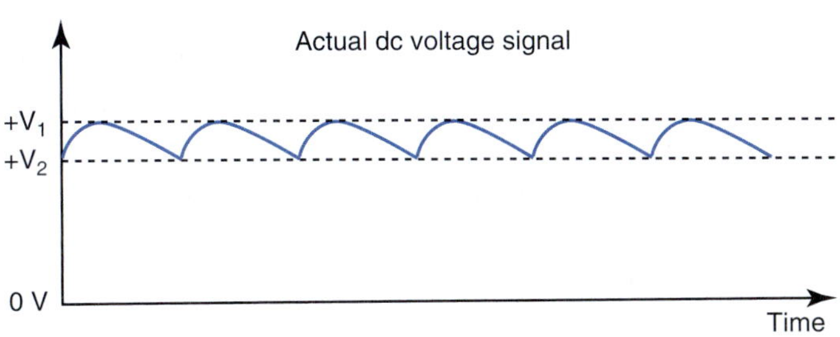

Goodheart-Willcox Publisher

Figure 6-10. Sample diagram of an actual dc voltage supply; in practice, the dc voltage signal will show some fluctuation.

to power our portable electric and electronic devices. The remainder of this chapter refers to dc power.

Because of the fluctuations that are often present in a dc power signal, the obtained measurement falls in between the

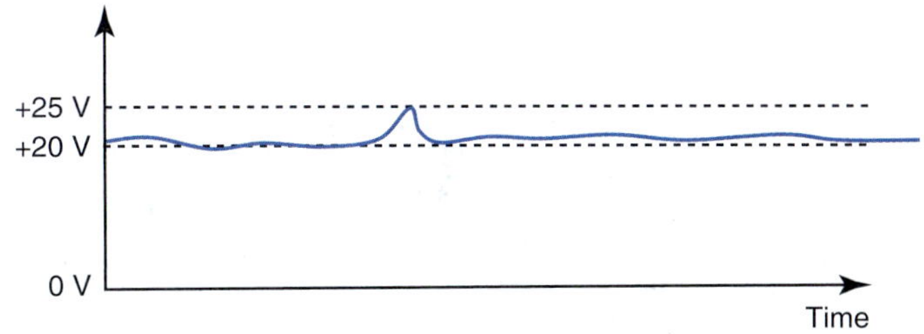

Goodheart-Willcox Publisher

Figure 6-11. The average voltage of this dc power supply is close to 20 V.

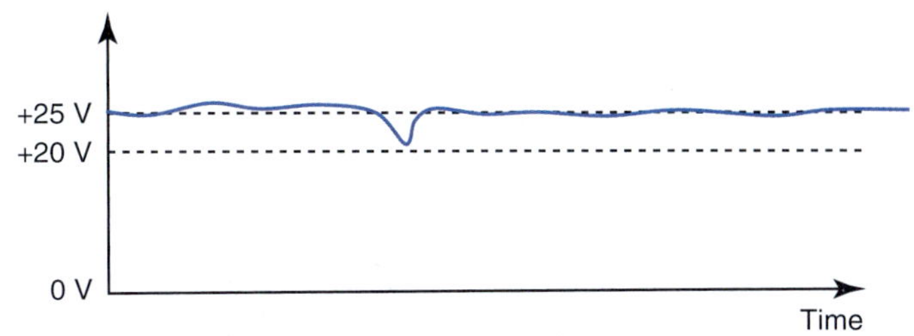

Goodheart-Willcox Publisher

Figure 6-12. The average voltage of this dc power supply is close to 25 V.

minimum and maximum voltages. The obtained dc voltage reading is referred to as the ***average dc voltage***. The average dc voltage is a weighted average that takes into account the amount of time that a particular voltage is being supplied to the circuit when compared to other voltages. The average voltage of the dc waveform illustrated in **Figure 6-11** is much closer to 20 V than 25 V, because, for the majority of the time, the voltage is very close to 20 V. Similarly, the average voltage of the dc waveform in **Figure 6-12** is much closer to 25 V than 20 V.

6.2 Current

The ***current*** in an electric circuit is the rate at which electrons flow through the circuit. Current is measured in amperes, or amps. We use the letter A to represent amps, while the letter *I*, which stands for the intensity of current, is used in mathematical calculations and formulas. For example, the expression $I = 3$ A is interpreted as the current is 3 A.

The unit ***ampere (A)*** refers to the rate of current flow, or more specifically, the number of electrons that pass through a circuit conductor per unit time. A related unit is the ***coulomb***, which is the unit for electric charge. One ampere is the equivalent of a 1-coulomb charge of electricity passing any given point in a circuit in one second. A 1-coulomb charge is equivalent to the charge of 6.241×10^{18} (6,241,000,000,000,000,000) electrons. Therefore, when 6.241×10^{18} electrons pass through a conductor in one second, a current of 1 A is flowing.

Referring back to our kitchen faucet example, an amp can be looked at as the water-equivalent of the amount of water that flows from the faucet, expressed in gallons of water per minute, or gpm. The amount of water flowing through the faucet is determined, in part, by the amount of water pressure in the piping arrangement.

The amount of current that flows in a circuit is determined by the voltage being supplied to the circuit as well as the ease with which the current can flow through the circuit. An increase in the voltage supplied to the circuit increases the amount of current in the circuit. As this pressure increases, the number of electrons pushed through the circuit per unit time increases, thereby increasing the current. On the other hand, a decrease in the voltage supplied to the circuit decreases the amount of current in the circuit.

6.3 Resistance

In addition to changes in the supplied voltage, the electron flow in an electric circuit is also affected by circuit elements and components that act as obstructions to the free flow of electrons. Collectively, the term ***resistance*** is used to describe any material that opposes current flow.

Resistance and current are inversely related. As the circuit resistance, or opposition to current flow, increases, electron flow decreases. As circuit resistance decreases, electron flow increases. Consider a blockage in a water pipe. The blockage causes resistance to the flow and, as a result, the rate of water flow through the pipe is reduced. Pipe diameter can also be compared with electrical resistance. A larger pipe (comparable to lower electrical resistance) allows more water to flow than a smaller diameter pipe (comparable to higher electrical resistance). In all cases, greater resistance results in less flow.

Good conductors of electricity have low resistance, while poor conductors, called insulators, have high resistance. In addition, good conductors of electricity are typically good conductors of heat, while poor conductors of electricity are typically poor conductors of heat. **Figure 6-13** provides examples of both good conductors and poor conductors of electricity.

Good Conductors
- Copper
- Aluminum
- Gold
- Platinum
- Silver

Poor Conductors
- Wood
- Plastic
- Paper
- Air
- Rubber

Goodheart-Willcox Publisher

Figure 6-13. Examples of good conductors and poor conductors of electricity.

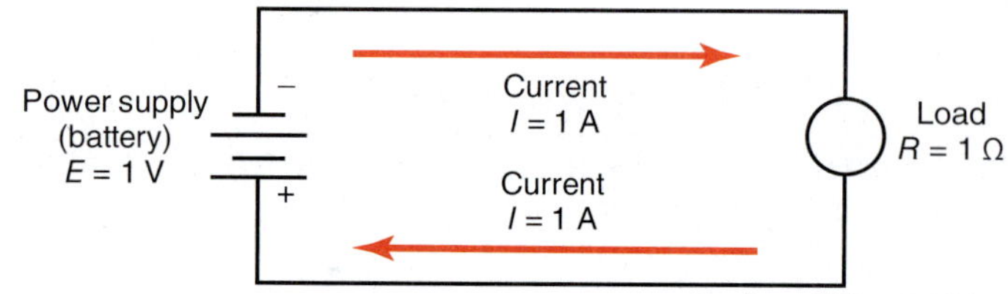

Goodheart-Willcox Publisher

Figure 6-14. Circuit representation of a current of 1 A, a 1-V battery, and a load of 1 Ω.

Resistance is measured in ohms, which is indicated by the Greek letter omega (Ω). The letter R is used to represent resistance in mathematical calculations and formulas. For example, the expression $R = 5\ \Omega$ is interpreted as the resistance is equal to 5 Ω. One ***ohm*** (Ω) is defined as the amount of resistance that will allow one ampere of electric current to flow in a circuit when a potential of 1 V is supplied to it, **Figure 6-14**.

6.4 Power

When voltage is applied to an electric circuit and current flows through the circuit, electric power is consumed. In science, power is defined as the rate at which work is done. In an electric circuit, ***power*** is the amount of electrical energy converted or transferred in a given amount of time. The amount of power consumed by an electric circuit depends on a number of factors, including the type of load that is being powered.

In its most general terms, power (P) is expressed in ***watts (W)***, where 1 W of power is transferred through a circuit when 1 V causes a current of 1 A to flow.

The term kilowatt (kW) might be familiar to many, especially for those who closely examine their energy bill. The charges on the typical energy bill are based on the amount of electrical energy used, expressed in kilowatt-hours, where 1 kW is equal to 1000 W, **Figure 6-15**. For example, if a 1000-W appliance operates for one hour, 1 kilowatt-hour of energy is used. Similarly, if a 1000-W appliance operates for two hours, or if a 2000-W appliance operates for one hour, 2 kilowatt-hours of energy are used. The cost of the energy used each month is determined by the number of kilowatt-hours used, multiplied by the cost per kilowatt-hour, plus any taxes, fees, and regulatory surcharges.

6.5 Ohm's Law and Watt's Law

To gain a better comprehension of the relationships that exist among voltage, current, resistance, and power, the concept of the electric circuit needs to be understood. All electrical

Common SI System Unit Prefixes

Prefix	Name	Abbreviation	Exponential Representation	Numerical Representation
milli	milliwatt	mW	1×10^{-3}	1 mW = 0.001 W
base unit	watt	W	1×10^{0}	1 W
kilo	kilowatt	kW	1×10^{3}	1 kW = 1000 W
mega	megawatt	MW	1×10^{6}	1 MW = 1,000,000 W
giga	gigawatt	GW	1×10^{9}	1 GW = 1,000,000,000 W

Goodheart-Willcox Publisher

Figure 6-15. Common abbreviations and exponential representations for power measured in watts.

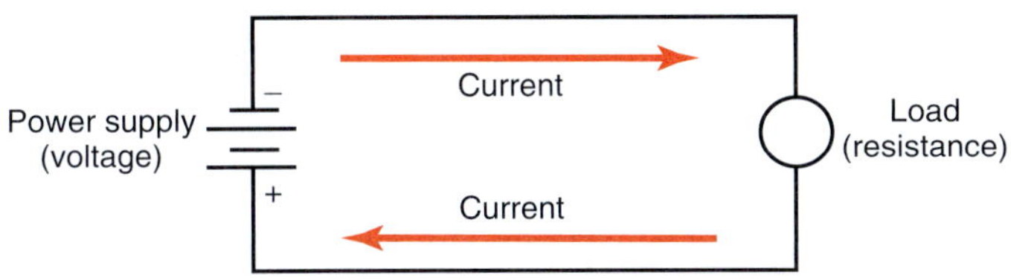

Goodheart-Willcox Publisher

Figure 6-16. A basic series circuit. Batteries supply dc voltage, resulting in current flow in one direction. Series circuits provide a single path for current to flow through the circuit.

circuits must include a power supply, a load, and a complete path to function, **Figure 6-16**.

The power supply is the means by which electrical pressure, or voltage, is supplied to the circuit. The power supply can be a dc (direct current) power supply or an ac (alternating current) power supply. Batteries are a common dc power supply, and generators and transformers are common ac power supplies. In a circuit with a dc power supply, current flows in one direction. In a circuit with an ac power supply, the direction of current flow alternates.

The ***load*** is the device that consumes electrical power (and provides resistance). A circuit may contain a single load or multiple loads. Loads can take the form of appliances, light bulbs, motors, electric heating elements, or any other power-consuming device. Although most circuits use some type of switching device to control the operation of the load, discussion of this greatly simplified circuit does not include a control device.

A predictable relationship exists among voltage, current, and resistance. To introduce this relationship at its most basic and fundamental level, the following relationships exist in a dc circuit:

- The current increases as the voltage increases.
- The current increases as the resistance decreases.
- The current decreases as the voltage decreases.
- The current decreases as the resistance increases.
- The voltage supplied to a circuit is equal to the circuit resistance multiplied by the circuit current.

Ohm's Law

The relationship that exists among voltage, current, and resistance is known as ***Ohm's law*** and is expressed in its simplest form as:

$$E = I \times R$$

where:

E = Voltage, measured in volts (V)
I = Current, measured in amperes (A)
R = Resistance, measured in ohms (Ω)

This formula states that the voltage supplied to a circuit is equal to the circuit resistance multiplied by the circuit current. Ohm's law can also be expressed as follows:

$$I = \frac{E}{R}$$

$$R = \frac{E}{I}$$

The first formula above states that the current flowing in a circuit is equal to the circuit voltage divided by the resistance of the circuit. The second formula states that the resistance of the circuit is equal to the circuit voltage divided by the current flowing in a circuit.

It is often easier to think of Ohm's law in a pictorial sense as opposed to memorizing its three basic forms, **Figure 6-17**. In the figure, a triangle is divided into three sections, with the top section representing voltage, E. The bottom two sections

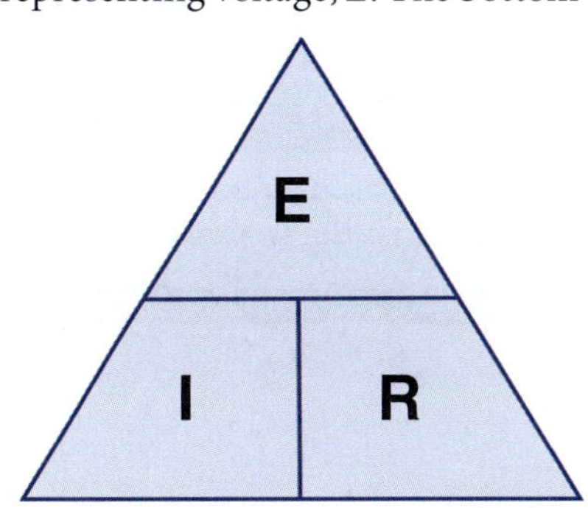

Goodheart-Willcox Publisher

Figure 6-17. The Ohm's law triangle is a helpful tool in remembering the formulas needed to solve problems.

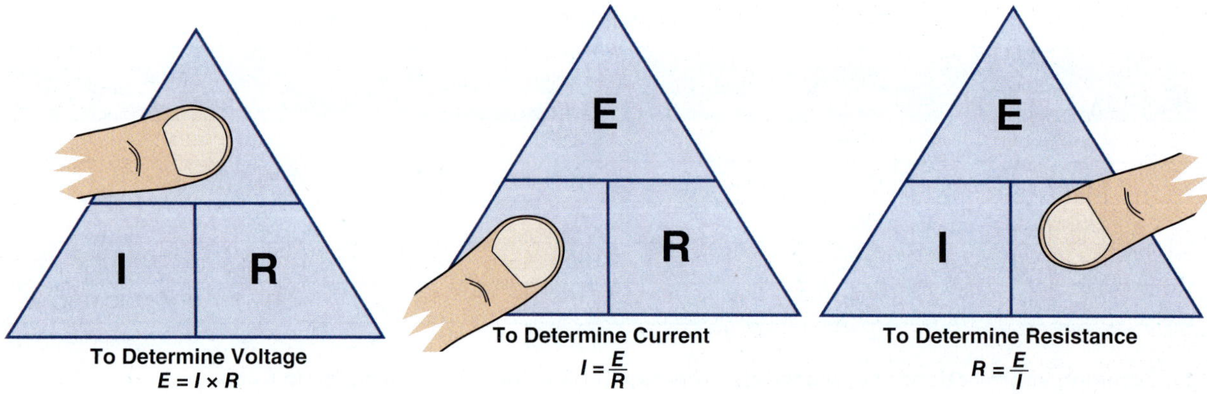

Goodheart-Willcox Publisher

Figure 6-18. To find the formula to solve for the unknown value, you can cover the unknown value on an Ohm's law triangle like the one shown here. For example, if the resistance (R) and current (I) of a circuit are known, and voltage (E) needs to be calculated, you will need to cover the E to reveal the formula to calculate voltage ($I \times R = E$).

represent current, I, and resistance, R. You can use the Ohm's law triangle to determine the correct formula to use to obtain the value needed, **Figure 6-18**. Refer to **Example 6-1** on how to calculate the amount of current flowing through a current.

Example 6-1

Consider the following circuit, which has a 10-V battery that is powering a 5-Ω load. Determine the amount of current flowing in this circuit.

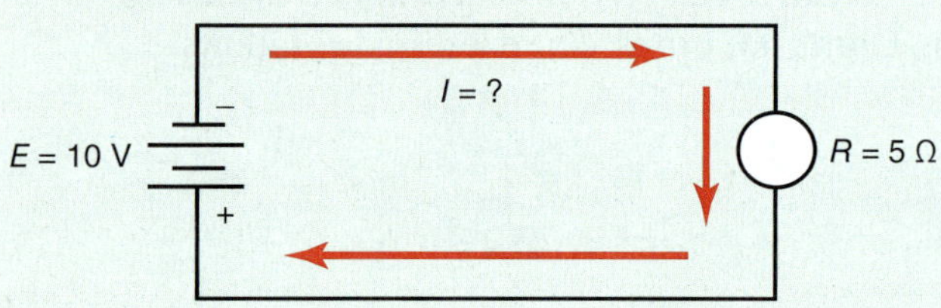

Goodheart-Willcox Publisher

✓ Solution

Using the Ohm's law triangle, determine the formula needed to find the amount of current (I).

$$I = \frac{E}{R}$$

$$I = \frac{10\ \text{V}}{5\ \Omega}$$

$$I = 2\ \text{A}$$

Now It's Your Turn 6-1

The following circuit has a 20-V battery and has 4 A of current. Using Ohm's law, determine the circuit resistance.

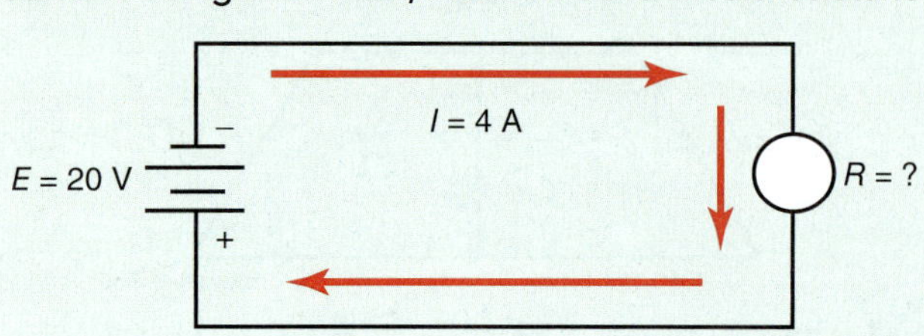

Goodheart-Willcox Publisher

Watt's Law

The relationship that exists among power, current, and voltage in an electric circuit is known as ***Watt's law***. Just as there are three different forms of Ohm's law, there are also three different forms of Watt's law:

$$P = I \times E$$

$$I = \frac{P}{E}$$

$$E = \frac{P}{I}$$

where:

P = Power, measured in watts (W)
I = Current, measured in amperes (A)
E = Voltage, measured in volts (V)

Similar to Ohm's law, Watt's law can be illustrated using a triangle, **Figure 6-19**. The use of the Watt's law triangle is described in **Figure 6-20**.

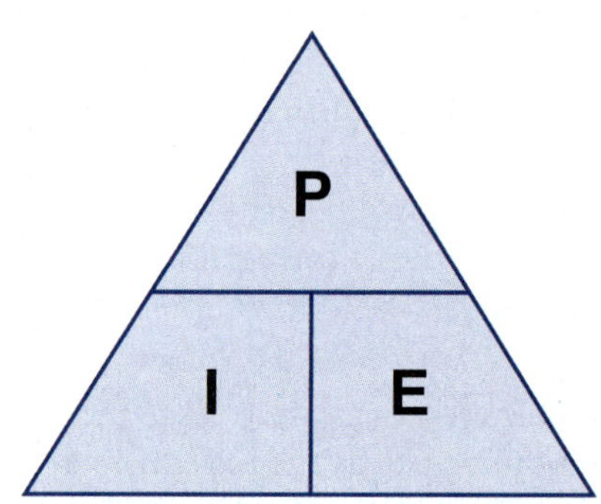

Goodheart-Willcox Publisher

Figure 6-19. As with Ohm's law, the Watt's law triangle can be used to help remember the formulas for solving the unknown value.

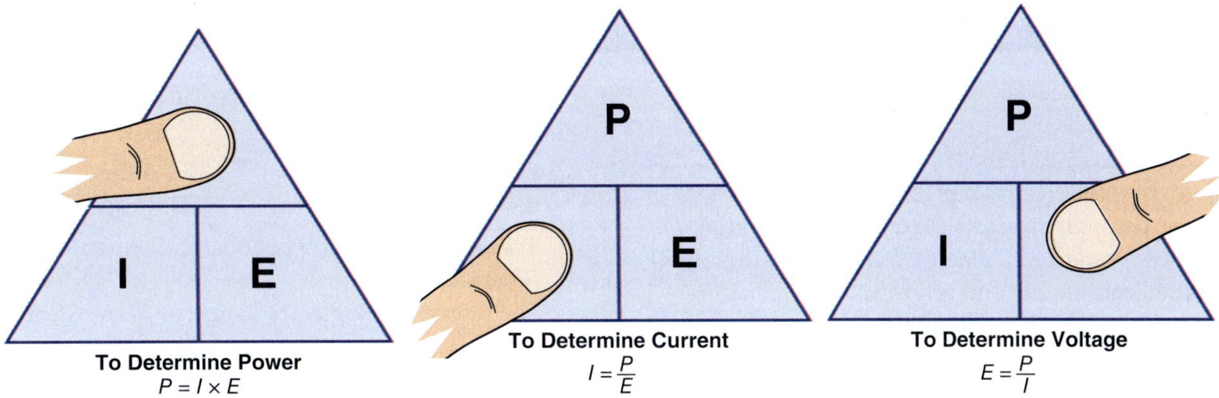

Goodheart-Willcox Publisher

Figure 6-20. Using the Watt's law triangle is similar to using the Ohm's law triangle to find the formula to solve for an unknown value. For example, when calculating power, cover the *P* in the triangle to reveal the formula $P = I \times E$.

The electrical characteristics of voltage, current, resistance, and power can be calculated in a number of different ways. In fact, there are three formulas for each of these four electrical quantities. A PIRE wheel is another tool you can use to recall the formulas that involve watts (*P*), amperes (*I*), ohms (*R*), and volts (*E*), **Figure 6-21**. Refer to **Example 6-2** on how to calculate power consumed by a circuit.

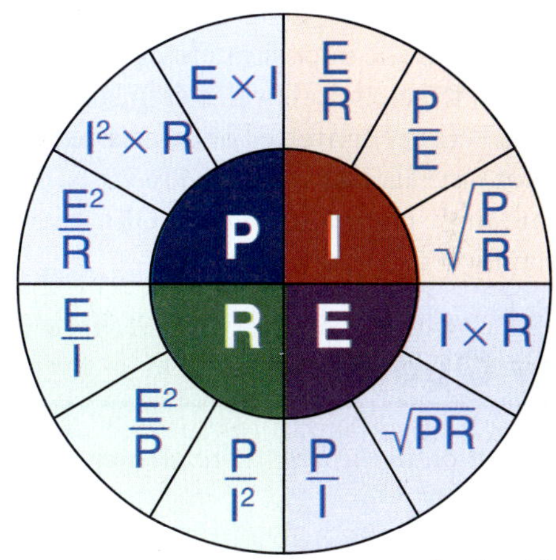

Goodheart-Willcox Publisher

Figure 6-21. A PIRE wheel shows formulas associated with Ohm's law and Watt's law. It is possible to calculate power, current, voltage, or resistance if any two of the other factors are known quantities.

Example 6-2

Calculate the power consumed by the circuit shown in **Example 6-1**.

✓ Solution

This circuit has a supply voltage of 10 V and a current of 2 A. Using the $P = I \times E$ form of Ohm's law, the power consumed by this circuit can be calculated as follows:

$$P = I \times E$$

$$P = 2\ \text{A} \times 10\ \text{V}$$

$$P = 20\ \text{W}$$

The power consumed by this circuit can also be calculated using the $P = I^2 \times R$ form of the power formula, as shown in **Figure 6-21**. Using the $P = I^2 \times R$ form of Ohm's law, the power consumed by this circuit can be calculated as follows:

$$P = I^2 \times R$$

$$P = (2\ \text{A})^2 \times 5\ \Omega$$

$$P = 20\ \text{W}$$

A third way to calculate the power consumed by this circuit uses the $P = \frac{E^2}{R}$ form of the power formula, as shown in **Figure 6-21**. The power consumed by this circuit can be calculated as follows:

$$P = \frac{E^2}{R}$$

$$P = \frac{(10\ \text{V})^2}{5\ \Omega}$$

$$P = 20\ \text{W}$$

Now It's Your Turn 6-2

If the voltage in **Example 6-1** is increased to 20 V, calculate the power consumed by the circuit using the three forms of the power formula.

CHAPTER 6 REVIEW

Summary

- Voltage, E, is the electrical pressure that pushes electrons through an electrical circuit. A potential difference is required if electric current is to flow. (6.1)
- Grounded conductors and ungrounded conductors carry current during the normal operation of an electrical circuit. (6.1)
- Grounding conductors are intended to carry current only when a fault condition is present. (6.1)
- Common voltages are alternating current, ac, and direct current, dc. (6.1)
- Current, I, refers to the flow of electrons through an electrical circuit. (6.2)
- Resistance, R, is the opposition to current flow. (6.3)
- Power (P) in the electrical field is defined as the amount of electrical energy converted or transferred in a given amount of time; usually measured in watts. (6.4)
- There are three common forms of Ohm's law and three common forms of Watt's law. (6.5)

Review Questions

Use the information in the chapter to answer the following questions.

Measuring Your Potential

1. The difference in electrical charge between two points in a circuit is known as _____. (6.1)
   A. voltage
   B. current
   C. resistance
   D. power
2. On a typical 120-V wall outlet, the ungrounded conductor is _____. (6.1)
   A. green
   B. grey
   C. black
   D. white
3. *True or False?* In direct current, the polarity of the voltage continuously changes from positive to negative. (6.1)
4. Of the following, which statement is accurate? (6.2)
   A. As the voltage supplied to a circuit increases, the resistance of the circuit will decrease.
   B. As the resistance of a circuit increases, the current flow through the circuit will increase.
   C. An increase in the voltage supplied to the circuit will increase the amount of current flow in the circuit.
   D. A decrease in the voltage supplied to the circuit will increase the amount of current flow in the circuit.
5. Resistance is measured in _____. (6.3)
   A. volts
   B. amps
   C. ohms
   D. watts
6. *True or False?* A good conductor of electricity has low resistance. (6.3)
7. The rate at which electrical energy is transferred through an electrical circuit is known as _____. (6.4)
   A. power
   B. resistance
   C. efficiency
   D. Ohm's law
8. Of the following, which is the formula used to determine voltage when the resistance and current of a circuit is known? (6.5)
   A. $E = I \times R$
   B. $E = \frac{I}{R}$
   C. $I = \frac{E}{R}$
   D. $I = E \times R$
9. The formula $P = I \times E$ is used to determine _____. (6.5)
   A. amps
   B. watts
   C. volts
   D. ohms

Plug-In

1. How many kilowatt-hours of energy are used if a 1500-W appliance is operated for 30 minutes? (6.4)
   A. 0.75 kilowatt-hours
   B. 1.0 kilowatt-hours
   C. 1.5 kilowatt-hours
   D. 3.0 kilowatt-hours
2. If 2 A of current flow through a 10-Ω load, the voltage supplied to the circuit will be closest to _____. (6.5)
   A. 10 V
   B. 15 V
   C. 20 V
   D. 25 V
3. If a 50-Vdc power supply is connected to a 100-Ω load, the current flowing in the circuit will be closest to _____. (6.5)
   A. 0.5 A
   B. 1.0 A
   C. 2.0 A
   D. 5.0 A
4. If 8 A of electricity is used at 110 V, how much power is consumed? (6.5)
   A. 880 W
   B. 880 Ω
   C. 13.75 A
   D. 0.07 W
5. If an appliance that is rated at 550 W is connected to a 117-V source, how much current will this appliance use? (6.5)
   A. 4.7 A
   B. 0.21 A
   C. 4.7 Ω
   D. 64 A

Making the Connection

1. Draw a simple circuit that uses a 10-V battery to power a 2-Ω load. Calculate the amperage of the circuit. Calculate the apparent power consumed by the circuit. (6.5)

CHAPTER 7 Mathematical Manipulation and Conversions

R. MACKAY PHOTOGRAPHY, LLC/Shutterstock.com

SPARKING DISCUSSION

Discuss practical applications of math and measurements that you have performed or observed in your personal, professional, and educational experience.

LEARNING OUTCOMES

After completing this chapter, you will be able to:

7.1 Use the PEMDAS acronym to perform mathematical operations in the correct order.

7.2 Calculate equations while correctly applying associative, distributive, and commutative properties.

7.3 Convert units of weight, time, distance, energy, power, and temperature in both the SI system (metric) and US Customary systems.

7.4 Convert from one unit of measurement to another to perform calculations.

TECHNICAL TERMS

additive inverse
associative law
British thermal unit (Btu)
Celsius scale
commutative law
distributive law
equation
exponent
Fahrenheit scale
multiplicative inverse
PEMDAS
power
reciprocal
SI system
US Customary system
work

Introduction

Whether installing, designing, or troubleshooting, every successful individual working in the electrical industry has a relatively strong grasp of math and how to manipulate numbers. Of course, not everyone needs to be a master of calculus, geometry, algebra, and trigonometry, but being able to navigate basic mathematical expressions to solve simple problems is a skill that most employers find extremely desirable.

This chapter provides information on how to change the configuration of mathematical expressions to suit the needs of a particular situation. In addition to discussing the manipulation of mathematical formulas and expressions, this chapter also addresses important conversion factors that can be used effectively and efficiently to convert units of weight, time, distance, energy, power, and temperature in both the metric and US Customary systems.

7.1 Order of Mathematical Operations

Taking the fundamentals of addition, subtraction, division, and multiplication to the next level makes it possible to solve equations in the electrical field. Providing insight into how mathematical equations can be manipulated will ultimately make your jobs easier. Consider the following mathematical expression:

$$3 + 5 \times 6$$

There are two simple operations, namely addition and multiplication. Depending on the order in which the operation is completed, two different answers can be calculated. If the addition is completed first ($3 + 5 = 8$), the result will be 48 (8×6). However, if the multiplication is completed first, $5 \times 6 = 30$, the result will be 33 ($3 + 30$).

Only one of these resulting values can be correct, but how do you determine which operation takes precedence? Luckily, a set of such rules exists. The acronym ***PEMDAS*** can be used to help remember the proper order for completing multiple operations in formulas and equations, **Figure 7-1**.

- **Parentheses.** Any operation inside parentheses are completed first.
- **Exponents.** Any number containing exponents are completed second.
- **Multiplication/division.** Multiplication and division are completed in left to right order.
- **Addition/subtraction.** Addition and subtraction are completed in left to right order.

Shedding Light

There is a lot of information to commit to memory when determining the proper order to calculate a mathematical expression. The acronym PEMDAS provides the correct order from the highest priority (parentheses) to the lowest (addition and subtraction). The acronym PEMDAS is often committed to memory by using the sentence *Please excuse my dear Aunt Sally,* where the first letter of each word comprises the PEMDAS acronym.

Expressions in Parentheses

When it comes to any mathematical expression, any operations that appear within a set of parentheses are completed first. Consider the expression $(6 + 15) \div 3$. By performing the addition first, as the parentheses indicate, the expression simplifies to $21 \div 3$, which equals 7. A similar calculation is provided in **Example 7-1**.

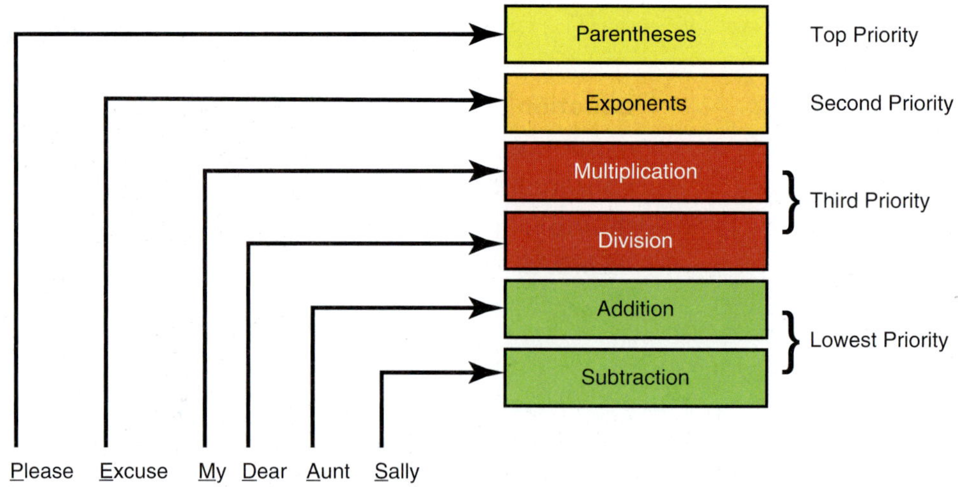

Goodheart-Willcox Publisher

Figure 7-1. Acronym to help remember the correct order to solve mathematical calculations.

Example 7-1

Calculate the following equation:

$$5 \times (3 + 7) \times (10 \div 5) = ?$$

✓ Solution

First, solve the expressions within the parentheses:

$$5 \times (3 + 7) \times (10 \div 5) = ?$$

$$5 \times 10 \times 2 = ?$$

Then working left to right:

$$5 \times 10 \times 2 = ?$$

$$50 \times 2 = 100$$

Now It's Your Turn 7-1

What is the value of $100 \div (20 \div 5) \div (10 \div 2)$?

Exponents

Once all of the operations within parentheses are calculated, the terms with exponents are calculated next. Consider the expression $5 + 4^2$. One might be tempted to add the 5 and 4 together before squaring, but this would be incorrect.

Math Sidebar

An ***exponent*** is a number or symbol denoting the power to which another number is to be raised, which is the number of times the value is multiplied by itself. For example, the square of 4 is written as 4^2 and read as four squared. It is also called four to the second power. The value 4^2 is equal to 4×4, or 16, and the value of 4^3 is equal to $4 \times 4 \times 4$, or 64.

To evaluate this expression correctly, the value of 4^2 must be determined first ($4 \times 4 = 16$). Then the value 16 is added to 5 to obtain the correct answer of 21. Note: The expression $5 + 4^2$ is not the same as $(5 + 4)^2$. As we have just seen, $5 + 4^2$ is equal to 21. The expression $(5 + 4)^2$ would be evaluated by first completing the operation within the parentheses. So, the expression $(5 + 4)^2$ becomes $(9)^2$, or simply 9^2, which equals 81. A similar calculation is provided in **Example 7-2**.

Multiplication and Division

Once the operations contained within parentheses and those involving exponents are completed, the operations of multiplication and division take priority. Neither multiplication nor division takes priority over the other. They are simply performed in order, as they appear from left to right.

Consider the following expression:

$$6 \times 4 + 8 \div 2 = ?$$

Example 7-2

What is the value of the following equation?

$$2^3 + 5^2 + 7 + 2^2 = ?$$

✓ Solution

Since there are no parentheses in this expression, the terms with exponents are calculated first.

$$2^3 + 5^2 + 7 + 2^2 = ?$$

where:

$2^3 = 8$

$5^2 = 25$

$2^2 = 4$

The expression then becomes $8 + 25 + 7 + 4$, which equals 44.

Now It's Your Turn 7-2

Calculate the value of the following equation.

$$2^3 + 5^2 + 9^2 = ?$$

In this case, first multiply 6 and 4, then divide 8 by 2. To easily visualize the original expression, it is often beneficial to place the multiplication and division operations inside sets of parentheses when they are separated by subtraction or addition operations. Doing so changes the appearance of the expression slightly. The expression then becomes:

$$(6 \times 4) + (8 \div 2) = ?$$

$$(24) + (4) = ?$$

$$24 + 4 = 28$$

A similar calculation is provided in **Example 7-3**.

Example 7-3

What is the value of the following equation?

$$5 \times 6 + 3 \times 9 + 4 \div 2 = ?$$

✓ Solution

First, simplify the appearance of the expression by placing parentheses around the expression to highlight the multiplication and division operations:

$$(5 \times 6) + (3 \times 9) + (4 \div 2) = ?$$

$$(30) + (27) + (2) = ?$$

$$30 + 27 + 2 = 59$$

Continued

Now It's Your Turn 7-3

Calculate the value of the following equation.

$$5 \times 4^2 \div 8 = ?$$

Math Sidebar

When the operations in an expression are the same level of priority, they are completed in order from left to right. In a situation where the only operations in a mathematical expression are multiplication and division, the operations can be calculated in any order and the correct answer will still be found.

Addition and Subtraction

The operations with the lowest priority are addition and subtraction. Calculations involving addition and subtraction are completed after all of those involving parentheses, exponents, multiplication, and division have been completed. As with multiplication and division, neither addition nor subtraction takes priority over the other, as these operations are performed in order, from left to right.

Consider the following expression:

$$4 - 2 + 1 = ?$$

Working left to right, first calculate $4 - 2$, which equals 2, and then add 1, which will result in the answer of 3. However, if the addition is done first, the expression becomes $4 - 3$, or 1, which is incorrect. To avoid calculation errors, the addition and subtraction operations must be performed from left to right. A similar calculation is provided in **Example 7-4**.

7.2 Algebraic Expressions

Sometimes, mathematical expressions are not presented in the most convenient form, based on the requirements of the application. In this situation, expressions may need to be reformatted or reconfigured. When working with mathematical expressions, you must do so without changing the value of the expression. As was established previously, performing mathematical operations in the incorrect order can result in an incorrect result. In a similar manner, when numerical expressions are manipulated incorrectly, computational errors are likely to be made.

To minimize errors when manipulating mathematical expressions, keep the following in mind: Whatever is done on one side of an equation must also be done on the other side. The term ***equation*** is used to represent any sequence of numbers and/or variables that contains an equal sign. Here are some key examples:

- If a number is added to, or subtracted from, one side of an equation, the same number must be added to, or subtracted from, the other side of the equation. For example, if $5X = 10$, then the expression $5X + 2 = 10 + 2$ must also be true.
- If one side of an equation is either multiplied or divided by a number or expression, the other side of the expression must also be either multiplied or divided by the same number or expression. For example, if $5Y = 10$, then the expression $2 \times 5Y = 2 \times 10$ must also be true.
- If one side of the equation is squared, the other side must also be squared. For example, if $5 + X = 9$, then the expression $(5 + X)^2 = 9^2$ must also be true.
- If the square root of one side of an equation is taken, the square root of the other side of the equation must also be taken. For example, if $5 + X = 9$, then the expression $\sqrt{(5 + X)} = \sqrt{9}$ must also be true.

Example 7-4

Calculate the value of the following expression.

$$2 \times (5 + 2)^2 + (2 + 1)^2 - 12 \div 3 = ?$$

✓ Solution

The first step is to solve the expressions appearing within the parentheses:

$$2 \times (5 + 2)^2 + (2 + 1)^2 - 12 \div 3 = ?$$

$$2 \times (7)^2 + (3)^2 - 12 \div 3 = ?$$

Next, the exponents are evaluated:

$$2 \times (7)^2 + (3)^2 - 12 \div 3 = ?$$

$$2 \times (49) + (9) - 12 \div 3 = ?$$

Then, perform the multiplication and division operations, in order, from left to right:

$$(2 \times 49) + 9 - (12 \div 3) = ?$$

$$(98) + 9 - (4) = ?$$

Finally, the addition and subtraction operations are performed from left to right:

$$98 + 9 - 4 = 103$$

Now It's Your Turn 7-4

Calculate the following expression.

$$5 \times 6 - 4 + 9 - 6 \div 2 = ?$$

Example 7-5 provides a calculation showing how to solve this type of equation.

Example 7-5

If the left side of the expression 2Y = 8 is multiplied by 5, what must the new expression look like if the value of the expression is to remain the same?

✓ Solution

To prevent the value of the expression from changing, if the left side of the equation is multiplied by 5, then the right side must also be multiplied by 5. Multiplying 2Y by 5 yields a product of 10Y, and multiplying 8 by 5 yields a product of 40. So, mathematically, the values of 2Y = 8 and 10Y = 40 are the same.

Now It's Your Turn 7-5

If the left side of the expression 2Y = 8 is divided by 2, what must the new equation look like if the value of the expression is to remain the same?

In addition to the four basic rules regarding the manipulation of terms in an equation, there are three laws that may be used to further manipulate mathematical expressions, as needed, to simplify calculations:

- Associative law of addition and multiplication
- Commutative law of addition and multiplication
- Distributive law

Associative Law for Addition and Multiplication

The concepts governed by the associative law for addition and the associative law for multiplication were discussed earlier, but are presented here in their formal formats, along with the technical names that are associated with them.

The ***associative law*** states that the order in which numbers are added or multiplied can be changed without changing the value of the expression. This law only applies to addition and multiplication. It does not apply to subtraction and division.

In general terms, the associative law for addition can be represented as follows:

$$(A + B) + C = A + (B + C)$$

The associative law for multiplication can be represented as follows:

$$(A \times B) \times C = A \times (B \times C)$$

Commutative Law for Addition and Multiplication

The ***commutative law*** states that the position of numbers being added or multiplied can be changed without changing the value of the expression. This law only applies to addition and multiplication. It does not apply to subtraction and division.

In general terms, the commutative law for addition can be represented as follows:

$$A + B = B + A$$

The commutative law for multiplication is represented as follows:

$$A \times B = B \times A$$

Although the underlying concept of the commutative law is intuitive, one very popular application of the commutative law is less apparent. The commutative law is commonly used to calculate the percentage of a number. The general form of the commutative law when used with percentages is as follows:

$$A\% \times B = B\% \times A$$

Although calculating common percentages such as 10%, 25%, or 50% will often not require any manipulation, the commutative law may greatly simplify the calculating of less commonly encountered percentages. Using the commutative law applied to percentages, the expression 16% of 50 can be rewritten as 50% of 16. After applying the commutative law to percentages, it is easier to see that 16% of 50 is the same as 50% of 16 or one-half of 16, which is 8. You can practice using the commutative law to calculate percentages in Example 7-6.

Example 7-6

What is 12% of 25?

✓ Solution

Without a calculator, this could take some time as you would need to manually multiply the numbers 0.12, which represents 12% as a decimal, and 25 together. However, using the commutative law for percentages, the expression can be manipulated to read 25% of 12. Since 25% is a percentage that is commonly used by many, as it represents one quarter (1/4), the expression becomes more easily manageable. One quarter of 12, which is 12 divided by 4, equals 3.

Now It's Your Turn 7-6

Without using a calculator, what is 16% of 75?

Distributive Law

The ***distributive law*** states that when a number is multiplied by an addition (or subtraction) operation contained within a set of parentheses, the expression can be restated as the sum (or difference) of the number multiplied by each term contained within the parentheses. The basic form of the distributive law, when used for addition, is provided here:

$$A \times (B + C) = (A \times B) + (A \times C)$$

When the distributive law is used for subtraction, the formula looks like this:

$$A \times (B - C) = (A \times B) - (A \times C)$$

Note: The distributive law does not work for division. Refer to **Example 7-7** on how to solve an equation using the distributive law.

Example 7-7

What is the value of $5 \times (6 + 3)$?

✓ Solution

There are two ways to arrive at the correct answer. The first way is to use the correct order of operations, where the expression within the parentheses, 6 + 3, is calculated first. Once this is done, the expression becomes 5×9, which equals 45.

The second way is to use the addition form of the distributive law:

$$A \times (B + C) = (A \times B) + (A \times C)$$

$$5 \times (6 + 3) = (5 \times 6) + (5 \times 3)$$

$$5 \times 9 = 30 + 15$$

$$45 = 45$$

Now It's Your Turn 7-7

Using the distributive law for subtraction, find the value of $7 \times (4 - 1)$?

Shedding Light

There is no one right way to arrive at the correct answer, as long as the rules of mathematical calculations are followed. Depending on the numbers involved and your level of comfort and confidence, you should choose the method that works best for you. Remember, it is all about getting to the right answer.

7.3 Measurement and Conversions

When working in the electrical industry, further manipulation of mathematical expressions is often required. It is one thing to see an expression not formatted or configured as desired, but quite another when the values are expressed in units that are either not familiar or inappropriate for the application. This section provides useful information that will allow for the easy conversion among units commonly used to express temperature, time, weight, and distance.

Temperature

Although the most common temperature scale used in the United States is the ***Fahrenheit scale***, it may be necessary to express temperatures using the ***Celsius scale***. Some tables and charts that are used by those in the electrical trades provide Celsius temperatures, while others provide both Fahrenheit and Celsius values. However, since all charts and tables are not configured as such, it is beneficial to know how to convert between these two common temperature scales.

To convert degrees Celsius (°C) to degrees Fahrenheit (°F):

$$°F = [(°C + 40) \times 9 \div 5] - 40$$

$$°F = (°C \text{ x } 9 \div 5) + 32$$

To convert degrees Fahrenheit (°F) to degrees Celsius (°C):

$$°C = [(°F + 40) \times 5 \div 9] - 40$$

$$°C = (°F - 32) \times 5 \div 9$$

Notice that two different formulas are provided to convert Fahrenheit to Celsius and to convert Celsius to Fahrenheit. The conversions that appear first may be different from the ones you likely learned in school, namely $°F = (1.8 \times °C) + 32$ and $°C = (°F - 32) \times 5 \div 9$. These formulas provide alternative methods for performing the calculations, while the results will be the same, as you will notice in **Example 7-8**. For some, using these formulas is easier, as these formulas are very similar to each other.

Time

When estimating energy usage, labor costs, person-hours, or any one of a number of other essential calculations that must be made, it is often required to change the units of time. It would not be feasible, for example, to refer to a year as 525,600 minutes. However, referring to a year having 8760 (365×24) hours is common, as this value may be used to determine the annual use of an electrical appliance or fixture. **Figure 7-2** provides conversion factors that can be used to change the units of time as required for a particular application, project,

Example 7-8

What is the Fahrenheit equivalent of 50°C?

✓ Solution

There are two formulas to use to convert from degrees Celsius to degrees Fahrenheit:

$$°F = [(50°C + 40) \times 9 \div 5] - 40$$
$$°F = [90 \times 9 \div 5] - 40$$
$$°F = [810 \div 5] - 40$$
$$°F = 162 - 40$$
$$°F = 122$$

or:

$$°F = (°C \times 9 \div 5) + 32$$
$$°F = (50 \times 9 \div 5) + 32$$
$$°F = (50 \times 1.8) + 32$$
$$°F = 90 + 32$$
$$°F = 122$$

Both methods determine that 50°C is equivalent to 122°F.

Now It's Your Turn 7-8

Convert 50°F to degrees Celsius.

Time Equivalents

Unit	Equivalents
1 second	1000 milliseconds 0.01667 minute 0.000278 hour
1 minute	60 seconds 0.1667 hour 0.000694 day
1 hour	3600 seconds 60 minutes 0.04167 day
1 day	1440 minutes 24 hours 0.1429 week 0.00274 year
1 week	10,080 minutes 168 hours 7 days 0.0192 year
1 year	8760 hours 365 days 52 weeks

Goodheart-Willcox Publisher

Figure 7-2. Equivalents to change units of time.

or calculation. Use this conversion table to calculate the problem presented in **Example 7-9**.

Example 7-9

If a lighting fixture is operated for a total of 4 hours per week, what percentage of the week is the lighting fixture in operation?

✓ Solution

There are many ways to reach this answer, but the easiest and most straightforward method would be to use the conversion factor between hours and weeks. Using the conversion table in **Figure 7-2** as a reference, there are 168 hours in a week. Setting up the conversion, the expression is written as follows:

$$4 \text{ hr/week} \times \frac{1 \text{ week}}{168 \text{ hr}} \times 100\% = ?$$
$$\frac{4 \text{ hr}}{168 \text{ hr}} \times 100\% = ?$$
$$0.0238 \times 100\% = 2.38\%$$

Notice that the units of hours in the numerator of 4 hr/week will cancel the units of hours in the denominator of 1 week/168 hr and the units of week in the denominator of 4 hr/week will cancel the units of week in the numerator of 1 week/168 hr. As the example is calling for a percentage and not an answer in terms of hours or weeks, it is appropriate for all of the working units to cancel each other out. With these cancelations made, the expression becomes 4 divided by 168, or 0.0238. Multiplying this result by 100% formats the answer as a percentage.

Now It's Your Turn 7-9

An electrician needs to replace 30 duplex receptacles on a job. It takes the electrician 20 minutes to replace each receptacle. How many hours will it take the electrician to complete the task?

Weight and Distance

In the electrical trades, weight and distance conversions are not very common. In the United States, you will encounter two systems of measurements on a daily basis: ***US Customary system*** and the ***SI system***, or *metric system*. Knowing how to do these conversions will be beneficial when they are needed. For example, many of the charts and tables provided by the *NEC* list metric units first. **Figure 7-3** and **Figure 7-4** provide charts that may be helpful when such calculations are needed. These charts will come in handy when working through the problem in **Example 7-10**.

Know the Code

Section 90.9 in the *NEC* covers the units of measurement used in the *NEC*. The *NEC* uses a dual system of units that contains the metric unit listed first and the US Customary unit (i.e., inch-pound) immediately following in parentheses.

Weight Measurement Conversion

To Convert from US Customary:	Multiply by:	To Get SI:
ounces	28.349	grams
ounces	0.028349	kilograms
pounds	453.59	grams
pounds	0.4536	kilograms
To Convert from SI:	**Multiply by:**	**To Get US Customary:**
grams	0.03527	ounces
grams	0.002205	pounds
kilograms	35.274	ounces
kilograms	2.2046	pounds

Goodheart-Willcox Publisher

Figure 7-3. Weight conversion factors.

Linear Measurement Conversion

To Convert from US Customary:	Multiply by:	To Get SI:
inches	25.4	millimeters
inches	2.54	centimeters
inches	0.0254	meters
feet	304.8	millimeters
feet	30.48	centimeters
feet	0.3048	meters
feet	0.0003048	kilometers
To Convert from SI:	**Multiply by:**	**To Get US Customary:**
millimeters	0.03937	inches
centimeters	0.3937	inches
meters	39.37	inches
millimeters	0.003281	feet
centimeters	0.03281	feet
meters	3.281	feet
kilometers	3281	feet

Goodheart-Willcox Publisher

Figure 7-4. Distance conversion factors.

Power

A great deal of the work performed in the electrical trades revolves around power, heat, and energy. Knowing how to convert among commonly used electrical units is an important skill. Refer to the table provided in **Figure 7-5** for common conversions used for power, heat, and energy.

Example 7-10

An electrician needs to suspend a lighting fixture that weighs 45 lb. The hanger that the electrician wants to use to support the fixture is rated in kilograms. How many kilograms does the light fixture weigh?

✓ Solution

Referring to the table in **Figure 7-3**, 1 kg is equal to 2.205 lb, rounding up slightly. The conversion between pounds and kilograms is set up as follows:

$$45 \text{ lb} \times \frac{1 \text{ kg}}{2.205 \text{ lb}}$$

The pounds units cancel each other out as they appear in both the numerator and denominator. Dividing 45 by 2.205 gives us 20.41 and the units are kilograms. You can also convert pounds to kilograms by using the following conversion:

$$1 \text{ lb} \times 0.4536 \text{ kg/lb} = 1 \text{ kg}$$

$$45 \text{ lb} \times 0.4536 \text{ kg/lb} = 20.41 \text{ kg}$$

Now It's Your Turn 7-10

An electrician needs 140 meters of THHN 12 gauge hook-up wire. A new spool of wire contains 500 feet of wire. Does the electrician have enough wire on the spool?

The terms work and power are related in the sense that ***power*** is the rate at which work is done. ***Work*** is commonly expressed in units of foot-pounds (ft-lbs) and refers to the equivalent of moving an object having a certain weight a specified distance. See **Example 7-11**. Power and work are further discussed in Chapter 15, *Single-Phase Alternating Current Power.*

Power is expressed in the units of ft-lb/min, or horsepower (hp). As shown in **Figure 7-5**, 1 hp equals 33,000 ft-lb/min. Horsepower is also defined as 746 W. Both ft-lb/min and horsepower have a time element associated with them. The min term in ft-lb/min references the period of time during which the work, expressed in ft-lb, is done. Horsepower is a

Example 7-11

How much work is done by moving a 500-lb object a distance of 20 ft?

✓ Solution

Work can be calculated by multiplying the weight of an object in pounds by the distance moved, in feet. To solve this problem:

$$500 \text{ lb} \times 20 \text{ ft} = 10{,}000 \text{ ft-lbs}$$

Energy Equivalents

Energy Measurement	Equivalents
1 Btu	778 ft-lb 1054.8 joules (J)
1 horsepower (hp)	33,000 ft-lb/min 550 ft-lb/sec 746 W 2545.6 Btu/hr 42.42 Btu/min 1.014 hp (metric)
1 horsepower-hour (hp-hr)	1 hp for 1 hr 1,980,000 ft-lb 746 W/hr 0.746 kWh 2545.6 Btu
1 watt (W)	3.414 Btu/hr
1 kilowatt (kW)	1000 W 1.34 hp
1 kilowatt hour (kWh)	1 kW for 1 hr 1000 W/hr

Goodheart-Willcox Publisher

Figure 7-5. Heat, energy, and power conversion factors.

measurement of power, which is a function of current and voltage. Current, by its very definition, refers to the quantity of electrons passing any given point in an electric circuit in one second, so power also includes a time factor, although it might not be immediately evident. Refer to **Example 7-12** for how to calculate power.

Example 7-12

How much power, in hp, is used when a 500-lb object is moved a distance of 20 ft in 30 seconds?

✓ Solution

In this example, 10,000 ft-lbs (500 lb × 20 ft) of work is being done in 30 seconds, or 0.5 minutes. Since one horsepower is the equivalent of doing 33,000 ft-lb of work in one minute, the calculation can be set up as follows:

$$\frac{10{,}000 \text{ ft-lb}}{0.5 \text{ min}} \times \frac{1 \text{ hp}}{33{,}000 \text{ ft-lb/min}}$$

Both the units of ft-lb and the units of minutes cancel each other out, leaving only the units of hp. Numerically, the expression becomes:

$$\frac{10{,}000}{0.5} \times \frac{1}{33{,}000}$$

$$\frac{20{,}000}{33{,}000} = 0.61 \text{ hp}$$

Math Sidebar

When the denominator of a fraction is another fraction, the denominator of the fraction in the denominator becomes the numerator in the main fraction. Consider the expression 1 divided by 1/4. In this case, the 1/4 is the denominator of the fraction, and 4 is the denominator of this fraction. The 4 moves to the numerator, making the result 4.

$$\frac{1}{1/4} = \frac{1 \times 4}{1} = \frac{4}{1} = 4$$

Expressed verbally, the expression is asking how many times one-quarter goes into one or, for a more graphic example, how many quarters are there in a dollar? The answer to this question is four.

In another example, two divided by 1/5 becomes 10. The five moves up to the numerator so the expression becomes 2 × 5 divided by 1, which equals 10.

Another way to manipulate expressions such as these is to change the operation of division into a multiplication operation. When changing division to multiplication, the inverse of the expression in the denominator, the dividend, is used. The expression 2 ÷ (1/5) can, therefore, be written as 2 × (5/1), which is 10.

$$2 \div \frac{1}{5} = \frac{2}{1} \div \frac{1}{5}$$

$$\frac{2}{1} \times \frac{5}{1} = \frac{10}{1}$$

$$\frac{10}{1} = 10$$

Whenever possible, avoid multiple division operations, such as dividing a fraction by another fraction, in a single calculation. Dividing by a fraction is equal to multiplying by the reciprocal of the fraction. Dividing 1/2 by 1/8 is the same as multiplying 1/2 and 8/1 together.

Heat and Energy

The foot-pound (ft-lb) is also directly related to the ***British thermal unit (Btu)***, which is a unit of heat energy. By definition, the Btu is the amount of heat energy required to raise the temperature of 1 lb of water 1°F. We know that 1 Btu is the equivalent of 778 ft-lb, and 1 ft-lb is equal to 0.001285 Btu. Neither the ft-lb nor the Btu make any reference to time. The ft-lb refers only to the amount of work done and not how much time it took for the work to be done. Similarly, the Btu refers only to the amount of heat transferred and not the duration of the heat transfer. Refer to **Example 7-13** for how to calculate heat.

Example 7-13

How much heat, in Btu, is generated by moving a 500-lb object a distance of 20 ft?

✓ Solution

This value can be converted to Btu as follows:

$$10{,}000 \text{ ft-lb} \times \frac{1 \text{ Btu}}{778 \text{ ft-lb}}$$

With the ft-lb units in both the numerator and denominator, they cancel each other out. By dividing 10,000 by 778, the result is 12.85 Btu.

Unlike the Btu, which refers only to the amount of heat generated or transferred, the unit Btu/hr refers to the amount of heat generated or transferred per unit of time, namely the hour. One watt of electric power, for example, is the equivalent of 3.412 Btu/hr. Similarly, 1 watt-hour is equal to 3.412 Btu. Refer to **Example 7-14**.

Example 7-14

How much would it cost to continuously operate a 3500-W load for a period of one week if the cost of energy is $0.15/kWh?

✓ Solution

To begin, the total number of hours that the load is operating must be calculated. With a week being seven days and each day having 24 hours, the total number of hours in the week is calculated as:

$$\frac{7 \text{ days}}{1 \text{ week}} \times \frac{24 \text{ hour}}{1 \text{ day}}$$

Notice that the day units cancel out and what is left is the units hour/week. By multiplying 7 and 24, it can be seen that there are 168 hours in a week.

From this, it is known that the electric load will operate for 168 hours. Multiplying 3500 W by 168 hours results in 588,000 Wh. However, we pay for energy by the kilowatt-hours. Converting to kilowatt-hours is accomplished as follows:

$$588{,}000 \text{ Wh} \times \frac{1 \text{ kWh}}{1000 \text{ Wh}}$$

The units of Wh cancel out, and what is left is the units of kWh. Dividing 588,000 by 1000 equals 588 kWh. Knowing the price per kWh, the final calculation becomes:

$$588 \times \$0.15 = \$88.20$$

The total cost to operate the load continuously for one week is $88.20.

Shedding Light

The unit of watt-hour (Wh) is used to calculate your monthly energy bill. If you examine your recent electric bill, you will notice you are billed by the number of kilowatt-hrs (kWh) used during the billing cycle. One kWh is equal to 1000 Wh. The amount of your bill is the number of kWh used multiplied by the cost per kWh, plus any service-related fees, taxes, and regulatory charges.

7.4 Converting Units

When working with linear measurements, electricians often convert measurements back and forth between feet and inches. This is a useful skill that can make addition, subtraction, multiplication, and division of linear measurements easier.

Performing operations in the correct order is very important. Equally important to this, however, is ensuring that the units are correct. If you add five apples to five apples, the result will be ten apples. What will happen if you add two apples to three oranges? Apple-oranges is not an acceptable unit. The resulting numbers, as well as the units, must be correct. Consider the electrician who needs to calculate the number of feet of conduit that will be required for a particular installation. The required lengths are 48 inches, 6 feet, 10 feet, and 36 inches. If not careful, the electrician might simply add 48, 6, 10, and 36 to arrive at an answer of 100. The question becomes, "100 what?" Does the electrician need 100 feet of conduit or 100 inches of conduit? The answer is neither, as the 100 has absolutely no meaning, since the units (some expressed in inches, while others expressed in feet) were added together.

To have meaning, the units must be consistent. Knowing that 48 inches is the same as 4 feet, and that 36 inches is the same as 3 feet, the calculation can be restated as:

$$4' + 6' + 10' + 3' = 23'$$

A similar calculation is provided in **Example 7-15**.

When changing the units of an expression, calculation errors are often made when division and multiplication operations are used improperly. To help ensure that errors are not made, pay careful attention to the units, the actual numbers, and calculation operations.

Shedding Light

The terms Btu and ft-lb are used in the US Customary system. In the metric system, the units of joules and calories are used. The units of joules and calories are often found in engineering documents, but they are sometimes found in the specification sheets on some end-user products, especially if they are manufactured abroad.

Example 7-15

What is the value of 10′8″ – 3′11″?

✓ Solution

This result cannot be calculated in the manner used in **Example 7-14** because 11″ cannot be subtracted from 8″. However, there are two other ways to perform this calculation. The first way would be to convert the measurements to inches.

First convert 10′8″ into inches: 10′ is the same as 120″ and adding 8″ gives the equivalent of 128″. Then convert 3′11″ to inches: 3′ is the same as 36″ and adding 11″ gives the equivalent of 47″. The expression now becomes:

128″– 47″ = 81″

This answer can now be converted back to feet by dividing the result by 12″, giving a result of 6′ with a remainder of 9, so the answer is 6′9″.

The second way to calculate this value is to first manipulate the 10′8″ expression. By borrowing 1′ from the 10′ and adding it, in the form of 12″ to the inches, this term becomes 9′8″ + 12″, or 9′20″. Now, the expression becomes 9′20″ – 3′11″. Now, subtracting the feet from the feet and the inches from the inches, the expression becomes a combination of (9′ – 3′) and (20″ – 11″), which equals 6′9″.

Now It's Your Turn 7-15

Solve the following equation:

9′8″– 41″ = ?

Math Sidebar

If an expression is both multiplied and divided by another term, the result is the original term. When dividing a number, value, or variable by the same number, value, or variable, the result will always be one. The same concept holds true for the units used to identify the actual numerical values.

Consider the situation when converting 5000 W to horsepower. From the table in **Figure 7-5**, 746 W equals 1 hp, but how is this factor used to make the conversion? To ensure accuracy, the units can be incorporated into the calculation process as follows:

$$5000\text{ W} \times \frac{1\text{ hp}}{746\text{ W}}$$

By setting the expression up in this manner, the 5000 W term is in the numerator as is the 1 hp term, while the 746 W term is in the denominator. The units of watts, being in both the numerator and the denominator, cancel each other out, leaving only the units of horsepower. Dividing 5000 by 746 yields a result of 6.7 hp. The same result can be obtained by dividing as follows:

$$\frac{5000\text{ W}}{746\text{ W/hp}}$$

Although dividing yields the same result, the formatting can become confusing as the units of the two expressions are themselves fractions or ratios. In this calculation, you can instead set it up as a multiplication operation instead of a division operation: multiply 5000 W by 1 hp/746 W.

Example 7-16 provides additional practice on converting measurements to their equivalent.

Example 7-16

Calculate the following equation:

33,000 ft-lb/min + 1 hp + 746 W = ?

✓ Solution

The first step is converting the various measurements to an equivalent. Using **Figure 7-5** as a reference, each of the terms in this expression is the equivalent of 1 hp:

33,000 ft-lb/min = 1 hp
746 W = 1 hp

As a result, the expression can be rewritten as:

1 hp + 1 hp + 1 hp = 3 hp

Now It's Your Turn 7-16

Calculate the following equation:

66,000 ft-lb/min + 1 hp + 2238 W = ?

Math Sidebar

Sometimes, the terms inverse and reciprocal are used interchangeably, but they are not the same thing. The term inverse refers to opposite. There are two common mathematical inverses, namely the ***additive inverse*** and the ***multiplicative inverse***. The additive inverse is the number that, when added to another number, will result in zero. Therefore, the additive inverse of +5 is –5, as the sum of these two numbers is zero. The multiplicative inverse is a fraction that, when multiplied by another fraction, will yield a result of 1. Therefore, the multiplicative inverse of 5 is 1/5, as the product of these two values is 1. Similarly, the multiplicative inverse of 2/7 is 7/2, as the product of these two fractions is 1. The term ***reciprocal*** refers only to the multiplicative inverse of a value.

Throughout your career in the electrical trades, you will encounter a number of prefixes. You are already familiar with several of the most common prefixes, such as kilo and mega.

Electricians will encounter kilowatts and megawatts when asked to calculate power usage on large and small scales. Solar panel usage and electric cars are becoming more popular every day. This type of equipment is rated in watts. As you move further into alternating current (ac) theory, you will also see the use of abbreviations such as kVA or kilo-volt-amps. You can also expect to encounter microfarads for capacitance and millihenries for inductance. Both units of measurement are essential in the study of capacitive and inductive circuits. The three forces that are behind the theory of alternating current are resistance, inductive reactance (X_L), and capacitive reactance (Xc). The manipulation of the metric system to properly apply electrical knowledge is necessary.

Figure 7-6 provides a more comprehensive list of prefixes, along with their symbols and their exponential and decimal representations.

Common Prefixes Used in the Electrical Industry

Prefix	Symbol	Exponent	Decimal
pico	p	10^{-12}	0.000000000001
nano	n	10^{-9}	0.000000001
micro	μ	10^{-6}	0.000001
milli	m	10^{-3}	0.001
centi	c	10^{-2}	0.01
deci	d	10^{-1}	0.1
deka	da	10^{1}	10
hecto	h	10^{2}	100
kilo	k	10^{3}	1000
Mega	M	10^{6}	1,000,000
Giga	G	10^{9}	1,000,000,000
Tera	T	10^{12}	1,000,000,000,000

Goodheart-Willcox Publisher

Figure 7-6. Comprehensive list of prefixes that are commonly used for linear measurement, weight, or volume.

Summary

- To reduce errors, perform calculations in the correct order. (7.1)
- The acronym PEMDAS is used to remember the order of operations as follows: parentheses, exponents, multiplication, division, addition, subtraction. (7.1)
- Any mathematical operation performed on one side of an equation must also be performed on the other side. (7.2)
- The associative law for addition and multiplication states that these calculations can be performed in any order. (7.2)
- The commutative law for addition and multiplication states that the order of numbers being added or the order of numbers being multiplied can be reversed. (7.2)
- The distributive law of addition states that multiplying a number, *A*, by a sum of numbers is the same as adding together the numbers that result from multiplying *A* by each number in the sum. (7.2)
- The distributive law of subtraction states that multiplying a number, *A*, by the difference between two numbers is the same as taking the difference between the numbers that result from multiplying *A* by each number in the subtraction operation. (7.2)
- The units of ft-lb are used to quantify work, while ft-lb/min are used to quantify the rate at which work is done. (7.3)
- One horsepower is the term used to indicate when work is done at the rate of 33,000 ft-lb/min. (7.3)
- Energy cost is often calculated as the number of kw-hr used multiplied by the cost per kw-hr. (7.3)
- Watt-hours can be converted to Btu by multiplying by 4.312 Btu/watt-hour. (7.3)
- The units of ft-lb and Btu are used in the US Customary system, while joules and calories are used in the metric system. (7.3)
- To help ensure error-free calculations, make certain that the units are converted when necessary; note when units will cancel out in performing a calculation. (7.4)

Review Questions

Use the information in the chapter to answer the following questions.

Measuring Your Potential

1. When calculating mathematical equations, the lowest priority mathematical operations are _____. (7.1)
   A. any operation inside parentheses
   B. any number containing exponents
   C. multiplication and division
   D. addition and subtraction
2. *True or False?* Multiplication and division operations are performed, in order, from left to right. (7.1)
3. Which of the following expressions has the same numerical value as $5 + 3 \times 7 - 5 + 8 \div 2$? (7.1)
   A. $8 \times 7 - 13 \div 2$
   B. $5 + 21 - 5 + 4$
   C. $5 + 3 \times 2 + 4$
   D. $5 + 3 - 2 + 16$
4. *True or False?* The expression $5 \times (3 + 2)$ is the same as $(5 \times 3) + (5 \times 2)$. (7.1)
5. *True or False?* It is acceptable to multiply one side of an equation by a number without multiplying the other side of the equation by the same number. (7.2)
6. The Celsius equivalent of 40°F is closest to _____. (7.3)
   A. –10°C
   B. 4°C
   C. 10°C
   D. 24°C
7. Of the following, which is the longest distance? (7.3)
   A. 5 meters
   B. 6 yards
   C. 200 inches
   D. 500 centimeters
8. Doing work at the rate of 99,000 ft-lb/min is the equivalent of _____. (7.3)
   A. 2 hp
   B. 3 hp
   C. 4 hp
   D. 5 hp

9. _____ is expressed in the units of foot-pounds (ft-lbs). (7.3)
   A. Power
   B. Horsepower
   C. Work
   D. Distance
10. Operating a 500-W load for 10 hours uses _____ of energy. (7.3)
   A. 5 kw-hr
   B. 50 kw-hr
   C. 500 kw-hr
   D. 5000 kw-hr
11. *True or False?* Multiplying a number by its reciprocal will always result in a value of 1. (7.4)
12. The additive inverse of +2 is _____. (7.4)
   A. +1/2
   B. −1/2
   C. −2
   D. +2
13. What is the value of 10′8″ + 3′11″? (7.4)
   A. 7′ 3″
   B. 13′ 9″
   C. 14′ 7″
   D. 15′ 3″

Plug-In

1. What is the value of $7 - (9 - 7)^2 + 3 - 2$? (7.1)
   A. −2
   B. 0
   C. 4
   D. 6
2. What is the value of $5 + 3^2 - 4 - 1$? (7.1)
   A. 7
   B. 9
   C. 11
   D. 13
3. What is the value of $20 - 5 - 8 + 6$? (7.1)
   A. 1
   B. 13
   C. 21
   D. 29
4. An electric furnace produces 18,766 Btu/h of heat energy. How many watts of electric power will this heater consume? (7.3)
   A. 5.5 W
   B. 5500 W
   C. 6403 W
   D. 64,030 W
5. An electrician needs to replace 500 light bulbs in an office. It will take three minutes to replace each bulb in an LED lamp. The company pays the electrician $35/hr to do the job. If the company's budget for the electrician's labor for this job is $700, the actual labor cost will _____. (7.3)
   A. exceed the budget by $175
   B. be $175 under budget
   C. exceed the budget by $350
   D. be $350 under budget

Making the Connection

1. How can the accuracy of a mathematical expression be evaluated without having to completely go through the calculations again? (7.1, 7.4)
2. How can accurate calculations on the job result in increased profitability for the contractor? (7.3)

CHAPTER

8 Electrical Tools and Testing Instruments

Avocado_studio/Shutterstock.com

SPARKING DISCUSSION

If dull drill bits and saw blades are less likely to cut the user of the tool, how can they be less safe to use?

LEARNING OUTCOMES

After completing this chapter, you will be able to:

8.1 Explain the importance of inspecting all tools and equipment prior to use.

8.2 Identify various types of hand tools used by electricians.

8.3 List the different types of power tools commonly used in the electrical trade.

8.4 Demonstrate the proper use of electrical testing equipment.

TECHNICAL TERMS

- adjustable wrench
- box wrench
- cable cutter
- chalk line
- circular saw
- clamp-on ammeter
- combination wrench
- conduit bender
- diagonal-cutting pliers
- digital multimeter (DMM)
- fish stick
- fish tape
- hacksaw
- hammer drill
- hex-key wrench
- jab saw
- linesman pliers
- locking pliers
- metallic cable cutter
- multipurpose wiring tool
- needle-nose pliers
- non-contact voltage tester
- nut driver
- open-end wrench
- plumb bob
- power analyzer
- power drill
- pump pliers
- reciprocating saw
- screwdriver
- skinning knife
- socket wrench
- soft-face hammer
- straight claw hammer
- tape measure
- teeth per inch
- torpedo level
- torque screwdriver
- torque wrench
- true-RMS meter
- utility knife
- wire stripper

Introduction

Those working in the electrical trades rely on basic hand and power tools as well as testing equipment to perform everyday tasks. To stay safe on the job, it is important for electricians to know what a particular tool is used for, how to properly use it, and know the limitations of the tool. To ensure tools are used correctly, it is important to read and understand all instructions before attempting to use them. If you are unsure how to use a tool, do not use it. Improper use of tools is a leading cause of personal injury.

This chapter describes some of the basic tools commonly used by those working in the electrical trades. Since not every tool you may encounter is covered here, be sure to consult the manufacturer's literature and follow guidelines established by national safety organizations, such as OSHA.

8.1 Tool Safety

When purchasing tools, keep safety and comfort in mind. Working on and around electrical circuits, fixtures, and appliances pose the potential for electric shock. The majority of hand tools are not insulated for shock protection. Although the rubber and plastic handles found on hand tools do provide some shock protection, they are for comfort only. Proper voltage-rated tools will generally have an orange-colored handle, with a symbol of two overlapping triangles, **Figure 8-1**.

As you will be working with tools constantly, you should make sure they are comfortable. Uncomfortable tools can lead to wrist and other injuries; therefore, it is a great idea to handle tools before purchasing them to make certain that they are comfortable. High-quality tools are more expensive, but they tend to last longer and have more features, such as an ergonomically desirable design.

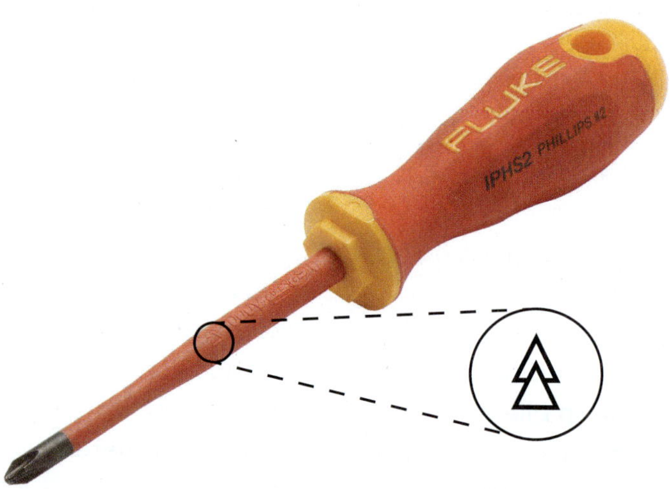

Courtesy Fluke Corporation

Figure 8-1. When working on or near electrical systems, insulated tools offer some protection against an accidental shock. The screwdriver shown here indicates that it is certified with a voltage rating of 1000 V.

Safety Alert

Never use a tool to perform a task that it was not intended to do. Doing so can result in damage to the tool as well as severe personal injury. Never take chances by using the incorrect tool.

Prior to using a tool, carefully inspect it to confirm that it is safe to use. Avoid using tools with damaged handles or damaged power cords. If there is doubt in your mind about the integrity of a tool, it is far better to err on the side of caution and not use it.

When using tools, be sure to keep the following in mind:

- Use tools only for the tasks they are intended to perform. For example, never use a screwdriver as a chisel or a wrench as a hammer. Doing so can cause severe personal injury.
- Never exceed the limits of a tool. This can cause a tool's handle to shatter.
- Make certain that you know how to properly use a tool. If you are unsure about how to use a particular tool, do not use it or ask for instructions for its proper use.
- If available, read the instructions supplied with the tool. If not provided or otherwise unavailable, then manufacturer's websites often have instruction manuals and guides for most of their products.
- Never apply excessive pressure or force to a drill, saw, or other cutting tool.
- When cutting or drilling, be sure to use sharp blades and bits. Sharp cutting surfaces cut better and faster than dull ones and require much less pressure or force to do their job. Applying less pressure on the tool reduces the possibility of falling into the tool or breaking the tool and sustaining an injury.

Safety Alert

Always wear the appropriate pieces of personal protective equipment (PPE) to help prevent personal injury. Be sure to refer to Chapter 3, *Workplace Safety*, for detailed information regarding various pieces of PPE and their uses.

Tool Belts

Electricians spend a great deal of time walking from one location to another on the jobsite. To avoid excessive walking and climbing ladders, electricians often carry their tools in a tool belt, **Figure 8-2**. Tool belts are extremely useful as they serve as an additional pair of hands, carrying the tools required for a specific task. Tools can be organized as the electrician wishes, with each tool having its own place in a separate pouch. It is also important to carry the appropriate number of tools. Carrying more tools than necessary will add extra weight to your load, which can lead to health issues in the long term.

Lyudmila Zavyalova/Shutterstock.com

Figure 8-2. Being organized means you will spend less time fumbling and searching for the right tool, and less time walking or climbing down a ladder to get the right tool.

8.2 **Hand Tools**

Electricians are called on to perform a multitude of tasks and must be proficient with the variety of tools necessary for this wide array of work. Tools that are used in the electrical field are generally divided into two categories: hand tools and power tools. The following section discusses some of the hand tools that are commonly used in the electrical trades.

Screwdrivers

Regardless of the trade, the most common hand tool is the ***screwdriver***. Screwdrivers come in a variety of configurations and are used to turn screws with different slotted or grooved head configurations, **Figure 8-3**. The two most common types of screwdriver tips are the Phillips head and the slotted. A screwdriver with a square tip, also called a Robertson tip, is another screwdriver commonly used by electricians, **Figure 8-4**. This type of screwdriver is used on circuit breakers, switches, and receptacles. Multitip screwdrivers are used for multiple applications as they have interchangeable tips.

Safety Alert

Only use screwdrivers to tighten and loosen screws. A screwdriver should never be used as a chisel, pry bar, or any other tool. Screwdriver handles should only be turned by hand, so never use pliers or other tools to grab or turn the handle. Never strike the handle of a screwdriver with a hammer.

Nut Drivers

Nut drivers are used to turn hex-head screws. Nut driver handles are often color-coded, making it easy to identify the different sizes. Similar to screwdrivers, there are shaft options for nut drivers. Shafts can be long, short, or even hollow to allow longer screws to extend into the shaft, **Figure 8-5**.

Common Screw Heads

Head	Type
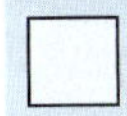	Square
	Hex
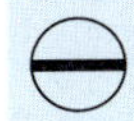	Slotted
	Phillips
	Frearson
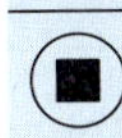	Robertson/square recess
	Torx®
	Security Torx®
	Hex socket
	Security hex socket
	Tri-wing
	Pozidriv
	Spanner
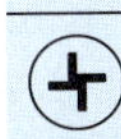	Torq-set
	Spline
	One-way
	Clutch

Goodheart-Wilcox Publisher

Figure 8-3. Screwdrivers come in a variety of different tip styles and heads.

Safety Alert

Insulated tools are required when working on energized circuits. The comfort grips on standard tools are not designed to protect against electrical shock. Be sure to inspect the insulation on all voltage-rated hand tools prior to use. Using a tool with damaged insulation can result in electric shock.

Goodheart-Wilcox Publisher

Figure 8-4. A square tip screwdriver is ideal when working on a circuit breaker.

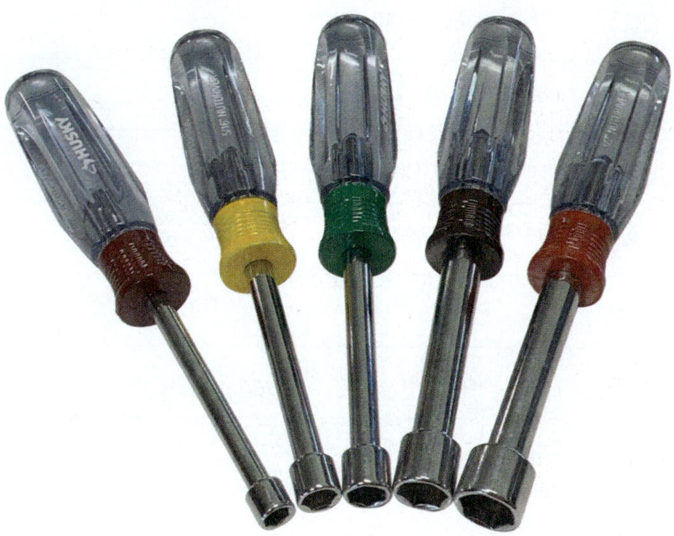

Goodheart-Wilcox Publisher

Figure 8-5. Nut drivers tighten and loosen nuts and hex-head fasteners; these are color-coded by size.

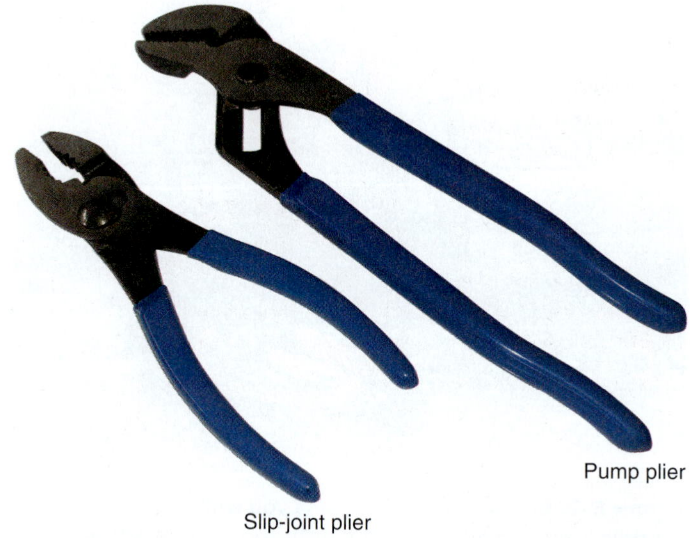

Goodheart-Wilcox Publisher

Figure 8-6. Used to hold and tighten connectors, slip-joint pliers and pump pliers have sliding jaws that lock in place at specific intervals to adjust for different nuts, bolts, or other items.

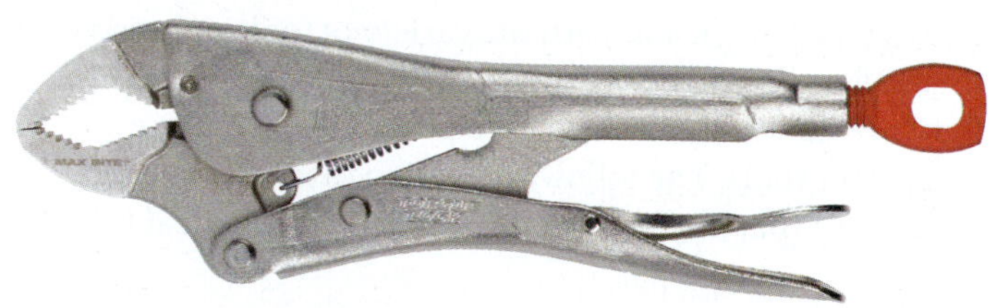

Milwaukee Tools

Figure 8-7. Locking pliers can be used as a handheld vise or clamp.

Image courtesy of Ideal Industries, Inc.

Figure 8-8. Needle-nose pliers have a long, thin nose that makes them useful for gripping or retrieving objects in tight spaces. They are also used to cut, hold, and pull conductors.

Pliers

Pliers are another often-used tool in an electrical worker's tool belt. Although pliers are used primarily to hold and twist, some are configured to cut and allow access into tight spaces. Common types of pliers include the following:

- General purpose, slip-joint pliers
- Pump pliers
- Locking pliers
- Needle-nose pliers
- Diagonal-cutting pliers
- Linesman pliers

General purpose and ***pump pliers*** are of the slip-joint variety and are used for holding and twisting, **Figure 8-6**. The position of the jaws on these types of pliers can be adjusted to provide a better grip on the material being secured. ***Locking pliers***, as the name implies, have the ability to lock the jaws in place, holding materials together, as a vise would, **Figure 8-7**.

Three specialty pliers commonly used in the electrical trades are the needle-nose pliers, side-cutting pliers, and linesman pliers. ***Needle-nose pliers***, with their long, thin jaws, are used to reach tight spaces that conventional pliers would not be able to reach, **Figure 8-8**. ***Diagonal-cutting pliers***, also referred to as side-cutters, are easily identified by

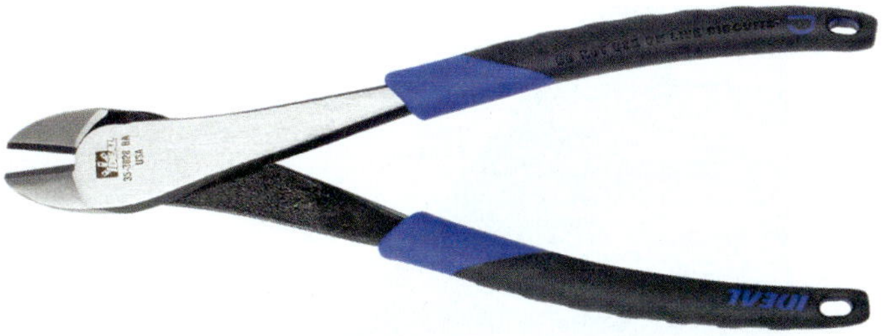

Image courtesy of Ideal Industries, Inc.

Figure 8-9. Diagonal-cutting pliers are used for cutting electrical cables and conductors.

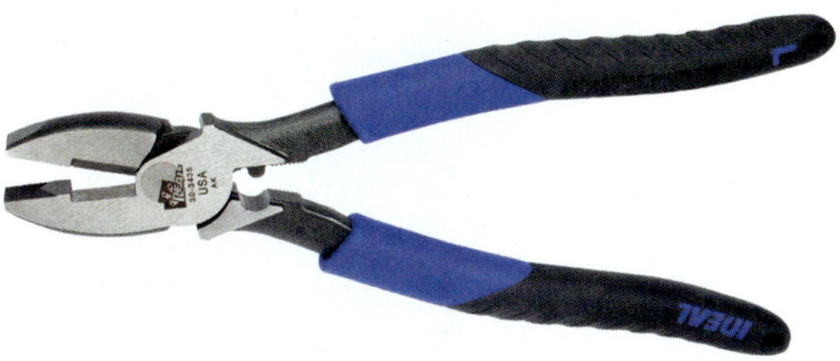

Image courtesy of Ideal Industries, Inc.

Figure 8-10. Linesman pliers are heavy pliers that are used to cut cables, conductors, and small screws as well as pull and hold conductors.

their angled cutting edges, **Figure 8-9**. This type of pliers is used to cut wire, small screws, or other thin metals close to a surface. ***Linesman pliers*** are very versatile, as they typically have wider and larger jaws than conventional pliers and can be used to cut, bend, and twist wires together, **Figure 8-10**.

Wrenches

Wrenches are used primarily to turn bolts with hex heads. Some wrench types, such as pipe wrenches, are used to tighten and loosen connections between pipe sections and fittings. Wrenches commonly used by electricians are adjustable wrenches, combination wrenches, socket wrenches, and torque wrenches.

Adjustable wrenches are a staple in most workers' tool belt as they take up less space than a complete run of individually sized wrenches. For proper use, the wrench should be pulled on, with force being exerted on the stationary jaw of the tool, **Figure 8-11**.

The ***combination wrench*** is a two-headed wrench that combines an open-end wrench and a box wrench. The ***open-end wrench*** is popular because it has an open end that can be positioned onto a hex head from the side. However, the open-end wrench should not be used for final tightening and, because of its open end, is more prone to slipping. The ***box wrench*** provides more torque than an open-end wrench. The box wrench completely encircles the fastener, which makes it less likely to slip. If a hex head can be reached and accessed with a box wrench, it is recommended that the box wrench is used instead of an open-end or adjustable wrench. See **Figure 8-12**.

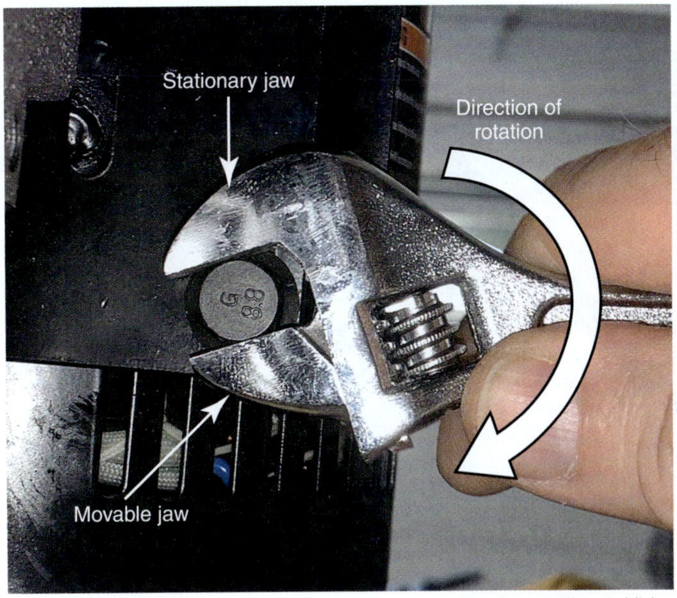

Goodheart-Willcox Publisher

Figure 8-11. Adjustable wrenches should be pulled on with force acting on the stationary jaw.

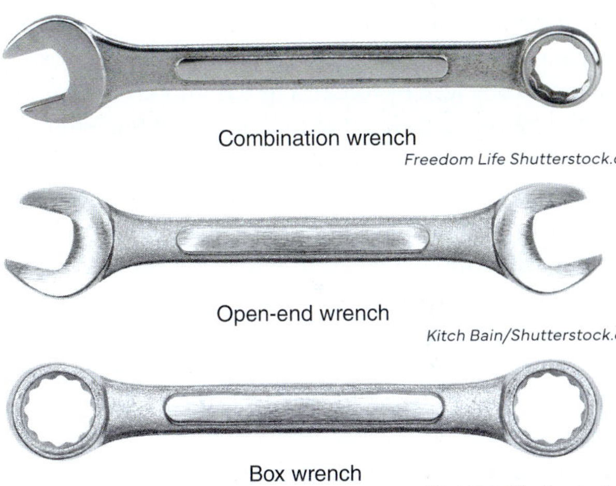

Freedom Life Shutterstock.com

Kitch Bain/Shutterstock.com

Kitch Bain/Shutterstock.com

Figure 8-12. The wrenches shown here are used to tighten and loosen bolts and nuts.

Socket wrenches are used in conjunction with a ratcheting handle, **Figure 8-13**. The socket wrench is also known as a ratchet set or socket set. Individual sockets attach to the handle, either directly or with an extension or universal joint. Extensions and universal joints allow the user to access hard-to-reach hex heads. The ratcheting handle on most socket wrench sets is reversible, so it can be used to either tighten or loosen nuts and bolts.

The sockets themselves come in two different configurations: 6-point and 12-point. The 6-point socket is better if more torque is required. It provides less slip, has more contact surface, and reduces the likelihood of stripping a bolt. The overall wall thickness of a 6-point socket increases its overall

Goodheart-Willcox Publisher

Figure 8-13. Socket wrenches have an assortment of head sizes to fit a variety of fasteners. Most sets come with extenders.

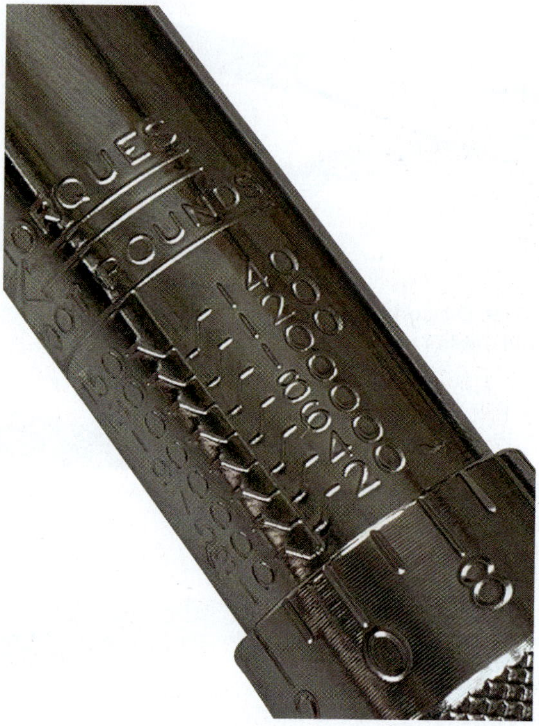

Goodheart-Willcox Publisher

Figure 8-14. The amount of torque required is set by rotating the calibrated scale on the handle of a torque wrench.

strength when compared to a 12-point. The advantage of a 12-point socket is its ability to more easily engage with the bolt head; a benefit when the bolt head is hard to see or must be located and engaged by touch. The likelihood of slip is greater with the 12-point socket due to less overall bolt contact.

The ***torque wrench*** and ***torque screwdriver*** are important tools to have on hand when installing electrical equipment. When tightening nuts and bolts, the torque wrench is used to ensure that the correct amount of force is used. This force is known as torque, which is measured in foot-pounds and inch-pounds. Using too much force can cause a nut or bolt to break, while using too little force can cause a nut or bolt to come loose. The basic, click adjustable torque wrench is easily identified by its unique handle, **Figure 8-14**. The amount of torque required for a particular application is set by rotating the handle to the desired position on the calibrated scale. With the correct socket connected to the wrench, the wrench is moved in the desired direction until a click is heard. At this point, the preset amount of torque has been applied. Over time, torque wrenches will need to be calibrated to ensure accuracy. Dropping or banging a torque wrench can bring the tool out of calibration, so be sure to handle your tools, especially this one, carefully. Torque screwdrivers are similar to the ratcheted-style torque wrench in that they will slip at the preset torque but are used on smaller-sized screws.

Know the Code

Follow the manufacturer's recommendations on how often a torque wrench or screwdriver requires recalibration. The *National Electrical Code NEC* requires that all electrical terminations be made to the manufacturer's required torque.

Unlike traditional wrenches, ***hex-key wrenches***, also referred to as hex keys, Allen wrenches, and Allen keys, are used to turn bolts and screws that have hex-shaped recesses in their heads, **Figure 8-15**. Some hex-key wrenches have ball-shaped tips, which allow the tool to be used at an angle. This option is especially useful in hard-to-reach situations. Hex-key bits are also used with socket sets and torque wrenches.

Hammers

Hammers used in the electrical trades are available in different sizes and types for various work purposes. The electrician's ***straight claw hammer*** is commonly used in electrical work. It has a longer nose than a traditional hammer and has a smooth striking face. The ***soft-face hammer*** is especially useful when soft or finished surfaces need to be gently hammered.

Measuring Devices

No electrician's tool belt is without a basic tape measure. ***Tape measures***, often called a tape, are available in different lengths and widths. Although shorter tape measures are smaller, it is recommended that a heavy-duty, wide tape measure with a length of 20 feet or more be used. Wider tape measures are more durable and will last longer. They are also easier to extend out longer distances without bending or collapsing. Some tape measures have magnetic tips, making it easier for one person to take measurements.

Goodheart-Willcox Publisher

Figure 8-15. Hex-key wrenches come in a wide variety of configurations including individual L-shaped wrenches, T-handle wrenches, and foldout multi-wrench sets.

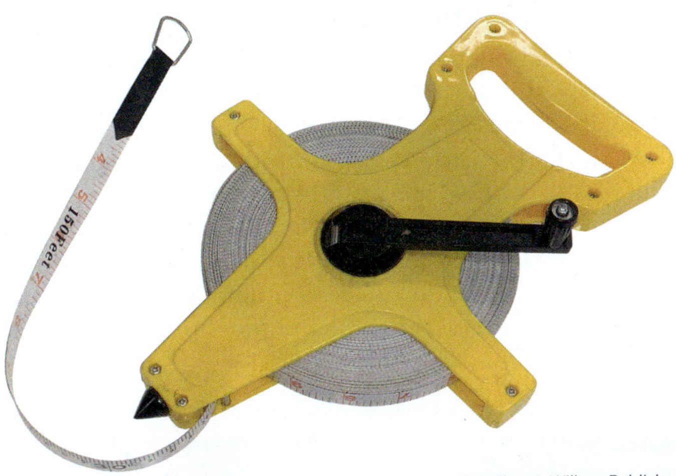

Goodheart-Willcox Publisher

Figure 8-16. An open reel field tape made of plastic can measure distances of over 300 feet.

When longer length measurements are needed, plastic tapes can be used, **Figure 8-16**. Plastic tapes are used to measure 100 feet or more and are rolled up easily and quickly with the aid of the handle on the device. Laser measuring tools can also be used to measure distances quickly and accurately.

Saws

In the electrical trades, various types of saws are used. A ***jab saw***, also called a drywall saw, is used to cut holes in drywall, **Figure 8-17**. A ***hacksaw*** is used to cut materials such as electrical metallic tubing (EMT) and PVC conduit, **Figure 8-18**.

Image courtesy of Ideal Industries, Inc.

Figure 8-17. A jab saw, also known as a drywall saw, is commonly used by electricians to cut drywall or thinner wood paneling for device boxes.

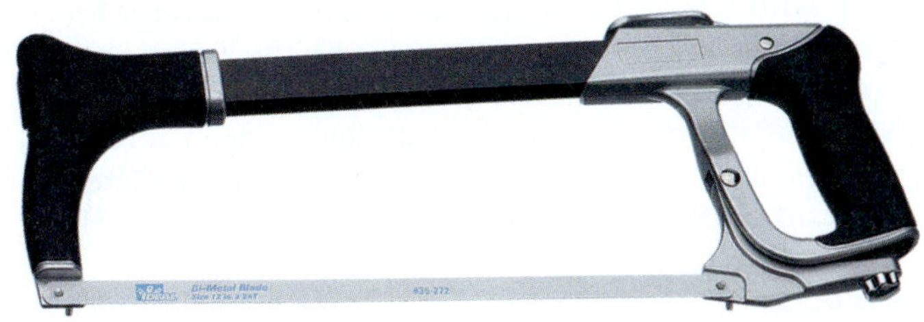

Image courtesy of Ideal Industries, Inc.

Figure 8-18. Hacksaws can be used to cut metal, large copper piping, and PVC conduit. They are available in manual with internal blade storage or fuel-powered.

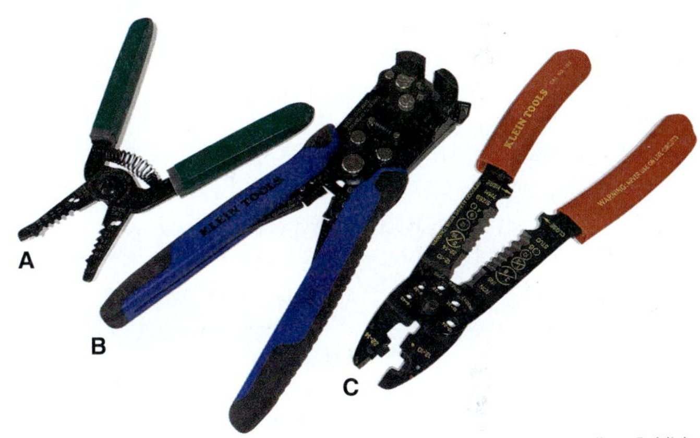

Goodheart-Willcox Publisher

Figure 8-19. Wire strippers and crimping tools are used to strip insulation, cut cables, and crimp assorted fittings. A—Manual wire stripper. B—Automatic wire stripper. C—Multipurpose wiring tool.

Wiring Tools

In the electrical trades, wiring tools are specifically designed to strip insulation from conductors, cut cables and conductors, and crimp assorted fittings and solderless connectors on conductors, **Figure 8-19**. Two commonly used wiring tools are wire strippers and crimping tools. ***Wire strippers*** are used to remove the insulation from wires without damaging the conductor itself. There are manual versions and automatic versions of the wire stripper. When using the manual wire stripper, the proper slot, which corresponds to the gauge or thickness of the conductor, must be selected. A ***multipurpose wiring tool*** is an all-purpose tool that is used to cut and strip wire in addition to crimping solderless connectors to stranded wires. Many electricians use a crimping tool to crimp connectors.

Safety Alert

Solderless connectors are not commonly crimped onto solid conductors, and stranded conductors are often not secured under screw terminals. To connect a stranded conductor to a screw terminal, a solderless connector must be connected to the conductor first. Then, the solderless connector can be secured under the screw terminal. Solderless connectors need to be rated by the manufacturer for use with solid conductors. Check with the listing and labeling of the connector to see if this is permitted. Stranded conductors can be installed under a screw terminal if the manufacturer approves this use. Often a circuit breaker will have a screw and plate assembly that is intended to accept both solid and stranded conductors.

Knives

Electricians often carry two types of knives with them: a utility knife and a skinning knife. The ***utility knife,*** like the one shown in **Figure 8-20**, is used for general cutting. The ***skinning knife*** has a hawkbill blade and is primarily used to remove insulation from wires and cables.

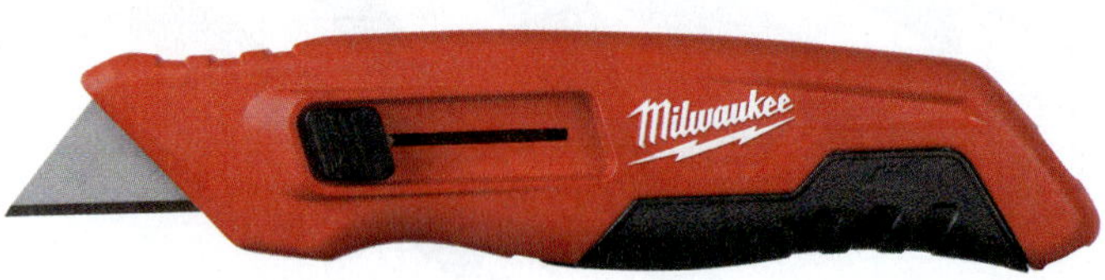

Milwaukee Tools

Figure 8-20. Electricians use a utility knife with a retractable blade to remove insulation from a wire.

Safety Alert

Always make certain that knife blades are sharp. Sharp blades cut easier and require less force, and as a result are safer to use than dull blades.

Wire Fishing Tools

Fish tapes and fish sticks are used to help electricians pull wire through suspended ceilings, walls, under raised floors, and through conduit, **Figure 8-21**. The end of a tape or stick is inserted into the end location for the wire and fed into the conduit or passage until it reaches the starting point. The wire or cable is then attached to the tape, which is pulled back through the conduit, along with the wire or cable.

Fish tapes are available in metallic and nonmetallic varieties. The flexible tape is stored in a reel. ***Fish sticks***, also called fish rods and fish poles, come in kits containing multiple rods that can be screwed together to assemble the length needed. Different tips, including a light, can be attached to the stick to adapt for different applications.

Conduit Bending Tools

When installing conduit, through which conductors or wires will run, it is often necessary to bend the conduit. Smaller electrical conduit, such as electric metallic tube (EMT), intermediate metallic conduit (IMC), and rigid metallic conduit (RMC), can be bent by using a ***conduit bender***. This tool has reference guides along its side, so it can be used to create an accurate and uniform bend. Conduit benders can be used to bend conduit while simultaneously using a foot lever and hand pressure, **Figure 8-22**.

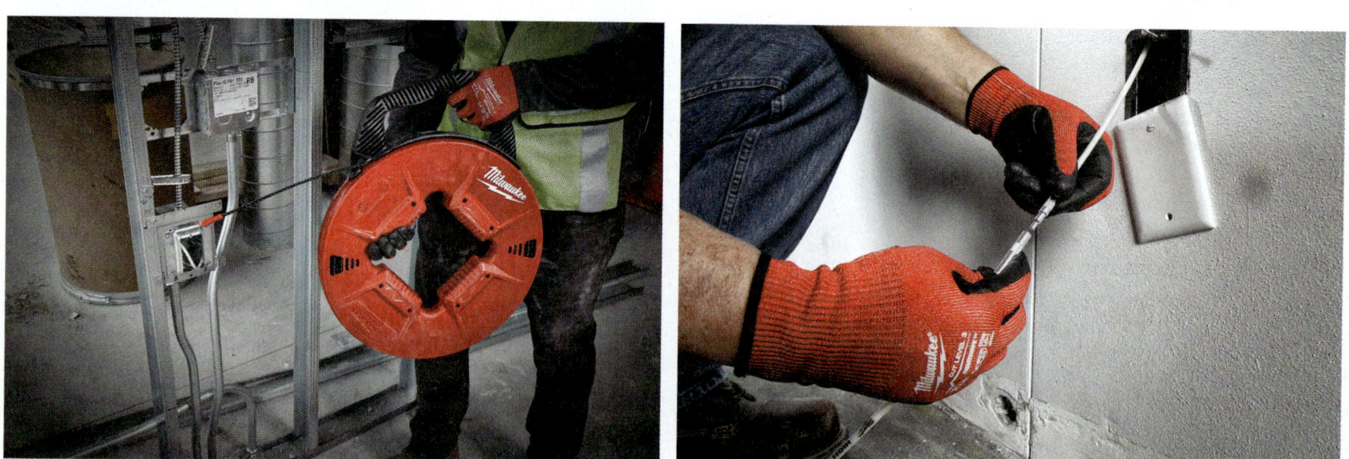

Milwaukee Tools

Figure 8-21. Fish tapes and fish sticks are used to insert wire and cable through voids in walls and ceilings or pull conductors through a conduit. A—Metallic and nonmetallic fish tapes are available in lengths from 25 feet to over 200 feet. B—Sections of fish stick come in 4 to 5 feet rods that can be assembled to form the length needed.

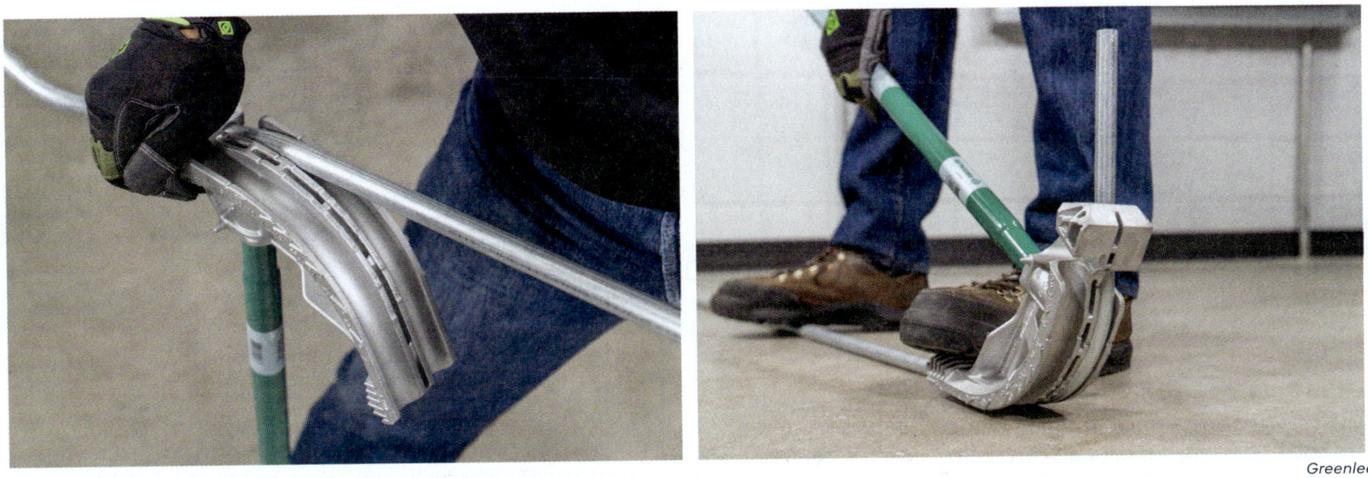

Greenlee©

Figure 8-22. When using a conduit bender, use both hand and foot pressure. If this is not done, the conduit will not bend correctly.

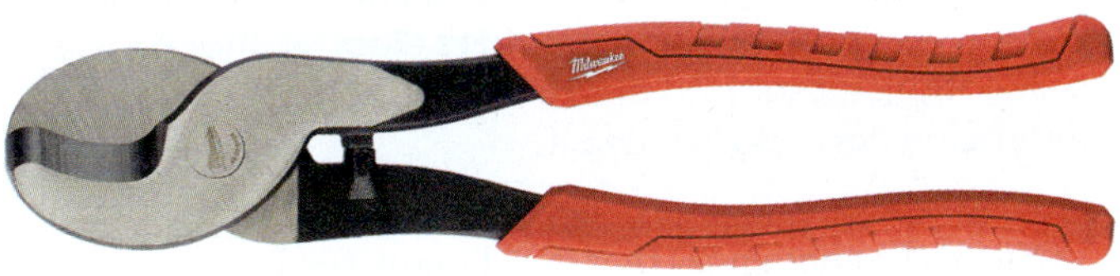

Milwaukee Tools

Figure 8-23. Cable cutters are used to cut non-armored cables.

Handheld benders will bend conduit as large as 1-1/4″ EMT and 1″ RMC or IMC.

Cable Cutters

Different types of ***cable cutters*** are available based on the types of materials that are to be cut. Some materials, referred to as non-armored cables, have vinyl or plastic casings, while other materials, referred to as armored cables, have metallic casings. Armored cables are commonly referred to as metal-clad, or MC, cables. Non-armored cables can be easily cut using a cutter that resembles a pair of pliers, **Figure 8-23**. This tool has rounded cutting edges to effectively cut round cables without affecting the shape or performance of the cable. A ***metallic cable cutter*** will cut across the ribbon, thus eliminating the sharp points, **Figure 8-24**.

Know the Code

When installing Type AC (armor clad) metallic shielded cable, be sure to insert a bushing or its equivalent into the end of the cable before connecting it to a box, connector, or fitting. This ensures that the conductors do not become damaged from contact with the sharp edges of the cut metallic cable. Remember to follow the manufacturers' installation, listing, and labeling instructions. Refer to *NEC Article 320* for specific rules. Type MC (metal-clad) cable does not require a bushing. Refer to *NEC Article 330* for specific installation requirements.

Goodheart-Willcox Publisher

Figure 8-24. A metallic shielded cable cutter helps eliminate sharp edges when cutting a metallic conduit.

Levels, Squares, Plumb Bobs, and Chalk Lines

When installing electric fixtures and lines, it is important that they are level and plumb to ensure a professional-looking job. The tools that assist the installer in accomplishing this important task are the level, the square, the plumb bob, and the chalk line.

The combination square and torpedo level are shown in **Figure 8-25**. These levels have floating bubbles that indicate when a section of conduit or a fixture is level. A bubble located in the center of the scale indicates that the installation is level. In addition to leveling, the combination square can also be used to ensure accurate and consistent 45- and 90-degree angles.

Torpedo levels are designed specifically for electrical metal conduit bending. They have a 30-degree angle bubble designed to aid in the bending of all types of conduit. In addition, electricians will often use specialty levels to help keep all conduit bends on the same flat plane.

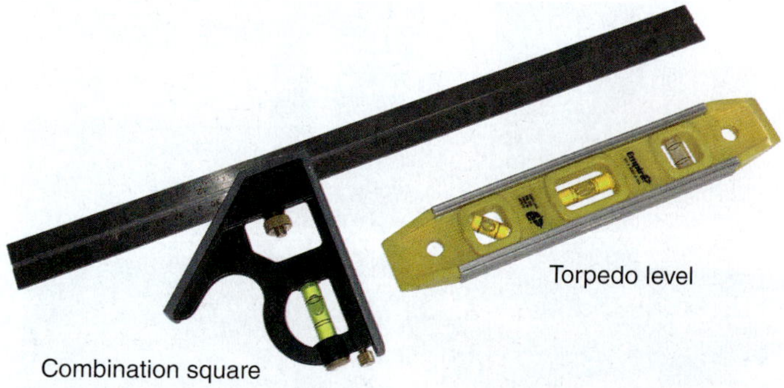

Goodheart-Willcox Publisher

Figure 8-25. A torpedo level or a combination square can be used to ensure that an item is level, meaning it is parallel with the floor or ground, or if the item is plumb, meaning it is perpendicular to the floor or ground.

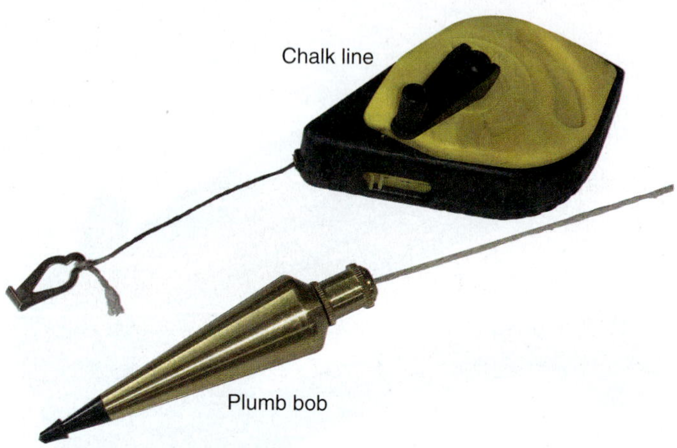

Goodheart-Willcox Publisher

Figure 8-26. Electricians use chalk lines and plumb bobs to help align items and ensure they are installed straight.

Chalk lines and plumb bobs are string-based tools that are used to ensure straight and plumb installations, **Figure 8-26**. The ***chalk line*** is comprised of a string that is housed in a chalk-filled container. As the string is pulled from the container, it is coated with chalk. By holding the string close to a surface in the desired position and then snapping the string, a straight chalk line is drawn on the surface.

The ***plumb bob*** is a useful tool that helps ensure that an installation is vertical. By holding the end of the string and allowing the plumb bob to freely hang, a vertical reference line can be established. Chalk lines and plumb bobs are being replaced with handheld laser tools that do the job more accurately and quickly.

8.3 Power Tools

Today's power tools are available in both corded and battery-powered (cordless) models. In either case, technological advances have paved the way for smaller, lighter-weight tools that have much more power and torque than those used only a few years ago. Ergonomic handles and configurations make modern power tools more comfortable and safer to use.

Shedding Light

On the jobsite, extension cords are often used to bring electric power to remote locations. Since these cords are often subjected to rough handling and possible damage, always inspect all extension cords carefully prior to use. When inspecting an extension cord, the entire length of the cord should be checked for cuts, gouges, or anything else that might compromise the integrity of the cord. Extension cords used on the job are required to have three-prong (grounding) plugs. The two flat prongs carry current to and from the plugged-in device, while the rounded prong is the ground prong, which provides electric shock protection. If the third prong is missing from an extension cord, do not use it.

One of the challenges that faced many field workers in the past was the amount of time required for a battery-powered tool to become fully charged. With today's battery technologies, batteries are holding their charge much longer than before, making these tools a more widely accepted option. The electrician's most commonly used power tools are drills and saws.

Safety Alert

Use only double-insulated two-prong or properly grounded three-prong power tools that are connected to properly grounded power supplies and circuits. Double-insulated power tools have no exposed metal parts that can present a shock hazard.

Power Drills

Every electrician uses a drill, but the application will determine the type needed. For general drilling and fastening, a battery-powered drill will often do the job, **Figure 8-27**. For heavier applications, corded tools are often used. ***Power drills*** can be used to create small or large holes in a variety of materials including wood, steel, and masonry. Always be sure to use the right drill and drill bit for the job. A ***hammer drill*** with masonry bits is used to drill holes into concrete or brick. To cut larger holes in soft steel sheet metal, a hole drill can be used, **Figure 8-28**.

Drills can be fitted with driver bits to install and remove fasteners. Various drill tip sets that include different sizes of slotted, Philips, star drive, square drive, and nut drivers are available to expand the applications for which the drill can be used, **Figure 8-29**.

Goodheart-Willcox Publisher

Figure 8-27. Battery-powered drills are commonly used for quickly loosening and tightening screws with a variety of head configurations.

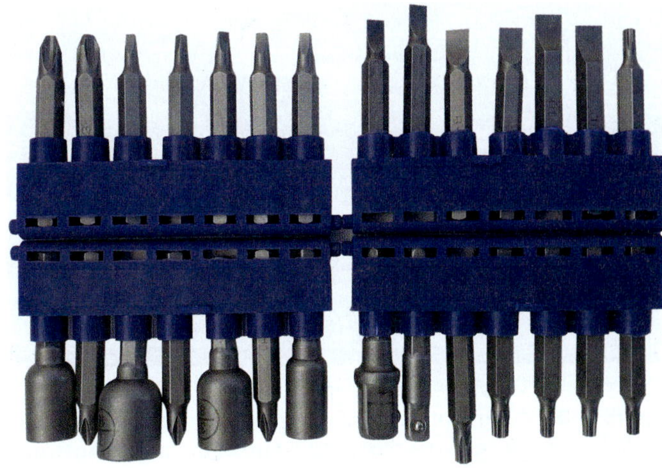

Goodheart-Willcox Publisher

Figure 8-29. A drill bit set provides a variety of drill tips to select from for any given project.

Goodheart-Willcox Publisher

Figure 8-30. A reciprocating saw is a power-operated saw in which the blade moves back and forth to make a cut.

Greenlee©

Figure 8-28. Electricians use hole drills to cut through thinner materials such as plywood and soft steel sheet metal used in many electrical cabinets.

Safety Alert

As with all cutting and boring tools, make certain that drill bits are sharp. Sharp bits work better, cut faster, require less force, and, as a result, are safer to use than dull bits. Always let the tool do the work. To avoid personal injury, never lean on or force a drill to cut faster.

Electric Saws

As with drills, the application determines the type of electric saw needed. One of the most common power saws used by electricians is the battery-powered reciprocating saw, **Figure 8-30**. A ***reciprocating saw*** can be used to cut wood, metal, or a number of other materials. The cutting blade is selected based on the material to be cut and the quality of the cut being made. Cut quality is determined by looking at the number of teeth on a saw blade, also known as the ***teeth per inch***. Blades with more teeth per inch (tpi) create finer cuts. Blades with fewer teeth per inch can be used for a rougher cuts. Metal blades, for example, have very fine teeth, while blades used for cutting wood have larger, deeper teeth, **Figure 8-31**.

Another electric saw commonly used by the electrician is the circular saw, **Figure 8-32**. The ***circular saw*** uses a

Goodheart-Willcox Publisher

Figure 8-31. When cutting a piece of wood, the ideal saw blade has fewer teeth per inch (tpi). The blade at the bottom, used to cut metal, has a greater tpi.

Goodheart-Willcox Publisher

Figure 8-32. A battery-powered circular saw uses a disk-shaped blade with spinning, rotary motion to cut various materials.

round blade and is primarily used to make straight cuts in sheet materials such as wood. Circular saws are also used to cut sections of lumber, such as two-by-fours. When using a metal blade, a circular saw can be used for cutting conduit, uni-strut, and various other materials commonly used in electrical applications.

Other powered saws commonly used in the electrical trades include band saws and chop saws. Small and larger corded and battery powered band saws are widely used for cutting all types of conduit and other material. Chop saws with a metal cutting blade are used when a square cut on conduit and other pieces of electrical material is needed.

8.4 Electrical Testing Equipment

In the field, electricians often test electric circuits to know if voltage is present or to know exactly how much voltage

Courtesy Fluke Corporation

Figure 8-33. Almost every electrician carries a non-contact voltage tester to quickly test for voltage. They are not foolproof, but if used properly, they will help prevent electrical shock.

is present. Electricians often need to determine whether power is being supplied to a fixture to determine if the problem lies within the fixture or with the wiring that carries power to it. If an electric load is operating, the electrician might need to measure the amount of electric current flowing through the device. When an intermittent problem persists, the electrician might need to use more sophisticated pieces of test equipment to observe the quality of a power signal over long periods of time. In any event, the job at hand dictates what pieces of test instrumentation are needed.

Non-Contact Voltage Tester

In some instances, the electrician needs to know if voltage is present at a particular point in an electric circuit. When such is the case, a ***non-contact voltage tester*** can be used, **Figure 8-33**. By positioning the device near a conductor, the user is alerted if voltage is present. Depending on the type of device used, there might be a light that indicates that the device is operational, while another light or buzzer will illuminate or sound to indicate the presence of voltage. Other pieces of electrical test instrumentation have a non-contact voltage-testing feature incorporated into them. For instance, the multimeter in **Figure 8-34** has a built-in non-contact voltage reader. When such a feature is activated, the meter will detect the presence of voltage when the instrument is brought in close contact with a circuit.

Digital Multimeter

A ***digital multimeter (DMM)***, **Figure 8-35**, allows the user to measure a wide range of electrical characteristics such as

Courtesy Fluke Corporation

Figure 8-34. A non-contact voltage testing feature on a multimeter. When such a feature is activated, the meter will detect the presence of voltage when the instrument is brought in close contact with a circuit.

Courtesy Fluke Corporation

Figure 8-36. A clamp-on ammeter is similar to a digital multimeter but also includes jaws that measure the current.

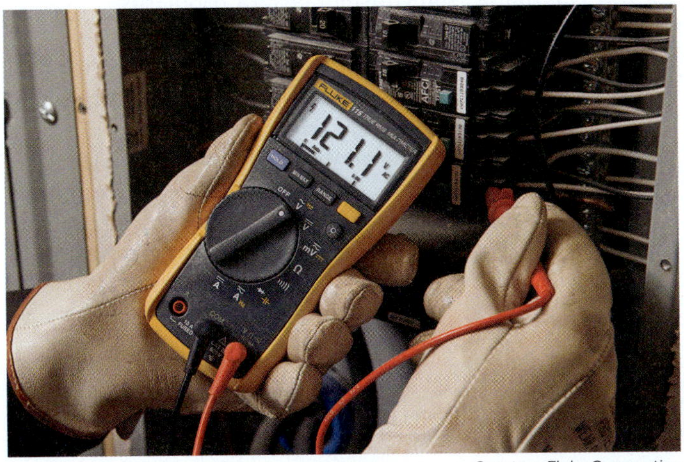

Courtesy Fluke Corporation

Figure 8-35. A digital multimeter is an all-in-one device that measures voltage, resistance, and current.

voltage, current, resistance, and continuity. Some meters have the ability to test electrical devices such as capacitors and diodes.

A digital multimeter that is classified as a ***true-RMS meter*** has the ability to accurately measure the characteristics of pure sine waves as well as nonsinusoidal waveforms. True-RMS meters calculate the equivalent dc values of the ac power signal being tested. This is especially useful when evaluating a power supply whose characteristics are not be known or might be problematic.

Safety Alert

Be sure to properly store all pieces of test instrumentation and protect them from falls. Damaged instrumentation, including test leads, can provide inaccurate readings that can lead to severe personal injury or death. Whenever using any meter, the accepted practice is to test on a known live circuit before using on the circuit to be tested. When testing is completed, check again on the known live circuit. This does not happen often, but test leads or the meter may malfunction between tests.

Clamp-On Ammeter

A ***clamp-on ammeter***, **Figure 8-36**, has all or most of the features of the digital multimeter, but also has the ability to measure the current flow through a conductor by simply clamping the operable jaws of the meter around the conductor to be tested. In some cases, it might be very difficult or not possible, to clamp the jaws of the ammeter around a conductor. In cases such as this, there are ammeters that have an open fork that can be positioned over a conductor to obtain current readings.

In situations where it might be unsafe to be close to the point at which a current reading is to be taken, such as within the arc-flash zone, app-based instruments can be used. These instruments, such as the one shown in **Figure 8-37**, allow the electrician to position a meter to take a current reading with the circuit is de-energized. Then, once at a safe distance, the

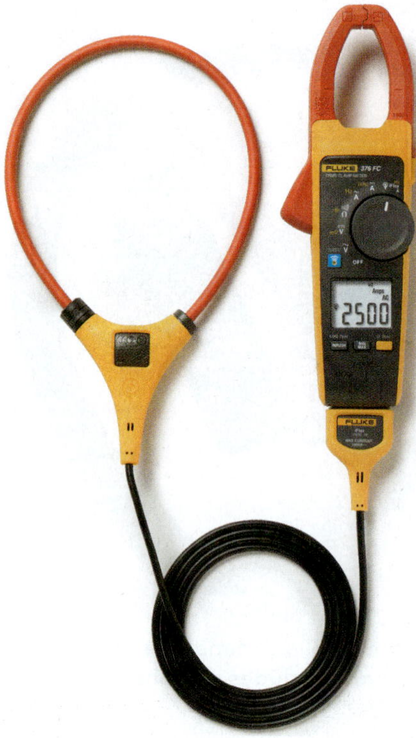

Courtesy Fluke Corporation

Figure 8-37. This wireless clamp meter allows an electrician to safely take a current reading in an arc-flash zone.

electrician can energize the circuit and monitor the current by means of an app on a smart device.

Power Analyzers

Power analyzers are extremely powerful pieces of test equipment as they calculate and display mechanical power and efficiency, **Figure 8-38**. Some models have the ability to perform these functions without requiring mechanical sensors. Power analyzers have the ability to identify problems with the power signal such as voltage dips, voltage swells, voltage transients, harmonics, and voltage unbalance. These power problems often go undetected and are often the cause for unexplainable, intermittent circuit problems.

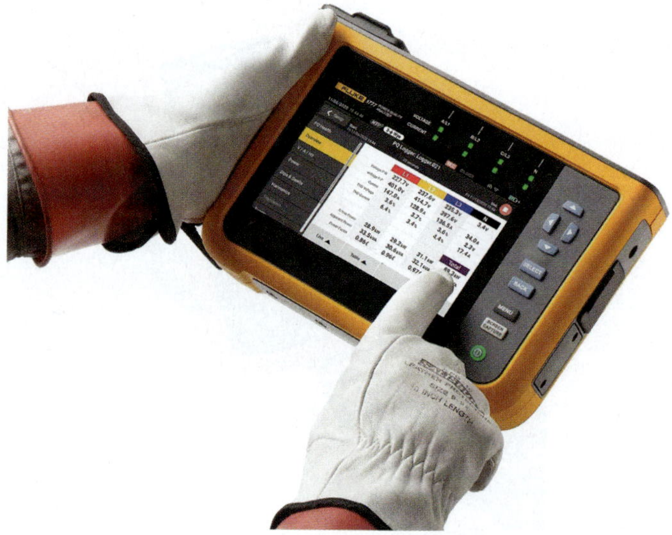

Courtesy Fluke Corporation

Figure 8-38. Power analyzers are used to measure the flow of power in an electrical system. This equipment can calculate mechanical power and efficiency.

Shedding Light

As a result of technological advances, the cost of high-quality test instrumentation has dropped significantly, while the reliability of these instruments has increased exponentially. Armed with the latest high-tech tools and instruments, properly trained electricians are able to identify power system problems that were once almost undetectable by field service personnel.

CHAPTER 8 REVIEW

Summary

- Tools must be used only to perform their intended job. Never exceed the limits of a tool. (8.1)
- Be sure to inspect all tools for damage prior to use. Damaged tools should not be used. Wearing a tool belt with the right tools for the job is extremely helpful. (8.1)
- Screwdrivers have many different tips with the most popular being the slotted and the Phillips. Star and square bits are becoming more popular along with specialty bits used on circuit breakers, light switches. and receptacles. (8.2)
- Nut drivers are used to tighten and loosen screws and bolts with hex-shaped heads and are often color-coded to easily identify the different sizes. (8.2)
- Primarily used to hold and twist, some pliers are configured to cut and allow access into tight spaces. (8.2)
- Wrenches are primarily used to turn bolts with hex heads. Some wrenches are used to tighten and loosen connections between pipe sections and fittings. (8.2)
- Commonly used hammers are the electrician's straight claw hammer and soft-faced hammer. (8.2)
- Electricians use a variety of saws such as the jab saw and hacksaw. (8.2)
- Wiring tools are specifically designed to strip insulation from conductors, cut cables and conductors, and crimp assorted fittings on conductors Fish tapes, poles, and sticks are used to snake wires and cables through conduits, dropped ceilings, walls, and raised floors. (8.2)
- Chalk lines, plumb bobs, levels, and squares are used to ensure straight, vertical, and level installations. Electronic measuring tapes and electronic laser levels will do the same job more accurately and quicker. (8.2)
- Be sure to use the right power drill and saw for the job. (8.3)
- Drill bits and cutting blades should be sharp as they cut better, require less force, and are safer to use. (8.3)
- Non-contact voltage testers are used to establish the presence of electrical potential. (8.4)
- Digital multimeters are commonly used to measure ac voltage, dc voltage, resistance, and continuity. (8.4)
- Clamp-on ammeters measure current in a conductor without direct contact. (8.4)
- Power analyzers are used to evaluate power supplies and identify potential problems such as voltage dips, voltage swells, voltage transients, harmonics, and voltage unbalance. (8.4)

Review Questions

Use the information in the chapter to answer the following questions.

Measuring Your Potential

1. Screwdrivers are intended to _____. (8.2)
   A. tighten and loosen screws
   B. pry heavy objects
   C. chisel wood
   D. scrape concrete
2. Bolts with hex-shaped heads are most likely to be turned with a _____. (8.2)
   A. nut driver or a screwdriver
   B. nut driver or a torque wrench
   C. nut driver or a socket wrench
   D. socket wrench or a screwdriver
3. *True or False?* Side-cutting pliers are used to cut thin metal close to a surface. (8.2)
4. A combination wrench is made up of an open-end wrench and a(n) _____. (8.2)
   A. pipe wrench
   B. Allen wrench
   C. socket wrench
   D. box wrench
5. The torque wrench is used to ensure proper twisting force, expressed in _____. (8.2)
   A. foot-pounds
   B. foot-inches
   C. pounds per square inch
   D. pounds per square foot
6. The purpose of a ball tip on a hex-key wrench is to _____. (8.2)
   A. allow it to turn round-headed screws and bolts
   B. allow it to be used at an angle
   C. make the end of the tool less sharp and safer
   D. make the end magnetic
7. The tool used to gently tap easily-damaged surfaces is the _____. (8.2)
   A. claw hammer
   B. soft-face hammer
   C. ball peen hammer
   D. jack hammer

8. The saw most likely used to create a hole in a wall to mount a duplex receptacle is the _____. (8.2)
   A. hole saw
   B. jab saw
   C. wood saw
   D. hacksaw
9. A crimping tool is used to _____. (8.2)
   A. secure stranded wires under screw terminals
   B. connect solderless connectors to stranded wires
   C. remove insulations from wires
   D. secure solid wires under screw terminals
10. A skinning knife is most likely intended to _____. (8.2)
   A. cut wire to size
   B. cut metallic shielded cable
   C. remove rust from metallic conduit
   D. remove insulation from a wire or conductor
11. *True or False?* Fish tapes are commonly used to snake wires through conduit. (8.2)
12. A conduit bender is most likely used on which of the following? (8.2)
   A. EMT and IMC
   B. EMT and BX
   C. IMC and Romex
   D. Romex and BX
13. To avoid leaving sharp, pointed edges when cutting electrical cables, it is recommended to use a _____. (8.2)
   A. cable cutter
   B. diagonal-cutting pliers
   C. hacksaw
   D. metallic shielded cable cutter
14. The chalk line is used to ensure a _____. (8.2)
   A. vertical installation
   B. straight installation
   C. horizontal installation
   D. level installation
15. *True or False?* A hammer drill with masonry bits is used to drill holes into brick. (8.3)
16. A _____ can be carried in an electrician's pocket and used to test if voltage is present. (8.4)
   A. digital multimeter
   B. true-RMS meter
   C. noncontact voltage tester
   D. power analyzer
17. A digital multimeter can be used to measure all of the following electrical characteristics *except* _____. (8.4)
   A. ac voltage
   B. resistance
   C. power
   D. continuity
18. A clamp-on ammeter senses the _____ to determine the amount of current flowing through a conductor. (8.4)
   A. voltage supplied to a circuit
   B. resistance of the circuit
   C. amount of power consumed
   D. strength of the generated magnetic field

Plug-In

1. You are installing a luminaire in the ceiling over a dining room table. Of the following, which tool would you use to ensure the luminaire is aligned vertically to the table? (8.2)
   A. Chalk line
   B. Plumb bob
   C. Measuring tape
   D. Fish tape
2. When working on the main service panel where the conductors attach to the main breaker, a(n) _____ should be used to tighten the hex head fasteners to the manufacturer's requirements. (8.2)
   A. torque wrench
   B. hex-key wrench
   C. box wrench
   D. adjustable wrench
3. Of the following, which saw and saw blade is ideal to cut through metal? (8.3)
   A. A reciprocating saw with a blade with a high tpi.
   B. A reciprocating saw with a blade with a low tpi.
   C. A circular saw with a blade with a with a low tpi.
   D. A hacksaw with a high tpi.
4. Before testing the resistance of a conductor, it is suggested to test for voltage. Of the following, which tool can be used to indicate the presence or absence of voltage as well as measure volts, values of resistance in ohms, and current in amps? (8.4)
   A. Noncontact voltage tester
   B. Clamp-on ammeter
   C. Digital multimeter
   D. Voltmeter

Making the Connection

1. Who would be most likely to use a power analyzer on a job? What events might have led to requiring the use of this instrument? (8.4)
2. Why is using a wire stripper better than using a pair of linesman pliers to remove the insulation from a conductor? (8.2)

SECTION

3 Electric Circuits and Circuit Evaluation Techniques

Eakrin Rasadonyindee/Shutterstock.com

CHAPTER

9 The Electric Circuit

Chaikom/Shutterstock.com

SPARKING DISCUSSION

Why do some loads require a heavier gauge conductor than other loads? Discuss this in relation to residential loads such as kitchen appliances, computers, or car chargers.

LEARNING OUTCOMES

After completing this chapter, you will be able to:

9.1 Understand the importance of an appropriate power supply; recognize ac and dc symbols in wiring diagrams.

9.2 Describe and list examples of resistive loads, inductive loads, and capacitive loads.

9.3 Discuss the importance of a complete conductive path in an electric circuit.

9.4 Describe types of external condition change that may cause an automatic switch to flip.

9.5 Understand the difference between operational controls and safety controls.

TECHNICAL TERMS

armature
automatic reset device
capacitive load
control device
electromechanical control device
electronic control device
inductive load
load
manual reset device
operational control
resistive load
safety control

Introduction

The basic electric circuit has three components: a power source, a device to consume the power, and a complete conductive path. In addition to these required elements, a circuit also normally includes a method to start, stop, adjust, or otherwise control the operation of the device being powered by the circuit. In addition, electrical circuits generally include an overcurrent protective device to protect the circuit in the event of unintended current flow.

In the simple electric circuit shown in **Figure 9-1**, a battery provides voltage to the circuit that is used to control the operation of a light bulb. The operation of the light bulb is controlled by a manually-operated switch. Wires connect the power supply to the switch and the light. In addition, a conductive path is provided between the switch and the light. When the switch is in the ON position, current flows through the circuit, powering the light. When the switch is in the OFF position, no current flows through the circuit, and the light will not illuminate.

This chapter discusses four elements of a typical circuit: the power source, the load, the conductive path, and the switch. On a day-to-day basis, electricians work with these elements. Electricians install the controls for the power source in the form of an electrical service, transformer, generator, or solar inverter. They size and install conductors between a power source and a load to create a conductive path. They select and install switches, circuit breakers, and manual disconnects to provide a means to disconnect a power source from loads. Properly knowing how these four elements interconnect, where they are located, and how they are sized is important to every electrician.

9.1 Power Source

The power source provides voltage, or electrical potential, to a circuit. Two common power sources are batteries and generators. Batteries supply direct current (dc) that travels in one direction and does not change polarity. DC batteries are used to energize many wireless tools and accessories. The batteries used must be properly sized and rated for the application.

Most of the power supplied to buildings is alternating current (ac) produced by generators. Unlike dc power, the polarity of the voltage continuously changes from negative to positive in ac power. AC power is used to energize the appliances in our homes, our lighting circuits, as well as our heating and air-conditioning systems. AC power can be supplied at different voltages and different phase configurations; therefore, the power requirements for a particular application must be known before connecting the system to a power source. In wiring diagrams, power sources are often identified by symbols similar to the ones shown in **Figure 9-2**.

9.2 Circuit Loads

In an electric circuit, a ***load*** is any circuit component that consumes power or energy. Common electrical loads include lighting fixtures, air-conditioning and heating equipment, computers, kitchen appliances, fans, and phone chargers. In more general terms, a load is any device or component that converts electrical energy into light, heat, or some form of useful motion. Electrical loads differ in how they consume power in an electric circuit and in what kind of power output they produce. There are three basic types of loads: resistive, inductive, and capacitive. Resistive and inductive loads

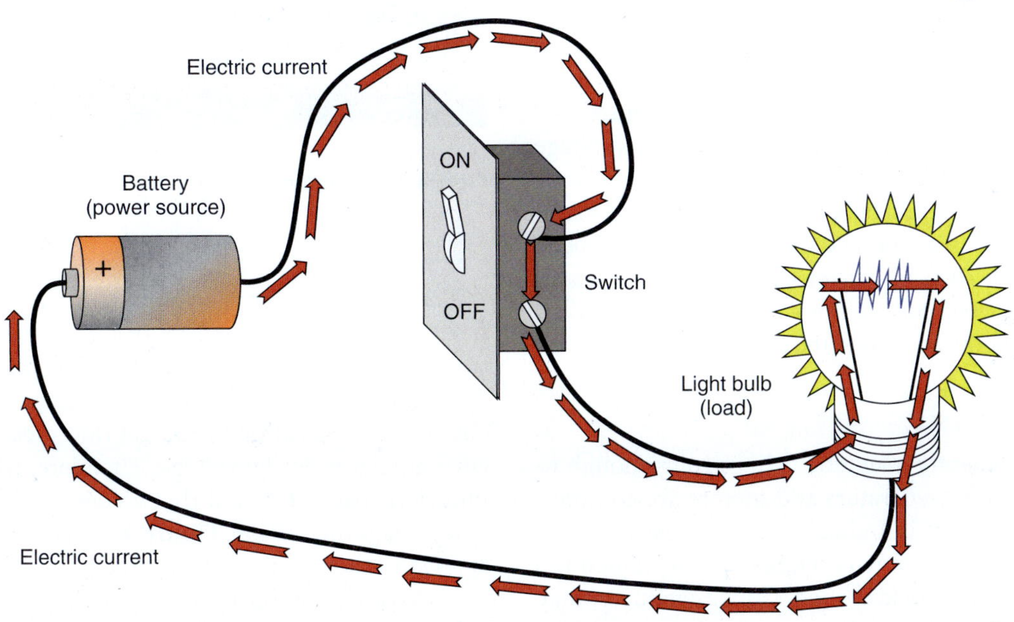

Goodheart-Willcox Publisher

Figure 9-1. In a basic electric circuit, when the switch is in the ON position, current flows through the circuit, causing the light bulb to illuminate.

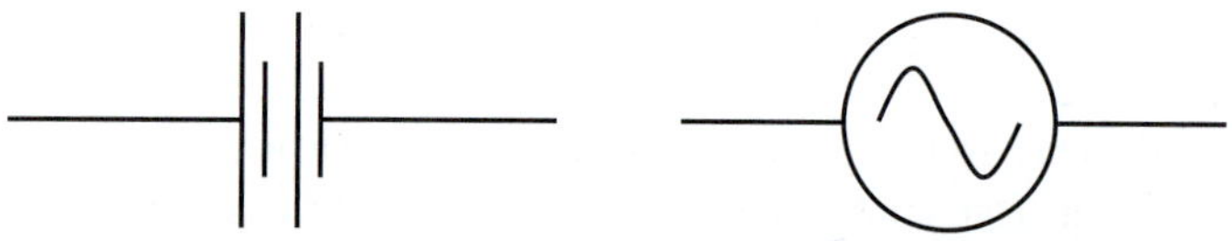

Goodheart-Willcox Publisher

Figure 9-2. Commonly used wiring diagram symbols for dc and ac power supplies.

Stenko Vlad/Shutterstock.com

Figure 9-3. Resistive loads primarily generate light and/or heat. The toaster shown here is an example of a resistive load.

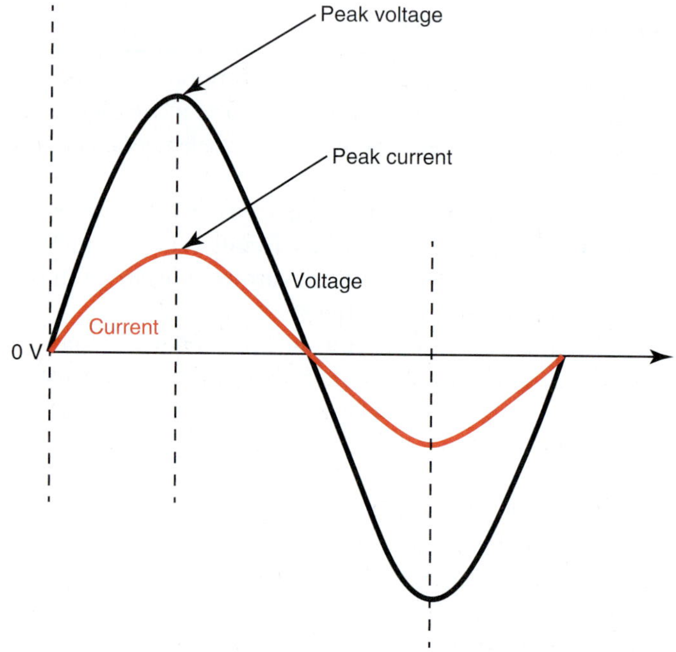

Goodheart-Willcox Publisher

Figure 9-4. In circuits containing only resistive loads, the circuit's current and voltage are in-phase with each other.

Goodheart-Willcox Publisher

Figure 9-5. Windings of a motor are inductive loads.

produce outputs that are useful to the end user. Capacitive loads are used to help control power and make the power source more efficient.

Resistive Loads

Resistive loads generate primarily light and heat. Electric heaters, incandescent light bulbs, coffee makers, and toaster ovens are common examples of resistive loads, **Figure 9-3**. Resistive loads do not cause a shift between a circuit's voltage and current, **Figure 9-4**.

In addition to the heat and light produced by resistive loads, a small magnetic field is also produced. The generated magnetic field is not due to any particular characteristics of resistive loads, but any time an electric current flows through a conductor, a magnetic field is produced around the conductor.

Inductive Loads

Inductive loads generate magnetic fields strong enough to perform useful work. Generators and motors are common examples of inductive loads. Inductive loads are easily identified as they contain coils of wire, **Figure 9-5**. Although the strength of the magnetic field around a single conductor may be weak, with each of the small magnetic forces inside the coils all pointing in the same direction, the concentrated overall magnetic field is much stronger, **Figure 9-6.**

When voltage is applied to an inductive load, a magnetic field is created. This magnetic field opposes the flow of electrons, or current flow. As the applied voltage increases, the opposition force is overcome and electrons flow. In an ac circuit, this opposition to current flow creates a delay in the current relative to the voltage. Therefore, when referring to inductive loads, it is said that voltage leads the current, or current lags the voltage, **Figure 9-7**. In either case, voltage is applied to the circuit and, at some point after that, current starts to flow. The concept of current flow being restricted by the effects of a magnetic field is called inductive reactance. This concept is further explained in Chapter 17, *Inductors and Inductance*.

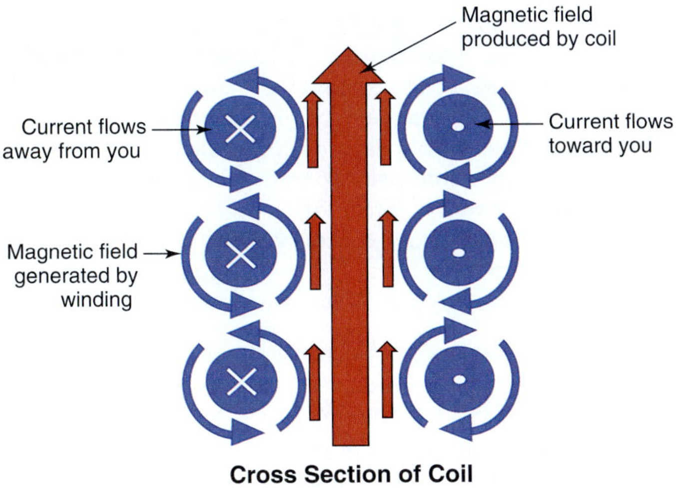

Goodheart-Willcox Publisher

Figure 9-6. Inside a coil of wire, the strength of the magnetic field is the sum of individual forces all acting in the same direction.

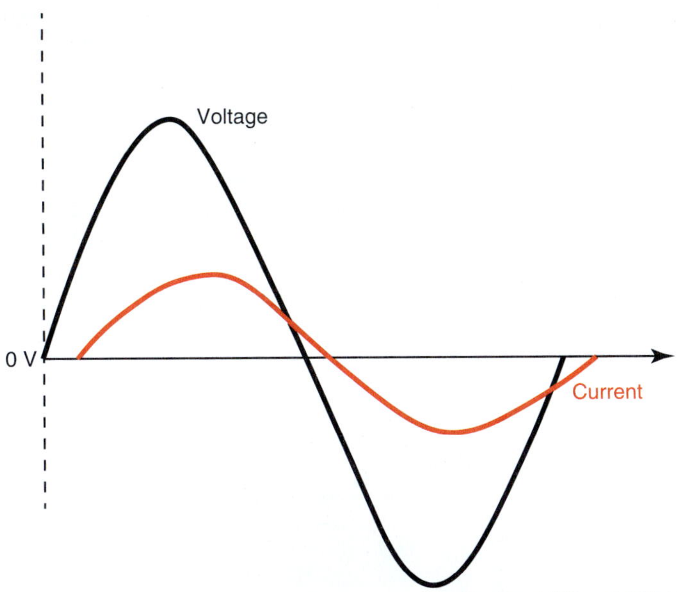

Goodheart-Willcox Publisher

Figure 9-7. In circuits containing inductive loads, the circuit's current lags the voltage by 30 electrical degrees.

Capacitive Loads

Unlike resistive and inductive loads, capacitive loads are not used to perform end-use functions. Instead, they are used to help control power quality and make its use more efficient. ***Capacitive loads*** are loads that store and release electrical energy.

In its simplest form, a capacitor is constructed as two conductive plates separated by an insulating material, **Figure 9-8**. The insulating material that separates the two conductive plates is called a dielectric. The electrical symbol for a capacitor is drawn as two lines, one straight and one curved. In some wiring diagrams, capacitors are represented by two straight, parallel lines, **Figure 9-9**.

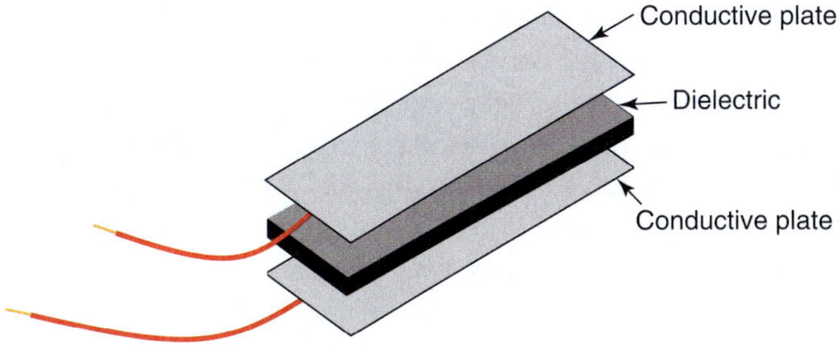

Goodheart-Willcox Publisher

Figure 9-8. A capacitor is made up of two conductive plates separated by a dielectric.

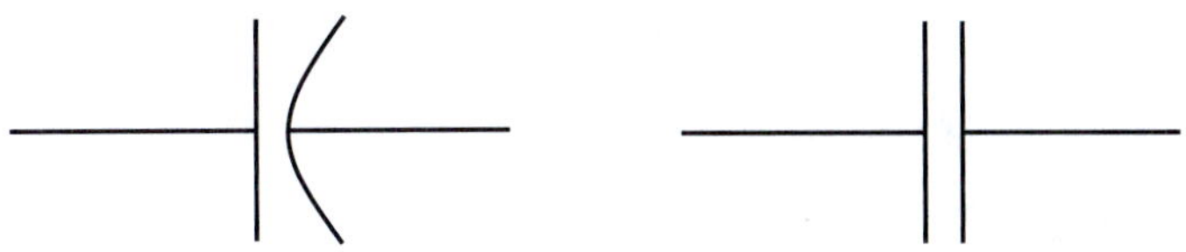

Goodheart-Willcox Publisher

Figure 9-9. Capacitor symbols that are used in circuit diagrams.

Shedding Light

The concept of a circuit's current lagging behind its voltage (or voltage leading the current) can be equated to pushing a chair across a carpeted floor. By gently pushing on the chair, a force is being exerted on the chair, but the chair does not move. There is opposition to movement resulting from friction that exists between the chair and the carpet. As the force exerted on the chair increases, the resistance to movement (friction) is eventually overcome, and the chair starts to move across the floor. In this example, the movement of the chair across the floor starts at some point in time after the pushing force was initiated. The movement of the chair can be equated to the flow of electrons, while the force being exerted on the chair can be equated to the voltage applied to the circuit. The greater the voltage, the greater the push.

In a dc circuit, current flows as the capacitor charges. When enough electrons have flowed so that the charge between the plates is equal to the electrical potential of the power source, current stops flowing and the capacitor is said to be charged. If the charged capacitor is connected to a load, it discharges, providing the source voltage to a load for a limited amount of time.

In an ac circuit, the function of a capacitor is to continually store and release electrical energy. As current flows in one direction, one plate charges and the other plate discharges. When the direction of current changes, so does the behavior of the plates, with the previously charged plate discharging and the previously discharged plate charging. This capacitive load causes current to lead the voltage, **Figure 9-10**, offsetting potential effects caused by inductive loads.

Capacitors are available in many different sizes, **Figure 9-11**. Small capacitors are commonly found on solid-state printed circuit boards, while larger capacitors can be found in appliances, such as furnaces and air-conditioning systems.

The resistance to current flow in a capacitor is called capacitive reactance. This is discussed in further detail in Chapter 18, *Capacitors and Capacitance.*

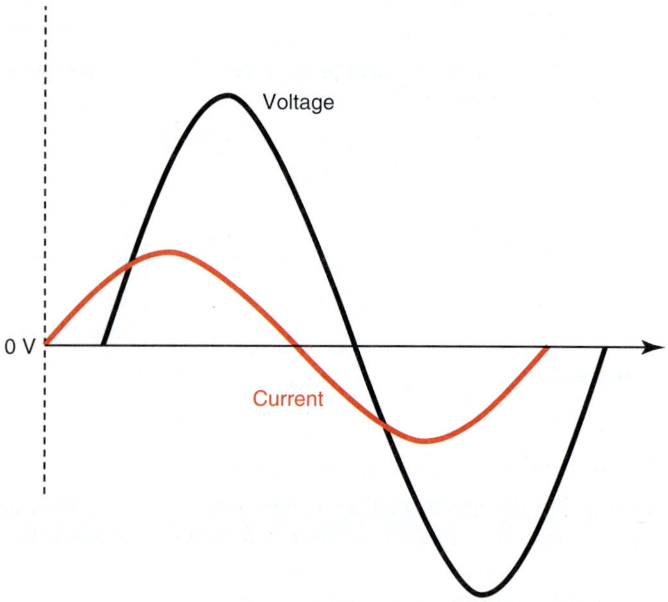

Goodheart-Willcox Publisher

Figure 9-10. A capacitor helps reduce the phase angle created by inductive loads, causing the voltage to lag behind current.

Shedding Light

Because of their ability to help increase circuit efficiency, capacitors are found in many electrical circuits and electronic devices. Industries and applications that use capacitors include telecommunications, information technologies, manufacturing, motors, motor-starting circuits, televisions, smartphones, and many other electronic devices.

Combination Loads and Circuits

Most electric circuits are not purely resistive, purely inductive, or purely capacitive. For the most part, circuits have multiple loads of different types. Some motor circuits, for example, contain windings that are made up of coils of wire (inductive loads) and capacitors (capacitive loads) that aid in starting the motor or improving its operating efficiency.

Shedding Light

To help remember when current leads the voltage and when current lags the voltage, the expression "ELI the ICE man" often helps. The "L" in ELI represents inductance and the "C" in ICE represents capacitance. In ELI, the voltage, *E*, comes before the current, *I*, so the voltage leads the current in an inductive circuit. In ICE, the current, *I*, comes before the voltage, *E*, so the current leads the voltage in a capacitive circuit.

A Goodheart-Willcox Publisher B Goodheart-Willcox Publisher

Figure 9-11. Capacitors vary in size. A—A small capacitor on a solid-state printed circuit board. B—A larger capacitor from an air-conditioning system.

It is often beneficial to have both inductive and capacitive loads in an electric circuit. Since inductive loads cause current to lag voltage, and capacitive loads cause voltage to lag current, the combined effect of these two load types is an increase in efficiency. A brief summary of how the three circuit loads work is provided in **Figure 9-12**.

Circuit Loads

Type of Load	Primarily Generates	Also Generates	Effect on Voltage-Current Relationship
Resistive	Light and/or heat	Magnetic field	None
Inductive	Magnetic field	Heat	Voltage leads current (ELI)
Capacitive	Stored electrical energy	Wasted heat	Current leads voltage (ICE)

Goodheart-Willcox Publisher

Figure 9-12. Summary of circuit load types.

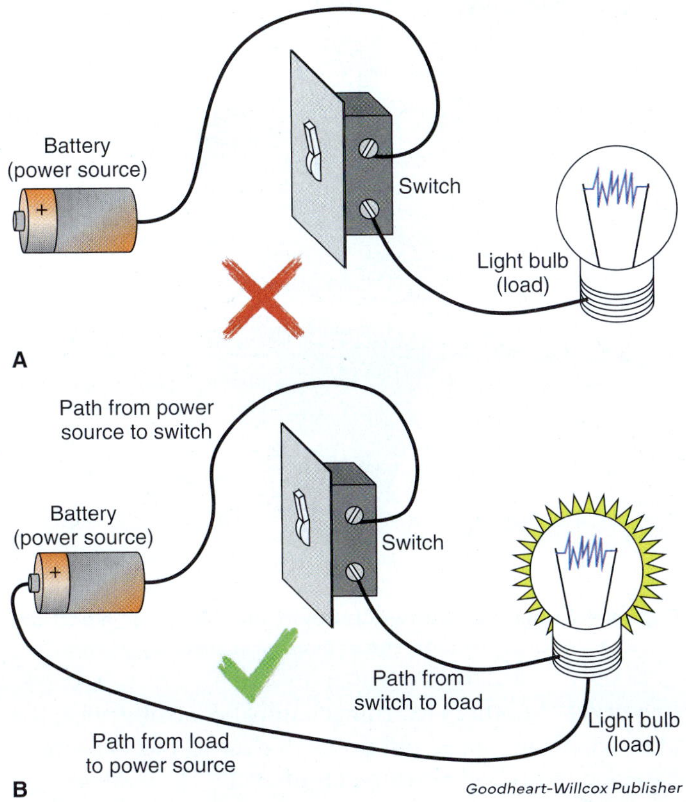

Goodheart-Willcox Publisher

Figure 9-13. A complete conductive path consists of a power supply to the switch, a path from the switch to the load, and a path from the load back to the power supply. A—Without a complete conductive path through a circuit, current will not flow. B—When a complete path exists, current will flow.

9.3 Complete Conductive Path

If voltage is supplied to the circuit loads, there must be a network of interconnecting wires that supplies voltage to the load, carries the circuit's current, and completes the circuit by carrying current back to the power supply. A complete path includes electrical connections between the power source and the switching devices, between the switching devices and the load being powered, and between the load and the power supply. When a complete current path exists, there will be a complete loop that includes the power supply, circuit switches, and the controlled load, **Figure 9-13**.

9.4 Circuit Controls and Switches

If a load, such as a pump, motor, light bulb, or heater, was simply connected to a power supply, the device would operate continuously, **Figure 9-14**. This would be inconvenient and will result in a large waste of energy. Imagine if all the lights in your house were on 24 hours a day, seven days a week. What if your car engine ran all the time? There needs to be some way to start and stop (energize and de-energize) the loads that need to be powered, so they operate at the right time, or when we need them to operate.

To ensure that circuit loads are energized and de-energized at the right time and operate safely while doing so, a control device, such as a switch, fuse/circuit breaker, or relay, is used. ***Control devices*** open and close sets of electrical contacts to energize or de-energize the circuits that power the main circuit loads. Control devices can operate in a number of different ways, depending on how the main load is to function.

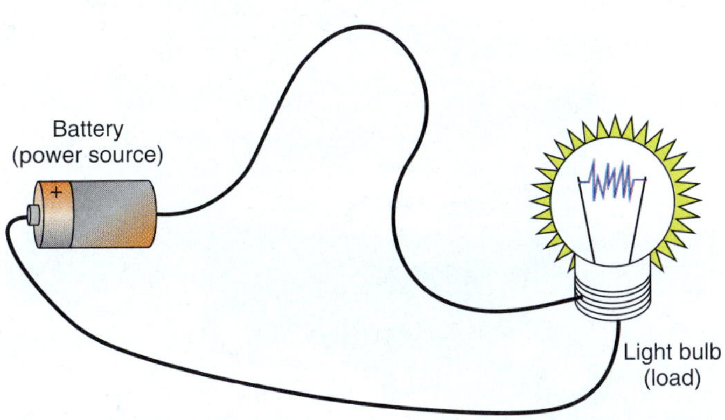

Goodheart-Willcox Publisher

Figure 9-14. In this scenario, the light bulb is connected directly to a battery. Without a switch, the light bulb will illuminate continuously.

Switches

Switches operate either manually or automatically. The position of manually operated switches must be physically changed by the user to either energize or de-energize a circuit. The wall switches in your house are examples of manually operated control devices that are used to turn the lights on and off in various areas in the living space. Automatic switches respond to some external force or condition to energize or de-energize circuits. Automatic switches can operate based on sensed conditions such as the following:

- Time
- Temperature
- Motion
- Light
- Sound
- Fluid level
- Air movement
- Pressure
- Fluid flow
- Electric current flow
- Voltage
- Resistance

Goodheart-Willcox Publisher

Figure 9-16. The coil and movable contacts on a general-purpose control relay.

Electromechanical Control Devices

Electromechanical control devices have a linkage or physical connection that causes sets of electrical contacts to open and close, **Figure 9-15**. The general-purpose relay opens and closes sets of mechanically controlled electrical contacts when a magnetic field is generated by a small coil of wire contained within the device, **Figure 9-16**. The part of the relay that moves to open and close the electrical contacts is called the ***armature***. Another electromechanical control device, the pressure switch, opens and closes an electrical set of contacts based on the sensed pressure, **Figure 9-17**. These devices are connected to pressurized containers or systems to ensure that the internal pressure limit is not exceeded. For example, an air compressor uses pressure controls to maintain the pressure in the air tank. Other electromechanical control devices include float switches and fluid flow switches.

Goodheart-Willcox Publisher

Figure 9-17. A pressure switch opens and closes an electrical set of contacts based on the sensed pressure.

Goodheart-Willcox Publisher

Figure 9-15. A general-purpose control relay is an example of an electromechanical control device.

Goodheart-Willcox Publisher

Figure 9-18. Photocells are used to automatically turn on and off outside lighting.

Chad Robertson Media/Shutterstock.com

Figure 9-19. A garage door safety sensor will stop action if the beam is crossed.

Electronic Control Devices

Control devices that rely on electronic circuitry and sensors to open and close sets of electrical contacts are referred to as ***electronic control devices***. The benefit of electronic control devices is the lack of, or very few, moving parts to fail. An example of this type of control device is a photocell, **Figure 9-18**. In a common dusk-to-dawn photocell, which is used to control outdoor lighting, the contacts are open during the day, preventing the light from turning on. As the sun sets, the amount of sensed light decreases, and the contacts close, turning the light on.

Operational and Safety Control Devices

In addition to the external conditions and forces that cause switches to open and close, there are two important classifications that are used to describe switches and their intended purpose. These switch classifications are operational controls and safety controls.

Some switches open and close sets of electrical contacts during the normal operation of the circuit. These switching devices are referred to as ***operational controls***. The controls found on a washing machine, the thermostat on an electric oven or central air-conditioning system, or the photocells that control outdoor lighting are all examples of operational controls.

Some switches are designed to change the position of their electrical contacts only when an unsafe or undesirable condition is present. These switching devices are referred to as ***safety controls***. The internal temperature sensor in your smartphone or the sensors on your automatic garage door opener are examples of safety controls.

For example, the sensors on an automatic garage door opener create a safety circuit, **Figure 9-19**. When there is no obstruction in the garage door's path, the light beam between the two sensors closes a circuit that allows the garage door to operate normally. However, if an obstruction blocks the light stream, the downward operation of the garage door is halted and the door raises to the completely open position. This direction-reversal process is a safety measure activated only by an obstruction indicated by the broken light beam.

Shedding Light

A thermostat in a central air-conditioning system is a great example of an operational control. Once the thermostat is set to the desired temperature, the air-conditioning system operates until the desired temperature is reached. At this point, the system cycles off. As the space temperature rises, the air-conditioning system cycles on to lower the temperature once again. The thermostat senses the temperature and cycles the system on and off to maintain the conditioned area at the correct temperature.

Automatic and Manual Reset Control Devices

Controls are also often classified as either automatic reset devices or manual reset devices. ***Automatic reset devices*** are designed to reset themselves automatically. Once the sensed condition that caused the switch to change its position has reached a desired or acceptable level, the contacts of the switch automatically return to their previous position.

Manual reset devices require that once the position of the contacts change, the switch be reset manually. Manual reset devices are commonly used as safety devices. Once the sensed condition has reached an unsafe, undesirable, or unacceptable level, the contacts of the switch will move to de-energize the load. Even if the condition returns to a safe level, a service technician, electrician, or maintenance team member will need to reset the device to restore operation. Once it has been established that the safety device was responsible for the shutdown, the cause for the safety event will need to be identified and corrected before restoring power to the circuit.

CHAPTER 9 REVIEW

Summary

- A basic electric circuit has three component parts: a power source, a device to consume the power, and a complete conductive path as well as a method to start, stop, adjust, or otherwise control the operation of the device being powered. (9.1)
- A power source provides voltage to a circuit. (9.1)
- Loads can be classified as resistive, inductive, or capacitive. (9.2)
- Resistive loads produce primarily light and/or heat. (9.2)
- Inductive loads produce primarily magnetic fields. (9.2)
- Capacitive loads store and release electrical energy into the circuit. (9.2)
- In circuits with only inductive loads, voltage leads the current. (9.2)
- Capacitors cause the current to lead voltage. (9.2)
- Capacitive and inductive loads are often used together to increase circuit efficiency. (9.2)
- A complete conductive path is needed to supply voltage to the circuit load. (9.3)
- Switches and controls are used to energize and de-energize circuit loads. (9.4)
- Electromechanical control devices have a linkage or physical connection that opens and closes the electrical contact. (9.4)
- Electronic control devices rely on electronic circuitry and sensors to open and close an electrical contact. (9.4)
- Operational controls open and close their contacts during the normal operation of a system. (9.4)
- Safety controls are typically normally-closed devices that open their contacts when unsafe or undesired conditions are present. (9.4)

Review Questions

Use the information in the chapter to answer the following questions.

Measuring Your Potential

1. A circuit's required elements are the _____. (9.1)
   A. power source, switch, and load
   B. load, conductive path, and switch
   C. load, conductive path, switch, and power supply
   D. power source, conductive path, and load
2. *True or False?* A standard battery is an example of an alternating current power source. (9.1)
3. In a circuit with only resistive loads, the _____. (9.2)
   A. voltage leads the current
   B. current leads the voltage
   C. voltage and current are in phase with each other
   D. current lags the voltage
4. Resistive loads primarily produce _____. (9.2)
   A. light and heat
   B. light and magnetic fields
   C. heat and motion
   D. magnetic fields and motion
5. What is primarily produced by an inductive load? (9.2)
   A. Light
   B. Magnetic fields
   C. Heat
   D. Motion
6. Which of the following best describes a capacitor? (9.2)
   A. Two insulating plates separated by a conductive material
   B. Two insulating plates separated by a third layer of insulating material
   C. Two conductive plates separated by an insulating material
   D. Two conductive plates separated by a third conductive material
7. Of the following, which best describes a complete conductive path? (9.3)
   A. Starts at the power supply and ends at the load
   B. Starts at the power supply and ends at the switch
   C. Starts at the switch and ends at the load
   D. Starts and ends at the power supply
8. Switches are used to _____. (9.4)
   A. start electrical devices at the correct time
   B. stop electrical devices at the correct time
   C. modulate the operation of electrical devices
   D. start, stop, and modulate the operation of electrical devices
9. Which type of control opens and closes during the normal operation of a circuit or system? (9.4)
   A. Safety device
   B. Electromechanical device
   C. Operational device
   D. Electronic device
10. A manual reset device is most commonly used as a(n) _____. (9.4)
    A. safety device
    B. electromechanical device
    C. operational device
    D. electronic device

Plug-In

1. Of the following, which is a common example of resistive loads? (9.2)
   A. Coffee makers
   B. Generators
   C. Air conditioners
   D. Transformers
2. In an electric clothes dryer, the electric heating element is a(n) _____ load. (9.2)
   A. capacitive
   B. electromechanical
   C. inductive
   D. resistive
3. In an electric clothes dryer, the motor that rotates the drum is a(n) _____ load. (9.2)
   A. capacitive
   B. electromechanical
   C. inductive
   D. resistive
4. *True or False?* ELI is best described as current lags the voltage. (9.2)
5. Thermostats are examples of a(n) _____ control devices. (9.5)
   A. electromechanical
   B. electronic
   C. automatic reset
   D. operational

Making the Connection

1. Explain why further investigation is required if it is determined that a safety switch has opened to de-energize a circuit or system. (9.4)
2. Circuit breakers are devices that must be reset manually. Since an over-current condition often occurs for only a short period of time, would it be more beneficial to have circuit breakers reset automatically? Why or why not? (9.4)

CHAPTER

10 Series Circuits

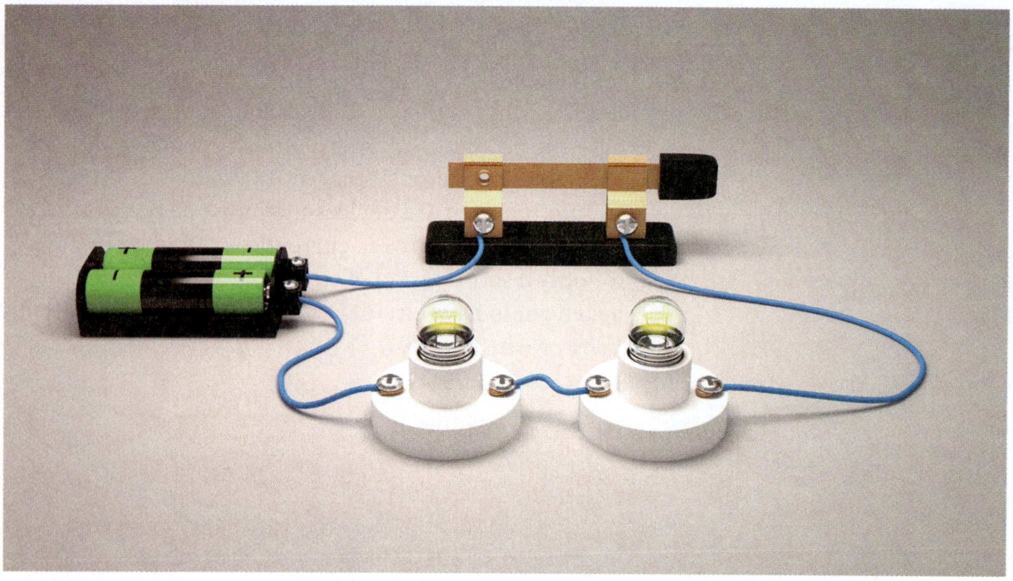

haryigit/Shutterstock.com

SPARKING DISCUSSION

If the voltage supplied to a series circuit is doubled, the power consumption of the circuit also doubles. Do you agree or disagree with this statement?

LEARNING OUTCOMES

After completing this chapter, you will be able to:

10.1 List the properties of a series circuit.
10.2 Calculate total current, resistance, and voltage in a series circuit.
10.3 Apply Ohm's law and power formulas to series circuits.
10.4 Explain the function of switches in series circuits.

TECHNICAL TERMS

rating
series circuit
terminal
voltage drop

Introduction

In Chapter 9, *The Electric Circuit*, the fundamental components of a basic electric circuit (power source, loads, switches, and conductive paths) were discussed. Armed with the knowledge of the roles these components play in electric circuits, our discussion will now turn to electric circuit configurations, the most basic of which is the series circuit.

In this chapter, you will learn the rules that govern the behavior of series circuits. Examples of each rule are provided, including mathematical calculations. It is important to understand the concepts presented in this chapter, as all future chapters build on this basic information.

10.1 Characteristics of Series Circuits

The term ***series circuit*** is used to describe an electric circuit that has only one possible path for electrons to follow. In a series circuit, all electrons follow the same path, and the intensity of electric current flow is the same at all points in the circuit. The series circuit is often compared to a winding, one-lane mountain pass road. There are no on-ramps or off-ramps—just one well-defined path that all vehicles must follow. If an obstruction blocks the road, the flow of traffic stops. Similarly, if a wire breaks in a series circuit, the flow of electrons stops. If the road surface is excessively rough, or if there are hairpin turns, traffic slows as a result of this opposition, or resistance, to flow. In a series circuit, electrical resistance results in a reduction of current flow.

Simple series circuits are easily identified as they have three common features:

- Most circuit components have two electrical terminals to which conductors or wires are connected.
- The output of one circuit component is connected, via conductors or wires, to the input of another circuit component.
- There is only one option when tracing out the path of current flow.

Aside from the current-carrying conductors, there are three components found in a basic series circuit: a power source, a load, and a switch. In the most basic series circuit, each of these components has two terminals to which the current-carrying conductors are connected, **Figure 10-1**.

Terminals are the connection points between circuit components and the wires or conductors that are used to connect them to each other. The basic series circuit is easy to wire as the output of one component is connected to the input of the next. The terms output and input refer to the two connection points. The order in which the components are connected does not matter. Electrically speaking, the wiring arrangement in **Figure 10-2A** is the same as that in **Figure 10-2B**.

When a series circuit is properly wired, there is one complete path that passes through each circuit component, and there is no more than one conductor or wire connected to any one end or terminal on any of the individual circuit components. When tracing the complete conductive path, there is only one option, and the path ends at the same component, although on the opposite side, where the tracing began. There are no "forks in the road" where current can split between two or more possible paths. Using basic electrical symbols instead of pictorial versions of the electrical components, the series circuits shown in **Figure 10-3** are electrically the same.

10.2 Series Circuit Calculations

There are three rules to calculate values of voltage, current, resistance, and power in any series circuit.

- Current flow is the same at all points in a series circuit.
- Total circuit resistance is equal to the sum of all circuit resistances.
- Total circuit voltage is equal to the sum of all voltage drops in the circuit.

In most real-world series circuits, there is a single load wired in series with one or more switches or control devices. Unless specifically designed and intended to be connected in series, loads are not connected in this manner. The following

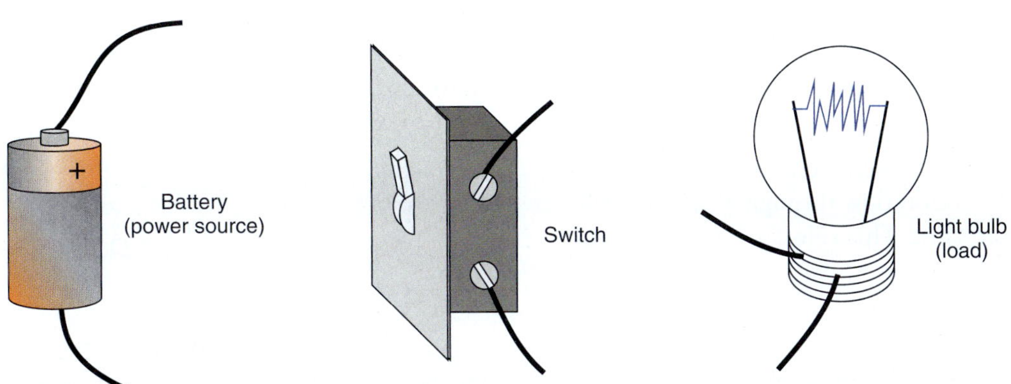

Goodheart-Willcox Publisher

Figure 10-1. In a series circuit, each component has two terminals that are used to connect with another component.

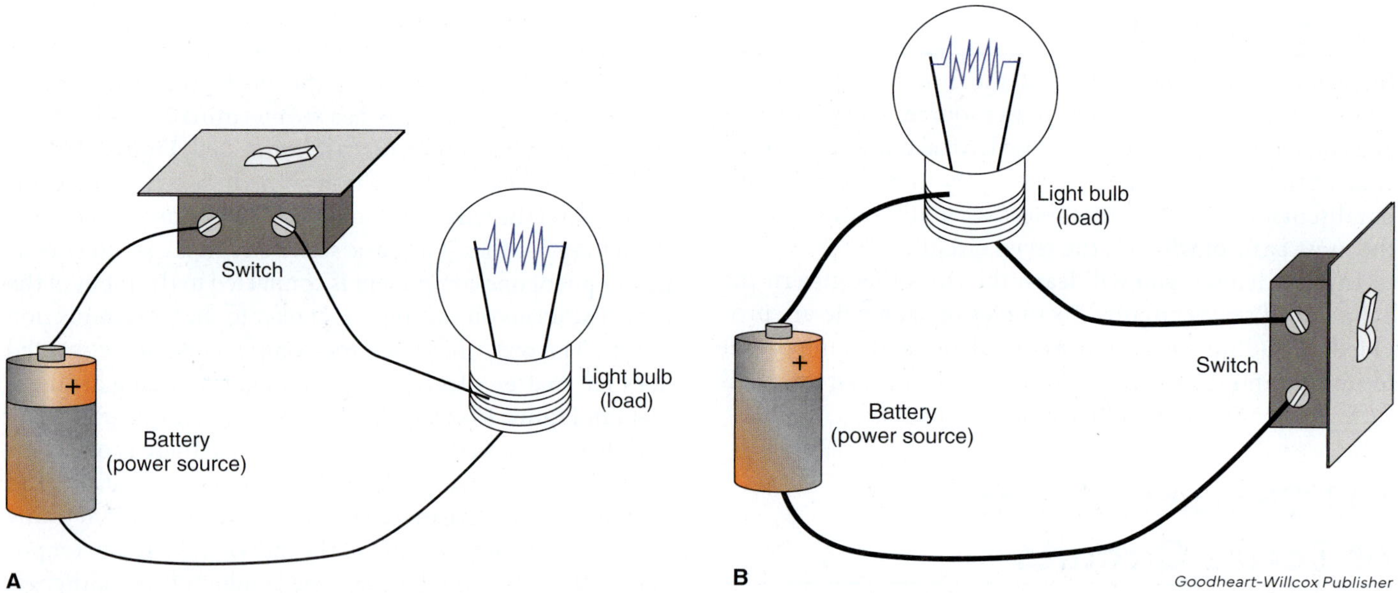

Goodheart-Willcox Publisher

Figure 10-2. In a basic series circuit, changing the order of the circuit components does not change the operation of the circuit. A—The positive terminal of the battery is connected to the switch, the switch is connected to the light bulb, and the light bulb is connected to the negative terminal of the battery. B—The relative positions of the switch and light bulb are changed with no effect on the operation of the circuit.

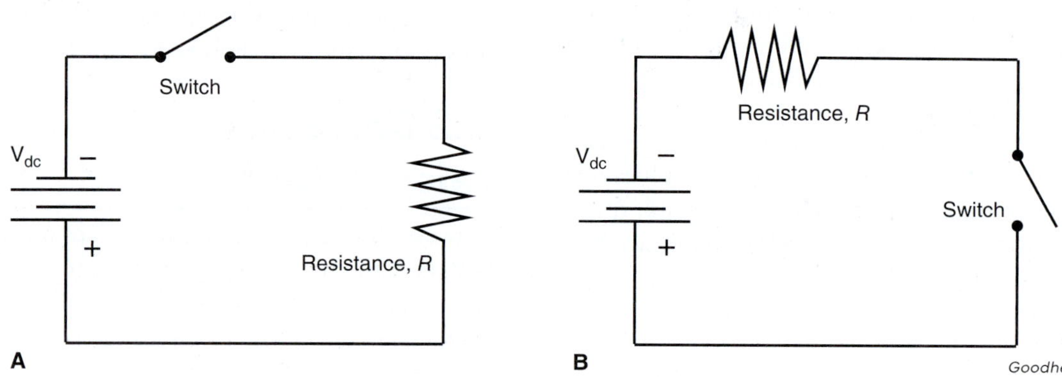

Goodheart-Willcox Publisher

Figure 10-3. Representation of the circuits shown in **Figure 10-2** using wiring diagram symbols.

examples are intended to provide you with an understanding of the laws that govern series circuits and the opportunity to perform some basic calculations. *Note:* You will need to be familiar with Ohm's law and the power formulas to help solve these equations. Refer to Chapter 6, *Electrical Characteristics*, where both are discussed in detail.

Calculating Current Flow

Current flow is the same at all points in a series circuit. With only one possible path for electric current to flow in a series circuit, the rate of electron flow through all components in a series circuit is the same. This rule can be expressed as:

$$I_T = I_1 = I_2 = I_3 = \ldots = I_N$$

In this expression, I represents current, I_T represents the total circuit current, while I_1, I_2, and I_3 represent the current flow through circuit components 1, 2, 3, etc. The I_N term indicates the current flow through the last circuit component, meaning no matter how many components are present in the circuit, the current will be the same at all points in the circuit. Wires and conductors are also referred to as circuit components since current must also flow through them to get to the circuit's loads and switches.

Referencing **Figure 10-4**, the amount of current flowing in the circuit can be calculated by using the current form of Ohm's law:

$$I = \frac{E}{R}$$

$$I = \frac{20\ \text{V}}{5\ \Omega}$$

$$I = 4\ \text{A}$$

A current of 4 A flows through all circuit elements, including the switch, the load, and all conductors used to connect the components. A clamp-on ammeter can be used to measure the current flow in a circuit. When taking a current reading

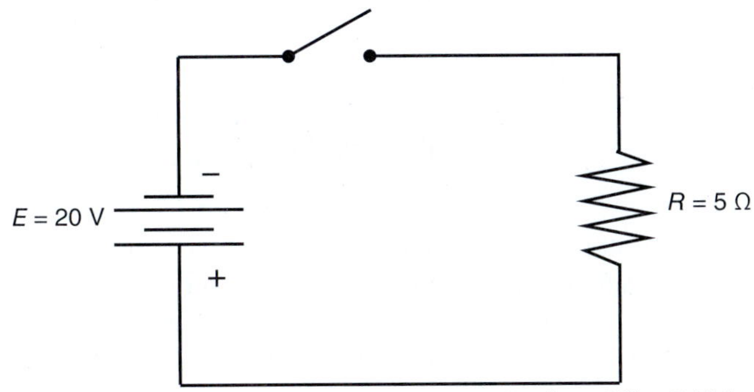

Figure 10-4. Wiring diagram showing a series circuit with a 20-V power supply and a 5-Ω load.

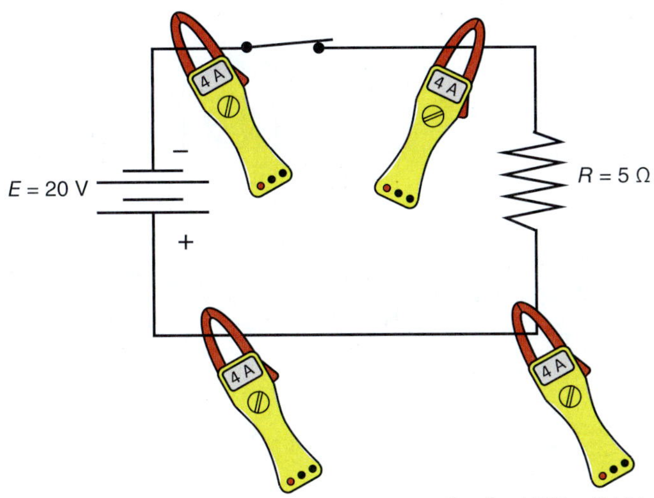

Figure 10-5. In a series circuit, measuring the current at any point in the closed circuit produces the same result.

anywhere in a series circuit, the same current value will be measured at each point, **Figure 10-5**. An additional current flow calculation is provided in **Example 10-1**.

When constructing a series circuit, all circuit components, including circuit protection devices, switches, and conductors, must be rated at or above the expected voltage and current. A device's ***rating*** indicates the voltage and current limits that the device can be safely subjected to.

Safety Alert

Switches and conductors must be rated and sized for the expected circuit current, which is determined by the load that is being powered. Fuses and circuit breakers are sized to protect the circuit wiring, not the load being powered. The ratings of fuses and circuit breakers should match the ampacity, or current-carrying capability, of the circuit conductors.

Know the Code

In *Table 210.24, Section 240.4 (B), (C) & (D), Tables 310.16 & 310.17,* the *NEC* provides an overview of wiring sizing and circuit protection guidelines based on the expected current flow in a circuit.

Example 10-1

What is the expected current flow in a 120 V circuit that is powering a resistive load that consumes 2400 W of power?

✓ Solution

Using the current form of the power formula, the expected current through this circuit can be calculated as follows:

$$I = \frac{P}{E}$$

$$I = \frac{2400\ \text{W}}{120\ \text{V}}$$

$$I = 20\ \text{A}$$

With the expected circuit current of 20 A, all switches, conductors, and circuit protection devices must be sized to handle at least 20 A. Regardless of where in the circuit amperage measurements are taken, the current will always be the same.

Now It's Your Turn 10-1

What is the expected current flow through an electric cooktop that consumes 6000 W of power and is supplied with 240 V?

Calculating Total Circuit Resistance

In a series circuit, total resistance is equal to the sum of all circuit resistances. With only one possible path for electric current to take in a series circuit, all circuit current flows through all components in the circuit. Since circuit resistance opposes current flow, the total resistance of a series circuit is the equivalent of the sum of the individual resistances in the circuit. This rule can be expressed as follows:

$$R_T = R_1 + R_2 + R_3 + \ldots + R_N$$

In this expression, R represents resistance, R_T represents the total circuit resistance, while R_1, R_2, and R_3 represent the resistance of circuit components 1, 2, 3, etc. The R_N term indicates the resistance of the last circuit component. **Example 10-2** provides a sample calculation to determine the total resistance in a series circuit.

Example 10-2

Calculate the total resistance for the following circuit.

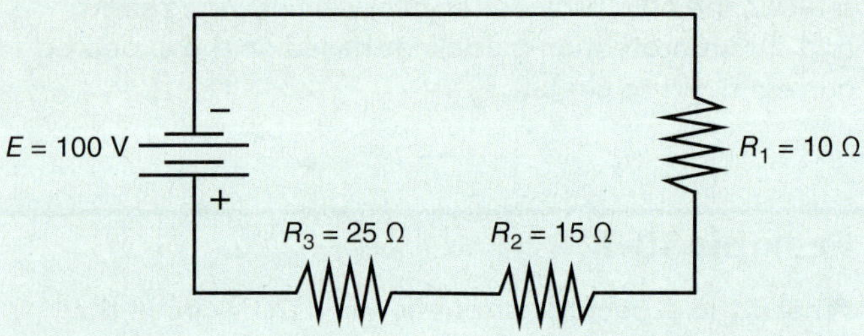

Goodheart-Willcox Publisher

✓ Solution

This 100-V series circuit has three resistances (10 Ω, 15 Ω, and 25 Ω). As total resistance is equal to the sum of all resistances, it can be calculated as follows:

$$R_T = R_1 + R_2 + R_3$$

$$R_T = 10\ \Omega + 15\ \Omega + 25\ \Omega$$

$$R_T = 50\ \Omega$$

Now It's Your Turn 10-2

Calculate the total resistance of the following circuit.

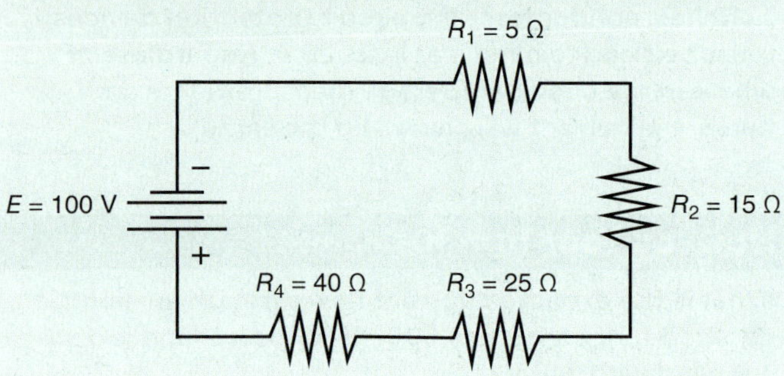

Goodheart-Willcox Publisher

Shedding Light

In real-world electric circuits, the conductors that carry electric current and even switches in the closed position do offer some resistance to current flow. The total resistance of a series circuit is, therefore, the sum of the resistance of the load being powered, plus the resistance of the conductors in the circuit, plus any resistance measured across the closed switches in the circuit.

Calculating Total Circuit Voltage

Total circuit voltage is equal to the sum of all voltage drops in the circuit. In a series circuit, each circuit component that has a measurable resistance will have a measurable voltage across it. This measurable voltage is referred to as a ***voltage drop***. The sum of the voltage measurements across each circuit component will be equal to the total voltage being supplied to the circuit. This rule can be expressed as follows:

$$E_T = E_1 + E_2 + E_3 + \ldots + E_N$$

In this expression, E represents voltage, E_T represents the total circuit voltage, while E_1, E_2, and E_3 represent the voltage measured across each of the circuit components 1, 2, 3, etc. The E_N term indicates the measured voltage across the last component in the series circuit. Refer to **Example 10-3** on how to calculate total circuit voltage in a series circuit.

Example 10-3

Calculate the voltage being supplied to the following series circuit.

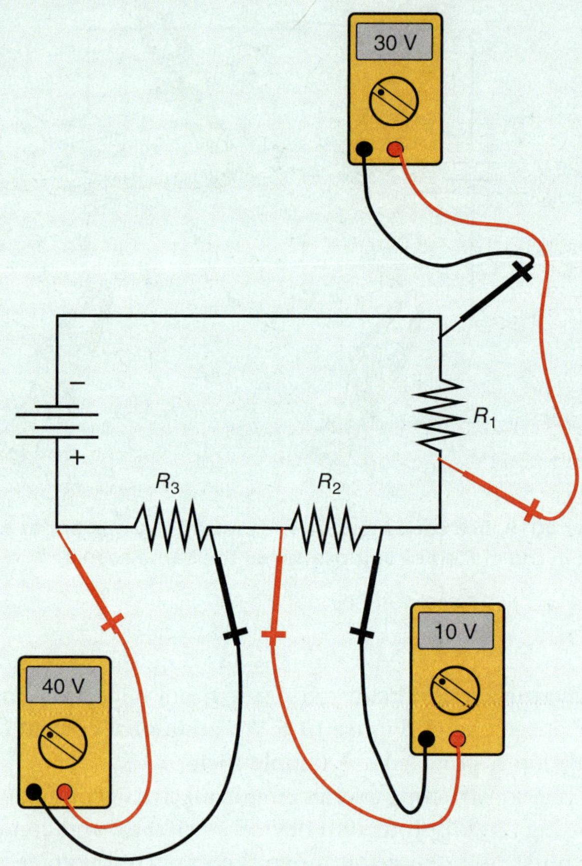

Goodheart-Willcox Publisher

✓ Solution

This circuit has three resistances and the measured voltage across each of them is 30 V, 10 V, and 40 V. Since the sum of the voltage measurements across each circuit component equals the total voltage being supplied to the circuit, the following can be used to determine the circuit's supply voltage:

$$E_T = E_1 + E_2 + E_3$$

$$E_T = 30\text{ V} + 10\text{ V} + 40\text{ V}$$

$$E_T = 80\text{ V}$$

Now It's Your Turn 10-3

Calculate the measured voltage across R_2 for the following series circuit.

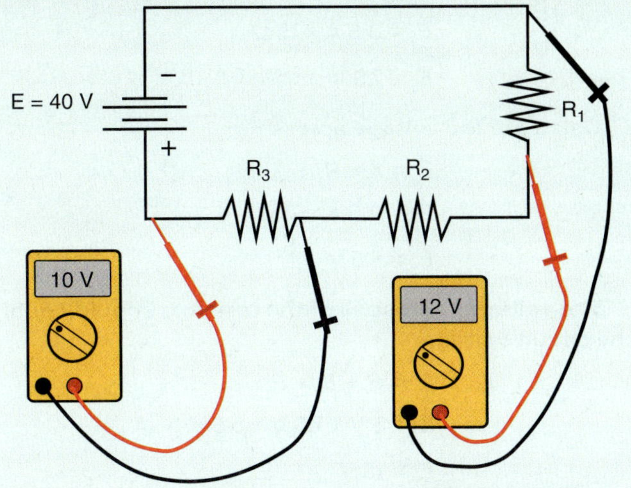

Goodheart-Willcox Publisher

Calculating Voltage Drop Across a Resistor

One method to determine the voltage drop across each of the resistances in a series circuit is to use the voltage form of Ohm's law:

$$E = I \times R$$

Refer to the circuit shown in **Figure 10-6**. In order to use the voltage form of Ohm's law to determine the voltage across each of the two resistances, first determine the total circuit current as follows:

$$I = \frac{E}{R}$$

$$I_T = \frac{E}{R_T}$$

$$I_T = \frac{80 \text{ V}}{(R_1 + R_2)}$$

$$I_T = \frac{80 \text{ V}}{(15\ \Omega + 25\ \Omega)}$$

$$I_T = \frac{80 \text{ V}}{40\ \Omega}$$

$$I_T = 2 \text{ A}$$

With the total circuit current of 2 A, the voltage form of Ohm's law ($E = I \times R$) can be used to determine the measured voltage across each of the resistances in the series circuit. Since the current is the same at all points in a series circuit, each of the two circuit's resistances (R_1 and R_2) has 2 A of current flowing through it. Expressed differently, the total current in a series circuit, I_T, is the same as the current flowing through R_1, denoted I_1, and the current flowing through R_2, denoted I_2, **Figure 10-7**.

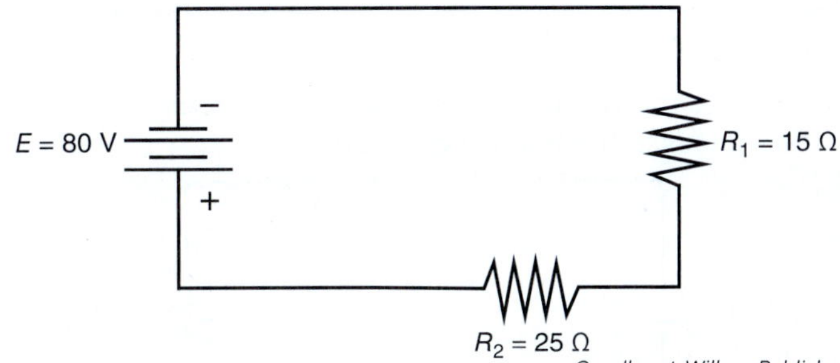

Goodheart-Willcox Publisher

Figure 10-6. An 80-V series circuit where R_1 equals 15 Ω and R_2 equals 25 Ω.

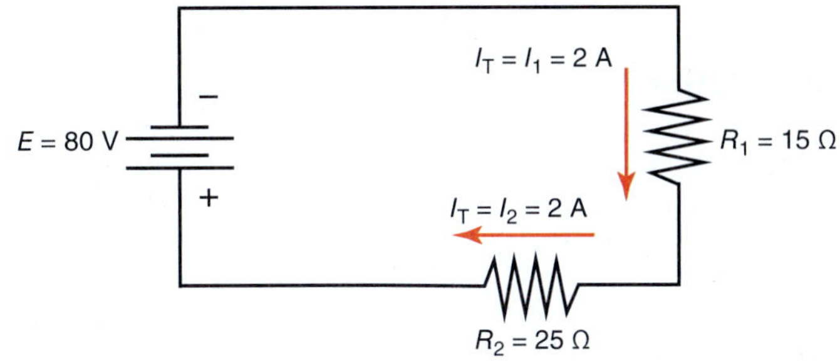

Goodheart-Willcox Publisher

Figure 10-7. The circuit current is 2 A, which is the same as the current flow through R_1 and R_2.

The voltage form of Ohm's law can now be used for each individual resistance as follows:

$$E_1 = I_1 \times R_1$$
$$E_2 = I_2 \times R_2$$

As previously calculated, I_T, I_1, and I_2 are all 2 A. In addition, the values for R_1 and R_2 are known. First, calculate E_1 as follows:

$$E_1 = I_1 \times R_1$$

$$E_1 = 2 \text{ A} \times 15\ \Omega$$

$$E_1 = 30 \text{ V}$$

Then, calculate E_2 as follows:

$$E_2 = I_2 \times R_2$$

$$E_2 = 2 \text{ A} \times 25\ \Omega$$

$$E_2 = 50 \text{ V}$$

Figure 10-8 shows the voltage readings across the circuit resistances. The total voltage supplied, which equals the sum of the measured voltages across the circuit components, is calculated as follows:

$$E_T = E_1 + E_2$$

$$E_T = 30 \text{ V} + 50 \text{ V}$$

$$E_T = 80 \text{ V}$$

Example 10-4

Calculate the voltage across each of the three resistances in the following circuit.

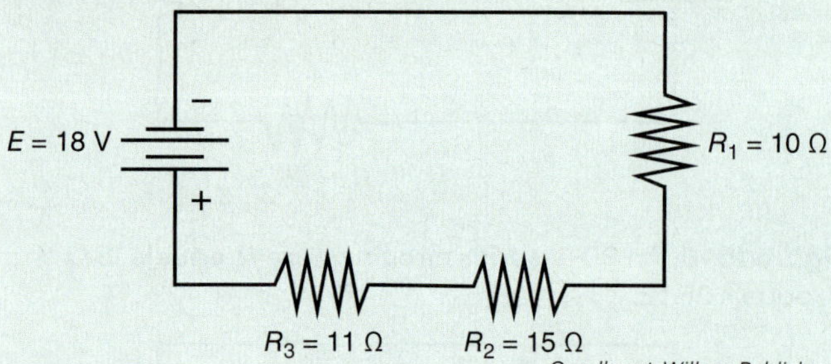

Goodheart-Willcox Publisher

✓ Solution

In order to determine the voltage across a resistor, first calculate the circuit current. To do so, begin by determining the total circuit resistance:

$$R_T = R_1 + R_2 + R_3$$
$$R_T = 10\ \Omega + 15\ \Omega + 11\ \Omega$$
$$R_T = 36\ \Omega$$

Then calculate the circuit current:

$$I_T = \frac{E_T}{R_T}$$
$$I_T = \frac{18\ \text{V}}{36\ \Omega}$$
$$I_T = 0.5\ \text{A}$$

The current through each resistor is the same as the total circuit current, so $I_1 = I_2 = I_3 = I_T = 0.5$ A. Calculate the voltage across R_1:

$$E_1 = I_1 \times R_1$$
$$E_1 = 0.5\ \text{A} \times 10\ \Omega$$
$$E_1 = 5\ \text{V}$$

Calculate the voltage across R_2:

$$E_2 = I_2 \times R_2$$
$$E_2 = 0.5\ \text{A} \times 15\ \Omega$$
$$E_2 = 7.5\ \text{V}$$

Calculate the voltage across R_3:

$$E_3 = I_3 \times R_3$$
$$E_3 = 0.5\ \text{A} \times 11\ \Omega$$
$$E_3 = 5.5\ \text{V}$$

The voltages across all of the resistors should equal the circuit voltage:

$$E_T = E_1 + E_2 + E_3$$
$$E_T = 5\ \text{V} + 7.5\ \text{V} + 5.5\ \text{V}$$
$$E_T = 18\ \text{V}$$

Now It's Your Turn 10-4

Calculate the voltage across each of the three resistances in the following circuit.

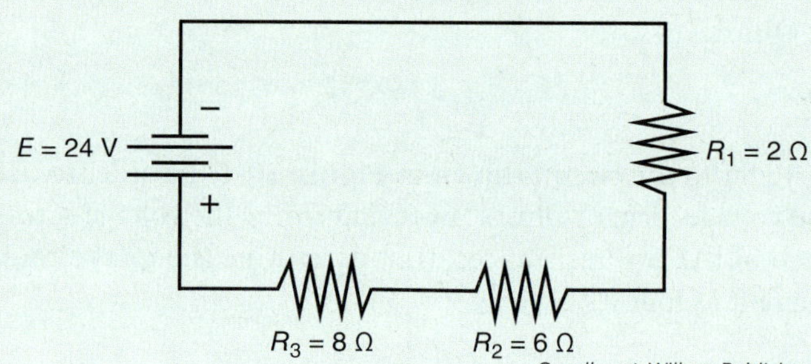

Goodheart-Willcox Publisher

Refer to **Example 10-4** for additional practice on calculating voltage across three resistances.

Alternative Methods of Calculating Voltage Drop

The voltage measured across each series circuit component is proportional to the resistance of the component. Consider the circuit shown in **Figure 10-9**. In this circuit, there are two circuit loads with resistances of 10 Ω and 20 Ω. With the value of R_2 being twice that of R_1, the measured voltage across R_2 will be twice the voltage measured across R_1. The sum of the voltages across R_1 and R_2 can be expressed as:

$$E_T = E_1 + E_2$$

In this expression, E_T is the voltage supplied to the circuit, E_1 is the voltage measured across R_1, and E_2 is the voltage measured across R_2. Since E_2 is two times E_1 (because the resistance value of R_2 is two times that of R_1), E_2 can be replaced by the term $2E_1$. Thus, the expression now becomes:

$$E_T = E_1 + E_2$$
$$E_T = E_1 + 2E_1$$
$$E_T = 3E_1$$

The value of E_T is provided in the circuit. Substituting 60 V for E_T changes the equation as follows:

$$E_T = 3E_1$$
$$60\ \text{V} = 3E_1$$

Next, divide both sides of the equation by three to solve the equation for E_1:

$$60\ \text{V} = 3E_1$$
$$\frac{60\ \text{V}}{3} = \frac{3E_1}{3}$$
$$E_1 = 20\ \text{V}$$

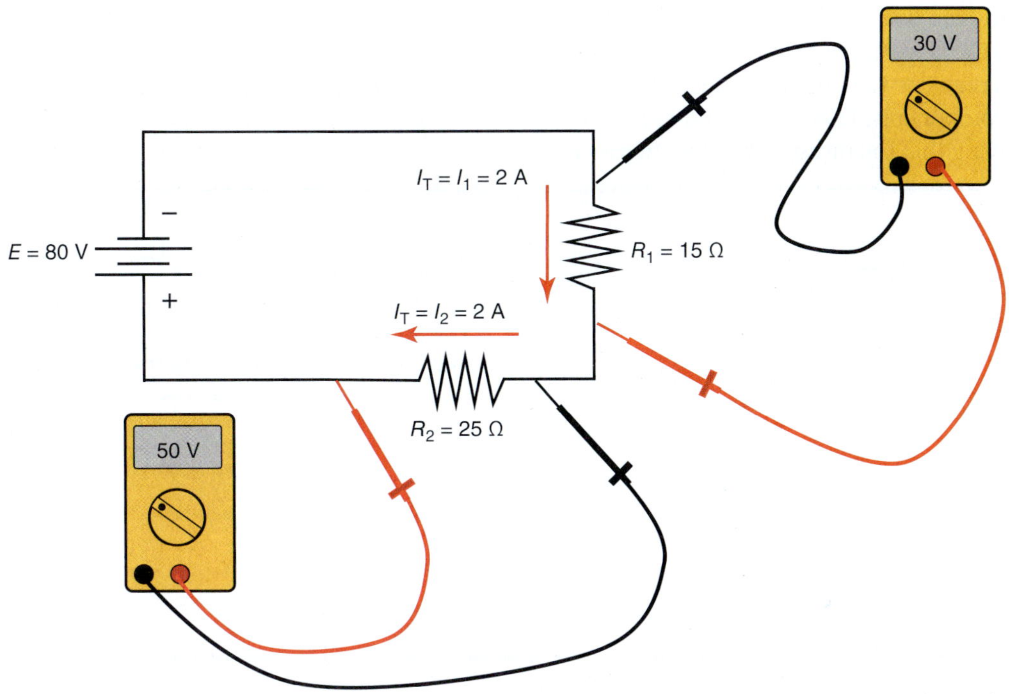

Goodheart-Willcox Publisher

Figure 10-8. Total voltage of 80 V is equal to the sum of voltages across circuit components; in this case, voltage of 30 V across R_1 and voltage of 50 V across R_2 equals total voltage of 80 V.

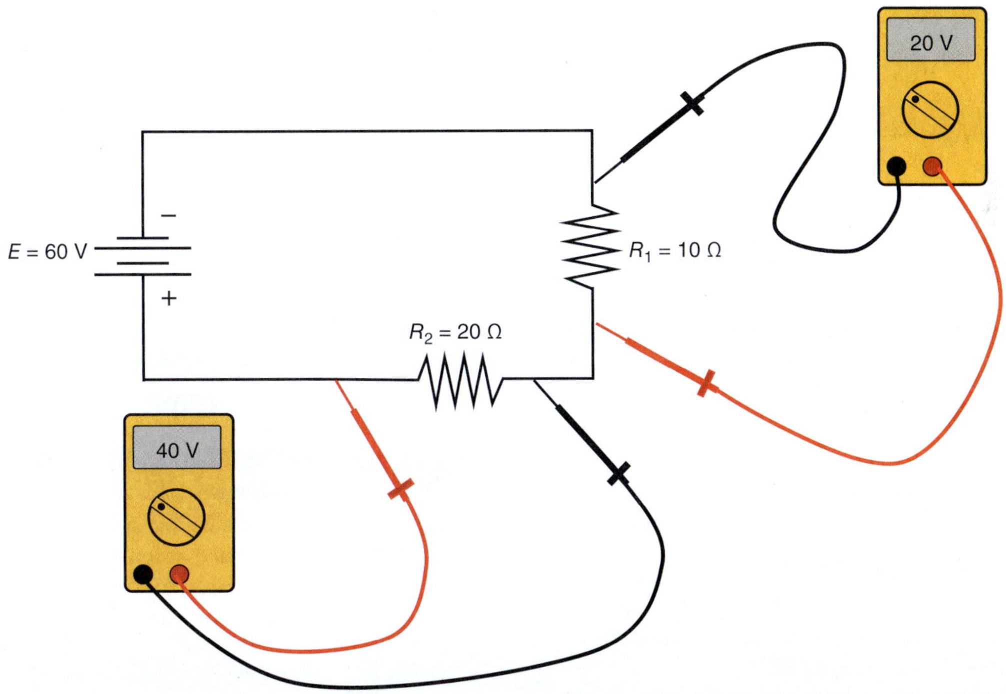

Goodheart-Willcox Publisher

Figure 10-9. A 60-V series circuit where R_1 equals 10 Ω and R_2 equals 20 Ω.

Knowing that the measured voltage across R_1 is 20 V, the voltage across R_2 can be determined. Since the total voltage supplied to the circuit is equal to the sum of the measured voltages across the circuit components, E_2 equals 40 V (60 V – 20 V). Alternatively, knowing that the measured voltage across R_2 is twice the voltage measured across R_1, doubling E_1 results in the same answer.

Example 10-5 and **Example 10-6** provide additional methods to calculate voltage drop.

Example 10-5

Calculate the voltage across each of the three resistances in the following circuit.

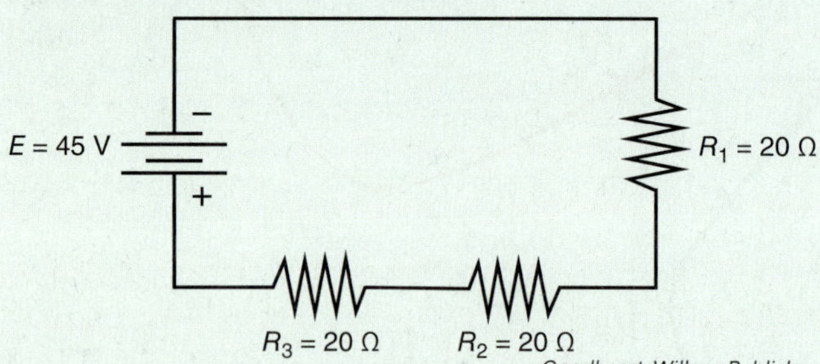

Goodheart-Willcox Publisher

✓ Solution

Since the three resistances are all equal, the measured voltage across each of them will be the same. With a supply voltage of 45 V, the measured voltage across each resistance will be one-third of 45 V, or 15 V.

Example 10-6

Calculate the voltage being supplied to the following circuit.

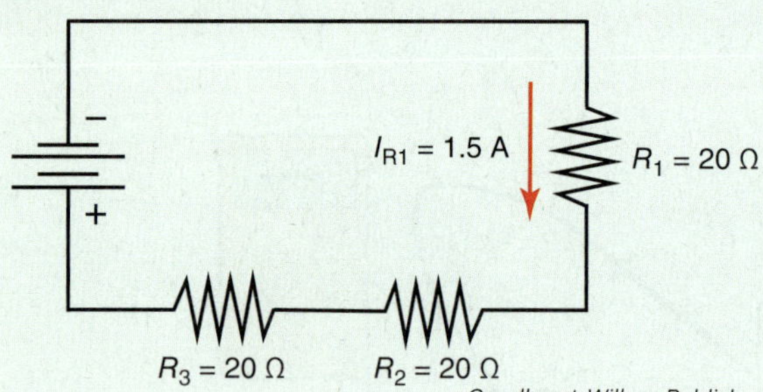

Goodheart-Willcox Publisher

✓ Solution

The circuit current is 1.5 A. Using the formula of Ohm's law, the measured voltage across R_1 can be calculated as follows:

$$E_1 = I_1 \times R_1$$

$$E_1 = 1.5\ \text{A} \times 20\ \Omega$$

$$E_1 = 30\ \text{V}$$

Since the three resistances are all equal, the measured voltage across each of them will be 30 V as well. With a measured voltage of 30 V across each of the three resistances, the voltage supplied to the circuit will be three times 30 V, or 90 V.

Alternatively, the voltage supplied to the circuit can be determined by calculating the total resistance of the circuit as follows:

$$R_T = R_1 + R_2 + R_3$$

$$R_T = 20\ \Omega + 20\ \Omega + 20\ \Omega$$

$$R_T = 60\ \Omega$$

The circuit can then be simplified by combining the three series resistances into a single resistance with a value of 60 Ω, as shown in the following image.

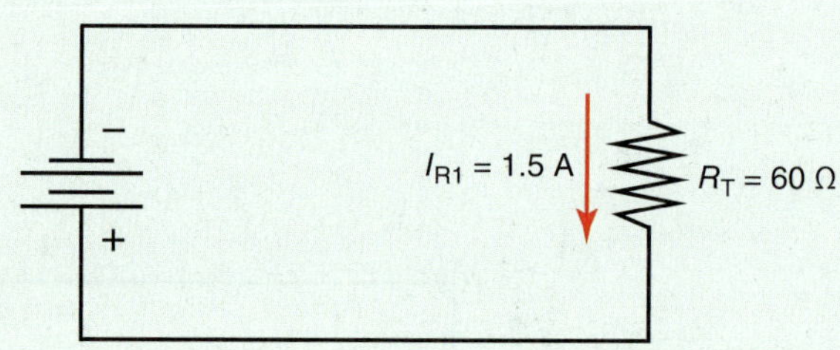

Goodheart-Willcox Publisher

Next, the voltage form of Ohm's law can be used to solve for the supply voltage as follows:

$$E_T = I_T \times R_T$$

$$E_T = 1.5\ \text{A} \times 60\ \Omega$$

$$E_T = 90\ \text{V}$$

Now It's Your Turn 10-6

Calculate the voltage being supplied to the following circuit. (Note: There are at least three different ways to calculate this value. How many ways can you find?)

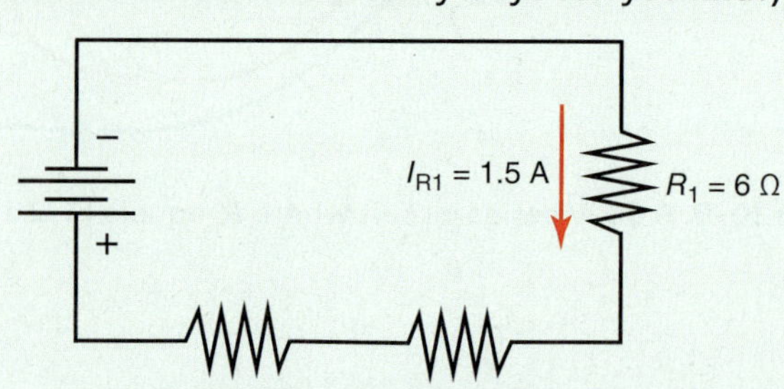

Goodheart-Willcox Publisher

10.3 **Calculating Power Consumption**

Using the rules of series circuits, the power consumption of a series circuit with multiple resistances can be calculated using the following forms of the power formula:

$$P = I^2 \times R$$
$$P = \frac{E^2}{R}$$
$$P = I \times E$$

Consider the circuit shown in **Figure 10-10**. To calculate the power consumed by the circuit, begin by determining the total circuit resistance. The total circuit resistance can be found by adding 8 Ω and 12 Ω to get 20 Ω.

Now the power can be calculated:

$$P = \frac{E^2}{R}$$
$$P = \frac{(80\ \text{V})^2}{20\ \Omega}$$
$$P = \frac{6400}{20}$$
$$P = 320\ \text{W}$$

The following is another method to determine the circuit current:

$$I_T = \frac{E_T}{R_T}$$
$$I_T = \frac{80\ \text{V}}{20\ \Omega}$$
$$I_T = 4\ \text{A}$$

Then the power consumed by the circuit can be calculated as follows:

$$P = I^2 \times R$$
$$P = (4\ \text{A})^2 \times 20\ \Omega$$
$$P = 16 \times 20$$
$$P = 320\ \text{W}$$

Alternately, the power consumed by the circuit can be calculated by using the following equation:

$$P = I \times E$$
$$P = 4\ \text{A} \times 80\ \text{V}$$
$$P = 320\ \text{W}$$

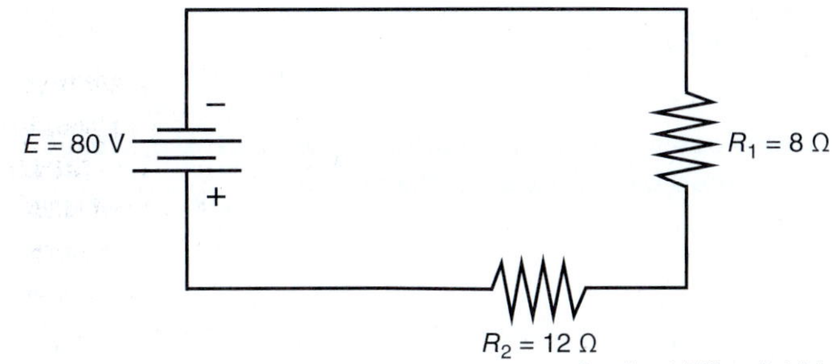

Figure 10-10. An 80-V series circuit where R_1 equals 8 Ω and R_2 equals 12 Ω.

As you can see, using any of the three equations yields the same result.

Since the individual resistances are known, along with the current flowing through each one, the power consumed by each of the resistances in this series circuit can also be calculated. Considering each of the resistances individually, the power consumed by R_1 can be determined as follows:

$$P_1 = I^2 \times R_1$$
$$P_1 = (4\ \text{A})^2 \times 8\ \Omega$$
$$P_1 = 16 \times 8$$
$$P_1 = 128\ \text{W}$$

To calculate the power consumed by R_2, the following equation can be used:

$$P_2 = I^2 \times R_2$$
$$P_2 = (4\ \text{A})^2 \times 12\ \Omega$$
$$P_2 = 16 \times 12$$
$$P_2 = 192\ \text{W}$$

For example, When the individual results are added together (128 W + 192 W), notice that the sum (320 W) is the same as the result when the total circuit resistance was used.

Additional practice on calculating power consumed in a series circuit is provided in **Example 10-7**.

10.4 **Use of Switches**

Switches are used to start and stop the flow of current in a circuit. In the most basic series circuit, one switch is used, **Figure 10-11**. When the switch is in the closed position, power can pass freely between the power source and the

Example 10-7

Referring to the following circuit, calculate the power consumed by the circuit as well as the power consumed by each load in the circuit.

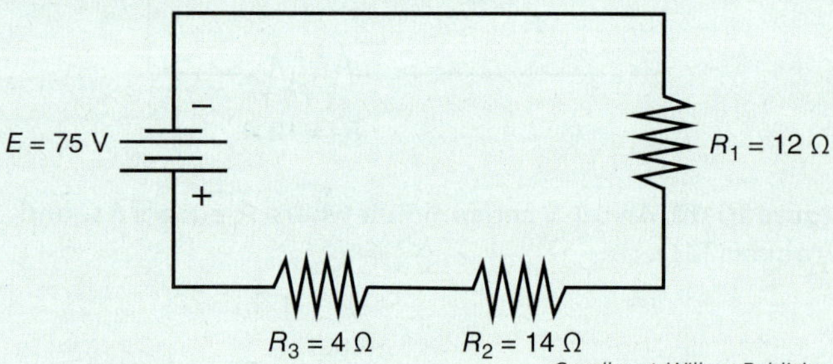

Goodheart-Willcox Publisher

✓ Solution

To calculate the total power consumption of the circuit, first determine the total circuit resistance as follows:

$$R_T = R_1 + R_2 + R_3$$

$$R_T = 12\ \Omega + 14\ \Omega + 4\ \Omega$$

$$R_T = 30\ \Omega$$

Next, calculate the power consumed by the circuit as follows:

$$P = \frac{E^2}{R_T}$$

$$P = \frac{(75\text{ V})^2}{30\ \Omega}$$

$$P = 187.5\text{ W}$$

The power consumed by the circuit can also be calculated if the circuit current is known, along with total resistance. In this instance, with a voltage of 75 V and a total resistance of 30 Ω, divide 75 V by 30 Ω, which will give you a circuit current of 2.5 A. Substituting the known values into the following equation provides the same result as above:

$$P = I^2 \times R$$

$$P = (2.5\text{ A})^2 \times 30\ \Omega$$

$$P = 187.5\text{ W}$$

The circuit current is constant throughout the circuit ($I_T = I_1 = I_2 = I_3$), so the power consumed by R_1 can be calculated as follows:

$$P_1 = I^2 \times R_1$$

$$P_1 = (2.5\text{ A})^2 \times 12\ \Omega$$

$$P_1 = 6.25 \times 12$$

$$P_1 = 75\text{ W}$$

Then calculate the power consumed by R_2:

$$P_2 = I^2 \times R_2$$

$$P_2 = (2.5\text{ A})^2 \times 14\ \Omega$$

$$P_2 = 6.25 \times 14$$

$$P_2 = 87.5\text{ W}$$

Next, calculate the power consumed by R_3:

$$P_3 = I^2 \times R_3$$

$$P_3 = (2.5\text{ A})^2 \times 4\ \Omega$$

$$P_3 = 6.25 \times 4$$

$$P_3 = 25\text{ W}$$

These individual power calculations could also be performed using the $P = E^2/R$ or $P = I \times E$ formulas.

Finally, you can check the accuracy of these three results by adding them together and comparing the final result to the result obtained for the total circuit's power consumption. These two figures should be the same.

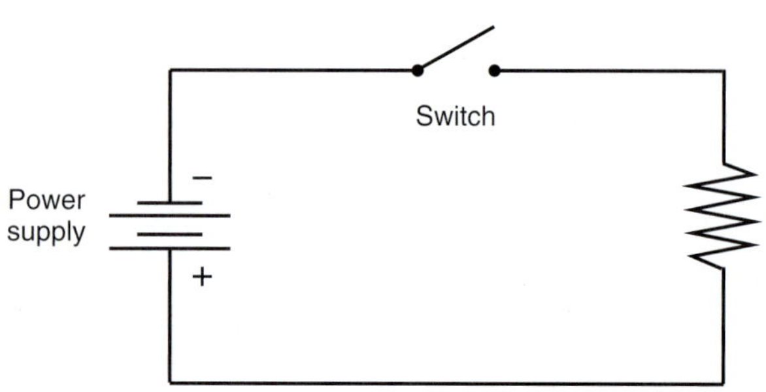

Goodheart-Willcox Publisher

Figure 10-11. A wiring diagram of a simple series circuit with a power supply, an open switch, and one load.

load, **Figure 10-12**. When closed, the voltage supplied to the load is equal to the voltage at the power supply. When a switch is in the closed position, the voltage drop across the switch will equal 0 V.

A voltage reading of 0 V across a switch does not necessarily mean that there is no electrical potential present, nor does it mean that electric current is not flowing. In the circuit shown in **Figure 10-13A**, for example, there is potential present and there is current flowing. The voltage reading provides the difference in potential between the two points in the electric circuit. In this circuit, the switch

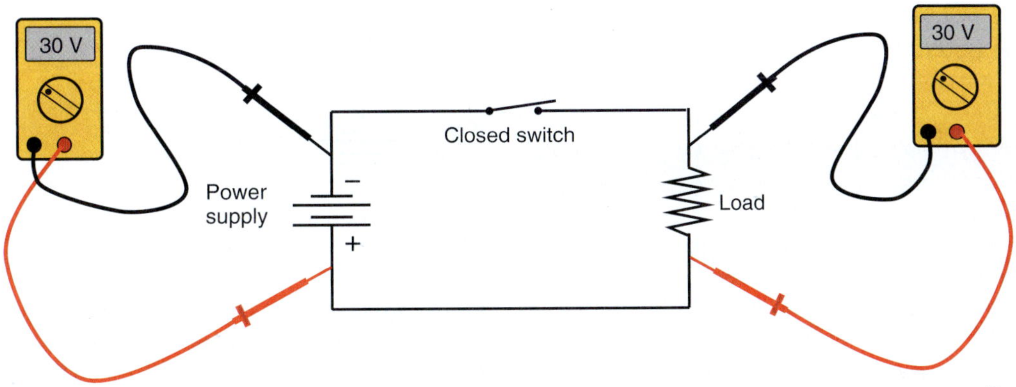

Goodheart-Willcox Publisher

Figure 10-12. When the switch is closed, the voltage across the load is equal to the voltage supplied to the circuit. As shown here, the measured voltage at both the power supply and the load is 30 V.

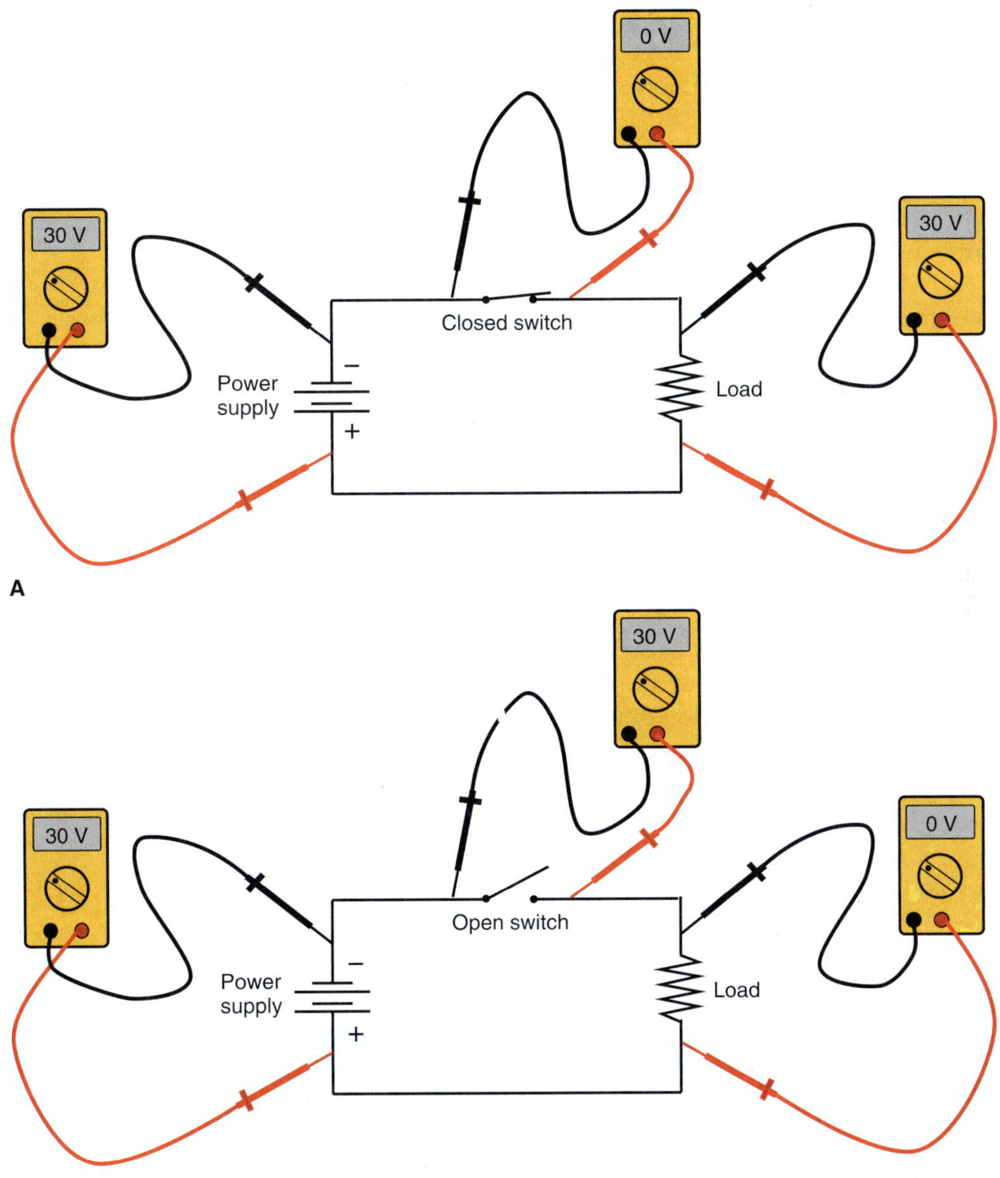

Goodheart-Willcox Publisher

Figure 10-13. Before working on any circuit, confirm that the circuit is de-energized. A—Although the voltage across a closed switch is always zero, current flow can still be present. B—When the switch is open, the voltage across the switch is equal to the voltage supplied to the circuit.

is closed, and the potential on both sides of the switch is 30 V. Thus, the potential difference between 30 V and 30 V is 0 V.

Safety Alert

Obtaining a voltage reading of 0 V across a switch does not mean that the circuit is de-energized. Be sure to measure the voltage between each side of the switch and ground or neutral (grounded conductor) to confirm that the circuit is de-energized. The voltage measured across a closed switch will always be 0 V, so be careful!

When the switch is in the open position, the load is no longer energized and the measured voltage across the switch is 30 V, **Figure 10-13B**. The voltage at the power supply is still 30 V.

Notice that when the switch is closed, there is no voltage drop across the switch and power is supplied to the load. However, when the switch is open, there is voltage drop across the switch and no voltage is supplied to the load.

In some applications, multiple switches are wired in series with each other. This arrangement is used when multiple conditions must be met in order for the load to become energized. An example of such a circuit is an outdoor light that illuminates only when it is dark outside and when motion is detected. In this case, one switch will be a photocell that causes a set of electrical contacts to close when the level of sensed light is low. The other switch will be a motion sensor that causes a set of electrical contacts to close when motion is sensed, **Figure 10-14**. The light will only turn on when both conditions, namely darkness and motion, are sensed. **Figure 10-15** illustrates all possible switch combinations for our photocell/motion example.

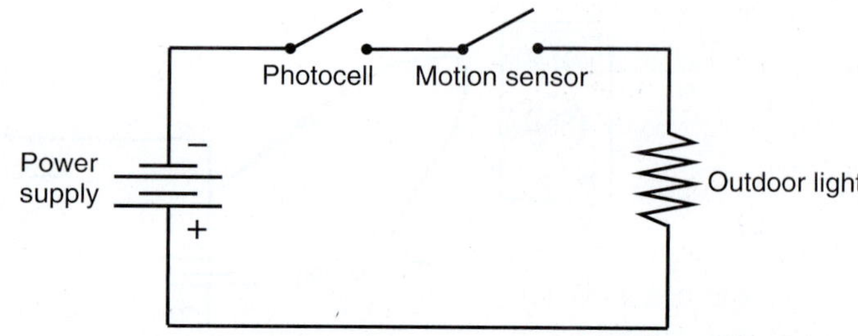

Goodheart-Willcox Publisher

Figure 10-14. The use of two switches in a series circuit configuration allows a light to illuminate only when motion is sensed and there is no light acting on the photocell.

	PHOTOCELL SENSING LIGHT (SWITCH OPEN)	PHOTOCELL NOT SENSING LIGHT (SWITCH CLOSED)
MOTION NOT SENSED (SWITCH OPEN)	LIGHT NOT ENERGIZED (A)	LIGHT NOT ENERGIZED (C)
MOTION SENSED (SWITCH CLOSED)	LIGHT NOT ENERGIZED (B)	LIGHT ENERGIZED (D)

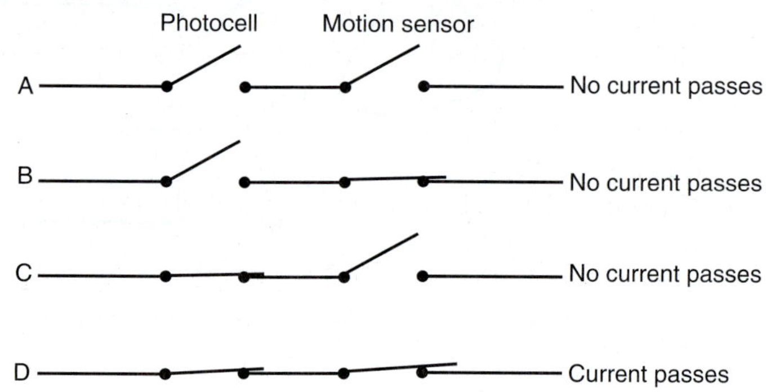

Goodheart-Willcox Publisher

Figure 10-15. All possible switch position combinations for photocell/motion sensor example.

CHAPTER 10 REVIEW

Summary

- In a series circuit there is only one path for electric current to follow. (10.1)
- A break anywhere in a series circuit will prevent current from flowing through the entire circuit. (10.1)
- In a basic series circuit, the output of one component is connected to the input of the next. (10.1)
- With only one possible path for electric current to flow in a series circuit, the rate of electron flow through all components in a series circuit is the same. (10.2)
- Total resistance in a series circuit is equal to the sum of all circuit resistances. (10.2)
- The voltage supplied to a series circuit is equal to the sum of all voltage drops in the circuit. (10.2)
- In a series circuit, all components must be rated to handle the voltage and current that they will be exposed to. (10.2)
- The total power in a series circuit is equal to the sum of the power drawn by each of the energy-consuming devices. (10.3)
- The power consumption of a series circuit increases as the circuit resistance decreases. (10.3)
- Switches are used to start and stop the flow of current in a circuit. (10.4)
- Multiple switches are sometimes connected in series with each other. (10.4)

Review Questions

Use the information in the chapter to answer the following questions.

Measuring Your Potential

1. Of the following, which is not a common feature of a simple series circuit? (10.1)
   A. Most circuit components have two electrical terminals.
   B. The output of one circuit component is connected via a conductor to the input of another circuit.
   C. There is one option for the path of current flow.
   D. There are two components in a series circuit.
2. The connection points between circuit components and the wires or conductors that are used to connect them to each other are known as _____. (10.1)
   A. ground
   B. terminals
   C. switches
   D. circuit breakers
3. *True or False?* When constructing a series circuit, all circuit components should be rated at or above the expected amperage draw of the load being powered. (10.2)
4. The rate of electron flow through all components in a series circuit is equal _____. (10.2)
   A. across all components in the circuit
   B. to the largest resistance in the circuit
   C. to the sum of all voltage drops in the circuit
   D. to the smallest resistance in the circuit
5. In a series circuit, I_T equals _____. (10.2)
   A. $I_1 + I_2 + \ldots I_N$
   B. $I_1 = I_2 = I_3 \ldots = I_N$
   C. $I_N - I_1 - I_2$
   D. $E_T \times R_T$
6. If four resistances with values of 10 Ω, 10 Ω, 15 Ω, and 20 Ω are wired in series with each other, the total resistance will be _____. (10.2)
   A. 45 Ω
   B. 55 Ω
   C. 65 Ω
   D. 75 Ω
7. In a series circuit, E_T equals _____. (10.2)
   A. $E_1 - E_2$
   B. $E_T \times R_T$
   C. $E_1 + E_2 + E_3 + \ldots + E_N$
   D. E_N
8. What is the power consumption of a 50-V series circuit with one 20 Ω load and one 30 Ω load? (10.3)
   A. 25 W
   B. 50 W
   C. 500 W
   D. 2500 W
9. *True or False?* In a series circuit, the voltage measured across an open switch will be equal to the voltage supplied to the circuit. (10.4)
10. In a series circuit with two switches wired in series with each other, power will pass to the load only when _____. (10.4)
    A. the switch closest to the power supply is closed
    B. the switch closest to the load is closed
    C. both switches are in the open position
    D. both switches are in the closed position

Plug-In

1. What is the expected current flow in a 50-V circuit that is powering a 5-Ω load? (10.2)
   A. 5 A
   B. 10 A
   C. 55 A
   D. 250 A

2. What is the total resistance of the following circuit? (10.2)

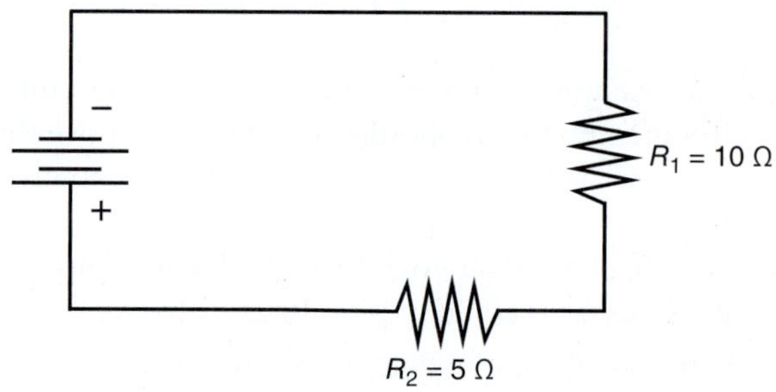

Goodheart-Willcox Publisher

A. 15 A
B. 15 Ω
C. 150 Ω
D. 2 Ω

3. Four 10 Ω resistances wired in series with each other will have the same resistance as _____ resistances wired in series with each other. (10.2)
   A. two 20 Ω
   B. five 9 Ω
   C. six 7 Ω
   D. three 15 Ω
4. A 50-V series circuit with one 20-Ω load and one 30-Ω load will operate with a total resistance of_____. (10.2)
   A. 25 Ω and a current of 1 A
   B. 50 Ω and a current of 1 A
   C. 25 Ω and a current of 2 A
   D. 50 Ω and a current of 2 A
5. A 50-V series circuit with one 20-Ω load and one 30-Ω load will operate with a voltage drop of_____. (10.2)
   A. 20 V across the 20-Ω load and 30 V across the 30-Ω load
   B. 30 V across the 20-Ω load and 20 V across the 30-Ω load
   C. 50 V across both the 20-Ω load and the 30-Ω load
   D. 25 V across both the 20-Ω load and the 30-Ω load
6. The power consumption of a 100-V series circuit with one 20-Ω load and one 30-Ω load is the same as the power consumption of a 20-V circuit with _____ load(s). (10.3)
   A. one 2-Ω
   B. two 10-Ω
   C. two 5-Ω
   D. one 10-Ω

Use the following circuit to solve for the unknown values in questions 7 – 9.

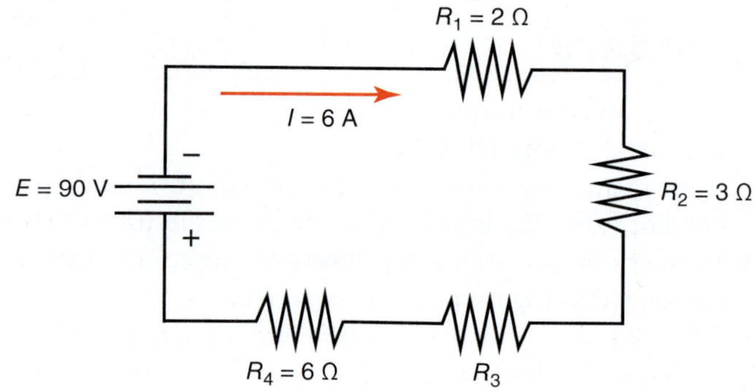

Goodheart-Willcox Publisher

7. What is the value of R_3? (10.2)
   A. 4 Ω
   B. 12 Ω
   C. 15 Ω
   D. 24 Ω
8. What is the value of E_1? (10.2)
   A. 4 V
   B. 12 V
   C. 15 V
   D. 24 V
9. What is the value of P_2? (10.3)
   A. 72 W
   B. 108 W
   C. 144 W
   D. 216 W

Making the Connection

1. Identify two real-world applications that utilize series-connected switches and explain the situation. (10.1)
2. Why is the measured voltage across a closed switch always very close to zero? (10.2)
3. Why might it be difficult to change the position of a manually operated switch in a series circuit? (10.4)

CHAPTER

11 Parallel Circuits

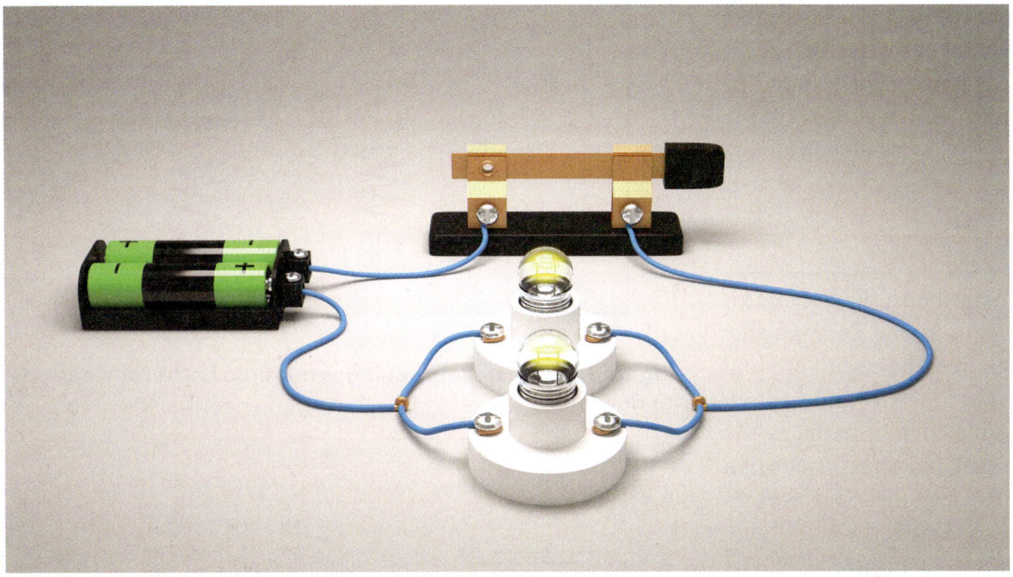

haryigit/Shutterstock.com

SPARKING DISCUSSION

Identify a real-world application that utilizes parallel-connected switches and explain the situation.

LEARNING OUTCOMES

After studying this chapter, you will be able to:

11.1 List the properties of parallel circuits.

11.2 Calculate voltage, current, and total resistance in parallel circuits.

11.3 Apply Ohm's law and power formulas to parallel circuits to calculate power consumption.

11.4 Explain why switches are sometimes connected in parallel with each other.

TECHNICAL TERMS

branch

node

parallel circuit

short circuit

Introduction

In Chapter 10, *Series Circuits*, the most basic of all electric circuit configurations is covered. Part of that discussion includes the rules that outline the behavior and properties of series circuits, as well as how to manipulate and evaluate these properties. This chapter concentrates on another common electric circuit configuration, the parallel circuit. This chapter discusses the rules that govern the behavior of parallel circuits and provides numerous examples of each rule, including mathematical calculations. The concepts presented in this chapter are critically important, as most electric circuits encountered in the field are configured as parallel circuits.

11.1 Parallel Circuits

The term ***parallel circuit*** is used to describe an electric circuit that has multiple paths for electrons to follow. In a parallel circuit, the voltage supplied to each path of the circuit is equal to the voltage supplied to the entire circuit. The term ***branch*** is used to describe each of the possible current paths in the circuit. In a simple parallel circuit, each branch consists of a separate load and, in many cases, the switches and devices that control the operation of that load, **Figure 11-1**.

Consider the parallel circuit shown in **Figure 11-2**. All points along the red lines of the circuit are connected directly to the negative terminals of each load and are at the same voltage or potential. This is because there are no loads or switches in that line that would otherwise affect the measured potential along the line. This assumes the wire offers no resistance to current flow. In addition, all points along the blue line of the circuit are connected directly to the positive terminal of the battery and the positive terminals of the loads. All blue lines are at the same voltage or potential. With the potential being the same at all points along the top, as well as being the same at all points along the bottom, voltage readings taken between any point on the top line and any point on the bottom line will provide the same reading, **Figure 11-3**. With this circuit

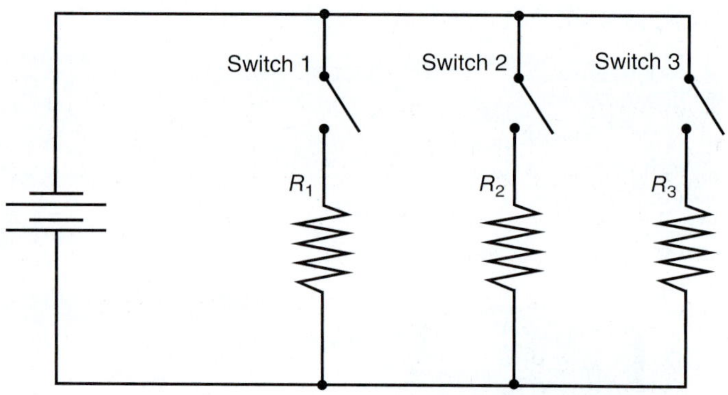

Goodheart-Willcox Publisher

Figure 11-1. A parallel circuit has multiple paths for electrons to flow.

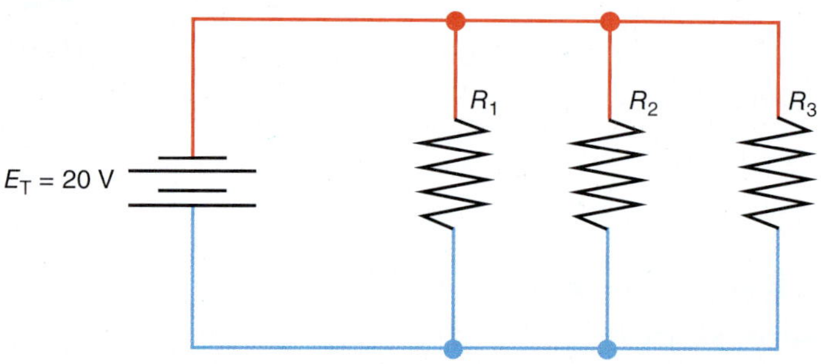

Goodheart-Willcox Publisher

Figure 11-2. In this parallel circuit, all points along the red line are electrically the same, as are all points along the blue line.

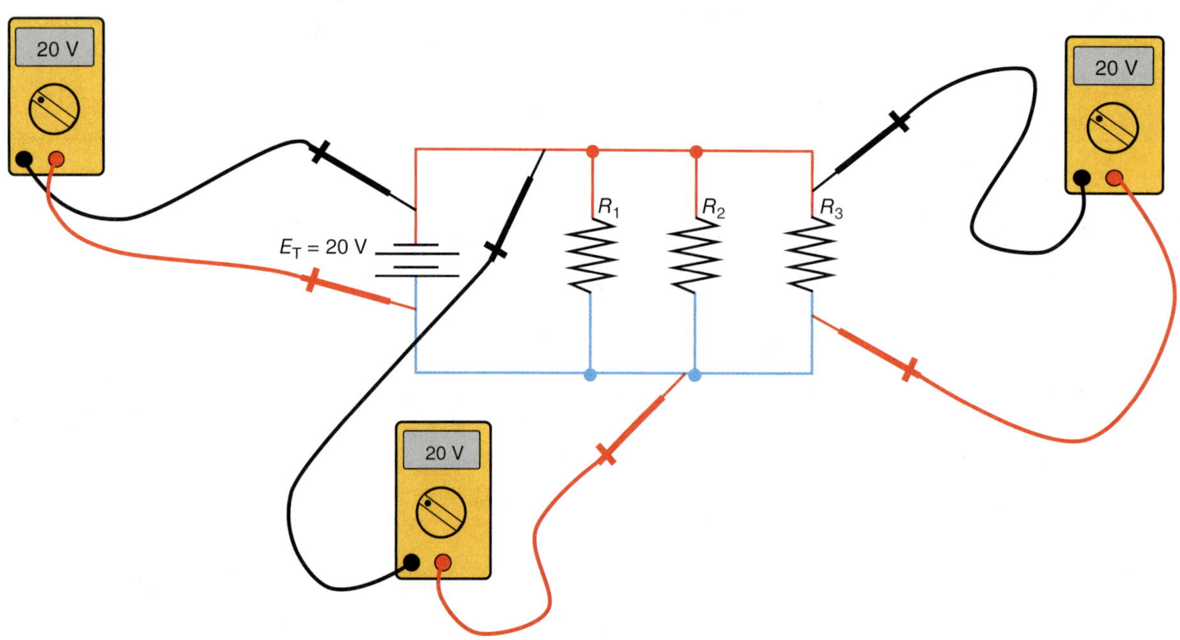

Goodheart-Willcox Publisher

Figure 11-3. Any voltage reading between the red and blue lines will be the same.

configuration, each of the three branches (powering R_1, R_2, and R_3) is being supplied with the same voltage, which is equal to that of the power source.

11.2 Calculating Parallel Circuit Values

Electricians must understand the rules associated with voltage, current, resistance, and power in a parallel circuit. Similar to the rules used for calculating series circuits covered in Chapter 10, *Series Circuits*, there are three rules used to calculate circuit values in parallel circuits:

- The voltage supplied to each branch of a parallel circuit is the same throughout the circuit.
- The total circuit current equals the sum of the current flows through each of the branches in the circuit.
- The total resistance of a parallel circuit is always less than the resistance of the lowest value resistance in the circuit.

The following examples help provide an understanding of the laws that govern parallel circuits and an opportunity to perform basic calculations. You will need to be familiar with Ohm's law and the power formulas to help solve these equations. Refer to Chapter 6, *Electrical Characteristics*, where both are discussed in detail.

Calculating Voltage

The first parallel circuit rule states that the voltage supplied across any branch of a parallel circuit is the same across the entire circuit. Just like loads connected in series, parallel loads share the same power supply. In a series circuit, however, the voltage supplied to the circuit is divided among the circuit loads, while each branch in a parallel circuit is supplied with a voltage that is equal to the total circuit voltage. This rule can be expressed as follows:

$$E_T = E_1 = E_2 = E_3 = \ldots = E_N$$

In this expression, E represents voltage, E_T represents the power supply voltage, while E_1, E_2, and E_3 represent the voltage supplied to circuit branches 1, 2, 3, etc. The E_N term refers to the voltage supplied across the last circuit branch.

Examining the parallel circuit shown in **Figure 11-4**, a 30-V power supply is powering two branches: one containing load R_1 and the other containing load R_2. Since the voltage supplied to each branch of a parallel circuit is equal to the circuit's supply voltage, the voltage supplied to both R_1 and R_2 is 30 V.

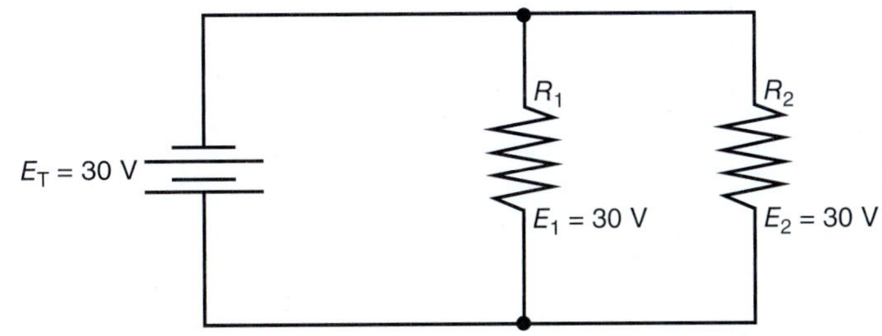

Goodheart-Willcox Publisher

Figure 11-4. In a parallel circuit, all branches are supplied with the same voltage, which is equal to the voltage at the power supply.

Calculating Total Branch Current

In a parallel circuit, each branch has a measurable current flowing through it. The current flowing through each branch is determined by the voltage supplied to the circuit and the resistance of the branch. The sum of the current flows through each branch is equal to the total current flowing through the circuit as measured at either the power supply or along any conductor before the first branch. This rule can be expressed as:

$$I_T = I_1 + I_2 + I_3 + \ldots + I_N$$

In this expression, I represents current, I_T represents the total circuit current at the power supply, while I_1, I_2, and I_3 represent the current flow through circuit branches 1, 2, 3, etc. The I_N term represents the current flowing across the last branch in the parallel circuit. Refer to **Example 11-1** and **Example 11-2** on how to calculate current flow in a parallel circuit.

Example 11-1

Calculate the total amount of current flow in the following parallel circuit.

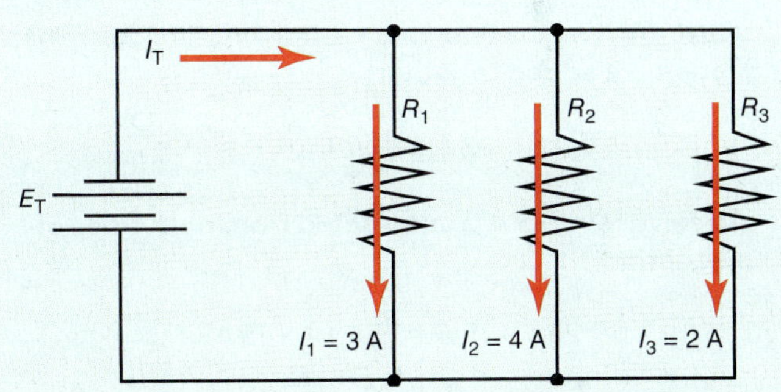

Goodheart-Willcox Publisher

✓ Solution

This circuit has three branches. Since the sum of the current flows through each circuit branch will be equal to the total current flow in the circuit, the following formula can be used to determine the circuit's total current:

$$I_T = I_1 + I_2 + I_3$$

$$I_T = 3\text{ A} + 4\text{ A} + 2\text{ A}$$

$$I_T = 9\text{ A}$$

Example 11-2

What is the current flow through the second branch in the following parallel circuit?

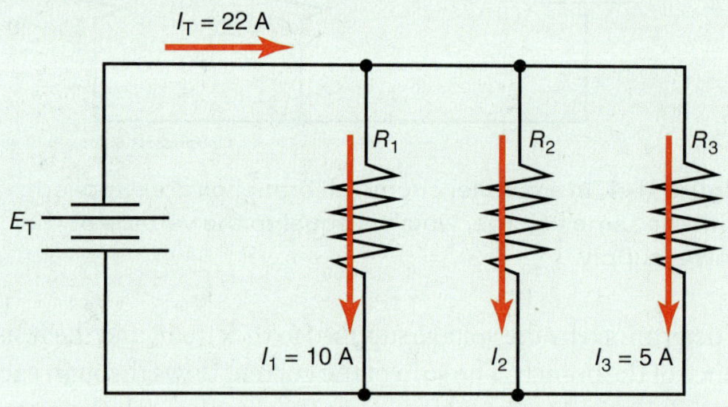

Goodheart-Willcox Publisher

✓ Solution

This circuit has three branches. The current flow through branch one is 10 A, and the current flow through branch three is 5 A. The total current flow in the circuit is a known value of 22 A. Since the sum of the current flows through each circuit branch will be equal to the total current flow in the circuit, the following formula can be used to determine the current flow through the second branch:

$$I_T = I_1 + I_2 + I_3$$

Substituting the known values leaves I_2 as the only unknown value:

$$22\ \text{A} = 10\ \text{A} + I_2 + 5\ \text{A}$$

Next, the two current values on the right side of the equation are added together:

$$22\ \text{A} = 15\ \text{A} + I_2$$

To solve for I_2, 15 A is subtracted from both sides of the equation:

$$22\ \text{A} - 15\ \text{A} = 15\ \text{A} + I_2 - 15\ \text{A}$$

$$I_2 = 7\ \text{A}$$

If the voltage supplied to a parallel circuit is known, the current flow through each branch with a known resistance can be calculated using the current form of Ohm's law. See **Example 11-3**.

In addition to the total circuit current being equal to the sum of the branch currents, the current flow through each parallel circuit branch will be inversely proportional to the resistance of each branch. As the resistance of a circuit branch increases, current flow through that branch decreases.

Example 11-3

Using Ohm's law, calculate the current flow through each branch in the following parallel circuit.

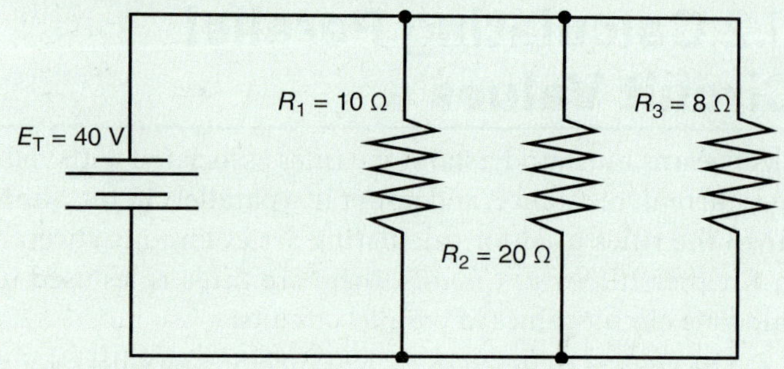

Goodheart-Willcox Publisher

✓ Solution

This 40-V parallel circuit has three branches with resistances of 10 Ω, 20 Ω, and 8 Ω. Using the current formula of Ohm's law, first determine the current through R_1:

$$I_1 = \frac{E_T}{R_1}$$

$$I_1 = \frac{40\ \text{V}}{10\ \Omega}$$

$$I_1 = 4\ \text{A}$$

Next, determine the flow through R_2:

$$I_2 = \frac{E_T}{R_2}$$

$$I_2 = \frac{40\ \text{V}}{20\ \Omega}$$

$$I_2 = 2\ \text{A}$$

Then determine the flow through R_3:

$$I_3 = \frac{E_T}{R_3}$$

$$I_3 = \frac{40\ \text{V}}{8\ \Omega}$$

$$I_3 = 5\ \text{A}$$

Since the sum of the current through each circuit branch will be equal to the total current flow in the circuit, the total circuit amperage can now be calculated as follows:

$$I_T = I_1 + I_2 + I_3$$

$$I_T = 4\ \text{A} + 2\ \text{A} + 5\ \text{A}$$

$$I_T = 11\ \text{A}$$

Consider the parallel circuit shown in **Figure 11-5**. In this circuit, there are two branches with resistances of 5 Ω and 10 Ω. With the value of R_2 being twice that of R_1, the current flow through R_1 will be twice the current flowing

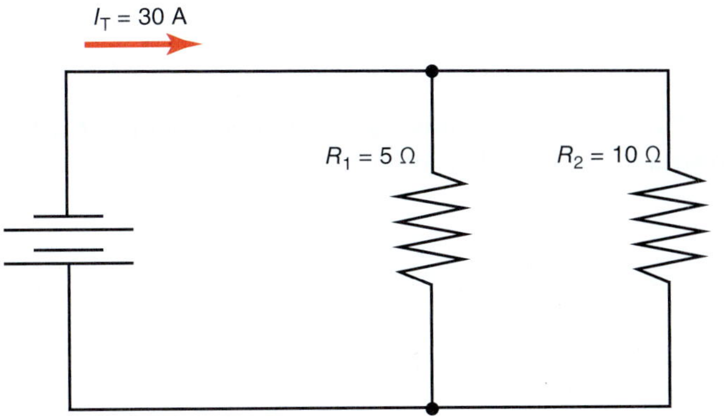

Goodheart-Willcox Publisher

Figure 11-5. Parallel circuit with branch resistances of 5 Ω and 10 Ω. With the value of R_2 being twice that of R_1, the current flow through R_1 will be twice the current flowing through R_2.

through R_2. The sum of the current flows through R_1 and R_2 can be expressed as follows:

$$I_T = I_1 + I_2$$

In this expression, I_T is the total current flow through the circuit, while I_1 and I_2 represent current flowing through branches one and two, respectively. Since I_1 is two times I_2 (because the resistance value of R_2 is two times that of R_1), I_1 can be replaced by the term $2I_2$. The equation then becomes:

$$I_T = I_1 + I_2$$
$$I_T = 2I_2 + I_2$$
$$I_T = 3I_2$$

Substituting 30 A for I_T changes the equation to:

$$I_T = 3I_2$$
$$30 \text{ A} = 3I_2$$

Dividing both sides of the equation by 3 solves the equation for I_2:

$$\frac{30 \text{ A}}{3} = \frac{3I_2}{3}$$
$$I_2 = 10 \text{ A}$$

Knowing that the measured current flowing through branch 2 is 10 A, the current flowing through branch 1 can be determined. Since the total current flowing in the circuit is equal to the sum of the measured current flows in each branch, I_1 will equal 20 A (30 A – 10 A). Alternately, knowing that the measured current through branch 1 must be twice the current flowing through branch 2 (because the resistance of branch 1 is one-half that of branch 2), doubling I_2 would yield the same

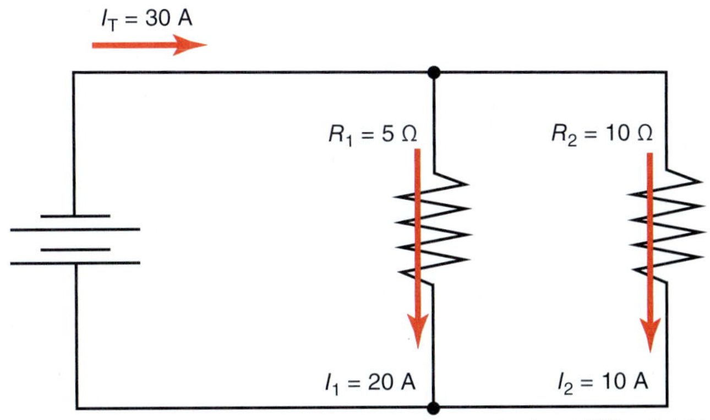

Goodheart-Willcox Publisher

Figure 11-6. The resistance of R_2 is twice that of R_1 so the current flow through R_2 is one-half of that through R_1.

20 A answer, **Figure 11-6**. For additional practice on calculating current, see **Example 11-4**.

Example 11-4

Calculate the current flowing through each branch of the following parallel circuit.

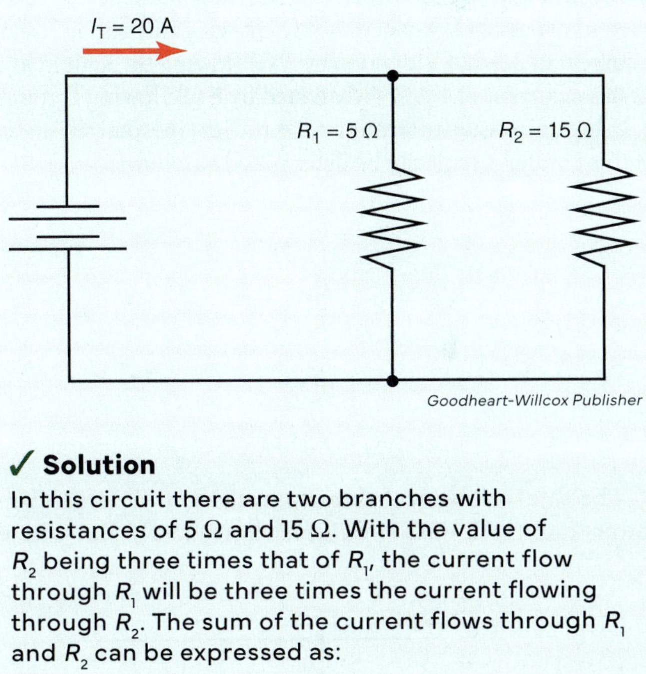

Goodheart-Willcox Publisher

✓ Solution

In this circuit there are two branches with resistances of 5 Ω and 15 Ω. With the value of R_2 being three times that of R_1, the current flow through R_1 will be three times the current flowing through R_2. The sum of the current flows through R_1 and R_2 can be expressed as:

$$I_T = I_1 + I_2$$

Since I_1 is three times I_2 (because the resistance value of R_2 is three times that of R_1), I_1 can be replaced by the term $3I_2$.

$$I_T = I_1 + I_2$$
$$I_T = 3I_2 + I_2$$
$$I_T = 4I_2$$

Continued

Next, substitute 20 A for I_T:

$$I_T = 4I_2$$
$$20\text{ A} = 4I_2$$

Then divide both sides of the equation by 4:

$$\frac{20\text{ A}}{4} = \frac{4I_2}{4}$$
$$I_2 = 5\text{ A}$$

Knowing that the measured current flowing through branch 2 is 5 A, the current flowing through branch 1 can now be determined. With the current through branch 1 being three times the current flowing through branch 2, tripling I_2 provides a current of 15 A flowing through R_1.

Calculating Total Circuit Resistance

The third rule for a parallel circuit states that the total resistance of a parallel circuit is always less than the resistance of the lowest value resistance in the circuit. Consider the 24-V parallel circuit shown in **Figure 11-7**. This circuit has three branches: R_1 has a current of 6 A (24 V divided by 4 Ω) flowing through it; R_2 has a current of 8 A (24 V divided by 3 Ω) flowing through it; and R_3 has a current of 4 A (24 V divided by 6 Ω) flowing through it. Using the resistance formula of Ohm's law, the total resistance of this parallel circuit can be determined as follows:

$$R = \frac{E}{I}$$
$$R_T = \frac{E_T}{I_T}$$
$$R_T = \frac{24\text{ V}}{18\text{ A}}$$
$$R_T = 1.33\ \Omega$$

The total resistance of this parallel circuit (1.33 Ω) is lower than the lowest circuit resistance (3 Ω). But why is this the case? With more current paths available, electron flow is divided among the available paths, lowering the resistance to flow.

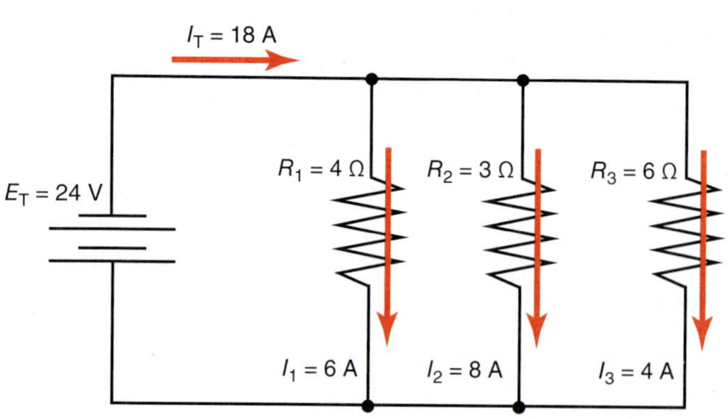

Goodheart-Willcox Publisher

Figure 11-7. In this parallel circuit, as in any parallel circuit, the total resistance is less than the resistance of the lowest-valued resistance in the circuit.

The reciprocal formula can be used to calculate the total resistance of a parallel circuit:

$$\frac{1}{R_T} = \frac{1}{R_1} + \frac{1}{R_2} + \frac{1}{R_3} + \ldots + \frac{1}{R_N}$$

In this expression, R_T is the total circuit resistance, while R_1, R_2, and R_3 represent branch resistances 1, 2, and 3. R_N represents the resistance across the last parallel circuit branch. See **Example 11-5**.

Math Sidebar

A reciprocal of a number is equal to one divided by that number. For example, the reciprocal of 2 is 1 divided by 2. Refer to Chapter 7, *Mathematical Manipulation and Conversions*, where calculating with reciprocals is explained.

Example 11-5

Determine the total resistance of the following parallel circuit.

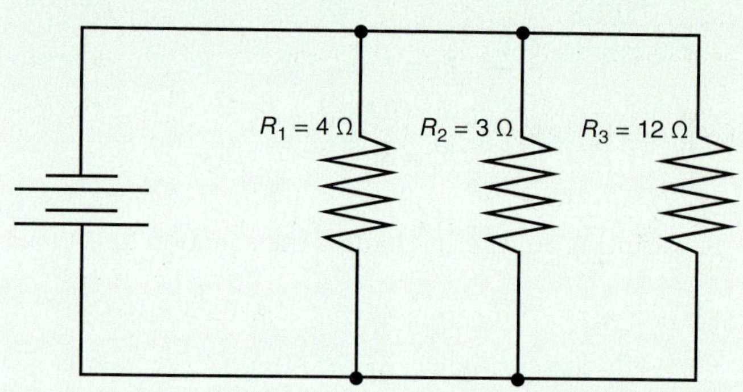

Goodheart-Willcox Publisher

✓ Solution

The total circuit resistance can be calculated by using the reciprocal method:

$$\frac{1}{R_T} = \frac{1}{R_1} + \frac{1}{R_2} + \frac{1}{R_3}$$

$$\frac{1}{R_T} = \frac{1}{4\ \Omega} + \frac{1}{3\ \Omega} + \frac{1}{12\ \Omega}$$

To add the fractions on the right side of the equation you will need to find the lowest common denominator. In this example, the lowest common denominator is 12. To convert the fractions to their lowest common denominator, multiply the 1/4 by 3/3 and then 1/3 by 4/4 as follows:

$$\frac{1}{R_T} = \frac{1}{4\ \Omega} \times \frac{3}{3} + \frac{1}{3\ \Omega} \times \frac{4}{4} + \frac{1}{12\ \Omega}$$

$$\frac{1}{R_T} = \frac{3}{12\ \Omega} + \frac{4}{12\ \Omega} + \frac{1}{12\ \Omega}$$

Then add the numerators:

$$\frac{1}{R_T} = \frac{8}{12\ \Omega}$$

Cross-multiplying this expression:

$$8R_T = 12\ \Omega$$

Finally, divide both sides of the equation by 8 to solve the expression for R_T:

$$8R_T = 12\ \Omega$$

$$\frac{8R_T}{8} = \frac{12\ \Omega}{8}$$

$$R_T = 1.5\ \Omega$$

Math Sidebar

You can also solve equations using a scientific calculator. For example, the total resistance of the circuit used in **Example 11-5** can be calculated by pressing the following keys on a scientific calculator:

4 1/x + 3 1/x + 12 1/x =

Goodheart-Willcox Publisher

Once you find the sum, then push the following key:

1/x

Goodheart-Willcox Publisher

Did you get the final answer that was noted in **Example 11-5**? Note: Some calculators may require a different key stroke or press the equal key at the end.

Alternative Calculation for Two Parallel Resistances

If the parallel circuit being evaluated has only two branches, the following formula can be used:

$$R_T = \frac{R_1 \times R_2}{R_1 + R_2}$$

Because of its format, this method of calculating the total resistance of two parallel resistances with different values is often referred to as the *product over sum formula*. This method is explained in **Example 11-6**.

Example 11-6

Calculate the total resistance for the following parallel circuit.

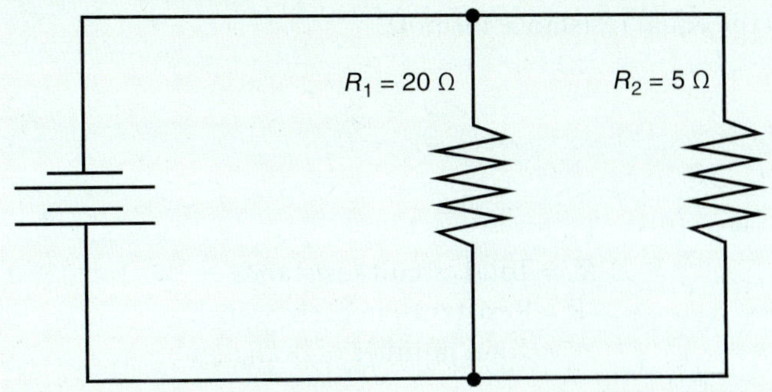

Goodheart-Willcox Publisher

✓ Solution

With only two resistances in this parallel circuit, the product over sum formula can be used to calculate the total resistance of the circuit. Substituting the known values for R_1 and R_2, the total resistance can be determined as follows:

$$R_T = \frac{R_1 \times R_2}{R_1 + R_2}$$

$$R_T = \frac{20\ \Omega \times 5\ \Omega}{20\ \Omega + 5\ \Omega}$$

$$R_T = \frac{100\ \Omega^2}{25\ \Omega}$$

$$R_T = 4\ \Omega$$

You can also get the same answer by using a scientific calculator by pressing the following keys:

20 1/x + 5 1/x =

Goodheart-Willcox Publisher

Then, press the 1/x key again. Did you get 4 Ω?

Now It's Your Turn 11-6

Using the product over the sum formula, calculate the total resistance of the circuit shown in the following parallel circuit.

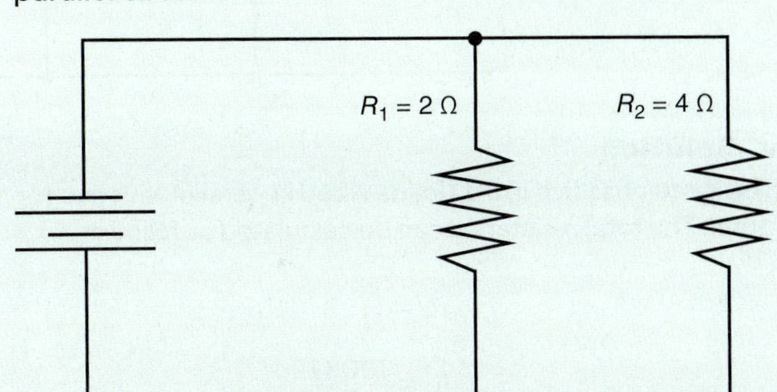

Goodheart-Willcox Publisher

Alternative Calculation for Equal Branch Resistances

Another useful formula that can be used when multiple resistances of the same value are wired in parallel with each other is the equal resistance formula:

$$R_T = \frac{R}{N}$$

where:

R_T = Total circuit resistance
R = Resistance in each branch
N = Total number of branches

Thus, when two resistances of the same value are wired in parallel with each other, the total resistance is equal to one-half of the value of one of the resistances. If the parallel circuit has three resistances with the same value, the total circuit resistance is R divided by 3. If the parallel circuit has four resistances with the same value, the total circuit resistance is R divided by 4. The denominator is always equal to the number of same-valued resistances in the parallel circuit. Refer to **Example 11-7** on how to calculate total resistance using this method.

To summarize, when calculating multiple resistances in parallel, three common formulas are used:

Reciprocal Formula (works in all cases):

$$\frac{1}{R_T} = \frac{1}{R_1} + \frac{1}{R_2} + \frac{1}{R_3} + \ldots + \frac{1}{R_N}$$

Product over Sum Formula (works with two branches):

$$R_T = \frac{R_1 \times R_2}{R_1 + R_2}$$

Equal Resistance Formula (works when all branches have the same resistance):

$$R_T = \frac{R}{N}$$

Calculating Current Flow

Unlike series circuits, where the current flow is the same at all points in the circuit, the measured current in a parallel circuit is different at various points in the circuit. Referring to the circuit in **Figure 11-8**, the current flow through each of the three circuit branches is different, as is the measured current at the power supply. As discussed earlier in this chapter, the current measured at or close to the power supply is the sum of the measured currents in each of the circuit's branches. This is also true of the current on the return path. Current returning to the power supply is equal to the current leaving the power supply. In **Figure 11-8**, the 15 A measurement (total circuit current) is the sum of the 4 A, 5 A, and 6 A current in the branches. What would the current be in the conductor that connects branch 1 and branch 2?

To help answer this question, refer to **Figure 11-9**. Point A is referred to as a node. In general, a ***node*** is a point in a circuit where at least two conductors or wires are connected. Typically, the term node is used when three or more conductors or components are connected in a singular location. On some wiring diagrams and schematics, a node is represented by a black dot.

Example 11-7

Calculate the total resistance for the following circuit.

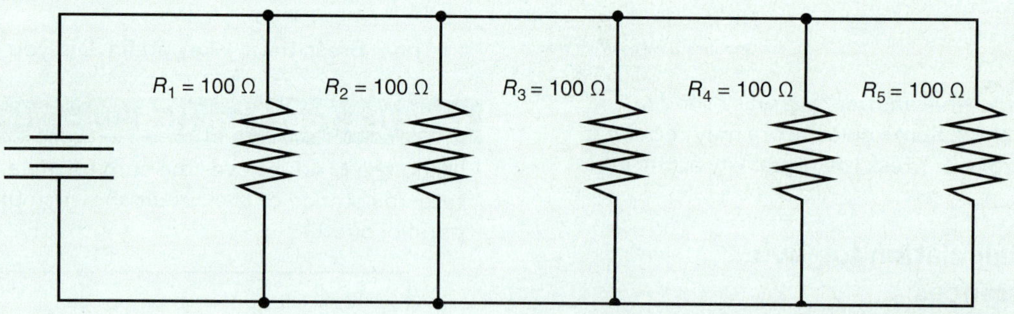

Goodheart-Willcox Publisher

✓ Solution

This circuit has five 100-Ω loads wired in parallel with each other. The total resistance can be calculated as follows:

$$R_T = \frac{R}{N}$$

$$R_T = \frac{100\ \Omega}{5}$$

$$R_T = 20\ \Omega$$

Now It's Your Turn 11-7

Calculate the total resistance of a four-branch parallel circuit where each branch has a resistance of 40 Ω.

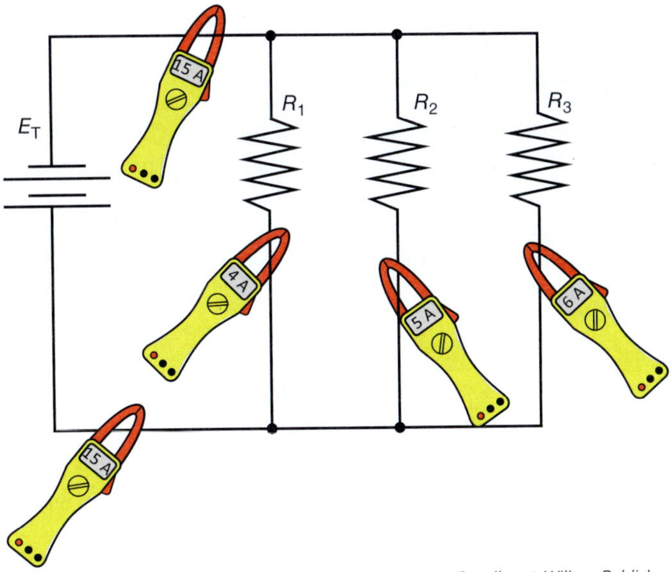

Goodheart-Willcox Publisher

Figure 11-8. The total circuit current is equal to the sum of the branch currents. In addition, note that the total current leaving the power supply is the same as the current returning to the power supply.

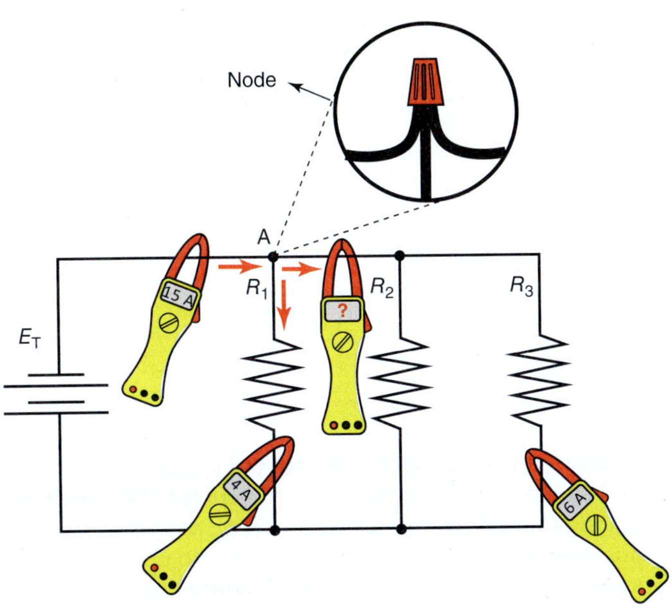

Goodheart-Willcox Publisher

Figure 11-9. A node consisting of three connected conductors. Current entering and leaving node A is known; this can be used to calculate current between branch 1 and branch 2.

The amount of current entering a node is equal to the amount of current leaving the node. This concept is referred to as Kirchhoff's Current Law. Thus, if you add all of the individual current values flowing into the node, the total will be the same as the sum of all the current values flowing out of the node.

Referring back to **Figure 11-9**, a current of 15 A is flowing into node A, while a current of 4 A is flowing out of node A toward R_1. Since the total current flowing out of the node must equal the total current flowing into the node (15 A), the current in the conductor between branch 1 and branch 2 is 11 A (15 A – 4 A). Since the 15-A flow is directed into the node and the 4-A flow is directed away from the node, the 11-A flow is also directed away from the node, as indicated in the **Figure 11-9**.

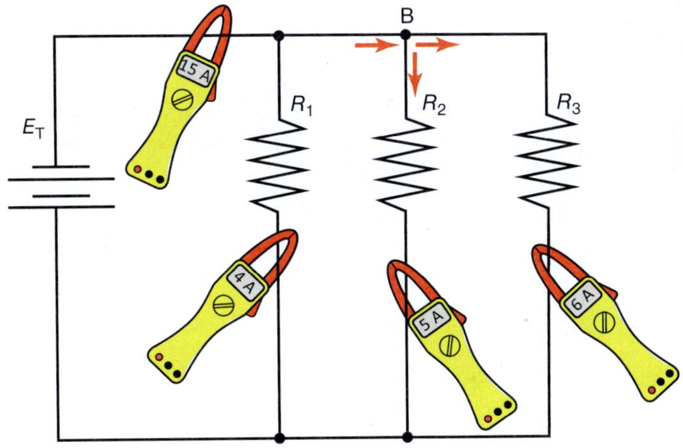

Goodheart-Willcox Publisher

Figure 11-10. When evaluating the current entering and leaving node B in this circuit, recall that the current entering and leaving a node is the same. This fact can also be used to calculate current between branch 1 and branch 2.

You can also use the measurements at node B in **Figure 11-10** to determine the current flowing between branch 1 and branch 2. The total current flowing out of node B is 11 A (5 A to branch 2 plus 6 A to branch 3), so the current flowing into the node through the conductor between branch 1 and branch 2 is also 11 A.

Example 11-8 provides additional practice on calculating current.

Example 11-8

Referring to the following circuit, calculate the current reading on meter B.

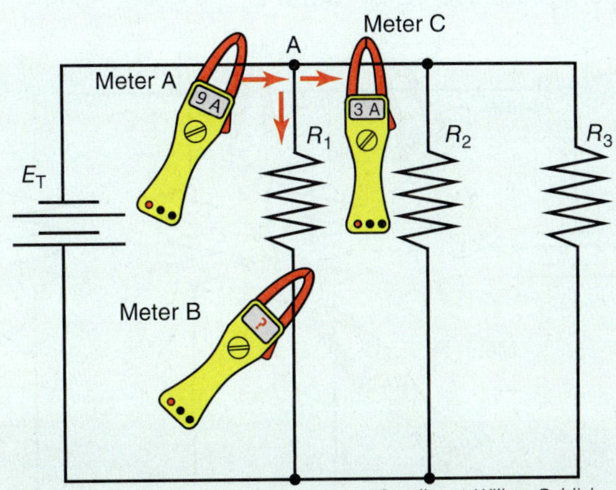

Goodheart-Willcox Publisher

✓ Solution

The inflow, as measured on meter A, is 9 A, and one outflow, as measured on meter C, is 3 A. The current flowing out of the node must equal the current flowing into the node (9 A), so meter B would display a value of 6 A (9 A – 3 A).

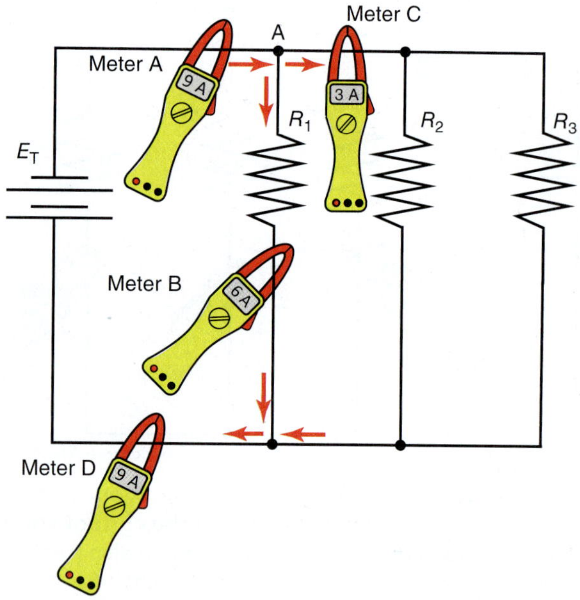

Goodheart-Willcox Publisher

Figure 11-11. The current flow returning to the power supply is the same as the current flow from the power supply.

The current flow returning to the power supply must also be the same as the current flowing from the power supply to the circuit loads. For example, the measured current on meter D in **Figure 11-11** must also be 9 A. In addition, the current flowing from a load is the same as the current flowing into the load. How to calculate current flow returning to the power supply is explained in **Example 11-9**.

11.3 Calculating Power Consumption

Using the rules of parallel circuits, the power consumption of a parallel circuit can be calculated using the power formulas that are discussed in detail in Chapter 6, *Electrical Characteristics*.

The formulas used to calculate power consumption are as follows:

$$P = I \times E$$
$$P = I^2 \times R$$
$$P = \frac{E^2}{R}$$

These formulas are also provided in the power wheel shown in **Figure 11-12**.

Consider the circuit shown in **Figure 11-13**. This 40-V circuit powers two parallel loads with resistances of 20 Ω and 5 Ω, resulting in a total circuit resistance of 4 Ω. With the circuit voltage and resistance known, the circuit's power consumption can be calculated as follows:

$$P = \frac{E^2}{R}$$
$$P = \frac{(40\text{ V})^2}{4\ \Omega}$$
$$P = \frac{1600}{4}$$
$$P = 400\text{ W}$$

Example 11-9

Referring to the following circuit, determine the amperage measured by meter E.

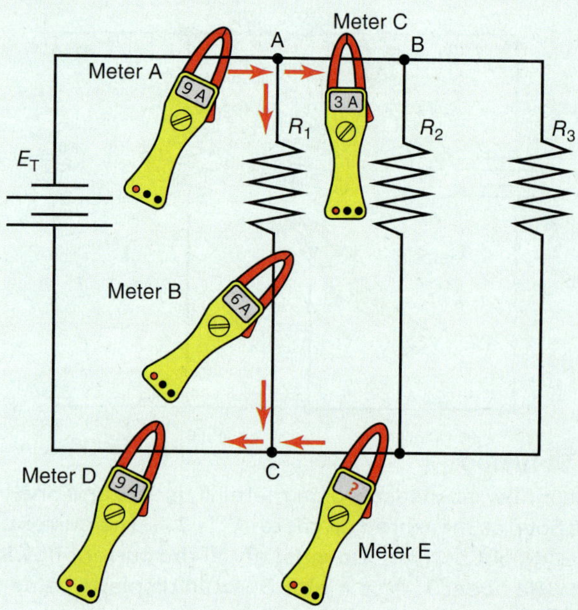

Goodheart-Willcox Publisher

✓ Solution

The amperage displayed on meter E can be determined in two ways.

The first method is to evaluate the inflows and outflows at node C. In this case, the current flowing out of node C is 9 A, so the total current flowing into node C must also be 9 A. The current flowing into node C from branch 1 is 6 A. Therefore, meter E would display a value of 3 A (9 A – 6 A).

The second way to determine the amperage reading on meter E is to observe that the current flowing toward branch 2 and branch 3 is 3 A (shown on meter C). Since the current leaving a load is equal to the current entering the load, the current measured on meter E must be the same as the current measured on meter C.

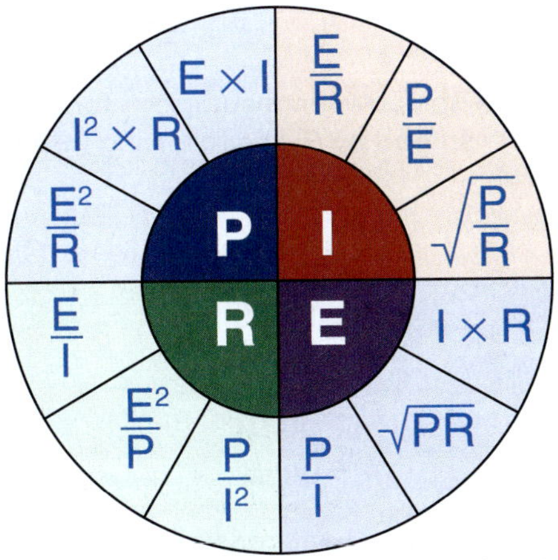

Goodheart-Willcox Publisher

Figure 11-12. The power wheel is a useful tool to calculate unknown values.

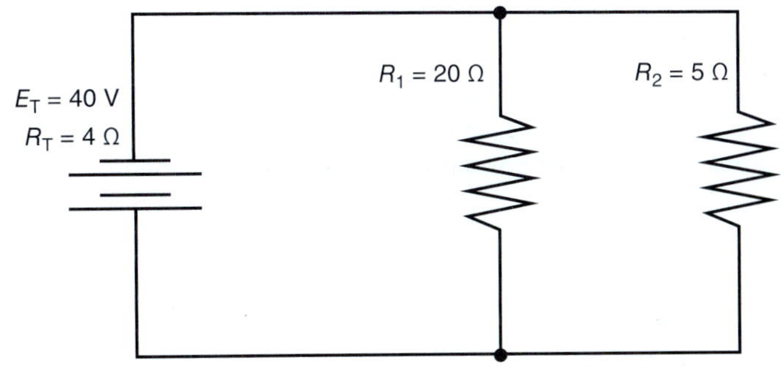

Goodheart-Willcox Publisher

Figure 11-13. The amount of power used in a parallel circuit can be calculated using the power formula of Ohm's law.

Example 11-10 and **Example 11-11** provide scenarios on how to calculate power consumed by a circuit.

Example 11-10

Referring to the following circuit, calculate the amount of power consumed by the circuit and the power consumed by each load.

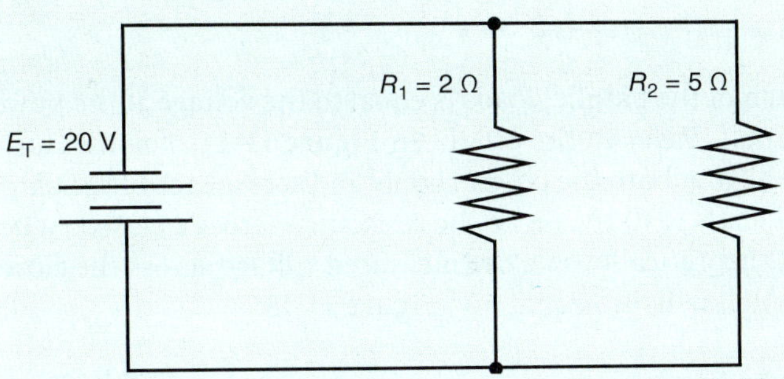

Goodheart-Willcox Publisher

✓ Solution

To determine this circuit's power consumption, either the total circuit current or the total circuit resistance must be known. The total circuit resistance can be calculated as follows:

$$R_T = \frac{R_1 \times R_2}{R_1 + R_2}$$

$$R_T = \frac{2\ \Omega \times 5\ \Omega}{2\ \Omega + 5\ \Omega}$$

$$R_T = \frac{10\ \Omega^2}{7\ \Omega}$$

$$R_T = 1.43\ \Omega$$

With a known supply voltage of 20 V and a known total resistance of 1.43 Ω, the circuit's power consumption can then be calculated as follows:

$$P_T = \frac{E^2}{R}$$

$$P_T = \frac{(20\ \text{V})^2}{1.43\ \Omega}$$

$$P_T = \frac{400}{1.43}$$

$$P_T = 279.7\ \text{W} \approx 280\ \text{W}$$

Power consumed by the individual loads can be calculated using the same power formula. The power consumed by R_1 is calculated as follows:

$$P_1 = \frac{(E_1)^2}{R_1}$$

$$P_1 = \frac{(20\ \text{V})^2}{2\ \Omega}$$

$$P_1 = \frac{400}{2}$$

$$P_1 = 200\ \text{W}$$

Next, the power consumed by R_2 is calculated as follows:

$$P_2 = \frac{(E_2)^2}{R_2}$$

$$P_2 = \frac{(20\ \text{V})^2}{5\ \Omega}$$

$$P_2 = \frac{400}{5}$$

$$P_2 = 80\ \text{W}$$

Notice that, when the individual results for P_1 and P_2 are added together, the sum is the same as the result that was obtained for total power in the circuit (P_T).

Example 11-11

Referring to the following circuit, calculate the total power consumption as well as the value of R_1 and the amount of power consumed by each circuit load.

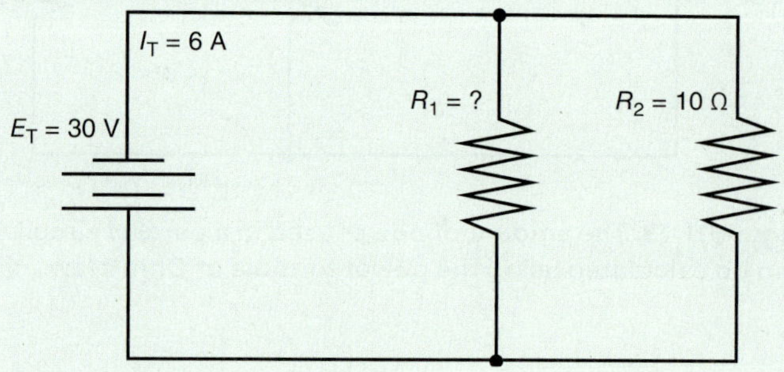

Goodheart-Willcox Publisher

✓ Solution

Since the total circuit voltage and amperage are known, the following formula can be used to calculate the circuit's power consumption as follows:

$$P_T = I_T \times E_T$$

$$P_T = 6\ \text{A} \times 30\ \text{V}$$

$$P_T = 180\ \text{W}$$

To calculate the power consumption for P_2 and then the power consumed by R_1, the following formulas can be used:

$$P_2 = \frac{(E_2)^2}{R_2}$$

$$P_2 = \frac{(30\ \text{V})^2}{10\ \Omega}$$

$$P_2 = \frac{900\ \text{V}}{10\ \Omega}$$

$$P_2 = 90\ \text{W}$$

The circuit's total power consumption is 180 W, and 90 W are consumed by R_2. Since the power consumed by the individual loads is equal to the total power consumed by the circuit, the power consumed by R_1 is also 90 W (180 W – 90 W). Therefore, the resistance of R_1 is the same as R_2, or 10 Ω.

11.4 Use of Switches

In the most basic parallel circuit, one switch is used to control multiple loads. When the switch is in the closed position, power can pass freely between the power source and the parallel-connected loads. When closed, the voltage supplied to each of the parallel loads is equal to the voltage at the power supply. Refer to the circuit in **Figure 11-14**. The measured voltage at both the power supply and across each of the parallel loads is 100 V. Since the resistance across a closed switch is, theoretically, 0 Ω, the measured voltage across the closed switch will also equal 0 V, **Figure 11-15**.

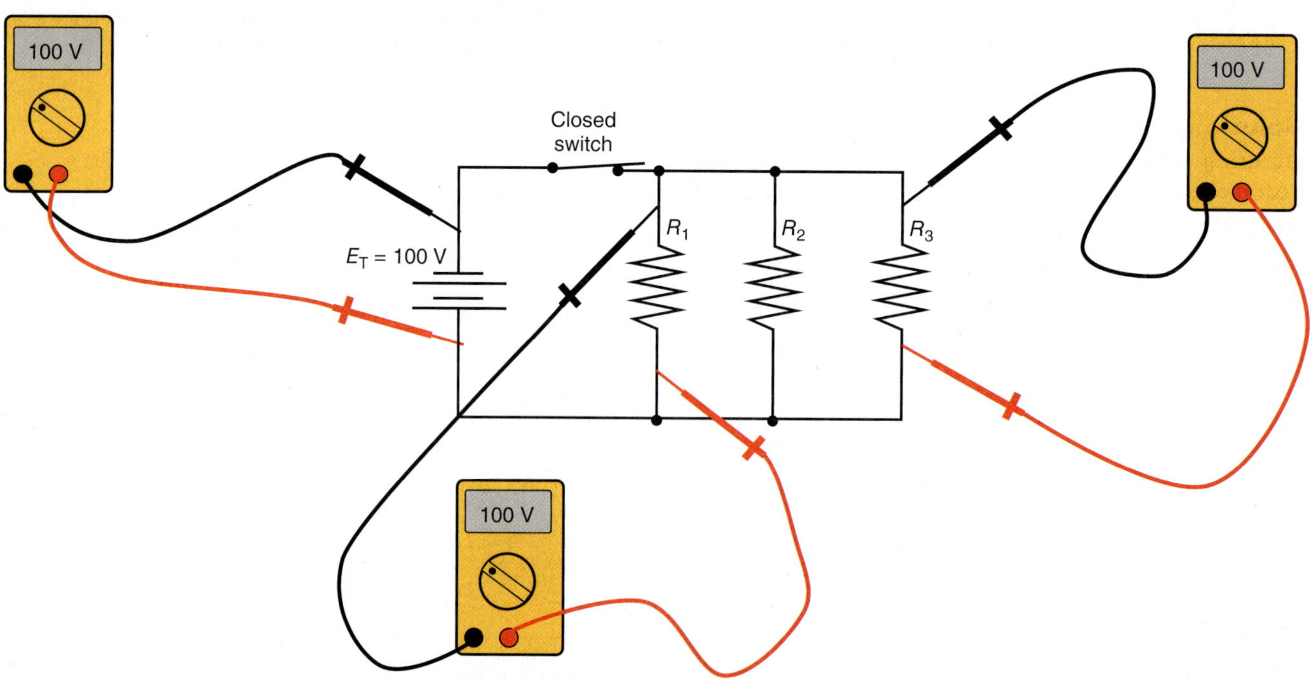

Goodheart-Willcox Publisher

Figure 11-14. In a parallel circuit, a single switch can be used to control multiple loads. When the switch is closed, the main circuit voltage is supplied to all circuit branches.

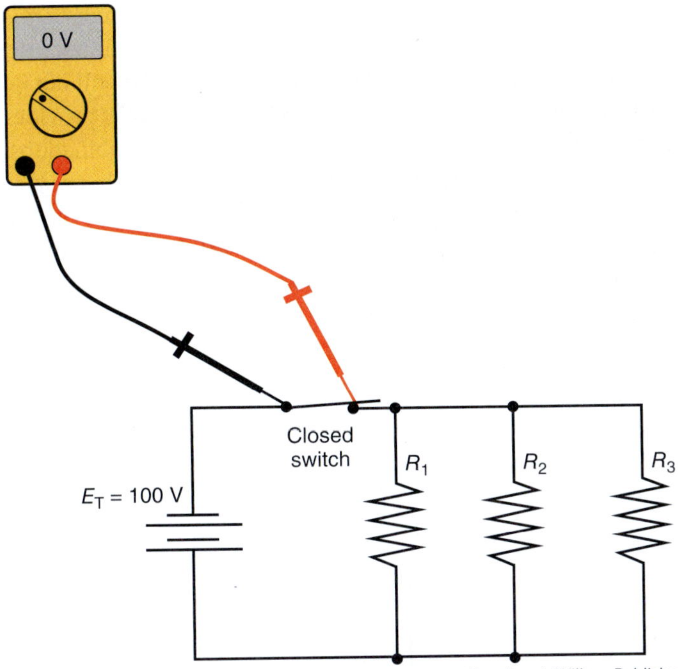

Goodheart-Willcox Publisher

Figure 11-15. Ideally, the resistance across a closed switch will be 0 Ω. Therefore, the measured voltage across a closed switch should be 0 V. However, caution should still be taken as the circuit may still be energized.

Safety Alert

Obtaining a voltage reading of 0 V across a switch does not mean that the circuit is de-energized. Be sure to measure the voltage between each side of the switch and ground to confirm that the circuit is de-energized. The voltage measured across a closed switch will *always* be 0 V, so be careful!

In parallel circuits, every branch must have a load. A branch cannot comprise only a switch, **Figure 11-16.** If the only component in a branch is a switch, a short circuit will occur when the switch closes. A ***short circuit*** occurs when a current path has little or no resistance, resulting in a dangerous and damaging huge flow of current. When a short circuit condition exists, circuit protection devices such as properly sized fuses and circuit breakers will open to de-energize the circuit.

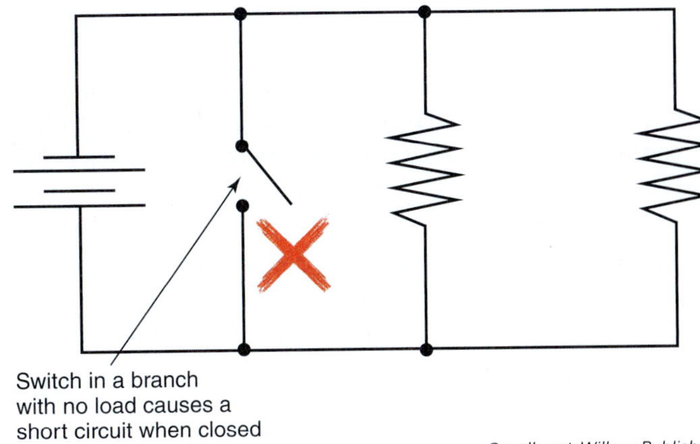

Goodheart-Willcox Publisher

Figure 11-16. All parallel circuit branches must have a load. Switches cannot be the only component in a parallel circuit branch. If this switch is closed, a dead short will occur, creating a short circuit.

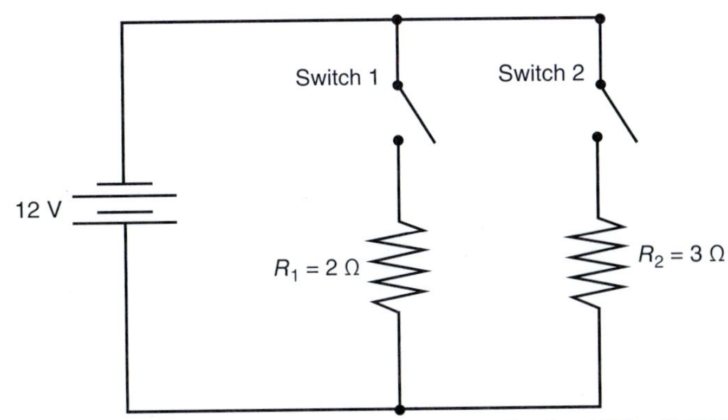

Goodheart-Willcox Publisher

Figure 11-17. Two switches in a parallel circuit, each controlling the one load that is wired in series with it.

In many applications, each branch has its own switch, **Figure 11-17.** In this configuration, each of the two loads is controlled independently of the other. When both switches are in the open position, neither load is energized. When both switches are in the closed position, both loads are energized. When only one switch is closed, the load wired in series with the closed switch is energized, while the other load is not. The table shown in **Figure 11-18** provides circuit and power

Possible Switch Combinations

Switch 1	Switch 2	R_1	R_2	Total Circuit Current	Total Circuit Power
Open	Open	De-energized	De-energized	0	0
Open	Closed	De-energized	Energized	4 A	48 W
Closed	Open	Energized	De-energized	6 A	72 W
Closed	Closed	Energized	Energized	10 A	120 W

Goodheart-Willcox Publisher

Figure 11-18. All possible switch combinations for the circuit shown in **Figure 11-17.**

information for the circuit shown in **Figure 11-17** with all possible switch position combinations.

In some applications, multiple switches are wired in parallel with each other, **Figure 11-19**. This arrangement is used when a load is to be energized when any one of multiple switches are closed.

Consider the situation where there is an outdoor light that illuminates either when motion is sensed or when it is raining. In this case, one switch will be a motion sensor that causes a set of electrical contacts to close when motion is sensed and the other will be a moisture sensor that closes its contacts when moisture is sensed. With the switches connected in parallel, the circuit load will be energized if either switch is closed. The load will not be energized when both switches are open. **Figure 11-20** provides circuit information for all possible switch position combinations.

When switches are wired in parallel with each other and at least one switch is closed, a voltage reading taken across either switch will be 0 V. The only time there will be a measurable voltage across a set of parallel switches is when all switches are in the open position.

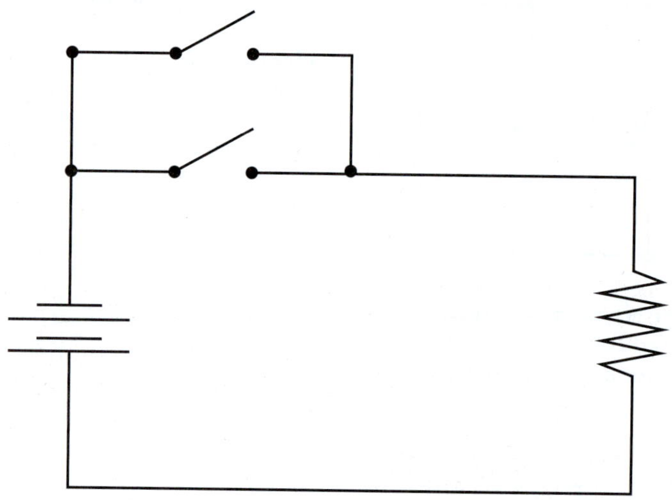

Goodheart-Willcox Publisher

Figure 11-19. A simple parallel circuit with two switches connected in parallel with each other. If either switch is closed, the circuit load will be energized.

	MOISTURE SENSOR DRY (SWITCH OPEN)	MOISTURE SENSOR WET (SWITCH CLOSED)
MOTION NOT SENSED (SWITCH OPEN)	LIGHT NOT ENERGIZED (A)	LIGHT ENERGIZED (C)
MOTION SENSED (SWITCH CLOSED)	LIGHT ENERGIZED (B)	LIGHT ENERGIZED (D)

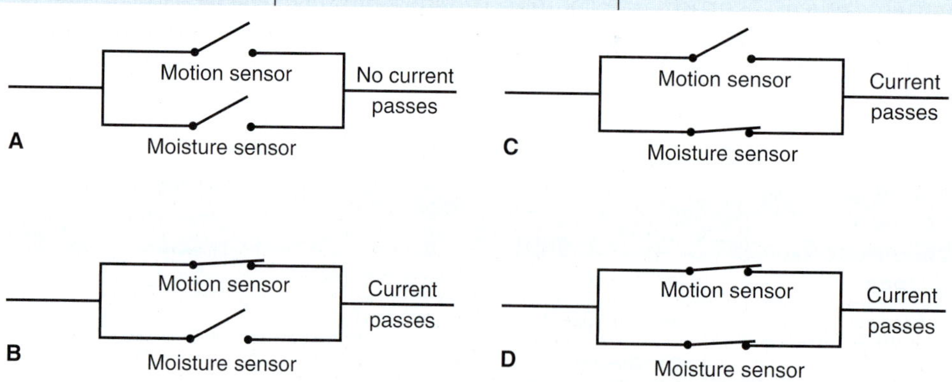

Goodheart-Willcox Publisher

Figure 11-20. All possible switch combinations for parallel-connected motion and moisture sensors.

CHAPTER 11 REVIEW

Summary

- In a parallel circuit there are multiple paths for electric current to follow. (11.1)
- Each path in a parallel circuit is called a branch. (11.1)
- A break in one branch of a parallel circuit will not affect current from flowing through the remainder of the circuit. (11.1)
- The total circuit current is the sum of the current flows through each individual branch. (11.2)
- The amount of current entering a node must equal the amount of current leaving the node. (11.2)
- The voltage supplied to a parallel circuit is the same voltage supplied to all branches in the circuit. (11.2)
- The total resistance in a parallel circuit is always lower than the lowest resistance in the circuit. (11.2)
- The power consumption of a parallel circuit increases as the circuit resistance decreases. (11.3)
- Multiple switches are sometimes connected in parallel with each other. (11.4)

Review Questions

Use the information in the chapter to answer the following questions.

Measuring Your Potential

1. In a parallel circuit, the voltage measured across any branch will be _____. (11.1)
   A. equal to the voltage measured across the largest resistance
   B. equal to the voltage supplied across the entire circuit
   C. equal to the voltage measured across the lowest branch resistance
   D. always less than resistance value resistance
2. If four resistances with values of 5 Ω, 10 Ω, 20 Ω, and 20 Ω are wired in parallel with each other, the total resistance will be equal to _____. (11.2)
   A. 2.0 Ω
   B. 2.5 Ω
   C. 3.0 Ω
   D. 4.0 Ω
3. If four 20 Ω resistances are wired in parallel with each other, the total resistance will be equal to _____. (11.2)
   A. 5 Ω
   B. 10 Ω
   C. 15 Ω
   D. 20 Ω
4. Two 100 Ω resistances wired in parallel with each other will have the same resistance as two _____ with each other. (11.2)
   A. 10 Ω resistances wired in series
   B. 15 Ω resistances wired in parallel
   C. 25 Ω resistances wired in series
   D. 40 Ω resistances wired in parallel
5. A 60-V parallel circuit with one 20-Ω load and one 5-Ω load will operate with a total resistance of _____. (11.2)
   A. 15 Ω and a current of 4 A
   B. 15 Ω and a current of 15 A
   C. 4 Ω and a current of 4 A
   D. 4 Ω and a current of 15 A
6. What is the power consumption of a 50-V parallel circuit with one 20-Ω load and one 5-Ω load? (11.3)
   A. 200 W
   B. 475 W
   C. 625 W
   D. 850 W
7. In a parallel circuit with two switches wired in parallel with each other, power will not pass to the load when _____. (11.4)
   A. the switch closest to the power supply is closed
   B. the switch closest to the load is closed
   C. both switches are in the open position
   D. both switches are in the closed position

Plug-In

1. Determine the total resistance for the following circuit. (11.2)

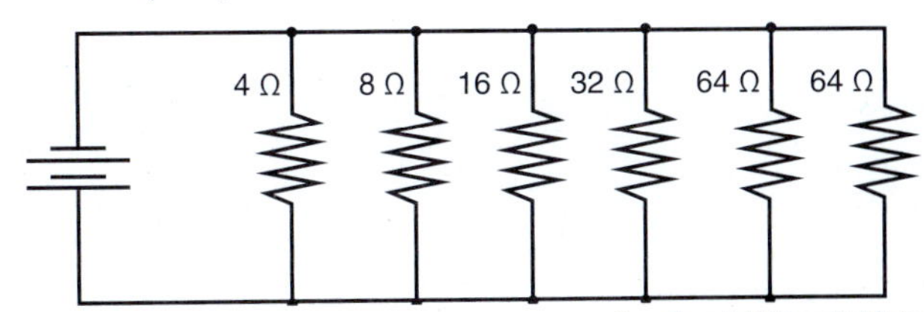

Goodheart-Willcox Publisher

   A. 2 Ω
   B. 2.5 Ω
   C. 7 Ω
   D. 8.5 Ω
2. The power consumption of a 100-V parallel circuit with one 10-Ω load and one 40-Ω load is the same as the power consumption of a 50-V circuit with two _____ loads wired in parallel with each other. (11.3)
   A. 4-Ω
   B. 8-Ω
   C. 12-Ω
   D. 15-Ω

Refer to the following circuit to answer questions 3 through 5.

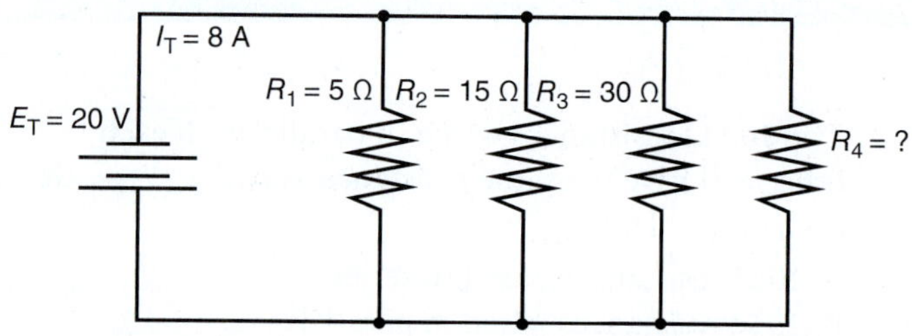

Goodheart-Willcox Publisher

3. The resistance of R_4 is _____. (11.2)
   A. 5 Ω
   B. 6 Ω
   C. 8 Ω
   D. 10 Ω
4. The I through R_1 equals _____. (11.2)
   A. 3 A
   B. 4 A
   C. 6 A
   D. 8 A
5. The total power of the circuit is _____. (11.3)
   A. 160 W
   B. 180 W
   C. 200 W
   D. 208 W

Making the Connection

1. Two lighting fixtures are wired in parallel with each other and each has its own switch. Both lights are illuminated. If one fixture is switched off the brightness of the other fixture will increase. Do you agree or disagree with this statement? Be able to provide support for your answer. (11.4)
2. As the diameter of a conductor increases, the resistance per unit foot of the conductor will decrease. Do you agree or disagree with this statement? Be able to provide support for your answer. (11.2)

CHAPTER

12 Combination Circuits

Quality Stock Arts/Shutterstock.com

SPARKING DISCUSSION

Would you consider the hotel key card shown here as part of a combination circuit? Why or why not?

LEARNING OUTCOMES

After completing this chapter, you will be able to:

12.1 Define and identify a combination circuit.
12.2 Determine the order in which series and parallel circuit rules are applied to combination circuits.
12.3 Determine the power consumption of combination circuit components and total current and power.
12.4 Evaluate combination switch arrangements.

TECHNICAL TERMS

AND operation
combination circuit
control circuit design
logic theory
OR operation

Introduction

In previous chapters, the fundamental concepts of series and parallel circuits are discussed as well as the rules that apply to each. In many applications and installations, the typical circuits an electrician encounters are combination circuits, meaning the circuit consists of both series and parallel elements. For example, the circuit breaker and circuit conductor at the panelboard are series elements. This energized conductor splits into parallel branches to feed parallel-connected loads. The return conductors then leave the loads and recombine as a single-series element back to the panelboard.

This chapter discusses combination circuits, how to easily identify them, and how to evaluate them. As the content in this chapter builds on the concepts covered in Chapter 10, *Series Circuits* and Chapter 11, *Parallel Circuits,* it might be useful to refer to those chapters as needed.

12.1 Identifying Combination Circuits

Circuits that contain both series-connected as well as parallel-connected elements are commonly referred to as ***combination circuits***. Combination circuits are also referred to as series-parallel circuits. In a series circuit, there is only one possible path for electrons to follow, while in a parallel circuit, there are multiple paths for current. Combination circuits have both series-connected elements as well as parallel-connected elements. In the circuit shown in **Figure 12-1**, loads R_1 and R_2 are located between points A and B. For current to pass from point A to point B, current must pass through both R_1 and R_2, meaning that these two loads are wired in series with each other. Between points B and C, loads R_3 and R_4 are connected. For current to pass from point B to point C, current can pass through R_3 or R_4, meaning that these two circuit loads are wired in parallel with each other.

When looking at a simple wiring diagram, it is relatively easy to identify the series and parallel components by inspection. This is not the case in all circuits. Asking one basic question, and using the circuit diagram to answer it, can help determine which circuit components are wired in series with each other and which are wired in parallel. This question is, "Can power be supplied to any other circuit load without passing through this load?" If the answer to this question is "no," the load is wired in series with the power supply. If the answer is "yes," then ask "Which one(s)?" That answer indicates loads that are wired in parallel with each other. The answers to the question can be categorized as follows:

- If all answers are no, then the circuit is a series circuit.
- If all answers are yes, and each response to "Which one(s)?" includes all circuit loads, then the circuit is a parallel circuit. (Each load can bypass all of the other loads.)
- If some answers are no and some answers are yes, then the circuit is a combination circuit.

This process can be summarized by the flowchart provided in **Figure 12-2**. This flowchart will help you as you navigate through **Example 12-1**, **Example 12-2**, **Example 12-3**, and **Example 12-4**.

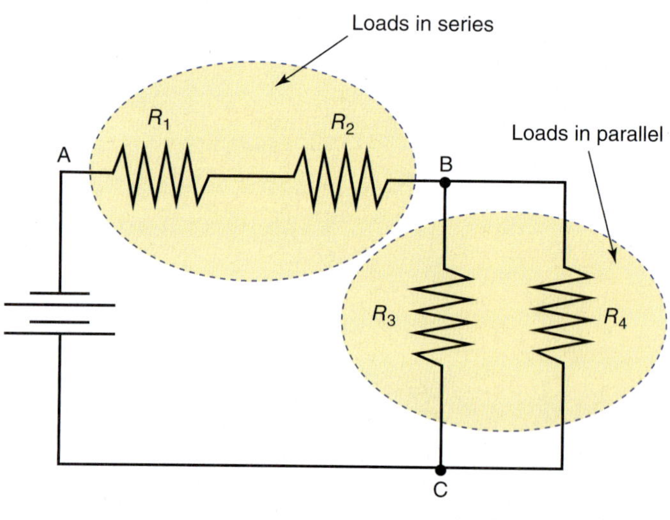

Figure 12-1. In this combination circuit, R_1 and R_2 are wired in series with each other, while R_3 and R_4 are wired in parallel with each other.

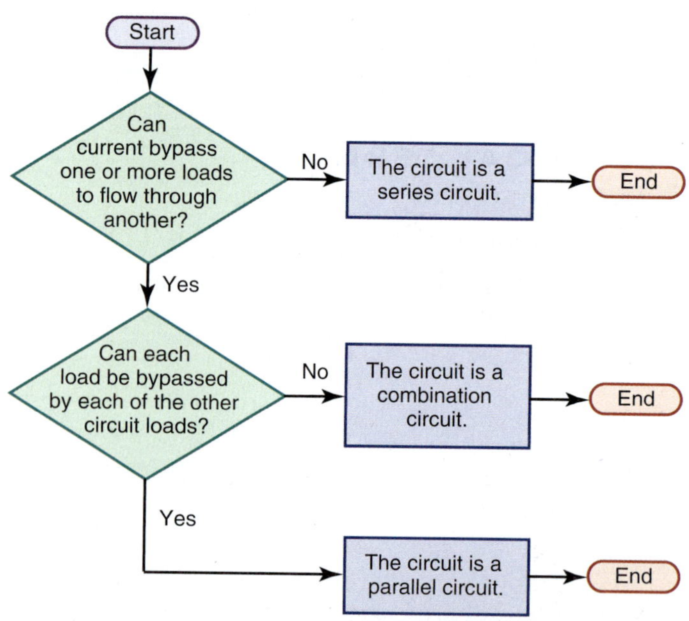

Figure 12-2. Using this flowchart can help you identify series, parallel, and combination circuits.

Example 12-1

Determine if the following circuit is a series circuit, a parallel circuit, or a combination circuit.

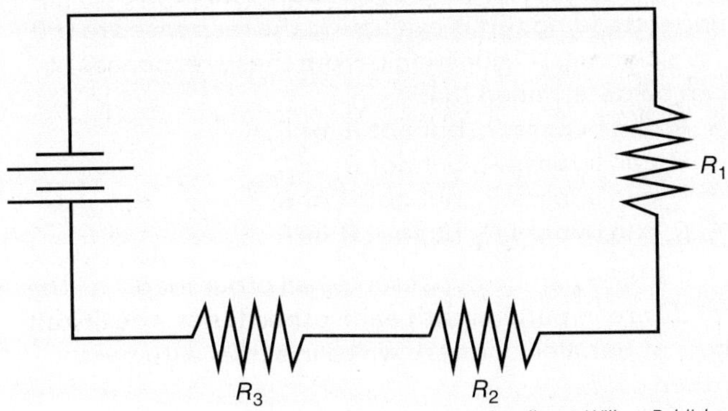

Goodheart-Willcox Publisher

✓ Solution

Asking if power can be supplied to any other circuit loads without passing through any other circuit loads yields the following results:

- Can power be supplied to any other circuit load without passing through R_1? *In this circuit, the answer is no.*
- Can power be supplied to any other circuit load without passing through R_2? *In this circuit, the answer is no.*
- Can power be supplied to any other circuit load without passing through R_3? *In this circuit, the answer is no.*

Since the answer to all three questions above is no, this is a series circuit.

Example 12-2

Determine if the following circuit is a series circuit, a parallel circuit, or a combination circuit.

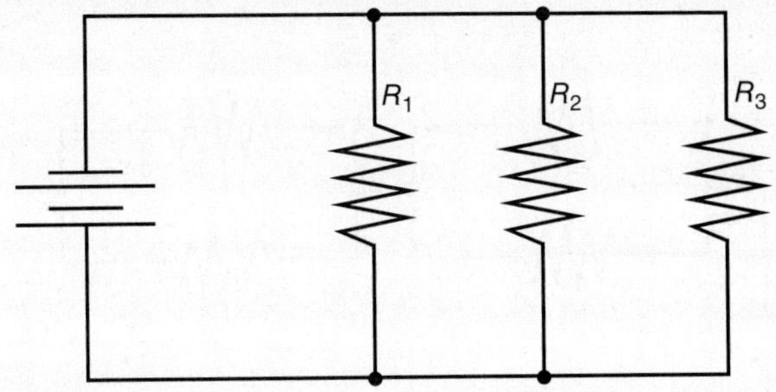

Goodheart-Willcox Publisher

✓ Solution

Asking if power can be supplied to any other circuit loads without passing through any other circuit loads yields the following results:

- Can power be supplied to any other circuit load without passing through R_1? *In this circuit, the answer is yes.*
- Which one(s)? R_2 *and* R_3
- Can power be supplied to any other circuit load without passing through R_2? *In this circuit, the answer is yes.*
- Which one(s)? R_1 *and* R_3
- Can power be supplied to any other circuit load without passing through R_3? *In this circuit, the answer is yes.*
- Which one(s)? R_1 *and* R_2

Since the answer to all three questions is yes, then this is a parallel circuit. Based on the answers to the which ones questions, all three of the loads are connected in parallel with each other. Although this might sound obvious, it is not always the case.

Example 12-3

Determine if the following circuit is a series circuit, a parallel circuit, or a combination circuit.

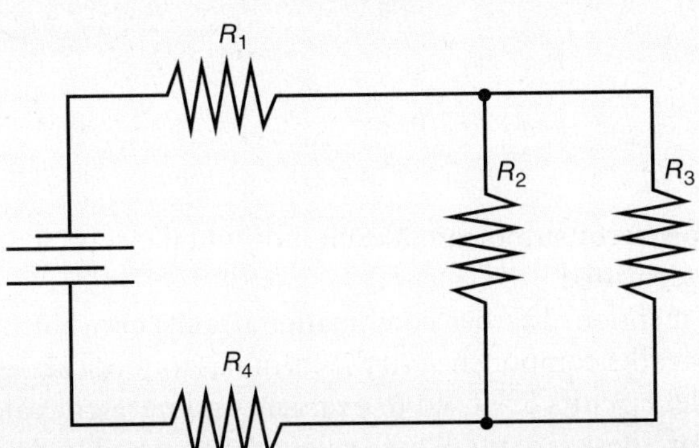

Goodheart-Willcox Publisher

✓ Solution

Asking if power can be supplied to any other circuit loads without passing through any other circuit loads yields the following results:

- Can power be supplied to any other circuit load without passing through R_1? *In this circuit, the answer is no.*
- Can power be supplied to any other circuit load without passing through R_2? *In this circuit, the answer is yes.*
- Can power be supplied to any other circuit load without passing through R_3? *In this circuit, the answer is yes.*
- Which one(s)? R_4
- Can power be supplied to any other circuit load without passing through R_4? *In this circuit, the answer is no.*

Since the answers to the questions referring to R_1 and R_4 are no, R_1 and R_4 are connected in series with each other as well as in series with the power supply. Since the answers to the questions referring to R_2 and R_3 are yes, these loads are connected in parallel. They are wired in parallel with each other since R_3 can bypass R_2 and R_2 can bypass R_3, as indicated by the responses to the questions. Since some answers are no and some answers are yes, then this is a combination circuit.

Example 12-4

Determine if the following circuit is a series circuit, a parallel circuit, or a combination circuit. Be careful, this one is tricky!

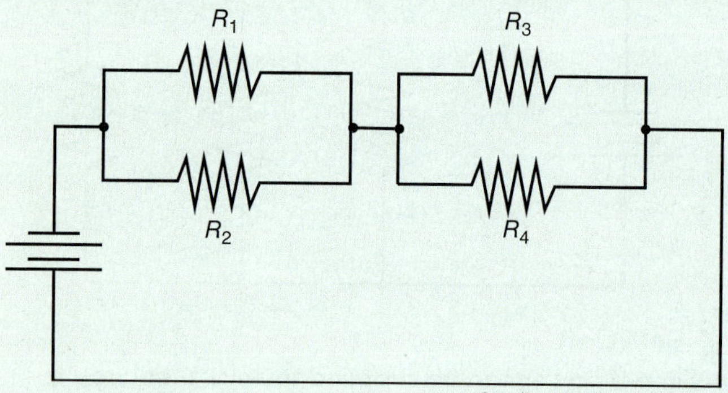

Goodheart-Willcox Publisher

✓ Solution

The following circuit is made up of two sets of parallel loads wired in series with each other. Both R_1 and R_2 are wired in parallel with each other, as are R_3 and R_4. However, these two parallel pairs are connected in series with each other. Asking if power can be supplied to any other circuit loads without passing through any other circuit loads yields the following results:

- Can power be supplied to any other circuit load without passing through R_1? *The answer is yes. Then ask yourself "which ones," and the answer is R_2.*
- Can power be supplied to any other circuit load without passing through R_2? *The answer is yes. Then ask yourself "which ones," and the answer is R_1.*
- Can power be supplied to any other circuit load without passing through R_3? *The answer is yes. Then ask yourself "which ones," and the answer is R_4.*
- Can power be supplied to any other circuit load without passing through R_4? *The answer is yes. Then ask yourself "which ones," and the answer is R_3.*

Since the answer to all three questions is yes, one might conclude that this circuit is a parallel circuit. However, it is a combination circuit. The key to understanding this is evaluating the responses to the "Which one(s)?" questions. From these responses, it can be determined that:

- R_1 can bypass R_2, but not R_3 or R_4.
- R_2 can bypass R_1, but not R_3 or R_4.
- R_3 can bypass R_4, but not R_1 or R_2.
- R_4 can bypass R_3, but *not* R_1 or R_2.

If a load can be bypassed by all other loads, all the loads are in parallel with each other. If not, the linked sets of parallel loads are wired in series with each other.

Now It's Your Turn 12-4

Evaluate the following circuit and determine if the circuit is a series circuit, a parallel circuit, or a combination circuit.

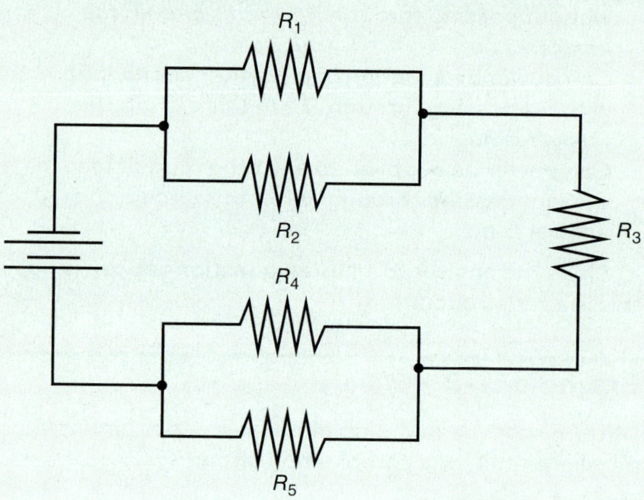

Goodheart-Willcox Publisher

12.2 **Evaluating Combination Circuits**

Calculations within a combination circuit can be made by continually evaluating portions of the circuit that contain loads connected either in series or parallel, until only a single equivalent resistance remains. As a portion of the circuit is evaluated, it is replaced with the resulting equivalent resistance. Once the final equivalent circuit resistance has been determined, the total circuit current and power consumption can be calculated. From this point, the power consumption of each individual circuit load can be determined.

As mentioned earlier, combination circuits have both series and parallel components, so the formulas for both series and parallel circuits are used to evaluate and perform calculations on these circuits. These rules are discussed in detail in Chapter 10, *Series Circuits* and Chapter 11, *Parallel Circuits*. The rules for series and parallel circuits are summarized in **Figure 12-3**.

Example 12-5 and **Example 12-6** provide the steps to calculate circuit load and power consumption.

Example 12-5

Determine the current flow and power consumption of the following circuit.

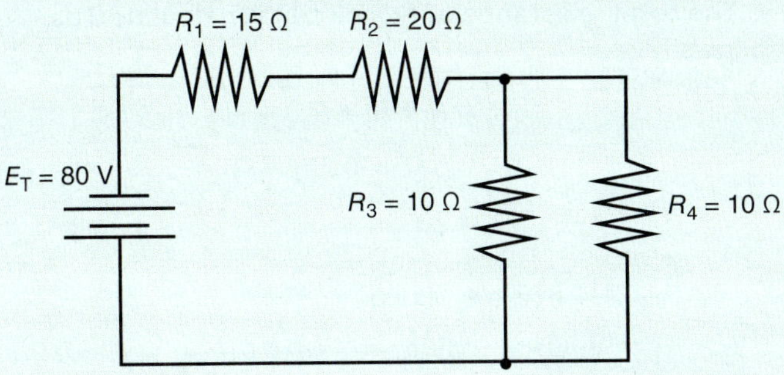

Goodheart-Willcox Publisher

✓ Solution

From an initial inspection of the circuit, loads R_1 and R_2 are wired in series with each other, and in series with these two loads is the parallel combination of loads R_3 and R_4.

The first step in simplifying this circuit is to evaluate the parallel combination of loads R_3 and R_4 and replace this combination with a single resistance. Since the value of both R_3 and R_4 is 10 Ω, the following formula can be used:

$$R_{3,4} = \frac{R}{N}$$

$$R_{3,4} = \frac{10\ \Omega}{2}$$

$$R_{3,4} = 5\ \Omega$$

The notation $R_{3,4}$ represents the equivalent resistance of R_3 and R_4. The parallel combination of loads R_3 and R_4 can, therefore, be replaced with a single 5 Ω load. The following circuit is now a simple series circuit.

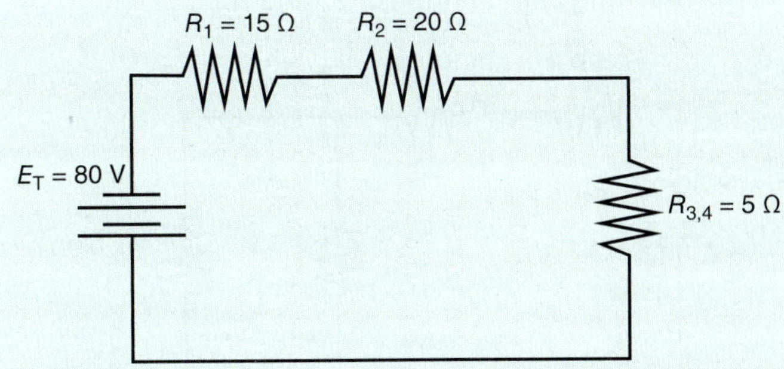

Goodheart-Willcox Publisher

Next, calculate the total resistance of this series circuit as follows:

$$R_T = R_1 + R_2 + R_{3,4}$$

$$R_T = 15\ \Omega + 20\ \Omega + 5\ \Omega$$

$$R_T = 40\ \Omega$$

Now the current flow through the circuit can be calculated by using Ohm's law:

$$I_T = \frac{E_T}{R_T}$$

$$I_T = \frac{80\ \text{V}}{40\ \Omega}$$

$$I_T = 2\ \text{A}$$

Finally, the power consumption of this circuit can be calculated using the basic form of the power formula:

$$P = I \times E$$

$$P = 2\ \text{A} \times 80\ \text{V}$$

$$P = 160\ \text{W}$$

Example 12-6

Determine the total resistance of the following circuit.

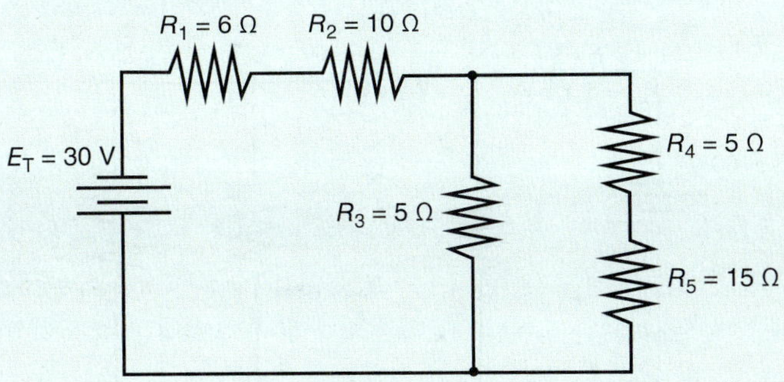

Goodheart-Willcox Publisher

✓ Solution

From an initial inspection of the circuit, this circuit closely resembles the one found in **Example 12-5**. Loads R_1 and R_2 are wired in series with each other and in series with a two-branch parallel combination comprising loads R_3, R_4, and R_5. However, in this parallel combination, series-connected loads R_4 and R_5 are nested within the parallel circuit.

The first step in simplifying this circuit is to evaluate the series combination of loads R_4 and R_5 and replace this combination with a single resistance. Since the equivalent value of resistances connected in series is the sum of the resistance values, loads R_4 and R_5 can be

Continued

replaced with a single 20 Ω resistance, where $R_{4,5}$ refers to the resulting combination of R_4 and R_5.

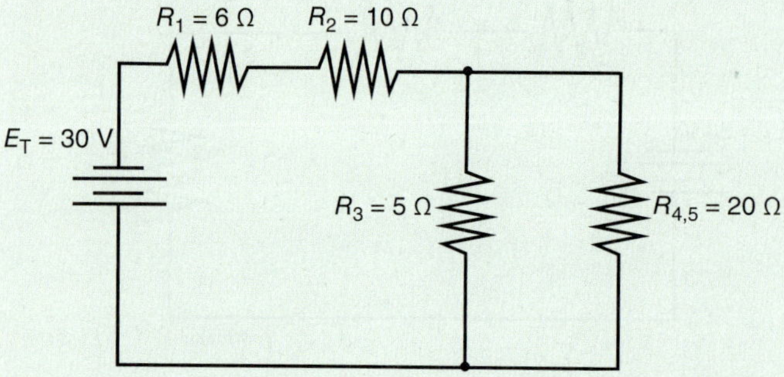

Goodheart-Willcox Publisher

The next step in simplifying this circuit is to combine the two parallel resistances, R_3 and $R_{4,5}$. Use the product over sum formula for combining two resistances with different values as follows:

$$R_{3,4,5} = \frac{R_3 \times R_{4,5}}{R_3 + R_{4,5}}$$

$$R_{3,4,5} = \frac{5\ \Omega \times 20\ \Omega}{5\ \Omega + 20\ \Omega}$$

$$R_{3,4,5} = \frac{100\ \Omega^2}{25\ \Omega}$$

$$R_{3,4,5} = 4\ \Omega$$

The combination of R_3 and $R_{4,5}$ can, therefore, be replaced with $R_{3,4,5}$, which has a value of 4 Ω. The circuit has now become a series circuit with three resistances with values of 6 Ω, 10 Ω, and 4 Ω.

The total resistance of this circuit is calculated as follows:

$$R_T = R_1 + R_2 + R_{3,4,5}$$

$$R_T = 6\ \Omega + 10\ \Omega + 4\ \Omega$$

$$R_T = 20\ \Omega$$

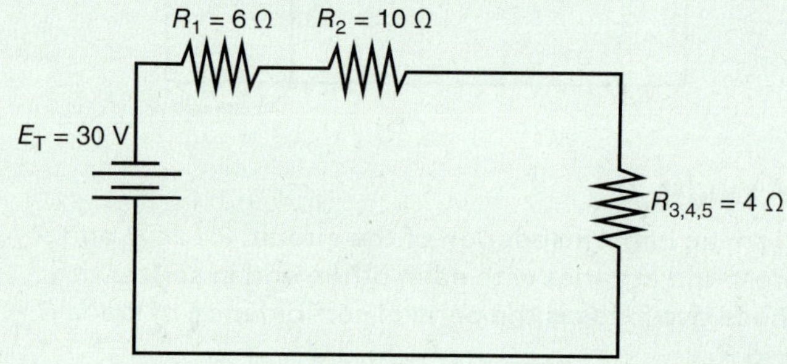

Goodheart-Willcox Publisher

Now It's Your Turn 12-6

Determine the total circuit current and total power consumption for the circuit shown in **Example 12-6**.

Series and Parallel Circuit Rules

Series Circuit Rules

$$I_T = I_1 = I_2 = I_3 = \ldots = I_N$$

$$R_T = R_1 + R_2 + R_3 + \ldots + R_N$$

$$E_T = E_1 + E_2 + E_3 + \ldots + E_N$$

Parallel Circuit Rules

$$E_T = E_1 = E_2 = E_3 = \ldots = E_N$$

$$I_T = I_1 + I_2 + I_3 + \ldots + I_N$$

Multiple Resistances of Different Values

$$\frac{1}{R_T} = \frac{1}{R_1} + \frac{1}{R_2} + \frac{1}{R_3} + \ldots + \frac{1}{R_N}$$

Two Resistances of Different Values

$$R_T = R_1 \times \frac{R_2}{R_1} + R_2$$

Multiple Resistances of the Same Value

$$R_T = \frac{R}{N}$$

Goodheart-Willcox Publisher

Figure 12-3. Rules for series and parallel circuits are used when evaluating combination circuits.

The same strategies used in the previous examples can be used no matter how complicated a circuit may appear. Consider the circuit shown in **Figure 12-4**, which is more intricate than those previously discussed. Before reading any further, identify what you would consider the first step in simplifying this circuit.

If your first step would be to combine loads R_4 and R_5, you are correct. R_1 and R_2 might look like they are wired in series

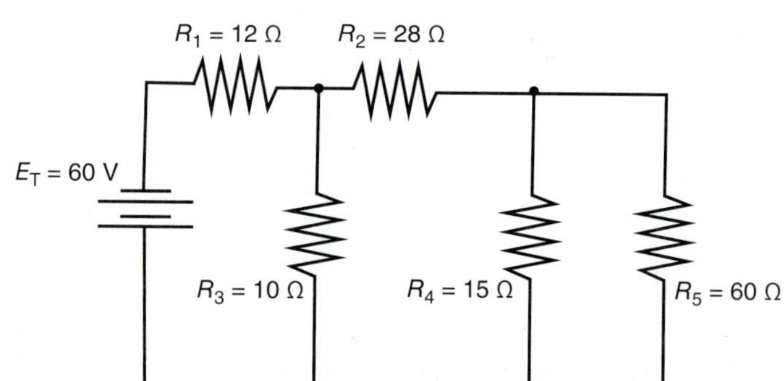

Goodheart-Willcox Publisher

Figure 12-4. How would you simplify this circuit? Referring back to **Figure 12-2** helps answer the question, "What is the first step in simplifying a circuit?"

with each other, but R_2, along with R_3, is actually part of a parallel portion of the circuit. R_3 and R_4 might look like they are wired in parallel with each other and can be combined, but R_2 is in between them, complicating the issue.

Looking at the parallel combination of R_4 and R_5, the equivalent resistance can be calculated as follows:

$$R_{4,5} = \frac{R_4 \times R_5}{R_4 + R_5}$$

$$R_{4,5} = \frac{15\ \Omega \times 60\ \Omega}{15\ \Omega + 60\ \Omega}$$

$$R_{4,5} = \frac{900\ \Omega^2}{75\ \Omega}$$

$$R_{4,5} = 12\ \Omega$$

These two resistances can be replaced by a single 12-Ω load, $R_{4,5}$, **Figure 12-5**. From this point, the two series-connected resistances, R_2 and $R_{4,5}$, can be represented by a single resistance, $R_{2,4,5}$, which has a value of 40 Ω, **Figure 12-6**. The 10-Ω load and the 40-Ω load are now connected in parallel and can be combined using the product over sum formula as follows:

$$R_{2,3,4,5} = \frac{R_3 \times R_{2,4,5}}{R_3 + R_{2,4,5}}$$

$$R_{2,3,4,5} = \frac{10\ \Omega \times 40\ \Omega}{10\ \Omega + 40\ \Omega}$$

$$R_{2,3,4,5} = \frac{400\ \Omega^2}{50\ \Omega}$$

$$R_{2,3,4,5} = 8\ \Omega$$

The combination of the four resistances, R_2, R_3, R_4, and R_5, results in an equivalent resistance of 8 Ω, **Figure 12-7**. The total circuit resistance for this equivalent series circuit is 20 Ω (12 Ω + 8 Ω). The total circuit current can now be calculated as follows:

$$I_T = \frac{E_T}{R_T}$$

$$I_T = \frac{60\ \text{V}}{20\ \Omega}$$

$$I_T = 3\ \text{A}$$

Next, the power consumption of this circuit is calculated by using the basic form of the power formula:

$$P = I \times E$$

$$P = 3\ \text{A} \times 60\ \text{V}$$

$$P = 180\ \text{W}$$

Another practice problem is provided in **Example 12-7**.

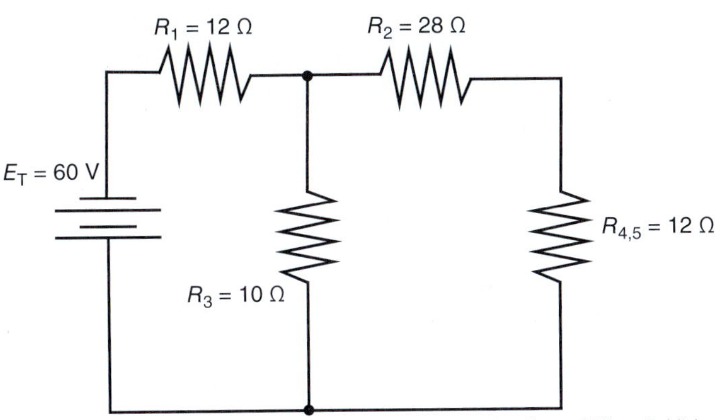

Goodheart-Willcox Publisher

Figure 12-5. Loads R_4 and R_5 are replaced with a single resistance, $R_{4,5}$.

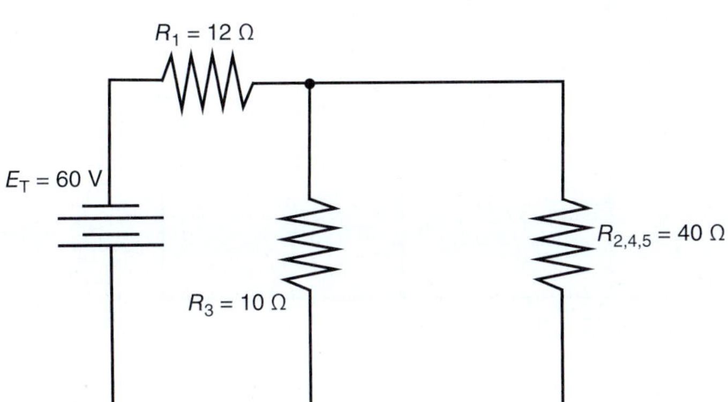

Goodheart-Willcox Publisher

Figure 12-6. Loads R_2 and $R_{4,5}$ are replaced with a single resistance, $R_{2,4,5}$.

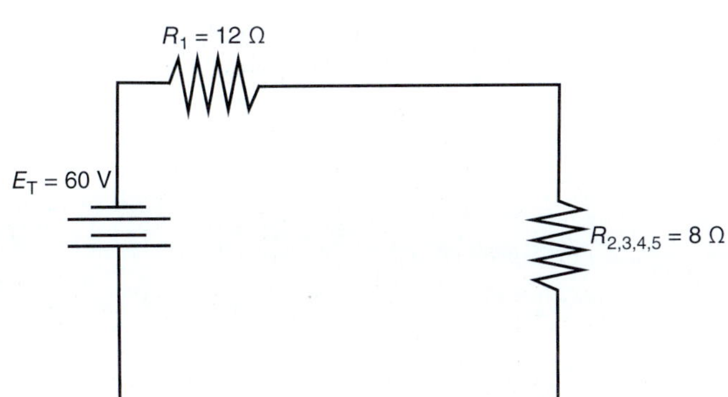

Goodheart-Willcox Publisher

Figure 12-7. Loads R_3 and $R_{2,4,5}$ are replaced with a single resistance, $R_{2,3,4,5}$.

Example 12-7

Determine the current flow and power consumption of the following circuit.

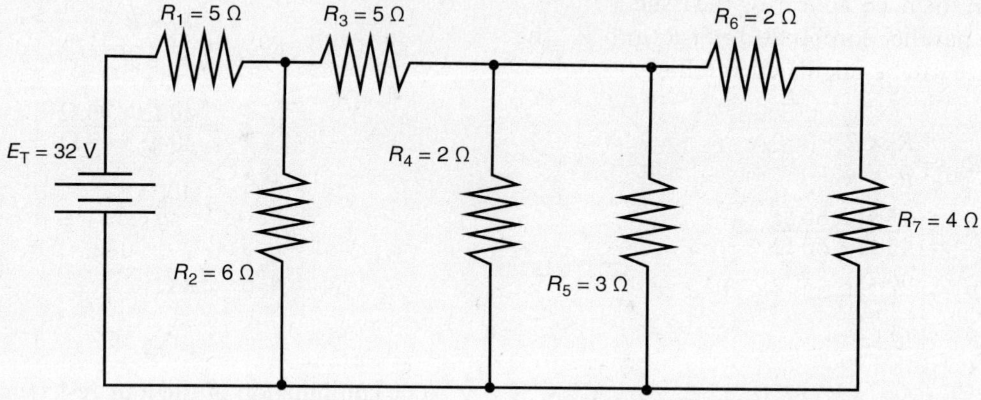

Goodheart-Willcox Publisher

✓ Solution

The first step in simplifying this circuit is to evaluate the series combination of loads R_6 and R_7 and replace this combination with a single resistance. Since the equivalent value of resistances connected in series is the sum of the resistance values, loads R_6 and R_7 can be replaced with a single 6 Ω resistance.

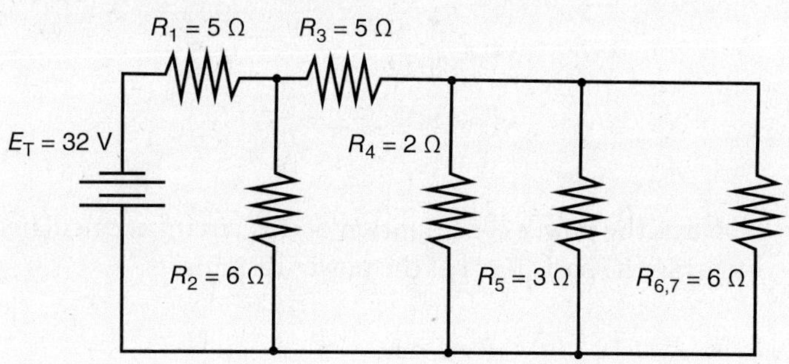

Goodheart-Willcox Publisher

Next, use the resistance formula for multiple resistances with different values to combine the three parallel resistances (R_4, R_5, and $R_{6,7}$) as follows:

$$\frac{1}{R_{4,5,6,7}} = \frac{1}{R_4} + \frac{1}{R_5} + \frac{1}{R_{6,7}}$$

$$\frac{1}{R_{4,5,6,7}} = \frac{1}{2\ \Omega} + \frac{1}{3\ \Omega} + \frac{1}{6\ \Omega}$$

In this example we continue solving the equation by converting fractions to have a common denominator, but this equation can also be solved using a scientific calculator as explained in Chapter 11, *Parallel Circuits*.

$$\frac{1}{R_{4,5,6,7}} = \frac{3}{3} \times \frac{1}{2\ \Omega} + \frac{2}{2} \times \frac{1}{3\ \Omega} + \frac{1}{6\ \Omega}$$

$$\frac{1}{R_{4,5,6,7}} = \frac{3}{6\ \Omega} + \frac{2}{6\ \Omega} + \frac{1}{6\ \Omega}$$

$$\frac{1}{R_{4,5,6,7}} = \frac{3+2+1}{6\ \Omega}$$

$$\frac{1}{R_{4,5,6,7}} = \frac{6}{6\ \Omega}$$

$$\frac{1}{R_{4,5,6,7}} = \frac{1}{1\ \Omega}$$

$$R_{4,5,6,7} = 1\ \Omega$$

As shown in the following circuit, the combination of R_4, R_5, and $R_{6,7}$ is replaced with $R_{4,5,6,7}$, which has a value of 1 Ω.

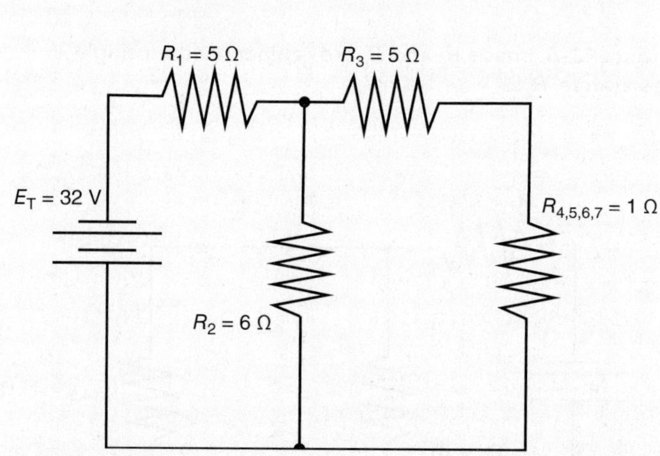

Goodheart-Willcox Publisher

By combining R_3 with $R_{4,5,6,7}$, the combined load will now have a value of 6 Ω, as shown.

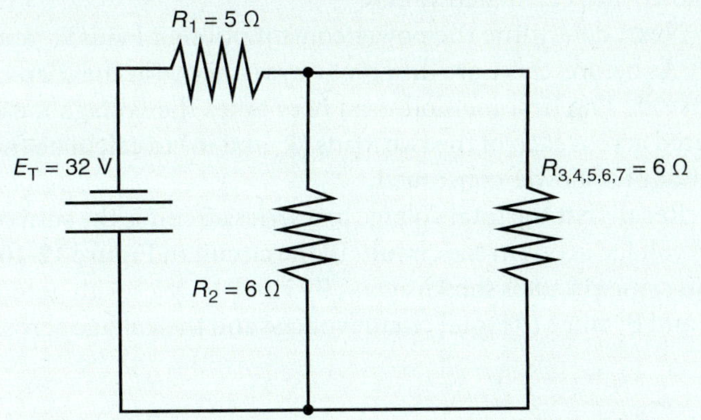

Goodheart-Willcox Publisher

Next, combine the two parallel, 6 Ω loads, which results in an equivalent resistance of 3 Ω. As shown in the following simplified circuit, this now becomes a series circuit with two resistances with values of 5 Ω and 3 Ω.

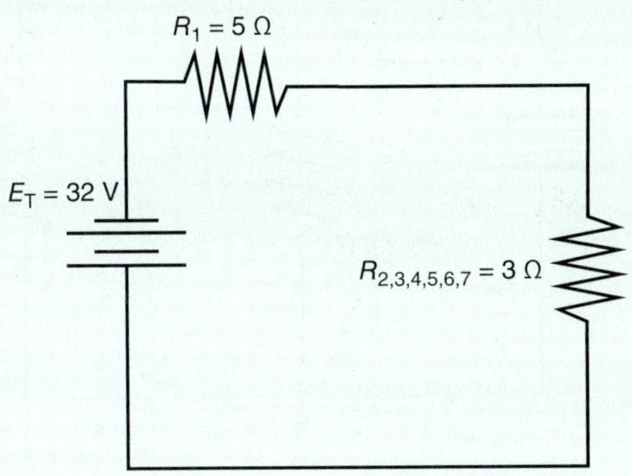

Goodheart-Willcox Publisher

Combine these two resistances to obtain the total circuit resistance (R_T) of 8 Ω.

The total circuit current can now be calculated as follows:

$$I_T = \frac{E_T}{R_T}$$

$$I_T = \frac{32\ \text{V}}{8\ \Omega}$$

$$I_T = 4\ \text{A}$$

The power consumption of this circuit can now be calculated as follows:

$$P = I \times E$$

$$P = 4\ \text{A} \times 32\ \text{V}$$

$$P = 128\ \text{W}$$

Now It's Your Turn 12-7

Evaluate the following circuit and determine the circuit's total current flow and the power consumption.

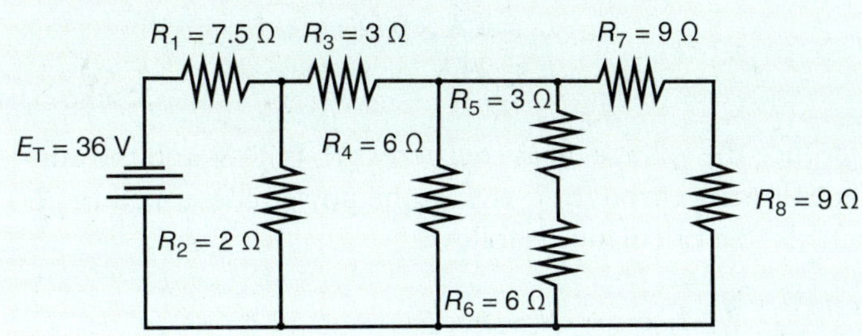

Goodheart-Willcox Publisher

12.3 Power Consumption of Individual Components

Now that you are familiar with calculating complex circuits, you are ready to calculate the power consumption of individual circuit loads. Although these calculations are not typically performed by an electrician in the field, evaluating circuits at a deeper level will help improve the level of understanding of how electric circuits function.

Chapter 10, *Series Circuits* and Chapter 11, *Parallel Circuits*, discuss how to calculate the power consumption of various series and parallel circuits as well as the power consumption of each of the loads in the circuit. The same concepts apply to combination circuits; however, the rules for both series and parallel circuits will be used as the need for each applies.

Consider the simple combination circuit shown in **Figure 12-8**. This circuit has a 6-Ω load, R_1, wired in series with the parallel combination of R_2 and R_3, each of which has a resistance of 8 Ω. Using the methods discussed earlier in this chapter, R_T is calculated as 10 Ω and I_T is calculated as 4 A.

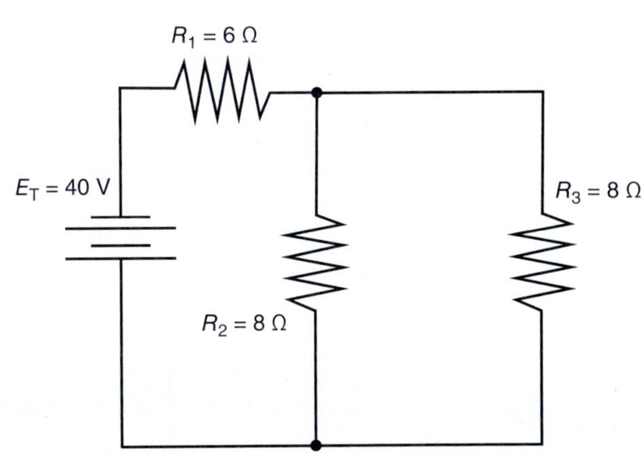

Goodheart-Willcox Publisher

Figure 12-8. A simple combination circuit. With total current, total voltage, and after calculating total resistance, total power can be calculated.

With the total circuit current, voltage, and resistance known, the total power consumption of the circuit can be calculated as follows:

$$P_T = I_T \times E_T$$
$$P_T = 4\text{ A} \times 40\text{ V}$$
$$P_T = 160\text{ W}$$

Now that the total circuit current flowing through R_1 and the resistance value is known, the power consumed by R_1 can be calculated as follows:

$$P_1 = (I_2)^2 \times R_1$$
$$P_1 = (4\text{ A})^2 \times 6\ \Omega$$
$$P_1 = 16 \times 6$$
$$P_1 = 96\text{ W}$$

The power consumed by R_1 can also be calculated by using the formula $P = I \times E$, but the voltage across R_1 would need to be calculated first. Since the current flowing through R_1 and the resistance value of R_1 is known, the voltage form of Ohm's law can be used to calculate the voltage across R_1 as follows:

$$E_1 = I_1 \times R_1$$
$$E_1 = 4\text{ A} \times 6\ \Omega$$
$$E_1 = 24\text{ V}$$

Since the voltage measured across R_1 is 24 V and the current flowing through R_1 is 4 A, the power consumed by R_1 can also be calculated as follows:

$$P_1 = I_1 \times E_1$$
$$P_1 = 4\text{ A} \times 24\text{ V}$$
$$P_1 = 96\text{ W}$$

As you can see, there are several different ways to arrive at the correct answer. **Figure 12-9** shows the circuit with the known and calculated values.

Next, determine the power consumption of loads R_2 and R_3. As before, there are different ways to arrive at the correct answer. The first method used here takes the voltage measured across each of the two loads (R_2 and R_3) to calculate the amount of power consumed.

Recall that the total voltage in a series circuit is the sum of all voltage drops in the circuit. In the circuit in **Figure 12-10**, this same circuit is shown again, R_1 wired in series with points A and B. Since the total circuit voltage and the voltage across

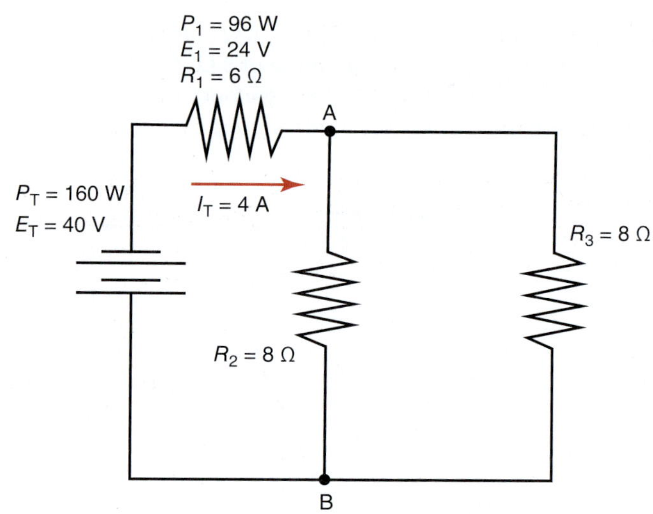

Goodheart-Willcox Publisher

Figure 12-9. Circuit from **Figure 12-8** with power and current data included.

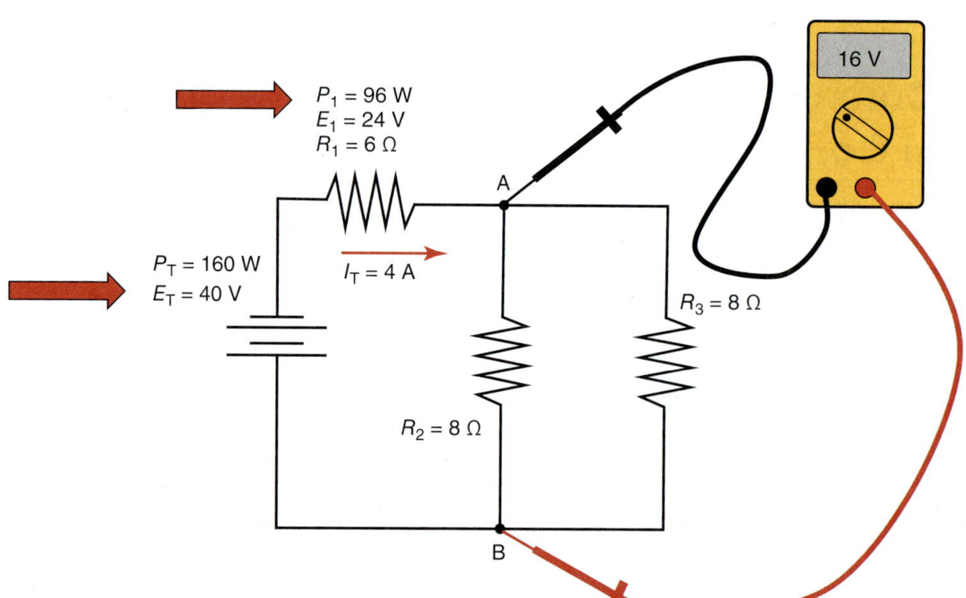

Goodheart-Willcox Publisher

Figure 12-10. A voltage of 16 V is measured between points A and B.

R_1 are known, the measured voltage between reference points A and B can be calculated as follows:

$$40\ \text{V} - 24\ \text{V} = 16\ \text{V}$$

Power consumed by R_2 can be calculated as follows:

$$P_2 = \frac{(E_2)^2}{R_2}$$

$$P_2 = \frac{(16\ \text{V})^2}{8\ \Omega}$$

$$P_2 = \frac{256\ \text{V}}{8\ \Omega}$$

$$P_2 = 32\ \text{W}$$

Since the resistance of R_3 is the same as the resistance of R_2 and the voltage across R_3 is the same as the voltage across R_2, the power consumed by R_3 is also 32 W.

The power consumed by R_2 and R_3 can also be calculated by using the amount of current flowing through each load. Since 16 V is measured across R_2 and R_3, the current flowing through both R_2 and R_3 is 2 A (16 V divided by 8 Ω). The power consumed by R_2 (and thus, R_3) can then be calculated as follows:

$$P_2 = (I_2)^2 \times R_2$$

$$P_2 = (2\ \text{A})^2 \times 8\ \Omega$$

$$P_2 = 4\ \text{A} \times 8\ \Omega$$

$$P_2 = 32\ \text{W}$$

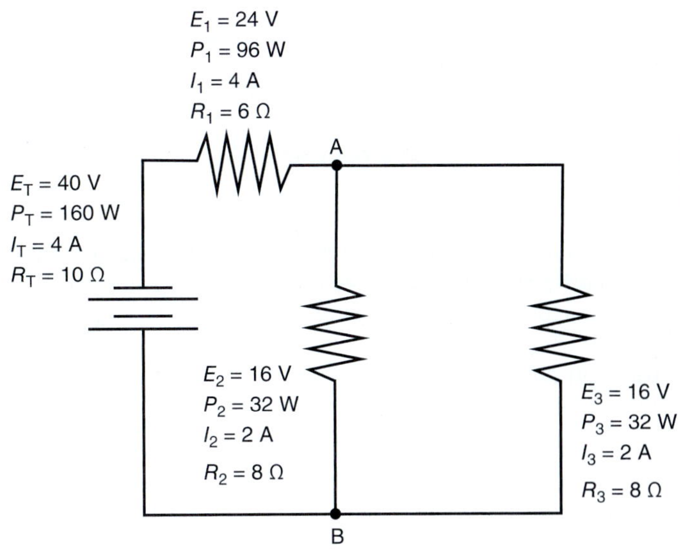

Goodheart-Willcox Publisher

Figure 12-11. Circuit from **Figure 12-8** with all power, current, voltage, and resistance data.

The current flow through R_2 and R_3 (each) can also be determined by evaluating the inflows and outflows at node A. With 4 A of current flowing into node A, 4 A must flow out. Since the values of R_2 and R_3 are the same, the current flow through each of these loads must be the same. With 4 A of current dividing equally between R_2 and R_3, the current flow through each is 2 A. A summary of the circuit values is provided in **Figure 12-11**. **Example 12-8** provides another circuit to calculate power consumption using this method.

Example 12-8

Determine the power consumption of each load in the following circuit.

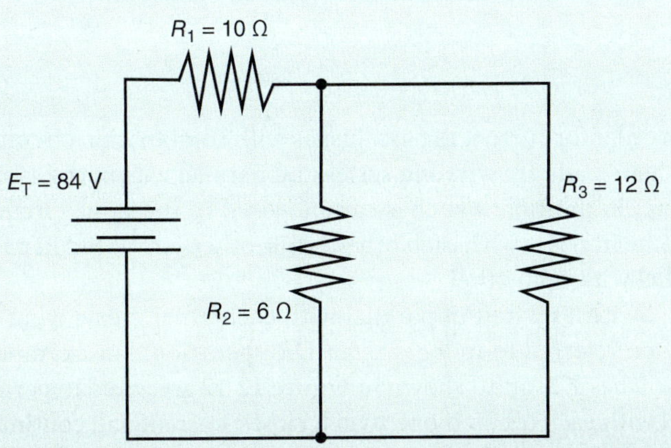

Goodheart-Willcox Publisher

✓ Solution

The first step in evaluating this circuit is to combine loads R_2 and R_3. Looking at the parallel combination of R_2 and R_3, the equivalent resistance is calculated by using the product over sum formula for resistance as follows:

$$R_{2,3} = \frac{R_2 \times R_3}{R_2 + R_3}$$

$$R_{2,3} = \frac{6\ \Omega \times 12\ \Omega}{6\ \Omega + 12\ \Omega}$$

$$R_{2,3} = \frac{72\ \Omega^2}{18\ \Omega}$$

$$R_{2,3} = 4\ \Omega$$

R_2 and R_3 are now replaced by a single resistance, $R_{2,3}$, which has a value of 4 Ω. The total circuit resistance (R_T) is 14 Ω (10 Ω + 4 Ω).

With this equivalent circuit, the total circuit current can be calculated as follows:

$$I_T = \frac{E_T}{R_T}$$

$$I_T = \frac{84\ \text{V}}{14\ \Omega}$$

$$I_T = 6\ \text{A}$$

Continued

Next, the total power consumption of the circuit can be calculated as follows:

$$P_T = I_T \times E_T$$

$$P_T = 6\ A \times 84\ V$$

$$P_T = 504\ W$$

Since R_1 is in series with the power supply, the current flowing through R_1 is equal to the total circuit current. The measured voltage across R_1 is 60 V ($E_1 = 6\ A \times 10\ \Omega$).

Knowing the resistance of R_1, the current flowing through the circuit, and the voltage measured across it, the power consumed by R_1 can now be calculated:

$$P_1 = I_T \times E_1$$

$$P_1 = 6\ A \times 60\ V$$

$$P_1 = 360\ W$$

The voltage applied to each of the two parallel loads (R_2 and R_3) can be found by subtracting the measured voltage across R_1 from the total voltage supplied to the circuit. Thus, E_2 and E_3 are 24 V (84 V – 60 V).

For R_2, supplying 24 V across the 6 Ω load, power consumption can be calculated as follows:

$$P_2 = \frac{(E_2)^2}{R_2}$$

$$P_2 = \frac{(24\ V)^2}{6\ \Omega}$$

$$P_2 = \frac{576\ V}{6\ \Omega}$$

$$P_2 = 96\ W$$

For R_3, supplying 24 V across the 12 Ω load, power consumption is then calculated as follows:

$$P_3 = \frac{(E_3)^2}{R_3}$$

$$P_3 = \frac{(24\ V)^2}{12\ \Omega}$$

$$P_3 = \frac{576\ V}{12\ \Omega}$$

$$P_3 = 48\ W$$

To check your calculations, the power consumption of all loads should equal the total power consumption of the circuit:

$$P_T = P_1 + P_2 + P_3$$

$$504\ W = 360\ W + 96\ W + 48\ W$$

$$504\ W = 504\ W$$

Now It's Your Turn 12-8

Determine the total power consumption as well as that of each load in the following circuit. Then, verify your answers by adding the individual power consumptions together and comparing this sum to the total power consumption of the circuit.

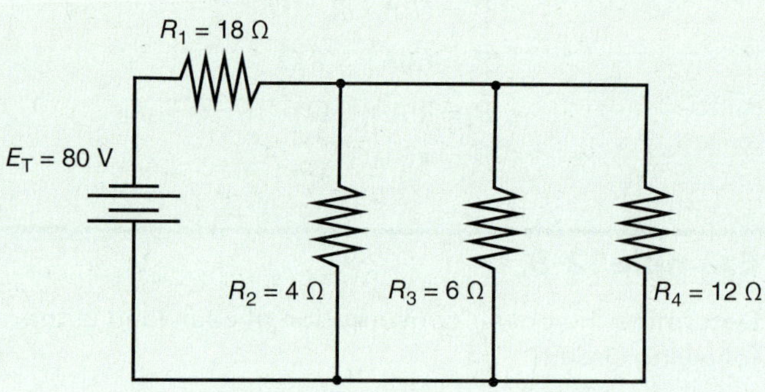

Goodheart-Willcox Publisher

12.4 Combination Switch Arrangements

Just as there are rules regarding the manipulation of series, parallel, and combination circuits, there are also rules that apply to switches connected in series and parallel. Although there are no mathematical relationships or formulas, there are rules of logic that apply to switches. ***Logic theory*** refers to a set of basic operations (such as AND and OR) that can be used to manipulate or evaluate binary elements, which have two positions, such as ON/OFF switches. These rules make up the fundamentals of computer programs and algorithms that control the operation of most systems, appliances, and devices.

When switches are wired in series with each other, both switches must be in the closed position in order for current to pass. When switches are wired in parallel with each other, any one or more of the switches must be in the closed position in order for current to pass. Just as with combination circuits, where loads are wired in series and parallel within the same circuit, multiple switches can be wired in the same circuit, some in series with each other, while others are wired in parallel with each other.

Switches wired in parallel with each other create what is often referred to in logic as an ***OR operation***. For example, switches 1, 2, and 3 shown in **Figure 12-12** are wired in parallel with each other. If one switch closes, current will continue to flow through that closed switch to the load and back to the power source. The OR operation states that switch 1 OR switch 2 OR switch 3 will complete the circuit, energizing the load.

Switches wired in series with each other create what is often referred to as an ***AND operation***. In **Figure 12-13**, switches 1, 2, and 3 are all wired in series with each other. If any one of the switches opens, current will not flow through the circuit. The only condition that will complete the circuit

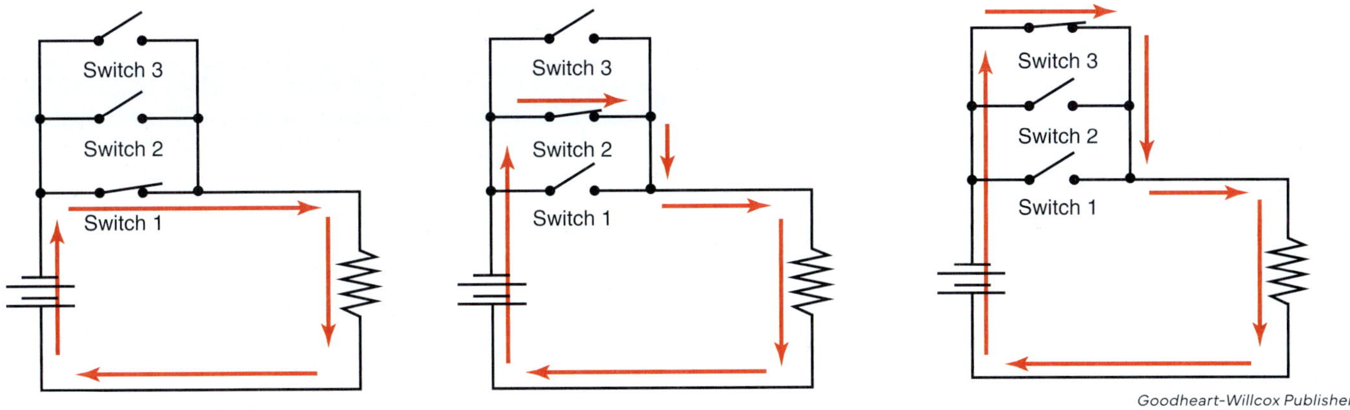

Figure 12-12. When switches are connected in parallel, closing any one switch will complete the circuit and energize the load.

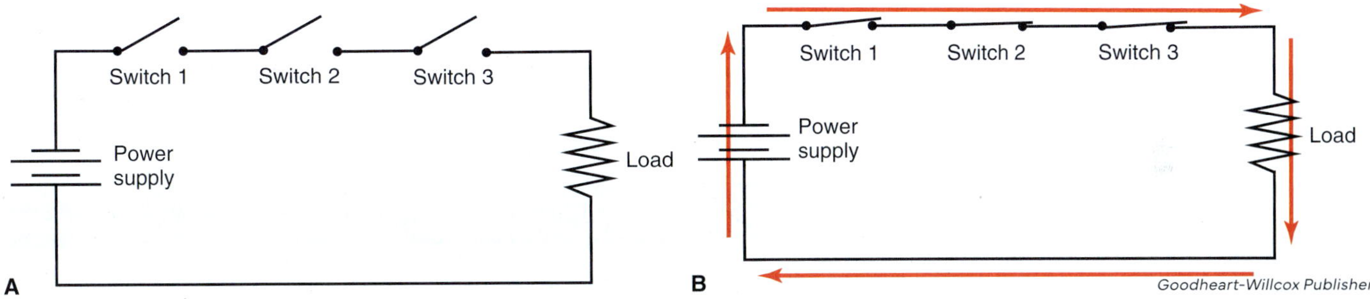

Figure 12-13. A circuit with three series-connected switches. A—When all series-connected switches are open, current will not flow through the circuit. B—All series-connected switches must be closed for current to flow through the circuit.

is when all series-connected switches are closed. The AND operation states that switch 1 AND switch 2 AND switch 3 will, when closed, complete the circuit, energizing the load. Remember, if any one of the series-connected switches opens, current will not flow and the load will not become energized.

Consider the combination switch arrangement shown in **Figure 12-14**. In this circuit, switches 1 and 2 are wired in parallel with each other, but both switches are wired in series with switch 3. From a logic standpoint, switch 3 must be closed as well as either switch 1 or switch 2 in order for power to pass to the load. If switch 3 opens for whatever reason, the load cannot become energized regardless of the positions of switches 1 and 2.

A summary of all possible switch combinations is shown in the chart. Notice that switches 1 and 2 are both in the open position on lines 1 and 2 so the load will not become energized even when switch 3 closes. On each of lines 3 through 8, either switch 1 or switch 2 is closed, or both are closed, so the position of switch 3 determines whether or not the load is energized.

Switches can be configured in many different ways to create the desired output of a system. Collectively, the process of designing a circuit to create the desired circuit output is called ***control circuit design***.

Shedding Light

In electrical construction, the majority of switches and their associated loads are connected in parallel with each other. Some common series-connected switches that serve an individual circuit include the following:

- Three-way and four-way switching, which is typically used for lights.
- Motor control installations where multiple emergency power off (EPO) switches are series connected. In this arrangement, if any one of the switches is open, the circuit is de-energized.
- Circuit breakers and disconnect switches are series-connected switching devices.

The switch arrangement shown in **Figure 12-15** incorporates two parallel switches, 1 and 2, that are wired in series with two additional switches, 3 and 4. In order for the circuit load to become energized, either one or both of switches 1 and 2 must be closed and, in addition, both switches 3 and 4 must be closed.

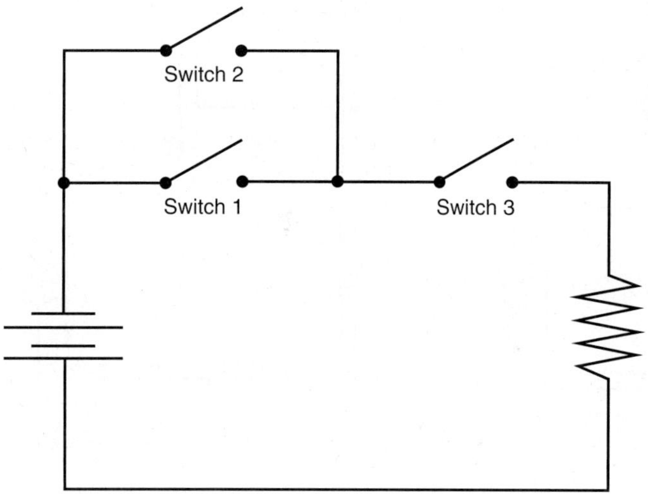

Possible Switch Combinations

Line	Switch 1	Switch 2	Switch 3	Is Load Energized?
1	Open	Open	Open	No
2	Open	Open	Closed	No
3	Open	Closed	Open	No
4	Open	Closed	Closed	Yes
5	Closed	Open	Open	No
6	Closed	Open	Closed	Yes
7	Closed	Closed	Open	No
8	Closed	Closed	Closed	Yes

Goodheart-Willcox Publisher

Figure 12-14. In this combination circuit, switch 1 and switch 2 are wired in parallel with each other, but are wired in series with switch 3.

Possible Switch Combinations

Line	Switch 1	Switch 2	Switch 3	Switch 4	Is Load Energized?
1	Open	Open	Open	Open	No
2	Open	Open	Open	Closed	No
3	Open	Open	Closed	Open	No
4	Open	Open	Closed	Closed	No
5	Open	Closed	Open	Open	No
6	Open	Closed	Open	Closed	No
7	Open	Closed	Closed	Open	No
8	Open	Closed	Closed	Closed	Yes
9	Closed	Open	Open	Open	No
10	Closed	Open	Open	Closed	No
11	Closed	Open	Closed	Open	No
12	Closed	Open	Closed	Closed	Yes
13	Closed	Closed	Open	Open	No
14	Closed	Closed	Open	Closed	No
15	Closed	Closed	Closed	Open	No
16	Closed	Closed	Closed	Closed	Yes

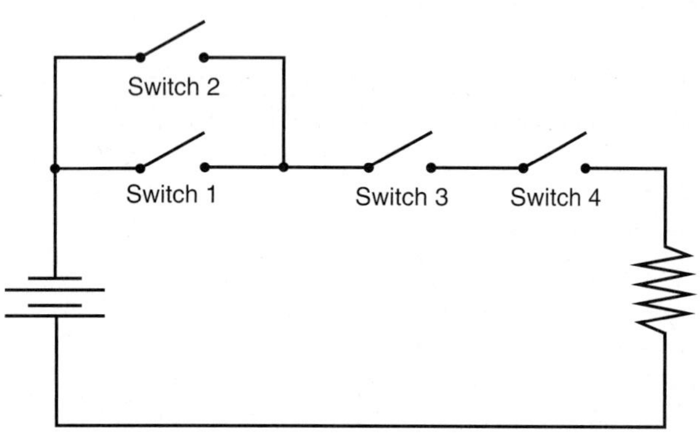

Goodheart-Willcox Publisher

Figure 12-15. In this combination circuit, switches 1 and 2 are parallel switches that are wired in series with switches 3 and 4.

If either switch 3 or 4 is in the open position, the load cannot become energized, regardless of the positions of switches 1 and 2. A chart showing all possible switch combinations, as well as the outcome, whether or not the load is energized, is provided in the table. Although there are 16 possible switch configurations, there are only four combinations where both switches 3 and 4 are closed. Of these four possible combinations that allow for load operation, only three will energize the load. This is because on line 4 neither switch 1 nor switch 2 is in the closed position, so no power can pass to the load.

CHAPTER 12 REVIEW

Summary

- A combination circuit includes both series- and parallel-connected components. (12.1)
- Both series circuit and parallel circuit rules must be used to evaluate combination circuits. (12.2)
- Grouped series or parallel loads are combined into single-valued resistances until only one single resistance remains. (12.2)
- The total circuit current and power are calculated once a single equivalent resistance remains. (12.3)
- Switches can be arranged in combination-type circuits, with both series- and parallel-connected devices. (12.4)
- An AND operation is often used to describe multiple switches connected in series with each other. (12.4)
- An OR operation is often used to describe multiple switches connected in parallel with each other. (12.4)

Review Questions

Use the information in the chapter to answer the following questions.

Measuring Your Potential

1. *True or False?* A combination circuit is made of series and parallel resistances. (12.1)
2. When calculating an unknown value in a combination circuit, _____. (12.2)
   A. use series circuit rules for all sections that are connected in series
   B. use parallel circuit rules for all sections that are connected in series and connected in parallel
   C. calculate only the parallel combination of loads and replace with a single resistance
   D. simplify the circuit by evaluating the loads connected either in series or parallel, until only a single equivalent resistance remains
3. *True or False?* If three loads are connected in parallel with each other, the total equivalent resistance will always be lower than the lowest resistance. (12.2)
4. If two parallel-connected 10 Ω resistances are wired in series with a 15 Ω resistance, the total resistance will be equal to _____. (12.2)
   A. 20 Ω
   B. 25 Ω
   C. 30 Ω
   D. 35 Ω
5. If two series-connected 10 Ω resistances are wired in parallel with a 5 Ω resistance, the total resistance will be equal to _____. (12.2)
   A. 1 Ω
   B. 4 Ω
   C. 10 Ω
   D. 25 Ω
6. When two switches are wired in series with each other, _____ position to allow current to pass. (12.4)
   A. both switches must be in the closed
   B. both switches must be in the open
   C. one or more of the switches must be in the closed
   D. one or more of the switches must be in the open
7. On evaluation of the following circuit, it can be determined that this is a _____. (12.1)

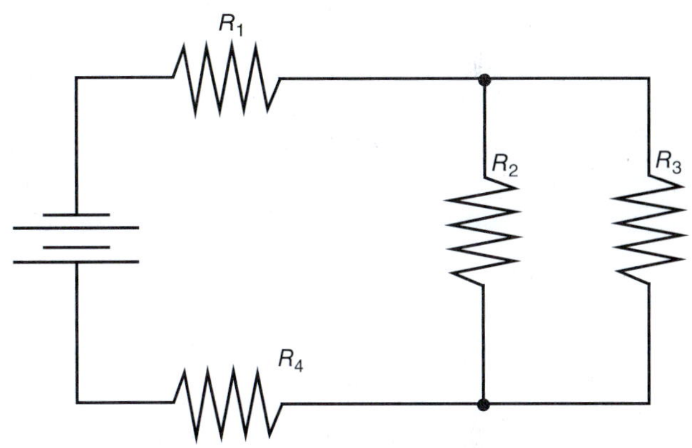

Goodheart-Willcox Publisher

   A. series circuit, as power cannot be supplied to any other circuit load without passing through R_1
   B. parallel circuit, as power can be supplied to any other circuit load without passing through R_2
   C. parallel circuit, as power cannot be supplied to any other circuit load without passing through R_2
   D. combination circuit, as power cannot be supplied to any other circuit load without passing through R_1 and R_4, but can supply power through R_2 and R_3

Plug-In

Use the following circuit to answer questions 1 through 5.

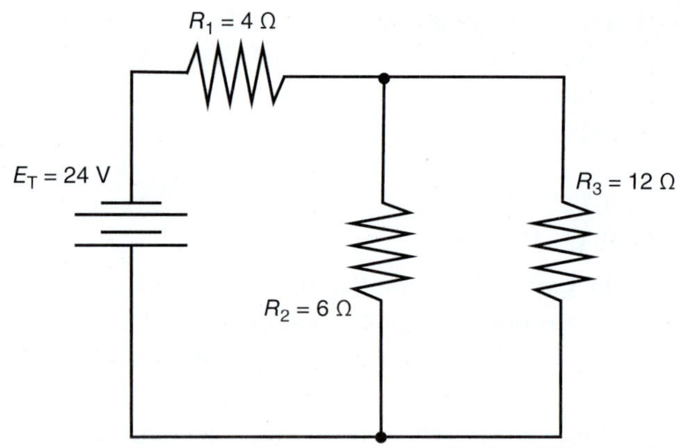

Goodheart-Willcox Publisher

1. Loads R_2 and R_3 can be replaced with a single load with a resistance of _____. (12.2)
   A. 1 Ω
   B. 2 Ω
   C. 3 Ω
   D. 4 Ω
2. The measured voltage across R_1 will be _____. (12.2)
   A. 6 V
   B. 12 V
   C. 18 V
   D. 24 V
3. The power consumption of this circuit is _____. (12.3)
   A. 24 W
   B. 48 W
   C. 72 W
   D. 96 W
4. The power consumption of R_2 is _____. (12.3)
   A. 12 W
   B. 24 W
   C. 36 W
   D. 48 W
5. The power consumption of R_3 is _____. (12.3)
   A. 12 W
   B. 24 W
   C. 36 W
   D. 48 W

Use the following circuit to answer questions 6 through 10.

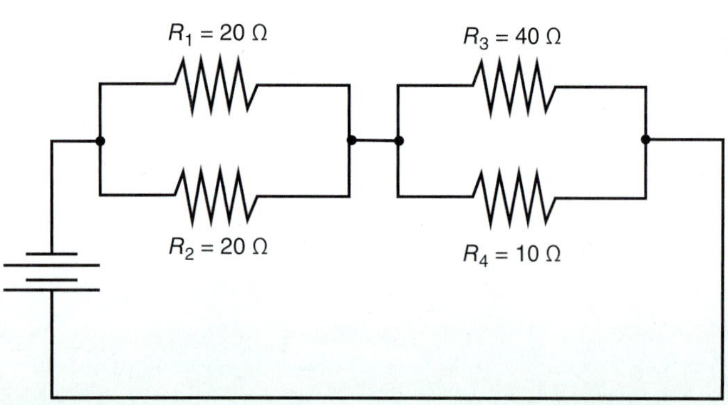

Goodheart-Willcox Publisher

6. The total circuit resistance is _____. (12.2)
   A. 9 Ω
   B. 18 Ω
   C. 27 Ω
   D. 36 Ω
7. If the total power consumption of the circuit is 200 W, what is the voltage supplied to the circuit? (12.3)
   A. 15 V
   B. 30 V
   C. 45 V
   D. 60 V
8. Which of the following statements is true? (12.2)
   A. The measured voltage across R_1 is greater than the measured voltage across R_2.
   B. The measured voltage across R_3 is greater than the measured voltage across R_4.
   C. The measured voltage across R_3 is less than the measured voltage across R_2.
   D. The measured voltage across R_1 is less than the measured voltage across R_4.
9. Which load consumes the most power? (12.3)
   A. R_1
   B. R_2
   C. R_3
   D. R_4
10. If load R_3 experiences a burned internal wire creating an open circuit through that load, which of the following statements is true? (12.2)
    A. The current flow through the entire circuit will increase.
    B. The power consumption of the circuit will increase.
    C. The total circuit resistance will increase.
    D. The total circuit voltage will increase.

Making the Connection

1. Why might it be difficult to determine which switch is open when multiple switches feeding the same load are wired in parallel with each other? (12.4)
2. In practical terms for an electrician, what type of circuit is typical: series, parallel, or combination? What impact will be likely if more loads are added? (12.2)

CHAPTER 13 Current, Voltage, and Superposition Laws and Theorems

Shaiith/Shutterstock.com

SPARKING DISCUSSION

Which are more reliable, theorems or laws? Explain your opinion.

LEARNING OUTCOMES

After completing this chapter, you will be able to:

13.1 Explain the difference between a law and a theorem.
13.2 Evaluate electric circuits using Kirchhoff's voltage law and Kirchhoff's current law.
13.3 Evaluate electric circuits using Thevenin's theorem.
13.4 Evaluate electric circuits using Norton's theorem.
13.5 Evaluate electric circuits using the superposition theorem.

TECHNICAL TERMS

current source
Kirchhoff's current law
Kirchhoff's voltage law
law
Norton current
Norton equivalent circuit
Norton resistance
Norton's theorem
superposition theorem
theorem
Thevenin equivalent circuit
Thevenin resistance
Thevenin's theorem
Thevenin voltage
Venn diagram
voltage source

Introduction

Simple series and parallel circuits can often be easily evaluated using Ohm's law. In the real world, however, most circuits are more complicated than those discussed earlier in this textbook.

This chapter will explain how laws and theorems are used to simplify the process of evaluating various electric circuits. These laws and theorems include Kirchhoff's voltage law, Kirchhoff's current law, Thevenin's theorem, Norton's theorem, and the superposition theorem. As each is discussed, detailed step-by-step examples are provided to guide you along the path to a better understanding of these important concepts.

You do not use every tool in your toolbox every time you have to complete a task. However, you need to have your tools ready, and you have to know how to use them for when you do need them. It is far better to have a tool ready and not need it than to need it and not have it readily available. The same holds true for electrical laws and theorems. As an electrician, you might not utilize all the concepts that are presented in this chapter daily, but you will become a stronger, more successful electrician if you have a solid understanding of the rules that electricity and all electric circuits follow.

13.1 Laws and Theorems

Although the terms law and theorem are often used interchangeably, there are technical differences between the two. ***Laws***, as referred to in the electrical discipline, are mathematical expressions that express relationships among electrical quantities. Ohm's law represents the relationship that exists among voltage, current, and resistance. Referring to **Figure 13-1**, if the resistance of a circuit is doubled, the current flow through the circuit will drop by a factor of two, as long as the voltage remains constant.

Theorems, on the other hand, are proven by a natural progression of logical, previously established statements. For example, consider the following two generic statements, which have previously been proven to be true:

- All *A* is *B*
- All *B* is *C*

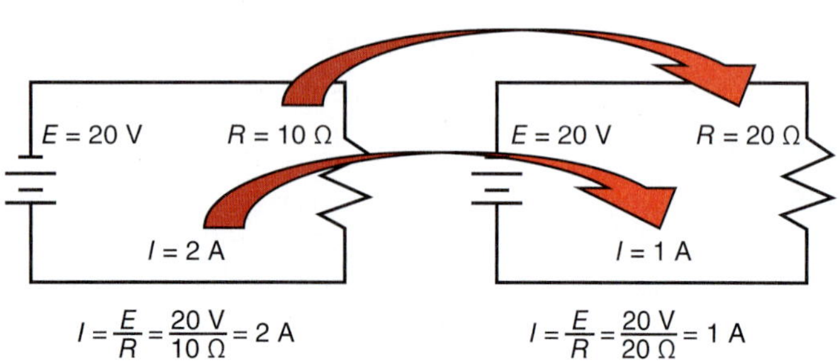

Figure 13-1. Doubling the resistance causes the current to drop by a factor of two.

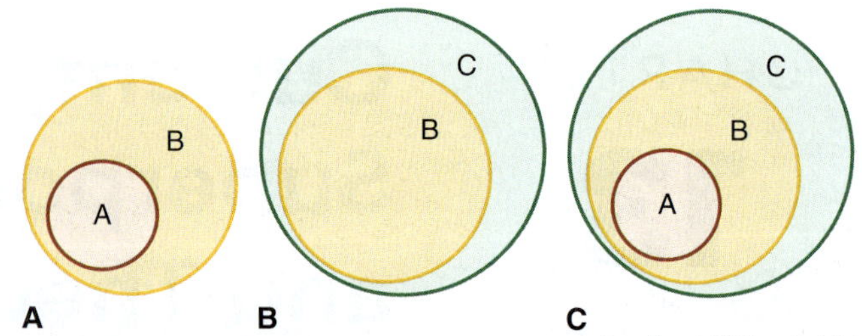

Figure 13-2. Venn diagrams are often used to show subsets, meaning a set that is contained within another set. A—All members of set *A* are also members of set *B*. B—All members of set *B* are also members of set *C*. C—Since all members of *A* are in *B* and all members of *B* are in *C*, then all members of *A* are also members of set *C*.

Since both statements are accepted to be true, it must also be accepted that the statement all *A* is *C* is also true. Graphically, a Venn diagram can be used to illustrate this concept. ***Venn diagrams*** are graphical representations of the relationships that exist among different groups, or sets, of mathematical and/or logic-related statements. For example, the Venn diagrams shown in **Figure 13-2** help illustrate these statements further.

13.2 Kirchhoff's Current Law and Voltage Law

Kirchhoff's current law is used to evaluate current flow in an electric circuit. This law states that the mathematical sum of all current flows into and out of any circuit node, connection, or point must equal zero. Kirchhoff's current law is an extended formal statement of the concept first introduced in Chapter 11, *Parallel Circuits*, that the sum of the inflow current values must always equal the sum of the outflow values.

Kirchhoff's current law assigns different polarities (positive or negative) to current inflows and outflows. The mathematical sum of the inflows to and outflows from a node will be zero if all the inflows have the same polarity and all the outflows have a polarity that is opposite to that of the inflows. The following is an example of the general form of Kirchhoff's current law for a node with three inflows and three outflows:

$$\text{INFLOW}_1 + \text{INFLOW}_2 + \text{INFLOW}_3 + \text{OUTFLOW}_1 + \text{OUTFLOW}_2 + \text{OUTFLOW}_3 = 0$$

In its simplest form, consider the portion of an electric circuit shown in **Figure 13-3**. Two current-carrying conductors, A and B, are connected to each other at the node. When the circuit is energized, electric current flows

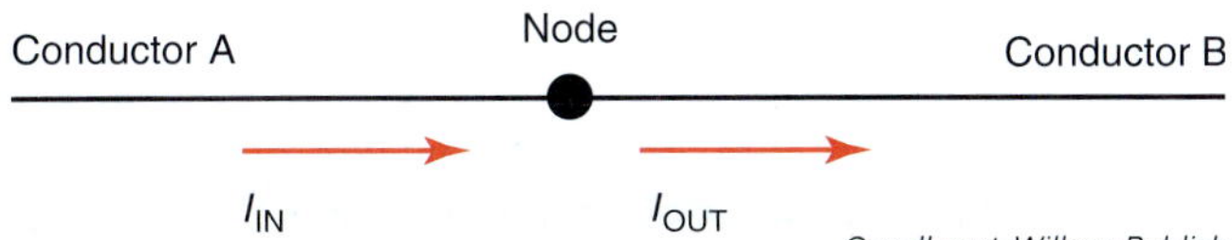

Goodheart-Willcox Publisher

Figure 13-3. In a node, the current entering the node is equal to the current leaving the node.

toward the node in one conductor and away from the node in the other. Since the node is an electrical connection, no electrical energy is used, stored, or otherwise consumed by the node. The node serves as a pass-through point. If the current flowing through conductor A toward the node is expressed as a positive value and the current flowing through conductor B away from the node is expressed as a negative value, the relationship between I_{IN} and I_{OUT} can be expressed as follows:

$$I_{IN} + I_{OUT} = 0$$

Example 13-1, **Example 13-2**, and **Example 13-3** use Kirchhoff's current law to determine current flow.

Shedding Light

It is important to note that there is no such thing as a negative current. The positive and negative signs assigned to the current values are used to identify the direction of the current flow relative to the node. The positive and negative signs are assigned based on the direction of current flow, with all inflows assigned the same polarity (positive), and all outflows assigned the opposite polarity (negative).

Kirchhoff's Current Law for Three-Branch Circuits

Kirchhoff's current law holds true for electric circuits regardless of the number of branches or current paths in the circuit. Consider the three-branch parallel circuit shown

Example 13-1

Referring to the following portion of a two-branch electric circuit, determine the current flow through R_2 if the total current (I_T) is 8 A and the current flow through R_1 (I_1) is 3 A.

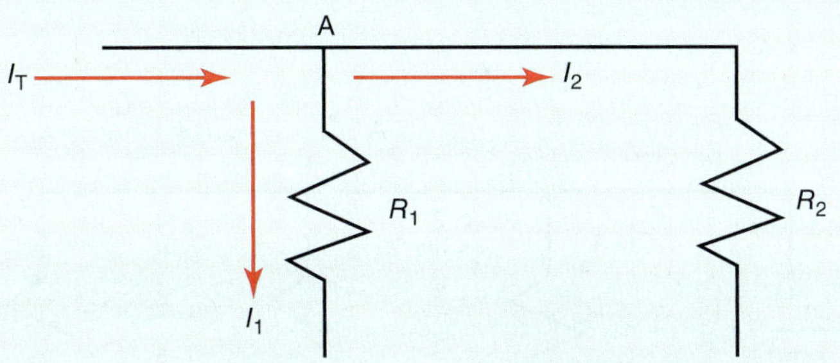

Goodheart-Willcox Publisher

✓ Solution

In this circuit, the total current (I_T) is 8 A and is flowing into node A. At point A, the total current splits into I_1 and I_2 to feed resistances R_1 and R_2, respectively. The total current flowing into the node will be assigned a positive polarity, while the outflows, I_1 and I_2, will be assigned negative polarities. Using Kirchhoff's current law, the mathematical relationship that exists among I_1, I_2, and I_T is expressed as:

$$I_T + I_1 + I_2 = 0$$

Next, substitute the known values for I_T (an inflow with a positive polarity) and I_1 (an outflow with a negative polarity) as follows:

$$8\text{ A} + (-3\text{ A}) + I_2 = 0$$

Then solve for I_2:

$$5\text{ A} + I_2 = 0$$

$$I_2 = -5\text{ A}$$

The current flowing through R_2 is 5 A. The negative polarity associated with I_2 indicates that the current of 5 A is flowing away from the node.

Now It's Your Turn 13-1

Determine the direction and amount of current flowing through conductor C in the following diagram.

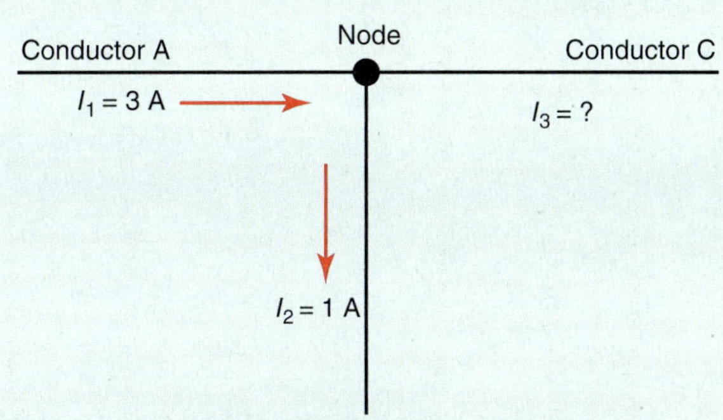

Goodheart-Willcox Publisher

Example 13-2

Referring to the portion of the electric circuit shown, determine the value of I_3 (the current flowing out of node B).

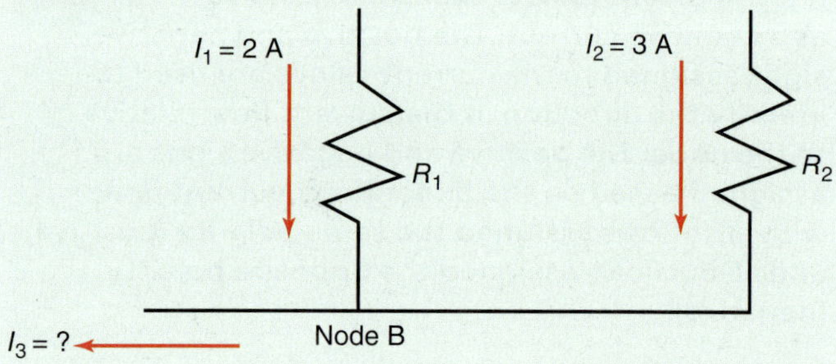

Goodheart-Willcox Publisher

✓ Solution

In this circuit, the value of I_1, the current flowing through R_1, is 2 A, and the value of I_2, the current flowing through R_2, is 3 A. Using the basic form of Kirchhoff's current law, substitute the known current values for I_1 and I_2 as follows:

$$I_1 + I_2 + I_3 = 0$$

$$2\text{ A} + 3\text{ A} + I_3 = 0$$

$$5\text{ A} + I_3 = 0$$

$$I_3 = -5\text{ A}$$

Since the currents I_1 and I_2 flow into node B, and I_3 flows out of node B, a positive polarity has been assigned to currents I_1 and I_2, and a negative polarity to current I_3. The result of −5 A for I_3 indicates that this current flows away from node B.

Now It's Your Turn 13-2

Determine the direction and amount of current flowing through conductor C in the following diagram.

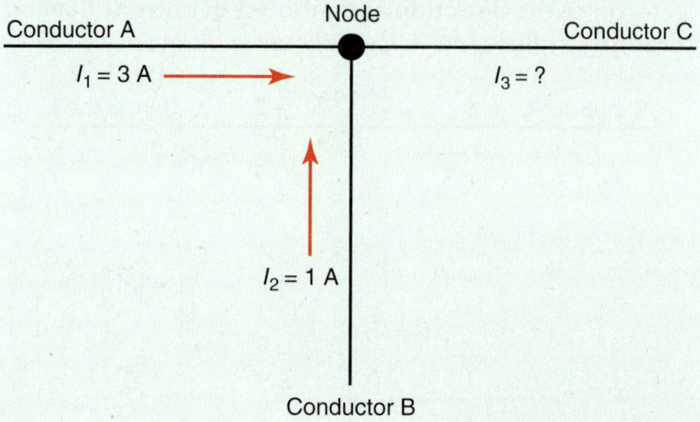

Goodheart-Willcox Publisher

Example 13-3

Referring to the portion of an electric circuit shown in the following diagram, determine the value of I_2, the current flowing through R_2.

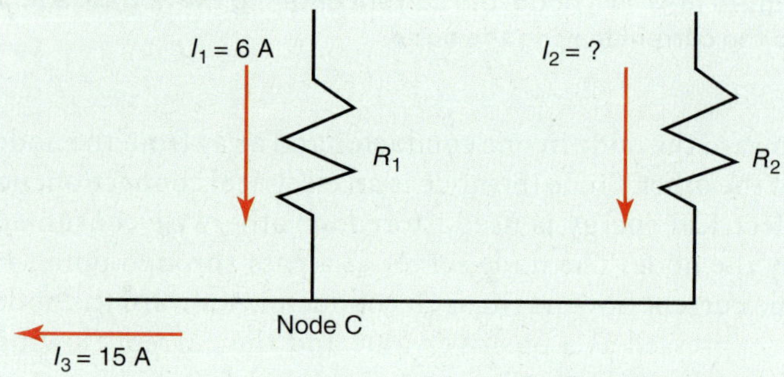

Goodheart-Willcox Publisher

✓ Solution

This circuit has two inflows to node C, I_1 and I_2, and one outflow from node C, I_3. Since I_1 and I_2 are inflows to node C, they will be assigned a positive polarity and I_3, an outflow from node 3, will be assigned a negative polarity. Using the basic form of Kirchhoff's current law, substitute the known current values for I_1 and I_3 as follows:

$$I_1 + I_2 + I_3 = 0$$

$$6\text{ A} + I_2 + (-15\text{ A}) = 0$$

$$I_2 + (-9\text{ A}) = 0$$

$$I_2 = 9\text{ A}$$

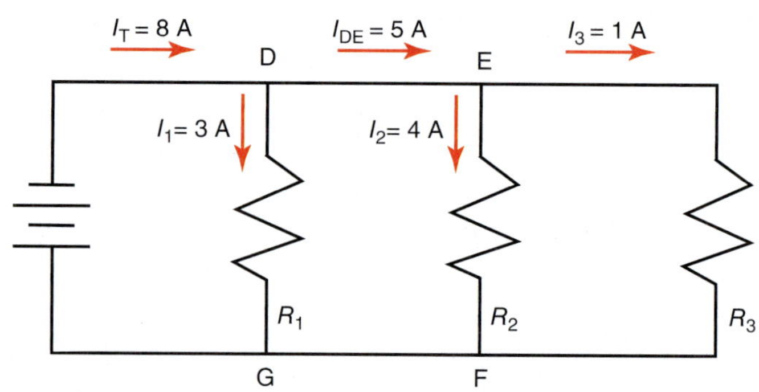

Goodheart-Willcox Publisher

Figure 13-4. A three-branch parallel circuit used to demonstrate Kirchhoff's current law.

in **Figure 13-4**. In the top-left corner of this circuit, a current of 8 A (I_T) flows from the power source to node D. At node D, two outflows can also be observed: a current flow of 3 A (I_1) flowing to load R_1, and a current flow (I_{DE}) of 5 A flowing toward node E. Mathematically, Kirchhoff's current law at node D can be expressed as:

$$I_T + I_1 + I_{DE} = 0$$

Assigning a positive polarity to inflow I_T and negative polarities to outflows I_1 and I_{DE}, the following expression verifies that Kirchhoff's current law holds true:

$$I_T + I_1 + I_{DE} = 0$$
$$8\text{ A} + (-3\text{ A}) + (-5\text{ A}) = 0$$
$$8\text{ A} + (-8\text{ A}) = 0$$
$$0 = 0$$

To verify Kirchhoff's current law at node E, refer to **Example 13-4**.

Electrically speaking, two points in an electric circuit are the same if they are connected to each other by only an uninterrupted conductor with negligible resistance. Based on this, nodes D and E in **Figure 13-4** are electrically the same and represent a single node. The same holds true for nodes F and G. Therefore, the circuit shown in **Figure 13-4** and **Figure 13-5** can be redrawn as shown in **Figure 13-6**.

Using Kirchhoff's current law to evaluate node H, this node has one inflow and three outflows. Assigning a positive polarity to the inflow to node H and negative polarities to the outflows from the node, the relationship can be expressed as follows:

$$I_T + I_1 + I_2 + I_3 = 0$$
$$8\text{ A} + (-3\text{A}) + (-4\text{ A}) + (-1\text{ A}) = 0$$
$$8\text{ A} + (-8\text{ A}) = 0$$
$$0 = 0$$

To verify Kirchhoff's current law at node J, refer to **Example 13-5**.

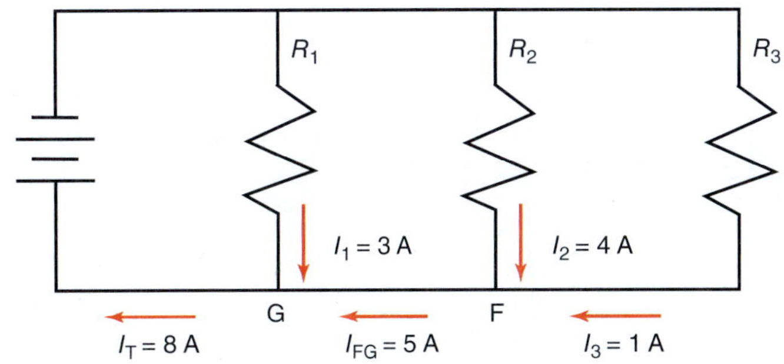

Goodheart-Willcox Publisher

Figure 13-5. The same circuit from **Figure 13-4** with the current inflows and outflows identified at nodes F and G.

Example 13-4

Referring to **Figure 13-4**, create a mathematical expression and verify Kirchhoff's current law at node E.

✓ Solution

At node E, one inflow and two outflows can be observed. The inflow is current I_{DE}, which is the same as the 5 A outflow from node D. The two outflows from node E are I_2, which is the 4 A current flow to load R_2, and I_3 is the 1 A current flow to load R_3. Kirchhoff's current law at node E can be expressed as:

$$I_{DE} + I_2 + I_3 = 0$$

Assigning a positive polarity to inflow I_{DE} and negative polarities to outflows I_2 and I_3, the following expression verifies that Kirchhoff's current law holds true:

$$I_{DE} + I_2 + I_3 = 0$$
$$5\text{ A} + (-4\text{ A}) + (-1\text{ A}) = 0$$
$$5\text{ A} + (-5\text{ A}) = 0$$
$$0 = 0$$

Now It's Your Turn 13-4

Referring to **Figure 13-5**, create mathematical expressions that verify Kirchhoff's current law at nodes F and G.

Example 13-5

Referring to **Figure 13-6**, create a mathematical expression and verify Kirchhoff's current law at node J.

✓ Solution

Node J has three inflows and one outflow. The relationship can be expressed as follows:

$$I_T + I_1 + I_2 + I_3 = 0$$
$$(-8\text{ A}) + 3\text{ A} + 4\text{ A} + 1\text{ A} = 0$$
$$-8\text{ A} + 8\text{ A} = 0$$
$$0 = 0$$

Now It's Your Turn 13-5

Referring to the following circuit, determine the value of current flow I_2 using Kirchhoff's current law. Check your result by performing calculations at both nodes K and L.

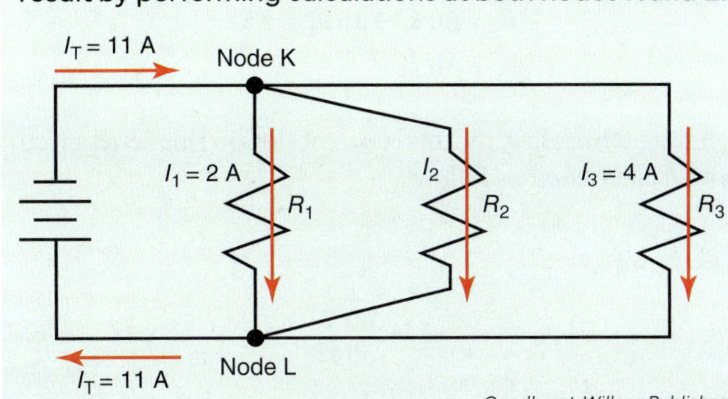

Goodheart-Willcox Publisher

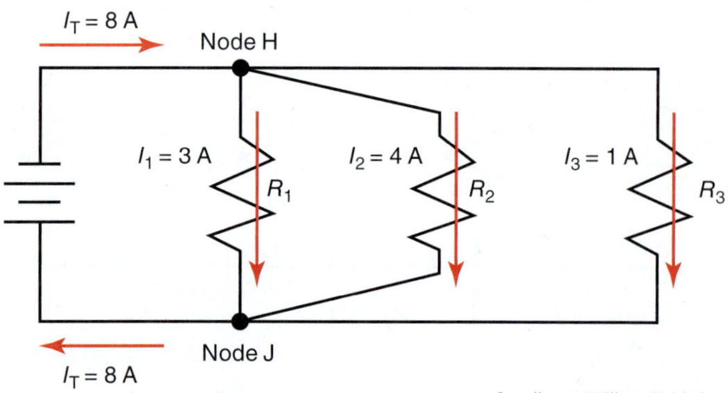

Goodheart-Willcox Publisher

Figure 13-6. Circuit from **Figures 13-4** and **Figure 13-5** redrawn with combined nodes. Notice that nodes D and E have been combined to form a single node (H) and nodes F and G have been combined to form a single node (J).

Kirchhoff's Voltage Law

Kirchhoff's voltage law provides a means to evaluate the voltage drops that are present in a complete closed-circuit path. Kirchhoff's voltage law states that the sum of all voltage drops, including power supplies and loads, in a closed-circuit path must equal zero. The formula is shown below:

$$E_1 + E_2 + E_3 + E_4 = 0$$

In Kirchhoff's voltage law, each voltage term represents the voltage as measured across each circuit component, be it a power supply, power-passing device, or a power-consuming device.

Just as Kirchhoff's current law assigned polarities to current flows based on their direction, Kirchhoff's voltage law assigns polarity to each voltage drop in a circuit. Circuit loads are assigned a negative polarity at the entry point of current and positive polarity at the exit point of current, **Figure 13-7**.

To illustrate the workings of Kirchhoff's voltage law, consider the simple series circuit shown in **Figure 13-8**. The total resistance of a series circuit is the sum of all the resistances in the circuit. The total resistance for the circuit shown in **Figure 13-8** can be calculated as follows:

$$R_T = R_1 + R_2 + R_3$$
$$R_T = 10\ \Omega + 15\ \Omega + 5\ \Omega$$
$$R_T = 30\ \Omega$$

Using Ohm's law, the total current flow in this series circuit can be calculated as follows:

$$I_T = \frac{E_T}{R_T}$$
$$I_T = \frac{60\text{ V}}{30\ \Omega}$$
$$I_T = 2\text{ A}$$

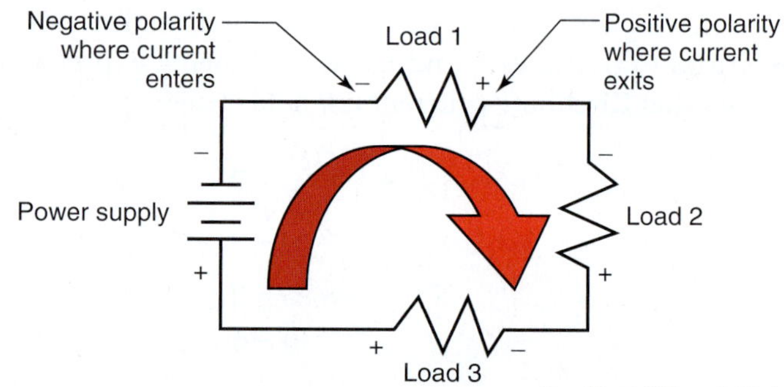

Goodheart-Willcox Publisher

Figure 13-7. Polarities assigned to power sources and circuit loads. The red arrow points in the direction of current flow.

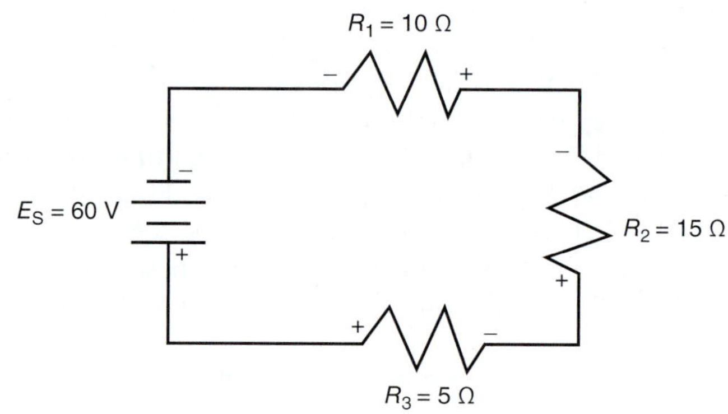

Goodheart-Willcox Publisher

Figure 13-8. A series circuit with three resistances.

Next, the voltage drop across each resistance can be determined by multiplying the circuit current and the value of each resistance together. The voltage drop across R_1 can be found as follows:

$$E_1 = I_T \times R_1$$
$$E_1 = 2\text{ A} \times 10\ \Omega$$
$$E_1 = 20\text{ V}$$

The voltage drop across R_2 can then be calculated as follows:

$$E_2 = I_T \times R_2$$
$$E_2 = 2\text{ A} \times 15\ \Omega$$
$$E_2 = 30\text{ V}$$

Finally, the voltage drop across R_3 can be calculated as follows:

$$E_3 = I_T \times R_3$$
$$E_3 = 2\text{ A} \times 5\ \Omega$$
$$E_3 = 10\text{ V}$$

Figure 13-9 is the same circuit shown in **Figure 13-8** with the total current and voltage drop across each circuit resistance added.

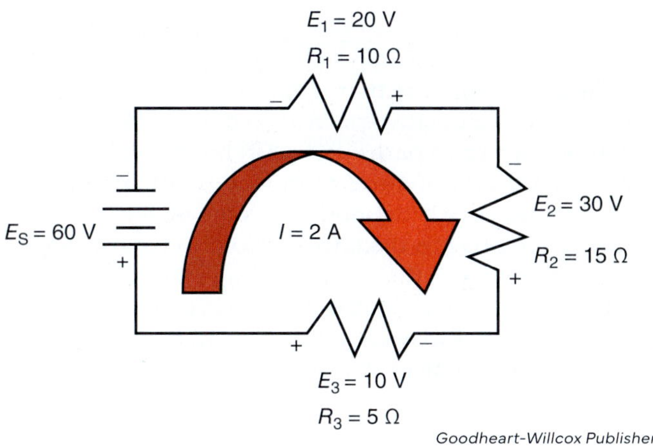

Figure 13-9. Same circuit as **Figure 13-8** with voltage drops, polarities, and direction of current flow indicated.

To verify Kirchhoff's voltage law, the voltage values across each relevant circuit element are added together, making certain that the proper polarity is associated with each. In this circuit, E_S represents the supply voltage, while E_1, E_2, and E_3 represent the measured voltages across R_1, R_2, and R_3, respectively. If Kirchhoff's voltage law holds true, the sum of these voltages will equal zero. To verify the total voltage values, move in a clockwise path around the circuit. Assign polarity to sources and loads based on the polarity first encountered following the path direction (clockwise in this case). When moving in a clockwise path, starting at the upper left corner of the wiring diagram, the negative polarity end of a load is encountered before the positive polarity end, while the positive polarity end of the power source is encountered before the negative polarity end. Use Kirchhoff's voltage law as follows:

$$E_S + E_1 + E_2 + E_3 = 0$$

$$60\text{ V} + (-20\text{ V}) + (-30\text{ V}) + (-10\text{ V}) = 0$$

$$60\text{ V} - 20\text{ V} - 30\text{ V} - 10\text{ V} = 0$$

$$60\text{ V} - 60\text{ V} = 0$$

$$0 = 0$$

Since the sum of the voltage values around the closed path is zero, Kirchhoff's voltage law is verified.

Refer to **Example 13-6** to determine voltage drop by using Kirchhoff's voltage law.

Shedding Light

Based on Ohm's law, if the voltage drop for two loads is the same and the current flow through the loads is the same, the resistances of the two loads must be the same.

Example 13-6

For the following circuit, use Kirchhoff's voltage law to determine the voltage across R_3.

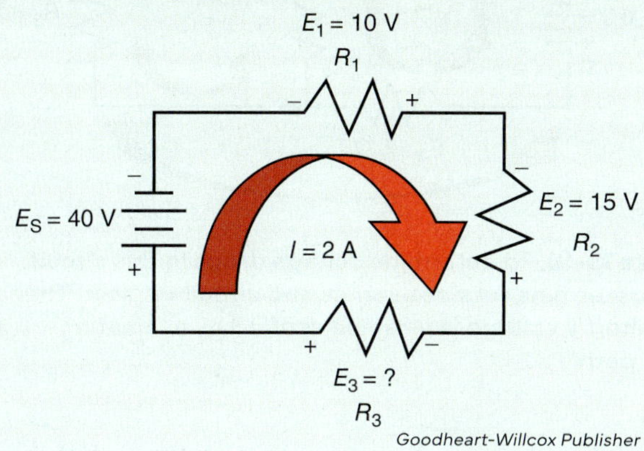

✓ Solution

Since the sum of the voltage values across each circuit element must be zero, a mathematical expression can be set up to solve for the one unknown value, E_3. Using Kirchhoff's law, start with the power supply and move in a clockwise path around the circuit. Moving in a clockwise direction, the supply voltage, E_S, has a positive polarity, while the voltage drops across the three circuit loads have negative polarities. Substitute the known voltage values as follows:

$$E_S + E_1 + E_2 + E_3 = 0$$

$$40\text{ V} + (-10\text{ V}) + (-15\text{ V}) + E_3 = 0$$

$$40\text{ V} - 25\text{ V} + E_3 = 0$$

$$15\text{ V} + E_3 = 0$$

$$E_3 = -15\text{ V}$$

The voltage reading across R_3 is 15 V. The negative sign indicates the polarity of R_3 with respect to the Kirchhoff's voltage law calculation. Since the voltage drops across R_2 and R_3 are both 15 V, R_2 and R_3 must have the same resistance.

Kirchhoff's Voltage Law in a Parallel Circuit

Previous examples of Kirchhoff's voltage law examined simple series circuits with only one possible current path. Our attention now turns to the process of evaluating a circuit with multiple paths. Consider the circuit shown in **Figure 13-10**. This circuit has three paths that can be evaluated using Kirchhoff's voltage law.

Path 1 moves in a clockwise direction through the power supply, R_1, R_4, and R_3. The power supply has a positive polarity and the three circuit loads have negative polarity.

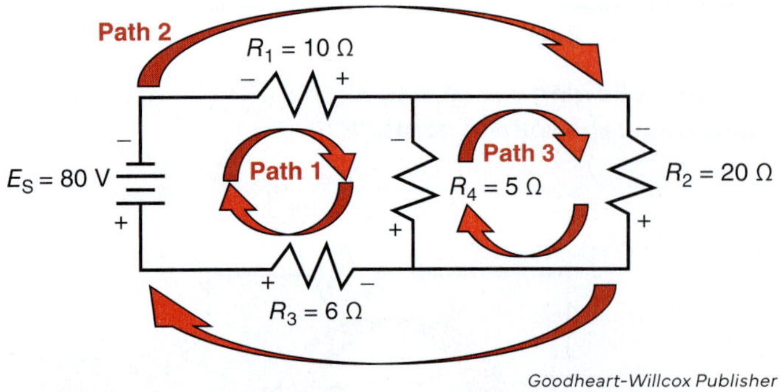

Figure 13-10. To determine voltage drops in this circuit, first determine total resistance and circuit current. Then Kirchhoff's voltage law can be verified by evaluating each path.

Path 2 moves in a clockwise direction through the power supply, R_1, R_2, and R_3. The power supply has a positive polarity and the three circuit loads have negative polarity.

Shedding Light

In **Figure 13-10**, note that R_4 has a negative polarity in path 1 but a positive polarity in path 3. When evaluating circuit paths using Kirchhoff's voltage law, the polarity of a source or load is not absolute. Instead, polarity is dependent on the device's position relative to the chosen direction of rotation through the path. Kirchhoff's voltage law holds true whether the path is evaluated in a clockwise or counterclockwise direction. Changing direction simply reverses all of the polarities.

Path 3 moves in a clockwise direction through R_2 and R_4. Polarity is determined by the end of the load first encountered based on the direction of rotation in the path (clockwise in this case), so R_2 has a negative polarity and R_4 has a positive polarity.

Refer to **Example 13-7** to evaluate a circuit using Kirchhoff's voltage law.

Example 13-7

Verify Kirchhoff's voltage law for the three paths identified in the circuit in **Figure 13-10**.

✓ Solution

To evaluate the circuit using Kirchhoff's voltage law, the voltage drops across each load must be known. Determining the voltage drop across each load requires the total circuit resistance and current values. The total resistance of the circuit can be found by first combining R_2 and R_4 into an equivalent resistance, $R_{2,4}$:

$$R_{2,4} = \frac{(R_2 \times R_4)}{(R_2 + R_4)}$$

$$R_{2,4} = \frac{(20\ \Omega \times 5\ \Omega)}{(20\ \Omega + 5\ \Omega)}$$

$$R_{2,4} = \frac{100\ \Omega}{25\ \Omega}$$

$$R_{2,4} = 4\ \Omega$$

Then calculate the total resistance of the circuit:

$$R_T = R_1 + R_{2,4} + R_3$$

$$R_T = 10\ \Omega + 4\ \Omega + 6\ \Omega$$

$$R_T = 20\ \Omega$$

With the total circuit resistance calculated and the supply voltage known, the total circuit current can be calculated as follows:

$$I_T = \frac{E_T}{R_T}$$

$$I_T = \frac{80\ \text{V}}{20\ \Omega}$$

$$I_T = 4\ \text{A}$$

Using Ohm's law, the voltage drop across each resistance can be determined by multiplying the circuit current and the value of each resistance together. First, calculate the voltage drop across R_1:

$$E_1 = I \times R_1$$

$$E_1 = 4\ \text{A} \times 10\ \Omega$$

$$E_1 = 40\ \text{V}$$

Then calculate the voltage drop across R_3:

$$E_3 = I \times R_3$$

$$E_3 = 4\ \text{A} \times 6\ \Omega$$

$$E_3 = 24\ \text{V}$$

Finally, calculate the voltage drop across the equivalent resistance, $R_{2,4}$:

$$E_{2,4} = I \times R_{2,4}$$

$$E_{2,4} = 4\ \text{A} \times 4\ \Omega$$

$$E_{2,4} = 16\ \text{V}$$

Since 16 V is applied to the $R_2 - R_4$ parallel combination, a voltage drop of 16 V will be present across each of the two resistances.

Kirchhoff's voltage law can now be verified by evaluating each of the three paths, **Figure 13-11**. First, verify path 1 (clockwise through the power supply, R_1, R_4, and R_3) as follows:

$$E_S + E_1 + E_4 + E_3 = 0$$

$$80\text{ V} + (-40\text{ V}) + (-16\text{ V}) + (-24\text{ V}) = 0$$

$$80\text{ V} - 80\text{ V} = 0$$

$$0 = 0$$

Next, verify path 2 (clockwise through the power supply, R_1, R_2, and R_3) as follows:

$$E_S + E_1 + E_2 + E_3 = 0$$

$$80\text{ V} + (-40\text{ V}) + (-16\text{ V}) + (-24\text{ V}) = 0$$

$$80\text{ V} - 80\text{ V} = 0$$

$$0 = 0$$

Then, verify path 3 (clockwise through R_2 and R_4) as follows:

$$E_2 + E_4 = 0$$

$$-16\text{ V} + 16\text{ V} = 0$$

$$0 = 0$$

Now It's Your Turn 13-7

Determine the value of the supply voltage E_S in the following circuit.

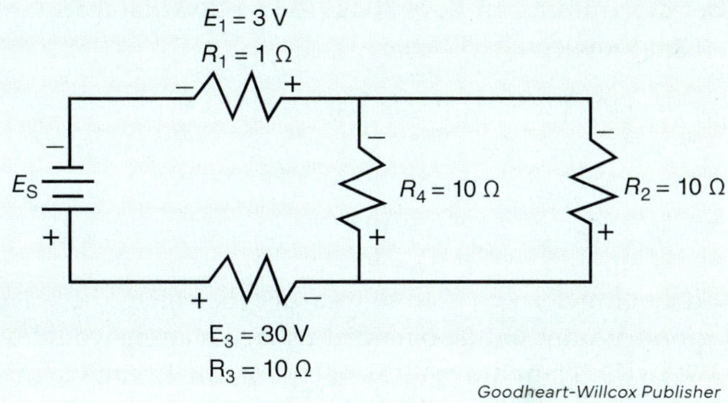

Goodheart-Willcox Publisher

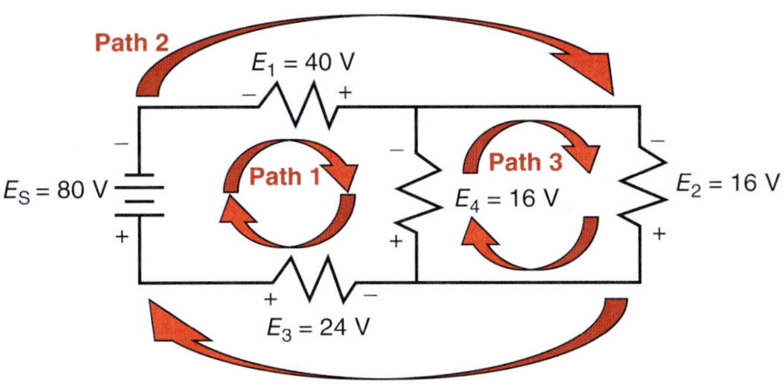

Goodheart-Willcox Publisher

Figure 13-11. Same circuit as **Figure 13-10**. Kirchhoff's law is verified by evaluating path 1, path 2, and finally path 3. Circuit with voltage drop information added.

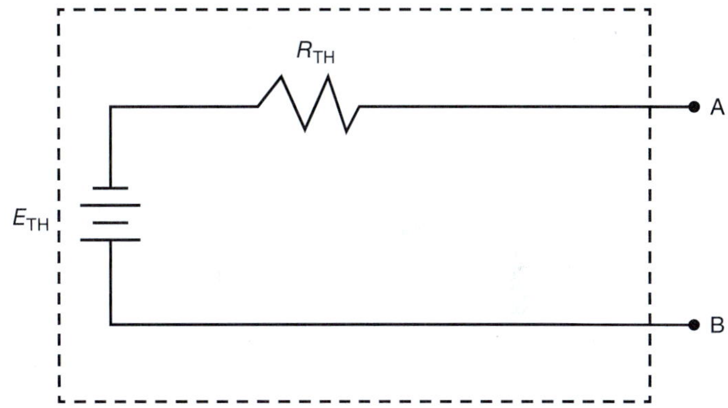

Goodheart-Willcox Publisher

Figure 13-12. A Thevenin equivalent circuit is made up of a voltage source and a series-connected resistance.

13.3 Thevenin's Theorem

Thevenin's theorem is used to simplify circuits for evaluation and design purposes. ***Thevenin's theorem*** states that any linear circuit, made up of one or more power supplies and one or more resistances, can be reduced to a simple series circuit consisting of a single voltage source and one resistor. A ***voltage source*** is a device, such as a battery, that has the ability to provide continuous constant voltage to a circuit. This single-voltage-source, single-resistor circuit reduction is referred to as the ***Thevenin equivalent circuit***. In a Thevenin equivalent circuit, shown in **Figure 13-12**, the equivalent voltage is identified as the Thevenin voltage (E_{TH}) and the equivalent circuit resistance is identified as the Thevenin resistance (R_{TH}). When evaluated between points A and B, a Thevenin equivalent circuit is equivalent to the original, unsimplified circuit.

Thevenin's theorem can be equated to the black box system concept, meaning a system that is only observable at its input and output terminals. Evaluation of the system is made solely by measurements taken at the output terminals. A Thevenin equivalent circuit provides the same outputs as the original circuit, but is a simplified version, or simulation, of the original. For example, if taking a reading across two terminals of an unknown circuit yields a resistance reading of 4 Ω, the actual circuit could be one 4-Ω resistor, two 8-Ω resistances connected in parallel, two 2-Ω resistances connected in series, or any other possible combination that results in a 4 Ω reading.

Thevenin Voltage

Thevenin voltage is an ideal value that is measured between the circuit's two output terminals, when no external components are connected to them. Consider the circuit shown in **Figure 13-13**. Although the voltage supplied to the circuit is 60 V, the Thevenin voltage is not equal to the 60 V being supplied to the circuit. Thevenin voltage is equal to the voltage measured across the 20-Ω circuit load, as the output terminals are positioned across this load.

Thevenin voltage (E_{TH}), which is also the voltage drop across R_2 in the circuit shown in **Figure 13-13**, can be found after determining an R_T of 30 Ω (10 Ω + 20 Ω) and an I_T of 2 A (60 V divided by 30 Ω):

$$E_{TH} = I_T \times R_2$$

$$E_{TH} = 2\text{ A} \times 20\ \Omega$$

$$E_{TH} = 40\text{ V}$$

If a voltmeter was used to measure the voltage between terminals A and B, the obtained voltage reading would be 40 V. In the Thevenin equivalent circuit, the circuit's actual 60-V power source is replaced with a 40-V ideal power source.

Thevenin Resistance

Thevenin resistance (R_{TH}) is the resistance that would be measured between the circuit's output terminals if all voltage sources were replaced with short circuits and all current sources were replaced with open circuits. A ***current source*** is a device that maintains circuit current at a constant rate, regardless of the circuit loads or other circuit conditions. To determine the Thevenin resistance for the circuit shown in **Figure 13-13**, the 60-V power source must be replaced with a short circuit, **Figure 13-14**. In this circuit, there is no current source, so no additional circuit modifications need to be made.

With the voltage source replaced with a short circuit, the Thevenin resistance, which is the equivalent resistance of the circuit, can be calculated. In this example, the resulting circuit is a 20-Ω resistance connected in parallel with a 10-Ω resistance. Using the formula for determining the equivalent resistance of two resistors connected in parallel, R_{TH} can be calculated:

$$R_{TH} = \frac{R_1 \times R_2}{R_1 + R_2}$$

$$R_{TH} = \frac{20\ \Omega \times 10\ \Omega}{20\ \Omega + 10\ \Omega}$$

$$R_{TH} = \frac{200\ \Omega^2}{30\ \Omega}$$

$$R_{TH} = 6.67\ \Omega$$

The process used to calculate the Thevenin resistance will vary from circuit to circuit.

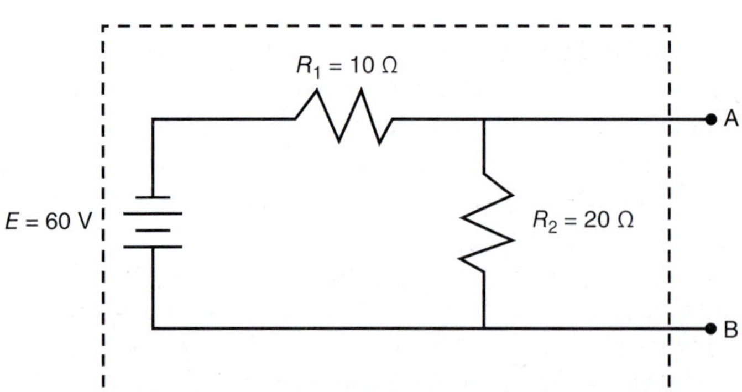

Goodheart-Willcox Publisher

Figure 13-13. Thevenin voltage is not always equal to the circuit's supply voltage.

Thevenin Equivalent Circuit

Combining the information obtained in the previous two sections, the Thevenin equivalent circuit can be constructed. The Thevenin equivalent circuit, with a Thevenin voltage of 40 V and a Thevenin resistance of 6.67 Ω is shown in **Figure 13-15**.

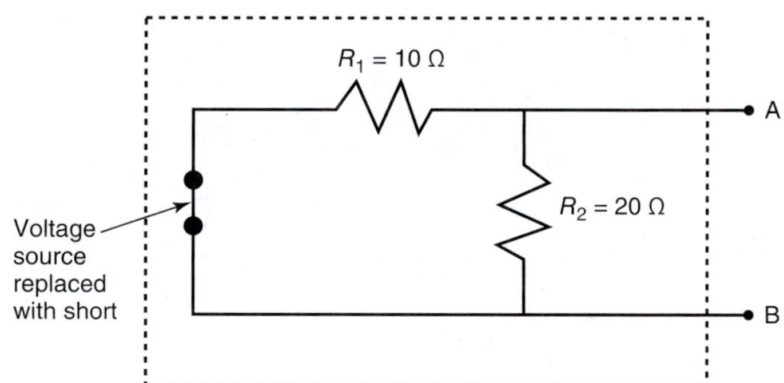

Goodheart-Willcox Publisher

Figure 13-14. To determine the Thevenin resistance of the circuit shown in **Figure 13-13**, the voltage source is replaced with a short circuit.

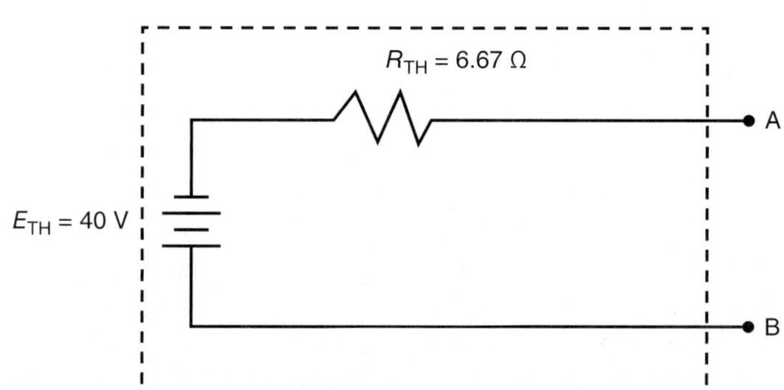

Goodheart-Willcox Publisher

Figure 13-15. The Thevenin equivalent circuit for the circuit presented in **Figure 13-13**.

13.4 Norton's Theorem

Similar to Thevenin's theorem, Norton's theorem serves to simplify otherwise complex circuits. Instead of the single power source and the series-connected resistance found in the Thevenin equivalent circuit, ***Norton's theorem*** reduces a circuit to a single current source with a single resistance connected in parallel to it. The general appearance of a ***Norton equivalent circuit*** is provided in **Figure 13-16**. The current source is represented in this diagram by the circle-shaped component with the directional arrow contained within it. As with the Thevenin equivalent circuit, the output terminals of the Norton equivalent circuit are the same as those of the original circuit and, in our example, are identified as terminals A and B. The ***Norton current*** is the current value obtained when the voltage measured between terminals A and B is divided by the Norton resistance value.

Norton Resistance

Consider the circuit shown in **Figure 13-17**. ***Norton resistance***, which is the resistance used to create the Norton equivalent circuit, is calculated in the manner that was used to determine the resistance in the Thevenin equivalent circuit: voltage sources are replaced with short circuits and current sources are replaced with open circuits, **Figure 13-18**. Norton resistance, R_N, for this circuit is calculated as follows:

$$R_N = \frac{R_1 \times R_2}{R_1 + R_2}$$

$$R_N = \frac{5\ \Omega \times 20\ \Omega}{5\ \Omega + 20\ \Omega}$$

$$R_N = \frac{100\ \Omega}{25\ \Omega}$$

$$R_N = 4\ \Omega$$

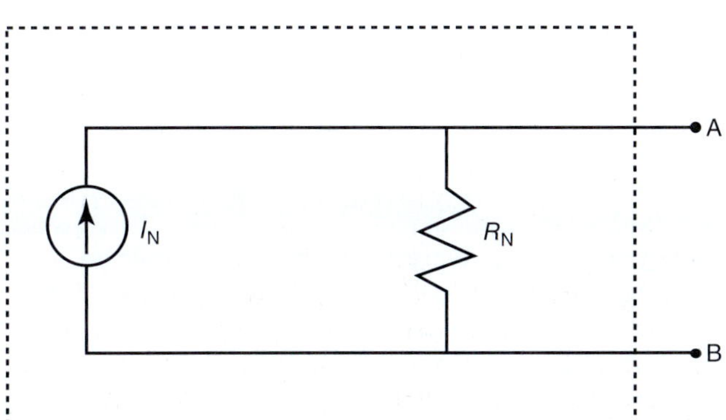

Goodheart-Willcox Publisher

Figure 13-16. The Norton equivalent circuit is made up of a current source and a parallel-connected resistance.

Norton Current

There are two methods that can be used to calculate the Norton current. The first method involves the following three steps:

1. Create a short circuit between terminals A and B.
2. Determine the resistance of this new circuit.
3. Divide this resistance into the original circuit's supply voltage.

The general formula for this calculation is:

$$I_N = \frac{V_{ORIGINAL}}{R_{SHORT}}$$

where:

I_N = Norton current
$V_{ORIGINAL}$ = Voltage supplied to the original circuit
R_{SHORT} = Circuit's resistance with the output terminals shorted

For the circuit shown in **Figure 13-17**, a short circuit is created between terminals A and B, **Figure 13-19**. With the short circuit in place, the 20-Ω resistance is effectively removed from the circuit, leaving only the 5 Ω.

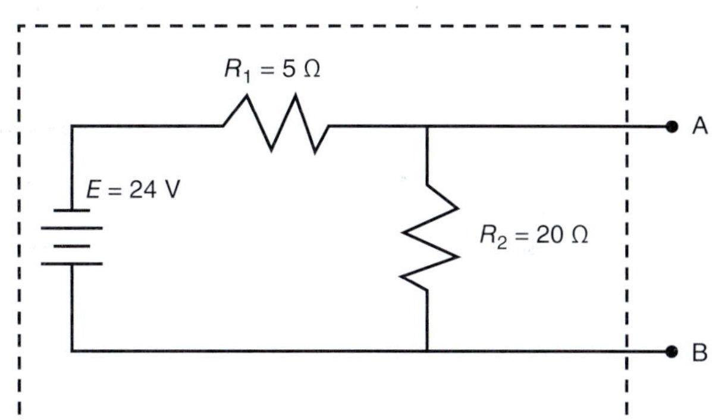

Goodheart-Willcox Publisher

Figure 13-17. A voltage divider circuit.

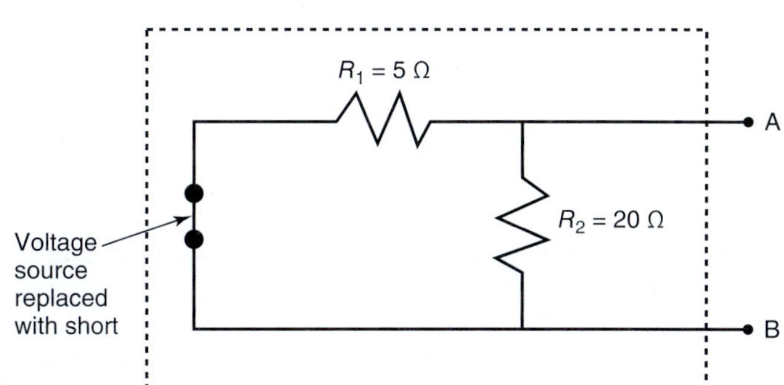

Goodheart-Willcox Publisher

Figure 13-18. Same circuit as **Figure 13-17**, with the 24-V voltage source replaced by a short circuit to calculate Norton's resistance.

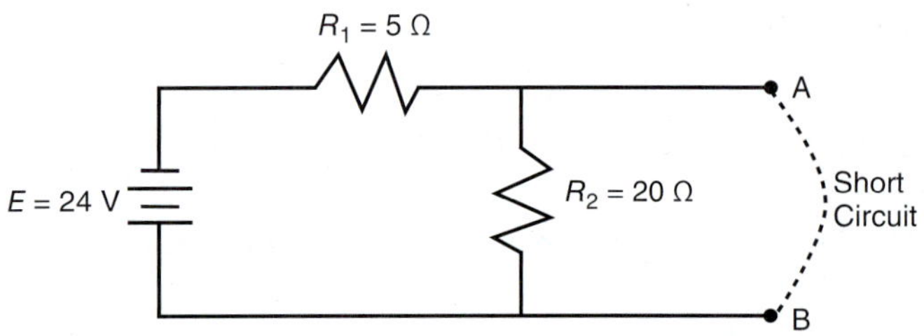

Figure 13-19. Same circuit as **Figure 13-17**, creating a short circuit between terminals A and B.

Shedding Light

Placing a short circuit across any circuit load reduces the resistance of that portion of the circuit to zero. With the resistance of a short circuit being very close to 0 Ω, the product over sum formula for two resistances in parallel can be used. From **Figure 13-19**, the resistance of the 20-Ω load in parallel with the 0-Ω short circuit is calculated as follows:

$$R = \frac{20\ \Omega \times 0\ \Omega}{20\ \Omega + 0\ \Omega}$$

$$R = \frac{0\ \Omega}{20\ \Omega}$$

$$R = 0$$

The resistance for the AB-shorted circuit is 5 Ω. To calculate the Norton current (I_N), the circuit voltage is then divided by the newly calculated circuit resistance, as follows:

$$I_N = \frac{V_{ORIGINAL}}{R_{SHORT}}$$

$$I_N = \frac{24\ V}{5\ \Omega}$$

$$I_N = 4.8\ A$$

The second method that can be used to calculate the Norton current involves dividing the open circuit voltage, as measured between terminals A and B, by the Norton resistance.

Shedding Light

If you are determining Thevenin and Norton equivalents for the same circuit, the open circuit voltage measured between terminals A and B is the Thevenin voltage, which is used to determine the Norton current.

The open circuit voltage, as measured between terminals A and B, is the same as the voltage drop across R_2. After calculating R_T as 25 Ω (5 Ω + 20 Ω) and I_T as 0.96 A (24 V divided by 25 Ω), the voltage between terminals A and B is calculated:

$$E_{AB} = I_T \times R_2$$

$$E_{AB} = 0.96\ A \times 20\ \Omega$$

$$E_{AB} = 19.2\ V$$

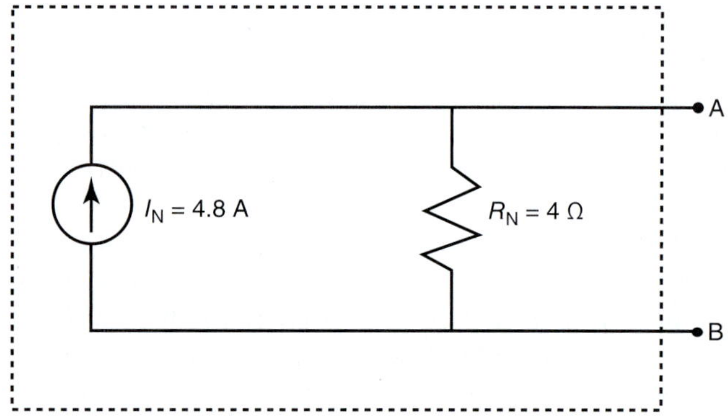

Figure 13-20. The Norton equivalent circuit for the circuit presented in **Figure 13-17.**

To calculate the Norton current, use the following formula:

$$I_N = \frac{V_{AB}}{R_N}$$

$$I_N = \frac{19.2\ V}{4\ \Omega}$$

$$I_N = 4.8\ A$$

As you can see, either method can be used to determine the Norton current, and the result will be the same.

Norton Equivalent Circuit

Combining the information obtained in the previous two sections, the Norton equivalent circuit can be constructed. A Norton equivalent circuit reduces a circuit to a single current source with a single parallel-connected resistance. This circuit has a Norton current of 4.8 A and a Norton resistance of 4 Ω, **Figure 13-20**.

13.5 Superposition Theorem

The ***superposition theorem*** is a tool that can be used to simplify and evaluate circuits with multiple voltage or current sources. When the superposition theorem is used, the circuit is evaluated multiple times, with only one source being active each time. As each circuit is evaluated with a single voltage source, the remaining voltage sources in the circuit are replaced with a single conductor, or short circuit.

Shedding Light

As with any circuit-related calculations, always be aware of the direction of current flow to ensure that the polarities are correct. In **Example 13-8**, the direction of current flow in R_4 was the same for both source one and source two so the two current values were simply added together. When the directions of flow are different, one value will be positive, while the other will be negative.

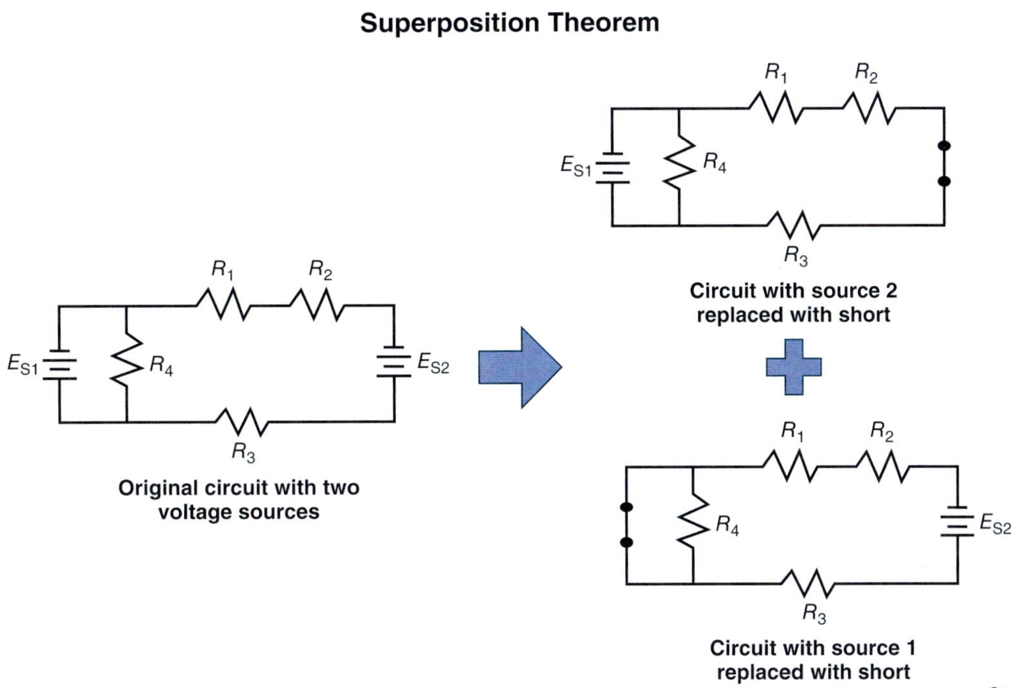

Figure 13-21. An electric circuit with two voltage sources, E_{S1} and E_{S2}. The superposition theorem can be used to make calculations with source two replaced with a short, then make calculations with source one replaced with a short. Next, add these values.

As an example, refer to the circuit shown in **Figure 13-21**. To use the superposition theorem, this circuit is first evaluated with voltage source E_{S1} as the active voltage source and voltage source E_{S2} replaced with a short circuit. Then, the circuit is evaluated again, this time with voltage source E_{S2} as the active voltage source and voltage source E_{S1} replaced with a short circuit.

The current flow through each circuit conductor, load, or component is then determined by adding up the individual current flows through those devices that are attributed to each individual voltage source.

Refer to **Example 13-8** for another scenario using the superposition theorem.

Example 13-8

Determine the current flow through R_4 in the following circuit.

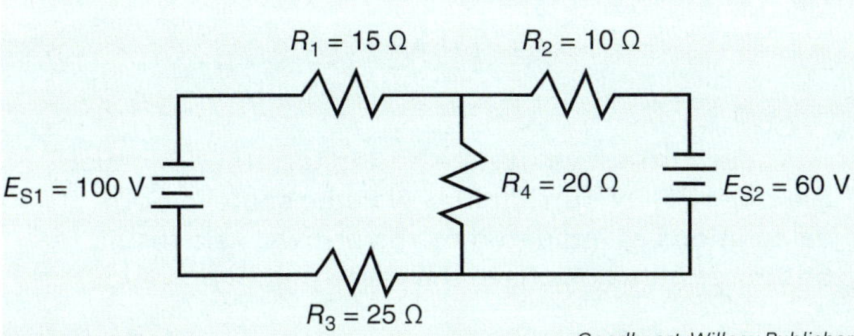

✓ Solution

The first step is to evaluate E_{S1} as the only power source by shorting out the voltage source E_{S2}. The arrows in the following circuit indicate the direction of current flow created by voltage source E_{S1}. Notice the direction of current flow through R_4 is from top to bottom.

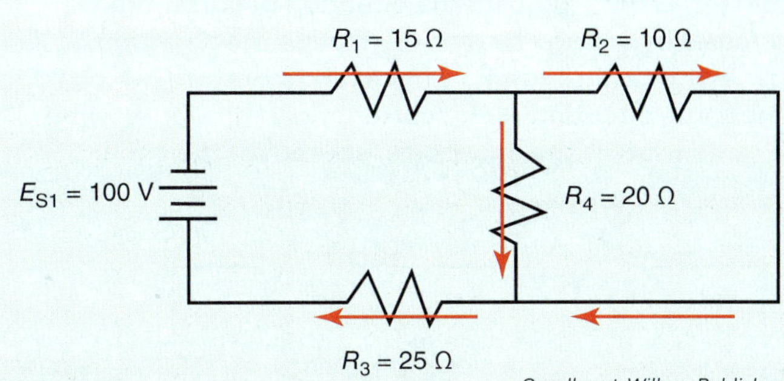

In this configuration, R_2 and R_4 are now connected in parallel with each other. This parallel combination is connected in series with both R_1 and R_3. The equivalent resistance $R_{2,4}$ can be calculated as follows:

$$R_{2,4} = \frac{10\ \Omega \times 20\ \Omega}{10\ \Omega + 20\ \Omega}$$

$$R_{2,4} = \frac{200\ \Omega^2}{30\ \Omega}$$

$$R_{2,4} = 6.67\ \Omega$$

Continued

With R_{TS1} representing the total circuit resistance with E_{S1} as the only active voltage source and $R_{2,4}$ representing the equivalent resistance of R_2 in parallel with R_4, the total resistance of the circuit can be calculated:

$$R_{TS1} = R_1 + R_{2,4} + R_3$$

$$R_{TS1} = 15\ \Omega + 6.67\ \Omega + 25\ \Omega$$

$$R_{TS1} = 46.67\ \Omega$$

Next, the total circuit current with E_{S1} as the only voltage source (I_{TS1}) can be calculated:

$$I_{TS1} = \frac{E_{S1}}{R_{TS1}}$$

$$I_{TS1} = \frac{100\ \text{V}}{46.67\ \Omega}$$

$$I_{TS1} = 2.143\ \text{A}$$

Using Ohm's law, the voltage drop across the $R_{2,4}$ parallel branches can then be determined:

$$E_{2,4} = I_{TS1} \times R_{2,4}$$

$$E_{2,4} = 2.143\ \text{A} \times 6.67\ \Omega$$

$$E_{2,4} = 14.3\ \text{V}$$

The current flow through R_4 can then be determined:

$$I_4 = \frac{E_{2,4}}{R_4}$$

$$I_4 = \frac{14.3\ \text{V}}{20\ \Omega}$$

$$I_4 = 0.71\ \text{A}$$

The next step in the process is to examine the circuit as if E_{S2} is the only power source. This is done by shorting out power source E_{S1}, as shown in the following circuit. The red arrows indicate the direction of current flow created by voltage source E_{S2}. Notice the direction of current flow through R_4 is from top to bottom, which is the same direction as the current flow through R_4 when E_{S1} was used as the only power source for the circuit.

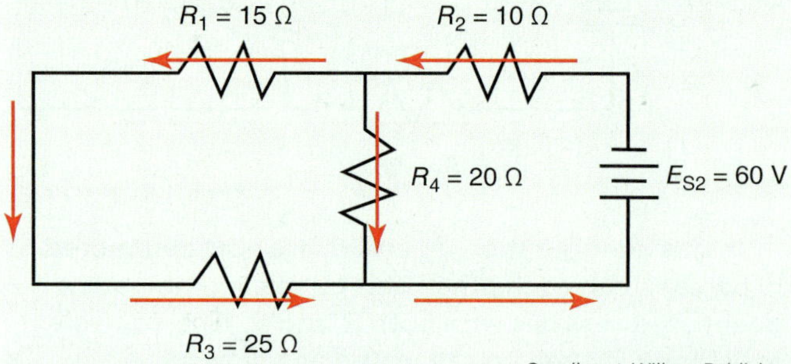

Goodheart-Willcox Publisher

In this circuit, series-connected resistances R_1 and R_3 have an equivalent resistance ($R_{1,3}$) of 40 Ω (15 Ω + 25 Ω). The equivalent resistance of parallel-connected $R_{1,3}$ and R_4 is then calculated:

$$R_{1,3,4} = \frac{40\ \Omega \times 20\ \Omega}{40\ \Omega + 20\ \Omega}$$

$$R_{1,3,4} = 13.33\ \Omega$$

With R_{TS2} representing the total circuit resistance with E_{S2} as the only active voltage source and $R_{1,3,4}$ representing the equivalent resistance of $R_{1,3}$ in parallel with R_4, the total resistance of the circuit can be calculated:

$$R_{TS2} = R_2 + R_{1,3,4}$$

$$R_{TS2} = 10\ \Omega + 13.33\ \Omega$$

$$R_{TS2} = 23.33\ \Omega$$

Then, the total circuit current with E_{S2} as the only voltage source (I_{TS2}) can be calculated:

$$I_{TS2} = \frac{E_{S2}}{R_{TS2}}$$

$$I_{TS2} = \frac{60\ \text{V}}{23.33\ \Omega}$$

$$I_{TS2} = 2.57\ \text{A}$$

Using Ohm's law, the voltage drop across the $R_{1,3,4}$ parallel branches can then be determined:

$$E_{1,3,4} = I_{TS2} \times R_{1,3,4}$$

$$E_{1,3,4} = 2.57\ \text{A} \times 13.33\ \Omega$$

$$E_{1,3,4} = 34.26\ \text{V}$$

Next, the current flow through R_4 can then be determined using Ohm's law:

$$I_4 = \frac{E_{1,3,4}}{R_4}$$

$$I_4 = \frac{34.26\ \text{V}}{20\ \Omega}$$

$$I_4 = 1.71\ \text{A}$$

The total current flowing through R_4 can now be determined by adding the I_4 value obtained when E_{S1} was the only voltage source to the value obtained when E_{S2} was the only voltage source:

$$I_4 = 0.71\ \text{A} + 1.71\ \text{A}$$

$$I_4 = 2.42\ \text{A}$$

These current values are added together because the direction of current flow was the same in both calculations as indicated by the arrows. As a result, both current values have the same polarity.

Now It's Your Turn 13-8

Using the superposition theorem, determine the current flow through R_3 in the following circuit.

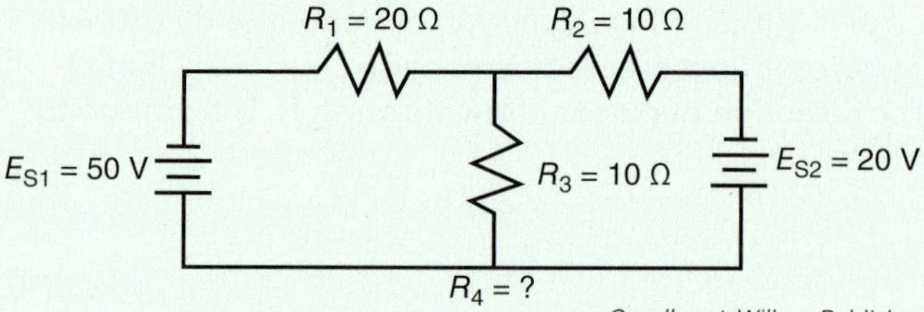

Goodheart-Willcox Publisher

CHAPTER 13 REVIEW

Summary

- Electrical laws and theorems can be used to simplify and evaluate electric circuits. (13.1)
- Kirchhoff's current law states that the mathematical sum of all current flows into and out of any circuit node, connection, or point must equal zero. (13.2)
- Kirchhoff's voltage law provides a means to evaluate the voltage drops that are present in a complete closed-circuit path. (13.2)
- Thevenin's theorem simplifies a complex circuit into a simple circuit with a single voltage source and a single resistance wired in series with it. (13.3)
- Norton's theorem simplifies a complex circuit into a simple circuit with a single current source and a single resistance wired in parallel with it. (13.4)
- The superposition theorem is used to evaluate circuits with more than one power source. Power sources can be any combination of current sources and voltage sources. (13.5)

Review Questions

Use the information in the chapter to answer the following questions.

Measuring Your Potential

1. A(n) _____ illustrates that all members of set A are also members of set B. (13.1)

   A. expression
   B. theorem
   C. Venn diagram
   D. law

2. The mathematical sum of all current flows into and out of any circuit node, connection, or point must equal zero is known as _____. (13.2)

   A. Kirchhoff's current law
   B. Kirchhoff's voltage law
   C. Norton's theorem
   D. Superposition theorem

3. _____ provides a means to evaluate the voltage drops that are present in a complete, series circuit. (13.2)

   A. Kirchhoff's current law
   B. Kirchhoff's voltage law
   C. Norton's theorem
   D. Superposition theorem

4. Of the following, this is used to simplify and evaluate circuits with multiple voltage or current sources. (13.3)

   A. Thevenin's theorem
   B. Ohm's law
   C. Norton's theorem
   D. Kirchhoff's current law

5. *True or False?* If three wires are connected together to form a node, and two of the wires each carry 2 A of current into the node, the third wire must carry 4 A of current out of the node. (13.2)

6. *True or False?* To calculate the Thevenin resistance, voltage sources are removed by replacing them with a short circuit. (13.3)

Use the following circuit to answer Questions 7 through 10.

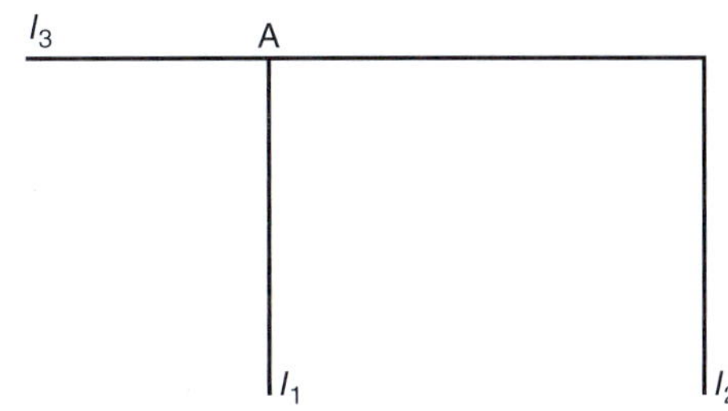

Goodheart-Willcox Publisher

7. If I_3 represents a current of 3 A flowing into node A and I_2 represents a current of 1 A flowing out of node A, then I_1 represents a current of _____. (13.2)

   A. 2 A flowing into node A
   B. 2 A flowing out of node A
   C. 4 A flowing into node A
   D. 4 A flowing out of node A

8. If I_3 represents a current of 3 A flowing into node A and I_2 represents a current of 1 A flowing into node A, then I_1 represents a current of _____. (13.2)

   A. 2 A flowing into node A
   B. 2 A flowing out of node A
   C. 4 A flowing into node A
   D. 4 A flowing out of node A

9. If I_3 represents a current of 2 A flowing out of node A and I_2 represents a current of 1 A flowing out of node A, then I_1 represents a current of _____. (13.2)

   A. 1 A flowing into node A
   B. 1 A flowing out of node A
   C. 3 A flowing into node A
   D. 3 A flowing out of node A

10. If I_3 represents a current of 3 A flowing out of node A and I_2 represents a current of 2 A flowing out of node A, then I_1 represents a current of _____. (13.2)

    A. 5 A flowing into node A
    B. 5 A flowing out of node A
    C. 1 A flowing into node A
    D. 1 A flowing out of node A

Plug-In

Use the following circuit to answer Questions 1 and 2.

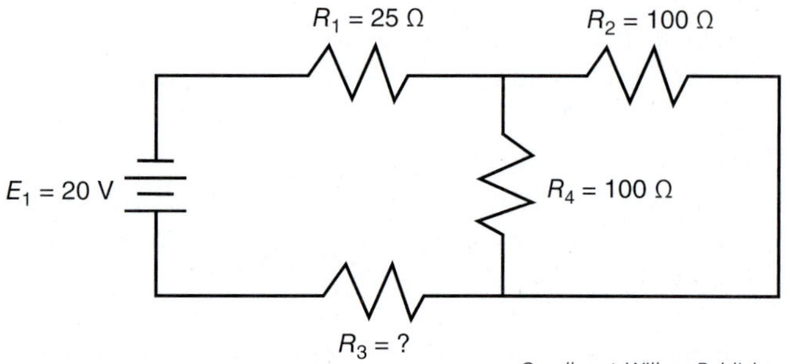

Goodheart-Willcox Publisher

1. If the total circuit current is 0.2 A, what must the voltage drop across R_3 be for Kirchhoff's voltage law to hold? (13.2)
   A. 5 V
   B. 10 V
   C. 15 V
   D. 20 V
2. If the total circuit current is 0.2 A, what is the value of R_3? (13.2)
   A. 5 Ω
   B. 15 Ω
   C. 25 Ω
   D. 35 Ω

Use the following circuit to answer Questions 3 through 8.

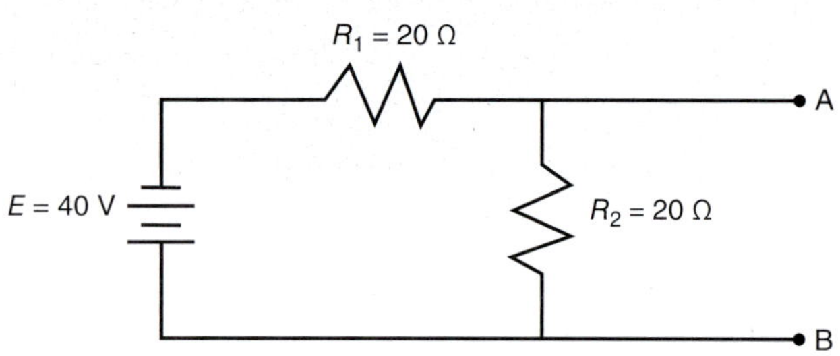

Goodheart-Willcox Publisher

3. The Thevenin voltage for the circuit is _____. (13.3)
   A. 10 V C. 30 V
   B. 20 V D. 40 V
4. The Thevenin resistance for the circuit is _____. (13.3)
   A. 10 Ω C. 20 Ω
   B. 15 Ω D. 30 Ω
5. The Thevenin equivalent circuit is made up of a voltage source in _____. (13.3)
   A. series with a current source
   B. parallel with another voltage source
   C. series with a resistance
   D. parallel with a resistance
6. The Norton current for the circuit is _____. (13.4)
   A. 1 A C. 3 A
   B. 2 A D. 4 A
7. The Norton resistance for the circuit is _____. (13.4)
   A. 10 Ω C. 20 Ω
   B. 15 Ω D. 30 Ω
8. The Norton equivalent circuit is made up of a current source in _____. (13.4)
   A. series with another current source
   B. parallel with a voltage source
   C. series with a resistance
   D. parallel with a resistance

Making the Connection

1. Create the Norton equivalent circuit for the following circuit; calculate both Norton resistance and Norton current. (13.4)

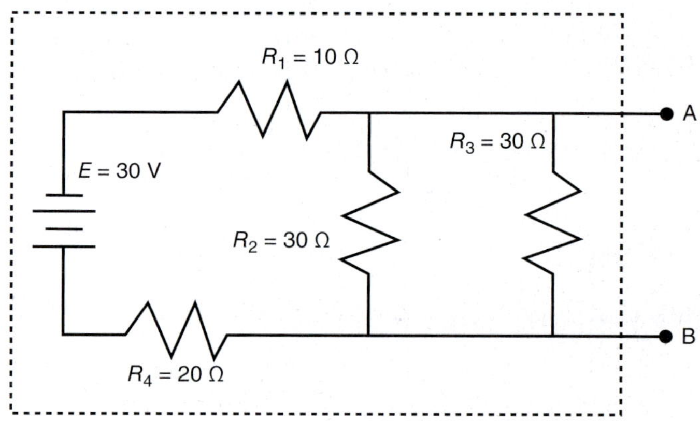

Goodheart-Willcox Publisher

2. Referring to the following circuit, determine the current flowing through R_4. (13.5)

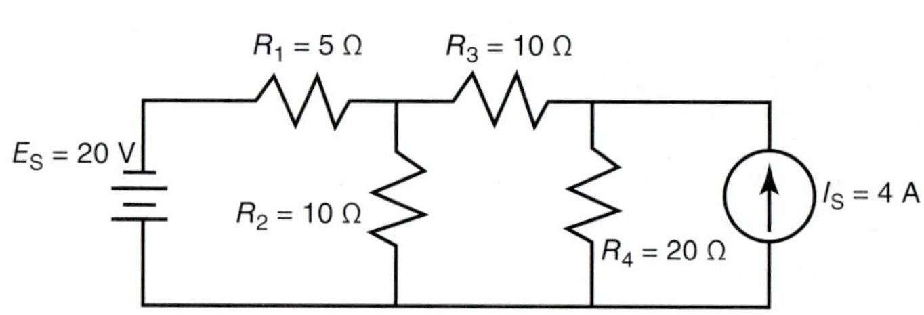

Goodheart-Willcox Publisher

SECTION

4

AC Fundamentals

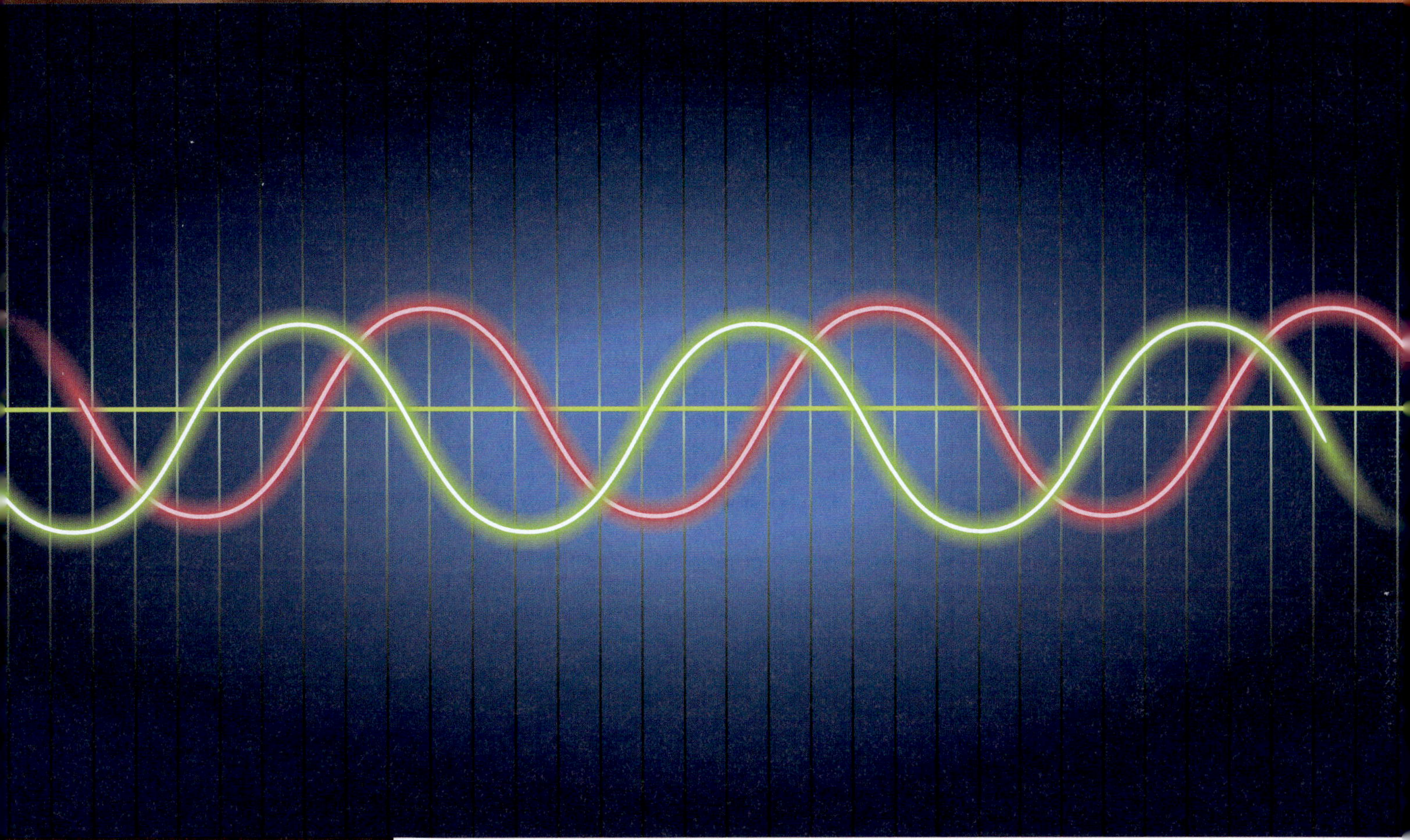

Marisha/Shutterstock.com

CHAPTER

14 Angles, Triangles, and Vectors

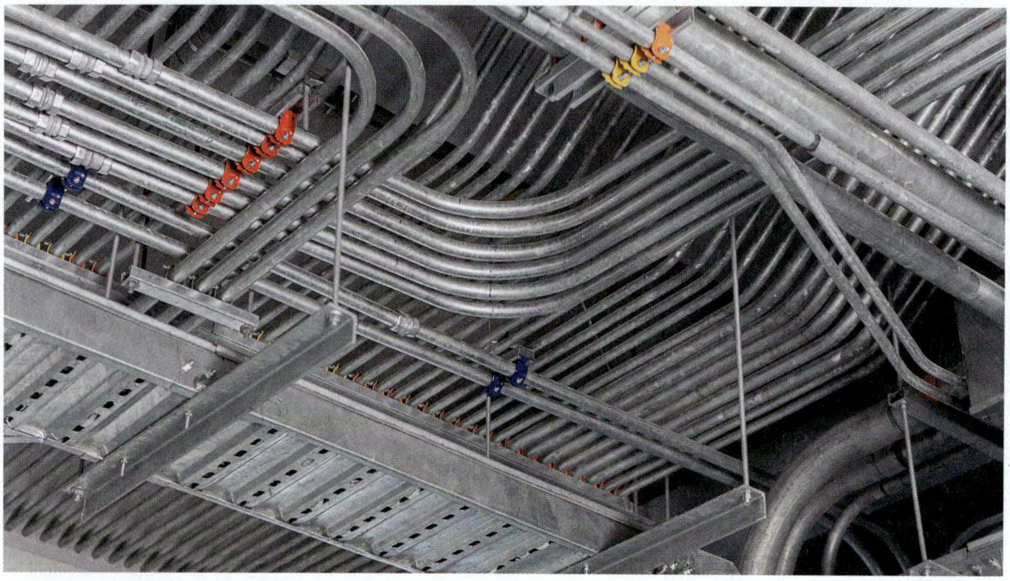

SB7/Shutterstock.com

SPARKING DISCUSSION

Geometry, algebra, and trigonometry are used to solve problems involving the flow of ac electricity and to measure efficiently the current flow at the load. Describe a practical jobsite scenario where an understanding of math is important to being a successful electrician.

LEARNING OUTCOMES

After completing this chapter, you will be able to:

14.1 Differentiate between supplementary and complementary angles.

14.2 Describe the properties of four common triangle types.

14.3 Use the Pythagorean theorem to solve for the unknown side of a triangle.

14.4 Use basic trigonometry to solve for sides and angles of right triangles.

14.5 Perform vector calculations.

TECHNICAL TERMS

acute angle
arccosine (arccos)
arcsine (arcsin)
arctangent (arctan)
complementary angle
cosine
equilateral triangle
hypotenuse
interior angle
inverse cosine
inverse sine
inverse tangent
isosceles right triangle
isosceles triangle
obtuse angle
Pythagorean theorem
Pythagorean triple
reflex angle
resultant vector
right angle
right triangle
scalene triangle
sine
straight angle
supplementary angle
tangent
vector
vector head
vector tail
vertex

Introduction

Having chosen to obtain a career in the electrical trades, it is very likely that you have the ability to think spatially and are able to easily visualize shapes and pictures. These are valuable skills to have as we begin our discussion of geometry and trigonometry.

Geometry and trigonometry are important elements in the study, application, and installation of electrical systems and components. To become proficient in complex conduit bending, for example, the use of basic algebra, geometry, and trigonometry is essential. Conduit bends involving offsets and concentric 90° bends require not only basic math and algebra but bending measurements derived from the use of trigonometry. Every job will have exposed lighting, and most will have concrete-concealed or buried conduit that may require an electrician to use basic algebra, geometry, and trigonometry. AC electrical theory relies heavily on trigonometry to help illustrate the expanding and contracting magnetic field and determine the actual amount of power, in watts, that is utilized by a load or multiple loads. Further, in the study of alternating current, trigonometry holds the key to tying together the opposite acting forces of inductance and capacitance.

We all get excited when we purchase a new tool for our toolbox and cannot wait to put it to use. Trigonometry is no different. It can strengthen your understanding of electrical theories. If you think of it as a tool that can be used to help you better understand the system or circuit you are working on, you will become more effective at your job. It is the goal of this chapter to present the material in a manner that stimulates the advancement of your electrical education. Consider it a first step in putting this shiny new tool in your toolbox.

14.1 Angles

In their simplest form, angles are formed when two lines intersect at a single point. An angle is the measurement of the opening between these two lines, **Figure 14-1**. Angles are measured in degrees. The smallest angle is 0° and the largest angle is 360°, **Figure 14-2**.

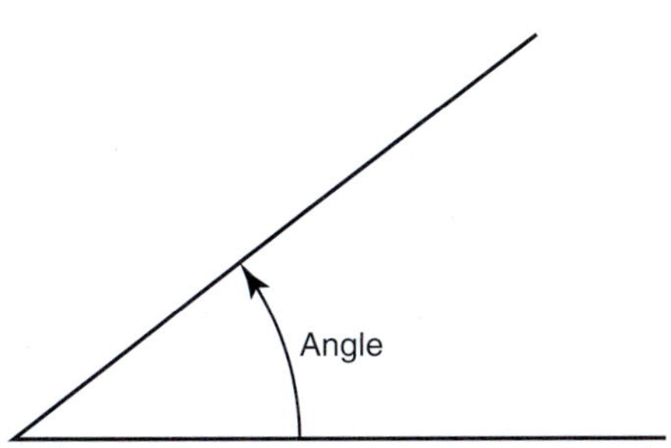

Goodheart-Willcox Publisher

Figure 14-1. An angle is the measurement of the opening between two intersecting lines.

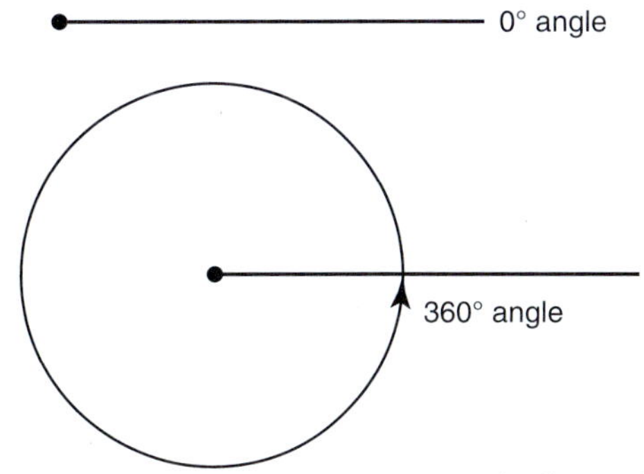

Goodheart-Willcox Publisher

Figure 14-2. In a 0° angle, the opening between the two intersecting lines is nonexistent. A 360° angle consists of a complete circular path around a point that starts and ends at the same point.

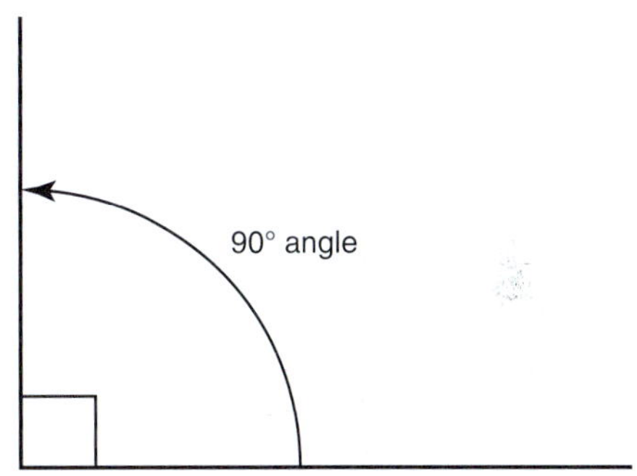

Goodheart-Willcox Publisher

Figure 14-3. A right angle is 90°. The small square drawn at the intersection point indicates that the angle is exactly 90°.

The most common angle is the right angle. A ***right angle*** is formed when two lines intersect each other in a perpendicular fashion. This means that these two lines intersect at an angle of 90° with each other. A small square is often drawn at the intersection point between these two lines to indicate that the two lines form a right angle, **Figure 14-3**.

Other common angles include straight angles, acute angles, obtuse angles, and reflex angles, **Figure 14-4**. When two lines intersect each other to form a straight line, the angle formed between these two lines is 180°. The angle formed between these two lines is called a ***straight angle***. ***Acute angles*** are formed when two lines intersect to create an angle that is less than 90°. Angles that are greater than 90° but less than 180° are called ***obtuse angles***. ***Reflex angles*** are formed when two lines intersect to create an angle that is greater than 180° but less than 360°.

Other angles include complementary and supplementary angles. ***Complementary angles*** refer to any two angles whose sum equals 90°. For example, the angles shown in **Figure 14-5** are complementary angles as the two angles (30° and 60°) complement each other, meaning the two angles equal 90°. ***Supplementary angles*** refer to any two angles whose sum equals 180°. The angles shown in **Figure 14-6** are supplementary angles. The 130° angle and 50° angle are supplements of each other since 130° plus 50° equals 180°.

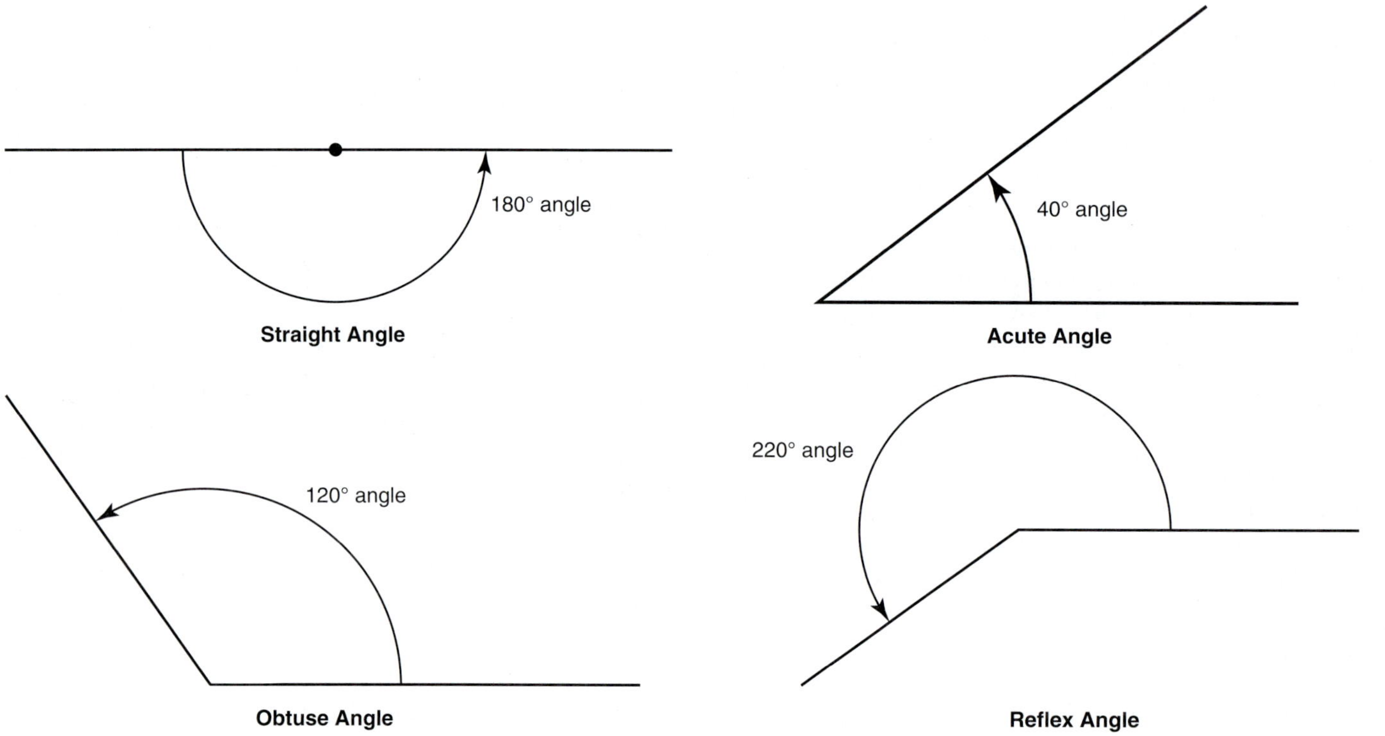

Goodheart-Willcox Publisher

Figure 14-4. The four common angles include straight angles, acute angles, obtuse angles, and reflex angles.

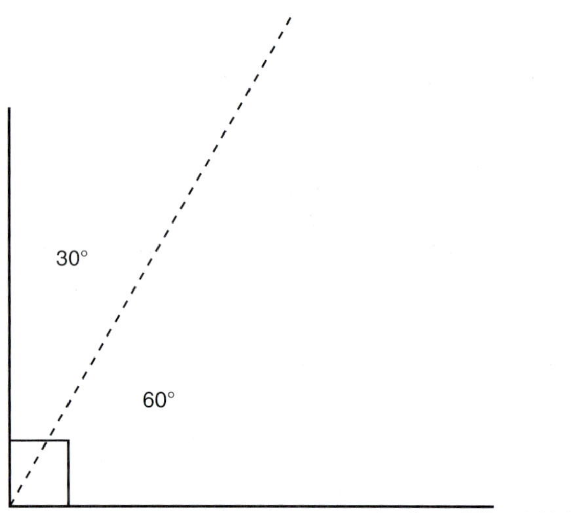

Goodheart-Willcox Publisher

Figure 14-5. Complementary angles have two angles that equal 90°.

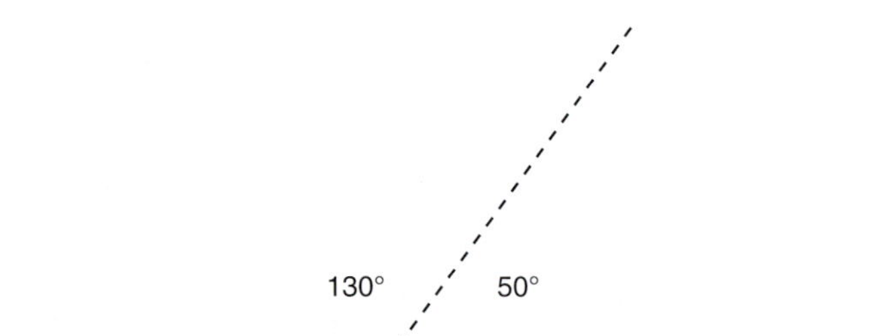

Goodheart-Willcox Publisher

Figure 14-6. Supplementary angles have two angles that equal 180°.

14.2 Triangles

Triangles are one of the most basic geometric shapes. If you wish to gain a solid understanding of vectors, forces, directional magnetic fields, and other important electricity-related topics and concepts, having a solid grasp of the characteristics and properties of triangles is important. The five important properties that all triangles have in common are illustrated in **Figure 14-7**. These properties include the following:

- All triangles have three sides.
- All triangles have three vertices, or corners. Each corner of a triangle is called a ***vertex***.
- All triangles have three interior angles. An ***interior angle*** of a triangle is an angle formed by two adjacent sides of a triangle.
- The sum of all three interior angles of a triangle must add up to 180°. Refer to **Example 14-1**.
- The sum of the lengths of any two sides of a triangle must always be greater than the length of the third side. Refer to **Example 14-2**.

Math Sidebar

In mathematics, the vertices of triangles are normally labeled as *A*, *B*, and *C*, with *C* being the right angle in a right triangle. The sides of a triangle are labeled *a*, *b*, and *c*, with letters corresponding to the opposite angle. For example, angle *A* is opposite side *a*, and side *c* is the hypotenuse in a right triangle.

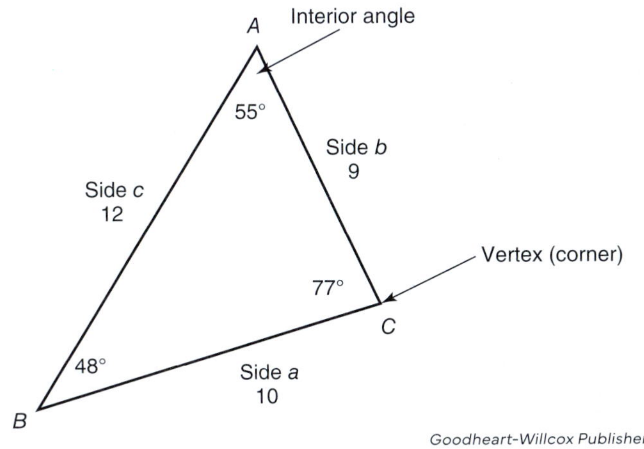

Figure 14-7. A triangle consists of three sides, three corners, and three interior angles; the sum of the three interior angles adds up to 180°.

Example 14-1

If two of the interior angles of a triangle are 30° and 70°, what is the measurement of the third interior angle?

✓ Solution

The sum of all three interior angles of a triangle must add up to 180°. If two of the angles are known, the third angle can be found by subtracting the two known angles from 180 to obtain the value of the third angle:

$$180° - 30° - 70° = 80°$$

Now It's Your Turn 14-1

If two of the interior angles of a triangle are 40° and 80°, determine the measurement of the third interior angle.

Example 14-2

Can a triangle be formed with three-line segments of lengths 9, 11, and 22?

✓ Solution

The sum of the lengths of any two sides of a triangle will always be greater than the length of the third side. In this example, the sum of the two shorter sides is 20 (9 + 11), which is less than the length of the longest side (22). Therefore, these three-line segments cannot form a triangle. You may want to visualize or even attempt to create a scaled drawing of the triangle with the lengths provided.

Now It's Your Turn 14-2

Can a triangle be formed with three line segments of lengths 9, 8, and 7?

Right Triangle

The term right is used to describe a 90° angle. A right angle is a 90° angle, and a ***right triangle*** is a triangle that has one 90° angle. The right triangle is the most useful triangle configuration, as two of its sides are perpendicular to each other just like the X and Y axes on an x-y coordinate scale, **Figure 14-8**. The longest side of a right triangle is found opposite the 90° angle and is called the ***hypotenuse***.

There are two special right triangles that are commonly encountered. These two triangles are the 30-60-90 triangle and the 45-45-90 triangle. The interior angles of a 30-60-90 triangle are 30°, 60°, and 90°, **Figure 14-9**. The lengths of the sides of this triangle have a ratio of 1 to $\sqrt{3}$ to 2 opposite the 30°, 60°, and 90°angles, respectively. When a 30-60-90 triangle is encountered, along with at least one side length,

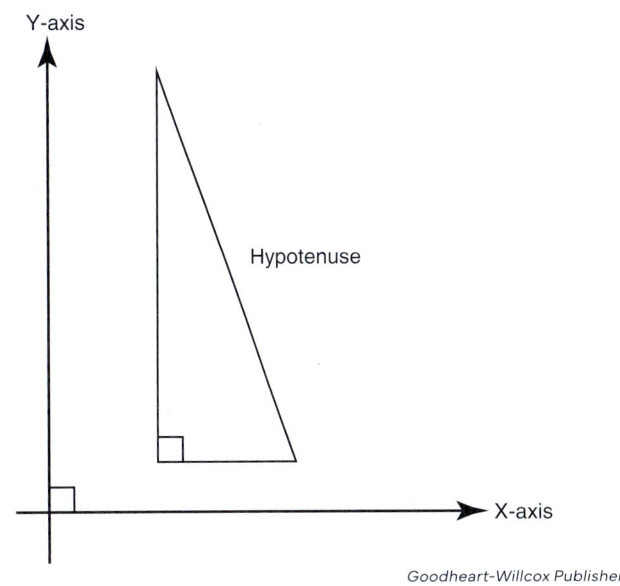

Figure 14-8. A right triangle has two perpendicular sides that form a 90° angle.

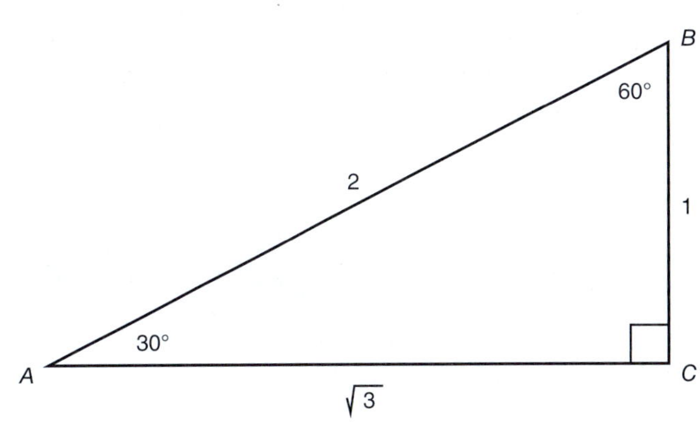

Figure 14-9. A 30-60-90 triangle consists of a 30°, a 60°, and a 90° angle. If one side length is known, the other two can be determined.

the remaining sides can be immediately determined using the 1, $\sqrt{3}$, 2 relationship or ratio:

$$\text{unknown length} = \text{known length} \times \frac{\text{unknown ratio value}}{\text{known ratio value}}$$

Example 14-3 shows how to calculate an unknown side of a 30-60-90 triangle.

Example 14-3

Determine the lengths of sides *b* and *c* for the following 30-60-90 triangle.

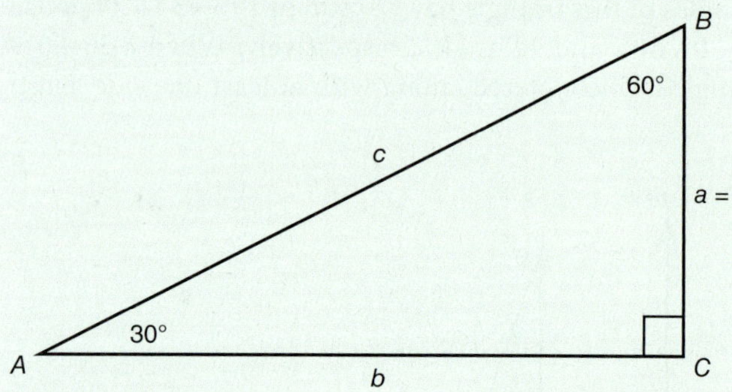

Goodheart-Willcox Publisher

✓ Solution

First, find the unknown length *b*. The ratio of *a* to *b* is 1 to $\sqrt{3}$, and known length *a* is 5:

$$\text{unknown length} = \text{known length} \times \frac{\text{unknown ratio value}}{\text{known ratio value}}$$

$$b = 5 \times \frac{\sqrt{3}}{1}$$

$$b = 8.66$$

Find the unknown length *c*. The ratio of *a* to *c* is 1 to 2, and known length *a* is 5:

$$c = 5 \times \frac{2}{1}$$

$$c = 10$$

Now It's Your Turn 14-3

Referring to the following triangle, determine the lengths of sides *a* and *b*.

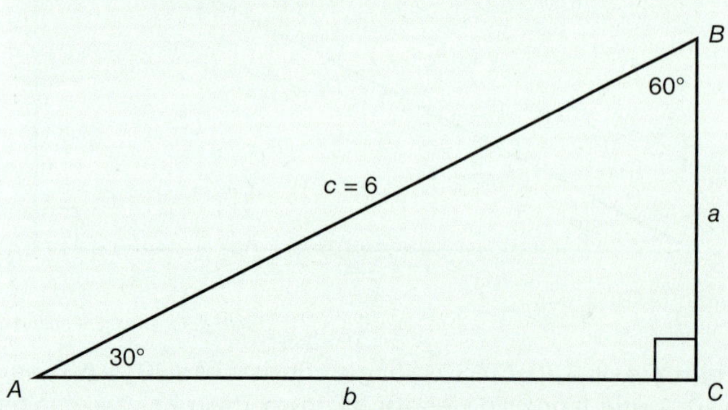

Goodheart-Willcox Publisher

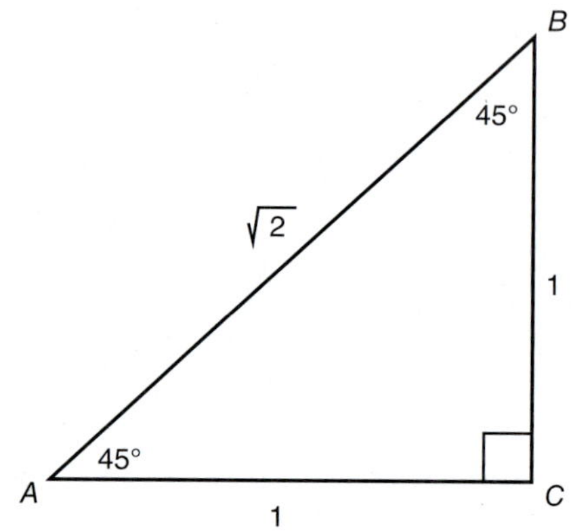

Goodheart-Willcox Publisher

Figure 14-10. A 45-45-90 triangle consists of a 45°, a 45°, and a 90° angle. If one side length is known, the other two can be determined.

As the name implies, the angles of a 45-45-90 triangle are 45°, 45°, and 90°. The lengths of the sides of this triangle have a ratio of 1 to 1 to $\sqrt{2}$ opposite the 45°, 45°, and 90° angles, respectively, **Figure 14-10**. When a 45-45-90 triangle is encountered, along with at least one side length, the remaining sides can be determined using the 1, 1, $\sqrt{2}$ relationship or ratio, using the equation presented earlier:

$$\text{unknown length} = \text{known length} \times \frac{\text{unknown ratio value}}{\text{known ratio value}}$$

Example 14-4 provides steps to calculate unknown sides of a 45-45-90 triangle.

Scalene Triangle

A ***scalene triangle*** is a triangle that has three sides all with different lengths, and three angles all with different measurements. A scalene triangle is not symmetrical, **Figure 14-11**. Some right triangles may also be scalene triangles, while others are not.

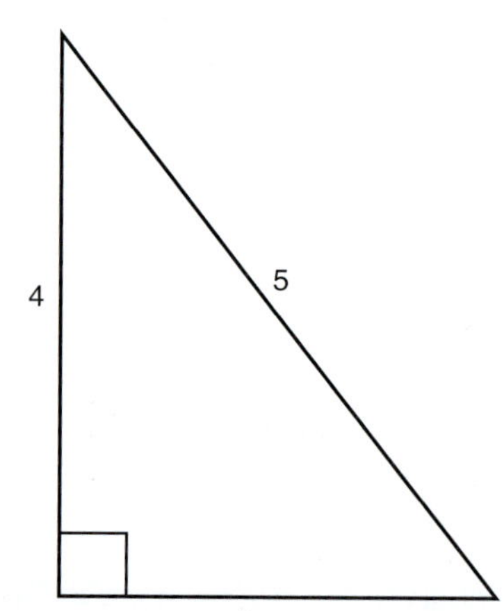

Goodheart-Willcox Publisher

Figure 14-11. A triangle that has no equal sides or angles is known as a scalene triangle.

Example 14-4

Referring to the following 45-45-90 triangle, determine the lengths of sides *b* and *c*.

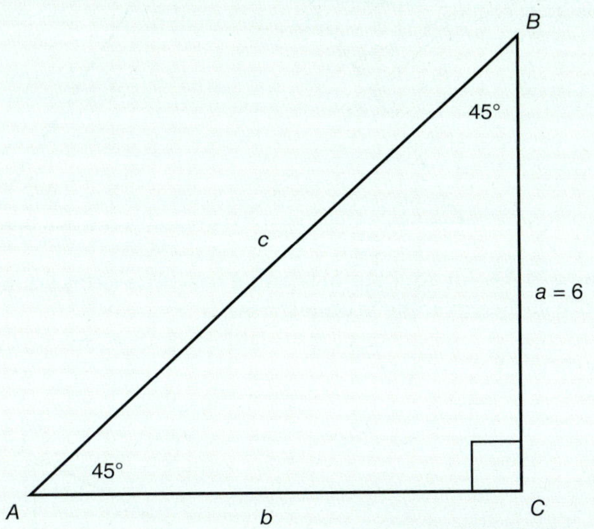

Goodheart-Willcox Publisher

✓ Solution

Both legs of a 45-45-90 triangle are the same length (1 to 1 ratio), so because the length of side *a* is 6, the length of side *b* is also 6.

The ratio of a leg to the hypotenuse of a 45-45-90 triangle is 1 to $\sqrt{2}$, so the length of side *c* can be calculated as follows:

$$\text{unknown length} = \text{known length} \times \frac{\text{unknown ratio value}}{\text{known ratio value}}$$

$$c = 6 \times \frac{\sqrt{2}}{1}$$

$$c = 8.49$$

Now It's Your Turn 14-4

Referring to the following triangle, determine the lengths of sides *a* and *b*.

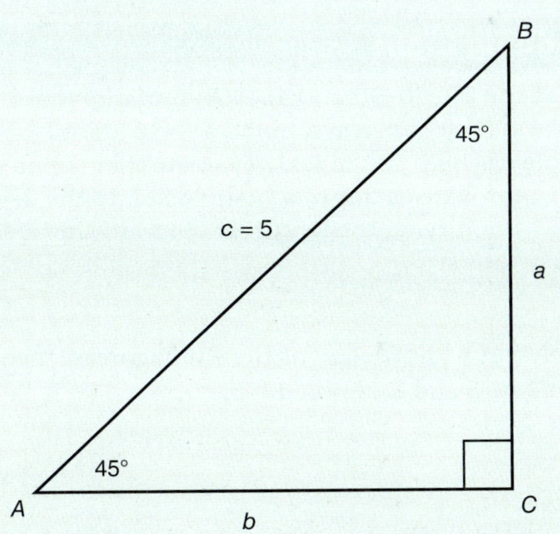

Goodheart-Willcox Publisher

Isosceles Triangle

An ***isosceles triangle*** is a triangle that has two equal sides and two equal angles. The two angles opposite the equal sides will always be the equal angles. Isosceles triangles are symmetrical as indicated by a line that cuts the unique angle in half, **Figure 14-12**. Some isosceles triangles may also be right triangles; however, they can never be scalene triangles as two sides are the same length. An ***isosceles right triangle*** will always have one 90° angle and two 45° angles, **Figure 14-13**. The isosceles right triangle is also a 45-45-90 right triangle.

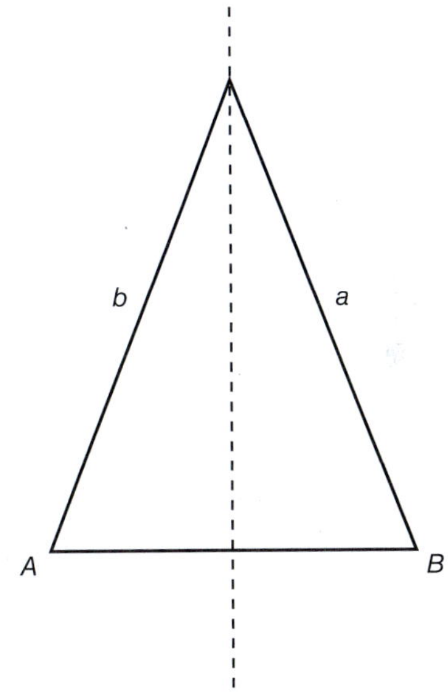

Goodheart-Willcox Publisher

Figure 14-12. An isosceles triangle has two equal sides and two equal angles; it is symmetrical as shown by the dotted line.

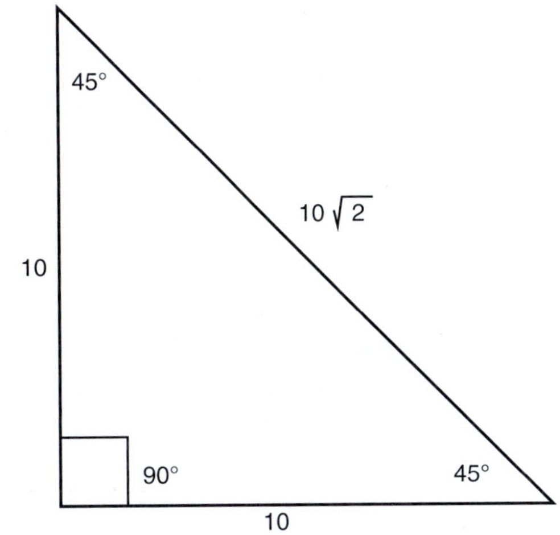

Goodheart-Willcox Publisher

Figure 14-13. An isosceles right triangle is also known as a 45-45-90 right triangle.

Equilateral Triangle

An ***equilateral triangle*** is a triangle that has three equal sides and three equal angles, **Figure 14-14**. All three angles in an equilateral triangle are equal to 60°. An equilateral triangle is unique in the sense that it cannot be a right triangle, a scalene triangle, or an isosceles triangle.

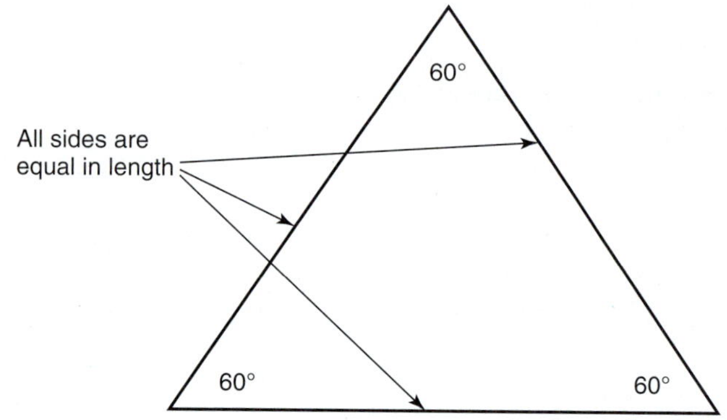

Figure 14-14. An equilateral triangle has all its sides equal in length and each angle is equal to 60°.

14.3 Pythagorean Theorem

The Pythagorean theorem is a powerful tool that can be used effectively to evaluate right triangles. The ***Pythagorean theorem*** states that the square of the longest side of the triangle, or hypotenuse, c, is equal to the sum of the squares of the other two sides, a and b, of the triangle. Using the Pythagorean theorem, the length of one side of a right triangle can be found if the lengths of the other two sides are known. The Pythagorean theorem is written as follows:

$$a^2 + b^2 = c^2$$

Math Sidebar

The formula for the Pythagorean theorem can be modified to determine the length of any side of the triangle that is unknown.

When the hypotenuse c is unknown, use:

$$c^2 = a^2 + b^2$$

When side a is unknown, use:

$$a^2 = c^2 - b^2$$

When side b is unknown, use:

$$b^2 = c^2 - a^2$$

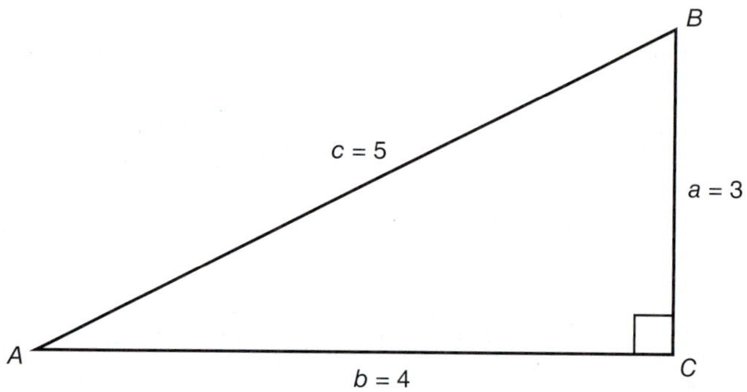

Figure 14-15. The right triangle shown here is also known as a 3-4-5 Pythagorean triple.

Consider the right triangle shown in **Figure 14-15**. The hypotenuse of the triangle is 5 units in length, while the other two sides are 3 and 4 units in length. From the Pythagorean theorem, it can be seen that:

$$a^2 + b^2 = c^2$$
$$3^2 + 4^2 = 5^2$$
$$(3 \times 3) + (4 \times 4) = (5 \times 5)$$
$$9 + 16 = 25$$
$$25 = 25$$

The triangle shown in **Figure 14-15** is known as a 3-4-5 triangle. This is one of two common triangles referred to as a ***Pythagorean triple***. A Pythagorean triple is a right triangle in which the length of each side is an integer, or whole number.

Shedding Light

Just as the 30-60-90 and 45-45-90 triangles are special in their own ways, being able to identify a 3-4-5 triangle will help you quickly evaluate them once you spot them. If two sides of a triangle fit into the 3-4-5 format, identifying the length of the third side will be an easy task.

Apply your knowledge of the Pythagorean theorem in **Example 14-5** and **Example 14-6**.

Example 14-5

Determine the length of side *a* for the following triangle.

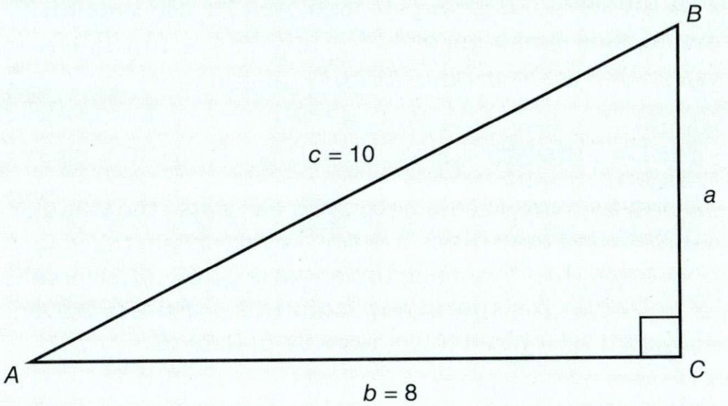

Goodheart-Willcox Publisher

✓ Solution

As side *a* is unknown, you will need to use the following variation of the Pythagorean theorem:

$$a^2 = c^2 - b^2$$
$$a^2 = 10^2 - 8^2$$
$$a^2 = 100 - 64$$
$$a^2 = 36$$
$$\sqrt{a^2} = \sqrt{36}$$
$$a = 6$$

The length of side *a* is 6 units in length. This triangle is a 3-4-5 triangle. Another way to solve this example is by inspection. The hypotenuse (*c*) is 2 times 5, side *b* is 2 times 4, so side *a* must be 2 times 3, which results in 6!

Example 14-6

Determine the length of side *b* for the following triangle.

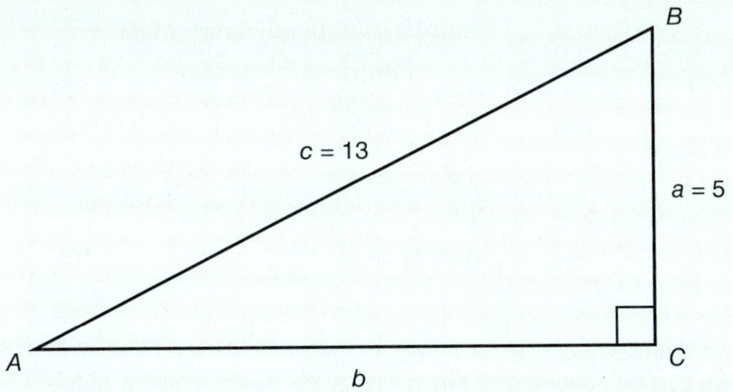

Goodheart-Willcox Publisher

✓ Solution

Using Pythagorean theorem to solve for side *b*:

$$b^2 = c^2 - a^2$$
$$b^2 = 13^2 - 5^2$$
$$b^2 = 169 - 25$$
$$b^2 = 144$$
$$\sqrt{b^2} = \sqrt{144}$$
$$b = 12$$

The length of side *b* is 12 units in length. The triangle in this example is another Pythagorean triple. This triple is a 5-12-13 triangle.

Now It's Your Turn 14-6

Referring to the following triangle, determine the length of the hypotenuse *c*.

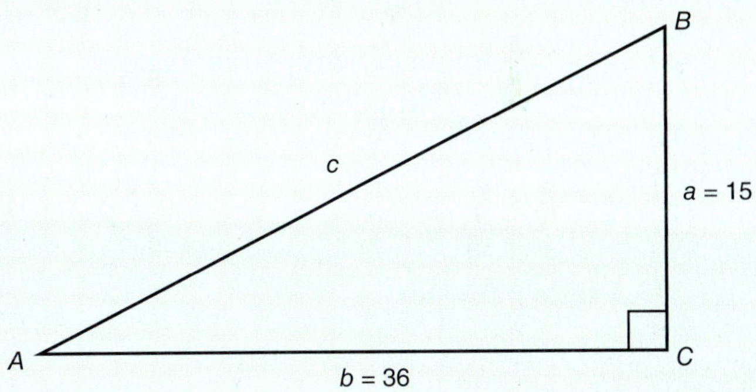

Goodheart-Willcox Publisher

14.4 Trigonometric Functions

There are three common relationships that exist between angles in a right triangle and the lengths of the sides of the triangle. These relationships, called trigonometric functions, are important to understand if you intend to work in electrical trades.

Shedding Light

Trigonometry will help you fully comprehend how alternating current works, how the opposite-acting forces of inductive reactance and capacitive reactance react with each other, and how to determine the amount of useful power delivered to a load.

Trigonometric Values

Angle (Degree)	Cosine	Sine	Tangent
0	1.000	0.000	0.000
15	0.966	0.259	0.268
30	0.866	0.500	0.577
45	0.707	0.707	1.000
60	0.500	0.866	1.732
75	0.259	0.966	3.732
90	0.000	1.000	Undefined

Goodheart-Willcox Publisher

Figure 14-16. Sine, cosine, and tangent values for some common angles.

The three trigonometric functions are sine, cosine, and tangent. The sine, cosine, and tangent values for some commonly used angles are provided in **Figure 14-16**.

Sine

The ***sine*** of an angle in a right triangle refers to the ratio of the length of the side opposite the angle divided by the length of the hypotenuse of the triangle. Consider the right triangle shown in **Figure 14-17**. This triangle has sides of lengths 5 (*a*), 12 (*b*), and 13 (*c*). The side opposite angle A is side *a*, which has a length of 5. The hypotenuse of the triangle, side *c*, has a length of 13. The sine of angle A can then be expressed as follows:

$$\sin A = \frac{\text{length of side opposite angle } A}{\text{length of hypotenuse}}$$

$$\sin A = \frac{a}{c}$$

$$\sin A = \frac{5}{13}$$

$$\sin A = 0.3846$$

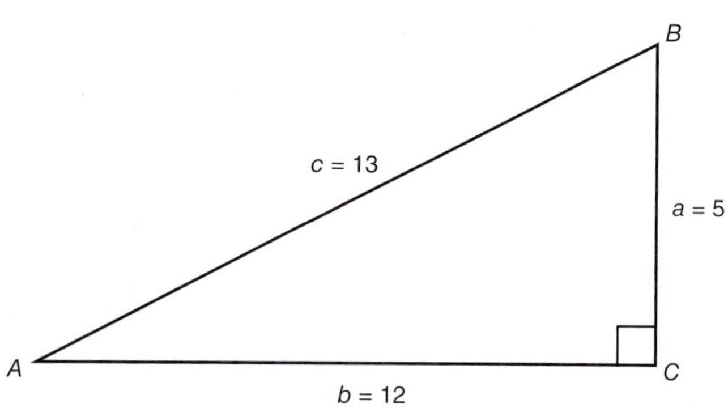

Goodheart-Willcox Publisher

Figure 14-17. Angle *A* can be calculated by dividing the length of the side opposite angle *A* by the length of the hypotenuse, and then taking the inverse sine of that value.

A term associated with the sine function is the ***inverse sine*** function. The inverse sine function symbol is written as $\sin^{-1}$. The inverse sine symbol can be translated as "the angle whose sine is." Since we know from the previous example that the sine of angle *A* is 0.3846, we can also state that the $\sin^{-1}$ 0.3846 equals angle *A*. The inverse sine function is also referred to as the ***arcsine (arcsin)*** function.

Math Sidebar

Using a scientific calculator, you can enter the value of 0.3846 and then press the $\sin^{-1}$ key to obtain the value for angle *A*. In this case, the angle is 22.62°. Conversely, if you enter the known angle of 22.62 and press the sin key, you will obtain the sine value of 0.3846.

Cosine

The ***cosine*** of an angle in a right triangle refers to the ratio of the length of the side adjacent to the angle divided by the hypotenuse of the triangle. To calculate the cosine for **Figure 14-17**, the side adjacent to angle *A* is side *b*, which has a length of 12. The hypotenuse of the triangle, side *c*, has a length of 13. The cosine of angle *A* can then be expressed as follows:

$$\cos A = \frac{\text{length of side adjacent angle } A}{\text{length of hypotenuse}}$$

$$\cos A = \frac{b}{c}$$

$$\cos A = \frac{12}{13}$$

$$\cos A = 0.9231$$

Just as with the inverse sine function, the ***inverse cosine*** function can be used to determine an angle given a cosine value. The inverse cosine function is written as $\cos^{-1}$. The inverse cosine symbol can be translated as "the angle whose cosine is." Since we know that the cosine of angle *A* is 0.9231, we can also state that the $\cos^{-1}$ 0.9231 equals angle *A*. The inverse cosine function is also referred to as the ***arccosine (arccos)*** function.

Math Sidebar

Using a scientific calculator, you can enter the value of 0.9231 and then press the $\cos^{-1}$ key to obtain the value for angle *A*. Referring to **Figure 14-17**, the angle is 22.62°. Conversely, if you enter the known angle of 22.62 and press the cos key, you will obtain the cosine value of 0.9231.

Tangent

The ***tangent*** of an angle in a right triangle refers to the ratio of the length of the side opposite the angle divided by the length of the side that is adjacent to the angle. Referring to the right triangle in **Figure 14-17**, the side opposite to angle *A* has a length of 5, while the side adjacent to the angle has a length of 12. The tangent of angle A can then be expressed as follows:

$$\tan A = \frac{\text{length of side opposite angle } A}{\text{length of side adjacent angle } A}$$

$$\tan A = \frac{a}{b}$$

$$\tan A = \frac{5}{12}$$

$$\tan A = 0.4167$$

Just as with the previous functions, the ***inverse tangent*** is written as $\tan^{-1}$ and is used to determine an angle corresponding to a given tangent value. Since we know that the tangent of angle A is 0.4167, we can also state that the $\tan^{-1} 0.4167$ is equal to angle A. The inverse tangent function is also referred to as the ***arctangent (arctan)*** function. You can verify that angle A is 22.62° by using the $\tan^{-1}$ key on a scientific calculator.

Shedding Light

To make it easier to remember the mathematical formulas for sine, cosine, and tangent functions, a popular acronym is used: SOHCAHTOA or SOH-CAH-TOA. This acronym can be broken down into three-letter chunks:

SOH stands for sine, opposite, and hypotenuse:

$$\text{Sine (sin)} = \frac{\text{Opposite (O)}}{\text{Hypotenuse (H)}}$$

CAH stands for cosine, adjacent, and hypotenuse:

$$\text{Cosine (cos)} = \frac{\text{Adjacent (A)}}{\text{Hypotenuse (H)}}$$

TOA stands for tangent, opposite, and adjacent:

$$\text{Tangent (tan)} = \frac{\text{Opposite (O)}}{\text{Adjacent (A)}}$$

14.5 Vectors

A ***vector*** is a directional line segment that has magnitude and direction. Magnitude can refer to length or, in the study of electricity, the strength of a force, **Figure 14-18**. The magnitude of a vector, even when used to identify the strength of a force, is indicated by the length of the line segment. A vector

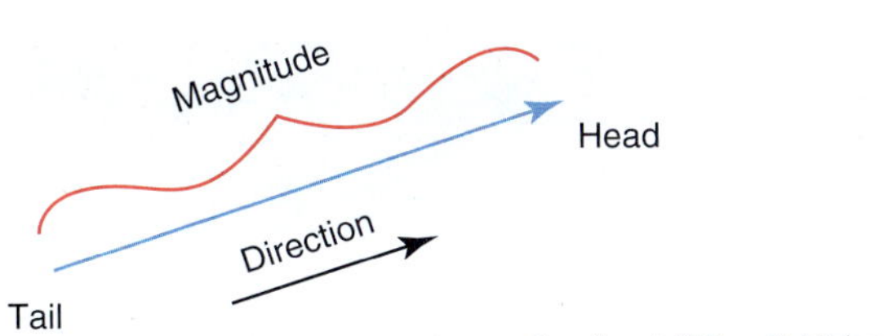

Figure 14-18. A vector indicates both magnitude and direction. An arrowhead on the line shows the direction of the force, and the length of the vector shows the magnitude.

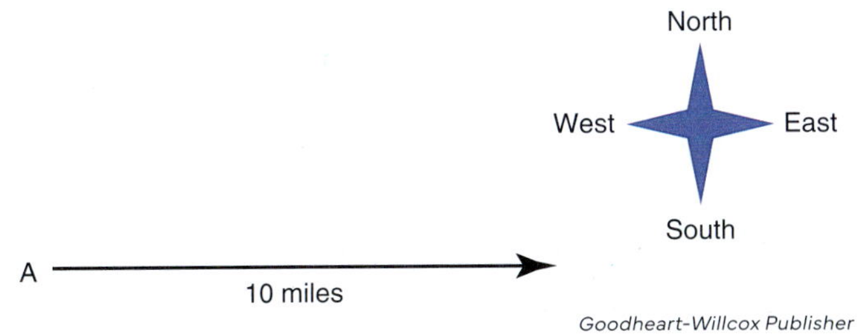

Figure 14-19. A vector representing a driving distance.

has two ends, which are identified as the vector tail and the vector head. The ***vector tail*** indicates the starting point of the vector, while the ***vector head*** indicates the ending point of the vector. The vector head is indicated by the arrowhead.

Shedding Light

Driving due east, a distance of ten miles from reference point A, can be represented by a vector ten units in length pointing east, **Figure 14-19**.

Adding Vectors

In many instances, there are multiple vectors acting on a system or object. To determine the result of multiple vectors acting on a system or object, the vectors must be added together. This is accomplished by connecting the head of one vector to the tail of another. This process continues until all vectors have been accounted for. The ***resultant vector***, which is the cumulative effect of all the vectors in the system, is drawn with its tail located at the tail of the first vector and its head located at the head of the final vector, **Figure 14-20**.

Just like in basic addition, where 3 + 5 + 8 will yield the same result as 5 + 8 + 3, vectors can be added in any order. Consider the following three vectors on an x-y coordinate system:

- Vector A—A three-pound force pushing in the +x direction.
- Vector B—A four-pound force pushing in the +y direction.
- Vector C—A five-pound force pushing in the –x direction.

Showing two of the six possible combinations of adding the three vectors, the resultant vector will be the same, regardless of the order in which the vectors are added, **Figure 14-21**.

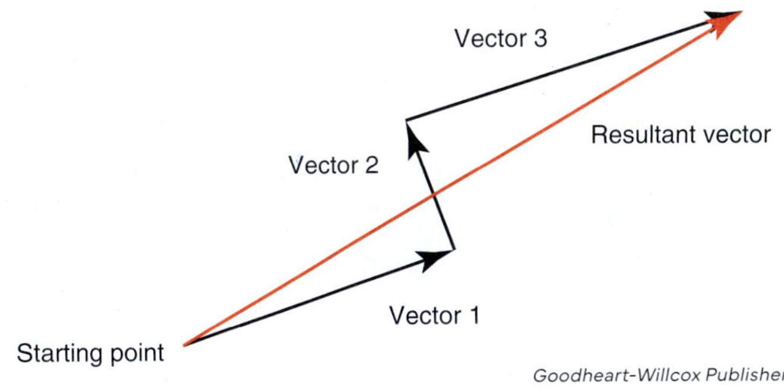

Figure 14-20. To add vectors, connect the vectors in a tail-to-head fashion. When adding two or more vectors, the outcome is the resultant vector, shown here in red.

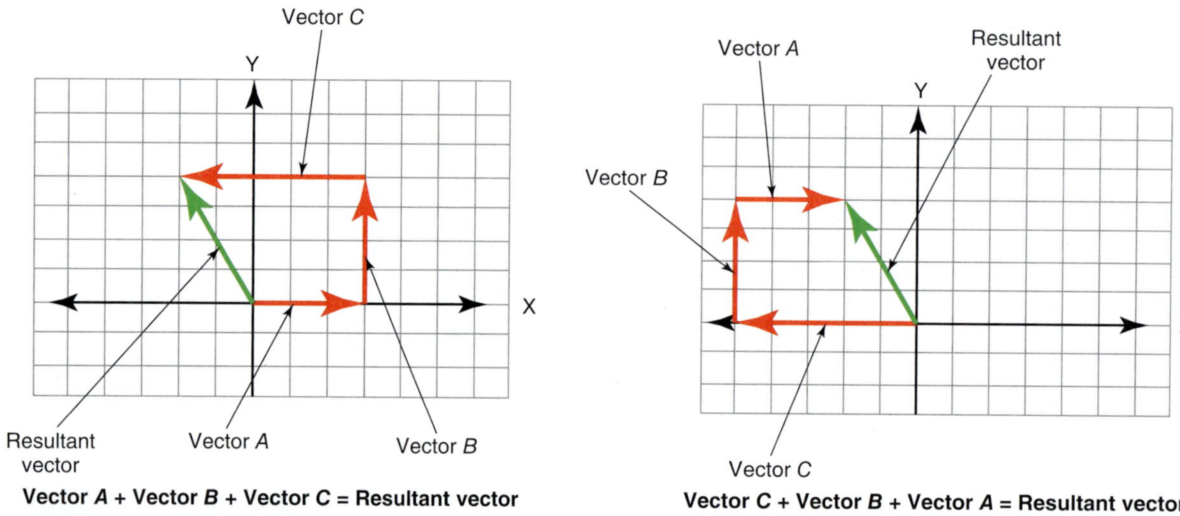

Goodheart-Willcox Publisher

Figure 14-21. The order of vector addition is not important, as the resultant vector is the same. Vectors *A*, *B*, and *C* are added in two different sequences, but the two resultant vectors are identical.

Shedding Light

Adding vectors is very similar to the rules of the popular "claw" game, found in many video arcades or local casual restaurants.

Goodheart-Willcox Publisher

In this game, the player first moves the claw in a left-to-right direction to reach the item of choice, indicated by the dashed line 1 in the following image. Then, the player moves the claw in the front-to-back direction, indicated by the dashed line 2, to arrive at the final claw position.

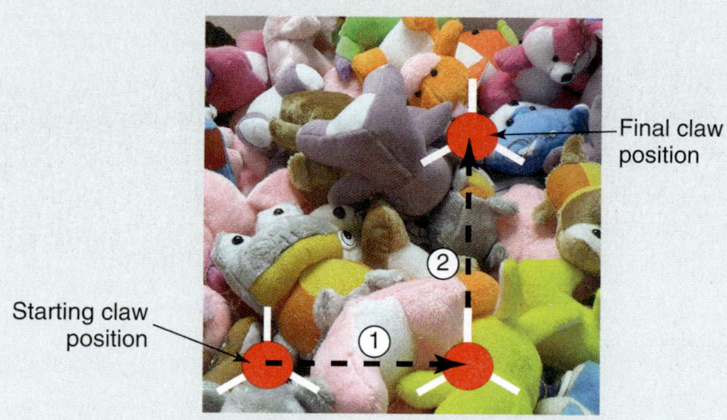

Goodheart-Willcox Publisher, chubbyrisa/Shutterstock.com

By connecting the start point to the final claw location point, a vector is created. This vector shown in the following figure is the sum of the left-to-right vector and the front-to-back vector.

Goodheart-Willcox Publisher, chubbyrisa/Shutterstock.com

CHAPTER 14 REVIEW

Summary

- Angles are formed when two lines intersect at a single point. (14.1)
- Angles of 90° are called right angles. (14.1)
- A straight angle is an angle that is equal to 180°. (14.1)
- Acute angles are less than 90°. (14.1)
- Angles that are greater than 90° but less than 180° are called obtuse angles. (14.1)
- Angles that are greater than 180° but less than 360° are called reflex angles. (14.1)
- Complementary angles are any two angles whose sum equals 90°. (14.1)
- Two angles are considered to be supplementary if the sum of the two angles is equal to 180°. (14.1)
- Triangles have three sides, three vertices (corners), and three interior angles. The sum of all three interior angles adds up to 180°, and the sum of the lengths of any two sides of a triangle will be greater than the length of the third side. (14.2)
- A right triangle has one 90° angle. (14.2)
- A scalene triangle has three sides with three different lengths and three angles with three different measurements. (14.2)
- An isosceles triangle has two equal sides and two equal angles. (14.2)
- An equilateral triangle has three equal sides and three equal angles. (14.2)
- The Pythagorean theorem can be used to evaluate right triangles. (14.3)
- The Pythagorean theorem states that the square of the hypotenuse is equal to the sum of the squares of the other two sides of a right triangle. (14.3)
- Three common trigonometric functions are sine, cosine, and tangent. (14.4)
- The sine of an angle in a right triangle is equal to the length of the side opposite the angle divided by the hypotenuse. (14.4)
- The cosine of an angle in a right triangle is equal to the length of the side adjacent to the angle divided by the hypotenuse. (14.4)
- The tangent of an angle in a right triangle is equal to the length of the side opposite the angle divided by the length of the side adjacent to the angle. (14.4)
- A vector is a line segment that has both direction and magnitude. (14.5)
- Vectors are added by connecting the head of one vector to the tail of another. (14.5)
- The resultant vector indicates the total result of all vectors acting on an object or system. (14.5)

Review Questions

Use the information in the chapter to answer the following questions.

Measuring Your Potential

1. Two angles that add up to 180° are known as _____ angles. (14.1)
   A. acute
   B. right
   C. supplementary
   D. complementary
2. *True or False?* The interior angles of a triangle must add up to 90°. (14.1)
3. *True or False?* A triangle can be formed with three-line segments of lengths 5, 7, and 10? (14.2)
4. A right triangle must contain one _____ angle. (14.2)
   A. 30°
   B. 45°
   C. 60°
   D. 90°
5. A(n) _____ triangle has no equal sides or equal angles. (14.2)
   A. isosceles
   B. equilateral
   C. obtuse
   D. scalene
6. The Pythagorean theorem is written as _____. (14.3)
   A. $a^2 \times b^2 = c^2$
   B. $a^2 + b^2 = c^2$
   C. $a^2 - b^2 = c^2$
   D. $a^2 \div b^2 = c^2$
7. The _____ of an angle is equal to the length of the opposite side of the triangle divided by the hypotenuse. (14.4)
   A. cosine
   B. sine
   C. tangent
   D. arcsine
8. The _____ is equal to the length of the adjacent side of the triangle divided by the hypotenuse. (14.4)
   A. cosine of an angle
   B. sine of an angle
   C. tangent of an angle
   D. arcsine of an angle

9. The cumulative effect of all the vectors in the system is called the _____. (14.5)
   A. magnified vector
   B. vector head
   C. resultant vector
   D. strength of force

Refer to the table shown in **Figure 14-16** *to answer questions 10 and 11.*

10. *True or False?* The sine of a 30° angle is 0.500. (14.4)
11. Of the following, which is true regarding $\sin^{-1}$ 0.00 and $\cos^{-1}$ 1.00? (14.4)
    A. They both represent an angle of 0° degrees.
    B. They both represent an angle of 90°.
    C. $\sin^{-1}$ 0.00 represents an angle of 90°, while $\cos^{-1}$ 1.00 represents an angle of 0°.
    D. $\sin^{-1}$ 0.00 represents an angle of 0°, while $\cos^{-1}$ 1.00 represents an angle of 90°.

Plug-In

1. If two sides of a triangle have lengths 9 and 13 inches, the third side can have a length of _____. (14.1)
   A. 2 inches
   B. 3 inches
   C. 4 inches
   D. 5 inches
2. If two of the interior angles of a triangle are 10° and 100°, the measurement of the third interior angle will be _____ (14.2)
   A. 10°
   B. 70°
   C. 90°
   D. 110°
3. If two sides of a right triangle have lengths 9 and 12 inches, the hypotenuse will have a length of _____. (14.2)
   A. 6 inches
   B. 10 inches
   C. 15 inches
   D. 20 inches
4. If the hypotenuse of a right triangle has a length of 26 inches and one side has a length of 10 inches, the remaining side must have a length of _____. (14.3)
   A. 10 inches
   B. 24 inches
   C. 36 inches
   D. 48 inches
5. If two sides of a right triangle are 8 and 9, which of the following is closest to the length of the hypotenuse? (14.3)
   A. 9
   B. 10
   C. 11
   D. 12
6. If the tangent of an angle is 1.000, what kind of triangle are you looking at? (14.4)
   A. A 45-45-90 right isosceles triangle.
   B. A 45-45-90 right equilateral triangle.
   C. A 30-60-90 right isosceles triangle.
   D. A 30-60-90 right equilateral triangle.
7. If two vectors, one pointing east and one pointing south, are added together, the resultant will point in which general direction? (14.5)
   A. Northeast
   B. Southeast
   C. Southwest
   D. Northwest

Making the Connection

1. Create four right triangles, identifying the lengths of all sides that follow the 3-4-5 triangle configuration. (14.2)
2. Use the Pythagorean theorem to verify the 1, 1, $\sqrt{2}$ relationship that exists in a 45-45-90 triangle. (14.3)
3. Draw a triangle that will help you understand why the tangent function is undefined at 90°. (14.4)

CHAPTER

15 Single-Phase Alternating Current Power

svetlaya_strana_omsk/Shutterstock.com

SPARKING DISCUSSION

Why do you think power losses in transmission lines are lower at higher voltages?

LEARNING OUTCOMES

After completing this chapter, you will be able to:

15.1 Explain the differences between ac power and dc power.
15.2 Compare ac and dc power transmission.
15.3 Define the terms rms voltage, average voltage, peak voltage, and peak-to-peak voltage as they apply to ac power.
15.4 Identify three non-sinusoidal waveforms.
15.5 Explain the concept of power factor as it applies to alternating current; define true power, apparent power, and reactive power.

TECHNICAL TERMS

amplitude
apparent power
average voltage
cycle
form factor
frequency
hertz (Hz)
in phase
instantaneous voltage
linear waveform
nonlinear waveform
non-sinusoidal waveform
peak-to-peak voltage
peak voltage
period
phase angle
power factor
reactive power
rms voltage
sawtooth waveform
sine wave
square waveform
triangular waveform
true power
volt-amperes (VA)
volt-amperes reactive (VAR)

Introduction

The vast majority of the electric power we use is alternating current (ac). The majority of the appliances, lighting fixtures, air conditioning systems, and electronic devices you find in your home are ultimately powered by alternating current. Even the wall receptacles you rely on every day provide ac power.

Unlike direct current (dc), where the voltage level is at a constant level and polarity, ac power continually changes its polarity and magnitude.

As a technician in the electrical trades, it is important to understand voltage values and how they apply to the work you do in the field. Having this knowledge will prepare you to work safely and will help you select the test instrument that best suits the needs of each job. In this chapter, you will gain insight into these voltages, how they relate to each other, and how to evaluate ac circuits.

15.1 Comparing AC to DC

In today's world, a large percentage of the dc power in use is produced chemically, such as in batteries and fuel cells, and from photovoltaic cells that are commonly used in solar panels. DC power can also be created by converting ac power using electronic circuits called rectifiers and filters, **Figure 15-1**. AC power is commonly generated by spinning coils of wire within a magnetic field. The force needed to turn these coils of wire can be provided by inexpensive and reliable sources such as flowing water, wind, and steam. As these coils of wire are rotated within a magnetic field, the strength and polarity of the generated voltage changes. This creates the ac waveform. The nature of the ac power signal makes it much easier to convert one ac voltage to another as compared to converting one dc voltage to another.

AC voltage levels can be easily increased or decreased by using mechanically simple and relatively inexpensive components called transformers. With advances in technology constantly being made, the efficiency of the transformers that are commonly used for power distribution can be as high as 99%. Although both ac and dc power are essential in today's world, a vast majority of generated power is ac. Additional information on the use of transformers is provided in Chapter 21, *Single-Phase Transformers*, and Chapter 22, *Three-Phase Transformers*.

In dc systems, the process of increasing and decreasing the voltage is much more complicated and, up until recently, has been very costly. Advancements in power electronics and semiconductor technologies, however, have made dc voltage conversions more popular and less expensive. In addition, the popularity of dc power transmission has greatly increased as the power generated by photovoltaic technologies is of the dc variety.

Shedding Light

In the late 19th century, there was an ac/dc dispute that later became known as the War of the Currents. This dispute was between Thomas Edison and Nikola Tesla. Edison was a proponent of dc, while Tesla believed that ac was the better choice. While the battle was heated, Tesla's ac power dominated. Today, the world is still predominantly powered by ac, but technological advances in dc are helping to bridge the gap.

15.2 AC Power Transmission

AC power can be easily transmitted at very high voltage and low current levels over long distances. This makes it easy to generate power in one location and ultimately use the power in a distant location, while experiencing only minimal losses during the transmission process. Transmitting voltages over long distances is more efficient at higher

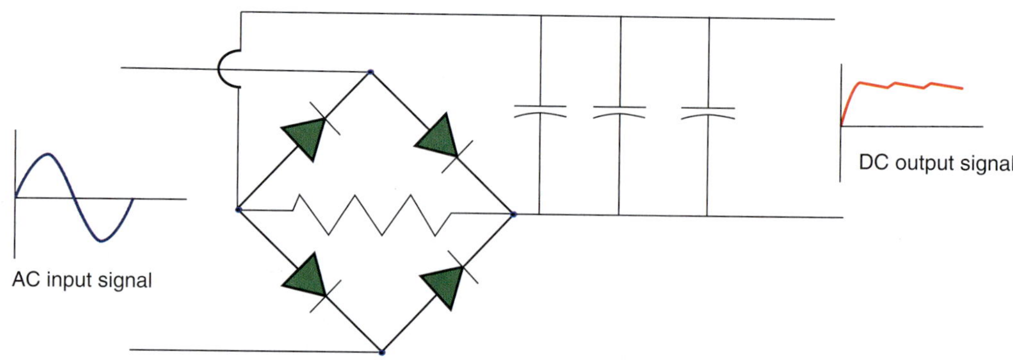

Goodheart-Willcox Publisher

Figure 15-1. DC power is created by converting ac power using ac-to-dc rectifier/filter circuits.

voltage levels. The basic power formula states that power is equal to the voltage multiplied by the current, commonly written as follows:

$$P = I \times E$$

If the voltage is increased, the current can be lowered while keeping the power unchanged. The power losses associated with voltage transmission are lower when the current is lower. For example, the power resulting from 5 A of current flow at a voltage of 20 V is the same as the power resulting from a 4 A current flow at a voltage of 25 V. In both cases, the calculated power is 100 W. At lower currents, the size of the current-carrying conductors can be smaller.

There are limits, however, to the voltage levels that can be transmitted. The higher the voltage, the greater the chance of personal injury, damage, or even death. At higher voltages, more insulation is required around the conductors. In addition, larger distances are needed between the conductors and, in many cases, the towers that support the transmission lines will need to be taller to keep them farther from the ground and from people.

When transmitting electricity, the number one goal is to lose as little power as possible during the process. Power transmission losses become larger as the transmission distance increases.

A large percentage of power transmission losses are resistive losses. Resistive power losses are often referred to as "I-squared-R" losses, where I is the current and R is the resistance. The formula used to calculate these losses is the same as the $P = I^2 \times R$ form of the power formula. Reducing the current flow through the transmission lines will greatly reduce the resistive power loss as this loss is proportional to the square of the current flow through the lines. Refer to **Example 15-1** and **Example 15-2**.

Example 15-1

Consider the scenario in the following image. A power plant supplies 240 Vdc at a current of 100 A to a residential home. The home is 2000 feet from the power plant and the resistance of the cable connecting the power plant to the home has a total resistance of 0.4 Ω. Determine the power loss associated with this transmission.

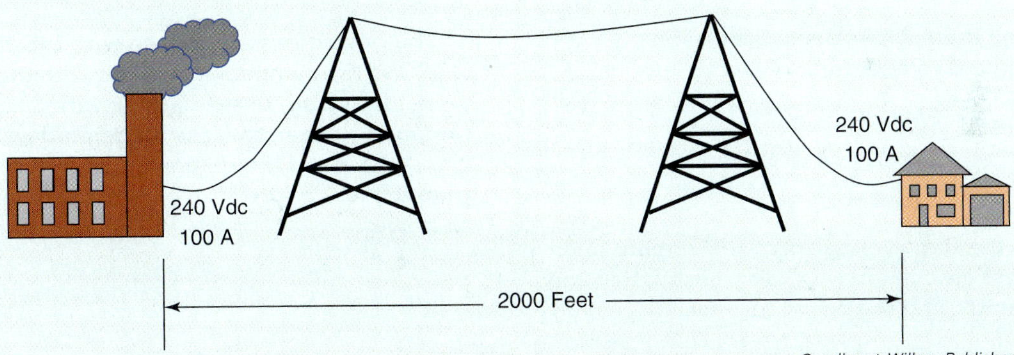

Goodheart-Willcox Publisher

✓ Solution

First, the voltage drop across the cable can be calculated using the voltage form of Ohm's law with the resistance of the conductor:

$$E = I \times R$$

where:

E = Voltage drop
I = Current flow through the cable
R = Resistance of the cable

$$E = 100\text{ A} \times 0.4\ \Omega$$

$$E = 40\text{ V}$$

The power loss associated with this transmission can then be calculated using the power formula:

$$P = I \times E$$

where:

P = Power loss
I = Current flow through the cable
E = Voltage drop through the cable

$$P = 100\text{ A} \times 40\text{ V}$$

$$P = 4000\text{ W}$$

The power loss associated with this transmission is 4000 W.

Example 15-2

Consider the scenario in the following image. A power plant supplies 240 Vac at a current of 100 A to a residential home. The home is 2000 feet from the power plant. In this example, the voltage is stepped up to 2400 V and the current is reduced to 10 A. Since the current flowing through the conductor is lower, the conductor can be smaller; therefore, its resistance will be greater. A value of 4 Ω will be used for this example. The power being transmitted is the same as **Exercise 15-1**.

Before stepping up:

$$P = I \times E$$

$$P = 100\ \text{A} \times 240\ \text{V}$$

$$P = 24{,}000\ \text{W}$$

After stepping up:

$$P = I \times E$$

$$P = 10\ \text{A} \times 2400\ \text{V}$$

$$P = 24{,}000\ \text{W}$$

Based on this information, determine the power loss associated with this transmission.

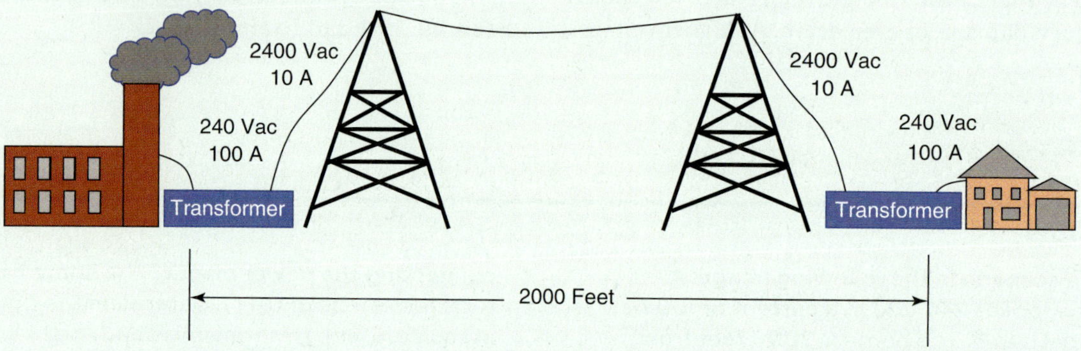

Goodheart-Willcox Publisher

✓ Solution

First, the voltage drop across the cable can be calculated using the voltage form of Ohm's law:

$$E = I \times R$$

$$E = 10\ \text{A} \times 4\ \Omega$$

$$E = 40\ \text{V}$$

Note: This is the same voltage drop in **Example 15-1**.

The power loss associated with this transmission can then be calculated using the power formula:

$$P = I \times E$$

$$P = 10\ \text{A} \times 40\ \text{V}$$

$$P = 400\ \text{W}$$

The power loss associated with this transmission is 400 W. By comparing this loss with the 4000 W power loss from **Example 15-1**, with all other factors remaining equal, the loss associated with the transmission of the dc power was ten times greater than the loss associated with the transmission of the ac power.

Shedding Light

Example 15-1 and **Example 15-2** illustrate why it is not economically feasible to use dc power. Distribution costs would be prohibitive. In the time of Tesla and Edison, it was not practical to take a higher dc voltage and step it down to a more manageable and useable lower voltage. Instead, transformers in an ac system could easily be installed to take the utility-generated higher voltages down to the needed levels for industry, residential, and light commercial use.

15.3 Anatomy of an Ideal Sine Wave

As mentioned earlier, alternating current is so named because the polarity of the voltage continuously changes from negative to positive. The ***sine wave***, as the name implies, is the graphical representation of the sine trigonometric function. An ideal uniform and symmetric ac waveform takes on the shape of a sine wave, **Figure 15-2**. Electrically speaking, the term ***cycle*** describes the repeating portion of a wave or signal. The cycle

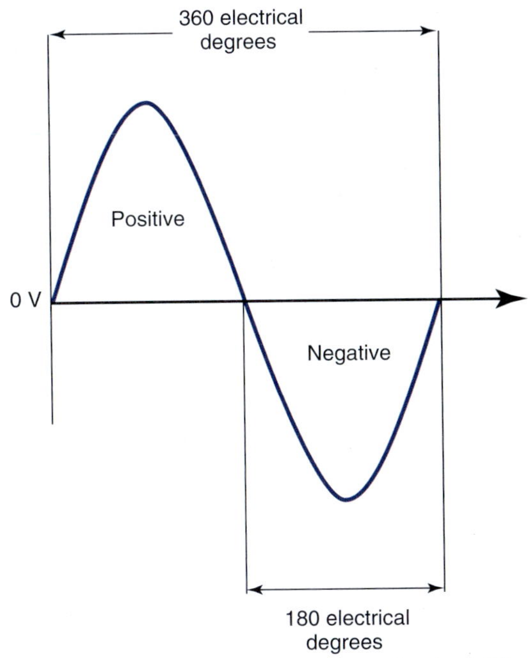

Goodheart-Willcox Publisher

Figure 15-2. One complete ac cycle takes 360 electrical degrees to complete.

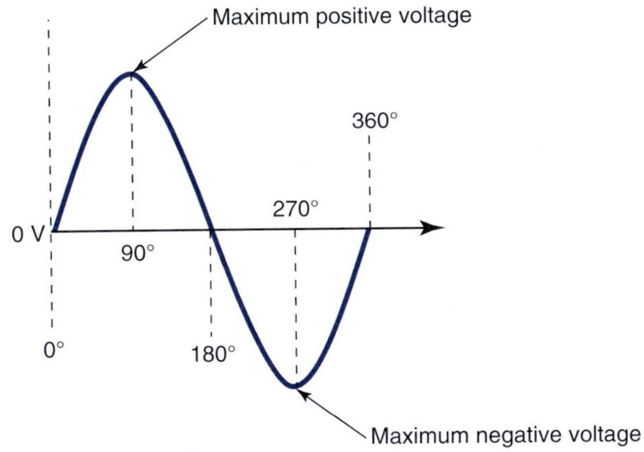

Goodheart-Willcox Publisher

Figure 15-3. Various reference points in an ac cycle.

is divided into two equal sections, one with a positive polarity and one with a negative polarity. The positive and negative portions of the cycle each occur for one-half of the cycle, or 180 electrical degrees. For example, the following describes the various points along the ac sine wave in **Figure 15-3**:

- The ac sine wave starts with a voltage of zero at a reference point of zero electrical degrees.
- The maximum positive voltage is reached at a point equal to one-quarter of the cycle, or 90 electrical degrees.
- The positive half of the sine wave ends at 180 electrical degrees, where the voltage is once again zero.

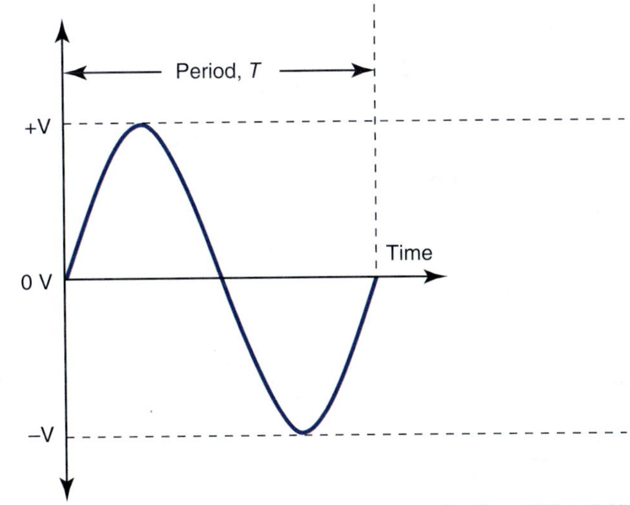

Goodheart-Willcox Publisher

Figure 15-4. The time to complete one cycle is called a period and is indicated by the letter *T*.

- The portion of the wave with a negative polarity starts at 180 electrical degrees.
- The maximum negative voltage is reached at a point halfway through the negative portion of the wave, or 270 electrical degrees.
- At 360 electrical degrees, one ac cycle is completed, and the voltage at this point is zero.

Period and Frequency

The time it takes to complete one cycle is called the ***period***. The period of a sine wave is identified by the letter *T*, **Figure 15-4**. The term ***frequency***, denoted *f*, is used to identify the number of cycles completed in one second. Frequency is expressed in units called ***hertz (Hz)***. In the United States, the standard frequency for ac power is 60 Hz, meaning that there are 60 cycles completed each second. In much of Europe, the standard frequency is 50 Hz. Frequency and period are mathematical inverses of each other: as one increases, the other decreases. Frequency and period are related by the following formulas:

$$T = \frac{1}{f}$$

$$f \times T = 1$$

$$f = \frac{1}{T}$$

For the standard power supply in your home or office, where the frequency of the power supply is 60 Hz, the period is equal to 1/60th of a second. Refer to **Example 15-3** on how to calculate the period and frequency of a sine wave.

Example 15-3

How long does it take for a 100 Hz sine wave to complete one cycle?

✓ Solution

Since the frequency of the sine wave is 100 cycles per second, the period can be found by taking the inverse of 100 (that is, 1/100), which is 0.01. It will, therefore, take 0.01 seconds to complete one cycle.

Now It's Your Turn 15-3

If the period of a sine wave is 0.02 seconds, calculate the frequency of the sine wave.

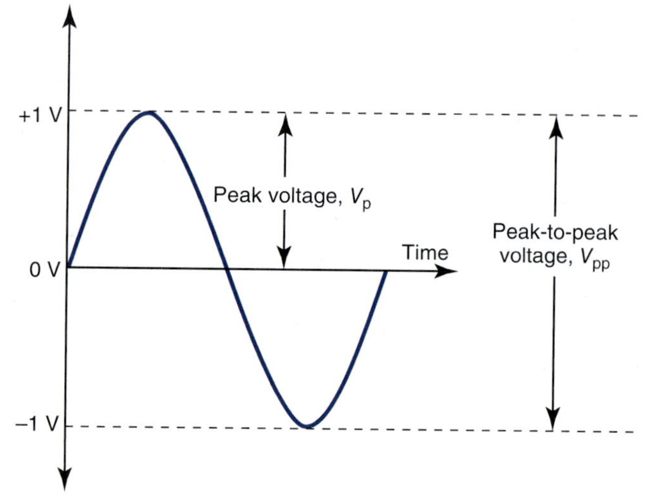

Figure 15-5. The peak voltage is measured from zero to the maximum voltage obtained in either the positive or negative portion of the wave. Peak-to-peak voltage is measured from the maximum positive voltage to the maximum negative voltage of the wave. In this waveform, the peak-to-peak voltage is 2 V (1 V minus –1 V).

Peak Voltage and Peak-to-Peak Voltage

Two other important terms used to describe the sine wave are peak voltage and peak-to-peak voltage, **Figure 15-5**. ***Peak voltage***, denoted V_P, represents the maximum voltage in either the positive or negative portion of the wave. Peak voltage is often referred to as the ***amplitude***, or maximum height of the waveform. ***Peak-to-peak voltage***, denoted V_{PP}, represents the voltage difference between the maximum positive voltage (+V) and the maximum negative voltage (–V) of the wave. For a perfectly symmetric waveform, peak-to-peak voltage is calculated as twice the peak voltage:

$$V_{PP} = 2 \times V_P$$

Peak-to-peak voltage is also calculated as the maximum positive voltage minus the maximum negative voltage:

$$V_{PP} = V_{MAX} - V_{MIN}$$

where:

V_{MAX} = Maximum positive voltage (maximum voltage value)

V_{MIN} = Maximum negative voltage (minimum voltage value)

This calculation holds true for waveforms that extend in both the positive and negative directions. When there is no negative portion of the wave, there is no maximum negative voltage value. In such cases, use the second formula: V_{PP} is equal to V_{MAX} minus V_{MIN}. Referring to the sine wave in **Figure 15-6**, for example, the peak-to-peak voltage is calculated as follows:

$$V_{PP} = V_{MAX} - V_{MIN}$$
$$V_{PP} = 20\text{ V} - 5\text{ V}$$
$$V_{PP} = 15\text{ V}$$

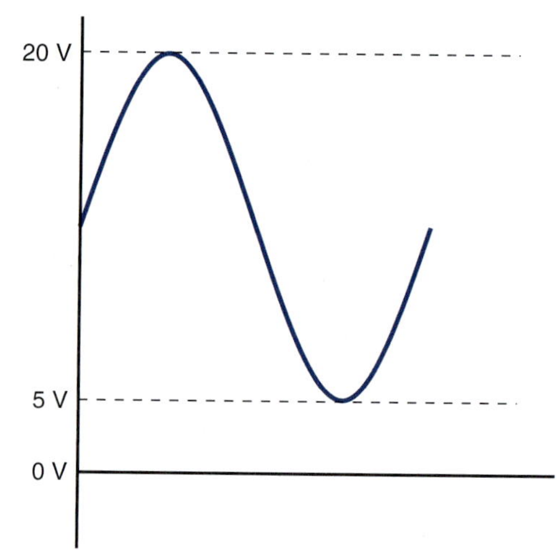

Figure 15-6. Peak-to-peak voltage can be calculated when there is no negative portion of the sine wave by using the two positive readings of the maximum voltage and the minimum voltage, shown here as 20 V and 5 V.

Refer to **Example 15-4** on how to calculate peak and peak-to-peak voltages.

Average Voltage

Since sine waves and other ac waveforms have voltage values that constantly change, obtaining an accurate and reliable voltage reading can be extremely difficult. This is why some manufacturers of electrical test equipment often rely on a signal's average voltage to obtain readings that can be converted into the value shown on a meter's display. It is important to note that the average voltage is not the value that is displayed by the meter.

Example 15-4

Determine the peak and peak-to-peak voltages for the following sine wave.

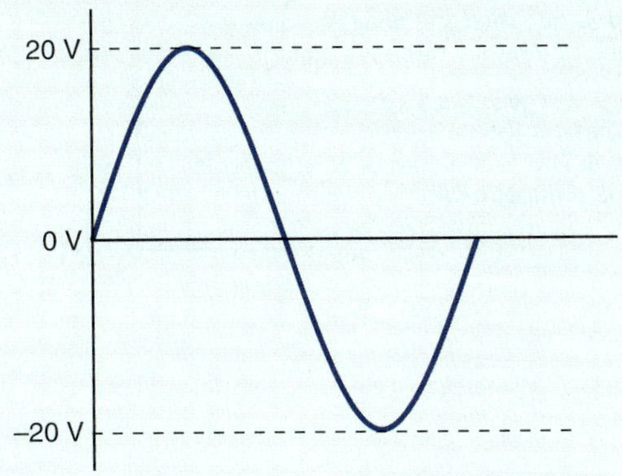

Goodheart-Willcox Publisher

✓ Solution

The peak voltage, or amplitude, is the greatest value in either the positive or negative direction. In this case, the peak voltage is 20 V. The peak-to-peak voltage is calculated as follows:

$$V_{PP} = V_{MAX} - V_{MIN}$$

$$V_{PP} = 20\text{ V} - (-20\text{ V})$$

$$V_{PP} = 40\text{ V}$$

Now It's Your Turn 15-4

Calculate the peak-to-peak voltage for the following sine waveform.

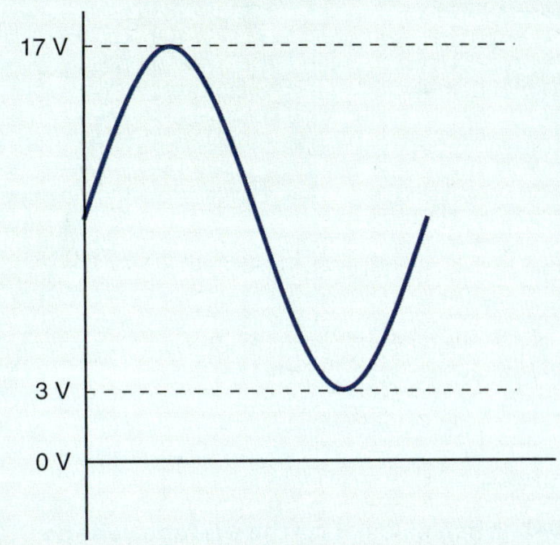

Goodheart-Willcox Publisher

The ***average voltage***, also referred to as the *mean voltage*, is calculated by taking the average of multiple, individual voltage readings. These individual readings are referred to as ***instantaneous voltage*** readings as they represent the voltage at the specific point in time the reading was obtained. Refer to **Example 15-5** on how to calculate an average voltage.

Math Sidebar

Subtracting a negative number from another number is the same as adding a positive number. The peak-to-peak voltage is equal to the difference between the maximum positive voltage and the maximum negative voltage, or maximum positive voltage minus maximum negative voltage. Subtracting V minus (–V) becomes adding V plus (+V). For example, if the maximum positive voltage is 1 V and the maximum negative voltage is –1 V, the peak-to-peak voltage is equal to 2 V.

Applying this same concept to an ideal sine wave would prove to be troublesome. Given the symmetrical nature of the sine wave, for every positive voltage there is a negative voltage to cancel it out. For example, if you add all the positive and negative values shown in **Figure 15-7** and divide the result by the number of values, the average will be zero. For a waveform that is symmetrical, therefore, the average voltage is calculated using only one-half of the cycle. Consider the sine wave shown in **Figure 15-8**. The positive voltage values on the left side of the waveform are measured as the voltage increases, while the voltages on the right side of the arc are measured as the voltage decreases. Using these numbers to calculate the average voltage will result in 12.74 V.

Example 15-5

Calculate the average voltage of the following measurements: 3 V, 8 V, 10 V, 12 V, 20 V, and 25 V.

✓ Solution

The average voltage is calculated by first adding all the numbers together, which equals 78 V. Next, divide the result by six, which represents how many numbers were added together.

$$\text{Average voltage} = \frac{78\text{ V}}{6}$$

$$\text{Average voltage} = 13\text{ V}$$

Average voltage in ideal sine waves can also be calculated by multiplying the peak value by the constant 0.637, as shown in the following formula:

$$V_{AVG} = V_P \times 0.637$$

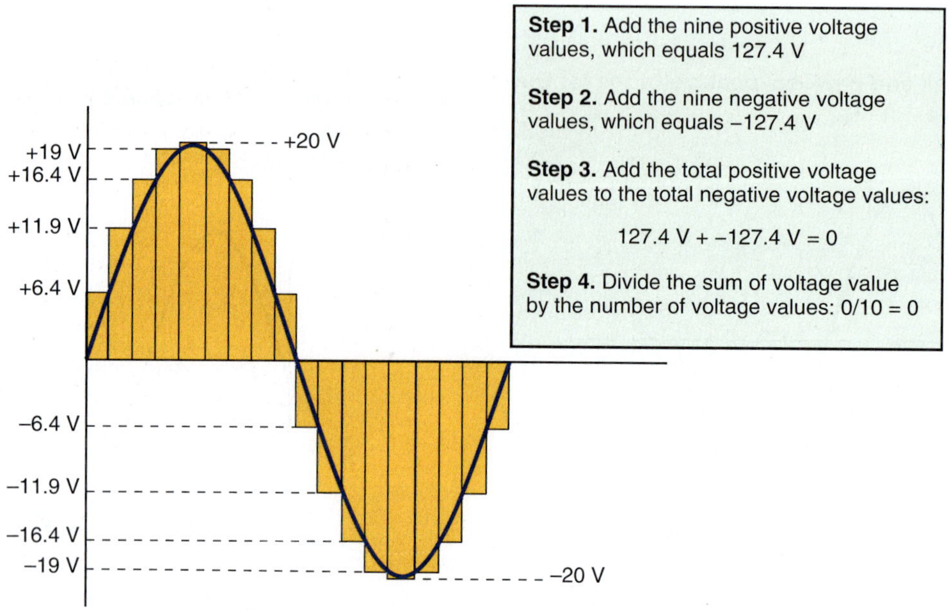

Goodheart-Willcox Publisher

Figure 15-7. When calculating the average voltage in an ideal sine wave, the answer will always equal zero.

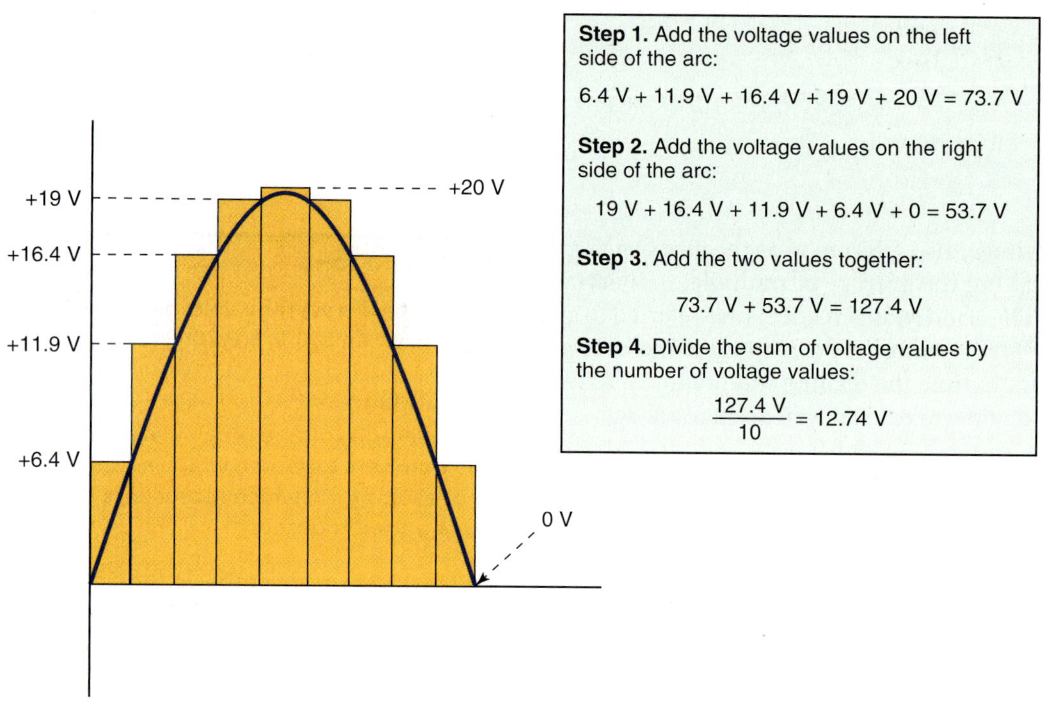

Goodheart-Willcox Publisher

Figure 15-8. When calculating the average voltage for a sine waveform, only consider the first half of the sine wave.

For example, the average voltage in **Figure 15-9** can be calculated as follows:

$$V_{AVG} = V_P \times 0.637$$

$$V_{AVG} = 20 \text{ V} \times 0.637$$

$$V_{AVG} = 12.74 \text{ V}$$

If the peak-to-peak voltage of an ideal sine wave is known, the average voltage can be determined using the following formula:

$$V_{AVG} = V_{PP} \times 0.3185$$

Refer to **Example 15-6** on how to determine the average voltage value of an ideal sine wave with a peak voltage. Refer to **Example 15-7** on how to determine the average voltage value of an ideal sine wave with a peak-to-peak voltage.

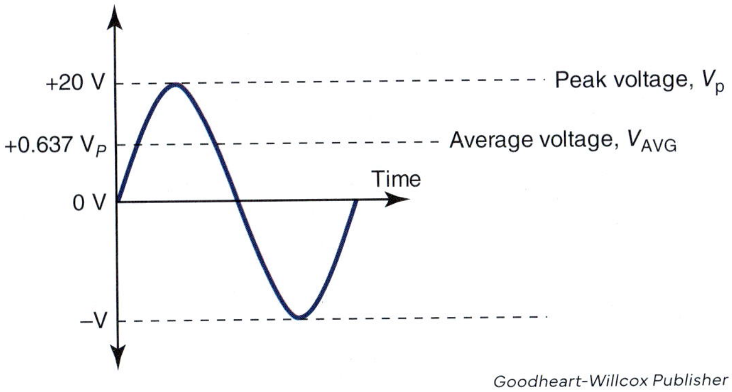

Figure 15-9. The average voltage for an ideal sine wave is calculated by multiplying the peak voltage by the constant 0.637.

Example 15-6

Calculate the average voltage value of an ideal sine wave with a peak voltage of 38 V.

✓ Solution

The average voltage value can be calculated as follows:

$$V_{AVG} = V_P \times 0.637$$

$$V_{AVG} = 38 \times 0.637$$

$$V_{AVG} = 24.2 \text{ V}$$

Now It's Your Turn 15-6

Calculate the average voltage value of an ideal sine wave with a peak voltage of 45 V.

Example 15-7

Calculate the average voltage value of an ideal sine wave with a peak-to-peak voltage of 88 V.

✓ Solution

Average voltage can be calculated as follows:

$$V_{AVG} = V_{PP} \times 0.3185$$

$$V_{AVG} = 88 \text{ V} \times 0.3185$$

$$V_{AVG} = 28.03 \text{ V}$$

Now It's Your Turn 15-7

Calculate the average voltage value of an ideal sine wave with a peak-to-peak voltage of 70 V.

RMS Voltage

Throughout the electrical trades and the *NEC*, ac voltage is expressed as an rms value. ***RMS voltage*** stands for the root-mean-square voltage and refers to the dc voltage equivalent of an ac power supply. The rms value is not an average voltage, but an effective voltage. An ac power supply with an rms voltage of 100 V, for example, will have the same effect as a 100 Vdc power supply. When making electrical calculations, such as power, amperage, and resistance, rms values are used. It is important to note that rms values do not apply to dc systems, as dc voltages are, for the most part, constant.

For an ideal sine wave, the rms voltage can be determined by multiplying the peak voltage by 0.707, as shown in the following formula.

$$V_{RMS} = V_P \times 0.707$$

Referring to the sine wave shown in **Figure 15-10**, the rms voltage can be calculated as follows:

$$V_{RMS} = V_P \times 0.707$$

$$V_{RMS} = 170 \text{ V} \times 0.707$$

$$V_{RMS} = 120 \text{ V}$$

Conversely, if the rms voltage value is known, the peak voltage of an equivalent ideal sine wave can be determined by multiplying the rms voltage by 1.414, the inverse of 0.707. Multiplying 120 V_{RMS} by 1.414, for example, will provide us with the peak voltage (170 V) we started with, as shown here:

$$V_P = V_{RMS} \times 1.414$$

$$V_P = 120 \text{ V} \times 1.414$$

$$V_P = 170 \text{ V}$$

Refer to **Example 15-8** on how to calculate peak voltage in an ideal sine wave.

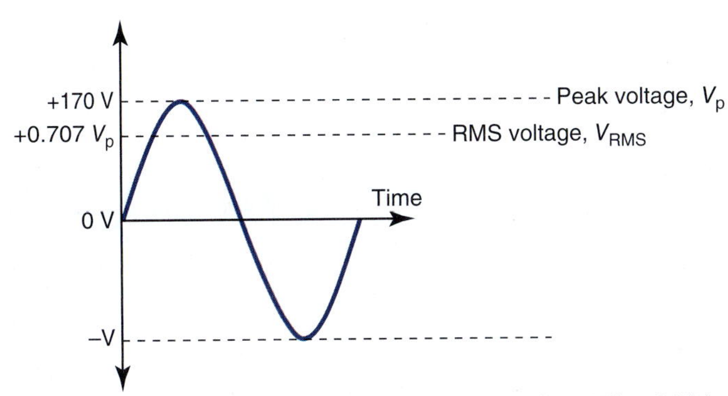

Figure 15-10. The rms voltage for an ideal sine wave is calculated by multiplying the peak voltage by 0.707.

Example 15-8

What is the peak voltage of an ideal sine wave if the rms voltage is 40 V?

✓ Solution

The peak voltage for an ideal sine wave can be calculated as follows:

$$V_P = V_{RMS} \times 1.414$$

$$V_P = 40\text{ V} \times 1.414$$

$$V_P = 56.6\text{ V}$$

Another way to calculate the peak value is to divide the rms voltage by the inverse of 1.414, which is 0.707.

$$V_P = \frac{V_{RMS}}{0.707}$$

$$V_P = \frac{40\text{ V}}{0.707}$$

$$V_P = 56.6\text{ V}$$

Since the peak voltage is one-half the peak-to-peak voltage, the rms value can be calculated if the peak-to-peak voltage is known, using the following formula:

$$V_{RMS} = 0.353 \times V_{PP}$$

In the formula above, the 0.353 represents one-half of 0.707. If the rms voltage is known, the peak-to-peak voltage can be calculated using the following formula, where 2.83 is approximately two times 1.414:

$$V_{PP} = 2.83 \times V_{RMS}$$

Average and RMS Voltages

As a review, average and rms voltage values are identified in **Figure 15-11** to illustrate their relationship to each other as well as to the peak voltage of a sine wave. Test instruments used in the electrical trades display ac voltages as rms values. These instruments calculate rms voltages by using either an averaging approach or a direct approach. Meters that use the averaging approach are called *averaging meters*, while those that determine the rms voltage directly are called true rms meters. Information on true rms meters and other digital multimeters is discussed in Chapter 8, *Electrical Tools and Testing Instruments.*

Averaging meters multiply the average voltage by 1.11 to obtain the rms voltage. This calculated rms value is what appears on the meter's display screen. The 1.11 value is called the ***form factor*** of a sine wave. The form factor defines the shape and proportion of a true, ideal sine wave. The relationship that exists between V_{AVG} and V_{RMS} for ideal sine waves is:

$$V_{RMS} = 1.11 \times V_{AVG}$$

If V_{RMS} is known, V_{AVG} can be calculated by using the following formula:

$$V_{AVG} = 0.9 \times V_{RMS}$$

This method of converting V_{AVG} to V_{RMS} is accurate when the waveforms are close to ideal sine waves, but not when the waveforms are not ideal or take on different shapes. When evaluating electronic circuits and loads, for example, true rms meters are strongly recommended.

A summary of the formulas that can be used to convert between V_{AVG}, V_{RMS}, V_P, and V_{PP} is provided in **Figure 15-12**.

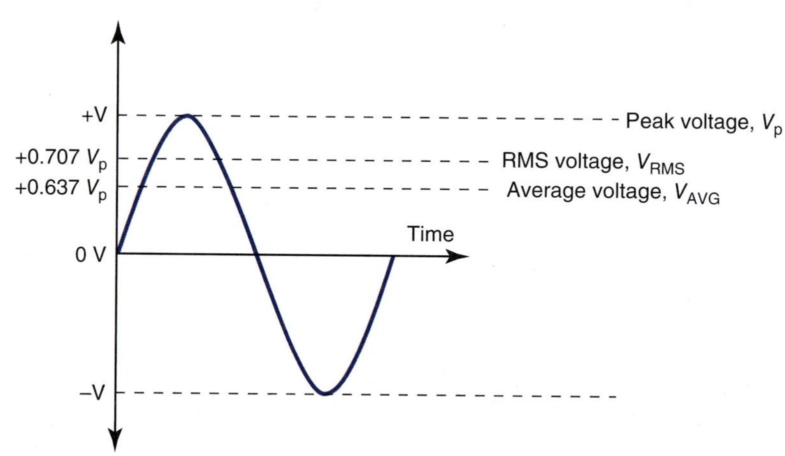

Goodheart-Willcox Publisher

Figure 15-11. Peak, rms, and average voltage comparison; rms is a higher value than average voltage.

Convert Peak, Peak-to-Peak, RMS, and Average Values

Convert:	To:	Multiply By:
V_P	V_{PP}	$V_P \times 2$
V_P	V_{AVG}	$V_P \times 0.637$
V_{PP}	V_{AVG}	$V_{PP} \times 0.3185$
V_P	V_{RMS}	$V_P \times 0.707$
V_{PP}	V_{RMS}	$V_{PP} \times 0.353$
V_{RMS}	V_{PP}	$V_{RMS} \times 2.83$
V_{RMS}	V_P	$V_{RMS} \times 1.414$
V_{AVG}	V_{RMS}	$V_{AVG} \times 1.111$
V_{RMS}	V_{AVG}	$V_{RMS} \times 0.9$
V_{AVG}	V_P	$V_{AVG} \times 1.57$

Goodheart-Willcox Publisher

Figure 15-12. Formulas to convert peak voltage (V_P), peak-to-peak voltage (V_{PP}), rms voltage (V_{RMS}), and average voltage (V_{AVG}).

15.4 Non-Sinusoidal Waveforms

Non-sinusoidal waveforms are waves that do not follow a typical sinusoidal shape. Non-sinusoidal waveforms have many applications in electronics and digital signals, including digital switching circuits, timing signals, electronic music, power supplies, and various types of modulation circuits. Common non-sinusoidal waveforms include triangular waveforms, sawtooth waveforms, and square (rectangle) waveforms.

Triangular Waveform

The ***triangular waveform*** is so named because its shape closely resembles a triangle, **Figure 15-13**. The triangular waveform is classified as a ***linear waveform*** because the voltage produced rises and falls at a constant rate with respect to time. A linear waveform is a waveform composed of linear segments. In a ***nonlinear waveform***, the rate of rise and fall is not constant, and the waveform has a curved shape.

For an ideal triangular waveform, the rise time is equal to the fall time, meaning that the time it takes for the voltage to increase from 0 V to its maximum is the same as the time it takes for the voltage to decrease from its maximum to 0 V. When plotted on a graph or observed on an oscilloscope, linear waves appear to be made up of straight lines or straight-line segments. Because the segments of these wave shapes are straight lines, triangular waveforms are easily evaluated.

Consider the triangular waveform shown in **Figure 15-14**. At time zero, the voltage generated by this triangular waveform is 0 V. As time passes, the peak voltage generated by

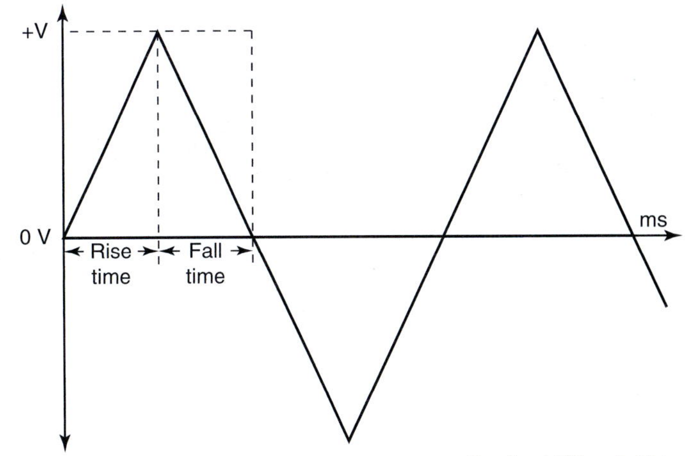

Goodheart-Willcox Publisher

Figure 15-13. Triangular waveforms are created when voltage increases and decreases at a constant rate.

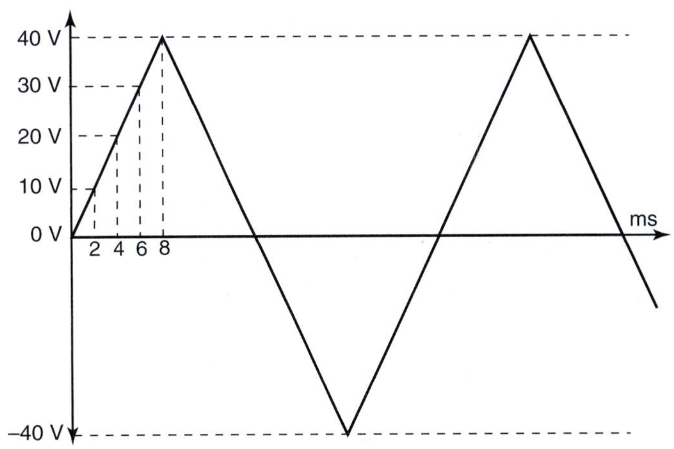

Goodheart-Willcox Publisher

Figure 15-14. In a triangular waveform, the voltage can be easily measured at various points since it increases at a constant rate.

this wave, 40 V, is reached at 8 milliseconds (ms). Because this voltage is increasing at a constant rate, we can determine that after 4 ms, which is half of the time it takes to reach its maximum value, the voltage generated by this waveform will be 20 V, which is one-half of the maximum voltage. It can also be predicted, and observed, that the generated voltage will be 10 V at 2 ms and 30 V at 6 ms.

Similarly, starting at 8 ms, the voltage generated by the wave will decrease at a constant rate until a voltage of 0 V is reached at 16 ms. The second half of the wave's cycle is a duplicate of the first half, but with a negative polarity, **Figure 15-15**. Just as with the traditional, ideal sine wave, once one cycle has been completed, it repeats itself continually, as long as the power source is energized.

Sawtooth Waveform

The ***sawtooth waveform*** resembles the cutting edge of a saw, **Figure 15-16**. The sawtooth pattern closely mirrors a triangular waveform but has some distinct differences. While

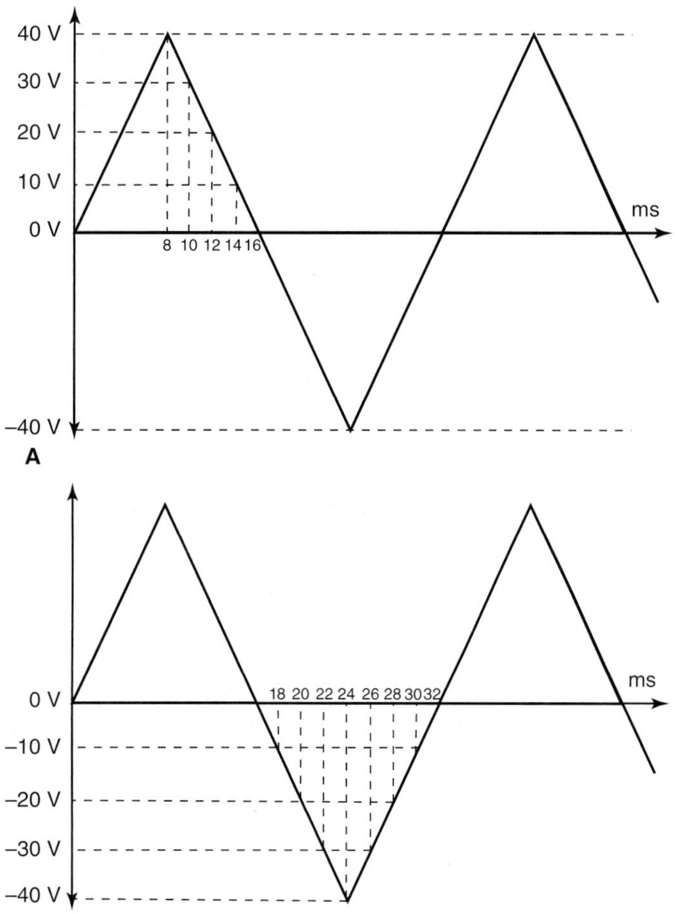

Goodheart-Willcox Publisher

Figure 15-15. In this triangle waveform, the voltage generated decreases at a constant rate. A—Voltage decreasing after reaching its peak voltage. B—The negative polarity portion of a triangular waveform.

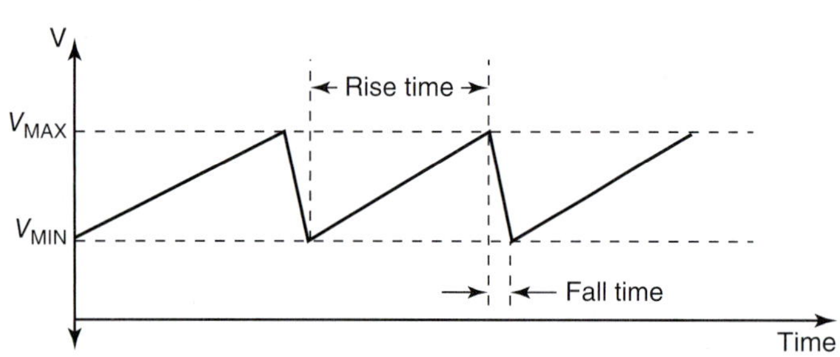

Goodheart-Willcox Publisher

Figure 15-16. Rise and fall (decay) times for a sawtooth waveform; rise times are significantly greater than fall times.

triangular waveforms are characterized as having equal rise and fall times, the sawtooth waveform is characterized as having a rise time that is significantly greater than the fall time. In an ideal sawtooth waveform, the fall time, also referred to as the *decay time*, is zero.

Square Waveform

Another non-sinusoidal waveform pattern is the square waveform, **Figure 15-17**. The ***square waveform***, also called

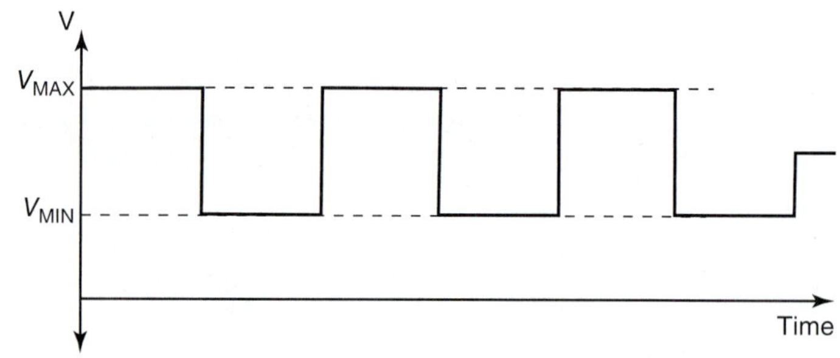

Goodheart-Willcox Publisher

Figure 15-17. Depending on the height and duration of the waves, a square waveform can resemble squares or rectangles.

a *rectangular waveform*, is characterized by its two distinct, fixed, maximum and minimum values. In an ideal square waveform, the transition time between the maximum and minimum values is zero. The square waveform is also commonly referred to as a *pulse train* or *pulse wave*.

15.5 Power in AC Circuits

Power in ac circuits can be expressed in three different ways:

- ***True power*** is the actual amount of power consumed by the circuit. Also called *active power* and *real power*, true power is expressed in units of watts (W).
- ***Apparent power*** is the power that appears to be present when voltage and current are measured separately. It is expressed in units of ***volt-amperes (VA)***.
- ***Reactive power*** is power consumed by an ac circuit that does not perform useful work. Reactive power is expressed in units of ***volt-amperes reactive (VAR)***.

In dc circuits and systems, true power and apparent power are the same. In ac circuits, however, true power is less than apparent power.

In dc circuits and systems, reactive power does not exist. In ac circuits, reactive power is responsible for the reduction in true power relative to apparent power. As the amount of reactive power increases, the true power decreases. As the amount of reactive power decreases, the true power increases. A summary of true power, apparent power, and reactive power is provided in **Figure 15-18**.

A Visual Representation

A visual representation is often used to describe the relationship that exists among apparent power, true power, and reactive power. Consider a person moving a box across the floor by pulling on a rope, **Figure 15-19**. In this image, the red arrow represents the total force being exerted and the blue arrow represents the horizontal force that is causing the box to move across the floor. The yellow arrow represents the vertical portion of the force being exerted on the rope that does

Summary of Power in AC Circuits

Power	Units	Presence in AC Circuits	Presence in DC Circuits
Apparent power	VA	Yes	Yes
True power	W	Yes	Yes
Reactive power	VAR	Yes	No

Goodheart-Willcox Publisher

Figure 15-18. Power in ac circuits can be expressed as apparent power, true power, or reactive power.

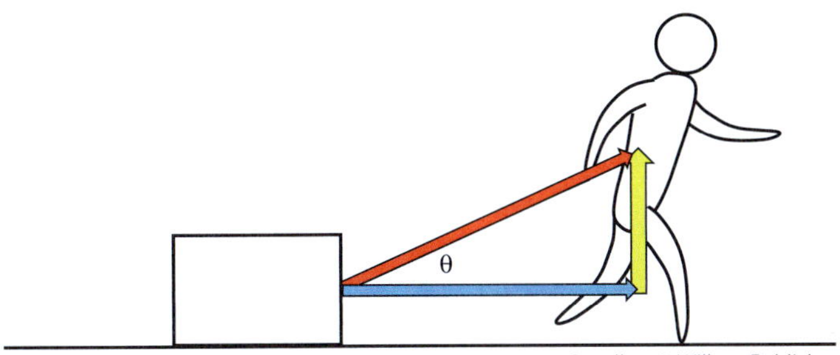

Goodheart-Willcox Publisher

Figure 15-19. Pulling a box across the floor helps create a visual representation of power.

not contribute to the moving of the box. The angle θ represents the angle between the lines that represent the total force and the useful force.

The force exerted by the person on the rope is greater than the horizontal force acting on the box, as evidenced by the difference in magnitude, or length, of the longer red and shorter blue force lines, or arrows. This indicates that there is another force, in the vertical direction, that does not contribute to the movement of the box. This force, which is at a 90-degree, or right angle to the horizontal force, is represented by the yellow arrow.

As the angle between the total force and the useful force increases, the amount of useful force decreases and the amount of wasted force increases. Consider the situation where the total force on the rope is exerted straight upward, **Figure 15-20**. In this case, since all the force is pulling up, there is no useful force moving the box in a horizontal direction. Therefore, the force is wasted. Similarly, if the total force is exerted in a horizontal direction, all the force is being used to move the box, and there is no wasted force, **Figure 15-21**. In this case, the total force is equal to the useful force.

When there is a useful force as well as a wasted force, the total force does not equal the mathematical sum of the useful and wasteful forces. Instead, the following two mathematical relationships exist:

$$(\text{Total force})^2 = (\text{Useful force})^2 + (\text{Wasted force})^2$$

$$\text{Cosine } \theta = \frac{\text{Useful force}}{\text{Total force}}$$

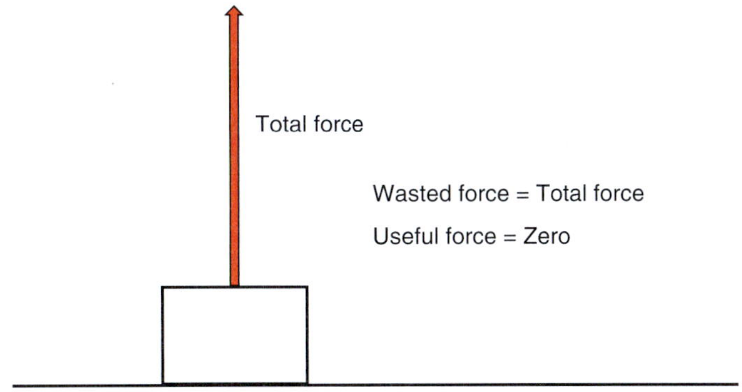

Goodheart-Willcox Publisher

Figure 15-20. The angle between the total force and the useful force will affect the amount of useful force. When the box is pulled straight up, all force is wasted.

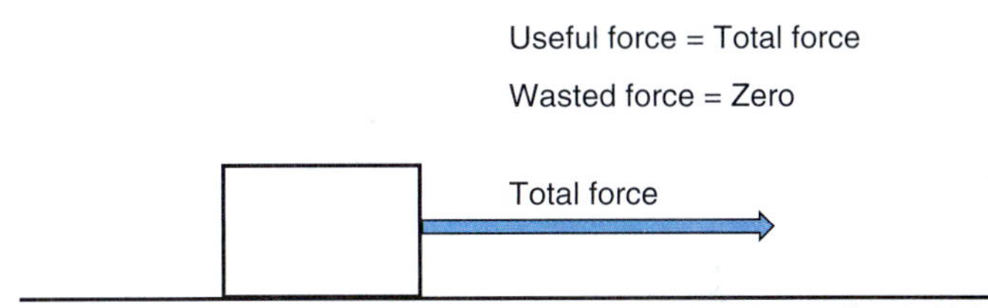

Goodheart-Willcox Publisher

Figure 15-21. When the force is pulled in a horizontal direction, the angle between the total force and the useful force is zero. As a result, all the force is directed in the useful direction.

The first formula listed is the Pythagorean theorem, which states that the square of the longest side of the triangle, or hypotenuse, is equal to the sum of the squares of the other two sides of the triangle.

The second formula listed is the definition of the cosine of an angle, which states that the cosine of an angle in a right triangle is equal to the length of the side adjacent to the angle divided by the length of the triangle's hypotenuse. Referring back to **Figure 15-19**, the side adjacent to angle θ is the useful force, while the hypotenuse of the triangle is the total force.

Power Factor and Phase Angle

Having discussed the concepts of useful force, total force, and wasted force, the transition to power becomes quite straightforward—simply replace the term force in our previous examples with power.

Power factor, denoted PF, is expressed as the ratio of true power to apparent power, where true power is the useful power and apparent power represents the total power, **Figure 15-22**. Power factor ranges from 0 to 1. For ac circuits, the basic power formula ($P = I \times E$) can be modified to account for the fact that the true power is less than the apparent power. This modified form of the power formula is provided here:

$$P = I \times E \times PF$$

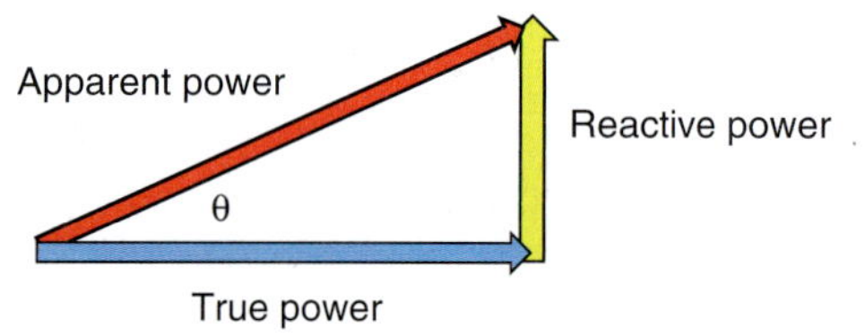

Goodheart-Willcox Publisher

Figure 15-22. Replacing force with the term power creates a power triangle consisting of apparent power, true power, and reactive power.

Refer to **Example 15-9** on how to calculate the power consumed.

Example 15-9

How much power is consumed by a 100 V circuit that draws 5 A of current if the power factor is 0.80?

✓ Solution

Using the modified form of the power formula, the power consumed is calculated as follows:

$$P = I \times E \times PF$$

$$P = 5\text{ A} \times 100\text{ V} \times 0.80$$

$$P = 400\text{ W}$$

The angle that exists between the true power and the apparent power is called the ***phase angle*** and is determined by the components found in the electric circuit. The phase angle is inversely related to the power factor. As the phase angle increases, the power factor decreases, and as the phase angle decreases, the power factor increases. When the current and voltage in an ac circuit reach their peak values at the same time, the phase angle is zero and the power factor is 1, its highest value, **Figure 15-23**.

When the current and voltage in an ac circuit reach their peak values at different times, the phase angle increases and the power factor drops, **Figure 15-24**. The phase angle is directly related to the number of electrical degrees that the voltage and current in an ac circuit are out of phase with each other.

Power factor is typically expressed either as a decimal or as a percentage. When expressed as a decimal, the formula for power factor is:

$$\text{PF} = \frac{\text{True power (W)}}{\text{Apparent power (VA)}}$$

Combining the formulas for power factor and the cosine of an angle, power factor can be expressed in decimal form, as the cosine of the phase angle:

$$\text{PF} = \frac{\text{True power (W)}}{\text{Apparent power (VA)}} = \cos\theta$$

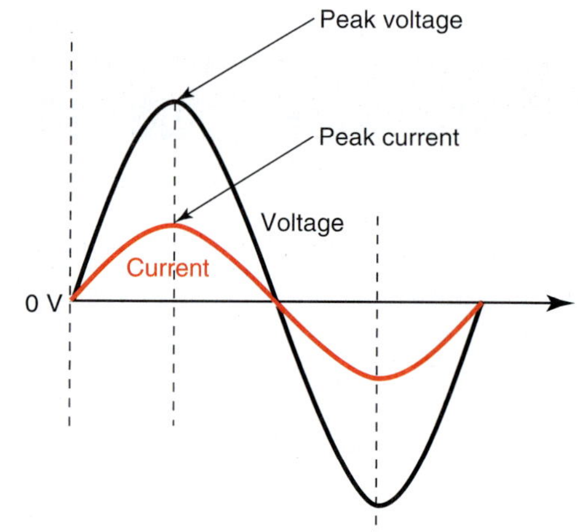

Goodheart-Willcox Publisher

Figure 15-23. Voltage and current are in phase with each other.

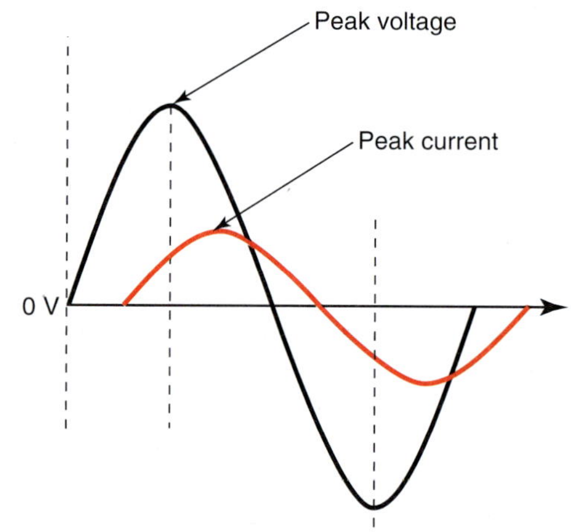

Goodheart-Willcox Publisher

Figure 15-24. Voltage and current are out of phase with each other.

When power factor is expressed as a percentage, the formula becomes:

$$\text{PF} = \frac{\text{True power}}{\text{Apparent power}} \times 100\%$$

or

$$\text{PF} = \cos\theta \times 100\%$$

A table of phase angles, cosines, and power factors is provided in **Figure 15-25**. In a dc circuit, for example, the phase angle is zero, so the true power is equal to the apparent power and the reactive power is zero. In this case, the power factor is equal to 1.0 or 100%. Refer to **Example 15-10** to practice calculating power factor.

Consider an electric circuit where the voltage and current are ***in phase*** with each other, meaning they reach their positive and negative peaks at the same time. When this occurs, active power and apparent power are the same. Assuming the true

Phase Angle, Cosine, and Power Factor

Phase Angle (Degrees)	Cosine	Power Factor
0	1.0	100%
10	0.985	98.5%
15	0.966	96.6%
20	0.940	94%
25	0.906	90.6%
30	0.866	86.6%
35	0.819	81.9%
40	0.766	76.6%
45	0.707	70.7%
50	0.643	64.3%
55	0.574	57.4%
60	0.500	50%
65	0.423	42.3%
70	0.342	34.2%
75	0.259	25.9%
80	0.174	17.4%
85	0.087	8.7%
90	0.00	0%

Goodheart-Willcox Publisher

Figure 15-25. Relationships among phase angle, cosine, and power factor.

Example 15-10

Determine the power factor for the following power triangle.

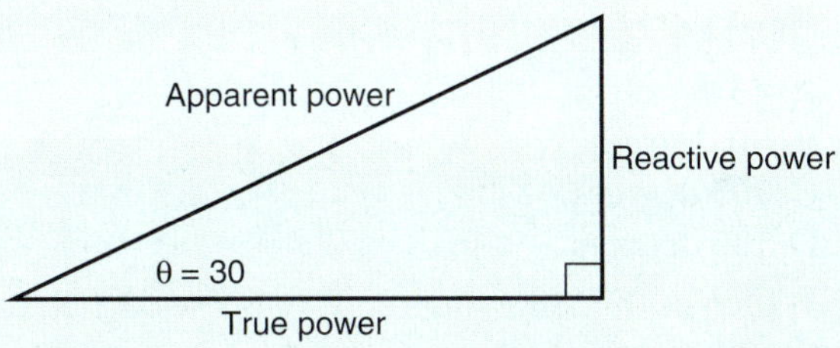

Goodheart-Willcox Publisher

✓ Solution

The phase angle between the true power and the apparent power is 30°. Referring to **Figure 15-25**, the power factor for the circuit represented by this power triangle is 86.6%.

Now It's Your Turn 15-10

Using the table in **Figure 15-25**, determine the power factor for the following power triangle.

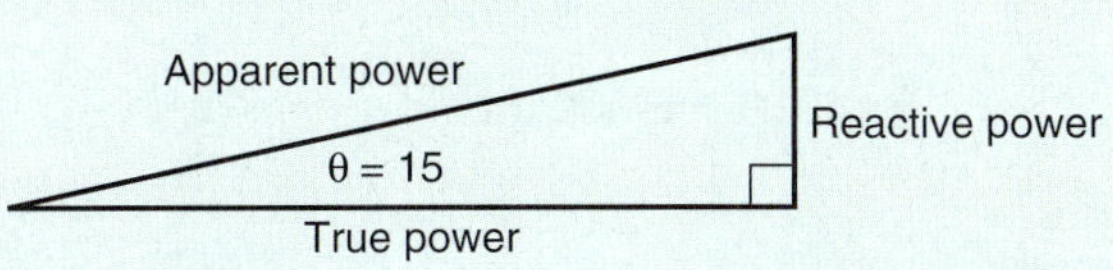

Goodheart-Willcox Publisher

power is 500 W and the apparent power is 500 VA, the power factor can be calculated as follows:

$$PF = \frac{\text{True power}}{\text{Apparent power}}$$

$$PF = \frac{500\ \text{W}}{500\ \text{VA}}$$

$$PF = 1.00$$

Using the same values, the power factor can also be expressed as 100%:

$$PF = \frac{\text{True power}}{\text{Apparent power}} \times 100\%$$

$$PF = \frac{500\ \text{W}}{500\ \text{VA}} \times 100\%$$

$$PF = 1.00 \times 100\%$$

$$PF = 100\%$$

Power Factor and Efficiency

The term efficiency applies to how well inputs are used to create or generate outputs. (Refer to Chapter 6, *Electrical Characteristics*, for further information on efficiency.) In the electrical industry, the term efficiency is often used to describe how effectively electrical energy is generated, used, or transmitted. Consider a power plant that burns natural gas to heat water to create the steam that turns the turbines that ultimately produce electricity. Not all the heat generated by burning fuel is transferred to the water being heated. Some heat is lost up the boiler's chimney. The more heat that is lost, the more waste there is and the lower the efficiency of the process.

Power factor is an expression of energy efficiency and is usually expressed as a percentage. The lower the percentage, the less efficient the power usage. Since power factor can be looked at as the cosine of the angle between true power and apparent power, the power factor will be greatest, or one, when the phase angle between active power and apparent power is zero. This is because the cosine of a zero-degree angle is one. The power factor will be lowest, or zero, when the phase angle between active power and apparent power is 90 degrees. This is because the cosine of a 90-degree angle is zero. The relationships that exist among phase angle, efficiency, true power, and power factor are illustrated in **Figure 15-26**.

Phase Angle, Efficiency, True Power, and Power Factor

Phase Angle	Efficiency	True Power	Power Factor
Low	High	High	High
High	Low	Low	Low

Goodheart-Willcox Publisher

Figure 15-26. Relationships among phase angle, efficiency, true power, and power factor.

CHAPTER 15 REVIEW

Summary

- The vast majority of power generated in the world today is ac power. (15.1)
- AC power has the property of continually changing its polarity and magnitude and can easily be transmitted at very high voltages over long distances. (15.2)
- Power losses are reduced as the current flow through the lines decreases. (15.2)
- AC power signals are most commonly represented by a sine wave. (15.3)
- The term frequency is used to describe the number of cycles that are completed each second. (15.3)
- The peak voltage is the maximum voltage of a waveform. (15.3)
- The peak-to-peak voltage is the difference between the maximum and minimum voltages of a waveform. (15.3)
- The average voltage is the average of multiple, individual voltage readings. (15.3)
- The rms voltage is the effective dc voltage of an ac power signal. (15.3)
- RMS voltage is calculated as the square root of the average of the squares of a series of voltage measurements. (15.3)
- Non-sinusoidal waveforms include the triangular, sawtooth, and square. (15.4)
- Power in ac circuits can take the form of apparent power (VA), true power (W), or reactive power (VAR). (15.5)
- Apparent power is the maximum amount of power supplied to a circuit. True or active power can be described as the power that is purposefully used by a circuit. Reactive power is the power consumed by an electric circuit that does not do useful work. (15.5)
- As the phase angle decreases, the power factor increases. (15.5)
- Power factor is calculated as the true power divided by the apparent power. (15.5)
- Electrically speaking, efficiency describes how effectively energy is generated, used, and/or stored. (15.5)

Review Questions

Use the information in the chapter to answer the following questions.

Measuring Your Potential

1. Common natural forces used to directly generate ac power include _____. (15.1)
   A. coal and natural gas
   B. wood and plastic
   C. flowing water and wind
   D. glass and plastic
2. AC voltage levels can be easily increased or decreased using _____. (15.1)
   A. rectifiers
   B. transistors
   C. relays
   D. transformers
3. Power losses are _____ of the current flow through the transmission lines. (15.2)
   A. equal to the square
   B. proportional to the square
   C. equal to the negative
   D. not equal to the square
4. *True or False?* For an ac waveform, the time required to complete one cycle is called the period. (15.3)
5. For an ideal sine wave, the voltage is at zero at _____. (15.3)
   A. 45 degrees
   B. 90 degrees
   C. 135 degrees
   D. 180 degrees
6. Which of the following represents the correct order, from lowest to highest, of the voltages associated with an ideal sine wave? (15.3)
   A. V_{AVG}, V_{RMS}, V_P
   B. V_{AVG}, V_{PP}, V_P
   C. V_{RMS}, V_{AVG}, V_P
   D. V_{RMS}, V_{AVG}, V_{PP}

7. For an ideal sine wave, the peak-to-peak voltage is equal to _____. (15.3)
   A. one-half of the peak voltage
   B. the square root of the peak voltage
   C. the square of the peak voltage
   D. two times the peak voltage
8. Which of the following waveforms is best described as a linear waveform with equal rise and fall times? (15.4)
   A. sinusoidal waveform
   B. triangular waveform
   C. square waveform
   D. sawtooth waveform
9. Which of the following is an accurate representation of ac power? (15.5)
   A. Power factor = Apparent power × True power
   B. Apparent power = True power + Reactive power
   C. Power factor = $\frac{\text{True power}}{\text{Apparent power}}$
   D. Apparent power = True power – Reactive Power
10. If the phase angle of an ac power supply is zero, which of the following is true? (15.5)
    A. Apparent power = True power
    B. Power Factor = True power
    C. Apparent power = Reactive power
    D. Power factor = Reactive power

Plug-In

1. The average voltage over a complete ideal sine wave cycle with a peak voltage of 30 V is _____. (15.3)
   A. 0 V
   B. 19.11 V
   C. 21.21 V
   D. 42.42 V
2. For an ideal sine wave, the voltage is at its maximum at _____. (15.3)
   A. 0 degrees
   B. 90 degrees
   C. 20 degrees
   D. 180 degrees
3. An ac power supply will have the same effect as a 50 Vdc power supply if its rms voltage is equal to _____. (15.3)
   A. 27 V
   B. 50 V
   C. 71 V
   D. 122 V
4. What is the frequency of an ac sine wave if its period is 1 ms? (15.4)
   A. 100 Hz
   B. 1000 Hz
   C. 10,000 Hz
   D. 100,000 Hz
5. With all other factors remaining equal, which of the following would result in the lowest power loss? (15.5)
   A. 1000 V at 100 A
   B. 100 V at 1000 A
   C. 500 V at 200 A
   D. 250 V at 400 A
6. If the phase angle of an ac power supply is 10 degrees, the power factor is closest to _____. (15.5)
   A. 0.707
   B. 0.819
   C. 0.906
   D. 0.985

Making the Connection

1. Explain why power transmission losses are lower at lower currents, even though the resistance of smaller conductors is higher. (15.2)
2. Explain why averaging-type digital multimeters might not be the best choice when evaluating electronic circuitry. (15.3)

CHAPTER

16 Three-Phase Alternating Current Power

Factory_Easy/Shutterstock.com

SPARKING DISCUSSION

Which is safer, three-phase power or single-phase power? Explain your position.

LEARNING OUTCOMES

After completing this chapter, you will be able to:

16.1 Explain how three-phase power is produced.

16.2 Describe the wye configuration of a three-phase alternating current power signal.

16.3 Describe the delta configuration of a three-phase alternating current power signal.

16.4 Calculate the voltage imbalance of a three-phase power signal.

16.5 Calculate true power in delta and wye systems with a given power factor.

TECHNICAL TERMS

current imbalance
delta configuration
high-leg delta configuration
line voltage
phase voltage
three-phase
voltage imbalance
wye configuration

Introduction

Unlike the single-phase or split-phase power that is found in most residences, most commercial and industrial structures rely on three-phase electric power. The popularity of three-phase systems stems from their ability to provide at least 50% more power with the addition of a third current-carrying conductor (from two to three) as compared to equivalent single-phase systems. Three-phase systems also deliver greater efficiency. This greater efficiency is obtained because the three-phase systems power being supplied to the load never falls to zero. With a 120° separation between any two of the three current-carrying conductors in a three-phase power system, voltage is always being applied to the controlled load. Even the single-phase and split-phase power we use in our homes and offices is derived from three-phase power systems.

The consistency and uniformity of three-phase signals are two of the factors that contribute to the higher efficiencies of these systems. When motors and other pieces of equipment are supplied with steady streams of constant power, they tend to operate more effectively and efficiently, saving energy. Therefore, whenever three-phase power is available, it is widely preferred over single-phase power. In this chapter you will learn about the two common three-phase power connections (wye and delta), how three-phase power is created, why it is so efficient, why those working in the electrical trades need to know about it, and how to perform power calculations on three-phase power systems.

16.1 Three-Phase Properties

The term ***three-phase*** refers to the simultaneous production of three evenly offset, single-phase, alternating current (ac) sine waves. Each of these three ac waveforms operates at the same frequency. The main difference is that each of the three sine waves reaches its instantaneous peak voltage value at different times. The three single-phase ac power supplies are separated from each other by 120 electrical degrees. Each 120° increment represents one-third of a complete cycle, as a complete cycle is made up of 360°. This phase separation has the effect of providing uniform, constant power over each cycle.

For example, in **Figure 16-1** the first sine wave (color-coded black) starts the first half of its waveform at zero electrical degrees in the positive direction. Exactly one third of a rotation later, at 120 electrical degrees, the next sine wave (color-coded red) starts the first half of its waveform, also in the positive direction. Following this sine wave by another 120 electrical degrees, another sine wave (color-coded blue) starts the first half of its waveform in the positive direction. Following the third sine wave at 360 electrical degrees, which is the equivalent of zero electrical degrees, this uniform cycle repeats as long as the system is operational.

Three-Phase Waveform

The following will help you visualize how a three-phase waveform is created. When a magnet is rotated between three equally-spaced conductors, **Figure 16-2**, its effect on the electrons in each conductor is observed. The north pole of the magnet has a negative charge, while the south pole of the magnet has a positive charge.

The space around the magnet can be divided further into 12, 30° sections, **Figure 16-3**. In this configuration, conductor A is positioned at 270°, conductor B is positioned at 30°, and conductor C is positioned at 150°. The magnet rotates in a clockwise direction, starting at 0°. The use of arrows with different lengths helps visualize the effect the magnet has on the electrons in each conductor at a specific position in its rotation. These arrows represent little effect, large effect, and maximum effect, **Figure 16-4**. The arrows will point in the direction of electron motion, with outward-pointing arrows representing a positive polarity, and inward-pointing arrows representing a negative polarity. The rotating magnetic field produces the sinusoidal waveforms in the three conductors.

When the south pole of the magnet aligns with 0°, there is no magnetic effect on the electrons in conductor A, **Figure 16-5**.

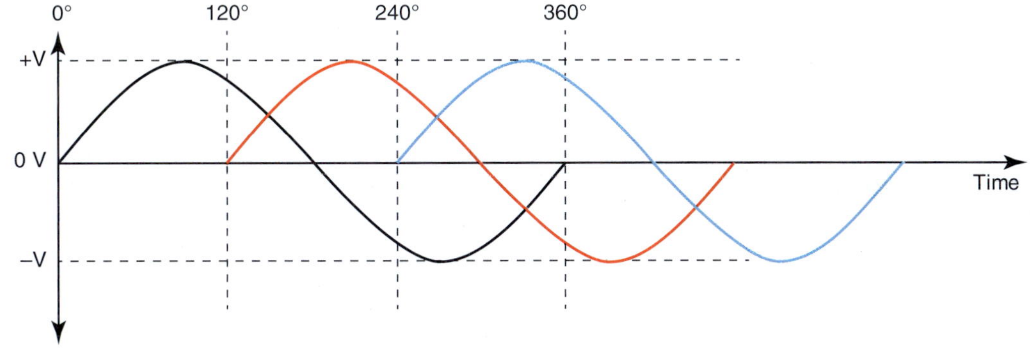

Goodheart-Willcox Publisher

Figure 16-1. In a three-phase ac system, each of the three single-phase waveforms is 120 electrical degrees out of phase with the next.

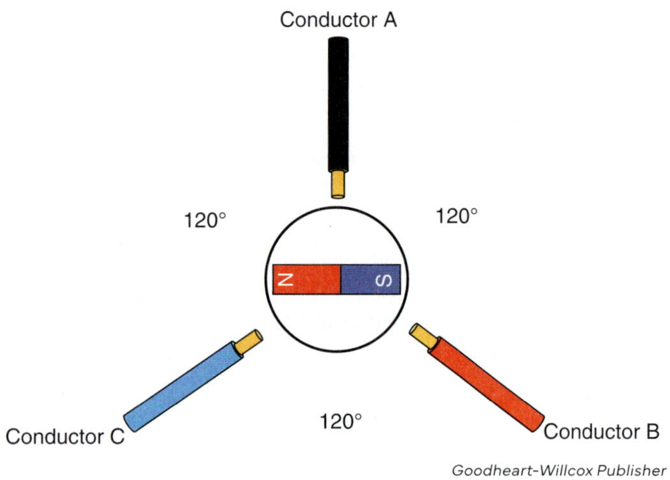

Figure 16-2. Three conductors positioned 120° apart from each other around a rotating magnet.

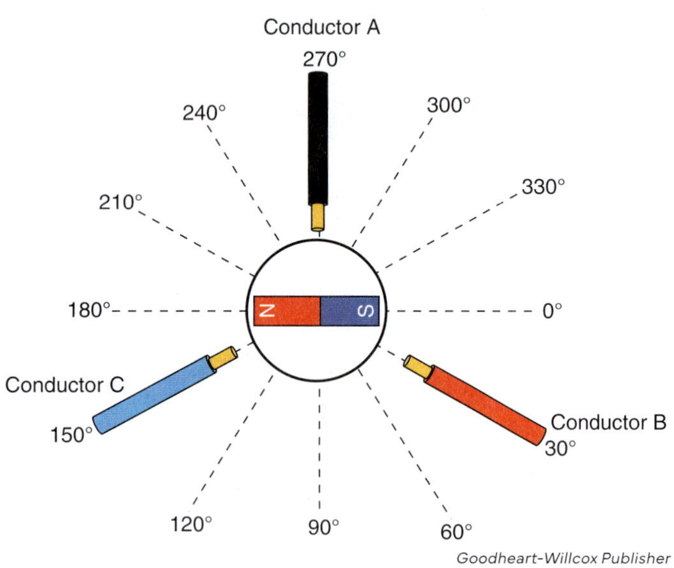

Figure 16-3. The complete circular path is divided into 30° sections.

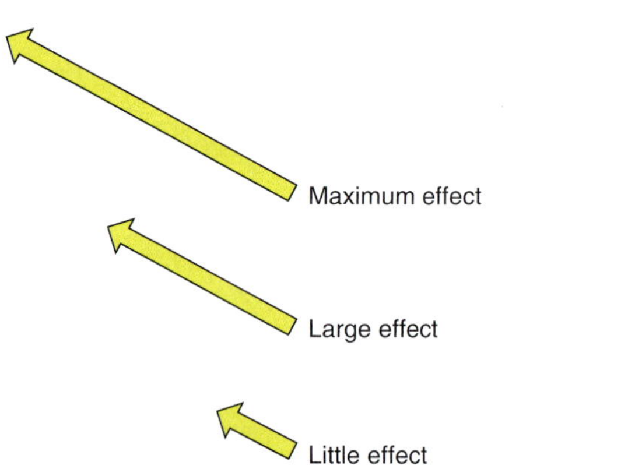

Figure 16-4. The length of the arrow represents the magnitude of the magnet's effect.

At this point, the magnet is perpendicular to this conductor and the effects of the north and south poles of the magnet on the conductor are equal and opposite. With the magnet in this position, the conductors B and C are both at the same angle to the magnet. However, conductor C is affected more by the north pole of the magnet, which is the negative side, while conductor B is affected more by the south pole, which is the positive side. The direction of the arrow for conductor C is pointing away from the center, indicating that electrons are repelled by the negative end of the magnet. The direction of the arrow for conductor B is pointing into the center, indicating that electrons are attracted to the positive end of the magnet. The sinusoidal waveforms in **Figure 16-5** highlight the condition at 0° of rotation: conductor A at zero, conductor C at a high positive value, and conductor B at a high negative value.

At 30° of rotation, there is a small negative magnetic effect on the electrons in conductors A and C. As illustrated in **Figure 16-6**, both conductors are exposed to more of the negative end of the magnet than the positive end. The effect of the magnet on the electrons in conductors A and C is the same in both direction and intensity. As can be seen in the highlighted portion of the waveform diagram in **Figure 16-6**, conductors A and C are at the same level and in the positive region of the diagram. Conductor B is in direct alignment with the positive end of the magnet and has the maximum interaction with it. This conductor is at its maximum negative value.

At 60° of rotation there is no magnetic effect on the electrons in conductor C, as the magnet is perpendicular to this conductor and the effects of the north and south poles of the magnet on the conductor are equal and opposite, **Figure 16-7**. The negative pole of the magnet is having more of an effect on conductor A than before, so there is more of a repulsive force in that conductor. The positive pole of the magnet is having less of an effect on conductor B than before, so the repulsive force in that conductor has dropped from the maximum level. Notice that the waveform corresponding to conductor C is at zero.

At 90° of rotation, there is a maximum negative magnetic effect on the electrons in conductor A, as the magnet is in perfect alignment with this conductor, **Figure 16-8**. This is the point where the repulsion force is the greatest, and the black conductor is at its peak positive value. The other two conductors are affected more by the positive pole of the magnet as the electrons in both conductors are attracted to the magnet as indicated by the inward-pointing arrows. Both of these conductors are shown in the negative portion of the waveform diagram highlighted in **Figure 16-8**.

As the magnet continues to rotate, the direction of the electron movement in each conductor changes, as well as the strength of the force acting on each conductor. After one revolution, the cycle is complete and another begins. The rising and falling occurs at the same frequency as the speed at which the magnet interacts with each conductor.

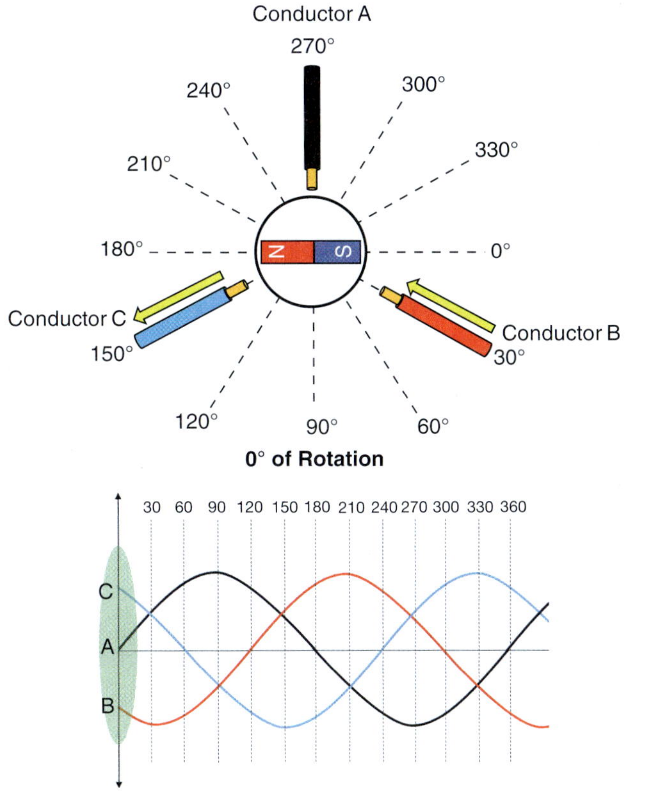

Goodheart-Willcox Publisher

Figure 16-5. Magnetic effect on the conductors with the south pole aligned at 0° of rotation.

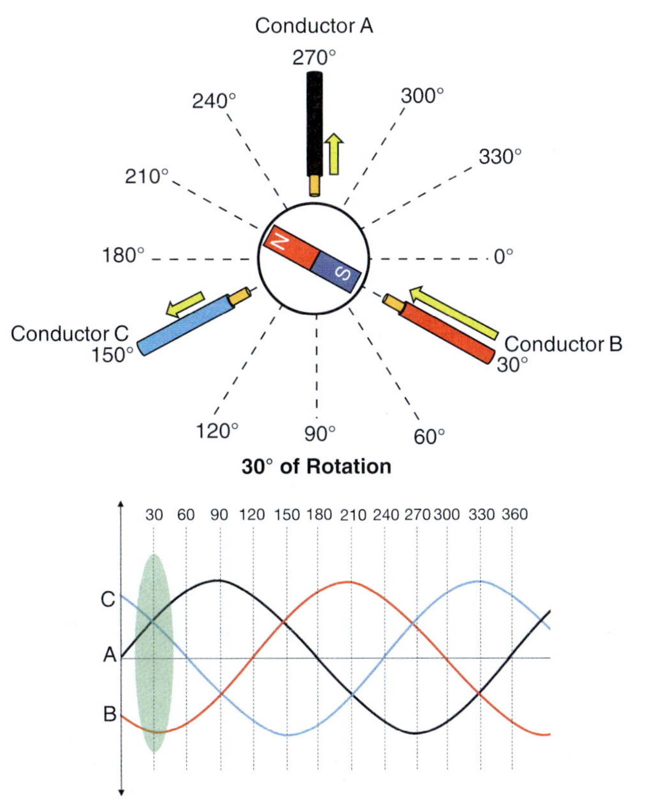

Goodheart-Willcox Publisher

Figure 16-6. Magnetic effect on the conductors with the south pole aligned at 30° of rotation.

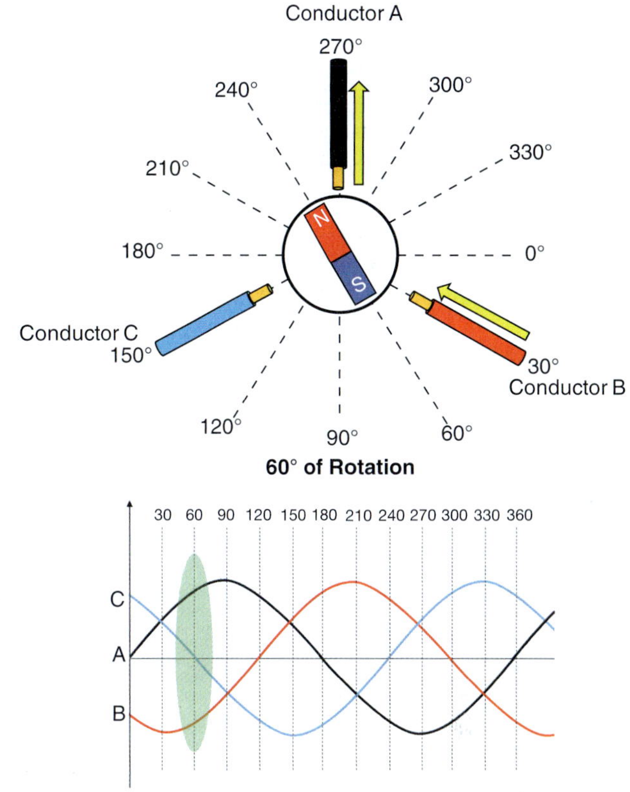

Goodheart-Willcox Publisher

Figure 16-7. Magnetic effect on the conductors with south pole aligned at 60° of rotation.

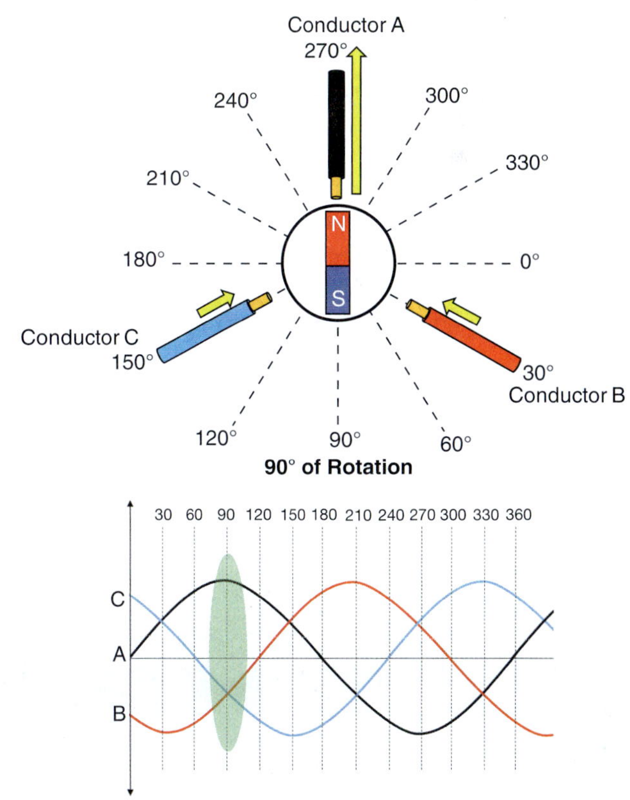

Goodheart-Willcox Publisher

Figure 16-8. Magnetic effect on the conductors with the south pole aligned at 90° of rotation.

16.2 Three-Phase Wye Configuration

The wye configuration of a three-phase voltage source is so named because of its similarity to the letter Y when drawn in schematic wiring diagrams, **Figure 16-9**. The ***wye configuration*** is characterized by a common connection point, referred to as a grounded neutral point, that joins one end of all three windings of the three-phase system together. Each winding is referred to as a phase and the ungrounded conductors connected to the windings are called *lines*. In a three-phase power system, the phases are referred to as phase A, phase B, and phase C and corresponding lines are referred to as A, B, and C, or L1, L2, and L3, respectively.

In a four-wire wye configuration, a neutral conductor is connected to the neutral point of the wye and grounded. A main bonding jumper (for services) or a system bonding jumper (for separately derived systems) is installed to create a point, or a bullseye, that is the common point to bond all metal enclosures, equipment grounding conductors, and the neutral to earth through a grounding electrode conductor. This bonding action makes the neutral the grounded conductor.

The voltage measured between any two ungrounded conductors (or phases) is referred to as ***line voltage*** or phase-to-phase voltage, while the voltage measured between any one ungrounded conductor (phase) and neutral or ground is referred to as ***phase voltage***, **Figure 16-10**. If a phase voltage of 120 V is measured between any one line and neutral or ground, the voltage measured between any two lines will be 208 V. The 208-V line voltage value is 1.732, or the square root of 3, times the phase voltage of 120 V. This is because the two phases are 120° out of phase with each other, **Figure 16-11**. When adding two phase-voltage vectors at an angle of 120° with each other, the resulting line-voltage vector will be equal to $\sqrt{3}$ times the phase voltage:

$$E_{\text{Line}} = \sqrt{3} \times E_{\text{Phase}}$$

If the line voltage is known, the phase voltage can be found using the following formula:

$$E_{\text{Phase}} = \frac{E_{\text{Line}}}{\sqrt{3}}$$

Shedding Light

You might expect that the voltage measured between any two lines would be 240 V, but this is not true! Remember that any two phases are 120° out of phase with each other, so the peak values of any two sine waves never line up, **Figure 16-12**.

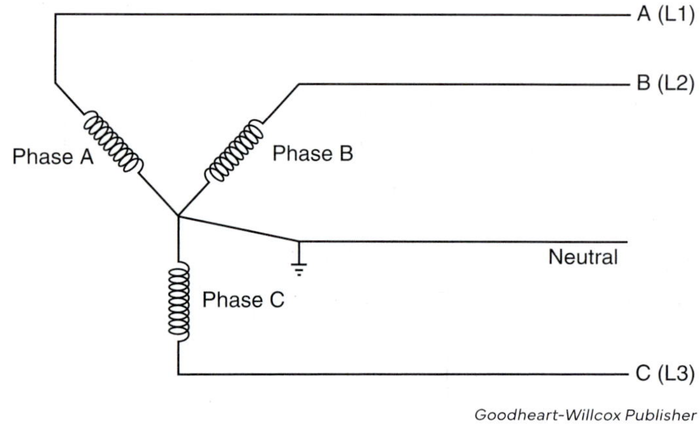

Figure 16-9. Three-phase, four-wire wye configuration with a grounded neutral point.

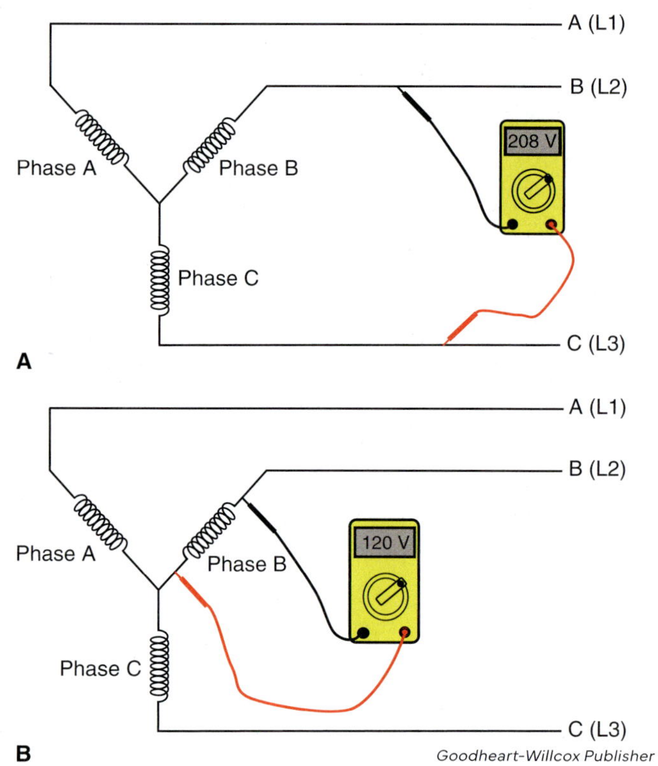

Figure 16-10. Measuring voltage in a wye configuration. A—Line voltage is measured between any two lines or across any two phases. B—Phase voltage is measured across any one phase.

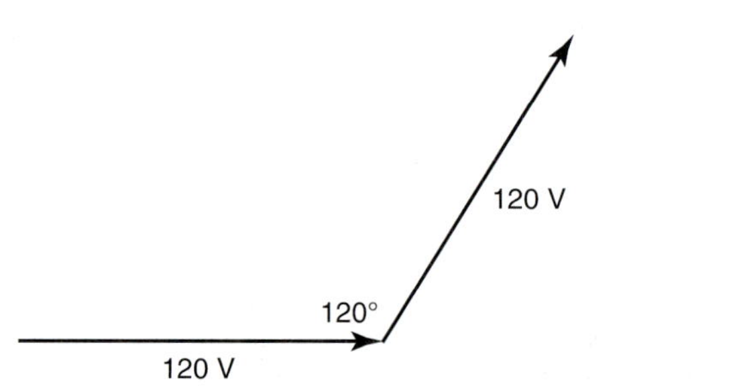

Figure 16-11. Adding two 120 V vectors at an angle of 120° with each other.

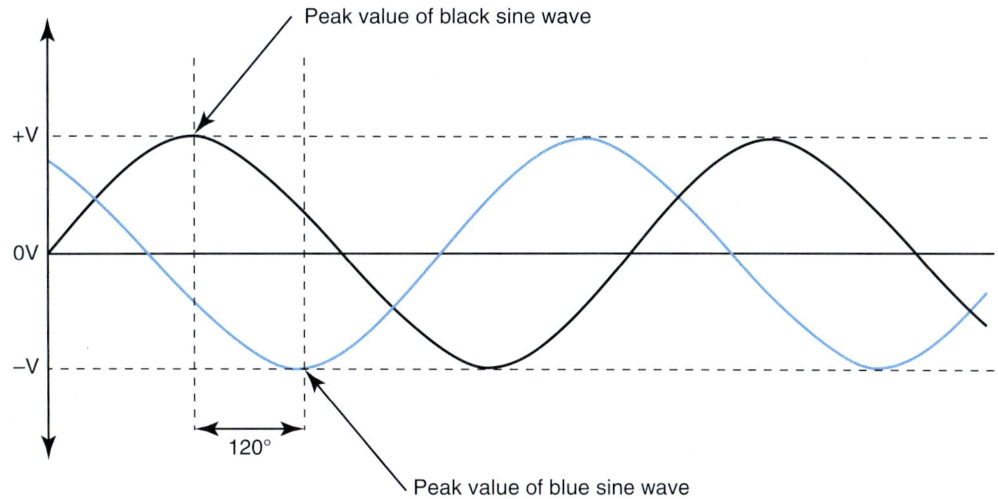

Goodheart-Willcox Publisher

Figure 16-12. The sine waves in a three-phase system do not reach their peaks at the same time.

The four-wire wye configuration provides various voltages for multiple applications:

- Single-phase nominal voltage of 120 V for general-purpose applications.
- Single-phase, line-to-line nominal voltage in the 208 V range for appliance and equipment applications.
- Three-phase nominal voltage in the 208 V range for heavy-duty equipment applications.

Refer to **Example 16-1** and **Example 16-2** on how to calculate voltage in a three-phase wye connection.

Example 16-1

If the phase voltage in a three-phase wye connection is 277 V, what is the expected line voltage?

✓ Solution

The line voltage can be calculated as follows:

$$E_{Line} = \sqrt{3} \times E_{Phase}$$

$$E_{Line} = 1.732 \times 277 \text{ V}$$

$$E_{Line} = 480 \text{ V}$$

Now It's Your Turn 16-1

If the phase voltage in a three-phase wye connection is 347 V, what is the expected line or phase-to-phase voltage?

Example 16-2

If the line voltage in a three-phase wye connection is 600 V, what is the expected phase voltage?

✓ Solution

The phase voltage can be calculated as follows:

$$E_{Phase} = \frac{E_{Line}}{\sqrt{3}}$$

$$E_{Phase} = \frac{600}{1.732}$$

$$E_{Phase} = 347 \text{ V}$$

Now It's Your Turn 16-2

If the line or phase-to-phase voltage in a three-phase wye connection is 400 V, what is the expected phase voltage?

Calculating Line Voltage

The $\sqrt{3}$ multiplier used to determine the line voltage when the phase voltage is known can be obtained by applying the sine function and the properties of triangles. For example, in the isosceles triangle in **Figure 16-13**, the resultant vector completes the triangle and connects the tail of one 120 V vector to the head of the other 120 V vector. The two 120 V vectors have the same magnitude. In an isosceles triangle, the angles opposite the equal sides must be equal as well. Since the angle between the vectors is 120°, then the remaining two angles must be 30° each.

By extending the horizontal vector to the right, to point C, and drawing a perpendicular line connecting point C to point B, triangle ACB is formed, **Figure 16-14**. Angle BAC is

60° since angle DAB and angle BAC are supplementary angles and their sum must be 180°.

Evaluating triangle ACB, the sine of angle BAC (60°) is equal to Y divided by 120 (opposite over hypotenuse). The height Y is determined as follows:

$$\begin{aligned} \text{Sin } 60° &= \frac{Y}{120} \\ Y &= 120 \times \text{Sin } 60° \\ Y &= 120 \times 0.866 \\ Y &= 103.92 \end{aligned}$$

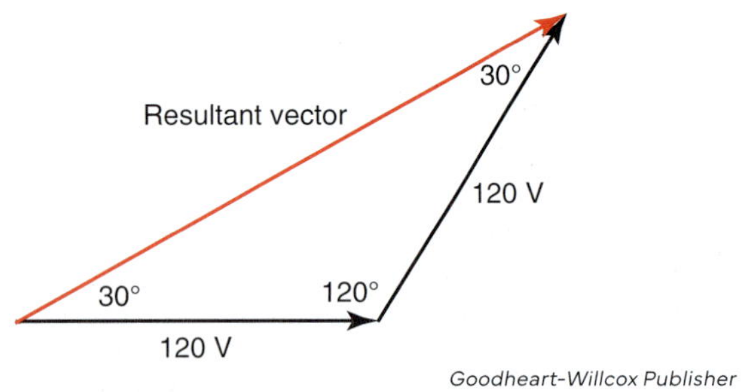

Figure 16-13. Angles in the isosceles triangle.

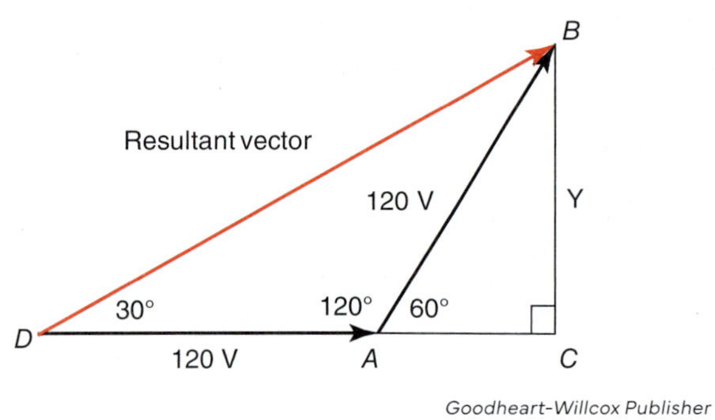

Figure 16-14. Creating a right triangle and solving for the height value Y.

Evaluating the triangle DCB in **Figure 16-14**, the sine of angle BDC (30°) is equal to Y divided by resultant vector. This is because the sine of an angle is equal to the length of the side opposite the angle divided by the triangle's hypotenuse.

$$\begin{aligned} \text{Sin } 30° &= \frac{\text{Opposite}}{\text{Hypotenuse}} \\ \text{Sin } 30° &= \frac{Y}{\text{Resultant Vector}} \\ \text{Resultant Vector} &= \frac{Y}{\text{Sin } 30°} \\ \text{Resultant Vector} &= \frac{103.92}{0.50} \\ \text{Resultant Vector} &= 207.84 \approx 208 \text{ V} \end{aligned}$$

When two 120 V vectors at an angle of 120° with each other, the resultant vector is 208 V. Dividing 208 V by 120 V provides the $\sqrt{3}$ multiplier value.

Current in a Three-Phase Wye Configuration

In a three-phase, four-wire wye-connected system, voltages at 208 V nominal will be measured between any two ungrounded conductors, or lines, **Figure 16-15**. In addition, at 120 V nominal will be measured between any ungrounded conductor (line) and the neutral grounded conductor or any grounding conductor, **Figure 16-16**. This is a main benefit of the wye-connection, as it can be used to power both three-phase and single-phase devices. With each of the three legs able to provide 120 V to individual branch circuits, the load on the entire three-phase system can be well balanced, thus increasing its efficiency and effectiveness.

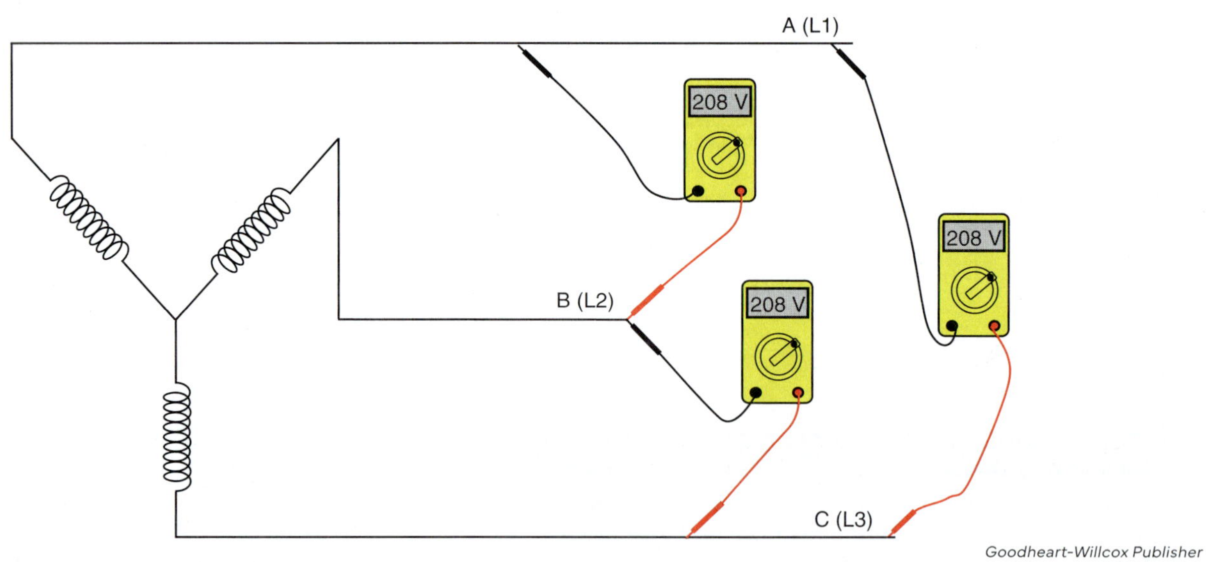

Figure 16-15. Obtaining 208 V readings between any two ungrounded conductors.

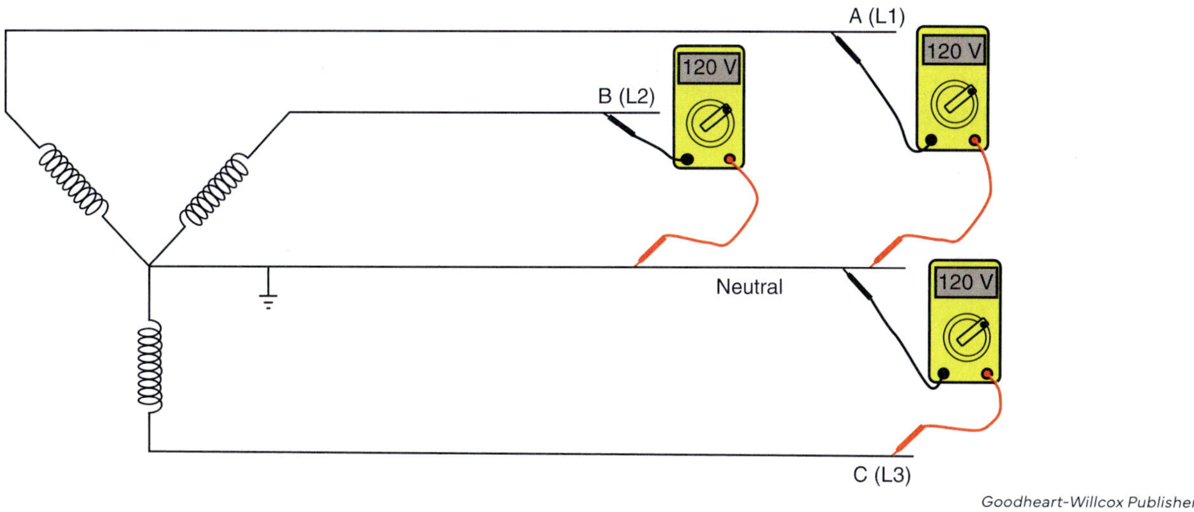

Figure 16-16. Obtaining 120 V readings between any ungrounded conductor and neutral.

Know the Code

Guidelines on which systems are to be grounded and how grounding is accomplished is explained in *Part II* of *Article 250, System Grounding*, of the *NEC*.

In a wye configuration, each line is connected to only one phase. Thus, each line and its corresponding phase are in series with each other. Since the current is the same at all points in a series circuit, the line and phase currents for each branch of a wye-connected three-phase system are the same, **Figure 16-17**. For example, if the current measured by meter A is 20 A, the current measured by meter B will also be 20 A. In general, for a wye-connected system, the following relationship between phase and line currents exists:

$$I_{\text{Phase}} = I_{\text{Line}}$$

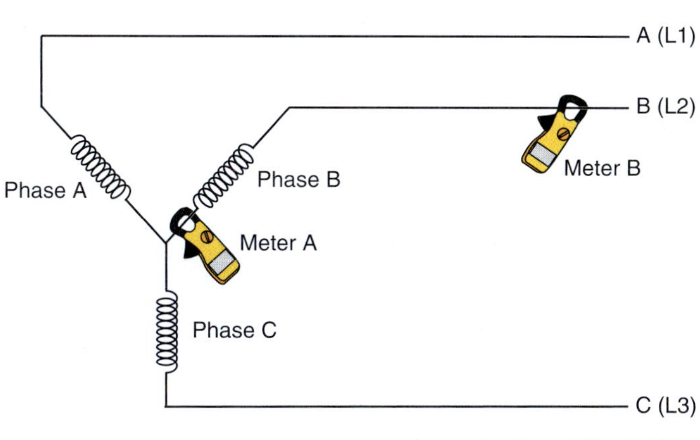

Figure 16-17. In a three-phase wye configuration, the amperage readings on meters A and B will be the same.

16.3 Three-Phase Delta Configuration

The schematic diagram for a ***delta configuration*** of a three-phase voltage source resembles the Greek letter delta, Δ, **Figure 16-18**. The three phases are connected to each other in a continuous loop where the outlet of one phase is connected to the inlet of the next. Unlike the wye connection, in a delta connection each line is connected to two phases. For example, Line 1 (L1) in **Figure 16-18** is connected to both phase A and phase B at point A.

When a delta-configured transformer is used for power distribution, there are three options concerning grounding and the potential installation/location of the grounded conductor:

- **Grounded conductor at the midpoint of one of the three-transformer windings.** This grounded conductor will also act as a neutral for two of the three phases.

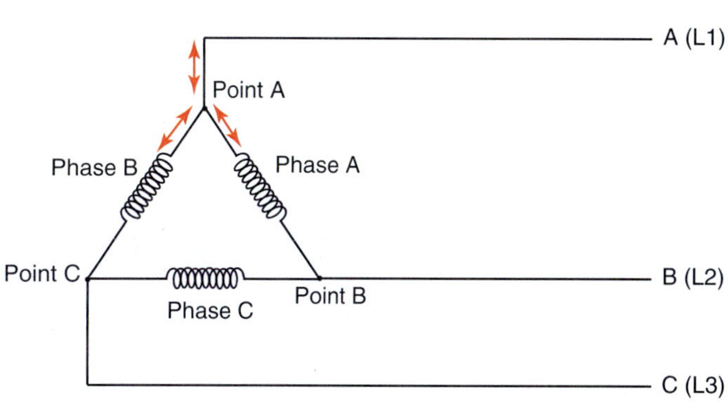

Figure 16-18. A delta configuration. At point A, Line 1 connects to phases A and B.

- **Corner-grounded B-phase system.** The B-phase conductor is grounded, making it a grounded conductor but not a neutral.
- **Ungrounded delta transformer.** This option is rarely used.

These choices will determine the voltage readings to the grounded conductor. Regardless of how a delta transformer is grounded or connected ungrounded, the voltage between phases or lines is the same or very close to the same.

The delta connection can provide various voltages for multiple applications:

- Single-phase voltage at 240 V nominal for appliance and equipment applications. This application uses any two ungrounded conductors from phase A, phase B, or phase C (L1, L2, or L3).
- Using a midpoint ground, single-phase voltage at 120 V nominal for small equipment and electronic applications. In this configuration, phase A or phase C to the neutral conductor supplies 120 V.
- Three-phase voltage at 240 V nominal for heavy-duty equipment applications; typically motors and welders. This application uses three ungrounded conductors, A-phase, B-phase, and C-phase (L1, L2, and L3).

In the delta configuration, a voltage reading taken across any two lines will be the same as a voltage reading taken across the phase located in between those two lines. In **Figure 16-19**, for example, meter A is measuring 240 V across phase A, while meter B is measuring 240 V between L1 and L2. Electrically speaking, the locations of the two black test leads, point 1 and point 2, are the same, as are the locations of the two red test leads, point 3 and point 4.

For a delta connection configured system, the phase voltage is equal to the line voltage:

$$E_{\text{Phase}} = E_{\text{Line}}$$

In some older commercial buildings, a modification of the delta configuration can be found. This modification is called the high-leg delta configuration. The high-leg delta configuration allows for four different voltages to be made available for use:

- Single-phase, 120 V for general purpose applications.
- Single-phase, 208 V nominal is available but is not typically used.
- Single-phase, 240 V nominal for appliance and equipment applications.
- Three-phase, 240 V nominal for heavy-duty equipment applications.

In the ***high-leg delta configuration***, a grounded conductor (neutral) is connected to the center of one of the three phases of the system, **Figure 16-20**. The addition of the grounded conductor does not change the voltage readings across each of the phases. The voltage readings across each phase will be 240 V nominal.

The addition of the grounded conductor now allows for the production of lower voltages. In **Figure 16-21**, the voltage measured from A to neutral is 120 V, the voltage measured from C to neutral is 120 V, and the voltage measured from B to neutral is 208 V.

The higher voltage reading between B and neutral is due to the path between B and neutral, **Figure 16-22A**. To move from B to neutral, current must pass through two different phases, which are 120° out of phase with each other. The path

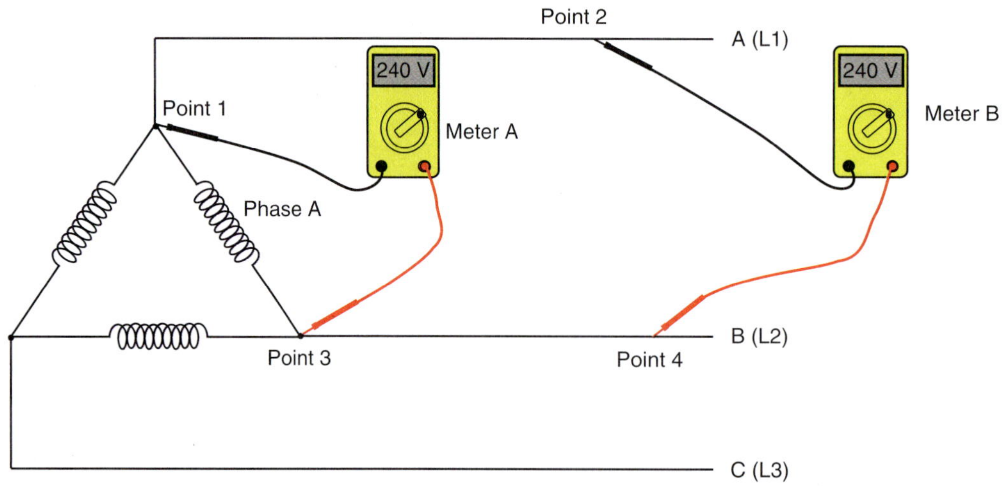

Goodheart-Willcox Publisher

Figure 16-19. In a delta configuration, phase and line voltage readings are the same across both lines.

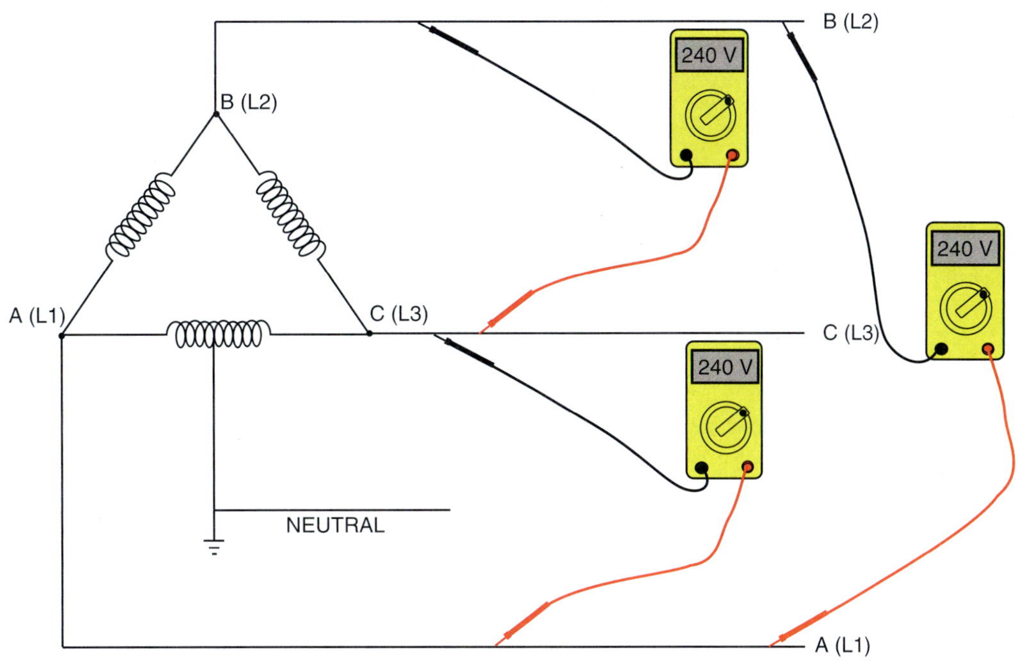

Goodheart-Willcox Publisher

Figure 16-20. In a high-leg delta connection, a grounded conductor is connected to the center of one of the three phases; line voltage readings across each phase are all 240 V.

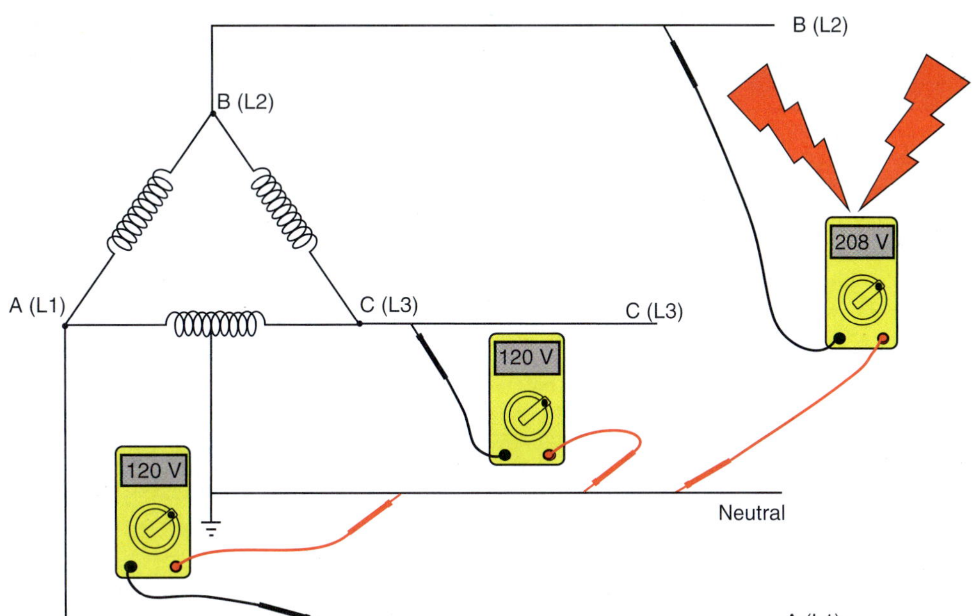

Goodheart-Willcox Publisher

Figure 16-21. Phase voltage measurements in a high-leg delta connection. The addition of the grounded conductor allows for the production of lower voltages.

from A to neutral, **Figure 16-22B**, and from C to neutral, **Figure 16-22C**, only involves one phase, so there is no need to add voltage vectors.

A graphical representation of the voltage readings in a high-leg delta connection is provided in **Figure 16-23**. The shaded space in this figure is an equilateral triangle and the high-leg voltage line divides this triangle into two equal 30-60-90 triangles. Triangles DCB and ACB in **Figure 16-14** are 30-60-90 triangles as well. These triangles have a special side-length relationship, namely 1, $\sqrt{3}$, 2. The shortest side will always be one-half as long as the middle side, while the hypotenuse, or longest side, will always be longer than the

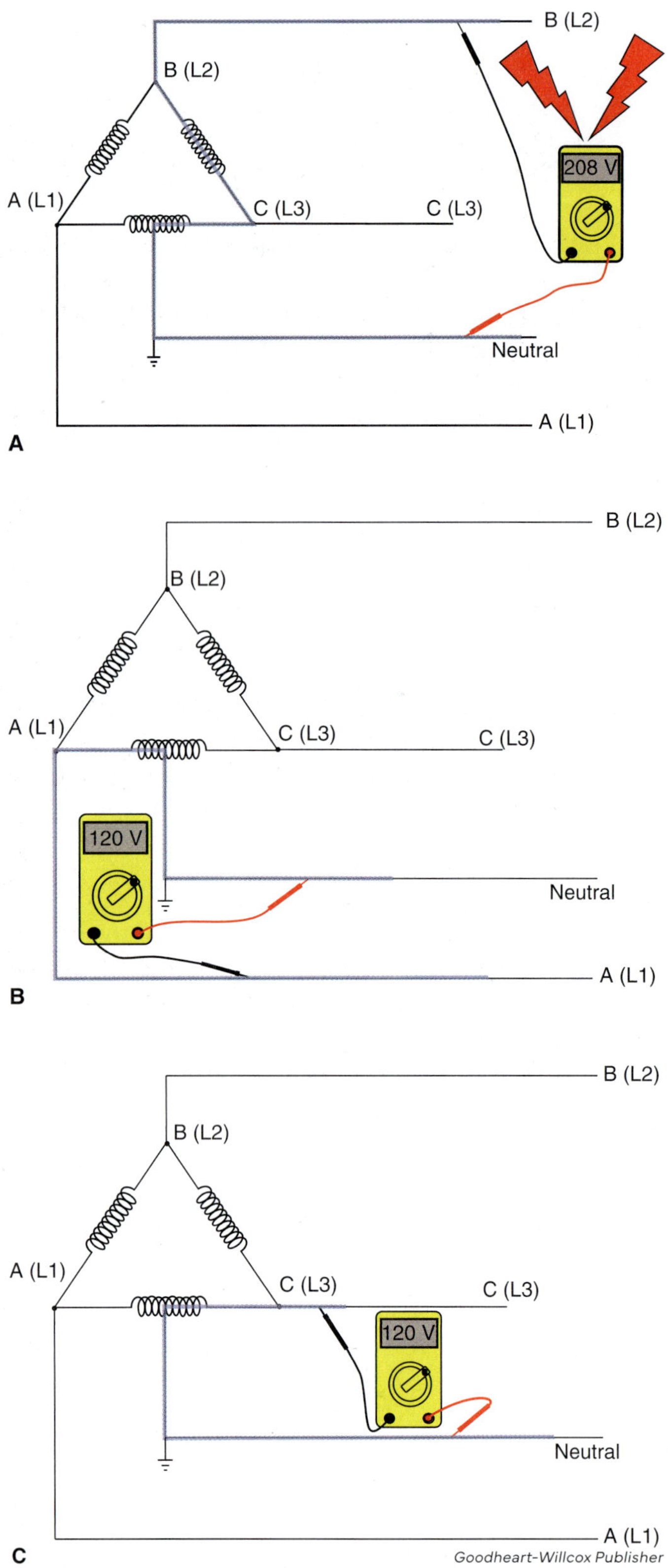

Goodheart-Willcox Publisher

Figure 16-22. When a path passes through two different phases, a high-leg delta connection is created. A—The path between L2 and neutral passes through only one phase. B—The path between L1 and neutral passes through two phases, which are 120° out of phase with each other, creating the high leg. C—The path between L3 and neutral passes through only one phase.

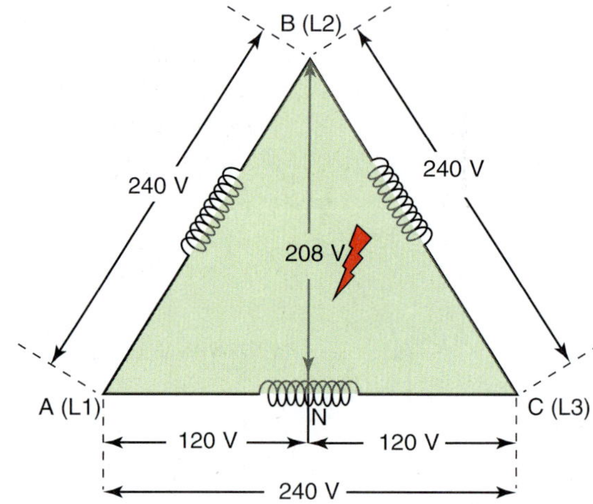

Goodheart-Willcox Publisher

Figure 16-23. Graphic representation of the voltages in a high-leg delta configuration.

shortest side by a factor of $\sqrt{3}$. The 120 V, 208 V, and 240 V values in each of the 30-60-90 triangles follow this same relationship. This configuration can be used to feed three-phase, 240 V loads, and two phases can be used to feed 120 V loads.

Current in a Three-Phase Delta Configuration

Referring to the delta connection in **Figure 16-24**, the following observations can be made:

- At Point A, line L1 is connected to Phase A and Phase C.
- At Point B, line L2 is connected to Phase A and Phase B.
- At Point C, line L3 is connected to Phase B and Phase C.

Recalling Kirchhoff's current law, the sum of the currents flowing into a node must equal the sum of the currents flowing out of the node. In the case of a three-phase system, where the phases are 120° out of phase with each other, the inflow and outflow currents must be treated as vectors. Calculating the line currents in a delta connection uses vector addition,

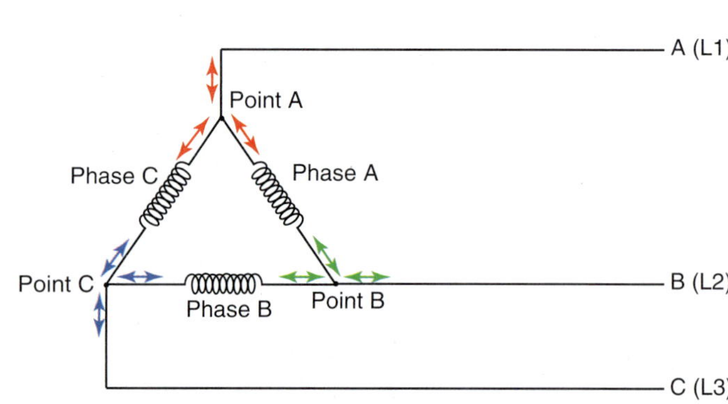

Goodheart-Willcox Publisher

Figure 16-24. In a delta connection, each line is connected to two phases.

which is the same method used to calculate the line voltage of a three-phase wye connection. As a result, the relationship that exists between the line and phase currents in a three-phase delta connection is:

$$I_{Line} = \sqrt{3} \times I_{Phase}$$

If the line current is known, the phase current in a delta configuration can be determined using the following formula:

$$I_{Phase} = \frac{I_{Line}}{\sqrt{3}}$$

Refer to **Example 16-3** and **Example 16-4** on how to calculate current in a delta connection.

Example 16-3

Referring to the following image, calculate the reading on meter B if the reading on meter A is 40 A.

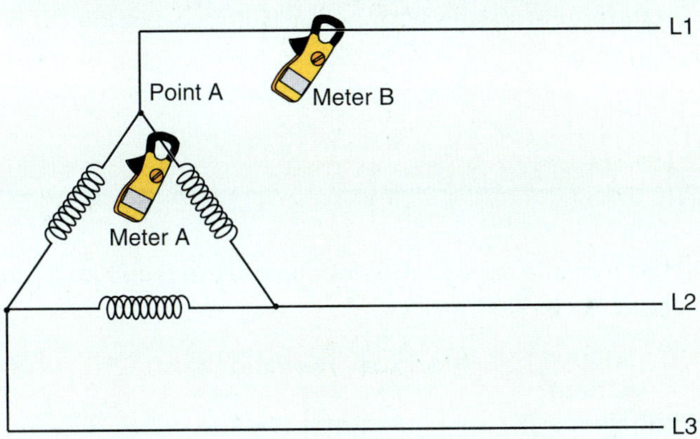

Goodheart-Willcox Publisher

✓ Solution

In this delta connection, meter A is measuring the phase current and meter B is measuring the line current. The line current can be calculated as follows:

$$I_{Line} = \sqrt{3} \times I_{Phase}$$

$$I_{Line} = 1.732 \times 40$$

$$I_{Line} = 69.28 \text{ A}$$

Shedding Light

Once energized, if a three-phase motor is turning in the wrong direction, changing the direction of rotation is easily accomplished. This can be done by switching the connection of any two lines feeding power to the motor. When energized again, the motor will rotate in the opposite direction.

Example 16-4

Referring to the image in **Example 16-3**, calculate the reading on meter A if meter B is reading 40 A.

✓ Solution

As meter A is measuring the phase current and meter B is measuring the line current, the phase current can be calculated as:

$$I_{Phase} = \frac{I_{Line}}{\sqrt{3}}$$

$$I_{Phase} = \frac{40}{1.732}$$

$$I_{Phase} = 23.1 \text{ A}$$

16.4 Voltage Imbalance

Voltage imbalance is an electrical characteristic that is used to quantify the equality, or inequality, that exists among the phases of a three-phase power system. Mathematically, it expresses the maximum difference between any one phase and the average voltage. It is desirable to keep this maximum difference, expressed as a percentage, as low as possible, especially since voltage imbalance is an indication of the "health" of the electrical system. As a generally accepted industry rule, a voltage imbalance below 2% is considered acceptable for the majority of applications.

Three-phase motors can be heavily impacted by a voltage imbalance. Even a low voltage imbalance of 1% can cause extra current flow and increased motor operating temperatures, thus decreasing a three-phase motor's efficiency and shortening life expectancy.

Know the Code

Depending on the codes, standards, and criteria being referenced, the suggested and allowable voltage imbalance levels will differ. Here are three examples:

- According to the American National Standards Institute (ANSI), in standard C84.1, electrical supply systems should operate with a maximum voltage imbalance of 3% under no-load conditions, when measured at the electric meter.
- Most utility companies in the United States set a voltage imbalance limit of 2.5%.
- Some manufacturers require that their motors operate with less than a 1% voltage imbalance if the motor warranty is to be valid.

To avoid confusion, it is important to understand the voltage imbalance guidelines that pertain to the specific application and project you are working on.

Voltage Imbalance Calculations

Voltage imbalance can be calculated using the following formula:

$$\text{Voltage Imbalance (\%)} = \frac{\text{Maximum Deviation from Average Voltage}}{\text{Average Voltage}} \times 100\%$$

The maximum deviation from the average voltage value comes from the phase voltage that differs the most when compared to the average voltage. Refer to **Example 16-5** on how to calculate voltage imbalance.

Math Sidebar

When calculating voltage imbalance, ignore negative signs for maximum deviation. Positive and negative values are unimportant. A voltage imbalance of −1% is electrically the same as a voltage imbalance of +1%. When calculating voltage deviation, a value of −9 V has the same effect as a voltage deviation of 9 V.

Possible Causes of Excessive Voltage Imbalance

There are a number of possible causes of excessive voltage imbalance. If an excessive voltage imbalance condition is observed, you might assume the problem lies with the utility company. Although this might be the case, there are other possible causes for the condition. Some potential causes for high voltage imbalance include:

- **Loose electrical connections.** If this is the case, extra heating and potential system damage will likely occur.
- **Worn or pitted electrical contacts on relays, contactors, or motor starters.** This will result in a higher impedance path, like a loose connection.
- **Uneven distribution of single-phase loads across the three phases.** Load balancing in a three-phase system is critical. This is accomplished by balancing the single-phase loads, such as lighting and general-purpose receptacles.
- **Faulty operation of power factor correction equipment.** Issues with power factor correction

Example 16-5

Calculate the voltage imbalance of a 460 V electric system using the following voltage readings:

Line 1 to Line 2 = 462 V
Line 2 to Line 3 = 448 V
Line 1 to Line 3 = 452 V

✓ Solution

The first step to calculate voltage imbalance is to obtain the average of the three voltage readings. The average of these three voltages is:

$$\frac{462\text{ V} + 448\text{ V} + 452\text{ V}}{3} = 454\text{ V}$$

Next, determine the largest voltage deviation, which is the difference between the voltage reading and the average voltage that was just calculated. The three voltage deviations are:

$$462 - 454 = 8\text{ V (Line 1 to Line 2)}$$

$$448 - 454 = -6\text{ V (Line 2 to Line 3)}$$

$$452 - 454 = -2\text{ V (Line 1 to Line 3)}$$

The largest voltage deviation is 8 V, so this is the value that will be used to calculate the voltage imbalance for this system. Plugging the known values for maximum voltage deviation and average voltage in the formula for voltage imbalance, the result is calculated as follows:

$$\text{Voltage Imbalance (\%)} = \frac{\text{Maximum Deviation from Average Voltage}}{\text{Average Voltage}} \times 100\%$$

$$\text{Voltage Imbalance} = \frac{8\text{ V}}{454\text{ V}} \times 100\%$$

$$\text{Voltage Imbalance} = 0.0176 \times 100\%$$

$$\text{Voltage Imbalance} = 1.76\%$$

This result falls within the acceptable range, since the accepted industry rule is an imbalance of under 2%.

Now It's Your Turn 16-5

Determine the voltage imbalance of an electrical system that has the following voltage readings:

Line 1 to Line 2 = 238 V

Line 2 to Line 3 = 227 V

Line 1 to Line 3 = 219 V

Is this voltage imbalance acceptable? Why or why not?

capacitor banks can contribute to increased voltage imbalance.

- **Electrical faults from one phase to ground.** These unintended pathways between a phase and ground can cause an increase in voltage imbalance if the current flow to ground is not high enough to trip the over-current protection device (OCPD).

Effects of Voltage Imbalance

When there is excessive voltage imbalance in a three-power system, the components and equipment connected to it can experience operational and performance-related problems. Some potential effects of high voltage imbalance include:

- Reduced performance
- Motor winding insulation overheating and damage
- Shortened equipment and motor life
- Motor controller failure or unreliable operation
- Solid-state component and circuit damage
- Increased mechanical stress and vibration
- Reduced motor torque
- Current flow fluctuations in individual phases
- Increases in energy costs

Just as voltage imbalance is a measure of the voltage deviation from the average voltage, ***current imbalance*** is a measure of the current deviation from the average current. Current imbalance can be caused by voltage imbalance. For example, each 1% of voltage imbalance in an induction motor increases the current imbalance by 6%.

Possible Remedies for Excessive Voltage Imbalance

Those working in the trades should be aware of voltage imbalance because it can be the cause of motor or other system failure. Although some sophisticated electronic systems feature voltage imbalance protection, it is also the responsibility of the equipment operators, as well as those who service and maintain equipment, to be aware of the potential problems that can arise. If a particular piece of equipment, for example, has experienced multiple motor or component failures, it is up to those who service the equipment to evaluate the cause for failure. Simply concluding that the motor overheated is not enough. Why the motor overheated is the question that needs to be answered.

If a system load, such as a motor, is suspected of causing excessive voltage imbalance, the load can be temporarily disconnected from the power source. With the load disconnected, the voltage imbalance can be recalculated. If the voltage imbalance drops to an acceptable level, the disconnected load is the cause of the problem. If, however, excessive voltage imbalance is still present, the system problem lies elsewhere.

Routine maintenance is key. Proper periodic maintenance of electrical equipment can help reduce, if not eliminate, excessive voltage imbalance that cannot be attributed to the utility company. By reducing voltage imbalance, the efficiency of the equipment will increase, and the operating costs associated with the equipment will fall. In addition, routine maintenance reduces system downtime and emergency repair costs.

Voltage Imbalance and Temperature Rise

One of the main effects of excessive voltage imbalance is an increase in motor temperature. To quantify the actual amount of temperature rise, expressed as a percentage, the following formula can be used to arrive at an approximate value:

$$\text{Temperature Rise (\%)} \approx 2 \times [\text{Voltage Imbalance (\%)}]^2$$

Refer to **Example 16-6** and **Example 16-7** on how to calculate temperature rise and motor temperature when there is a voltage imbalance.

Example 16-6

Calculate the temperature rise for a 3% voltage imbalance.

✓ **Solution**

The temperature rise value is calculated as follows:

$$\text{Temperature Rise (\%)} \approx 2 \times [\text{Voltage Imbalance (\%)}]^2$$

$$\text{Temperature Rise (\%)} \approx 2 \times (3\%)^2$$

$$\text{Temperature Rise (\%)} \approx 2 \times 9\%$$

$$\text{Temperature Rise (\%)} \approx 18\%$$

Now It's Your Turn 16-6

Calculate the approximate temperature rise percentage if the voltage imbalance is 2.5%.

The approximate motor temperature can be determined if two factors are known: the voltage imbalance and the motor's rated operating temperature at a voltage imbalance of 0%. The formula that can be used is:

$$\text{Motor Temperature} = (1 + \text{Temp Rise}) \times \text{Motor Temp}_{0\%}$$

In this formula, motor temperature represents the temperature of a motor with voltage imbalance and temperature rise is the calculated temperature rise due to voltage imbalance, which is expressed as a decimal. Motor temp$_{0\%}$ is the motor's rated operating temperature with no voltage imbalance.

Example 16-7

Under ideal conditions (with a voltage imbalance of 0%), a particular motor operates at a temperature of 125°F. If the voltage imbalance is 4%, what is the expected motor temperature?

✓ Solution

Remember that the temperature rise is calculated as two times the square of the voltage imbalance. In this example, the voltage imbalance is 4%, so the temperature rise is calculated as $2 \times (4)^2$, which equals 32%, or 0.32.

Next, calculate the expected motor temperature as follows:

$$\text{Motor Temperature} = (1 + \text{Temp Rise}) \times \text{Motor Temp}_{0\%}$$

$$\text{Motor Temperature} = (1 + 0.32) \times 125°\text{F}$$

$$\text{Motor Temperature} = 1.32 \times 125°\text{F}$$

$$\text{Motor Temperature} = 165°\text{F}$$

Now It's Your Turn 16-7

Calculate the expected temperature of a motor with an ideal operating temperature rating of 80°F if it is connected to a power supply that has a voltage imbalance of 5%.

Voltage Imbalance and Operating Costs

As the voltage imbalance of a power supply increases, the efficiency of the motors and any equipment powered by it decreases. As efficiency drops, the costs associated with operating the equipment rises. Mathematically, the cost of operating a piece of equipment is inversely related to the efficiency of the equipment.

When efficiency is expressed as a percentage, such as 93% efficient, the true meaning is that 93 units of useful output are produced for every 100 units of input. The expression 93% efficiency, therefore, can be written as 93/100. When calculating the effects of reduced efficiency on cost, the inverse of efficiency is used. In the case of our 93% efficient piece of equipment, the cost would be 100/93, or 1.075 times the cost of a similar piece of equipment operating at 100% efficiency. A formula that can be used to solve for operating cost is:

$$\text{Cost} = \text{Ideal Cost} \times \frac{100}{\text{Efficiency Percentage}}$$

Refer to **Example 16-8** and **Example 16-9** on how to calculate operating cost.

Example 16-8

If the cost to operate an ideal motor (100% efficient) for one month is $500, what would the cost be to operate the same motor for one month if its actual efficiency was 90%?

✓ Solution

The cost can be calculated as follows:

$$\text{Cost} = \text{Ideal Cost} \times \frac{100}{\text{Efficiency Percentage}}$$

$$\text{Cost} = \$500 \times \frac{100}{90}$$

$$\text{Cost} = \$555.56$$

Example 16-9

How much money can be saved if the efficiency of a motor is increased from 70% to 90%? At the present 70% efficiency, the cost to operate the motor is $6000 per year.

✓ Solution

To calculate the savings, the 100% efficiency cost would first need to be calculated. Since the present cost and efficiency values are known, they can be plugged into the cost formula as follows:

$$\text{Cost} = \text{Ideal Cost} \times \frac{100}{\text{Efficiency Percentage}}$$

$$\$6000 = \text{Ideal Cost} \times \frac{100}{70}$$

$$\$6000 = \text{Ideal Cost} \times 1.43$$

$$\text{Ideal Cost} = \frac{\$6000}{1.43}$$

$$\text{Ideal Cost} = \$4195.80$$

With the ideal cost and the desired efficiency known, these values can then be plugged into the formula to get the cost of operating the motor at 90%:

$$\text{Cost} = \text{Ideal Cost} \times \frac{100}{\text{efficiency percentage}}$$

$$\text{Cost} = \$4195.80 \times \frac{100}{90}$$

$$\text{Cost} = \frac{\$4195.80}{1.11}$$

$$\text{Cost} = \$4662$$

The cost savings will be equal to the $6000 operating cost at 70% efficiency minus the $4662 operating cost at 90%:

$$\text{Savings} = \$6000 - \$4662$$

$$\text{Savings} = \$1338$$

Now It's Your Turn 16-9

Due to an increase in voltage imbalance, the efficiency of a motor dropped from 83% to 67%. How much did the annual operating cost increase if the cost to operate the motor before the efficiency drop was $4700 per year?

Math Sidebar

When efficiency is expressed as a decimal, an efficiency of 87% would be written as 0.87. In this case, the efficiency could be described as a system that provides 0.87 units of useful output for every one unit of input. An efficiency of 0.87, therefore, can be written as 0.87/1. Just as before, when calculating the effects of reduced efficiency on cost, the inverse of efficiency is used. In the case of our 87% efficient piece of equipment, the cost would be 1/0.87, or 1.15 times the cost of a similar piece of equipment operating at 100% efficiency.

16.5 Three-Phase Power

Three-phase power will produce at least 50% more power with the addition of a third current-carrying conductor when compared to an equivalent single-phase power supply. When the power consumption of a single-phase system is calculated, the following formula is used:

$$\text{Power (Single-Phase)} = \text{Current} \times \text{Voltage} \times \text{Power Factor}$$

If the loads in the single-phase circuit were 100% resistive, the power factor is equal to 1, and can be calculated as follows:

$$\text{Power (Single-phase)} = \text{Current} \times \text{Voltage}$$

For three-phase power, there are two different formulas that can be used to calculate the apparent power, or volt-amperes (VA), of a circuit. The formula used depends on whether the phase values for current and voltage are used or if the line values for current and voltage are used. The two formulas are:

$$\text{Apparent Power (VA)} = 1.732 \times I_{\text{Line}} \times E_{\text{Line}}$$

$$\text{Apparent Power (VA)} = 3 \times I_{\text{Phase}} \times E_{\text{Phase}}$$

In most cases, the line voltage and current values are easier to obtain in the field, so the first formula listed is used more often than the second. Either formula will provide accurate VA calculation results, regardless of whether the three-phase system is configured as a wye or as a delta. Refer to **Example 16-10**.

Example 16-10

For the wye connection shown in in the following circuit, calculate the apparent power, in VA, using the line current and voltage values. Then, calculate the apparent power, in VA, using the phase current and voltage values.

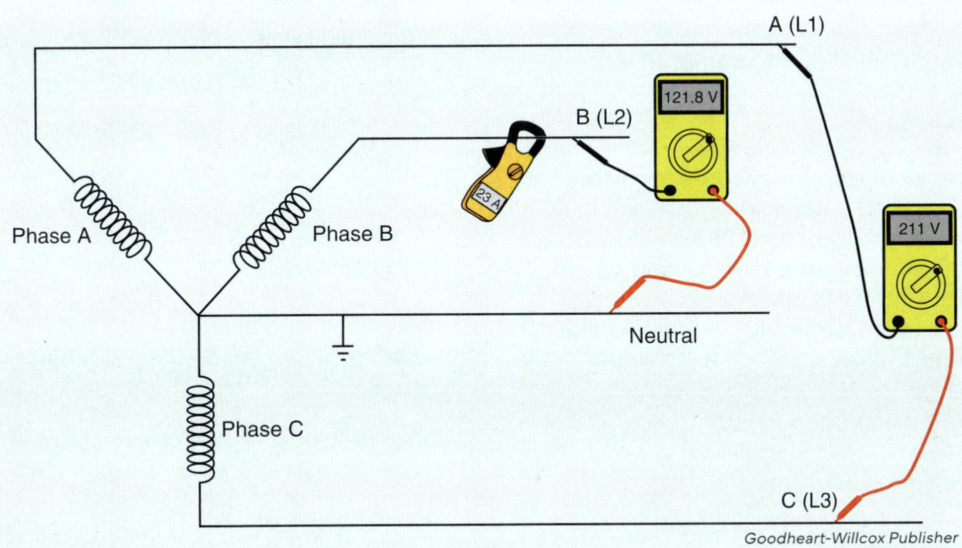

Goodheart-Willcox Publisher

Continued

✓ Solution

First, calculate the apparent power using the line current and voltage values. From the circuit, the line current is 23 A and the line voltage is 211 V. Using these values, the apparent power is calculated as:

$$VA = 1.732 \times I_{Line} \times E_{Line}$$

$$VA = 1.732 \times 23\ A \times 211\ V$$

$$VA = 8405\ VA$$

Next, calculate apparent power using the phase current and voltage values. From the circuit, the phase current is 23 A and the phase voltage is 121.8 V. Using these values, the apparent power is calculated as:

$$VA = 3 \times I_{Phase} \times E_{Phase}$$

$$VA = 3 \times 23\ A \times 121.8\ V$$

$$VA = 8404\ VA$$

Now It's Your Turn 16-10

Using the information provided in the delta connection shown in the following circuit, calculate the apparent power, in VA, using the phase current and voltage values. Then using the same circuit, calculate the apparent power, in VA, using the line current and voltage values.

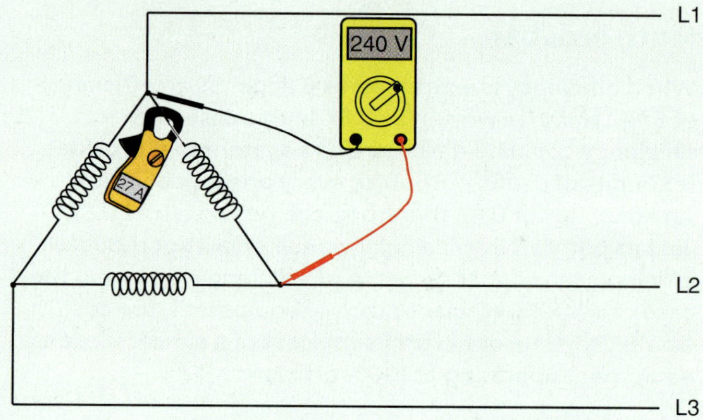

Goodheart-Willcox Publisher

In many electrical systems, the circuit loads are not 100% resistive, so the true power (active power) is less than the apparent power just discussed. The formulas for active power are similar to those previously introduced for apparent power, but include the power factor:

$$\text{True Power (W)} = 1.732 \times I_{Line} \times E_{Line} \times PF$$

$$\text{True Power (W)} = 3 \times I_{Phase} \times E_{Phase} \times PF$$

Either line values or phase values for the current and voltage can be used to calculate the true power of a three-phase system. Refer to **Example 16-11** and **Example 16-12**.

Math Sidebar

Although true or active power is measured in watts (W), the term kilowatt (kW) is commonly used to express the power of electric motors. To convert watts into kilowatts, just remember that 1 kW is equal to 1000 W, or 1 W is equal to 0.001 kW. For example, to convert 15,000 W into kW, you would need to use the following calculation:

$$15{,}000\ W \times 0.001 = 15\ kW$$

Example 16-11

What is the true or active power of a three-phase system with the following operating parameters?

Line current = 30 A
Line voltage = 230 V
PF = 0.85

✓ Solution

True power can be calculated as follows:

$$\text{True Power (W)} = 1.732 \times I_{Line} \times E_{Line} \times PF$$

$$\text{True Power (W)} = 1.732 \times 30\ A \times 230\ V \times 0.85$$

$$\text{True Power (W)} = 10{,}158.2\ W$$

$$10{,}158.2\ W = 10.16\ kW$$

Example 16-12

What is the true or active power of a three-phase system with the following operating parameters?

Phase current = 26 A
Phase voltage = 238 V
PF = 0.92

✓ Solution

True power can be calculated as:

$$\text{True Power (W)} = 3 \times I_{\text{Phase}} \times E_{\text{Phase}} \times \text{PF}$$

$$\text{True Power (W)} = 3 \times 26\ \text{A} \times 238\ \text{V} \times 0.92$$

$$\text{True Power (W)} = 17{,}078.9\ \text{W}$$

$$17{,}078.9\ \text{W} = 17.08\ \text{kW}$$

Now It's Your Turn 16-12

Using the information provided in the delta connection shown in the following circuit, calculate the true power, in watts, using the phase current and voltage values. Assign a power factor value of 0.87.

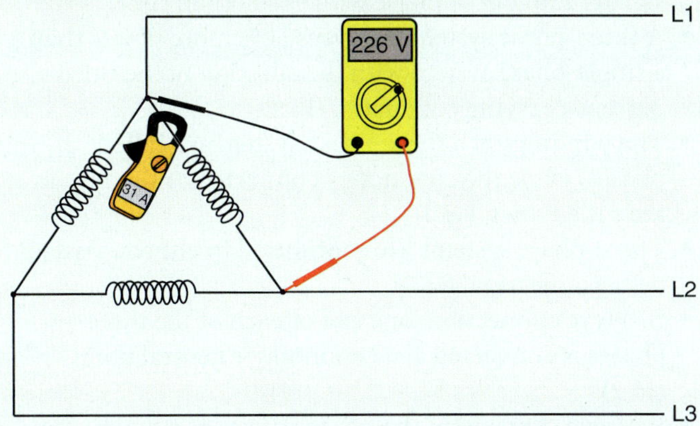

Goodheart-Willcox Publisher

Summary

- Three-phase power refers to the simultaneous generation of three single-phase alternating current signals 120° out of phase with each other. (16.1)
- A three-phase system produces 50% more power than a single-phase system with the addition of the third current-carrying conductor. (16.1)
- The windings of a three-phase system are called phases, while the conductors connected to the system are called lines. (16.1)
- Three-phase systems are configured in either a wye or delta connection. (16.1)
- In a wye connection, one end of each of the three phases is connected at a common, or neutral point. (16.2)
- In a wye connection the phase current is equal to the line current. (16.2)
- The wye connection can provide 120 V single-phase power as well as higher-voltage single-phase and three-phase power. (16.2)
- In a delta connection, the outlet of one of the three phases is connected to the inlet of the next, forming a complete loop. (16.3)
- In a delta connection the phase voltage is equal to the line voltage. (16.3)
- In a delta connection, the line current is equal to 1.732 times the phase current. (16.3)
- The delta connection provides higher-voltage single-phase and three-phase power. (16.3)
- The high-leg delta connection has the ability to provide 120 V single-phase power as well as higher-voltage, single-phase and three-phase power. (16.3)
- Voltage imbalance is used to evaluate the equality of the phases in a three-phase system. (16.4)
- Voltage imbalance can cause a number of problems including efficiency reduction, increased operating costs, component overheating, and premature device failure. (16.4)
- Voltage imbalance can be impacted by different amperage draws on each leg. An attempt should be made to balance the load among each of the three phases of a wye system. (16.4)
- The active power or true power (wattage) and apparent power (volt-amperes) of a three-phase system can be calculated using either the line current and voltage values, or the phase current and voltage values. (16.5)

Review Questions

Use the information in the chapter to answer the following questions.

Measuring Your Potential

1. Each of the phases of a three-phase system are separated by _____. (16.1)
   A. 60°
   B. 90°
   C. 120°
   D. 180°
2. *True or False?* In a working three-phase system, the total voltage supplied is never equal to zero. (16.1)
3. The windings of a three-phase system are referred to as _____. (16.2)
   A. phases
   B. lines
   C. grounded conductors
   D. ungrounded conductors
4. Including the GEC (grounding electrode conductor), how many conductors are found on a three-phase wye connection that provides 120 V as well as three-phase power? (16.2)
   A. 2
   B. 3
   C. 4
   D. 5
5. If the phase current on a three-phase wye connection is 27 A, what is the expected line current? (16.2)
   A. 15.6 A
   B. 27 A
   C. 46.8 A
   D. 54 A
6. Including the GEC (grounding electrode conductor), how many conductors are found on a three-phase delta connection? (16.3)
   A. 2
   B. 3
   C. 4
   D. 5
7. A voltage imbalance _____ is considered acceptable for many electrical systems. (16.4)
   A. below 4%
   B. below 2%
   C. above 1%
   D. below 1%

8. *True or False?* Voltage imbalance is caused primarily by motor overheating. (16.4)
9. Which of the following relationships is true regarding the apparent power calculation in a three-phase system? (16.5)
   A. Apparent power $= 3 \times \frac{I_{Line}}{E_{Line}}$
   B. Apparent power $= 1.732 \times I_{Phase} \times E_{Phase}$
   C. Apparent power $= 3 \times I_{Phase} \times E_{Phase}$
   D. Apparent power $= 1.732 \times \frac{I_{Phase}}{E_{Line}}$
10. *True or False?* Either line values or phase values for the current and voltage can be used to calculate the active power of a three-phase system. (16.5)

Plug-In

Use the following figure to answer questions 1–5.

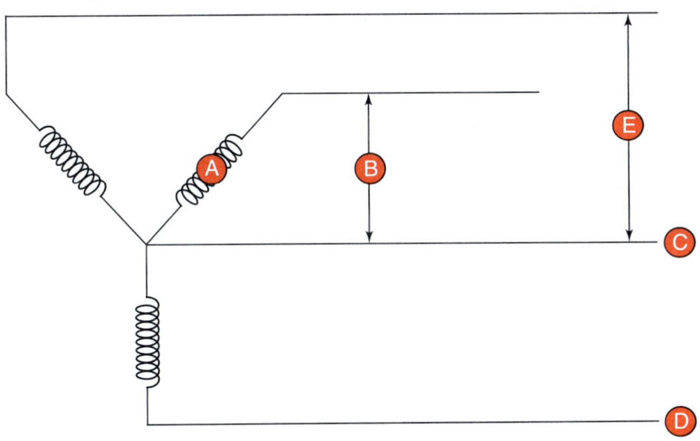

Goodheart-Willcox Publisher

1. The item labeled A is the _____. (16.2)
   A. underground conductor
   B. grounded conductor
   C. phase
   D. phase voltage
2. The item labeled B is the _____. (16.2)
   A. underground conductor
   B. grounded conductor
   C. phase
   D. phase voltage
3. The item labeled C is the _____. (16.2)
   A. underground conductor
   B. neutral point
   C. line voltage
   D. phase voltage
4. The item labeled D is the _____. (16.2)
   A. underground conductor
   B. grounded conductor
   C. line voltage
   D. phase voltage
5. The item labeled E is the _____. (16.2)
   A. underground conductor
   B. grounded conductor
   C. line voltage
   D. phase voltage

Use the following figure to calculate the answers for questions 6–8.

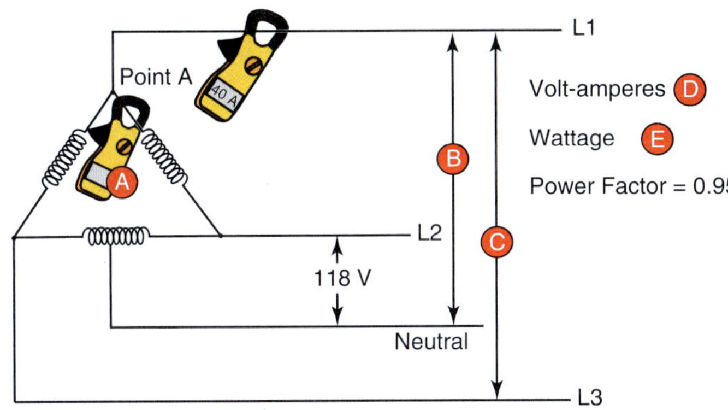

Goodheart-Willcox Publisher

6. What is the reading for the meter labeled A? (16.3)
   A. 23 A
   B. 30 A
   C. 40 A
   D. 63 A
7. What is the reading at the line labeled C? (16.3)
   A. 208 V
   B. 236 V
   C. 204 V
   D. 240 V
8. The wattage (labeled E) is _____. (16.3)
   A. 15,000 W
   B. 15,500 W
   C. 16,300 W
   D. 17,500 W

Making the Connection

1. Why is it important for a tradesperson working on a construction site to check the voltage of a power source before connecting and using a power tool? (16.4)
2. Why is it that a grounded conductor (neutral) cannot simply be connected to one of the points of the delta connection to create a neutral? (16.3)

SECTION 5

Inductance, Capacitance, Reactance, and Impedance

Courtesy Fluke Corporation

CHAPTER

17 Inductors and Inductance

KPixMining/Shutterstock.com

SPARKING DISCUSSION

In a purely inductive circuit, when the applied voltage is zero, the current should also be zero. Do you agree or disagree with this statement?

LEARNING OUTCOMES

After completing this chapter, you will be able to:

17.1 Explain how voltage is created by induction.
17.2 Evaluate inductors connected in parallel with each other.
17.3 Calculate the quality factor of an inductor.
17.4 Evaluate inductors connected in series with each other.
17.5 Evaluate the current and voltage in an inductive circuit.

TECHNICAL TERMS

air-core inductor
back electromotive force (bemf)
choke
counter electromotive force (cemf)
eddy current losses
henry (H)
hysteresis losses
inductance
inductive reactance
inductor
iron-core inductor
Lenz's law
powdered-iron core inductor
quality factor (Q)
reactors

Introduction

In our industry, electricians work with a wide variety of motors, transformers, relays, and solenoids. One thing these devices have in common is they all rely on coils of wire to operate. Electrical devices that contain coils are referred to as inductors.

Inductors serve many purposes in both electrical and electronic circuits. The operation of both motors and transformers is dependent on inductors. Inductors are used in some sensors, as well as filtering and tuning circuits. Some of the functions of inductors within a circuit include energy storage, current reduction, and voltage stabilization.

In this chapter, we will examine purely inductive circuits, which are circuits that contain only inductors. It is here that you will learn about the properties of inductors and inductance, the effects inductors have on the voltage and current in electric circuits, the calculations required to evaluate various inductor circuit connections, the quality factor of an inductor, and the process of calculating reactive power in an inductive circuit. In later chapters, additional circuit component types are discussed.

17.1 Inductance

The coils located in motors, transformers, relays, and solenoids are collectively referred to as ***inductors***, or inductive loads. When a coil is connected to a power source, a magnetic field is created when the circuit is energized. The strength of the magnetic field that is produced when current flows through a coil is much stronger than the field created when current flows through a straight conductor. This is because a coil magnifies and concentrates the effects of the magnetic field. This electromagnetic effect is fundamental to the operation of transformers, motors, and motor controls. The force created by these coils of wire is called inductance. The instant current flows through the inductor stops, the inductor will no longer be energized and magnetic field collapses.

Inductance is the measure of an inductor's ability to store energy, in the form of a magnetic field, for a given amount of current flow. Inductors use the effect of this concentrated electromagnetic force to perform work: some useful, some not. Because inductors have the ability to store energy in the form of a magnetic field, their behavior in electric circuits differs greatly from that of simple resistors.

Depending on the application, inductors can be referred to as chokes or reactors. When used to remove or filter unwanted ac ripples and fluctuations from a direct current (dc) power signal, inductors are called ***chokes***. When used in power transmission applications to maintain current and stabilize voltage, large inductors are called ***reactors***.

Lenz's Law

A discussion of inductors and inductance would not be complete without first examining an important underlying principle that explains, at least in part, the behavior of inductors in electric circuits. This principle is known as Lenz's law. ***Lenz's law*** states that, as the strength of a magnetic field changes, the induced current flows in a direction that is opposite to the change in magnetic field that originally created it. This concept is especially important in ac circuits. Since the polarity of the voltage in an ac circuit is constantly changing, the amount of current flow in the circuit is also constantly changing. As a result, the strength of the generated magnetic field is also constantly changing. When current flows through a coil of wire in an ac circuit, there is an additional force created by the coil that pushes to oppose it, **Figure 17-1**.

As an extension of Lenz's law, when an electromotive force (emf), or voltage, is applied to a coil, a secondary electromotive force is also generated. This secondary electromotive force is called ***counter electromotive force (cemf)*** or ***back electromotive force (bemf)***. Counter electromotive force pushes in a direction that is opposite to that of the original applied voltage or electromotive force. Inductance can also be described, therefore, as a measure of the force that opposes a change in current.

The counter electromotive force does not have the same magnitude as the original electromotive force, and is determined, in part, by the rate of current change in the circuit. In addition, the polarities of the counter electromotive force and electromotive force are opposites of each other. As a result, the two waveforms will be 180° out of phase with each other, **Figure 17-2**.

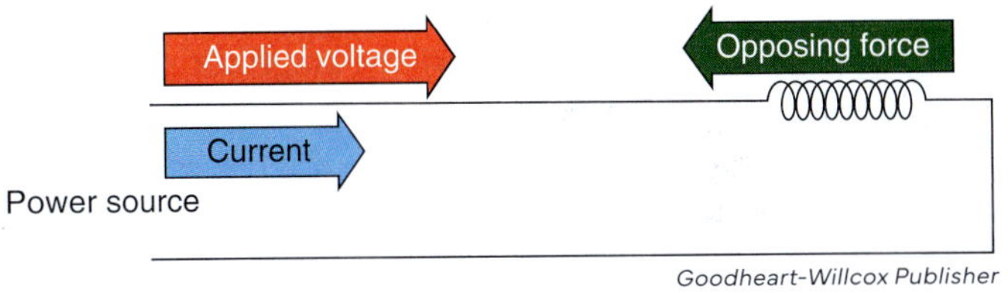

Figure 17-1. When energized, a coil produces a force that opposes the applied force.

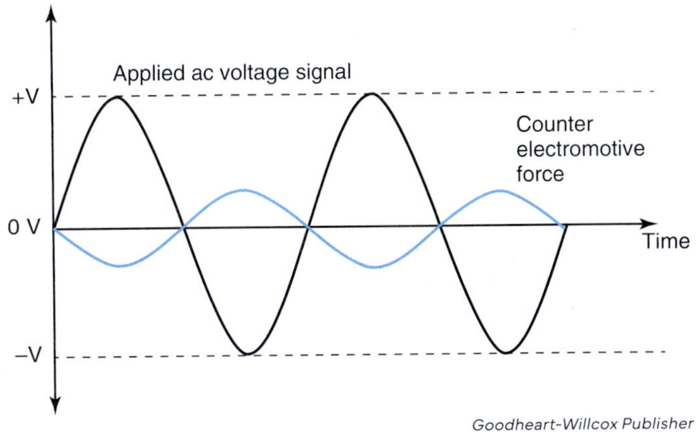

Figure 17-2. When the voltage of the applied ac signal reaches its positive peak, the counter electromotive force signal reaches its negative peak, and vice versa.

Quantities of Inductance

Name	Symbol	Decimal	Description	Fraction
henry	H	1.00	One	1
millihenry	mH	0.001	One one-thousandth	1/1000
microhenry	μH	0.000001	One one-millionth	1/1,000,000

Goodheart-Willcox Publisher

Figure 17-3. Inductance is measured in quantities of the henry.

Measuring Inductance

Inductance, identified by the letter *L*, is measured in henrys. One ***henry (H)*** is defined as the quantity of inductance that results when the rate of change of one ampere per second produces a counter electromotive force of 1 V. Depending on the application and the configuration of a circuit and its components, the quantities of a henry, a millihenry, and a microhenry can be used, **Figure 17-3.**

Rate of Change in a Sine Wave

It is important to stress that the magnitude of the generated counter electromotive force is proportional to the rate of current change, not the actual amount of current. If there is a constant flow of current, there is no change in current, so no counter electromotive force will be generated. As long as there is a change in the rate of current flow, a counter electromotive force will be produced. If the rate of current change is low, the induced voltage (counter electromotive force) will be low. If the rate of current change is high, the counter electromotive force will be high.

To help clarify the concept of rate of change, let us examine the rate of voltage change in a simple sine waveform, **Figure 17-4**. In this waveform, the maximum voltage is found at 90°, while a voltage of zero is found at 180°. The rate of voltage change can be visualized by the slope of the curve at any given point. Horizontal, or gently-sloped curves or lines, indicate smaller rates of change, while vertical, or steeper curves and lines, indicate larger rates. At 90°, the voltage is at its maximum, but the rate of voltage change is the lowest at this point. In the sine wave in **Figure 17-4A**, the highlighted area located at the top of the sine wave is nearly horizontal when compared to the rest of the waveform. At 180°, the voltage is at zero, but the slope of the sine wave at this point is very steep when compared to the rest of the waveform, **Figure 17-4B**.

Counter Electromotive Force and the Rate of Current Change

In an ac circuit, the magnetic field surrounding an inductor is constantly expanding and collapsing. As the magnitude of the circuit current increases, the magnetic field expands. When the magnetic field expands, the windings of the coil are cut by the magnetic field and a counter electromotive force is produced. This counter electromotive force then opposes the source voltage. When the current reaches its maximum value, either positive or negative, there is no longer a change in the rate of current flow. As a result, a counter electromotive force is no longer being created, **Figure 17-5**. With no counter electromotive force being produced, the circuit current is affected only by the resistance of the circuit and the voltage applied to the circuit.

If the power supply is turned off and voltage is no longer being supplied to the circuit, there will be no current flow through the circuit. This sudden change in current flow causes the magnetic field stored in the inductor to collapse. The collapse of the magnet field produces a counter

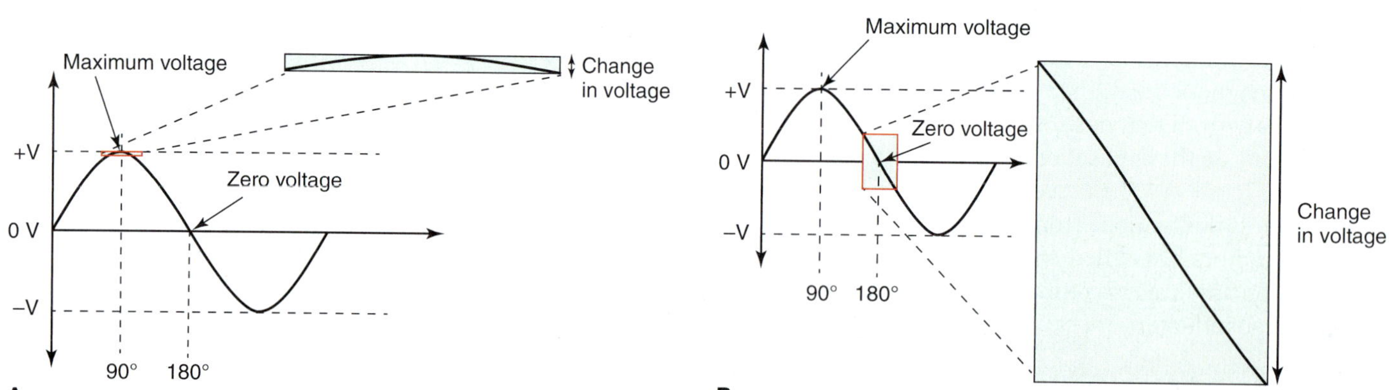

Goodheart-Willcox Publisher

Figure 17-4. Points at which an ideal sine wave reaches its maximum and zero values. A—When a sine wave reaches its peak values, the rate of change is close to zero. B—When a sine wave approaches a value of zero, the rate of change is maximized.

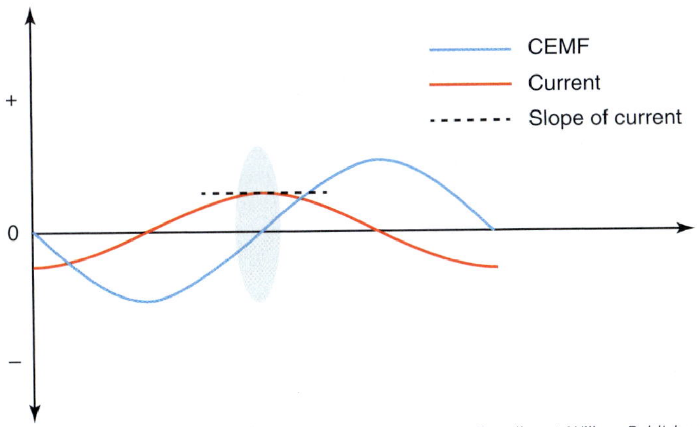

Figure 17-5. When the current reaches its maximum value, the slope of the line drawn parallel to the current line is horizontal, and the counter electromotive force drops to zero.

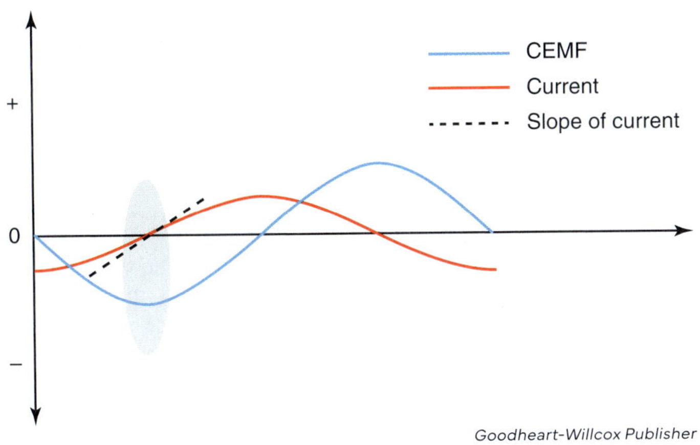

Figure 17-6. When the slope is steep, the rate of current change is the greatest and the counter electromotive force is at its maximum magnitude.

electromotive force. When the slope of the line drawn tangent to the current curve is steeply sloped, there is a large change in the rate of current flow and the counter electromotive force is at its maximum, **Figure 17-6**. The counter electromotive force produced when the magnetic field collapses has the opposite polarity as when the magnetic field was initially produced.

Voltage Leads Current in an Inductive Circuit

When voltage is applied to an inductive load, a counter electromotive force is created. In order for current to flow though the load, the voltage must first overcome the counter electromotive force. Once the counter electromotive force has been offset, current can flow through the circuit, **Figure 17-7**. Since current starts to flow sometime after voltage is supplied to the coil, it is said that the current lags behind the voltage, or that the voltage leads the current. This is a very important property of inductive circuits loads.

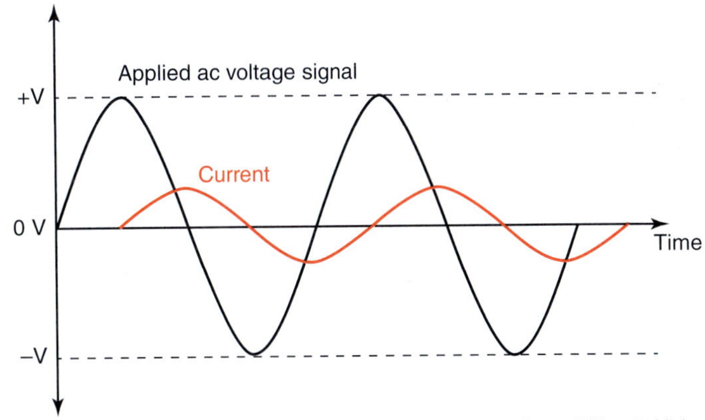

Figure 17-7. When voltage is applied to an inductive load, the voltage must overcome the counter electromotive force before current will start to flow. Current lags behind the voltage in an inductive circuit.

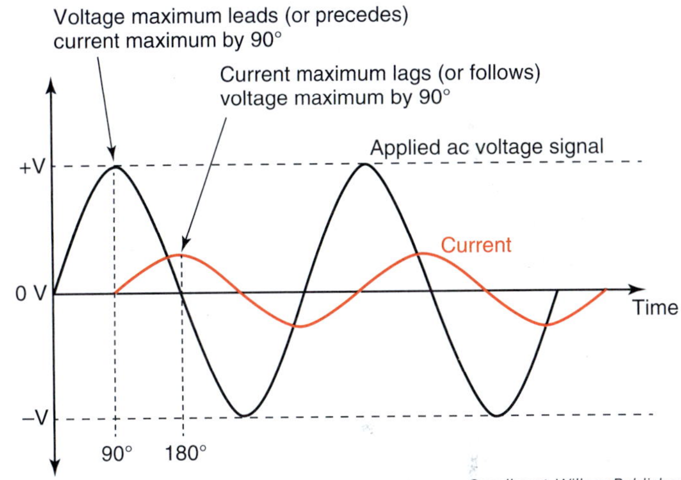

Figure 17-8. In a purely inductive circuit, the current lags behind voltage by 90°.

Shedding Light

The mnemonic "ELI the ICE man" is a helpful tool for remembering the relationship between voltage and current in inductive and capacitive circuits. The *L* in "ELI" represents inductive circuits, with the *E* (voltage) preceding, or leading, the *I* (current) in "ELI." Thus, "ELI" can help you remember that in inductive circuits, the voltage leads the current and the current lags, or follows, the voltage. The *C* in "ICE" represents capacitive circuits, in which the current (*I*) leads the voltage (*E*). Capacitive circuits are discussed in Chapter 18, *Capacitors and Capacitance*.

In a purely inductive circuit, the voltage leads the current by 90 electrical degrees, **Figure 17-8**. When the current is at its maximum, the voltage is zero. Similarly, when the current is zero, the voltage is at its maximum. A graphical representation among applied voltage, counter electromotive force, and current is provided in **Figure 17-9**.

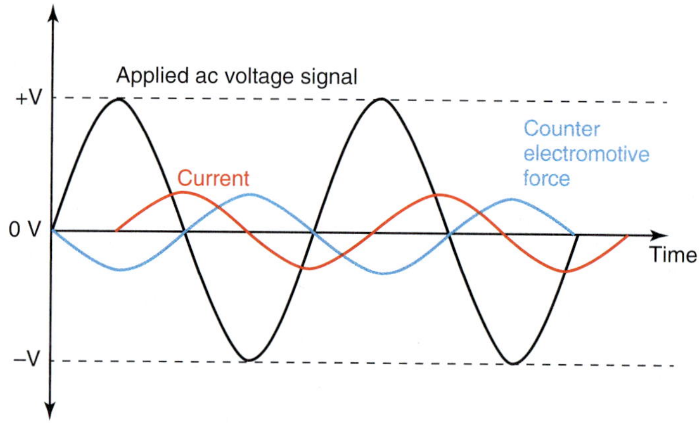

Goodheart-Willcox Publisher

Figure 17-9. Relationship among applied voltage, counter electromotive force, and current.

Shedding Light

Voltage leading the current (ELI) in an inductive circuit can be equated to pushing a chair across a plush carpet. When we apply pressure to the chair that is resting on the carpet, the chair does not immediately move. The initial force, voltage, that we apply to the chair is being used to overcome the friction, or resistance, that exists between the chair legs and the carpet. Once the resistance between the chair and carpet is overcome, then the chair starts to move. In this example, the pressure or force that we apply to the chair equates to the voltage that we apply to an inductive circuit, while the movement of the chair relates to the flow of electrons through the inductive circuit. The friction between the chair and the carpet represents the counter electromotive force.

Inductor as Both a Load and a Power Source

In an ac inductive circuit, inductors store and release energy at different points in the cycle. When the polarities of the voltage and current are opposite each other, the magnetic field collapses and the inductor acts as an energy source as it releases stored energy to the circuit. When the polarities of the voltage and current are the same, the magnetic field builds and the inductor acts as a load as it stores energy.

Shedding Light

Ideal inductors have no losses associated with them, so they store the same amount of energy as they release. In the real world, ideal inductors, or purely inductive circuits, do not exist.

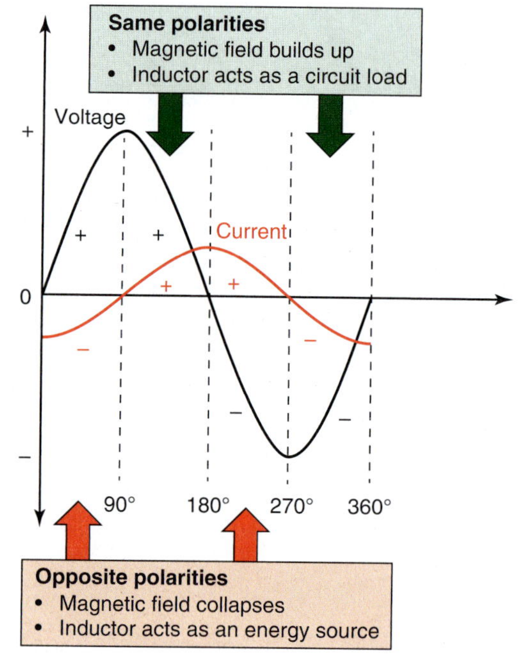

Goodheart-Willcox Publisher

Figure 17-10. Locations where the inductor acts as a load and where it acts as a power source.

Consider the voltage and current waveforms in **Figure 17-10**. In this purely inductive circuit, the voltage and current are 90° out of phase with each other. The voltage and current are both positive in the 90° to 180° section and are both negative in the 270° to 360° section. The voltage and current have opposite polarities in the 0° to 90° section and in the 180° to 270° section.

If an inductor is to store more energy, the current flow through it must increase. As can be seen in **Figure 17-10**, from 90° to 180° and from 270° to 360° the magnitude of the current is increasing toward its maximum value, in either the positive or negative direction. In these areas, the inductor is charging, as the amount of energy being stored in the inductor is increasing. When the current flow through an inductor decreases in magnitude, or moves away from its maximum value toward zero, the inductor discharges. This reduction of magnitude can be seen from 0° to 90° and from 180° to 270°.

Inductor Configuration and Wiring Diagram Symbols

The strength of an inductor depends on a number of factors including the current flow through the inductor, the core of the inductor, the frequency of the power signal, and the number of turns in the wire coil forming the inductor.

An ***air-core inductor*** is a coil of wire with a nonmetallic core located inside it, **Figure 17-11**. This nonmetallic core is commonly air or plastic. Since air and plastic do not conduct electricity well, air-core inductors produce the weakest opposition to current flow. In order to achieve the same inductance as other types of solid-core inductors, air-core inductors require more turns of wire.

Shedding Light

Radio speakers and radio transmitters often use air-core inductors. These devices reduce sound distortion and help to minimize harmonic vibrations that can be created when electromagnetic waves travel across them. The distortions and vibrations are low because there is electromagnetic interaction between the coil and the core.

Iron-core inductors are constructed as a conductor wound around a soft magnetic core, **Figure 17-12**. The core is often made of soft magnetic materials such as iron or ferrite. Because of the high magnetic permeability of these soft magnetic materials, iron-core inductors can produce larger amounts of inductance in a smaller space. The main drawback to iron-core inductors is that they experience more energy losses, especially at higher frequencies.

Powdered-iron core inductors have cores of soft magnetic materials, but the cores also have air gaps in them. The main benefit of powdered-iron core inductors is that they are able to store more energy when compared to other types of inductors. All inductors experience power losses, but powdered-iron core inductors are desirable for many applications because the magnetic losses associated with these inductors are lower than with other types of inductor. The electrical symbols for common types of inductors are provided in **Figure 17-13**.

Shedding Light

Common magnetic losses are hysteresis losses and eddy current losses. ***Hysteresis losses*** are associated with the charging and discharging of an inductor as the magnetic field builds up and collapses. These losses exist when the magnetic field builds up and collapses at different rates. As the difference between the rates of magnetic field build-up and collapse increases, the hysteresis losses increase. These types of losses are greater at higher frequencies and in metals that have a higher retentivity. ***Eddy current losses*** are the power losses that result when the induced electromotive force in an inductor causes circular currents to flow around the conductor. These types of losses occur in transformers and motors. Eddy currents can be reduced by using very thin insulating laminations that will allow the passage of the magnetic lines of flux but prohibit the current flow through the laminations.

KPixMining/Shutterstock.com

Figure 17-11. Air-core inductors are commonly used in radio speakers and radio transmitters.

Zigzag Mountain Art/Shutterstock.com

Figure 17-12. Iron-core inductors are commonly used in electrical power circuits.

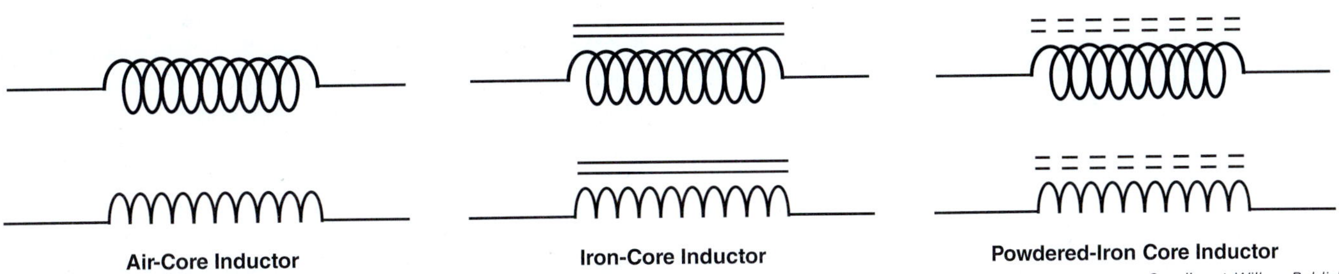

Goodheart-Willcox Publisher

Figure 17-13. Electrical symbols used for common inductor types.

Calculating Inductance

If the resistance of a coil is very low and can be disregarded, the inductance of a coil can be calculated using the following formula:

$$L = \frac{E}{6.28 \times I \times f}$$

where:

L = Inductance (H)
E = Voltage supplied to the circuit (V)
I = Current flow through the circuit (A)
f = Frequency of the power supply (Hz)
6.28 = Constant equal to 2π or 2×3.14

Refer to **Example 17-1** and **Example 17-2** on how to calculate the inductance of a coil.

Example 17-1

If a 190 Hz, 120 V signal is applied to a coil with negligible resistance, what is the inductance of the coil if the measured current through the coil is 0.4 A?

✓ Solution

The inductance of the coil can be calculated as follows:

$$L = \frac{E}{6.28 \times I \times f}$$

$$L = \frac{120 \text{ V}}{6.28 \times 0.4 \text{ A} \times 190 \text{ Hz}}$$

$$L = \frac{120}{477.28}$$

$$L = 0.25 \text{ H}$$

Now It's Your Turn 17-1

Calculate the inductance of a coil with negligible resistance that draws 0.8 A and is powered by a 120 V, 80 Hz signal.

Example 17-2

What is the expected current flow through a 0.6 H inductor with negligible resistance that is powered by a 120 V, 60 Hz signal?

✓ Solution

The expected current (I) can be calculated as follows:

$$I = \frac{E}{6.28 \times L \times f}$$

$$I = \frac{120 \text{ V}}{6.28 \times 0.6 \text{ H} \times 60 \text{ Hz}}$$

$$I = \frac{120}{226.08}$$

$$I = 0.53 \text{ A}$$

Now It's Your Turn 17-2

What is the frequency of a 130 V signal powering a 0.7 H coil with negligible resistance if the amperage draw of the coil is 0.3 A?

17.2 Inductive Reactance

As explained earlier, a counter electromotive force is created when voltage is applied to an inductor. This counter electromotive force acts to restrict current flow, which is why the current lags behind the voltage in an inductive circuit. The property of an inductor to limit current flow is called ***inductive reactance*** and is identified by the symbol X_L. Inductive reactance behaves like resistance and, as such, inductive reactance is measured in ohms. Inductive reactance is calculated using the following formula:

$$X_L = 2\pi \times f \times L$$

where:

X_L = Inductive reactance (Ω)
2π = Constant equal to 2×3.14 or 6.28
f = Frequency of the power supply (Hz)
L = Inductance (H)

At higher frequencies, the inductive reactance will rise, resulting in more opposition to current flow. An inductor with high inductance will also offer more opposition to current flow than an inductor with a lower inductance, when exposed to a power signal of the same frequency.

Refer to **Example 17-3** and **Example 17-4** on how to calculate inductive reactance.

Example 17-3

What is the inductive reactance of a 0.8 H coil if the frequency of the power supply is 60 Hz?

✓ Solution

Inductive reactance can be calculated as follows:

$$X_L = 2\pi \times f \times L$$

$$X_L = 6.28 \times 60 \text{ Hz} \times 0.8 \text{ H}$$

$$X_L = 301.4\ \Omega$$

Shedding Light

In an ac circuit, current flow through an inductor is limited by the counter electromotive force and the resistance of the conductor from which the inductor is created. If the conductor resistance is so low that it can be ignored, the inductive reactance is the only opposition to current flow in the circuit. Using Ohm's law, it can be determined that the inductive reactance, X_L is equal to E/I when, and only when, the conductor resistance is negligible.

Example 17-4

What is the frequency of the power supply if the inductive reactance of a 0.9 H coil is 200 Ω?

✓ Solution

The frequency of the power supply can be calculated as follows:

$$f = \frac{X_L}{2\pi \times L}$$

$$f = \frac{200\ \Omega}{6.28 \times 0.9\ \text{H}}$$

$$f = \frac{200\ \Omega}{5.652\ \text{H}}$$

$$f = 35.39\ \text{Hz}$$

Now It's Your Turn 17-4

Determine the inductance (L) of a coil if its inductive reactance (X_L) is 150 Ω when the frequency of the power supply is 80 Hz.

17.3 Quality Factor

The ***quality factor (Q)*** is the relationship that exists between the inductive reactance, X_L, and the resistance, R, in an inductor. The higher the inductive reactance and the lower the resistance, the higher the quality factor. The quality factor (Q) is calculated as follows:

$$Q = \frac{X_L}{R}$$

The inductive reactance and the resistance are both expressed in ohms, so the quality factor is a unitless quantity. If the quality factor of an ideal inductor with a resistance of 0 Ω was calculated, the result would be infinity. However, all inductors offer at least some resistance to current flow. Inductors classified as pure inductors typically have a quality factor of at least 10. By inspecting the quality factor formula, resistance is one of the major factors that affects the performance of an inductor.

The quality factor rates the overall performance and efficiency of an inductor. As the quality factor rises, the energy losses associated with the device fall. Of note, the cost of an inductor is directly proportional to its quality factor.

See **Example 17-5** to calculate resistance of an inductor. Refer to **Example 17-6** on how to calculate inductive reactance.

Shedding Light

For radio frequency applications, inductors with high quality factors are desirable. This is because communications channels typically have very narrow bandwidths, and the high quality factor allows the system to effectively block out signals that are not within the desired range. Some sound equipment, such as microphones and speakers, benefit from circuitry that utilizes inductors with low quality factors. These devices are expected to operate effectively over a wide range of frequencies and inductors with low quality factors help achieve that outcome.

Example 17-5

If the quality factor of an inductor is 15 and the inductive reactance of the inductor is 125 Ω, what is the resistance of the inductor?

✓ Solution

Resistance can be calculated by rearranging the quality factor formula as follows:

$$R = \frac{X_L}{Q}$$

$$R = \frac{125\ \Omega}{15\ \Omega}$$

$$R = 8.3\ \Omega$$

Example 17-6

Consider an inductor that has resistance of 15 Ω. What is the lowest value of the component's inductive reactance if the quality factor of the inductor is to be at least 20?

✓ Solution

Assuming a quality factor of 20 and a resistance of 15 Ω, the inductive reactance can be calculated by rearranging the quality factor formula to solve for X_L as follows:

$$X_L = Q \times R$$

$$X_L = 20 \times 15\ \Omega$$

$$X_L = 300\ \Omega$$

With an inductive reactance of 300 Ω, the quality factor will be 20. If the inductive reactance is higher than 300 Ω, the quality factor will be higher than 20. As a result, the lowest inductive reactance value that will ensure a quality factor of at least 20 is 300 Ω.

Now It's Your Turn 17-6

Consider an inductor that has an inductive reactance of 120 Ω. What is the highest value of the component's resistance if the quality factor of the inductor is to be at least 15?

Characteristics other than resistance affect the quality factor of an inductor. To determine the quality factor of an inductor, you will need to divide the inductive reactance by the resistance. From our earlier discussion on inductive reactance, it was determined that the expression $2\pi \times f \times L$ can be substituted for X_L in the quality factor formula:

$$Q = \frac{2\pi \times f \times L}{R}$$

From this formula, we can establish that the quality factor is affected by the resistance of the inductor, the inductance of the inductor, and the frequency of the power supply. See **Example 17-7**.

Example 17-7

What is the quality factor of a 1.2 H inductor that has a resistance of 20 Ω if it is connected to a power supply with a frequency of 60 Hz?

✓ Solution

Using the quality factor formula, the quality factor can be determined by substituting the known quantities for inductance, frequency, and resistance as follows:

$$Q = \frac{2\pi \times f \times L}{R}$$

$$Q = \frac{6.28 \times 60 \text{ Hz} \times 1.2 \text{ H}}{20 \ \Omega}$$

$$Q = \frac{452.16}{20}$$

$$Q = 22.61$$

17.4 Series-Connected Inductors

When inductors are wired in series with each other, the total inductance is equal to the sum of all of the individual inductance in the series connection. The general formula for series-connected inductors is as follows:

$$L_T = L_1 + L_2 + L_3 + \ldots + L_N$$

Where L_T is the total inductance and L_1 through L_N are the inductance values for each series-connected inductor. Refer to **Example 17-8** on how to calculate the inductance in a series circuit.

Example 17-8

What is the total inductance of a series circuit that has three inductors with values of 0.95 H, 0.75 H, and 1.1 H?

✓ Solution

The total inductance of a series circuit can be calculated as follows:

$$L_T = L_1 + L_2 + L_3$$

$$L_T = 0.95 \text{ H} + 0.75 \text{ H} + 1.1 \text{ H}$$

$$L_T = 2.8 \text{ H}$$

Now It's Your Turn 17-8

In the following circuit, determine the value of inductor L_2 if the total circuit inductance is 1.9 H.

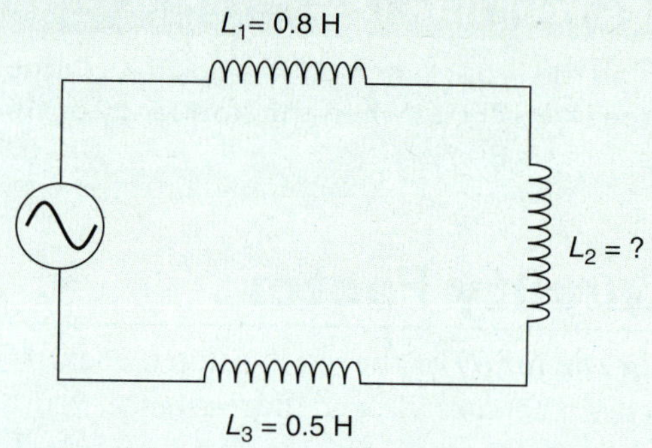

Goodheart-Willcox Publisher

17.5 Parallel-Connected Inductors

Just as with resistors wired in parallel with each other, the total inductance of a circuit that consists of inductors connected in parallel with each other is always lower than the lowest individual circuit inductance. The following general formula can be used to calculate the total inductance of a circuit with parallel-connected inductors:

$$\frac{1}{L_T} = \frac{1}{L_1} + \frac{1}{L_2} + \frac{1}{L_3} + \ldots + \frac{1}{L_N}$$

In this expression, L_T is the total circuit inductance, while L_1, L_2, and L_3 represent branch inductances 1, 2, and 3. L_N represents the inductance in the last parallel circuit branch. Refer to **Example 17-9** on how to calculate the inductance in a parallel circuit.

If the parallel circuit being evaluated has only two branches, the following formula can be used to find the total inductance of the parallel connection:

$$L_T = \frac{L_1 \times L_2}{(L_1 + L_2)}$$

This method of calculating the values of two inductors connected in parallel with each other is often referred to as product over sum. Refer to **Example 17-10**.

Example 17-9

Determine the total inductance of the following parallel circuit.

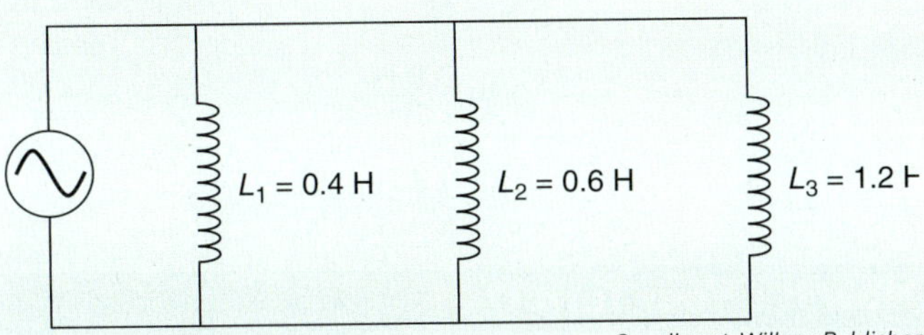

Goodheart-Willcox Publisher

✓ Solution

This three-branch parallel inductive circuit has inductance branches of 0.4 H, 0.6 H, and 1.2 H. The total circuit inductance can be calculated using the X-1 reciprocal key on your calculator or by using the following formula:

$$\frac{1}{L_T} = \frac{1}{L_1} + \frac{1}{L_2} + \frac{1}{L_3}$$

$$\frac{1}{L_T} = \frac{1}{0.4\ \text{H}} + \frac{1}{0.6\ \text{H}} + \frac{1}{1.2\ \text{H}}$$

This expression can be simplified using the reciprocal key (1/x or x^{-1}) on your calculator:

$$\frac{1}{L_T} = 5$$

Use the reciprocal key (1/x or x^{-1}) on your calculator again:

$$L_T = 0.2\ \text{H}$$

Example 17-10

Calculate the total inductance of the following parallel circuit.

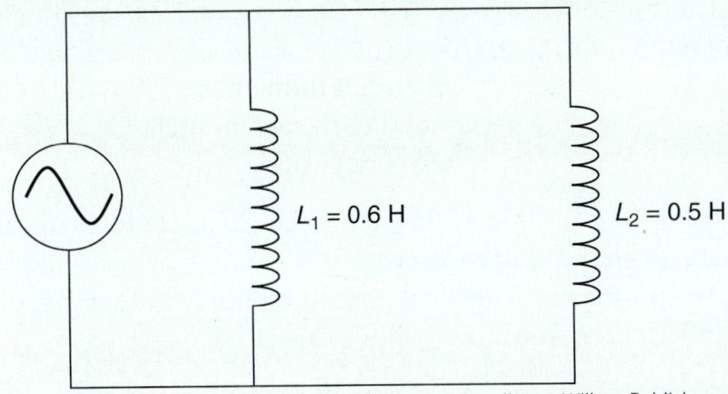

Goodheart-Willcox Publisher

✓ Solution

With only two inductors in this parallel circuit, the product over the sum formula can be used to calculate the total inductance of the circuit. Substituting the known values for L_1 and L_2, the total inductance can be determined as follows:

$$L_T = \frac{L_1 \times L_2}{(L_1 + L_2)}$$

$$L_T = \frac{0.6\ \text{H} \times 0.5\ \text{H}}{0.6\ \text{H} + 0.5\ \text{H}}$$

$$L_T = \frac{0.3\ \text{H}^2}{1.1\ \text{H}}$$

$$L_T = 0.27\ \text{H}$$

Now It's Your Turn 17-10

Using the product over the sum formula, calculate the total inductance of the circuit shown in the following parallel circuit.

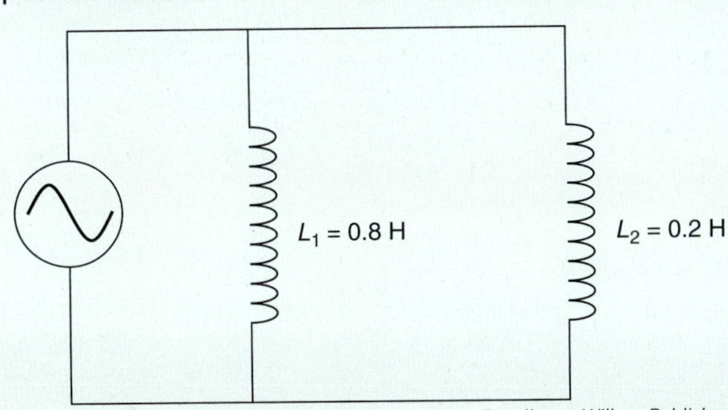

Goodheart-Willcox Publisher

When multiple inductors of the same value are wired in parallel with each other, a useful formula that can be used is:

$$L_T = \frac{L}{N}$$

where:

L_T = Total circuit inductance
L = Inductance in each branch
N = Total number of branches

Refer to **Example 17-11** on how to calculate total inductance using the above formula.

To summarize, when calculating multiple inductances in parallel, three common formulas are used:

Reciprocal Formula (works in all cases):

$$\frac{1}{L_T} = \frac{1}{L_1} + \frac{1}{L_2} + \frac{1}{L_3} + \dots + \frac{1}{L_N}$$

Product over Sum Formula (works with two branches):

$$L_T = \frac{L_1 \times L_2}{L_1 + L_2}$$

Equal Inductance Formula (works when all branches have the same resistance):

$$L_T = \frac{L}{N}$$

The same formula configurations that we used to evaluate series-connected and parallel-connected resistors apply to inductors as well. Series-connected inductors follow the rules for series-connected resistors and parallel-connected inductors follow the rules for parallel-connected resistors.

Example 17-11

Calculate the total inductance for the following parallel circuit.

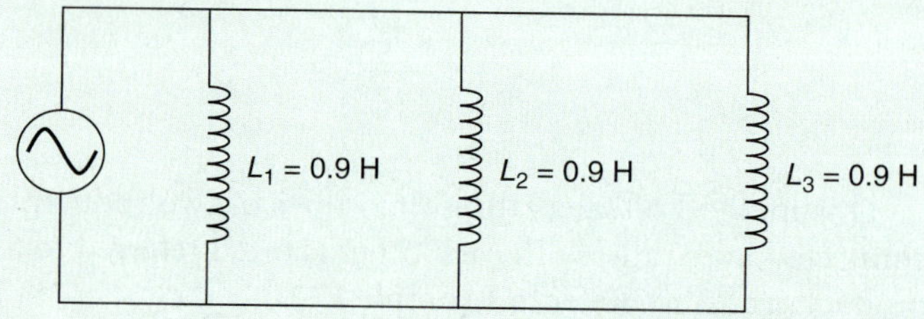

Goodheart-Willcox Publisher

✓ Solution

This circuit has three 0.9 H inductors wired in parallel with each other. The total inductance for this circuit can be calculated as follows:

$$L_T = \frac{L}{N}$$

$$L_T = \frac{0.9\text{ H}}{3}$$

$$L_T = 0.3\text{ H}$$

Now It's Your Turn 17-11

Calculate the total inductance of a four-branch parallel circuit where each branch contains a 0.8 H inductor.

CHAPTER 17 REVIEW

Summary

- Inductors store energy in the form of a magnetic field. (17.1)
- In operation, an inductor can act as both a power source and a load. (17.1)
- Inductance is a measure of the energy-storing capability of an inductor. (17.1)
- Common inductor types are air-core, iron-core, and powdered-iron core. (17.1)
- The counter electromotive force always has a polarity that is opposite to the polarity of the applied voltage. (17.1)
- Inductors generate a counter electromotive force when energized. (17.1)
- In an inductive circuit, the voltage leads the current ("ELI"). (17.1)
- Inductive reactance, X_L, is expressed in ohms and limits the current flow of a circuit. (17.2)
- Inductive reactance is a function of an inductor's inductance and the frequency of the power supply. (17.2)
- The quality factor of an inductor rates the overall performance and efficiency of the device. (17.3)
- The quality factor is directly proportional to the efficiency of an inductor. (17.3)
- The quality factor of an inductor is equal to the inductive reactance divided by the resistance. (17.3)
- Series-connected inductors follow the same rules as series-connected resistors. (17.4)
- Parallel-connected inductors follow the same rules as parallel-connected resistors. (17.5)

Review Questions

Use the information in the chapter to answer the following questions.

Measuring Your Potential

1. An inductor stores energy in the form of ______. (17.1)
   A. a magnetic field
   B. chemical energy
   C. a battery
   D. mechanical energy
2. The concept of Lenz's law revolves around the idea that a produced force ______. (17.1)
   A. pushes in the same direction as the original force
   B. is always greater than the original force
   C. pushes in the opposite direction from the original force
   D. is always equal to the original force
3. Which of the following statements is true regarding an applied electromotive force and the counter electromotive force generated by an inductor? (17.1)
   A. The electromotive force and counter electromotive force have the same magnitude.
   B. The electromotive force and counter electromotive force are 90° out of phase with each other.
   C. The electromotive force and counter electromotive force have different frequencies.
   D. The electromotive force and counter electromotive force are exactly 180° out of phase with each other.
4. In a purely inductive circuit, the voltage leads the current by ______. (17.1)
   A. 0°
   B. 45°
   C. 90°
   D. 180°
5. *True or False?* In an ac inductive current, inductors store and release energy different points in the cycle. (17.1)
6. Of the following, which inductor produces the weakest opposition to current flow? (17.1)
   A. Iron-core
   B. Air-core
   C. Powdered-iron core
   D. Both A and B
7. Of the following, which is the equation used to calculate inductance of a coil? (17.1)
   A. $I = \frac{E}{(6.28 \times L \times f)}$
   B. $L = \frac{E}{(6.28 \times I \times f)}$
   C. $X_L = 22\pi \times L \times f$
   D. $f = \frac{X_L}{2\pi \times L}$

8. Of the following, which is the formula used to calculate inductive reactance? (17.2)
   A. $L = \frac{X_L}{(6.28 \times f)}$
   B. $L = \frac{E}{6.28 \times I \times f}$
   C. $X_L = 2\pi \times f \times L$
   D. $f = \frac{X_L}{2\pi \times L}$
9. To increase the quality factor of an inductor, the _____. (17.3)
   A. physical size of the inductor should be decreased
   B. frequency of the power signal should be decreased
   C. resistance of the inductor should be decreased
   D. inductance of the inductor should be decreased
10. The inductive reactance and the resistance are both expressed in _____. (17.3)
    A. watts
    B. ohms
    C. hertz
    D. amperes
11. *True or False?* When inductors are wired in series, total inductance is calculated by adding all the individual inductance values. (17.4)
12. Of the following, which formula is used to calculate two parallel inductances with different values? (17.5)
    A. $\frac{1}{L_T} = \frac{1}{L_1} + \frac{1}{L_2} + \frac{1}{L_3}$
    B. $L_T = L_1 + L_2 + L_3$
    C. $L_T = \frac{(L_1 \times L_2)}{(L_1 + L_2)}$
    D. $L_T = \frac{L}{N}$

Plug-In

1. What is the inductive reactance of a 0.75 H inductor if it is connected to a 120 V, 60 Hz power supply? (17.2)
   A. 90 Ω
   B. 90 V
   C. 283 Ω
   D. 283 V
2. Which of the following inductors would be classified as a pure inductor? (17.3)
   A. An inductor with an X_L of 100 Ω and a resistance of 50 Ω.
   B. An inductor with an X_L of 150 Ω and a resistance of 10 Ω.
   C. An inductor with an X_L of 200 Ω and a resistance of 25 Ω.
   D. An inductor with an X_L of 250 Ω and a resistance of 125 Ω.
3. What is the quality factor of an inductor that has an inductive reactance of 125 Ω and a resistance of 25 Ω? (17.3)
   A. 0.2 Ω
   B. 5 Ω
   C. 100 Ω
   D. 150 Ω
4. If four 1.2 H inductors are wired in series with each other, the total inductance will be _____. (17.4)
   A. 0.3 H
   B. 0.6 H
   C. 1.2 H
   D. 4.8 H
5. If four 1.6 H inductors are wired in parallel with each other, the total inductance will be _____. (17.5)
   A. 0.4 H
   B. 0.8 H
   C. 1.6 H
   D. 6.4 H

Making the Connection

1. Explain why, in a purely inductive circuit, the current flow is zero when the applied voltage is at its highest level. (17.1)
2. Explain how an inductor can act as both a load and a power source. (17.1)
3. How will the quality factor of an inductor be affected if the frequency of the power signal is changed? (17.3)

SECTION 5

CHAPTER

18 Capacitors and Capacitance

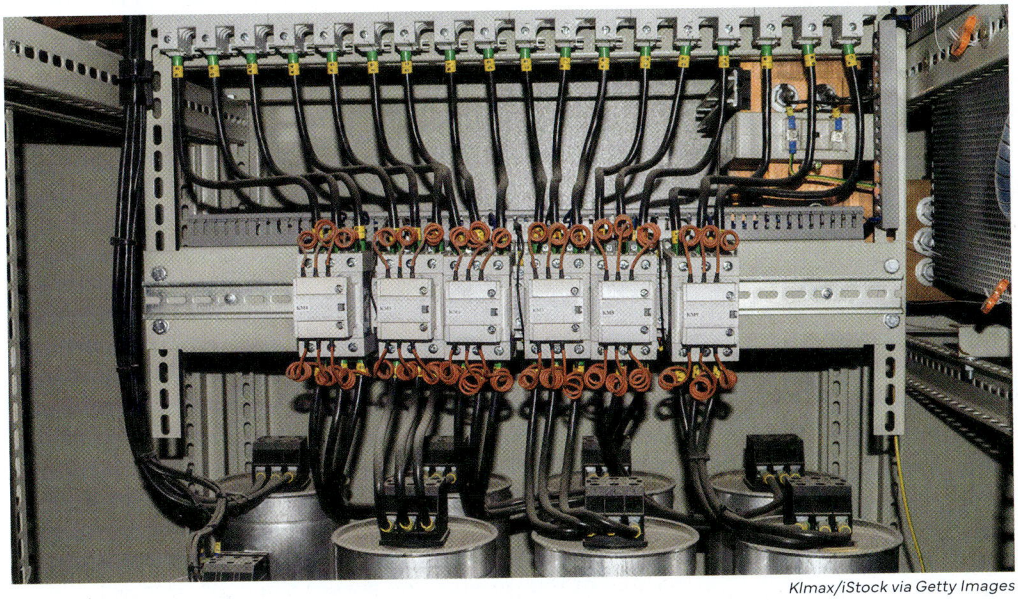

Klmax/iStock via Getty Images

SPARKING DISCUSSION

What would be the reason to add capacitors to a single-phase electric motor circuit?

LEARNING OUTCOMES

After completing this chapter, you will be able to:

18.1 Define the term capacitance as it applies to the electrical industry.
18.2 Explain the difference between a non-polarized and a polarized capacitor.
18.3 Explain how voltage and charge are affected by capacitance.
18.4 List the factors that determine and affect capacitance.
18.5 Describe the basic operation of a capacitor.
18.6 Demonstrate how to calculate the capacitive reactance of a capacitor.
18.7 Determine capacitor charge and discharge rate using the RC time constant.
18.8 Evaluate parallel-connected capacitors.
18.9 Evaluate series-connected capacitors.
18.10 Describe quality factor as it applies to capacitors.

TECHNICAL TERMS

anode
axial capacitor
bleed resistor
capacitance
capacitive reactance
capacitor
cathode
dielectric
dielectric constant
dielectric stress
electrolyte
electrolytic capacitor
electrostatic charge
leakage current
non-polarized capacitor
permittivity (ε)
polarized capacitor
radial capacitor
RC time constant
relative permittivity (ε_r)

Introduction

Capacitors can be found in nearly every segment of our industry. From the generation of electricity to its transmission and distribution and, ultimately its consumption, capacitors help ensure that the quality of the energy made available for consumption is as high as possible. Capacitors also help ensure that the equipment and appliances that rely on electricity to work do so as efficiently as possible.

In this chapter, we will examine purely capacitive circuits, which are circuits that contain only capacitors. You will learn about the properties of capacitors and capacitance, the effects capacitors have on the voltage and current in electric circuits, and the calculations required to evaluate various capacitor circuit connections.

18.1 Capacitors

Capacitors perform extremely useful and important functions in the electrical industry. However, most of the work done by these components is in a supportive role, to enhance the effectiveness and efficiency of an electric circuit or component.

Unlike an inductor, which opposes a change in current, a ***capacitor*** is an electric component that opposes a change in voltage. When referring to capacitors and their ability to oppose a change in voltage, the term capacitance is used. ***Capacitance*** is a measure of the ability of a material to hold an electric charge if a voltage is applied across it. When an electric charge in a capacitor is held, electrons are not flowing and the charge is said to be stationary, or electrostatic. The ***electrostatic charge***, also referred to as a *static charge*, is held until changes in external conditions permit the stored charge to be released.

Shedding Light

The electrical charge created when you walk across a carpeted floor is a static charge. This charge may be released when you touch a metal doorknob.

Although many consumer products that require electricity are classified as either inductive or resistive loads, capacitors often play an integral role in their operation. Some of the functions capacitors perform are improving power factor, storing an electric charge, smoothing dc power signals, filtering electrical signals, smoothing power supply outputs, stabilizing voltages, reducing unwanted or extraneous pulsations, and stabilizing current flow.

Basic Capacitor Function: A Water Flow Analogy

In *Chapter 6, Electrical Characteristics*, the concepts of voltage, current, and resistance were illustrated by a water pipe and a valve. When the valve is closed there is water pressure behind the valve, but the water does not flow. In this position, the closed valve represents resistance and the water pressure behind the valve represents voltage. When the valve is closed, there is no water flow, meaning there is no current. When the valve is open, the resistance is removed and water flows through the pipe. If the valve had only two positions, fully-open and fully-closed, there would be either full flow (when the valve is open) or no flow (when the valve is closed).

Safety Alert

Capacitors can store an electric charge even when not connected to a power supply or circuit. This stored charge can deliver a significant electric shock that can cause severe personal injury and, in extreme cases, death. Before working on circuits that contain capacitors, be sure to properly discharge them. To discharge a capacitor safely, it is recommended that a high-value (at least 100 Ω), high-wattage (at least 0.5 W) resistor be placed across the capacitor's terminals. Leaving this resistor in place for at least five seconds should be sufficient to ensure that the capacitor has discharged completely. Be sure to avoid coming in contact with the capacitor terminals when attempting to discharge a capacitor. The higher the resistance value of the resistor, the slower the capacitor will discharge. Be sure to check the manufacturer's literature for specific information on safely discharging capacitors.

Now consider the situation where the valve is constantly opening and closing to various positions at a randomly changing frequency. The water's flow rate at the outlet of the pipe will be just as erratic and unpredictable as the opening and closing of the valve, **Figure 18-1**. The maximum flow rates correspond to the time at which the valve is in the fully-open position. Similarly, there is no flow through the pipe when the valve is fully closed. Adding an additional

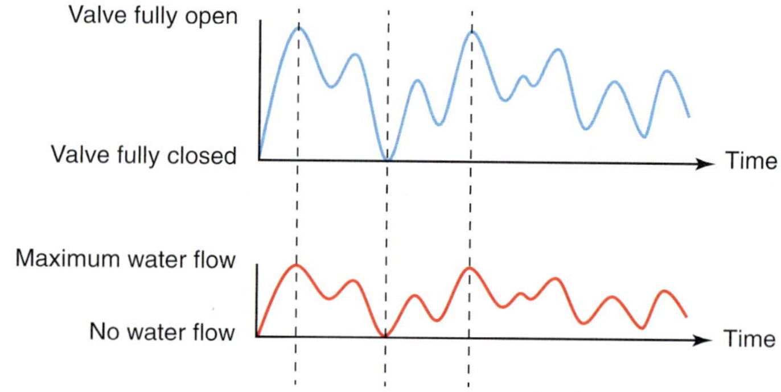

Figure 18-1. Given a valve opening and closing at randomly changing frequencies, the rate of flow is equally unpredictable.

component, a tank positioned at the outlet of the pipe, **Figure 18-2**, will help create a smoother, more constant and consistent flow rate.

In this configuration, the valve controls the flow of water into the tank, which has an outlet pipe located at its bottom. If the valve is closed, there is no water flow into the tank and, because the tank is empty, there is no flow from the tank. When the valve is opened, water will flow into the tank. As long as the flow rate into the tank is at least slightly greater than the flow rate out of the tank, the tank will begin to fill, even as water flows out of the tank from the bottom. With water accumulated in the tank, water will continue to flow from the bottom of the tank, even if the valve is closed, either fully or partially. Once some water has accumulated in the tank, the water output from the tank remains relatively consistent, even though the rate of water flow into the tank changes, **Figure 18-3**.

The capacitor performs the same basic function as the tank shown in **Figure 18-2**. As one of its functions, the capacitor has the ability to accept a varying voltage signal and convert it to a smooth, continuous voltage signal with a relatively uniform magnitude. The capacitor is able to store an electric charge, just as the tank was able to store water, and act as a voltage source, even when there is no voltage being supplied to it.

Shedding Light

Capacitors are commonly used in circuits that convert ac voltage to dc voltage. Passing a voltage signal through a capacitor, or bank of capacitors, is often the final step of the ac-to-dc conversion process. The capacitors help remove any variations in the voltage signal supplied to them and pass along to their output a consistent dc voltage signal, **Figure 18-4**.

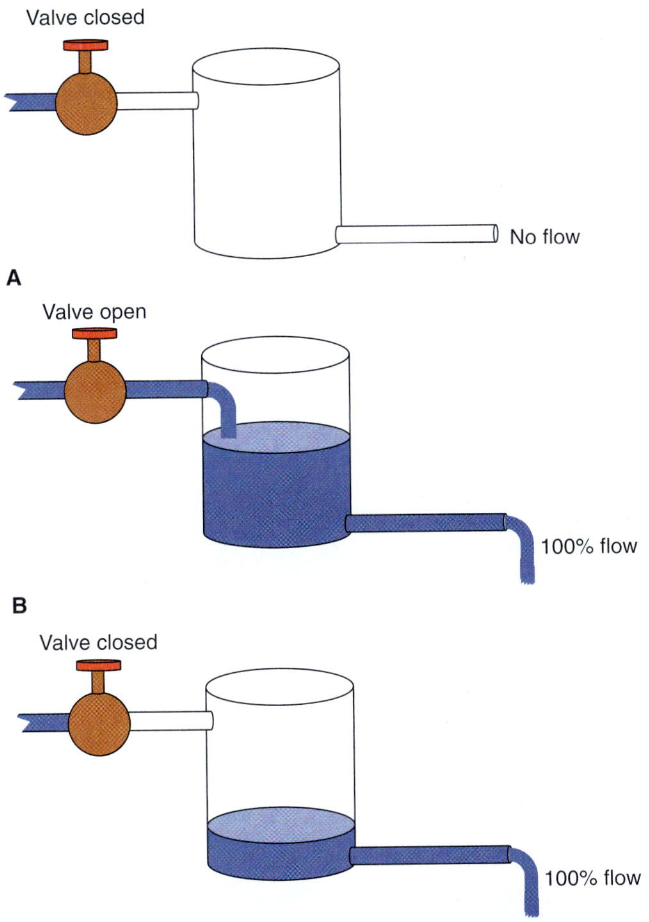

Goodheart-Willcox Publisher

Figure 18-2. Connecting a tank to the outlet of the water pipe. A—With the valve closed, no water flows into the tank. B—With the valve open, water flows into the tank, and some flows out. C—With water in the tank and the valve closed, water still flows from the tank.

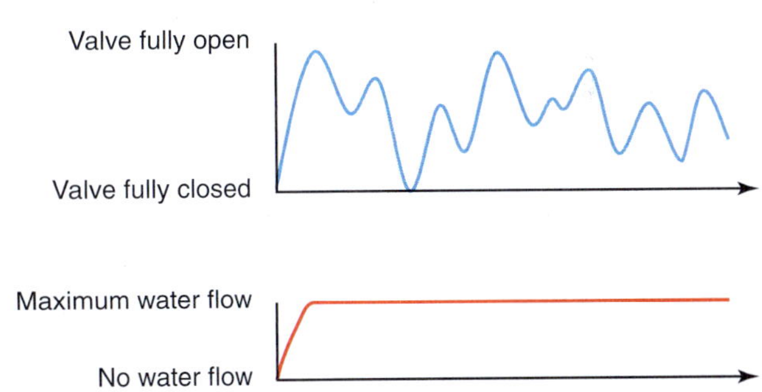

Goodheart-Willcox Publisher

Figure 18-3. The water flow out of the tank is constant, even though the valve opens and closes.

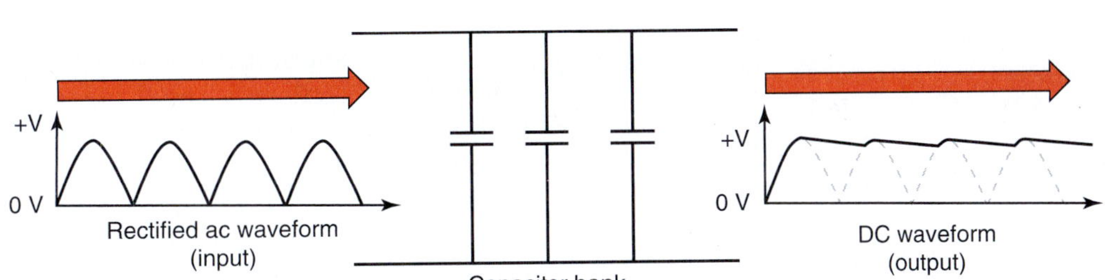

Goodheart-Willcox Publisher

Figure 18-4. Capacitors can be used to create a smooth, constant, and consistent voltage waveform.

18.2 Capacitor Construction

In its simplest form, a typical capacitor is constructed of two conductive metal plates separated by a layer of insulating material such as air, paper, plastic, or ceramic, **Figure 18-5**. This insulating material is called a ***dielectric***. Capacitors are identified as either polarized or non-polarized components. The electrical symbols for polarized and non-polarized capacitors are shown in **Figure 18-6**. A chart comparing the characteristics of polarized and non-polarized capacitors is provided in **Figure 18-7**.

Factors that Determine Capacitance

The design and construction of a capacitor determines the capacitance. There are three factors that are taken into account when a capacitor is fabricated. These factors include:

- **Area of the capacitor's conductive plates.** The greater the area of the plates, the greater the capacitance.
- **Distance between the plates.** The smaller the distance between the plates, the greater the capacitance.
- **Dielectric material.** Every material has a property called the *dielectric constant*. The greater the dielectric constant of the dielectric material, the greater the capacitance. (Dielectric constant is discussed later in this chapter.)

Non-Polarized Capacitors

A ***non-polarized capacitor***, also referred to as a *nonpolar capacitor*, is symmetric in its construction and, as such, has no polarity. Non-polarized capacitors can be connected in a circuit in either direction with no change in performance or possibility of component or circuit damage. Capacitors typically installed in appliances containing single-phase motors, such as air conditioning and heating systems, are non-polarized, **Figure 18-8**. With no polarity, the direction in which the capacitor is installed makes no difference. Non-polarized capacitors have directional symmetry, so the effects of current flow on the device in either direction are exactly the same.

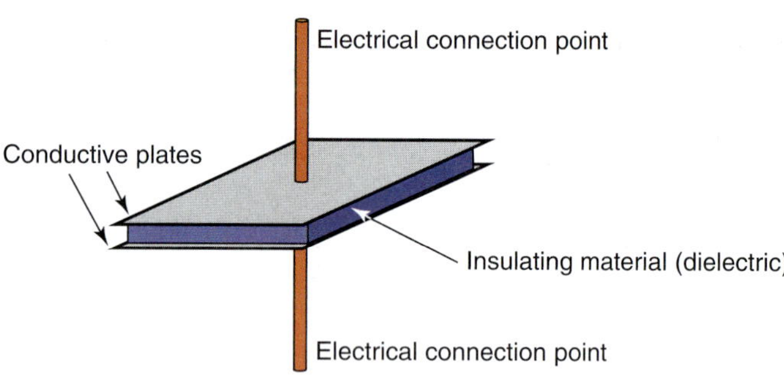

Goodheart-Willcox Publisher

Figure 18-5. Basic components of a capacitor.

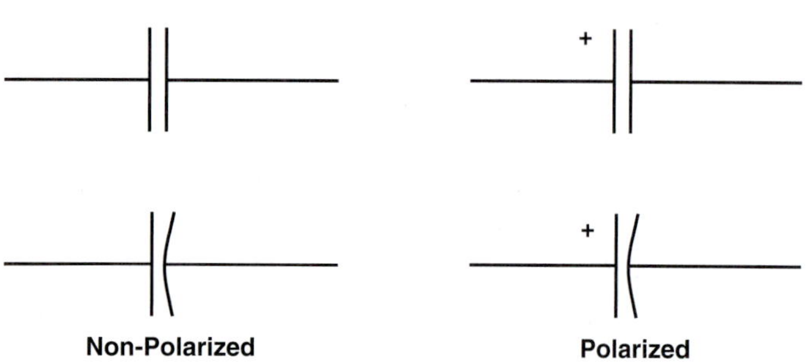

Goodheart-Willcox Publisher

Figure 18-6. Electrical symbols for capacitors.

Comparison of Polarized Capacitors to Non-Polarized Capacitors

Polarized Capacitors	Non-Polarized Capacitors
Have anode/cathode identification	Have no polarity identification
Used only in dc circuits	Used in both ac and dc circuits
Experience large leakage current	Experience low/no leakage current
Physically smaller in size	Physically larger in size
Asymmetrical construction	Symmetrical construction
Asymmetrical appearance	Symmetrical appearance
Less expensive	More expensive
Lower frequency limits	Higher frequency limits

Goodheart-Willcox Publisher

Figure 18-7. Characteristics of polarized and non-polarized capacitors.

Ratanapon Sangounsiritham/iStock via Getty Images

Figure 18-8. A non-polarized capacitor.

Polarized Capacitors

A ***polarized capacitor***, also referred to as a *polar capacitor*, is asymmetrical in its construction and has a distinct polarity associated with it. Polarized capacitors must be connected in a circuit in the proper orientation to prevent possible component and circuit damage, **Figure 18-9**.

Safety Alert

Connecting a polarized capacitor to a power supply in the wrong direction can result in premature component failure or circuit damage.

Polarized capacitors have marked terminals so the polarity can be easily identified. The positive (+) terminal is identified as the ***anode***, and the negative (–) terminal is identified as the ***cathode***. Nearly all polarized capacitors have only one terminal identified. When polarized capacitors are connected according to their marked polarity, they can achieve their rated maximum capacitance. In dc circuits, capacitors are polarized, so the positive side must always be at a higher potential than the negative side to ensure proper circuit operation. The electrical terminals on smaller capacitors are either on the same side of the capacitor or on opposite sides. Capacitors with their electrical connections positioned on opposite ends of the component are called ***axial capacitors*** and those with the electrical connections on the same side are referred to as ***radial capacitors***, **Figure 18-10**.

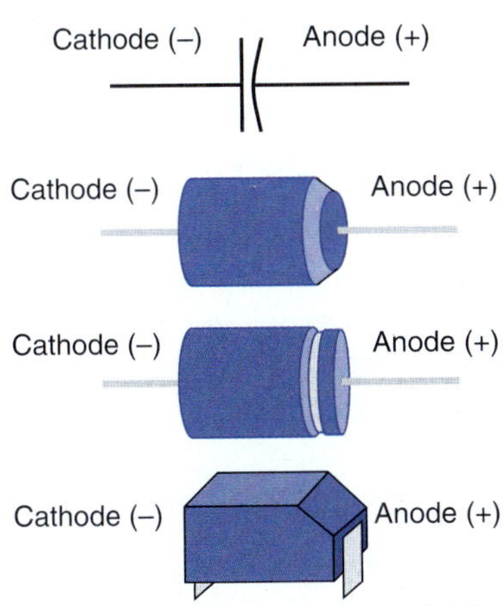

Goodheart-Willcox Publisher

Figure 18-9. Sample identification methods used to identify a capacitor's anode. Notice the lack of symmetry in the physical construction of a polarized capacitor.

Goodheart-Willcox Publisher

Figure 18-10. Electrical terminals are located on opposite ends of an axial capacitor and are located on the same side of a radial capacitor.

Shedding Light

To help remember which of the capacitor terminals is the anode and which is the cathode in wiring diagrams, it might be helpful to notice the curved shape of the cathode's electrical symbol. This curved shape closely resembles the letter *C*, which stands for cathode.

Shedding Light

A polarized capacitor can be replaced with a non-polarized capacitor, but the opposite is not always true. Be sure to carefully check the literature and all replacement guidelines to ensure that the replacement capacitor is installed correctly and that it is an acceptable replacement for the specific application.

Electrolytic Capacitor

Common types of polarized capacitors are the aluminum electrolytic capacitor and the tantalum electrolytic capacitor. An ***electrolytic capacitor*** is a capacitor type that has an electrolytic material as one of its conducting plates. An ***electrolyte*** is a liquid or pasty, gel-like substance that has a high concentration of ions, which are either positively or negatively charged atoms. The most common electrolytes are sodium, potassium, chloride, calcium, and phosphorus.

Electrolytic capacitors are made up of two metallic foil layers, a layer of oxide material, and an electrolytic layer, **Figure 18-11**. The electrolytic layer can be paper soaked with an electrolytic solution or can be a solid electrolytic material. Common materials used in electrolytic capacitors are aluminum and tantalum.

In electrolytic capacitors, the plates to which the electrical connections are made are constructed using pure metallic foil. The anode is etched to increase its surface area and increase the energy-storing capability of the capacitor while keeping the physical size of the device small. The anode is also coated with a thin layer of an oxide material, which is electrically bonded to the anode. Oxides are compounds that are made up of a base metal and oxygen. In aluminum electrolytic capacitors, the oxide material is aluminum oxide (Al_2O_3). In tantalum electrolytic capacitors, the oxide material is tantalum oxide (Ta_2O_5).

The oxide layer, which acts as the dielectric, separates the anode from the cathode. Since the anode and the electrolytic material are in contact with the oxide layer, the electrolytic material is the real cathode. The second plate, which is the electrical connection point, is referred to as the apparent cathode. In the electrical industry, the electrical connection points on a capacitor are typically referred to as the anode and cathode, ignoring reference to the outer foil layer (the apparent cathode) having an indirect connection to the dielectric.

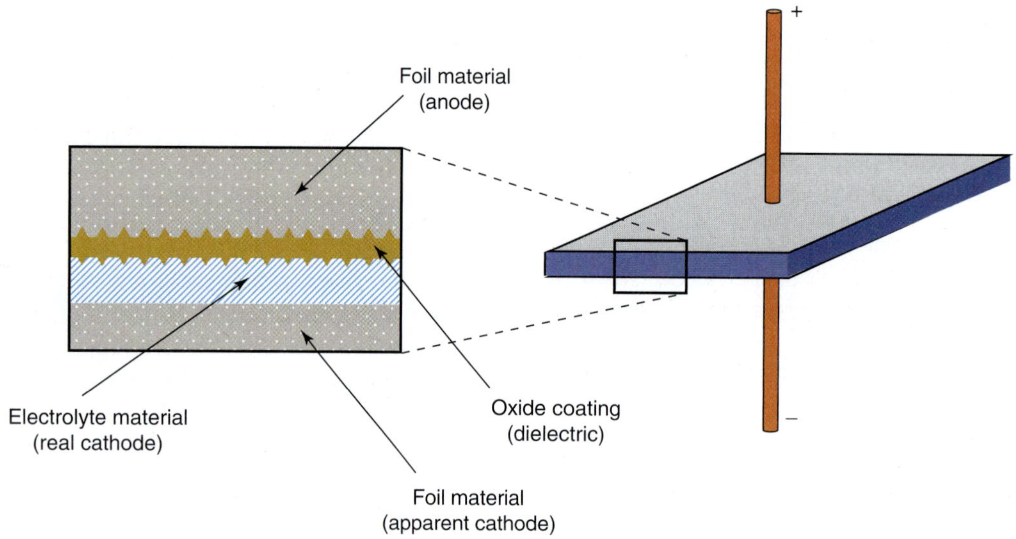

Figure 18-11. Cross-sectional view of an electrolytic capacitor.

Shedding Light

The question often arises about why polarized capacitors are used, given that non-polarized capacitors can be installed in any direction and are less likely to become damaged if not installed correctly. The polarizing of capacitors was, for the most part, an unintended consequence of creating capacitors with an oxide layer and an electrolytic material. The desire was to create devices with greater capacitances in smaller spaces. It was found that the oxide layer was more easily damaged when the polarity of the power supply was reversed. To maintain the structural and operational integrity of the electrolytic capacitors, polarities were assigned to the terminals on these devices.

Voltage, Temperature, and Leakage Current

An unintended consequence of the construction of an electrolytic capacitor is that when voltage is applied to it, a small amount of current passes through the oxide layer. This current is called ***leakage current*** and flows between the capacitor's anode and cathode, resulting in circuit losses. Excessive leakage current can cause damage to the capacitor. The amount of leakage current is determined by the potential difference, the dielectric material, and the measured resistance between the capacitor terminals. A higher potential difference or a lower dielectric resistance will result in higher leakage current.

Shedding Light

The factors that affect leakage current can be verified by referring to the form of Ohm's law, $I = E/R$. An increase in voltage or a decrease in resistance will cause an increase in current.

If a capacitor is exposed to higher temperatures, the capacitance of the device increases, but the leakage current will increase as well. If an increase in the voltage applied to an electrolytic capacitor causes the leakage current to increase, it is important to ensure that the rated capacitor voltage is never exceeded.

Shedding Light

In an ideal capacitor with a dielectric that insulates perfectly, there would be no leakage current at all. If this were the case, a capacitor would hold its electrostatic charge indefinitely. However, given the imperfect nature of insulating materials, even a relatively small amount of leakage current will cause a capacitor to lose its charge over time, even when not connected in a circuit.

Electrolytic Capacitor Comparison

Although aluminum and tantalum are both metals, there are significant differences that exist between electrolytic capacitors made from each base metal. Here are a few of these differences that should be taken into account when selecting between the two:

- Aluminum electrolytic capacitors can withstand higher voltages than tantalum electrolytic capacitors.
- Aluminum electrolytic capacitors have capacitance tolerances of up to 20%, while the average tolerance of a tantalum electrolytic capacitor is typically in the 10% range.
- Tantalum electrolytic capacitors are more sensitive to reverse voltages than aluminum electrolytic capacitors.

Safety Alert

Some capacitors are manufactured with a resistor connected between the terminals of the device. This resistor is called a ***bleed resistor***, **Figure 18-12**. The purpose of this resistor is to allow the capacitor to discharge safely, reducing the potential for electric shock, and to help prevent the capacitor from becoming damaged or destroyed in the event that its terminals come in contact with each other. It is important to never remove the bleed resistor from a capacitor.

Goodheart-Willcox Publisher

Figure 18-12. A bleed resistor is located on some capacitors to help release dielectric stress slowly and to allow the capacitor to discharge safely.

18.3 Capacitor Charge and Ratings

The charge of a capacitor, Q, is a function of the voltage, V, applied across it and the capacitance, C, of the capacitor. The relationship that exists among charge, voltage, and capacitance is expressed mathematically as follows:

$$Q = C \times V$$

$$C = \frac{Q}{V}$$

$$V = \frac{Q}{C}$$

The relationship that exists among charge (Q), voltage (V), and capacitance (C) can be graphically represented by a triangle similar to the one used to represent Ohm's law as well as the basic power formula, $P = I \times E$. The triangle is divided into three sections, with the top section representing charge, Q. The bottom two sections represent capacitance, C, and voltage, V. By covering the value that is to be obtained, the appropriate formula can be readily seen, **Figure 18-13**.

A capacitor's charge is expressed in terms of coulombs (C). One coulomb is equal to the charge on 6.241×10^{18} protons. The charge of one proton is, therefore, 1.6×10^{-19} coulomb ($1/6.241 \times 10^{18}$). Since the charge of an electron is opposite to that of a proton, an electron has a charge of -1.6×10^{-19} coulomb.

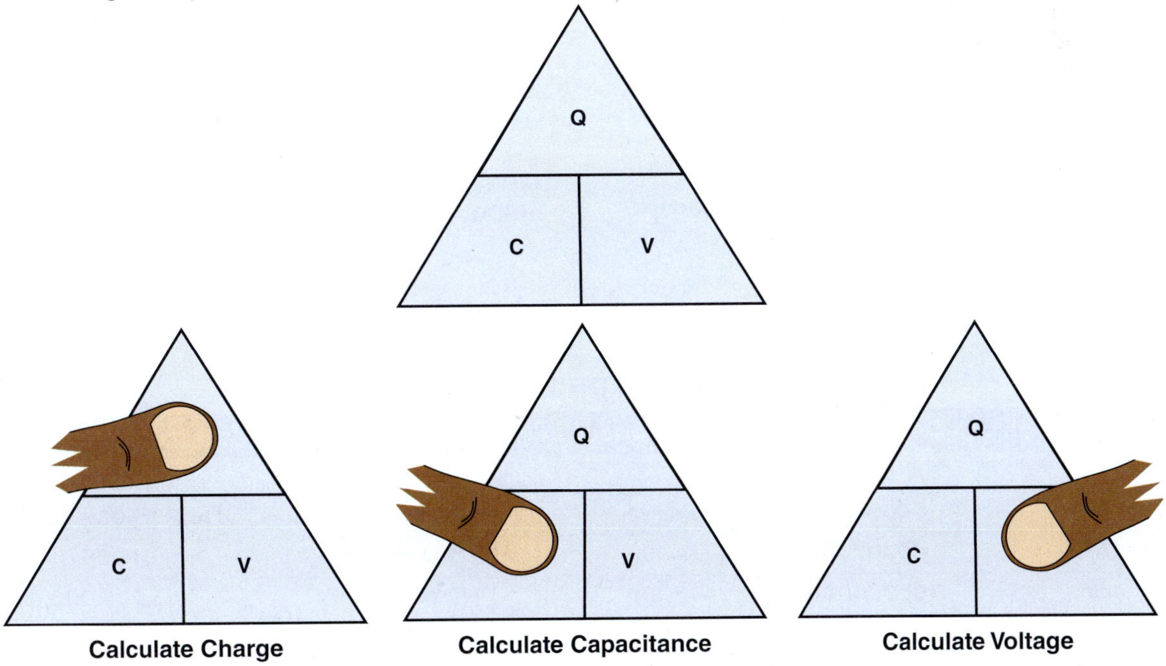

Goodheart-Willcox Publisher

Figure 18-13. If capacitance and voltage of a circuit are known, then charge can be calculated by covering the *Q* to reveal the *C* × *V* formula. When calculating capacitance, cover the *C* to reveal the formula of *Q* divided by *V*. When calculating voltage, cover the *V* to reveal the formula *Q* divided by *C*.

Shedding Light

Charge is used to describe current flow. One ampere is the rate of current equal to a charge of one coulomb of electrons (6.241×10^{18}) passing a given point in a circuit in one second.

Since a capacitor's charge, *Q*, is equal to the capacitor's capacitance multiplied by the voltage applied to the capacitor, when the voltage is zero, the capacitor's charge will also be zero. A sine waveform representing the capacitor's charge will be in phase with, and have the same frequency as, the sine waveform representing the applied voltage, **Figure 18-14**.

The unit of measurement for capacitance (*C*) is the farad (F), in honor of the English physicist and chemist, Michael Faraday. Since capacitance is calculated as charge, in coulombs, divided by voltage, the farad is expressed as coulombs per volt. One farad is a very large quantity of capacitance. As such, it is quite common to see capacitance expressed in much smaller units. The most common capacitance units are the microfarad (μF), the nanofarad (nF), and the picofarad (pF), **Figure 18-15**.

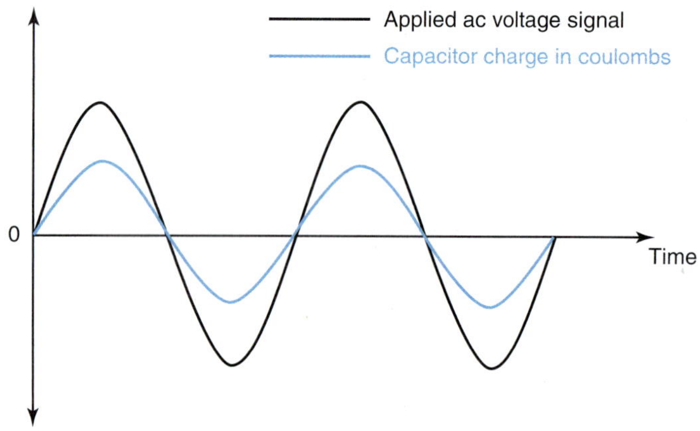

Goodheart-Willcox Publisher

Figure 18-14. Waveforms of applied voltage and capacitor charge.

Refer to **Example 18-1** and **Example 18-2** on how to calculate charge and capacitance of a capacitor.

Math Sidebar

In order to have correct units for calculations, it is sometimes necessary to convert among farads, microfarads, nanofarads, and picofarads. Until you get comfortable working with such small numbers, it is suggested that you work slowly and take incremental steps to arrive at the answer.

For example, consider the value 4500 pF, which is 4500×10^{-12} F. Moving the decimal point one place to the left brings the exponent one number closer to zero. So, 4500×10^{-12} is the same as 450×10^{-11}. The progression from picofarads to microfarads can be seen here:

$$4500 \times 10^{-12} = 450 \times 10^{-11}$$

$$450 \times 10^{-11} = 45 \times 10^{-10}$$

$$45 \times 10^{-10} = 4.5 \times 10^{-9}$$

$$4.5 \times 10^{-9} = 0.45 \times 10^{-8}$$

$$0.45 \times 10^{-8} = 0.045 \times 10^{-7}$$

$$0.045 \times 10^{-7} = 0.0045 \times 10^{-6}$$

Since the term micro means 10^{-6}, the expression 0.0045×10^{-6} represents 0.0045 μF.

To move from microfarads back to picofarads (from 10^{-6} to 10^{-12}), the decimal point needs to move six places to the right. At the same time, the exponent needs to be reduced by six. Moving the decimal point in 0.0045 six places to the right gives us 4500, while reducing the exponent 10^{-6} by six gives us 10^{-12}. With these changes, we can see that 0.0045 μF (0.0045×10^{-6} F) is the same as 4500 pF (4500×10^{-12} F).

Common Capacitance Units

Name	Symbol	Decimal	Exponent	Description	Fraction
Farad	F	1.00	1×10^{0}	One	1
Microfarad	μF	0.000001	1×10^{-6}	One-millionth	1/1,000,000
Nanofarad	nF	0.000000001	1×10^{-9}	One-billionth	1/1,000,000,000
Picofarad	pF	0.000000000001	1×10^{-12}	One-trillionth	1,000,000,000,000

Goodheart-Willcox Publisher

Figure 18-15. Expressions for capacitance units.

Example 18-1

What is a capacitor's charge if it has a capacitance of 275 μF and is connected to a 100-V power source?

✓ Solution

The capacitor's charge can be calculated as follows:

$$Q = C \times V$$

$$Q = 275\ \mu F \times 100\ V$$

Since capacitance is expressed in farads, this expression can be rewritten as follows:

$$Q = 275 \times 10^{-6}\ F \times 100\ V$$

$$Q = 0.000275\ F \times 100\ V$$

$$Q = 0.0275\ C$$

Now It's Your Turn 18-1

What is a capacitor's charge if it has a capacitance of 350 μF and is connected to a 240-V power source?

Example 18-2

What is a capacitor's capacitance if it has a charge of 0.05 C and is connected to a 50-V power source?

✓ Solution

The capacitor's capacitance can be calculated as follows:

$$C = \frac{Q}{V}$$

$$C = \frac{0.05\ C}{50\ V}$$

$$C = 0.001\ F$$

Since there are one million (1×10^6) μF in one farad, this answer can be expressed as follows:

$$0.001\ F \times (1 \times 10^6\ \mu F/F) = 1000\ \mu F$$

Now It's Your Turn 18-2

What is the voltage being supplied to a 500 μF capacitor if it has a charge of 0.04 C?

In addition to the capacitance rating, capacitors also have a voltage rating. The voltage rating of a capacitor represents the maximum voltage that can be applied to the capacitor without causing damage to the dielectric. The actual capacitance of a capacitor falls within a range specified by the device's rating and a tolerance, both of which are indicated on the capacitor's nameplate.

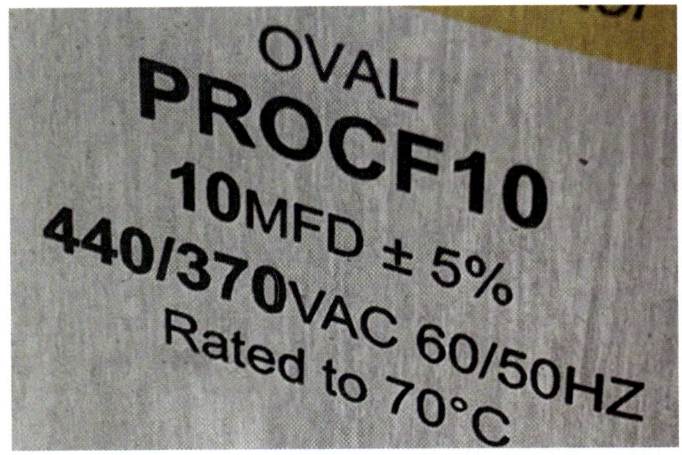

Goodheart-Willcox Publisher

Figure 18-16. The capacitor nameplate shown here indicates that the capacitance is 10 μF +/–5% and the rated voltage is 440/370 Vac. (For capacitors, microfarad is sometimes abbreviated as MFD.)

Shedding Light

If no tolerance is listed on a capacitor, it is assumed that the tolerance is 10%. On smaller capacitors, where it is not possible to list the capacitance, voltage, and tolerance, color-coded bands, dots, and stripes are used to identify the capacitor's characteristics.

Consider the capacitor label shown in **Figure 18-16**. For this capacitor, the capacitance rating is 10 μF. The capacitance rating includes a tolerance of +/–5%. Since 5% of 10 μF is 0.5 μF (10 μF × 0.05), the actual acceptable capacitance of the capacitor is the range of 9.5 μF to 10.5 μF (or 10 μF +/– 0.5 μF). The voltage rating of the capacitor is listed as 440/370 Vac. This means that this capacitor can be used to replace a capacitor with either a 370 Vac rating or a 440 Vac rating. Exposing this capacitor to ac voltages above 440 Vac can cause damage to the component.

Shedding Light

A capacitor with a higher voltage rating can be used to replace a capacitor with a lower voltage rating, but the reverse is not true. A capacitor rated at 440 Vac can safely replace the capacitor with a 370 Vac rating but a capacitor with a 370 Vac rating cannot be used to replace a capacitor with a 440 Vac rating. Remember that the voltage rating of a capacitor is not the operating voltage of the component. The rated voltage is the highest voltage to which the capacitor can safely be exposed without causing damage to the dielectric.

18.4 Permittivity, Relative Permittivity, and Dielectric Constant

When discussing capacitance, a great deal of focus needs to be placed on the properties of the dielectric material, meaning the substance that separates the capacitor's conductive plates.

One important factor that affects capacitance is the dielectric's ability to become polarized by an electric field. This factor is called ***permittivity*** and is identified by the lowercase Greek letter epsilon (ε). Permittivity is expressed in the units of farads per meter (F/m). The permittivity of a perfect vacuum, identified as ε_0, is used as a reference point to compare the permittivity levels of various materials. The term ***relative permittivity*** (ε_r) identifies the permittivity of a substance when compared to the permittivity of a perfect vacuum. The relative permittivity of a vacuum, therefore, is equal to one. The general expression for relative permittivity is:

$$\varepsilon_r = \frac{\varepsilon}{\varepsilon_0}$$

Since both ε and ε_0 are expressed in the units of F/m, the relative permittivity is a unitless quantity. In the electronics and semiconductor sectors of our industry, the term relative permittivity is often used, while in other sectors of our industry, the term ***dielectric constant*** is used. Dielectric constant is also a unitless quantity. Throughout the rest of this text, the term dielectric constant (κ) will be used when referring to the relative permittivity of a dielectric material. A partial list of the dielectric constants for some common dielectric materials is provided in **Figure 18-17**.

Dielectric Constants for Common Dielectric Materials

Material	Approximate Dielectric Constant
Vacuum	1.0
Air	1.0
Paper	1.5
Teflon™	2.0
Rubber	3.0
Glass	5.0
Porcelain	5.5
Silicon	12
Graphite	15
Water (pure)	80
Ceramic	1200

Goodheart-Willcox Publisher

Figure 18-17. The value of the dielectric constant of the dielectric material affects the capacitance of a capacitor. Dielectric materials with high dielectric constants hold an electric charge relatively well.

Shedding Light

Dielectric materials with low dielectric constants do not have the ability to hold an electric charge well. These materials typically make good insulators. In the communications sector of our industry, dielectric materials with low dielectric constants are often used to electrically isolate signal-carrying conductors from each other to reduce interference and improve transmission quality. When used in the production of capacitors, dielectric materials with high dielectric constants are desired. These materials hold an electric charge well, so the physical size of a capacitor can be reduced.

18.5 Basic Capacitor Operation

Having discussed the properties and construction of a capacitor, we will now examine its basic operation. Consider a capacitor connected across the terminals of a battery, **Figure 18-18**. When initially connected, some electrons from the negative terminal of the battery will shift to the capacitor plate that is connected to it. At the same time, some electrons from the opposite plate of the capacitor are attracted to the positive terminal of the battery. As a result, an electric charge is built up on one of the plates, while a net opposing charge is created on the other. The dielectric located between the two plates of the capacitor prevents the electrons from moving from one plate to the other. The process of building up a charge across the capacitor's plates is called *charging*. When the measured voltage across the capacitor reaches a potential that is equal to voltage across the battery, the capacitor is said to be charged, **Figure 18-19**. When charged, there is no additional electron shift from the battery to the capacitor. If the capacitor is disconnected from the circuit, the capacitor retains its charge.

A charged capacitor has the ability to act as a voltage source. A charged capacitor can supply voltage to a circuit, or part of

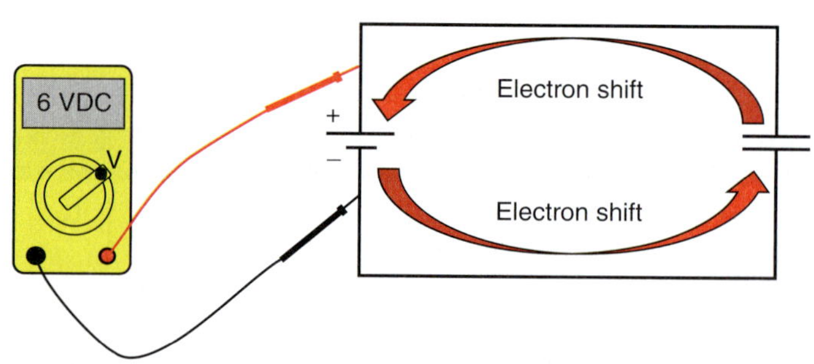

Goodheart-Willcox Publisher

Figure 18-18. When connecting a capacitor across the terminals of a battery, electrons shift from the battery's negative terminal to one of the capacitor's plates; an opposing charge is created on the other plate.

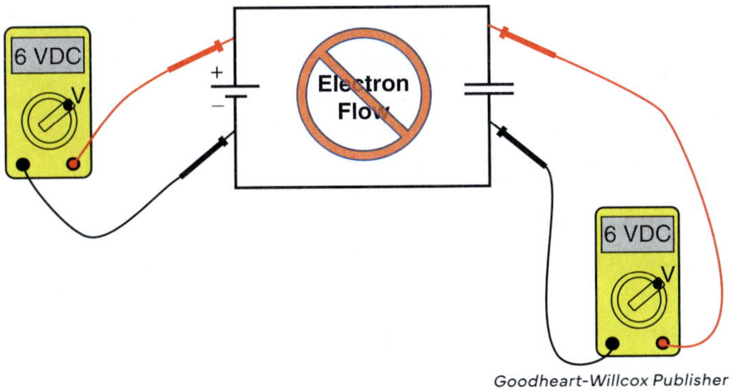

Figure 18-19. When the capacitor is fully charged, no current flows in the circuit. The capacitor retains its charge, even when disconnected from the voltage source.

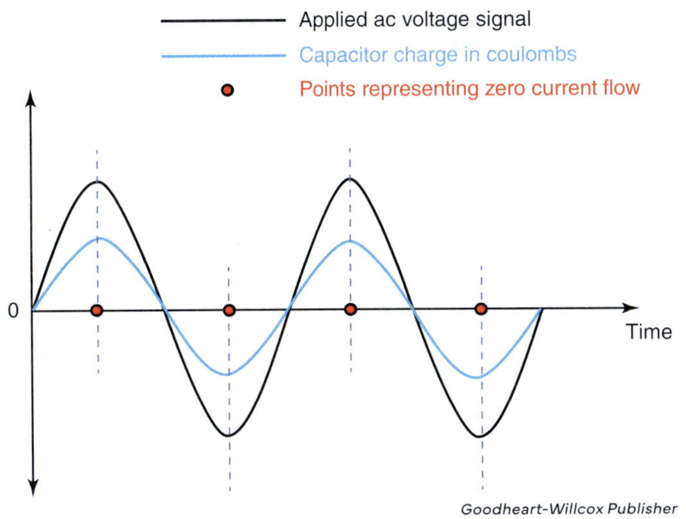

Figure 18-20. Applied voltage and capacitor charge reach maximum and zero together. When the capacitor is fully charged, there are points of no current flow.

a circuit, even if there is no external voltage source supplying potential. When acting as a voltage source, the capacitor is discharging. The magnitude of the stored charge will vary depending on the structure of the capacitor. When referring to capacitor charging and discharging, it is important to keep the following in mind:

- Electrons flow when a capacitor is either charging or discharging.
- When a capacitor is fully charged, there is no flow of electrons.

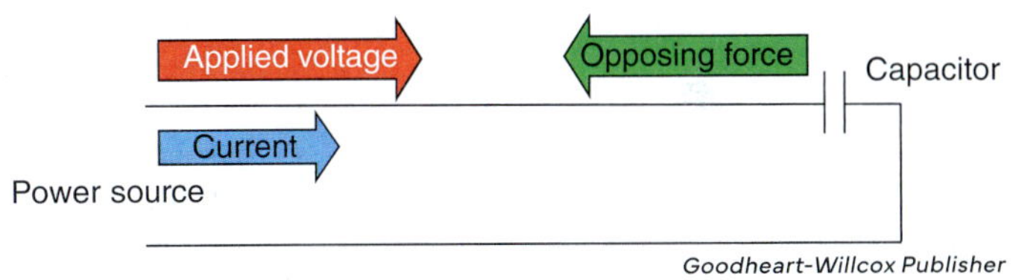

Figure 18-21. When voltage is applied to a capacitor, an opposing voltage is produced.

Current Leads the Voltage in a Capacitive Circuit

From our earlier discussion on the relationship that exists among capacitance, charge, and voltage, it was established, from the $Q = C \times V$ formula, when the voltage applied to a capacitor is zero, the capacitor's charge is also zero. Similarly, when the voltage applied to a capacitor is at its maximum, the capacitor's charge is also at its maximum. Refer back to **Figure 18-14** for the graphical representation of this relationship. It was also established, in the previous section of this chapter that, when the capacitor's charge is at its maximum, or when the capacitor is fully charged, there is no current flow, **Figure 18-20**.

In capacitive circuits, a counter force that opposes current flow is produced when voltage is applied to a capacitor, **Figure 18-21**. This counter force is similar to the counter electromotive force (cemf) that is generated when current flows through a coil of wire in an ac circuit. In a capacitive circuit, the counter force is a result of the electrostatic charge generated between the capacitor's plates. As an extension of this statement, when there is no voltage applied to a capacitor, there is no opposing force generated, and the current flow can be at its maximum level, **Figure 18-22**.

Since ideal ac current and voltage waveforms follow ideal sine wave patterns, the ac current waveform can be illustrated by connecting the points identified in **Figure 18-20** and **Figure 18-22**. It can be seen that in a purely capacitive circuit, which are circuits that contain only capacitors, the current leads the voltage by 90 electrical degrees, **Figure 18-23**. When the current is at its maximum, the voltage is zero. Similarly, when the current is zero, the voltage is at its maximum. A graphical representation among applied voltage, capacitor charge, and current is provided in **Figure 18-24**.

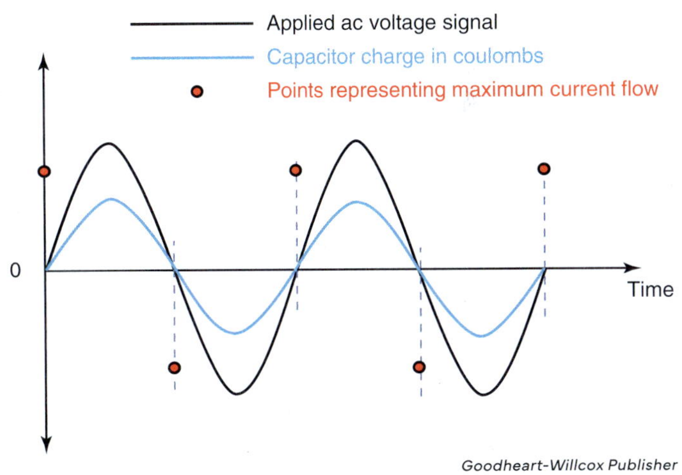

Figure 18-22. When no opposing voltage is present, current is at its maximum.

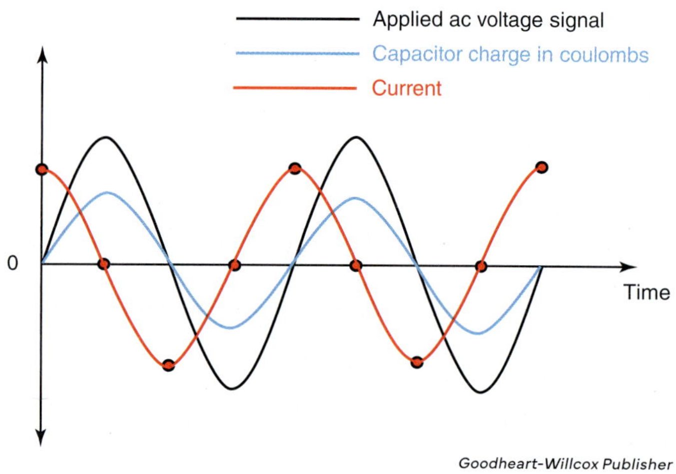

Figure 18-23. Connecting the points in **Figure 18-20** and **Figure 18-22** shows that the current leads the voltage by 90 electrical degrees in a purely capacitive circuit.

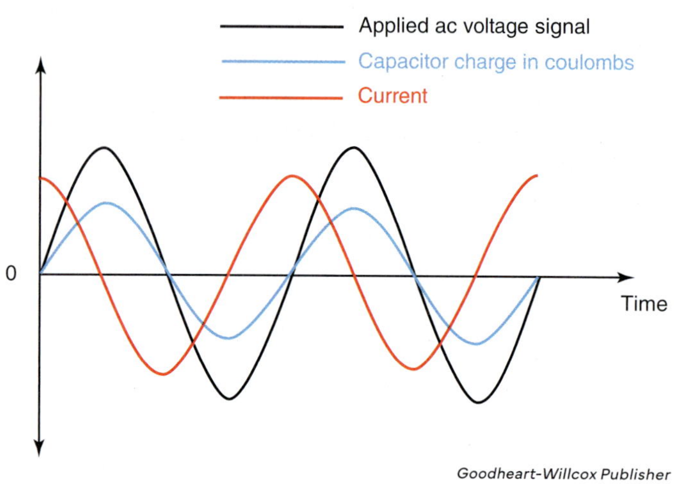

Figure 18-24. Graphical representation of applied voltage, current, and capacitor charge.

Shedding Light

The mnemonic "ELI the ICE man" is a helpful tool for remembering the relationship between voltage and current in capacitive and inductive circuits. The *C* in "ICE" represents capacitive circuits, with the *I* (current) preceding, or leading, the *E* (voltage) in "ICE." Thus, "ICE" can help you remember that in capacitive circuits, the current leads the voltage and the voltage lags, or follows, the current. The *L* in "ELI" represents inductive circuits, in which the voltage (*E*) leads the current (*I*). Inductive circuits are discussed in Chapter 17, *Inductors and Inductance.*

Dielectric Stress

When a capacitor is charged, one plate has a negative charge and the other has a positive charge. This difference in charge has an effect on the electrons in the capacitor's dielectric material, and has the potential to cause damage to the capacitor. If the dielectric becomes damaged, the capacitor will not be able to effectively store an electrical charge. In extreme situations, the capacitor can become shorted and will not be able to store an electric charge at all.

In a charged capacitor, the dielectric's electrons are attracted to the positive plate of the capacitor. This attraction causes the electron orbit to stretch in the direction of the positive plate, **Figure 18-25**. This stretching is referred to as ***dielectric stress.*** The amount of dielectric stress is determined by the potential difference across the capacitor. The larger the potential difference, the greater the dielectric stress. By ensuring that a capacitor is not exposed to voltages higher than their rated values, the amount of dielectric stress can be kept within acceptable limits.

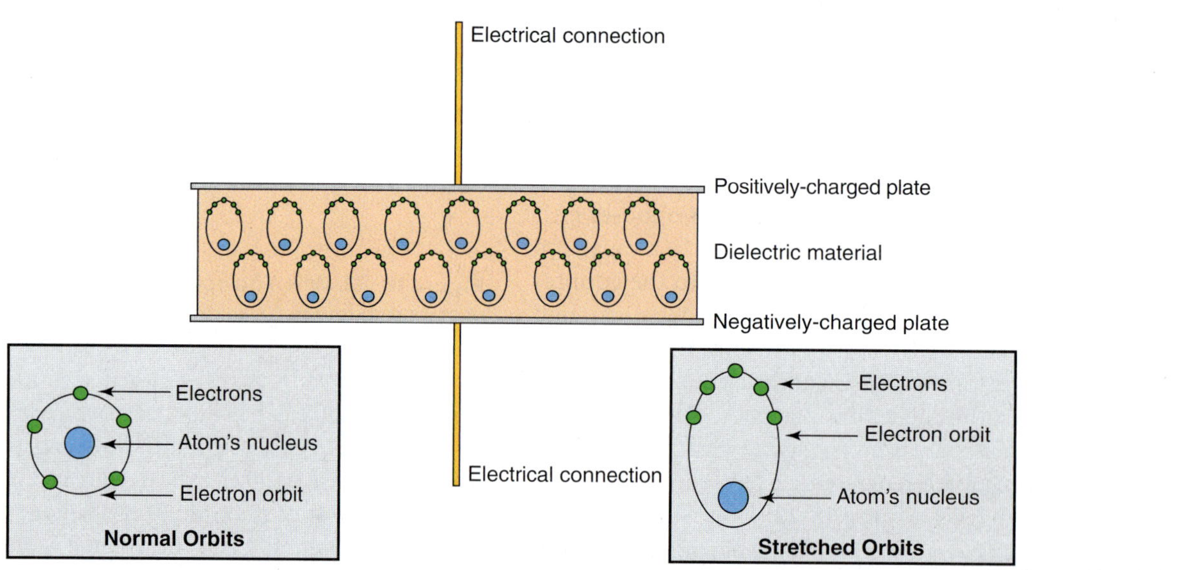

Figure 18-25. Voltage across a capacitor causes a stretching of a dielectric's electron orbits; this is called *dielectric stress.*

Shedding Light

Releasing dielectric stress too quickly can damage the dielectric material and shorten the life of the capacitor. Discharging a capacitor with a screwdriver, for example, will cause the immediate release of dielectric stress and can cause damage to the capacitor. The bleed resistor found on some capacitors helps release the dielectric stress slowly, allowing the capacitor to discharge safely.

18.6 Capacitive Reactance

In capacitive circuits, the electrostatic charge generated between the capacitor's plates produces a counter voltage that opposes current flow. In a purely capacitive ac circuit, the waveform for this opposition voltage is 180° out of phase with the applied voltage, **Figure 18-26**. A graphical representation among applied voltage, opposing voltage, capacitor charge, and current is provided in **Figure 18-27**. The opposition to current flow caused by this opposing voltage is called ***capacitive reactance*** and is identified by the symbol X_C. Capacitive reactance behaves like resistance and is measured in ohms. Capacitive reactance is calculated using the following formula:

$$X_C = \frac{1}{2\pi \times f \times C}$$

where:

2π = Constant equal to 2 × 3.14, or 6.28
f = Frequency of the power supply (in hertz)
C = Capacitance of the capacitor (in farads)

At higher frequencies, the capacitive reactance drops, resulting in less opposition to current flow. A capacitor with higher capacitance also offers less opposition to current flow than a capacitor with lower capacitance.

Refer to **Example 18-3, Example 18-4**, and **Example 18-5**.

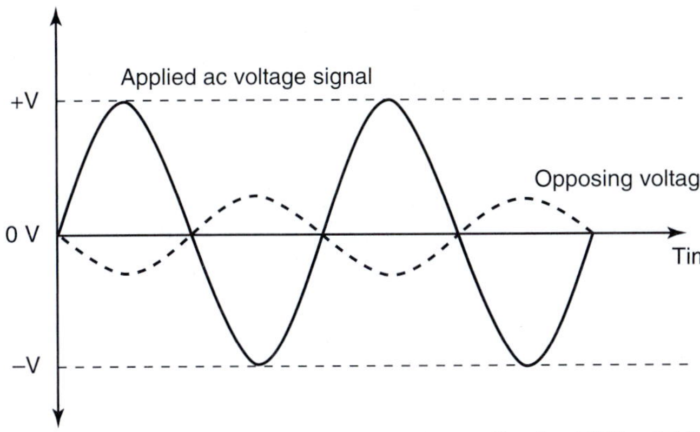

Figure 18-26. In a purely capacitive ac circuit, the opposing force is 180° out of phase with the applied voltage.

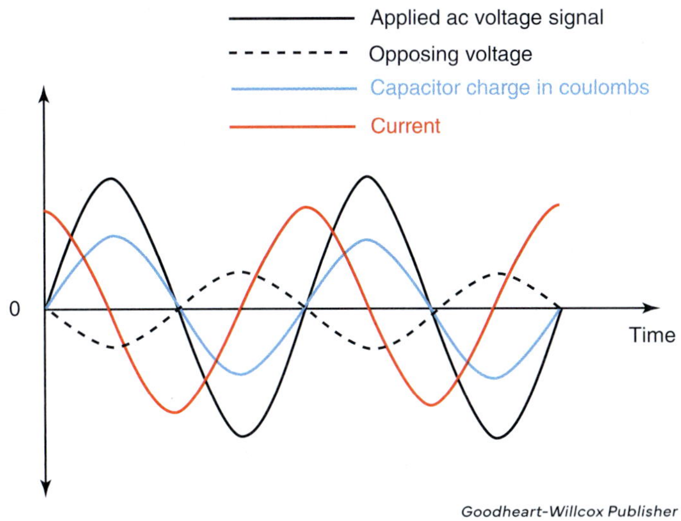

Figure 18-27. Graphical representation of applied voltage, opposing voltage, current, and capacitor charge.

Example 18-3

What is the capacitive reactance of a 120 µF capacitor if the frequency of the power supply is 60 Hz?

✓ Solution

The capacitive reactance formula requires that capacitance is expressed in units of farad. Thus, 120 µF must be converted to 120×10^{-6} F or 0.000120 F. (Since 120×10^{-6} is equal to 0.000120, either number can be used in the equation.)

$$X_C = \frac{1}{2\pi \times f \times C}$$

$$X_C = \frac{1}{6.28 \times 60 \text{ Hz} \times 0.00012 \text{ F}}$$

$$X_C = \frac{1}{0.0452}$$

$$X_C = 22.1\ \Omega$$

Example 18-4

What is the frequency of the power supply if the capacitive reactance of a 75 µF capacitor is 25 Ω?

✓ Solution

First, express 75 µF as 75×10^{-6} F, which is also 0.000075 F. To solve for frequency, rearrange the capacitive reactance formula as shown in the following:

$$f = \frac{1}{2\pi \times X_C \times C}$$

$$f = \frac{1}{6.28 \times 25\ \Omega \times 0.000075 \text{ F}}$$

$$f = \frac{1}{0.0118}$$

$$f = 84.75 \text{ Hz}$$

Example 18-5

What is the capacitance of a capacitor if it is connected to a power supply with a frequency of 100 Hz and has a capacitive reactance of 50 Ω?

✓ **Solution**

To solve for capacitance, you will need to modify the capacitive reactance formula as shown here:

$$C = \frac{1}{2\pi \times f \times X_C}$$

$$C = \frac{1}{6.28 \times 100 \text{ Hz} \times 50\ \Omega}$$

$$C = \frac{1}{31{,}400}$$

$$C = 31.8\ \mu\text{F}$$

Example 18-6

How much time will it take to charge a 150 μF capacitor if there is a 10,000 Ω (10 kΩ) resistor wired in series with it?

✓ **Solution**

The time constant can be calculated as follows:

$$\tau = R \times C$$

$$\tau = 10{,}000\ \Omega \times 0.00015 \text{ F}$$

$$\tau = 1.5 \text{ seconds}$$

Since it takes five time constants to fully charge or discharge a capacitor, the total charging time will be 5 multiplied by 1.5 seconds, or 7.5 seconds.

18.7 RC Time Constants

Capacitors charge and discharge at predictable rates that can be determined using the RC time constant. Identified by the lowercase Greek letter tau (τ), the RC time constant is equal to the capacitor's capacitance (in farads) multiplied by the series-connected resistance (in ohms). This expression is written as follows:

$$\tau = R \times C$$

Since this time constant is a function of resistance and capacitance, it is referred to as the ***RC time constant***. The RC time constant is expressed in seconds. The time required to fully charge or discharge a capacitor is five time constants.

Refer to **Example 18-6** on how to calculate the time constant of a capacitor.

Although the time constant varies depending on the series-connected resistance value and the capacitance of the capacitor, the percentage of charge gained or lost after each time constant is known, **Figure 18-28**.

When a capacitor is being charged, it reaches the following percentages of the total charge after each time constant:

- After one time constant, the capacitor reaches 63.2% of its total charge.
- After two time constants, the capacitor reaches 86.4% of its total charge.
- After three time constants, the capacitor reaches 95% of its total charge.
- After four time constants, the capacitor reaches 98.2% of its total charge.
- After five time constants, the capacitor reaches 100% of its total charge.

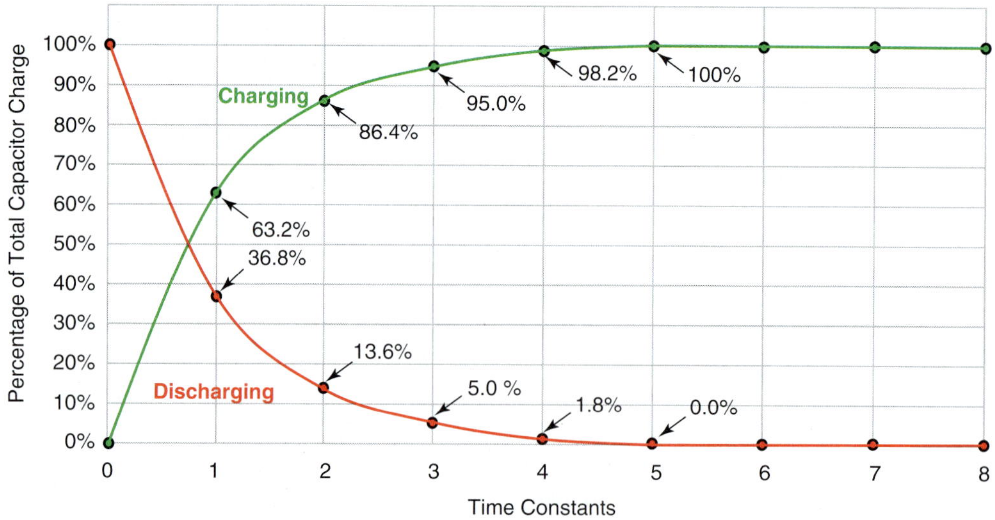

Goodheart-Willcox Publisher

Figure 18-28. Capacitor charging and discharging.

When a capacitor is being discharged, it loses the following percentages of the total charge after each time constant:

- After one time constant, the capacitor loses 63.2% of its total charge. This means that the capacitor has 36.8% (100% – 63.2%) of its total charge remaining.
- After two time constants, the capacitor loses 86.4% of its total charge. This means that the capacitor has 13.6% (100% – 86.4%) of its total charge remaining.
- After three time constants, the capacitor loses 95% of its total charge. This means that the capacitor has 5% (100% – 95%) of its total charge remaining.
- After four time constants, the capacitor loses 98.2% of its total charge. This means that the capacitor has 1.8% (100% – 98.2%) of its total charge remaining.
- After five time constants, the capacitor loses 100% of its total charge. The capacitor is completely discharged and has no charge remaining.

Example 18-7 provides how to calculate amount of time for a capacitor to discharge.

Example 18-7

Consider a fully charged 15 μF capacitor that has a 1000 Ω (1 kΩ) resistor wired in series with it. How long will it take for the capacitor to lose 95% of its total charge?

✓ Solution

First, convert 15 μF to 15×10^{-6} F, which is also 0.000015 F. Then calculate the RC time constant:

$$\tau = R \times C$$

$$\tau = 1000\ \Omega \times 0.000015\ \text{F}$$

$$\tau = 0.015\ \text{seconds}$$

Since it takes three time constants for a fully charged capacitor to lose 95% of its total charge, the time required will be three multiplied by 0.015 seconds, or 0.045 seconds.

18.8 Parallel-Connected Capacitors

When capacitors are wired in parallel with each other, the total capacitance is equal to the mathematical sum of all of the individual capacitances in the parallel connection. The general formula for parallel-connected capacitors is:

$$C_T = C_1 + C_2 + C_3 + \dots + C_N$$

In this equation, C_T is the total capacitance and C_1 through C_N are the capacitance values for each parallel-connected capacitor. Refer to **Example 18-8** on how to calculate total capacitance in a parallel-connected capacitor.

Shedding Light

Calculations for parallel-connected capacitors use equations similar to those for series-connected resistors and inductors, while calculations for series-connected capacitors use equations similar to those for parallel-connected resistors and inductors.

Example 18-8

What is the total capacitance of a parallel circuit that has three capacitors with values of 40 μF, 30 μF, and 50 μF?

✓ Solution

The total capacitance can be calculated as follows:

$$C_T = C_1 + C_2 + C_3$$

$$C_T = 40\ \mu\text{F} + 30\ \mu\text{F} + 50\ \mu\text{F}$$

$$C_T = 120\ \mu\text{F}$$

Now It's Your Turn 18-8

Referring to the following circuit, determine the value of capacitor C_2.

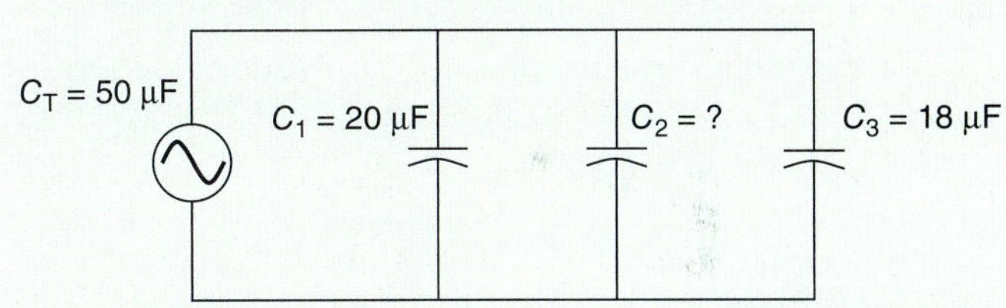

Goodheart-Willcox Publisher

18.9 Series-Connected Capacitors

The total capacitance of a circuit made up of capacitors connected in series with each other is always lower than the capacitance of the lowest individual capacitor in the circuit. Calculating the total capacitance of a circuit with series-connected capacitors uses the following general formula:

$$\frac{1}{C_T} = \frac{1}{C_1} + \frac{1}{C_2} + \frac{1}{C_3} + \dots + \frac{1}{C_N}$$

In this expression, C_T is the total circuit capacitance, while C_1 through C_N represent the series-connected capacitance values in the circuit. Refer to **Example 18-9** on how to calculate total capacitance in a series-connected capacitor.

Example 18-9

Determine the total capacitance of the following series circuit.

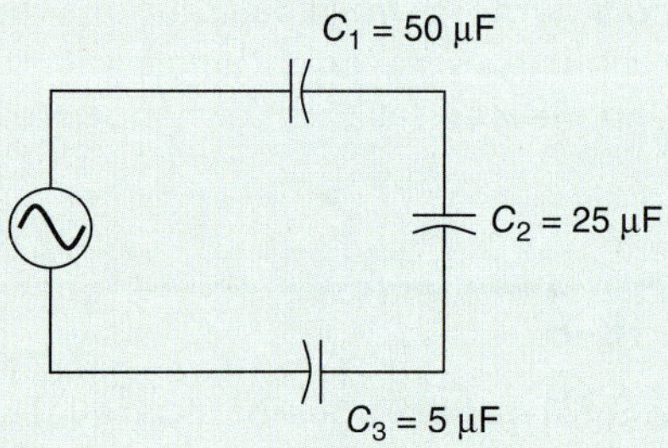

Goodheart-Willcox Publisher

✓ Solution

This series capacitive circuit has three capacitors with values of 25 µF, 50 µF, and 5 µF. The total circuit capacitance can be calculated using the following formula:

$$\frac{1}{C_T} = \frac{1}{C_1} + \frac{1}{C_2} + \frac{1}{C_3}$$

$$\frac{1}{C_T} = \frac{1}{50\ \mu F} + \frac{1}{25\ \mu F} + \frac{1}{5\ \mu F}$$

This expression can be simplified using the reciprocal key (1/x or x^{-1}) on your calculator:

$$\frac{1}{C_T} = 0.26$$

Use the reciprocal key (1/x or x^{-1}) on your calculator again:

$$C_T = 3.85\ \mu F$$

Math Sidebar

When working with the capacitive reactance formula earlier in the chapter, it was necessary to express capacitance values in units of farads, rather than microfarads or picofarads because the capacitive reactance formula includes variables that must be expressed in the specified units. When calculating total capacitance of parallel-connected or series-connected capacitors, any of the farad-based units (such as farads or microfarads) can be used, but all of the units must be the same. The total capacitance is expressed in the same unit as the individual capacitors.

If the series capacitive circuit being evaluated has only two capacitors, the following formula can be used to find the total capacitance of the series connection:

$$C_T = \frac{C_1 \times C_2}{C_1 + C_2}$$

This method of calculating the values of two capacitors connected in series with each other is often referred to as product over sum. Example 18-10 and Example 18-11 provide additional practice on how to calculate total capacitance in a series-connected capacitor.

Example 18-10

What is the total capacitance of the following series circuit?

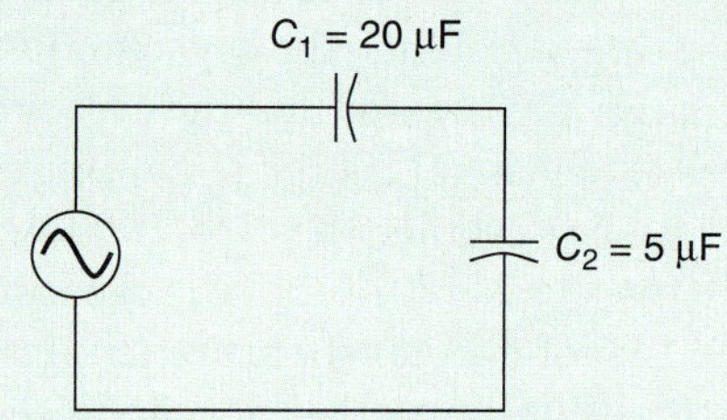

Goodheart-Willcox Publisher

✓ Solution

With only two capacitors in this series circuit, the product over the sum formula can be used to calculate the total capacitance of the circuit. The total capacitance can be determined as follows:

$$C_T = \frac{C_1 \times C_2}{C_1 + C_2}$$

$$C_T = \frac{5\ \mu F \times 20\ \mu F}{5\ \mu F + 20\ \mu F}$$

$$C_T = \frac{100\ \mu F^2}{25\ \mu F}$$

$$C_T = 4\ \mu F$$

Now It's Your Turn 18-10

Using the product over the sum formula, calculate the total capacitance for the following series circuit.

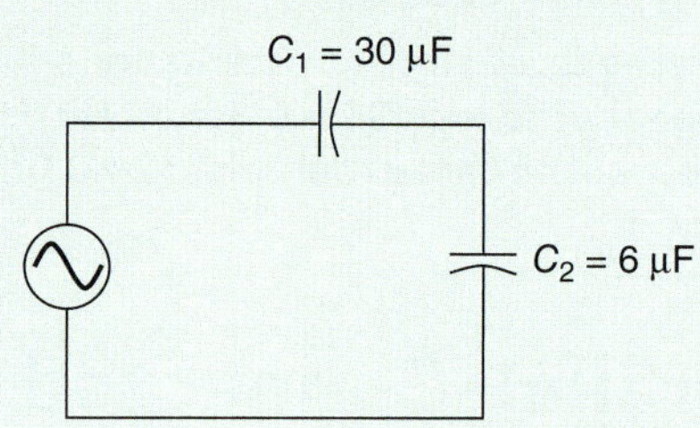

Goodheart-Willcox Publisher

When multiple capacitors of the same value are wired in series with each other, a useful formula that can be used is:

$$C_T = \frac{C}{N}$$

where:

C_T = Total circuit capacitance
C = Capacitance of each capacitor
N = Total number of series-connected capacitors

Example 18-11

Determine the total capacitance for a series circuit made up of three 75 μF capacitors.

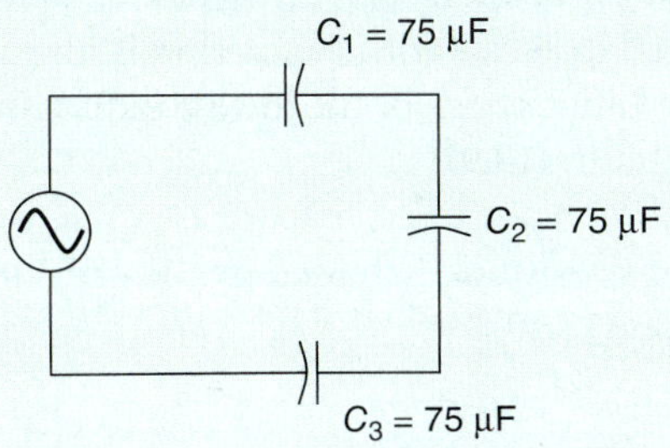

Goodheart-Willcox Publisher

✓ Solution

The total capacitance can be calculated as follows:

$$C_T = \frac{C}{N}$$

$$C_T = \frac{75\ \mu F}{3}$$

$$C_T = 25\ \mu F$$

Now It's Your Turn 18-11

Calculate the total capacitance of a series circuit made up of five 100 μF capacitors.

To summarize, when calculating multiple capacitors in series, three common formulas are used:

Reciprocal Formula (works in all cases):

$$\frac{1}{C_T} = \frac{1}{C_1} + \frac{1}{C_2} + \frac{1}{C_3} + \ldots + \frac{1}{C_N}$$

Product over Sum Formula (works with two series-connected capacitors):

$$C_T = \frac{C_1 \times C_2}{C_1 + C_2}$$

Equal Capacitance Formula (works when all capacitors have the same capacitance):

$$C_T = \frac{C}{N}$$

18.10 Capacitor Quality Factor

The quality factor (Q) of a capacitor is the relationship that exists between the capacitive reactance and the resistance of a capacitor. The lower the capacitive reactance and the higher the resistance, the higher the quality factor. The quality factor for a capacitor is calculated as follows:

$$Q = \frac{R}{X_C}$$

The steps to calculate the quality factor of a capacitor are highlighted in Example 18-12.

The capacitive reactance and the resistance are both expressed in ohms, so the quality factor is a unitless quantity. The quality factor rates the overall performance and efficiency of a capacitor. As the quality factor rises, the energy losses associated with the device fall. Here are some key things to keep in mind regarding the quality factor of a capacitor:

- If leakage current is low, the resistance of a capacitor will be very high and the quality factor will be high.
- As the capacitance or frequency increases, the capacitive reactance will decrease, and the quality factor will increase.

From these two statements it can be concluded that lower-valued capacitors will typically have lower quality factors than similar capacitors with higher capacitance ratings. In addition, exposing capacitors to higher voltages will increase the leakage current and dielectric stress, potentially damaging the dielectric and causing the resistance of the capacitor to drop. A drop in resistance will result in a lower quality factor. High-quality capacitors can have quality factors over 10,000 at high frequencies.

Example 18-12

What is the quality factor of a 555 μF capacitor with a resistance value of 2000 Ω if it is connected to a 100 Hz power source?

✓ Solution

The first step in calculating the quality factor is to determine the capacitive reactance, which requires expressing the 555 μF capacitance as 0.000555 F:

$$X_C = \frac{1}{2\pi \times f \times C}$$

$$X_C = \frac{1}{6.28 \times 100\ \text{Hz} \times 0.000555\ \text{F}}$$

$$X_C = \frac{1}{0.3485}$$

$$X_C = 2.87\ \Omega$$

Next, calculate the quality factor as follows:

$$Q = \frac{R}{X_C}$$

$$Q = \frac{2000\ \Omega}{2.87\ \Omega}$$

$$Q = 696.8$$

CHAPTER 18 REVIEW

Summary

- Capacitance is a measure of the ability of a material to hold an electrical charge. (18.1)
- A simple capacitor is made up of two conductive plates separated by an insulating material called a dielectric. (18.2)
- Polarized capacitors must be installed in the proper direction, while non-polarized capacitors can be installed in either direction. (18.2)
- The positive terminal on a capacitor is called the anode, while the negative terminal on a capacitor is called the cathode. (18.2)
- Electrolytic capacitors are polarized capacitors that use an electrolyte as the cathode material. (18.2)
- A capacitor's charge, Q, is equal to the capacitor's capacitance multiplied by the voltage applied to the capacitor. (18.3)
- Capacitance is measured in farads (F). Common capacitance ratings are in microfarads, nanofarads, and picofarads. (18.3)
- The voltage rating of a capacitor represents the maximum voltage that can be applied to the capacitor without causing damage to the dielectric. (18.3)
- Permittivity is the dielectric's ability to become polarized by an electric field. Relative permittivity is the permittivity of a substance when compared to the permittivity of a perfect vacuum. (18.4)
- Dielectric materials with low dielectric constants do not have the ability to hold an electric charge well. These materials typically make good insulators. (18.4)
- Electrons flow when a capacitor is either charging or discharging. When a capacitor is fully charged, there is no flow of electrons. (18.5)
- In a purely capacitive circuit, the current leads the voltage by 90 electrical degrees. (18.5)
- Capacitive reactance is the resistance to current flow that is produced by the electrostatic charge on the capacitor's plates. (18.6)
- The RC time constant is calculated as resistance multiplied by capacitance. (18.7)
- It takes five RC time constants to fully charge or fully discharge a capacitor. (18.7)
- When capacitors are wired in parallel, the total capacitance is the mathematical sum of all parallel-connected capacitors. (18.8)
- The total capacitance of a circuit made up of capacitors connected in series with each other is always lower than the capacitance of the lowest individual capacitor in the circuit. (18.9)
- The quality factor of a capacitor is equal to the capacitor's resistance divided by the capacitive reactance. (18.10)

Review Questions

Use the information in the chapter to answer the following questions.

Measuring Your Potential

1. A capacitor stores energy in the form of a(n) _____. (18.1)
   A. magnetic field
   B. electrostatic charge
   C. electromagnetic field
   D. electrokinetic charge
2. A(n) _____ can be connected in a circuit in either direction with no change in performance. (18.2)
   A. polarized capacitor
   B. non-polarized capacitor
   C. electrolytic capacitor
   D. bleed resistor
3. Of the following, which is *not* a factor that is used to determine capacitance? (18.2)
   A. Area of the capacitor's conductive plates
   B. Relative permittivity
   C. Distance between the plates
   D. Dielectric constant
4. Which of the following represents the lowest capacitance? (18.3)
   A. 1 μF
   B. 1000 pF
   C. 0.001 F
   D. 1,000,000 nF
5. Which of the following would result in an increase in a capacitor's capacitance? (18.4)
   A. Increasing the distance between the two conductive plates.
   B. Decreasing the area of the capacitor's plates.
   C. Replacing the dielectric with one with a higher dielectric constant.
   D. Exposing the capacitor to a lower temperature.

6. When a capacitor has a measurable voltage across it that is increasing, the capacitor is said to be _____. (18.5)
   A. charging
   B. fully charged
   C. discharging
   D. fully discharged
7. In a purely capacitive circuit, the current leads the voltage by _____. (18.5)
   A. 0°
   B. 45°
   C. 90°
   D. 180°
8. The stretching of the electron orbits within a dielectric material is called _____. (18.5)
   A. electromagnetic stress
   B. dielectric stress
   C. electromagnetic elasticity
   D. dielectric elasticity
9. Capacitive reactance is identified as _____. (18.6)
   A. X_R and is measured in volts
   B. X_C and is measured in ohms
   C. X_R and is measured in ohms
   D. X_C and is measured in volts
10. In a purely capacitive ac circuit, the waveform representing the capacitor's charge, in coulombs, is how far out of phase with the ac voltage applied to it? (18.6)
    A. 0°
    B. 45°
    C. 90°
    D. 180°
11. *True or False?* Capacitors charge and discharge at predictable rates. (18.7)
12. *True or False?* Lower-valued capacitors will typically have lower quality factors than similar capacitors with higher capacitance ratings. (8.10)

Plug-In

1. 8 µF = _____ F (18.3)
   A. 8,000,000
   B. 8000
   C. 0.008
   D. 0.000008
2. If four 80 µF capacitors are wired in parallel with each other, the total capacitance will be _____. (18.8)
   A. 20 µF
   B. 80 µF
   C. 160 µF
   D. 320 µF
3. If four 80 µF capacitors are wired in series with each other, the total capacitance will be _____. (18.9)
   A. 20 µF
   B. 80 µF
   C. 160 µF
   D. 320 µF
4. What is the total equivalent capacitance of a 5 µF capacitor wired in series with a 20 µF capacitor? (18.9)
   A. 2 µF
   B. 2.5 µF
   C. 4 µF
   D. 25 µF
5. Three capacitors, with values of 5 µF, 30 µF, and 60 µF, are wired in series with each other. The total equivalent capacitance will be closest to _____. (18.9)
   A. 5 µF
   B. 7 µF
   C. 50 µF
   D. 95 µF

Making the Connection

1. Explain why, in a purely capacitive circuit, the current flow is zero when the applied voltage is at its highest level. (18.3)
2. Explain how a capacitor can act as both a load and a power source. (18.5)
3. How will the quality factor of a capacitor be affected if the frequency of the power signal is changed? (18.10)

CHAPTER

19 Power in Capacitive and Inductive Circuits

drnadig/E+ via Getty Images

SPARKING DISCUSSION

A purely capacitive or purely inductive circuit generates reactive power, so the power factor of these circuits cannot be zero. Do you agree or disagree with this statement? Be prepared to provide support for your position.

LEARNING OUTCOMES

After completing this chapter, you will be able to:

19.1 Explain the concept of power as it applies to purely capacitive and purely inductive circuits.
19.2 Explain how reactive power is measured.
19.3 Calculate reactive power in a purely inductive circuit and a purely capacitive circuit.

TECHNICAL TERMS

apparent power (*S*)
power triangle
reactive power (*Q*)
true power (*P*)

Introduction

Having introduced capacitance and inductance in previous chapters, our focus now turns to the concepts of power and reactive power in purely capacitive and purely inductive alternating current (ac) circuits. In direct current (dc) circuits, power is calculated as the product of the circuit current multiplied by the circuit voltage. In ac circuits, however, the concept of power is more involved. In ac circuits, there are often three types of power: apparent power, true power, and reactive power.

Reactive power is a function of reactance, which is the opposition to current flow created in capacitive, inductive, and capacitive/inductive circuits. Capacitive reactance, X_C, is found in capacitive circuits. Inductive reactance, X_L, is primarily found in inductive circuits, although it does occur, in smaller quantities, in all ac circuits. This is due to the fact that a magnetic field is produced whenever current flows through a conductor. Capacitive reactance must be added by introducing capacitors to an ac system. Adding capacitors to the circuit helps increase the overall power factor of a system. Higher power factors help increase efficiency.

Inductive reactance and capacitive reactance have opposite effects on ac power. The electrical industry has found ways to add the appropriate amount of capacitance to a system or circuit to increase its power factor and efficiency. Transformers, fluorescent lighting, inductors, chokes, and asynchronous motors all increase the inductive reactance of a system. Underloaded three-phase motors are a huge contributor to reducing the power factor in commercial and industrial settings. In large factories, banks of capacitors might be installed to help correct objectionable power factors. As an electrician, knowing how to identify and help solve power factor problems is an important and valuable skill.

This chapter builds on previously discussed topics and concepts such as series-connected and parallel-connected inductors and capacitors, Ohm's law, and various forms of the power formula. You will perform circuit calculations on purely inductive and purely capacitive circuits to determine reactance and reactive power, along with other basic electric circuit characteristics including current and voltage.

19.1 Power in Purely Capacitive and Purely Inductive Circuits

Capacitors and inductors, classified as reactive devices, act as both circuit loads and voltage supplies in ac circuits. During half of the cycle these components store energy. During the other half of the cycle they release the stored energy to the circuit. As a result, the average power is, in effect, zero. However, capacitors and inductors have potential differences across them. As current flows through them, there is an element of power associated with them. This power, called *reactive power*, is often referred to as *lost*, *hidden*, or *phantom power*, because it performs no useful work in a circuit. Whenever possible, reactive power should be minimized. A circuit with a large amount of reactive power will have a lower power factor than a circuit with less reactive power.

The voltages and currents in dc circuits are, for the most part, constant. As a result, the power in the dc circuits is constant as well. In ac circuits, however, the instantaneous voltage and current values, which are the values at any given point in time, are constantly changing. Using constantly changing voltage and current inputs to calculate power results in a constantly changing result. Depending on the polarity of the instantaneous voltages and currents, as well as their polar relationships to each other, inductors and capacitors can act as both circuit loads and power sources. The function of the inductor or capacitor depends on the polarities of the current and voltage waveforms at various points in the cycle.

The polarity of the power waveform also relies on the polarities of the current and voltage waveforms. In a purely inductive or capacitive circuit, the power is positive when the polarities of the current and voltage are the same, either both positive or both negative. The power has a negative polarity when the current and voltage have opposite polarities, **Figure 19-1**. The frequency of the power waveform will be twice that of the current and voltage waveforms.

Math Sidebar

When the polarities of the current and voltage are the same, the product of the two values will always be positive. Multiplying two positive values together provides a positive result, as does multiplying two negative values together. For example, multiplying +5 and +3 equals +15, and multiplying −5 and −3 also equals +15. When current and voltage have opposite polarities, the product of the two values is negative. For example, multiplying +5 and −3 equals −15, and multiplying −5 and +3 also equals −15.

Power Polarity

Voltage	Current	Power
Positive (+)	Positive (+)	Positive (+)
Negative (−)	Negative (−)	Positive (+)
Positive (+)	Negative (−)	Negative (−)
Negative (−)	Positive (+)	Negative (−)

Goodheart-Willcox Publisher

Figure 19-1. Power has a negative polarity when current and voltage have opposite polarities; power has a positive polarity when current and voltage have the same polarities.

Purely Inductive Circuits

Consider the current and voltage waveforms of a purely inductive circuit shown in **Figure 19-2**. The voltage has a positive polarity from 0° to 180°, and a negative polarity from 180° to 360°. The current has a positive polarity from 90° to 270°, and a negative polarity from 0° to 90° and from 270° degrees to 360°. By examining this figure, the following can be concluded:

- Between 0° and 90°, the voltage and current polarities are opposite each other.
- Between 90° and 180°, the voltage and current polarities are the same.
- Between 180° and 270°, the voltage and current polarities are opposite each other.
- Between 270° and 360°, the voltage and current polarities are the same.

The circuit power has a positive polarity from 90° to 180° and from 270° to 360°. When the polarity of the power waveform is positive, the inductor acts as a circuit load and absorbs power from the circuit. The circuit power has a negative polarity from 0° to 90° and from 180° to 270°. When the polarity of the power waveform is negative, the inductor acts as a power supply and releases power back to the circuit.

The power waveform is illustrated in **Figure 19-3**. The power waveform for this purely inductive circuit is uniformly shaped with respect to the horizontal zero line, just as the voltage and current waveforms are. This indicates that an inductor releases the same amount of power to the circuit as it absorbs over time.

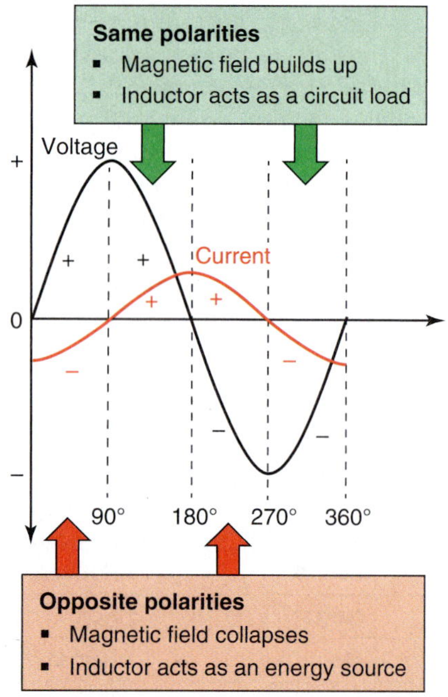

Goodheart-Willcox Publisher

Figure 19-2. When an inductor voltage and current have the same polarities, the inductor acts as a circuit load. When the inductor voltage and current have opposite polarities, the inductor acts as an energy source.

Purely Capacitive Circuits

Consider the current and voltage waveforms of a purely capacitive circuit shown in **Figure 19-4**. The voltage has a positive polarity from 0° to 180° and a negative polarity from 180° to 360°. The current has a positive polarity from 0° to 90° and from 270° to 360°, and a negative polarity from 90° to 270°.

By examining this figure, the following can be concluded:

- Between 0° and 90°, the voltage and current polarities are the same.
- Between 90° and 180°, the voltage and current polarities are opposite each other.
- Between 180° and 270°, the voltage and current polarities are the same.

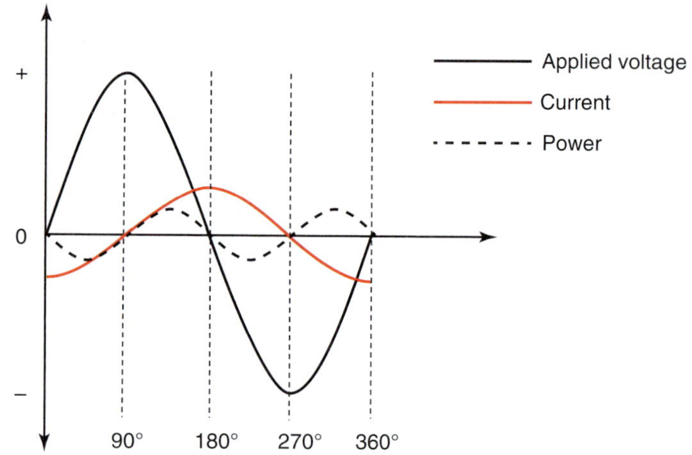

Goodheart-Willcox Publisher

Figure 19-3. Power waveform in a purely inductive circuit.

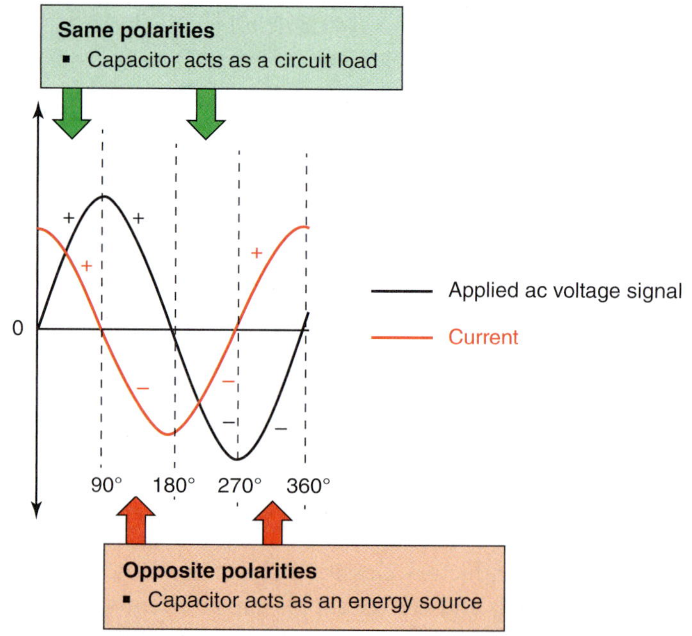

Goodheart-Willcox Publisher

Figure 19-4. When a capacitor voltage and current have the same polarities, the capacitor acts as a circuit load. When the capacitor voltage and current have opposite polarities, the capacitor acts as an energy source.

- Between 270° and 360°, the voltage and current polarities are opposite each other.

The circuit power has a positive polarity from 0° to 90° and from 180° to 270°. When the polarity of the power waveform is positive, the capacitor acts as a circuit load and absorbs power from the circuit. The circuit power has a negative polarity from 90° to 180° and from 270° to 360°. When the polarity of the power waveform is negative, the capacitor acts as a power supply and releases power back to the circuit.

The waveform of the power is illustrated in **Figure 19-5**. The power waveform for this purely capacitive circuit is uniformly shaped with respect to the horizontal zero line, just as the voltage and current waveforms are. This indicates that a purely capacitive circuit releases just as much power to the circuit as it absorbs over time.

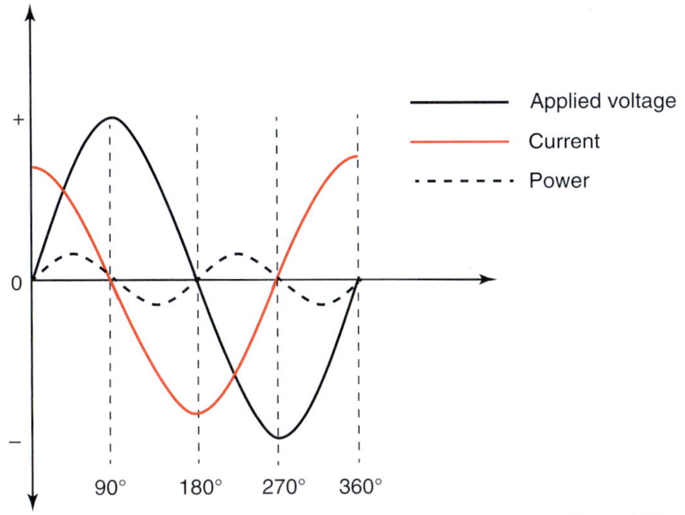

Goodheart-Willcox Publisher

Figure 19-5. Power waveform in a purely capacitive circuit.

Shedding Light

Since the power in a purely inductive or purely capacitive circuit is continually being absorbed and then returned to the circuit at the same rate, there is no true power associated with purely inductive or purely capacitive circuits. Another way of stating this is to say purely inductive or purely capacitive circuits have a power factor of zero.

19.2 Reactive Power

Reactive power refers to the energy that is stored and then released in the form of a magnetic field for inductors or an electrostatic charge for capacitors. ***Reactive power (Q)*** represents the power consumed by a circuit that does not actually perform any useful work. Because reactive power does not enhance circuit operation or function, reactive power is referred to as *wasted* or *wattless power.* Reactive power is produced whenever there is a phase shift between a circuit's voltage and current, **Figure 19-6**. Since a phase shift can only be experienced in ac circuits, reactive power is not found in dc circuits. In ideal, purely resistive ac circuits, there are no capacitors or inductors, so there is no phase shift. As a result, there is no reactive power associated with purely resistive ac circuits, **Figure 19-7**.

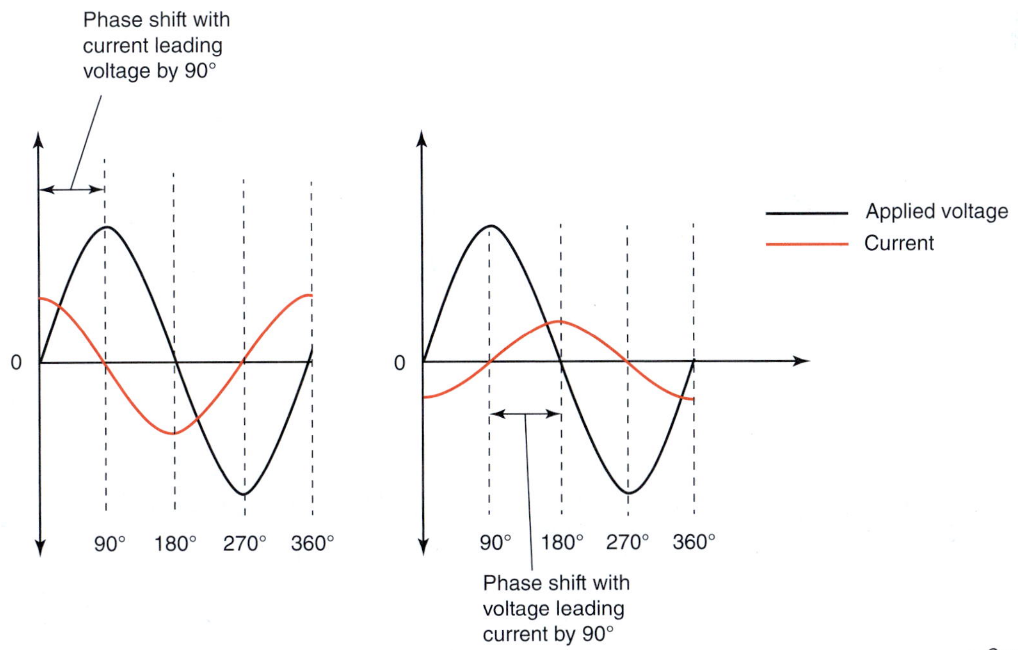

Goodheart-Willcox Publisher

Figure 19-6. Reactive power refers to the energy that is stored and then released in the form of a magnetic field for inductors or an electrostatic charge for capacitors. Reactive power causes a phase shift between voltage and current waveforms, as shown here.

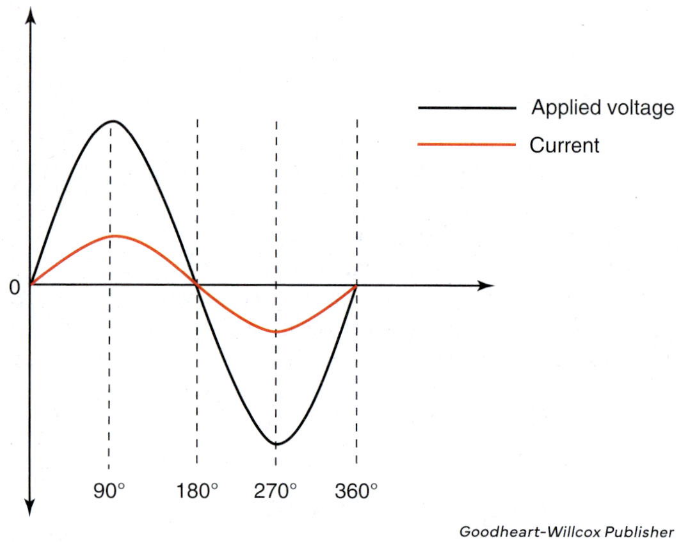

Figure 19-7. No reactive power is produced when the voltage and current are in phase with each other.

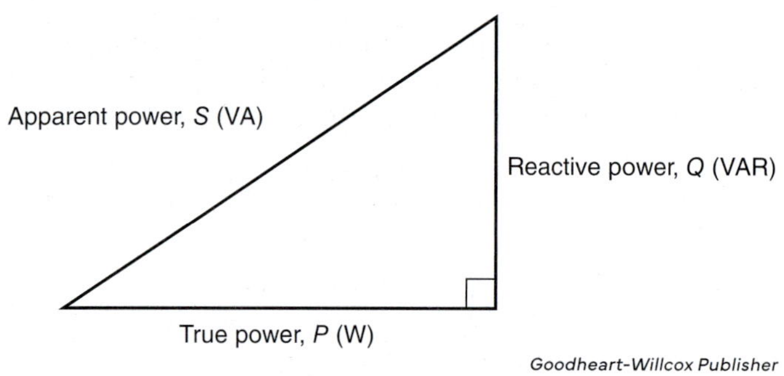

Figure 19-8. A power triangle is a useful tool to calculate true, reactive, and apparent power in an ac circuit.

Apparent power** (S)* is obtained when a circuit's applied voltage is multiplied by the circuit's amperage. Apparent power is expressed in volt-amperes, or VA. The apparent power of a circuit is only equal to the actual power consumed by a circuit when the voltage and current are in phase with each other. ***True power** (P)*, also known as real or *active power*, is the actual power consumed by the circuit. True power is expressed in watts (W) and is the product of the circuit's current, voltage, and power factor. In ac circuits where there is a phase shift between the current and voltage, the power factor will be less than one and the true power will be less than the apparent power. Apparent power, true power, and reactive power can be represented by a right triangle where apparent power is represented by the triangle's hypotenuse and true and reactive power are represented by the other two sides of the triangle. This right triangle is also known as the ***power triangle, **Figure 19-8.**

Units and Measurements

Although there are differences in the behavior of power in an inductive or capacitive circuit when compared to dc and resistive circuits, the power in an inductive or capacitive circuit can be calculated in a similar way to how power is calculated in purely resistive circuits. The main difference is that power in an inductive or capacitive circuit is referred to as *reactive power*, Q. Reactive power is expressed in units called volt-amperes reactive, VAR, or kilovolt-amperes reactive, kVAR, where 1 kVAR is equal to 1000 VAR. For VAR calculations, the basic power formulas are modified slightly:

- Instead of power (P), Q is used to represent reactive power. To represent reactive power in a purely inductive circuit, Q_L is used. To represent reactive power in a purely capacitive circuit, Q_C is used.
- The voltage is identified as E_L for the voltage supplied to an inductive circuit, or E_C for the voltage supplied to a capacitive circuit. The terms E or E_T are used to represent the total circuit voltage.
- The current flow through an inductor is identified as I_L and I_C represents the current flow in a capacitive circuit.
- Instead of resistance, R, the inductive reactance, either X_L or X_C, is used.

The three forms of the reactive power (Q) formula for purely inductive circuits are:

$$Q_L = I_L \times E_L$$
$$Q_L = I_L^2 \times X_L$$
$$Q_L = \frac{E_L^2}{X_L}$$

The three forms of the reactive power (Q) formula for purely capacitive circuits are:

$$Q_C = I_C \times E_C$$
$$Q_C = I_C^2 \times X_C$$
$$Q_C = \frac{E_C^2}{X_C}$$

The total reactive power of a circuit is equal to the sum of the individual reactive powers produced by each circuit component. This can be represented mathematically as follows:

$$Q_T = Q_1 + Q_2 + Q_3 + \ldots + Q_N$$

In this equation, Q_T is the total reactive power of a circuit and Q_1 to Q_N is the reactive power across all circuits.

19.3 Calculating Reactive Power

In order to calculate reactive power, the reactance of a purely inductive or purely capacitive circuit may need to be calculated. Chapter 17, *Inductors and Inductance* explains how to calculate inductive reactance, and Chapter 18, *Capacitors and Capacitance* explains how to calculate capacitive reactance.

As a reminder, the formula for inductive reactance, X_L, is as follows:

$$X_L = 2\pi f L$$

where:

X_L = Inductive reactance (Ω)
2π = Constant equal to 2×3.14, or 6.28
f = Frequency of the power supply (Hz)
L = Inductance of the circuit (H)

The formula for capacitive reactance, X_C, is as follows:

$$X_C = \frac{1}{2\pi \times f \times C}$$

where:

2π = Constant equal to 2×3.14, or 6.28
f = Frequency of the power supply (Hz)
C = Capacitance of the capacitor (F)

The following **Examples** will provide the steps on how to calculate reactive power of purely inductive circuits and purely capacitive circuits.

Example 19-1

What is the reactive power of a purely inductive circuit if the inductive reactance is 57 Ω and the voltage supplied to the circuit is 100 Vac?

✓ Solution

With the voltage supplied to the circuit and the inductive reactance known, the reactive power, Q_L, for this purely inductive circuit can be calculated as follows:

$$Q_L = \frac{E_L^2}{X_L}$$

$$Q_L = \frac{(100 \text{ V})^2}{57\ \Omega}$$

$$Q_L = \frac{10{,}000}{57}$$

$$Q_L = 175.4 \text{ VAR}$$

Now It's Your Turn 19-1

What is the reactive power of an inductor that has an applied voltage of 100 V and has a current of 1.2 A flowing through it?

Example 19-2

In the following 150 V, 60 Hz, 1.6 H purely inductive circuit, determine the reactive power and the total current flow.

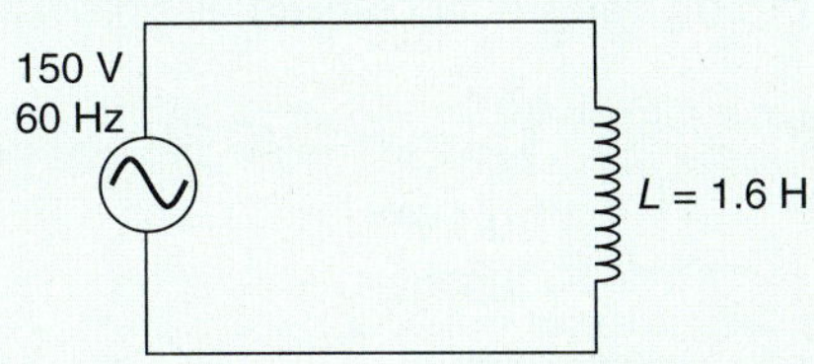

Goodheart-Willcox Publisher

✓ Solution

The first step is to determine the inductive reactance for the circuit as follows:

$$X_L = 2\pi \times f \times L$$

$$X_L = 6.28 \times 60 \text{ Hz} \times 1.6 \text{ H}$$

$$X_L = 602.88\ \Omega$$

With the voltage supplied to the circuit and the inductive reactance known, then the reactive power, Q_L, for this circuit can be calculated as follows:

$$Q_L = \frac{E^2}{X_L}$$

$$Q_L = \frac{(150 \text{ V})^2}{602.88\ \Omega}$$

$$Q_L = \frac{22{,}500}{602.88}$$

$$Q_L = 37.32 \text{ VAR}$$

Next, the current flow in the circuit can be calculated as follows:

$$I_L = \frac{E}{X_L}$$

$$I_L = \frac{150 \text{ V}}{602.88\ \Omega}$$

$$I_L = 0.25 \text{ A}$$

You can check the accuracy of your calculations by using the other forms of the Q_L formula. For example, using the formula $Q_L = I_L \times E_L$ and substituting the known values for I_L and E_L, the value for Q_L can be determined as follows:

$$Q_L = I_L \times E_L$$

$$Q_L = 0.25 \text{ A} \times 150 \text{ V}$$

$$Q_L = 37.5 \text{ VAR}$$

Similarly, using the formula $Q_L = I_L^2 \times X_L$ and substituting the known values for I_L and X_L, the value for Q_L can also be determined as follows:

$$Q_L = I_L^2 \times X_L$$

$$Q_L = (0.25 \text{ A})^2 \times 602.88\ \Omega$$

$$Q_L = 0.0625 \times 602.88\ \Omega$$

$$Q_L = 37.68 \text{ VAR}$$

Using any of the three formulas for Q_L will provide the same result, although small differences will be noticed due to rounding during the calculations.

Example 19-3

What is the reactive power of a purely capacitive circuit with a total capacitance of 35 μF if the frequency of the power supply is 80 Hz and the circuit current flow is 1.35 A?

✓ Solution

First, convert 35 μF to 0.000035 F. Then determine the capacitive reactance for the circuit as follows:

$$X_C = \frac{1}{2\pi \times f \times C}$$

$$X_C = \frac{1}{6.28 \times 80\text{ Hz} \times 0.000035\text{ F}}$$

$$X_C = \frac{1}{0.01758}$$

$$X_C = 56.9\ \Omega$$

With the circuit current and the capacitive reactance known, then reactive power, Q_C, for this circuit can be calculated as follows:

$$Q_C = I_C^2 \times X_C$$

$$Q_C = (1.35\text{ A})^2 \times 56.9\ \Omega$$

$$Q_C = 1.8225 \times 56.9$$

$$Q_C = 103.7\text{ VAR}$$

Now It's Your Turn 19-3

What is the reactive power of a capacitor with a capacitance reactance of 120 Ω if it is supplied with a voltage of 85 V?

Example 19-4

Referring to the following 100 V, 60 Hz capacitive circuit, determine the reactive power and the current flow of the circuit.

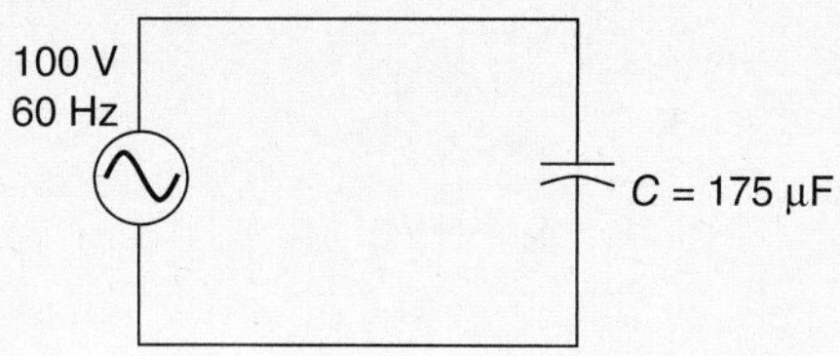

Goodheart-Willcox Publisher

✓ Solution

First, convert 175 μF to 0.000175 F. Then determine the capacitive reactance for the circuit as follows:

$$X_C = \frac{1}{2\pi \times f \times C}$$

$$X_C = \frac{1}{6.28 \times 60\text{ Hz} \times 0.000175\text{ F}}$$

$$X_C = \frac{1}{0.0659}$$

$$X_C = 15.17\ \Omega$$

With the voltage supplied to the circuit and the capacitive reactance known, then the reactive power Q_C for this circuit can be calculated as follows:

$$Q_C = \frac{E^2}{X_C}$$

$$Q_C = \frac{(100\text{ V})^2}{15.17\ \Omega}$$

$$Q_C = \frac{10{,}000}{15.17\ \Omega}$$

$$Q_C = 659.2\text{ VAR}$$

Next, the current flow in the circuit can be calculated as follows:

$$I_C = \frac{E}{X_C}$$

$$I_C = \frac{100\text{ V}}{15.17\ \Omega}$$

$$I_C = 6.59\text{ A}$$

Example 19-5

Determine the reactive power of the following capacitive 240 V, 60 Hz circuit.

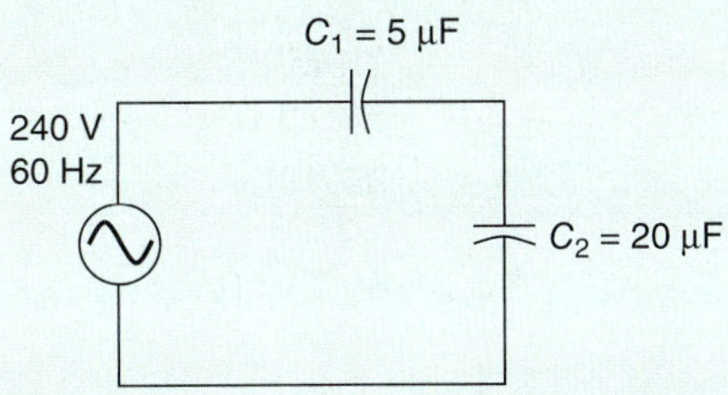

Goodheart-Willcox Publisher

✓ Solution

The first step to determine the reactive power is to determine the total capacitive reactance for the circuit. As this particular circuit has two capacitors wired in series with each other, the product over the sum formula can be used to calculate the total capacitive reactance as follows:

$$C_T = \frac{C_1 \times C_2}{C_1 + C_2}$$

$$C_T = \frac{5\ \mu F \times 20\ \mu F}{5\ \mu F + 20\ \mu F}$$

$$C_T = \frac{100\ \mu F^2}{25\ \mu F}$$

$$C_T = 4\ \mu F$$

Next, calculate reactive power as follows:

$$X_{CT} = \frac{1}{2\pi \times f \times C_T}$$

$$X_{CT} = \frac{1}{6.28 \times 60\ Hz \times 0.000004\ F}$$

$$X_{CT} = \frac{1}{0.00151}$$

$$X_{CT} = 662.3\ \Omega$$

With the voltage supplied to the circuit and the total capacitive reactance known, now the total reactive power Q_C for this circuit can be calculated as follows:

$$Q_C = \frac{E^2}{X_{CT}}$$

$$Q_C = \frac{(240\ V)^2}{662.3\ \Omega}$$

$$Q_C = \frac{57{,}600}{662.3}$$

$$Q_C = 87.0\ VAR$$

Now It's Your Turn 19-5

Determine the reactive power of the following 240 V, 60 Hz circuit.

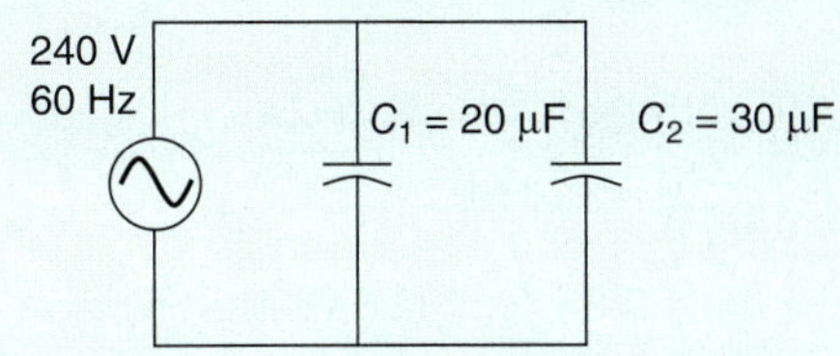

Goodheart-Willcox Publisher

Example 19-6

Referring to the 240 V, 60 Hz circuit from **Example 19-5**, determine the reactive power associated with each capacitor as well as the measured voltage across each capacitor. Note: In the circuit, $C_1 = 5\ \mu F$ and $C_2 = 20\ \mu F$.

✓ Solution

In **Example 19-5**, it was determined that the total capacitive reactance, X_{CT}, for this circuit is 662.3 Ω. The total circuit current can be determined as follows:

$$I_T = \frac{E}{X_{CT}}$$

$$I_T = \frac{240\ V}{662.3\ \Omega}$$

$$I_T = 0.36\ A$$

The capacitive reactance for each of the capacitors can be found using the capacitance reactance formula for each capacitor. For C_1, which has a capacitance of 5 µF, the capacitive reactance is calculated as follows:

$$X_{C1} = \frac{1}{2\pi \times f \times C_1}$$

$$X_{C1} = \frac{1}{6.28 \times 60\ Hz \times 0.000005\ F}$$

$$X_{C1} = \frac{1}{0.001884}$$

$$X_{C1} = 530.8\ \Omega$$

Continued

Now, calculate the capacitive reactance for C_2, as follows:

$$X_{C2} = \frac{1}{2\pi \times f \times C_2}$$

$$X_{C2} = \frac{1}{6.28 \times 60 \text{ Hz} \times 0.00002 \text{ F)}}$$

$$X_{C2} = \frac{1}{0.00754}$$

$$X_{C2} = 132.6\ \Omega$$

With the total circuit current and the individual capacitive reactance values known, the measured voltage across each capacitor can be calculated. For C_1, the voltage is calculated as:

$$E_{C1} = I_T \times X_{C1}$$

$$E_{C1} = 0.36 \text{ A} \times 530.8\ \Omega$$

$$E_{C1} = 191 \text{ V}$$

For C_1, the voltage is calculated as:

$$E_{C2} = I_T \times X_{C2}$$

$$E_{C2} = 0.36 \text{ A} \times 132.6\ \Omega$$

$$E_{C2} = 47.74 \text{ V}$$

The sum of these two voltage values is 238.74 V, which can be rounded up to 240 V. The difference between these two values is a result of rounding in the calculations.

Next, calculate reactive power. The reactive power for C_1 is calculated as follows:

$$Q_{C1} = \frac{E_{C1}^2}{X_{C1}}$$

$$Q_{C1} = \frac{(191 \text{ V})^2}{530.8\ \Omega}$$

$$Q_{C1} = \frac{36{,}481}{530.8}$$

$$Q_{C1} = 68.7 \text{ VAR}$$

Then the reactive power for C_2 is calculated as follows:

$$Q_{C2} = \frac{E_{C2}^2}{X_{C2}}$$

$$Q_{C2} = \frac{(47.74 \text{ V})^2}{132.6\ \Omega}$$

$$Q_{C2} = \frac{2279}{132.6\ \Omega}$$

$$Q_{C2} = 17.2 \text{ VAR}$$

Note: The reactive power for each capacitor can also be determined by using one of the following formulas:

$$Q_C = I_T^2 \times X_C$$

$$Q_C = I_T \times E_C$$

Now It's Your Turn 19-6

Referring to the 240 V, 60 Hz circuit from *Now It's Your Turn* in **Example 19-5**, determine the reactive power associated with each capacitor and the current in each branch. Note: In this circuit, C_1 = 20 μF and C_2 = 30 μF.

Example 19-7

Determine the reactive power of the following 200 V, 60 Hz purely inductive circuit.

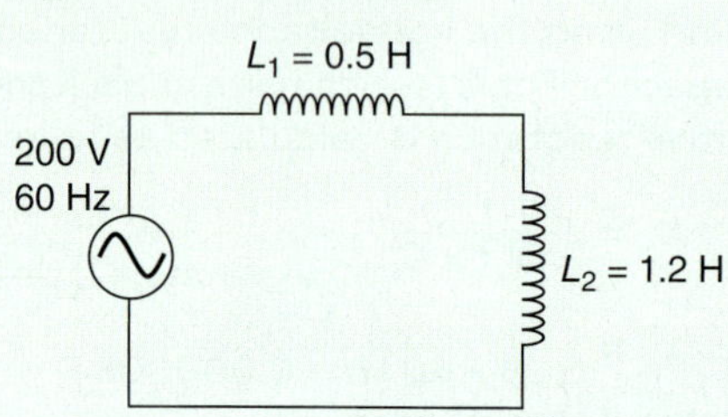

Goodheart-Willcox Publisher

✓ Solution

The first step to determine the reactive power is to determine the total inductive reactance for the circuit. For two inductors wired in series with each other, inductive reactance can be calculated as follows:

$$L_T = L_1 + L_2$$

$$L_T = 0.5 \text{ H} + 1.2 \text{ H}$$

$$L_T = 1.7 \text{ H}$$

Next, calculate the total inductive reactance as follows:

$$X_{LT} = 2\pi \times f \times L_T$$
$$X_{LT} = 6.28 \times 60\ \text{Hz} \times 1.7\ \text{H}$$
$$X_{LT} = 640.56\ \Omega$$

With the voltage supplied to the circuit and the total inductive reactance known, the total reactive power Q_L for this circuit can be calculated as follows:

$$Q_L = \frac{E^2}{X_{LT}}$$
$$Q_L = \frac{(200\ \text{V})^2}{640.56\ \Omega}$$
$$Q_L = \frac{40{,}000}{640.56}$$
$$Q_L = 62.44\ \text{VAR}$$

Now It's Your Turn 19-7

Determine the total reactive power of the following 200 V, 60 Hz circuit.

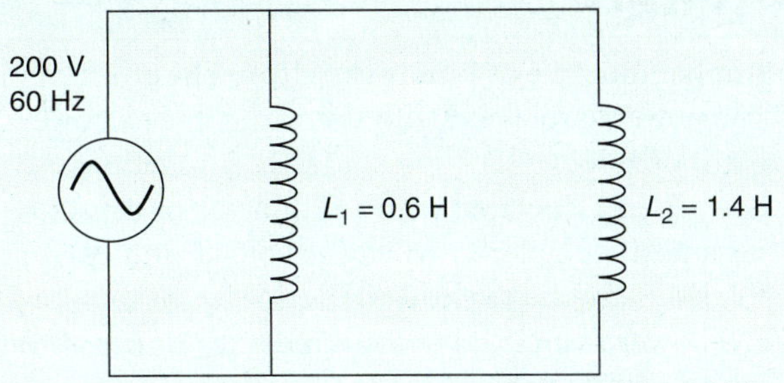

Goodheart-Willcox Publisher

Example 19-8

Referring to the 200 V, 60 Hz purely inductive circuit from **Example 19-7**, determine the reactive power associated with each inductor. Note: In this circuit, L_1 = 0.5 H and L_2 = 1.2 H.

✓ Solution

From **Example 19-7**, it was determined that the total inductive reactance for this circuit is 640.56 Ω. The total circuit current can be calculated as follows:

$$I_T = \frac{E}{X_{LT}}$$
$$I_T = \frac{200\ \text{V}}{640.56\ \Omega}$$
$$I_T = 0.312\ \text{A}$$

Next, the inductive reactance for each of the inductors will need to be determined. First, the inductive reactance for L_1 is calculated as follows:

$$X_{L1} = 2\pi \times f \times L_1$$
$$X_{L1} = 6.28 \times 60\ \text{Hz} \times 0.5\ \text{H}$$
$$X_{L1} = 188.4\ \Omega$$

Next, the inductive reactance for L_2 is calculated as follows:

$$X_{L2} = 2\pi \times f \times L_2$$
$$X_{L2} = 6.28 \times 60\ \text{Hz} \times 1.2\ \text{H}$$
$$X_{L2} = 452.16\ \Omega$$

With the total circuit current and the individual inductive reactance values known, the reactive power associated with each inductor can be calculated. The reactive power for L_1 is calculated as follows:

$$Q_{L1} = I^2 \times X_{L1}$$
$$Q_{L1} = (0.312\ \text{A})^2 \times 188.4\ \Omega$$
$$Q_{L1} = 0.097 \times 188.4$$
$$Q_{L1} = 18.33\ \text{VAR}$$

The reactive power for L_2 is calculated as follows:

$$Q_{L2} = I^2 \times X_{L2}$$
$$Q_{L2} = (0.312\ \text{A})^2 \times 452.16\ \Omega$$
$$Q_{L2} = 0.0973 \times 452.16$$
$$Q_{L2} = 44.00\ \text{VAR}$$

Now It's Your Turn 19-8

Referring to the 200 V, 60 Hz circuit from the *Now It's Your Turn* in **Example 19-7**, determine the reactive power associated with each inductor as well as the current flow through each inductor. Note: In this circuit, L_1 = 0.6 H and L_2 = 1.4 H.

CHAPTER 19 REVIEW

Summary

- Purely inductive or purely capacitive circuits have no true power associated with them and have a power factor of zero. (19.1)
- The power associated with an inductor or capacitor has a positive polarity when the voltage and current polarities are the same. (19.1)
- The power associated with an inductor or capacitor has a negative polarity when the voltage and current polarities are opposite. (19.1)
- The only power produced by purely inductive or purely capacitive circuits is reactive power. (19.2)
- The reactive power generated by a purely inductive or purely capacitive circuit is measured in units called volt-amperes reactive (VAR). (19.2)
- In order to calculate reactive power, the reactance of a purely inductive or purely capacitive circuit may need to be calculated or known. (19.3)

Review Questions

Use the information in the chapter to answer the following questions.

Measuring Your Potential

1. *True or False?* In a purely inductive circuit, the power is positive when the polarities of the current and voltage are both negative. (19.1)
2. Which of the following statements is true regarding the power in a purely inductive or a purely capacitive circuit? (19.1)
   A. The reactive power is equal to the voltage times the inductive reactance.
   B. The power waveform has a frequency that is one-half that of the voltage.
   C. The reactive power is equal to the current times the inductive reactance.
   D. When the polarity of the power waveform is positive, the inductor acts as a circuit load and absorbs power from the circuit.
3. In an inductor, _____ refers to the energy that is stored and then released in the form of a magnetic field. (19.2)
   A. power factor
   B. apparent power
   C. reactive power
   D. true power
4. In a power triangle, the hypotenuse represents _____. (19.2)
   A. apparent power
   B. true power
   C. reactive power
   D. power factor
5. A reactive power of 10.5 kVAR is equal to _____. (19.2)
   A. 105,000 VAR
   B. 10,500 VAR
   C. 1500 VAR
   D. 105 VAR
6. Of the following, which formula is used to calculate inductive reactance? (19.3)
   A. $X_L = 2\pi fL$
   B. $X_C = \frac{1}{2\pi fC}$
   C. $Q_T = Q_1 + Q_2 + Q_3$
   D. $Q_L = I_L \times E_L$
7. The formula used to calculate reactive power for a purely capacitive circuit is _____. (19.3)
   A. $Q_L = I_L \times E_L$
   B. $Q_T = Q_1 + Q_2 + Q_3$
   C. $Q_C = I_C \times E_C$
   D. $X_C = \frac{1}{2\pi fC}$

Plug-In

Refer to the following circuit to calculate the answers for questions 1 to 5.

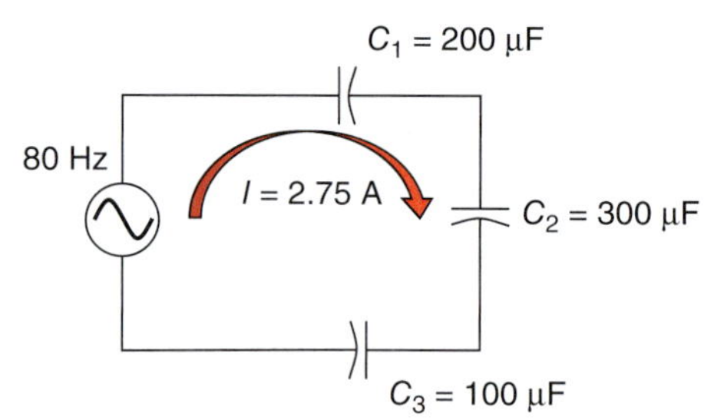

Goodheart-Willcox Publisher

1. The total equivalent capacitance for this circuit is closest to _____. (19.3)
   A. 35 μF
   B. 45 μF
   C. 55 μF
   D. 65 μF

2. The total capacitive reactance for this circuit is closest to _____. (19.3)
   A. 21 Ω
   B. 37 Ω
   C. 53 Ω
   D. 74 Ω
3. The total reactive power, Q_C, for this circuit is closest to _____. (19.3)
   A. 275 VAR
   B. 375 VAR
   C. 475 VAR
   D. 575 VAR
4. The supply voltage for this circuit is closest to _____. (19.3)
   A. 50 V
   B. 100 V
   C. 150 V
   D. 200 V
5. The reactive power associated with the 300 μF capacitor is closest to _____. (19.3)
   A. 75 VAR
   B. 150 VAR
   C. 225 VAR
   D. 300 VAR

Refer to the following circuit to calculate the answers for questions 6 to 9.

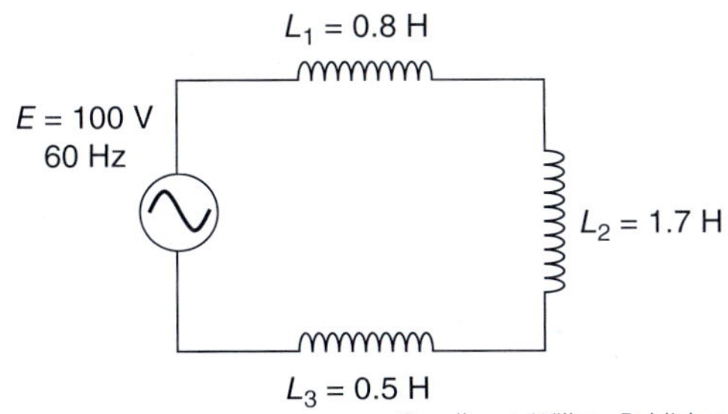

Goodheart-Willcox Publisher

6. The total inductive reactance for this circuit is closest to _____. (19.3)
   A. 0.25 H
   B. 1.0 H
   C. 3.0 H
   D. 6.5 H
7. The total reactive power for this circuit is closest to _____. (19.3)
   A. 4 VAR
   B. 9 VAR
   C. 15 VAR
   D. 21 VAR
8. The total circuit current is closest to _____. (19.3)
   A. 0.1 A
   B. 0.5 A
   C. 0.9 A
   D. 1.5 A
9. The reactive power associated with the 1.7 H inductor is closest to _____. (19.3)
   A. 1 VAR
   B. 5 VAR
   C. 10 VAR
   D. 20 VAR

Making the Connection

1. Why is the frequency of the power waveform in an inductive circuit twice that of the voltage or current waveform? (19.1)
2. Explain why purely capacitive and purely inductive circuits have no true power associated with them. (19.1)
3. Why does an increase in reactive power result in an increase in apparent power and a decrease in power factor? (19.1)
4. Explain how adding capacitors to otherwise inductive circuits could reduce the reactive power produced by the circuit. (19.2)

CHAPTER

20 Resistive, Capacitive, and Inductive Circuits

Amorn Suriyan/Shutterstock.com

SPARKING DISCUSSION

Talk about situations where the electrical contractor may be asked to help correct a bad power factor condition in an industrial occupancy. How would the electrical contractor determine the existing power factor? What steps could be taken to remedy this situation?

LEARNING OUTCOMES

After completing this chapter, you will be able to:

20.1 Define the term impedance as it is applied to the electrical industry.

20.2 Evaluate the impedance in series R-L, R-C, R-L-C, and L-C circuits.

20.3 Evaluate the impedance in parallel R-L, R-C, R-L-C, and L-C circuits.

20.4 Calculate the current flow in series R-L, R-C, R-L-C, and L-C circuits.

20.5 Calculate the voltage drops in series R-L, R-C, R-L-C, and L-C circuits.

20.6 Calculate the current flow in parallel R-L, R-C, R-L-C, and L-C circuits.

20.7 Calculate the power factor and phase angles of series and parallel R-L, R-C, and R-L-C circuits.

TECHNICAL TERMS

impedance (*Z*)
lagging phase angle
leading phase angle

Introduction

Up to this point, our discussions covered purely resistive, inductive, and capacitive circuits. Purely resistive circuits have resistance, but no reactance, while purely reactive circuits, either capacitive or inductive, have reactance, but no resistance. In purely capacitive circuits, the current leads the voltage by 90°, and in purely inductive circuits, the current lags the voltage by 90°. Unlike these ideal circuits, in the real world many of the circuits encountered include combinations of resistance, capacitance, and inductance.

Having practiced evaluating purely resistive, inductive, and capacitive circuits, you are now ready to evaluate more complicated circuits. The first important step in evaluating these circuits is to understand the concept of impedance, which is the total opposition to current flow in a circuit. In this chapter you will evaluate series and parallel combinations of resistive-inductive (R-L) circuits, resistive-capacitive (R-C) circuits, inductive-capacitive (L-C) circuits, and resistive-inductive-capacitive (R-L-C) circuits. Many of the concepts you will use to calculate these electrical characteristic values are the same ones that are discussed in previous chapters. It is recommended that you refer back to these chapters to review these concepts as a reference if needed.

20.1 Impedance, Resistance, and Reactance

Impedance (Z) represents the total opposition force to current flow in a circuit, and is a function of resistance and reactance. Resistance is a consistent force that acts to impede current flow, while reactance is a variable opposition to current flow. For this reason, resistors are referred to as nonreactive devices, while inductors and capacitors are classified as reactive devices.

Purely resistive circuits have resistance, but not reactance. Purely inductive and purely capacitive circuits have reactance, but no resistance. With impedance being a combination of resistance and reactance, all electrical components, devices, and circuits have impedance. Although the source of the current-opposing force can vary, impedance, reactance, and resistance are all expressed using the same units, namely ohms (Ω).

The method used to calculate impedance varies depending on the circuit components and how they are connected. Impedance calculations, however, have one thing in common—they are not performed by simply adding the reactance and resistance values. For resistive loads, the voltage (E_R) and the current are in phase with each other, **Figure 20-1**. For inductive loads, the current lags the voltage (E_L) by 90°, **Figure 20-2**. For capacitive loads, the current leads the voltage (E_C) by 90°, **Figure 20-3**. Although resistance and reactance are both expressed in ohms, they cannot be added to each other. A summary of resistance, reactance, and phase angles for resistors, inductors, and capacitors is provided in **Figure 20-4**.

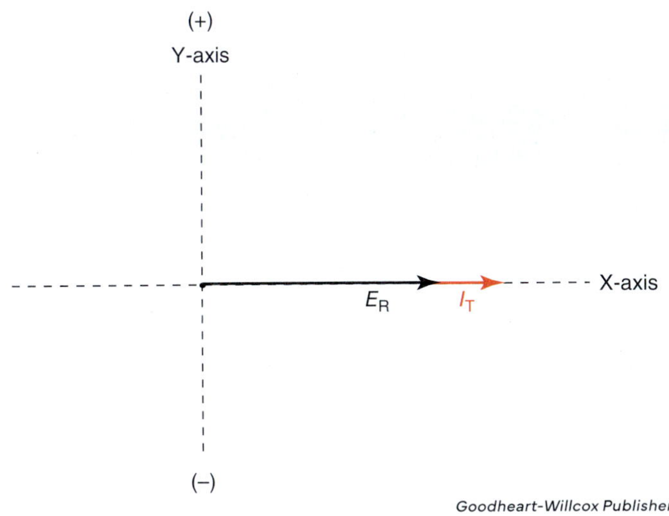

Figure 20-1. In a purely resistive circuit, the voltage and current are in phase with each other.

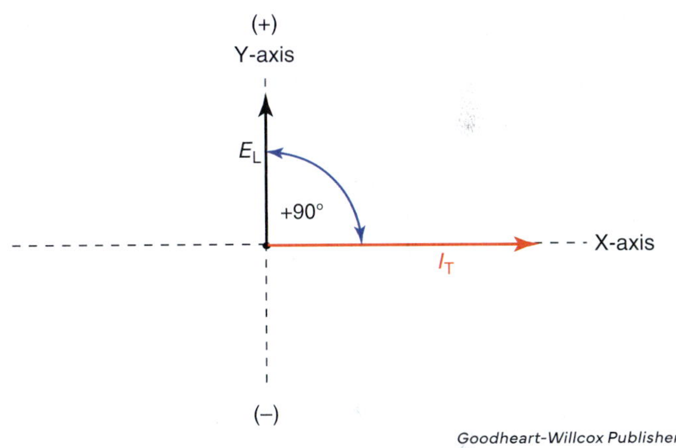

Figure 20-2. In a purely inductive circuit, the current lags the voltage by 90°.

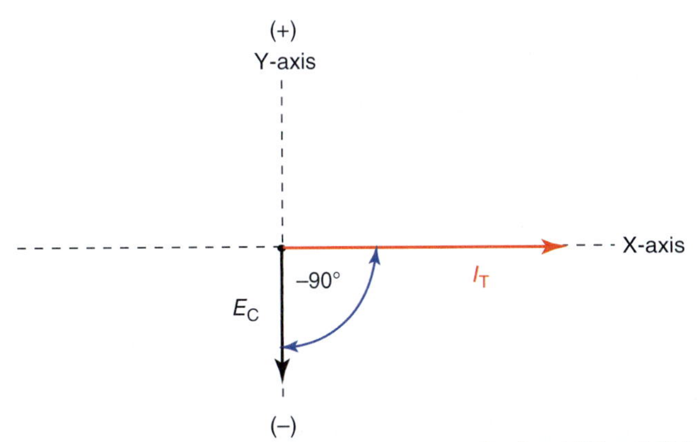

Figure 20-3. In a purely capacitive circuit, the current leads the voltage by 90°.

Resistor, Inductor, and Capacitor Properties

Circuit Element	Resistance	Reactance	Phase Angle
Resistor	R	0	0°
Inductor	0	$X_L = 2\pi \times f \times L$	+90°
Capacitor	0	$X_C = \frac{1}{2\pi} \times f \times C$	−90°

Goodheart-Willcox Publisher

Figure 20-4. Summary of resistance, reactance, and phase angles for resistors, inductors, and capacitors.

20.2 **Impedance in Series Circuits**

We will continue our discussion of impedance by examining simple series circuits made up of various combinations of resistors, inductors, and capacitors. The four basic circuit types include the following:

- Series resistive-inductive (R-L) circuits
- Series resistive-capacitive (R-C) circuits
- Series inductive-capacitive (L-C) circuits
- Series resistive-inductive-capacitive (R-L-C) circuits

These are the building blocks toward total understanding of the different factors impacting ac circuits and how changing these factors will impact the efficiency of a system. Inefficient systems with a less than desirable power factor are more costly to operate.

Impedance in Series R-L and R-C Circuits

In a series circuit made up of resistance and reactance, the resistance and the reactance are 90° out of phase with each other. In a series circuit with resistance and inductive reactance, the inductive reactance, X_L, points in the positive direction along the Y-axis, just as the leading voltage does in this type of circuit. In a series circuit with resistance and capacitive reactance, the capacitive reactance, X_C, points in the negative direction along the Y-axis, just as the lagging voltage does for this type of circuit, **Figure 20-5**.

In a series circuit made up of resistance and inductive reactance, the impedance is calculated using vector addition. Arranging the tail of the inductive reactance vector to the head of the resistance vector, the impedance vector can be found by connecting the tail of the resistance vector to the head of the inductive reactance vector, creating a right triangle, **Figure 20-6**.

As with any right triangle, the relationship among the lengths of the sides of the triangle can be expressed by the Pythagorean theorem. Right triangles and the Pythagorean theorem are explained in Chapter 14, *Angles, Triangles, and Vectors*. If the resistance and inductive reactance values are known, the impedance can be calculated. The hypotenuse of the right triangle represents the impedance (Z), and the other two sides of the triangle represent resistance (R) and inductive reactance (X_L). To calculate the impedance of a series R-L circuit, the Pythagorean theorem becomes:

$$Z = \sqrt{R^2 + X_L^2}$$

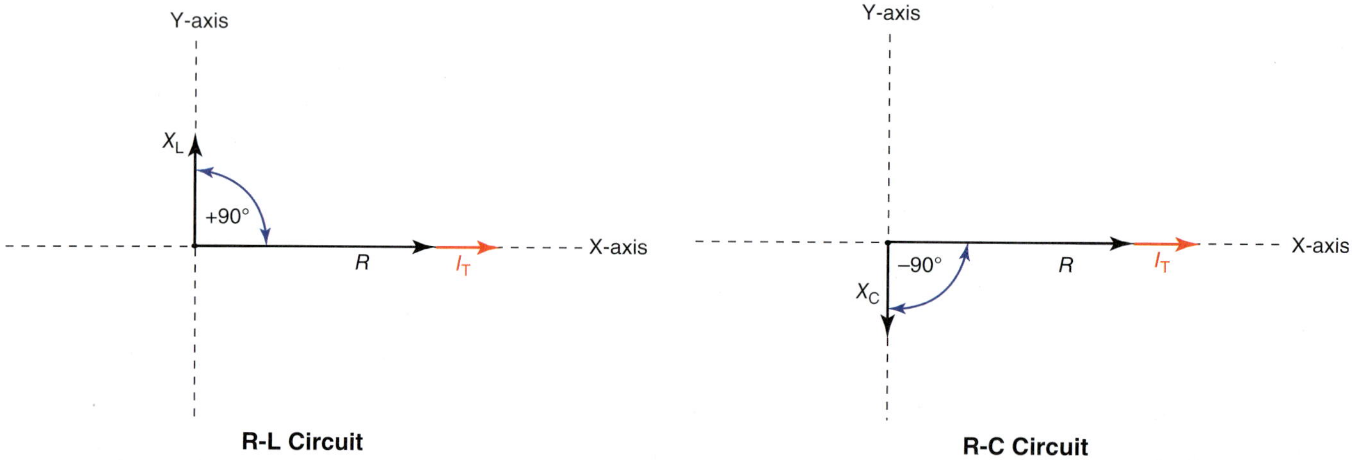

Goodheart-Willcox Publisher

Figure 20-5. In an R-L circuit, the inductive reactance leads the resistance by 90°; inductive reactance points in a positive direction. In an R-C circuit, the resistance leads the capacitive reactance by 90°; the capacitive reactance points in a negative direction.

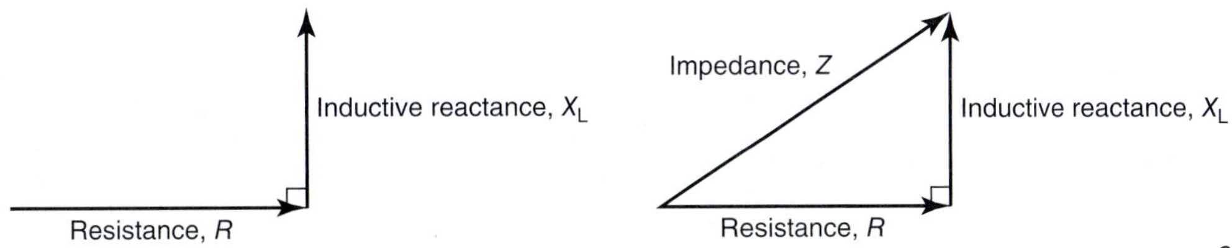

Goodheart-Willcox Publisher

Figure 20-6. In a series R-L circuit, the resistance and inductive reactance are added as vectors. The resultant of this vector addition is the impedance vector.

Refer to **Example 20-1** on how to use this formula to calculate the impedance of a series R-L circuit.

In a series circuit made up of resistance and capacitive reactance, the impedance is calculated using the same formula that was used to calculate the impedance of a series R-L circuit. The only difference is that the X_L for inductive reactance is replaced with X_C for capacitive reactance, as follows:

$$Z = \sqrt{R^2 + X_C^{\,2}}$$

Example 20-1

Calculate the impedance of the following series R-L circuit that has a resistor with a value of 45 Ω that is connected in series with an inductor that has an inductive reactance of 30 Ω.

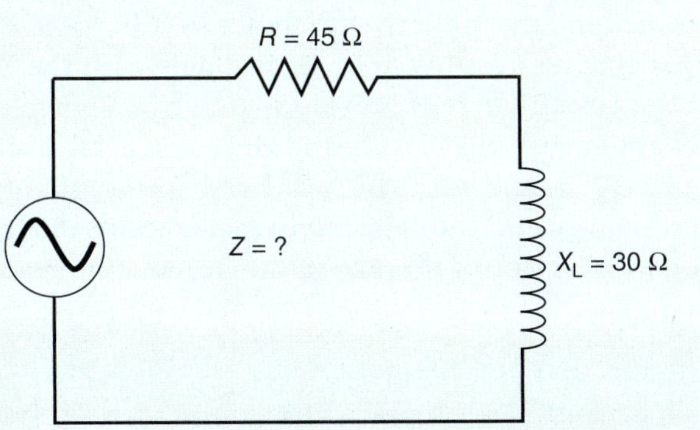

Goodheart-Willcox Publisher

✓ Solution

The circuit impedance can be calculated as follows:

$$Z = \sqrt{R^2 + X_L^{\,2}}$$

$$Z = \sqrt{(45\ \Omega)^2 + (30\ \Omega)^2}$$

$$Z = \sqrt{2025 + 900}$$

$$Z = \sqrt{2925}$$

$$Z = 54.1\ \Omega$$

Now It's Your Turn 20-1

What is the impedance of a circuit containing an inductor with an inductive reactance of 30 Ω that is wired in series with a 15 Ω resistance?

Refer to **Example 20-2** on how to calculate the impedance of a series R-C circuit.

Example 20-2

What is the impedance of a series R-C circuit that has a resistor with a value of 20 Ω connected in series with a capacitor that has a capacitive reactance of 14 Ω?

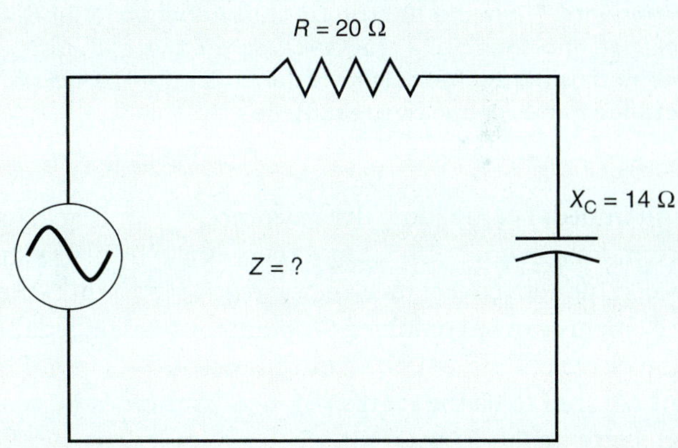

Goodheart-Willcox Publisher

✓ Solution

The circuit impedance can be calculated as follows:

$$Z = \sqrt{R^2 + X_C^{\,2}}$$

$$Z = \sqrt{20\ \Omega^2 + 14\ \Omega^2}$$

$$Z = \sqrt{400 + 196}$$

$$Z = \sqrt{596}$$

$$Z = 24.4\ \Omega$$

Now It's Your Turn 20-2

Calculate the impedance for a circuit containing a capacitor with capacitive reactance of 27.5 Ω that is wired in series with a 12 Ω resistance.

Impedance in Series L-C Circuits

An inductive-capacitive (L-C) circuit has only reactive components, namely inductors and capacitors. With no resistors in the circuit, there is no resistance. The only two elements of

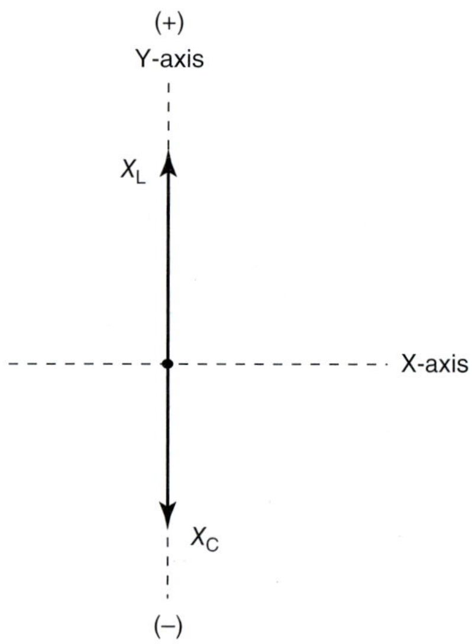

Goodheart-Willcox Publisher

Figure 20-7. In an L-C circuit there are only reactive components. There is a 180° phase shift between inductive reactance and capacitive reactance, so the two elements are combined using the formula impedance is equal to inductive reactance minus capacitive reactance.

circuit impedance are inductive reactance, X_L, and capacitive reactance, X_C, **Figure 20-7.** Since there is a 180° phase shift between inductive reactance and capacitive reactance, these two elements can be combined to obtain the total impedance. In the electrical industry, the impedance of L-C circuits is often referred to as the total reactance, as there is no resistance component. In this text, the term impedance will be used. The formula used to calculate the impedance of a series L-C circuit is:

$$Z = X_L - X_C$$

Refer to **Example 20-3** on how to calculate the impedance of a series L-C circuit.

Shedding Light

The expression $X_L - X_C$ is found in many equations for inductive and capacitive circuits. This expression represents the net reactance of the circuit. Since inductive capacitance and reactive capacitance are 180° out of phase, they in effect cancel each other, and the net reactance is the difference between the two. By convention, this difference is calculated by subtracting X_C from X_L. When this difference is positive, the inductive reactance is greater than the capacitive reactance, and the net reactance is inductive in the positive Y-axis direction. When the difference is negative, the capacitive reactance is greater and the net reactance is capacitive and in the negative Y-axis direction.

Example 20-3

What is the impedance of a series L-C circuit that has an inductive reactance of 28 Ω and a capacitive reactance of 18 Ω?

✓ Solution

The impedance for this L-C circuit is calculated as follows:

$$Z = X_L - X_C$$
$$Z = 28\ \Omega - 18\ \Omega$$
$$Z = 10\ \Omega$$

Now It's Your Turn 20-3

Calculate the impedance of a series L-C circuit that has an inductive reactance of 300 Ω and a capacitive reactance of 80 Ω.

Impedance in Series R-L-C Circuits

In a resistive-inductive-capacitive (R-L-C) circuit there is both inductive reactance and capacitive reactance. When capacitors and inductors are in the same circuit, it is important to note the polarities assigned to the capacitive reactance and the inductive reactance. As discussed earlier in this chapter, resistance is at a 0° angle with the X-axis and the circuit current, while inductive reactance is at a +90° angle with the X-axis and the capacitive reactance is at a –90° angle with the X-axis. Meaning, the inductive reactance and the capacitive reactance are 180° out of phase with each other, **Figure 20-8**. The inductive reactance

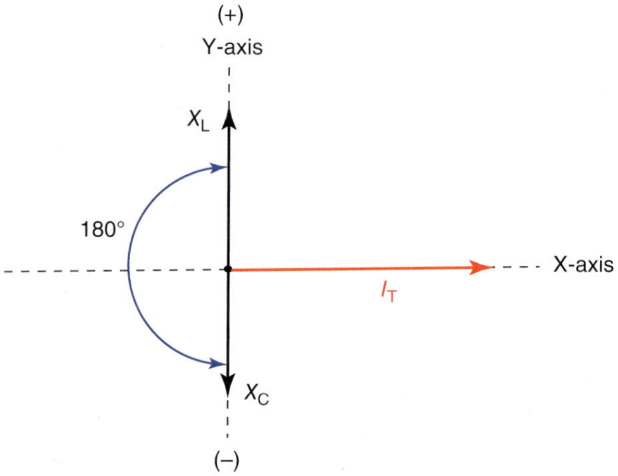

Goodheart-Willcox Publisher

Figure 20-8. Capacitive and inductive reactance are 180° out of phase with each other.

points in the positive direction along the Y-axis, while the capacitive reactance points in the negative direction along the Y-axis. Inductive reactance has a positive (+) polarity, while capacitive reactance has a negative polarity (–).

Shedding Light

In many impedance calculations, the capacitive reactance value is squared, so the end result will be the same, regardless of the polarity assigned to the capacitive reactance. However, in some calculations, for example when calculating the impedance of an R-L-C circuit, the negative polarity of the capacitive reactance is taken into account in the expression $X_L - X_C$.

The formula used to calculate the impedance of a series R-L-C circuit is similar to the one used to calculate the impedance of a series R-L or series R-C circuit. The main difference is that there are two opposing reactance values, namely X_L and X_C, that must be combined to obtain the net reactance of the circuit. By calculating the difference between X_L and X_C, the reactive portion of the circuit can be obtained. The formula that can be used to calculate the impedance of a series R-L-C circuit is:

$$Z = \sqrt{R^2 + (X_L - X_C)^2}$$

Refer to **Example 20-4** on how to calculate impedance of a series R-L-C circuit.

As a review, the following are the formulas used to calculate impedance of series R-L, R-C, L-C, and R-L-C circuits:

Series Resistive-Inductive (R-L):

$$Z = \sqrt{R^2 + X_L^2}$$

Series Resistive-Capacitive (R-C):

$$Z = \sqrt{R^2 + X_C^2}$$

Series Inductive-Capacitive (L-C):

$$Z = X_L - X_C$$

Series Resistive-Inductive-Capacitive (R-L-C):

$$Z = \sqrt{R^2 + (X_L - X_C)^2}$$

Example 20-4

Calculate the impedance of a series R-L-C circuit that has a resistance of 16 Ω, an inductive reactance of 22 Ω, and a capacitive reactance of 14 Ω.

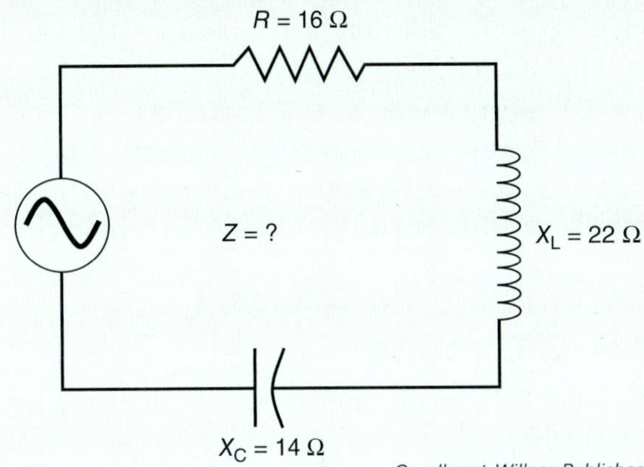

Goodheart-Willcox Publisher

✓ Solution

The impedance is calculated using the impedance formula for series R-L-C circuits:

$$Z = \sqrt{R^2 + (X_L - X_C)^2}$$

$$Z = \sqrt{(16\ \Omega)^2 + (22\ \Omega - 14\ \Omega)^2}$$

$$Z = \sqrt{(16\ \Omega)^2 + (8\ \Omega)^2}$$

$$Z = \sqrt{256 + 64}$$

$$Z = \sqrt{320}$$

$$Z = 17.9\ \Omega$$

Now It's Your Turn 20-4

Calculate the impedance of the following circuit.

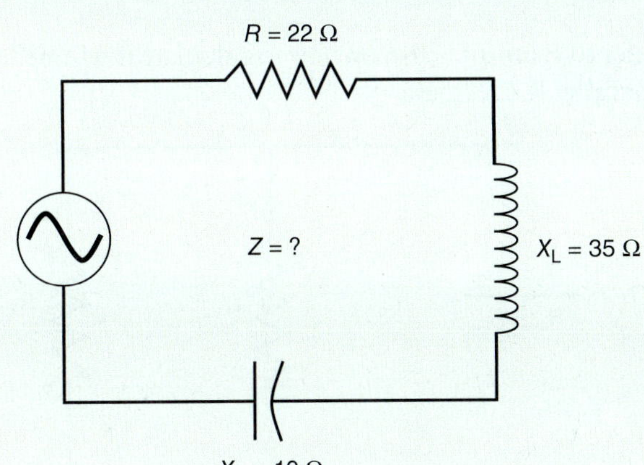

Goodheart-Willcox Publisher

20.3 Impedance in Parallel Circuits

Our discussion of impedance continues by moving on to simple parallel circuits that are made up of various combinations of resistors, inductors, and capacitors. The four basic circuit types include the following:

- Parallel resistive-inductive (R-L) circuits
- Parallel resistive-capacitive (R-C) circuits
- Parallel inductive-capacitive (L-C) circuits
- Parallel resistive-inductive-capacitive (R-L-C) circuits

Impedance in Parallel R-L and R-C Circuits

In a parallel R-L circuit, impedance is calculated using a variation of the product over the sum formula used to calculate the value of two resistances connected in parallel with each other, **Figure 20-9**.

In a parallel R-L circuit, however, the resistance and the inductive reactance are 90° out of phase with each other. Modifying the product over the sum formula to account for the 90° phase shift gives us the following formula:

$$Z = \frac{R \times X_L}{\sqrt{R^2 + X_L^2}}$$

Refer to **Example 20-5** on how to calculate the impedance of a parallel R-L circuit.

In a parallel circuit made up of resistance and capacitive reactance, the impedance is calculated using the same formula that was used to calculate the impedance of a parallel R-L circuit. The only difference is that the X_L for inductive reactance is replaced with X_C for capacitive reactance:

$$Z = \frac{R \times X_C}{\sqrt{R^2 + X_C^2}}$$

Refer to **Example 20-6** on how to calculate the impedance of a parallel R-C circuit.

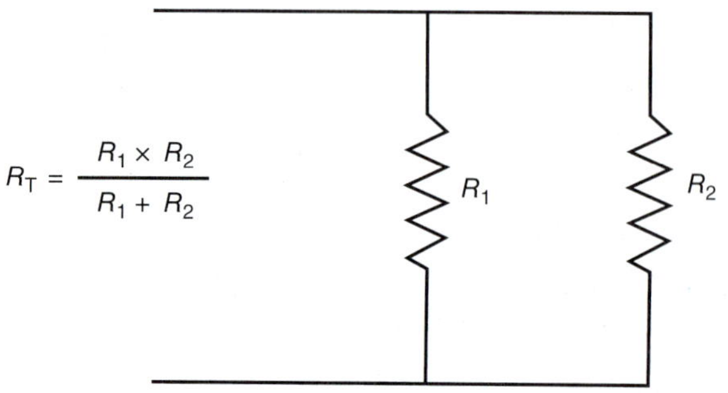

Goodheart-Willcox Publisher

Figure 20-9. Product over sum formula for two resistances connected in parallel with each other. For an R-L circuit, this formula is modified to account for inductive reactance.

Example 20-5

What is the impedance of a parallel R-L circuit that has a resistor with a value of 25 Ω connected in parallel with an inductor that has an inductive reactance of 18 Ω?

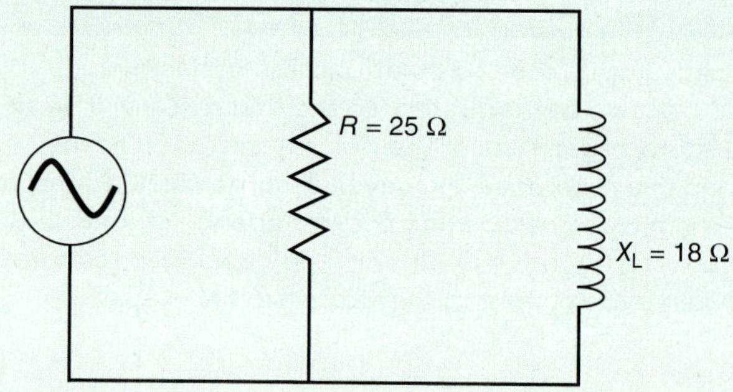

Goodheart-Willcox Publisher

✓ Solution

The impedance for this parallel R-L circuit can be calculated as follows:

$$Z = \frac{R \times X_L}{\sqrt{R^2 + X_L^2}}$$

$$Z = \frac{25\ \Omega \times 18\ \Omega}{\sqrt{(25\ \Omega)^2 + (18\ \Omega)^2}}$$

$$Z = \frac{450\ \Omega^2}{\sqrt{625\ \Omega^2 + 324\ \Omega^2}}$$

$$Z = \frac{450\ \Omega^2}{\sqrt{949\ \Omega^2}}$$

$$Z = \frac{450\ \Omega^2}{30.8\ \Omega}$$

$$Z = 14.6\ \Omega$$

Just as with parallel resistance calculations, the circuit impedance of a parallel circuit will always be lower than the ohm reading of any branch in the parallel circuit.

Now It's Your Turn 20-5

What is the impedance of a parallel R-L circuit if the inductive reactance is 19 Ω and the resistance is 35 Ω?

Example 20-6

What is the impedance of a parallel R-C circuit that has a resistor with a value of 12 Ω connected in parallel with a capacitor that has a capacitive reactance of 17 Ω?

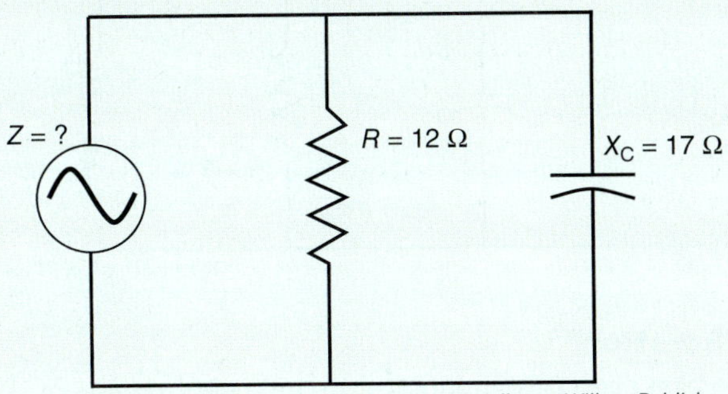

Goodheart-Willcox Publisher

✓ **Solution**

The circuit impedance can be calculated as follows:

$$Z = \frac{R \times X_C}{\sqrt{R^2 + X_C^2}}$$

$$Z = \frac{12\ \Omega \times 17\ \Omega}{\sqrt{(12\ \Omega)^2 + (17\ \Omega)^2}}$$

$$Z = \frac{204\ \Omega^2}{\sqrt{144\ \Omega^2 + 289\ \Omega^2}}$$

$$Z = \frac{204\ \Omega^2}{\sqrt{433\ \Omega^2}}$$

$$Z = \frac{204\ \Omega^2}{20.8\ \Omega}$$

$$Z = 9.8\ \Omega$$

Now It's Your Turn 20-6

What is the impedance of a parallel R-C circuit that has a 20 Ω resistor wired in parallel with 35 Ω of capacitive reactance?

Impedance in Parallel L-C Circuits

Inductive-capacitive (L-C) circuits have only reactive components, so the only two elements of circuit impedance are inductive reactance, X_L, and capacitive reactance, X_C. Since there is a 180° phase shift between inductive reactance and capacitive reactance, the magnitudes of these two values are polar opposites, meaning they have opposite polarities. Other than the polarity difference, the inductive reactance and capacitive reactance can be treated the same way as resistors wired in parallel with each other. The formula used to calculate the impedance of a parallel L-C circuit is as follows:

$$Z = \frac{X_L \times X_C}{X_L - X_C}$$

If the capacitive reactance is greater than the inductive reactance (that is, $X_C > X_L$), the negative value in the denominator causes the calculated impedance to have a negative value. This negative sign can be ignored when calculating the magnitude of the impedance. The negative result is simply an indication that the capacitive reactance is greater than the inductive reactance, and the net reactance points in the negative Y-axis direction.

Refer to **Example 20-7** on how to calculate the impedance of a parallel L-C circuit.

Example 20-7

What is the impedance of a parallel L-C circuit that has an inductive reactance of 10 Ω and a capacitive reactance of 50 Ω?

✓ **Solution**

The impedance for this circuit is calculated as:

$$Z = \frac{X_L \times X_C}{X_L - X_C}$$

$$Z = \frac{10\ \Omega \times 50\ \Omega}{10\ \Omega - 50\ \Omega}$$

$$Z = \frac{500\ \Omega^2}{-40\ \Omega}$$

$$Z = -12.5\ \Omega$$

The magnitude of the impedance is −12.5 Ω. The negative value indicates that X_C is greater than X_L, and the net reactance points in the negative Y-axis direction.

Now It's Your Turn 20-7

Calculate the impedance of a parallel L-C circuit that has an inductive reactance of 85 Ω and a capacitive reactance of 30 Ω.

Impedance in Parallel R-L-C Circuits

In a parallel R-L-C circuit, impedance is calculated using a two-step process that involves first calculating the net reactance (X) and then using the net reactance to calculate the circuit impedance. The equation for calculating net reactance in a parallel R-L-C circuit is identical to the equation for calculating impedance in a parallel L-C circuit:

$$X = \frac{X_L \times X_C}{X_L - X_C}$$

As discussed in the previous section, this calculation yields a negative value when capacitive reactance is greater than the inductive reactance (that is, $X_C > X_L$). The negative result is simply an indication that the capacitive reactance is greater than the inductive reactance, and the net reactance points in the negative Y-axis direction.

Shedding Light

The impedance in a parallel L-C circuit has no resistive component, so net reactance is the same as the impedance. In a parallel R-L-C, the net reactance is one component of the impedance, with resistance being the second component.

The net reactance is then used to calculate the impedance using the following formula:

$$Z = \frac{R \times X}{\sqrt{R^2 + X^2}}$$

Refer to **Example 20-8** on how to calculate the impedance of a parallel R-L-C circuit.

As a review, the following are the formulas used to calculate impedance of parallel R-L, R-C, L-C, and R-L-C circuits:

Parallel Resistive-Inductive (R-L):

$$Z = \frac{R \times X_L}{\sqrt{R^2 + {X_L}^2}}$$

Parallel Resistive-Capacitive (R-C):

$$Z = \frac{R \times X_C}{\sqrt{R^2 + {X_C}^2}}$$

Parallel Inductive-Capacitive (L-C):

$$Z = \frac{X_L \times X_C}{X_L - X_C}$$

Parallel Resistive-Inductive-Capacitive (R-L-C):

$$Z = \frac{R \times X}{\sqrt{R^2 + X^2}}$$

$$X = \frac{X_L \times X_C}{X_L - X_C}$$

Example 20-8

Calculate the impedance of the following parallel R-L-C circuit.

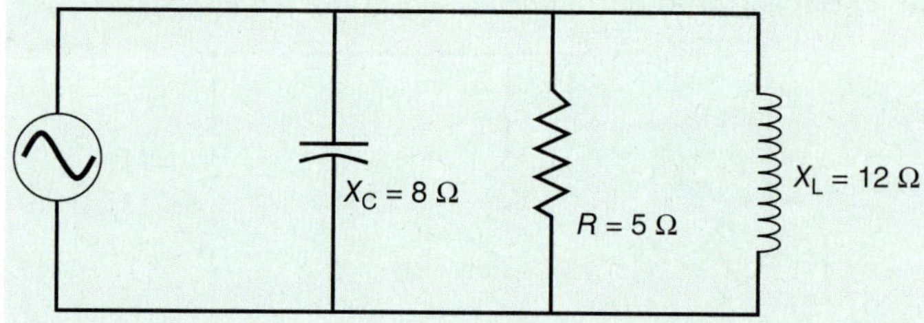

Goodheart-Willcox Publisher

✓ Solution

The first step is to calculate the net reactance:

$$X = \frac{X_L \times X_C}{X_L - X_C}$$

$$X = \frac{12\ \Omega \times 8\ \Omega}{12\ \Omega - 8\ \Omega}$$

$$X = \frac{96\ \Omega^2}{4\ \Omega}$$

$$X = 24\ \Omega$$

The net reactance is then used to calculate the impedance:

$$Z = \frac{R \times X}{\sqrt{R^2 + X^2}}$$

$$Z = \frac{5\ \Omega \times 24\ \Omega}{\sqrt{(5\ \Omega)^2 + (24\ \Omega)^2}}$$

$$Z = \frac{120\ \Omega^2}{\sqrt{25\ \Omega^2 + 576\ \Omega^2}}$$

$$Z = \frac{120\ \Omega^2}{\sqrt{601\ \Omega^2}}$$

$$Z = \frac{120\ \Omega^2}{24.5\ \Omega}$$

$$Z = 4.90\ \Omega$$

Now It's Your Turn 20-8

Calculate the impedance of the following parallel R-L-C circuit.

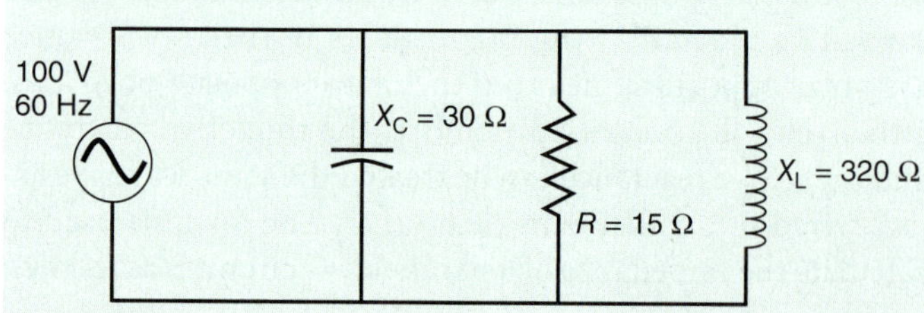

Goodheart-Willcox Publisher

Calculating Impedance with Ohm's Law

The previous sections provide specific equations and methods to calculate impedance in series and parallel R-L, R-C, L-C, and R-L-C circuits. Impedance can also be calculated using the following form of Ohm's law:

$$Z = \frac{E_T}{I_T}$$

where:

Z = Impedance (in ohms)
E_T = Total circuit voltage (in volts)
I_T = Total circuit current (in amps)

This form of Ohm's law is true for all circuits, and in some cases may provide an easier calculation to determine circuit impedance. Methods for calculating current flow in series and parallel R-L, R-C, L-C, and R-L-C circuits is covered in later sections of this chapter.

20.4 Current Flow in Series Circuits

In a series circuit, the current is the same at all points in the circuit and is a function of the circuit voltage and the total opposition to current flow. This is true for all series circuits, including R-L, R-C, L-C, and R-L-C circuits. The following key points regarding opposition to current flow have been previously discussed:

- In a purely resistive series circuit, the total circuit resistance, *R*, is equal to the sum of all the individual resistances in the circuit.
- In a circuit containing inductors, inductive reactance, X_L, is calculated using the following formula:

$$X_L = 2\pi \times f \times L$$

- In a circuit containing capacitors, capacitive reactance, X_C, is calculated using the following formula:

$$X_C = \frac{1}{2\pi \times f \times C}$$

- Impedance, *Z*, is a combination of the resistance and reactance found in a circuit. Impedance calculations vary depending on the circuit components and how they are connected to each other and the power supply.

Once the circuit impedance has been determined, this value can be substituted into Ohm's law to calculate the total current flow in a series circuit.

Refer to **Example 20-9** on how to calculate total current flow.

Refer to **Example 20-10** and **Example 20-11** on how to calculate the total current flow in a series L-C circuit.

Example 20-9

What is the total current flow in the circuit shown below?

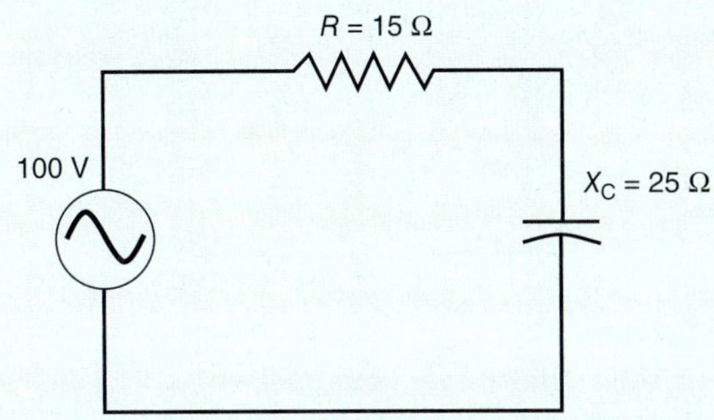

Goodheart-Willcox Publisher

✓ Solution

In this circuit, the supply voltage is known, as well as the resistance and reactance values. The first step in determining the total current flow in the circuit is to calculate the impedance for a series R-C circuit as follows:

$$Z = \sqrt{R^2 + X_C^{\,2}}$$

$$Z = \sqrt{(15\ \Omega)^2 + (25\ \Omega)^2}$$

$$Z = \sqrt{225\ \Omega^2 + 625\ \Omega^2}$$

$$Z = \sqrt{850\ \Omega^2}$$

$$Z = 29.2\ \Omega$$

The next step in determining the total current flow in the circuit is to use a modified form of Ohm's law, where the resistance, *R*, value is replaced with the impedance, *Z*, value. Then calculate for total circuit current (I_T) and by substituting the known values for the circuit voltage (E_T), and the impedance (*Z*), as follows:

$$E_T = I_T \times Z$$

$$I_T = \frac{E_T}{Z}$$

$$I_T = \frac{100\ \text{V}}{29.2\ \Omega}$$

$$I_T = 3.4\ \text{A}$$

Example 20-10

What is the total current flow in the series L-C circuit shown below?

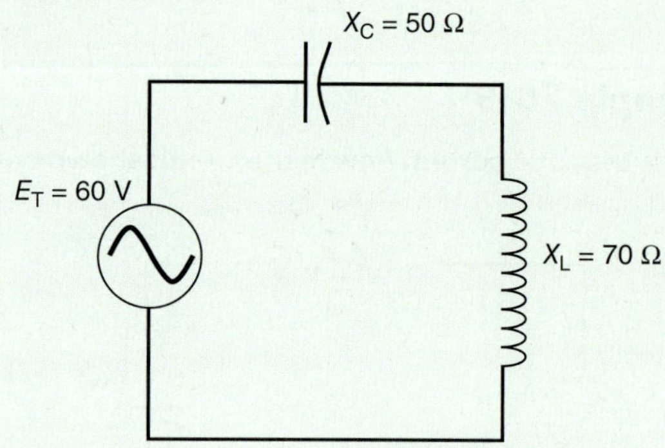

Goodheart-Willcox Publisher

✓ Solution

In this circuit, the supply voltage is known, as well as the inductive and capacitive reactance values. The first step in determining the total current flow in the circuit is to determine the impedance as follows:

$$Z = X_L - X_C$$

$$Z = 70\ \Omega - 50\ \Omega$$

$$Z = 20\ \Omega$$

The next step is to solve the $E_T = I_T \times Z$ for the current, I_T, and substitute the known values for the circuit voltage and the impedance as follows:

$$I_T = \frac{E_T}{Z}$$

$$I_T = \frac{60\text{ V}}{20\ \Omega}$$

$$I_T = 3\text{ A}$$

Example 20-11

Calculate the current flow in the following circuit.

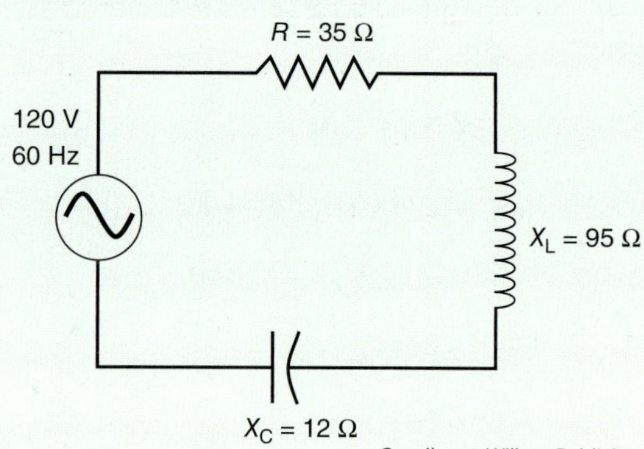

Goodheart-Willcox Publisher

✓ Solution

The first step is solving for the impedance for this series R-L-C circuit. With the resistance, capacitive reactance, and inductive reactance values known, they can be substituted into the impedance formula for a series R-L-C circuit as follows:

$$Z = \sqrt{R^2 + (X_L - X_C)^2}$$

$$Z = \sqrt{(35\ \Omega)^2 + (95\ \Omega - 12\ \Omega)^2}$$

$$Z = \sqrt{(35\ \Omega)^2 + (83\ \Omega)^2}$$

$$Z = \sqrt{1225\ \Omega^2 + 6889\ \Omega^2}$$

$$Z = \sqrt{8114\ \Omega^2}$$

$$Z = 90.1\ \Omega$$

With the total circuit voltage and impedance values known, the total circuit current can be calculated using a variation of Ohm's law, where impedance is used instead of resistance:

$$I_T = \frac{E_T}{Z}$$

$$I_T = \frac{120\text{ V}}{90.1\ \Omega}$$

$$I_T = 1.33\text{ A}$$

20.5 Voltage Drops in Series Circuits

In a purely resistive circuit, the total voltage supplied to the circuit is equal to the sum of the voltage drops across each circuit component. In series R-L, R-C, L-C, or R-L-C circuits, however, this rule does not apply. In these circuits, the voltage drop across each circuit component is found using variations of the $E = I \times R$ form of Ohm's law. The variations are as follows:

- Voltage across a resistor: $E_R = I \times R$
- Voltage across a capacitor: $E_C = I \times X_C$
- Voltage across an inductor: $E_L = I \times X_L$

In a series circuit, the current passing through each of these components (I) is the total circuit current (I_T). Notice that the voltage across a resistor is proportional to the resistance, the voltage across a capacitor is proportional to the capacitive reactance, and the voltage across an inductor is proportional to the inductive reactance.

Refer to **Example 20-12** on how to calculate voltage drop.

Example 20-12

Calculate the voltage drops across the resistor and the inductor in the following circuit.

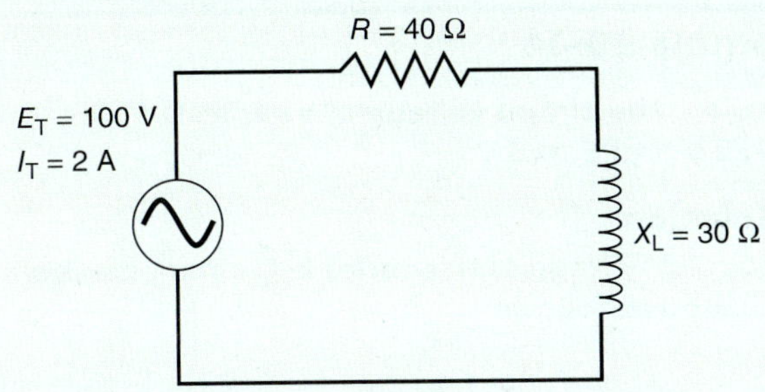

✓ Solution

In this series R-L circuit, the current is known as well as the resistance and the inductive reactance. The voltage drop across the resistor, E_R, is calculated as follows:

$$E_R = I \times R$$
$$E_R = 2\text{ A} \times 40\ \Omega$$
$$E_R = 80\text{ V}$$

The voltage drop across the inductor, E_L, is calculated as follows:

$$E_L = I \times X_L$$
$$E_L = 2\text{ A} \times 30\ \Omega$$
$$E_L = 60\text{ V}$$

In **Example 20-12**, the sum of the voltage drops is 140 V (80 V + 60 V), which is greater than the circuit's supply voltage of 100 V. The voltage in a series R-L or R-C circuit behaves in a similar manner to the impedance in these circuits. Just as impedance is calculated using either of the following formulas:

$$Z = \sqrt{R^2 + X_L^{\,2}}$$
$$Z = \sqrt{R^2 + X_C^{\,2}}$$

The supply voltage for series R-L or R-C circuits is calculated using the following formulas:

$$E_T = \sqrt{E_R^{\,2} + E_L^{\,2}}$$
$$E_T = \sqrt{E_R^{\,2} + E_C^{\,2}}$$

These formulas are used because of the phase angle between the resistive circuit component and the reactive circuit component, **Figure 20-10**. **Figure 20-11** shows the vector diagram for the circuit voltage in a series R-L circuit and in a series R-C circuit.

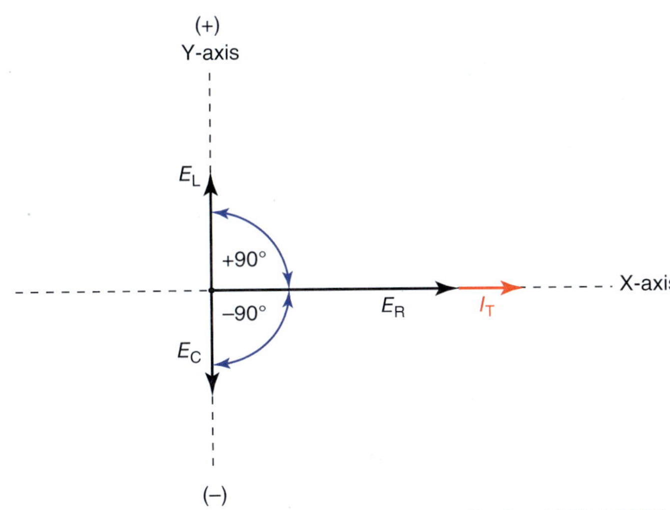

Goodheart-Willcox Publisher

Figure 20-10. Phase angle relationship among inductor voltage, capacitor voltage, and resistor voltage.

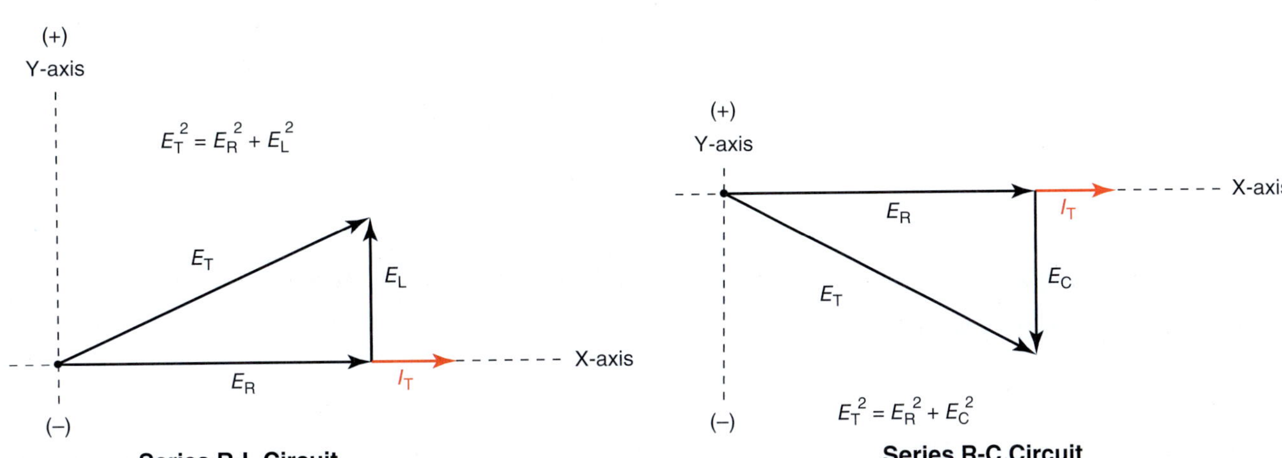

Goodheart-Willcox Publisher

Figure 20-11. Voltage vector diagram for the circuit voltage in a series R-L circuit and a series R-C circuit.

Refer to **Example 20-13** on how to calculate voltage drop and total circuit voltage in a series L-C circuit.

Example 20-13

Calculate the voltage drop across the capacitor, the voltage drop across the inductor, and the total circuit voltage in the following circuit.

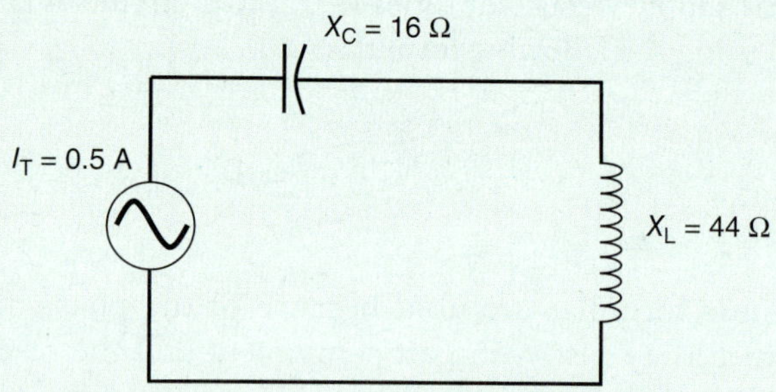

Goodheart-Willcox Publisher

✓ Solution

In this series L-C circuit, the current is known as well as the capacitive and inductive reactance. First, calculate the voltage drop across the capacitor, E_C, as follows:

$$E_C = I \times X_C$$

$$E_C = 0.5\text{ A} \times 16\ \Omega$$

$$E_C = 8\text{ V}$$

Then, calculate the voltage drop across the inductor, E_L, as follows:

$$E_L = I \times X_L$$

$$E_L = 0.5\text{ A} \times 44\ \Omega$$

$$E_L = 22\text{ V}$$

In this example we will use the circuit impedance to calculate the total circuit voltage. Impedance is calculated using the series L-C circuit impedance formula:

$$Z = X_L - X_C$$

$$Z = 44\ \Omega - 16\ \Omega$$

$$Z = 28\ \Omega$$

Using the $E_T = I_T \times Z$ formula, the circuit voltage is calculated as follows:

$$E_T = I_T \times Z$$

$$E_T = 0.5\text{ A} \times 28\ \Omega$$

$$E_T = 14\text{ V}$$

In a series L-C circuit, there is no resistive component. The total circuit voltage of a series L-C circuit is calculated in a manner that is similar to the $Z = X_L - X_C$ method of calculating the impedance of a series L-C circuit. In this case, the impedance and reactance values are replaced with voltage values as follows:

$$E_T = E_L - E_C$$

Refer to **Example 20-14**.

Example 20-14

Calculate the circuit voltage of a series L-C circuit if $E_L = 22$ V and $E_C = 8$ V.

✓ Solution

The circuit voltage of this series L-C circuit can be calculated as follows:

$$E_T = E_L - E_C$$

$$E_T = 22\text{ V} - 8\text{ V}$$

$$E_T = 14\text{ V}$$

Now It's Your Turn 20-14

Calculate the voltage drops across the capacitor and the inductor and the total circuit voltage in the following circuit.

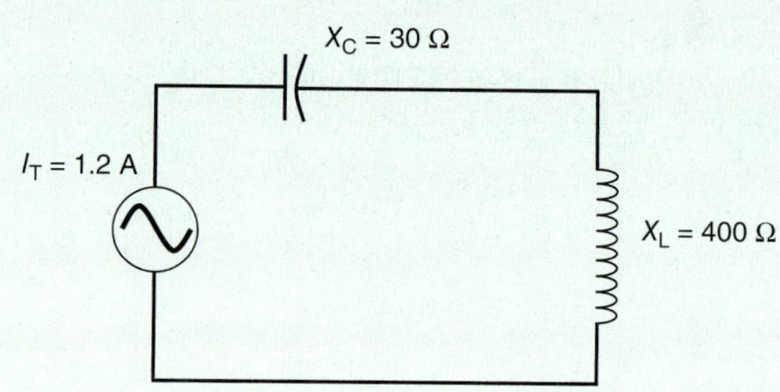

Goodheart-Willcox Publisher

Notice that that answer in **Example 20-13** for the total voltage is the same answer as in **Example 20-14**.

As a reminder, E_L and E_C are 180° out of phase with each other in a purely capacitive and inductive circuit. This results in the lower voltage being subtracted from the higher voltage. The resulting voltage will determine if the circuit is more capacitive or more inductive. Similar to the corresponding impedance calculations, this calculation yields a negative value when capacitive reactance is greater than the inductive reactance (that is, $X_C > X_L$). The negative result is simply an indication that the total circuit voltage is more capacitive and points in the negative Y-axis direction. A positive value occurs when inductive reactance is greater than capacitive reactance, the circuit is more inductive, and total circuit voltage points in the positive Y-axis direction.

As with series R-L, R-C, and L-C circuits, the voltage drop across each circuit component in a series R-L-C circuit can be found using Ohm's law. The circuit current is the same at all points in the circuit, so the same current value will be present in all three calculations. For the voltage across the resistance, E_R, the value of the resistance, R, will be used. For the voltage across the capacitor, E_C, the capacitive reactance, X_C, will be used. For the voltage across the inductor, E_L, the inductive reactance, X_L, will be used.

Refer to **Example 20-15**.

Example 20-15

Calculate the voltage drops across the resistor, inductor, and capacitor in the following circuit.

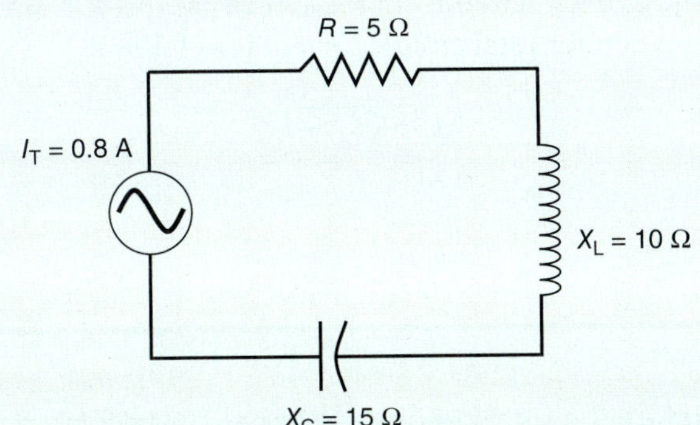

Goodheart-Willcox Publisher

✓ Solution

First, calculate the voltage across the resistor as follows:

$$E_R = I_T \times R$$
$$E_R = 0.8\ \text{A} \times 5\ \Omega$$
$$E_R = 4\ \text{V}$$

Then, calculate the voltage across the inductor as follows:

$$E_L = I_T \times X_L$$
$$E_L = 0.8\ \text{A} \times 10\ \Omega$$
$$E_L = 8\ \text{V}$$

Next, calculate the voltage across the capacitor as follows:

$$E_C = I_T \times X_C$$
$$E_C = 0.8\ \text{A} \times 15\ \Omega$$
$$E_C = 12\ \text{V}$$

The total circuit voltage of a series R-L-C circuit can be calculated in a manner that is similar to the method used to calculate the impedance of a series R-L-C circuit. In this case, the impedance and reactance values are replaced with voltage values:

$$E_T = \sqrt{E_R^2 + (E_L - E_C)^2}$$

Refer to **Example 20-16** on how to calculate total circuit voltage of a series R-L-C circuit.

Example 20-16

Calculate the total circuit voltage for the circuit shown in **Example 20-15**.

✓ Solution

Substituting the known values for E_R, E_L, and E_C, the total circuit voltage, E_T, is calculated as follows:

$$E_T = \sqrt{E_R^2 + (E_L - E_C)^2}$$
$$E_T = \sqrt{(4\ \text{V})^2 + (8\ \text{V} - 12\ \text{V})^2}$$
$$E_T = \sqrt{(4\ \text{V})^2 + (-4\ \text{V})^2}$$
$$E_T = \sqrt{16\ \text{V}^2 + 16\ \text{V}^2}$$
$$E_T = \sqrt{32\ \text{V}^2}$$
$$E_T = 5.66\ \text{V}$$

A summary of the voltage formulas that can be used on series R-L, R-C, L-C, and R-L-C circuits follows:

Series Resistive-Inductive (R-L):

$$E_T = \sqrt{E_R^2 + E_L^2}$$

Series Resistive-Capacitive (R-C):

$$E_T = \sqrt{E_R^2 + E_C^2}$$

Series Inductive-Capacitive (L-C):

$$E_T = E_L - E_C$$

Series Resistive-Inductive-Capacitive (R-L-C):

$$E_T = \sqrt{E_R^2 + (E_L - E_C)^2}$$

20.6 Current Flow in Parallel Circuits

In a purely resistive circuit, the total circuit current is equal to the sum of the individual branch currents. In parallel R-L, R-C, L-C, or R-L-C circuits, however, this rule does not apply.

The current flow through any individual branch of a circuit can be found using variations of Ohm's law. The variations are as follows:

- Current flow through a resistive parallel circuit branch:

$$I_R = \frac{E}{R}$$

- Current flow through a capacitive parallel circuit branch:

$$I_C = \frac{E}{X_C}$$

- Current flow through an inductive parallel circuit branch:

$$I_L = \frac{E}{X_L}$$

The total circuit current in a parallel R-L, R-C, L-C, or R-L-C circuit can be found using the following form of Ohm's law:

$$I_T = \frac{E}{Z}$$

Refer to **Example 20-17** on how to calculate the current flow in a parallel circuit.

Current Flow in Parallel R-L and R-C Circuits

In addition to using Ohm's law, the total circuit current in parallel R-L or R-C circuits can also be calculated using a formula similar to the one used to calculate the impedance of series R-L or R-C circuits. In the case of parallel R-L circuits, the equation for total circuit current is as follows:

$$I_T = \sqrt{I_R^2 + I_L^2}$$

Example 20-17

Calculate the current flow through each branch and the total circuit current in the following parallel R-L circuit.

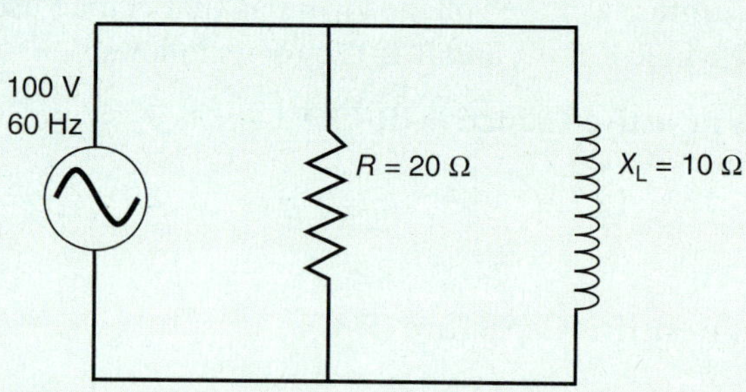

Goodheart-Willcox Publisher

✓ Solution

The current flow in the resistive branch in this parallel R-L circuit can be calculated as follows:

$$I_R = \frac{E}{R}$$

$$I_R = \frac{100\ \text{V}}{20\ \Omega}$$

$$I_R = 5\ \text{A}$$

The current flow in the inductive branch in this parallel R-L circuit can be calculated as follows:

$$I_L = \frac{E}{X_L}$$

$$I_L = \frac{100\ \text{V}}{10\ \Omega}$$

$$I_L = 10\ \text{A}$$

Before the total circuit current can be calculated, the impedance of this parallel R-L circuit must be calculated as follows:

$$Z = \frac{R \times X_L}{\sqrt{R^2 + X_L^2}}$$

$$Z = \frac{20\ \Omega \times 10\ \Omega}{\sqrt{(20\ \Omega)^2 + (10\ \Omega)^2}}$$

$$Z = \frac{200\ \Omega^2}{\sqrt{400\ \Omega^2 + 100\ \Omega^2}}$$

$$Z = \frac{200\ \Omega^2}{\sqrt{500\ \Omega^2}}$$

$$Z = \frac{200\ \Omega^2}{22.4\ \Omega}$$

$$Z = 8.9\ \Omega$$

The total circuit current can now be calculated as follows:

$$I_T = \frac{E}{Z}$$

$$I_T = \frac{100\ \text{V}}{8.9\ \Omega}$$

$$I_T = 11.2\ \text{A}$$

Refer to **Example 20-18** on how to calculate total current flow in a parallel R-L circuit.

Example 20-18

Recalculate the total circuit current from **Example 20-17** using the $I_R = 5$ A and $I_L = 10$ A values.

✓ Solution

Substitute the known values for I_R and I_L in the total circuit current formula:

$$I_T = \sqrt{I_R^2 + I_L^2}$$

$$I_T = \sqrt{(5\text{ A})^2 + (10\text{ A})^2}$$

$$I_T = \sqrt{25\text{ A} + 100\text{ A}}$$

$$I_T = \sqrt{125\text{ A}}$$

$$I_T = 11.2\text{ A}$$

Notice that this answer is the same as the amperage obtained in **Example 20-17** when Ohm's law formula was used.

Parallel R-C circuits use a similar formula, with the I_L^2 term replaced with I_C^2:

$$I_T = \sqrt{I_R^2 + I_C^2}$$

The relationship among currents in R-L and R-C circuits is shown graphically in **Figure 20-12.**

Current Flow in Parallel L-C Circuits

In a parallel L-C circuit, there is no resistive element. There is only inductive reactance and capacitive reactance. As with parallel R-C and R-L circuits, the current flow through a capacitive or inductive branch is calculated using one of the following two formulas:

$$I_L = \frac{E}{X_L}$$

$$I_C = \frac{E}{X_C}$$

Since there is no resistive component in a parallel L-C circuit, the formula used to calculate the total current in a parallel L-C circuit is similar to the formula used to calculate impedance in a series L-C circuit. The exception is that the impedance and reactance values are replaced with current values:

$$I_T = I_L - I_C$$

As with series L-C circuit impedance calculations, a negative result occurs when $I_C > I_L$, indicating a total circuit current pointing in the negative Y-axis direction.

Refer to **Example 20-19** on how to calculate current flow in a parallel L-C circuit.

Current Flow in Parallel R-L-C Circuits

As with parallel R-L, R-C, and L-C circuits, the current flow through a resistive, capacitive, or inductive branch is calculated using one of the following formulas:

$$I_R = \frac{E}{R}$$

$$I_L = \frac{E}{X_L}$$

$$I_C = \frac{E}{X_C}$$

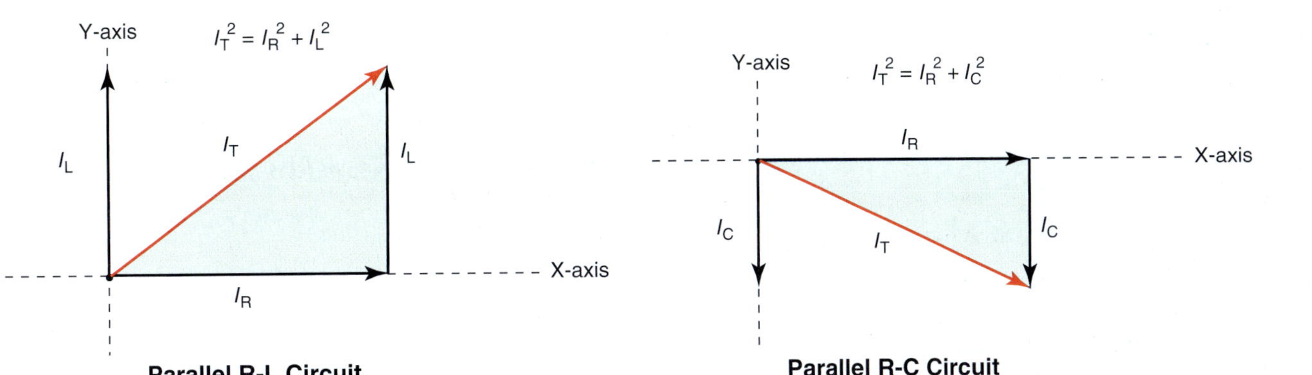

Figure 20-12. Vector representation of the relationship among I_T, I_R, and I_C in a parallel R-L or R-C circuit.

The total circuit current of a parallel R-L-C circuit can be calculated in a manner that is similar to the method used to calculate the impedance of a series R-L-C circuit. In this case, however, the impedance and reactance values are replaced with current values as follows:

$$I_T = \sqrt{I_R^2 + (I_L - I_C)^2}$$

Example 20-19

Calculate the current flow through each of the two circuit branches and the total circuit current for the following circuit.

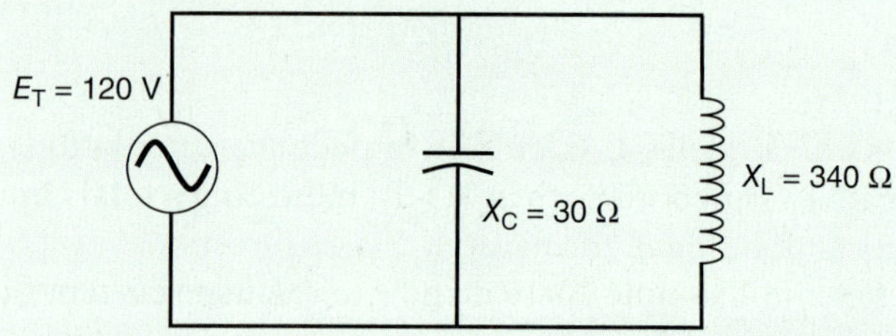

Goodheart-Willcox Publisher

✓ Solution

With the reactance of each branch known, along with the voltage supplied to each branch, the current flow through each branch can be calculated. The current flow through the capacitive branch is calculated as follows:

$$I_C = \frac{E}{X_C}$$

$$I_C = \frac{120\ \text{V}}{30\ \Omega}$$

$$I_C = 4.0\ \text{A}$$

Then the current flow through the inductive branch can be calculated as follows:

$$I_L = \frac{E}{X_L}$$

$$I_L = \frac{120\ \text{V}}{340\ \Omega}$$

$$I_L = 0.35\ \text{A}$$

The total circuit current of this parallel L-C circuit can now be calculated as follows:

$$I_T = I_L - I_C$$

$$I_T = 0.35\ \text{A} - 4\ \text{A}$$

$$I_T = -3.65\ \text{A}$$

The magnitude of the circuit current is 3.65 A. The negative sign indicates that the total circuit current points in the negative Y-axis direction.

Refer to **Example 20-20** on how to calculate current flow in a parallel R-L-C circuit.

Example 20-20

Using the total circuit current for parallel R-L-C circuits formula, calculate the total circuit current for the following circuit.

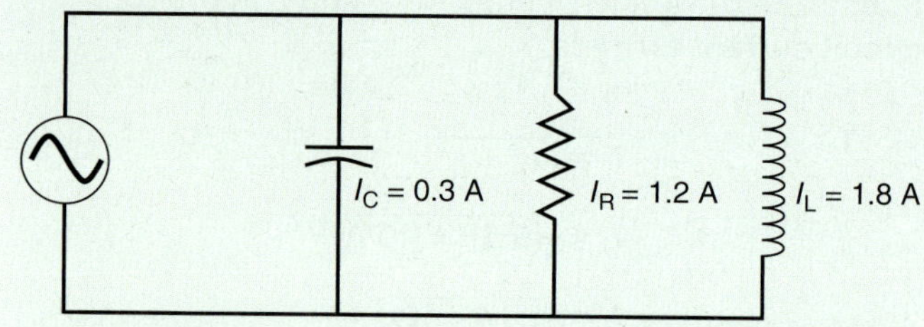

Goodheart-Willcox Publisher

✓ Solution

Substituting the known values for I_R, I_L, and I_C, the total circuit current can be calculated as follows:

$$I_T = \sqrt{I_R^2 + (I_L - I_C)^2}$$

$$I_T = \sqrt{(1.2\ \text{A})^2 + (1.8\ \text{A} - 0.3\ \text{A})^2}$$

$$I_T = \sqrt{(1.2\ \text{A})^2 + (1.5\ \text{A})^2}$$

$$I_T = \sqrt{1.44\ \text{A}^2 + 2.25\ \text{A}^2}$$

$$I_T = \sqrt{3.69\ \text{A}^2}$$

$$I_T = 1.92\ \text{A}$$

As a review, the current formulas that can be used on parallel R-L, R-C, L-C, and R-L-C circuits are as follows:

Parallel Resistive-Inductive (R-L):

$$I_T = \sqrt{I_R^2 + I_L^2}$$

Parallel Resistive-Capacitive (R-C):

$$I_T = \sqrt{I_R^2 + I_C^2}$$

Parallel Inductive-Capacitive (L-C):

$$I_T = I_L - I_C$$

Parallel Resistive-Inductive-Capacitive (R-L-C):

$$I_T = \sqrt{I_R^2 + (I_L - I_C)^2}$$

20.7 Power Factor and Phase Angle

In most real-world circuits, there is a combination of resistance and reactance. In the majority of these circuits, there is apparent power, true (active) power, and reactive power. Power factor will be revisited here in the context of R-L, R-C, L-C, and R-L-C circuits.

The term power factor relates to the percentage of a circuit's apparent power that is actually useful. Power factor is expressed as either a decimal or a percentage. Reactive power is, in essence, wasted power as it does no useful work in the circuit. Power factor is introduced in detail in Chapter 15, *Single-Phase Alternating Current Power.*

The power triangle is used to graphically relate true power to apparent power, **Figure 20-13**. The angle theta, θ, is located between the true power and the apparent power. The larger the angle θ, the more reactive power there is, and the lower the power factor. A lower power factor indicates a less efficient use of the power being consumed. The smaller the angle θ, the less reactive power there is, and the greater the power factor.

The cosine of angle θ represents the power factor, and the inverse cosine of the power factor represents the phase angle, or how many degrees out of phase the circuit components are. Power factor can be calculated in several ways:

For all circuits:

$$PF = \frac{\text{True power (W)}}{\text{Apparent power (VA)}}$$

$$PF = \cos\theta$$

For series circuits only:

$$PF = \frac{R}{Z}$$

$$PF = \frac{E_R}{E_T}$$

For parallel circuits only:

$$PF = \frac{Z}{R}$$

$$PF = \frac{I_R}{I_T}$$

How to calculate power factor and phase angle is provided in **Example 20-21**.

In circuits that have both capacitive and inductive components, the circuit may be more inductive than capacitive, or more capacitive than inductive, depending on the circuit connection and the relationship between the capacitive reactance and the inductive reactance. A table identifying the state of circuits containing both capacitive and inductive components is provided in **Figure 20-14.**

- **Series L-C or R-L-C circuit.** If the capacitive reactance, X_C, is greater than the inductive reactance, X_L, the overall circuit reactance is capacitive, and the phase angle is said to be leading. In this case, the current leads the voltage.
- **Parallel L-C or R-L-C circuit.** If the capacitive reactance, X_C, is greater than the inductive reactance, X_L, the overall circuit reactance is inductive and the phase angle is said to be lagging. In this case, the current lags the voltage.

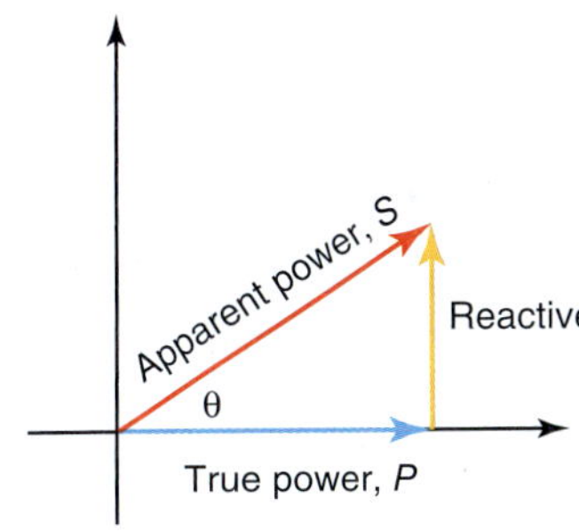

Figure 20-13. The power triangle is a graphic representation of true power, reactive power, apparent power, and power factor. The larger the phase angle θ, the smaller the power factor.

Shedding Light

In order for a circuit to produce true power and have a power factor, the circuit must include a resistive component. L-C circuits do not include a resistive component, so L-C circuits have no true power and no power factor, so none of the power factor equations are applicable to L-C circuits.

The terms ***lagging phase angle*** and ***leading phase angle*** refer to the circuit's power factor and relate to the phase angle between the current and voltage waveforms, as well as the angle between apparent power and true/active power. Whether the circuit has a lagging or leading phase angle does not affect the overall operation of the circuit. Instead, it indicates whether the circuit is more inductive than capacitive, or more capacitive than inductive.

In circuits that have only resistive and inductive components, the circuit will always be inductive, where the voltage leads the current, or the current lags behind the voltage. In circuits that have only resistive and capacitive components, the circuit will always be capacitive, where the current leads the voltage, **Figure 20-15.**

Example 20-21

Calculate the power factor and phase angle for the following circuit.

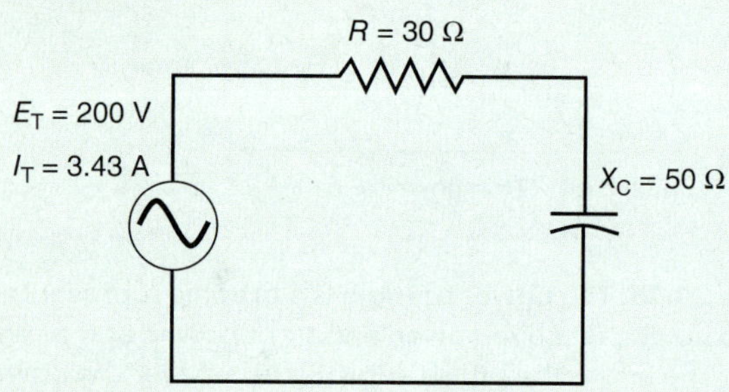

Goodheart-Willcox Publisher

✓ Solution

There are several ways to calculate the power factor for this circuit.

As an initial step, calculate the impedance for this series R-C circuit:

$$Z = \sqrt{R^2 + X_C^{\,2}}$$

$$Z = \sqrt{(30\ \Omega)^2 + (50\ \Omega)^2}$$

$$Z = \sqrt{900\ \Omega^2 + 2500\ \Omega^2}$$

$$Z = \sqrt{3400\ \Omega^2}$$

$$Z = 58.3\ \Omega$$

With the circuit impedance and resistance known, these values can then be used to calculate the power factor for a series circuit (PF) as follows:

$$\text{PF} = \frac{R}{Z}$$

$$\text{PF} = \frac{30\ \Omega}{58.3\ \Omega}$$

$$\text{PF} = 0.515\text{, or } 51.5\%$$

Since the power factor is the cosine of angle θ, the phase angle can be found by using the inverse cosine ($\cos^{-1}$) function as follows:

$$\text{phase angle, } \theta = \cos^{-1} 0.515$$

$$\text{phase angle, } \theta = 59.0°$$

The inverse cosine function can be interpreted as "the angle whose cosine is." The expression $\cos^{-1} 0.515$ can then be "translated" as the angle whose cosine is 0.515. On most scientific calculators, enter the value and then use the $\cos^{-1}$ button to find an inverse cosine value.

A second method of calculating the power factor in a series circuit is by dividing the voltage drop across the resistance by the total circuit voltage. For this series R-C circuit, the supply voltage is known and the voltage across the resistance, E_R, can be calculated as follows:

$$E_R = I \times R$$

$$E_R = 3.43\ \text{A} \times 30\ \Omega$$

$$E_R = 102.9\ \text{V}$$

With the circuit voltage and the voltage across the resistance known, these values can be used to calculate the power factor as follows:

$$\text{PF} = \frac{E_R}{E_T}$$

$$\text{PF} = \frac{102.9\ \text{V}}{200\ \text{V}}$$

$$\text{PF} = 0.515$$

From the previous solution, it is known that a power factor of 0.515 equates to a phase angle of 59.0°.

L-C and R-L-C Circuits

Circuit Type	Reactance	Capacitive / Inductive	Current	Phase Angle
Series Circuits	$X_L > X_C$	Inductive (ELI)	Current lags voltage	Lagging phase angle
Series Circuits	$X_C > X_L$	Capacitive (ICE)	Current leads voltage	Leading phase angle
Parallel Circuits	$X_L > X_C$	Capacitive (ICE)	Current leads voltage	Leading phase angle
Parallel Circuits	$X_C > X_L$	Inductive (ELI)	Current lags voltage	Lagging phase angle

Goodheart-Willcox Publisher

Figure 20-14. Relationship among X_L, X_C, and phase angle for L-C and R-L-C circuits.

R-L and R-C Circuits

Circuit Type	Resistance (R) and Impedance (Z)	Cos (Power Factor)	Capacitive/Inductive	Lead/Lag
R-L Series	$R < Z$	R/Z	Inductive (ELI)	Current lags voltage
R-L Parallel	$R > Z$	Z/R	Inductive (ELI)	Current lags voltage
R-C Series	$R < Z$	R/Z	Capacitive (ICE)	Current leads voltage
R-C Parallel	$R > Z$	Z/R	Capacitive (ICE)	Current leads voltage

Goodheart-Willcox Publisher

Figure 20-15. Relationship among *R*, *Z*, and whether the current leads or lag for R-L and R-C circuits.

CHAPTER 20 REVIEW

Summary

- Impedance is the total opposition force to current flow in a circuit. (20.1)
- Impedance is a function of circuit resistance and reactance. (20.1)
- The four basic series circuit types include resistive-inductive (R-L), resistive-capacitive (R-C), inductive-capacitive (L-C), and resistive-inductive-capacitive (R-L-C) circuits. (20.2)
- The method used to calculate series circuit impedance, voltage, and current will vary depending on the circuit components and how they are connected in the circuit. (20.2)
- The four basic parallel circuit types include resistive-inductive (R-L), resistive-capacitive (R-C), inductive-capacitive (L-C), and resistive-inductive-capacitive (R-L-C) circuits. (20.3)
- In a series circuit, the current is the same at all points in the circuit and is a function of the circuit voltage and the total opposition to current flow. (20.4)
- In series R-L, R-C, L-C, or R-L-C circuits, the voltage drop across each circuit component is found using variations of the $E = I \times R$ form of Ohm's law. (20.5)
- In parallel R-L, R-C, L-C, or R-L-C circuits, the current flow through any individual branch of the circuit can be found using variations of Ohm's law. (20.6)
- The power factor of a circuit is calculated as the cosine of the angle formed between a resistive circuit element and its total circuit counterpart. (20.7)

Review Questions

Use the information in the chapter to answer the following questions.

Measuring Your Potential

1. *True or False?* The method used to calculate impedance varies depending on the circuit components and how they connect in the circuit. (20.1)
2. In a series circuit made up of resistance and reactance, the resistance and reactance will be _____ out of phase with each other. (20.2)
   A. 90°
   B. 180°
   C. 270°
   D. 360°
3. Of the following, the formula used to calculate the impedance of a series R-L circuit is _____. (20.2)
   A. $Z = \sqrt{R^2 + X_L^2}$
   B. $Z = \sqrt{R^2 + X_C^2}$
   C. $Z = X_L + X_C$
   D. $Z = \sqrt{R^2 + (X_L - X_C)^2}$
4. In a circuit containing inductors and capacitors, the X_L and X_C are how far out of phase with each other? (20.2)
   A. 90°
   B. 180°
   C. 270°
   D. 360°
5. Of the following, the formula used to calculate the impedance of a parallel R-L circuit is _____. (20.3)
   A. $Z = \dfrac{R \times X}{\sqrt{R^2 + X^2}}$
   B. $Z = \dfrac{R \times X_L}{\sqrt{R^2 + X_L^2}}$
   C. $Z = \dfrac{R \times X_C}{\sqrt{R^2 + X_C^2}}$
   D. $Z = \dfrac{X_L \times X_C}{X_L - X_C}$
6. In series circuit, the first step to calculate the total current flow is to determine the circuit's _____. (20.4)
   A. current flow using the formula $E = I \times R$
   B. current flow using the formula $E = I \times Z$
   C. impedance using the formula $Z = \sqrt{R^2 + (X_L - X_C)^2}$
   D. impedance using the formula $Z = \sqrt{R^2 + X_C^2}$
7. In a series R-L circuit, what formula is used when determining the voltage drop across a capacitor? (20.5)
   A. $E_T = \sqrt{E_R^2 + (E_L - E_C)^2}$
   B. $E_T = E_L - E_C$
   C. $E_T = \sqrt{E_R^2 + E_L^2}$
   D. $E_T = \sqrt{E_R^2 + E_C^2}$
8. Of the following, the formula used to calculate the current flow through an inductive parallel circuit branch is _____. (20.6)
   A. $I_R = \dfrac{E}{R}$
   B. $I_C = \dfrac{E}{X_C}$
   C. $I_L = \dfrac{E}{X_L}$
   D. $I_T = \dfrac{E}{Z}$

9. The form of Ohm's law used to calculate the total circuit current in a parallel R-L, R-C, L-C, or R-L-C circuit is _____. (20.6)

   A. $I_R = \frac{E}{R}$

   B. $I_C = \frac{E}{X_C}$

   C. $I_L = \frac{E}{X_L}$

   D. $I_T = \frac{E}{Z}$

10. With the circuit impedance and resistance known in a series circuit, which of the following formulas can be used to calculate the power factor? (20.7)

    A. $\text{PF} = \frac{R}{Z}$

    B. $\text{PF} = \frac{I_R}{I_T}$

    C. $\text{PF} = \frac{Z}{R}$

    D. $Z = \sqrt{R^2 + X_C^{\,2}}$

Plug-In

Refer to the following circuit to calculate the answers for questions 1 to 5.

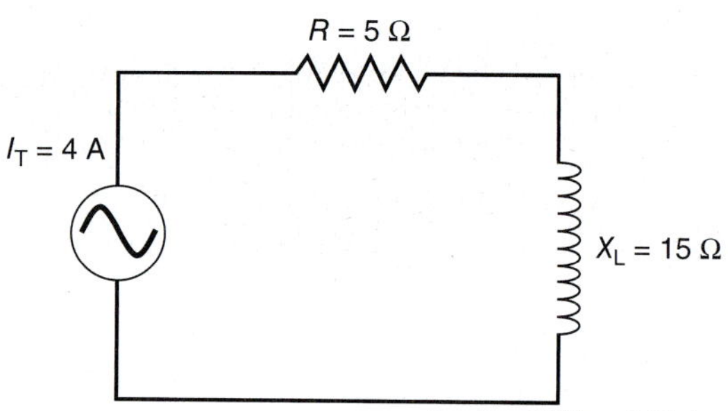

Goodheart-Willcox Publisher

1. The voltage across the resistor is closest to _____. (20.5)

   A. 5 V C. 15 V
   B. 10 V D. 20 V

2. The voltage across the inductor is closest to _____. (20.5)

   A. 50 V C. 70 V
   B. 60 V D. 80 V

3. The total circuit voltage is closest to _____. (20.5)

   A. 31.6 V C. 63.2 V
   B. 45.9 V D. 80.7 V

4. The power factor of this circuit is closest to _____. (20.7)

   A. 0.18 C. 0.61
   B. 0.32 D. 0.87

5. The phase angle for this circuit is closest to _____. (20.7)

   A. 52° C. 72°
   B. 62° D. 82°

Refer to the following circuit to calculate the answers for questions 6 to 12.

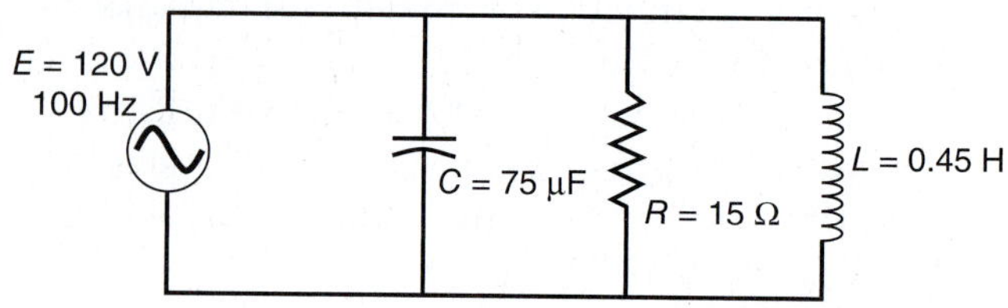

Goodheart-Willcox Publisher

6. The current flow through the capacitor is closest to _____. (20.6)

   A. 1.6 A C. 5.7 A
   B. 3.9 A D. 8.4 A

7. The current flow through the inductor is closest to _____. (20.6)

   A. 0.42 A C. 118.7 A
   B. 27.9 A D. 267.4 A

8. The current flow through the resistor is closest to _____. (20.6)

   A. 4 A C. 12 A
   B. 8 A D. 16 A

9. The total reactance in this circuit is closest to _____. (20.6)

   A. 200 Ω C. 300 Ω
   B. 260 Ω D. 320 Ω

10. The total impedance of this circuit is closest to _____. (20.3)

    A. 12.55 Ω C. 50.25 Ω
    B. 25.15 Ω D. 75.35 Ω

11. The power factor for this circuit is closest to _____. (20.7)

    A. 0.21 C. 0.84
    B. 0.67 D. 0.98

12. The phase angle for this circuit is closest to _____. (20.7)

    A. 33° C. 55°
    B. 44° D. 66°

Making the Connection

1. Explain why capacitive reactance is always a negative value. (20.2)
2. Explain why an increase in reactive power causes the efficiency of a circuit to decrease. (20.7)

SECTION

6 Transformers

WEG

CHAPTER

21 Single-Phase Transformers

P A/Shutterstock.com

SPARKING DISCUSSION

List examples of transformers you use throughout your typical day (at home, work, etc.) and describe the purpose or function of each example.

LEARNING OUTCOMES

After completing this chapter, you will be able to:

21.1 Explain the function of a transformer.
21.2 Describe the function of the primary and secondary windings of a transformer.
21.3 Evaluate single-phase step-up and step-down transformers.
21.4 Explain the function of an isolation transformer.
21.5 Compare the configurations of different multitap transformers.
21.6 Explain the function of a variable transformer.
21.7 Describe the function of a multiple-winding transformer.
21.8 List the advantages and disadvantages of an autotransformer.
21.9 Explain the function of buck transformers and boost transformers.
21.10 Compare the relationship among primary voltage, primary current, secondary voltage, and secondary current.
21.11 Evaluate and troubleshoot single-phase transformers.

TECHNICAL TERMS

air-core transformer
autotransformer
boost transformer
buck transformer
iron-core transformer
isolation transformer
laminations
multitap transformer
mutual induction
primary winding
secondary winding
step-down transformer
step-up transformer
turns ratio
variable transformer
volts-per-turn ratio

Introduction

In simple terms, a transformer is a device that changes the voltage levels in an electric circuit. From reducing a heating system's 240-V power supply to a control voltage of 24 V, to raising an electric generator's output voltage to facilitate power transmission over hundreds of miles, transformers can be found in nearly all electrical applications. For this reason, it is important that anyone working in the electrical trades be familiar with transformer types, configurations, installation, service, and troubleshooting.

This chapter introduces single-phase transformers and includes a discussion of how induction plays an important role in the functionality of most transformers. Without induction, transformers could not exist. Various configurations and applications of single-phase transformers are discussed, along with calculations that will enable you to evaluate these devices.

21.1 Transformer Applications

Before the discussion of transformer theory and operation, it would be beneficial to briefly discuss some of the many transformer-related applications that affect the electrical industry. Gaining insight into the roles that these all-important devices perform strengthens the argument that those working in the electrical trades need to have a solid understanding of transformers and the theories that direct and control their operation.

AC Voltage Regulation

Transformers can be used to increase or decrease the voltage of an alternating current (ac) power signal in a circuit or system. By controlling the magnitude of the voltage, energy consumption can be reduced, resulting in an increase in system efficiency. Many battery chargers, for example, rely on transformers to regulate the voltage being supplied to the battery to prevent current surges from the power supply to the battery. Uncontrolled current surges can cause damage to the battery being charged.

Processing and Manufacturing Facilities

In processing and manufacturing facilities, many different types of equipment can be found. It is not uncommon to encounter a wide range of voltages, often from 120 Vac to over 1000 Vac, that provide the required power to each type of equipment. Transformers can be used to provide the correct voltage to the various stations in the facility. These can include welding, product assembly, packaging, robotics stations, computer areas, and general function areas. Transformers associated with processing and manufacturing facilities must be properly maintained and serviced to help reduce unexpected system down time.

Laboratory Processes

In chemical and medical laboratories, it is often required that during an experiment or research project, chemical reactions be artificially driven or stimulated to achieve or speed up the desired results. This process, known as *electrolysis*, often relies on transformers to produce the high voltage that is needed to drive these chemical reactions. Without transformers to help speed the processes, experimentation, research, and results can be significantly delayed.

Impedance-Matching Applications

Transformers can be used to match the impedance of a power source to that of the load. When the load impedance and the power source impedance match, the efficiency of the energy transfer between the system components is greatly improved. In specialized applications, these transformers can be used as switches to start and stop the flow of current through a circuit. When used in conjunction with safety or control devices, the transformer can function to energize or de-energize a circuit when unsafe conditions are present or when the system parameters fall outside normal or desired operating ranges. Another impedance-matching application for transformers is the audio transformer, **Figure 21-1**. Audio transformers isolate and filter audio signals to improve sound quality.

Control Applications

In many appliances, such as air conditioning and heating systems, the operation of line voltage components such as fans and compressors is often controlled by the end user of

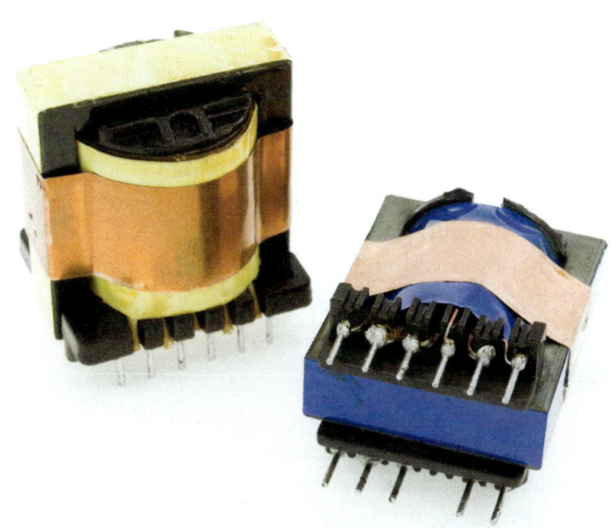

SeDmi/Shutterstock.com

Figure 21-1. Audio transformers are commonly found in amplifiers, speakers, and microphones.

Milwaukee Tools

Figure 21-2. A clamp-on ammeter uses a current transformer that allows the test instrument to obtain current readings when not physically connected to the circuit.

the appliance at the thermostat. To help ensure the safety of the operator of the system, as well as those around the thermostat, transformers are commonly used to convert the system's line voltage power supply to a safer low-voltage level, which is typically 24 V. By using a series of relays and other control devices, the low voltage, low amperage control signal produced by the transformer controls the operation of the line voltage system components.

Electrical Test Instrumentation

In the electrical trades, technicians rely heavily on test instruments such as ammeters and voltmeters. These test instruments often incorporate transformers to help increase their accuracy and reliability, **Figure 21-2**. Transformers also aid in the measuring and determining of a circuit's power factor, energy consumption, and efficiency.

Stabilizers

Transformers are commonly found in stabilizer circuits, which automatically increase or decrease circuit voltage or current to provide the circuit load with optimal operating values. By providing a load with a relatively constant and consistent power supply, the possibility of system or component failure is reduced.

Rectifiers

Transformers can also be used to convert ac power into dc power. This process is referred to as *rectification*. Although this is commonly achieved by using circuits constructed of diodes, resistors, and capacitors, specialized transformers can perform the same function. Rectifier transformers can be found in motor controls, heating and air-conditioning systems, as well as in high-voltage power transmission systems.

DC Blockers

Transformers can be used to block portions of a power signal. For example, a transformer can be used when the ac portion of the power supply is desired, but not the dc portion. The construction of the transformer, which is established at the design state, determines the signal-blocking characteristics of the device.

21.2 Basic Transformer Theory

Based on the intended function of a transformer, its configuration and construction will vary. However, the fundamental theory of operation for all transformers is, for the most part, the same.

Basic Transformer Construction

A simple transformer is constructed as two separate coils of wire, referred to as windings, **Figure 21-3**. One winding, called the ***primary winding***, is connected to the main power source. The second winding, called the ***secondary winding***, acts as the power source for a secondary circuit, which is electrically isolated from the main power source. When voltage is supplied to the primary winding, an electromagnetic field is produced around the winding. The secondary winding, being positioned close to the primary winding, experiences this magnetic field and, as a result, a voltage is induced across the secondary winding.

Transformer Turns Ratio

The relationship that exists between the voltage supplied to the primary winding of a transformer and the voltage produced at the secondary winding depends on the turns ratio. The ***turns ratio*** is the value obtained when the number of turns in the primary winding is divided by the number of turns in the secondary winding. The turns ratio is expressed as a relationship between two numbers such as 2:1. A turns ratio of 2:1 indicates that for every two turns in the primary

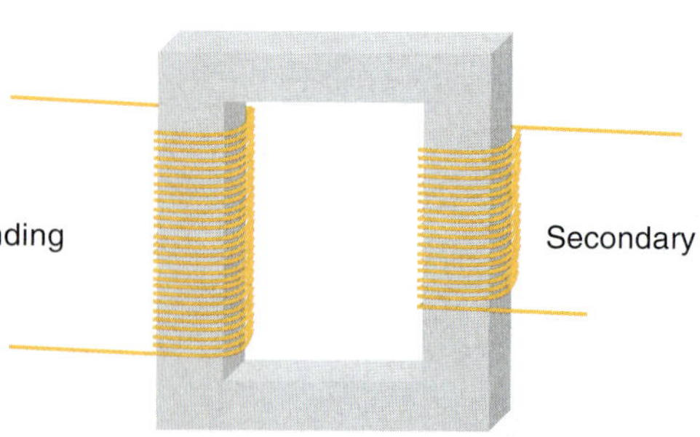

Goodheart-Willcox Publisher

Figure 21-3. A transformer is constructed with a primary winding connected to the main power source and a secondary winding that acts as the power source for a secondary circuit.

winding, there is one turn in the secondary winding. A transformer with a 2:1 turns ratio is said to have a "turns ratio of two to one." The formula for calculating the turns ratio is:

$$TR = \frac{N_P}{N_S}$$

where:

TR = Turns ratio
N_P = Number of turns in primary winding
N_S = Number of turns in secondary winding

Refer to **Example 21-1** and **Example 21-2** on how to calculate the turns ratio for a transformer.

Math Sidebar

The turns ratio is normally expressed as a ratio of whole numbers reduced to simplest form, such as 10:1 or 1:5. When the turns ratio is used in an equation, the ratio is converted to a fraction. For example, a turns ratio of 10:1 can be expressed as the fraction 10/1. This fraction can then be simplified to 10. A turns ratio of 1:5 can be expressed as 1/5 or 0.2.

Example 21-1

What is the turns ratio of a transformer that has 2500 turns in the primary winding and 500 turns in the secondary winding?

✓ Solution

The turns ratio of a transformer is calculated by dividing the number of turns in the primary winding by the number of turns in the secondary winding. Using the turns ratio formula, substitute the known values for N_P and N_S as follows:

$$TR = \frac{N_P}{N_S}$$

$$TR = \frac{2500}{500}$$

$$TR = 5:1$$

Dividing 2500 by 500 gives a result of 5, but the turns ratio must always be expressed as a relationship between two numbers. The result of 5 becomes 5:1.

Now It's Your Turn 21-1

What is the turns ratio of a transformer that has 6000 turns in the primary winding and 2000 turns in the secondary?

Example 21-2

What is the turns ratio of a transformer that has 1000 turns in the primary winding and 4000 turns in the secondary?

✓ Solution

Substituting the known values for N_P and N_S, the turns ratio is calculated as follows:

$$TR = \frac{N_P}{N_S}$$

$$TR = \frac{1000}{4000}$$

$$TR = 1:4$$

In this example, dividing N_P by N_S provides a result of 0.25 or 1/4. When the turns ratio is expressed, only whole numbers are used. To prevent using fractions or decimals, both sides of the ratio are multiplied by the lowest number that will give a result with whole numbers on both sides. A turns ratio of 0.25:1 becomes 1:4 when both sides of the ratio are multiplied by 4.

Now It's Your Turn 21-2

What is the turns ratio of a transformer that has 1000 turns in the primary winding and 8000 turns in the secondary?

Transformer Voltages

The voltage supplied to the primary winding of a transformer ultimately produces an induced voltage at the transformer's secondary winding. The relationship that exists between the transformer's primary and secondary voltages is determined by the turns ratio, and is expressed as follows:

$$\frac{N_P}{N_S} = \frac{E_P}{E_S}$$

where:

E_P = Voltage supplied to the primary winding
E_S = Voltage induced at the secondary winding

Since N_P divided by N_S is the turns ratio, the formula can also be stated as follows:

$$TR = \frac{E_P}{E_S}$$

Refer to **Example 21-3** on how to calculate measured voltage of a transformer.

Example 21-3

Calculate the measured voltage at the secondary winding of a transformer with a turns ratio of 5:1 and 120 V being supplied to the primary winding of the device.

✓ Solution

Substitute the known values into the turns ratio-voltage formula. The turns ratio of 5:1 is expressed as 5/1 or 5 to simplify the equation:

$$TR = \frac{E_P}{E_S}$$

Rearrange the equation to solve for secondary voltage:

$$E_S = \frac{E_P}{TR}$$

$$E_S = \frac{120\text{ V}}{5}$$

$$E_S = 24\text{ V}$$

When applied to primary and secondary voltages, the turns ratio can be thought of as the ratio of primary voltage to secondary voltage. For the 5:1 turns ratio in this example, we can say that for every 5 V applied to the primary winding, 1 V is induced in the secondary winding.

Now It's Your Turn 21-3

Calculate the turns ratio of a transformer that produces a voltage of 50 V at the secondary winding if 200 V is supplied to the transformer's primary winding.

Transformer Currents

The current flow rates through the primary and secondary windings are inversely proportional to the primary and secondary voltages. The relationship that exists among the primary voltage, secondary voltage, primary current, and secondary current can be expressed as:

$$\frac{E_P}{E_S} = \frac{I_S}{I_P}$$

By cross multiplying, we can see that:

$$E_P \times I_P = E_S \times I_S$$

Refer to **Example 21-4** on how to determine current flow through the secondary winding of a transformer.

Since the turns ratio of a transformer is equal to N_P divided by N_S as well as E_P divided by E_S, the turns ratio can also be used to establish the relationship between the primary and secondary current values as follows:

$$TR = \frac{I_S}{I_P}$$

Refer to **Example 21-5** on how to determine current flow through the primary winding of a transformer.

Example 21-4

Determine the current flow through the secondary winding of a transformer if the primary voltage is 80 V, the secondary voltage is 20 V, and the primary current is 2 A.

✓ Solution

Substitute the known values for the primary and secondary voltages, as well as the primary current, into the following formula:

$$E_P \times I_P = E_S \times I_S$$

Rearrange the equation to solve for secondary current:

$$I_S = E_P \times \frac{I_P}{E_S}$$

$$I_S = 80\text{ V} \times \frac{2\text{ A}}{20\text{ V}}$$

$$I_S = \frac{160\text{ VA}}{20\text{ V}}$$

$$I_S = 8\text{ A}$$

Example 21-5

Calculate the current in the primary winding of a transformer with a turns ratio of 3:1 and with a current of 6 A flowing in the secondary winding.

✓ Solution

Substitute the known values into the turns ratio-current formula:

$$TR = \frac{I_S}{I_P}$$

Rearrange the equation to solve for primary current. The turns ratio of 3:1 can be expressed as 3/1 or 3 to simplify calculations:

$$I_P = \frac{I_S}{TR}$$

$$I_P = \frac{6\text{ A}}{3}$$

$$I_P = 2\text{ A}$$

Now It's Your Turn 21-5

What is the turns ratio of a transformer that operates with a primary winding current of 3 A and a secondary winding current of 6 A?

Transformer Volt-Ampere Rating

As discussed in earlier chapters, apparent power is calculated as the voltage applied to a circuit multiplied by the circuit's current. Apparent power is expressed in volt-amperes (VA). In the $E_P \times I_P = E_S \times I_S$ formula, both sides of the equation are expressed in the units of volt-amperes. Transformers are rated in volt-amperes, as indicated on their nameplate, **Figure 21-4**.

Know the Code

Article 450 of the *NEC* provides requirements for transformer installations. *Article 450* requires transformer nameplates to include the following items:

- Name of manufacturer
- VA rating (generally expressed in kVA)
- Frequency
- Primary voltage and secondary voltage
- Impedance (for transformer 25 kVA and larger)
- Clearances (for transformers with ventilation openings)
- Type and amount of insulating liquid (when used)
- Temperature class for insulation (for dry-type transformers)

A transformer's VA rating can be obtained from either the primary voltage and current or the secondary voltage and current.

$$\text{VA rating} = E_P \times I_P = E_S \times I_S$$

Goodheart-Willcox Publisher

Figure 21-4. A nameplate of a transformer will provide information such as the transformer's VA rating.

Refer to **Example 21-6** and **Example 21-7** on how to calculate the VA rating of a transformer.

Example 21-6

Calculate the VA rating of a transformer with a primary voltage of 60 V, a secondary voltage of 15 V, and a primary current of 2 A.

✓ Solution

The VA rating of the transformer can be found by multiplying the primary voltage (E_P) by the primary current (I_P) as follows:

$$\text{VA rating} = E_P \times I_P$$

$$\text{VA rating} = 60 \text{ V} \times 2 \text{ A}$$

$$\text{VA rating} = 120 \text{ VA}$$

The VA rating in this example could also be calculated by first determining the secondary current and then multiplying it by the secondary voltage.

Now It's Your Turn 21-6

Calculate the VA rating of a transformer with a primary voltage of 25 V and a secondary voltage of 100 V if the primary current is 2 A.

Example 21-7

Calculate the maximum current in the secondary winding of a 40 VA transformer that is being used to power a 24 V control circuit of a residential air-conditioning system.

✓ Solution

The VA rating is obtained by multiplying a winding's voltage and its current together. Using the VA rating formula and substituting the known values for the VA rating and the secondary voltage, the current can be calculated as follows:

$$\text{VA rating} = E_S \times I_S$$

Rearrange the equation to solve for secondary current.

$$I_S = \frac{\text{VA rating}}{E_S}$$

$$I_S = \frac{40 \text{ VA}}{24 \text{ V}}$$

$$I_S = 1.67 \text{ A}$$

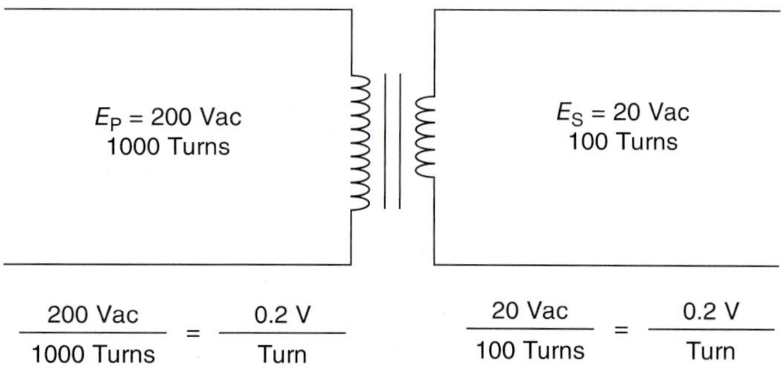

Goodheart-Willcox Publisher

Figure 21-5. The volts-per-turn ratio for this transformer is 0.2 V per turn. Since the secondary winding has 100 turns, the voltage at the secondary is 20 Vac (0.2 V/turn × 100 turns).

Shedding Light

The theoretical calculations presented in this chapter assume that the apparent power of the primary circuit is the same as that of the secondary circuit. However, in reality, losses occur as power transfers from the primary winding to the secondary winding. Common losses include copper losses and core losses caused by eddy currents, hysteresis, flux linkage, and core saturation. Overall, the loss range from primary to secondary is typically in the range of 1% to 5%, resulting in an overall efficiency of 95% to 99%.

Transformer Volts-per-Turn Ratio

The relationship between a transformer's primary and secondary voltages is determined by the turns ratio that exists between the primary and the secondary windings. The ***volts-per-turn ratio*** relates the primary and secondary voltages to the number of turns in the corresponding winding. A transformer's primary and secondary have the same volts-per-turn ratio. The volts-per-turn ratio can be expressed mathematically as follows:

$$\text{volts-per-turn ratio (V/turn)} = \frac{E_P}{N_P} = \frac{E_S}{N_S}$$

For example, to calculate the volts-per-turn ratio for the primary winding in **Figure 21-5**, first divide 200 V by 1000 turns, which equals 1:5. Since the expression provides the number of volts per winding turn, the second number in the ratio should be 1. Dividing both sides of the 1:5 ratio by 5, the result is 0.2 V per turn (1/5 : 5/5 = 0.2 V : 1 turn). If there are 100 turns in the secondary winding, the volts-per-turn ratio can be used to determine the secondary voltage as follows:

$$E_S = \text{volts-per-turn ratio} \times N_S$$
$$E_S = 0.2 \text{ V/turn} \times 100 \text{ turns}$$
$$E_S = 20 \text{ V}$$

Example 21-8 provides additional practice on calculating the number of turns in the winding of a transformer.

Example 21-8

With a primary voltage of 150 V, how many turns are in the primary winding of a transformer with a 5:1 turns ratio and a 0.5 V/turn ratio?

✓ **Solution**

First, substitute the known values into the volts-per-turn formula:

$$\text{volts-per-turn ratio (V/turn)} = \frac{E_P}{N_P}$$

Then, rearrange the equation to solve for number of turns.

$$N_P = \frac{E_P}{\text{volts-per-turn ratio}}$$
$$N_P = \frac{150 \text{ V}}{0.5 \text{ V/turn}}$$
$$N_P = 300 \text{ turns}$$

21.3 Single-Phase Step-Up and Step-Down Transformers

In the examples presented so far in this chapter, there were instances where the turns ratio was greater than one, such as 4:1 or 3:1, and other instances where the turns ratio was less than one, such as 1:3 or 1:4. When the turns ratio is greater than one, the voltage of the secondary winding is lower than the voltage of the primary winding. When the turns ratio is less than one, the voltage of the secondary winding is higher than the voltage of the primary winding. The turns ratio determines whether the transformer is stepping up the voltage or stepping down the voltage. Transformers that reduce voltage are called ***step-down transformers***, while transformers that increase voltage are called ***step-up transformers***.

Shedding Light

The majority of transformers encountered by residential, commercial, and industrial electricians are step-down transformers. The power supplied from the utility is generally provided at a relatively high voltage that must be "stepped down" to supply many or most of the circuits and loads. Step-up transformers will be encountered occasionally. For example, long wire runs for parking lot lighting may require voltage to be increased to offset voltage drop, or a supply voltage of 208 V may need to be increased to supply a single 480 V piece of equipment.

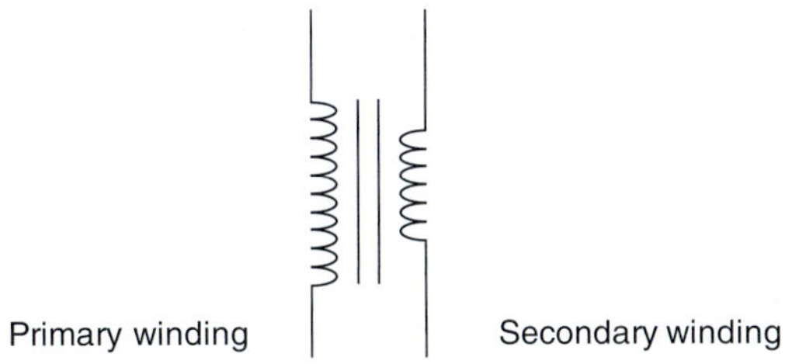

Goodheart-Willcox Publisher

Figure 21-6. Wiring diagram symbol for a step-down transformer.

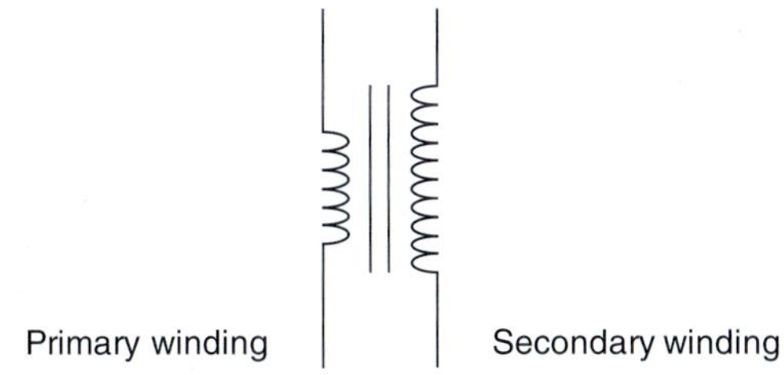

Goodheart-Willcox Publisher

Figure 21-7. Wiring diagram symbol for a step-up transformer.

Step-Down Transformers

Step-down transformers "step down" the voltage supplied to the primary winding to a lower voltage at the transformer's secondary winding. To accomplish this, the number of turns in the secondary winding is fewer than the number of turns in the primary winding, **Figure 21-6**.

Shedding Light

A welding transformer is an example of a step-down transformer that produces the high currents required to generate enough heat to melt the tip of the welding electrode. Although the voltage at the secondary winding of the welding transformer is in the 15 V to 45 V range, the amperage in the secondary can be in the 400 A to 600 A range. Current flowing between the electrode and the pieces being welded creates the heat that melts the filler material.

Step-down transformers are commonly found in appliances and systems where low-voltage, low-current control circuits are used to control the operation of higher voltage components that draw significantly more current. The electrical properties of a step-down transformer include the following:

- The primary voltage is greater than the secondary voltage.
- The secondary current is greater than the primary current.

Step-Up Transformers

Step-up transformers are designed to "step up" the primary voltage to a higher voltage at the transformer's secondary winding. This is accomplished by the transformer's secondary winding having more turns than the primary winding, **Figure 21-7**. The electrical properties of a step-up transformer include the following:

- The secondary voltage is greater than the primary voltage.
- The primary current is greater than the secondary current.

21.4 Isolation Transformers

An ***isolation transformer*** is a transformer that isolates one circuit from another. Isolation transformers can be used to separate circuits that operate at different voltages or to separate circuits that operate at the same voltage. The electrical symbol for an isolation transformer, **Figure 21-8**, shows that the two windings are electrically separated from each other. When used to separate circuits that operate at the same voltage, the turns ratio of the device is 1:1, **Figure 21-9**. The circuit connected to the secondary winding receives the same voltage as the primary but is not electrically connected to it. Isolation transformers are often used for safety reasons, as they create a barrier between the electrical components and wiring of different circuits.

The step-up and step-down transformers discussed earlier in this chapter are also technically isolation transformers, as there is a barrier between the primary and secondary windings. In these cases, however, the secondary voltage differs from the primary voltage. The barrier between the

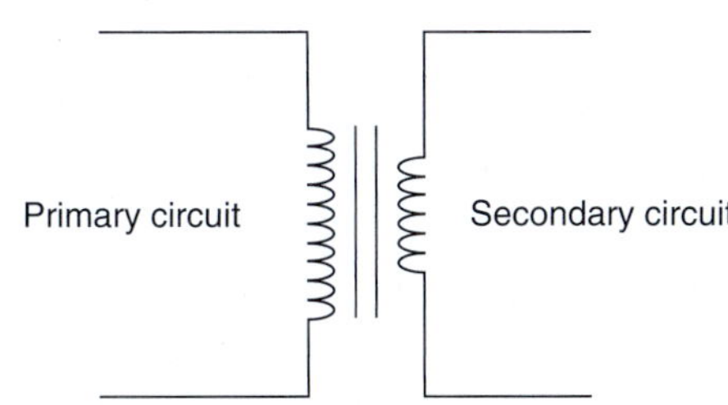

Goodheart-Willcox Publisher

Figure 21-8. Wiring diagram symbol for an isolation transformer.

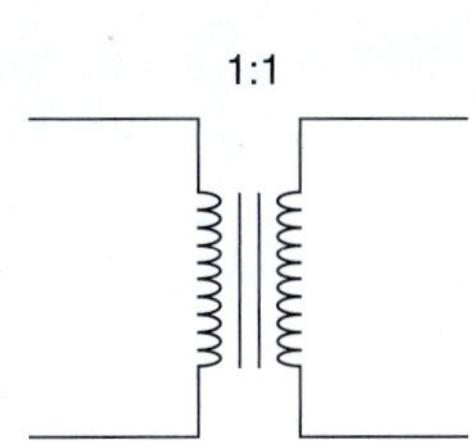

Goodheart-Willcox Publisher

Figure 21-9. Isolation transformer with a 1:1 turns ratio.

primary and secondary transformer windings is called the *core*. The transformer windings are wound around the core. As voltage is applied to the primary winding, current flows through the primary winding and creates a magnetic field. This magnetic field induces a voltage in the secondary winding, which is also formed around the core. The ability of the primary winding to induce a voltage in the secondary winding is referred to as ***mutual induction***.

Many transformers utilize cores made from soft iron and are called ***iron-core transformers***. The iron cores on these transformers are often constructed of many insulated layers, called ***laminations***, **Figure 21-10**. The laminations help reduce eddy current losses that result when an induced electromotive force causes circular currents to flow around a conductor. Eddy currents and eddy current losses are discussed further in Chapter 17, *Inductors and Inductance*. The number and size of the laminations, and hence the core, is determined by the power rating of the transformer. Some transformers utilize cores that are made from insulating materials. These transformers are called ***air-core transformers***. The electrical symbols for air-core and iron-core step-up and step-down transformers are provided in **Figure 21-11**.

Goodheart-Willcox Publisher

Figure 21-10. Transformer laminations minimize energy losses due to eddy currents and hysteresis.

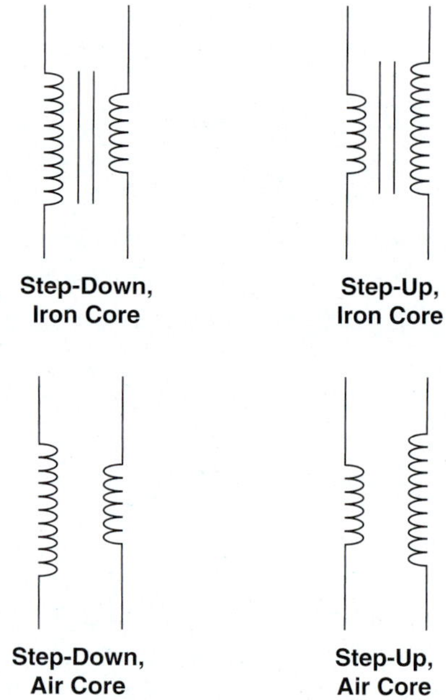

Goodheart-Willcox Publisher

Figure 21-11. Transformer wiring diagram symbols used for iron-core transformers and air-core transformers.

Shedding Light

For most mechanical applications, iron-core transformers are commonplace, as they are suitable for large loads that operate at low frequencies. Air-core transformers are the popular choice for high frequency applications, such as audio and radio frequency transmission applications. Air-core transformers operate with little to no signal loss and almost no signal distortion, which makes them desirable when transmission quality is to be maximized.

21.5 Multitap Transformer

In many cases, transformers are constructed with a single primary winding and a single secondary winding. In this configuration, the device is supplied with a single, known voltage, and induces a single, known voltage in the secondary winding. To accommodate a number of different voltages, either at the primary or secondary winding, some transformers are constructed with windings that have multiple lead wires connected to the primary or secondary winding. These transformers are called ***multitap transformers***.

Multitap Primary Windings

A multitap transformer has multiple leads connected to the primary winding at various points. The multitap transformer in **Figure 21-12** has four wires or leads connected to the primary winding and are color-coded black, white, red, and orange. There are two wires connected to the secondary winding which are color-coded yellow and blue. These two

Goodheart-Willcox Publisher

Figure 21-12. A transformer with multiple primary winding leads. In this configuration, the multitap transformer can be used to convert one of several primary voltages to a known secondary voltage.

secondary wires are intended to provide a constant 24 V signal to the circuit to which they will be connected.

The black wire is the common lead, which is always connected to the voltage source, **Figure 21-13**. Only one of the other three primary winding leads are connected to the voltage source. The lead to be connected is determined by the voltage being supplied to the transformer as follows:

- If the supply voltage is 120 V, the white wire is connected to the voltage source. The transformer functions as a 5:1 step-down transformer and steps the 120 V voltage down by a factor of 5 to 24 V.
- If the supply voltage is 208 V, the red wire is connected to the voltage source. The transformer functions as a 9:1 step-down transformer and steps the 208 V voltage down by a factor of 9 to 24 V.
- If the supply voltage is 240 V, the orange wire is connected to the voltage source. The transformer functions as a 10:1 step-down transformer and steps the 240 V voltage down by a factor of 10 to 24 V.

Safety Alert

When wiring any type of multitap transformer, it is important that all unused leads be capped separately to prevent damage to the component and reduce the risk of fire and severe personal injury.

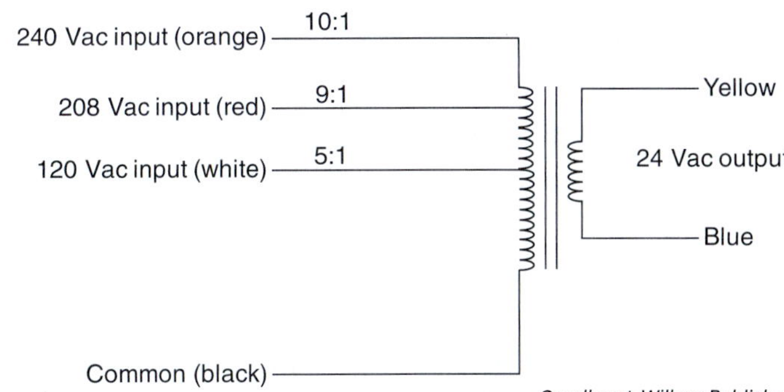

Goodheart-Willcox Publisher

Figure 21-13. Wiring diagram of a transformer with multiple primary winding leads.

Multitap Secondary Windings

Another configuration of a multitap transformer is to have multiple leads connected to the secondary winding at various points. The transformer contains a single primary winding and secondary winding that has been tapped at several points.

In the configuration illustrated in **Figure 21-14**, a single known voltage is supplied to the transformer's primary winding, but the secondary voltage can be varied, depending on which secondary leads are utilized. When the known voltage of 120 V is supplied to the primary winding, a secondary voltage of 12 V, 24 V, or 48 V can be induced in the secondary winding as follows:

- If the common lead and lead 1 are used, the secondary voltage will be 12 V.
- If the common lead and lead 2 are used, the secondary voltage will be 24 V.
- If the common lead and lead 3 are used, the secondary voltage will be 48 V.

As the number of secondary winding turns in the active electric circuit is increased, the transformer's turn ratio will decrease. As a result, with an unchanged volts-per-turn ratio, the transformer's secondary voltage will increase. The relationship that exists among the transformer's turn ratio, primary and secondary voltages, and the volts-per-turn ratio is provided in the table shown in **Figure 21-14**.

Multitap Switch Transformer

Just like the multitap transformers, the multitap switch transformer has multiple leads connected to various points on the transformer's secondary winding. The main difference between the multitap transformer and the multitap switch transformer is that the multitap switch transformer has all of its winding leads permanently connected to a multipole rotary switch, **Figure 21-15**. To change the transformer's secondary voltage, no wiring changes are necessary. All the user needs to do is rotate the multipole switch to the desired position to obtain the desired output voltage. For instance, by turning the rotary switch in a clockwise

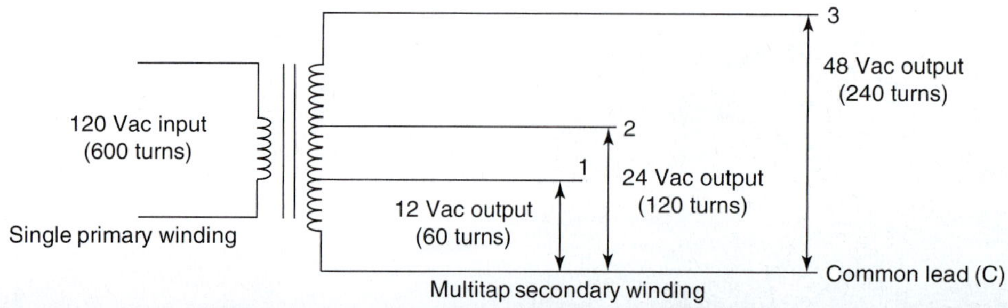

Secondary Lead Connection	Primary Voltage	Secondary Voltage	Secondary Turns	Transformer Turns Ratio	Secondary Volts/Turn Ratio
C-1	120 Vac	12 Vac	60	10:1	0.2:1
C-2	120 Vac	24 Vac	120	5:1	0.2:1
C-3	120 Vac	48 Vac	240	2.5:1	0.2:1

Goodheart-Willcox Publisher

Figure 21-14. A transformer with multiple secondary winding leads, showing a summary of transformer values.

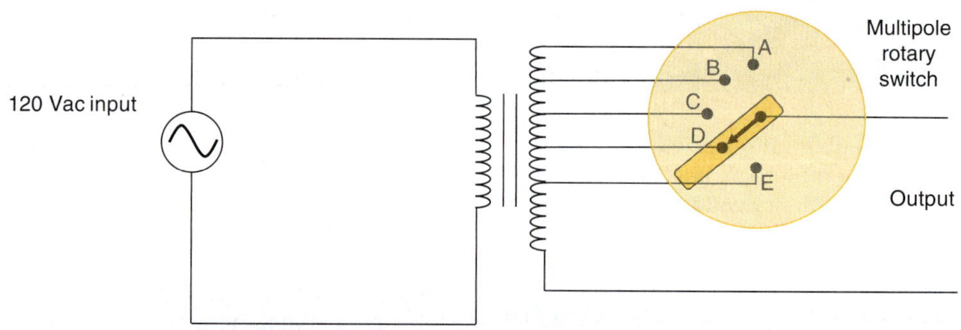

Input Voltage	Switch Position	Output Voltage	Turns Ratio
120 Vac	A	180 Vac	2:3
120 Vac	B	150 Vac	4:5
120 Vac	C	120 Vac	1:1
120 Vac	D	100 Vac	6:5
120 Vac	E	60 Vac	2:1

Goodheart-Willcox Publisher

Figure 21-15. Wiring diagram of a multitap switch transformer with available output voltage and turns ratio values for the various switch positions.

direction, the transformer's secondary (output) voltage increases by increasing the number of turns in the secondary winding and decreasing the transformer's turns ratio. Turning the rotary switch in a counterclockwise direction decreases the transformer's secondary (output) voltage by decreasing the number of turns in the secondary winding and increasing the transformer's turns ratio.

21.6 Variable Transformer

The ***variable transformer*** can be thought of as a multitap transformer with the number of taps, in essence, equal to the number of turns in the transformer's secondary winding. Instead of having a set number of leads or switch positions, the variable transformer is constructed with a sliding contact that moves along the surface of the secondary winding, **Figure 21-16**. The sliding contact is able to connect with the secondary winding at any point along its path. The variable transformer was commonly used to control the speed of model trains. For the transformer shown in **Figure 21-17**, more turns are introduced to the secondary winding when the dial is turned in the counterclockwise direction. This increases the voltage produced by the transformer at the secondary winding. Turning the dial in the clockwise direction

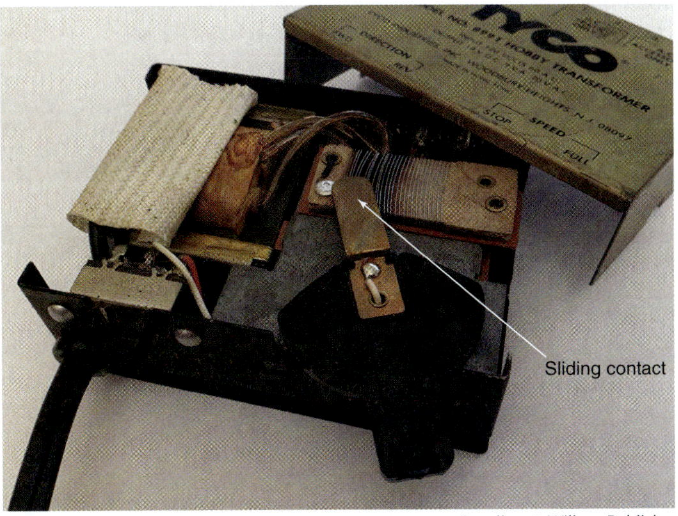

Goodheart-Willcox Publisher

Figure 21-16. The sliding contact in a variable transformer connects with the secondary winding at any point along its path.

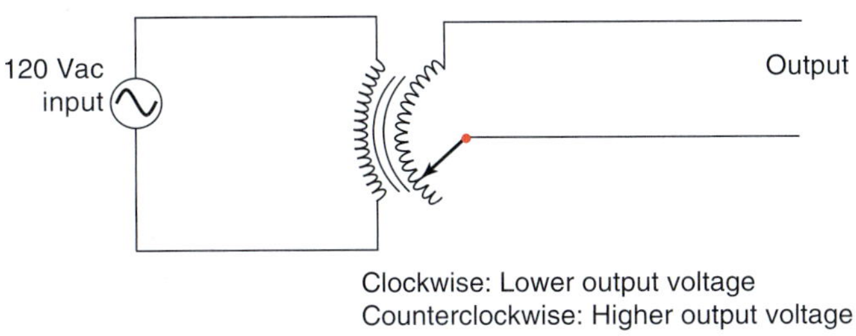

Goodheart-Willcox Publisher

Figure 21-17. Wiring diagram of a variable transformer.

reduces the number of turns in the secondary winding and, as a result, also reduces to measured voltage across the secondary winding.

21.7 Multiple-Winding Transformer

The multiple-winding transformer is constructed with multiple windings, **Figure 21-18**. In this type of transformer, the number of turns in each of the secondary windings can be different from the number of turns in each of the other windings. In addition, each of the secondary windings can act independently as a step-up or step-down transformer, or even as an isolation transformer with a turns ratio of 1:1. One of the many benefits of a multiple-winding transformer is that each of the secondary circuits is electrically isolated from the others.

When performing calculations involving a multiple-winding transformer, each primary-secondary winding combination is evaluated separately.

Refer to **Example 21-9** and **Example 21-10** on how to calculate the measured voltage for a multiple-winding transformer.

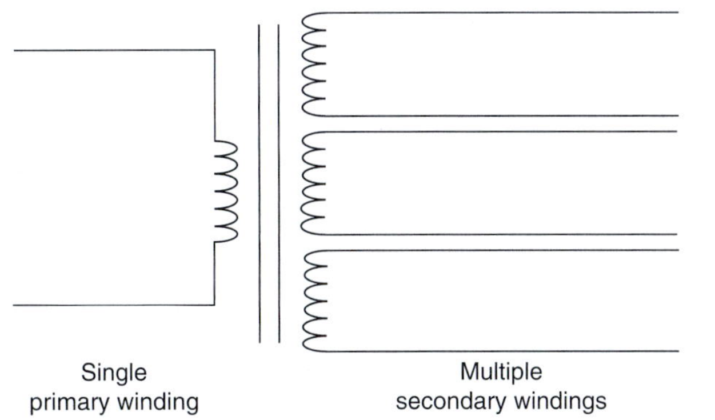

Goodheart-Willcox Publisher

Figure 21-18. A multiple-winding transformer with a single primary winding and multiple secondary windings.

Example 21-9

Referring to the following wiring diagram, calculate the measured voltage at each of the two secondary windings.

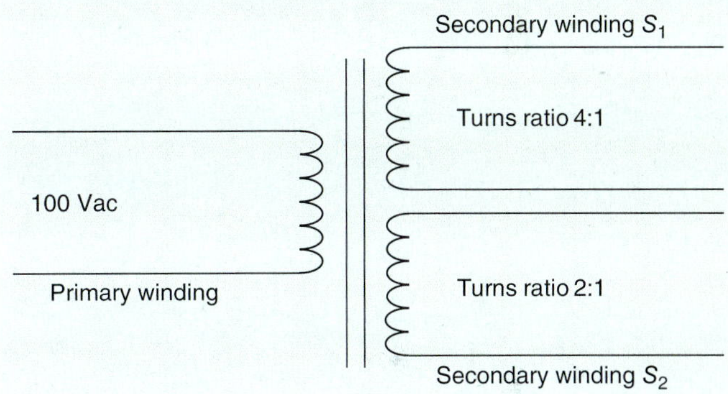

Goodheart-Willcox Publisher

✓ Solution

Treating each primary-secondary winding combination as a separate transformer, the voltage at secondary winding S_1 can be calculated using the turns ratio-voltage formula:

$$TR = \frac{E_P}{E_{S1}}$$

Rearrange the equation to solve for the secondary voltage:

$$E_{S1} = \frac{E_P}{TR}$$

$$E_{S1} = \frac{100 \text{ V}}{4}$$

$$E_{S1} = 25 \text{ V}$$

The voltage at the secondary winding S_2 is then calculated as follows:

$$E_{S2} = \frac{E_P}{TR}$$

$$E_{S2} = \frac{100 \text{ V}}{2}$$

$$E_{S2} = 50 \text{ V}$$

Example 21-10

Calculate the current flow through each of the secondary circuits in the following wiring diagram.

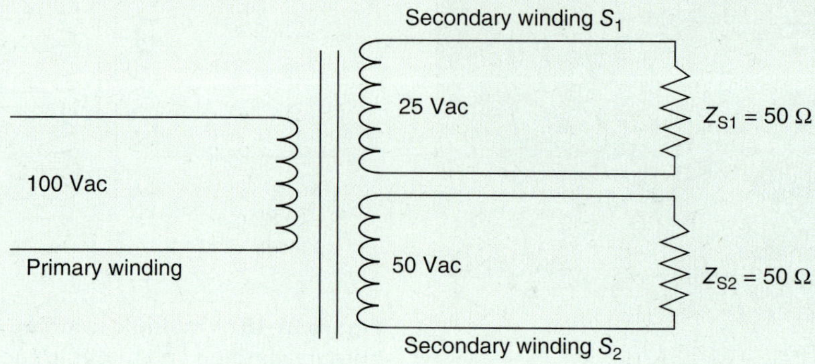

Goodheart-Willcox Publisher

✓ Solution

The current flow through secondary circuit S_1 can be calculated using Ohm's law:

$$I_{S1} = \frac{E_S}{Z_{S1}}$$

$$I_{S1} = \frac{25\text{ V}}{50\ \Omega}$$

$$I_{S1} = 0.5\text{ A}$$

The current in the S_2 circuit is represented by I_{S2} and is calculated as follows:

$$I_{S2} = \frac{E_{S2}}{Z_{S2}}$$

$$I_{S2} = \frac{50\text{ V}}{50\ \Omega}$$

$$I_{S2} = 1\text{ A}$$

Calculating the current flow in the primary winding of a multiple-winding transformer requires the use of a modified form of the $E_P \times I_P = E_S \times I_S$ formula. This formula, discussed earlier in this chapter, states that the apparent power (VA) of the primary is equal to the apparent power (VA) of the secondary. If each primary-secondary winding combination were evaluated separately, the apparent power of each secondary winding would be equal to the primary winding. For a transformer with two secondary windings, there would be twice as much apparent power at the secondary winding than at the primary winding. This, of course, would be greatly beneficial, but is impossible. The apparent power of the primary winding is, therefore, calculated as the sum of the apparent power of each of the secondary windings. For a multiple-winding transformer with two secondary windings, the formula would be as follows:

$$E_P \times I_P = (E_{S1} \times I_{S1}) + (E_{S2} \times I_{S2})$$

The following formula can be used to calculate the current flow in the primary winding of a multiple-winding transformer:

$$I_P = \frac{(E_{S1} \times I_{S1}) + (E_{S2} \times I_{S2})}{E_P}$$

Refer to **Example 21-11** on how to calculate current flow in the primary winding of a multiple-winding transformer.

Example 21-11

Referring to the following wiring diagram, calculate the current flow in the primary circuit.

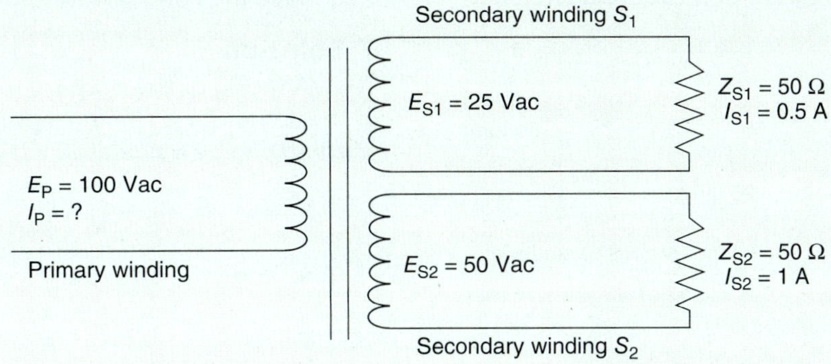

Goodheart-Willcox Publisher

✓ Solution

Substitute the known value into the formula for primary winding current in a two-secondary multiple-winding transformer:

$$I_P = \frac{(E_{S1} \times I_{S1}) + (E_{S2} \times I_{S2})}{E_P}$$

$$I_P = \frac{(25\ \text{V} \times 0.5\ \text{A}) + (50\ \text{V} \times 1\ \text{A})}{100\ \text{V}}$$

$$I_P = \frac{12.5\ \text{VA} + 50\ \text{VA}}{100\ \text{V}}$$

$$I_P = \frac{62.5\ \text{VA}}{100\ \text{V}}$$

$$I_P = 0.625\ \text{A}$$

21.8 Autotransformers

Unlike an isolation transformer, an ***autotransformer*** has only one winding. This single winding is connected to both the power leads that supply voltage to the transformer and the leads that supply power to the secondary circuit, **Figure 21-19**. An autotransformer can be used as either a step-down transformer or a step-up transformer.

One of the main advantages of autotransformers is that they are lighter and less expensive to produce than transformers constructed with two separate windings. The main disadvantage of autotransformers is that they do not isolate the primary and secondary circuits. The electrically connected primary and secondary circuits cause autotransformers to encounter more problems than isolation transformers. As a result, the use of autotransformers is rare. The calculations for autotransformers are the same as those used to evaluate isolation transformers. Just like isolation transformers, an autotransformer can be configured as a multitap transformer, a multitap switch transformer, or a variable transformer. Autotransformers cannot be configured as multiple-winding devices, as autotransformers share a single winding. Refer to **Example 21-12**, **Example 21-13**, and **Example 21-14**.

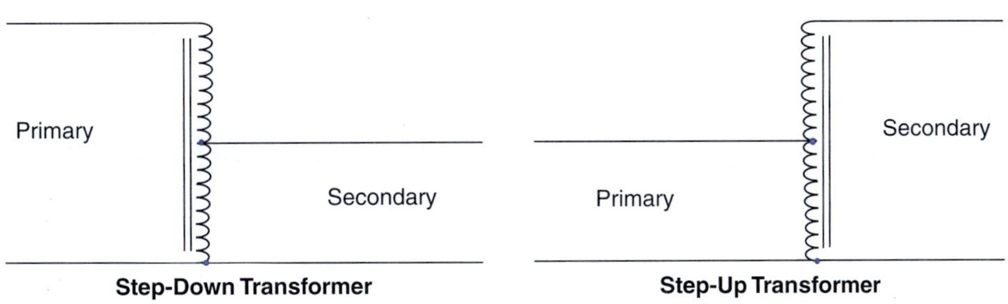

Goodheart-Willcox Publisher

Figure 21-19. Autotransformers can either be step-down transformers or step-up transformers.

Example 21-12

Calculate the voltage measured across leads C and 3 for the following wiring diagram of an autotransformer.

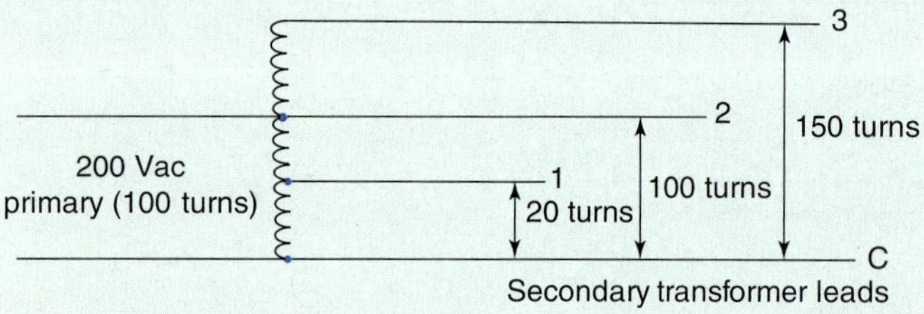

Goodheart-Willcox Publisher

✓ Solution

The measured voltage between leads C and 3 is determined using the turns ratio of the transformer. With 100 turns in the primary winding and 150 turns between leads C and 3, the turns ratio is 100:150, which can be expressed as 1:1.5, 2:3, 2/3, or 0.667.

With the turns ratio known, the secondary voltage can now be calculated using the turns ratio-voltage formula:

$$TR = \frac{E_P}{E_S}$$

Rearrange the equation to solve for secondary voltage:

$$E_S = \frac{E_P}{TR}$$

$$E_S = \frac{200\text{ V}}{0.667}$$

$$E_S = 300\text{ V}$$

Example 21-13

If the impedance between leads C and 3 in **Example 21-12** is 1.5 kΩ, what is the current flow through the circuit connected to the secondary winding?

✓ Solution

From **Example 21-12**, it was determined that the secondary voltage, E_S, is 300 V. Using Ohm's law, current is calculated as follows:

$$I = \frac{E}{Z}$$

$$I = \frac{300\text{ V}}{1500\ \Omega}$$

$$I = 0.2\text{ A}$$

Example 21-14

Using the results from **Example 21-12** and **Example 21-13**, determine the current flow through the primary winding.

✓ Solution

The VA rating of the primary must equal the VA rating of the secondary. The current flow in the primary winding can be calculated as follows:

$$E_P \times I_P = E_S \times I_S$$

Rearrange the equation to solve for primary current:

$$I_P = E_S \times \frac{I_S}{E_P}$$

$$I_P = 300\text{ V} \times \frac{0.2\text{ A}}{200\text{ V}}$$

$$I_P = \frac{60\text{ VA}}{200\text{ V}}$$

$$I_P = 0.3\text{ A}$$

Now It's Your Turn 21-14

For the autotransformer shown in the following wiring diagram, calculate the secondary voltage, secondary current, and primary current.

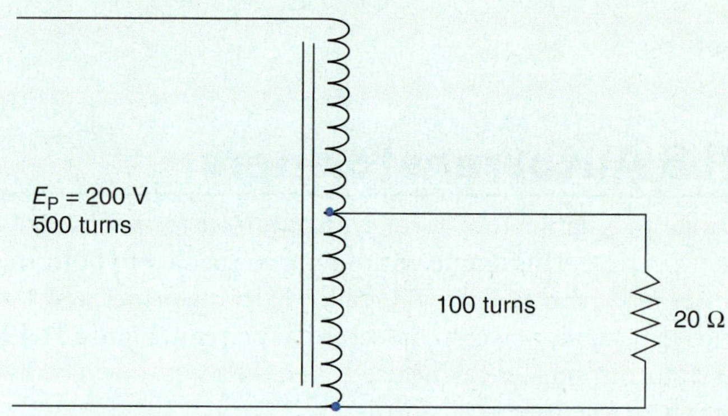

Goodheart-Willcox Publisher

21.9 Transformer Polarity

Electrical windings or coils shown in wiring diagrams, particularly those found in transformers, are marked with either a dot, a square, or an X to indicate their polarity, **Figure 21-20**. The transformer polarity markings on a transformer indicate two things:

- The manner in which the leads are connected to the transformer windings.
- The relative direction of the primary voltage when compared to the induced voltage in the transformer's secondary winding.

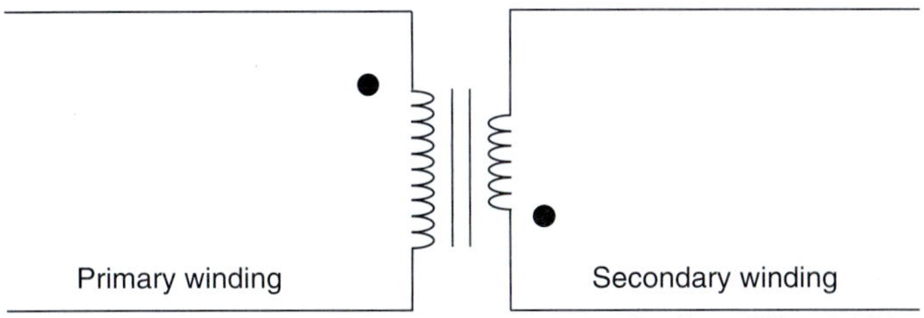

Goodheart-Willcox Publisher

Figure 21-20. Polarity markings on a transformer wiring diagram. The two dots indicate connection points on the transformer that have the same polarity.

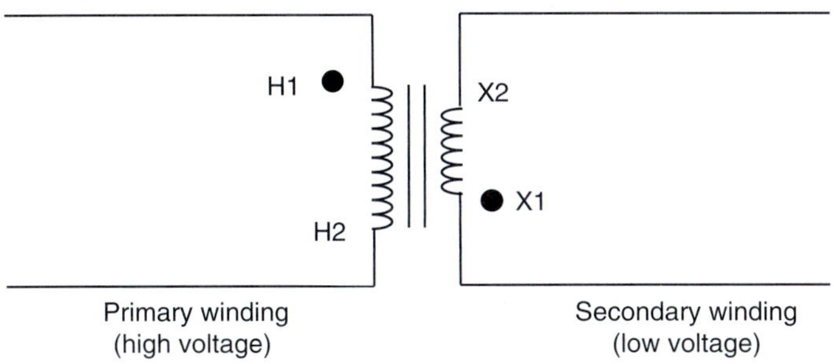

Goodheart-Willcox Publisher

Figure 21-21. The transformer polarity markings in this wiring diagram indicate that the H1 and X1 transformer leads have the same polarity.

Shedding Light

For technicians in the electrical trades, it is important to understand transformer polarities, especially when installing transformers in existing electrical systems. If the system has only one transformer, the polarity of the windings is less important, but should, under no circumstances, be ignored.

Transformer Lead Identification

The leads on transformers are typically marked for easy identification. High-voltage leads are typically identified as H1 and H2 and low-voltage leads are typically identified as X1 and X2. The H1 and X1 leads have the same polarity and will both have similar markings, **Figure 21-21**.

If a transformer does not have polarity markings on its leads or connection points, use the following procedure to identify the leads that have similar polarities.

1. Ensure that the transformer is not powered and is physically disconnected from the power source.
2. Position the transformer so it is being viewed from the low-voltage side. The left lead on the high-voltage side is, by industry conventions, the H1 lead. The right lead on the high-voltage side is the H2 lead, **Figure 21-22**.

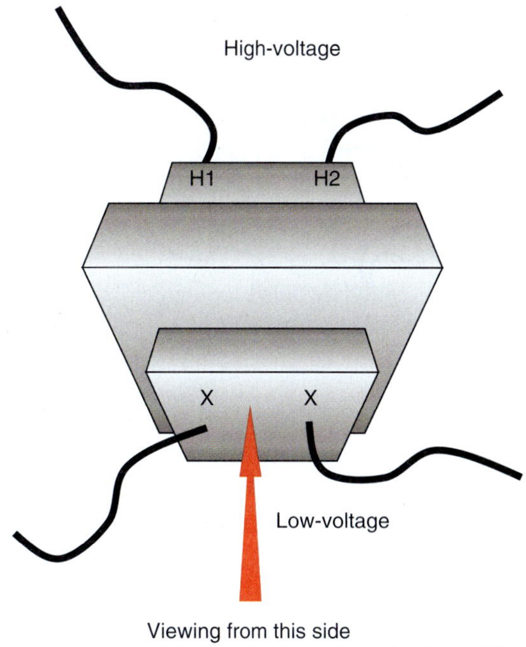

Goodheart-Willcox Publisher

Figure 21-22. When viewing the transformer from the low-voltage side, the left lead on the high-voltage side is the H1 lead and the right lead on the high-voltage side is the H2 lead.

3. Connect a jumper wire from the H1 lead to the X lead on the left side of the low-voltage transformer winding.
4. Connect a voltmeter between the H2 lead and the X lead on the right side of the low-voltage transformer winding.
5. Apply a known voltage to the H1/H2 leads of the transformer.
6. Measure the voltage between the H2 lead and the X lead on the right side of the low-voltage transformer winding.
7. Compare the measured voltage to the voltage supplied to the transformer.
8. If the measured voltage is greater than the voltage supplied to the transformer, then the X2 terminal is the left terminal on the low-voltage transformer winding, **Figure 21-23**.
9. If the measured voltage is less than the voltage supplied to the transformer, then the X2 terminal is the right terminal on the low-voltage transformer winding, **Figure 21-24**.

If the H1 and X1 terminal leads are on the same side of the transformer, the polarity markings will also be on the same side of the transformer. In this configuration, the voltage measured is lower than the voltage supplied to the transformer. Transformers configured in this manner are called ***buck transformers***, **Figure 21-25**. A buck transformer is also referred to as a *subtractive transformer*. In operation, the secondary winding of the buck transformer is being used to reduce, or buck, the main supply voltage.

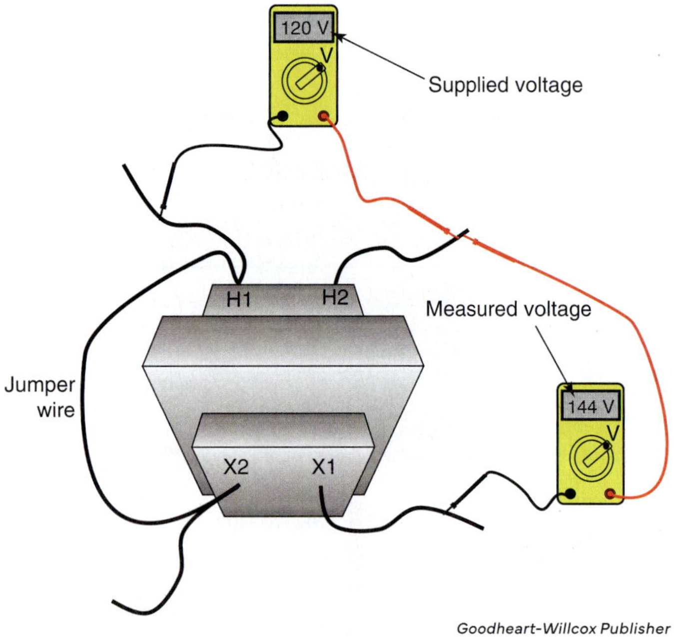

Figure 21-23. When the measured voltage is greater than the voltage supplied to the transformer, then the X2 terminal is the left terminal on the low-voltage transformer winding.

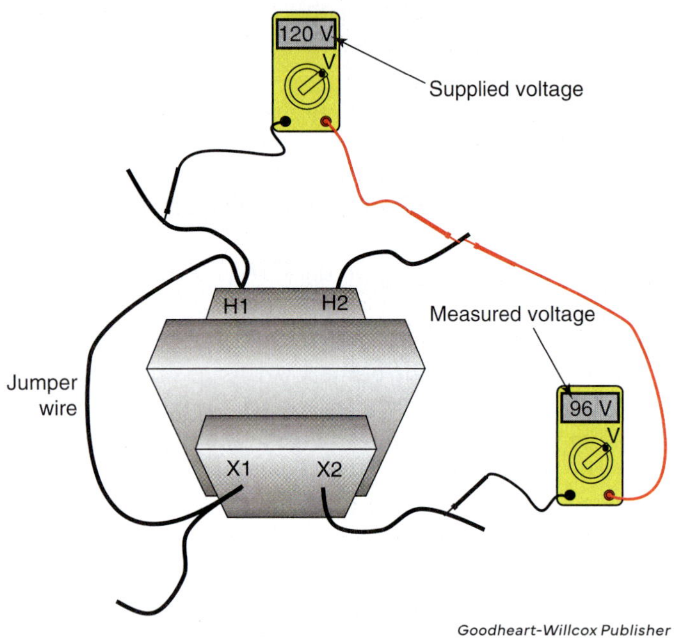

Figure 21-24. When the measured voltage is less than the voltage supplied to the transformer, then the X2 terminal is the right terminal on the low-voltage transformer winding.

Shedding Light

The buck and boost transformer concept is used in many applications to adjust the voltage supplied to equipment that relies on ac power. The boost transformer is used in step-up applications, while the buck transformer is used in step-down applications. For example, a boost transformer could be installed at the end of a long run of wire to increase voltage lost to voltage drop.

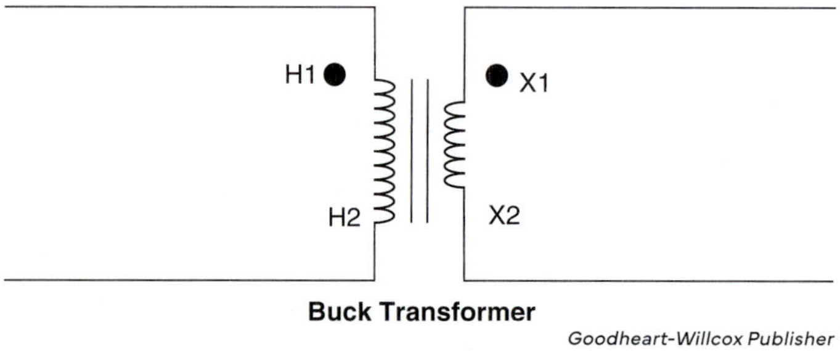

Figure 21-25. Terminal identification for a buck transformer.

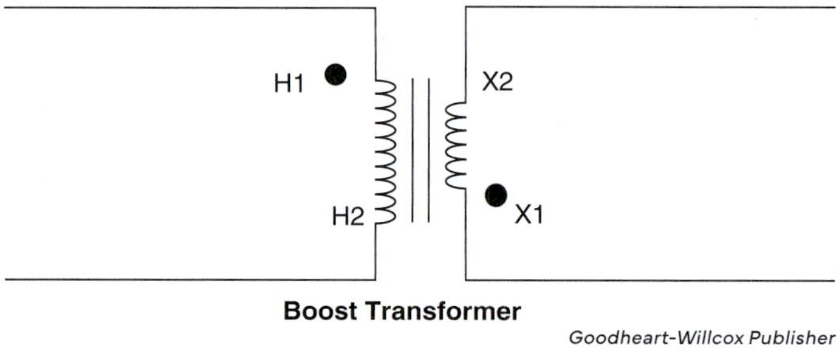

Figure 21-26. Terminal identification for a boost transformer.

If the H1 and X1 terminal leads are on opposite sides of the transformer, the polarity markings will also be on opposite sides of the transformer. In this configuration, the voltage measured is greater than the voltage supplied to the transformer. Transformers configured in this manner are called ***boost transformers***, **Figure 21-26**. A boost transformer is also referred to as an *additive transformer*. In operation, the secondary winding of the boost transformer is being used to increase, or boost, the main supply voltage.

Polarity and Current Flow

A general rule regarding a transformer's polarity markings and the direction of current flow through each winding is as follows: current will always flow into the polarity mark of one winding and flow out of a polarity mark on the other winding.

In a subtractive transformer, the polarity markings are directly across from each other, **Figure 21-27**. The current flows through the primary winding in a clockwise direction. In the primary winding, the current flows from the polarity marking toward the winding. In the secondary winding, current flows through the winding toward the polarity marking in a clockwise direction.

In an additive transformer, the polarity markings are positioned diagonally across from each other, **Figure 21-28**. In the primary winding, the current flows clockwise from the polarity marking toward the winding. In the secondary winding, current flows in a counterclockwise direction through the winding toward the polarity marking.

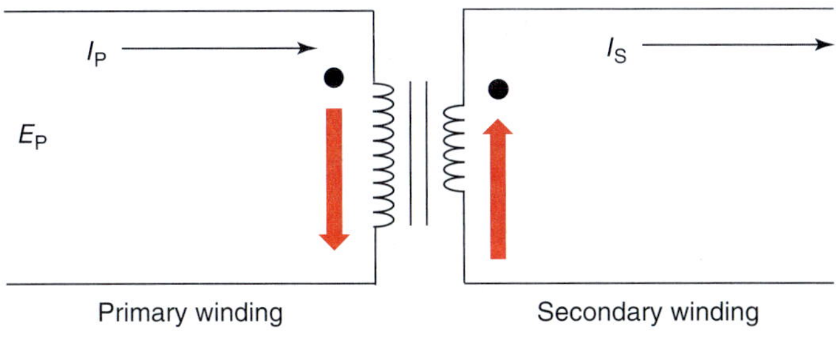

Figure 21-27. For subtractive transformers, the current flow directions in the primary and secondary windings are the same.

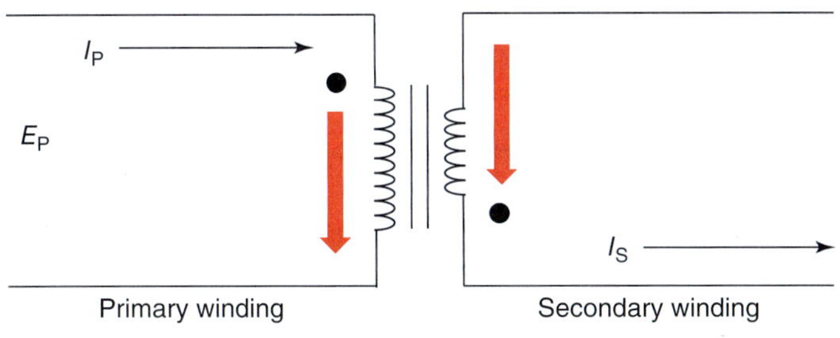

Figure 21-28. For additive transformers, the current flow directions in the primary and secondary windings are opposite to each other.

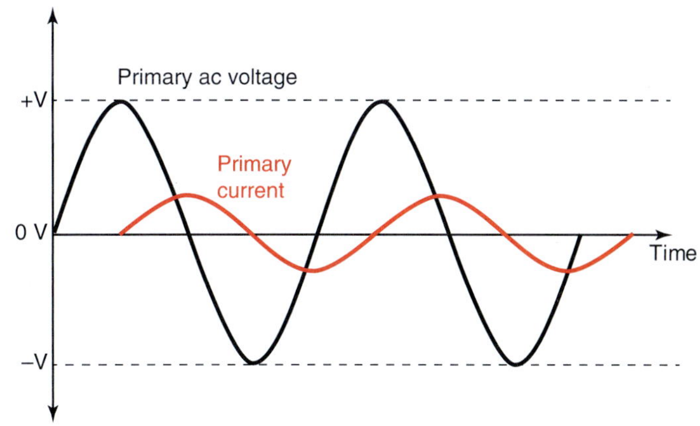

Figure 21-29. The primary current lags the primary voltage by 90°.

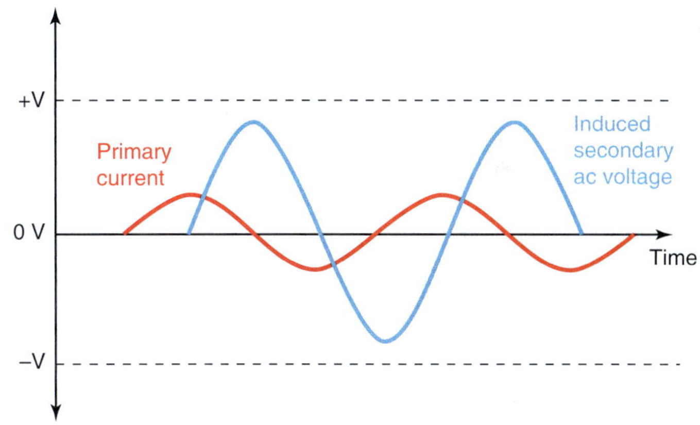

Figure 21-30. The secondary voltage lags the primary current by 90°.

21.10 Transformer Voltage and Current Waveforms

Transformers are inductive loads and, as such, the current lags the voltage by 90°, **Figure 21-29**. This current-lag condition is present in both the primary and secondary windings.

The current flow through the primary winding induces a voltage in the secondary winding. The magnitude of the induced voltage in the secondary winding is proportional to the rate of current change. As a result, there is a 90° offset between the primary current and the secondary voltage, **Figure 21-30**.

Shedding Light

When the rate of current change is the greatest, namely at 0° and 180°, the voltage is at its maximum. When the rate of current change is zero, at 90° and 270°, the voltage is zero. Chapter 17, *Inductors and Inductance*, discusses the rate of current change in inductors.

Since current lags the voltage by 90° in a purely inductive circuit, the secondary current lags the secondary voltage by 90°, **Figure 21-31**. The following summarizes the relationship that exists among the primary voltage, the primary current, the secondary voltage, and the secondary current:

- The primary current lags the primary voltage by 90°.
- The secondary voltage lags the primary current by 90°.
- The secondary current lags the secondary voltage by 90°.

The primary voltage and the secondary voltage are 180° out of phase with each other. The primary and secondary currents are also 180° out of phase with each other, **Figure 21-32**.

21.11 Transformer Troubleshooting

Single-phase transformers are easy to evaluate and troubleshoot. If voltage is supplied to the transformer's primary winding, voltage should be present at the secondary. A simple input-output troubleshooting flowchart can be used to evaluate the operation of a transformer, **Figure 21-33**.

If a transformer is defective, it is beneficial to know why the transformer failed. If the transformer failed due to unresolved external circuit or component problems, replacement transformers are likely to fail as well. Using a digital multimeter to test the windings of the transformer can help determine the cause for failure. Common transformer failures are open

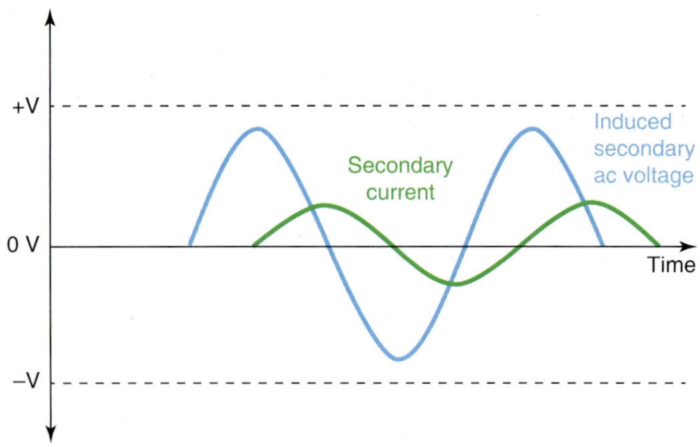

Figure 21-31. The secondary current lags the secondary voltage by 90°.

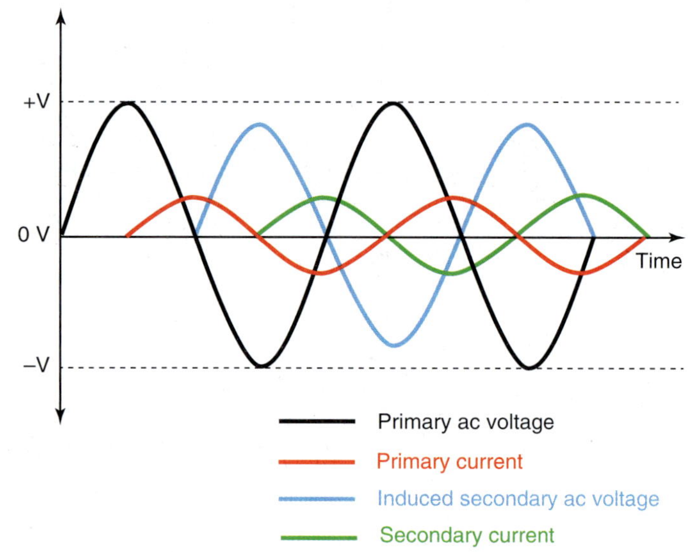

Figure 21-32. Graphical relationship among primary and secondary voltages and currents.

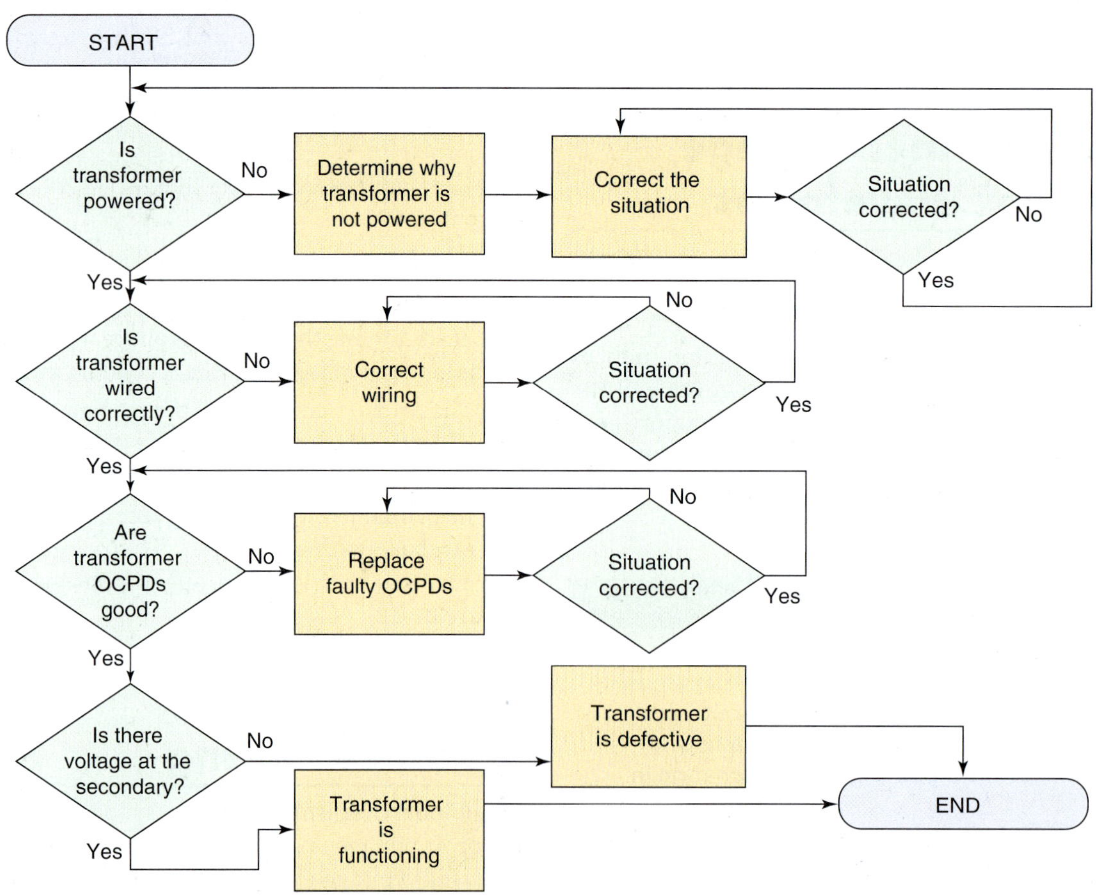

Figure 21-33. Troubleshooting flowchart for simple transformers.

windings, shorted windings, and grounded windings. An open, shorted, or grounded condition can occur in either the primary or secondary winding. Some larger transformers, such as those located outdoors, can become damaged by lightning strikes and other natural occurrences. With the aid of a digital multimeter set to measure resistance, in ohms, the integrity of a transformer can be easily tested. Troubleshooting steps are provided in **Procedure 21-1**, **Procedure 21-2**, and **Procedure 21-3**.

Safety Alert

Before using an ohmmeter to check the windings of a transformer or other electrical component, be sure that the device being tested is de-energized and all interconnecting wiring has been disconnected. This will help to reduce the possibility of damage to the component as well as the piece of test instrumentation.

Procedure 21-1

Testing for Open Transformer Windings

1. Turn the digital multimeter on and set the instrument to measure ohms.
2. Check that the transformer being tested is de-energized, lockout/tagout procedures have been followed, and the connection leads are all disconnected. On a good transformer, there should be measurable resistance across both the primary and secondary winding, **Figure 21-34**. If the resistance reading across either the primary or secondary winding indicates OL on the meter's display, the winding is open, **Figure 21-35**.
3. Verify that any transformer-mounted overcurrent protective devices (OCPDs) are not tripped or open.
4. If an OCPD is open, determine the cause for the condition, remedy the problem, and then replace or reset the OCPD. Follow *NEC* requirements for sizing transformer OCPDs.
5. Once this has been done, retest the transformer.

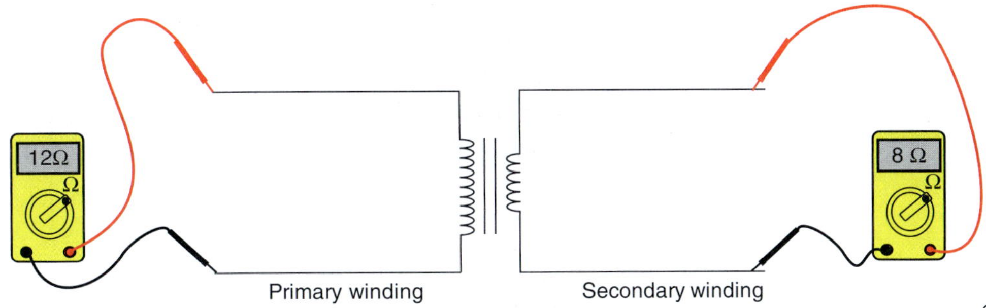

Goodheart-Willcox Publisher

Figure 21-34. Resistance measurements of a good transformer.

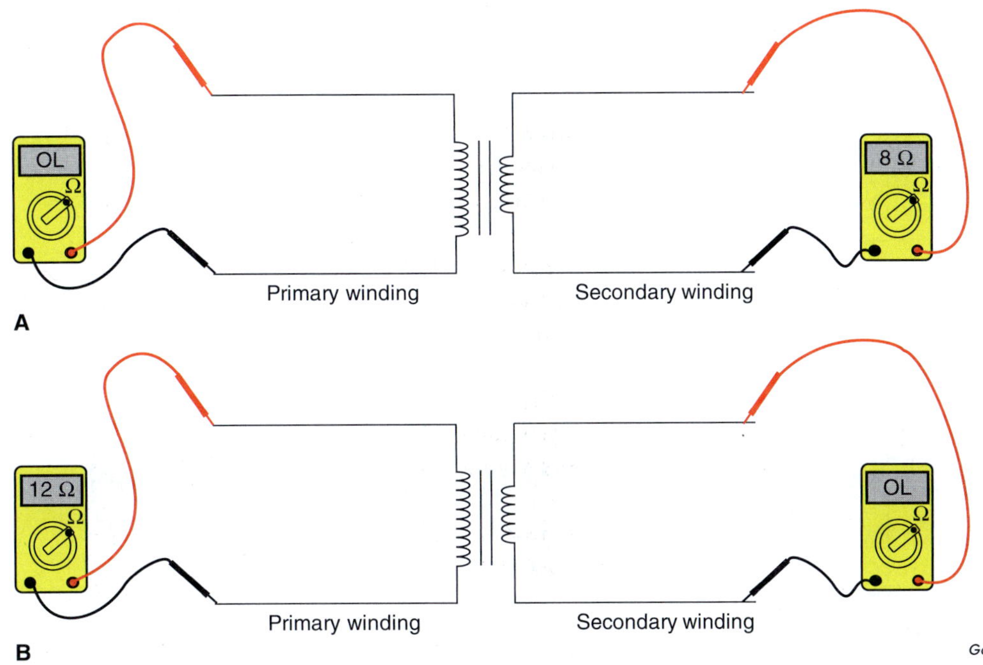

Goodheart-Willcox Publisher

Figure 21-35. Resistance measurements of an opened winding. A—Infinite resistance through the primary winding indicates a probable open primary winding. B—Infinite resistance through the secondary winding indicates a probable open secondary winding.

Procedure 21-2

Test for a Shorted Transformer Winding

1. Turn the digital multimeter on and set the instrument to measure ohms.
2. Ensure that the transformer being tested is de-energized, lockout/tagout procedures are followed, and the connection leads are all disconnected.
3. Measure the resistance across both the primary winding and the secondary winding. On a good transformer, there should be measurable resistance across both windings. If the resistance reading across either the primary or secondary winding is close to zero, or significantly lower than the normal or expected winding resistance, the winding is shorted, **Figure 21-36**.
4. If a shorted winding is found, locate the cause of this condition. This will help direct you to the original cause of the problem. For example, loose electrical connections can cause excessive component humming and overheating, which can lead to insulation breakdown. Insulation breakdown can cause shorted windings. When excessive heat is generated, windings can also be severed.
5. Measure the resistance between each primary winding lead and each secondary winding lead. There should be no electrical connection between the primary and secondary windings (except in autotransformers). There should be infinite resistance measured between any primary winding lead and any secondary winding lead, **Figure 21-37**. If there is measurable resistance between any primary winding lead and any secondary winding lead, **Figure 21-38**, there is a short circuit present in the transformer.

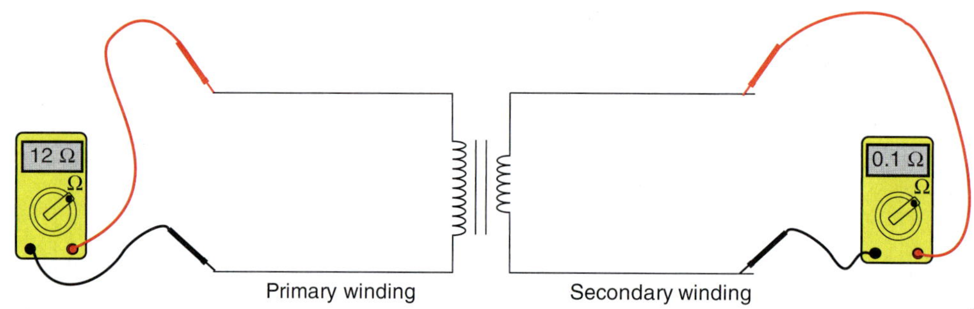

Goodheart-Willcox Publisher

Figure 21-36. Resistance measurements of a shorted secondary winding; the unexpectedly low resistance across the secondary winding indicates the problem.

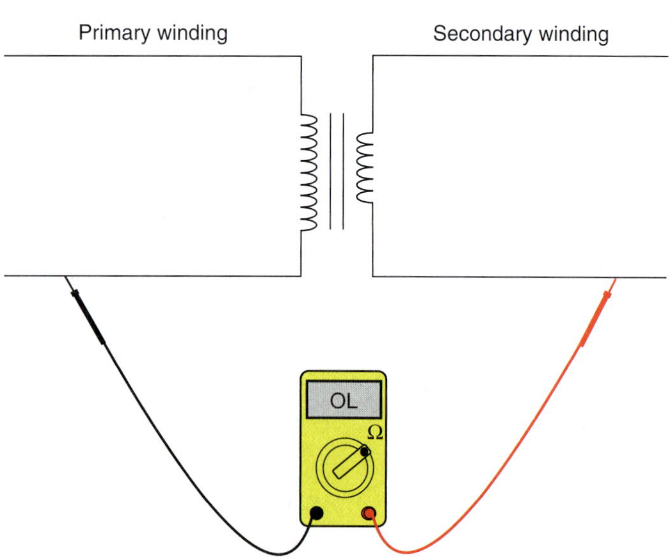

Goodheart-Willcox Publisher

Figure 21-37. Resistance measurements of a good isolation transformer.

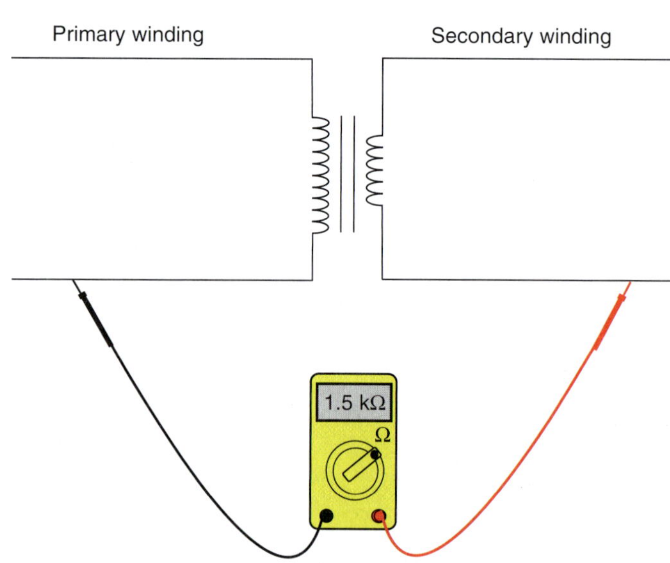

Goodheart-Willcox Publisher

Figure 21-38. Resistance measurements of a shorted isolation transformer.

Procedure 21-3

Test for a Grounded Transformer Winding

1. Turn the digital multimeter on and set the instrument to measure ohms.
2. Ensure that the transformer being tested is de-energized and that the connection leads are all disconnected.
3. Measure the resistance between each lead and ground. On a good transformer, there should be infinite resistance between any transformer lead and ground, **Figure 21-39**. If there is any measurable resistance between ground and any primary or secondary winding lead, the transformer is grounded, **Figure 21-40**.

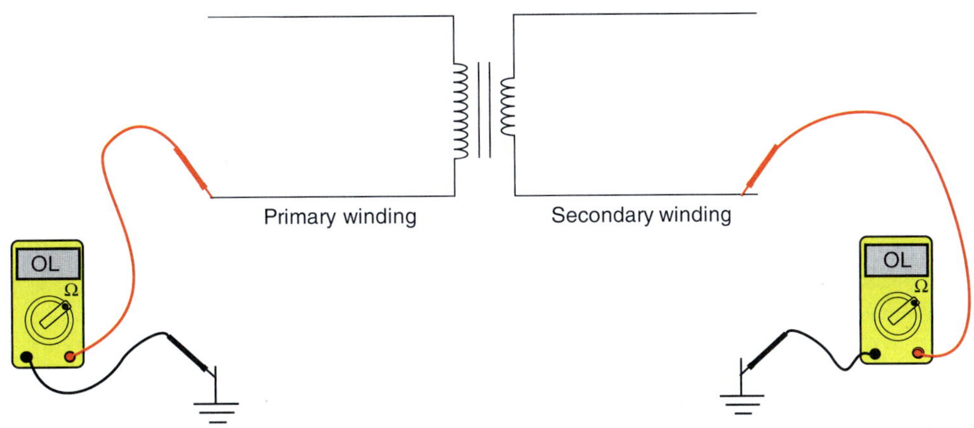

Goodheart-Willcox Publisher

Figure 21-39. Resistance measurements of a transformer with no grounds.

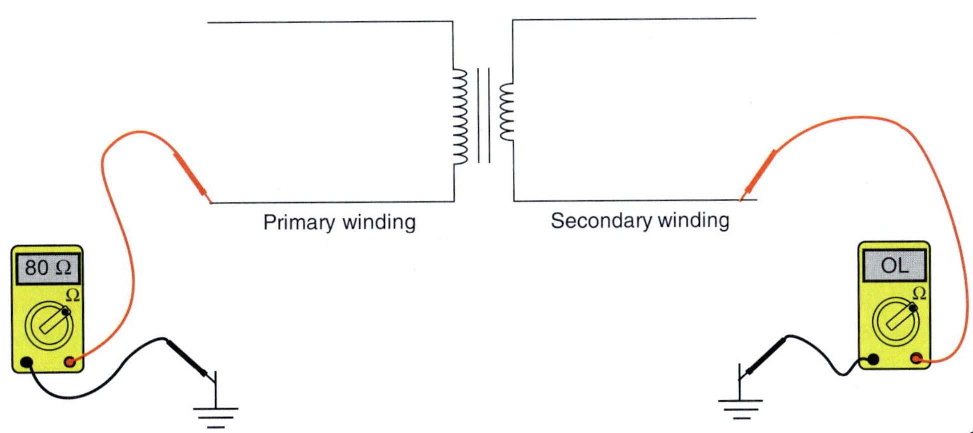

Goodheart-Willcox Publisher

Figure 21-40. Resistance measurements of a grounded transformer.

CHAPTER 21 REVIEW

Summary

- Transformers are responsible for changing voltage levels in an electric circuit. (21.1)
- Transformers are commonly found in appliances, test equipment, power distribution and transmission systems, stabilizers, rectifiers, and voltage regulation systems. (21.1)
- The basic transformer is made up of a primary winding and a secondary winding. (21.2)
- The relationship between the number of turns in the primary winding and the number of turns in the secondary winding is called the turns ratio. (21.2)
- The relationship between the current in the primary winding and the current in the secondary winding is the inverse of the turns ratio. (21.2)
- The relationship between a winding's voltage and the number of turns in the winding is called the volts-per-turn ratio. (21.2)
- A step-up transformer increases the voltage supplied to it, while a step-down transformer reduces the voltage supplied to it. (21.3)
- Isolation transformers have separate primary and secondary windings that are electrically isolated from each other. (21.4)
- Multitap transformers are configured with multiple lead wires connected to the primary or secondary winding. (21.5)
- Variable transformers have a sliding contact that moves along the surface of the secondary winding. (21.6)
- Multiple-winding transformers are connected with multiple windings: one single primary winding and multiple secondary windings, for example. (21.7)
- For calculations involving a multiple-winding transformer, the apparent power of the primary winding is calculated as the sum of the apparent power of each of the secondary windings. (21.7)
- Autotransformers have a single winding that serves as both the primary and secondary winding. (21.8)
- When connecting multiple transformers, the polarity of each transformer must be known to ensure proper connections. (21.9)
- A buck transformer is also referred to as a *subtractive transformer*. In operation, the secondary winding of the buck transformer is being used to reduce, or buck, the main supply voltage. (21.9)
- In a buck transformer, the winding current flows are opposite to each other. (21.9)
- A boost transformer is also referred to as an *additive transformer*. In operation, the secondary winding of the boost transformer is being used to increase, or boost, the main supply voltage. (21.9)
- In a boost transformer, the winding current flows are in the same direction. (21.9)
- Transformers are inductive loads and, as such, the current lags the voltage by 90°. (21.10)
- Transformer failures are commonly caused by open windings, shorted windings, and grounded windings. (21.11)

Review Questions

Use the information in the chapter to answer the following questions.

Measuring Your Potential

1. Of the following, which is *not* true regarding the use of transformers? (21.1)
   A. Electrolysis often relies on transformers to produce the high voltage needed to drive chemical reactions.
   B. Transformers are used to match the impedance of a power source.
   C. Rectifier transformers convert dc power to ac power.
   D. Transformers are used to convert a system's line voltage power supply to a safer lower level.
2. The two windings on a transformer are the _____. (21.2)
   A. auto winding and secondary winding
   B. self-winding and auto winding
   C. primary winding and secondary winding
   D. self-winding and primary winding
3. If a transformer has a primary voltage of 120 V and a secondary voltage of 24 V, which of the following is a possible combination of primary and secondary current values? (21.2)
   A. $I_P = 1$ A; $I_S = 5$ A
   B. $I_P = 5$ A; $I_S = 1$ A
   C. $I_P = 1.2$ A; $I_S = 2.4$ A
   D. $I_P = 2.4$ A; $I_S = 1.2$ A
4. What is the approximate turns ratio of a step-up ignition transformer that, when supplied 120 V, provides a spark of 10,000 V? (21.3)
   A. 85:1
   B. 1:85
   C. 170:1
   D. 1:170

5. If the primary voltage is 100 V and the secondary voltage is 20 V, the transformer is a _____. (21.3)
   A. step-down device with a turns ratio of 5:1
   B. step-down device with a turns ratio of 1:5
   C. step-up device with a turns ratio of 5:1
   D. step-up device with a turns ratio of 1:5
6. *True or False?* An isolation transformer has a single winding that is used for both the primary and secondary windings. (21.4)
7. A _____ has multiple leads connected to the primary winding at various points. (21.5)
   A. multitap transformer
   B. multitap switch transformer
   C. variable transformer
   D. multiple-winding transformer
8. Which of the following transformers has its winding leads permanently connected to a multipole rotary switch? (21.5)
   A. Multitap transformer
   B. Multitap switch transformer
   C. Variable transformer
   D. Multiple-winding transformer
9. The _____ has a sliding contact that moves along the surface of the secondary winding. (21.6)
   A. multitap transformer
   B. multitap switch transformer
   C. variable transformer
   D. multiple-winding transformer
10. Of the following, which is the formula used to calculate the current flow in a multiple-winding transformer with two secondary windings? (21.7)
    A. $TR = \frac{N_P}{N_S}$
    B. $\frac{N_P}{N_S} = \frac{E_P}{E_S}$
    C. $E_P \times I_P = (E_{S1} \times I_{S1}) + (E_{S2} \times I_{S2})$
    D. $\frac{E_P}{E_S} = \frac{I_S}{I_P}$
11. *True or False?* Autotransformers can be desirable because they are lighter than isolation transformers. (21.8)
12. *True or False?* Transformer polarity is commonly indicated by dots, triangles, or squares. (21.9)
13. *True or False?* In transformers, the current lags the voltage by 180°. (21.10)
14. A resistance reading of OL across either the primary or secondary winding of a multimeter indicates that the _____. (21.11)
    A. winding is open
    B. winding is shorted
    C. transformer has a short-circuit
    D. transformer is grounded

Plug-In

1. Calculate the turns ratio of a transformer that operates with a primary winding current of 6 A and a secondary winding current of 2 A. (21.2)
2. Calculate the VA rating of a transformer with a primary voltage of 250 V and a secondary voltage of 25 V if the primary current is 0.25 A. (21.2)
3. If the turns ratio of a step-down transformer is 6:1, determine the secondary voltage and current if the primary voltage is 120 Vac and the primary current is 0.5 A. (21.3)
4. Determine the turns ratio of the step-down transformer shown in the following wiring diagram. (21.3)

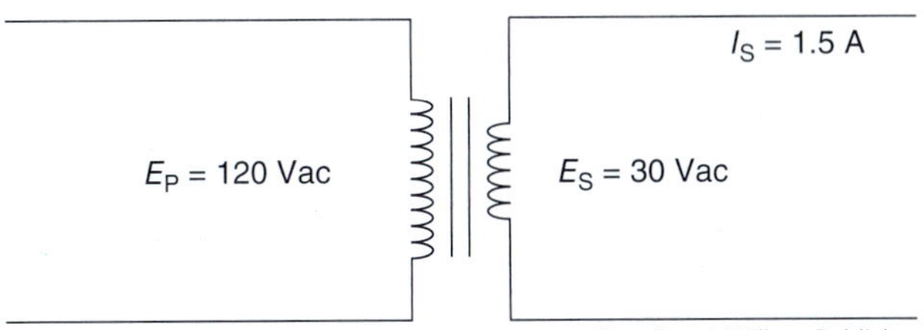

Goodheart-Willcox Publisher

5. If a transformer has a primary voltage of 120 Vac and a secondary voltage of 30 Vac, which of the following is true? (21.9)
   A. A connection voltage of 150 Vac will be measured between the primary and secondary if the transformer is a buck transformer.
   B. A connection voltage of 120 Vac will be measured between the primary and secondary if the transformer is a buck transformer.
   C. A connection voltage of 150 Vac will be measured between the primary and secondary if the transformer is a boost transformer.
   D. A connection voltage of 30 Vac will be measured between the primary and secondary if the transformer is a boost transformer.

Making the Connection

1. Explain some of the possible reasons that the VA calculation at the primary side of a transformer will differ from the VA calculation at the secondary side of a transformer. (21.2)
2. Explain why the VA calculation at the secondary side of a transformer and the VA calculation at the primary side of a transformer are equal when evaluating a "perfect" transformer. (21.7)

CHAPTER

22 Three-Phase Transformers

kaninw/Shutterstock.com

SPARKING DISCUSSION

Why are 480/277 V wye systems the primary service utilized by larger occupancies of all types?

LEARNING OUTCOMES

After completing this chapter, you will be able to:

22.1 Explain the function of a three-phase transformer.

22.2 List the voltage and current formulas for a delta and wye configuration.

22.3 Describe the difference between a wye connection and a delta connection.

22.4 Explain the wiring and function of an open-delta transformer, a T-connected transformer, and a Scott transformer.

22.5 Evaluate voltage and current on three-phase transformer systems.

TECHNICAL TERMS

center tap

delta-delta three-phase transformer

delta-wye three-phase transformer

open-delta transformer

Scott transformer

T-connected transformer

transformer bank

wye-delta three-phase transformer

wye-wye three-phase transformer

Introduction

Throughout the electrical industry, three-phase transformers are commonly found in power distribution, larger commercial applications, and industrial applications. Like single-phase transformers, three-phase transformers are used to increase or decrease voltage levels. The majority of power generated in the United States is three-phase power, so it stands to reason that step-up and step-down three-phase transformers play a major role in the power distribution process.

As a worker in the electrical trades, you will need to have a working knowledge and understanding of three-phase transformers. As this is an advanced topic, you will be tapping into some previously obtained knowledge, and expanding on this knowledge as you move forward in your studies.

Familiarize yourself with the material covered in Chapter 16, *Three-Phase Alternating Current Power,* and Chapter 21, *Single-Phase Transformers,* before you begin this chapter. The material in this chapter combines the theories and concepts covered in those two chapters along with your previously acquired electrical knowledge.

In this chapter, you will learn about different types of three-phase transformers, their configurations, electrical connections, and how to calculate electrical values as they apply to three-phase transformers. The benefits and drawbacks of each type of three-phase transformer configuration are also discussed.

22.1 Three-Phase Transformer Construction

In its simplest form, a three-phase transformer is created by combining three single-phase transformers in a variety of configurations. The term ***transformer bank*** is used to describe a three-phase transformer created when three individual, single-phase transformers are connected together. When a three-phase transformer is created from three separate single-phase transformers, close attention must be paid to ensure that the polarity of the connections is correct. In general, transformer banks are common for utilities involved in the transmission and generation of larger-scale power.

To help reduce the possibility of improper wiring and phasing, preassembled, three-phase transformers are used. They consist of three pairs of single-phase windings wound around a single laminated core, **Figure 22-1**. These preassembled transformers are often preferred because they are generally smaller, polarity-balanced, lighter, and, in most cases, less expensive than an equivalent transformer bank.

Shedding Light

In addition to primary and secondary voltage ratings and winding configurations, transformers are also rated by their apparent power (VA or kVA). The power rating of a transformer dictates the size of loads that can be safely powered. Some common sizes of three-phase transformers installed in large commercial and industrial occupancies include 15 kVA, 30 kVA, 45 kVA, 75 kVA, 112.5 kVA, 225 kVA, 300 kVA, and 500 kVA.

22.2 Three-Phase Transformer Windings

By their very nature, transformers are made up of primary and secondary windings. The primary windings of a transformer

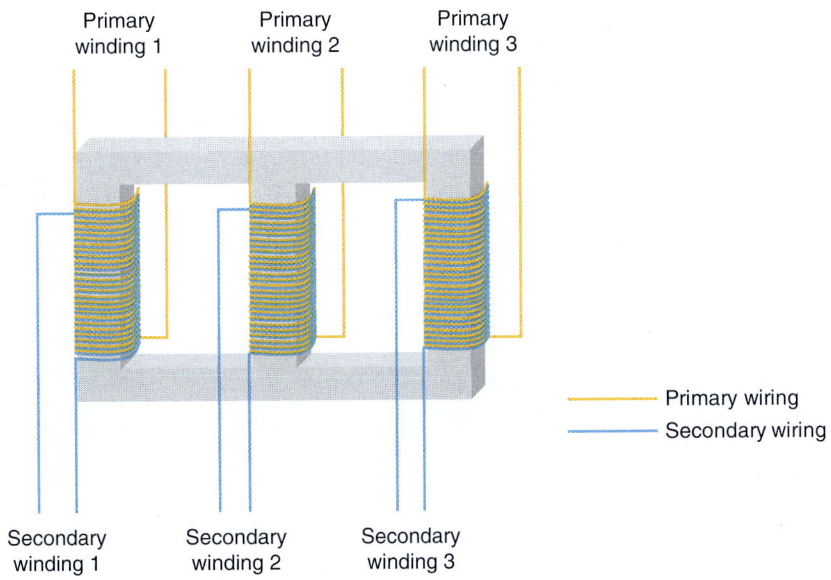

Goodheart-Willcox Publisher

Figure 22-1. A three-phase transformer consists of three single-phase transformers wound around a single core.

are connected to the system's main power supply. The secondary windings are found at the outlet of the transformer and act as the power supply for the circuitry connected to the transformer's secondary windings.

The windings in three-phase transformers are commonly configured as either a delta connection or a wye connection, **Figure 22-2**. The voltage and current formulas vary depending on the winding configuration. The sections that follow provide a review of these formulas.

Transformer nameplates were introduced in Chapter 21, *Single-Phase Transformers*. A nameplate includes critical information about the transformer, including information about the primary and secondary winding configurations and electrical connections, **Figure 22-3**.

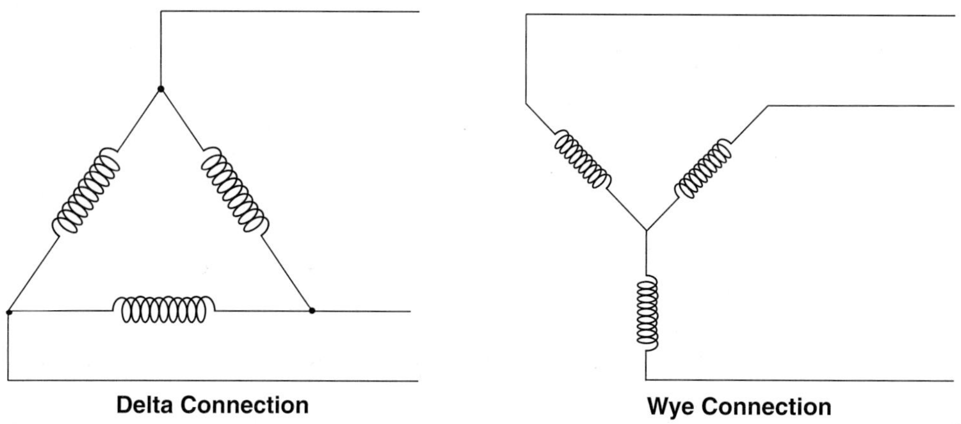

Goodheart-Willcox Publisher

Figure 22-2. Three-phase transformers are commonly configured as either a delta or wye connection.

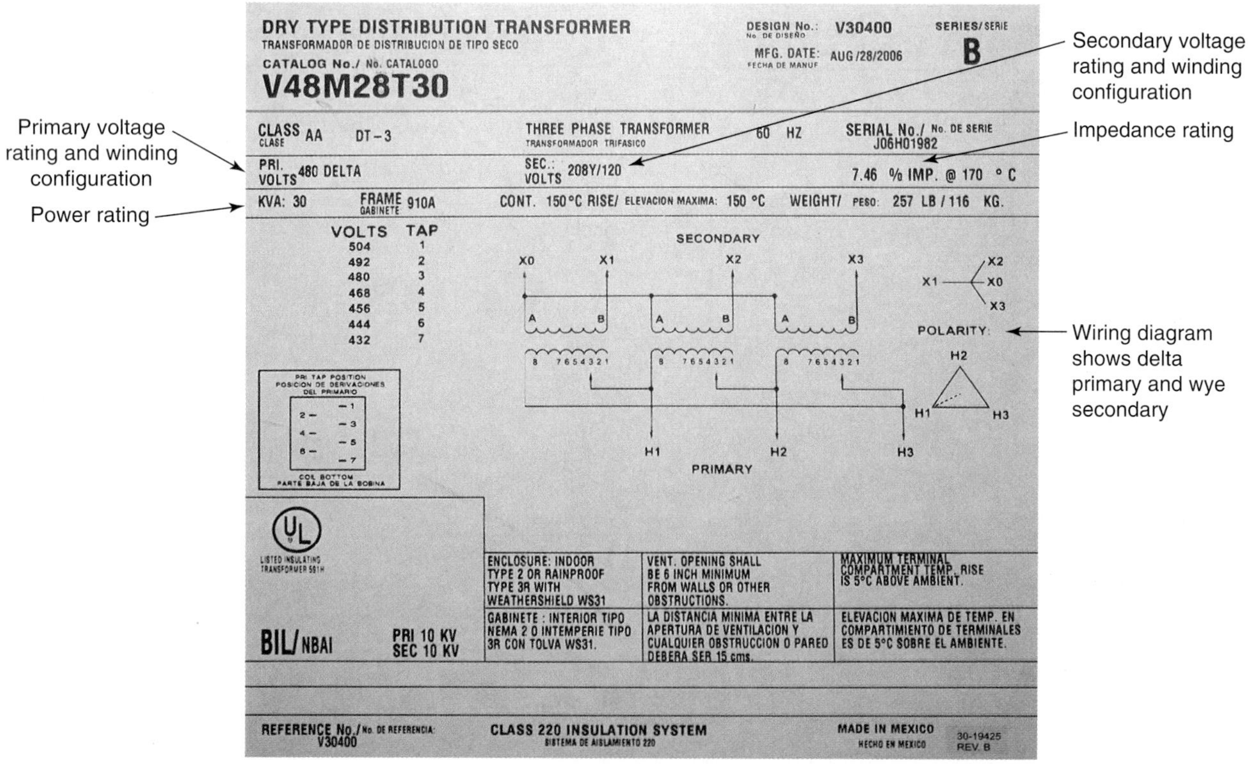

Goodheart-Willcox Publisher

Figure 22-3. This three-phase transformer nameplate provides critical technical characteristics of the transformer.

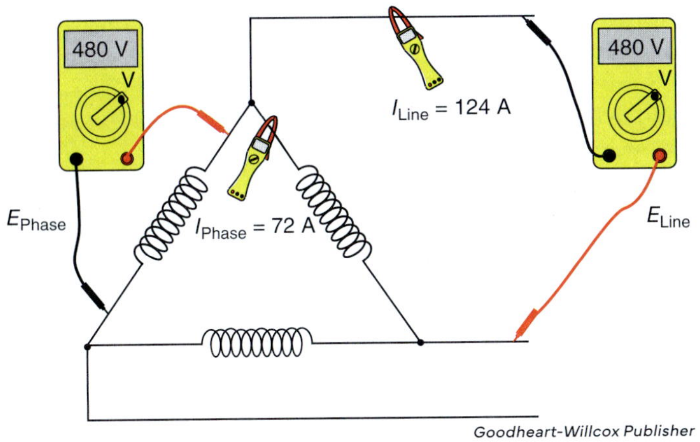

Goodheart-Willcox Publisher

Figure 22-4. Locations for voltage and current measurements on a delta connection. The phase voltage and line voltage are the same in a delta configuration.

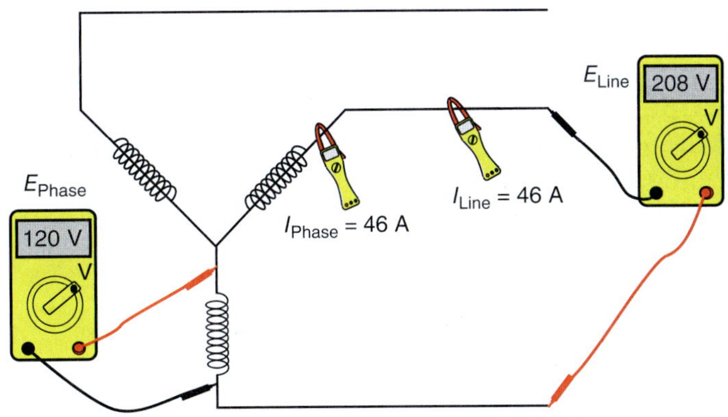

Goodheart-Willcox Publisher

Figure 22-5. Locations for voltage and current measurements on a wye connection. The phase current and line current are the same in a wye configuration.

Voltage and Current Formulas for a Delta Configuration

When windings are configured as a delta connection, the following properties apply:

- The phase voltage, E_{Phase}, is equal to the line voltage, E_{Line}.

$$E_{Phase} = E_{Line}$$

- The phase current, I_{Phase}, is equal to the line current, I_{Line}, divided by $\sqrt{3}$.

$$I_{Phase} = \frac{I_{Line}}{\sqrt{3}}$$

- The line current, I_{Line}, is equal to the phase current, I_{Phase}, multiplied by $\sqrt{3}$.

$$I_{Line} = I_{Phase} \times \sqrt{3}$$

The points at which these values are measured are identified in **Figure 22-4**. In this figure, from an electrical standpoint, the phase voltage, E_{Phase}, equals the line voltage, E_{Line}.

Voltage and Current Formulas for a Wye Configuration

When windings are configured as a wye connection, the following properties apply:

- The phase voltage, E_{Phase}, is equal to the line voltage, E_{Line}, divided by $\sqrt{3}$.

$$E_{Phase} = \frac{E_{Line}}{\sqrt{3}}$$

- The line voltage, E_{Line}, is equal to the phase voltage, E_{Phase}, multiplied by $\sqrt{3}$.

$$E_{Line} = E_{Phase} \times \sqrt{3}$$

- The phase current, I_{Phase}, is equal to the line current, I_{Line}.

$$I_{Phase} = I_{Line}$$

The points at which these values are measured are identified in **Figure 22-5**. From an electrical standpoint, the phase current, I_{Phase}, is equal to the line current, I_{Line}.

Some other useful formulas that will be used throughout this chapter include the following:

$$\text{Turns ratio } (TR) = \frac{E_{Phase(P)}}{E_{Phase(S)}}$$

$$TR = \frac{I_{Phase(S)}}{I_{Phase(P)}}$$

$$VA = \sqrt{3} \times I_{Line} \times E_{Line}$$

$$VA = 3 \times I_{Phase} \times E_{Phase}$$

where:

$E_{Phase(P)}$ = Phase voltage in a primary winding
$E_{Phase(S)}$ = Phase voltage in a secondary winding
$E_{Line(P)}$ = Line voltage in a primary winding
$E_{Line(S)}$ = Line voltage in a secondary winding
VA = Apparent power (in volt-amperes)

These formulas can be used on both wye connections and delta connections. Each of the two formulas provided for VA can be used on either the primary or secondary side of the transformer. Refer to **Appendix A** for a complete reference for these formulas, as well as others covered in this text.

22.3 **Common Three-Phase Transformers**

The following are common three-phase transformer configurations:

- Wye-wye transformer
- Delta-delta transformer
- Wye-delta transformer
- Delta-wye transformer

Wye-Wye Transformer

The ***wye-wye three-phase transformer***, also referred to as the *star-star configuration*, has both the primary and secondary windings configured as wye, or star, connections, **Figure 22-6**.

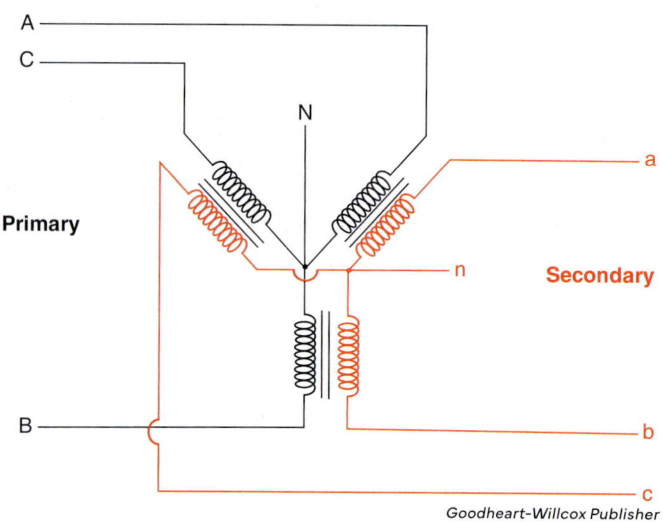

Figure 22-6. A wye-wye three-phase transformer with grounded conductors at the common connection point.

A schematic wiring diagram of the wye-wye connection is provided in **Figure 22-7**.

Some of the advantages of the wye-wye configuration include the following:

- There are no phase differences between the primary and secondary voltages.
- The phase voltage is lower than the line voltage, so fewer turns and less insulation is needed for the windings.
- The turns ratio is low.
- As they use less materials, they are lighter.
- Two different secondary voltages can be utilized: line-to-line and line-to-neutral.

The wye-wye three-phase transformer is the most economical option when the secondary load is balanced. When the neutral conductor is included and grounded, the effects of an unbalanced load can be minimized. Slightly unbalanced loads can cause signal distortion and interference. For this reason, when interference is a concern, especially in communication systems, the wye-wye configuration is not commonly used. With the grounded conductor in place, two different secondary voltages can be used, **Figure 28-8**.

A main drawback of individual transformers connected in a wye-wye configuration is that if any one of the three transformers fail, the transformer will no longer have the ability to provide three-phase power.

Delta-Delta Transformer

The ***delta-delta three-phase transformer*** has both the primary and secondary windings configured as delta connections, **Figure 22-9**. A schematic wiring diagram of the delta-delta connection is provided in **Figure 22-10**. On the

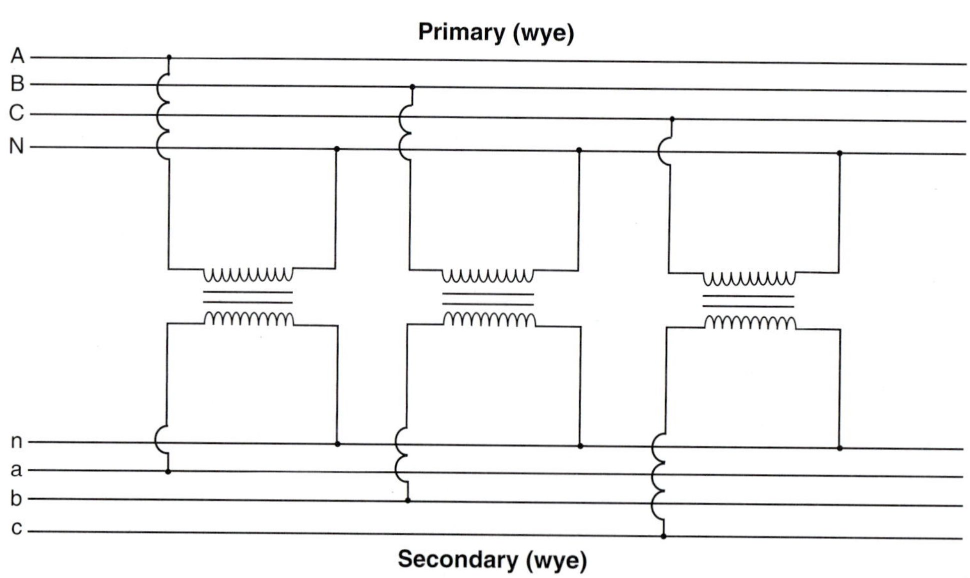

Figure 22-7. A wiring diagram of a wye-wye three-phase transformer.

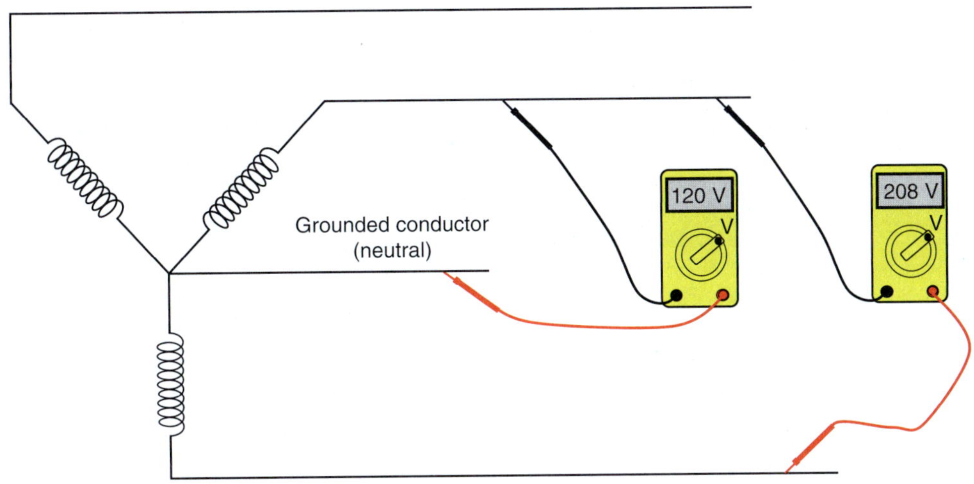

Goodheart-Willcox Publisher

Figure 22-8. Two different voltages can be obtained from a wye-wye connection with a connected grounded conductor.

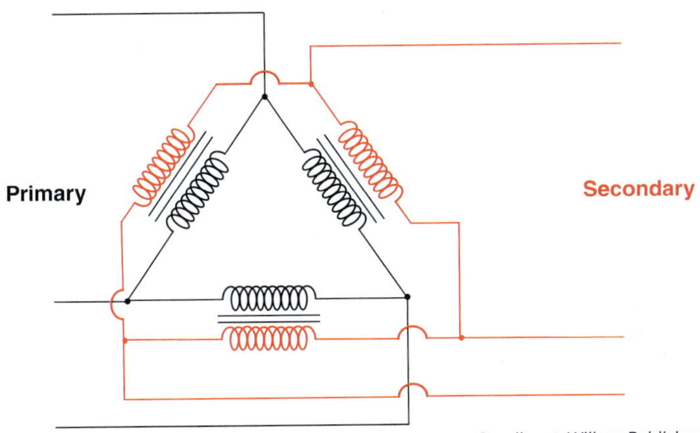

Goodheart-Willcox Publisher

Figure 22-9. A delta-delta three-phase transformer.

primary side of the transformer, each ungrounded conductor (line) is connected to one side of two different windings. This, in essence, connects the three windings in a continuous loop. The same wiring strategy is used on the secondary side.

Some of the advantages of the delta-delta configuration include the following:

- The phase currents are lower on delta-delta transformers than on wye-wye transformers, so the conductors used in the windings can have a smaller cross-sectional area.
- If one phase fails, the transformer will still provide three-phase power, but at a reduced output equal to 57.7% of the original capacity.

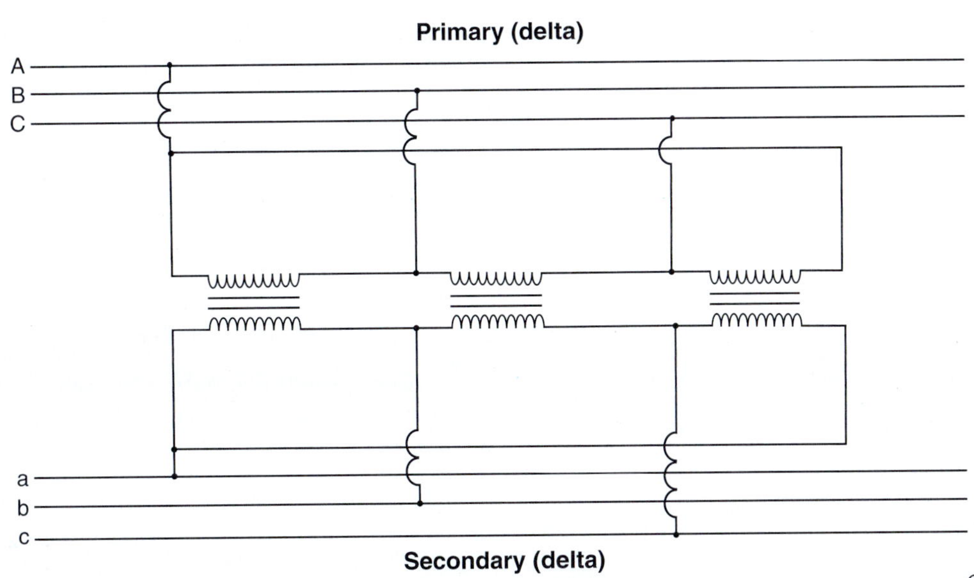

Goodheart-Willcox Publisher

Figure 22-10. A wiring diagram of a delta-delta three-phase transformer.

- This configuration can handle varying, unbalanced loads better than a wye-configured transformer.
- There is no phase displacement between the primary and secondary windings, meaning the secondary voltage is less likely to experience distortion.

Shedding Light

If a three-phase transformer is created by connecting three single-phase transformers together, it is important to ensure that all three transformers are phased correctly, meaning the polarity of all connected transformers must be correct. One way to check the polarity of the transformer connections on a transformer with a delta secondary is to take a voltage reading across two secondary winding leads before the final connection is made to close the delta. If the polarity of the connections is correct, the obtained voltage reading will be close to 0 V. If the polarity of a winding is incorrect, the voltage reading will be twice the expected value. If the voltage reading indicates an incorrect winding polarity, the connections will need to be checked and corrected. It is important to note that the incorrect winding polarity can be in either the transformer's primary or secondary winding.

One disadvantage of the delta-delta configuration is that the voltage applied to the windings is equal to the line voltage, increasing the insulation requirements for the transformer windings relative to a comparable wye configuration. Unlike transformers with a wye-connected secondary, the delta-delta transformer does not have the ability to provide multiple secondary voltages without the addition of a center tap. The concept of a center tap will be discussed later in this chapter when considering the open-delta configuration.

Two additional disadvantages of three-phase transformers with a delta-configured secondary are the following:

- Each transformer must be wound for the full-line voltage, requiring more conductive material and insulation.
- There is no grounded conductor or neutral to facilitate multiple voltages.

Delta-delta transformers are best suited for applications requiring a single secondary voltage for three-phase equipment, such as motors. Additionally, these transformers are most suited for applications with relatively small voltage changes between primary and secondary.

Refer to **Example 22-1** on how to evaluate a delta-delta transformer.

Example 22-1

For the 45 kVA three-phase, delta-delta transformer shown in the following figure, calculate:

- Primary winding phase voltage, $E_{Phase(P)}$
- Secondary winding phase voltage, $E_{Phase(S)}$
- Line voltage at the transformer's secondary winding, $E_{Line(S)}$
- Line current of the primary winding, $I_{Line(P)}$
- Phase current of the primary winding, $I_{Phase(P)}$
- Line current of the secondary winding, $I_{Line(S)}$
- Phase current of the secondary winding, $I_{Phase(S)}$

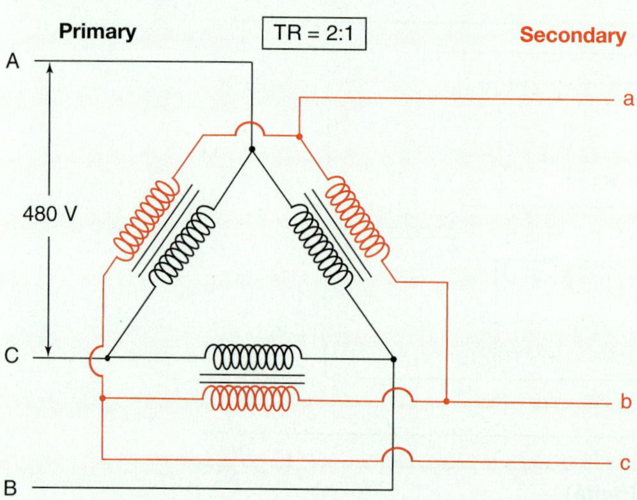

Goodheart-Willcox Publisher

✓ Solution

The formulas used to evaluate a delta-delta three-phase transformer are the same ones used to evaluate a single-phase transformer and a delta-configured three-phase power supply.

Step 1. Calculate the primary winding phase voltage.

For a delta connection, the phase voltage, which is the voltage measured across each individual winding, is equal to the line voltage. Since the primary winding's line voltage is 480 V, the primary winding's phase voltage is also 480 V.

Step 2. Calculate the secondary winding phase voltage.

The phase voltage at the transformer's secondary winding can be calculated using the turns ratio. Note that the turns ratio refers to phase voltages in the transformer.

$$TR = \frac{E_{Phase(P)}}{E_{Phase(S)}}$$

Rearrange the equation to solve for $E_{Phase(S)}$. The turns ratio of 2:1 is expressed as 2/1 or 2.

$$E_{Phase(S)} = \frac{E_{Phase(P)}}{TR}$$

$$E_{Phase(S)} = \frac{480\ V}{2}$$

$$E_{Phase(S)} = 240\ V$$

Step 3. Calculate the line voltage at the secondary winding.

Since the phase and line voltages are the same for a delta connection, the line voltage at the secondary winding is also 240 V.

Step 4. Calculate the line current of the primary winding.

Line current of the primary winding can be calculated by using the known values for VA and line voltage of the primary winding:

$$VA = \sqrt{3} \times E_{Line(P)} \times I_{Line(P)}$$

Rearrange the equation to solve for $I_{Line(P)}$. The turns ratio of 2:1 is expressed as 2/1 or 2.

$$I_{Line(P)} = \frac{VA}{\sqrt{3} \times E_{Line(P)}}$$

$$I_{Line(P)} = \frac{45\text{ kVA}}{\sqrt{3} \times 480\text{ V}}$$

$$I_{Line(P)} = \frac{45{,}000\text{ VA}}{831.4\text{ V}}$$

$$I_{Line(P)} = 54\text{ A}$$

Step 5. Calculate the phase current of the primary winding.

The phase current of the primary winding can be found by substituting the known values for primary line current into the formula for delta configurations as follows:

$$I_{Phase(P)} = \frac{I_{Line(P)}}{\sqrt{3}}$$

$$I_{Phase(P)} = \frac{54\text{ A}}{\sqrt{3}}$$

$$I_{Phase(P)} = 31\text{ A}$$

Step 6. Calculate the line current of the secondary winding.

Substituting the known values for VA and line voltage at the secondary winding, the line current of the secondary winding is calculated as follows:

$$VA = \sqrt{3} \times E_{Line(S)} \times I_{Line(S)}$$

Rearrange the equation to solve for $I_{Line(S)}$.

$$I_{Line(S)} = \frac{VA}{\sqrt{3} \times E_{Line(S)}}$$

$$I_{Line(S)} = \frac{45\text{ kVA}}{\sqrt{3} \times 240\text{ V}}$$

$$I_{Line(S)} = \frac{45{,}000\text{ VA}}{416\text{ V}}$$

$$I_{Line(S)} = 108\text{ A}$$

Step 7. Calculate the phase current of the secondary winding.

The phase current of the secondary winding can be found by using the known value for line current of the secondary winding. These values are substituted into the following formula for delta configurations:

$$I_{Phase(S)} = \frac{I_{Line(S)}}{\sqrt{3}}$$

$$I_{Phase(S)} = \frac{108\text{ A}}{\sqrt{3}}$$

$$I_{Phase(S)} = 62\text{ A}$$

Wye-Delta Transformer

The ***wye-delta three-phase transformer***, also referred to as the *star-delta configuration*, has its primary windings configured as a wye and its secondary windings configured as a delta, **Figure 22-11**.

Although this figure might appear to be configured as a wye-wye connection, by tracing a path through the three secondary windings, you will see that a continuous loop is created, which is indicative of a delta connection. A schematic wiring diagram of the wye-delta connection is provided in **Figure 22-12**. On the secondary side of the transformer, each ungrounded conductor (line) is connected to one side of two different windings, creating a continuous loop.

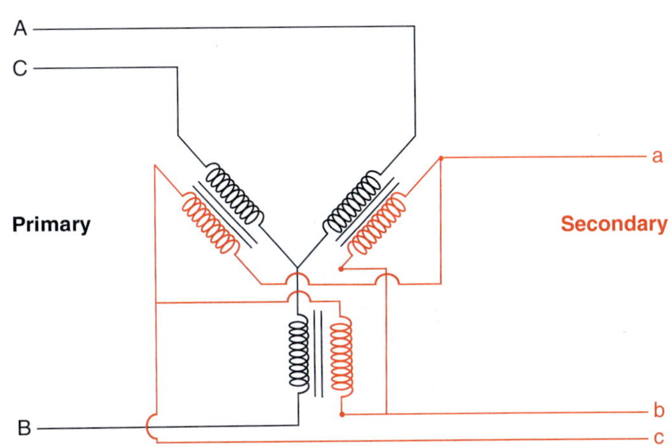

Figure 22-11. A wye-delta three-phase transformer.

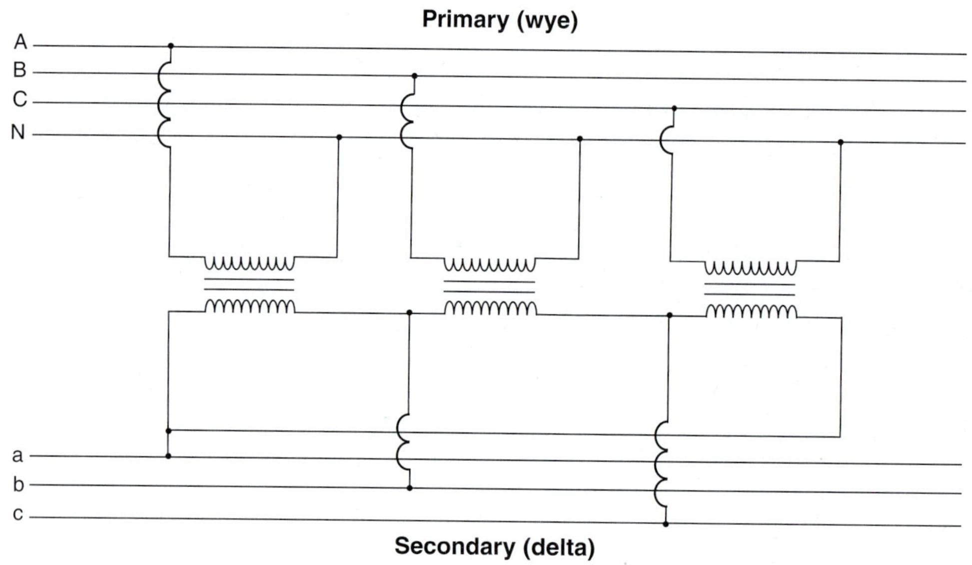

Goodheart-Willcox Publisher

Figure 22-12. A wiring diagram of a wye-delta three-phase transformer.

Some of the advantages of the wye-delta configuration include the following:

- Ability to utilize a line-to-grounded conductor voltage source at the primary winding.
- The phase voltage at the primary is lower than the primary line voltage, reducing the number of turns and the amount of insulation needed in the primary winding.
- The turns ratio is lower than that of a delta-delta configuration.
- Fewer turns in the primary winding are required, so less conductive material is used.
- There is a 30° phase shift between the primary and secondary currents, which reduces electrical noise and interference.

The wye-delta configuration has the benefits of the wye-wye connection without the disadvantage of increased possibility for unbalanced loads and increased signal interference. When interference is a concern, especially in communication systems, a transformer with a delta-configured secondary winding is recommended.

Two disadvantages of three-phase transformers with a delta-configured secondary are:

- Each transformer must be wound for the full-line voltage, requiring more conductive material and insulation.
- There is no grounded conductor or neutral to facilitate multiple voltages in the secondary circuit.

In addition, if any one of the three transformers fails in the wye-delta configuration, the transformer will no longer be able to provide three-phase power.

Like delta-delta transformers, wye-delta transformers are best suited for applications requiring a single secondary voltage for three-phase equipment, such as motors. The reduction of line voltage to phase voltage in wye-connected primary makes wye-delta transformers better suited for step-down applications.

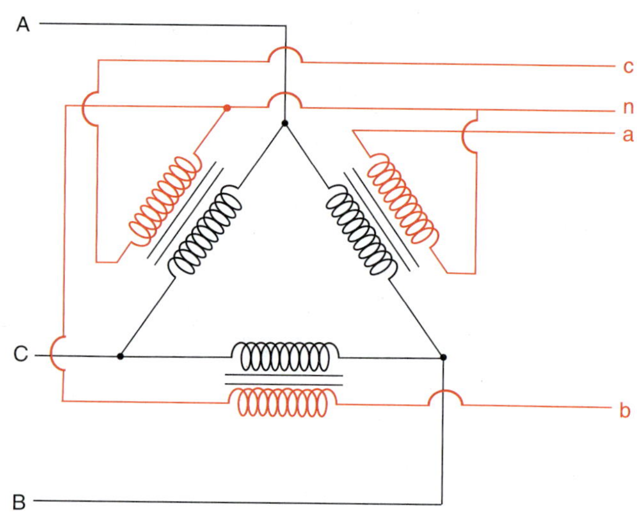

Goodheart-Willcox Publisher

Figure 22-13. A delta-wye three-phase transformer.

Delta-Wye Transformer

The delta-wye connection is the most popular transformer configuration. The ***delta-wye three-phase transformer***, also referred to as the *delta-star configuration*, has its primary windings configured as a delta and its secondary windings configured as a wye, **Figure 22-13**.

Although this figure might appear to be configured as a delta-delta connection, by tracing a path through the three secondary windings, you will see that a continuous loop through all windings is not possible. There is only one winding connected between the "n" terminal and each of the other terminals "a", "b", and "c". This confirms that the secondary windings are configured as a wye connection. A schematic wiring diagram of the delta-wye connection is provided in **Figure 22-14**.

Some of the advantages of the delta-wye configuration include the following:

- The ability to use a line-to-grounded conductor voltage source at the secondary winding.
- Lower amperage draws and heat generation.
- The phase voltage at the secondary is lower than the line voltage at the secondary.
- Fewer turns in the primary winding are required, so less conductive material is used.
- There is a phase shift between the primary and secondary currents, which reduces electrical noise and interference.
- Self-contained transformers including primary and secondary windings are widely available in many common sizes from 15kVA to 500kVA.

Even though the delta-wye configuration is the most popular configuration in the majority of large commercial and industrial applications, there are three primary drawbacks to this configuration:

- The turns ratio of the delta-wye configuration is greater than in any of the other three configurations discussed.
- Although rare, if any of the transformers fail, the entire transformer will fail.
- The phase shift that is present between the primary and secondary voltage and current signals may require isolation circuits to minimize circuit interactions.

Refer to **Example 22-2** on how to evaluate a delta-wye transformer.

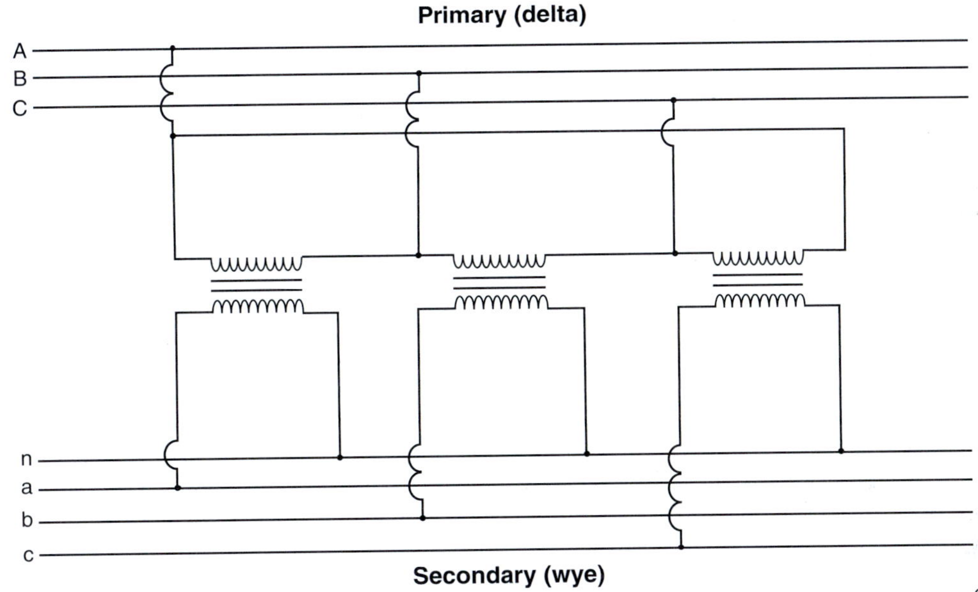

Figure 22-14. A wiring diagram of a delta-wye three-phase transformer.

Example 22-2

For the 75 kVA three-phase, delta-wye step-down transformer shown in the following figure, calculate the following:

- Primary winding phase voltage, $E_{Phase(P)}$
- Secondary winding phase voltage, $E_{Phase(S)}$
- Transformer's turns ratio, *TR*
- Line current of the primary winding, $I_{Line(P)}$
- Phase current of the primary winding, $I_{Phase(P)}$
- Line current of the secondary winding, $I_{Line(S)}$
- Phase current of the secondary winding, $I_{Phase(S)}$

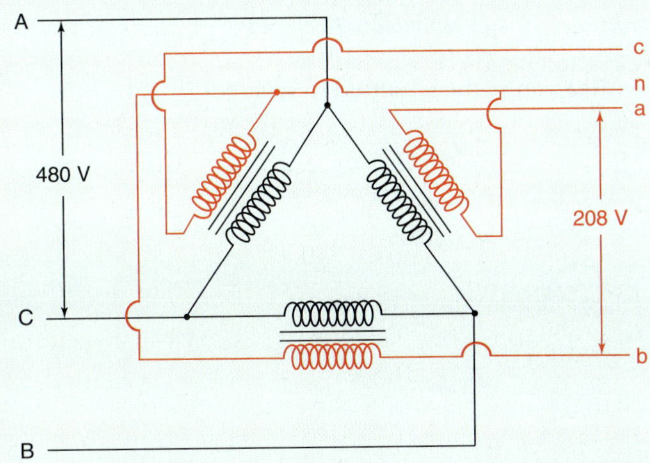

Continued

✓ Solution

Step 1. Calculate the primary winding phase voltage.

For the primary winding's delta connection, the phase voltage is equal to the line voltage. Since the primary winding's line voltage is 480 V, the primary winding's phase voltage is also 480 V.

Step 2. Calculate the secondary winding phase voltage.

To calculate the phase voltage at the transformer's secondary winding, substitute the known value for the line voltage of the secondary wiring into the following formula for wye configurations:

$$E_{Phase(S)} = \frac{E_{Line(S)}}{\sqrt{3}}$$

$$E_{Phase(S)} = \frac{208\text{ V}}{\sqrt{3}}$$

$$E_{Phase(S)} = 120\text{ V}$$

Step 3. Calculate the transformer's turns ratio.

Phase voltages must be used for turns ratio calculations:

$$TR = \frac{E_{Phase(P)}}{E_{Phase(S)}}$$

$$TR = \frac{480\text{ V}}{120\text{ V}}$$

$$TR = 4{:}1$$

Step 4. Calculate the line current of the primary winding.

To calculate the line current of the primary winding, substitute the known values for VA and line voltage of the primary winding, as follows:

$$VA = \sqrt{3} \times E_{Line(P)} \times I_{Line(P)}$$

Rearrange the equation to solve for $I_{Line(P)}$:

$$I_{Line(P)} = \frac{VA}{\sqrt{3} \times E_{Line(P)}}$$

$$I_{Line(P)} = \frac{75\text{ kVA}}{\sqrt{3} \times 480\text{ V}}$$

$$I_{Line(P)} = \frac{75{,}000\text{ VA}}{831\text{ V}}$$

$$I_{Line(P)} = 90\text{ A}$$

Step 5. Calculate the phase current of the primary winding.

To calculate the phase current of the primary winding, use the following formula for delta configurations:

$$I_{Phase(P)} = \frac{I_{Line(P)}}{\sqrt{3}}$$

$$I_{Phase(P)} = \frac{90\text{ A}}{\sqrt{3}}$$

$$I_{Phase(P)} = 52\text{ A}$$

Step 6. Calculate the line current of the secondary winding.

The line current of the secondary is found by using the known values for the secondary line voltage and the VA rating of the transformer as follows:

$$VA = \sqrt{3} \times E_{Line(S)} \times I_{Line(S)}$$

Rearrange the equation to solve for $I_{Line(S)}$.

$$I_{Line(S)} = \frac{VA}{\sqrt{3} \times E_{Line(S)}}$$

$$I_{Line(S)} = \frac{75\text{ kVA}}{\sqrt{3} \times 208\text{ V}}$$

$$I_{Line(S)} = \frac{75{,}000\text{ VA}}{360\text{ V}}$$

$$I_{Line(S)} = 208\text{ A}$$

Step 7. Calculate the phase current of the secondary winding.

Since the line current and the phase current are the same for a wye connection, the phase current for the secondary is also 208 A.

22.4 Other Three-Phase Transformer Configurations

The three-phase transformers covered up to this point are those constructed of three pairs of windings connected in a variety of configurations. It is also possible to produce three-phase power with only two pairs of windings. In addition, multiple single-phase or three-phase power signals can also be created at the secondary windings of a three-phase transformer. In this section, the open-delta transformer, the T-connected transformer, and the Scott transformer are discussed.

Open-Delta Transformer

The ***open-delta transformer*** creates a three-phase secondary signal using only two single-phase transformers. The open-delta transformer is so named because, upon inspection of the wiring diagram, it can be seen that one side of the traditional delta configuration is missing, creating what appears to be an open delta, **Figure 22-15**. Even with only two single-phase transformers, the open-delta configuration is still able to provide three-phase power at its output. All three legs will provide the same voltage output. Because of its appearance, the open-delta transformer is sometimes referred to as a V-connected transformer. Open-delta transformers are mainly used for lighter-duty applications as they only provide 57.7% of the output power (VA) supplied by a traditional three-winding delta transformer with the same phase voltage. This power reduction occurs because the line current in an open-delta configuration is equal to the phase current, **Figure 22-16**.

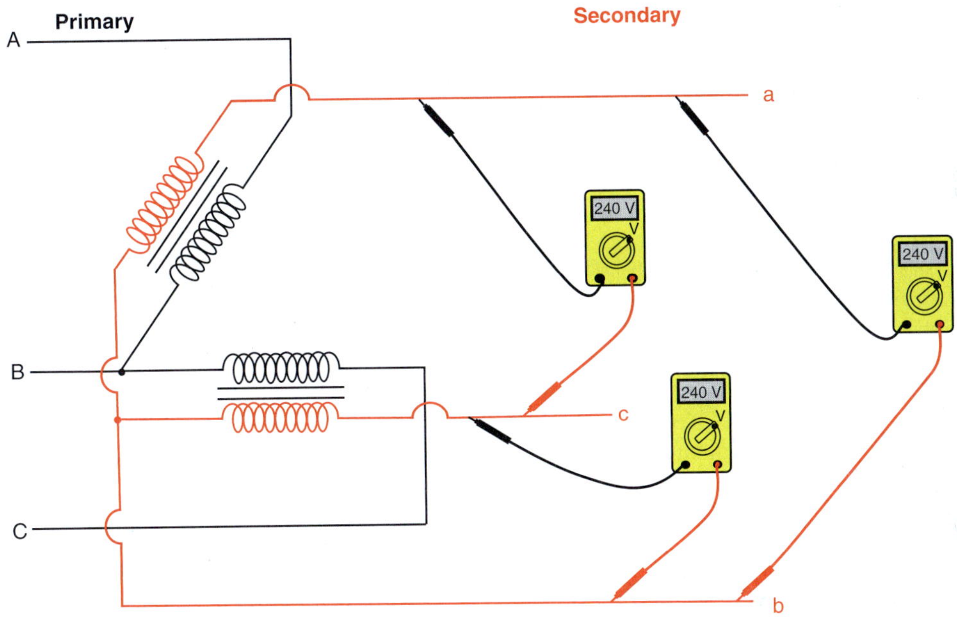

Goodheart-Willcox Publisher

Figure 22-15. An open-delta winding configuration.

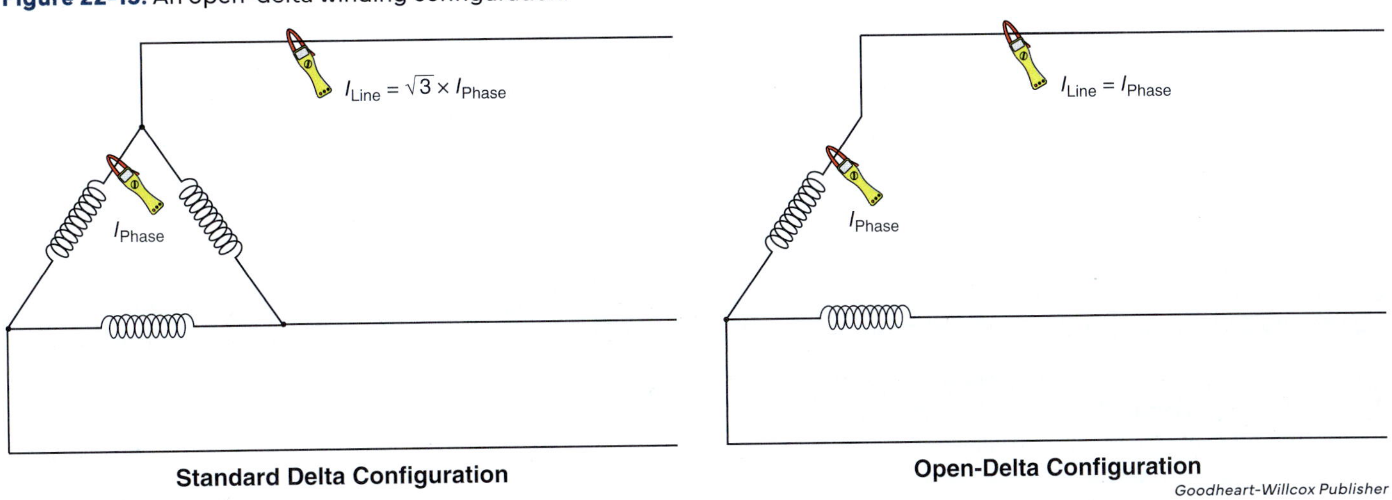

Goodheart-Willcox Publisher

Figure 22-16. In a standard delta configuration, line current is greater than phase current. In an open-delta configuration, the line current is the same as the phase current.

Shedding Light

If one transformer of a traditional delta transformer fails, the resulting transformer will be configured as an open-delta transformer. Although the failed transformer will still provide three-phase power, it will do so at a 43.3% reduction in output. At this reduced output, the transformer may or may not be able to satisfy the system or circuit requirements. In rare cases, a traditional delta transformer may be intentionally oversized by 43.3%, so it can continue to supply loads in an open-delta condition in the event that one phase fails or needs to be removed for maintenance.

Example 22-3 evaluates a delta transformer bank with a transformer failure.

Example 22-3

Consider a delta transformer that is made up of three 50 kVA single-phase transformers for a total of 150 kVA. If one of the three transformers fails, what is the expected capacity of the newly-formed open-delta transformer?

✓ Solution

The output of the open-delta transformer will be 57.7% of the output of the original 150 kVA transformer. Calculate the VA output as follows:

$$0.577 \times 150 \text{ kVA} = 86.6 \text{ kVA}$$

For a three-phase delta-connected transformer, the capacity, in kVA, is found by adding the kVA ratings of the three transformers. For example, the three 50 kVA transformers in **Example 22-3** were connected to form a closed-delta transformer with an output of 150 kVA. However, this is not the case with the open-delta configuration. For an open-delta transformer, the output of the transformer will be 86.6% of the sum of the two transformer output capacities.

Example 22-4 evaluates an open-delta transformer.

Example 22-4

What is the output of an open-delta transformer that is made up of two 150 kVA single-phase transformers?

✓ Solution

The sum of the capacities of the two single-phase transformers is 300 kVA. Multiplying this value by 0.866 (86.6%), gives a result of 259.8 kVA.

Now It's Your Turn 22-4

What would the VA rating be for each of two similar single-phase transformers if they were combined to form an open-delta transformer with an output capacity of 130 kVA?

Chapter 16, *Three-Phase Alternating Current Power*, introduced the high-leg delta configuration, in which a grounded conductor is connected to the midpoint of one of the windings. This grounded conductor is called a ***center tap***, and a high-leg delta configuration is also called a *four-wire delta configuration*. A center tap can also be used in an open-delta transformer, **Figure 22-17**.

The center tap allows the transformer to provide additional voltages. In addition to the higher voltages referenced earlier, the neutral allows for 120 V to be used if the secondary is configured as a 240 V delta connection. Some utilities use a system for distribution providing three-phase open delta for three-phase motors and the ability to generate 120 V for household use. The advantage of this system is distribution costs are lower for a three-phase system as the need for the third distribution wire is eliminated. This type of system may still be used in rural areas. In an open-delta system, only two of the three ungrounded conductors can be used in conjunction with the grounded conductor (neutral) to obtain 120 V. The voltage at the third ungrounded conductor, when measured with respect to ground, will be higher than the other two by a factor equal to the square root of three. For a 240-V system, this is 208 V on the high leg.

Know the Code

NEC Section 110.15 references high-leg markings and the required conductor markings on delta-configured systems. *Section 110.15* specifies that when the midpoint of a conductor is grounded, the conductor with the higher phase voltage to ground is to be "durably and permanently" color-coded orange. The code also allows for other effective marking methods to be used. The code also states that the high leg should be appropriately marked at any point in the system where an electrical connection is made if the grounded conductor (neutral) is present.

Safety Alert

Be very careful when working on or around three-phase systems, especially those that have orange-colored conductors. These conductors are called *high legs* and are at a potential, measured from line to neutral or ground, that is significantly higher than 120 V. Be sure to check the line-to-ground voltage before connecting any 120 V appliances or tools to help prevent equipment damage and personal injury. The panels used in these types of systems will have no single-phase, 120-V circuit breakers on the phase with the higher-leg voltage to ground.

T-Connected Transformer

The ***T-connected transformer*** is another configuration where two transformers are used to create three-phase power at the

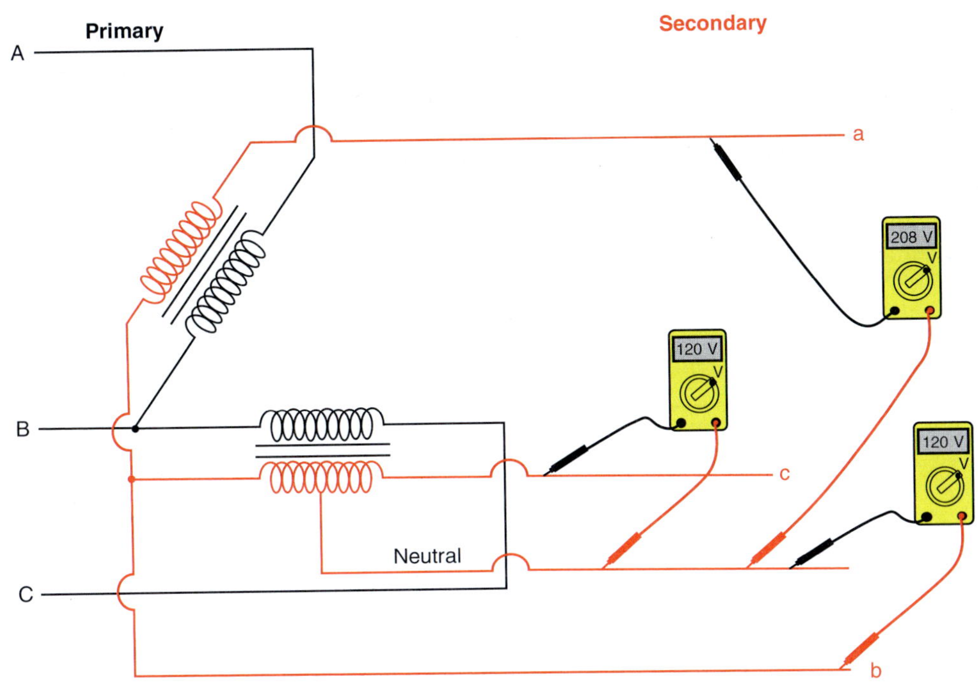

Goodheart-Willcox Publisher

Figure 22-17. A center tap in a 240 V open-delta configuration supplies two phase-to-neutral potentials of 120 V and one high-leg potential of 208 V.

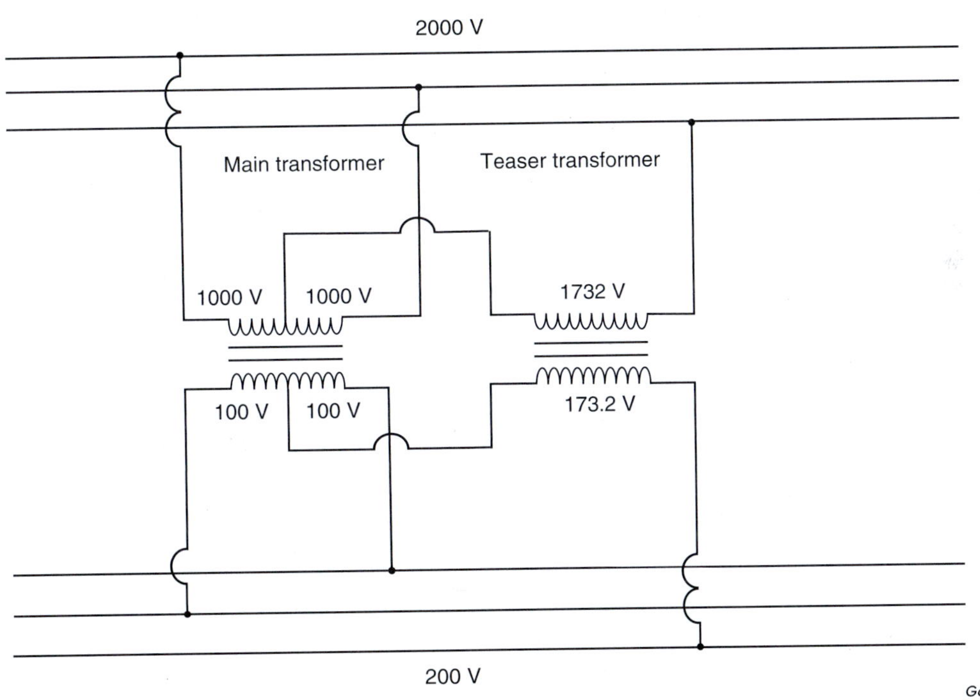

Goodheart-Willcox Publisher

Figure 22-18. Wiring diagram of a T-connected transformer.

secondary. This configuration utilizes one transformer with a center tap and a second transformer with an input voltage that is equal to 86.6% of the line voltage at the primary. The center-tapped transformer is called the *main transformer*, and the other transformer is called the *teaser transformer*, **Figure 22-18**. In this configuration, the output signals, one from the main transformer and one from the teaser transformer, are 90 electrical degrees out of phase with each other.

The voltage measured between any two of the secondary leads matches the phase voltage of the main transformer. The voltage value of the teaser transformer is designed to be 86.6% of the main transformer's voltage. This ensures that the combination of this winding with one-half of the main transformer provides the desired secondary voltage. Current flow through the secondary winding of the teaser transformer must also pass through one-half of the secondary winding of the main transformer, **Figure 22-19**, which is 90° out of phase with the teaser winding.

The T-connected transformer is produced as a single unit, where the individual transformers are wound to precise requirements and specifications. Both transformers are paired at the time of component manufacturing and are mounted in the same housing. As with other two-transformer configurations, the T-connected transformer results in a lighter, less expensive option than a three-winding transformer. As with the open-delta transformer, the T-connected transformer is best suited for lighter-duty applications.

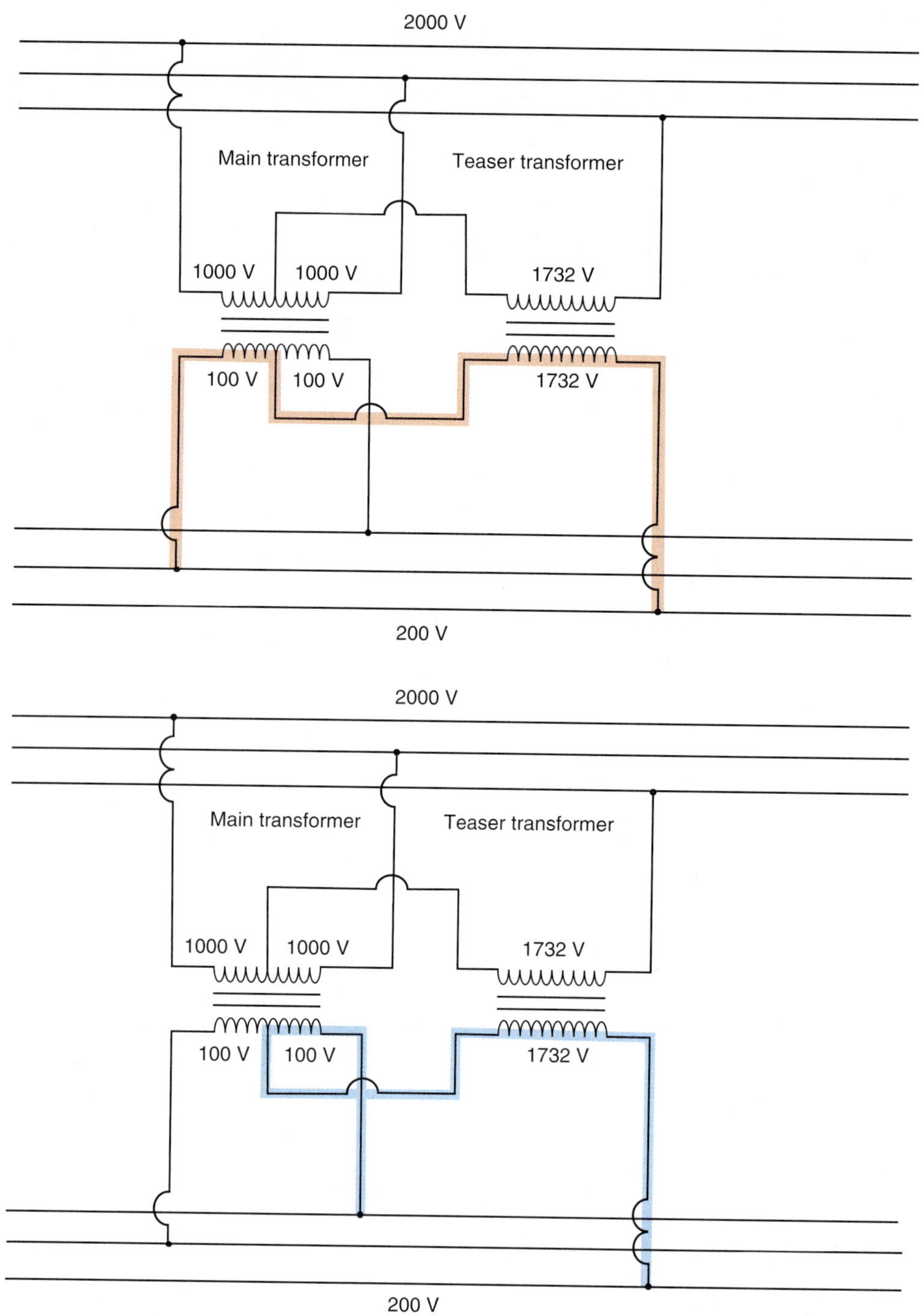

Goodheart-Willcox Publisher

Figure 22-19. Two current paths through the secondary windings of a T-connected transformer (highlighted). Current passes through half of the main transformer winding and then through the entire teaser transformer winding.

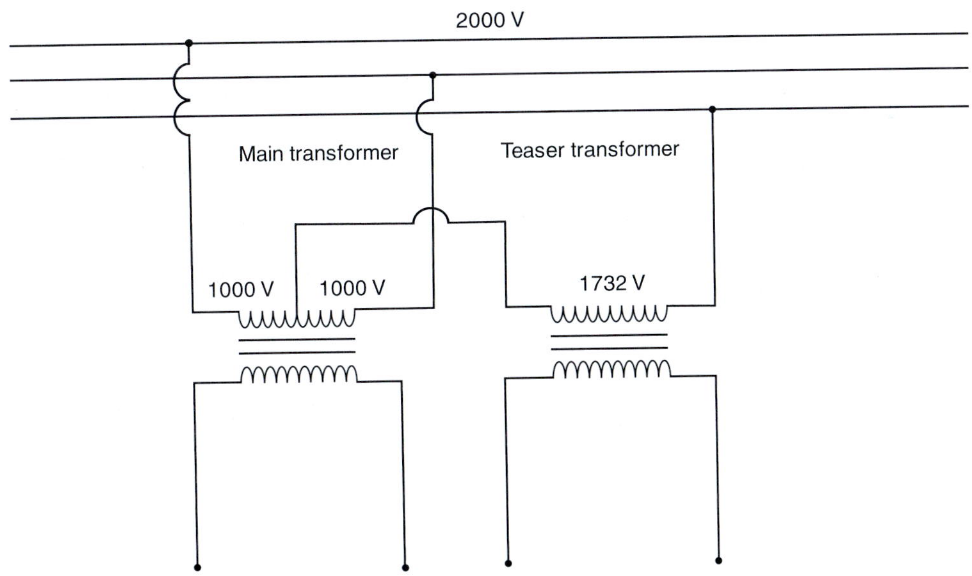

Goodheart-Willcox Publisher

Figure 22-20. Wiring diagram for a Scott transformer shows a T-connected primary and two single-phase secondary windings.

Safety Alert

To avoid potential overheating, operational irregularities, and system balancing problems, T-connected transformers should not be connected in parallel to conventional three-winding transformers.

Scott Transformer

The Scott transformer is similar to the T-connected transformer. The main difference is that the ***Scott transformer*** produces two separate single-phase signals at the transformer's secondary, **Figure 22-20**. These two secondary signals have the same voltage and are 90° out of phase, just as was the case with the T-connected transformer. The combination of the 1000-V vector and the 1732-V vector, connected at 90° relative to each other, produces the same 2000 V that is found at the primary of the main transformer. The Scott transformer arrangement is commonly used to power isolated single-phase circuits.

22.5 Calculating Three-Phase Transformer Loads

Up to this point, isolated three-phase transformers and their associated calculations were covered. Three-phase transformer configurations that have loads connected to their secondary windings may include the following connections:

- Wye-configured secondary load connected to a wye-wye transformer
- Delta-configured secondary load connected to a wye-wye transformer
- Wye configured secondary load connected to a delta-delta transformer
- Delta-configured secondary load connected to a delta-delta transformer
- Wye-configured secondary load connected to a wye-delta transformer
- Delta-configured secondary load connected to a wye-delta transformer
- Wye-configured secondary load connected to a delta-wye transformer
- Delta-configured secondary load connected to a delta-wye transformer

The following examples provide the steps on how to calculate values for two of these configurations. **Example 22-5** evaluates a delta-configured secondary load that is connected to a delta-delta transformer. This type of transformer could be installed in an industrial setting to power three-phase loads such as motors. **Example 22-6** evaluates a wye-configured secondary load that is connected to a delta-wye transformer. This configuration is common in larger commercial applications, industrial applications, and institutional applications such as schools and hospitals.

Example 22-5

The following circuit is an example of a delta-configured secondary load connected to a delta-delta transformer.

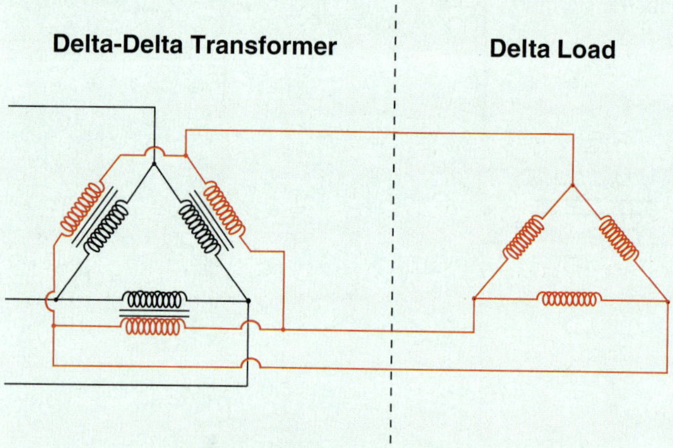

$E_{Line(P)} = 600$ V
$E_{Line(S)} = 240$ V
Load phase resistance = 20 Ω

Goodheart-Willcox Publisher

Using the values listed in the image, determine the following:

- $E_{Phase(P)}$
- $E_{Phase(S)}$
- Transformer turns ratio
- $E_{Line(Load)}$
- $E_{Phase(Load)}$
- $I_{Phase(Load)}$
- $I_{Line(Load)}$
- $I_{Line(S)}$
- $I_{Phase(S)}$
- $I_{Phase(P)}$
- $I_{Line(P)}$
- VA

✓ Solution

Step 1. Calculate the primary winding phase voltage.

For a delta connection, the phase voltage is equal to the line voltage, so the primary phase voltage, $E_{Phase(P)}$, is 600 V.

Step 2. Calculate the secondary winding phase voltage.

The secondary winding is also configured as a delta connection. In a delta connection, the phase voltage is equal to the line voltage, so the secondary phase voltage, $E_{Phase(S)}$, is 240 V.

Step 3. Calculate the transformer's turns ratio.

By substituting the known values for the primary phase voltage and the secondary phase voltage, the turns ratio is calculated as follows:

$$TR = \frac{E_{Phase(P)}}{E_{Phase(S)}}$$

$$TR = \frac{600 \text{ V}}{240 \text{ V}}$$

$$TR = 5:2 \text{ or } 2.5:1$$

Step 4. Calculate the line voltage supplied to the load.

The line voltage supplied to the load is the same as the line voltage at the transformer's secondary winding. Thus, the line voltage supplied to the load, $E_{Line(Load)}$, is 240 V.

Step 5. Calculate the phase voltage at the load.

For a delta connection, the phase voltage at the load is equal to the line voltage at the load. Therefore, the phase voltage at the load, $E_{Phase(Load)}$, is 240 V.

Step 6. Calculate the phase current at the load.

The phase current at the load is found using the current form of Ohm's law and substituting the load's phase voltage and the phase resistance, R_{Phase}.

$$I_{Phase(Load)} = \frac{E_{Phase(Load)}}{R_{Phase}}$$

$$I_{Phase(Load)} = \frac{240 \text{ V}}{20 \ \Omega}$$

$$I_{Phase(Load)} = 12 \text{ A}$$

Step 7. Calculate the line current at the load.

For a delta connection, the line current at the load is equal to $\sqrt{3}$ times the phase current at the load:

$$I_{Line(Load)} = I_{Phase(Load)} \times \sqrt{3}$$

$$I_{Line(Load)} = 12 \text{ A} \times 1.732$$

$$I_{Line(Load)} = 20.8 \text{ A}$$

Step 8. Calculate the line current at the secondary winding.

The line current leaving the transformer is the same line current that flows to the load. Therefore, the line current at the secondary winding, $I_{Line(S)}$, is 20.8 A.

Step 9. Calculate the phase current at the secondary winding.

For a delta connection, the phase current at the transformer's secondary winding is equal to the line current at the secondary winding divided by $\sqrt{3}$.

$$I_{Phase(S)} = \frac{I_{Line(S)}}{\sqrt{3}}$$

$$I_{Phase(S)} = \frac{20.8\text{ A}}{1.732}$$

$$I_{Phase(S)} = 12\text{ A}$$

Step 10. Calculate the phase current at the primary winding.

The phase current at the transformer's primary can be found using the turns ratio current formula:

$$TR = \frac{I_{Phase(S)}}{I_{Phase(P)}}$$

Rearrange the equation to solve for $I_{Phase(P)}$. The turns ratio of 2.5:1 is expressed 2.5.

$$I_{Phase(P)} = \frac{I_{Phase(S)}}{TR}$$

$$I_{Phase(P)} = \frac{12\text{ A}}{2.5}$$

$$I_{Phase(P)} = 4.8\text{ A}$$

Step 11. Calculate the line current at the primary winding.

For a delta connection, the line current is equal to the phase current multiplied by $\sqrt{3}$:

$$I_{Line(P)} = I_{Phase(P)} \times \sqrt{3}$$

$$I_{Line(P)} = 4.8\text{ A} \times 1.732$$

$$I_{Line(P)} = 8.3\text{ A}$$

Step 12. Calculate the transformer VA.

By substituting the known values for the line current and the line voltage, the VA is calculated as follows:

$$VA = \sqrt{3} \times I_{Line(P)} \times E_{Line(P)}$$

$$VA = 1.732 \times 8.3\text{ A} \times 600\text{ V}$$

$$VA = 8625\text{ VA or } 8.6\text{ kVA}$$

Example 22-6

The following circuit is an example of a wye-configured secondary load connected to a delta-wye transformer.

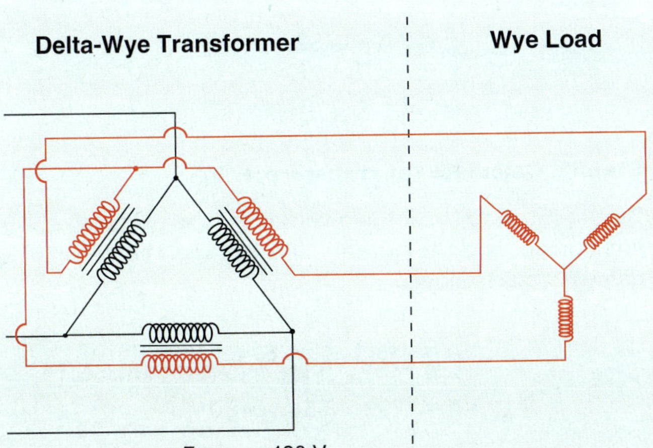

$E_{Line(P)} = 480$ V

$E_{Line(S)} = 208$ V

Load phase resistance = 15 Ω

Goodheart-Willcox Publisher

Using the values listed in the image, determine the following:

- $E_{Phase(P)}$
- $E_{Phase(S)}$
- Transformer turns ratio
- $E_{Line(Load)}$
- $E_{Phase(Load)}$
- $I_{Phase(Load)}$
- $I_{Line(Load)}$
- $I_{Line(S)}$
- $I_{Phase(S)}$
- $I_{Phase(P)}$
- $I_{Line(P)}$
- VA

✓ Solution

Step 1. Calculate the primary winding phase voltage.

For a delta connection, the phase voltage and the line voltage are the same. Thus, the primary phase voltage, $E_{Phase(P)}$, is 480 V.

Step 2. Calculate the secondary winding phase voltage.

The secondary winding is configured as a wye connection; therefore, the phase voltage of the

Continued

secondary is equal to the line voltage of the secondary winding divided by $\sqrt{3}$:

$$E_{Phase(S)} = \frac{E_{Line(S)}}{\sqrt{3}}$$

$$E_{Phase(S)} = \frac{208\text{ V}}{1.732}$$

$$E_{Phase(S)} = 120\text{ V}$$

Step 3. Calculate the transformer's turns ratio.

By substituting the known values for the primary phase voltage and the secondary phase voltage, the turns ratio is calculated as follows:

$$TR = \frac{E_{Phase(P)}}{E_{Phase(S)}}$$

$$TR = \frac{480\text{ V}}{120\text{ V}}$$

$$TR = 4{:}1$$

Step 4. Calculate the line voltage supplied to the load.

The line voltage supplied to the load is the same as the line voltage at the transformer's secondary winding. Therefore, the line voltage supplied to the load, $E_{Line(Load)}$, is 208 V.

Step 5. Calculate the phase voltage at the load.

For a wye connection, the phase voltage at the load is equal to the line voltage at the load divided by $\sqrt{3}$:

$$E_{Phase(Load)} = \frac{E_{Line(Load)}}{\sqrt{3}}$$

$$E_{Phase(Load)} = \frac{208\text{ V}}{1.732}$$

$$E_{Phase(Load)} = 120\text{ V}$$

Step 6. Calculate the phase current at the load.

The phase current at the load is found using the current form of Ohm's law and then substituting the load's phase voltage and the phase resistance (R_{Phase}).

$$I_{Phase(Load)} = \frac{E_{Phase(Load)}}{R_{Phase}}$$

$$I_{Phase(Load)} = \frac{120\text{ V}}{15\ \Omega}$$

$$I_{Phase(Load)} = 8\text{ A}$$

Step 7. Calculate the line current at the load.

For a wye connection, the line current at the load is equal to the phase current at the load. Therefore, the line current at the load, $I_{Line(Load)}$, is 8 A.

Step 8. Calculate the line current at the secondary winding.

The line current leaving the transformer is the same line current that flows to the load. Therefore, the secondary line current, $I_{Line(S)}$, is 8 A.

Step 9. Calculate the phase current at the secondary winding.

For a wye connection, the phase current is equal to the line current. Therefore, the phase current at the transformer's secondary winding, $I_{Phase(S)}$, is 8 A.

Step 10. Calculate the phase current at the primary winding.

The phase current at the transformer's primary can be found using the turns ratio current formula:

$$TR = \frac{I_{Phase(S)}}{I_{Phase(P)}}$$

Rearrange the equation to solve for $I_{Phase(P)}$. The turns ratio of 4:1 is expressed as 4/1 or 4.

$$I_{Phase(P)} = \frac{I_{Phase(S)}}{TR}$$

$$I_{Phase(P)} = \frac{8\text{ A}}{4}$$

$$I_{Phase(P)} = 2\text{ A}$$

Step 11. Calculate the line current at the primary winding.

For a delta connection, the line current is equal to the phase current multiplied by $\sqrt{3}$:

$$I_{Line(P)} = I_{Phase(P)} \times \sqrt{3}$$

$$I_{Line(P)} = 2\text{ A} \times 1.732$$

$$I_{Line(P)} = 3.5\text{ A}$$

Step 12. Calculate the transformer VA.

By substituting the known values for the primary line current and the primary line voltage, the VA is calculated as follows:

$$VA = \sqrt{3} \times I_{Line(P)} \times E_{Line(P)}$$

$$VA = 1.732 \times 3.5\text{ A} \times 480\text{ V}$$

$$VA = 2910\text{ VA or }2.9\text{ kVA}$$

CHAPTER 22 REVIEW

Summary

- Three-phase transformers are commonly used to increase or decrease the voltage of three-phase signals. (22.1)
- Three-phase transformers are made up of three individual single-phase transformers connected in a variety of configurations. (22.1)
- In a delta configuration, the phase and line voltages are the same. (22.2)
- In a delta configuration, the phase current is equal to the line current divided by $\sqrt{3}$. (22.2)
- In a delta configuration, the line current is equal to the phase current multiplied by $\sqrt{3}$.
- In a wye configuration, the line and the phase currents are the same. (22.2).
- In a wye configuration, the line voltage is equal to the phase voltage multiplied by $\sqrt{3}$. (22.2)
- In a wye configuration, the phase voltage is equal to the line voltage divided by $\sqrt{3}$. (22.2)
- Common three-phase transformer configurations are the wye-wye, wye-delta, delta-delta, and delta-wye. (22.3)
- Open-delta transformers use two transformers connected in a V configuration to create a three-phase secondary signal. (22.4)
- The T-connected transformer uses two single-phase transformers to produce a three-phase secondary signal. (22.4)
- The Scott transformer is used to produce two single-phase signals that are 90° out of phase with each other. (22.4)
- There are eight different three-phase transformer configurations that can have loads connected to their secondary windings. (22.5)

Review Questions

Use the information in the chapter to answer the following questions.

Measuring Your Potential

1. The term _____ best describes the three-phase transformer that is created when three individual, single-phase transformers are connected together. (22.1)
   A. center tap
   B. transformer bank
   C. open-delta
   D. T-connected
2. Referring to the electrical connections between the transformer's secondary windings and the load, which two readings will always be the same? (22.2)
   A. $I_{Phase(L)}$ and $I_{Line(S)}$
   B. $E_{Line(S)}$ and $E_{Line(L)}$
   C. $E_{Line(S)}$ and $E_{Line(P)}$
   D. $I_{Phase(S)}$ and $I_{Phase(L)}$
3. The turns ratio for any three-phase transformer will always be equal to _____. (22.2)
   A. $\frac{E_{Phase(P)}}{E_{Phase\ (S)}}$
   B. $\frac{E_{Phase(S)}}{E_{Phase(P)}}$
   C. $\frac{E_{Line(P)}}{E_{Line(S)}}$
   D. $\frac{E_{Line(S)}}{E_{Line(P)}}$
4. *True or False?* When a wye-configured load is connected to a transformer's wye-configured secondary winding, the $I_{Phase(S)}$, $I_{Line(S)}$, $I_{Line(L)}$, and $I_{Phase(L)}$ are all the same. (22.3)
5. *True or False?* If one of the three transformers of a wye-connected transformer fails, the transformer will no longer be able to produce a three-phase signal at the transformer's secondary. (22.3)
6. *True or False?* If one of the three transformers of a delta-connected transformer fails, the transformer will still produce a three-phase signal at the transformer's secondary. (22.3)
7. The center tap is best described as a(n) _____. (22.4)
   A. grounded conductor
   B. ungrounded conductor
   C. grounded insulator
   D. ungrounded insulator
8. According to the *NEC*, high-leg markings must be color-coded _____. (22.4)
   A. red
   B. blue
   C. yellow
   D. orange

9. A high leg is found on which of the following systems? (22.4)
   A. A wye connection with a grounded conductor at the center connection point.
   B. A wye connection with an ungrounded conductor at the center connection point.
   C. A delta connection with a center tap at the middle of a winding.
   D. A delta connection with a grounded conductor at one of the line terminals.
10. At the secondary of a properly operating Scott transformer, an electrician is most likely to find _____. (22.4)
   A. two single-phase signals that are in phase with each other
   B. two single-phase signals that are 90° out of phase with each other
   C. one three-phase low-voltage signal
   D. two three-phase low-voltage signals

Plug-In

1. The turns ratio of a delta-delta three-phase transformer with a primary line voltage of 200 V and a secondary line voltage of 50 V is _____. (22.5)
   A. 2:1
   B. 3:1
   C. 4:1
   D. 5:1
2. The turns ratio of a wye-delta three-phase transformer with a primary line voltage of 200 V and a secondary line voltage of 50 V is closest to _____. (22.5)
   A. 2.3:1
   B. 3.3:1
   C. 4.3:1
   D. 5.3:1
3. The VA output capacity of an open-delta transformer with two 100-kVA transformers will be _____. (22.5)
   A. 100 kVA
   B. 173.2 kVA
   C. 200 kVA
   D. 346.4 kVA

Refer to the following 50 VA three-phase, wye-wye transformer to answer questions 4 to 10.

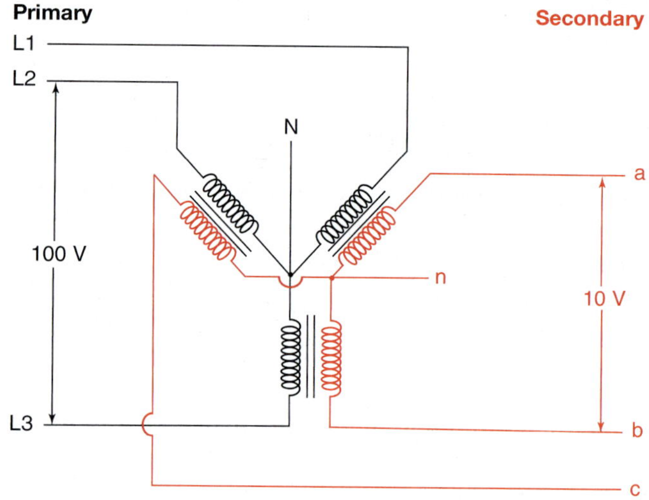

Goodheart-Willcox Publisher

4. Calculate the primary winding phase voltage. (22.5)
5. Calculate the secondary winding phase voltage. (22.5)
6. Calculate the transformer's turns ratio. (22.5)
7. Calculate the phase current of the primary winding. (22.5)
8. Calculate the line current of the primary winding. (22.5)
9. Calculate the line current of the secondary winding. (22.5)
10. Calculate the phase current of the secondary winding. (22.5)

Making the Connection

1. Explain why the turns ratio is lower on a wye-delta transformer than on a delta-delta transformer. Provide numerical examples as necessary to support your response. (22.3)
2. Explain why the two VA formulas yield the same result and can be used on both delta and wye connections. Provide numerical examples as necessary to support your response. (22.2)
3. Explain why the output capacity, in VA, of an open-delta configuration is only 86.6% of the mathematical sum of the two transformers that make up the connection. Compare this to the 100% output capacity obtained when the capacities of the three transformers that make up a delta connection are added together. Provide numerical examples as necessary to support your response. (22.4)

SECTION

7 Motor Fundamentals

WEG

CHAPTER

23 Motor Operation Basics

Courtesy Fluke Corporation

SPARKING DISCUSSION

Motors, engines, generators, and alternators take one form of energy and convert it into a different form. Identify some examples of each type of device. For each example, indicate the forms of energy being converted and the basic process or mechanism used to achieve the conversion.

LEARNING OUTCOMES

After completing this chapter, you will be able to:

23.1 Explain the differences between a motor and an engine, a generator, and an alternator.

23.2 Explain the differences between ac motors and dc motors.

23.3 Calculate synchronous and actual motor speeds.

23.4 Calculate slip as it applies to motors.

TECHNICAL TERMS

alternator
brush
commutator
engine
generator
motor
pole
slip
synchronous speed

Introduction

Of all electrical devices, motors rank as one of the most commonly encountered. It would be difficult to live a single day and not receive at least some benefit from the operation of a motor. It is far more likely that you have interacted, either directly or indirectly, with at least half a dozen of them before today's sunset.

Motors can be used to rotate fans, blowers, pumps, rotating drums, and a multitude of other items. Motors are also found in rotating and reciprocating power tools, elevators, escalators, dishwashers, microwave ovens, electric vehicles, computers, air-conditioning systems, and many other electrically powered items that rely on a rotating or linear action to accomplish a task. A motor used for any one of these applications will very likely appear to be at least somewhat different from a motor used for another application. This is mainly because there are so many different types and styles of motors.

In the United States, approximately 50% of all electricity produced is used to power electric motors. Because of their widespread use in so many different applications, it is important that individuals who are tasked with installing, troubleshooting, maintaining, and operating motors have a strong working knowledge of how they operate. It is the responsibility of a service technician, for example, to determine not only if a motor failed, but why the motor failed. If the cause of motor failure is not determined and a defective motor is simply replaced, it is likely that the new replacement motor will also fail. This chapter discusses the basics of motor operation and serves as a foundation for the chapters that follow.

23.1 Motor Terminology

In casual conversation, it is quite common to hear the terms motor and engine used interchangeably but, in fact, they are very different. In addition, the terms generator and alternator are also often misused. As we begin our discussion of electric motors, let us first define each of these four components and describe some of the differences that exist among them.

Motors versus Engines

A ***motor*** is a machine that converts electrical energy into mechanical energy. During the conversion from electrical energy to mechanical energy, magnetic fields are generated. The mechanical energy that is produced is in the form of a rotational force that can be used directly, indirectly, or converted to a linear, reciprocating action for various additional applications.

An ***engine*** also produces mechanical energy, but it is the energy source that differs between motors and engines, **Figure 23-1**. Engines, such as those found in gas-burning vehicles, rely on the combustion of a fuel to create mechanical energy. This allows our automobiles to move. The combustion of fuel, for example, creates combustion gases that push pistons connected to a crankshaft, causing it to turn. The rotating crankshaft provides the mechanical energy to propel the vehicle.

One easy way to tell the difference between a motor and an engine is that an engine has cylinders, pistons, connecting rods, and a crankshaft, while electric motors have windings, **Figure 23-2**.

Motors versus Alternators and Generators

Alternators and generators are, in essence, the polar opposite of motors. While motors use electrical energy to produce mechanical energy, a ***generator*** is a machine that converts mechanical energy into electrical energy, **Figure 23-3**. An ***alternator***, also called an *ac generator*, is a generator that produces ac electrical energy.

An alternator in a car, for example, relies on the combustion of fuel to turn a belt to provide the rotational mechanical energy that is converted to ac electrical energy. A gas-fired backup generator uses an engine to provide rotating mechanical energy that is then converted to electrical energy. A detailed discussion on alternators and generators is provided in Chapter 34, *Generator Fundamentals*, Chapter 35, *DC Generators*, and Chapter 36, *AC Generators*.

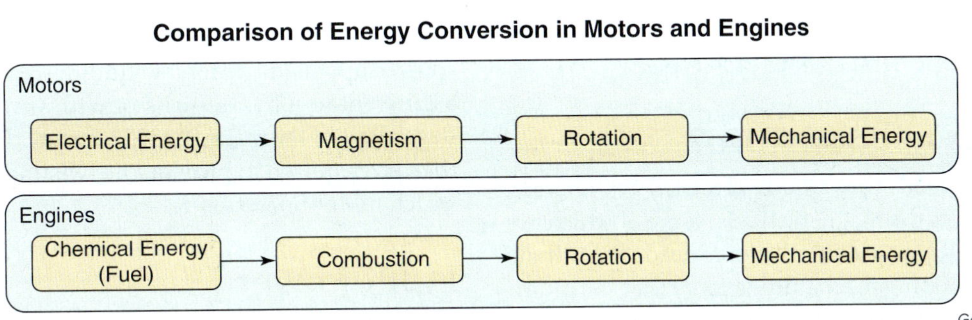

Goodheart-Willcox Publisher

Figure 23-1. In an electric motor, electrical energy is converted to mechanical energy through the production of magnetic fields. In an engine, chemical energy is converted to mechanical energy through combustion.

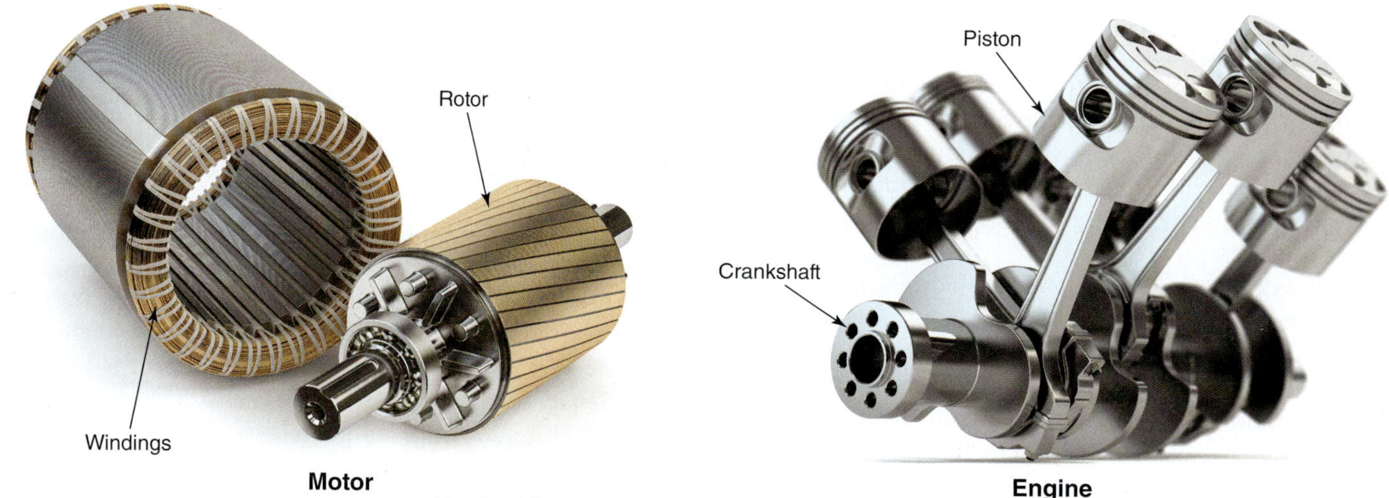

Maxx-Studio/Shutterstock.com

KsanderDN/Shutterstock.com

Figure 23-2. Comparison of a motor and an engine. A motor uses magnets, windings, a rotor, and a stator to convert electrical energy to mechanical energy. An engine uses mechanical components such as pistons and a crankshaft to convert chemical energy to mechanical energy.

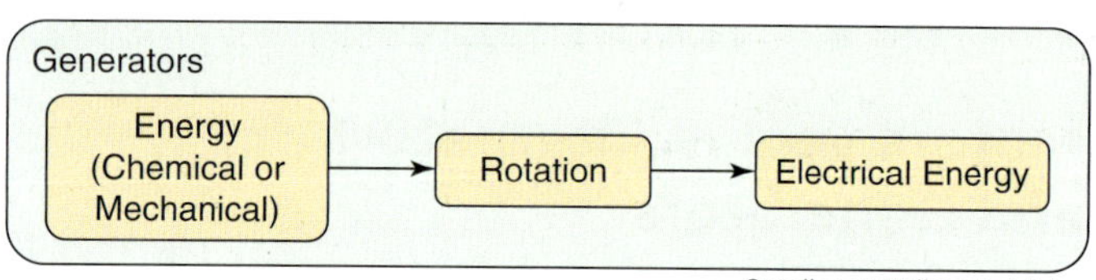

Goodheart-Willcox Publisher

Figure 23-3. Process flow diagram of a generator.

23.2 AC and DC Motors

Although motors are commonly classified by the job they are intended to perform and the environments in which they are intended to operate, the basic operational goal of a motor is the same—to convert electrical energy into mechanical energy. How motors make this energy conversion varies, in part, by the type of power supply to which the motor is connected. Power supplies are most generally classified as either ac or dc. Similarly, motors are also classified as either ac or dc devices. Both ac and dc motors are constructed differently, operate differently, and use different methods to convert electrical energy into mechanical energy. The following provides a general overview of the differences between ac and dc motors, and then more specific details about each type.

What Makes a Motor Turn?

Electric motors are loads that require electricity to operate. More specifically, it is the magnetic fields, some of which are created by electricity, that provide the force needed to initiate and maintain the rotating mechanical energy that the motor produces. A single magnetic field, however, is not enough to generate the rotating force needed to start a motor. Imagine a sailboat floating on a lake with no wind whatsoever acting on its sails. Unless there is a force, namely wind, acting on the sail, the boat will remain motionless. If there is no difference in pressure or force across the sail, there is no motion. Picture two groups of people engaged in an age-old test of group strength, a tug-of-war. The flag, which hangs midway between the two teams, will move from one side to the other as the net force exerted by the two teams changes or varies. If the force exerted by each team is the same, the net force will be zero, and the flag will remain motionless. The lack of flag motion does not indicate that there is no force or pressure acting on the flag. In this case, the lack of motion is due to the fact that the forces acting on the flag are equal and opposite, so they cancel each other out. The same concept can be easily applied to a motor. There needs to be more than a magnetic field. There needs to be an imbalance in magnetic fields which creates forces that attract and repel in order for a motor to turn.

Different types of motors produce magnetic fields differently. Some rely on permanent magnets, while others use electromagnets. Some use ac power, while some use dc power. Some motors rely on a single power source supplied to a single set of windings, while others rely on a single power source supplied to different windings, each constructed with a different number of turns or with wire of different gauge. Regardless of the method used to create these magnetic fields, they are required to provide the twisting force, or torque, on which motor operation relies.

Input Power

In order for a motor to operate, the polarity of the power being supplied to the motor must be continually changing. If a change in polarity does not occur, there can be no continuous, uninterrupted rotation. AC power supplies, represented most commonly by sine waves, provide a signal that changes its direction two times during each cycle, or 120 times per second for a 60 Hz signal, **Figure 23-4**. AC motors can function directly when supplied with the proper ac signal.

DC motors, on the other hand, are supplied with a direct current, or dc power signal. DC power signals do not change polarity. In order for a dc motor to operate, there must be a means to change the polarity of the unidirectional power supply. This can be accomplished with a ***commutator***, **Figure 23-5,** or electronic circuitry that accomplishes the commutation task. Commutation is the process of reversing the polarity of a power supply to change the direction of current flow through a circuit. DC motors are, therefore, designed to accept dc power and commutate the signal to facilitate motor operation. In a dc motor, there is no permanent physical connection between the main circuitry and the rotating windings. Instead, spring-loaded ***brushes*** are used to ensure a consistent electrical connection between the power source and the rotating windings. A detailed description of dc motors and how they operate is provided in Chapter 27, *DC Motors*.

Magnetic Fields

In ac motors, the alternating power source creates an imbalance in magnetic fields that allows the motor to operate. When the stationary portion (stator) of the motor is energized, the alternating current creates a rotating magnetic field. The rotating portion (rotor) of the motor is located

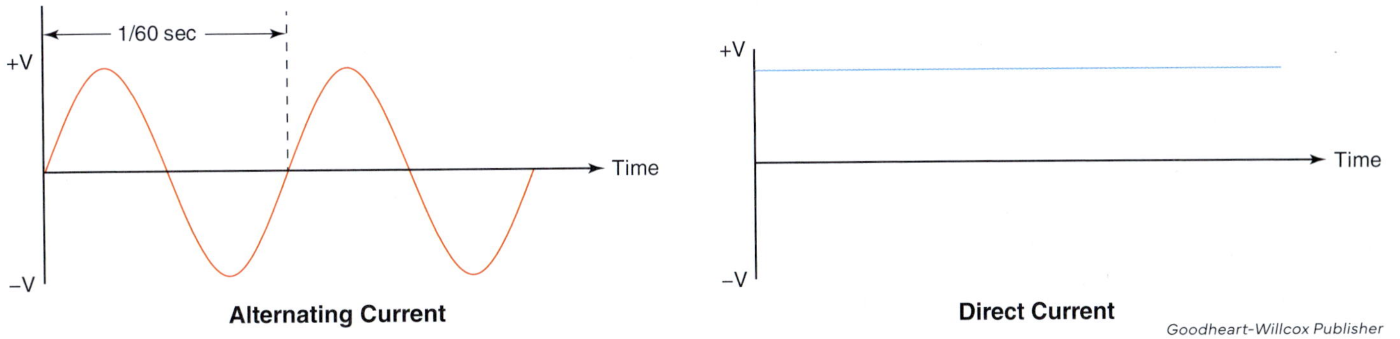

Goodheart-Willcox Publisher

Figure 23-4. Comparison of alternating current and direct current. A 60 cycle-per-second ac power signal represented by a sine wave. A dc power signal does not change polarity and is represented by straight horizonal line.

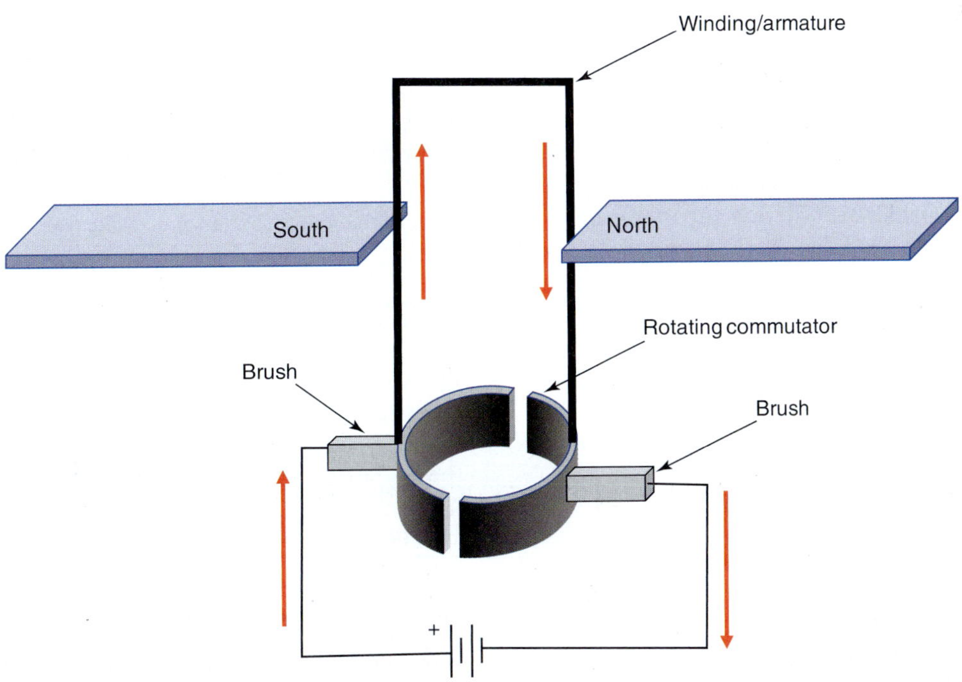

Goodheart-Willcox Publisher

Figure 23-5. A rotating commutator and brush assembly.

within the stator. Although not electrically connected to it, the rotor is acted upon by the generated magnetic field. This creates a force, or torque, that causes the rotor to turn. The rotation of the rotor generates an additional rotating magnetic field. Although there is one power source to the ac motor, magnetic fields of different strengths are produced.

In a dc motor, the power supplied to the motor does not alternate, so two different magnetic fields are required. A set of permanent magnets produces a nonrotating magnetic field, and a dc power source is supplied to the motor windings. The interaction between the magnetic fields created by the permanent magnet and the dc current flow through the windings produces the motor's rotation.

Controlling Motor Speed

The speed of an ac motor is controlled, in part, by the frequency of the power supply applied to the motor. If increased motor speed is desired, the frequency of the power supply can be increased. Conversely, if decreased motor speed is to be achieved, the frequency of the power supply can be decreased. Calculating the speed of an ac motor is discussed later in this chapter.

Since dc motor power signals are not cyclical, they do not have a frequency. Instead, the speed of dc motors is controlled by varying the voltage and/or current that is applied to the motor's windings, or by adjusting the strength of the magnetic field in which the windings are positioned. Adjusting the strength of the magnetic field is possible if an externally controlled and generated power source is used to create the magnetic field instead of using permanent magnets.

Side-by-Side Efficiency and Performance Comparison

When deciding between using an ac motor or a dc motor, it is important to weigh as many factors as possible to ensure that the best decision is reached. Two of the major, and most obvious, points to consider are whether the motor meet the needs of the project, and how much the motor will cost to purchase and install. Other things to consider include:

- Speed of response to load changes
- Motor efficiency
- Physical size
- Starting torque
- Level of electromagnetic interference
- Starting/stopping speed
- Speed control
- Level of required maintenance
- Use in explosive conditions
- Power consumption
- Speed of rotation reversal

The importance of each item on this list will vary depending on the particular project and the intentions of the project owner. For example, if a building owner needs to replace a motor prior to selling the building, then the initial cost of the motor over the efficiency of the motor will be more of a concern. In this case, a less-expensive motor with higher operating costs would be more appealing than a higher-priced, high-efficiency model. Motor location might also be a factor in some cases. Consider a motor that is to be installed in a hard-to-access location. A motor that requires less maintenance would be more desirable than one that requires more frequent or substantive maintenance. If a motor is to be installed in close proximity to communications equipment, electromagnetic interference created by the motor will be a concern.

A chart comparing ac and dc motors is provided in **Figure 23-6.** The entries in the chart are general in nature and there will likely be other factors that might be considered when choosing between an ac and a dc motor.

Factors that Affect the Speed of an AC Motor

The speed of an ac motor is determined by three factors:

- The frequency of the power supply
- The number of polarity changes per cycle of the power supply
- The number of motor poles

The frequency of the power supply and the number of polarity changes per cycle are established by the power supply, not the motor. The number of poles is determined by the motor construction. The term ***pole*** represents the total number of positive and negative (north and south) magnetic

Comparison of AC and DC Motors

Motor Quality/ Characteristic	AC Motors	DC Motors
Initial cost	Lower	Higher
Response to load changes	Slower	Faster
Efficiency	Lower	Higher
Physical size	Larger	Smaller
Starting torque	Lower	Higher
Electromagnetic interference	Higher	Lower
Starting/stopping speed	Slower	Faster
Speed control	Harder	Easier
Maintenance	Less	More
Use in explosive conditions	Yes	No
Power consumption	Lower	Higher
Speed of rotation reversal	Slower	Faster

Goodheart-Willcox Publisher

Figure 23-6. Comparison of ac and dc motor characteristics.

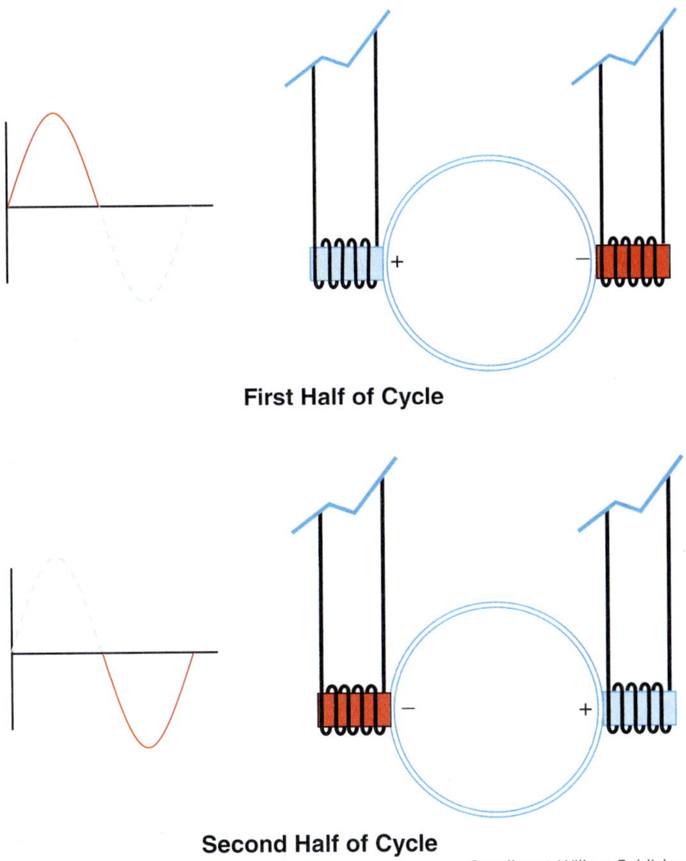

Goodheart-Willcox Publisher

Figure 23-7. Configuration of a two-pole ac motor. During the first half of an ac cycle, one pole has a positive polarity, while the other has a negative polarity. During the second half of an ac cycle, the pole's polarities are reversed.

fields. Two common motor pole configurations are two-pole and four-pole.

The two-pole motor configuration has two magnetic fields, one positive and one negative. These two magnetic fields change their polarity as the polarity of the power supply changes. As an ac power signal flows through the motor, one magnetic field will be positive and the other will be negative for one-half of the ac cycle. For the second half of the cycle, the charges will reverse, **Figure 23-7**.

To illustrate how the number of poles affects the speed of an ac motor, consider the simplified diagrams in **Figure 23-8** and **Figure 23-9**. **Figure 23-8** illustrates a two-pole motor configuration, which causes the motor to rotate 180° during each half cycle. **Figure 23-9** shows a four-pole motor configuration that causes the motor to rotate 90° during each half cycle. Therefore, four polarity changes of the power supply are needed to cause a four-pole motor to make one complete rotation.

In more specific detail, consider **Figure 23-8**, in which a magnet is suspended between the two poles. The magnet is free to rotate and, as such, has positioned itself so its negatively charged pole is aligned with the motor's positively charged pole, and the magnet's positively charged

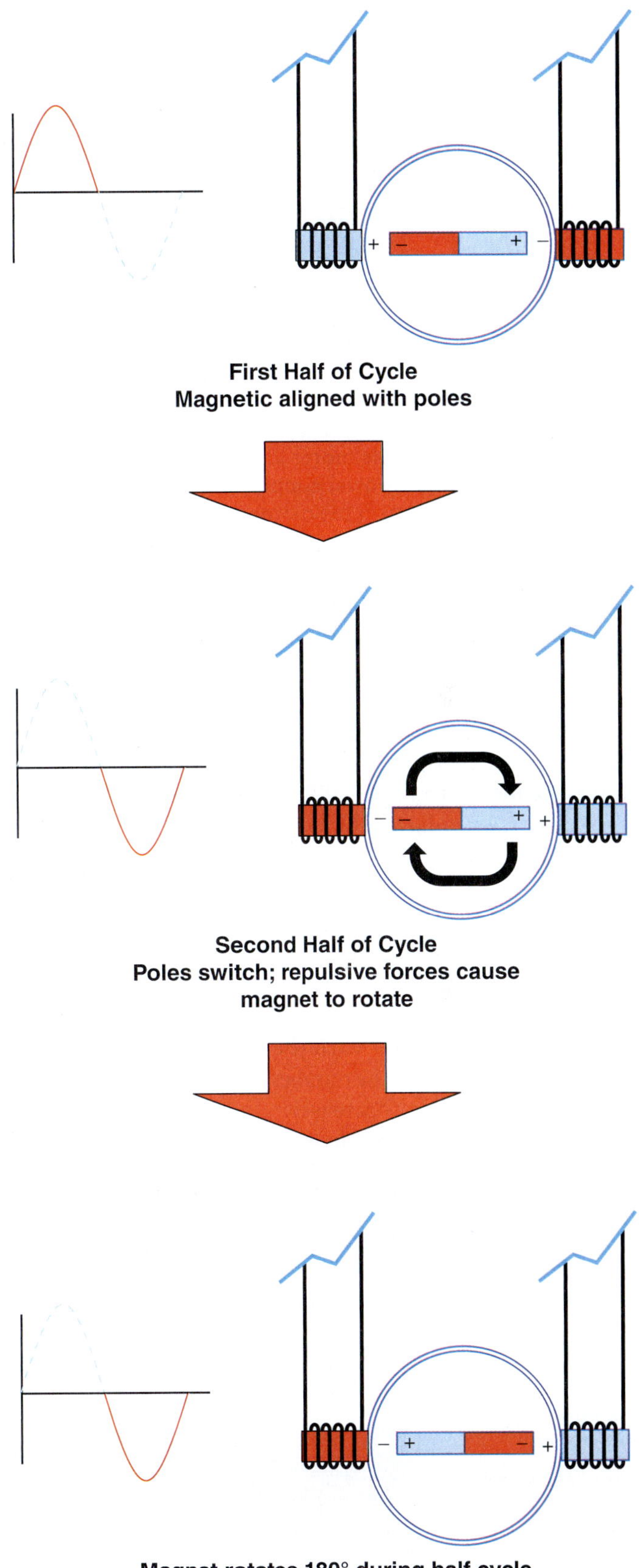

Goodheart-Willcox Publisher

Figure 23-8. In a two-pole motor configuration, two polarity changes of the power supply are needed to cause one complete revolution in the motor.

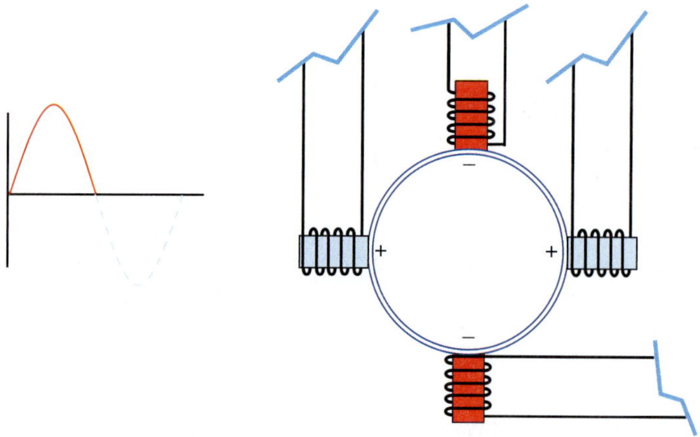

Figure 23-9. In a four-pole motor configuration, four polarity changes of the power supply are needed to cause one complete revolution in the motor.

pole is aligned with the motor's negatively charged pole. In the second half of the ac power signal, the polarity of the motor's poles changes. At this point, the suspended magnet experiences repulsive forces, negative-negative and positive-positive alignment, which causes it to rotate. The suspended magnet will rotate 180° until it comes to rest. From this scenario, in a two-pole motor, each polarity change will cause the motor to rotate 180° degrees, or one-half of a revolution. Two polarity changes are needed to cause a full revolution of the motor.

A four-pole motor configuration has two negative magnetic fields and two positive magnetic fields, **Figure 23-9.** As with the two-pole configuration, the polarity of the magnetic fields changes as the polarity of the power supply changes. As the polarities of the motor's poles change, the motor will rotate 90° to align itself with the polarities of the motor windings. In general, as the number of motor poles increases, the speed of the motor decreases.

23.3 Synchronous Speed Calculations for AC Motors

Motor speed is expressed in the units of rpm, which stands for revolutions per minute or rotations per minute. The term used to express the calculated, or ideal, motor speed is ***synchronous speed***. To calculate the synchronous speed of an ac motor, use the following formula:

$$N_S = 120 \times \frac{f}{p}$$

where:

N_S = Synchronous speed (in rpm)
120 = Constant for two polarity changes per cycle
f = Frequency of supplied current (in Hz)
p = Number of poles

For example, for a 60 Hz ac power signal, this formula becomes:

$$N_S = 120 \times \frac{60}{p}$$

$$N_S = \frac{7200}{p}$$

Refer to **Example 23-1** and **Example 23-2** on how to calculate synchronous speed.

Shedding Light

It is important to remember that the formula 7200 divided by the number of poles only applies to motors that are being powered by a 60 Hz signal. If the frequency of the power supply is different, this formula cannot be used!

Example 23-1

What is the synchronous speed of a four-pole motor connected to an ac power supply with a frequency of 60 Hz?

✓ Solution

The synchronous motor speed for this four-pole motor connected to an ac power frequency of 60 Hz is calculated as follows:

$$N_S = \frac{7200}{p}$$

$$N_S = \frac{7200}{4}$$

$$N_S = 1800 \text{ rpm}$$

Example 23-2

What is the synchronous speed of a four-pole motor connected to an ac power supply with a frequency of 80 Hz?

✓ Solution

With an 80 Hz power supply, the motor speed is calculated by using the following formula:

$$N_S = 120 \times \frac{f}{p}$$

$$N_S = 120 \times \frac{80 \text{ Hz}}{4}$$

$$N_S = \frac{9600}{4}$$

$$N_S = 2400 \text{ rpm}$$

Now It's Your Turn 23-2

What is the synchronous speed of a two-pole motor connected to an ac power supply with a frequency of 80 Hz?

23.4 Motor Slip

The actual speed at which a motor turns is lower than its calculated synchronous speed. The term ***slip*** is used to refer to the difference between a motor's synchronous speed and its actual speed. There will always be forces acting on a motor's rotational components, causing it to turn at a lower speed. These forces include friction, inductive reactance, and the load on the motor.

Shedding Light

Although slip seems to be a negative effect of ac motors, slip is needed. If a motor operated at its synchronous speed, there would be no induction. As a result, the ac motor would produce no torque. If the motor turns at the same speed as the rotating magnetic fields, there is no net change in magnetic field and, as a result, there is no induction. Slip results in the creation of enough induction and torque to help overcome friction and other losses when the motor is operated without a mechanical load connected to it.

Motor slip, denoted by the capital letter S, can be expressed as a percentage or as a numerical value, depending on the application. When expressed as a numerical value, slip is calculated as follows:

$$S = N_S - N$$

where:

S = Slip (in rpm)
N_S = Synchronous motor speed (in rpm)
N = Actual motor speed (in rpm)

When expressed as a percentage, percentage slip is calculated as follows:

$$\%S = \frac{N_S - N}{N_S} \times 100\%$$

where:

$\%S$ = Percentage slip
N_S = Synchronous motor speed (in rpm)
N = Actual motor speed (in rpm)

Example 23-3 calculates the slip for a two-pole motor, and **Example 23-4** calculates the percentage of a motor slip.

Example 23-3

Calculate the motor slip for a two-pole motor connected to a 60 Hz power supply if the motor's actual speed is 3450 rpm.

✓ Solution

The synchronous speed for this motor is calculated using the simplified formula for a 60 Hz power supply as follows:

$$N_S = \frac{7200}{P}$$

$$N_S = \frac{7200}{2}$$

$$N_S = 3600 \text{ rpm}$$

With an actual speed of 3450 rpm, the motor slip is then calculated as follows:

$$S = N_S - N$$

$$S = 3600 \text{ rpm} - 3450 \text{ rpm}$$

$$S = 150 \text{ rpm}$$

Example 23-4

Calculate the percentage of motor slip for the motor in **Example 23-3**, which has a synchronous motor speed of 3600 rpm and an actual motor speed of 3450 rpm.

✓ Solution

The percentage slip is calculated as follows:

$$\%S = \frac{N_S - N}{N_S} \times 100\%$$

$$\%S = \frac{3600 \text{ rpm} - 3450 \text{ rpm}}{3600 \text{ rpm}} \times 100\%$$

$$\%S = \frac{150 \text{ rpm}}{3600 \text{ rpm}} \times 100\%$$

$$\%S = 0.0417 \times 100\%$$

$$\%S = 4.17\%$$

Now It's Your Turn 23-4

Calculate the motor slip, expressed as both a numerical value and a percentage, for a four-pole motor connected to a 50 Hz power supply if the motor's actual speed is 1455 rpm.

CHAPTER 23 REVIEW

Summary

- A motor is a machine that converts electrical energy into mechanical energy. (23.1)
- Engines and motors both produce mechanical energy, but engines typically rely on fuel combustion as their energy source. (23.1)
- Alternators and generators rely on various forms of mechanical energy to create electrical energy. (23.1)
- Motors, in their simplest form, are classified as either ac or dc machines. (23.2)
- Motors require an imbalance in magnetic fields to create the twisting force needed to turn them. (23.2)
- AC motors are supplied with a power signal that changes its polarity two times every cycle. (23.2)
- DC motors rely on commutators or electronic circuitry to reverse the polarity of the power signal being supplied to them. (23.2)
- Deciding factors that help determine if an ac or dc motor is to be used include cost, desired speed of motor response, motor efficiency, motor size, starting torque, starting or stopping speed, electromagnetic interference levels, power consumption requirements, and level of required maintenance. (23.2)
- The speed of an ac motor is determined by three factors: the frequency of the ac power signal, the number of motor poles, and the number of polarity changes per cycle. (23.2)
- The speed of a dc motor is controlled by varying the voltage and/or current supplied to the motor, or by adjusting the strength of the magnetic field in which the motor windings are located. (23.2)
- The term synchronous motor speed represents the calculated speed of the motor. (23.3)
- Slip is the difference between a motor's synchronous speed and its actual speed. (23.4)
- Slip can be expressed as a numerical value or as a percentage. (23.4)

Review Questions

Use the information in the chapter to answer the following questions.

Measuring Your Potential

1. A(n) _____ converts electrical energy into mechanical energy. (23.1)
   A. motor
   B. engine
   C. alternator
   D. generator
2. A generator uses _____ to provide mechanical energy that is then converted to electrical energy. (23.1)
   A. a turning belt
   B. an engine
   C. combustion of a fuel
   D. magnetic fields
3. Of the following, which is a benefit of an ac motor over a dc motor? (23.2)
   A. AC motors are less expensive than dc motors.
   B. AC motors respond to load changes faster than dc motors.
   C. AC motors are smaller than dc motors.
   D. AC motors have higher starting torque than dc motors.
4. Of the following, which is a benefit of a dc motor over an ac motor? (23.2)
   A. DC motors are less expensive than ac motors.
   B. DC motors can be used in explosive locations, while ac motors cannot.
   C. DC motors can reverse their direction of rotation faster than ac motors.
   D. DC motors require less maintenance than ac motors.
5. *True or False?* Many dc motors rely on a commutator to reverse the polarity of the power supplied to the motor. (23.2)
6. As the number of poles increases, the speed of an ac motor _____. (23.2)
   A. increases
   B. decreases
   C. remains the same
   D. will vary
7. The term used to express the calculated, or ideal, motor speed is _____. (23.3)
   A. rotations per minute
   B. synchronous speed
   C. commutator
   D. slip
8. *True or False?* The synchronous speed of an eight-pole motor connected to an ac power supply with a frequency of 100 Hz is 1500 rpm.? (23.3)
9. The term slip refers to the _____. (23.4)
   A. difference between a motor's actual speed and its calculated speed
   B. difference between a motor's amperage rating and its actual amperage draw
   C. situation when the motor's amperage draw rises above its nameplate current rating
   D. situation when the voltage supplied to the motor exceeds its nameplate voltage rating

Plug-In

1. The actual speed of a four-pole ac motor connected to a 60 Hz power supply is closest to which of the following if the motor is operating with a 6% slip? (23.4)
   A. 1492 rpm
   B. 1592 rpm
   C. 1692 rpm
   D. 1792 rpm
2. Of the following, which is the fastest motors? (23.4)
   A. Four-pole, 100 Hz, 3% slip
   B. Two-pole, 60 Hz, 5% slip
   C. Four-pole, 70 Hz, 4% slip
   D. Two-pole, 50 Hz, 5% slip
3. Of the following, which is the slowest motor? (23.4)
   A. Four-pole, 100 Hz, 3% slip
   B. Two-pole, 60 Hz, 5% slip
   C. Four-pole, 70 Hz, 4% slip
   D. Two-pole, 50 Hz, 5% slip

Making the Connection

1. Explain why the speed of an ac motor decreases as the number of motor poles increases. Provide as much proof as you can to support your response. (23.3)
2. Explain where the constant 7200 comes from when determining the speed of an ac motor. Under what conditions will this constant change? (23.3)

CHAPTER

24 Motor Anatomy, Types, and Construction

Maxx-Studio/Shutterstock.com

SPARKING DISCUSSION

Explore the websites of a few motor manufacturers, taking note of the various categories and classifications the manufacturer uses to organize their motors. Compare and contrast the various categories you find.

LEARNING OUTCOMES

After completing this chapter, you will be able to:

24.1 Describe the function of the common components of a motor.

24.2 Explain the difference between an open motor, a totally enclosed motor, a hazardous location motor, a flame-proof motor, and an explosion-proof motor.

TECHNICAL TERMS

axial load
bearing
end bell
explosion-proof (XP) motor
flame-proof (FLP) motor
flat
hazardous location (HAZ) motor
hub
key
keyway
National Electrical Manufacturers Association (NEMA)
open drip-proof (ODP) motor
radial load
rotor
shaft
stator
totally enclosed air-over (TEAO) motor
totally enclosed blower-cooled (TEBC) motor
totally enclosed fan-cooled (TEFC) motor
totally enclosed nonventilated (TENV) motor
totally enclosed washdown (TEWD) motor
totally enclosed water-cooled (TEWC) motor

Introduction

With motors being among the most common electrical devices encountered by those working in the electrical trades, it is important to have a solid understanding of them. To gain a better understanding of electric motors, this chapter provides you with an overview of the basic parts of a motor, as well as some popular types of motors, their intended applications, and how they are differentiated from each other. This chapter lays the foundation needed for the following chapters: Chapter 25, *Motor Drives*, and Chapter 26, *Motor Installation and Replacement*.

24.1 Anatomy of a Motor

There are many individual, interdependent components that, collectively, make up a motor. These components must work together seamlessly in order for the motor to operate as effectively and efficiently as possible. Some of these parts contribute to the creation of rotation, while others transfer this mechanical energy to pumps, fans, and other controlled devices. Some parts allow for smoother, more efficient, and quieter motor operation, while others facilitate the connecting, or mounting, of the motor to the larger system of which the motor is a part. Regardless of an individual component's role in ensuring effective and efficient motor operation, the failure of one element will, undoubtedly, result in the reduced operating performance of the entire motor assembly. Some basic motor components are shown in **Figure 24-1**. Motor parts that are specific to certain motor types, such as commutators and brushes in a dc motor, are covered in Section 8, *Motor Functionality*.

Shaft

When all is said and done, it is the rotation of the motor's shaft that allows the fan, pump, or other devices connected to the motor to function. The ***shaft*** is the all-important link between the electrical and magnetic portions of the motor itself and the mechanical work that the motor is helping facilitate. The shaft is physically connected to the rotor, or rotating portion of the motor.

Because they serve as the connection point to fans, pumps, and other components, motor shafts are subject to physical damage. Depending on the motor's application, shafts can be manufactured from many different materials including brass, aluminum, and titanium, but most larger applications use a variety of steel alloys.

A motor shaft can be connected to many driven devices, including fan blades, gears, pulleys, sprockets, and other shafts. The mechanical methods used to connect these driven devices to a shaft include setscrews, pins, keys and keyways, hubs, bushings, and couplings.

Flat

A shaft ***flat*** is formed when a portion of the round shaft is ground or machined to form a flat surface. Motor shafts can have one or more flats machined into them, **Figure 24-2**. When a motor shaft has one flat, the shaft is often referred to as a *D-shaft*, referring to its resemblance to the letter *D* when viewed from the end. Some shafts have two flats. Depending on the application and the motor manufacturer, the two flats can be adjacent to each other or parallel to each other.

Flats are used in conjunction with setscrews. The flat should line up with the setscrew tapping on the ***hub***, or mounting portion, of the controlled device, **Figure 24-3**. When tightened, the setscrew contacts the flat portion of the shaft, securing the device in place and reducing the chances of slippage.

Keyway

A ***keyway*** is a two-part channel that is machined into both the motor shaft and the hub. Half of the channel is

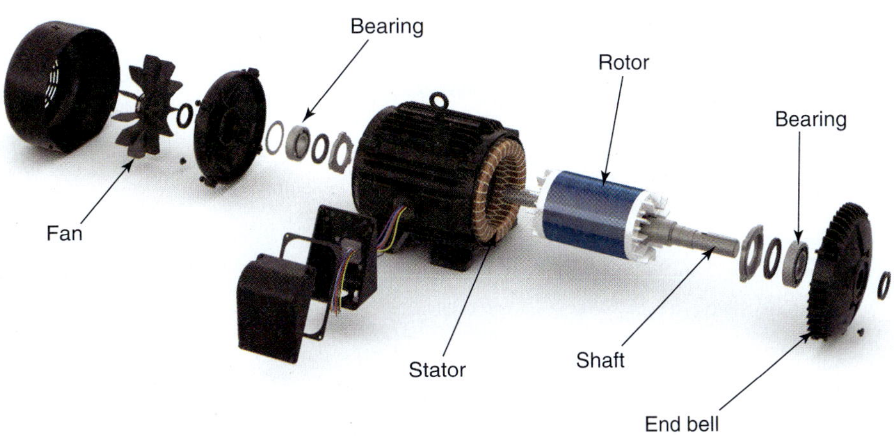

ABB Motors and Mechanical

Figure 24-1. Most motors contain the same basic components.

machined into the shaft, while the other half is machined into the hub. When the controlled device is mounted on the shaft, the two half-channels must be aligned in order to accept the key. The ***key*** is a rectangular prism, or brick-shaped pin, that is inserted into the keyway to help provide a secure connection between the controlled device and the shaft. The key helps prevent slippage of the entire assembly. The key is held in place with one or more setscrews, **Figure 24-4**.

Several types of keys and keyways are used for different applications, **Figure 24-5**. The most common type of keyway is the parallel keyway, which accepts a flat, square-shaped or rectangular-shaped key. The Woodruff keyway uses a semicircular key. In some cases, the motor shaft is partially cross-drilled with a round hole. This allows the tightened setscrew to extend into the shaft itself. Motor shafts that are completely cross-drilled are used in conjunction with a cotter pin or other type of connecting pin.

Shedding Light

Just because a fan, pump, or other component is not rotating, even when the proper voltage is being supplied to the motor, do not immediately conclude that the motor is defective. Inspecting the connection between the motor shaft and the controlled device may prove invaluable. If the setscrew or other connector has come loose, it is possible that the motor will operate correctly, but the shaft is simply rotating within the controlled device, not turning it. In addition, if the controlled device is jammed, or otherwise stuck, the motor may not rotate when energized. If possible, check motor operation with the controlled device disconnected. If the motor operates when isolated, but not when connected to the driven component, do not condemn the motor!

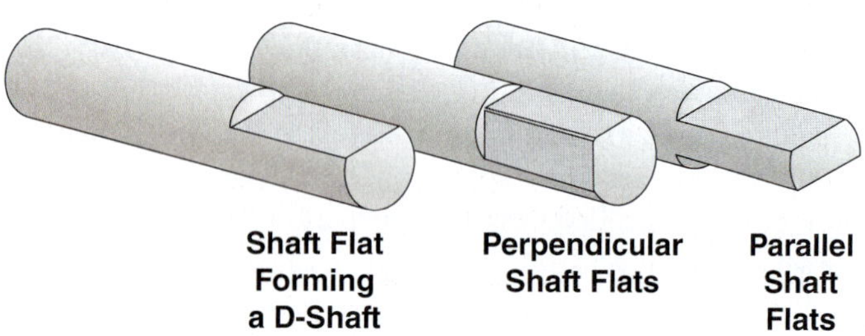

Goodheart-Willcox Publisher

Figure 24-2. Flats are machined into motor shafts to provide a flat surface for mechanical connections.

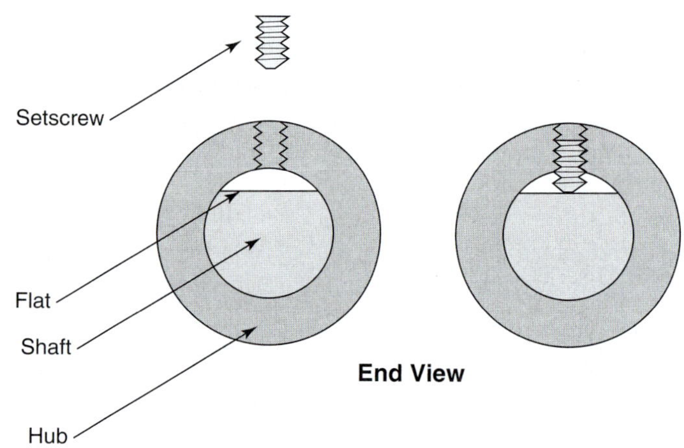

Goodheart-Willcox Publisher

Figure 24-3. A hub is connected to a shaft by tightening a setscrew against the flat.

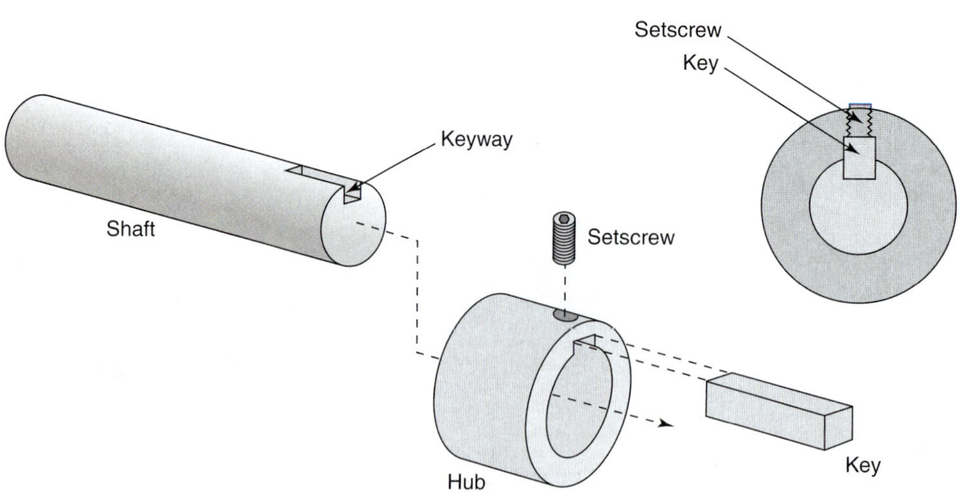

Goodheart-Willcox Publisher

Figure 24-4. This hub is secured to the motor shaft by inserting a key into the keyway and then tightening a setscrew against the key.

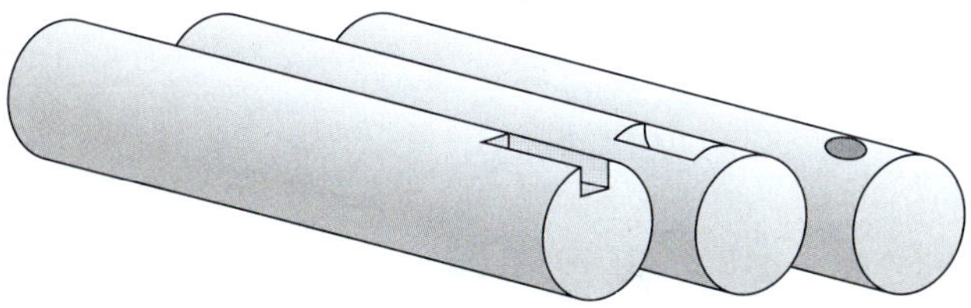

Goodheart-Willcox Publisher

Figure 24-5. Various types of keys and keyways are used for different applications.

BVNIMA/Shutterstock.com

Figure 24-6. Many laminations are tightly stacked to form the body of a rotor.

Rotor

The ***rotor*** is the portion of an electric motor that rotates. The motor shaft is connected directly to the rotor. Since the shaft and rotor are physically connected to each other, the speed at which the rotor spins determines the speed at which the motor shaft rotates. The rotor is often constructed as a series of copper or aluminum bars set in place on iron or steel laminations. Laminations are very thin slices of steel that are individually coated and stacked, **Figure 24-6**.

The rotor is not electrically connected to the motor's power circuits, but functions in a manner that can be equated to the secondary winding of a transformer. Basic transformers have two windings, a primary winding and a secondary winding. When voltage is supplied to the primary winding that is electrically connected to the power supply, a magnetic field is produced, and voltage is induced in the secondary winding. The stator produces the motor's magnetic field, while the rotor is positioned within the stator.

Stator

The ***stator*** is the stationary part of an electric motor. The stator, **Figure 24-7**, is the portion of the motor that is electrically connected to the motor's power source. Stators house the motor's windings that are responsible for creating the magnetic field that ultimately causes the motor's rotor to turn. The stator's windings can be configured and constructed differently based on the size and application of the motor. The most common stator windings are constructed as insulated copper coils composed of many wire turns. The strength of the generated magnetic field is proportional to the number of turns in the stator winding coils.

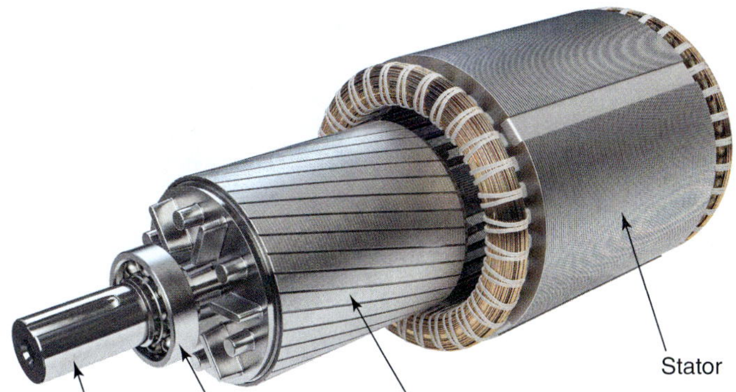

Maxx-Studio/Shutterstock.com

Figure 24-7. The stator is the stationary part of the motor that is connected to the motor's power source. The rotor turns within the stator.

ABB Motors and Mechanical

Figure 24-8. Motor bearings reduce the friction between the rotor and the main body of the motor.

Bearings

Motor ***bearings***, **Figure 24-8**, are located within an electric motor's housing and are instrumental in ensuring the efficient operation of an electric motor. The main purpose of a motor bearing is to reduce the friction that might otherwise affect the ability of the motor's shaft to rotate. By reducing the friction between the motor's rotor and the main body of the motor, losses are reduced, increasing the power, or work potential, of the motor. The two main classifications of bearings are the ball bearing and the sleeve bearing, **Figure 24-9**. Ball bearings rely on carefully machined spheres of steel to create two mating, low-friction surfaces. A sleeve bearing is a lubricated cylindrical surface in which a shaft rotates.

One main consideration when selecting a bearing type is the type and intensity of the loads the bearings will be subjected

Figure 24-9. Two common bearing types are ball bearings and sleeve bearings. A ball bearing consists of an inner race, which fits tightly to the shaft; rolling elements (balls) held in place by a cage; and an outer race, which is connected to the motor enclosure. The sleeve bearing shown here is part of a pillow block, a housing used in mechanical assemblies to support shafts.

Characteristic Comparison for Ball and Sleeve Bearings

Characteristic	Ball Bearing	Sleeve Bearing
Can carry radial loads	Yes	Yes
Can carry axial loads	Yes	No
Size	More compact	Longer
Heat generated	Less	More
Lubrication	Requires less	Requires more
Operating speed	Lower speeds	Wider range of speeds
Cost	Less expensive	More expensive
Replacement	More difficult	Easier
Friction losses	Lower	Higher
Noise	Typically louder	Typically quieter

Goodheart-Willcox Publisher

Figure 24-10. Comparison of ball bearing and sleeve bearing characteristics.

to. ***Radial loads*** push on the motor shaft in directions that are perpendicular to the shaft axis, while ***axial loads*** push in directions that are parallel to the axis of the shaft. Axial loads are also called *thrust loads*. Radial loads may be caused when belts pull on driven pulleys or gears push against one another. Axial loads are caused by the weight of the rotor in vertical motor applications and are common in pump applications, due to fluid pressure on an impellor transferring force to the shaft. Different types of bearings are designed to handle these loads differently. Some of the most common characteristics of ball bearings and sleeve bearings are provided in **Figure 24-10**.

End Bells

End bells are the part of an electric motor that support the bearings. End bells are located on both ends of a motor,

Figure 24-11. Each end of a motor is capped by an end bell, as shown in this cutaway of a totally enclosed fan-cooled (TEFC) motor. Bearings are housed in the center of the end bells.

Figure 24-11. In addition, they act to protect the ends of the motor windings from physical damage. End bells can be solid, or they can have holes or louvers to allow air to move freely across the motor windings. End bells are also commonly known as *shields*, *end shields*, or *end brackets*.

Frame

One of the ways motors are identified is by their frame size or frame number. The frame number mainly refers to the mounting characteristics and measurements between various parts of the motor. Frame numbers are established by the ***National Electrical Manufacturers Association (NEMA)*** and follow a coding system that is used to identify motor measurements. Some of the basic NEMA motor dimensions are shown in **Figure 24-12**. Refer to *Appendix C* for more information on motor frame numbers and dimensions.

One common motor measurement is the D measurement. This is the measurement, in inches, from the center of the motor shaft to the bottom of the motor mount.

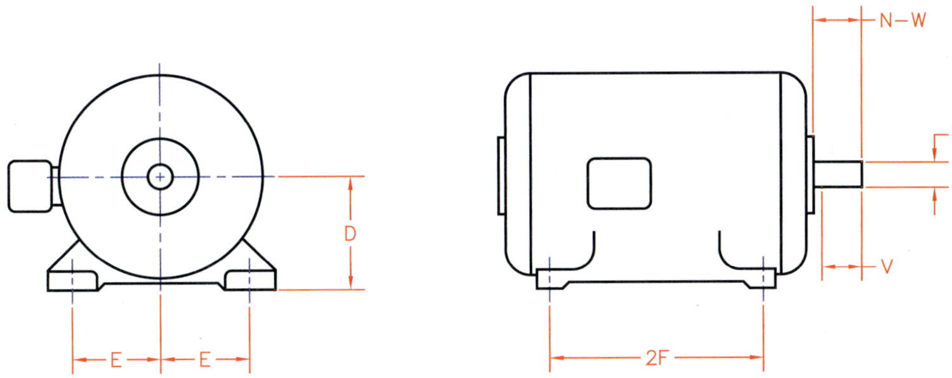

Basic NEMA Motor Dimensions

NEMA Letter	Description of Dimension
D	Centerline of shaft to bottom of feet
E	Centerline of shaft to mounting holes in feet
2F	Distance between centerlines of mounting holes in feet (side view)
N-W	Length of shaft extension from shoulder
U	Shaft diameter
V	Length of shaft available for coupling

Goodheart-Willcox Publisher

Figure 24-12. NEMA standards define a set of motor dimensions to ensure that motor features are consistently described by manufacturers and designers. The dimensions shown here are some of the basic dimensions included in the standards.

For motors with a two-digit frame number, the D measurement is found by dividing the frame number by 16. For example, the D measurement for a 48 frame motor is calculated by dividing 48 by 16, resulting in a D measurement of 3″. For a 56 frame motor, the D measurement is 3.5″ (56 divided by 16).

For frame numbers that have three or more digits, the D measurement is equal to the first two frame number digits divided by 4. **Example 24-1** provides additional practice in calculating the D measurement of a motor.

Example 24-1

What is the D measurement for a motor with a frame number of 182?

✓ Solution

For a 182 frame motor, the D measurement is found by dividing the first two digits of 182 by 4, so the D measurement is 4.5″.

Now It's Your Turn 24-1

What is the D measurement for a motor with a frame number of 2110?

Motor frame numbers are often followed by a frame letter, or suffix. The frame letter provides more information about the intended application for a particular frame number, or additional frame measurement criteria. Frame numbers that have an "M" or "N" suffix, for example, are intended for use as oil burner motors, while frame numbers that have a "G" suffix are intended for gasoline pump motors. Refer to *Appendix C*, for more information on motor frame letters and their intended applications.

Shedding Light

When replacing a motor, be sure to match the frame number of the replacement motor to that of the motor being changed out. Failing to do this can result in an additional trip back to the supply house or an improperly selected, applied, or installed motor.

24.2 Motor Types Classified by Enclosure and Cooling Method

Motors are designed and constructed based on their intended use. In addition, the conditions the motor will be subjected to must be considered. Proper motor selection is a detailed process that, when executed properly, will help ensure problem-free and reliable operation. Some common motor classifications based on enclosure and cooling method include:

- Open drip-proof (ODP)
- Totally enclosed fan-cooled (TEFC)
- Totally enclosed blower-cooled (TEBC)
- Totally enclosed water-cooled (TEWC)
- Totally enclosed nonventilated (TENV)
- Totally enclosed washdown (TEWD)
- Totally enclosed air-over (TEAO)
- Hazardous location (HAZ)
- Flame-proof (FLP)
- Explosion-proof (XP)

Shedding Light

When selecting, replacing, installing, or servicing a motor, make certain that all manufacturer's instructions, recommendations, safety guidelines and procedures are followed. Failure to do so can result in premature motor failure, personal injury, or property damage.

Open Drip-Proof Motors

Open drip-proof (ODP) motors allow air to flow freely through them, but do not allow dripping water to enter the enclosure, **Figure 24-13**. A louver-type arrangement prevents water from dripping inside the enclosure. However, if the motor is not mounted level, or according to manufacturer's guidelines, water may enter. As an ODP motor is an open-type motor, it is not recommended for installation or use in wet locations. Open motors rely on airflow through them to keep the motor windings cool. Under certain conditions, water can enter the motor's housing. In the event water does enter the motor, many manufacturers have provided small drain holes located at the bottom of the housing to allow water to drain.

Totally Enclosed Fan-Cooled Motors

Totally enclosed fan-cooled (TEFC) motors, unlike open-type motors, do not allow external air to flow into the motor enclosure and across the windings for cooling purposes, **Figure 24-14**. Instead, TEFC motors have a small fan attached to the rear end of the shaft. This fan moves air across the motor when operating to help keep it cool. TEFC motors are the most commonly used motors in industrial settings.

Even though TEFC motors are totally enclosed, moisture in the form of condensation may accumulate inside the motor's housing. To allow any condensation to drain from the motor, small weep or drain holes are provided at the bottom of the motor. When installing any motor, ensure that all drain holes are positioned at the bottom of the motor, **Figure 24-15**. Failure to do so can result in water accumulation in the housing and premature motor failure.

Totally Enclosed Blower-Cooled Motors

In a TEFC motor, the cooling fan is attached to the motor shaft, so the fan rotates at the same speed as the motor. In low-speed applications, the lower fan speed limits the amount of airflow over the motor and may not be sufficient to cool the motor. ***Totally enclosed blower-cooled (TEBC) motors*** are similar to TEFC motors, except the fan is driven by a separate motor, rather than being attached to the motor shaft,

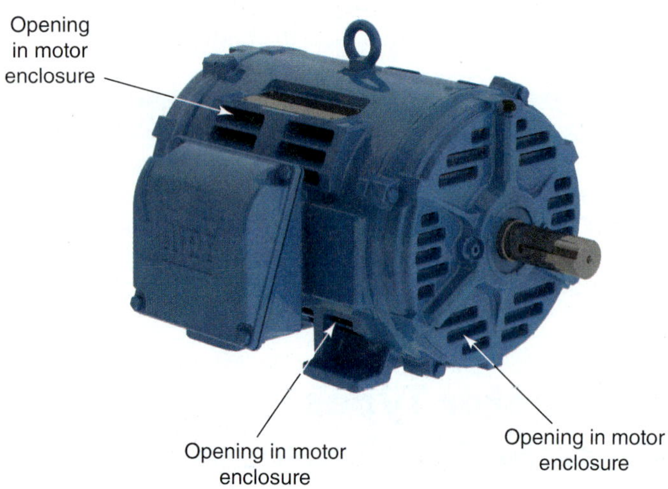

WEG

Figure 24-13. An open drip-proof (ODP) motor has openings in the motor housing to allow external airflow into the motor. Openings are shielded from above to prevent water striking the motor at an angle up to 15° from vertical.

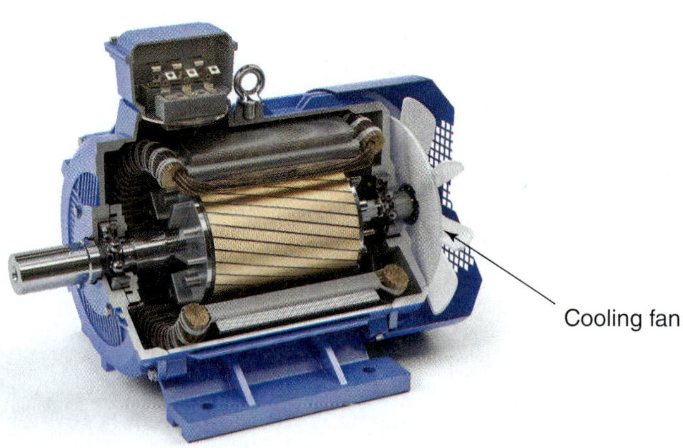

Maxx-Studio/Shutterstock.com

Figure 24-14. Totally enclosed fan-cooled (TEFC) motors have a cooling fan connected to the motor shaft. When the motor is running, the fan circulates air over the outside of the motor enclosure. The end bell between the fan and the stator and rotor prevents the airflow from entering the interior of the motor.

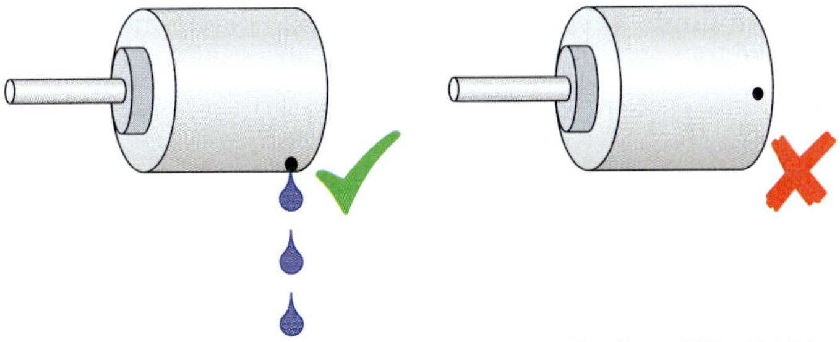

Goodheart-Willcox Publisher

Figure 24-15. Proper motor installation and positioning of weep holes ensures that any condensation in the motor can drain.

ABB Motors and Mechanical

Figure 24-16. A totally enclosed blower-cooled (TEBC) motor has a separate motor driving the cooling fan.

Figure 24-16. This allows the fan speed to be greater than the motor speed, ensuring adequate airflow to cool the motor.

Totally Enclosed Water-Cooled Motors

All air-cooled motors use the air around the motor to cool the motor. When motors are located in a high-temperature area, the hot air around the motor cannot cool the motor as well as cooler air. ***Totally enclosed water-cooled (TEWC) motors*** do not rely on the surrounding air for cooling, **Figure 24-17**. Instead, the motor housing includes channels through which water flows. The water flows into the motor housing, absorbs heat from the motor, and then flows out of the housing.

Totally Enclosed Nonventilated Motors

Totally enclosed nonventilated (TENV) motors are similar to TEFC motors, but they do not have a fan to cool the motor, **Figure 24-18**. As this type of motor relies on the surrounding air to provide cooling, it is important to carefully evaluate the expected ambient conditions. This will help ensure that the motor will operate properly without overheating. These motors can be used in areas where a fan may be subject to clogging, such as nonhazardous locations that are moist or dirty.

WEG

Figure 24-17. This totally enclosed water-cooled (TEWC) motor has a water channel cast in its frame. Water circulates through the channels, absorbing heat and cooling the motor.

WEG

Figure 24-18. A totally enclosed nonventilated (TENV) motor has no cooling fan as part of its construction.

Totally Enclosed Washdown Motors

Totally enclosed washdown (TEWD) motors are designed to be installed in wet environments, including environments that require periodic, thorough cleaning of motors, **Figure 24-19**. There are different types of washdown motors available, each being able to withstand varying degrees of water exposure. Be sure to select the type of motor properly to ensure continued reliable operation.

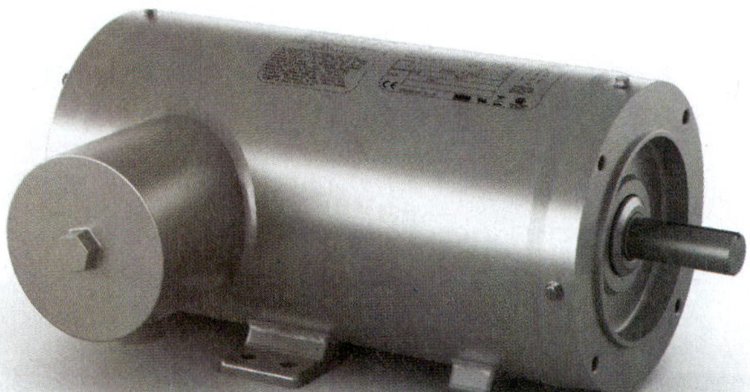

ABB Motors and Mechanical

Figure 24-19. This totally enclosed washdown (TEWD) motor has a stainless-steel enclosure and can be used in wet environments such as food processing facilities.

WEG

Figure 24-20. Totally enclosed air-over (TEAO) motors are sometimes installed in a forced-air duct system.

Shedding Light

The United States Food and Drug Administration (FDA) has strict standards that must be met in food processing plants. These standards include the periodic deep cleaning of every exposed surface in the plant. This includes motors. Motors selected for use in food-processing equipment must be able to withstand constant exposure to large amounts of water, which often has chemical cleaning agents dissolved in it.

Totally Enclosed Air-Over Motors

Totally enclosed air-over (TEAO) motors are often referred to as dust-tight motors, **Figure 24-20**. These motors are similar in construction to TEFC motors but do not use an integral fan to keep them cool. Instead, TEAO motors are often installed within an airstream, such as in an HVAC system's duct system.

Hazardous Location Motors

Hazardous location (HAZ) motors are classified by a number of factors including the type of hazard that the motor is likely to be exposed to. These factors include environmental exposure potential, maximum temperature limits, and guidelines specific to the safe containment of certain substances and materials. When selecting a motor for use in potentially hazardous locations, ensure that all criteria are carefully evaluated.

Know the Code

Details about hazardous locations can be found in *Article 500, Hazardous (Classified) Locations, Classes I, II, and III, Divisions 1 and 2* of the *NEC*. *Article 501* covers Class I locations that include flammable gases, flammable vapors, or combustible liquid-produced vapors such as propane and acetylene. *Article 502* covers Class II locations that include combustible dust such as coal, grain, and metal. *Article 503* covers Class III locations that include combustible fibers or flyings. *Division 1* refers to constant or expected hazards, while *Division 2* refers to intermittent or unexpected hazards. Alternative classifications to these division methods are covered in the "zone methods" in *Article 505* and *Article 506*.

Flame-Proof Motors

Flame-proof (FLP) motors are a type of hazardous location motor designed to prevent flames produced inside the motor housing from traveling to their surroundings, **Figure 24-21**. These motors are also intended to contain any explosions that might occur within the motor's housing. Since FLP motors are used in areas in which hazardous or dangerous chemicals and substances are found, the goal is to prevent an internal motor failure from causing a more catastrophic event should flames or sparks penetrate the housing. Flame-proof motors are rated for different gas groups. When selecting a motor, ensure the motor is rated for the appropriate application. Common gas groups include the following:

- Gas Group IIA: Propane or gases of equivalent hazard
- Gas Group IIB: Ethylene or gases of equivalent hazard
- Gas Group IIC: Acetylene, hydrogen, or gases of equivalent hazard

Flame-proof motors are similar to explosion-proof motors, but they adhere to different standards. The International Electrotechnical Commission (IEC) sets the standards for flame-proof motors.

Figure 24-21. Flame-proof (FLP) motors are rated for specific hazardous locations.

Figure 24-22. Explosion-proof (XP) motors are rated for specific hazardous locations. This motor is equipped with a wireless sensor that monitors the health and performance of the motor.

Explosion-Proof Motors

Explosion-proof (XP) motors are a type of hazardous location motor designed to withstand and contain an explosion that might occur within the motor's casing, **Figure 24-22**. Similar to flame-proof motors, explosion-proof motors are designed to minimize the possibility of a larger problem involving the potentially combustible nature of the motor's surroundings. One way to help prevent a devastating event from occurring in a chemical environment is to ensure that the motor's normal operating temperature is well below the ignition temperature or flashpoint of the substances that surround the motor's housing. Some gases can have a lower-than-expected flashpoint.

Explosion-proof motors adhere to different sets of standards than flame-proof motors do. Explosion-proof motors adhere to United States standards organizations, including NEMA, the American National Standards Institute (ANSI), and the Underwriters Laboratories (UL). These sets of standards are typically stricter than those set forth by the IEC. As a result, explosion-proof motors typically undergo more extensive testing and have an overall heavier construction. In many cases, specialized tools are required to access the housing in an effort to maintain the integrity of the motor shell. To ensure compliance with the strict standards set by these organizations, explosion-proof motors are not manufactured using gasketing materials in the motor's housing. All connections between the motor's housing and any electrical junction boxes are strictly metal-to-metal. Explosion-proof motors are also equipped with breathers and drains. Drains allow for the removal of any condensation that might accumulate within the motor casing, while breathers allow air to enter the housing to equalize pressure.

Summary

- The motor shaft is connected to the rotor and turns at the same speed as the rotor. (24.1)
- Flats are formed on the motor shaft to help connect the driven component to the shaft. Keyways and keys are used in larger applications to secure the driven component to the shaft. (24.1)
- The rotor is the portion of a motor that rotates. The stator is the stationary portion of a motor that houses the windings. Bearings are used to reduce rotational friction of the rotor. (24.1)
- Motors are equipped with end bells to protect the motor windings and support the bearings. Common bearing types are ball bearings and sleeve bearings. (24.1)
- Radial loads push in a direction that is perpendicular to the shaft. Axial (thrust) loads push in a direction that is parallel to the shaft. (24.1)
- The motor frame number identifies a motor's mounting characteristics and measurements. (24.1)
- Motor types are classified by enclosure and cooling method. (24.2)
- Open drip-proof (ODP) motors rely on airflow through them to keep the motor windings cool. (24.2)
- A totally enclosed fan-cooled (TEFC) motor has a small fan attached to the shaft to move air across the exterior of the motor when operating. (24.2)
- A totally enclosed blower-cooled (TEBC) motor has a cooling fan driven by a separate, dedicated motor. (24.2)
- A totally enclosed water-cooled (TEWC) motor has channels in the motor housing through which water flows. The water flows into the motor housing, absorbs heat from the motor, and then flows out of the housing. (24.2)
- A totally enclosed nonventilated (TENV) motor is totally enclosed and does not have a fan. This type of motor relies on the surrounding air to provide cooling. (24.2)
- A totally enclosed washdown (TEWD) motor is intended for use in wet environments or areas where the motor can be subjected to regular, thorough cleaning. (24.2)
- A totally enclosed air-over (TEAO) motor is a dust-tight motor installed within an airstream to provide cooling. (24.2)
- Hazardous location (HAZ) motors are classified by environmental exposure, maximum temperature limits, and guidelines specific to the safe containment of certain substances. (24.2)
- Flame-proof (FLP) and explosion-proof (XP) motors are hazardous location motors. (24.2)

Review Questions

Use the information in the chapter to answer the following questions.

Measuring Your Potential

1. Which of the following two components are in direct physical contact with each other? (24.1)
   A. The rotor and the shaft
   B. The stator and the hub
   C. The rotor and the stator
   D. The stator and the shaft
2. *True or False?* A motor shaft with one flat is referred to as a D-shaft. (24.1)
3. A keyway is formed between grooves in the _____. (24.1)
   A. hub and shaft
   B. hub and flat
   C. flat and bearing
   D. bearing and shaft
4. *True or False?* The motor shaft rotates at the same speed as the motor's stator. (24.1)
5. Thin slices of steel that are individually coated and stacked are called _____. (24.1)
   A. layers
   B. laminations
   C. elevations
   D. extrusions
6. The _____ is electrically connected to the motor's power source. (24.1)
   A. stator
   B. rotor
   C. shaft
   D. hub
7. *True or False?* Ball bearings are better at handling axial thrust loads than sleeve bearings. (24.1)

8. Radial thrust loads act in a direction that is _____. (24.1)
   A. parallel to the motor's shaft
   B. perpendicular to the motor's shaft
   C. parallel to the motor's rotor
   D. perpendicular to the motor's rotor
9. A motor's end bells protect the _____. (24.1)
   A. stator windings and support the bearings
   B. rotor windings and support the motor shaft flats
   C. shaft and support the bearings
   D. stator windings and the motor shaft flats
10. The motor type that is equipped with a shaft-mount fan is the _____ motor. (24.2)
    A. totally enclosed fan-cooled
    B. totally enclosed air-over
    C. totally enclosed nonventilated
    D. totally enclosed washdown
11. *True or False?* Totally enclosed nonventilated motors rely on the movement of the surrounding air to cool them. (24.2)
12. The motor type that is best suited for use in wet locations is the _____. (24.2)
    A. open drip-proof motor
    B. totally enclosed washdown motor
    C. totally enclosed air over motor
    D. flame-proof motor
13. The two motor types that are most similar to each other are the _____. (24.2)
    A. open drip-proof motor and the totally enclosed washdown motor
    B. flame-proof motor and the explosion-proof motor
    C. hazardous location motor and the totally enclosed washdown motor
    D. totally enclosed washdown motor and the totally enclosed air-over motor

Plug-In

1. Electrician A states that a motor with a frame number of 48M is an oil burner motor with a 3″ D dimension. Electrician B states that a motor with a frame number of 48N is gasoline pump motor with a 3″ D dimension. Who is correct? (24.1)
   A. Electrician A only
   B. Electrician B only
   C. Both electrician A and electrician B
   D. Neither electrician A nor electrician B
2. Electrician A states that a motor with a frame number of 120M is an oil burner motor with a 7.5″ D dimension. Electrician B states that a motor with a frame number of 142N is an oil burner motor with a 3″ D dimension. Who is correct? (24.1)
   A. Electrician A only
   B. Electrician B only
   C. Both electrician A and electrician B
   D. Neither electrician A nor electrician B
3. Which of the following organizations provides standards for flame-proof motors? (24.2)
   A. The American National Standards (ANSI)
   B. The Underwriters Laboratories (UL)
   C. The National Electrical Manufacturers Association (NEMA)
   D. The International Electrotechnical Commission (IEC)

Making the Connection

1. Explain why thrust loads should be considered when choosing a motor. How will varying thrust load types affect this decision? (24.1)
2. Explain why shafts on motors used for higher-torque applications are not completely round. (24.1)

CHAPTER

25 Motor Drives

EvijaF/Shutterstock.com

SPARKING DISCUSSION

On the job, you have encountered a motor that has failed. Your coworker suggests that you simply replace the motor, check its operation, and leave the job. Do you agree or disagree with your coworker's suggestion? Provide support for your position.

LEARNING OUTCOMES

After completing this chapter, you will be able to:

25.1 Describe the basic function of a motor drive.
25.2 Explain how energy can be saved by using a motor drive.
25.3 Explain how pulse width modulation (PWM) is used to convert a dc power source to an ac voltage signal.
25.4 Describe the construction and function of a transistor.
25.5 Explain how a variable frequency drive (VFD) is used to change the speed of an ac motor.
25.6 Describe constant torque, variable torque, and constant power applications.

TECHNICAL TERMS

affinity laws
base
bipolar junction transistor (BJT)
collector-emitter junction
converter
dc bus
drive system
full-wave bridge rectifier
insulated-gate bipolar transistor (IGBT)
inverter
metal oxide semiconductor field effect transistor (MOSFET)
motor drive
pulse width modulation (PWM)
saturation voltage
transistor
variable frequency drive (VFD)

Introduction

Historically, motors were designed and intended to rotate at a single speed, as indicated on the motor's nameplate. These older motors had two speeds—on and off. Running a motor, pump, fan, or compressor at full speed whenever it cycled on subjected the motor to excessive wear and tear. In addition, a great deal of energy was wasted. Today, motors have the ability to operate safely at multiple speeds. By varying the speed at which a motor operates, the work done can match the requirements of the application. This saves energy and, as a result, saves money.

With motors being one of the most utilized electrical components in our industry, it is important that you understand the methods and strategies that are used to control them.

This chapter provides the theories and concepts used in motor drives. The content provided is based on material and concepts covered earlier in this text, so be sure to refer back as needed to maximize your understanding.

25.1 Motors are Not Motor Drives

It is important to understand that motors and motor drives are not the same. A motor is the device that produces the force, either rotational or linear, that powers an apparatus or machine. A ***motor drive*** is an electronic device that controls the operational characteristics of a motor, such as motor speed, torque, position, and direction of rotation. A motor drive controls the operation of the motor by modifying the electrical power waveform supplied to the motor. Specifically, a drive can control the voltage and frequency of the power, continuously changing both characteristics to achieve the desired motor output. Most motor drives are discreet, stand-alone devices, but in some cases a driver may be attached to the motor as a single assembly, **Figure 25-1**. The combination of a motor and a motor drive forms a ***drive system***.

Most motor drives can be categorized as one of four types:

- **DC motor drives.** DC drives convert ac power into dc power to supply a dc motor.
- **AC motor drives.** AC drives convert ac power into a dc signal, and then convert the dc signal back into ac power to supply an ac motor.
- **Servo motor drives.** Servo drives are specially designed to control servo motors. Servo motors rely on one or more feedback signals and are used for specialized applications where precise control of speed, torque, and/or position are required.
- **Stepper motor drives.** Stepper drives have functions specifically tailored to control stepper motors. Stepper motor drives convert digital, electronic signals into a mechanical force that causes the motor to turn in fixed, predetermined rotational increments.

Shedding Light

Motor drive names are inconsistent throughout the electrical industry. AC drives and dc drives may be referred to as adjustable speed drives, adjustable frequency drives, variable speed drives (VSDs), variable frequency drives (VFDs), inverters, dc converters, and power converters. An inverter is nearly always an ac drive, because an inverter is a device that converts dc to ac.

When compared with a motor starter, which provides a single, constant power supply to a motor operating at a single speed, a motor drive provides several significant benefits:

- **Improved process control.** Motor speed can be reduced during periods when the driven equipment does not need to operate at 100%.
- **Energy efficiency.** Operating at reduced motor speeds, matching motor torque to load, and ramping up motor speed on start-up all reduce the total amount of electricity consumed by a motor.
- **Motor and driven-equipment longevity.** Ramp-up and ramp-down during starting and stopping and matching torque to load reduce mechanical stress on the motor and the driven equipment (such as pumps, compressors, fans, and blowers).

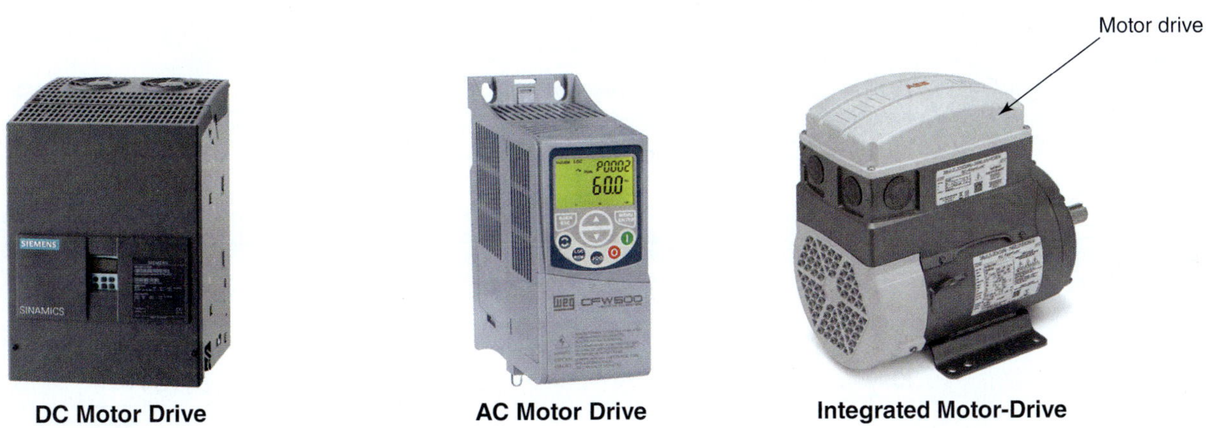

© Siemens AG 2025, All rights reserved. WEG ABB Motors and Mechanical

Figure 25-1. Motor drives are most commonly stand-alone devices, but a motor drive can be integrated with the motor.

These benefits are discussed in greater detail throughout this chapter.

In operation, motor drives react to inputs received from a control module that processes data from a motor feedback loop and from the system's operating parameters, **Figure 25-2**. Consider a pump that provides water for a production process that requires different amounts of water at different stages of the manufacturing process. If the pump provides water at a constant rate of 2000 gallons per minute, water will be wasted whenever the process requires anything less than 2000 gallons per minute. Similarly, not enough water will be provided if the process requires more than 2000 gallons of water per minute. By using a motor drive, the speed of the motor controlling the pump can vary, and the pump's flow rate will match that of the process. As a result, the amount of water wasted is reduced, as is the amount of wasted electrical energy.

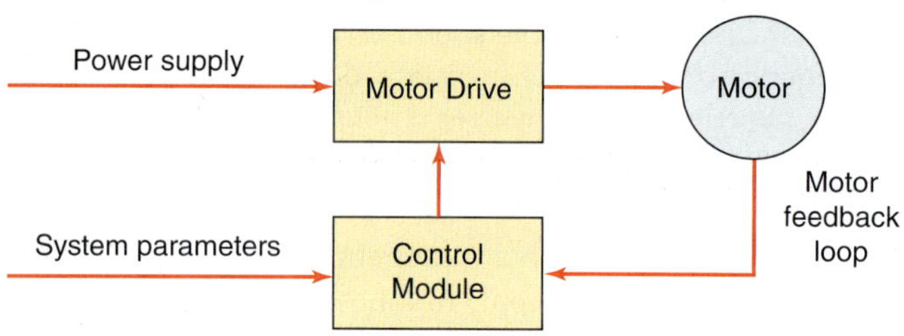

Goodheart-Willcox Publisher

Figure 25-2. Process diagram of a motor drive.

Early Generation Drive Theory

To set the stage for understanding more recent developments, an early generation method to vary the speed of a dc motor is discussed here. Further information on dc motors is explained in Chapter 27, *DC Motors*.

Varying the voltage supplied to a dc motor causes the speed of the motor to change. **Figure 25-3** illustrates the basic concept of varying the voltage supplied to a constant-resistance load in a dc circuit using a variable resistor. By varying the value of the variable resistor in this circuit, the voltage supplied to the 3 Ω load will change. As the resistance of the variable resistor increases from 0 Ω to 15 Ω, the voltage across the 3 Ω load decreases from 12 V to 2 V.

One main drawback of using a variable resistor to adjust the voltage supplied to the load is the loss of power associated with the load. As the table in **Figure 25-3** implies, as the resistance of the variable resistor increases, it consumes more power, and the power loss increases.

One alternative option to vary the voltage supplied to a load is to use a switch to start and stop current flow through the circuit. When the switch is open, there is no current flow and, as a result, no voltage is supplied to the load. When the switch is closed, voltage to the load is provided. By adjusting the switch's on and off times, the average voltage supplied to the motor can be changed as the needs of the system change.

Consider the dc waveform shown in **Figure 25-4**. A 10 Vdc voltage signal is supplied at time zero and continues for

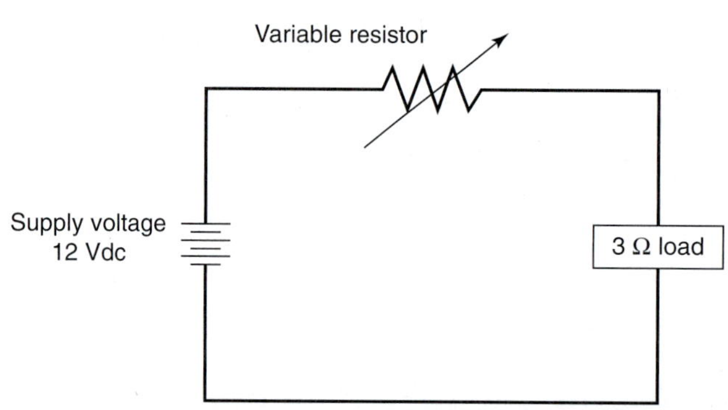

Power Supply (Vdc)	Variable Resistance Value (Ω)	Variable Resistance Voltage (V)	Load Resistance (Ω)	Load Voltage (V)	Circuit Current (A)	Circuit Power (W)	Load Power (W)	Power Loss (%)
12	0	0	3	12	4	48	48	0
12	0.6	2	3	10	3.33	40	33.3	16.7
12	1.5	4	3	8	2.67	32	21.4	33.3
12	3	6	3	6	2	24	12	50
12	6	8	3	4	1.33	16	5.3	66.9
12	15	10	3	2	0.67	8	1.34	83.4

Goodheart-Willcox Publisher

Figure 25-3. Voltage at a load can be controlled with a series-connected variable resistor. The chart lists voltage at the load and power loss for various variable resistor settings.

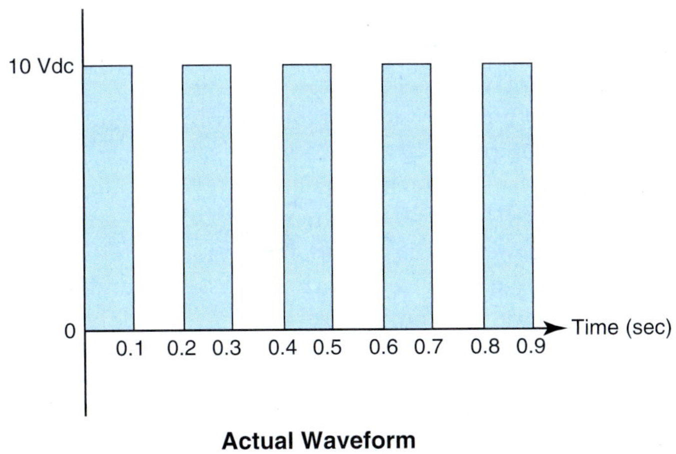

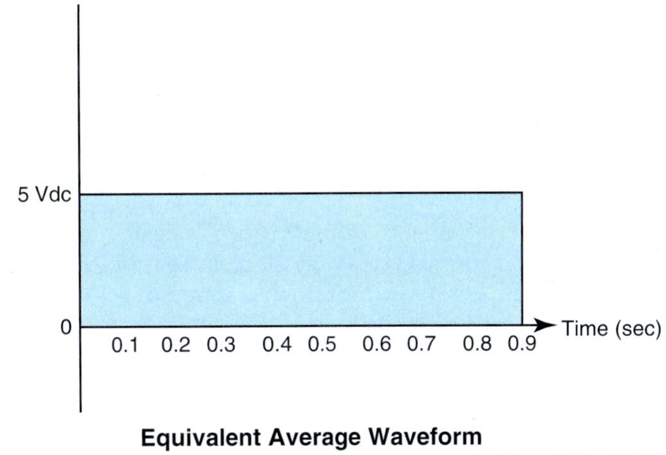

Goodheart-Willcox Publisher

Figure 25-4. Pulses of dc voltage can be represented as an equivalent average waveform.

0.1 second. After 0.1 seconds, the voltage signal stops for 0.1 seconds. At 0.2 seconds, the 10 Vdc signal is once again provided for 0.1 seconds. This on-and-off cycle repeats itself indefinitely. With the signal being on and off for the same time interval, namely 0.1 second, it can be established that the average voltage being supplied will be one-half of the 10 Vdc signal, or 5 Vdc. By varying the on and off times of the dc voltage signal, the average voltage supplied to the load can be changed as needed. This underlying concept, referred to as pulse width modulation, or PWM, is discussed in more detail later in this chapter.

Safety Alert

Although motor drives allow for motor operation at speeds above the motor's nameplate rating, it is usually not desirable to do so. Not only is power consumption greatly increased, but the motor can overheat and even break apart if it runs at unsafe speeds. Most motors can be safely operated at speeds up to 110% of their rated speed, which is 10% over the nameplate speed. If you notice a motor operating at speeds that greatly exceed the nameplate rating, you should investigate the application to determine if there is another motor that would be better suited for the purpose. Overheating of the motor will shorten its life expectancy.

Shedding Light

To help reduce the possibility of motor damage from operating at excessive speeds, programmable motor drives are often equipped with a maximum frequency setting. In the United States, where the standard supply signal frequency is 60 Hz, the motor drive can be programmed to limit the frequency of the power signal supplied to the motor to 66 Hz, which is 10% higher than the 60 Hz standard.

25.2 Energy-Saving Benefits of Varying Motor Speed

In an age where energy consumption and efficiency are of utmost importance, it seems logical to vary the speed of a motor to match the needs of the system. Intuitively, you might think that running a motor at one-half of its design speed would use one-half of the power. In actuality, the amount of power consumed is much less, as demonstrated through the use of affinity laws.

Affinity laws are a series of mathematical formulas and expressions that approximate the performance of fans and pumps. Consider the situation where fan speed and horsepower change in response to changes in system or load requirements. This relationship between motor speed and power consumption is expressed as:

$$\frac{P_I}{P_F} = \left(\frac{\text{rpm}_I}{\text{rpm}_F}\right)^3$$

where:

P_I = Initial power
P_F = Final power
rpm_I = Initial motor speed
rpm_F = Final motor speed

From this formula, a cubic (or power of three) relationship exists between the power consumption ratio of a fan or blower and its speed ratio. Refer to **Example 25-1** on how to calculate power consumption.

Variable Speed Motor Drive Applications

Variable speed drives provide smoother and more consistent operation when compared to a fixed, constant speed

Example 25-1

How much power is consumed by a motor that operates at 50% of its 1000 rpm rated speed if, at 1000 rpm, the motor consumes 5000 W?

✓ Solution

In this case, P_I is equal to 5000 W, rpm_I is equal to 1000 rpm, and rpm_F is equal to 500 rpm. To find P_F, which is the power consumption at a speed of 500 rpm, use the following formula:

$$\frac{P_I}{P_F} = \left(\frac{rpm_I}{rpm_F}\right)^3$$

First, simplify the motor speed ratio expression:

$$\left(\frac{rpm_I}{rpm_F}\right)^3 = \left(\frac{1000}{500}\right)^3 = 2^3 = 8$$

Rearrange the equation to solve for P_F and substitute 8 for the speed ratio expression:

$$P_F = \frac{P_I}{8}$$

$$P_F = \frac{5000 \text{ W}}{8}$$

$$P_F = 625 \text{ W}$$

Based on this calculation, we can see that when the motor speed is reduced by 50%, the power consumption of the motor drops by 87.5%, to only 12.5% of the motor's full speed power consumption. The calculation is provided here:

$$\text{Half Speed Power Consumption (\%)} = \left(\frac{\text{Half Speed Power}}{\text{Full Speed Power}}\right) \times 100\%$$

$$\text{Half Speed Power Consumption (\%)} = \frac{625 \text{ W}}{5000 \text{ W}} \times 100\%$$

$$\text{Half Speed Power Consumption (\%)} = 0.125 \times 100\%$$

$$\text{Half Speed Power Consumption (\%)} = 12.5\%$$

This represents an 87.5% reduction in power consumption.

alternative. Imagine for a moment that the speed control for your car had two options: full speed and full stop. In this situation, you would be exposed to sudden starts, sudden stops, and as a result, a lot of wasted fuel and, very likely, an overall feeling of discomfort. The sudden starts and stops are also very likely to put undue stress on the mechanical systems of your car and contribute to excessive wear and tear. By slowly varying the rates at which the vehicle accelerates and decelerates, a more even and comfortable ride will be experienced, as well as an increase in fuel efficiency and a likely decrease in the frequency of required vehicle repairs.

Now, consider the situation where a submersible pump is located at the bottom of a drilled well. If the pump is actuated by a fixed-speed motor, the motor's pump operation will transition between on and off, depending on the desired or required conditions as established by the system's control strategy. Continually cycling the pump on and off will result in frequent surges of electric current needed to start the motor. This results in electrical, mechanical, hydraulic, and thermal stresses on and within the motor as well as on its control equipment and the pump itself. Since the pump, wiring, and piping are likely located many feet below the ground's surface, any mechanical or electrical failure in the pump will be expensive and time-consuming to repair. If an adjustable or variable speed drive is used, the pump will have the ability to operate continuously, but it will do so at a speed that gradually and smoothly increases and decreases, ramping up and down to match the required water flow rate.

Fixed-Speed Blower Application

Fans, blowers, and pumps consume a large part of the electrical energy used by commercial and industrial motors. Early generation fans and pumps were typically driven by single-speed motors, even though the process requirements were variable. A commonly used strategy was to vary the fluid flow rate to the process by installing a damper or valve arrangement at the outlet of the fan, blower, or pump.

Consider the situation where a centrifugal blower is used to provide airflow for a particular application. The airflow for this application will vary based on a certain set of criteria, but the blower is driven by a constant speed motor. In this configuration, the maximum delivered airflow, in cfm (cubic feet per minute), will often, by design, be greater than the maximum amount of air required by the process. This is to ensure that enough air is supplied to the process. The excess air is typically delivered to a dump zone, which is often an unused, unoccupied, unheated, or uncooled area, **Figure 25-5**. In this fixed-speed blower system, airflow to the process can be controlled by a multi-vein opposing damper arrangement. The dampers feed air to the process and dump zones, as required.

As the amount of required airflow for the process decreases, the process air damper closes, while the dump zone damper opens, **Figure 25-6**. The air flowing into the dump zone represents the difference between the total amount of air delivered by the blower and the amount of air required by the process. In this configuration, the amount of air delivered by the blower is the same regardless of how much air is actually being used by the process and how much air is wasted. At a constant airflow rate and a constant motor speed, the amount of energy used is also constant. A more effective and efficient method of controlling airflow to the process would be to directly regulate

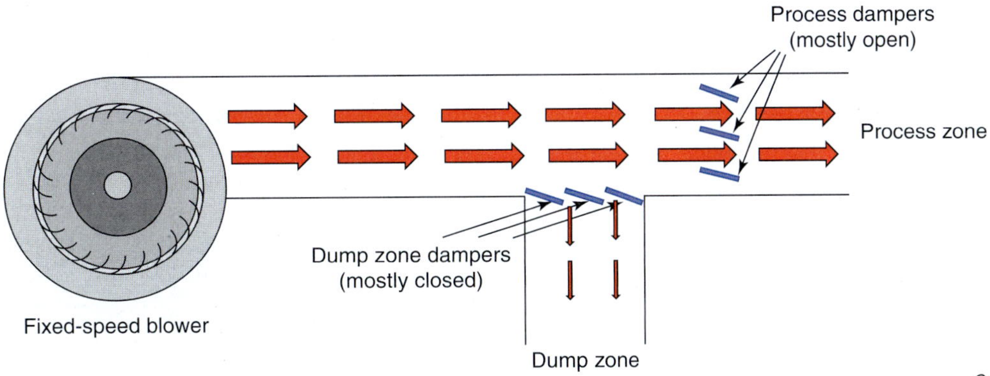

Goodheart-Willcox Publisher

Figure 25-5. A fixed-speed blower delivers constant airflow, regardless of system need. In this configuration, excess air escapes into the dump zone.

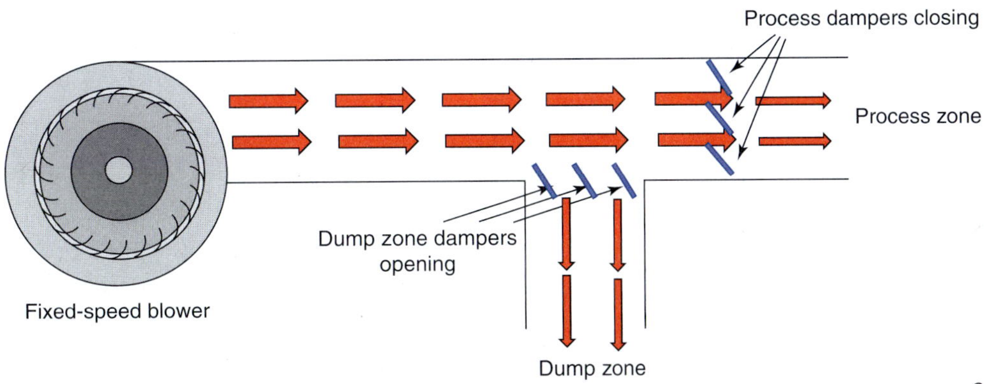

Goodheart-Willcox Publisher

Figure 25-6. As less air is needed by the process, more air and energy are wasted.

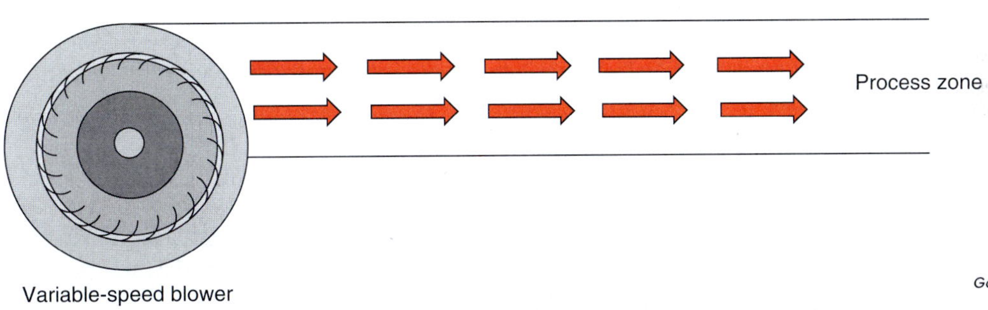

Goodheart-Willcox Publisher

Figure 25-7. A variable-speed blower delivers only the airflow that is needed.

the speed of the motor so the blower only discharges the required amount of air. This eliminates the need for the dump zone and, in addition, little to no energy is wasted, **Figure 25-7**.

According to affinity laws, as the motor speed drops, the amount of energy saved increases exponentially. If, in this scenario, the speed of a variable-speed motor is reduced to 60% of its design speed, the motor's power consumption will be reduced to approximately 21.6% of the power that would have been consumed at full speed.

The table and graph in **Figure 25-8** show the percentage of power consumption when compared to the percentage of motor speed.

25.3 Pulse Width Modulation

Pulse width modulation (PWM) is a digital control technique commonly used to create an artificial ac power signal from a series of dc power signals. The output voltage is determined by the amount of time a dc voltage pulse signal is applied and how much time elapses between dc voltage pulses. Applying a dc voltage for a short period of time results in a lower output voltage. Applying a dc voltage for a longer period of time results in a higher output voltage.

Examples earlier in this chapter evaluated equally spaced dc voltage pulses, each with the same duration. To create an ac voltage signal, or one that closely resembles an ac voltage

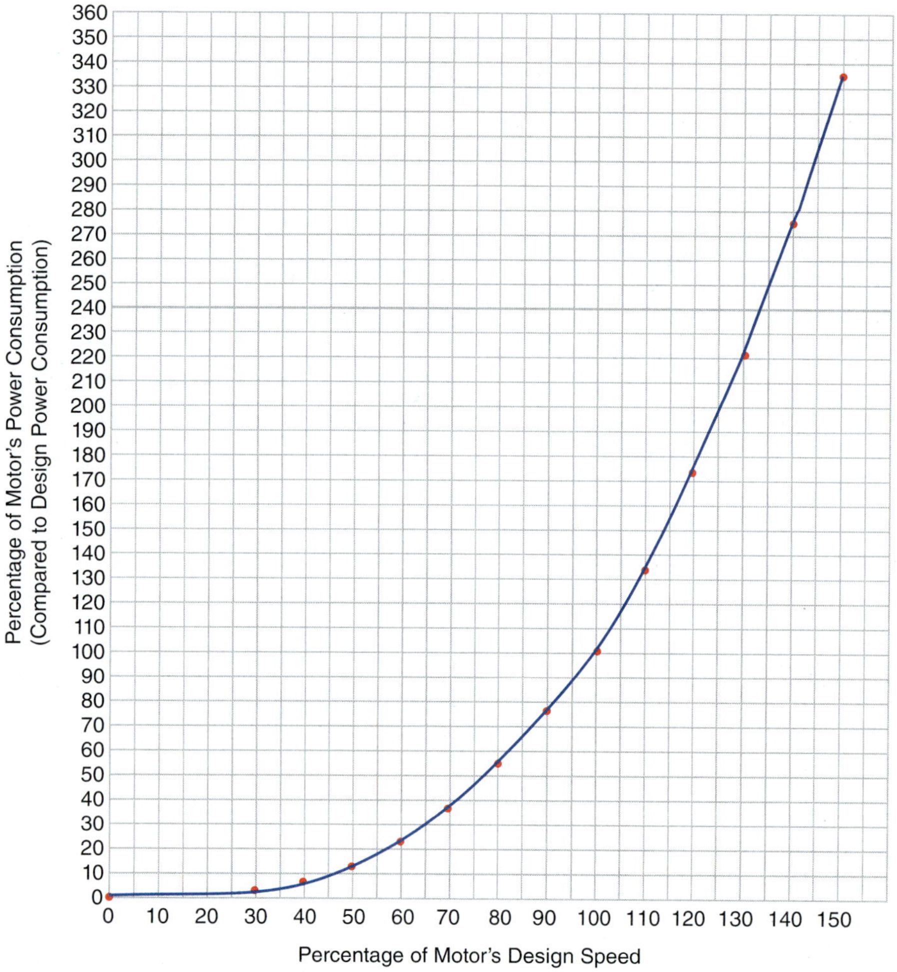

Effect of Motor Speed Change on Power Consumption

Percentage of Motor's Design Speed (%)	Percentage of Power Consumption (%)
30	2.7
40	6.4
50	12.5
60	21.6
70	34.3
80	51.2
90	72.9
100	100
110	133.1
120	172.8
130	219.7
140	274.4
150	336.2

Goodheart-Willcox Publisher

Figure 25-8. Comparison of motor speed and power consumption, relative to motor speed and power ratings. Reducing motor speed by 50% reduces power consumption by 87.5%, while increasing motor speed by 50% increases power consumption by 336%.

signal, the polarity, duration, and spacing between the dc voltage pulses will vary.

Shedding Light

The creation of artificial ac power signals using pulse width modulation relies on complex electronic circuitry which is often proprietary. In most cases, it is not necessary for those working in the electrical trades to understand the specifics of the circuitry involved. However, it is important to possess a fundamental understanding of how the motors we service are controlled.

Changing the Pulse Spacing

The average voltage produced with respect to the magnitude of the individual voltage pulses relies on the space, or time lapse, between the pulses, **Figure 25-9**. A large time lapse between pulses results in an average voltage that is significantly lower than the magnitude of the individual pulses. As the time lapse between pulses decreases, the average resulting voltage is significantly higher, and much closer to the magnitude of the individual pulses.

Changing the Pulse Width

Changing the width of the pulse also affects the average voltage produced. Wider pulses result in a higher output voltage, while narrower pulses produce a lower output voltage. Combining pulses of different widths with different pulse spacings or time gaps results in an output voltage signal that varies in magnitude. For example, the wider, closely-spaced pulses in **Figure 25-10** produce a higher voltage signal. The narrower voltage pulses that are spaced further apart from each other produce a lower voltage. The total effect of these individual pulses produces an output voltage signal that closely resembles the first half of an ac sine wave.

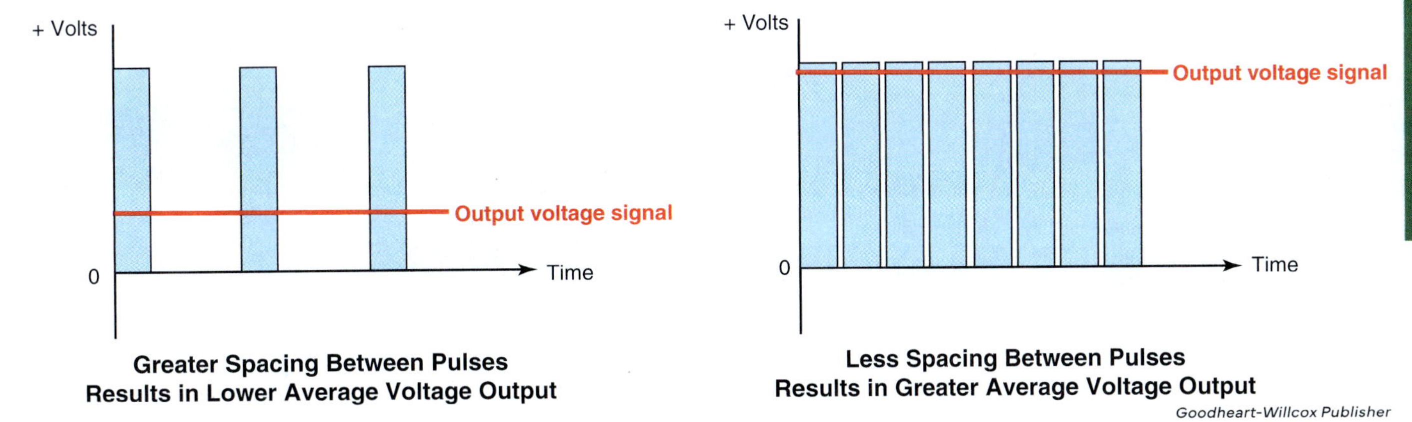

Figure 25-9. The effects of pulse spacing on average voltage output.

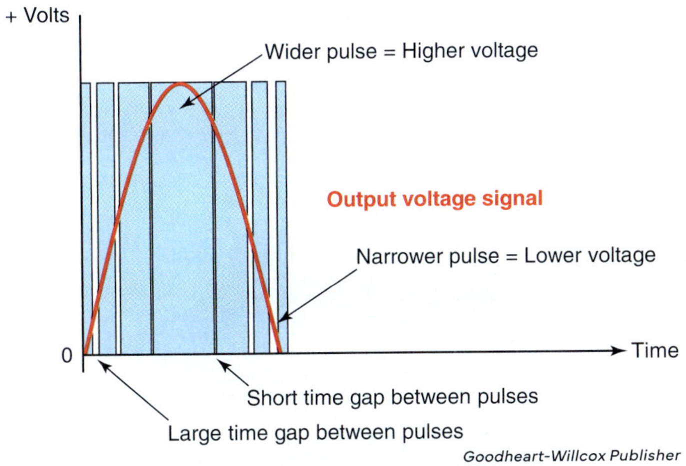

Figure 25-10. Pulse width and spacing between pulses affects the magnitude of the output voltage.

Changing the Pulse Polarity

If the polarity and magnitude of the dc voltage pulses are the same, the average voltage signal can be accurately represented by a horizontal line. However, if the polarity of the pulses is reversed at periodic intervals, a square wave is produced. If the spacing between the individual pulses is small, the magnitude of the square wave will be close to that of the original voltage pulses. If the spacing between the individual pulses is large, the magnitude of the square wave will be much lower than that of the original voltage pulses, **Figure 25-11**.

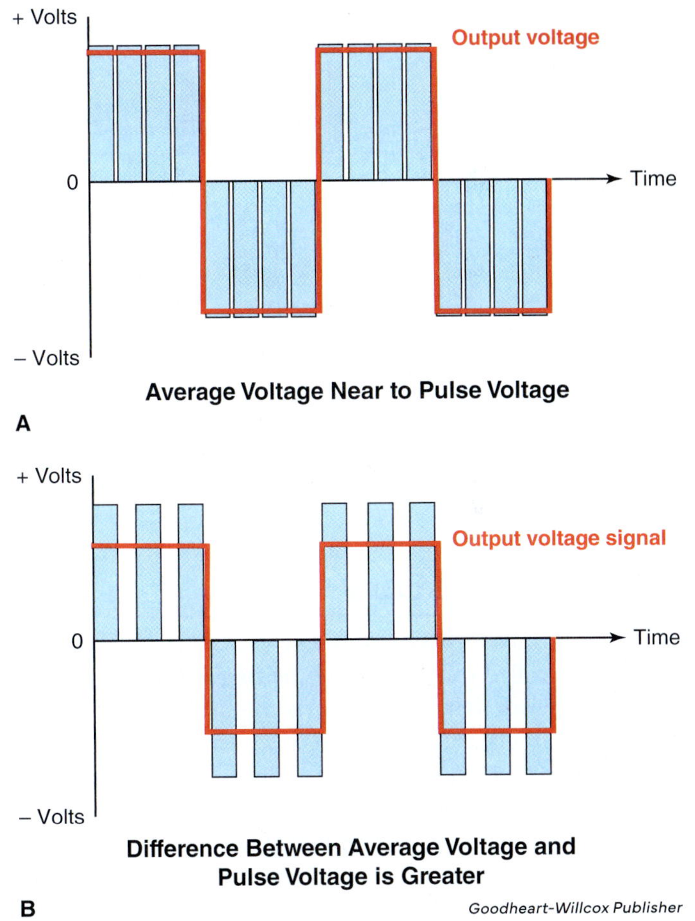

Figure 25-11. The difference between the pulse voltage and the average voltage in the output waveform is proportional to the amount of time being pulses. A—Relatively less time between pulses. B—Relatively more time between pulses.

Changing the Pulse Width, Polarity, and Spacing

By combining the effects of changing the width, polarity, and spacing of the individual voltage pulses, an average output signal that closely resembles an ac waveform can be created, **Figure 25-12**. By controlling the timing of the polarity changes and the duration of the pulses, the frequency of the signal sent to the motor can be changed, thereby changing the speed at which the motor operates. At a higher frequency, the motor speed increases, while at a lower frequency, the motor speed decreases, **Figure 25-13**.

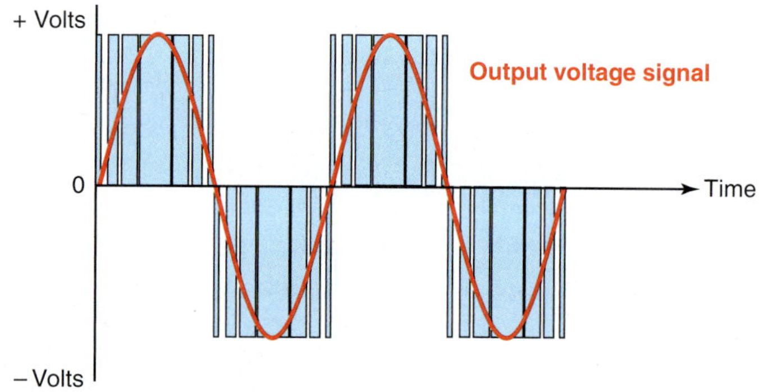

Goodheart-Willcox Publisher

Figure 25-12. Adjusting the pulse widths and spacing simulates a sine wave.

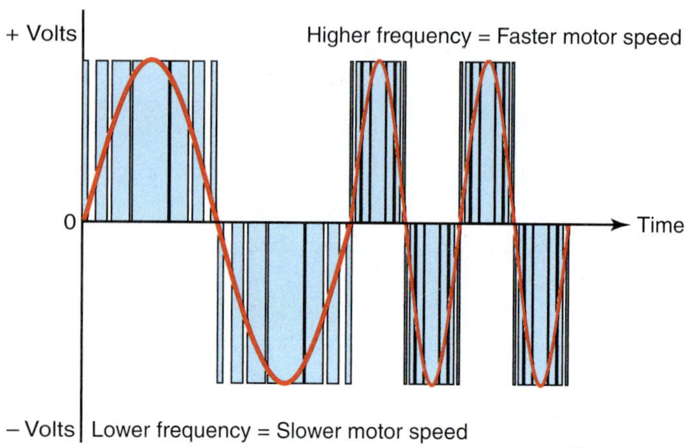

Goodheart-Willcox Publisher

Figure 25-13. Adjusting the timing and spacing of the polarity changes increases or decreases frequency. When a higher frequency is created, motor speed increases.

25.4 Transistors

Most newer motor drive systems rely on electronic switching devices called transistors to function. A ***transistor*** is an electronic switch that, just like its mechanical and electromechanical counterparts, opens and closes an electrical current path, **Figure 25-14**. A mechanical switch relies on an external force, such as someone physically flipping a switch to change the position of its contacts. An electromechanical switch, such as a relay, **Figure 25-15**, relies on the flow of current to create a magnetic field that causes sets of electrical contacts to either open or close.

Basic Transistor Operation

A basic transistor functions in a manner similar to that of the relay. A transistor has three terminals that are identified as the base, the collector, and the emitter, **Figure 25-16**.

The main current-carrying path on the transistor is located between the collector and the emitter. This path is referred to as the ***collector-emitter junction***. The ***base*** controls the

Zagach Letters/Shutterstock.com

Figure 25-14. Transistors, similar to the ones shown here, open and close an electrical current path.

Goodheart-Willcox Publisher

Figure 25-15. A general purpose relay relies on the flow of current to create a magnetic field that opens and closes a current path.

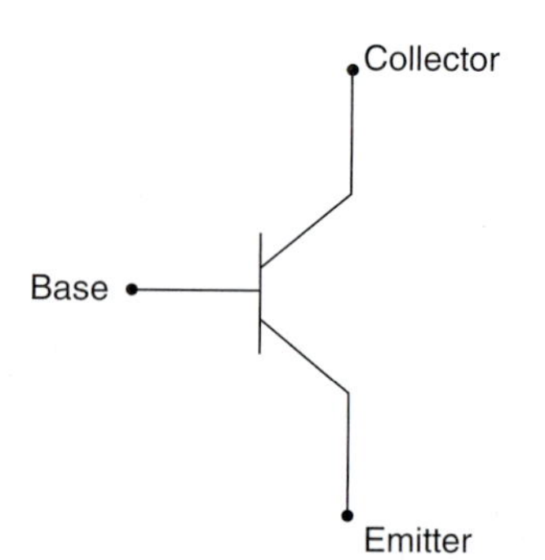

Goodheart-Willcox Publisher

Figure 25-16. Schematic symbol of a simple, generic transistor.

opening and closing of this junction. When the appropriate control voltage is supplied to the base, the collector-emitter junction acts as a closed switch and current passes freely through the collector-emitter junction, **Figure 25-17**. When no voltage is applied to the base, the collector-emitter junction acts as an open switch and no current can pass through. The base of a transistor performs the same basic function as the coil on a relay, and the collector-emitter junction functions in a manner similar to the relay's set of electrical contacts.

Unlike the general purpose relay that relies on the physical opening and closing of sets of electrical contacts to operate, transistors have no moving parts. With no moving parts, the transistor is able to shift between the conducting and nonconducting states many thousands of times per second. Since transistors have no moving parts, they generally have a longer operational life than that of electromechanical relays and contactors.

Transistors in Motor Drives

Transistors play an important role in amplifying an electrical signal. The transistor commonly found in electronic motor drives and other motor control circuits is the ***insulated-gate bipolar transistor (IGBT)***. The IGBT combines the most advantageous characteristics of the ***metal oxide semiconductor field effect transistor (MOSFET)*** and the ***bipolar junction transistor (BJT)***, **Figure 25-18**. The MOSFET is a voltage-controlled transistor, while the BJT is a current-controlled transistor.

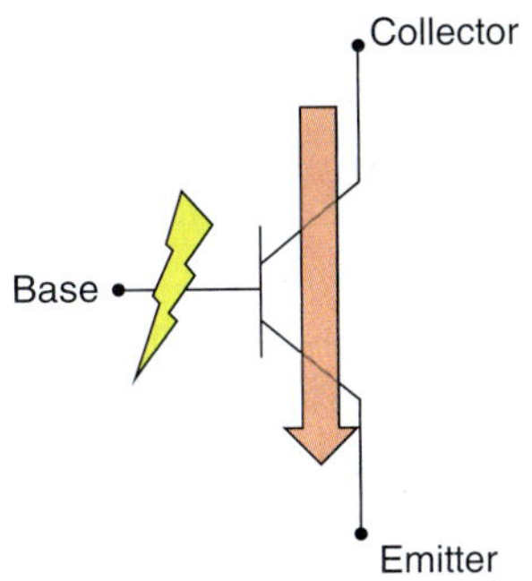

Goodheart-Willcox Publisher

Figure 25-17. Current can pass through the collector-emitter junction when the base is energized.

The MOSFET, BJT, and IGBT all have a control voltage that causes the main current channel to open and close. This control voltage is applied to the third terminal, which is not part of the main current channel. On the MOSFET and the IGBT, this terminal is called the *gate*. On the BJT, this terminal is called the *base*. In addition, the collector and emitter terminals on a MOSFET transistor are called the drain and source, respectively. Although the terminal names are different depending on the transistor type, their basic functions are the same.

The IGBT has a very high switching speed, which is a major advantage over the MOSFET. Both the IGBT and MOSFET transistors have isolated gates, which are so named because the gates are physically isolated from the collector-emitter junctions. The electrical symbols for the IGBT and MOSFET transistors clearly show that the gates are isolated from the rest of the transistor. By isolating the gate from the collector-emitter junction, the power losses of the transistor are much lower when compared to those of traditional transistors.

The IGBT also has a low saturation voltage, which is a major advantage over the BJT. ***Saturation voltage*** refers to the minimum gate voltage that creates a near resistance-free current path between the collector and the emitter. The low saturation voltage means that only a small voltage on the gate is required to maintain the low-resistance collector-emitter path through the transistor. By combining these three transistor characteristics, the IGBT becomes a desirable semiconductor switching device for use in motor drives. Because of this hybrid-type design, the IGBT experiences low failure rates, lower input energy losses, and can be used effectively for higher voltage and higher-frequency applications.

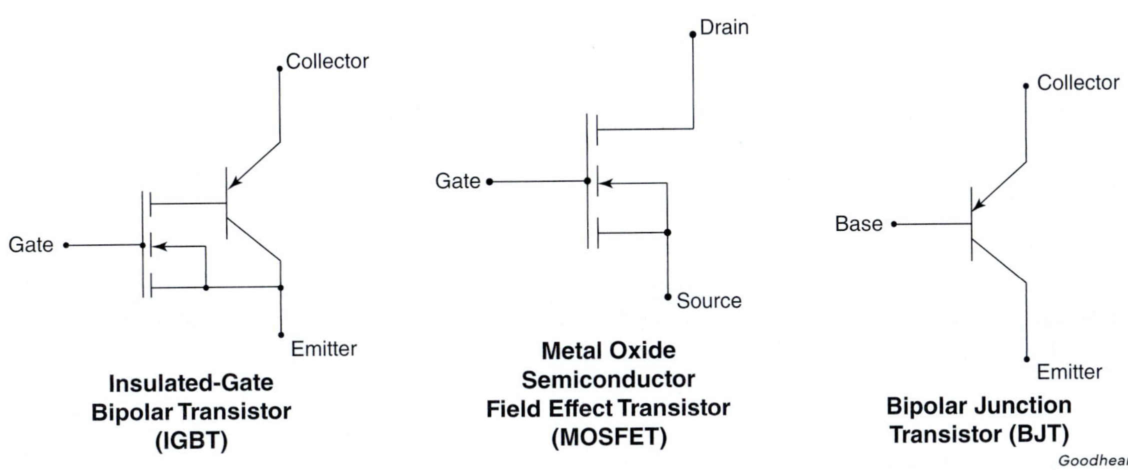

Goodheart-Willcox Publisher

Figure 25-18. An IGBT combines features of MOSFETs and BJTs. In IGBTs and MOSFETs, the base is called the gate. IGBTs and MOSFETs have insulated gates. Notice that the collector and emitter terminals on the MOSFET transistor are identified as the drain and the source, respectively.

25.5 Variable Frequency Drives

As you recall from Chapter 23, *Motor Operation Basics*, the speed of a motor is determined by the frequency of the power signal supplied to the motor and the number of motor poles:

$$N_S = \frac{120 \times f}{p}$$

where:

N_S = Synchronous speed (in rpm)
120 = Constant for two polarity changes per cycle
f = Frequency of input power signal (in Hz)
p = Number of poles

Changing the speed of a motor, therefore, requires that either the number of motor poles or the frequency of the power supply is changed. Changing the number of motor poles is not an easy task, as the motor wiring will typically need to be altered. Changing the frequency of the power supply is relatively easy to accomplish using a variable frequency drive.

A ***variable frequency drive (VFD)*** is an electronic motor drive that has the ability to control, monitor, and adjust the speed of an induction or synchronous motor by changing the frequency of the signal that provides power to the motor. By controlling the frequency of the motor's power signal, the motor's speed can be gradually increased or decreased to closely match the requirements of the system. This process has several benefits:

- Lessens mechanical and electrical stress on the motor and the driven devices.
- Reduces the frequency of an emergency and unexpected system failure and shutdown.
- Reduces maintenance and repair costs.
- Extends the life of the motor and the components driven by it.
- Reduces the energy costs associated with motor operation.

Shedding Light

VFDs are used in a wide range of applications including small appliances, pumps, fans, blowers, and compressors. Electric drive systems provide increased reliability and efficient operation. Advances in manufacturing processes and power electronics technology have paved the way for reduced VFD costs and improved performance.

The VFD's output signal, which is transferred to the motor, is created by external inputs to the circuitry that helps determine the frequency that will deliver optimal motor speed. Transforming the original input signal to the VFD to one signal that has the ability to vary the frequency of the motor's input signal is a three-step process:

1. The ac input signal is converted to a rectified dc signal.
2. The dc signal is smoothed and filtered.
3. An ac signal is created (inverted) from the dc signal based on system requirements.

Converting the AC Input Signal

The first section of the VFD is the converter. A ***converter*** is an electronic circuit that converts an ac signal to a dc signal. The typical converter is made up of four diodes and an output resistor. A diode is best described as an electronic check valve that only permits electric current to flow in one direction. The diodes and the resistor are arranged to form a bridge. When configured in this manner, the circuit operates as a full-wave bridge rectifier. A ***full-wave bridge rectifier*** converts the complete ac cycle into a pulsating dc signal. By comparing the waveforms of the input and output signals in **Figure 25-19**, it can be seen that the polarity of the negative portions of the input signal is changed to a positive polarity at the converter's output.

First Half-Cycle

During the first half-cycle, the polarity of the input ac sine wave is positive. In **Figure 25-20**, for example, the current flows from the ac voltage source in a clockwise direction and follows a path through diode D2. Since diodes only permit current flow in one direction, the direction of the diode's arrow represents the direction in which current is permitted

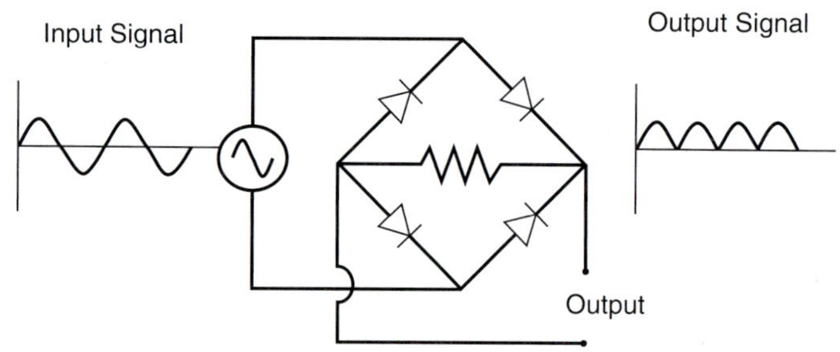

Goodheart-Willcox Publisher

Figure 25-19. Wiring diagram of a full-wave bridge rectifier.

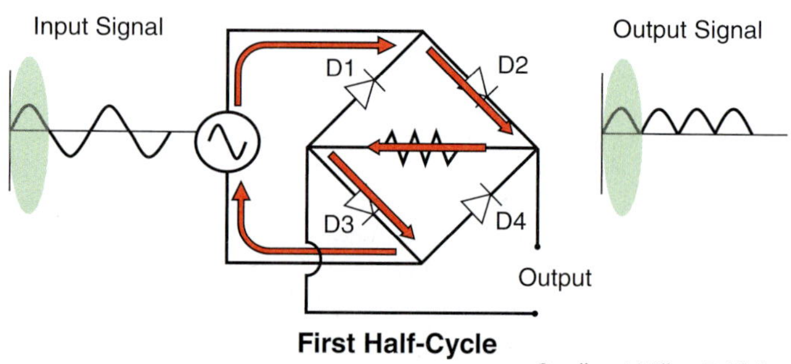

Goodheart-Willcox Publisher

Figure 25-20. First half-cycle of a converter section of a VFD.

to flow. Current will not flow through diode D1, as the arrow is pointing in the opposite direction. After flowing through D2, current is directed through the output resistor. Current will not flow through diode D4, as the arrow is pointing in the opposite direction. After flowing through the output resistor, current is directed through D3 and then back to the voltage source. Note: The potential at the outlet of the output resistor is lower than the potential at the D1-D2 junction, so current will not be directed through D1. By following the path as indicated, the flow through the output resistor is in the right-to-left direction.

Second Half-Cycle

During the second half-cycle, the polarity of the input ac sine wave is negative, **Figure 25-21**. During this portion of the cycle, current flows from the ac voltage source in a counterclockwise direction and follows a path through diode D4. Current will not flow through diode D3 at this point. After flowing through D4, current is directed, once again, through the output resistor. Current will not flow through diode D2. After flowing through the output resistor, current is directed through D1 and then back to the voltage source. The direction of current flow through the output resistor is, once again, in the right-to-left direction. The output resistor is experiencing current flow in the right-to-left direction regardless of the polarity of the original input signal. As a result, the polarity of the output waveform is always positive.

Smoothing the DC Signal

The next section of the VFD is called the dc bus. A ***dc bus*** is a bank of capacitors connected in parallel with each other and is responsible for filtering and smoothing the waveform, **Figure 25-22**. The input of the dc bus is the rectified signal that is present at the outlet of the bridge rectifier. The voltage supplied to the dc bus causes the capacitors to charge and discharge. When the capacitors discharge, the potential at the dc bus outlet is higher than the potential of the signal at the inlet of the bus. As a result, the signal at the outlet of the dc bus represents a more consistent dc voltage, **Figure 25-23**. There will always be a small ripple at the output of the dc bus, as it would be impossible to produce a 100% pure dc signal.

Creating an Artificial AC Signal

The final step in the transformation process is sending the smoothed dc voltage signal from the outlet of the dc bus to the inverter. The ***inverter*** is the part of the VFD that is responsible for changing the dc signal into an ac signal. The inverter can be made up of two, four, or six transistors, arranged in pairs. Each pair of transistors simulates a single-phase, ac voltage waveform.

A two-transistor inverter is illustrated in **Figure 25-24**. To start the process of creating an ac voltage waveform, the gate on transistor 1 is energized, causing the collector-emitter junction to close, starting the flow of current. From zero to 60 electrical degrees, voltage across the collector-emitter junction increases to its maximum. From 60 to 120 electrical degrees, the maximum voltage is present at the output. At 120 electrical degrees, the gate on transistor 1 is de-energized, causing the collector-emitter junction to open, stopping the flow of current, **Figure 25-25**. The voltage at the output of the inverter drops to zero at 180 electrical degrees.

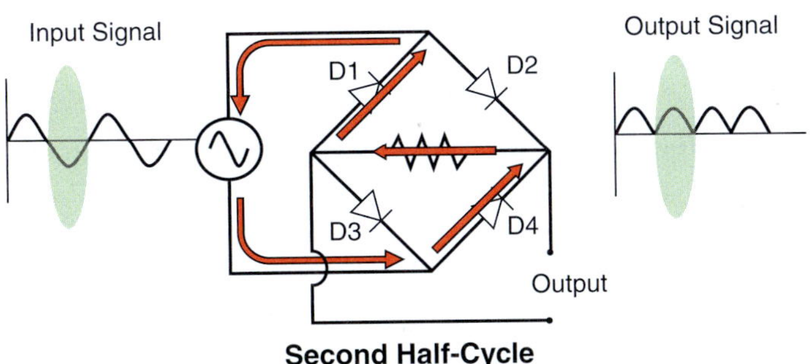

Goodheart-Willcox Publisher

Figure 25-21. Second half-cycle of a converter section of a VFD.

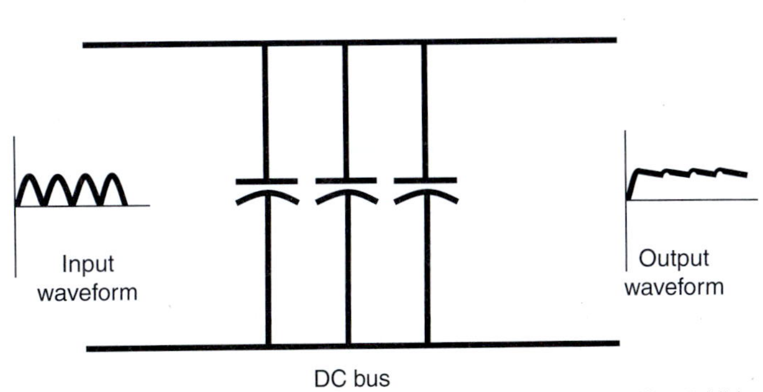

Goodheart-Willcox Publisher

Figure 25-22. Wiring diagram of a dc bus.

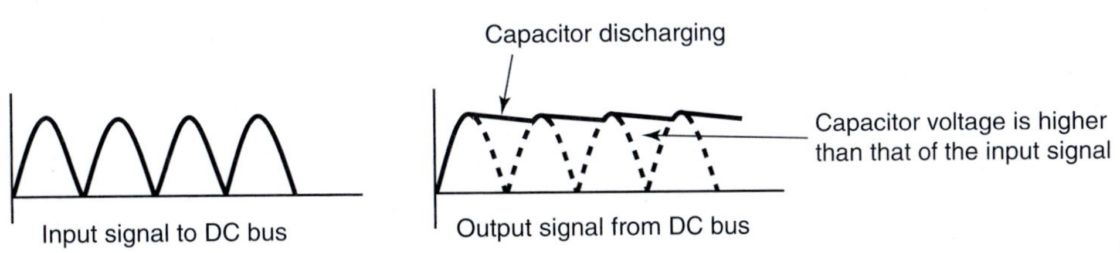

Goodheart-Willcox Publisher

Figure 25-23. The input and output signals on a dc bus. Notice that the waveform is smoothed in the output signal.

Once the voltage has reached zero, the gate on transistor 2 is energized, causing the collector-emitter junction to close, starting the flow of current. However, the voltage transmitted through the inverter through transistor 2 has a negative polarity. From 180 to 240 electrical degrees, the magnitude of the voltage increases to its maximum with a negative polarity. From 240 to 300 electrical degrees, the collector-emitter junction is fully open, and the maximum

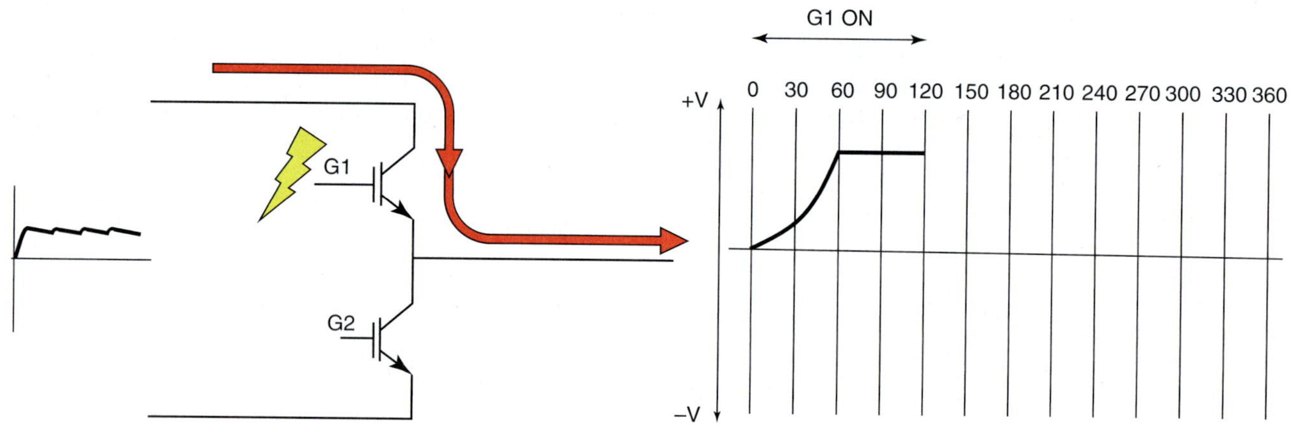

Goodheart-Willcox Publisher

Figure 25-24. The inverter portion of a variable frequency drive. Transistor 1 conducts between zero and 120 electrical degrees.

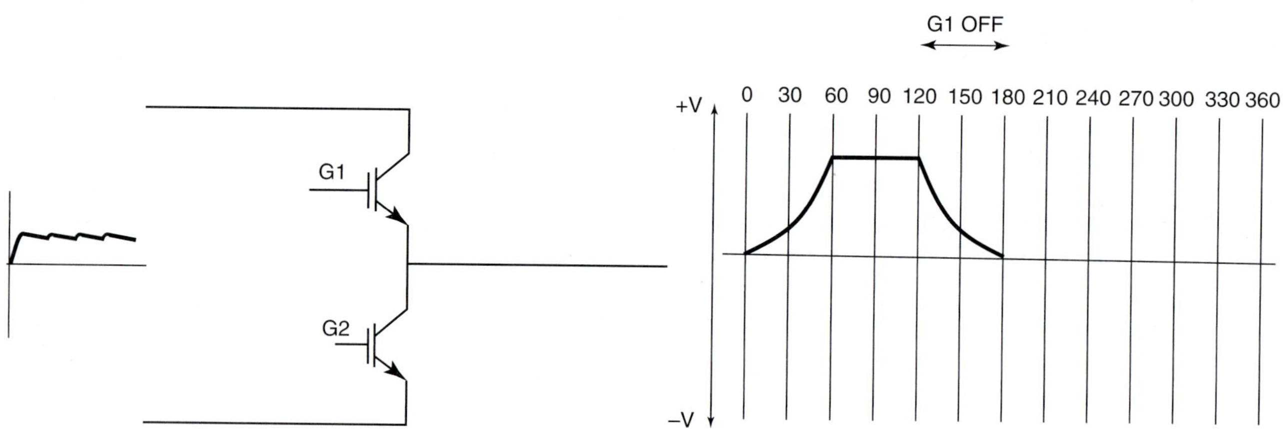

Goodheart-Willcox Publisher

Figure 25-25. Transistor 1 is de-energized at 120 electrical degrees and voltage output drops to zero at 180 electrical degrees.

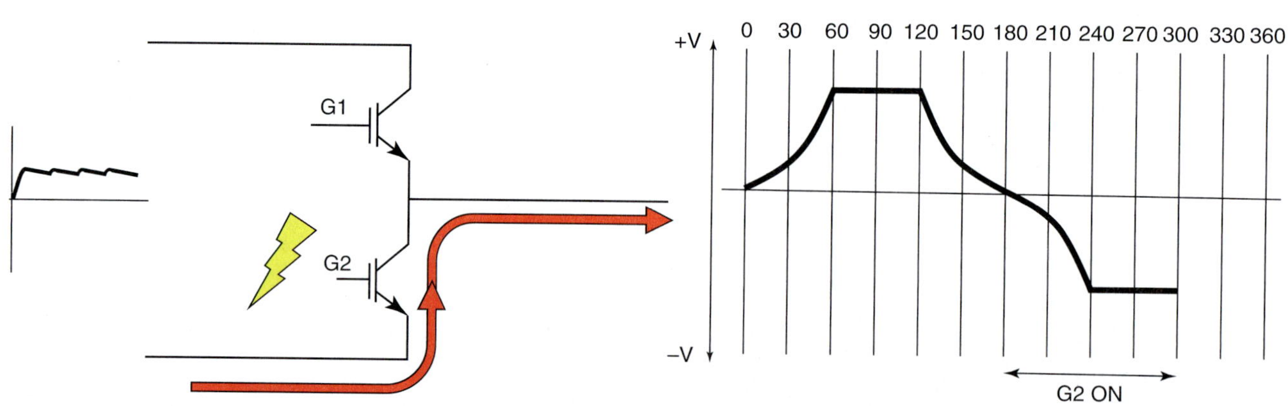

Goodheart-Willcox Publisher

Figure 25-26. Transistor 2 conducts between 180 and 300 electrical degrees. Notice the voltage increases to its maximum negative voltage at 240 degrees, and maximum negative voltage is present at the output.

negative voltage is present at the output, **Figure 25-26**. At 300 electrical degrees, the gate on transistor 2 is de-energized, with the voltage at the output of the inverter reaching zero at 360 electrical degrees, **Figure 25-27**.

Combining all three steps of the transformation process, the circuit will closely resemble the diagram in **Figure 25-28**. The waveform that is created does not perfectly recreate an ac sine wave, but it is close enough to allow for acceptable motor performance. When two transistor pairs are used, a 240 Vac single-phase supply can be produced, **Figure 25-29**. When three transistor pairs are used, a three-phase supply can be produced, **Figure 25-30**.

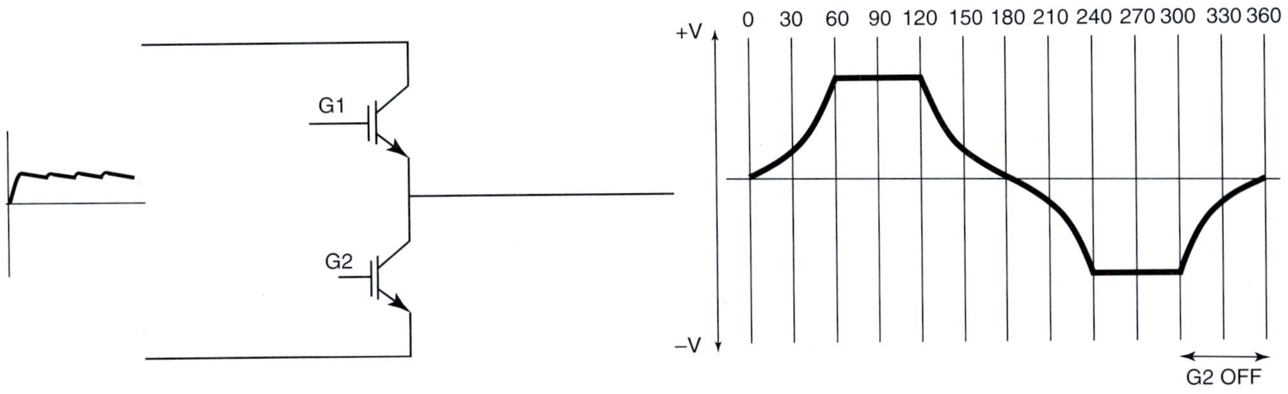

Goodheart-Willcox Publisher

Figure 25-27. Transistor 2 de-energizes at 300 electrical degrees.

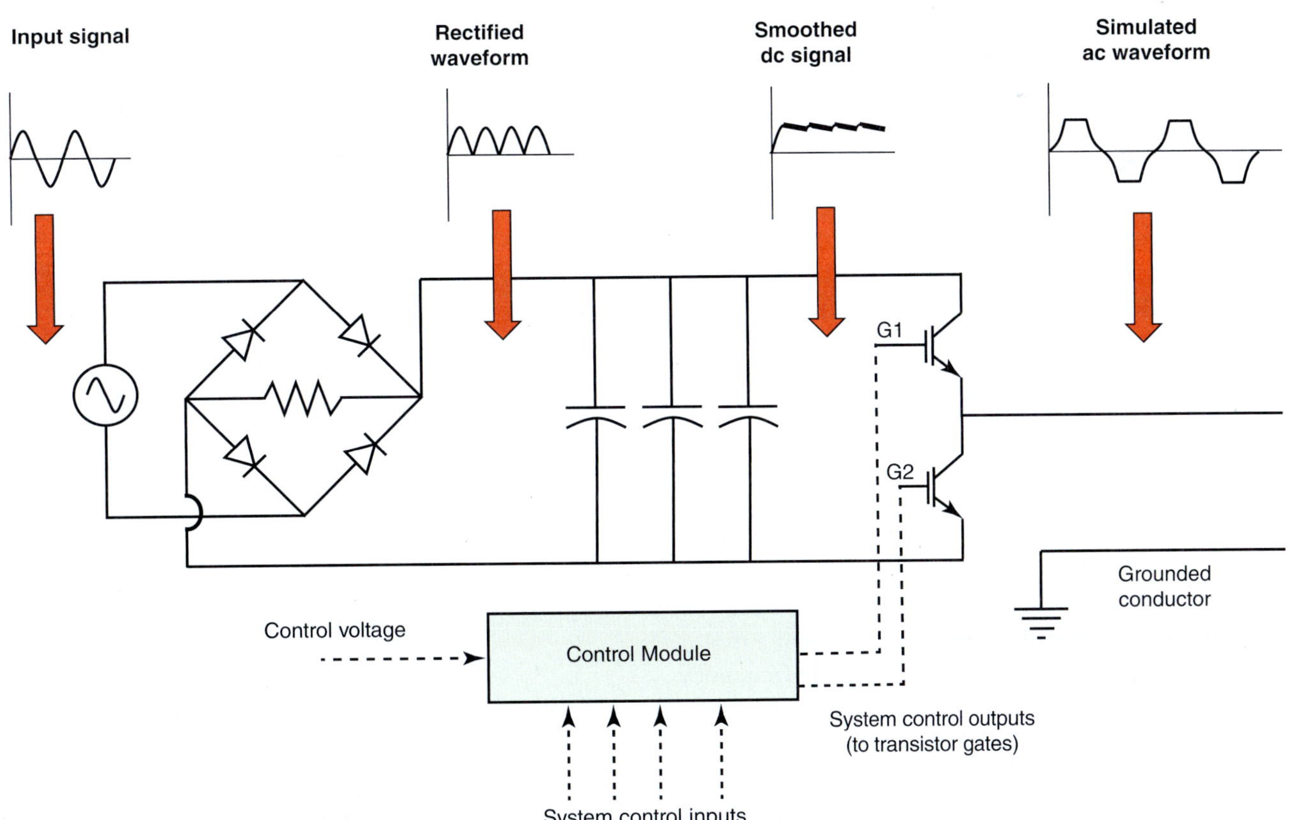

Goodheart-Willcox Publisher

Figure 25-28. Complete wiring diagram of the three stages of a variable frequency drive.

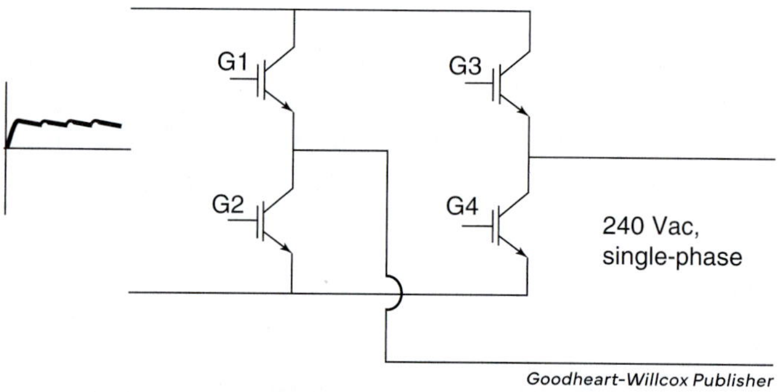

Figure 25-29. Two transistor pairs can create a 240 Vac single-phase voltage supply.

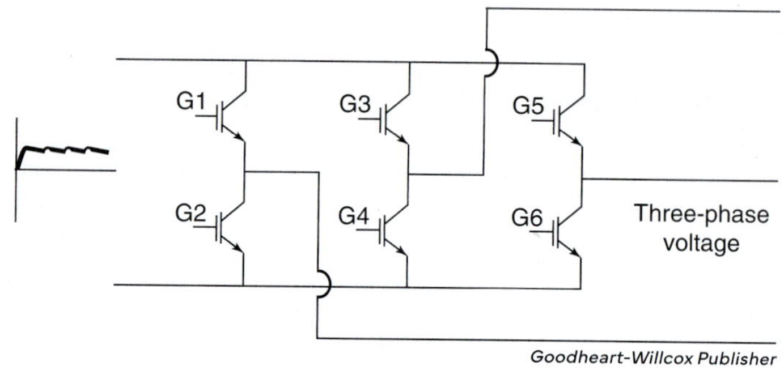

Figure 25-30. Three transistor pairs can create a three-phase voltage supply.

Shedding Light

As with any electrical device or component, it is important to take some precautions to ensure the continued safe and efficient operation of a VFD. In the field, electricians should ensure that the devices are kept dust-free, as dust can cause an overheated condition and reduced VFD performance. As with other electrical devices, VFDs should be kept dry and, as part of routine maintenance, all electrical connections should be checked to ensure they are tight. Loose electrical connections in a VFD can lead to overheating, erratic motor operation, intermittent motor stoppages, or nuisance fault generation.

25.6 Motor Drives and Torque

Some motor drives have the ability to control the torque that is produced by the motor. The amount of torque produced is determined by the voltage-to-frequency (V/Hz) ratio. As the name implies, the V/Hz strategy is based on the ratio of the voltage supplied to the motor to the frequency of the input power signal. The torque of a motor is proportionally related to the V/Hz ratio. As the V/Hz ratio increases, the motor's torque also increases. As the voltage-to-frequency ratio decreases, the motor's torque also decreases, **Figure 25-31**. For applications where the motor torque is to remain constant, the power consumed by the motor can be varied by making proportional changes to the supply voltage and the frequency of the supply signal to keep the V/Hz ratio constant. Refer to **Example 25-2**.

Example 25-2

Consider a motor that is powered by a 240 Vac, 60 Hz power supply. What is the frequency of the power supply needed for the motor to operate at 200 Vac if it is to provide the same amount of torque?

✓ **Solution**

If the torque is to remain the same, the voltage-to-frequency ratio must also remain the same. For this motor, the voltage-to-frequency ratio is calculated as:

$$\text{V/Hz Ratio} = \frac{\text{Supply Voltage}}{\text{Power Supply Frequency}}$$

$$\text{V/Hz Ratio} = \frac{240 \text{ Vac}}{60 \text{ Hz}}$$

$$\text{V/Hz Ratio} = 4{:}1 \text{ or } 4$$

If the voltage is changed to 200 Vac, the new frequency is calculated as follows:

$$\text{V/Hz Ratio} = \frac{\text{Supply Voltage}}{\text{Power Supply Frequency}}$$

Rearrange the equation to solve for power supply frequency:

$$\text{Power Supply Frequency} = \frac{\text{Supply Voltage}}{\text{V/Hz Ratio}}$$

$$\text{Power Supply Frequency} = \frac{200 \text{ Vac}}{4}$$

$$\text{Power Supply Frequency} = 50 \text{ Hz}$$

For constant voltage applications, the torque of a motor can be changed by increasing or decreasing the frequency of the power signal sent to the motor. At a constant voltage, an increase in power signal frequency will decrease the V/Hz ratio, lowering the motor's torque. Similarly, a decrease in power signal frequency at a constant voltage will increase the V/Hz ratio, increasing the motor's torque. With constant voltage being delivered to the motor, the power consumed by the motor can be held constant while changing the motor's torque. Refer to **Example 25-3**.

Torque requirements are typically classified as either constant torque or variable torque. Some applications require constant power, regardless of motor speed. Constant power systems are a variation of variable torque control systems.

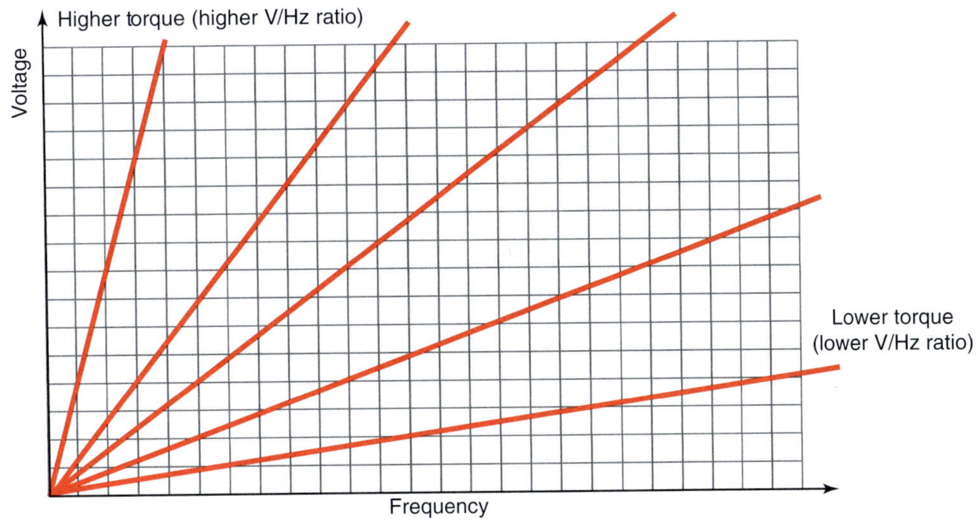

Goodheart-Willcox Publisher

Figure 25-31. The amount of torque produced is determined by the voltage-to-frequency ratio.

Example 25-3

Consider a motor with a V/Hz ratio of 3:1 that is powered by a 300 V signal. If the power consumption of the motor is to remain constant, what will the frequency of the power supply to the motor be if the V/Hz ratio is 5:1? Will this result in an increase or decrease in motor torque? Why or why not?

✓ Solution

This is a constant voltage application. With a V/Hz ratio of 3:1 and a voltage of 300 V, the frequency of the power supply at the motor is 100 Hz (300 V divided by 100 Hz equals 3).

With a V/Hz ratio of 5:1 and a voltage of 300 V, the frequency of the power supply at the motor is 60 Hz (300 V divided by 60 Hz equals 5).

Since the frequency of the power signal has decreased from 100 Hz to 60 Hz, the V/Hz ratio increases and the motor's torque increases as well.

Constant Torque

In constant torque applications, the amount of torque produced is the same regardless of the motor's speed. This means that the slope of the torque characteristic line must remain the same at all motor speeds. Therefore, if the frequency is changed to alter the motor's speed, the voltage supplied to the motor must also change to keep the V/Hz ratio the same. Refer to **Example 25-4**.

The power consumption of the motor is directly proportional to the motor's speed. So, if a motor speed is increased by 50%, the power consumption will also increase by 50% to keep the torque constant. Similarly, if a motor speed is decreased by 50%, the power consumption will also decrease by 50% to keep the torque constant, **Figure 25-32**.

If the motor speed is to increase:

- The frequency of the power supply will need to increase.
- An increase in frequency requires an increase in voltage to keep the V/Hz ratio constant.
- An increase in voltage results in an increase in power consumption.

If the motor speed is to decrease:

- The frequency of the power supply will need to decrease.
- A decrease in frequency requires a decrease in voltage to keep the V/Hz ratio constant.
- A decrease in voltage results in a decrease in power consumption.

Variable Torque

Variable torque applications typically require the use of motors that deliver less torque at lower speeds and more torque at higher speeds. The relationship between torque and speed is expressed by another affinity law:

$$\frac{T_F}{T_I} = \left(\frac{rpm_F}{rpm_I}\right)^2$$

where:

T_F = Final torque
T_I = Initial torque
rpm_F = Final motor speed
rpm_I = Initial motor speed

From this equation, if the speed of a motor is reduced by a factor of 2, the motor torque will be reduced by a factor

Example 25-4

Consider a motor that provides a constant torque for a conveyor belt application. The motor is powered by a 240 V, 60 Hz signal. The motor drive reduces the frequency of the power signal to 50 Hz in order to slow the motor. To maintain the same amount of torque, what must the voltage being supplied to the motor be adjusted to?

✓ Solution

For a constant torque application, the V/Hz ratio must remain the same. The original V/Hz ratio is calculated as:

$$V/Hz = \frac{240\text{ V}}{60\text{ Hz}}$$

$$V/Hz = 4$$

The voltage required at 50 Hz is calculated using the constant V/Hz ratio of 4:

$$V/Hz = \frac{\text{Voltage}}{\text{Frequency}}$$

Rearrange the equation to solve for voltage.

$$\text{Voltage} = V/Hz \times \text{Frequency}$$

$$\text{Voltage} = 4 \times 50\text{ Hz}$$

$$\text{Voltage} = 200\text{ V}$$

Now It's Your Turn 25-4

A motor for a constant torque application is being powered by a 208 V, 60 Hz signal. If the voltage supplied to the motor is 240 V, what must the frequency of the motor's power supply be adjusted to?

of 4, which is the square of 2. Refer to **Example 25-5** and **Example 25-6**.

Motors have four different associated torque values:

- **Locked-rotor torque.** Locked-rotor torque is the amount of twisting force that is required to get the motor's shaft turning from a completely stopped position. If the driven load (such as a pump, blower, fan, or compressor) has more rotational resistance than the motor's locked-rotor torque can overcome, the motor will not be able to start and will trip any motor or circuit protection devices that are in place.
- **Pull-up torque.** Pull-up torque is the amount of torque required to bring the driven load up to its desired speed. If the amount of torque produced is lower than the pull-up torque, the driven load may rotate, but at a speed that is lower than required.
- **Breakdown torque.** Breakdown torque represents the maximum amount of torque that can be produced before the motor is unable to maintain the driven load at the desired or required speed. This is the point at which the motor will stall.
- **Full-load torque.** Full-load torque refers to the torque level at 100% of the motor's rated power, voltage, and current, as indicated on the motor's nameplate. This value should always be higher than the motor's actual operating conditions, as no motor should run at 100% of its capacity to achieve the desired output.

Constant Power

Constant horsepower loads, such as lathes, grinders, and similar pieces of equipment and machinery, require more

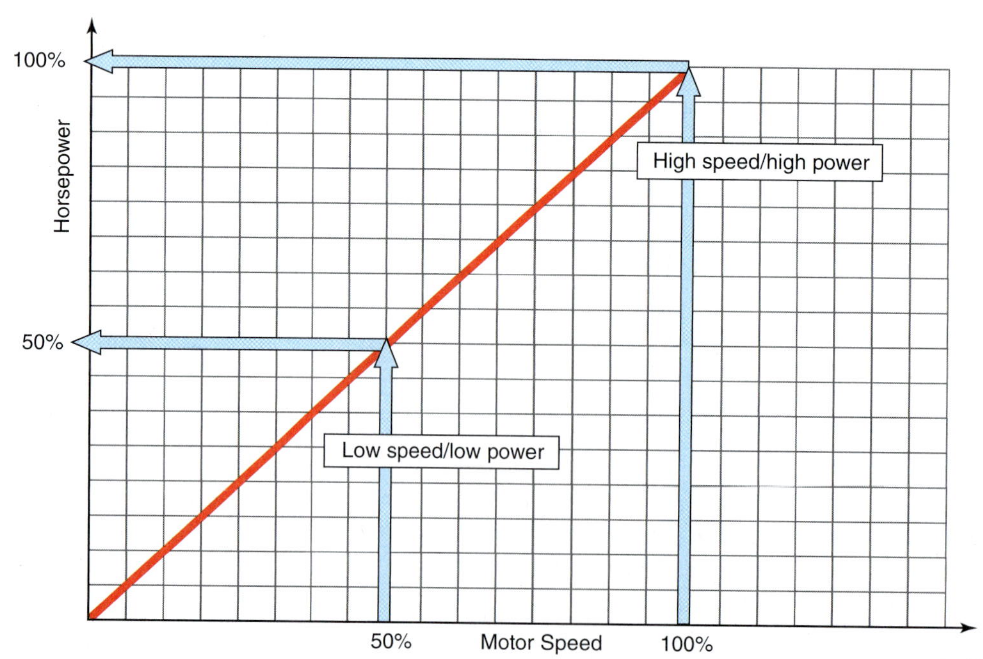

Goodheart-Willcox Publisher

Figure 25-32. Constant torque chart plotting motor speed against horsepower.

Example 25-5

If the speed of a variable torque motor is to be reduced from 1500 rpm to 750 rpm, how will the torque be affected?

✓ Solution

Use the torque-speed affinity law:

$$\frac{T_F}{T_I} = \left(\frac{rpm_F}{rpm_I}\right)^2$$

Simplify the motor speed expression:

$$\left(\frac{rpm_F}{rpm_I}\right)^2 = \left(\frac{750}{1500}\right)^2 = (0.5)^2 = 0.25$$

With the ratio of T_F to T_I being 0.25, the resulting torque is one-fourth that of the initial torque. As the speed decreases by a factor of 2, the required torque decreases by a factor of 4.

Example 25-6

If the speed of a variable torque motor is to be increased from 500 rpm to 2000 rpm, how will the torque be affected?

✓ Solution

Using the torque-speed affinity law:

$$\frac{T_F}{T_I} = \left(\frac{rpm_F}{rpm_I}\right)^2$$

$$\frac{T_F}{T_I} = \left(\frac{2000}{500}\right)^2$$

$$\frac{T_F}{T_I} = 4^2$$

$$\frac{T_F}{T_I} = 16$$

With the ratio of T_F to T_I being 16, the resulting torque is 16 times that of the initial torque. As the speed increases by a factor of 4, the required torque increases by a factor of 16.

torque at low speeds than at high speeds. The relationship, therefore, between torque and speed is inversely proportional, **Figure 25-33**. As the speed of the motor increases, the required torque decreases. As the speed of the motor decreases, the required torque increases.

Constant power applications are not prime candidates for use in conjunction with variable frequency drives. Since the power consumption of a constant power application remains the same regardless of the motor's speed or torque, there is no energy savings to be realized by using a VFD.

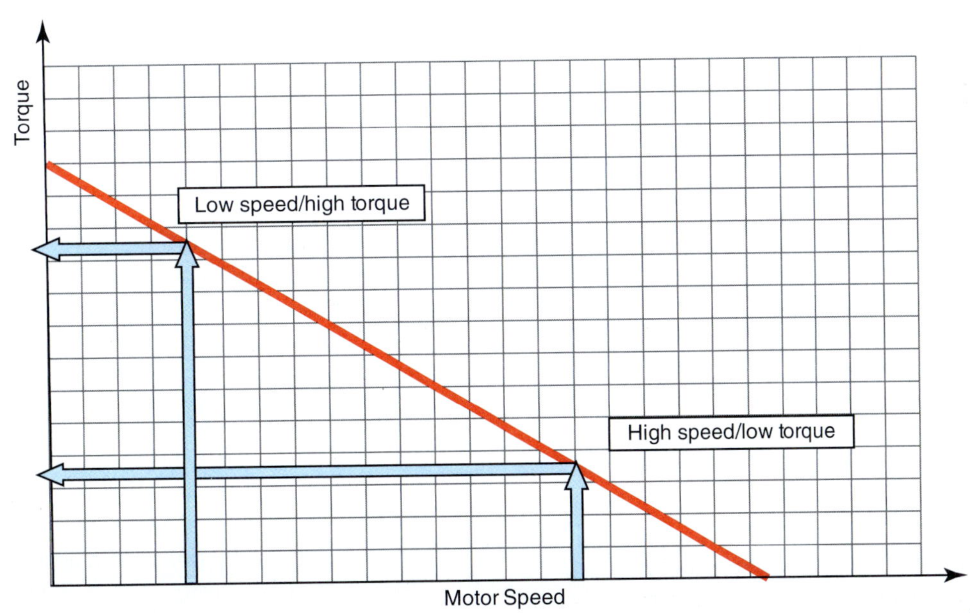

Goodheart-Willcox Publisher

Figure 25-33. Constant power chart plotting motor speed against torque.

CHAPTER 25 REVIEW

Summary

- Motor drives are designed to control the speed and/or torque of a motor. (25.1)
- A drive system is the combination of a motor and a motor drive. (25.1)
- Motor drives are used primarily to reduce energy costs and prolong the useful life of a motor. (25.2)
- Motor speed is proportional to power consumption, but not directly. A cubic (or power of three) relationship exists between the power consumption and speed of the motor. (25.2)
- Affinity laws are mathematical expressions that approximate the performance of fans and pumps over their complete operating ranges. (25.2)
- Pulse width modulation is a design strategy that creates an artificial ac power signal from a series of dc power signals. (25.3)
- Transistors are commonly used in the electronic circuits of motor drives. (25.4)
- The transitor most commonly used in motor drives is the insulated-gate bipolar transistor (IGBT). (25.4)
- Transistors function in a manner similar to relays, where a small control voltage controls current flow in a larger, main path. (25.4)
- A variable frequency drive, VFD, is used to change the speed of a motor by varying the frequency of the power supply that powers the motor. (25.5)
- There are three main parts of a variable frequency drive used to vary the frequency: the converter, the dc bus, and the inverter. (25.5)
- A converter utilizes a bridge circuit to convert an ac signal to a dc signal. (25.5)
- An inverter converts a dc signal to an artificial ac signal. (25.5)
- Torque applications are typically classified as either constant torque or variable torque. (25.6)
- In a constant torque application, power consumption is proportional to motor speed. (25.6)
- Four common types of torque are locked-rotor torque, pull-up torque, breakdown torque, and full-load torque. (25.6)
- Constant power applications will not see a financial benefit from using a VFD or other motor drive system. (25.6)

Review Questions

Use the information in the chapter to answer the following questions.

Measuring Your Potential

1. A drive system is the combination of a _____. (25.1)
   A. power supply and a motor drive
   B. motor and a motor drive
   C. motor and a circuit protector
   D. power supply and a motor
2. *True or False?* To make a motor run more efficiently, you should run it at one-half its design speed, which will result in the motor using one-half of its power. (25.2)
3. In a PWM application, a square wave can be created by _____. (25.3)
   A. increasing the space between the dc voltage pulses
   B. decreasing the space between the dc voltage pulses
   C. increasing the width of the dc voltage pulses
   D. changing the polarity of the dc voltage pulses
4. Which of the following two transistor terminals perform similar functions? (25.4)
   A. Base and gate
   B. Base and collector
   C. Gate and emitter
   D. Collector and emitter
5. The main current path in a transistor is between which two terminals? (25.4)
   A. Base and emitter
   B. Collector and emitter
   C. Gate and collector
   D. Base and gate
6. The purpose of the converter in a variable frequency drive is to change a(n) _____. (25.5)
   A. higher ac voltage signal to a lower ac voltage signal
   B. higher dc voltage signal to a lower dc voltage signal
   C. ac voltage signal to a dc voltage signal
   D. dc voltage signal to an ac voltage signal
7. The purpose of the inverter in a variable frequency drive is to change a(n) _____. (25.5)
   A. higher ac voltage signal to a lower ac voltage signal
   B. higher dc voltage signal to a lower dc voltage signal
   C. ac voltage signal to a dc voltage signal
   D. dc voltage signal to an ac voltage signal

8. The dc bus in a variable frequency drive is made up primarily of _____ (25.5)
   A. capacitors
   B. variable resistors
   C. transistors
   D. relays
9. If the motor speed in a constant torque application increases by 25%, the power consumed will _____. (25.6)
   A. decrease by 25%
   B. increase by 25%
   C. decrease by 50%
   D. increase by 50%
10. The twisting force that initiates a motor's rotation from the fully stopped position is called _____. (25.6)
    A. acceleration torque
    B. full-load torque
    C. locked rotor torque
    D. breakdown torque
11. *True or False?* For a constant power application, as the motor speed increases, the torque requirement also increases. (25.6)

Plug-In

1. How much power is consumed by a motor that operates at 150% of its 1000 rpm rated speed if, at 1000 rpm, the motor consumes 5000 W? (25.2)
2. List precautions you can take in the field to prevent a VFD from overheating. (25.5)
3. For a constant torque application, what will the new voltage have to be if the V/Hz ratio is 5:1 and the frequency is changed to 80 Hz? (25.6)
   A. 200 V
   B. 300 V
   C. 400 V
   D. 500 V

Making the Connection

1. Explain why a variable frequency drive is not typically recommended as an energy-saving measure for constant power applications. (25.6)
2. Explain why it is typically not recommended to run a motor at speeds that are in excess of its nameplate rating. (25.5)
3. A commercial electrical customer is looking to make changes to reduce the amount of energy that is being consumed. These energy-saving measures include installing variable frequency drives on all of their equipment, including the motor on a constant horsepower conveyor belt system. The customer is always right, so you do it. Do you agree or disagree with your customer's decision to install a variable frequency drive on the conveyor belt system? Provide support for your position. (25.5)

CHAPTER

26 Motor Installation and Replacement

Siemens AG 2025, All rights reserved

SPARKING DISCUSSION

Installing a motor that has a much larger power rating than what is needed will help ensure that the motor will not have to work as hard and will last longer. Do you agree or disagree with this statement? What alternative argument can be made to this claim?

LEARNING OUTCOMES

After completing this chapter, you will be able to:

- **26.1** Explain the importance of following the motor manufacturer's instructions.
- **26.2** Interpret information on a motor's nameplate.
- **26.3** Identify why a motor failed and determine if it can be repaired or needs to be replaced.
- **26.4** Explain how to properly remove a motor.
- **26.5** Determine the correct method to properly lift a motor.
- **26.6** Describe how to properly mount a motor.
- **26.7** Determine how to align drive and driven pulleys.
- **26.8** List the steps to check for proper belt tension.
- **26.9** Explain how to connect a motor electrically in accordance with local and national electrical codes and manufacturer recommendations.
- **26.10** List the steps to properly start up and put a motor back into operation.

TECHNICAL TERMS

angular misalignment
offset misalignment
thread engagement

Introduction

Although your primary role as an electrician is to service, install, and maintain electrical circuits, you might be called upon to install or replace electrical components, such as motors. The installation or replacement of a motor can be a time-consuming and often tedious process, but proper installation is essential if the motor is to operate efficiently and reliably.

This chapter focuses on the tasks required to properly install a motor. These tasks include selecting a suitable motor replacement, disconnecting the motor, both mechanically and electrically, properly lifting the old motor out, setting the new motor in place, mounting and securing the motor, aligning pulleys and couplings, and, of course, making the electrical connections and checking the motor's operation once the installation has been completed.

26.1 Safety Considerations

It would be impossible to cover all scenarios and situations regarding motor installation and replacement in this chapter. With that in mind, it is important that you adhere to, and demonstrate, best industry practices with regards to the safe handling of motors, motor controllers, and motor circuits. The National Electrical Manufacturers Association (NEMA), along with their MG-2 standard *Safety Standard and Guide for Selection, Installation, and Use of Electric Motors and Generators*, is a valuable resource for anyone in the electrical trades who is involved with the installation, replacement, or selection of motors or generators. In addition, standards and guidelines published by the Occupational Safety and Health Administration (OSHA) must also be studied and followed during the removal and installation of motors and their associated controls.

Safety Alert

Working on or around electric motors is potentially dangerous. Although the potential for accidents can be greatly reduced by proper selection and installation, hazardous situations cannot be completely eliminated. When installing or servicing equipment, be sure to always follow the guidelines and recommendations of the equipment manufacturer. Although best practices apply to all situations and circumstances, manufacturers often provide information that is specific to their product. Thoroughly reading and becoming familiar with the manufacturer's literature will help save time and help you perform your job more effectively. Once you have completed your work, it is recommended that the manufacturer's literature remain with the equipment. This way, those who follow you will have access to this important information as well.

Your personal safety and the safety of those around you is especially important when electric motors are involved. A safe motor environment starts with proper design, application, and selection of the motor and its associated components, operational controls, and safety devices. The following is a list of safety-related items that you should be aware of when performing motor-related work:

- Ensure that the motor has been properly selected to match the needs of the system and the type of load that will ultimately be driven.
- Follow motor instructions to help ensure safe and proper mounting, wiring, and operation of the motor.
- Mount guards over or around rotating parts such as pulleys, couplings, blowers, fans, and shaft extensions to help prevent accidental, unintended contact with them. This will greatly reduce the chance of sustaining an injury.
- Before servicing a motor, verify that the motor drive or control circuitry does not have the ability to supply voltage to the motor or its control system when the main circuit power is off.
- Ensure that all power leads at the motor read 0 V with respect to each other and with respect to ground.
- Lock out and tag out all power to the motor before servicing a motor. This will help prevent the accidental and unintentional energizing of the motor's power or control circuits.
- Avoid contact with all energized circuits and rotating parts.
- Avoid bypassing any safety or protective devices.
- Avoid contact with capacitors unless they have been safely and completely discharged.
- Avoid prolonged exposure to any motor or piece of mechanical equipment that operates with high noise levels without wearing proper ear protection.

Although many of the items listed pertain to many aspects of the work you will perform in the field, it is important to continually stress them to help ensure safety on the job. Other safety guidelines to follow when working in the electrical field are discussed in Chapter 3, *Workplace Safety*.

26.2 Selecting a Motor

The number one concern when selecting a motor replacement is to check that the motor is properly matched to the application for which it will be used. An exact motor replacement is always best, but this is not always possible. If an exact motor replacement is not available, do your best to match the replacement to the original as closely as possible. This involves not only matching the electrical specifications of the replacement motor to those of the old one, but also the shaft size, frame size, motor enclosure type, and, if possible, the motor's mounting configuration.

Motor Enclosure

The type of motor enclosure is determined by the location of the motor, the conditions under which the motor is to operate, the substances that the motor is likely to be exposed to, and the motor's cooling requirements. The following are common motor enclosure types:

- Explosion-proof (XP)
- Flame-proof (FLP)
- Hazardous location (HAZ)
- Open drip-proof (ODP)
- Totally enclosed air-over (TEAO)
- Totally enclosed fan-cooled (TEFC)
- Totally enclosed nonventilated (TENV)
- Totally enclosed washdown (TEWD)

Installing a motor with an enclosure type that is not consistent with the motor's intended application can cause premature motor failure, equipment damage, or severe personal injury. Chapter 24, *Motor Anatomy, Types, and Construction*, provides information on various motor enclosure types and their applications.

Motor Nameplate

In most cases, the motor's nameplate will provide you with most, if not all of the information you will need to identify a suitable replacement motor, **Figure 26-1**. The motor's nameplate typically provides the following information:

- Manufacturer
- Model number
- Serial number
- Number of phases for ac motor
- Rated voltage
- Full-load current
- Frequency
- Horsepower (hp)
- Speed (rpm)
- Rotation
- Frame size
- Enclosure type
- Temperature rise
- Locked-rotor current (or locked rotor kVA per hp)
- Service factor (SF)

The motor's model number is one of the most valuable pieces of information you can have about a motor or any other piece of equipment or machinery. By accessing the motor manufacturer's website or published literature, you can search by the model of the motor to gain access to information about the motor's features and operating characteristics. The serial number is also an important piece of information, especially if it is suspected that the motor might be under warranty. By providing the serial number to the manufacturer, the date of production can be determined, as well as other important details about the motor.

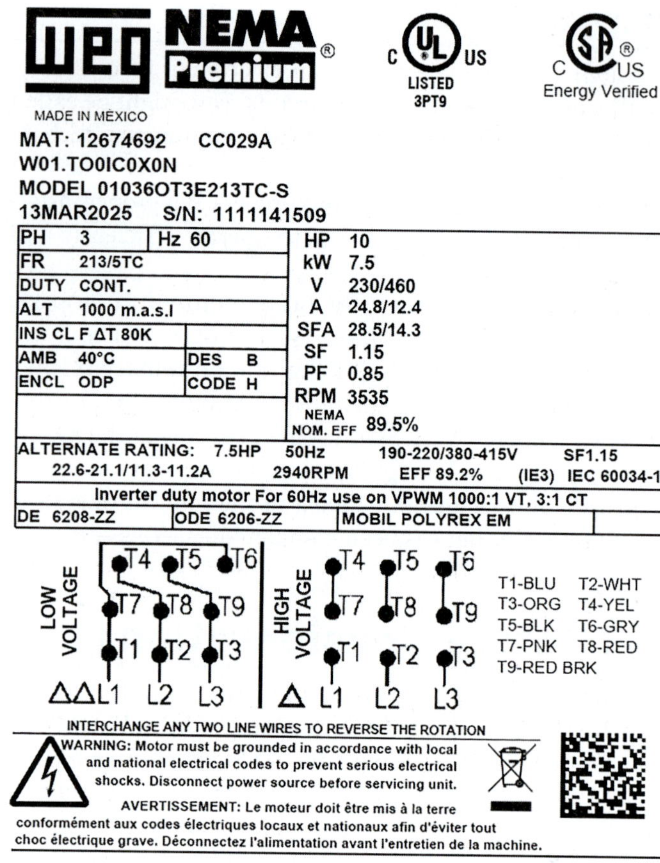

WEG

Figure 26-1. A motor nameplate provides essential information about a motor, including, but not limited to, model and serial numbers, the motor's rated voltage, frequency, horsepower, and speed.

The nameplate voltage and frequency should correspond to the voltage and frequency of the power supply. A motor will typically operate satisfactorily at a supply voltage within 10% (plus or minus) of the voltage indicated on the nameplate and a frequency that is within 5% (plus or minus) of the nameplate frequency. The combined deviations of voltage and frequency should not exceed 10%.

Shedding Light

If you cannot find the nameplate on the motor, refer to the mechanical equipment specifications to locate the power ratings and other important motor information. The mechanical specifications can often be found on the project's drawings or plans. In some cases, the motor's model number is also specified on these documents.

Example 26-1 and **Example 26-2** provide information on how to determine whether or not a motor will likely function properly when powered by the stated input signal.

The horsepower (hp) rating of a motor, where 746 W equals 1 hp, can be expressed a number of different ways. For larger motors, the horsepower rating is expressed as a whole number,

Example 26-1

A motor is rated at 230 Vac/60 Hz. The actual power supply provides a 207 Vac/60 Hz signal. Will this motor operate satisfactorily when powered by this signal?

✓ Solution

Since the actual frequency is equal to the rated frequency, we only need to determine if the voltage supplied to the motor is within 10% of the nameplate voltage. The voltage deviation is calculated as follows, where the initial value indicates the nameplate/rated voltage and the final value indicates the actual voltage being supplied to the motor:

$$\text{Percentage Change} = \frac{\text{Final Value} - \text{Initial Value}}{\text{Initial Value}} \times 100\%$$

$$\text{Percentage Change} = \frac{207 - 230}{230} \times 100\%$$

$$\text{Percentage Change} = \frac{-23}{230} \times 100\%$$

$$\text{Percentage Change} = -10\%$$

The combined deviations of voltage and frequency do not exceed 10%. Although the total deviation is at the very top of the acceptable range, the motor should operate properly.

such as 5 or 10. For fractional horsepower motors, this rating is expressed as a fraction, such as 1/3 or 1/2. In many instances, small fractional horsepower motors are rated in watts. Refer to **Example 26-3**.

The speed, expressed in rotations per minute (rpm), and direction of rotation, expressed as either clockwise (CW) or counterclockwise (CCW), provide valuable information about how fast the motor turns and in what direction. Motor speed indicates the maximum full-load speed at which the motor can safely operate with minimal damage potential. If the required motor speed is higher than the rpm rating on the nameplate, another motor type should be selected.

Motor rotation refers to motors with a single shaft. Motors with two shafts can operate in either the clockwise direction or the counterclockwise direction, depending on how they are mounted. In most cases, rotation refers to the direction the motor's shaft will spin when observed from the shaft end, **Figure 26-2**. Some motor manufacturers will further clarify the direction of motor rotation by adding the letters *SE* to the rotation notation, which indicates "shaft end." Motors with two shafts are marked as CWLE or CCWLE. The additional letters *LE* refer to "lead end." The lead end of a motor is the end where the motor leads protrude from the motor's housing, **Figure 26-3**. A double-shafted motor identified as CWLE will rotate in the clockwise direction when the motor is observed from the lead end. Similarly, a double-shafted motor identified as CCWLE will rotate in the counterclockwise direction when the motor is observed from the lead end.

Example 26-2

A motor is rated at 230 Vac/60 Hz. The actual power supply provides a 212 Vac/65 Hz signal. Will this motor operate satisfactorily when powered by this signal?

✓ Solution

To determine if the power supplied to the motor is acceptable, with a total combined deviation of no more than 10% of the nameplate voltage and the frequency, you will need to calculate the voltage and frequency deviations. First, calculate the voltage deviations as follows:

$$\begin{matrix}\text{Percentage} \\ \text{Voltage Change}\end{matrix} = \frac{\text{Final Value} - \text{Initial Value}}{\text{Initial Value}} \times 100\%$$

$$\begin{matrix}\text{Percentage} \\ \text{Voltage Change}\end{matrix} = \frac{212 - 230}{230} \times 100\%$$

$$\begin{matrix}\text{Percentage} \\ \text{Voltage Change}\end{matrix} = \frac{-18}{230} \times 100\%$$

$$\begin{matrix}\text{Percentage} \\ \text{Voltage Change}\end{matrix} = -7.8\%$$

Next, calculate the frequency deviation as follows:

$$\begin{matrix}\text{Percentage} \\ \text{Frequency Change}\end{matrix} = \frac{\text{Final Value} - \text{Initial Value}}{\text{Initial Value}} \times 100\%$$

$$\begin{matrix}\text{Percentage} \\ \text{Frequency Change}\end{matrix} = \frac{65 - 60}{60} \times 100\%$$

$$\begin{matrix}\text{Percentage} \\ \text{Frequency Change}\end{matrix} = \frac{5}{60} \times 100\%$$

$$\begin{matrix}\text{Percentage} \\ \text{Frequency Change}\end{matrix} = 8.3\%$$

The combined deviations of voltage and frequency is 16.1% (7.8% + 8.3%) and exceeds the 10% maximum. You might think that, since adding −7.8% and +8.3% results in a sum of 0.5%, this result is acceptable, but this is not true. When calculating the total deviations, we ignore the positive and negative signs and simply add the magnitude of the deviation percentages.

Example 26-3

A motor that is rated at 2/3 hp can also be referred to a motor as having what wattage rating?

✓ Solution

Since 1 hp is equal to 746 W, a 2/3 hp motor can be referred to as a 500 W motor:

$$\frac{2}{3\ \text{hp}} \times \frac{746\ \text{W}}{\text{hp}} = 497.3\ \text{W} \approx 500\ \text{W}$$

Now It's Your Turn 26-3

What is the approximate horsepower rating of a 250 W motor?

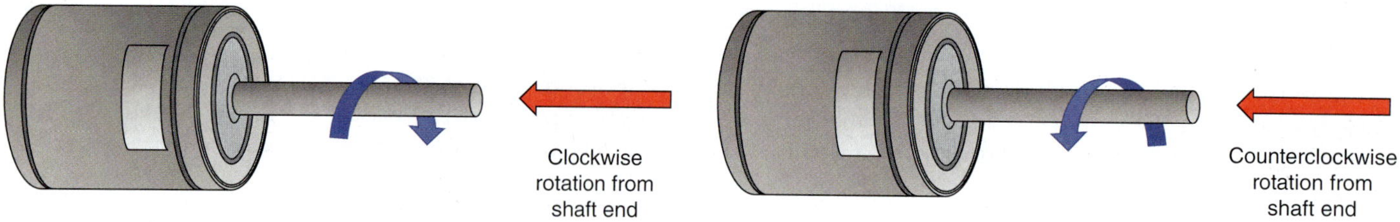

Figure 26-2. Observing motor direction of rotation from the shaft end.

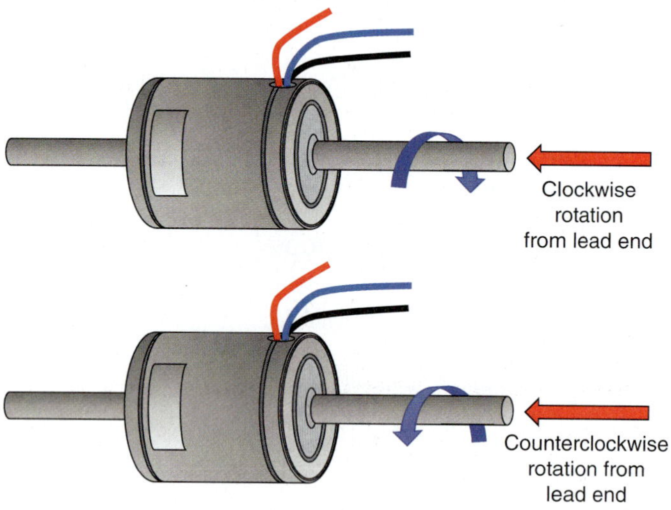

Figure 26-3. Observing motor direction of rotation from the lead end.

The frame number refers to a motor's mounting characteristics and measurements. Frame numbers are established by NEMA and are consistent from manufacturer to manufacturer. Chapter 24, *Motor Anatomy, Types, and Construction*, provides additional information on motor frame numbers and important motor measurements.

Shedding Light

Before uncrating, moving, or installing a motor, thoroughly inspect the motor. Signs of damage, such as dents, scratches, or other imperfections are indications that there might be something wrong with the motor. If possible, attempt to turn the shaft by hand to ensure that it turns freely. Also, keep all the literature supplied with the motor. This includes any warning or safety-related tags or stickers, lubrication information, and installation and maintenance manuals.

26.3 Troubleshooting Motor Failure

Motor failure does not always mean that the motor needs to be replaced. When a motor fails to operate, it is important to determine why the motor failed, and if the problem can be resolved by either repairing the motor or by correcting external conditions that have caused the motor to malfunction. Repair is much more common on larger multi-horsepower motors than on smaller motors. It would, for example, not be economically practical to replace the bearings on a 1/2 hp motor, as the cost of repair would likely exceed the cost of a new replacement motor.

Some commonly encountered motor problems that should be investigated before condemning a motor include the following:

- **Motor has open windings.** This will likely require repair or replacement.
- **Motor stalls after starting causing the overloads to open.** Too much of a load or misalignment of the load could be the cause.
- **Motor fails to start.** A blown fuse or open circuit breaker could be the cause as could an open overload.
- **Motor does not reach desired speed.** Excessive load on the motor, improper supply voltage, or an improperly selected motor are likely causes for this condition.
- **Excessive motor vibration or noise.** Loose motor mountings or the motor being out of alignment with the driven load could be the cause.
- **Voltage imbalance, also called voltage unbalance.** The actual applied voltage should be within 10% of the voltage rating. However, a balanced voltage between the phases in a three-motor is also important. Each 1% of voltage variation between phases can result in a 6% to 10% current imbalance. This will likely result in the motor overheating.

If addressing the items in this section fails to correct the problem, the motor should be removed and either repaired or replaced.

Open Motor Windings

If one or more motor windings are open, an internal overload or protection device might be open. If the motor is warm, the thermal protector might be open. In this situation, it is recommended that the motor be allowed to completely cool down. Once cool, the integrity of the motor windings can be rechecked. If the windings now test good, the cause for motor overheating must be determined and corrected.

Common causes for motor overheating include the following:

- Insufficient airflow across and/or through the motor
- Voltage imbalance
- Current imbalance
- Excessive surrounding (ambient) air temperature
- Motor operating in an overloaded condition
- Loss of one voltage phase (three-phase motors)
- Bearing damage
- Excessive friction in driven device
- Excessive load

Once the cause of open motor windings and/or motor overheating has been identified and corrected, the motor should be able to restart safely. Once restarted, motor operation should be monitored to ensure that the problem has been resolved. If the motor has cooled and the windings are still open without a resolution to the problem being identified, then the motor should be removed.

Motor Stalls after Starting

If a motor starts when energized but stalls after a period of time, the following can be checked before opting to replace the motor:

- Loss of one voltage phase (three-phase motors)
- Wrong application
- Excessive load on the motor
- Low supply voltage
- Loose electrical connection
- Power failure
- Blown fuse or tripped circuit breaker or protector

Motor Does Not Reach Design Speed

If the motor runs but does not reach its desired speed, or it takes too long for the motor to reach its design speed, the following items should be checked before concluding that the motor is defective:

- Incorrect motor application
- Excessive load on motor start-up
- Reduced supply voltage
- Loss of one voltage phase (three-phase motors)
- Damaged or loose rotor
- Damaged bearings on motor or on driven load

Motor Fails to Start

If the motor fails to start, it is recommended that the following items be checked before condemning the motor:

- Blown fuses
- Tripped circuit breakers
- Open overload on motor starter
- No supply voltage
- Mechanical motor failure
- Overloaded motor
- Open thermal overload or protector
- Improper supply voltage or loss of a phase in a three-phase system

Excessive Motor Vibration or Noise

If the motor vibrates excessively or makes a great deal of noise, the following items should be checked before concluding that the motor needs to be replaced:

- Misalignment of drive and driven components
- Defective bearings
- Loss of voltage phase (three-phase motors)
- Excessive motor shaft play
- Motor not mounted properly
- Loose motor mounts
- Unbalanced driven load
- Driven device striking or rubbing against housing or shield

Safety Alert

Before disconnecting any motor wiring, make certain that all power and control circuits are completely de-energized and that any power-storing devices, such as capacitors, are properly and completely discharged. It is also important to note the existing connections of the ungrounded or "hot" conductors energizing a three-phase motor. Correct motor rotation depends on this. If any two conductors are connected to phase that are not the same as the original installation, the motor will rotate in reverse of the intended direction. This is of great importance when connected to exhaust fans, pumps, and compressors. Incorrect rotation will result in damage to equipment.

26.4 Removing a Motor

When removing a motor from a piece of equipment, it is important to document the conditions as they exist prior to disconnecting or removing any wires, mounting hardware, driven components, belts, pulleys, couplings, or motor guards. If the motor is being removed and replaced with a motor of the same model, marking the footprint of the original motor mount will save time later on. By positioning the new motor in the exact position as the original, shaft alignment with the driven system component is almost guaranteed. In addition, before loosening motor mounts, take an accurate center-to-center measurement between the motor shaft and the driven shaft. If the existing pulleys are to be reused, taking this measurement helps ensure that proper belt tension will be maintained. This, of course, assumes that the initial belt tension was correct.

Be sure to document all electrical connections as they presently exist. This can be done by using wire markers,

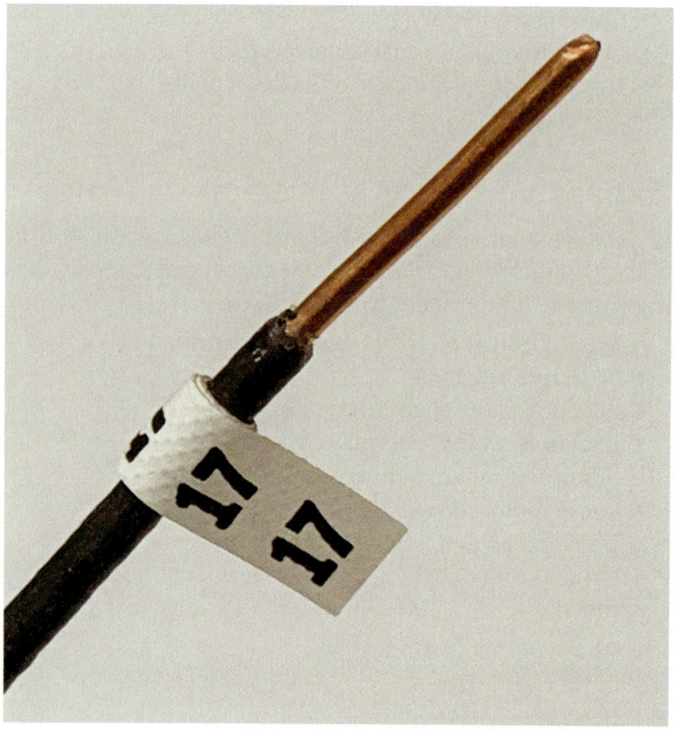

Goodheart-Willcox Publisher

Figure 26-4. Wire-marking tags like the one shown here are used to indicate the location of electrical connections.

Photo courtesy of Jason Obrzut

Figure 26-5. A motor shaft can be damaged when a pulley is not properly removed.

Figure 26-4, taking notes, and/or taking a picture of the wiring connections. It is good practice to loosen the motor mounts before removing the belts from pulley arrangements. By allowing the motor to shift, belt tension will be reduced, making belt removal easier.

Safety Alert

Be sure to keep your fingers away from the area where belts come in contact with pulleys. Severe personal injury can result if fingers get caught between a belt and pulley. Also, never attempt to stop a rotating driven device by hand. Always allow it to come to a full stop on its own. Never attempt to stop a belt-driven load by grabbing the belt. Once again, fingers can get caught under the belt, causing severe personal injury.

Once any belts have been removed from the pulley arrangement, the pulley can be removed from the motor. First, loosen any setscrews that are securing the pulley to the motor's shaft. Once the setscrews have been loosened or removed, then the pulley can be removed from the motor shaft. If the pulley cannot be removed easily, apply some penetrating oil to help loosen the pulley. Under no circumstances should you ever strike a motor shaft with a hammer, as it can become damaged, making the pulley more difficult to remove, **Figure 26-5**. Never pull directly on a driven device such as a blower wheel. Pulling on the wheel can take it out of balance, creating excessive amounts of noise and putting an unbalanced load on the motor bearings. If a pulley or blower wheel needs to be pulled with force from a motor shaft, then it is suggested to use a wheel puller that is designed specifically for this purpose.

Once the motor has been disconnected electrically and mechanically, it can be lifted and removed.

26.5 Lifting a Motor

Depending on the size of the motor, lifting methods will differ greatly. Fractional horsepower motors can often be lifted by a single individual, while larger, multi-horsepower motors are lifted using hoists and other heavy equipment.

Motors that require a hoist to be lifted can have either one or two lifting eyebolts, **Figure 26-6**. For motors equipped with one eyebolt, the angle between the shaft of the eyebolt and the vertical lifting line should not exceed 30°, **Figure 26-7**. For motors equipped with two eyebolts, the angle between the shaft of one eyebolt and either of the individual lifting lines should not exceed 45°, **Figure 26-8**.

Safety Alert

Keep fingers, feet, and all body parts out from under the motor when lifting it. It is far better to prepare for a motor that might fall by staying clear of the area in case an accident occurs. Motors can be easily replaced, body parts cannot.

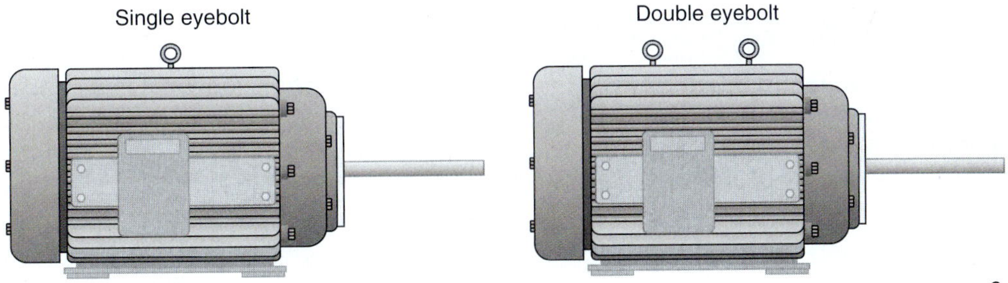

Goodheart-Willcox Publisher

Figure 26-6. A hoist is required when lifting motors with one or two eyebolts.

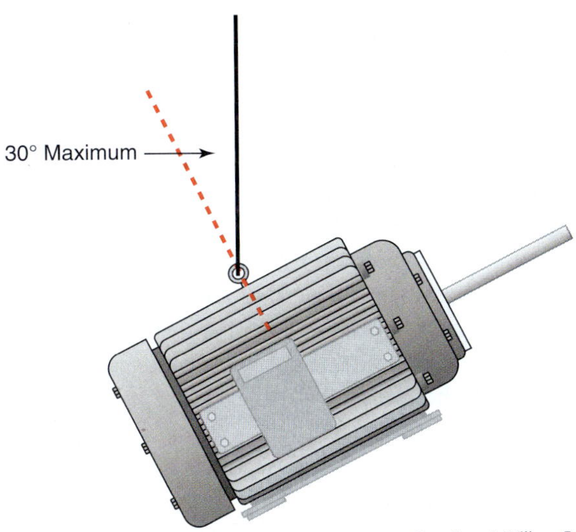

Goodheart-Willcox Publisher

Figure 26-7. The maximum lifting angle for a motor with one eyebolt is 30° between the vertical lifting line and the shaft of the eyebolt.

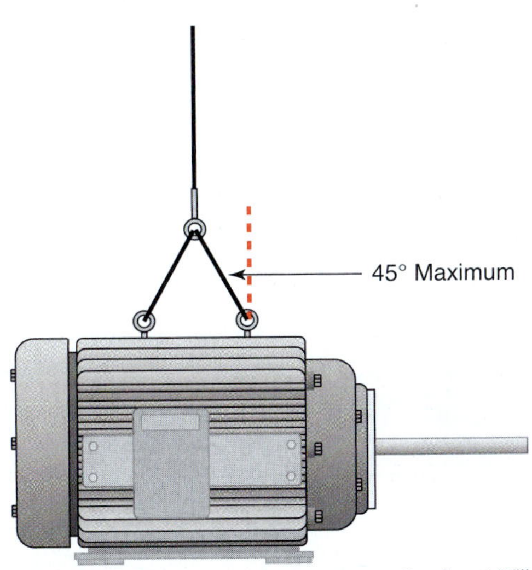

Goodheart-Willcox Publisher

Figure 26-8. The maximum lifting angle for a motor with two eyebolts is 45° between the shaft of one eyebolt and either of the individual lifting lines.

It is important to note that eyebolts or lifting lugs are intended to lift only the motor. Nothing should be attached to the motor shaft when it is being lifted. This includes pumps, fans, blowers, or any other piece of driven equipment. The location of the eyebolts is determined based on the motor's center of gravity. Lifting the motor and a connected device as a single unit can create a weight imbalance issue that can cause serious problems. In addition, the combined weight of the motor and driven equipment may exceed the weight-supporting capability of the eyebolts. If the motor must be lifted with a driven component connected to it, another method of lifting the motor, such as a forklift or crane, should be used.

Safety Alert

Motor eyebolts are often not a permanent part of the motor. Before attempting to lift a motor, make certain that the eyebolt is not cross-threaded and is fully threaded into the motor's mounting hole. Failure to check the integrity of the eyebolt before lifting can result in severe personal injury or damage to the motor should it fall.

Safety Alert

Never use the motor shaft as a means to lift, reposition, or shift a motor. Bearing damage can result.

Avoid using motor-mounted eyebolts when temperatures are below 0°F. At low temperatures, eyebolts can become brittle and can shatter if they are struck, either intentionally or accidentally. If a motor must be hoisted when the temperature is low, it is recommended that another lifting method be used.

26.6 Mounting a Motor

Motors must be mounted securely on a level, rigid base to ensure proper operation. Poorly or improperly mounted motors have the potential to vibrate excessively. Excessive vibration can cause motor or bearing damage, as well as excessively high noise levels. In addition, a rigid foundation helps ensure proper alignment between the motor shaft and that of the driven load.

Shedding Light

The International Organization for Standardization, ISO, in their ISO-20816-1 standard, establishes the general conditions and procedures for the measurement and evaluation of motor vibration. In this standard, vibrations on the nonrotating parts of motors are evaluated.

When mounting a motor, be sure to use bolts with the largest bolt diameter that will fit through the holes in the base of the motor. When selecting bolts, be sure to consider the bolt's hardness, length, and diameter. The bolt should be long enough to pass completely through the motor's mounting foot, the mounting base or platform, any washers, and of course, the securing nut or nuts, **Figure 26-9**. The ***thread engagement***, which refers to the amount of the bolt that is physically threaded into the securing nuts, should be at least equal to the diameter of the bolt, **Figure 26-10**.

Shedding Light

Hardened bolts that are not typical hardware grade should be used. For example, grade 8 hardened bolts are more durable and resistant to wear and tear than typical hardware grade bolts. Using bolts with a finer thread will provide a tighter and better holding force. The use of softer bolts will be more likely to fail.

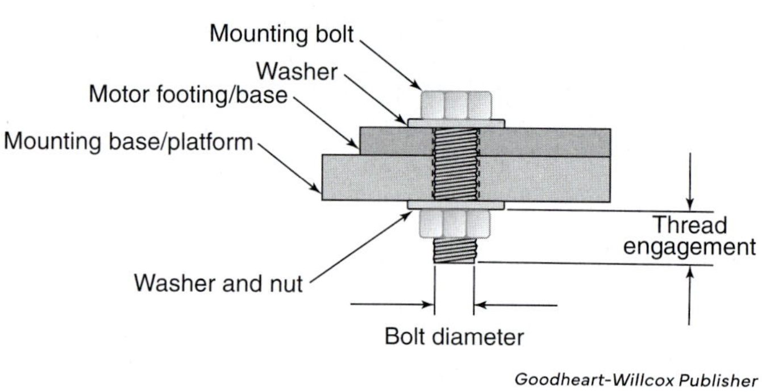

Figure 26-9. When selecting a mounting bolt, the bolt should be long enough to pass completely through the motor's mounting foot, the mounting base, washers, and through the securing nut or nuts.

When mounting the motor, it is strongly recommended that the motor shaft and the shaft of the driven load are in perfect alignment before tightening the mounting bolts. Do not rely on flexible couplings between the drive and driven components to compensate for any alignment problems, however slight they might be. To help align the drive and driven shafts, it may be necessary to shim the motor's base to obtain the desired height. Once the motor and driven components are aligned, the mounting bolts can be tightened. Tightening the mounting bolts to the manufacturer's torque recommendations is always a good practice to follow.

26.7 Pulley Positioning and Alignment

Once the motor is securely in place and the drive shaft is in perfect alignment with the shaft of the driven component, attention can then turn to the drive and driven pulleys. Before securing any pulleys, couplings, or other connection components to the motor shaft, it is important to verify that the motor specifications match those of the power supply. Any motor selection problems should be identified before connecting anything to the motor's shaft. Failure to do so will very likely leave permanent marks and indentations on the shaft that will make the returning and exchanging of the motor a much more difficult process.

The drive and driven pulleys must be in perfect alignment. Alignment should be checked with a straightedge or laser alignment tool, **Figure 26-11**. There should be no signs of offset misalignment or angular misalignment. ***Offset misalignment***, also referred to as *parallel misalignment*, occurs when the drive and driven shaft are parallel to each other, but the pulleys are not in alignment. Offset misalignment is easily corrected by repositioning either the drive or driven pulley on the shaft until both pulleys are properly aligned.

Angular misalignment occurs when the drive and driven shaft are not parallel to each other. To repair angular misalignment problems, either the motor or driven component must be repositioned to align both shafts. In some cases, pulley-positioning problems can be a combination of offset

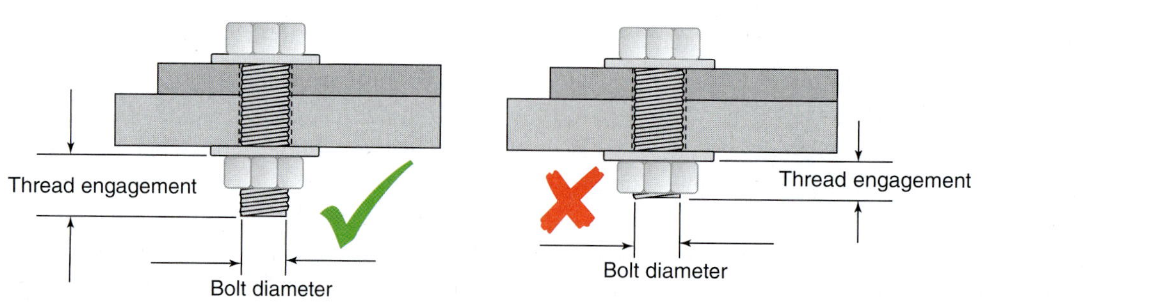

Figure 26-10. The minimum thread engagement should be equal to the diameter of the bolt.

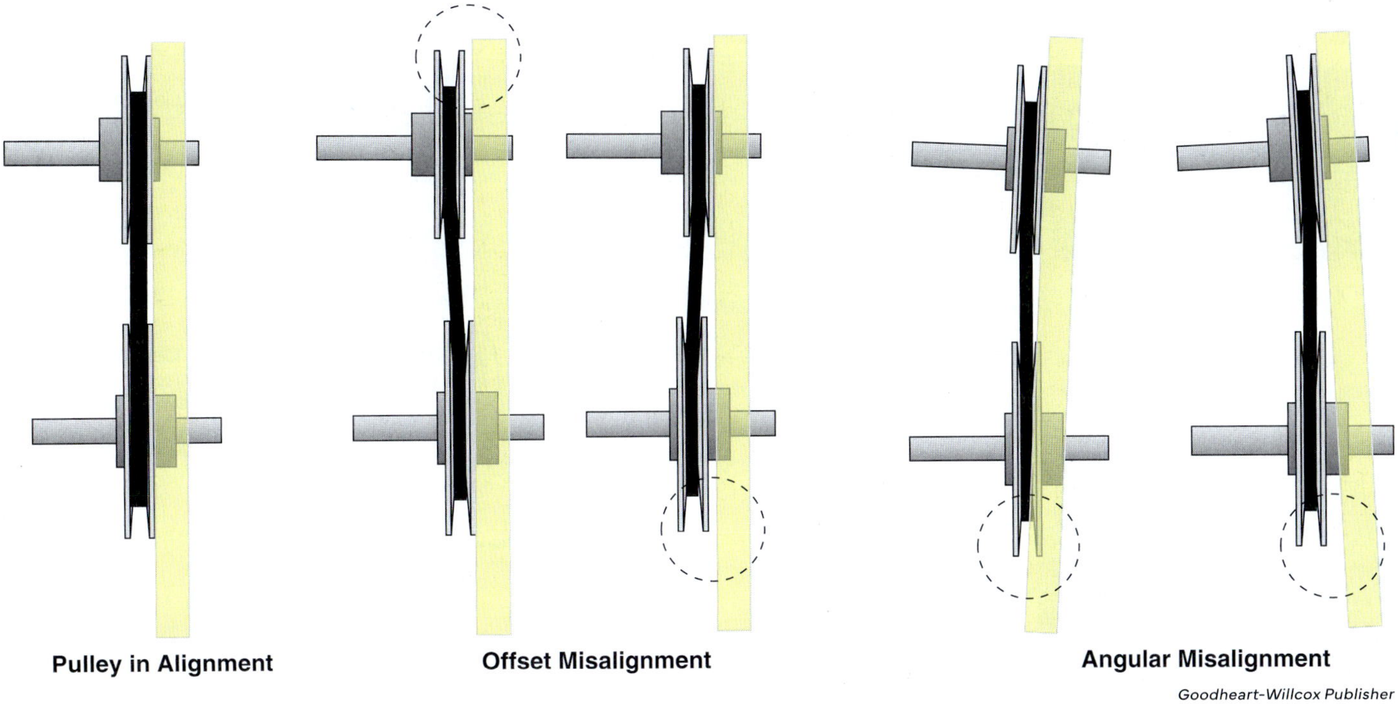

Figure 26-11. Before securing any pulley or drive and driven shaft, use a straightedge to verify alignment. The drive and driven pulleys must be in perfect alignment

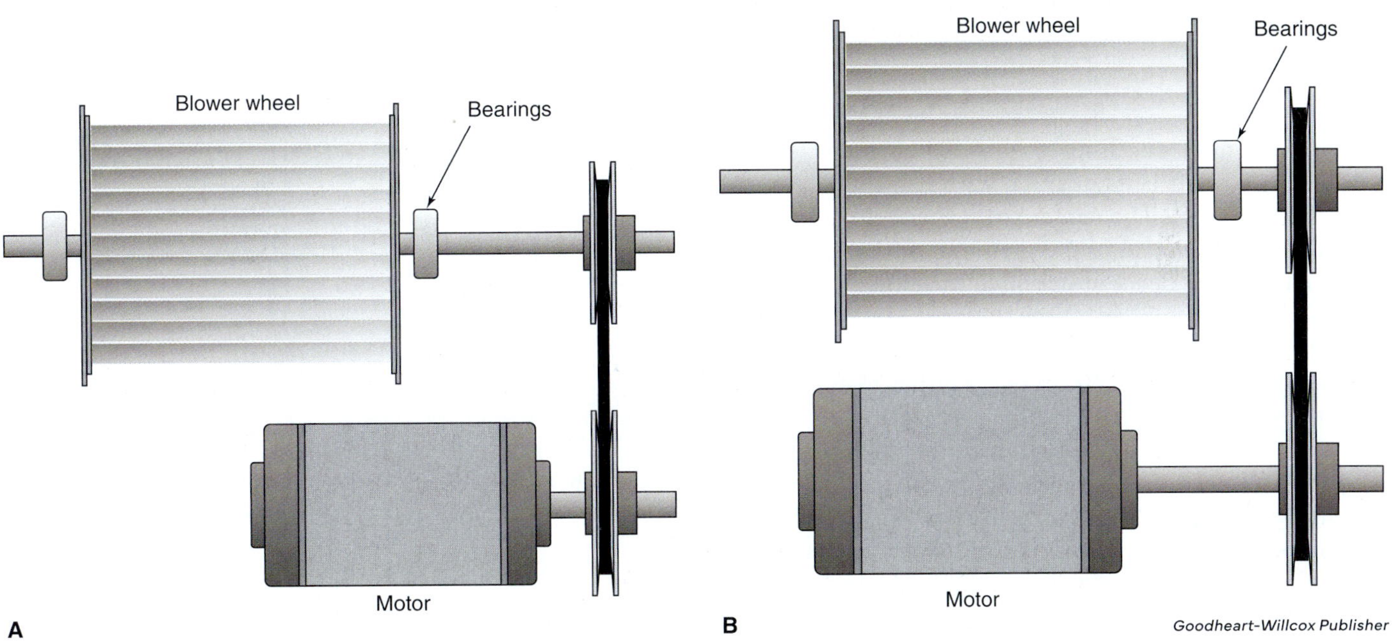

Figure 26-12. Less bearing stress contributes to longer bearing life. A—The distance between the pulley and the bearings is greater than it needs to be, putting additional, unwanted stress on the bearings. B—The distance between the pulley and the bearings is as short as is practical, putting less stress on the bearings.

and angular misalignment. If both situations are present, the pulleys must be repositioned on the shafts, but only after the angular misalignment situation has been corrected.

If the pulleys are not properly aligned, the belts can become damaged. In some cases, when belt tension is too loose, the belts can come off the pulleys altogether.

When positioning pulleys on a shaft, the distance between the pulley and the bearing, either on the drive side or the driven side, should be as short as practical. The closer the pulleys are mounted to the motor bearings or the driven component bearings, the less stress there will be on the bearings, **Figure 26-12**.

Figure 26-13. The belt center line should not extend past the end of the shaft.

Another important thing to consider when positioning a pulley on a shaft is its position with respect to the end of the motor shaft. The center point of a belt should not extend beyond the end of the motor shaft, **Figure 26-13**. If the belt extends beyond the end of the motor shaft, then the stability of the pulley on the shaft is reduced. In addition, this can cause excessive vibration, cause damage to the pulley and the belt, and cause the pulley to come loose from the shaft.

Before installing belts on the pulleys or otherwise making physical connections between the motor and the driven load, it is important to verify that the direction of motor rotation is correct. This can be accomplished by briefly applying power to the motor leads and observing the direction of rotation. This must be done carefully to ensure that all safety precautions are taken to prevent accidental contact with live conductors or the rotating motor shaft.

Shedding Light

For three-phase motors, interchanging any two of the three line voltage leads will change the rotation of the motor, **Figure 26-14**.

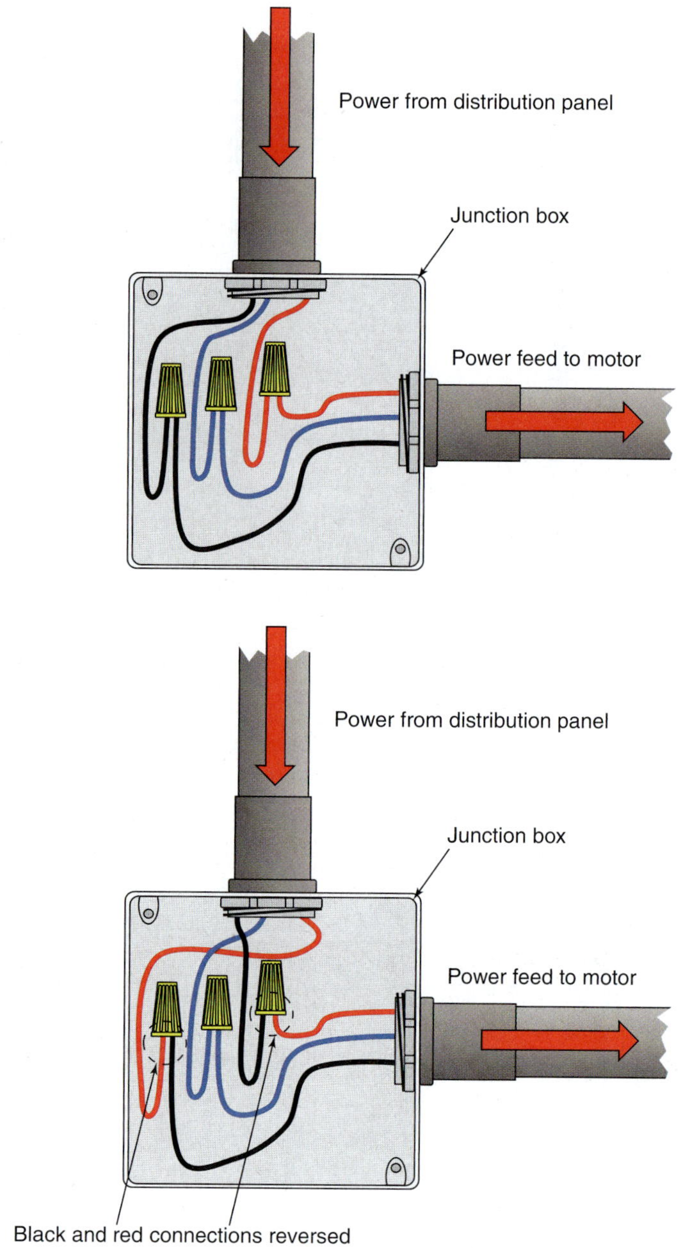

Figure 26-14. Changing the direction of rotation of a three-phase motor.

26.8 Belt Tension

Belts, when properly installed, should be tight enough to prevent slippage, but loose enough to avoid excessive pressure or strain on the drive or driven bearings. A belt tension gauge is used to ensure the tension falls within the manufacturer's recommended range. There are several other specialized tools that can be used to evaluate belt tension, such as sonic belt tension meters, belt tension frequency meters, and laser belt tensioning meters. The belt tension gauge in **Figure 26-15** is

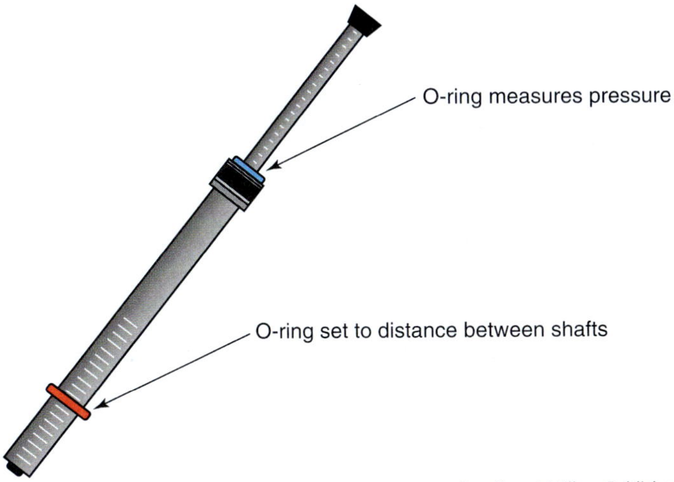

Goodheart-Willcox Publisher

Figure 26-15. The belt tension gauge shown here is equipped with two O-rings. The larger O-ring is set to the center-to-center distance between the two shafts and the pressure is read from the smaller O-ring located at the top of the gauge.

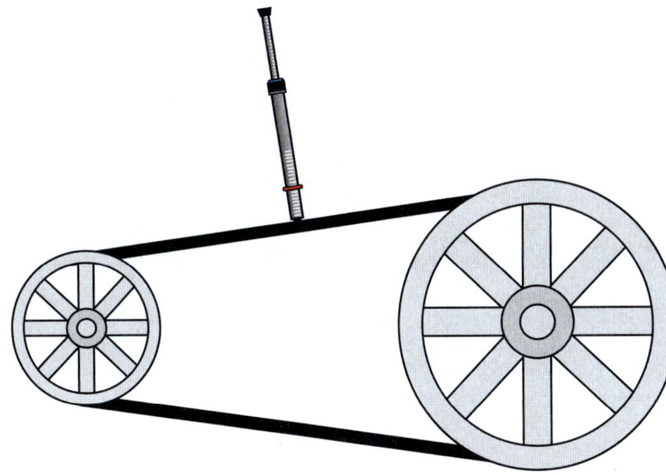

Goodheart-Willcox Publisher

Figure 26-16. Positioning the belt tension gauge.

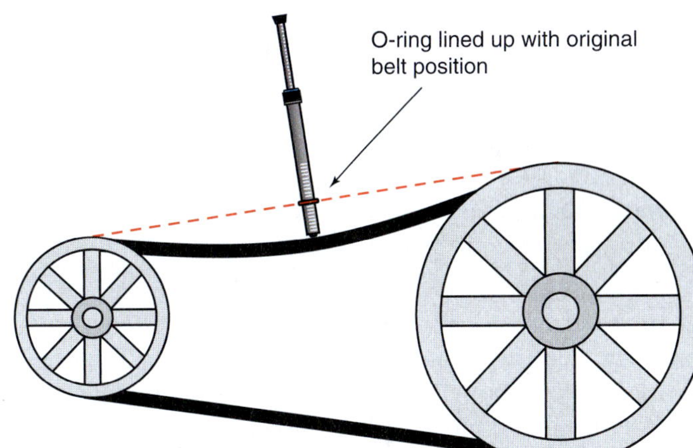

Goodheart-Willcox Publisher

Figure 26-17. Final position of the belt tension gauge.

configured so that a smaller, spring-loaded handle is pushed into the larger chamber. Steps to check belt tension are provided in **Procedure 26-1**.

Safety Alert

Rotating parts such as couplings and pulleys should be guarded to prevent personal injury.

Procedure 26-1

Using a Belt Tension Gauge to Check Belt Tension

1. Ensure the machine is powered OFF and locked out.
2. Visually inspect the belt for signs of damage, wear, or misalignment.
3. Determine the correct tension by reviewing the manufacturer's specifications.
4. Measure the center-to-center distance between pulleys along a line parallel to the belt.
5. Position the larger O-ring on the belt tension gauge at the value of the distance measurement.
6. Position the belt tension gauge perpendicular to the belt, with the foot of the gauge resting on the belt at the center of its span, **Figure 26-16**.
7. Deflect the belt by pushing the gauge handle until the large O-ring lines up with the original belt position, **Figure 26-17**. This action pushes the smaller O-ring upwards.
8. Read the tension displayed by the smaller O-ring, then compare your reading to the manufacturer's chart to determine if the belt tension is correct.

Evaluating belt tension should not be overlooked. Excessive belt tension shortens belt life and can damage bearings by pulling the shaft out of alignment, **Figure 26-18**. When replacing a motor, it is good field practice to inspect all bearings for excessive wear or damage.

26.9 Electrical Connections

Once the motor has been properly mounted, secured, and connected to the driven component, the permanent electrical connections can be made. Connecting the motor to the power supply should be done in accordance with the wiring diagram accompanying the motor and other instructions supplied by the motor manufacturer, as well as following local and national electrical codes.

Before energizing any motor, be sure to double-check the electrical wiring to ensure proper connections, including

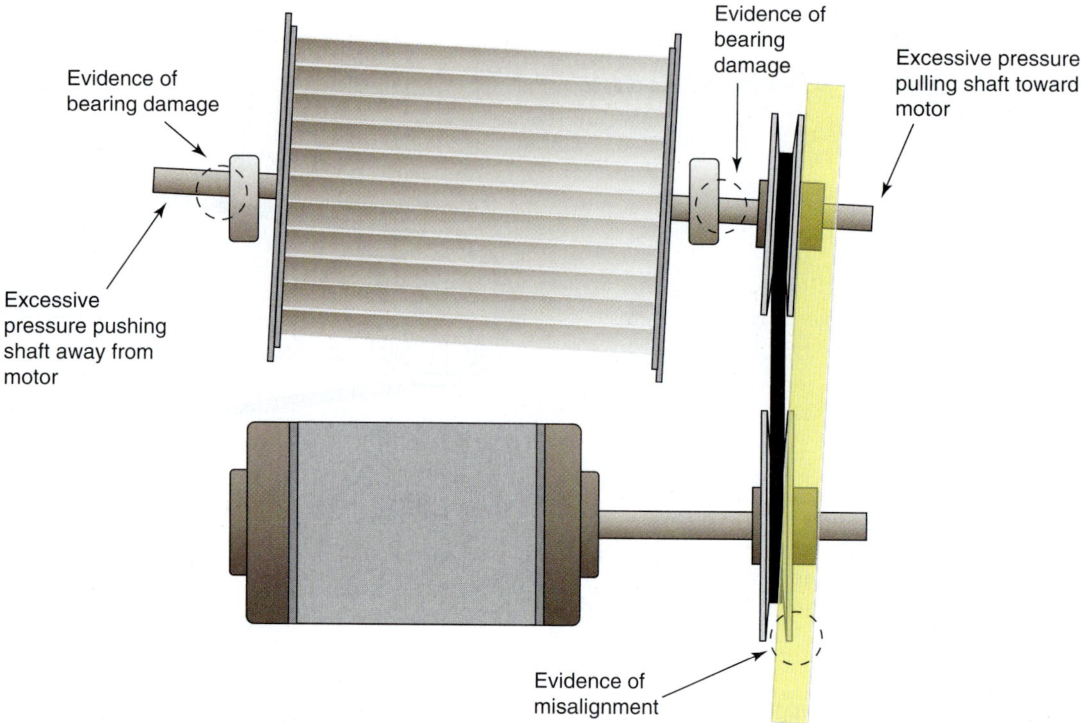

Goodheart-Willcox Publisher

Figure 26-18. Excessive stress on the bearings can result in bearing damage and misalignment of the shaft.

connections that ensure proper rotation. Failure to connect the motor correctly can cause damage to the motor. For example, motors that operate at different voltages by altering the electrical connections must be wired properly to avoid damage that can result if the motor is not wired for the voltage being supplied.

Shedding Light

Do not be overly concerned if the branch conductors and the motor leads are not the same size. Branch circuit conductors are sized based on a motor's full-load, or expected circuit current, as determined by the *NEC*. Some motor leads are smaller than branch circuit conductors. This is because, depending on the motor configuration and enclosure, internal wiring can be cooled by air circulating across and/or through the motor. This allows these conductors to carry currents that are greater than the *NEC* rating for same-size branch circuit conductors.

Since motor leads and power circuit conductors may be different sizes, wire-to-wire connections should be made with connectors that can be used effectively on conductors

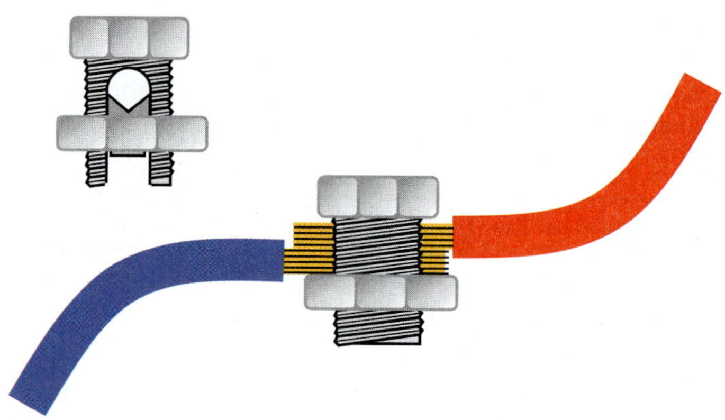

Goodheart-Willcox Publisher

Figure 26-19. A split-bolt connector is used to bond copper wires together to complete an electrical circuit.

of different sizes. One such connector is a split-bolt connector, **Figure 26-19**. Once the connection is made, be sure to insulate the connection to help ensure no contact is made with the enclosure or other conductors. A combination of rubber tape wrapped around the connection and vinyl tape wrapped around the rubber tape is acceptable.

Know the Code

NEC Article 310, Conductors for General Wiring, provides detailed information on conductor sizing for general wiring applications with more specific information on proper wire sizing for motors provided in *Article 430*.

NEC Article 430 covers motors, motor control circuits, motor circuit conductor protection, motor overloads, motor controllers, and grounding as it applies to motors. Be sure to reference the appropriate sections of *Article 430* to ensure compliance with the *NEC*. In addition, be aware of the local codes in your area as they pertain to the installation, wiring, replacing, and protecting of motors and their associated control and power circuits.

After checking the connection, ensure that the motor is properly grounded. The grounding method used must be in accordance with the *NEC* and any applicable local codes. Motor frames and metal exteriors are required to be grounded. When making electrical connections, you are required to make certain there is a permanent ground connection between the ground point, the motor's housing, and the frame. The conductors that are used as grounding wires are sized based on the size of the branch-circuit short-circuit and ground-fault protective device, the fuses, or circuit breaker. *NEC Section 250.122 (D)* provides specific requirements for motor circuits.

Know the Code

NEC Sections 430.241 through *430.245* address the grounding of exposed metal parts of motors, motor frames, and controllers that are not intended to carry electric current. When questions arise regarding the grounding of motors or other motor-related components or controllers, these sections should be referenced. Information regarding metallic, non-current-carrying equipment parts can be found in *Section 250.22*.

Once all electrical connections have been made, be sure to replace all motor terminal box covers as well as other junction box covers.

26.10 Motor Start-Up

Before starting a motor for the first time, check the following items:

- Turn the shaft by hand, if possible, to make certain that it rotates freely.
- Check that all electrical connections are consistent with the motor's wiring requirements.
- Disconnect the motor from its load during the initial start to make sure that the motor rotates in the correct direction, using a phase rotation tester. During initial motor start-up, it is good practice to quickly start and stop the motor, called bumping, to confirm rotation is correct.
- Remove any hex keys or other loose parts from the pulleys or driven components.
- Replace all the terminal box covers.
- Ensure that all motor, pulley, belt, and coupling guards or shields are in place and do not make contact with any rotating parts of the equipment.

Once the motor is up and running, be sure to monitor its operation. Pay attention to any unusual sounds that are coming from the motor or the driven components and address accordingly. Measuring the motor's amperage draw and comparing it to the nameplate amperage rating will also alert you to any potential problems. To help ensure proper operation of the motor and the driven components, it is good practice to monitor the bearing temperatures of the motor and driven components. A bearing temperature rise of more than 3.5°F per minute indicates that there might be a problem. If such a temperature rise is observed, you should turn the motor off and investigate the situation. Common causes of excessive temperature rise include improper shaft alignment, pulley misalignment, and excessive belt tension.

CHAPTER 26 REVIEW

Summary

- While working on electric motors, potential for accidents can be greatly reduced by proper selection and installation of an appropriate motor. (26.1)
- When selecting a replacement motor, make certain that the new motor matches the application for which it is to be used. (26.2)
- The majority of the information needed to locate a suitable motor replacement can be found on the old motor's nameplate. (26.2)
- The voltage supplied to a motor should be within 10% of the nameplate value. (26.2)
- The frequency of a motor's power supply should be within 5% of the nameplate value. (26.2)
- The combined deviations of voltage and frequency should not exceed 10%. (26.2)
- The rotation of a single-shafted motor is determined by looking at the motor from the shaft end. (26.2)
- The rotation of a double-shafted motor is often determined by looking at the motor from the lead end. Motors with two shafts can operate in either the clockwise direction or the counterclockwise direction, depending on how they are mounted. (26.2)
- The direction of rotation of a three-phase motor can be reversed by switching any two power lead connections. (26.2)
- Before replacing a motor, it is strongly recommended that the cause of the motor failure be identified. (26.3)
- Before disconnecting any parts of the motor, document conditions as they presently exist and document all electrical connections using wire markers or by taking pictures. (26.4)
- If replacing an old motor with an exact replacement, mark the footprint of the original motor. (26.4)
- Motor lifting methods vary. Fractional horsepower motors can be lifted by a single person, but multi-horsepower motors require either a team or hoists. (26.5)
- Motors that require a hoist to be lifted can have either one or two lifting eyebolts. (26.5)
- When mounting a motor, use bolts with the largest bolt diameter that will fit through the holes in the motor's base. (26.6)
- The use of hardened bolts with a higher rating will result in a more secure and safe installation. Softer bolts will be more likely to fail. (26.6)
- The drive and driven shafts must be in perfect alignment to help ensure proper operation of the system. (26.6)
- The distance between a pulley and the bearings should be as short as practical. (26.7)
- Angular misalignment occurs when the drive and driven shafts are not aligned. (26.7)
- Offset misalignment occurs when the drive and driven pulleys are not aligned. (26.7)
- Belt tension should be properly evaluated to prevent premature belt and bearing failure. (26.8)
- Double-check all power wiring and ground connections to ensure they correspond to the motor's wiring diagram for the intended voltage and rotation. (26.9)
- Make sure all connections are properly torqued. (26.9)
- Before starting a motor, make certain that all safety guards are in place and that all terminal and motor box covers are securely in place. (26.10)
- Once a motor is replaced, its operation should be observed to verify the installation was successful. (26.10)

Review Questions

Use the information in the chapter to answer the following questions.

Measuring Your Potential

1. Which of the following is *not* a personal safety item to be aware of when working on a motor? (26.1)
   A. Adhere to NEMA's best practice guidelines and standards.
   B. Follow all manufacturer's guidelines and instructions.
   C. Install guards over pulleys and other rotating parts.
   D. Attempt to manually turn the shaft on an energized motor.
2. If a motor needs to be replaced and there is no nameplate on the motor, it is best to _____. (26.2)
   A. call NEMA for assistance
   B. refer to the project plans for motor specifications
   C. send pictures of the motor to the manufacturer to be identified
   D. rely on the motor's physical size and shaft diameter to determine its horsepower rating

3. If the letters CCWLE are indicated on a motor's nameplate, which of the following statements is correct? (26.2)
   A. The motor has one shaft and will rotate in the clockwise direction when looking at the end of the motor's shaft.
   B. The motor has one shaft and will rotate in the counterclockwise direction when looking at the end of the motor's shaft.
   C. The motor has two shafts and will rotate in the counterclockwise direction when looking at the end of the motor shaft on the lead end.
   D. The motor has two shafts and will rotate in the clockwise direction when looking at the end of the motor shaft on the lead end.
4. Before uncrating a motor, it is good field practice to _____. (26.2)
   A. inspect for damage and rotate the shaft by hand, if possible
   B. inspect for damage and energize the motor to determine rotation
   C. rotate the shaft by hand, if possible, and energize the motor to determine rotation
   D. lift the motor by the shaft to check the bearings
5. A 1/3 horsepower motor can also be identified as a motor of what approximate wattage? (26.2)
   A. 150 W
   B. 200 W
   C. 250 W
   D. 300 W
6. Common causes for intermittent motor stalling may include _____. (26.3)
   A. excessive load on motor, loose electrical connection, and/or blown fuse
   B. overloaded motor, open thermal overload, and/or incorrect motor application
   C. defective bearings, unbalanced drive load, and/or loss of voltage phase
   D. insufficient motor airflow, bearing damage, and/or loss of one voltage phase
7. After you ensure that the electrical conductors are not energized and the main circuits are locked out/tagged out, the next step you should perform before removing a motor from a piece of equipment is _____. (26.4)
   A. document the conditions as they presently exist
   B. remove belts from the pulley arrangement
   C. loosen any setscrews that are securing the pulley to the motor's shaft
   D. disconnect all conductors from the motor's electrical terminals
8. When lifting a motor using one eyebolt, _____. (26.5)
   A. the angle created between the eyebolt shaft and vertical should not be less than 30°
   B. the angle created between the eyebolt shaft and vertical should not be less than 45°
   C. the driven device can be safely lifted along with the motor
   D. ensure the eyebolt connection to the motor is loose
9. Thread engagement refers to the _____. (26.6)
   A. length of a bolt used to mount a motor
   B. distance the securing nut is threaded onto a bolt
   C. diameter of the nut used to secure a motor in place
   D. number of bolt threads per inch
10. Before connecting a motor to its driven load, you should _____. (26.7)
    A. briefly power the motor to check its rotation
    B. jump out the motor's safety devices
    C. ensure that the covers on all electrical boxes are removed
    D. remove all guards from rotating components
11. *True or False?* Overtightening a belt reduces belt slippage and prolongs the life of the bearings. (26.8)
12. The portion of the *NEC* that covers motors, motor control circuits, motor circuit conductor protection, motor overloads, motor controllers, and motor grounding is _____. (26.9)
    A. *Article 230*
    B. *Article 320*
    C. *Article 340*
    D. *Article 430*

Plug-In

1. Which of the following sets of voltage and frequency deviations is most likely to be deemed acceptable? (26.2)
   A. +5% voltage deviation and a −6% frequency deviation
   B. −4% voltage deviation and a +5% frequency deviation
   C. −8% voltage deviation and a −6% frequency deviation
   D. +7% voltage deviation and a +7% frequency deviation
2. A motor that does not reach its design speed may result from a(n) _____. (26.3)
   A. overloaded motor
   B. blown fuse
   C. damaged or loose rotor
   D. insufficient motor airflow
3. A bearing temperature of more than 3.5°F per minute indicates that there might be a problem with the motor. Explain what steps you should take to address this condition, and list possible causes for this excessive rise in temperature. (26.10)

Refer to the following nameplate to complete questions 4 to 6.

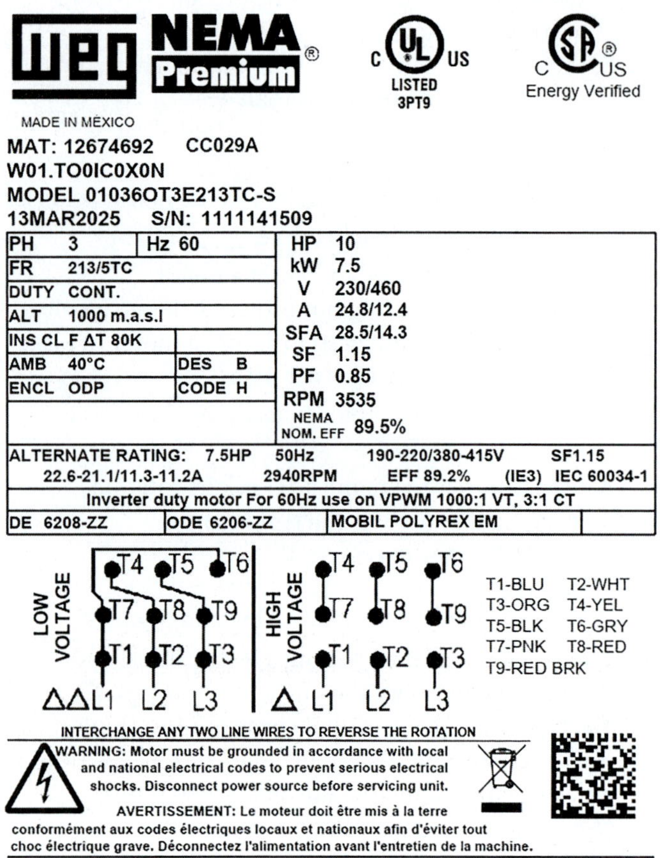

WEG

4. This nameplate is from a _____ motor. (26.2)
   A. direct current
   B. single-phase ac
   C. three-phase ac
   D. Cannot be determined from the nameplate.
5. When this two-pole motor is supplied by 230 V, what is the percentage slip? (26.2)
   A. 1.8%
   B. 2.0%
   C. 18.3%
   D. Cannot be determined from the nameplate.
6. The motor is a(n) _____ motor. (26.2)
   A. explosion-proof
   B. open drip-proof
   C. totally enclosed fan-cooled
   D. Cannot be determined from the nameplate.

Making the Connection

1. Explain why it is important to determine the cause for motor failure. (26.3)
2. Explain how improper pulley alignment can cause premature bearing failure. (26.7)
3. Explain why motor horsepower is a major factor in selecting a replacement motor. (26.2)

SECTION

8

Motor Functionality

Ingo Bartussek/Shutterstock.com

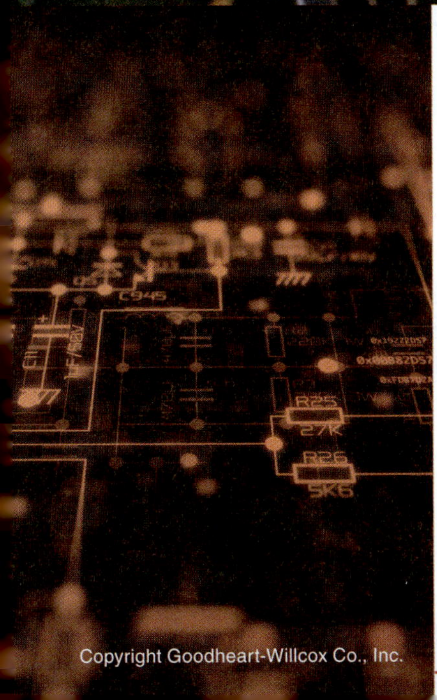

CHAPTER

27 DC Motors

KPixMining/Shutterstock.com

SPARKING DISCUSSION

When considering cost, a motor replacement should be selected based only on its first-cost standpoint. Do you agree or disagree with this statement? Provide details on why a first-cost standpoint might be or might not be the best option.

LEARNING OUTCOMES

After completing this chapter, you will be able to:

27.1 List the advantages of using dc motors over ac motors.
27.2 Describe how to use the right-hand rule for motors to determine the direction of force applied to a conductor.
27.3 Explain the function of commutators and brushes as they apply to dc motors.
27.4 Describe the differences among the four common types of series dc motor speed controls.
27.5 Explain the operation of a shunt dc motor speed control.
27.6 Explain the differences between cumulatively and differentially compounded dc motors.
27.7 Explain the construction of a permanent magnet dc motor.
27.8 Describe the two dc power sources of a separately excited dc motor.
27.9 Describe how the efficiency of a dc motor is determined.

TECHNICAL TERMS

compound dc motor
core losses
mechanical losses
permanent magnet dc motor
series dc motor
shunt dc motor
torque
windage losses
winding losses

Introduction

Direct current motors (also known as dc motors) are a popular choice for many applications, especially those that are portable and rely on battery power. Two desirable properties of dc motors are high starting torque and the ability to control motor speed accurately over the entire range of operating conditions. Although these properties are especially desirable in larger applications such as locomotives and elevators, dc motors are also very effective for smaller scale applications including consumer electronic devices, fans, blowers, and pumps.

After reading this chapter, you will have a better understanding of the relationship that exists among the direction of current flow, the direction of the stationary magnetic field, and the direction of the generated torque or twisting force as these factors relate to dc motors. You will also have a better understanding of the construction of a basic dc motor and its operation. Specific types of dc motors are examined, as well as the criteria that should be used when selecting the best type of dc motor to use for a particular application.

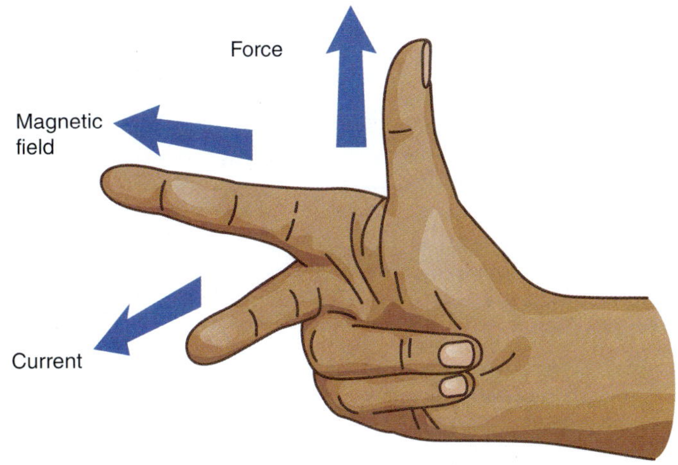

VectorMine/Shutterstock.com

Figure 27-1. The right-hand rule for motors indicates the direction of motion (force), magnetic field, and current. With your fingers positioned as shown here, your index finger indicates the direction of the magnetic field (from north to south). Your middle finger indicates the direction of current flow (from positive to negative). Your thumb indicates the force, or the direction in which the conductor is moving.

27.1 Overview of DC Motors

Generally speaking, direct current power is more efficient than alternating current power. This extends to many aspects of the electrical industry, including lighting circuits and equipment as well as motors. End users and those who specify or choose motors are increasing the use of dc as their desired power source when it is a viable option. This is mainly because of the higher efficiencies of dc motors and the vast improvements made to dc technologies over the past few decades. The most significant advances in dc power are in the areas of long-distance power transmission and ac-to-dc power conversion.

In addition, keep in mind that dc motors are valued for their high starting torque and ability to control motor speed accurately. These points are discussed further for several dc motor types.

27.2 Right-Hand Rule for Motors

Just as with any other motor, dc motors rely on the attraction and repulsion of magnetic forces to operate. The law of charges states that like charges and poles repel each other, while opposite charges and poles attract each other.

When a current-carrying conductor is positioned within a magnetic field, a directional force is created. The right-hand rule for motors provides an easy-to-understand and easy-to-visualize relationship that exists among the directions of current flow, the generated magnetic field, and the force that is created. To use the right-hand rule for motors, the thumb, forefinger, and middle finger of the right hand are positioned so they are at 90° angles to each other. The index finger represents the direction of the magnetic field, the middle finger represents the direction of current flow, and the thumb represents the direction of the conductor's motion, **Figure 27-1**.

The right-hand rule for motors can also be used to determine the direction of current flow when a conductor is passed through a magnetic field. In this case, the direction of the magnetic field and the direction of the force (movement of conductor) are known. Consider the situation where two magnets are positioned as illustrated in **Figure 27-2**. The direction of the magnetic field is from left to right. A conductor cuts through these lines of magnetic flux from the bottom to the top, so the direction of force is upward. Current flow is at a 90° angle to both of these.

Shedding Light

The right-hand rule for motors applies when the electron flow convention for direct current direction (negative to positive) is in use, as it is in this text. When conventional current flow (positive to negative) is being used, the right-hand rule for motors is replaced by Fleming's left-hand rule for motors. When using either the right-hand or left-hand rules to determine electrical behavior, be sure to use the appropriate current flow direction.

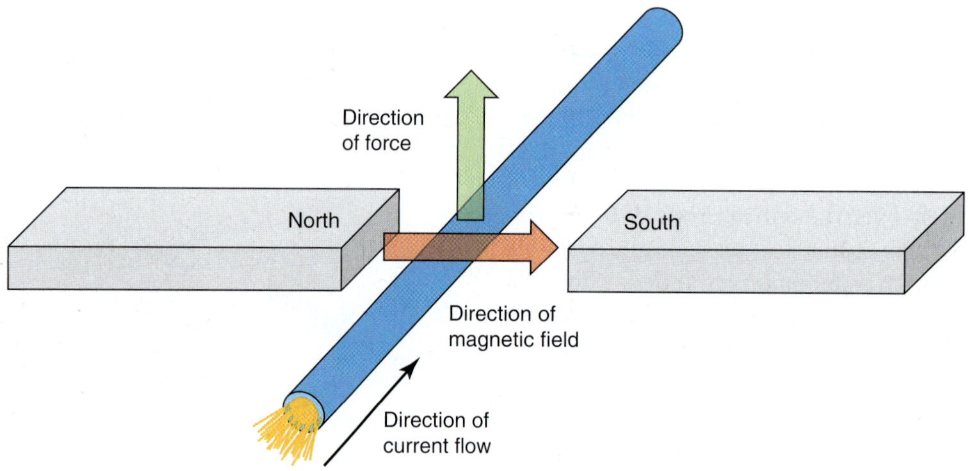

Goodheart-Willcox Publisher

Figure 27-2. Direction of current flow can be determined if a conductor is passed through a magnetic field; current flow is at a 90° angle to both the direction of force and the direction of the magnetic field.

27.3 Basic DC Motor Construction and Operation

The most basic dc motor is made up of a stator, a rotor, a commutator, brushes, and a shaft, **Figure 27-3**. As with any motor, the stator is the stationary portion of the motor, while the rotor refers to the portion of the motor that is connected directly to the motor's rotating shaft. **Figure 27-4** provides a skeletal view of the dc motor's basic components.

The coils of wire that rotate within the motor's magnetic field are collectively referred to as the *armature*. DC motors operate on the basic principle of a rotating armature within a magnetic field. The commutator and the brushes are the motor components that are responsible for keeping the motor turning in one direction. Motor rotation in a single direction is required if the motor is to work.

The magnetic field is produced by permanent magnets or windings installed in the stator. These windings in the stator are called *field windings*. A dc motor might have a series field winding, a shunt (parallel) field winding, or both a series field winding and a shunt field winding. Common types of dc motors are shown in **Figure 27-5**.

In our discussion of dc motors, a common labeling strategy will be used to identify the connecting terminals on the armature, series field winding, and the shunt field winding. On the armature, the terminal connections are identified as A_1 and A_2. The series field winding terminal connections are identified as S_1 and S_2, while the shunt field winding terminals are identified as F_1 and F_2, **Figure 27-6**.

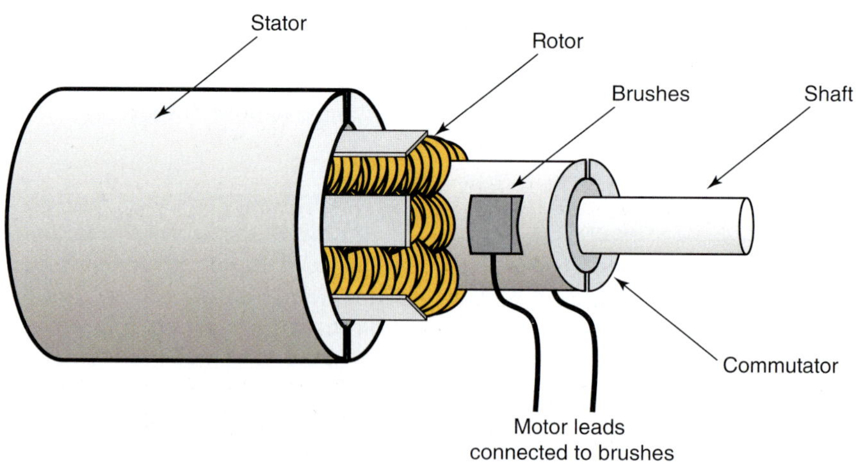

Goodheart-Willcox Publisher

Figure 27-3. A cutaway view of a simple dc motor.

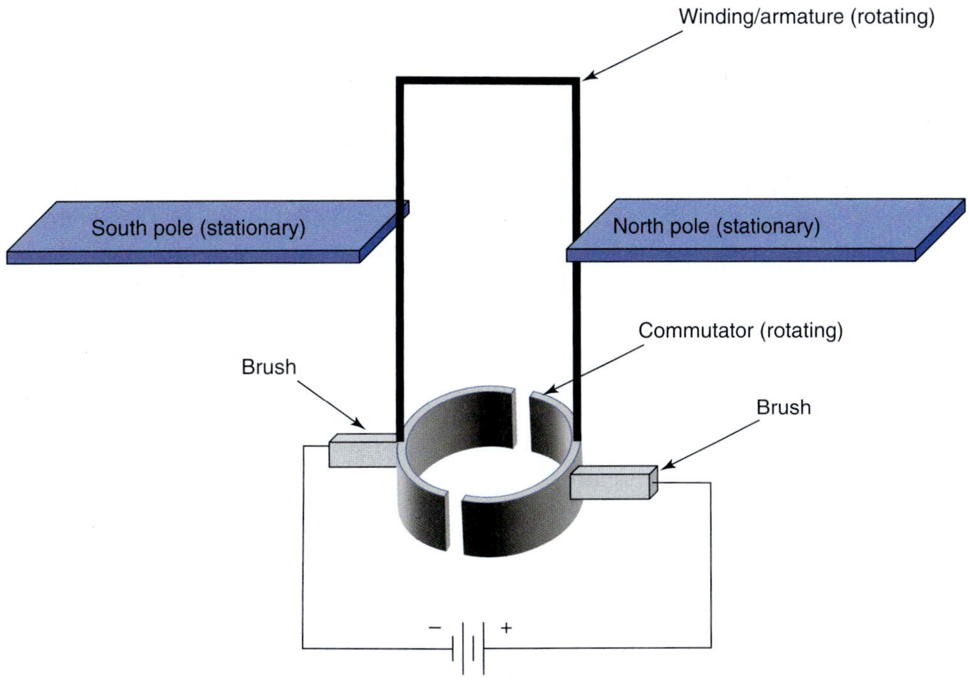

Goodheart-Willcox Publisher

Figure 27-4. In this representation of a dc motor, the stationary magnets (both the south and north pole) are part of the stator. The motor windings and the commutator, along with the shaft, make up the dc motor's rotor.

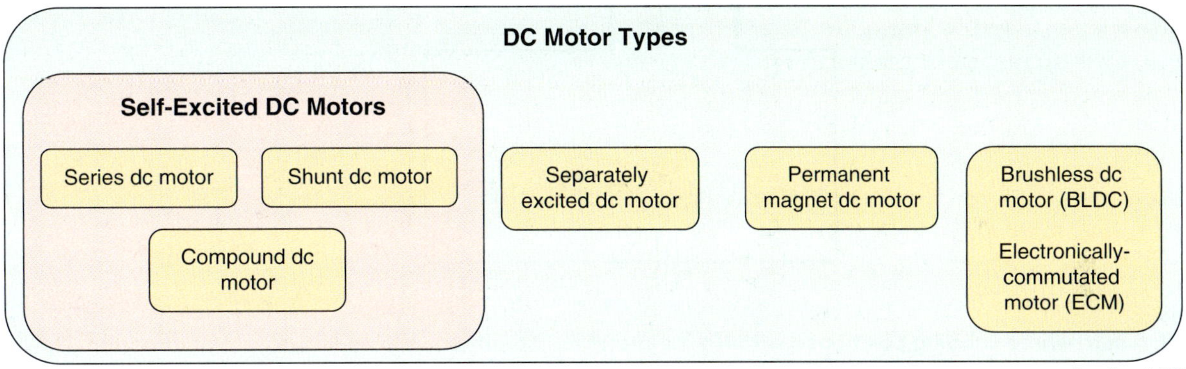

Goodheart-Willcox Publisher

Figure 27-5. DC motors are categorized into these six types. Brushless dc motors are discussed in Chapter 30, *Electronically Commutated Motors*.

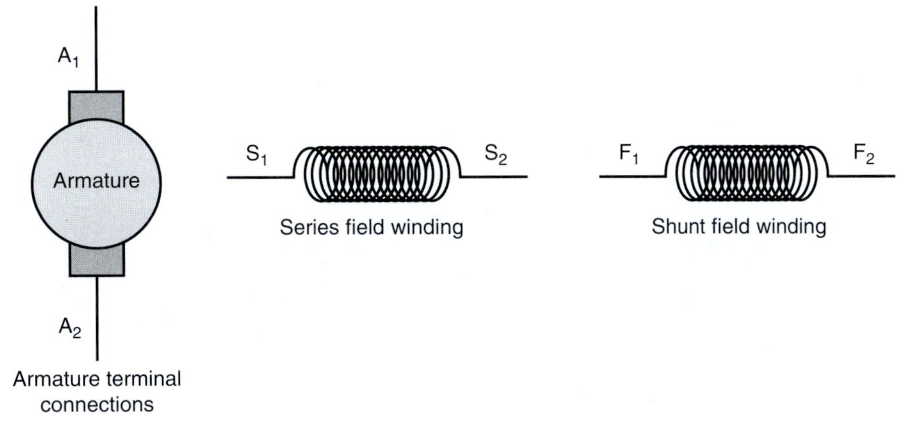

Goodheart-Willcox Publisher

Figure 27-6. The armature, series winding, and shunt winding have unique markings. The armature leads are labeled A_1 and A_2, the series field winding leads are labeled S_1 and S_2, while the shunt field winding leads are labeled F_1 and F_2.

Commutators and Brushes

To keep a dc motor rotating in only one direction, a commutator is used. The commutator is a type of rotary switch that continually reverses the polarity of the voltage being supplied to the motor's armature. This constant polarity reversal produces constant torque in the same direction, allowing the motor to turn in one direction when the motor is energized. Since the polarity of an alternating current power supply is constantly changing, commutators are not found on ac motors.

The commutator is an integral part of a dc motor. Consider the circuit shown in **Figure 27-7**. This diagram shows a winding positioned within a magnetic field oriented in the north-to-south direction. This winding is hard-wired or permanently connected to the power supply terminals. Current flows through the circuit in a clockwise direction. Using the right-hand rule for motors, it can be determined that the direction of force on the left side, closest to the south pole of the magnet, is directed downward. The direction of force on the right side, closest to the north pole of the magnet, is directed upward. With the force on the left pushing down and the force on the right pushing up, the winding will rotate 180° in a counterclockwise direction, **Figure 27-8**. After making one-half of a full rotation, or

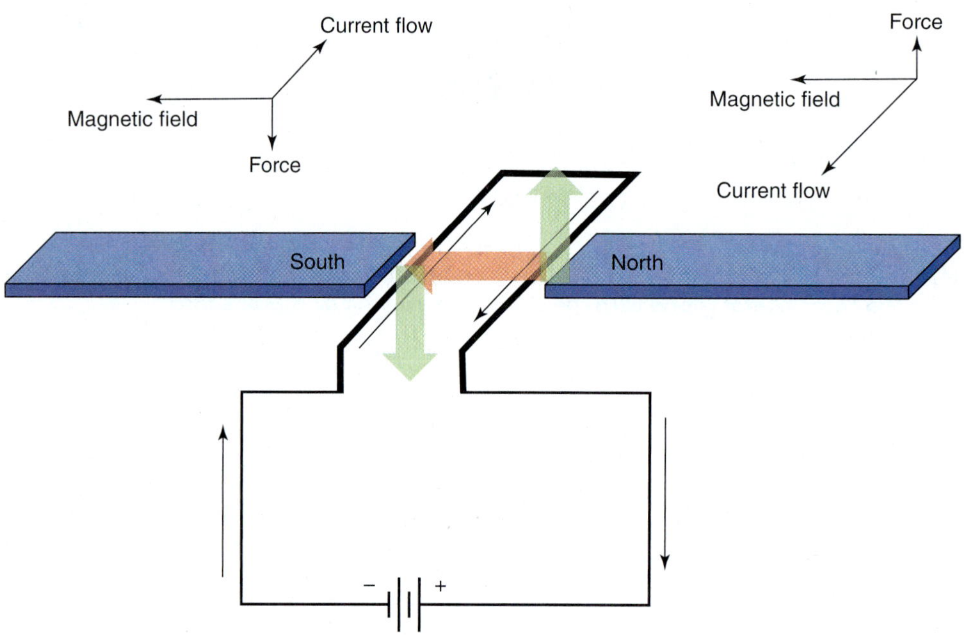

Goodheart-Willcox Publisher

Figure 27-7. A hard-wired winding positioned within a magnetic field oriented in the north-to-south direction. The black arrows indicate the direction of current flow through this circuit. The directions of the forces acting on the left and right portions of the coil are indicated by the right-hand rule for motors.

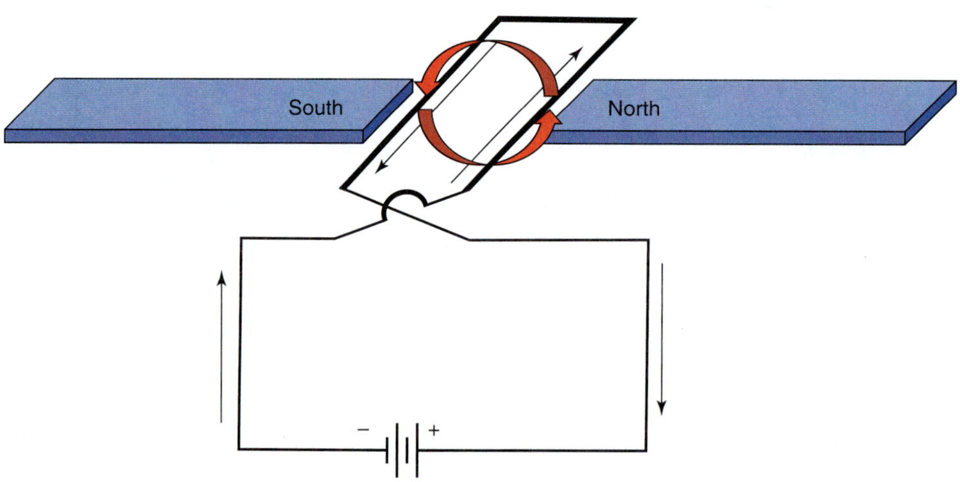

Goodheart-Willcox Publisher

Figure 27-8. With the forces acting in the indicated directions, the winding rotates 180° in the counterclockwise direction.

180°, the direction of rotation reverses, as demonstrated once again by the right-hand rule for motors. On the side of the winding closest to the south pole, the force is in the upward direction, while on the north pole side of the winding, the force is in the downward direction, **Figure 27-9**. As a result, the winding will rotate in the clockwise direction, **Figure 27-10**. Without a commutator, therefore, the armature would continually rotate 180° counterclockwise, then 180° clockwise, rendering the motor, for the most part, useless.

Commutators are connected to the armature or windings of the motor, **Figure 27-11**. A simple commutator has two sections and is referred to as a split-ring commutator. The segments of the split ring are attached to the opposite ends of the coil or winding. For armatures with multiple coils, the commutator will have multiple segments, two for each coil or winding. For example, an armature with two coils will utilize a four-segment commutator, **Figure 27-12**.

Although the commutator is permanently connected to the motor windings, it is not permanently connected to the motor's

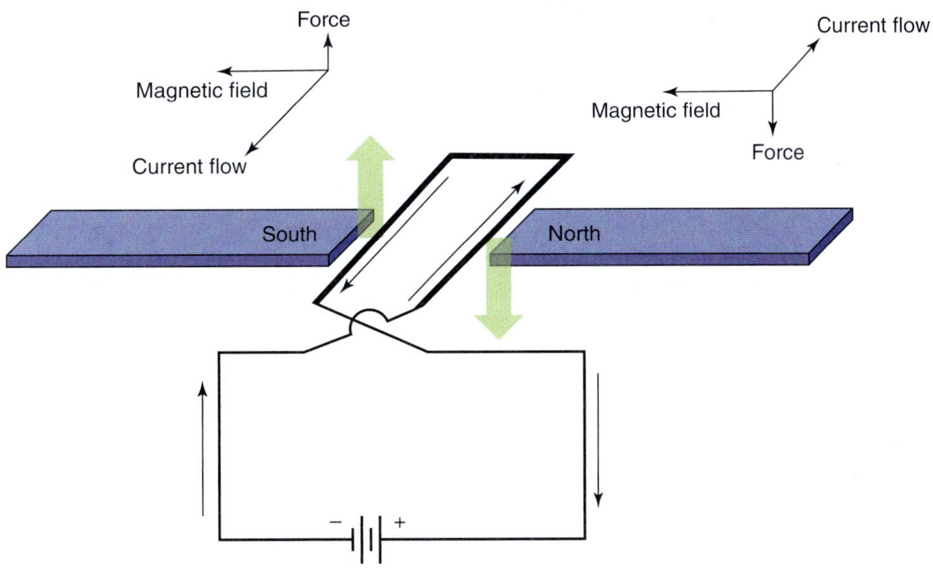

Goodheart-Willcox Publisher

Figure 27-9. Using the right-hand rule for motors, the directions of the generated forces are determined. Since the current flow directions are different at the left and right sides of the arrangement, the generated forces are in different directions as well.

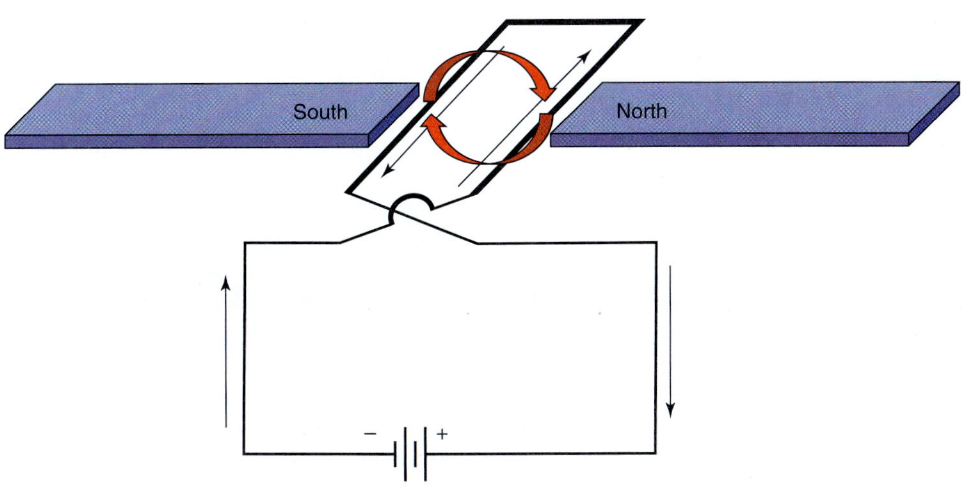

Goodheart-Willcox Publisher

Figure 27-10. With the forces acting in the indicated directions, the winding now rotates 180° in the clockwise direction.

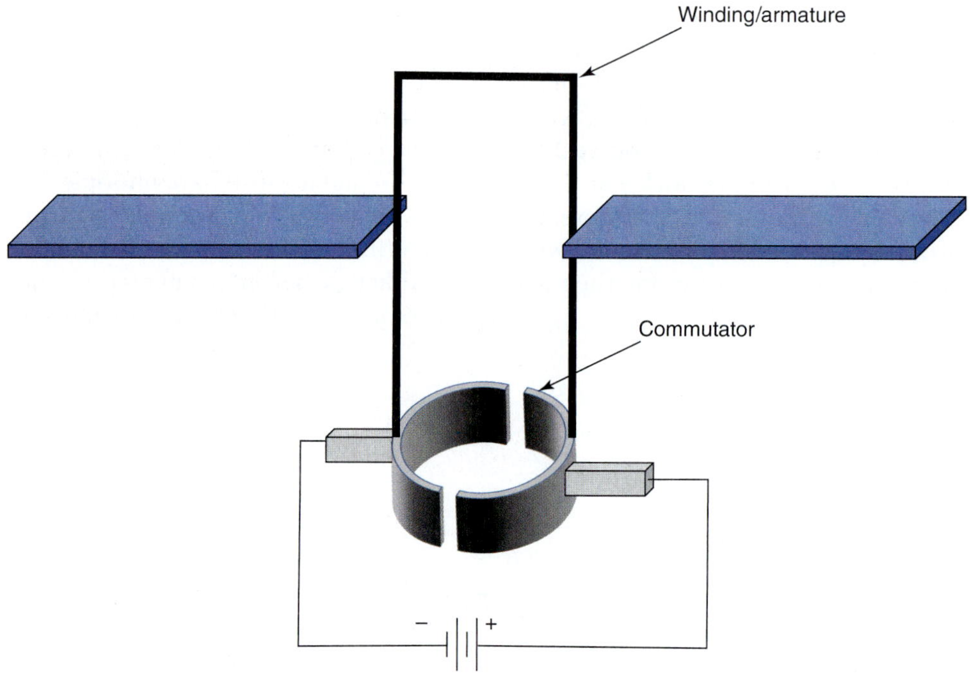

Goodheart-Willcox Publisher

Figure 27-11. Commutator permanently connected to the motor winding or armature. This simple commutator has two sections and is known as a split-ring commutator.

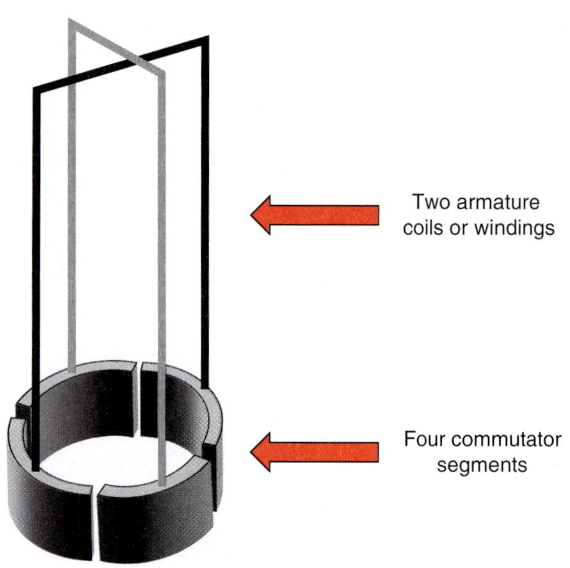

Goodheart-Willcox Publisher

Figure 27-12. The number of commutator segments is twice the number of armature windings or coils.

power supply. Spring-loaded connection points, called *brushes*, are positioned on each side of the commutator, **Figure 27-13**. These brushes make contact with the commutator as it rotates. By using the commutator/brush arrangement, the direction of current flow will always be in the same direction and the armature will continue to rotate in only one direction.

To illustrate the operation of a commutator and brushes, we will examine the end-view of the motor winding/commutator arrangement shown in **Figure 27-14A**. The ends of the commutator have been color-coded for clarification. The left (red) side of the commutator is connected to the negative side of the power source, while the right (blue) side is connected to the positive side of the power source. As previously discussed, the left side of the armature will be pushed down, and the right side will be pushed up, causing the armature to rotate in the counterclockwise direction, **Figure 27-14B**. As the armature continues to rotate, the red side remains in contact with the left brush and the negative side of the power source, while the right side remains in contact with the brush on the right side, **Figure 27-14C**. Once the armature rotates past the vertical position, the brushes now make contact with the opposite portion of the armature, **Figure 27-14D**. Now, the blue side of the commutator is connected to the negative side of the power source, while the red side is connected to the positive side of the power source. It is this constant switching of polarity that allows the motor to keep running in only one direction.

Shedding Light

One category of dc motors does not use brushes and a commutator. Brushless dc motors (BLDC), also called *electronically commutated motors (ECM)*, replace the brushes and commutator with an electronic circuit. The electronic circuit modifies the current supplied to the motor to duplicate the polarity-changing behavior of the brushes and commutator. These motors are discussed in Chapter 30, *Electronically Commutated Motors*.

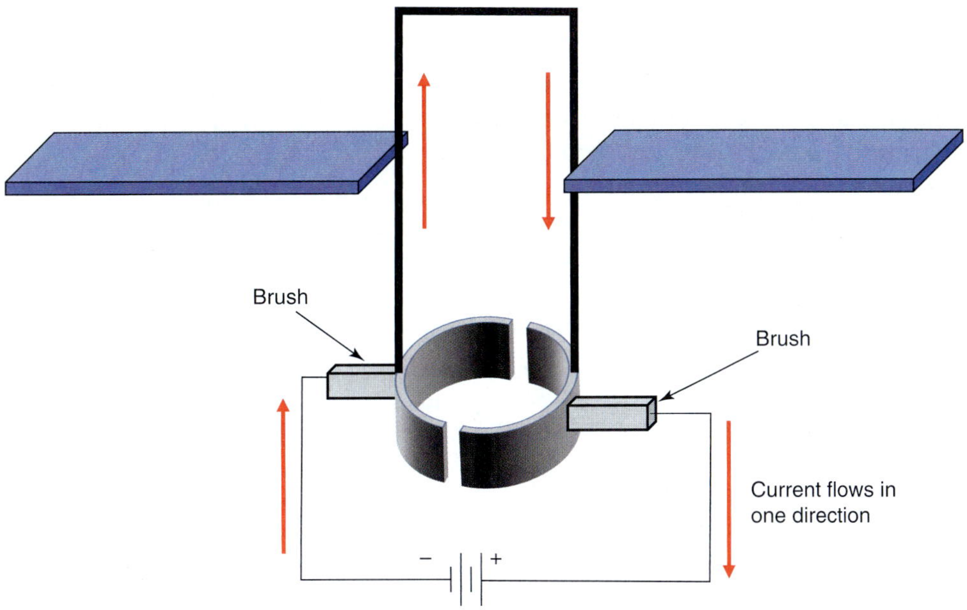

Figure 27-13. Spring-loaded brushes make contact with the commutator. As a result, the current flow will be in the same direction.

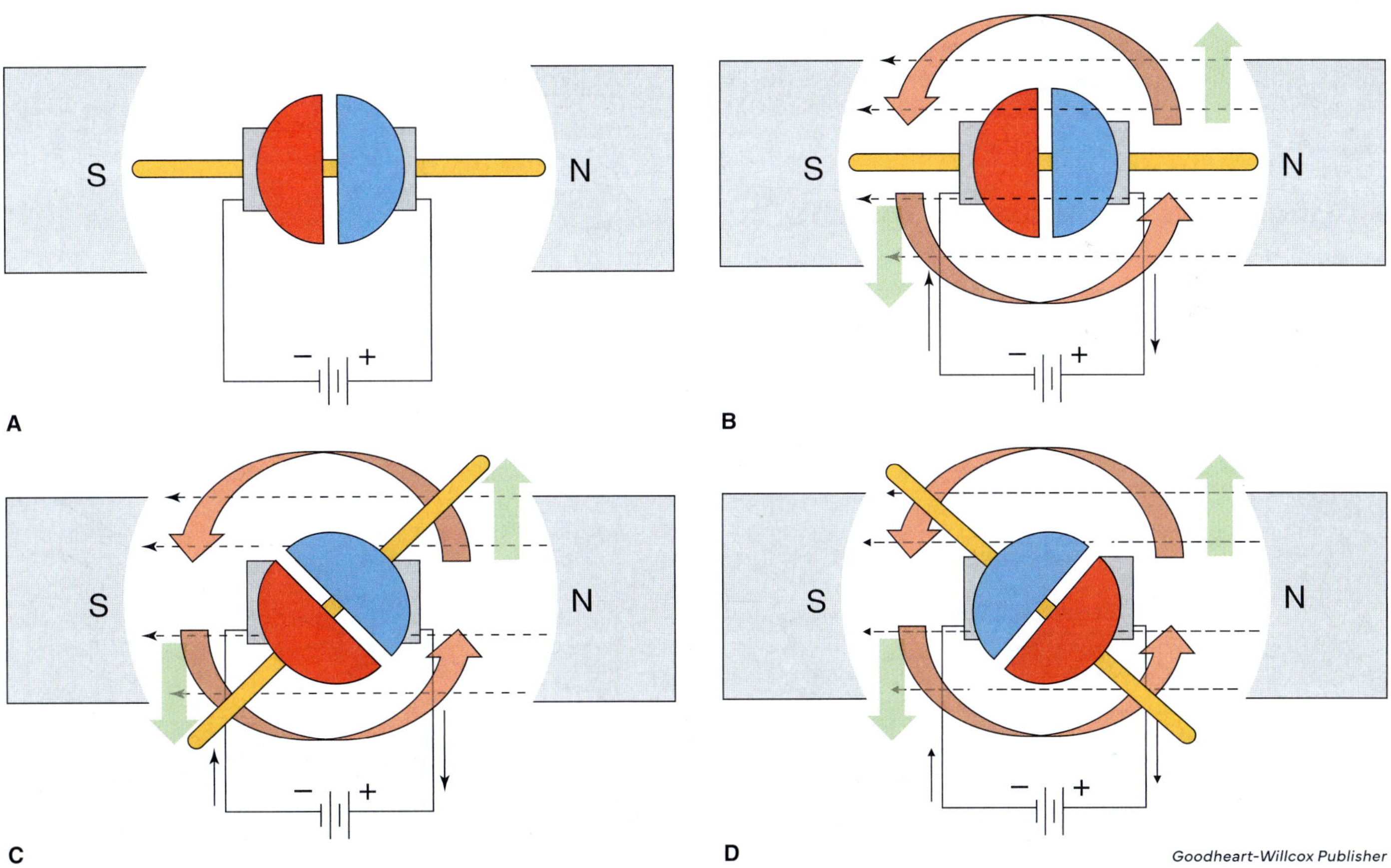

Figure 27-14. As current flows in the same direction, the armature will rotate in one direction. A—End view of a commutator and brush arrangement. The left side of the commutator is making contact with the negative side of the power supply; the right side is making contact with the positive side of the power supply. B—Armature turning in the counterclockwise direction. C—Before reaching the vertical position, the left side of the armature is still making contact with the negative side of the power supply. D—After passing the vertical position, the polarity of the power supplied to the armature reverses.

As the commutator rotates, the brushes will periodically pass over the commutator gaps and no voltage is applied to the winding. Even though no voltage is being supplied to the winding, momentum propels the armature forward until the brushes, once again, make contact with the commutator segments. On motors with multiple windings, force is still being generated in other windings that continues to rotate the armature.

Direction of Motor Rotation

Even when an imbalance in magnetic fields is present, it is the interaction between these magnetic fields that determines the direction in which a motor will rotate. When a conductor moves within a magnetic field, electrons within the conductor move. The direction in which the electrons move is determined by the direction of the magnetic field and the motion path the conductor takes within the magnetic field. A practical extension of this concept is the right-hand rule for motors. By using this rule, the direction of the generated ***torque***, which is the twisting or rotating thrust or force, can be determined. The direction of the generated torque determines the motor's directional rotation.

27.4 Series DC Motors

As the name implies, ***series dc motors*** are wired with the field winding in series with the armature, **Figure 27-15**. In this configuration, the current flowing through the armature and field winding will always be the same.

Applications of Series DC Motors

Series dc motors are generally used where high starting torque is required and when there is very little chance of operating under no-load conditions. Because of the high starting torque characteristic of the series dc motor, it is a popular choice for use in the following applications:

- Air compressors
- Elevators
- Power tools
- Home appliances

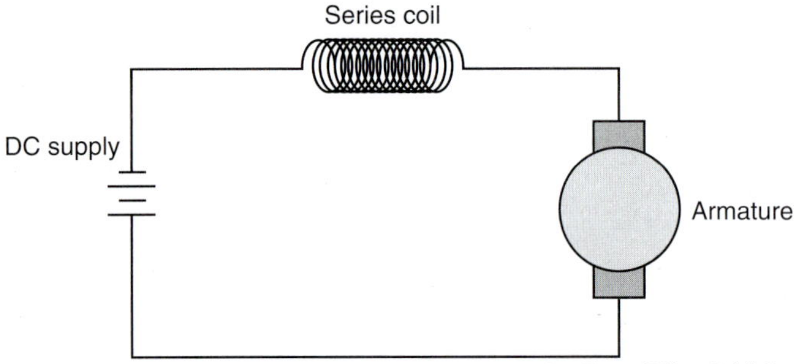

Goodheart-Willcox Publisher

Figure 27-15. In a series dc motor, the winding is wired in series with the armature. The current flowing through both the winding and the armature will be the same.

Series DC Motor Speed Control

The speed of a series dc motor is inversely related to the load on the motor. When the load on the motor is low, the motor speed is high. Similarly, when the load on the motor is high, the motor speed decreases. A decrease in motor speed causes a reduction in the counter-electromotive force (cemf). With lower cemf, the current flow through the motor (armature and field winding) increases. An increase in current leads to an increase in the amount of torque produced. **Figure 27-16** summarizes the functionality of a series dc motor with respect to the load on the motor. As the motor load increases, torque and current increase, while motor speed and cemf decrease. When the load on a series dc motor is low, the current and torque are low, while the speed and cemf are high.

Series dc motors should not be operated when there is no load on them. When the motor load is low, the higher motor speed can cause an overheating condition and physical damage to the motor. To help reduce the possibility of having a series dc motor operate under no-load conditions, it is strongly recommended that these motors only be used for direct-drive applications. Using series dc motors in conjunction with belts, linkages, chains, or other non-direct connection methods increases the possibility of motor damage in the event the drive-driven connection fails.

There are four common methods used to control the speed of a series dc motor:

- Armature resistance control
- Armature diverter control
- Tapped field control
- Field diverter control

Armature Resistance Control

The armature resistance motor control method is the most common method used to control the speed of series dc

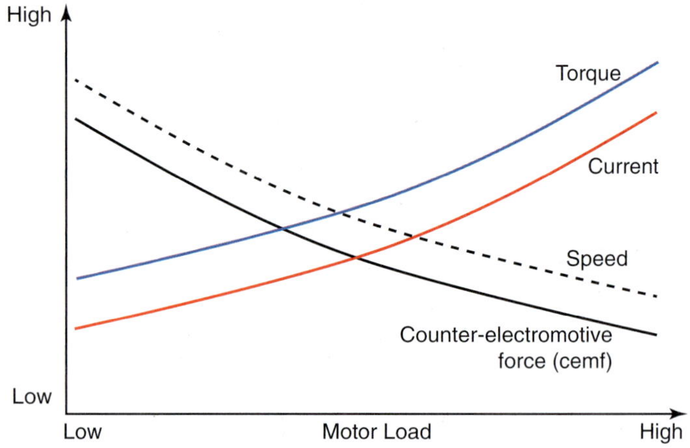

Goodheart-Willcox Publisher

Figure 27-16. Graphical representation of the functionality of a series dc motor with respect to the load on the motor. As the load on the motor increases, the motor's torque and current increase, while the motor's speed and cemf decrease.

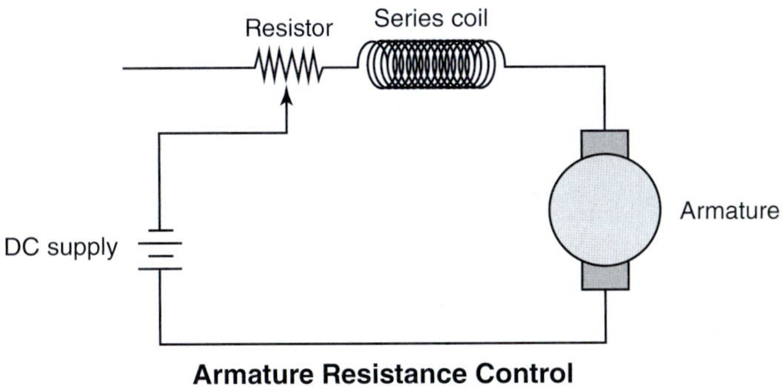

Goodheart-Willcox Publisher

Figure 27-17. Wiring diagram of an armature resistance speed control circuit for a series dc motor.

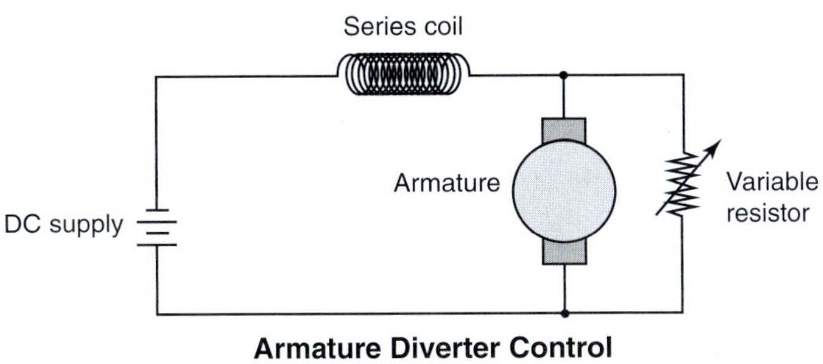

Goodheart-Willcox Publisher

Figure 27-18. Wiring diagram of an armature diverter speed control circuit for a series dc motor.

motors. The armature resistance control strategy uses a resistor that is wired in series with the series field winding, **Figure 27-17.** This control strategy is used when motor speeds below the motor's normal design speed are desired. A switch is used to control the amount of additional resistance that is wired in series with the series field winding. If no additional resistance is placed in series with the series field winding, the motor will operate at a speed that is close to the motor's normal design speed. As the amount of added series resistance increases, the voltage supplied to the armature and the cemf decreases. With motor speed being directly proportional to both the voltage and cemf, a reduction in these two factors causes the speed of the motor to drop below the motor's normal design speed.

Armature Diverter Control

The armature diverter speed control strategy uses a variable resistor wired in parallel with the armature, **Figure 27-18.** This resistor diverts current from the armature, lowering the armature current. When the diverter resistance is highest and acts as an open switch, the diverted current is close to zero, and all of the current is directed through the armature. This is when the armature current is at its maximum.

Assuming the load on the motor is constant, a reduction in the armature current requires that the flux increases to maintain the load torque at a constant level. Since speed and flux are inversely related to each other, an increase in flux results in a reduction of motor speed. The higher the resistance of the diverter resistor, the lower the flux and the higher the motor speed. The interactions among diverter resistance, armature current, flux, torque, and motor speed for a constant load application are shown in **Figure 27-19.** Using the armature diverter speed control strategy allows the motor to operate at speeds up to and including the normal design speed of the motor. This strategy does not allow for the motor to operate at speeds that exceed the normal design speed.

Tapped Field Control

The tapped field motor speed control strategy uses a tap to bypass a portion of the series field winding, **Figure 27-20.** This strategy is used when motor speeds in excess of the motor's normal design speed are desired. When no portion of the series field winding is bypassed, the motor operates at its normal design speed. As the number of turns in the series field winding being bypassed increases, the amount of flux being produced decreases. Since motor speed and flux are inversely proportional to each other, this reduction in flux results in an increase in motor speed.

Field Diverter Control

The field diverter speed control strategy uses a variable resistor that is wired in parallel with the series field winding,

Effects of Diverter Resistance

Diverter Resistance	Armature Current	Flux	Torque	Motor Speed
High	High	Low	Constant	High
Low	Low	High	Constant	Low

Goodheart-Willcox Publisher

Figure 27-19. Relationship among diverter resistance, armature current, flux, torque, and motor speed.

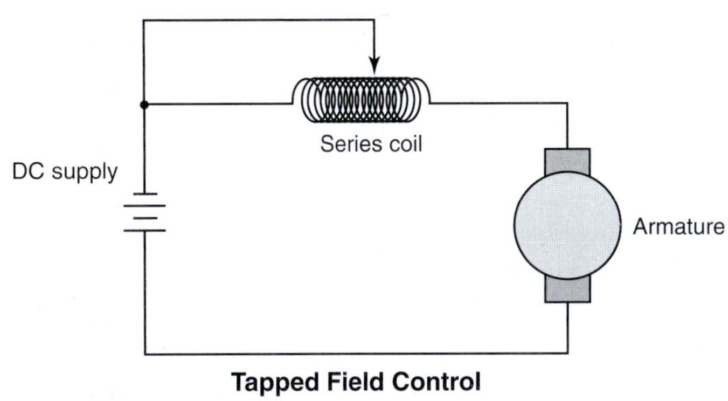

Goodheart-Willcox Publisher

Figure 27-20. Wiring diagram of a tapped field control circuit for a series dc motor.

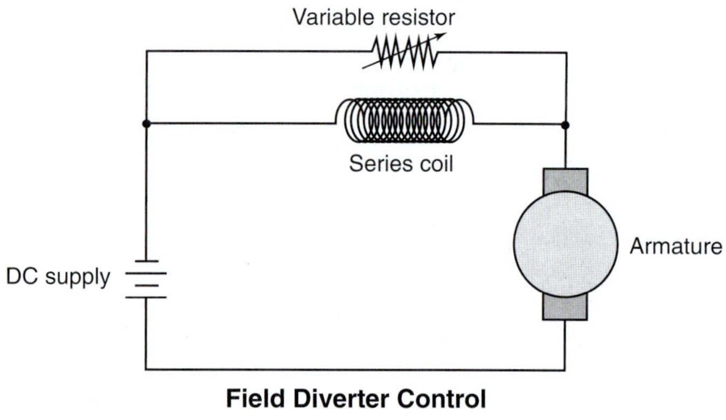

Goodheart-Willcox Publisher

Figure 27-21. Wiring diagram of a field diverter control circuit for a series dc motor.

Figure 27-21. This control strategy is used when the motor needs to operate at speeds that are greater than the motor's normal design speed. When the resistance of the parallel-connected resistor is infinite, the resistor acts like an open switch, so all circuit current flows through the series field winding. In this condition, the motor operates at its normal design speed. As the resistance of the field diverter resistor drops, current is diverted from the series field winding to flow through the diverter resistor path, so less current flows through the series field winding. With less current flowing through the series field winding, the series field strength is weakened. A reduction in current flow through the series field winding causes a reduction in the amount of flux generated. Since the motor speed is inversely related to the amount of flux produced, this flux reduction causes the speed of the motor to increase.

27.5 Shunt DC Motors

Shunt dc motors are wired with the field winding in parallel with the armature, **Figure 27-22**. In this configuration, different amounts of current flow through the armature and field winding. The parallel winding is energized when the motor is energized, even when the commutator gaps are aligned with the brushes.

Shunt dc motors are generally used when relatively constant speed is needed. Some common applications for shunt dc motors include the following:

- Fabrication machines such as lathes
- Centrifugal pumps
- Conveyor belts
- Blowers and fans

Shunt DC Motor Speed Control

There are two common methods used to control the speed of a shunt dc motor. These methods are flux control and armature resistance control.

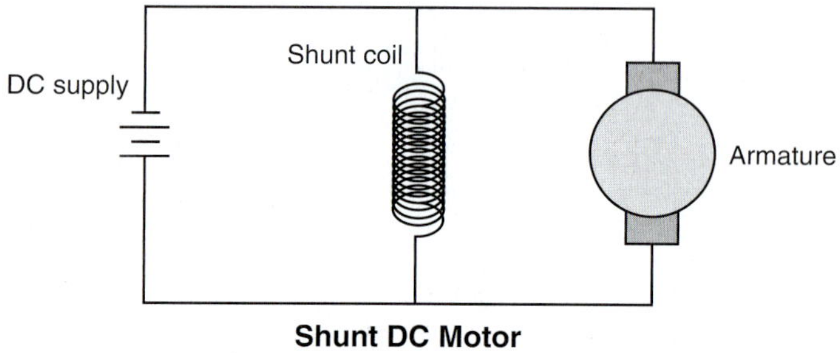

Goodheart-Willcox Publisher

Figure 27-22. In a shunt dc motor, the winding is wired in parallel with the armature. The amount of current flow through the winding and the armature will be different.

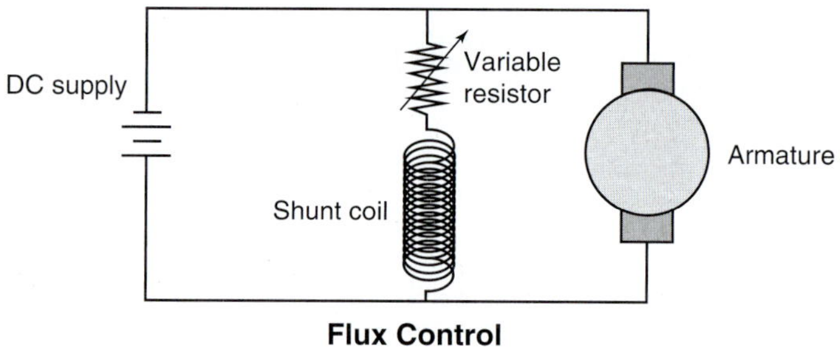

Goodheart-Willcox Publisher

Figure 27-23. The flux speed control method uses a variable resistor wired in series with the shunt field winding.

Flux Control

The flux speed control strategy uses a variable resistor wired in series with the shunt field winding and in parallel with the motor's armature, **Figure 27-23**. When the value of the variable resistor is close to zero, the motor operates at a speed close to its normal design speed. As the value of the variable resistor increases, the shunt field current decreases, and the shunt field flux decreases as well. This decrease in shunt field flux results in an increase in motor speed.

The advantages of the flux control strategy are that this method is simple and inexpensive. The drawbacks to this motor speed strategy include the following:

- Wasted power due to the series resistance
- Reduced motor efficiency
- Inability to maintain constant motor speed when motor load changes

Armature Resistance Control

The armature resistance speed control strategy uses a variable resistor wired in series with the armature and in parallel with the shunt field winding, **Figure 27-24**. This control strategy allows for motor speeds equal to or lower than the motor's normal design speed. By changing the

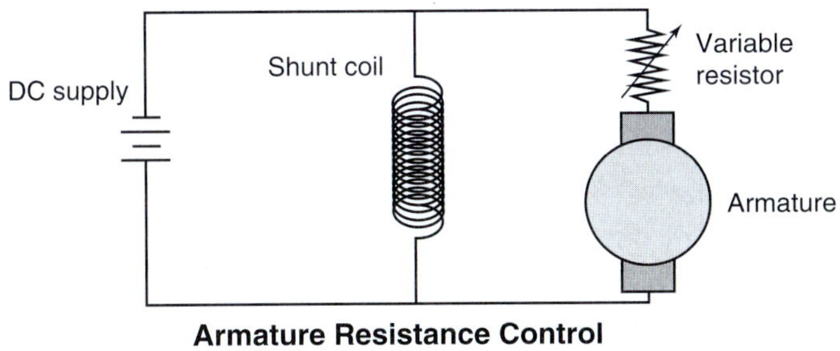

Goodheart-Willcox Publisher

Figure 27-24. The armature resistance speed control strategy uses a variable resistor wired in series with the armature and in parallel with the shunt field winding.

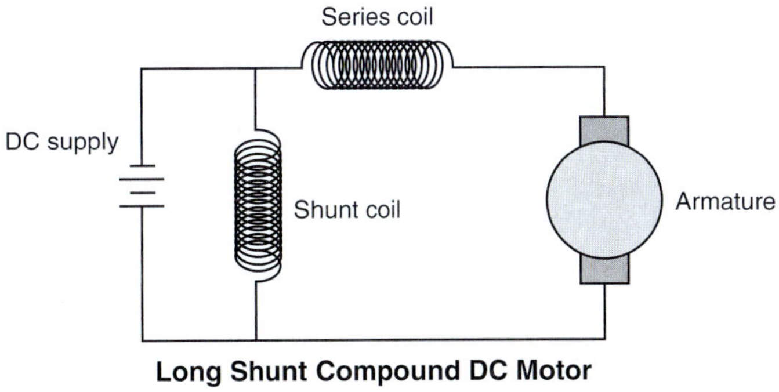

Goodheart-Willcox Publisher

Figure 27-25. In a long shunt compound dc motor, the shunt winding is connected in parallel with both the series winding and the armature.

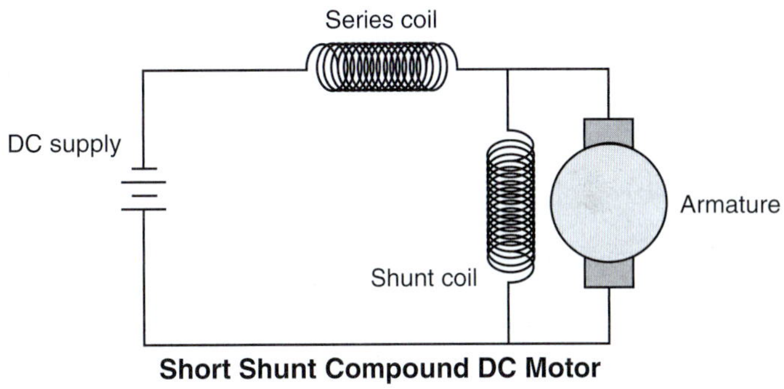

Goodheart-Willcox Publisher

Figure 27-26. In a short shunt compound dc motor, the shunt winding is wired in parallel with only the armature winding.

value of the series resistance, the measured voltage across the armature changes. When the series resistance is close to zero, the voltage across the armature is at its highest level, and the motor's operating speed will be very close to the motor's design speed. Similarly, the voltage across the armature is lowest when the series resistance is at its highest value. As the resistance of the series resistor increases, the armature voltage and the cemf decrease, and the motor speed decreases.

27.6 Compound DC Motors

A compound dc motor is a cross between the series dc motor and the shunt dc motor. Unlike the series dc motor and the shunt dc motor, the compound dc motor has two field windings. One winding is wired in series with the armature coil and one is wired in parallel with it. By incorporating properties of the series and shunt varieties of dc motors, the ***compound dc motor*** can operate with high torque at low speeds, a major benefit of the series dc motor, and can be precisely speed-controlled, a major benefit of the shunt dc motor. Compound dc motors are popular choices for applications where high starting torque and relatively constant speed are desired, such as elevators and conveyor belts.

Long and Short Shunt

There are two common wiring configurations for compound dc motors: the long shunt and the short shunt. The long shunt compound dc motor is the most popular. In a long shunt compound dc motor, the shunt winding is connected in parallel with both the series winding and the armature, **Figure 27-25**. This configuration is what allows for superior motor speed control.

In a short shunt compound dc motor, the shunt winding is wired in parallel with only the armature winding, **Figure 27-26**. When wired in this configuration, the entire circuit current flows through the series winding. After passing through the series winding, the current splits to flow through the shunt winding and the armature. The short shunt configuration allows for high starting torque and is best suited for constant speed applications.

Cumulative and Differential Compounding

In addition to the long and short shunt wiring configurations, the series and shunt windings of compound dc motors can be connected differently with respect to their polarities. When current passes through the series and shunt windings in the same direction, the compound motor is said to be cumulatively compounding, **Figure 27-27**. In this configuration, the shunt field flux produced by the shunt winding assists or enhances the effect of the main field flux produced by the series winding.

When the current passed through the series and shunt windings is in opposite directions, the compound motor is said to be differentially compounding, **Figure 27-28**. In this configuration, the flux created by the shunt field winding causes a reduction in the total amount of produced flux. As a result, differentially compounded motors are much less common than cumulatively compounded motors.

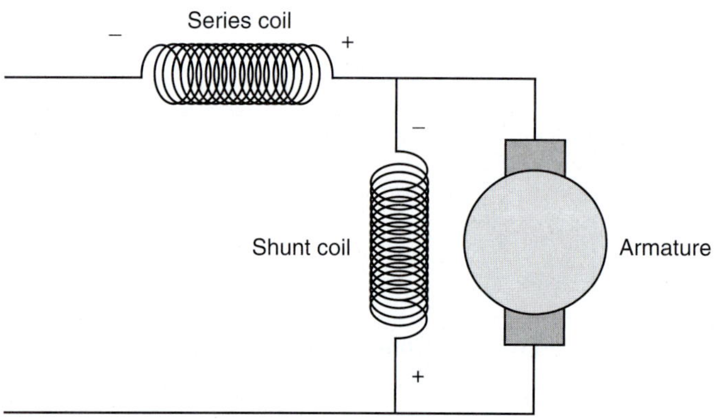

Goodheart-Willcox Publisher

Figure 27-27. A cumulatively compounded dc motor is wired so current passes through the series and shunt windings in the same direction. The field produced by the shunt winding enhances or adds to the main field produced by the series winding.

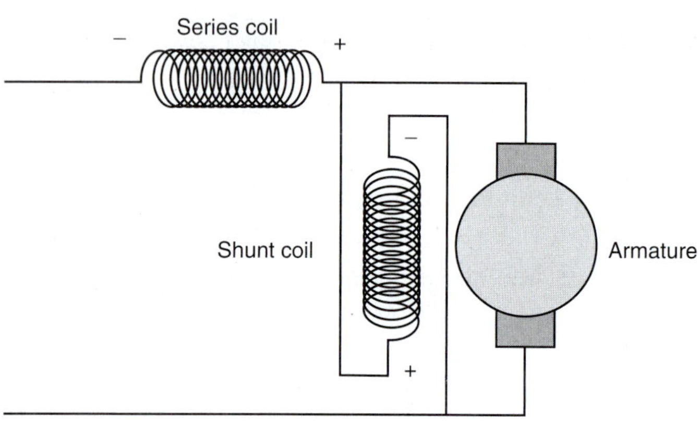

Goodheart-Willcox Publisher

Figure 27-28. A differentially compounded dc motor is wired so current passes through the series and shunt windings in opposite directions. The field produced by the shunt winding offsets some of the main field produced by the series winding, resulting in a weaker net field.

The cumulative compounding connection provides higher motor starting torque but less speed regulation. The differential compounding connection offers more precise motor speed control and is typically used for applications that require constant motor speed.

Applications of Compound DC Motors

The compound dc motor's ability to provide high starting torque and precise speed regulation makes these motors a great choice for many larger applications. For example, compound dc motors perform well in elevators due to their extreme load changes. With their high starting torque and good speed control, compound dc motors are ideal in many applications, including the following:

- Elevators, due to their ability to perform well when there are extreme load changes
- Manufacturing shears and punches, due to their high starting torque and good speed control
- Steel rolling, due to their high starting torque and ability to function well with heavy-duty loads
- Cutting machinery, due to their ability to drive heavy loads
- Mixing equipment, due to their high starting torque

27.7 Permanent Magnet DC Motors

Permanent magnet dc motors reduce or, in some cases, completely eliminate the need for external currents to produce magnetic fields. Permanent magnets are used to create the field flux that interacts with the rotor.

Permanent magnet dc motors are typically physically smaller, lighter, and more efficient than other dc motors with similar operating characteristics. Permanent magnet dc motors are a popular choice for applications that require smaller motors, typically less than 2 or 3 hp.

Permanent magnet dc motors operate with high starting torque and have the ability to control speed well. These motors are popular choices for many lighter-duty applications such as the following:

- Clothes-washing machines
- Windshield wipers
- Automobile starters
- Personal computer disc drives
- Toys
- Wheelchairs
- Blowers in heating and air-conditioning equipment

27.8 Separately Excited DC Motors

Series dc motors, shunt dc motors, and compound dc motors are categorized as self-excited dc motors. In a self-excited dc motor, the source that supplies power to the armature coils also supplies power to the field windings.

Separately excited dc motors are supplied by two separate dc power sources. One source supplies the armature coils, and a second source supplies the field windings, **Figure 27-29**.

The field winding in a separately excited dc motor is similar to the field winding in a shunt dc motor. Speed control methods are also similar to those for a shunt dc motor.

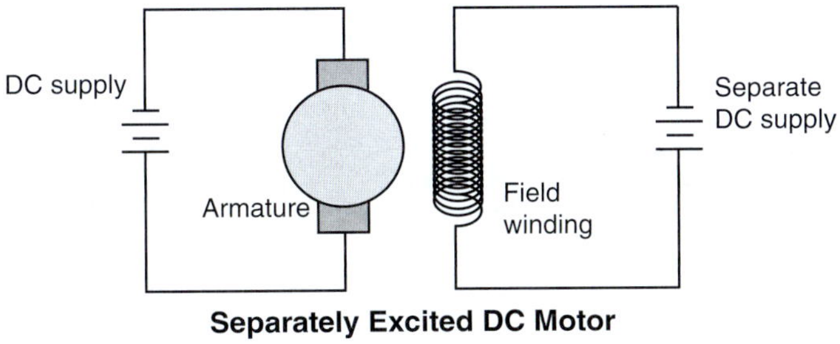

Goodheart-Willcox Publisher

Figure 27-29. A separately excited dc motor uses a second dc power source to supply the field windings.

Supplying the field winding with an independent power source provides precise control of the current. The current in the field windings can remain constant while the current in the armature coils varies.

27.9 Efficiency Losses in DC Motors

No machine is 100% efficient, and dc motors are no exception. Common sources of efficiency reductions in dc motors include core losses, winding losses, and mechanical losses. Cumulatively, these losses amount to wasted energy and should be as low as possible to help ensure higher efficiency levels.

Core Losses

Core losses, also referred to as *iron losses* or *armature losses*, are the combined effects of hysteresis losses and eddy current losses. Hysteresis losses result from the changing magnetic field polarity, while eddy current losses result from the relative motion between a conductor and a magnetic field. When the armature of the dc motor rotates, a small electric current is produced in the armature. This is an induced current that acts to oppose the supply current flowing through the armature. This opposition to current flow results in hysteresis losses. In addition, when the armature rotates, an electromotive force is also induced. This induced voltage also causes an additional opposition current to be produced. This opposition current, which is the result of the induced voltage, produces eddy current losses. Although core losses vary, they are relatively constant, regardless of motor speed or load, when compared to other losses experienced by dc motors.

Winding Losses

Winding losses, also known as *copper power losses*, result from the motor's winding resistance. Unlike core losses, which are relatively constant, winding losses vary depending on the amount of current flowing through the motor. Winding losses are commonly referred to as I^2R losses. In the power formula ($P = I^2R$), power is a function of the square of the current. Because of this, power losses are a function of the square of the motor current. By doubling the motor current, a fourfold increase in power loss ($2^2 = 4$) results. Increasing the motor current by a factor of three causes the power losses to increase by a factor of nine ($3^2 = 9$). Under low-load conditions when motor current is low, the winding losses are also low.

Shedding Light

When a dc motor is operating under high-load conditions, the power losses increase, and the motor temperature also increases. As motor temperature increases, coil resistance increases and motor torque decreases. These changes result in increased energy losses. Increased motor temperatures can be addressed by increasing the amount of air passing over the motor. One problem created by increasing airflow over the motor is increased wind resistance, which also contributes to efficiency losses. Energy losses due to wind resistance are commonly referred to as ***windage losses***, which fall under a broader energy loss classification known as mechanical losses.

Mechanical Losses

Mechanical losses result from the actual operation of the motor. The most common type of mechanical loss is due to friction. In dc motors, friction is typically present in two locations: the motor bearings and the surface where the motor's brushes and commutator make contact with each other. Another type of mechanical losses comes from the resistance of the air that surrounds the motor. Mechanical losses are directly proportional to the speed of the motor. At low speeds, the mechanical losses associated with the dc motor are low. As the speed of the motor increases, the mechanical losses also increase. Mechanical losses account for about 20% of total motor losses.

CHAPTER 27 REVIEW

Summary

- DC motors are desirable because of their high starting torque and their ability to accurately control motor speed. (27.1)
- Most sustainable and renewable energy sources are completely compatible with dc power. (27.1)
- The right-hand rule for motors can be used to illustrate the directional relationship that exists among current, force, and magnetic field. (27.2)
- A simple dc motor is made up of a stator, a rotor, a commutator, brushes, and a shaft. (27.3)
- The commutator is a rotary switch that continually reverses the polarity of the motor's supply voltage. (27.3)
- The direction of rotation of a dc motor is determined by the direction of the generated torque. (27.3)
- DC motors with only series windings are called series dc motors. (27.4)
- The speed of a series dc motor is commonly controlled using an armature resistance control, and armature diverter control, a tapped field control, or a field diverter control. (27.4)
- DC motors with only shunt windings are called shunt dc motors. (27.5)
- The speed of a shunt dc motor is commonly controlled using a flux control or an armature resistance control. (27.5)
- DC motors with both series and shunt windings are called compound dc motors. (27.6)
- Compound dc motors can be connected as either cumulatively compounded or differentially compounded. (27.6)
- Permanent magnet motors rely on permanent magnets to create the field flux that interacts with the motor. Some permanent magnet motors also use an additional, external current source. (27.7)
- Separately excited dc motors are supplied by two separate dc power sources: one source supplies the armature coils, and a second source supplies the field windings. (27.8)
- Losses in dc motors come in the form of core losses, winding losses, and mechanical losses. In some cases, increasing airflow over a motor will create windage losses. (27.9)

Review Questions

Use the information in the chapter to answer the following questions.

Measuring Your Potential

1. *True or False?* A dc motor is desirable for use on portable applications that rely on battery power. (27.1)
2. *True or False?* When using the right-hand rule for motors, the thumb points in the direction of current flow. (27.2)
3. *True or False?* The brushes on a dc motor are in direct contact with the commutator. (27.3)
4. Together, the commutator and brushes work to _____. (27.3)
   A. keep the motor turning in only one direction
   B. maintain the desired supply voltage to the motor
   C. align the motor's internal magnets
   D. prevent the motor from turning too fast
5. The commutator is responsible for _____. (27.3)
   A. connecting the power supply to the motor windings
   B. connecting the brushes to the power supply
   C. reversing the polarity of the power supplied to the windings
   D. reversing the polarity of the power supplied to the brushes
6. A series dc motor equipped with a variable resistor wired in series with the series field winding is using which motor control strategy? (27.4)
   A. Armature resistance control
   B. Armature diverter control
   C. Tapped field control
   D. Field diverter control
7. A series dc motor equipped with a variable resistor wired in parallel with the armature is using which motor control strategy? (27.4)
   A. Armature resistance control
   B. Armature diverter control
   C. Tapped field control
   D. Field diverter control

8. *True or False?* A shunt winding is a winding that is wired in parallel with the motor's armature. (27.5)
9. *True or False?* Cumulatively compounded dc motors are more widely used than differentially compounded dc motors. (27.6)
10. *True or False?* A permanent magnet dc motor relies on permanent magnets, but may use another dc power supply. (27.7)
11. Which of the following dc motors has two separate dc power sources? (27.8)
    A. Series dc motor
    B. Shunt dc motor
    C. Separately excited dc motor
    D. Compound dc motor
12. Hysteresis losses are the result of the _____. (27.9)
    A. relative motion between a conductor and a magnetic field
    B. reversal of magnetic field polarity
    C. changing direction of the motor
    D. generation of friction between the commutator and the brushes
13. The losses associated with the resistance of the motor's windings are often referred to as _____. (27.9)
    A. winding losses
    B. windage losses
    C. eddy current losses
    D. mechanical losses
14. Increased airflow over a motor can result in which type of losses? (27.9)
    A. Mechanical losses
    B. Hysteresis losses
    C. Core losses
    D. Windage losses

Plug-In

1. Without a commutator, a motor will _____. (27.3)
    A. rotate in the wrong direction all the time
    B. not rotate at all
    C. rotate in the wrong direction half the time
    D. never turn off
2. Eddy current losses in a motor are the result of the relative motion between _____. (27.9)
    A. a conductor and a magnetic field
    B. two magnets
    C. the commutator and the brushes
    D. the armature and the commutator
3. Using the right-hand rule for motors to analyze the following figure, determine the direction of current flow, magnetic field, and force both the north and south poles. (27.2)

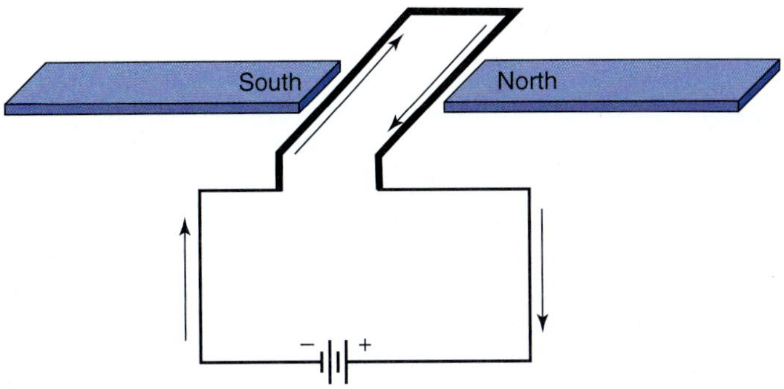

Goodheart-Willcox Publisher

Refer to the following diagram to answer questions 4 and 5.

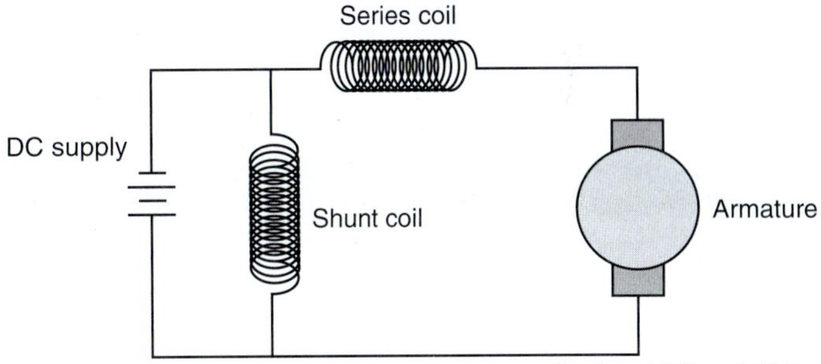

Goodheart-Willcox Publisher

4. The diagram best represents which of the following types of dc motors? (27.6)
    A. Compound dc motor
    B. Series dc motor
    C. Shunt dc motor
    D. Permanent magnet dc motor
5. Give two examples of where you might find this type of dc motor. (27.6)

Making the Connection

1. Explain how commutators and brushes work together to ensure that a motor turns in only one direction. (27.3)
2. Explain the relationship that exists among current, flux, and motor speed. (27.2)

CHAPTER

28 Single-Phase AC Motors

Wachirawit kamai/Shutterstock.com

SPARKING DISCUSSION

In most dc motors, power is supplied to both the armature (rotor) and field (stator) windings. In single-phase ac motors, power is supplied to only the stator. Why do dc motors require power to be supplied to the rotor and the stator, but ac motors do not?

LEARNING OUTCOMES

After completing this chapter, you will be able to:

28.1 Describe the construction and operation of a shaded-pole motor.
28.2 Discuss the winding configuration and operation of a split-phase motor.
28.3 Explain the operation of an induction-start, induction-run (ISIR) motor.
28.4 Explain the operation of a capacitor-start, induction-run (CSIR) motor.
28.5 Explain the operation of a permanent-split capacitor (PSC) motor.
28.6 Explain the operation of a capacitor-start, capacitor-run (CSCR) motor.
28.7 Explain how to change the speed of a multispeed motor.
28.8 Explain how to change the direction of rotation of a single-phase motor.
28.9 Describe the two wiring configurations of a dual-voltage, single-phase motor.

TECHNICAL TERMS

capacitor-start, capacitor-run (CSCR) motor
capacitor-start, induction-run (CSIR) motor
centrifugal switch
current magnetic relay (CMR)
dual-voltage, single-phase motor
induction motor
induction-start, induction-run (ISIR) motor
multispeed motor
permanent-split capacitor (PSC) motor
positive temperature coefficient (PTC) relay
potential magnetic relay (PMR)
run capacitor
run winding
shaded-pole motor
shading coil
split-phase motor
start capacitor
start winding

Introduction

Although three-phase power is more efficient than single-phase power, single-phase motors are found in nearly all aspects of every segment of every industry, especially in residential and light commercial applications. The fans, pumps, and blowers found in lighter-duty heating, air-conditioning, and refrigeration equipment, as well as motors found in a wide range of small appliances, are often of the single-phase variety. The power supply for most single-phase ac motors is 120 V, 208 V, or 240 V. Residential and light commercial fossil fuel heating systems typically use 120-V fans and blowers, while residential and light commercial air-conditioning compressors commonly use 240-V motors. Usually, single-phase, 120/240-V power is supplied to most residential areas and many small commercial installations. This choice of voltage distribution necessitates equipment designed to operate at 120/240 V. Many motors are able to operate at a wide range of voltages depending on how the motor is connected to the source voltage.

In this chapter, you will learn about common types of single-phase motors, their basic modes of operation, as well as common applications for each type. You will also be briefly introduced to starting components and devices that are used to facilitate and enhance single-phase motor operation. Additional information about these starting components and devices, including their configurations, characteristics, and operational details and ratings, is provided in Chapter 31, *Single-Phase Motor Starting and Protection Devices.*

ROMBIX STUDIO/Shutterstock.com

Figure 28-1. A shaded-pole motor is an example of a simple, single-phase ac induction motor.

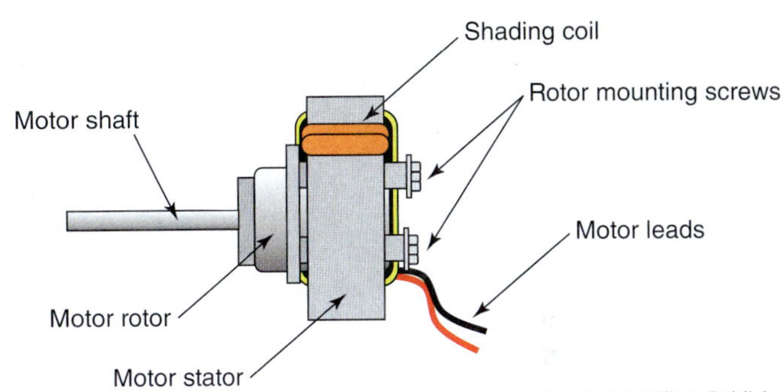

Goodheart-Willcox Publisher

Figure 28-2. Cross-sectional view of a shaded-pole motor. This type of motor is easily identified by the thick shading coils that create the imbalance in magnetic fields required to make the motor turn.

28.1 Shaded-Pole Motors

The ***shaded-pole motor*** is one of the simplest of all ac induction motors, **Figure 28-1**. This motor type is inexpensive to construct and purchase and does not require the use of any starting components, relays, or devices to operate. Its simple construction and lack of starting devices combine to create a reliable motor with a very long, useful service life.

Shaded-Pole Motor Construction

The simple shaded-pole motor is constructed with a squirrel-cage rotor positioned within a stator field, **Figure 28-2**. Just as with any other motor type, the shaded-pole motor operates on the principle of differences in magnetic field strength to create the rotating force required to turn the motor. To create an imbalance in magnetic fields, loops of thick copper wire or wide copper bands are positioned around portions of the motor's stationary winding. These loops or bands have very low resistance and are commonly referred to as ***shading coils***, **Figure 28-3**.

Shaded-Pole Motor Operation

When ac voltage is initially applied to the shaded-pole motor, the current flow through the windings results in the creation of a magnetic field. This magnetic field acts on both the rotor and the shading coils, inducing current flow. Because the shading coils have very low resistance, the induced current flow through the coils is quite large. In the portion of the ac waveform where current is increasing, the current flow through the shading coils opposes the change in magnetic flux, so no force in the shaded area is able to act on the motor's rotor, **Figure 28-4**. This creates the force imbalance that causes the motor to turn.

In the portion of the ac waveform where current is decreasing, the strength of the magnetic field produced by the windings is also decreasing. However, the shading coils again oppose this change, resulting in a stronger magnetic field in the portions of the rotor shielded by the shading coils, **Figure 28-5**.

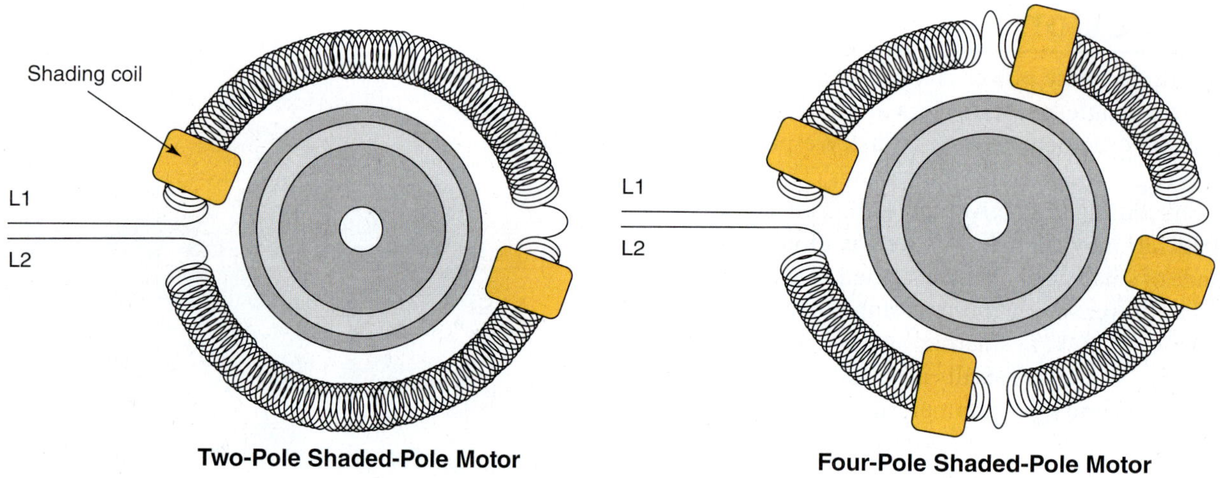

Goodheart-Willcox Publisher

Figure 28-3. Shaded-pole motors are commonly configured as either two-pole or four-pole devices.

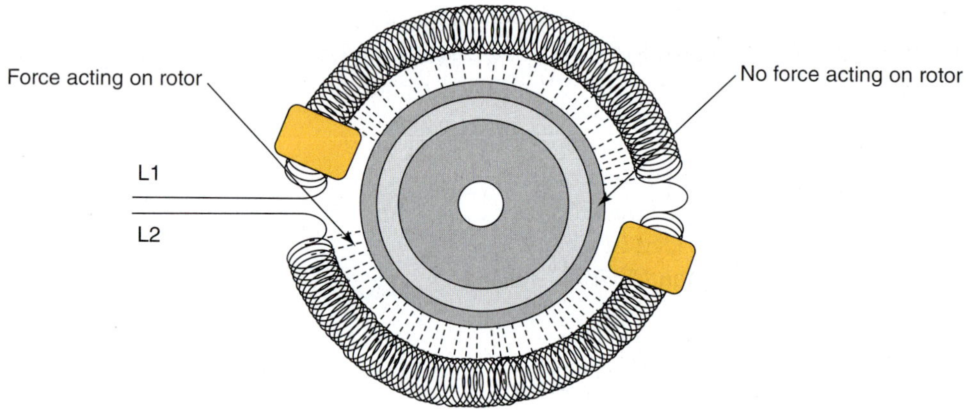

Goodheart-Willcox Publisher

Figure 28-4. When current is induced in the shading coil, a magnetic field is produced that prevents the main magnetic field from acting on the rotor.

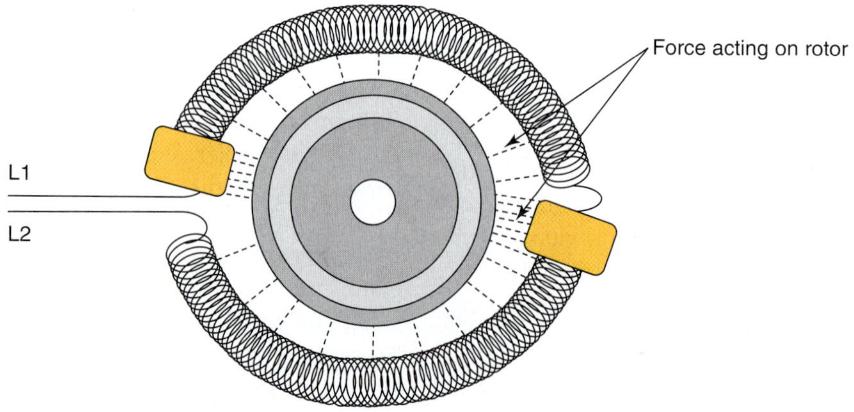

Goodheart-Willcox Publisher

Figure 28-5. When the main magnetic field is collapsing, the shading coils oppose the reduction, resulting in stronger magnetic fields affecting the portions of the rotor aligned with the shading coils.

The shading coils create variations in the net magnetic field that is acting on the rotor. These variations in the net magnetic field are sufficient to start the motor.

Shaded-Pole Motor Applications

Shaded-pole motors are fractional horsepower, low-torque motors used for light-duty applications. The efficiency of this type of motor is very low, but for the intended applications this is not a drawback. These motors are used for direct-drive applications, where small fan blades, blower wheels, and other light-duty driven loads are connected directly to the motor shaft. See **Figure 28-1**. Shaded-pole motors are not used in conjunction with larger pulley-driven or other high-torque belt-driven applications.

Because of their low horsepower ratings, in many instances, shaded-pole motors are rated in watts (W) instead of horsepower (hp). For example, a small shaded-pole motor that is used to turn a small propeller-type fan blade in a household refrigerator is rated at about 6 W, or about 1/125 hp. Typically, shaded-pole motors are available in sizes up to 1/20 horsepower.

Shaded-Pole Motor Direction of Rotation

Unlike other common motor types, changing the direction of rotation of a shaded-pole motor cannot be done by changing the wiring connections. Instead, the location of the shading coils on the windings determines the motor's direction of rotation. To reverse the rotation of a shaded-pole motor, the motor needs to be disassembled and reassembled with the rotor rotated 180°, **Figure 28-6**.

Not all shaded-pole motors can have their rotational direction reversed. Before attempting to disassemble the motor, carefully inspect the motor and refer to the manufacturer's literature to determine if the motor can be taken apart to change its rotational direction, and how to properly do so. In order for the motor to operate properly in both the clockwise and counterclockwise direction, the rotor must align perfectly with the stator in both positions. Improper motor disassembly and reassembly can result in the motor failing to operate, operating in an overheated condition, or failing prematurely.

28.2 Split-Phase Motors

Motor rotation occurs when a magnetic field produced in the stator reacts with the magnetic field generated in the rotor. For any ac motor in which the rotor has no direct connection to a power supply, an induced current will generate a magnetic field. When motor rotation relies on induction, the motor is referred to as an ***induction motor***. Induction motors are classified as asynchronous motors, meaning that the motor's actual speed is lower than the motor's calculated, or synchronous, speed. ***Split-phase motors*** are induction motors that have two distinct windings connected in parallel with each other. These windings are a high-resistance start winding and a low-resistance run winding, **Figure 28-7**. The ***start winding*** is typically wound with many turns of thinner wire, while the ***run winding*** is wound with fewer turns of thicker wire. The term split-phase refers to the fact that, in most cases, the current flow through the motor is shared, or split, between the start winding and the run winding. Split-phase, single-phase, fractional horsepower motors are commonly used for medium torque applications, such as fans, blowers, and pumps. These motors typically operate with moderate efficiency. Fractional-horsepower motors, which are less than 1 hp, do not consume extremely large amounts of energy.

When voltage is supplied to a split-phase motor, the measured voltage across the start winding is the same as the measured voltage across the run winding, **Figure 28-8**. The current flow through each of the windings, however, is not the same since the winding resistances are different. More current flows through the run winding because it has lower resistance than the start winding. In **Figure 28-9**, for example, the total current draw of the motor is 4 A, while the current flow through the start winding is 1 A and the current flow through the run winding is 3 A. It is often beneficial to visualize the start and run windings as two inductive loads connected in parallel with each other, **Figure 28-10**.

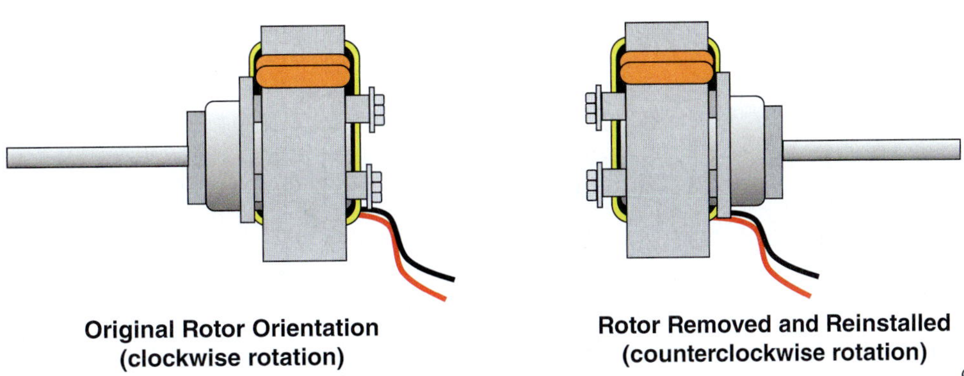

Figure 28-6. Some shaded-pole motors can have the direction of rotation reversed by removing the rotor from the motor, rotating it 180°, and then reinstalling it facing the other direction.

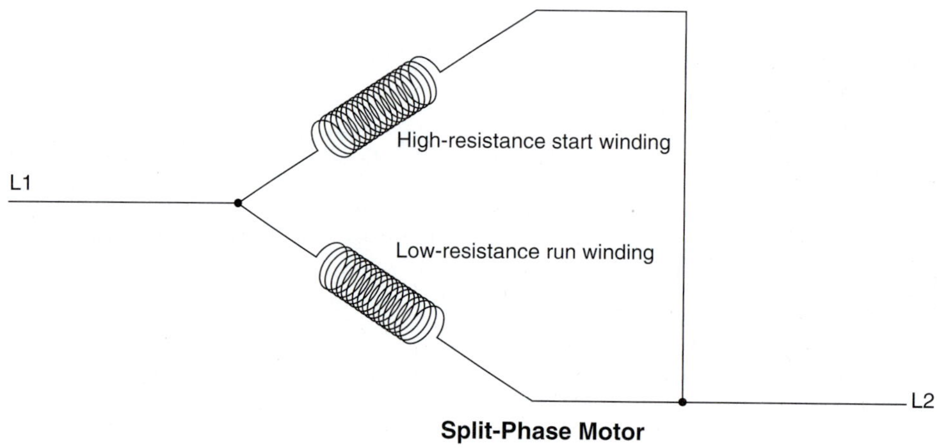

Goodheart-Willcox Publisher

Figure 28-7. Split-phase motors have a high-resistance start winding and a low-resistance run winding.

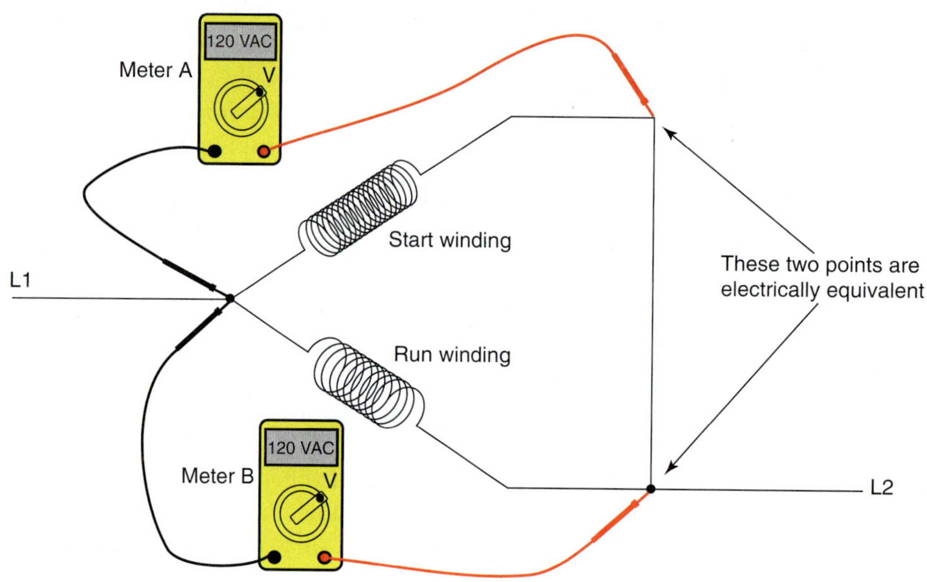

Goodheart-Willcox Publisher

Figure 28-8. When the start winding and the run winding are supplied with the same line voltage, the measured voltage across the start winding is the same as the measured voltage across the run winding.

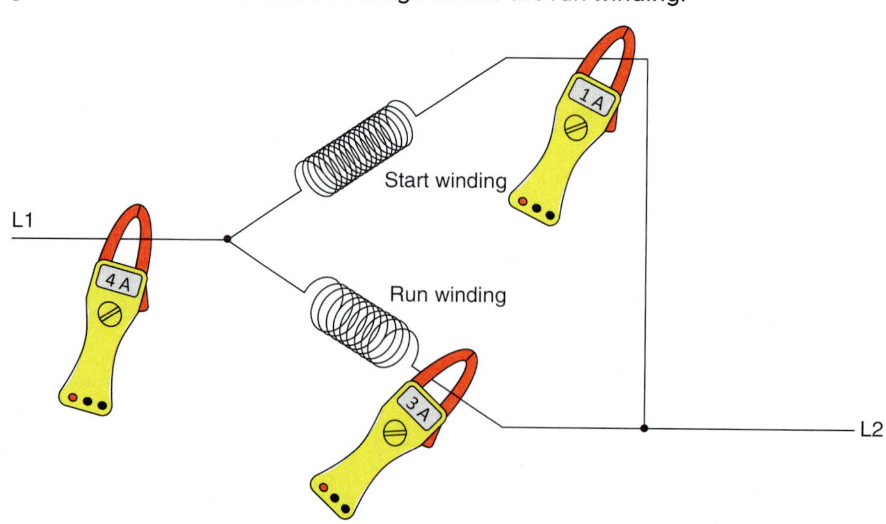

Goodheart-Willcox Publisher

Figure 28-9. The total motor current is split unevenly between the start and run windings. With a lower resistance, the current flow is greater through the run winding.

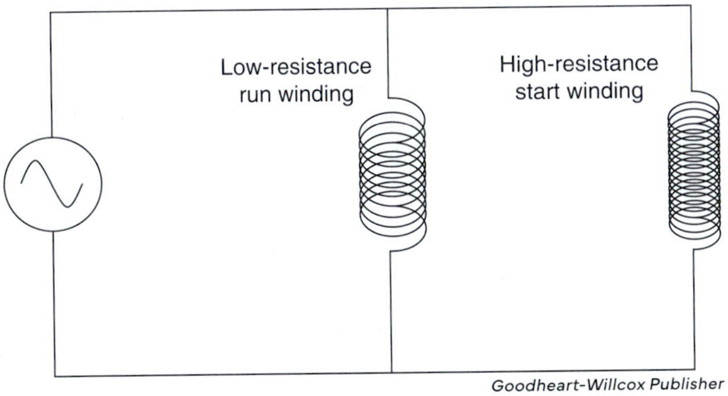

Figure 28-10. The windings of a split-phase motor can be easily visualized as two inductive loads connected in parallel with each other.

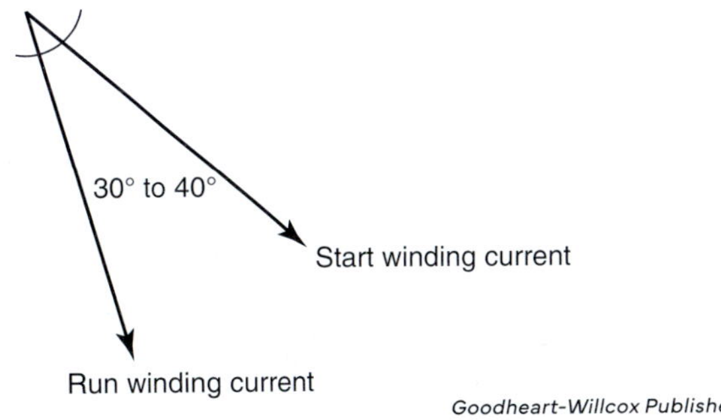

Figure 28-12. In the ISIR motor, there is a 30° to 40° phase angle between the run winding current and the start winding current.

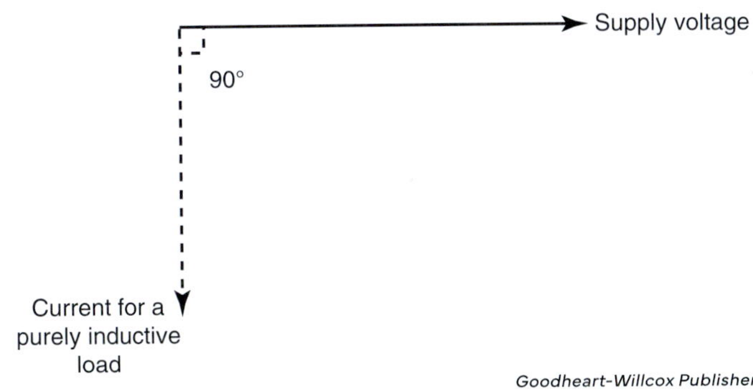

Figure 28-13. In a purely inductive circuit, the voltage and current are 90° out of phase with each other.

28.3 Induction-Start, Induction-Run Motors

The ***induction-start, induction-run (ISIR) motor*** uses a high resistance start winding, a low resistance run winding, and a starting device, **Figure 28-11**. The starting device is used to remove the start winding from the active electric circuit once the motor is up and running at a speed that is close to the motor's design speed.

Operation of the ISIR motor relies on the out-of-phase condition that is created because of the resistance and inductance differences associated with the start and run windings. Since the two windings have different resistance and inductance values, there is a phase angle between the current flows through the two windings. This phase angle varies from motor to motor but is typically between 30° and 40°, **Figure 28-12**. The current that flows through the motor's two windings is also out of phase with the voltage supplied to the motor. The voltage and current of a purely inductive circuit are 90° out of phase with each other, **Figure 28-13**. Since motor windings have both inductance and resistance, the phase angle between the currents and the voltage will always be less than 90°, **Figure 28-14**. A phase angle of less than 90° results in a motor with lower starting torque.

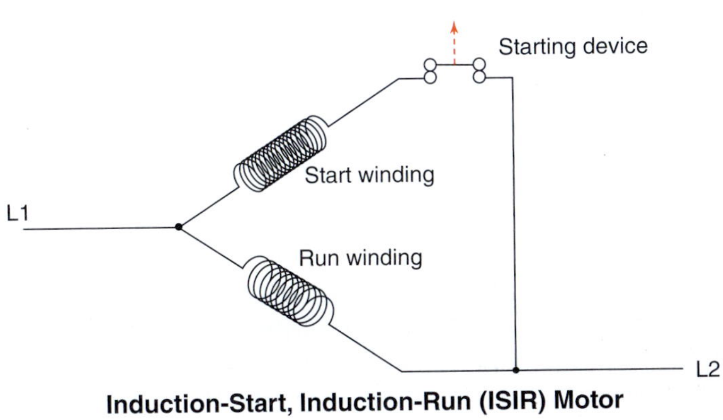

Figure 28-11. The ISIR motor uses a high-resistance start winding, a low-resistance run winding, and a starting device. The starting device is used to remove the start winding from the active electric circuit once the motor is running at a speed near design speed.

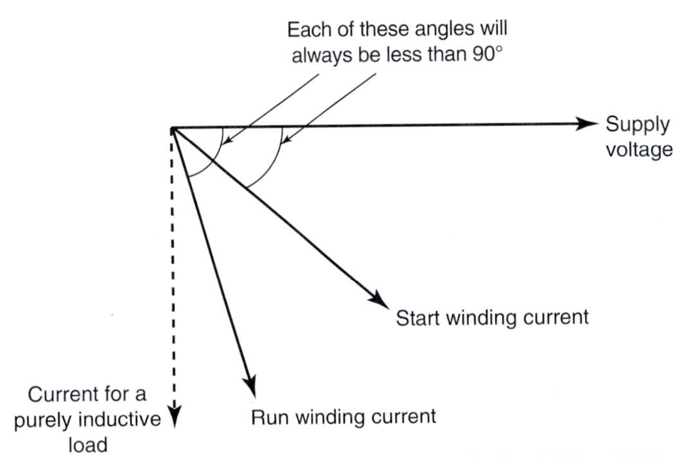

Figure 28-14. In an ISIR motor, the phase angle between the voltage and the start winding current will always be less than 90°. The angle between the voltage and the run winding current will also always be less than 90°.

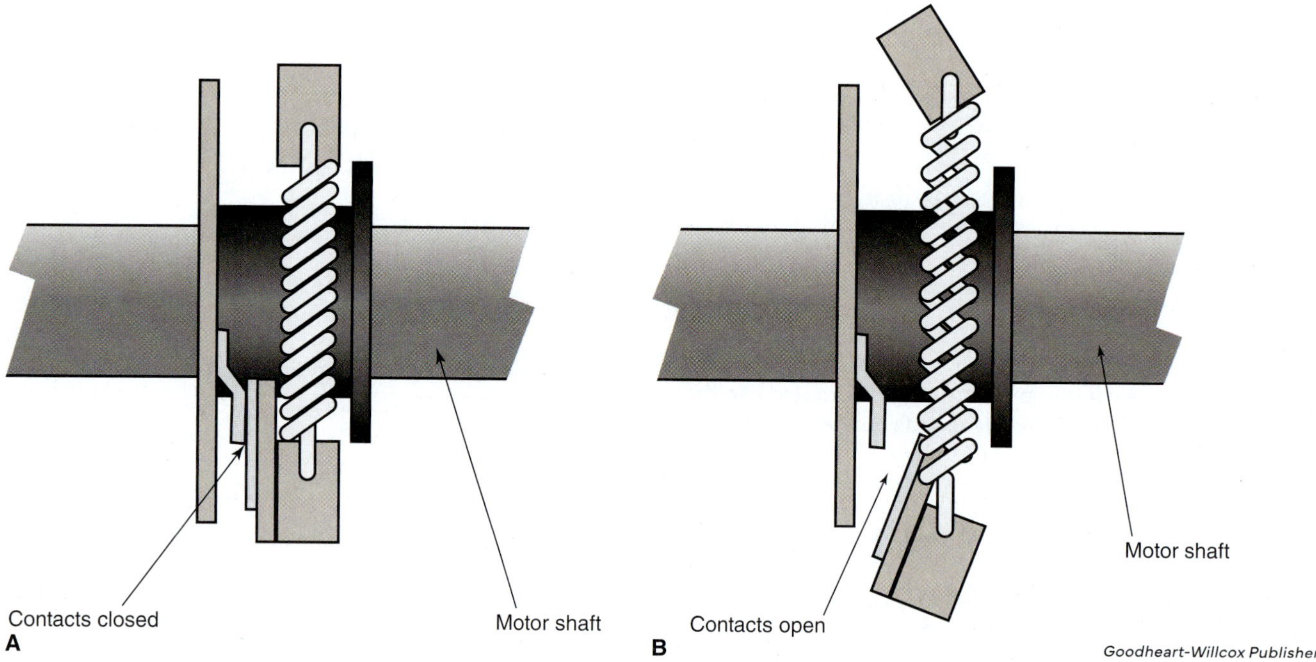

Figure 28-15. In an ISIR motor, the centrifugal switch is mounted to the motor's shaft. A—When the motor is not operating, or operating at a low speed, the contacts on the centrifugal switch are closed. B—When the motor is rotating close to its operating speed, the contacts on the centrifugal switch are pushed open by centrifugal force, removing the start winding from the active circuit.

Removing the Start Winding

A difference between the two magnetic fields is only needed to start the ISIR motor. Once the motor has started and is operating at a speed of at least 70% of its rated speed, the motor can continue to operate with only the run winding in the active electric circuit. The turning motor shaft and rotor will create the needed magnetic field imbalance to keep the motor operating.

The start winding is often removed from the circuit using a device such as the ***centrifugal switch***. The centrifugal switch is mounted to the motor's shaft. The position of the switch's contacts is determined by the speed of the motor. The switch is connected in series with the motor's start winding. When the motor is stopped, or operating at a low speed, the contacts are closed and the start winding remains in the active electric circuit. Once the motor reaches about 70% of its rated speed, centrifugal force pushes the switch's contacts open, removing the start winding from the active circuit, **Figure 28-15**. When the motor is de-energized, the motor slows down and the contacts on the centrifugal switch close. With the switch's contacts closed, both windings will be in the active circuit when motor operation is next called for.

A ***positive temperature coefficient (PTC) relay*** can also be used to remove the start winding from the circuit. Just like the centrifugal switch, it is wired in series with the start winding, **Figure 28-16**. The PTC relay is a solid-state device that operates on the principle that current flow generates heat. The PTC relay reacts to an increase in temperature by increasing its resistance. When the sensed temperature is low, the device's resistance is low, and the PTC acts like a closed switch and passes current freely. When the motor is initially energized, both the start and run windings are in the circuit and the motor starts. As the motor continues to operate, current flow through the start winding causes the PTC to heat up. This increases its resistance, and, after a brief period of time, the PTC has a high enough resistance to act as an open switch, effectively removing the start winding from the circuit. Once the motor is de-energized, the PTC cools down and its resistance once again drops. Motors that utilize PTC starting devices must remain off for at least a few minutes between calls for motor operation. This will allow the PTC device to cool down enough to once again act like a closed circuit.

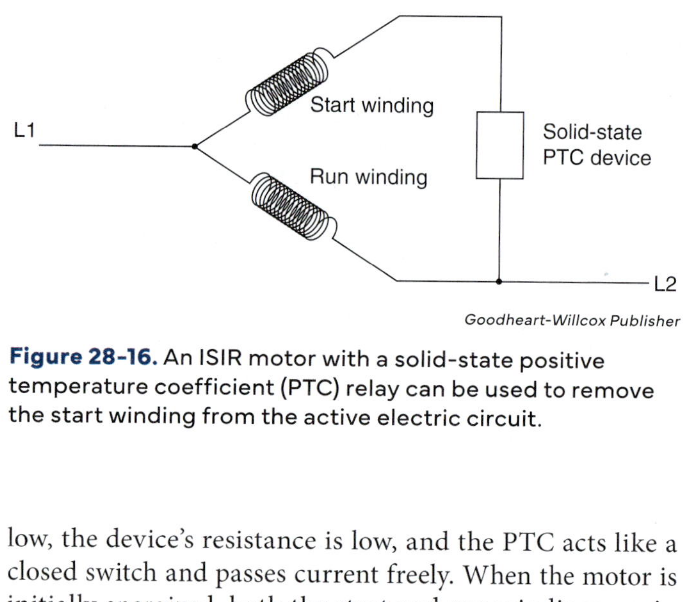

Figure 28-16. An ISIR motor with a solid-state positive temperature coefficient (PTC) relay can be used to remove the start winding from the active electric circuit.

ISIR Motor Applications

Induction-start, induction-run motors operate with low starting torque, so they are intended for use in light-duty and medium-duty applications, including the following:

- General-purpose woodworking and shop tools, such as drills, band saws, grinders, and sanders
- Household appliances, such as vacuum cleaners, mixers, and blenders
- Low-horsepower fans, blowers, pumps, and compressors

28.4 Capacitor-Start, Induction-Run Motors

The ***capacitor-start, induction-run (CSIR) motor*** closely resembles the split-phase, induction-start, induction-run motor. Just like the ISIR motor, the CSIR motor has two distinct windings (the start winding and the run winding) and a means to remove the start winding from the active electric circuit. Unlike the ISIR motor, the CSIR motor uses a capacitor wired in series with the start winding and in parallel with the run winding, **Figure 28-17**. This capacitor is called a ***start capacitor***. Depending on the motor's specifications, the start capacitor's capacitance can range from 20 μF to about 1500 μF. As the name implies, the start capacitor's function is to aid in motor starting and to increase the starting torque. The start capacitor is intended to be in the active circuit for a very short time, and only when the motor is initially energized. Once the motor has started, the start capacitor is removed from the circuit, along with the start winding. The same device is used to remove the start winding and the start capacitor from the circuit. Once the start winding and the start capacitor have been removed from the circuit, the motor will continue to run with only the run winding energized, as in the case of the ISIR motor.

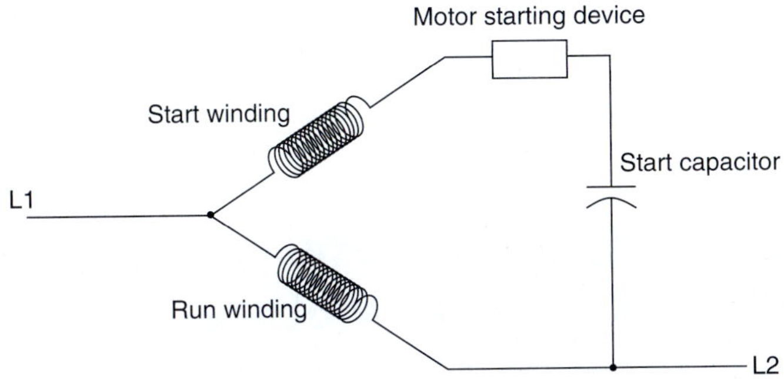

Goodheart-Willcox Publisher

Figure 28-17. The CSIR motor uses a starting device and a start capacitor, both of which are wired in series with the start winding and in parallel with the run winding.

Removing the Start Winding and the Start Capacitor

The start winding and start capacitor are typically removed from the active circuit based on the current flow through the motor's run winding. This is done by a device called the ***current magnetic relay (CMR)***. The CMR is made up of a normally open set of contacts and a low-resistance coil. The CMR's coil is wired in series with the run winding, and the normally open contacts are wired in series with the start winding and the start capacitor, **Figure 28-18**. The contacts of the CMR are held open by gravity; therefore, the relay must be in its correct orientation to operate properly.

When the motor is initially energized, the CMR's contacts are open. In this position, the start winding and start capacitor are not in the active electric circuit. Current flows only through the run winding, **Figure 28-19**. As a result, the motor cannot start as there is no imbalance in the magnetic field. Not being able to start, the motor's current draw is equal to its locked rotor amperage (LRA). LRA is the amperage draw of an energized

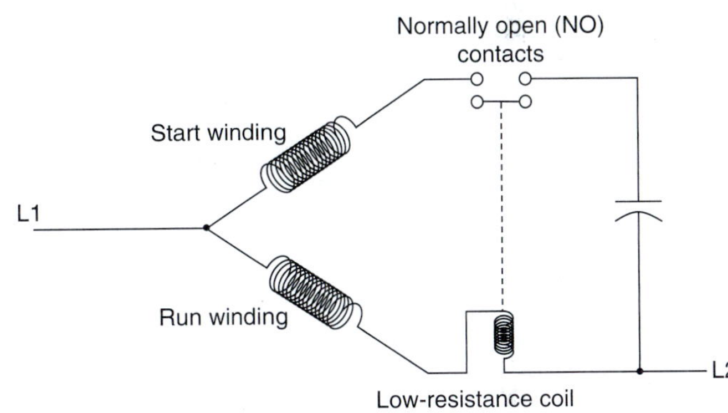

Goodheart-Willcox Publisher

Figure 28-18. The current magnetic relay in a CSIR motor has a low-resistance coil wired in series with the run winding and a normally open set of contacts wired in series with the start winding and the start capacitor.

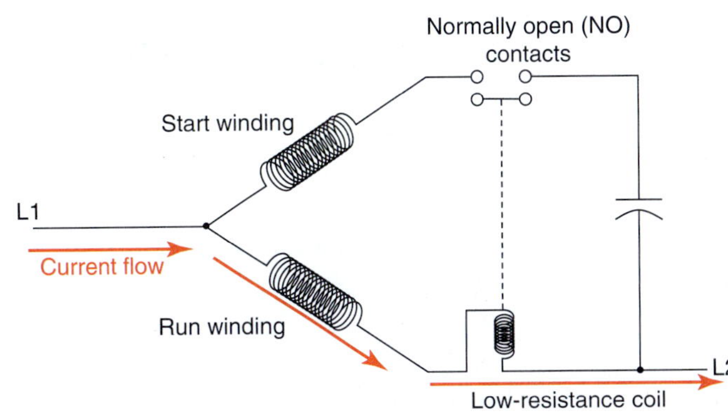

Goodheart-Willcox Publisher

Figure 28-19. In a CSIR motor, the contacts of the current magnetic relay are normally open, so only the run winding is energized when power is initially supplied to the motor.

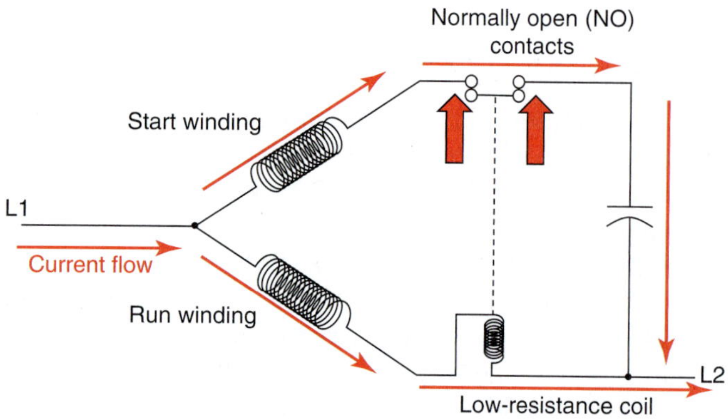

Goodheart-Willcox Publisher

Figure 28-20. When the starting current is sufficiently high in a CSIR motor, there is enough force to close the current magnetic relay's contacts and add the start winding and start capacitor to the active electric circuit.

motor that has its shaft immobilized or cannot otherwise turn. Since there is no imbalance in the magnetic field and the shaft cannot turn, the LRA is the amount of current flowing through the only complete circuit, which is through the run winding and the coil of the CMR. The high current flow through the CMR's coil causes the normally open contacts to close, adding the start winding and the start capacitor to the circuit, **Figure 28-20**. This allows the motor to start. Once started, the current draw drops to the motor's normal running load amperage (RLA). This drop in current reduces the strength of the magnetic field generated by the CMR's coil and its strength is no longer strong enough to keep the contacts of the CMR closed. As a result, gravity pulls the relay's contacts open. The motor continues to operate with only the run winding energized.

CSIR Motor Applications

Once operating, the CSIR and ISIR motors are very similar to each other as they both operate with only the run winding in the active electric circuit. The main difference is that the CSIR motor, with its start capacitor, is desirable for applications that require more starting torque than shaded-pole or induction-start, induction-run motors can provide. CSIR motors are not intended for high-torque applications, such as belt-driven systems, but can be effectively used for medium-duty, direct-drive purposes. Applications for which the CSIR motor is often used include the following:

- Direct-drive pumps, fans, and blowers
- Air-conditioning and refrigeration compressors
- Furnaces
- Spas and hot tubs

28.5 Permanent-Split Capacitor Motors

The ***permanent-split capacitor (PSC) motor*** is very efficient, while being one of the simplest types of single-phase motors.

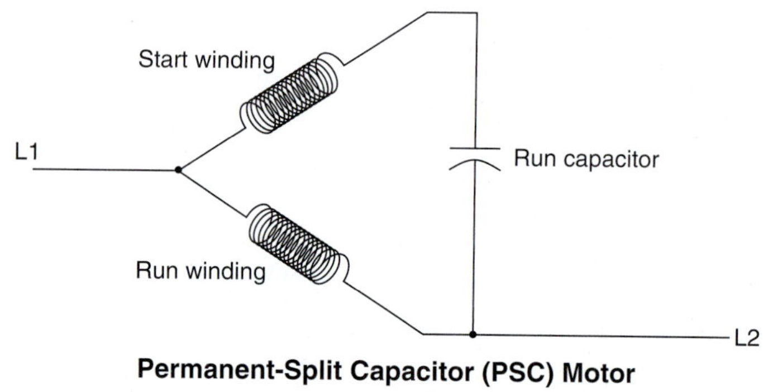

Goodheart-Willcox Publisher

Figure 28-21. Wiring diagram of a PSC motor. The run capacitor is connected in series with the start winding and in parallel with the run winding.

The PSC motor has a start winding, a run winding, and a capacitor that is intended to be in the active circuit whenever the motor is energized. This capacitor, called a ***run capacitor***, is connected in series with the start winding and in parallel with the run winding, **Figure 28-21**. Run capacitors have a much lower microfarad rating than start capacitors. Depending on the motor's specifications and design, run capacitors typically range from 4 μF to about 75 μF.

PSC motors do not use a starting relay or device. The start winding, the run winding, and the run capacitor are in the active electric circuit whenever the motor is energized. Although the PSC motor is very efficient and requires no moving parts to start the motor, the motor's starting torque is very low. As a result, the PSC motor can be used for only fractional horsepower applications that require low starting torque. The PSC motor has a very soft start, meaning that it starts slowly and operates quietly. Since the PSC motor starts slowly, it is a popular choice when the motor is located in or near an occupied location where high noise levels are not desirable. The PSC motor is commonly found on air-conditioning systems and heating system blower assemblies.

Shedding Light

PSC motors are specially designed to have the start winding energized whenever the motor is energized, so be sure to use and wire the motor according to the manufacturer's recommendations and guidelines.

28.6 Capacitor-Start, Capacitor-Run Motors

Capacitor-start, capacitor-run (CSCR) motors are a cross between the capacitor-start, induction-run (CSIR) motor and the permanent-split capacitor (PSC) motor. CSCR motors use a starting relay, a run capacitor, and a start capacitor. The start capacitor provides motor-starting assistance, while the run capacitor helps create an efficient phase angle between the current and voltage when the motor is running. As with other motor types discussed in this chapter, the start capacitor is removed from the circuit once the motor has started, and the run capacitor is in the active circuit whenever the motor is energized. The start and run capacitors are both wired in series with the start winding but are in parallel to one another. A starting relay is responsible for removing the start capacitor from the circuit once the motor has started, **Figure 28-22.**

When the motor is initially energized, the total capacitance provided to assist motor starting is the sum of the two parallel-connected capacitances. For example, a CSCR motor with a 15 µF run capacitor and a 100 µF start capacitor would have a total circuit capacitance of 115 µF on motor start-up. This increased capacitance creates a greater phase angle between the run and start windings, giving the motor higher starting torque. Once the motor is up and running, the start capacitor is removed from the circuit. However, unlike the CSIR motor, the start winding and the run capacitor stay in the circuit. In this configuration, the CSCR motor resembles a PSC motor once the start capacitor has been removed, **Figure 28-23.**

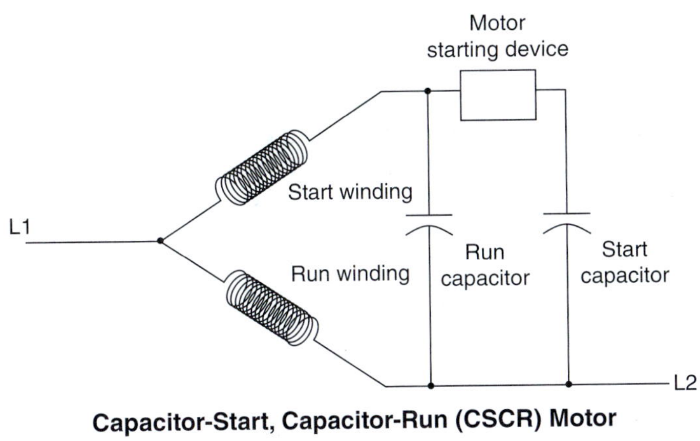

Goodheart-Willcox Publisher

Figure 28-22. In addition to the start and run windings, a CSCR motor uses a starting relay, a run capacitor, and a start capacitor. A starting relay removes the start capacitor from the circuit once the motor has started.

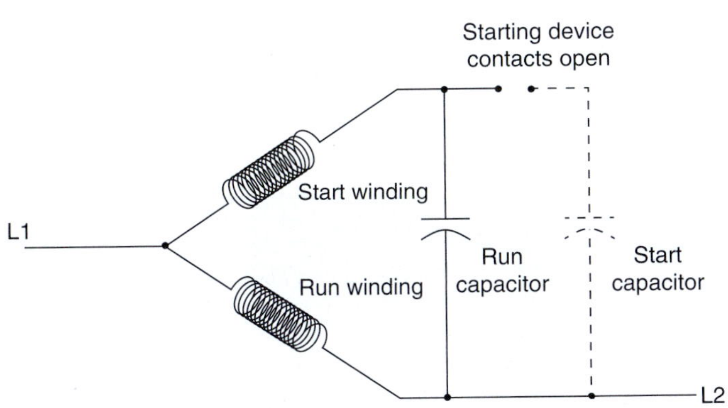

Goodheart-Willcox Publisher

Figure 28-23. With the start capacitor removed from the circuit, the CSCR motor resembles a PSC motor.

Removing the Start Capacitor

On a CSCR motor, the start capacitor is often removed from the circuit using a ***potential magnetic relay (PMR).*** The PMR has a high-resistance coil and a normally closed set of contacts, **Figure 28-24.** The coil is connected between terminals 2 and 5, while the normally closed contacts are connected between terminals 1 and 2.

The coil of the PMR is connected in parallel with the start winding, while the normally closed contacts are wired in series with the start capacitor, **Figure 28-25.** The voltage sensed by the coil of the PMR causes the relay contacts to open and close. When the motor is initially energized, the voltage sensed by the PMR coil is equal to the line voltage supplied to the motor's start winding. This voltage is not high enough to cause the normally closed contacts on the relay to open. As the motor speed increases, the voltage sensed by the PMR coil is the combination of the line voltage supplied to the motor's start winding and a voltage that is induced by the motor's turning rotor. As the motor's speed increases, the induced voltage and the voltage sensed by the PMR coil increases. At a predetermined voltage, the relay's normally closed contacts open, and the motor continues to operate as a PSC motor.

If at any time during motor operation the motor experiences a high load, the motor might slow down. If this should occur, the voltage sensed by the PMR coil will drop, causing the normally closed contacts to close once again, putting the start capacitor back into the circuit. By reintroducing the start capacitor into the circuit, the motor will be able to respond to the increased load by creating additional torque. Once the motor reaches its desired speed once again, the relay's contacts will open, and the motor will continue to operate as a PSC motor.

CSCR Motor Applications

With the exception of three-phase motors, the CSCR motor is one of the most efficient traditional motor types in use today. The CSCR motor combines a great deal of starting torque with the high operating efficiency of a PSC motor. It is frequently used for larger belt-driven applications and other applications

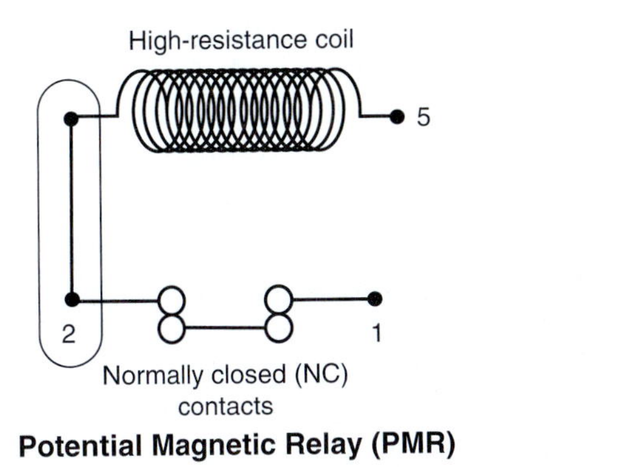

Goodheart-Willcox Publisher

Figure 28-24. A PMR has a high-resistance coil and a normally closed set of contacts.

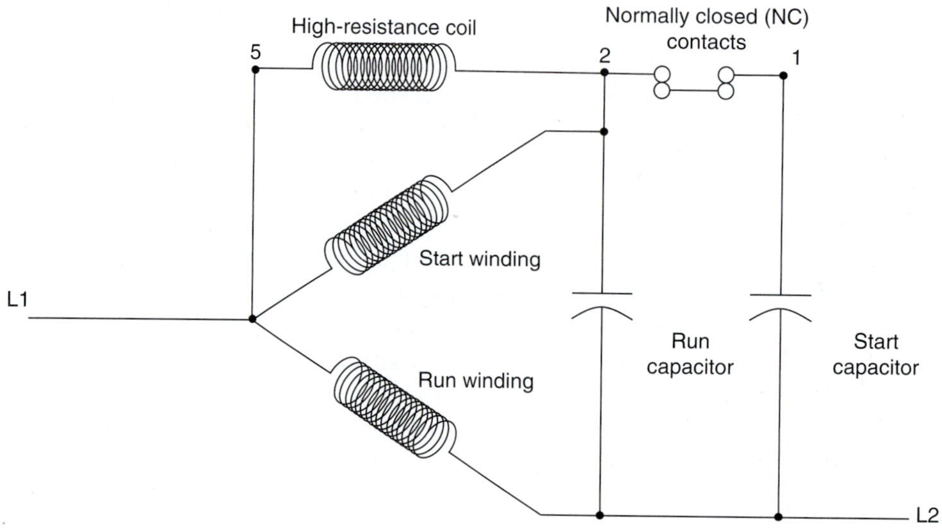

Goodheart-Willcox Publisher

Figure 28-25. The coil of the PMR is wired in parallel with the start winding, and the normally closed contacts are wired in series with the start capacitor.

that require more starting torque and where three-phase power is not readily available. Belt-driven motor and blower assemblies offer a great deal of resistance when they initially start because the resistance of the drive pulleys, belts, driven pulleys, driven device, and driven bearings must all be overcome. Once the motor starts and the initial resistance has been overcome, keeping the motor turning does not require the same amount of torque. Other applications for which the CSCR motor is often used include the following:

- Pumps, fans, and blowers
- Air compressors
- Refrigerant compressors
- Conveyor belt systems
- Medical equipment

28.7 Multispeed Motors

Single-phase motors are often configured as ***multispeed motors***, which are motors that can run at more than one speed. A multispeed PSC motor, for example, is easily identified as there are many colored wires extending out from the motor casing, **Figure 28-26**. The color-coding can vary from one motor manufacturer to another, but a commonly used color-coding configuration is as follows:

- Black: High speed
- Blue: Medium speed
- Red: Low speed
- White or Yellow: Line voltage
- Brown: Capacitor
- Green: Ground

The speed of the motor can be changed by switching the wiring connections. By switching the wiring connections, inductance is either added to the circuit or removed from the circuit. As the amount of added inductance decreases, the speed of the motor increases. When more inductance is wired into the circuit, the motor speed decreases. Consider the wiring diagram for a multispeed PSC motor in **Figure 28-27**. If line voltage is connected between the black power lead and L2, the motor will operate at high speed. If line voltage is connected between the red power lead and L2, the motor will operate at low speed.

In some cases, the speed of a motor can be changed using a selector switch, **Figure 28-28**. By changing the position of the selector switch, the amount of run winding inductance can be changed. As more run winding inductance is added to the active circuit, the motor will operate at a lower speed. If the selector switch is set to low, the entire run winding is in the active circuit and the motor will operate at its lowest speed. If the selector switch is set to high, only a small portion of the run winding is in the circuit and, as a result, the motor will operate at its highest speed. Selector switches and other control devices are often used in conjunction with multispeed motors as this enables the end user of the motor to change speeds without having to make wiring changes.

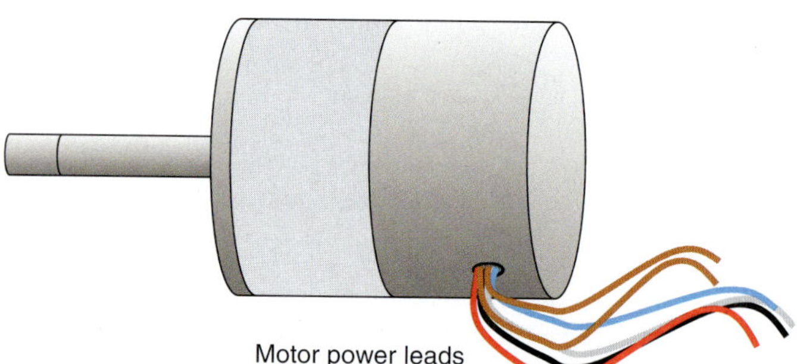

Goodheart-Willcox Publisher

Figure 28-26. Color-coded power leads on a multispeed motor.

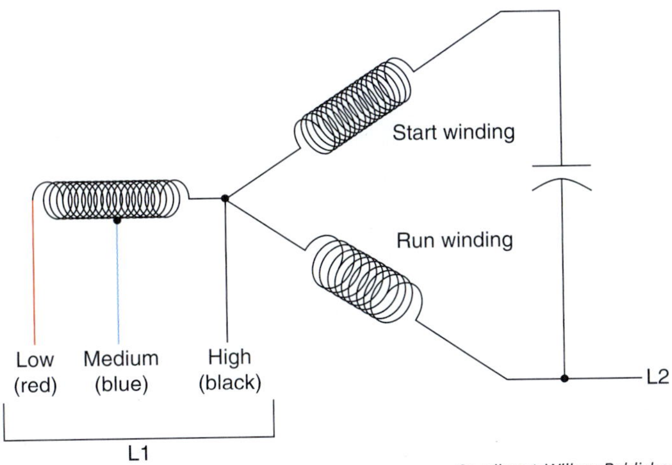

Goodheart-Willcox Publisher

Figure 28-27. In a multispeed motor, the speed of the motor can be changed by connecting power to different leads on the motor.

28.8 Single-Phase Motor Direction of Rotation

Some single-phase motors are manufactured to rotate in only one direction, while others have the ability to rotate in either the clockwise or counterclockwise direction. The direction of rotation of a single-phase motor is determined by the direction of current flow through the motor's start winding.

By reversing the start winding connections, the direction of current flow through the start winding is reversed, changing the motor's pole charges, **Figure 28-29**. This will cause the motor to rotate in the opposite direction. On motors that can have their direction of rotation reversed, there will often be two loops of colored wire extending from the motor's housing. Each loop will be a different color and will have

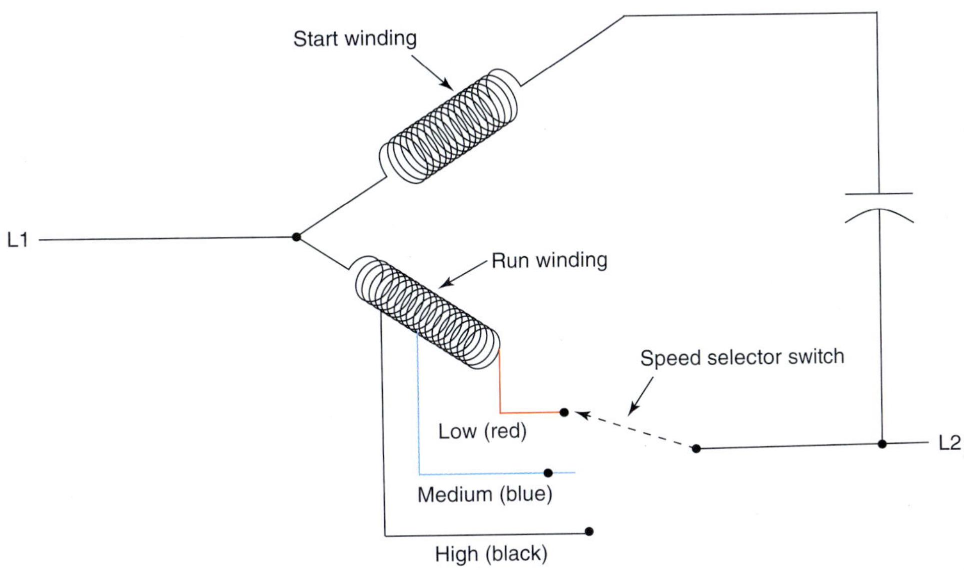

Goodheart-Willcox Publisher

Figure 28-28. In a PSC motor, the selector switch can be used to change the speed of a motor. The selector switch connects different motor speed leads to the power source.

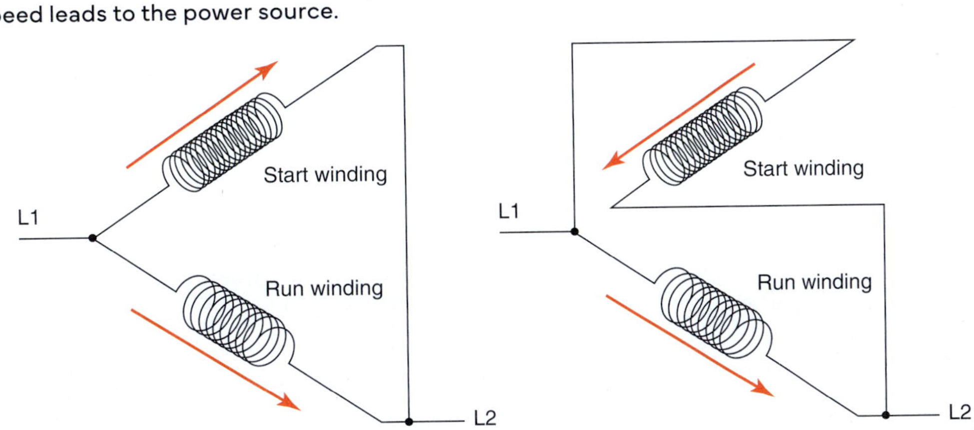

Goodheart-Willcox Publisher

Figure 28-29. The direction of rotation of some single-phase motors can be changed by reversing the direction of current flow through the start winding. The two motors shown in this diagram will rotate in opposite directions because the direction of current flow through the start windings is different.

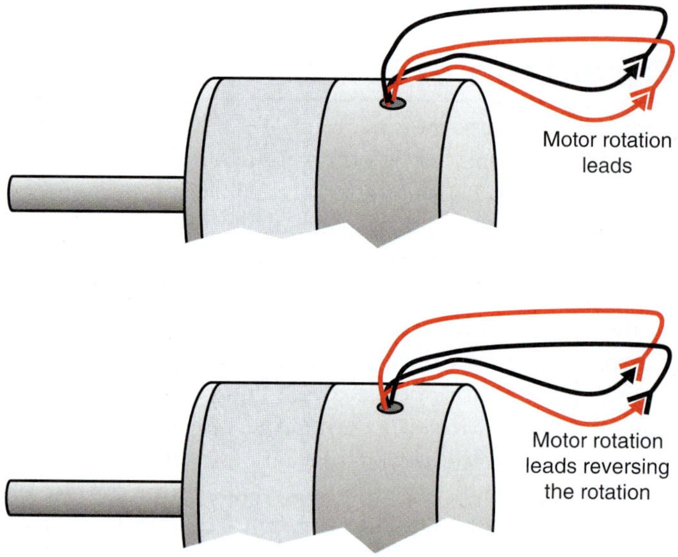

Goodheart-Willcox Publisher

Figure 28-30. Reversing the connection on the loops of colored wires of a single-phase motor will change the direction of the motor's rotation.

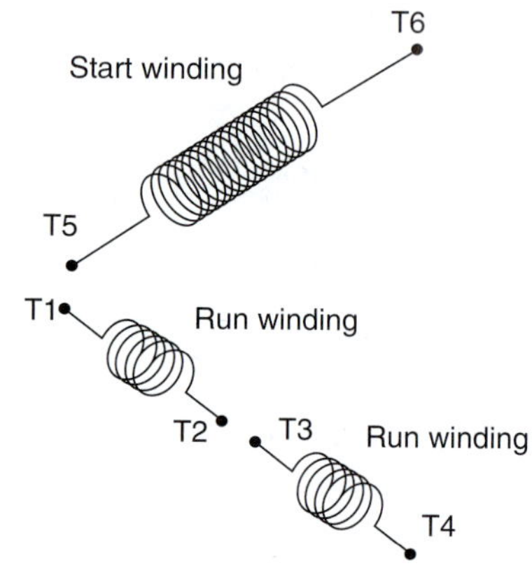

Goodheart-Willcox Publisher

Figure 28-31. Terminal numbering for a dual-voltage, single-phase motor with one start winding and two run winding segments.

Shedding Light

Be sure to check the wiring diagram for the specific motor you are working on, as terminal identifications can vary from one motor manufacturer to the next.

a solderless connector, which joins two sections of wire to create the loop. To change the direction of rotation, the loop connections are switched, so each loop is now made up of two wires, each with a different color, **Figure 28-30**. By switching these connections, the direction of current flow through the start winding is reversed. It is important to note that different motor manufacturers can utilize loops of wire with different color combinations.

28.9 Dual-Voltage, Single-Phase Motors

Some single-phase motors can operate at one of two different voltages. These motors are referred to as ***dual-voltage, single-phase motors***. Typically, a dual-voltage single-phase motor can operate at either 120 V or 240 V and is commonly manufactured in one of two ways:

- Two run windings and one start winding
- Two run windings and two start windings

Two Run Windings and One Start Winding Configuration

The two run winding and one start winding configuration uses two low-resistance run windings of equal resistance and one high-resistance start winding. In this configuration, the winding connection terminals for the dual-voltage motor are typically identified as illustrated in **Figure 28-31**.

When a 120 V/240 V dual-voltage motor is to be powered by a 240-V power supply, it is said to be wired for high-voltage operation. When wired to operate at high voltage, the run windings are connected in series with each other, and the start winding is wired in parallel with one of the run windings, **Figure 28-32**. With 240 V being supplied to the motor, approximately 120 V will be measured across any of the three windings, as the two run winding segments are wired in series with each other and effectively share the 240 V. In addition, one of the run winding segments is wired in series with the start winding. Therefore, the circuit's line voltage is shared between these windings. The larger the difference between the run winding resistance and the start winding resistance, the closer the voltage readings across the run winding segments will be to each other.

When a 120 V/240 V dual-voltage motor is to be powered by a 120-V power supply, it is said to be wired for low-voltage operation. When wired to operate at low voltage, the run windings and the start winding are all connected in parallel with each other, **Figure 28-33**. With 120 V being supplied to the motor, the same 120 V will be measured across any of the three windings. As a reminder, the amperage will be doubled when the voltage to the motor is cut in half.

Regardless of the voltage supplied to the motor, each winding segment will have approximately 120 V supplied to it. The electrical connections for a 120 V/240 V dual-voltage single-phase motor, when configured with two run windings and one start winding, are summarized in **Figure 28-34**.

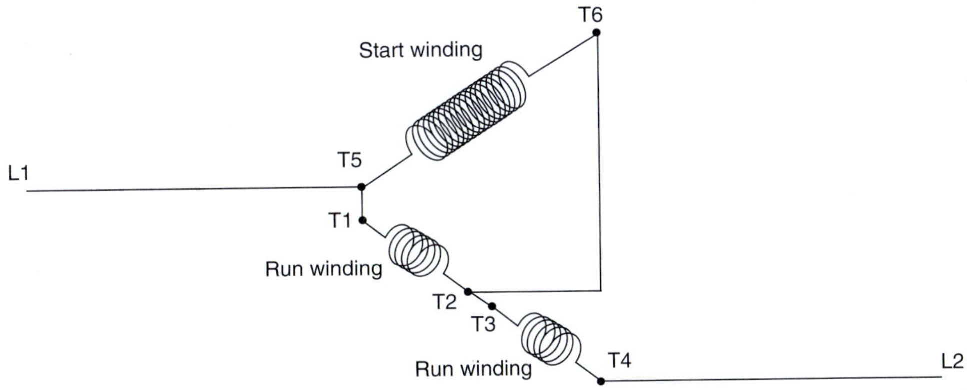

Goodheart-Willcox Publisher

Figure 28-32. High-voltage wiring diagram for a dual-voltage, single-phase motor with one start winding and two run winding segments.

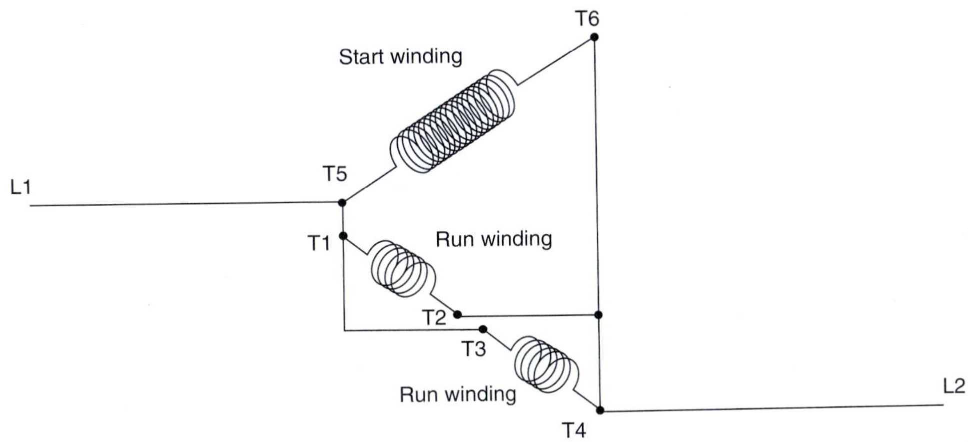

Goodheart-Willcox Publisher

Figure 28-33. Low-voltage connection diagram for a dual-voltage, single-phase motor with one start winding and two run winding segments.

Wiring Terminal Connections
Two Run Windings and One Start Winding

High Voltage (240 V)	Low Voltage (120 V)
L1 – T1 – T5	L1 – T1 – T3 – T5
L2 – T4	L2 – T2 – T4 – T6
T2 – T3 – T6	

Goodheart-Willcox Publisher

Figure 28-34. Wiring terminal connections for a dual-voltage, single-phase motor with two run windings and one start winding.

Two Run Windings and Two Start Windings Configuration

The two run windings and two start windings configuration uses two low-resistance run windings of equal resistance and two high-resistance start windings, also of equal resistance. For this configuration, the winding connection terminals for the dual-voltage motor are identified in **Figure 28-35**.

When this motor is wired to operate at 240 V, or high voltage, the run windings are connected in series with each other,

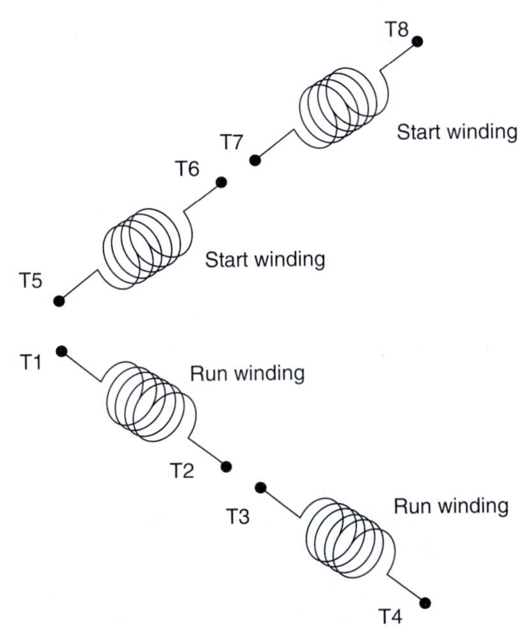

Goodheart-Willcox Publisher

Figure 28-35. Terminal numbering for a dual-voltage, single-phase motor with two start winding segments and two run winding segments.

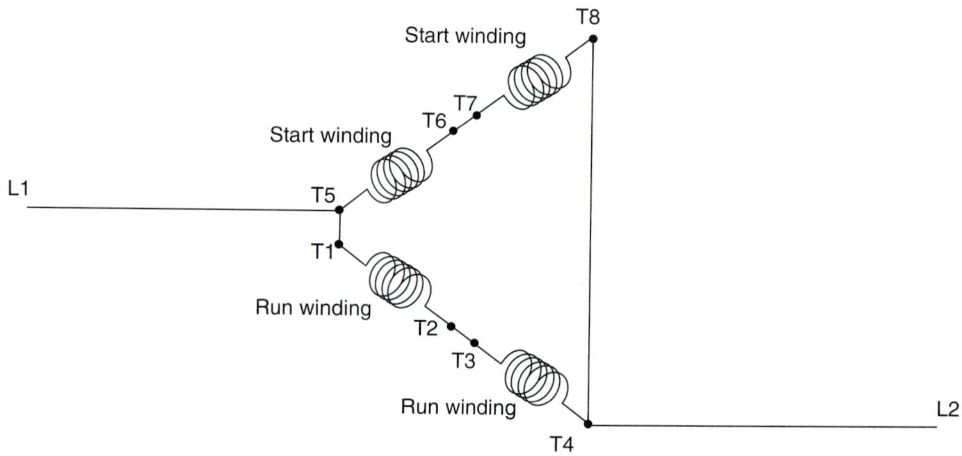

Goodheart-Willcox Publisher

Figure 28-36. High-voltage connection diagram for a dual-voltage, single-phase motor with two start windings and two run winding segments.

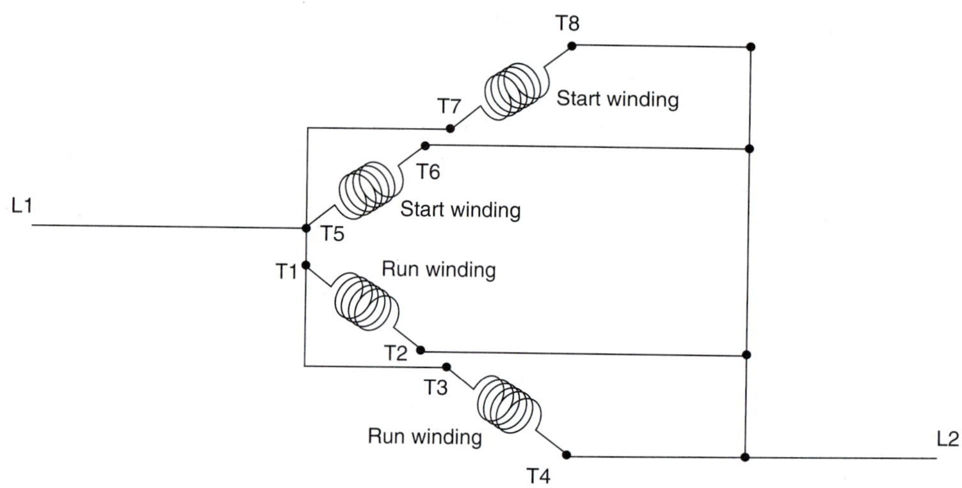

Goodheart-Willcox Publisher

Figure 28-37. Low-voltage connection diagram for a dual-voltage, single-phase motor with two start windings and two run winding segments.

as are the start windings, **Figure 28-36**. With 240 V being supplied to the motor, approximately 120 V will be measured across any of the four winding segments. With the two run winding segments, each with the same resistance, wired in series, they share the 240-V power signal equally. The same holds true for the two start winding segments.

When this motor is wired to operate at 120 V, or low voltage, all four winding segments are connected in parallel with each other, **Figure 28-37**. With 120 V being supplied to the motor, 120 V will be measured across each of the four winding segments. The electrical connections for a 120 V/240 V dual-voltage, single-phase motor configured with two run windings and two start windings are summarized in **Figure 28-38**.

Wiring Terminal Connections
Two Run Windings and Two Start Windings

High Voltage (240 V)	Low Voltage (120 V)
L1 – T1 – T5 L2 – T4 – T8 T6 – T7 T2 – T3	L1 – T1 – T3 – T5 – T7 L2 – T2 – T4 – T6 – T8

Goodheart-Willcox Publisher

Figure 28-38. Wiring terminal connections for a dual-voltage, single-phase motor with two run winding segments and two start winding segments.

CHAPTER 28 REVIEW

Summary

- The shaded-pole motor does not require any starting components and is a reliable motor with a very long, useful service life. (28.1)
- The shaded-pole motor uses shading coils to create the imbalance in the magnetic field needed to initiate and maintain motor operation. (28.1)
- Shaded-pole motors are used for small fractional-horsepower, direct-drive applications. (28.1)
- Shaded-pole motors are often rated in watts instead of horsepower. (28.1)
- Split-phase motors are induction motors that have two distinct windings connected in parallel with each other. (28.2)
- Induction-start, induction-run (ISIR) motors use a start winding, a run winding, and a starting device. The starting device, often a centrifugal switch, removes the start winding from the circuit once the motor nears its design speed. (28.3)
- ISIR motors are intended for use on light-duty and medium-duty applications that do not require a great deal of starting torque. (28.3)
- Capacitor-start, induction-run (CSIR) motors use a start winding, a run winding, a start capacitor, and a starting device. The starting device removes the start winding and start capacitor from the circuit after the motor starts. (28.4)
- CSIR motors have more starting torque than ISIR motors. (28.4)
- The current magnetic relay (CMR) is commonly used to remove the start winding of a CSIR motor from the circuit once the motor starts. (28.4)
- The permanent-split capacitor (PSC) motor has a run winding, a start winding, and a permanently installed run capacitor. (28.5)
- The PSC motor has low starting torque but high operating efficiency. (28.5)
- The PSC motor is a soft-start motor and is a popular choice when the noise level is a concern. (28.5)
- Capacitor-start, capacitor-run (CSCR) motors use a start winding, a run winding, a start capacitor, a run capacitor, and a starting device. The starting device removes the start capacitor from the circuit after the motor starts. (28.6)
- On start-up of the CSCR motor, the run capacitor and the start capacitor are connected in parallel with each other. (28.6)
- On a CSCR motor, the start capacitor is often removed from the circuit using a potential magnetic relay (PMR). (28.6)
- CSCR motors are used for applications that require a great deal of starting torque and a high running efficiency, such as belt-driven applications. (28.6)
- The speed of a multispeed motor can be changed by switching the wiring connections or using a selector switch. (28.7)
- Some single-phase motors can be made to rotate in either a clockwise or counterclockwise direction. The direction of rotation of a single-phase motor can be changed by reversing the start winding lead, which changes the direction of current flow through the start winding. (28.8)
- Dual-voltage, single-phase motors typically operate at either 120 V or 240 V. These motors are commonly manufactured with two run windings and one start winding or with two run windings and two start windings. (28.9)

Review Questions

Use the information in the chapter to answer the following questions.

Measuring Your Potential

1. Two common operating voltages for single-phase ac motors are _____. (28.1)
   A. 120 V and 24 V
   B. 120 V and 240 V
   C. 24 V and 240 V
   D. 240 V and 277 V
2. Shaded-pole motors are typically rated in _____. (28.1)
   A. watts
   B. amperes
   C. ohms
   D. volts
3. On most shaded-pole motors, the direction of rotation is changed by _____. (28.1)
   A. rotating the rotor 180°
   B. changing the motor's power lead connections
   C. grounding the motor's line voltage power lead
   D. changing the motor's shading coil connections
4. In a single-phase motor, the highest resistance will be measured across the _____. (28.2)
   A. closed contacts of the current magnetic relay
   B. closed contacts of the potential magnetic relay
   C. run winding
   D. start winding

5. The PTC motor starting device operates on the fact that an increase in resistance _____. (28.3)
   A. increases current
   B. decreases current
   C. increases voltage
   D. decreases voltage
6. The single-phase motor starting device that opens and closes its contacts based solely on the speed of the motor is the _____. (28.3)
   A. centrifugal switch
   B. positive temperature coefficient device
   C. current magnetic relay
   D. potential magnetic relay
7. The induction-start, induction-run motor uses _____. (28.3)
   A. only start and run windings
   B. a start winding, a run winding, and a motor starting device
   C. two run windings and a run capacitor
   D. a start winding, a run winding, a run capacitor, and a motor starting device
8. The phase angle between the run winding current and the start winding current is typically in the range of _____. (28.3)
   A. 0° to 20°
   B. 30° to 40°
   C. 40° to 60°
   D. 70° to 90°
9. The single-phase motor starting device that operates on the current flow through the run winding is the _____. (28.4)
   A. PTC
   B. CMR
   C. PMR
   D. ISR
10. The single-phase motor that operates with a run capacitor and no starting relay is the _____. (28.5)
    A. shaded-pole motor
    B. capacitor-start, capacitor-run motor
    C. permanent-split capacitor motor
    D. capacitor-start, induction-run motor
11. The single-phase motor starting device that operates on the measured voltage across the start winding is the _____. (28.6)
    A. PTC
    B. CMR
    C. PMR
    D. ISR
12. When a capacitor-start, capacitor-run motor is operating in its steady-state condition, the _____. (28.6)
    A. run capacitor is no longer in the active circuit
    B. current magnetic relay's coil will have a high resistance
    C. potential magnetic relay's coil will have a low resistance
    D. motor's active circuit resembles that of a PSC motor
13. On a multispeed PSC motor, the _____ lead is most likely the high-speed lead. (28.7)
    A. blue
    B. red
    C. black
    D. green
14. The direction of rotation of a single-phase motor is determined by the _____. (28.8)
    A. polarity of the power source
    B. direction of current flow through the run winding
    C. direction of current flow through the start winding
    D. electrical connections on the run capacitor

Plug-In

1. Changing the direction of rotation of a _____ motor cannot be done by changing the wiring connection. (28.1)
   A. capacitor-start, induction-run
   B. multispeed
   C. shaded-pole
   D. dual-voltage single phase
2. A low-horsepower fan is an example of a(n) _____ motor. (28.3)
   A. induction-start, induction-run
   B. capacitor-start, capacitor-run
   C. capacitor-start, induction-run
   D. permanent-split capacitor
3. A capacitor-start, capacitor-run motor is commonly found in _____. (28.6)
   A. general-purpose woodworking tools
   B. medical equipment
   C. household blenders
   D. furnaces

Making the Connection

1. Explain what will likely happen if the contacts of a centrifugal switch are stuck in the open position when the motor is initially energized. (28.3)
2. Explain what will happen if a dual-voltage motor is wired to operate at low voltage, but is connected to a high-voltage power supply. (28.9)

CHAPTER 29 Three-Phase AC Motors

Courtesy Fluke Corporation

SPARKING DISCUSSION

An electrician states that the voltage supplied to a three-phase motor is a factor that affects the motor's operating speed. Do you agree or disagree with this statement? Why or why not?

LEARNING OUTCOMES

After completing this chapter, you will be able to:

29.1 Explain the operation of a three-phase ac motor.

29.2 Calculate the synchronous speed of a three-phase motor.

29.3 Calculate the motor slip of a three-phase motor.

29.4 Evaluate the winding resistance in a wye-connected three-phase motor and a delta-connected three-phase motor.

29.5 Describe how the winding arrangements differ between a single-voltage and a dual-voltage three-phase motor.

29.6 Explain how to change the direction of rotation of a three-phase motor.

TECHNICAL TERMS

motor and phase rotation indicator

Introduction

Three-phase ac motors are widely used in various industrial and commercial applications because of their high efficiency levels and reliability over a wide range of loads and applications. They also offer excellent speed control capabilities and require a minimal amount of maintenance.

Unlike single-phase motors, which are commonly used in residential and light commercial applications, three-phase motors are designed for more demanding tasks that require higher starting and operating torque and a higher overall power output. In addition, a major distinction between three-phase motors and single-phase motors is that three-phase motors do not require the use of starting and operational devices such as start and run capacitors.

In this chapter, the operation of three-phase motors is covered. Various winding configurations, the concept of a rotating magnetic field, motor speed calculations, and direction of motor rotation are explained in the context of three-phase motors.

29.1 Three-Phase Motor Winding Polarity

As with any ac motor, the operation of a three-phase motor relies on a rotating magnetic field. What differentiates three-phase motors from single-phase motors is the fact that three-phase motors are supplied with three-phase power. Three-phase power is comprised of three individual single-phase power supplies that are 120° out of phase with each other. Three-phase power refers to the simultaneous production of three evenly offset, single-phase waveforms, each with the same frequency, **Figure 29-1**. Each phase of the three-phase power supply is connected to the stator windings of the three-phase motor. Each stator phase winding is often made up of numerous pole pieces. **Figure 29-2**, for example, illustrates a two-pole, three-phase motor, where each stator phase winding has two pole pieces. As the polarity of each phase of the power supply changes, so does the polarity of each pole piece. These pole pieces are evenly distributed or arranged in the motor's stator to create an even, high-torque force that helps make three-phase motors very efficient.

The pole pieces in a particular stator phase winding are wound so that they will always have opposite polarities. For example, if the A-1 pole piece acts as a south pole, the A-2 pole piece will act as a north pole, **Figure 29-3A**. If the direction of current flow through the A-1 pole piece changes, and the A-1 pole piece acts a north pole, then the A-2 pole piece will act as a south pole, **Figure 29-3B**. The same holds true for the B-1/B-2 and C-1/C-2 pole piece pairs. In **Figure 29-3C**, the B-1 and B-2 pole pieces have opposite polarities, as do the C-1 and C-2 pole pieces.

The B-1 and C-1 pole pieces are both wound in the same direction with respect to each other, so they will have the same polarity if the direction of current flow through each of these pole pieces is the same. For example, in **Figure 29-4** the direction of current flow for both the B and C phases is in the clockwise, or positive (+) direction. As a result, the polarity of the B-1 pole piece and the C-1 pole piece will be the same. At this point in time, **Figure 29-4** indicates that the C-1 and B-1 pole pieces are acting as magnetic north poles. If the directions of current flow through the B-1 and C-1 pole pieces is in opposite directions, the polarities of these two pole pieces will be

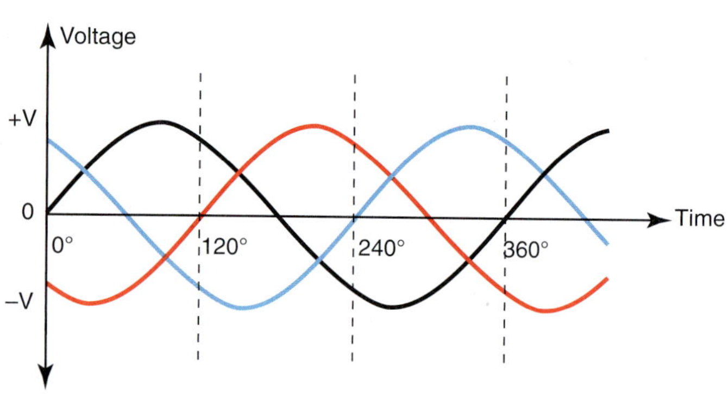

Goodheart-Willcox Publisher

Figure 29-1. A three-phase power supply is made up of three single-phase waveforms, each with the same frequency, positioned 120 electrical degrees out-of-phase with each other.

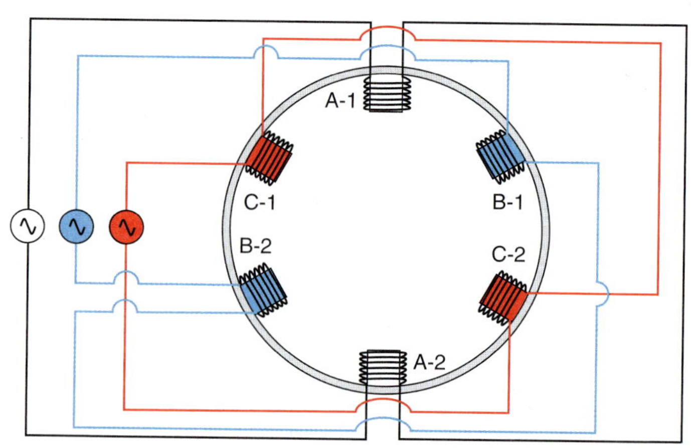

Goodheart-Willcox Publisher

Figure 29-2. Each phase winding has multiple pole pieces. In this example, phase A has pole pieces A-1 and A-2, phase B has pole pieces B-1 and B-2, and phase C has pole pieces C-1 and C-2.

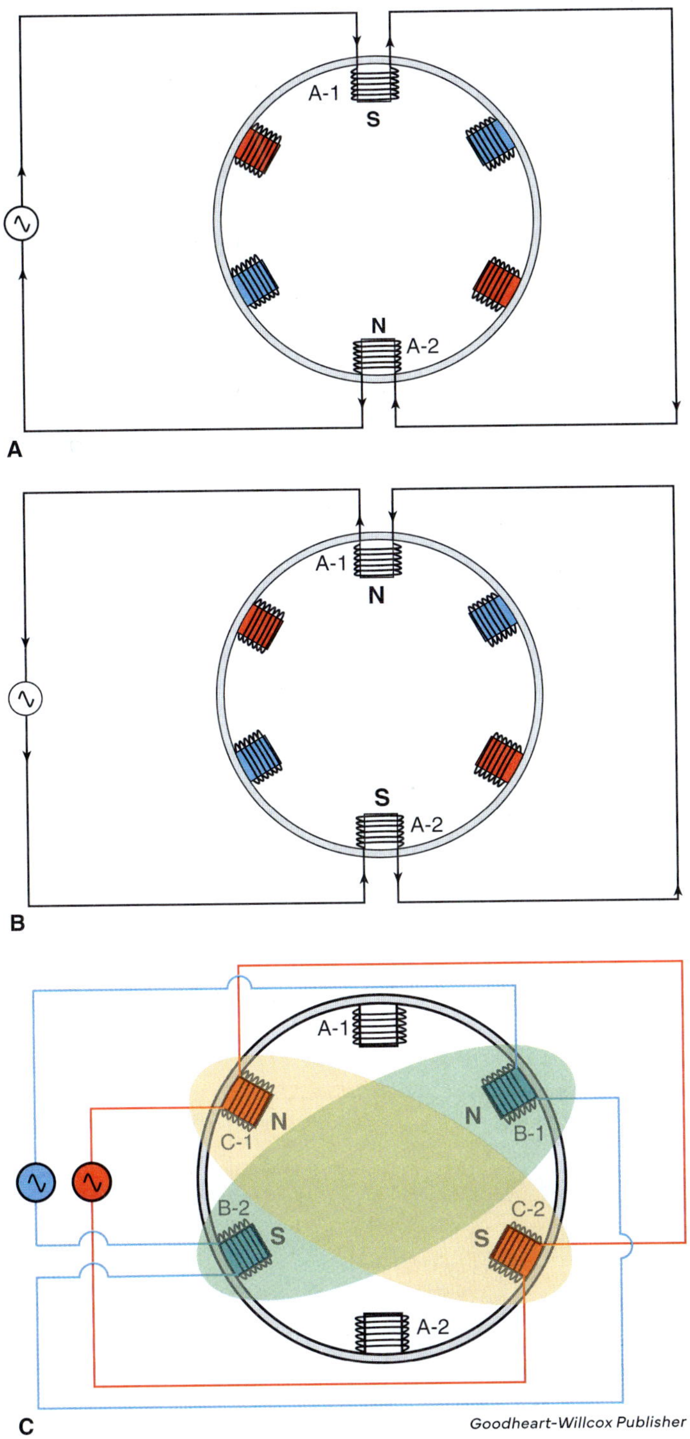

Goodheart-Willcox Publisher

Figure 29-3. Pole pieces A-1 and A-2 have opposite polarities. A—In the example shown, pole piece A-1 is acting as a south pole, while pole piece A-2 is acting as a north pole. B—When the polarity of the power signal reverses, so do the polarities of the pole pieces. With the direction of current reversed, pole piece A-1 acts as a north pole, while pole piece A-2 acts as a south pole. C—All sets of pole pieces will have opposite polarities.

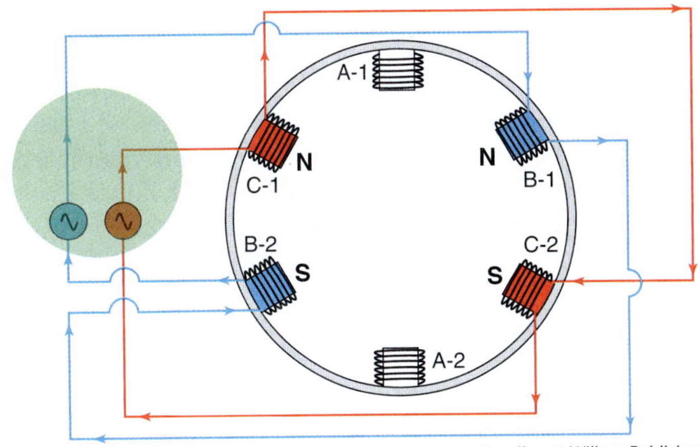

Goodheart-Willcox Publisher

Figure 29-4. Phase B and phase C are wound in the same direction. As a result, the B-1 and C-1 pole pieces will have the same polarity when both phases have the same polarity.

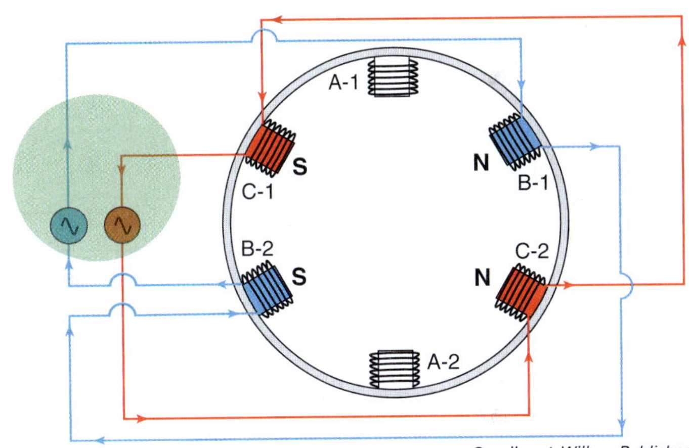

Goodheart-Willcox Publisher

Figure 29-5. When the phase currents for phase B and phase C are in the opposite direction, then B-1 and C-1 pole pieces will have opposite polarities.

opposite. In **Figure 29-5**, the current flow through the B phase is in the clockwise direction, while the current flow through the C phase is in the counterclockwise direction. As a result, the polarities of the B-1 and C-1 pole pieces are opposite to each other. The B-1 pole piece is acting as a north pole, while the C-1 pole piece is acting like a south pole.

Although the B-1 and C-1 pole pieces are both wound in the same direction with respect to each other, they are wound in the opposite direction with respect to the A-1 pole piece. In this configuration, if the direction of current flow through all three phases is the same, then the B-1 and C-1 polarities will be the same, but the polarity of the A-1 pole piece will be different, **Figure 29-6**. If, however, the direction of current flow through the A-1 pole piece is reversed while not changing the direction of current flow through the B-1 and C-1 pole pieces, the polarity of the A-1, B-1, and C-1 pole pieces will all be the same, **Figure 29-7**.

Magnetic Field Rotation

Basic three-phase motor operation relies on the interaction among rotating magnetic fields and the conductors that carry ac current through the motor's windings. The changing pole piece polarities follow a uniform path that facilitates effective and efficient rotor rotation. To illustrate pole alignment and the rotating magnetic fields, we will examine the lines

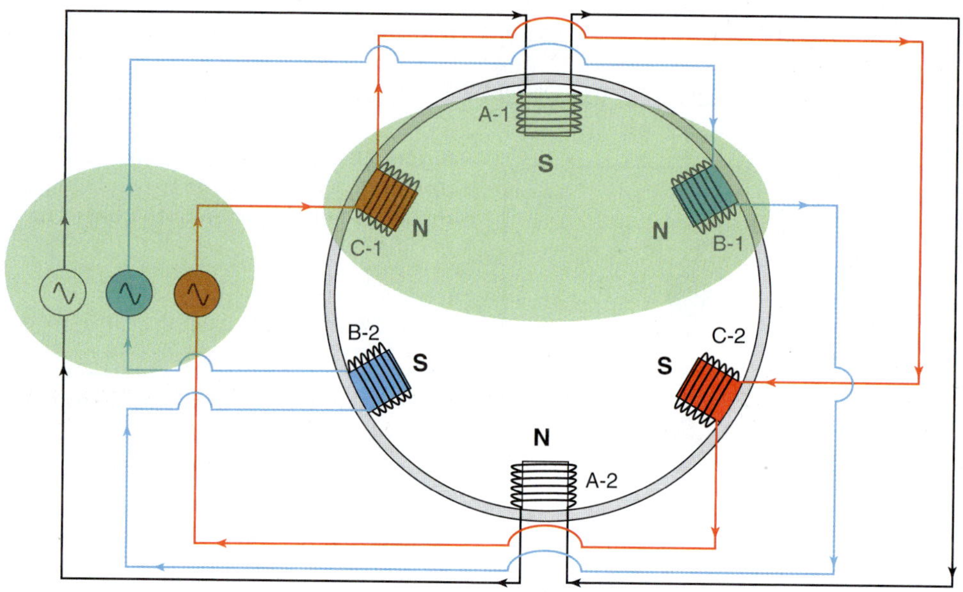

Goodheart-Willcox Publisher

Figure 29-6. Phase A is wound in a direction that is opposite to the direction of phases B and C. If the polarity of phase A is the same as that of phase B or phase C, the polarity of the phase A pole pieces will be opposite that of the phase B or phase C pole pieces.

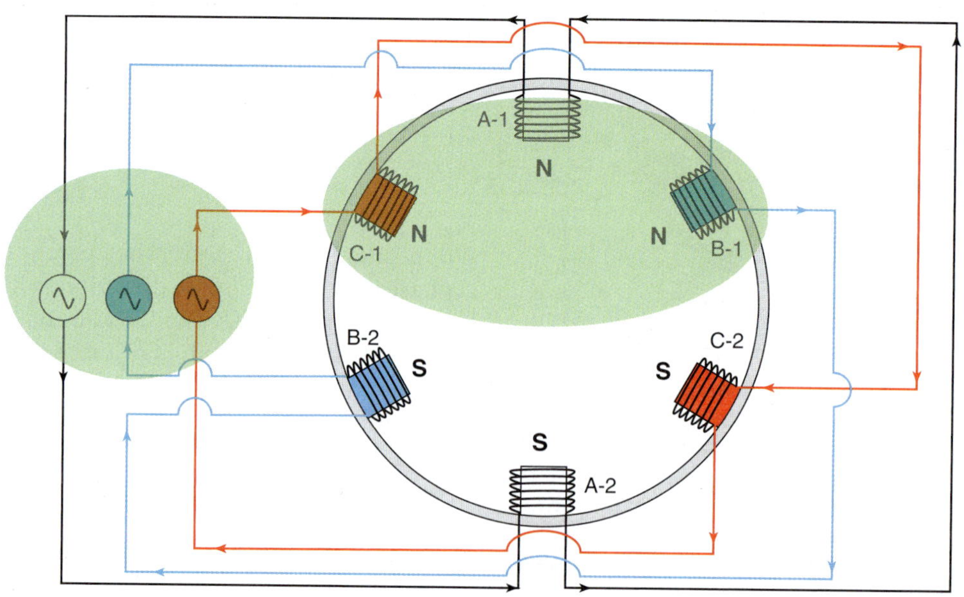

Goodheart-Willcox Publisher

Figure 29-7. If the A-1, B-1 and C-1 pole pieces all have the same polarity, then the polarity of phase B and phase C will be the same, and the polarity of phase A will be opposite to that of phase B and phase C.

of force that are produced during a complete three-phase ac cycle. A single, 360-degree cycle can be broken down into six, 60-degree segments for further understanding.

The zero electrical degree point represents the start of a cycle for Phase A. At this point, the polarity of the B phase is positive, and the polarity of the C phase is negative. The A phase is at a potential of 0 V and, therefore, is not producing any magnetic field, **Figure 29-8**. Since the potential of the A phase is zero and no magnetic field is being produced by that phase, it is represented by a dotted line and no polarities are indicated for the A-1 and A-2 pole pieces. The current flow through the B phase is in the clockwise direction, indicating a positive polarity. If the B phase has a positive polarity, then the B-1 pole piece will act as a north pole. The current flow through the C phase is in the counterclockwise direction, indicating a negative polarity. If the C phase has a negative polarity, then the C-1 pole piece will act as a south pole. As a result, the B-2 pole piece will act as a south pole, and the C-2 pole piece will act as a north pole. With the direction of force acting on the motor's rotor in the north-to-south direction, the forces is directed to the left. See **Figure 29-8.**

At the 60 electrical degree point, the polarity of the A phase is positive, and the polarity of the C phase is negative. The B phase is at a potential of 0 V and, therefore, is not producing any magnetic field, **Figure 29-9**. Since the potential of the B phase is zero and no magnetic field is being produced by that phase, it is represented by a dotted line and no polarities are indicated for the B-1 and B-2 pole pieces. The current flow through the A phase is in the clockwise direction, indicating a positive polarity. With a positive polarity, the A-1 pole piece acts as a south pole. The A phase is wound in a direction that is opposite in polarity to that of the B and C phases. The current flow through the C phase is in the counterclockwise direction, indicating a negative polarity. With a negative polarity, the C-1 pole piece will act as a south pole. The A-2 and C-2 pole pieces will both act as north poles. With the direction of force acting on the motor in the north-to-south direction, the directional force is directed up and to the left. See **Figure 29-9**.

As the cycle continues, the polarity of the poles and the direction of force continue to change based on the polarity of the waveform as illustrated in **Figure 29-10**.

Observing the positions of the arrows in **Figures 29-11**, it can be concluded that the magnetic field is rotating in the clockwise direction. The motor's rotor rotates in the same direction as the magnetic field, so the motor's shaft is also turning in that direction.

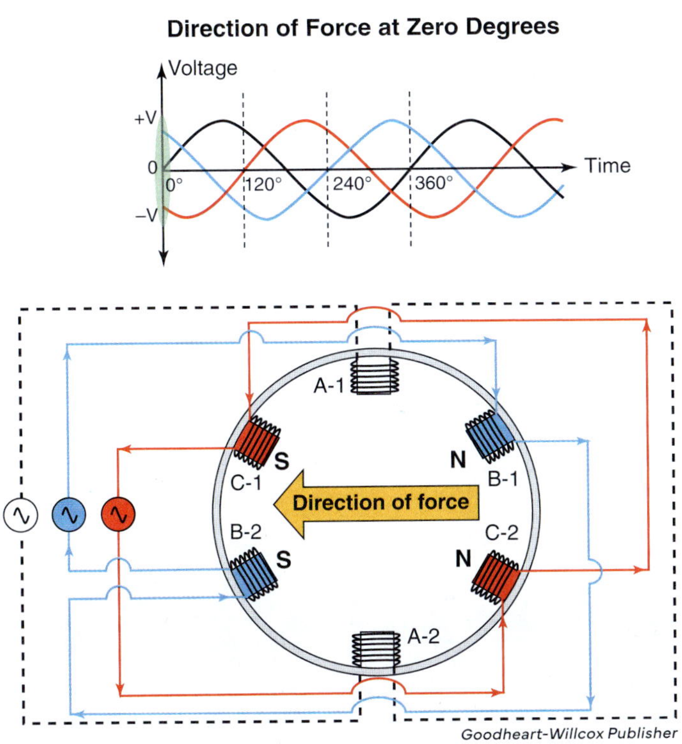

Figure 29-8. At zero electrical degrees, phase A has no effect on the force. The B-1 and C-2 pole pieces act as north poles, and the B-2 and C-1 poles act as south poles.

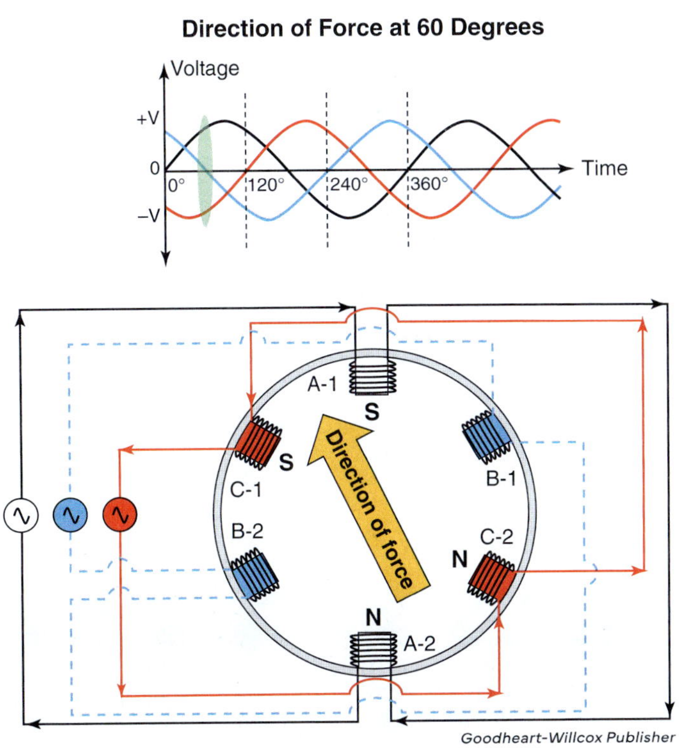

Figure 29-9. At 60 electrical degrees, phase B has no effect on the force. The A-1 and C-1 pole pieces act as south poles, and the A-2 and C-2 poles act as north poles.

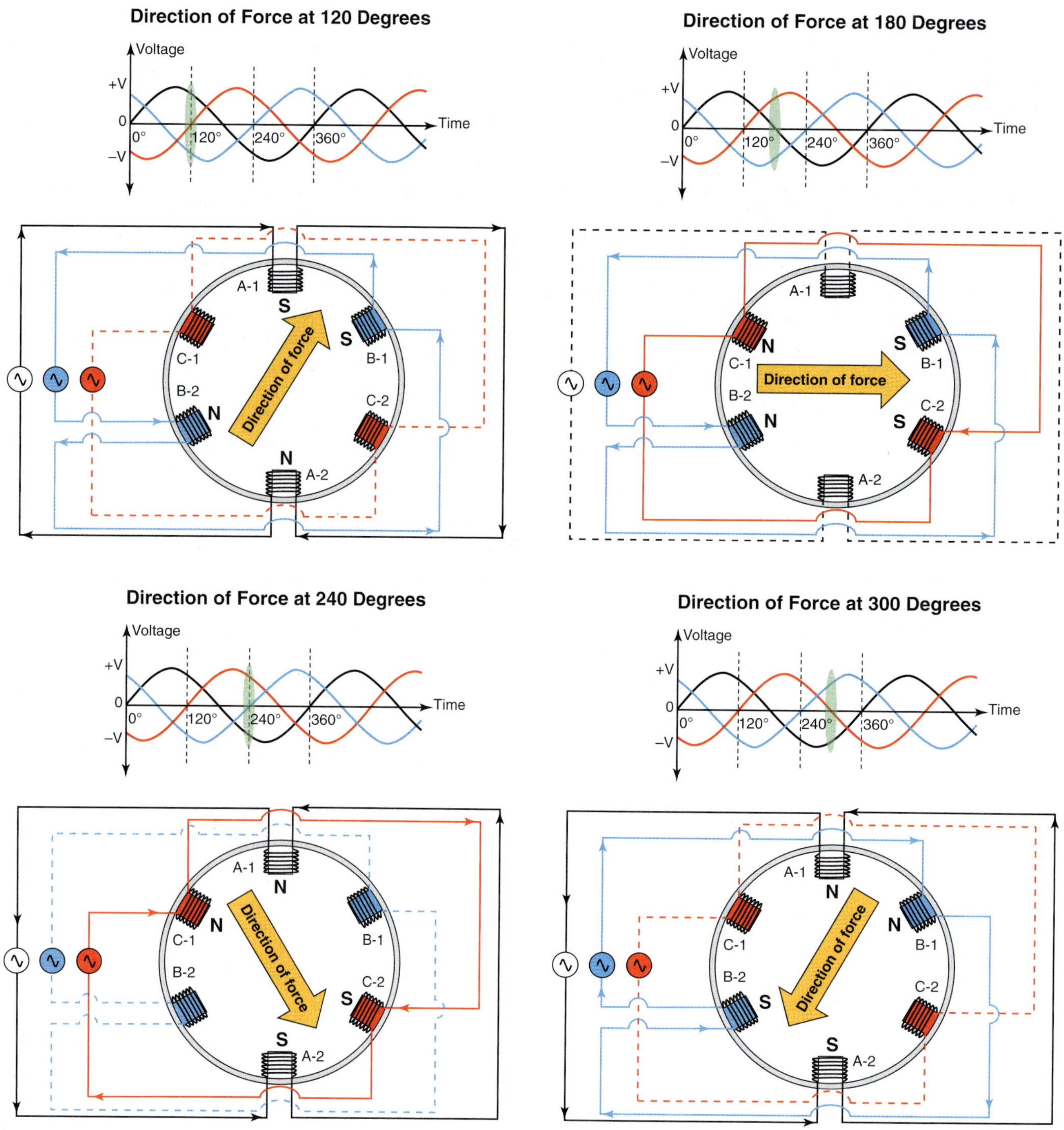

Goodheart-Willcox Publisher

Figure 29-10. The polarity of the poles and direction of magnetic force change as the cycle of the electrical waveform progresses.

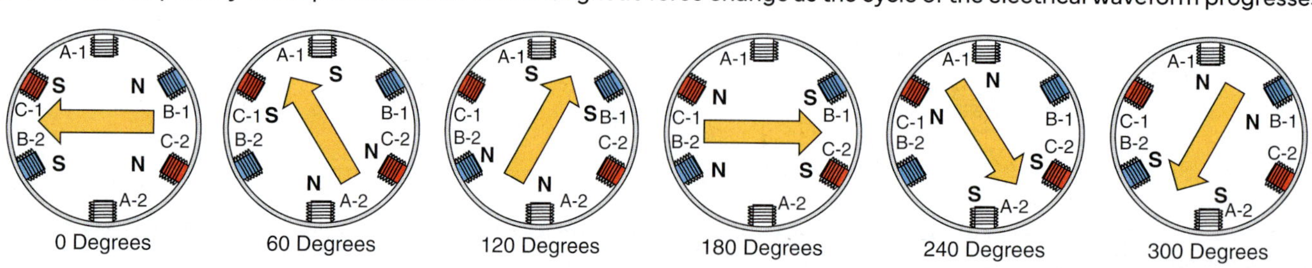

Goodheart-Willcox Publisher

Figure 29-11. The force direction in the 60-degree segments shows that the direction of force rotates in a clockwise direction.

29.2 Three-Phase Motor Speed

A motor's synchronous speed is the motor's ideal, or calculated, speed. The synchronous speed of a three-phase motor is expressed in the units of rotations per minute, or rpm. Motor speed is a function of two factors: the frequency of the power supply and the number of pole pieces per phase. In addition to these two factors, a conversion factor of 120 is incorporated into the motor's synchronous speed calculation. This constant takes into account two unit conversions to obtain the units of rotations per minute. The first unit conversion factor, 60 seconds/minute, changes seconds into minutes. The second conversion factor, two polarity changes/cycle, is used to convert from signal frequency to motor rotation speed. The formula used to calculate motor speed is as follows:

$$N_S = \frac{120 \times f}{P}$$

where:

N_S = Synchronous speed (in rpm)
120 = Constant for two polarity changes per cycle
f = Frequency of supplied current (in Hz)
P = Number of poles

To calculate synchronous speed in a three-phase motor, see **Example 29-1** and **Example 29-2**.

Example 29-1

What is the synchronous speed of a four-pole, three-phase motor if the power supply has a frequency of 60 Hz?

✓ Solution

Use the synchronous speed formula:

$$N_S = \frac{120 \times f}{P}$$

$$N_S = \frac{120 \times 60\text{ Hz}}{4}$$

$$N_S = \frac{7200}{4}$$

$$N_S = 1800\text{ rpm}$$

Example 29-2

If the speed of the motor in **Example 29-1** must be increased to 2700 rpm, what change must be made to the frequency of the power supply?

✓ Solution

Since we know that the motor is a four-pole motor and the synchronous speed is to be 2700 rpm, the required frequency can be found using the synchronous speed formula:

$$N_S = \frac{120 \times f}{P}$$

Rearrange the equation to solve for *f*.

$$f = \frac{N_S \times P}{120}$$

$$f = \frac{2700\text{ rpm} \times 4}{120}$$

$$f = \frac{10{,}800}{120}$$

$$f = 90\text{ Hz}$$

Now It's Your Turn 29-2

How many pole pieces per phase does a three-phase motor have if it operates at a synchronous speed of 1500 rpm when supplied with a 100 Hz power supply?

29.3 Three-Phase Motor Slip

The speed at which a motor turns is always lower than the synchronous, or calculated, motor speed. This difference between a motor's calculated and actual speed is called motor slip. Refer to Chapter 23, *Motor Operations Basics*, where motor slip is first explained.

Motor slip can be expressed as a numerical value in rotations per minute, rpm, or as a percentage. Whether expressed as a numerical value or a percentage, motor slip should be no more than 5% of the synchronous motor speed. The numerical value for slip is calculated as follows:

$$S = N_S - N$$

where:

S = Slip (in rpm)
N_S = Synchronous motor speed (in rpm)
N = Actual motor speed (in rpm)

Refer to **Example 29-3** and **Example 29-4** to calculate motor slip using this formula.

Example 29-3

Calculate the slip of a three-phase motor if the motor's actual speed is 1725 rpm and its synchronous speed is 1800 rpm.

✓ Solution

The motor slip is then calculated as follows:

$$S = N_S - N$$

$$S = 1800 \text{ rpm} - 1725 \text{ rpm}$$

$$S = 75 \text{ rpm}$$

Example 29-4

Calculate the slip of a two-pole, three-phase motor if the frequency of the power supply is 60 Hz and the motor's actual speed is 3450 rpm.

✓ Solution

The solution is determined by first calculating the synchronous speed of the motor as follows:

$$N_S = \frac{120 \times f}{P}$$

$$N_S = \frac{120 \times 60 \text{ Hz}}{2}$$

$$N_S = \frac{7200}{2}$$

$$N_S = 3600 \text{ rpm}$$

The motor slip is then calculated as follows:

$$S = N_S - N$$

$$S = 3600 \text{ rpm} - 3450 \text{ rpm}$$

$$S = 150 \text{ rpm}$$

When expressed as a percentage, slip is calculated as follows:

$$\%S = \frac{N_S - N}{N_S} \times 100\%$$

where:

$\%S$ = Percentage slip
N_S = Synchronous motor speed (in rpm)
N = Actual motor speed (in rpm)

To calculate the percentage slip for a three-phase ac motor, see **Example 29-5**.

Example 29-5

Calculate the percentage slip for a three-phase ac motor that operates with a synchronous speed of 900 rpm and actual speed of 860 rpm.

✓ Solution

The percentage of motor slip is calculated by substituting the known values into the expression as follows:

$$\%S = \frac{N_S - N}{N_S} \times 100\%$$

$$\%S = \frac{900 - 860}{900} \times 100\%$$

$$\%S = \frac{40}{900} \times 100\%$$

$$\%S = 4.44\%$$

Now It's Your Turn 29-5

Calculate the actual speed of a two-pole, three-phase motor if its synchronous speed is 3600 rpm and the motor is operating with a 3.5% motor slip.

29.4 Three-Phase Motor Winding Configurations

Unlike single-phase motors that have two windings, simple three-phase motors have three windings. All three windings in a three-phase motor have the same resistance. In Chapter 16, *Three-Phase Alternating Current Power*, the concepts of three-phase wye and delta connections are discussed as they apply to three-phase power. The same wye and delta configurations can be found in three-phase motors.

Wye-Connected Three-Phase Motors

Recall that the wye configuration is so named because of its similarity to the letter *Y* when drawn in wiring diagrams. In a simple three-phase motor, the three windings are joined together at a common connection point. When energized, each of the three ungrounded line voltage conductors (L1, L2, and L3) will be connected to each of the unique winding terminals, or connection points, on the motor, **Figure 29-12**.

Wye-Connected Winding Resistance

Although the resistance of the windings on a three-phase motor is the same, the measured resistance across any two motor leads of a properly operating motor will not be equal

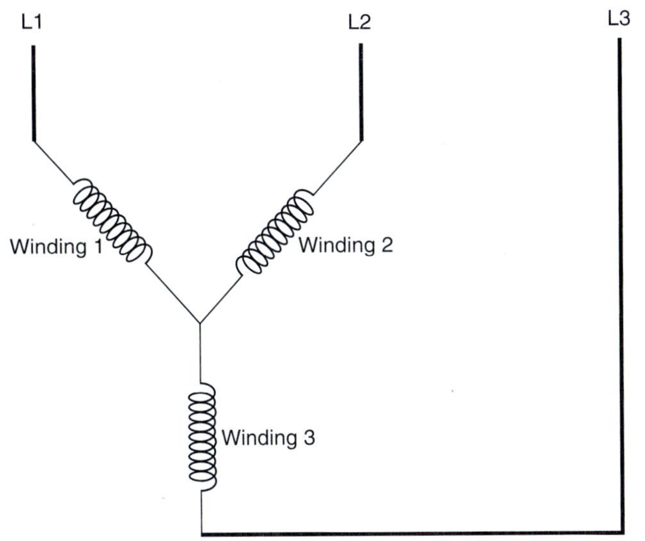

Goodheart-Willcox Publisher

Figure 29-12. Wiring for a wye-connected three-phase motor. Each ungrounded line voltage conductor (L1, L2, and L3) is connected to one of the motor leads.

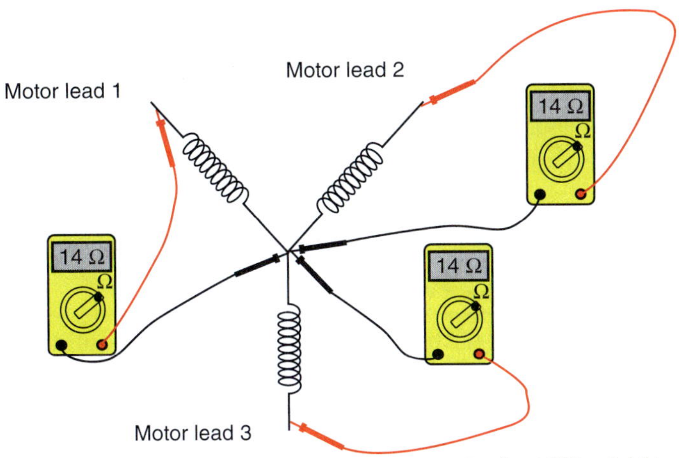

Goodheart-Willcox Publisher

Figure 29-13. A wye-connected three-phase motor with individual winding resistances of 14 Ω.

to the actual resistance of the motor's winding. Consider a simple three-phase motor with three motor leads and individual winding resistances of 14 Ω, **Figure 29-13**. If you were to take an actual resistance reading between any two of the motor leads, the meter's display would register a resistance that is twice the actual resistance of a single motor winding, **Figure 29-14**. The conductive path that is being evaluated with the ohmmeter is comprised of two of the motor windings connected in series with each other, **Figure 29-15**. When evaluating resistance in a wye-connected three-phase motor, each obtained resistance reading is actually being taken through two windings, not one.

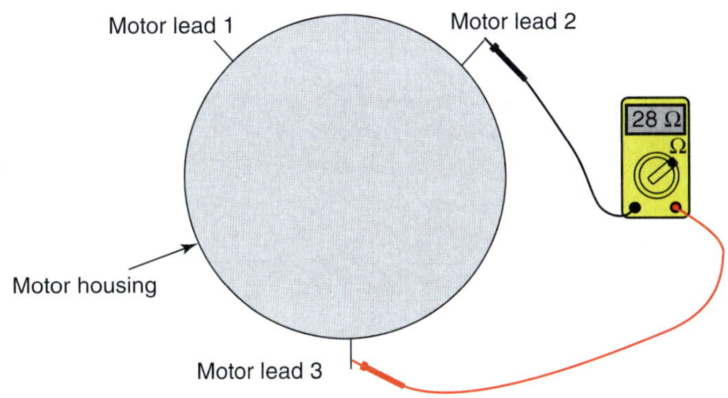

Goodheart-Willcox Publisher

Figure 29-14. In a wye-connected three-phase motor, the measured resistance between any two of the motor's three leads will be twice the resistance of an individual winding.

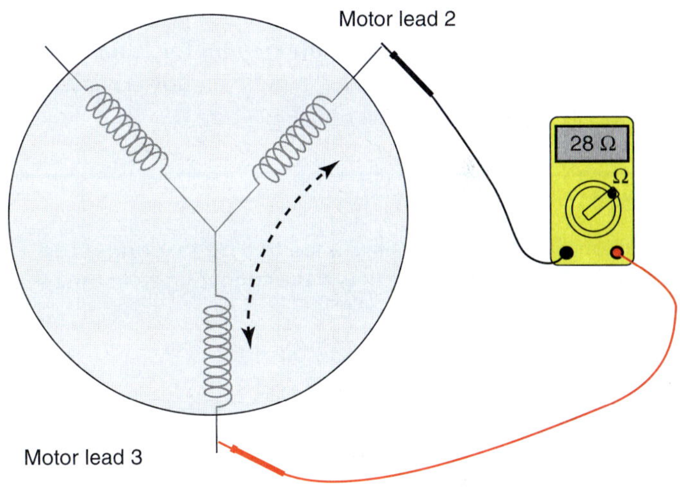

Goodheart-Willcox Publisher

Figure 29-15. The conductive path between any two of the motor's three leads includes two of the motor's windings.

Delta-Connected Three-Phase Motors

The delta wiring configuration resembles the Greek letter delta, Δ. The three phases are connected to each other to form a continuous loop, where the outlet of one phase is connected to the inlet of the next. Each phase of the power supply connects to two phases of the motor, **Figure 29-16**.

Delta-Connected Winding Resistance

As in the case of the wye-connected windings of a three-phase motor, the measured resistance across any two motor leads will not be equal to the actual resistance of the motor's winding. In the case of a delta-configured three-phase motor, the measured resistance between any two motor leads will be equal to two-thirds of the resistance of one of the windings. If the resistance of an individual motor winding is 15 Ω, the

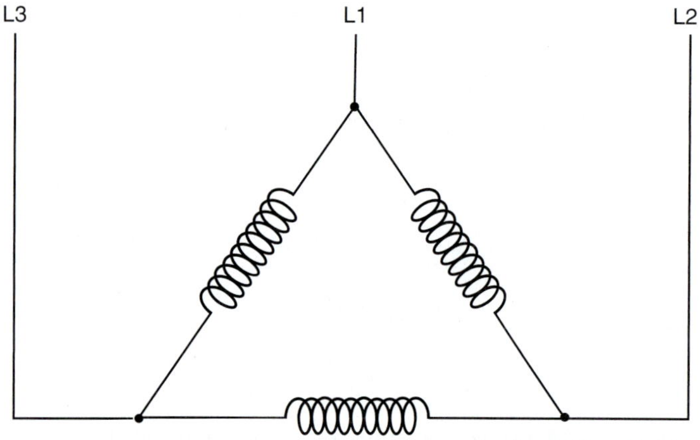

Goodheart-Willcox Publisher

Figure 29-16. Each ungrounded line voltage conductor (L1, L2, and L3) is connected to one of the motor leads at each corner of the delta configuration.

measured resistance between any two motor leads will be 10 Ω. **Example 29-6** provides a sample on how to calculate the actual resistance.

Example 29-6

Determine the actual resistance of a delta-connected three-phase motor winding if the following resistance readings are obtained.

- Resistance between Leads 1 and 2: 10 Ω
- Resistance between Leads 1 and 3: 10 Ω
- Resistance between Leads 2 and 3: 10 Ω

✓ Solution

Referring to the following figure, a random value of *x* represents the resistance of each of the motor's three windings. It can also be seen in this figure that there are two parallel paths through which the ohmmeter is measuring resistance: path A and path B.

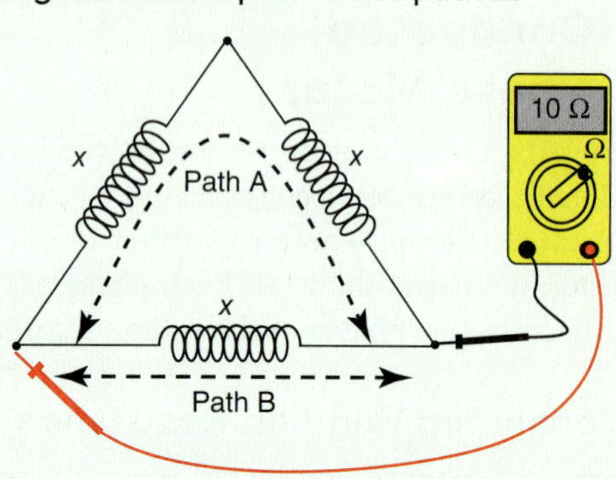

Goodheart-Willcox Publisher

Path A contains two resistors connected in series, so the total resistance of path A is determined by adding the two resistances. Thus, the resistance of path A is $x + x$, or $2x$. The resistance of path B is x. The total resistance of two resistances connected in parallel with each other is given by the following product over sum formula:

$$R_T = \frac{R_1 \times R_2}{R_1 + R_2}$$

Substitute the value of $2x$ (path A) for R_1 and x (path B) for R_2, as follows:

$$R_T = \frac{2x \times x}{2x + x}$$

$$R_T = \frac{2x^2}{3x}$$

$$R_T = \frac{2x}{3}$$

$$R_T = 0.667x$$

Since the measured, or total, resistance is 10 Ω, this value is substituted for R_T as follows:

$$R_T = 0.667x$$

$$10\ \Omega = 0.667x$$

Solving for x, the resistance of an individual motor winding is calculated as follows:

$$10\ \Omega = 0.667x$$

$$x = \frac{10\ \Omega}{0.667}$$

$$x = 15\ \Omega$$

29.5 Dual-Voltage Three-Phase AC Motors

A large number of three-phase motors have the ability to operate at one of two voltages, depending on how they are wired. These motors, called dual-voltage motors, have two windings per phase, for a total of six windings, instead of the single winding phases that were introduced earlier in this chapter, **Figure 29-17**.

Dual-voltage three-phase motors are commonly configured to operate at 208/230 V or 460 V. The 208 V or 230 V operating voltage is referred to as low voltage, while the 460 V operating voltage is referred to as high voltage. Typically, these motors have nine leads extending from the motor's housing and are numbered in a standard manner. However, in some cases, dual-voltage three-phase motors are constructed with 12 motor leads extending from the motor housing.

Wye-Connected Dual-Voltage Motors

For ac motors manufactured to NEMA standards, terminals are marked as T1, T2, T3, etc. For a wye-connected dual-voltage motor, the numbering starts at the outermost edge and follows a clockwise direction, moving inward to the next available unnumbered motor lead. The numbering starts with

T1 and continues inward until the last lead, T9, is identified, **Figure 29-18**.

For high-voltage operation, the windings in each phase are wired in series with each other to create one large wye connection. In the wye-connected dual-voltage motor in **Figure 29-19**, motor lead T4 is connected to lead T7, lead T5 is connected to lead T8, and lead T6 is connected to lead T9. Each leg of the three-phase power supply is connected to one

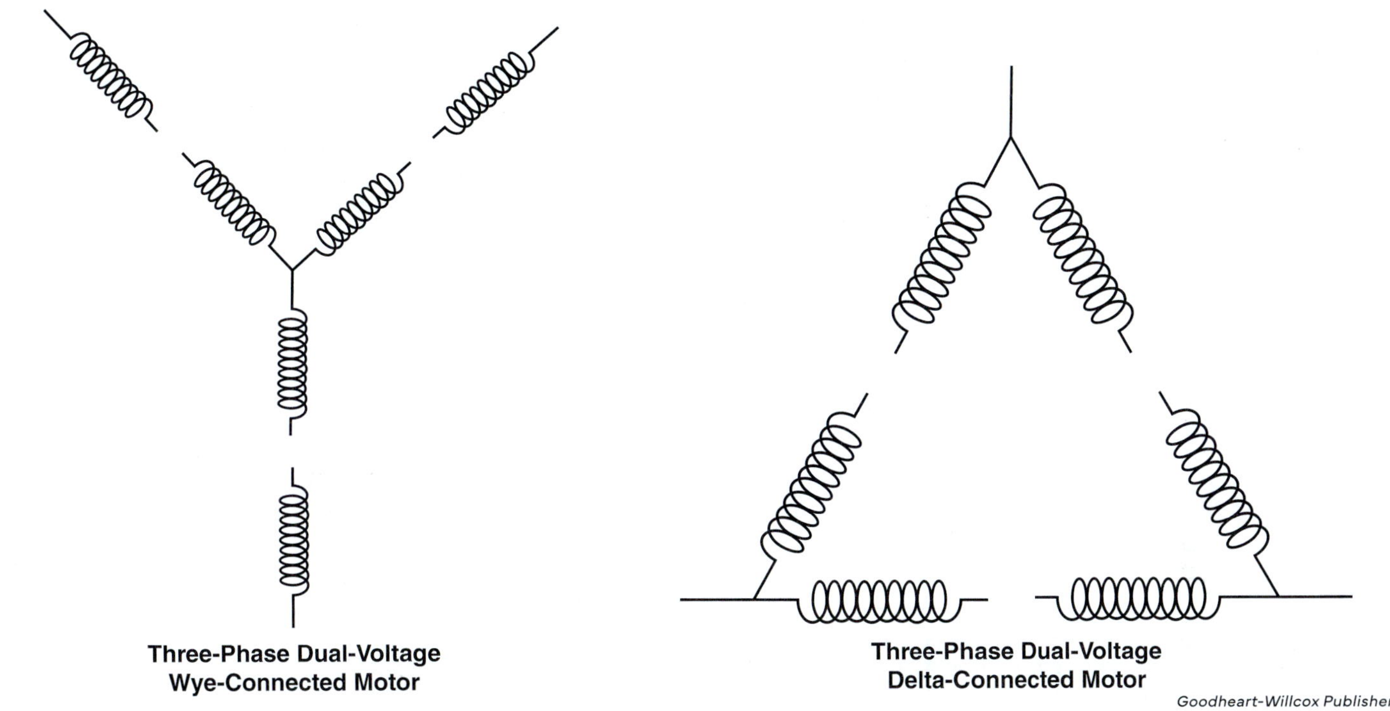

Figure 29-17. Winding configurations for dual-voltage motors.

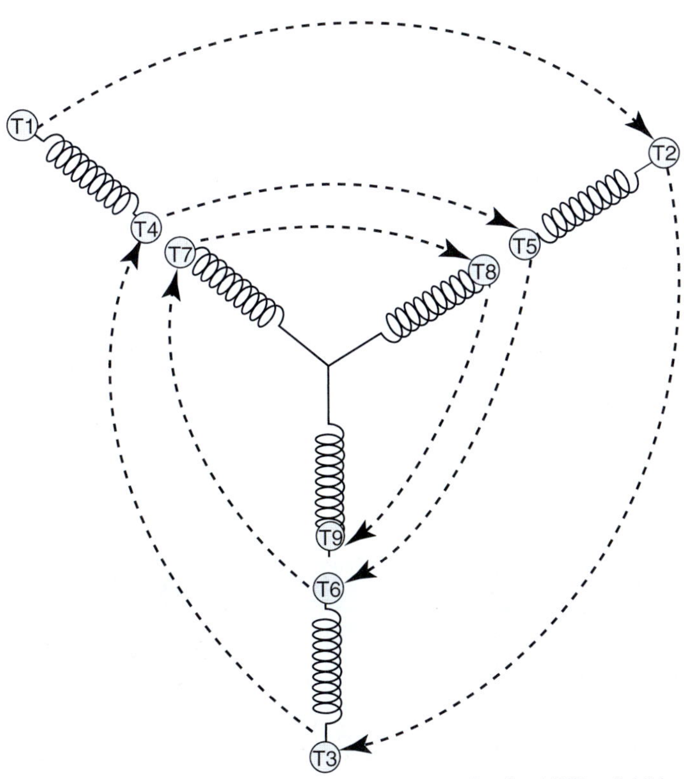

Figure 29-18. The numbering for a three-phase dual-voltage wye-connected motor follows an inward-spiraling path.

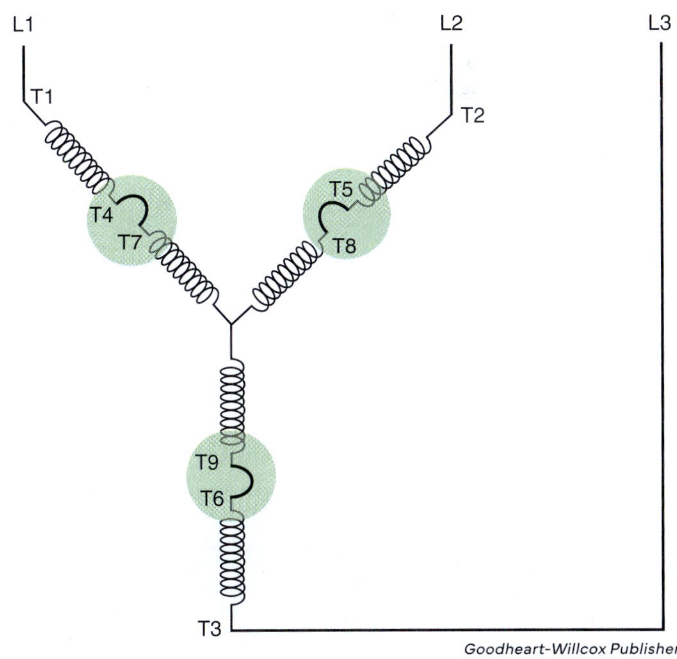

Figure 29-19. For high-voltage operation, leads T4 and T7, T5 and T8, and T6 and T9 are connected together to form one large wye connection. The ungrounded line voltage conductors (L1, L2, and L3) are connected to motor leads T1, T2 and T3.

of the remaining motor leads. A simplified wiring diagram for a wye-connected dual-voltage motor connected to operate at high voltage is provided in **Figure 29-20**.

For low-voltage operation, the windings in each phase are wired in parallel with each other, creating two smaller wye connections. This is accomplished by first connecting motor leads T4, T5, and T6 together, **Figure 29-21**. Once connected in this manner, two smaller wye connections are formed, **Figure 29-22**. Each leg of the three-phase power supply is connected to two of the remaining six motor leads, **Figure 29-23**; L1 is connected to motor leads T1 and T7, L2 is connected to leads T2 and T8, and L3 is connected to leads T6 and T9. **Figure 29-24** shows a simplified connection diagram. The wiring diagram for a wye-connected dual-voltage

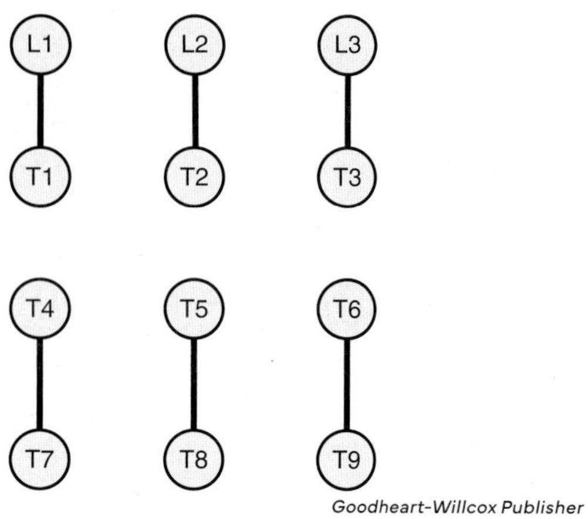

Figure 29-20. A simplified connection diagram for a nine-wire, wye-connected dual-voltage three-phase motor that is to operate at high voltage.

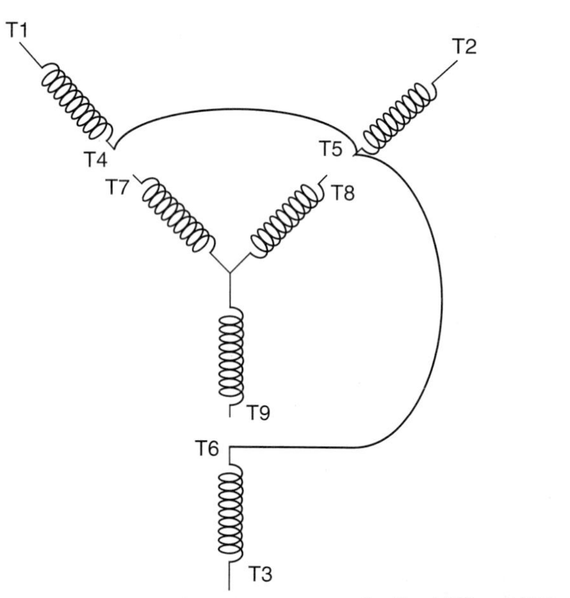

Figure 29-21. For low-voltage operation of a wye-connected dual-voltage three-phase motor, leads T4, T5, and T6 are connected together.

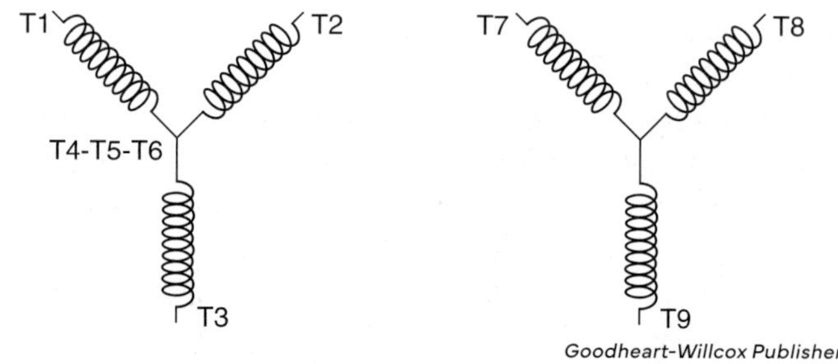

Figure 29-22. For low-voltage operation of a wye-connected dual-voltage three-phase motor, two smaller wye configurations are formed. The center of one of the small wyes is created by the connection of leads 4, 5, and 6.

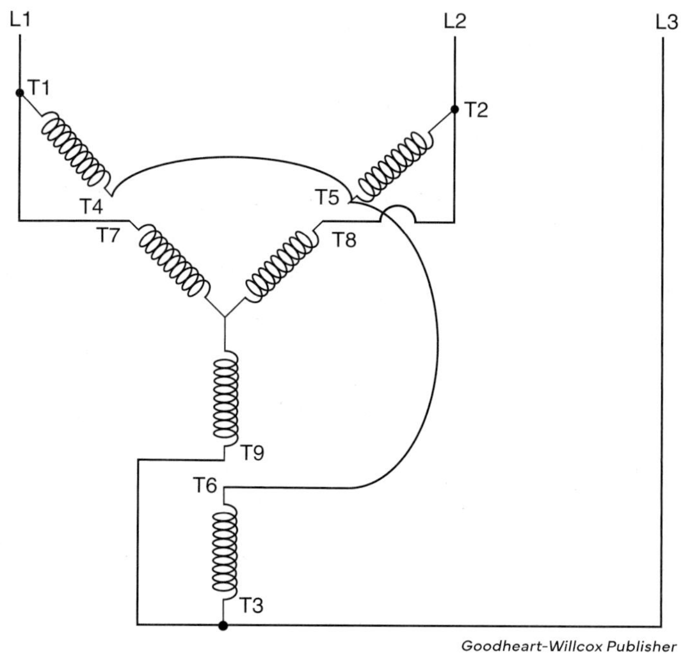

Figure 29-23. Wiring diagram of a nine-wire, wye-connected dual-voltage three-phase motor that is to operate at low voltage.

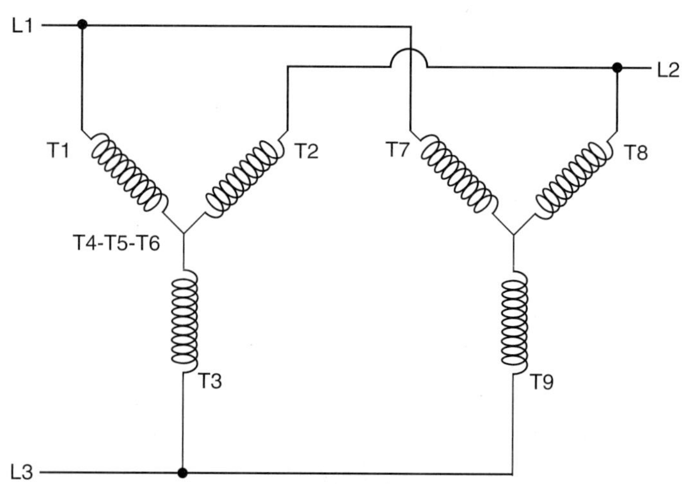

Figure 29-24. Simplified wiring diagram of a nine-wire, wye-connected dual-voltage three-phase motor that is wired to operate at low voltage.

motor connected to operate at low voltage is provided in **Figure 29-25**.

In a wye-connected dual-voltage motor with 12 motor leads extending from the motor housings, the three windings connected to the T7, T8, and T9 motor leads are separated. This is illustrated in **Figure 29-26**. These three additional motor leads are numbered T10, T11, and T12. The numbering sequence follows the same inward-spiraling format that was used with the 9-lead, wye-connected motor.

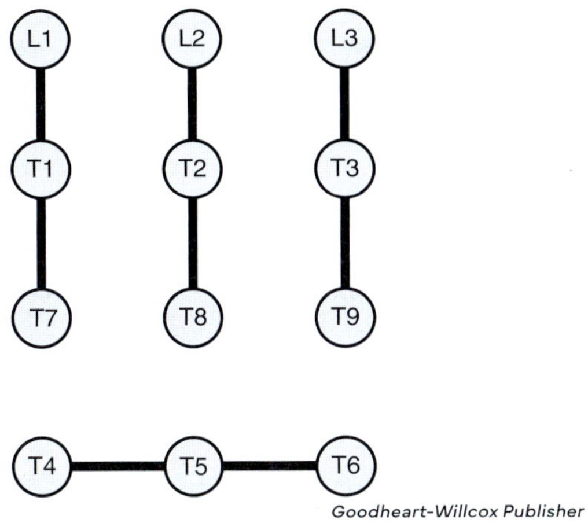

Figure 29-25. Connection diagram for a nine-wire, wye-connected dual-voltage three-phase motor that is wired for low voltage operation.

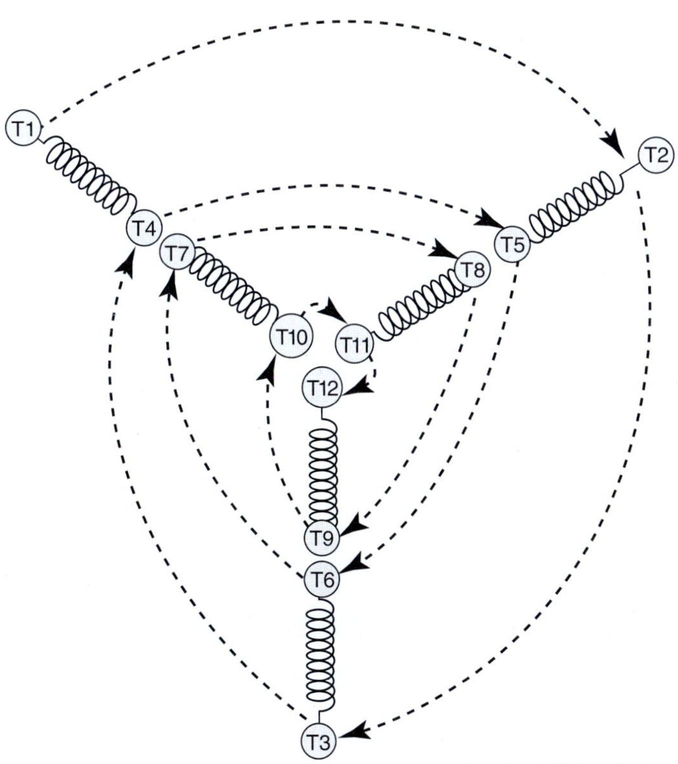

Figure 29-26. Motor lead numbering for a 12-wire, wye-connected dual-voltage three-phase motor.

Shedding Light

Three-phase motors manufactured to IEC standards use a terminal marking convention different from that of NEMA motors. On IEC motors, the terminals for each phase are marked as U, V, and W. Thus, on wye-connected IEC motor, leads T1 through T9 on a NEMA motor would correspond to leads U1, V1, W1, U2, V2, W2, U3, V3, W3, respectively.

Delta-Connected Dual-Voltage Motors

For the delta-connected dual-voltage motor, the motor lead numbering follows a similar pattern to that used in the wye configuration. In the wye configuration, each of the power phases is connected, either directly or indirectly to three motor leads:

- Phase 1: Connected to motor leads T1, T4, and T7
- Phase 2: Connected to motor leads T2, T5, and T8
- Phase 3: Connected to motor leads T3, T6, and T9

The delta wiring uses the same motor lead triplets as the wye connection. However, instead of circling around the windings, as shown in Figure 29-18, the terminals are assigned their numbers by following a path through the windings. Starting at reference point A in **Figure 29-27** and moving in a clockwise direction through the windings, the motor leads are labeled T1, T4, and T7. Then, continuing in a clockwise direction from reference point B, the numbering continues with T2, T5, and T8. The remaining three leads are labeled, in a clockwise direction, as T3, T6, and T9.

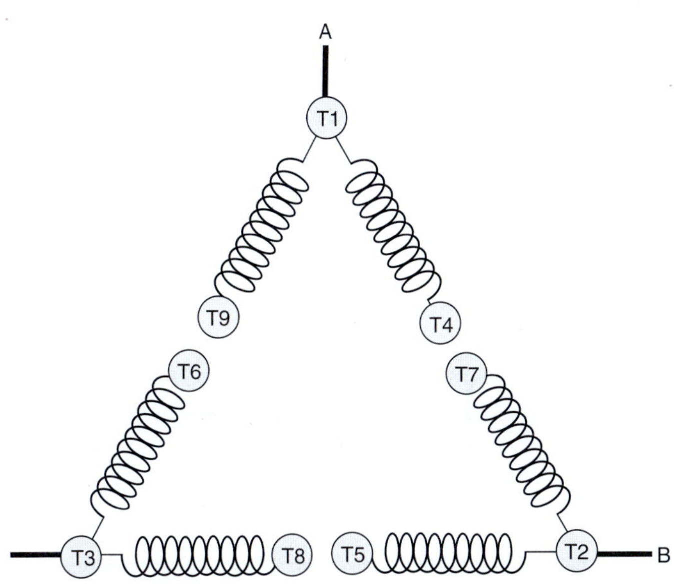

Figure 29-27. Numbering of the three phases for a wye-connected dual-voltage three-phase motor.

For high-voltage operation, the windings in each phase are wired in series with each other, to create one large delta connection. For example, in **Figure 29-28** motor lead T4 is connected to lead T7, lead T5 is connected to lead T8, and lead T6 is connected to lead T9. Each leg of the three-phase power supply is connected to one of the corners of the delta connection, namely leads T1, T2, and T3. A simplified wiring diagram for a delta-configured dual-voltage motor connected to operate at high voltage is provided in **Figure 29-29**. Note: This is the same diagram for wye-connected dual-voltage motor operating at high voltage.

For low-voltage operation, the windings in each phase are wired in parallel with each other. Each leg of the three-phase power supply is connected to one side of the two winding segments in each phase that the power leg is connected to. For example, L1 is connected to motor lead T1, as well as motor leads T6 and T7, **Figure 29-30**.

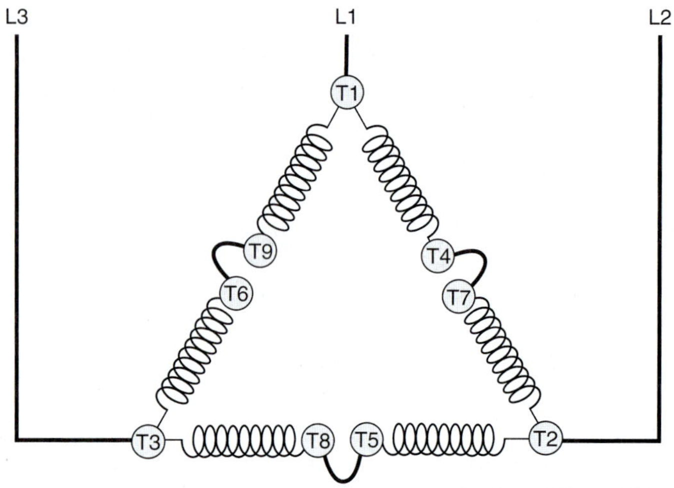

Figure 29-28. For high-voltage operation, leads T4 and T7, T5 and T8, and T6 and T9 are connected together to form one large delta connection. The ungrounded line voltage conductors (L1, L2, and L3) are connected to motor leads T1, T2, and T3.

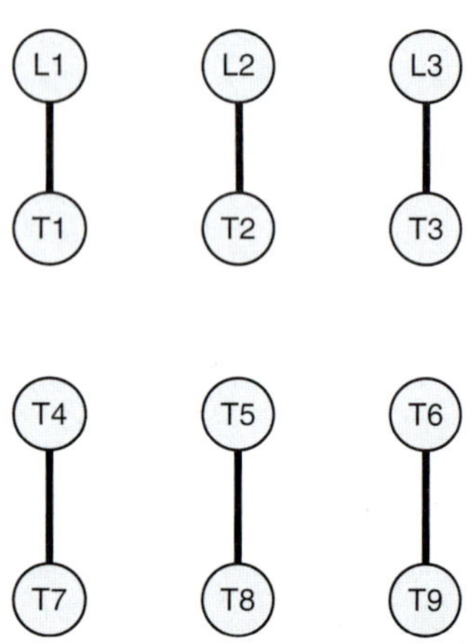

Figure 29-29. Simplified connection diagram for a nine-wire, delta-connected dual-voltage three-phase motor wired to operate at high voltage.

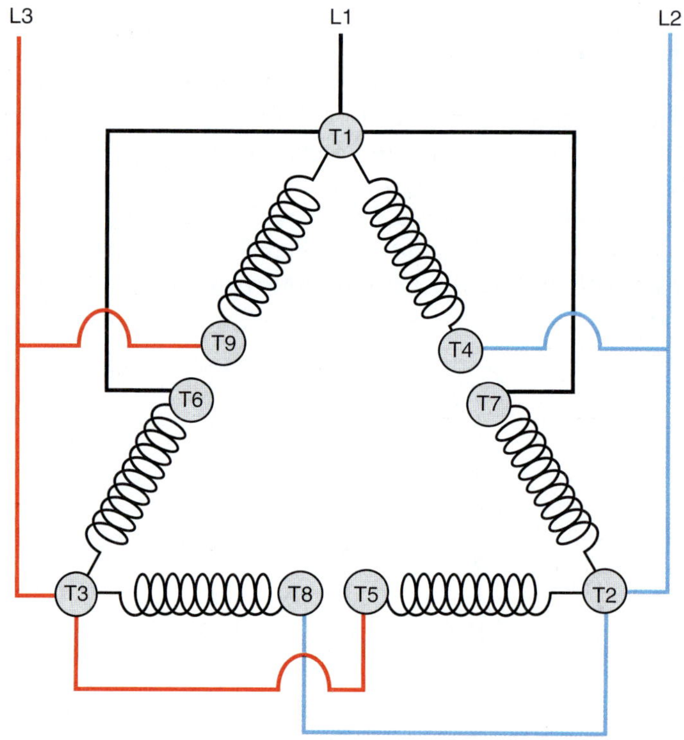

Figure 29-30. Wiring diagram showing the completed connections to facilitate low-voltage operation.

L2 is connected to motor leads T2, T4, and T8, while L3 is connected to motor leads T3, T5, and T9. For the windings connected between L1 and L2, a simplified connection diagram is provided in **Figure 29-31**. A modified ladder diagram for the low-voltage delta connection is provided in **Figure 29-32**. A connection diagram is provided in **Figure 29-33**.

For delta-connected dual-voltage three-phase motors with 12 motor leads extending from the motor housings, the two windings that are connected to each of the motor leads T1, T2, and T3 are separated. These three additional motor leads are numbered T10, T11, and T12. The numbering sequence for these three additional motor leads is illustrated in **Figure 29-34**.

29.6 Three-Phase Motor Direction of Rotation

The direction of rotation of a three-phase motor is determined by the order in which the individual motor phases are wired and how the motor is connected to the power supply.

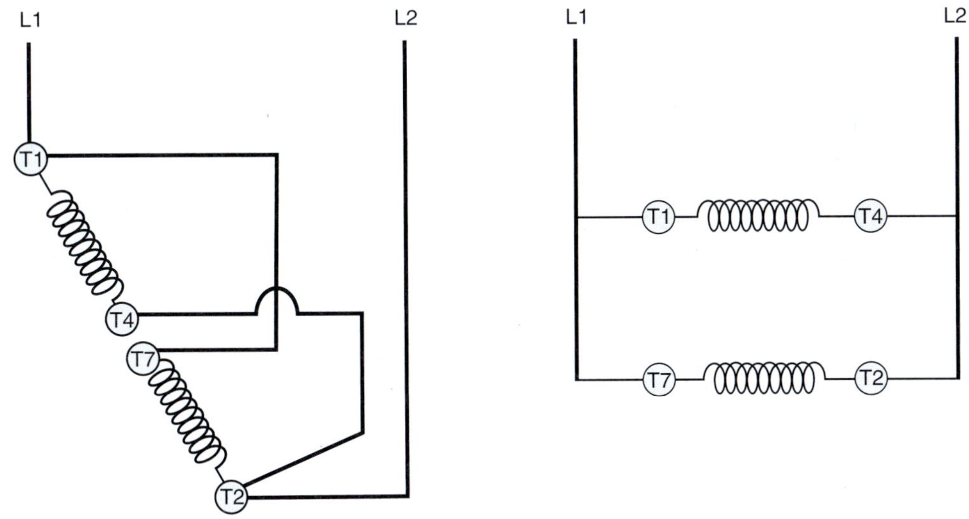

Goodheart-Willcox Publisher

Figure 29-31. Simplified diagram showing the connections for the two windings connected between L1 and L2.

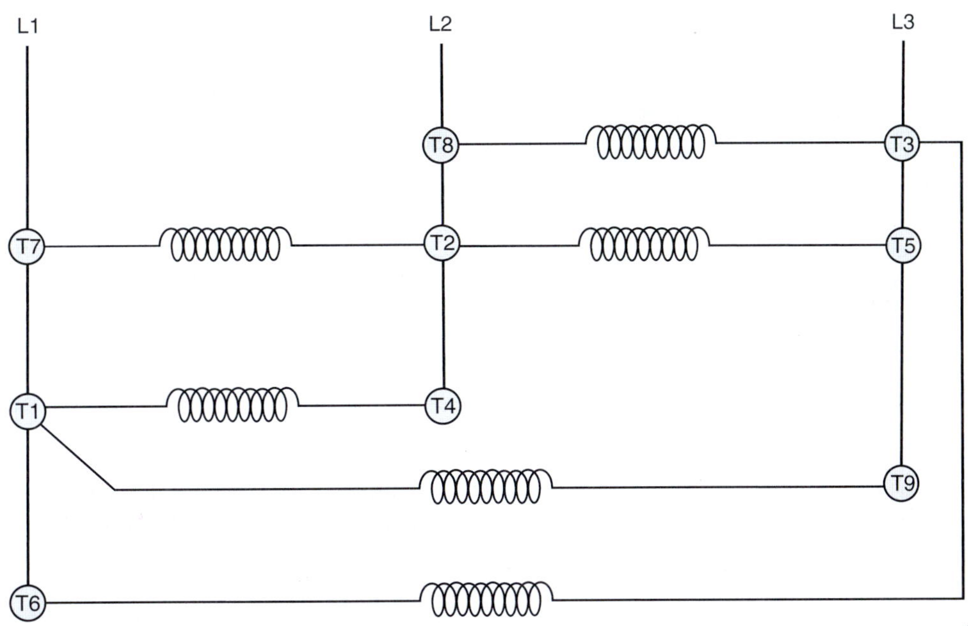

Goodheart-Willcox Publisher

Figure 29-32. Ladder diagram showing the completed connections to facilitate low-voltage operation of a delta-connected dual-voltage motor.

The direction of motor rotation can be determined in one of three ways:

- Briefly energize, or bump, the motor prior to mounting or connecting it to the driven load and observe the motor's direction of rotation.
- Mount and connect the motor to the driven load and then briefly energize the motor to observe the direction of rotation of the driven load.
- Use a ***motor and phase rotation indicator*** to evaluate the rotary field direction of the power supply and the intended motor connections to ensure that, when energized, the motor rotates in the desired direction, **Figure 29-35**.

If the three-phase motor is not rotating in the desired direction, then it might be necessary to change the direction of rotation. The direction of rotation is reversed by changing the connections of any two power leads. See **Procedure 29-1**.

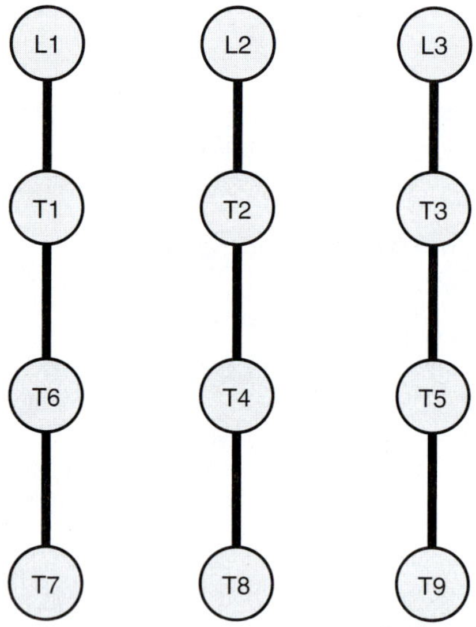

Goodheart-Willcox Publisher

Figure 29-33. Connection diagram for a nine-wire, delta-connected dual-voltage three-phase motor that is to operate at low voltage. Note that the numbering of the leads in the diagram are not in sequential order.

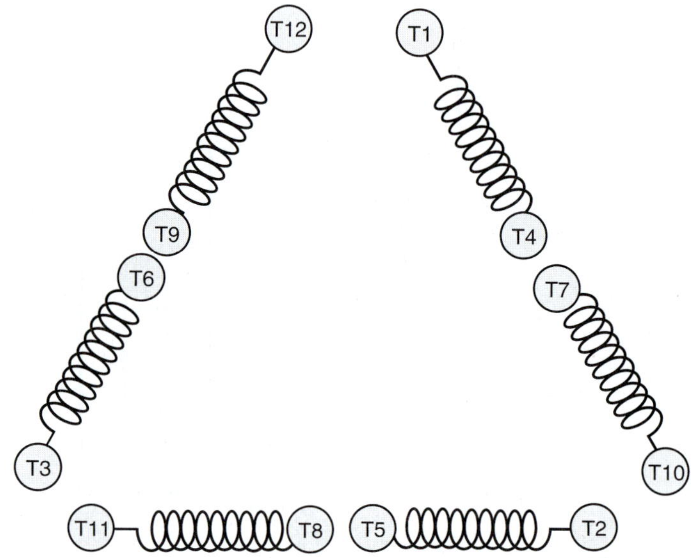

Goodheart-Willcox Publisher

Figure 29-34. Motor lead numbering for a 12-wire, delta-connected dual-voltage three-phase motor.

Courtesy Fluke Corporation

Figure 29-35. A motor and phase rotation indicator can be used to determine phase sequence and motor rotation direction. Meter probes are connected to the phase voltage terminals to measure phase sequence. Some noncontact meters can determine motor rotation by simply being placed on the motor enclosure.

Safety Alert

Never check motor rotation direction by connecting to a driven load that cannot be safely driven in the wrong direction. If rotating the driven load may cause damage to the equipment or system, use one of the other methods to determine motor rotation direction.

Procedure 29-1

Changing the Direction of Rotation in a Three-Phase Motor

1. Ensure that the main power supply is disconnected. This involves not only de-energizing the motor's control circuit but also turning off any circuit breakers that power the motor.
2. Identify the three power phases, often identified as L1, L2, and L3.
3. Identify the three motor leads, often identified as T1, T2, and T3.
4. Label the three power leads and the three motor leads. Or, with your cell phone, take a picture of the connections if the wires are color coded.
5. Change the positions of any two power leads or motor leads.
6. Make certain that the electrical connections are tight.
7. Restore power to the motor circuit.
8. Briefly energize and de-energize (bump) the motor and observe its direction of rotation.
9. Once the correct rotation has been verified, energize the motor. Monitor its performance by checking the amperage and listening for any unusual noises or vibrations.

CHAPTER 29 REVIEW

Summary

- A rotating magnetic field allows three-phase motors to operate. (29.1)
- Three-phase motors are supplied with three-phase power, which is comprised of three individual single-phase power supplies that are 120° out of phase with each other. (29.1)
- The calculated, or ideal, motor speed is called the motor's synchronous speed. (29.2)
- A motor's synchronous speed is determined by the frequency of the power supply and the number of poles per phase. (29.2)
- Motor slip is the difference between a motor's calculated and actual speed. (29.3)
- Three-phase motors have three windings that have the same resistance.
- These motors can be wye-connected or delta-connected. (29.4)
- Resistance measured between motor leads depends on whether the motor is wye-connected or delta-connected. (29.4)
- The windings of a dual-voltage three-phase motor are wired in series for high-voltage applications. (29.5)
- The windings of a dual-voltage three-phase motor are wired in parallel for low-voltage applications. (29.5)
- Some three-phase motors have nine leads extending from the motor housing, while others have 12 leads. (29.5)
- The direction of rotation of a three-phase motor can be changed by changing any two motor lead connections. (29.6)

Review Questions

Use the information in the chapter to answer the following questions.

Measuring Your Potential

1. *True or False?* Three-phase power refers to the simultaneous production of three evenly-offset, single-phase waveforms, each with the same frequency. (29.1)
2. When calculating a motor's synchronous speed, a conversion factor of _____ is incorporated into the calculation. (29.2)

   A. 120
   B. 240
   C. 360
   D. 480
3. The synchronous speed of a three-phase motor is determined by the _____. (29.2)

   A. number of motor pole pieces per phase and the frequency of the power supply
   B. voltage supplied to the motor and the frequency of the power supply
   C. number of motor pole pieces per phase and the amperage draw of the motor
   D. voltage supplied to the motor and the amperage draw of the motor
4. *True or False?* Motor slip is calculated by subtracting the actual motor speed from the synchronous speed. (29.3)
5. Motor slip should be _____. (29.3)

   A. no more than 5% of the synchronous motor speed
   B. no more than 10% of the synchronous motor speed
   C. between 5% and 10% of the actual motor speed
   D. no more than 5% of the actual motor speed
6. *True or False?* When evaluating a three-winding, wye-connected three-phase motor, each obtained resistance reading is being taken through two windings. (29.4)
7. Two common operating voltages for dual-voltage three-phase motors are _____. (29.5)

   A. 115 V and 230 V
   B. 120 V and 240 V
   C. 200 V and 400 V
   D. 208 V and 460 V
8. On a 12-wire, three-phase dual-voltage motor, how many separate stator windings are found within the motor housing? (29.5)

   A. 3
   B. 6
   C. 9
   D. 12
9. A _____ can be used to help ensure that a three-phase motor turns in the proper direction. (29.6)

   A. motor rotation sensor
   B. clamp-on ammeter
   C. phase indicator
   D. motor and phase rotation indicator

Plug-In

1. If the A-1 and B-1 pole pieces on a three-phase motor have positive polarities, the _____. (29.1)
   A. A-2 pole piece will have a positive polarity and the B-2 pole piece will have a negative polarity
   B. A-2 pole piece will have a negative polarity and the B-2 pole piece will have a positive polarity
   C. C-1 pole piece will have a negative polarity and the C-2 pole piece will have a positive polarity
   D. A-2 and B-2 pole pieces will both have a negative polarity
2. The synchronous speed of an eight-pole, three-phase motor that is powered by a 240 V/50 Hz signal will be closest to _____ if the amperage draw is 45 A. (29.2)
   A. 750 rpm
   B. 1200 rpm
   C. 1400 rpm
   D. 1600 rpm
3. The percentage slip for a motor that has a synchronous speed of 3600 rpm and an actual speed of 3450 rpm is closest to _____. (29.3)
   A. 4.2%
   B. 4.35%
   C. 4.5%
   D. 4.65%

Making the Connection

1. Describe the two factors that determine the synchronous speed of a three-phase motor. (29.2, 29.3)
2. Explain what will happen if a dual-voltage three-phase motor is wired to operate at low voltage but is connected to a high-voltage power supply. (29.5)
3. Explain why the measured resistance values of wye-connected and delta-configured windings differ from each other, even if the individual winding resistances are the same. (29.4)

CHAPTER

30 Electronically Commutated Motors

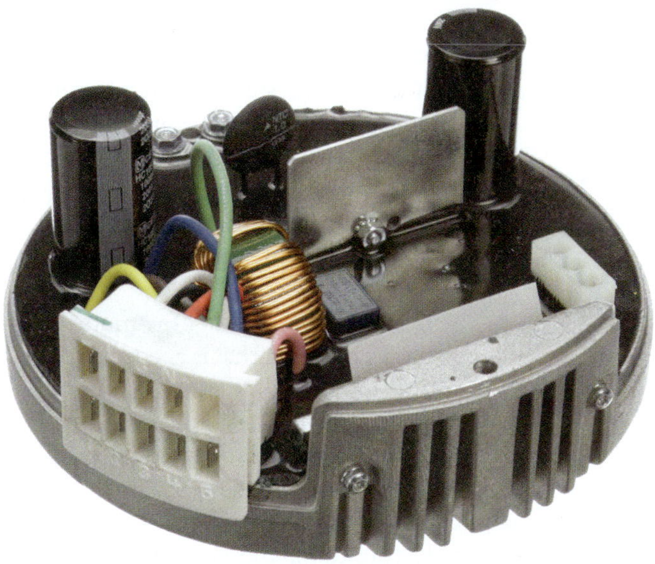

Image provided by Regal Rexnord. All rights reserved.

SPARKING DISCUSSION

It is often stated that electronically commutated motors are simply three-phase motors. Do you agree or disagree with this statement?

LEARNING OUTCOMES

After completing this chapter, you will be able to:

30.1 List three conditions commonly controlled and maintained by an electronically commutated motor (ECM).

30.2 Identify the components and configurations of an ECM.

30.3 Explain the operation of diodes and transistors in electronic devices.

30.4 Explain how rectifiers and inverters can be used to create a dc signal.

30.5 Describe how the speed of an ECM can be changed.

30.6 List general troubleshooting steps to determine if an ECM has malfunctioned.

TECHNICAL TERMS

depletion region
doping
electronically commutated motor (ECM)
forward bias
load matching
N-type material
P-type material
rectifier
reverse bias
soft-start motor
threshold voltage

Introduction

Early generation electrical machines and motors were typically single-speed devices. Regardless of the load on the system or its operational requirements, energy was often wasted as these single-speed components provided more power than was needed for the particular application. Technological advances in electric motor design paved the way for the introduction of electronically commutated motors, commonly referred to as ECMs.

This chapter covers the theories underlying how ECMs operate, as well as how previously discussed concepts are adapted to apply to electronically commutated motors. This chapter also includes a discussion on ECM evaluation, where you will learn how to isolate the cause of failure if an ECM does not perform as intended.

30.1 What is an ECM?

An ***electronically commutated motor (ECM)*** is a motor that uses electronic components and circuitry, rather than brushes, to control the polarity of the power signal supplied to the motor windings, **Figure 30-1**. ECMs are high-efficiency motors that are used primarily to control the flow of fluids, such as air, in a variety of applications. While the ECM is classified as a dc three-phase brushless motor and operates as a three-phase motor, it is actually powered by a single-phase ac power supply. The motor's single-phase power supply is first converted to a dc signal, which is then manipulated to create an artificial three-phase signal that ultimately powers the motor. Presently, ECMs are commonly used for applications of less than 1 hp. As technologies advance, this is very likely to change.

ECMs are major components in newer-generation high-efficiency and modulating systems and machines. Unlike the conventional induction motors discussed earlier in this text, ECMs have the ability to adjust their operation based on external factors and conditions that act upon them. Depending on their design, ECMs have the ability to modulate torque, speed, and/or the amount of fluid moved to provide an output consistent with the needs of the system.

Figure 30-1. An electronically commutated motor relies on electronic circuitry to supply an appropriately configured power signal to help ensure the motor matches the needs of the system.

The ability of ECMs to modulate their output is made possible by the introduction of a frequency drive, an electronic control module, and a microprocessor. The ECM's microprocessor can be programmed to ensure the motor responds to, and maintains, the system's desired operating conditions. By modulating motor speed and/or torque to match the needs of the system, a great deal of energy can be saved. In air-conditioning applications, for example, consistent comfort levels can be maintained, even when external conditions change, by varying the amount of air being moved by the system.

The process of controlling the motor to provide only the output energy or power that is needed is called ***load matching***. In addition, load matching reduces, and in some cases eliminates, frequent motor starts and stops. As a result, the amount of wear and tear on the motor is reduced. Most electrical and mechanical motor losses occur during motor start-up and shut-down.

Unlike traditional induction motors that experience high inrush currents on start-up, ECMs are classified as ***soft-start motors***. Soft-start motors gradually ramp up their speed, further reducing mechanical stress on the motor bearings and the amount of heat generated within the motor. Reductions in start-up or inrush current reduce energy consumption and help to extend the useful life of the motor. Another benefit of soft-start motors is quieter operation.

Shedding Light

ECMs are sometimes referred to as EC motors or brushless dc (BLDC) motors.

30.2 ECM Components and Configurations

Generally speaking, the ECM has three main control elements: a programmable microprocessor, a frequency drive, and an electronic control module. In most cases, ECMs are constructed as two separate components: a motor and a control module. These two components are connected mechanically and electrically to create a complete motor assembly. The motor comprises stator windings and a permanent magnet rotor. The control module houses the microprocessor and connectors. Each motor manufacturer carefully designs their motors and control modules to ensure that they work together seamlessly.

There are three common configurations of electronically commutated motors. The ECM with an integrated motor

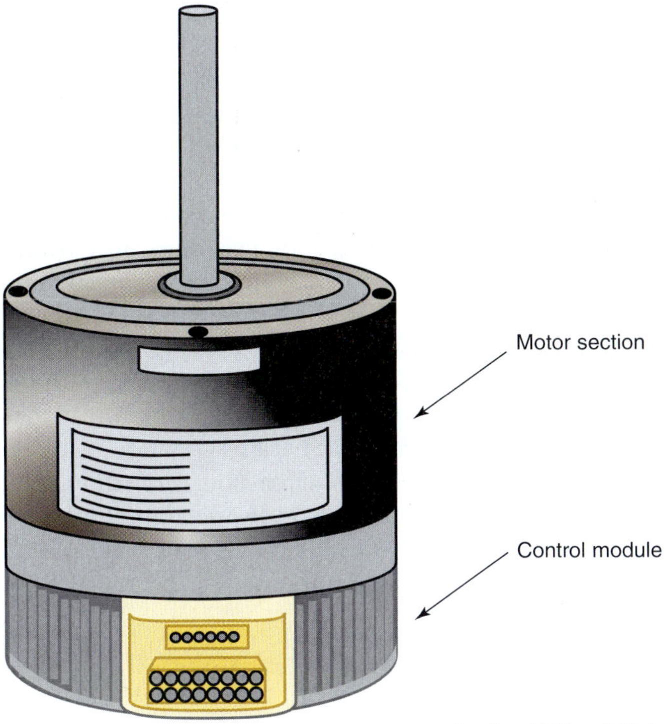

Figure 30-2. A one-piece electronically commutated motor with an integrated control module.

control is constructed as a single unit, with the control module incorporated into the motor casing, **Figure 30-2**. If the ECM needs to be replaced for any reason, the entire unit, including the motor and control module, is replaced as it encompasses both portions of the motor assembly. Some ECMs have a detachable motor control module, **Figure 30-3**. In this configuration, the motor and control module are constructed as separate units and are bolted together for installation. A third, less common, configuration of ECM motors uses separate motor and control modules that are not intended to be bolted together, **Figure 30-4**. This option is a popular choice for applications and installations that have space restrictions that prevent the larger assembly from being installed.

30.3 Underlying Electronic Concepts of ECMs

Unlike most motors discussed in this text, ECMs use various fundamental electronic components and devices. These components rely on basic semiconductor and electronic theories and concepts to function. The concept of semiconductors, originally introduced in Chapter 4, *Building Blocks of Electricity and Electronics*, will be revisited here, along with the following electronics-based concepts:

- Doping
- Diodes
- Transistors

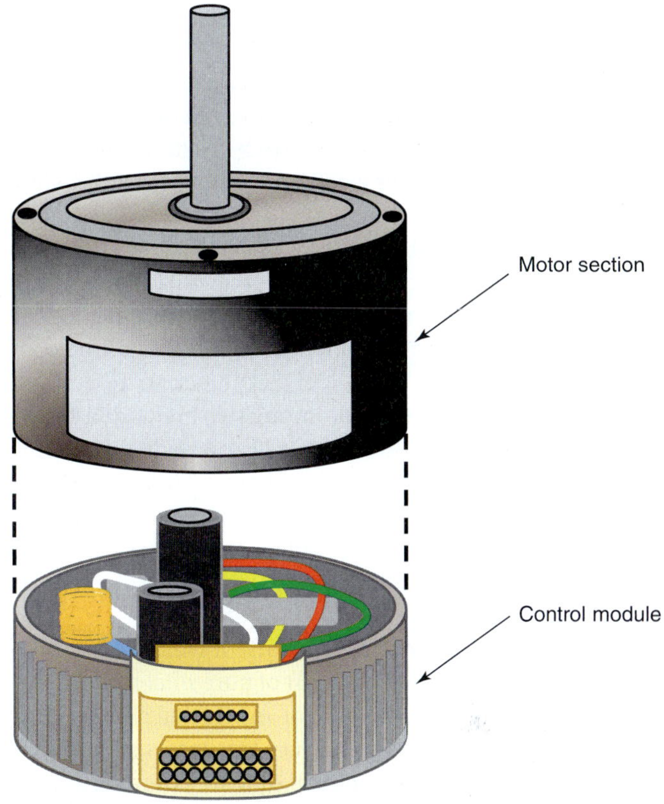

Figure 30-3. A two-piece electronically commutated motor assembly. The motor and the control module are bolted together before installation.

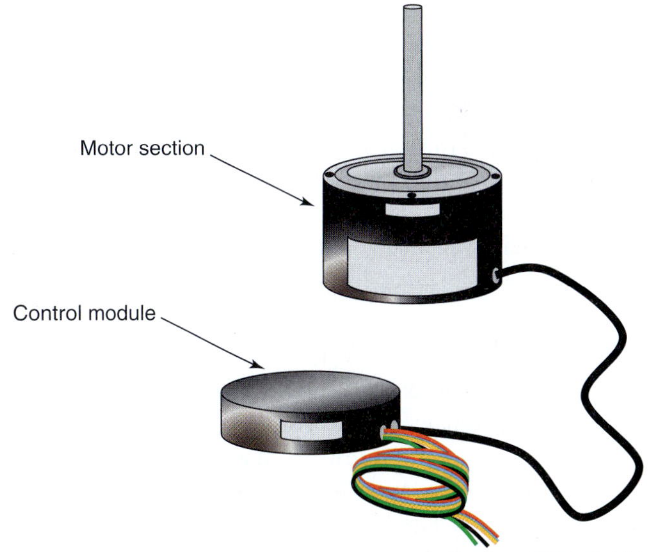

Figure 30-4. An electronically commutated motor with a remotely controlled module. The control module is designed to be located remotely from the motor.

Semiconductors and Doping

Semiconductor materials are the fundamental building blocks for the electronic devices commonly found on printed circuits or solid-state control modules. Most equipment, machinery, and appliances today, regardless of how simple they might appear to be, use solid-state control boards, devices, and modules. Common materials used to create semiconductor materials are silicon and germanium. Silicon, the second most abundant element in the earth's crust, is commonly used in the production of electronic chips and solid-state circuit boards.

Despite the availability and common use of silicon in electronic circuitry, pure silicon is not a good conductor of electricity. In order to conduct electricity, electrons must be free to move from atom to atom. In its pure crystalline configuration, silicon does not allow this to occur. To create a conductor out of pure silicon, impurities are added to the material in a process called ***doping***. (For a review of atomic structure, it may be useful to revisit Chapter 4, *Building Blocks of Electricity and Electronics*. Atomic structure and the capacity of an atom's energy shells to hold electrons are explained in that chapter. In particular, the structure of a silicon atom is explained.)

There are two types of doping, both of which are used to create semiconductive materials out of pure silicon. One doping method results in the creation of an electron deficiency in the material, while the other produces an excess of electrons in the material. The two materials that are commonly used in the doping process are boron and phosphorus.

P-Type Material

The term ***P-type material*** is used to describe doped silicon that has a deficiency of electrons. The *P* in P-type refers to the net positive charge that results from a lack of negatively charged electrons. P-type material is commonly created using boron as the impurity, or doping agent. The boron atom has five electrons: two in the first shell and three electrons in the outermost shell. By doping pure silicon with boron, holes are created. A hole represents a space that would have normally been occupied by an electron. It is this lack of electrons that create the positively charged P-type material.

N-Type Material

The term ***N-type material*** is used to describe doped silicon that has an excess of electrons. The *N* in N-type refers to the net negative charge that results from an excess of negatively charged electrons. N-type material is commonly created using phosphorus. The phosphorus atom has 15 electrons: two in the first shell, eight electrons in the next shell, and five electrons in the outermost shell. By doping pure silicon with phosphorus, an excess of electrons is present in the material.

Diodes

The diode is one of the simplest and most straightforward of the electronic devices. The diode is a circuit component that allows electricity to flow in only one direction. The diode is constructed of one layer of P-type material bonded to a layer of N-type material. The electrical symbol for the diode closely resembles an arrow, **Figure 30-5**. The two terminals or connection points on the diode are called the anode and the cathode. The cathode is the side of the diode that is made up of N-type material, while the anode is made up of P-type material.

Consider a simple series dc lighting circuit that consists of a battery, a light bulb, and a diode, **Figure 30-6**. In the diagram, the battery's negative terminal is connected to the diode's cathode. When the circuit is energized, the higher concentration of electrons at the battery's negative terminal causes electrons to move from the battery toward the negatively charged cathode. Electrons flowing from the battery repel the excess electrons in the negatively charged cathode, pushing them to the anode side of the diode. From the anode, electrons flow through the light bulb and continue toward the positive side of the battery. Since current is able to flow

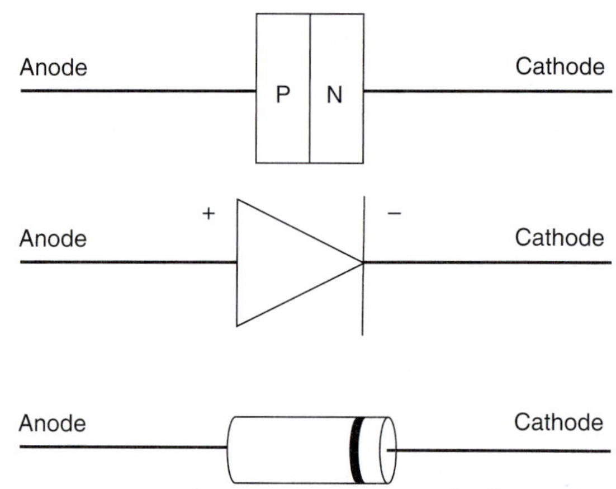

Goodheart-Willcox Publisher

Figure 30-5. Various representations of a diode. The anode is the diode's positive terminal, while the cathode is the diode's negative terminal. The cathode is typically the marked or banded terminal or connection point on the diode.

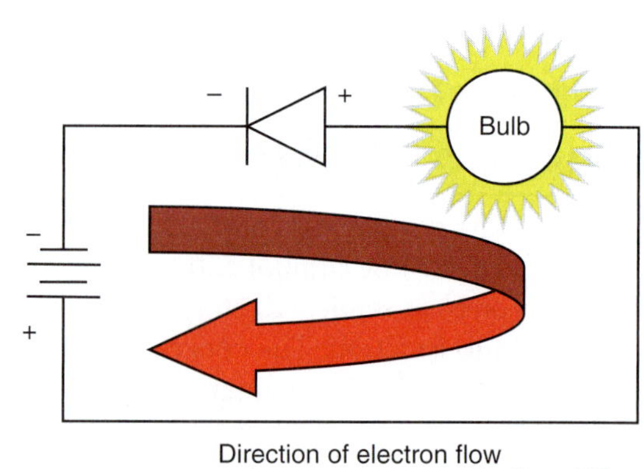

Goodheart-Willcox Publisher

Figure 30-6. When the diode's cathode is connected to a battery's negative terminal, the diode is said to be forward biased and can act as a closed switch.

through the diode, the diode is acting like a closed switch, and the light bulb will illuminate. When a diode acts as a closed switch, the diode is said to be ***forward biased***.

In **Figure 30-7**, the orientation of the diode has been reversed. The battery's negative terminal is connected to the diode's positively-charged anode. In this case, electrons flow from the negative terminal on the battery to the diode's anode. Electrons flowing from the battery fill the holes in the anode that were created by the boron doping in the P-type material. By filling the holes in the anode's P-type material, electrons cannot flow through the diode to the cathode. At the same time, excess electrons in the cathode will be attracted to the positive terminal on the battery, **Figure 30-8**. As a result, a barrier is created at the P-N junction that electrons cannot cross. This junction is called the ***depletion region***. Since current cannot pass through the diode, the diode acts like an open switch and the light bulb will not illuminate. When the diode acts like an open switch, the diode is said to be ***reverse biased***.

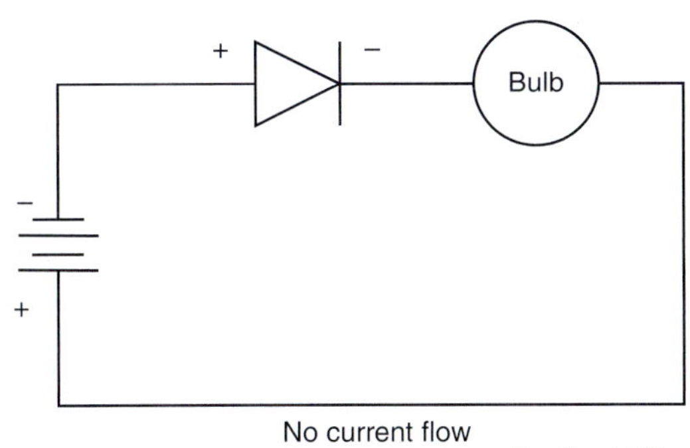

Goodheart-Willcox Publisher

Figure 30-7. When the diode's anode is connected to a battery's negative terminal, the diode is said to be reverse biased and will act as an open switch.

Shedding Light

When a diode is manufactured, a layer of P-type material is connected to a layer of N-type material. When connected, some electrons from the N-type material travel across this connection point and fill the holes in the P-type material. This creates negatively charged ions on the P-type side of the diode and positively charged ions on the N-type side of the diode. In this configuration, the depletion region is formed, and electrons cannot flow freely from the N-type side of the diode to the P-type side when electrons are externally introduced to the diode's N-type side.

Ideal diodes are either 100% forward biased (ON), or 100% reverse biased (OFF). However, in the real world, this is not the case. Even when a diode is forward biased, there will be a measurable voltage across the device. The voltage drop across a diode depends on the type of diode being used. The voltage drop across a silicon diode, for example, will be about 0.7 V, **Figure 30-9**. This is referred to as ***threshold voltage***. In order for the diode to turn on, or become functionally forward biased, the voltage across it must be at least 0.7 V. If the voltage across the diode is less than the threshold voltage, the diode will be reverse biased and the diode will act as an open switch. Once the voltage across the diode reaches 0.7 V, the diode will begin to conduct. **Figure 30-10** illustrates the relationship between current and voltage with respect to the threshold voltage. The current flow through the ideal diode will be zero whenever the voltage across the diode is less than 0.7 V, and will allow unrestricted current flow whenever the voltage across the diode is greater than 0.7 V.

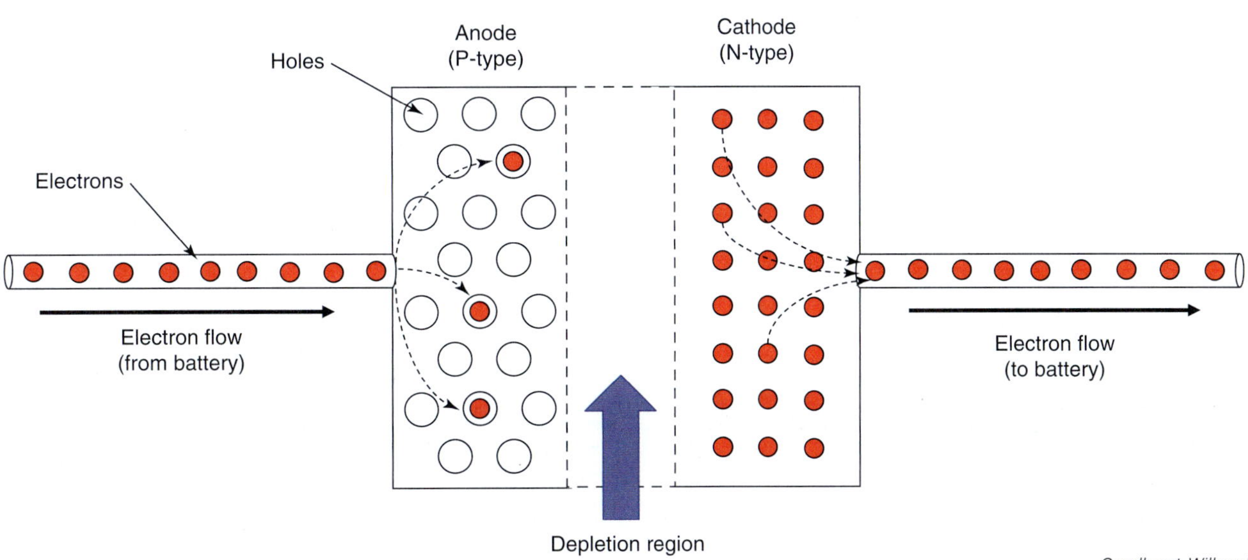

Goodheart-Willcox Publisher

Figure 30-8. A depletion region is created when electrons from the negatively charged side of the power source fill the holes in a diode's anode, and when excess electrons in the diode's cathode flow toward the positively charged side of the power source.

> **Shedding Light**
>
> Different diodes have different threshold voltages. The threshold voltage of a germanium diode, for example, will have a threshold voltage of about 0.4 V, while a light-emitting diode, or LED, will have a threshold voltage of about 1.7 V.

Diodes in AC Circuits

Although previous examples refer to diode use in dc circuits, diodes also function in ac circuits. Since the polarity of an ac signal changes two times each cycle, the diode will be forward biased for half of the cycle and reverse biased for the other half. Consider the ac circuit in **Figure 30-11**. During the first half of the cycle, the diode is forward biased and current flows through the circuit, **Figure 30-11A**. The voltage polarity in the first half of the signal is positive and is connected to the anode, which is the diode's positive terminal. This results in the light bulb illuminating. The voltage supplied to the light bulb will be slightly lower than the circuit's supply voltage, given the threshold voltage of 0.7 V that must be overcome to turn the diode on.

During the second half of the cycle, the diode is reverse biased and no current flows through the circuit, **Figure 30-11B**. When the diode is reverse biased, the light bulb will not illuminate. This is due to the voltage polarity in the second half of the signal being negative and connected to the positive terminal of the diode—namely, the anode. In this configuration, the diode is off and acts like an open switch.

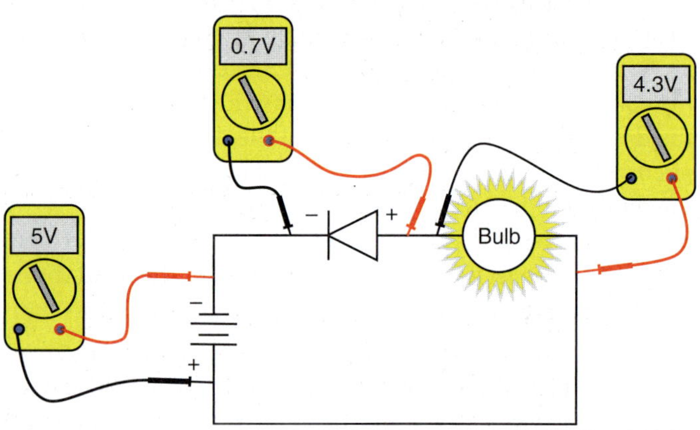

Goodheart-Willcox Publisher

Figure 30-9. A silicon diode has a threshold voltage of about 0.7 V. In order for the diode to become forward biased, the voltage across the diode must exceed the threshold voltage.

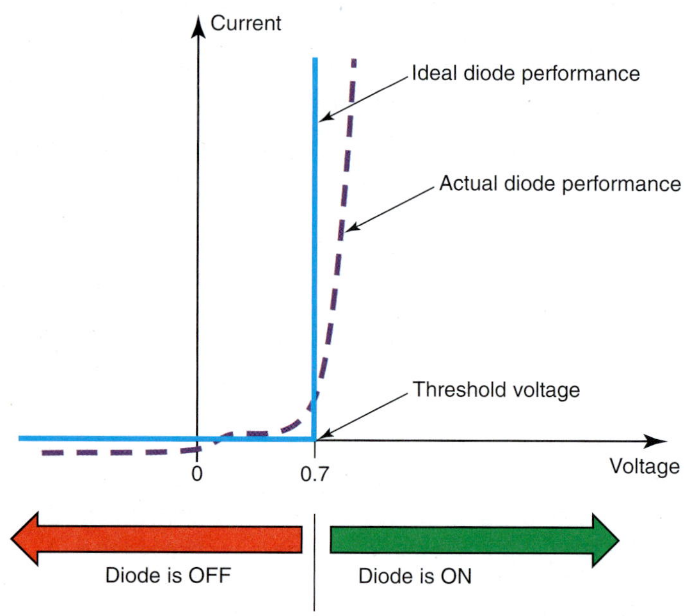

Goodheart-Willcox Publisher

Figure 30-10. Ideal and actual diode performance before and after voltage is supplied. Ideally, the diode is either off or on. In reality, some current may flow through the device before the threshold voltage is reached (dotted line).

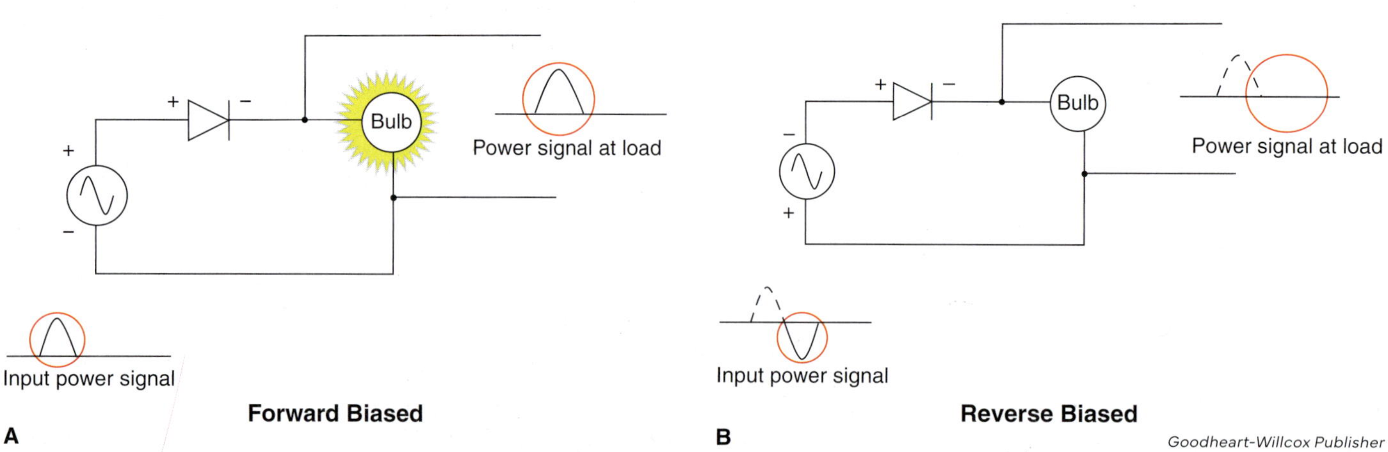

Goodheart-Willcox Publisher

Figure 30-11. A diode in an ac circuit. A—In the first half of the ac signal cycle, the diode is forward biased and acts like a closed switch, energizing the circuit load. B—In the second half of the ac signal cycle, the diode is reverse biased and acts like an open switch, which de-energizes the circuit load.

Testing Diodes

A diode can be tested using an ohmmeter, but it must first be removed from the circuit. A properly operating diode will conduct electricity in only one direction. Measuring across the device, the resistance should be close to 0 Ω when connected in one direction, and close to infinity when the test leads are reversed. If the resistance is close to 0 Ω when connected in both directions, the diode is defective, as it is acting like a closed switch, regardless of polarity. Similarly, if the resistance is high when connected in both directions, the diode is defective, as it acts like an open switch, regardless of polarity.

Transistors

Like the diode, the transistor is a common electronic device used for control and switching purposes. The transistor, unlike the diode, has the ability to amplify a signal, which can then be used to control the operation of other circuits and devices. Transistors operate in a manner that is similar to that of a general-purpose relay. For example, a simple relay consists of a coil and a set of electrical contacts. When voltage is supplied to the relay's coil, a magnetic field is produced, and the position of the electric contacts changes: normally closed contacts open, while normally open contacts close. When voltage is no longer supplied to the coil, the contacts return to their original, normal position.

Transistors have three terminals identified as the collector, the base, and the emitter, **Figure 30-12**. They are configured with three layers of alternating semiconductive material. Transistors can be made up of either two layers of P-type material separated by a layer of N-type material, or two layers of N-type material separated by a layer of P-type material, **Figure 30-13**. A transistor comprised of two layers of P-type material and one layer of N-type material is known as a PNP transistor, while a transistor comprised of two layers of N-type material and one layer of P-type material is known as an NPN transistor. A transistor is, in effect, two diodes connected in series with each other with a center tap connection, **Figure 30-14**. The electrical symbols for PNP and NPN transistors are provided in **Figure 30-15**. Note that the terminals are the same for both PNP and NPN transistors. The difference between the two symbols is the direction in which the arrow on the symbol is pointing. The direction of the arrow indicates the polarity of the power supply, which

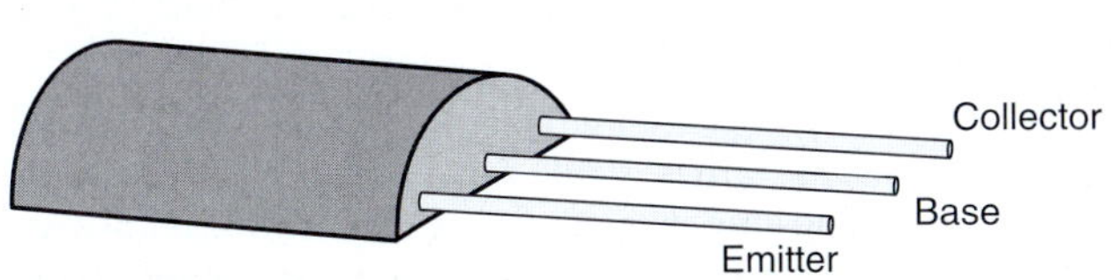

Goodheart-Willcox Publisher

Figure 30-12. A transistor has three terminals: the collector, the base, and the emitter.

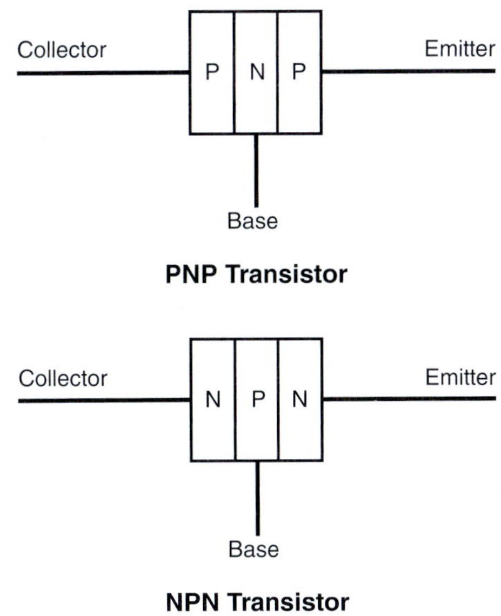

Goodheart-Willcox Publisher

Figure 30-13. The basic transistor is configured with three alternating layers of P-type and N-type material. A PNP transistor has two outer layers of P-type material separated by a layer of N-type material. An NPN transistor has two outer layers of N-type material separated by a layer of P-type material.

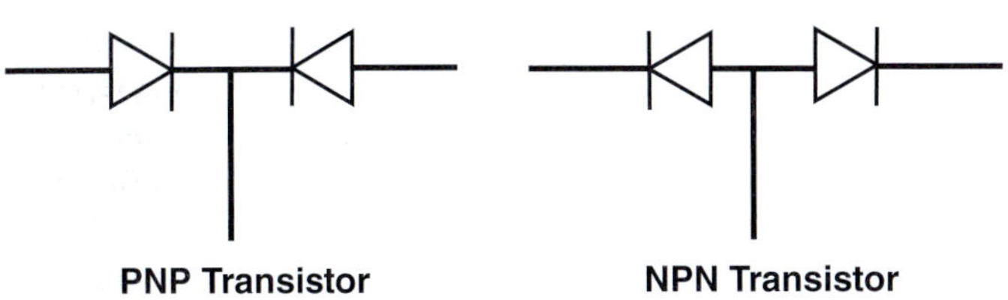

Goodheart-Willcox Publisher

Figure 30-14. The transistor is, in effect, two diodes connected in series with each other. A center tap conductor is connected between the two diodes.

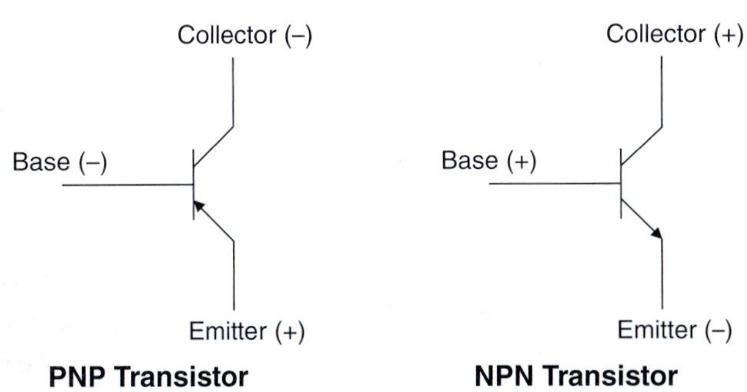

Goodheart-Willcox Publisher

Figure 30-15. The electrical symbols for the PNP and NPN transistor. The arrows here always point to the negative polarity. On the PNP transistor the arrow points to the base and collector, because they both have a negative polarity. The arrow points toward the emitter, which has a negative polarity, on the NPN transistor.

is opposite for PNP and NPN transistors. The arrow symbol always points to the negative polarity. For PNP transistors, the arrow points toward the base and the collector, which both have a negative polarity. In the case of the NPN transistor, the arrow is pointing toward the emitter, which has a negative polarity.

The current signal that controls the operation of the transistor flows between the base and emitter terminals. This control current is referred to as the base-emitter current. When a control voltage is supplied across the base and emitter terminals, referred to as the base-emitter voltage, current will flow between the base and the emitter, **Figure 30-16.** The transistor, like the diode, has a threshold voltage associated with it. As long as the base-emitter voltage is greater than the transistor's threshold voltage, the transistor will open a conductive path between the collector and emitter. In this condition, the transistor is said to be conducting, **Figure 30-17.**

By energizing and de-energizing the base of the transistor, the path between the collector and emitter can be opened and closed. Since transistors are electronic switches with no moving parts, the opening and closing of the collector-emitter path can occur thousands of times per second. This is not possible with mechanical switches.

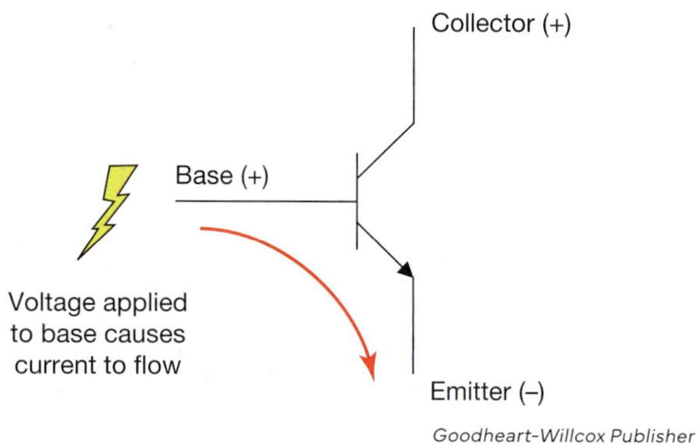

Figure 30-16. The transistor's control voltage is applied to the base, causing current to flow between the base and the emitter.

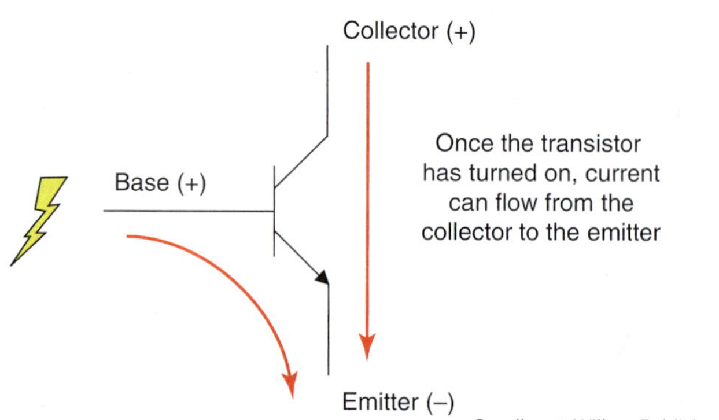

Figure 30-17. When the transistor's control voltage exceeds the threshold voltage, a current path between the collector and emitter is created.

30.4 Creating a Three-Phase Signal

Powering the rotational portion of the ECM is a multistep process that uses electronic circuits made up of multiple diodes and transistors. Diodes, transistors, and the circuitry on which ECM functionality and operation are based is discussed in Chapter 25, *Motor Drives.*

The following are the individual steps to provide power to the ECM:

- Single-phase ac power is supplied to the motor's controller.
- The ac power signal is rectified to a dc signal.
- The dc signal is inverted to a three-phase, ac signal.
- The three-phase signal is supplied to the motor's stator windings, allowing the motor to operate.

Of these four steps, two require the use of electronic circuitry to manipulate one signal to create another. The electronic circuits used are converters and inverters.

Rectifiers

A dc rectifier is used to convert an ac power signal to a dc signal. The circuit used to alter an ac power supply, and ultimately create dc power, is called a ***rectifier.*** In ECMs, an array of diodes is used.

The diode arrangement shown in **Figure 30-18** is called a full-wave bridge rectifier. This example of a full wave rectifier is similar to the rectifier found in some motor drives and uses

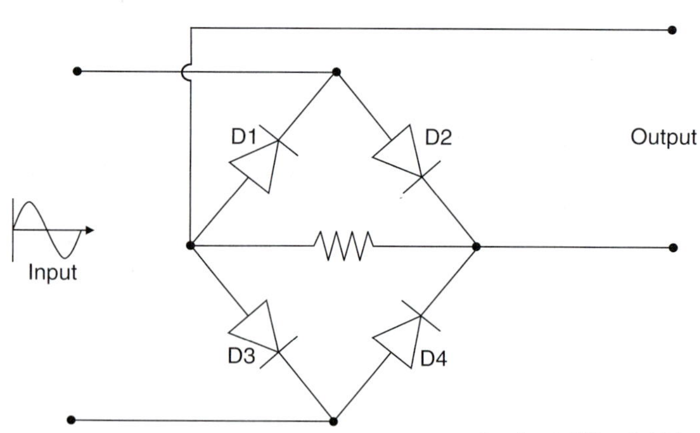

Figure 30-18. Diagram of a full-wave bridge rectifier. In this four-diode arrangement, each set of two diodes is forward biased during one half of the input power signal and reverse biased during the other half of the cycle. The output of the circuit is obtained across the resistor positioned within the diode array.

four diodes, configured as two sets of two diodes. Each set of diodes is forward biased during one half of the input power signal and reverse biased during the other half of the cycle. Diodes D1 and D4 are paired and have the same polarity. Diodes D2 and D3 are also paired and have the same polarity. The rectifier circuit's output is obtained across a resistor in the rectifier circuit.

During the first half of the ac sine wave signal, the input power signal is positive and current flows through the rectifier as indicated by the blue arrows in **Figure 30-19A**. During this half of the cycle, diodes D2 and D3 are forward biased, while diodes D1 and D4 are reverse biased. During the second half of the ac sine wave signal, the input power signal is negative and current flows through the rectifier as indicated by the red arrows in **Figure 30-19B**. During this half of the cycle, diodes D1 and D4 are forward biased, while diodes D2 and D3 are reverse biased. Regardless of the polarity of the input power signal, the output resistor is only exposed to a positive signal, **Figure 30-20**. This rectifier has allowed the positive portion of the signal to pass through the resistor as positive (blue arrow over the resistor in **Figure 30-19A**) and changed the negative portion of the signal to positive as well (red arrow over the resistor in **Figure 30-19B**).

Although the waveform at the outlet of the rectifier is always positive, the voltage is constantly changing. The output waveform can be manipulated to create a more well-defined

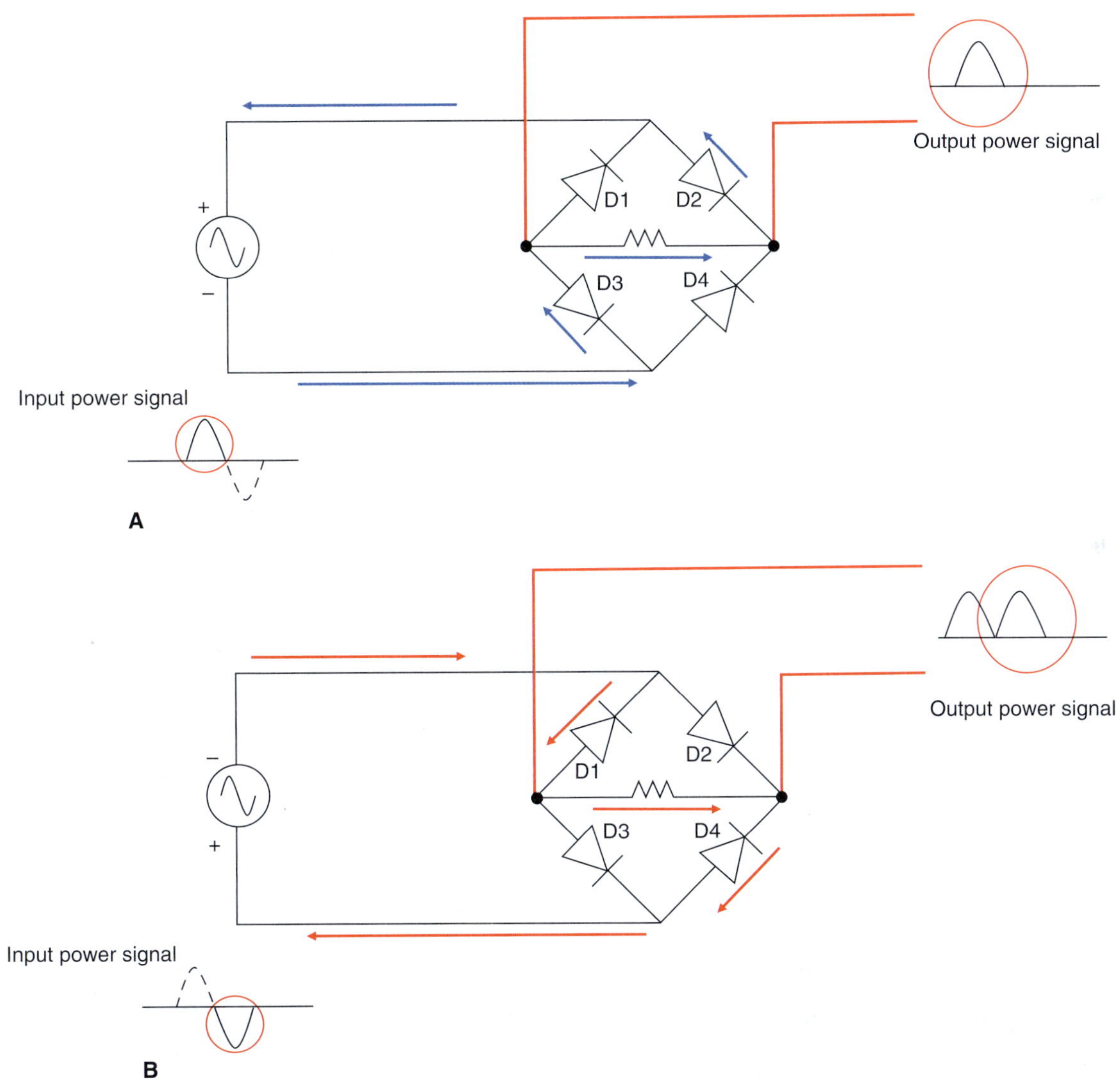

Goodheart-Willcox Publisher

Figure 30-19. Two half cycles of a full-wave bridge rectifier. A—Diodes D2 and D3 are forward biased, and current flows through the rectifier as indicated by the blue arrows. The input signal is positive and the output signal is positive as indicated by the arrow across the resistor. B—During the second half of the cycle, diodes D1 and D4 are forward biased, creating current flow as shown by the red arrows. Current flow through the resistor is shown by the red arrow there, creating a positive output signal.

and consistent voltage signal by passing the signal through a bank of capacitors. The waveform at the outlet of the capacitor bank has a ripple-shaped pattern, **Figure 30-21**. This output waveform is created when the capacitors continually charge and discharge. A wiring diagram that combines the full-wave rectifier and signal-smoothing circuits is illustrated in **Figure 30-22**.

When voltage is initially supplied to a capacitor, the capacitor charges. Once fully charged, the capacitor begins to discharge and the voltage across the capacitor begins to

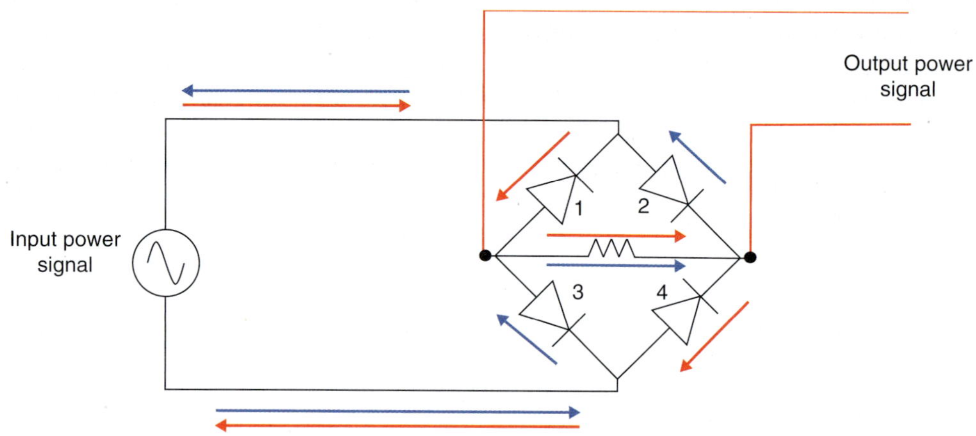

Goodheart-Willcox Publisher

Figure 30-20. Regardless of the polarity of the input signal, the polarity at the output terminals is always positive. Notice that the direction of current flow through the output resistor is in the same direction regardless of the input signal's polarity.

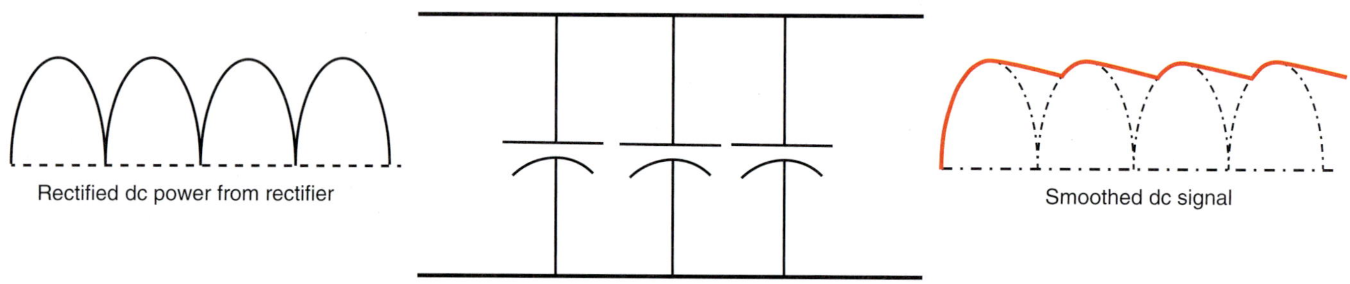

Goodheart-Willcox Publisher

Figure 30-21. The rectified dc waveform is uneven. Passing the pulsing dc signal through a bank of capacitors helps smooth the dc signal, creating a ripple-shaped pattern.

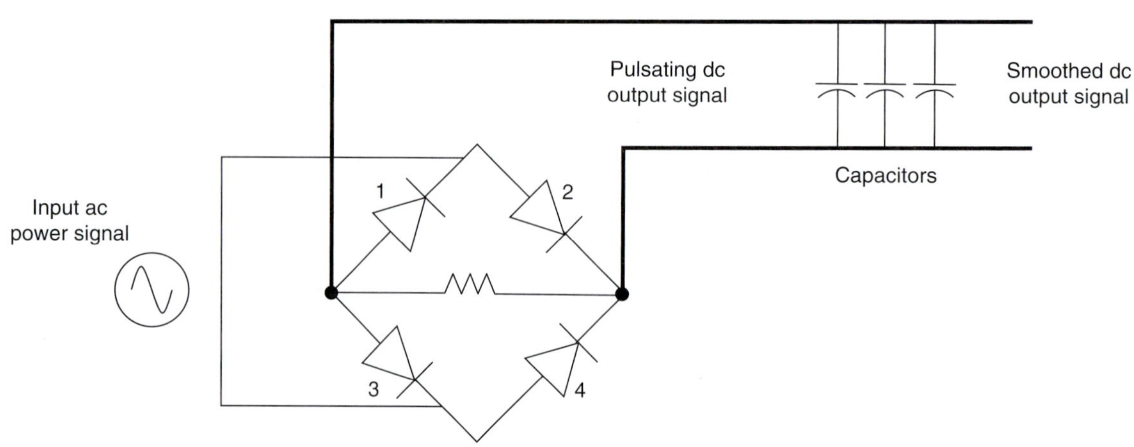

Goodheart-Willcox Publisher

Figure 30-22. Wiring diagram showing the combination of the full-wave bridge rectifier and the signal-smoothing circuits (the capacitors).

drop, **Figure 30-23**. While discharging, the voltage across the capacitor is lower than the maximum circuit voltage, but higher than the instantaneous voltages of the original pulsating dc waveform. The second half of the pulsating waveform supplies a voltage to the capacitor that is higher than the voltage across the capacitor. This will cause the capacitor to charge once again, **Figure 30-24**. The charging and discharging process continues as long as voltage is supplied to the rectifier and signal-smoothing circuits.

Inverters

The inverter changes the smoothed-out dc signal to a three-phase ac signal. It is the inverter that produces a motor's power signal at a frequency that closely matches the particular application and set of operating conditions. The speed of a motor is determined, in part, by the frequency of the power supply and the number of motor poles. Since the number of motor poles is, for the most part, unchangeable, frequency is the main factor that controls motor speed. The inverter, therefore, is mainly responsible for controlling, maintaining, or modulating the frequency of the power signal and, as a result, motor speed. A common inverter used for ECM applications is the six-step voltage-controlled inverter. The input to the inverter circuitry is the output of the dc converter.

The six-step voltage-controlled inverter is constructed with six transistors, two for each of the three phases of power that will ultimately be supplied to the motor. Each of the two transistors in any given phase is responsible for facilitating current flow for one-half of the cycle. One-half of the cycle will have a positive polarity, while the other half-cycle will have a negative polarity. In the six-step voltage-controlled inverter in **Figure 30-25**, transistors 1 and 4 combine to create one power phase, transistors 2 and 5 combine to create another power phase, and transistors 3 and 6 combine to create the third power phase.

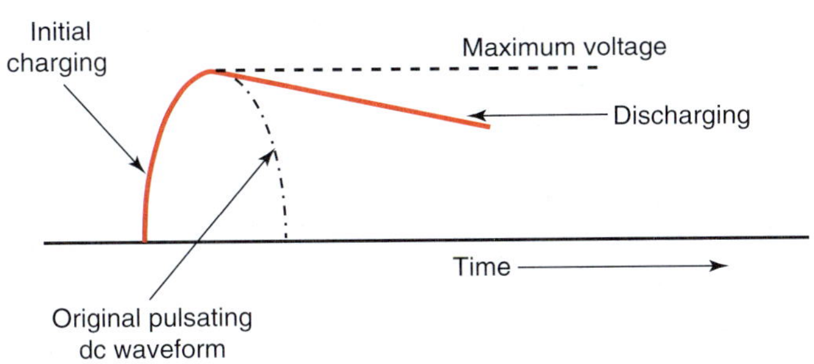

Figure 30-23. Once a capacitor is fully charged, it will begin to discharge. While discharging, the voltage across the capacitor will drop, but the voltage is still greater than the instantaneous voltage of the original pulsating dc waveform.

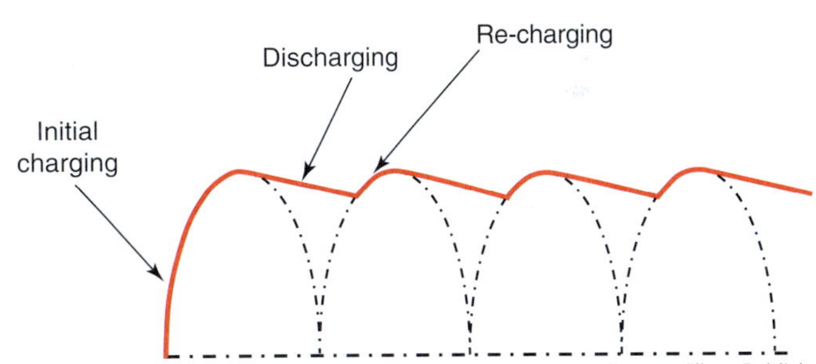

Figure 30-24. Subsequent voltage pulses cause the capacitor to begin charging again. The result is a ripple-shaped, dc waveform.

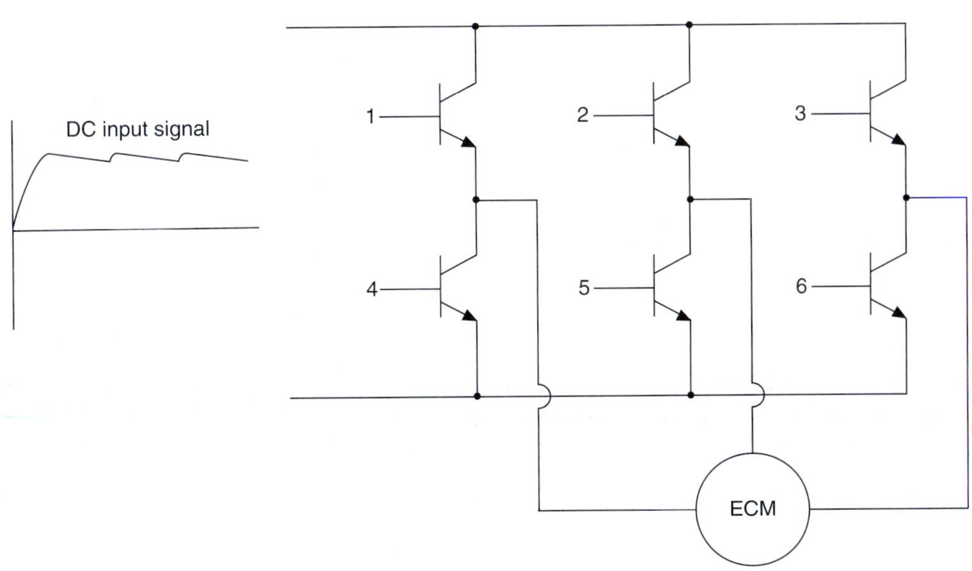

Figure 30-25. In a six-step inverter, three pairs of transistors are responsible for creating three phases of power for the motor. These transistors operate in pairs: transistors 1 and 4 are a pair, as are transistors 2 and 5, as well as 3 and 6.

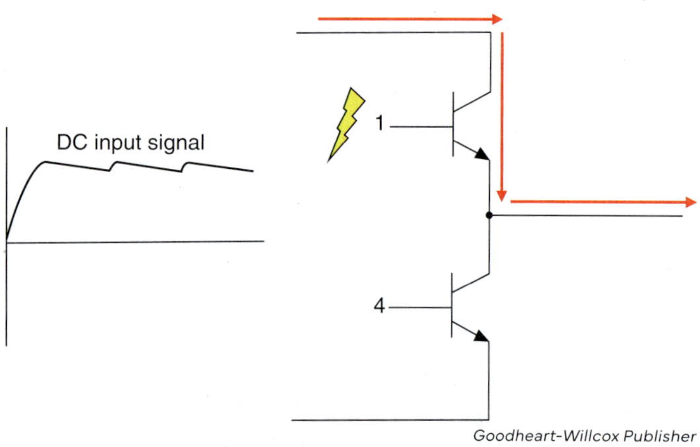

Figure 30-26. First section of the six-step inverter made up of two transistors. When a sufficient control voltage is supplied to the base of transistor 1, current flows through the collector-emitter junction. The voltage at the output has a positive polarity.

In the portion of a six-step voltage-controlled inverter illustrated in **Figure 30-26**, two transistors, identified as transistors 1 and 4, control one phase of the three-phase ECM power signal being generated. At the start of the cycle, a control voltage signal is supplied to the base of transistor 1, and the transistor begins to conduct current. Because the direction of current flow is in the same direction as indicated by the arrow on the NPN transistor, the voltage at the outlet of the inverter circuit has a positive polarity. As long as the voltage at the base exceeds the transistor's threshold voltage, the collector-emitter junction will allow current to flow. In **Figure 30-27A**, the output waveform is created when transistor 1 is conducting. When the control voltage signal is removed from the base of transistor 1, the collector-emitter junction will open, and current flow will stop. This causes the output voltage to drop to zero, **Figure 30-27B**.

Shortly after the base of transistor 1 is de-energized, a control voltage signal is supplied to the base of transistor 4, **Figure 30-28**. Transistor 4 now begins to conduct current. Because the direction of current flow is in a direction that is opposite to the direction indicated by the arrow on the NPN transistor, the voltage at the outlet of the inverter circuit has a negative polarity, **Figure 30-29A**. As long as there is voltage at the base, the collector-emitter junction will allow current to flow. When the control voltage signal is removed from the base of transistor 4, the collector-emitter junction will open, and current flow will stop, **Figure 30-29B**. The output signal created by the combination of transistors 1 and 4 closely resembles a sine wave, **Figure 30-30**.

The single-phase ac waveform created by transistor combinations 2 and 5, **Figure 30-31**, and 3 and 6, **Figure 30-32** is the same as the waveform created by the combination of transistors 1 and 4 with one exception. The differences are that the other two waveforms are generated 120° and 240° out-of-phase with the first. Together, the three individual

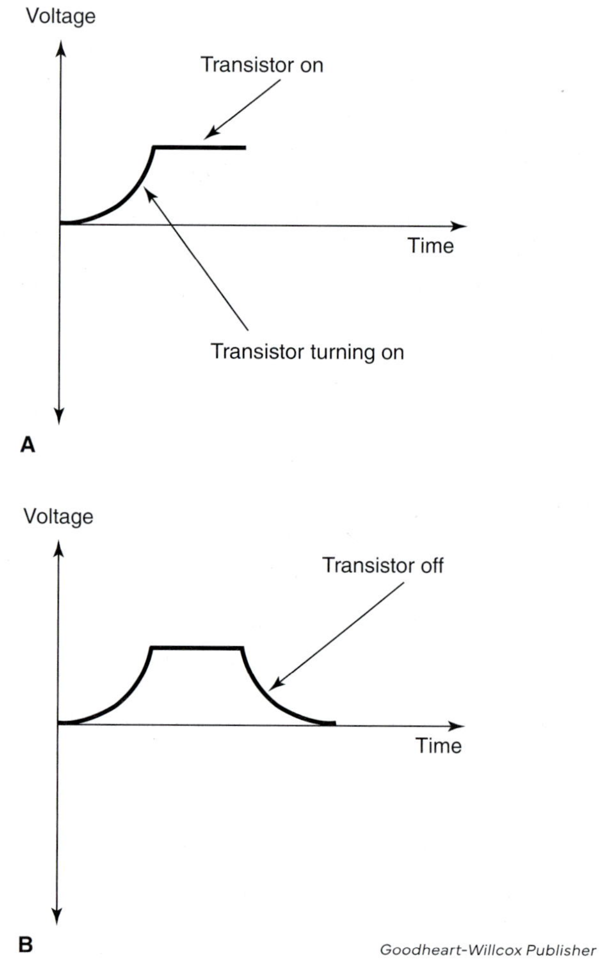

Figure 30-27. Same transistor pair as in **Figure 30-26**. A—As long as voltage at the base exceeds the transistor's threshold voltage, the collector-emitter junction allows current to flow. B—When voltage is removed from transistor 1, the junction opens, and current flow stops. Output voltage drops to zero.

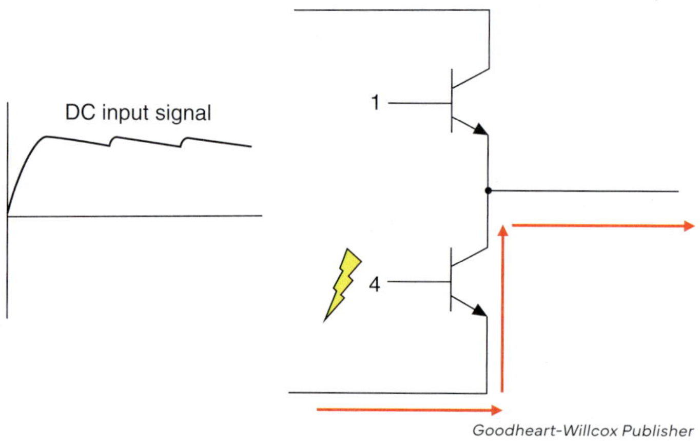

Figure 30-28. When a sufficient control voltage is supplied to the base of transistor 4, current flows through that transistor's collector-emitter junction. The voltage at the outlet of the inverter circuit has a negative polarity.

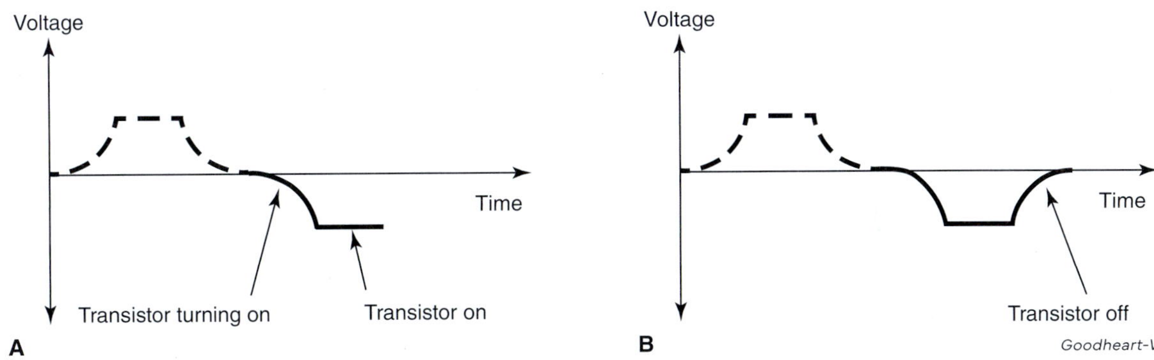

Goodheart-Willcox Publisher

Figure 30-29. Same transistor pair as in **Figure 30-27**, with transistor 4 conducting current. A—The voltage at the output will have a negative polarity while this transistor is conducting. B—With the signal removed from the base of transistor 4, the collector-emitter junction opens and the voltage returns to zero.

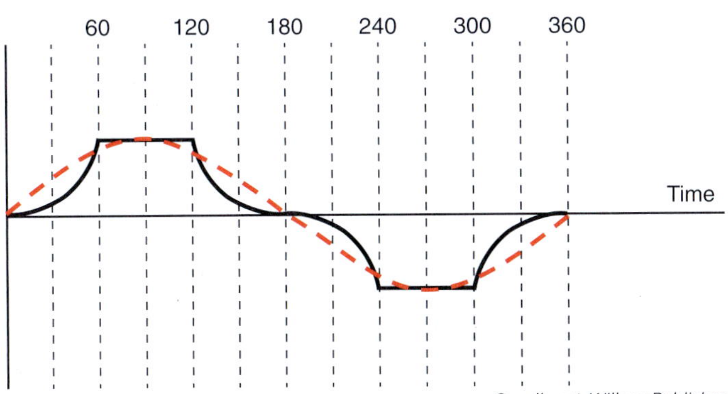

Goodheart-Willcox Publisher

Figure 30-30. The combination of the two signals created by transistors 1 and 4 produces a waveform that closely resembles a sine wave.

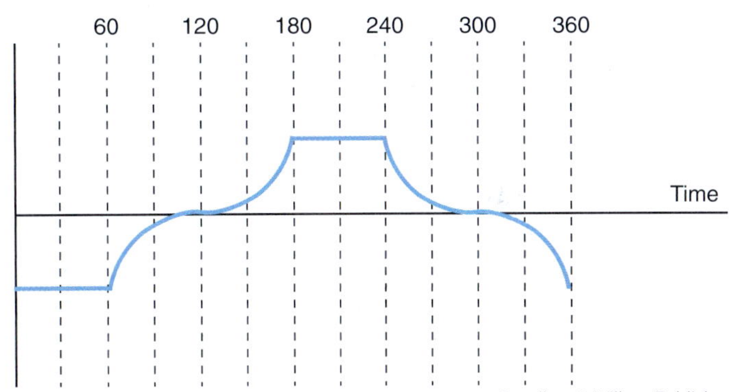

Goodheart-Willcox Publisher

Figure 30-31. The waveform created by transistors 2 and 5. This waveform is the same as the one produced by transistors 1 and 4 but is 120° out of phase with that waveform. Notice that this waveform has its positive polarity starting at 120°, which is 120° after the start of the waveform shown in **Figure 30-30**.

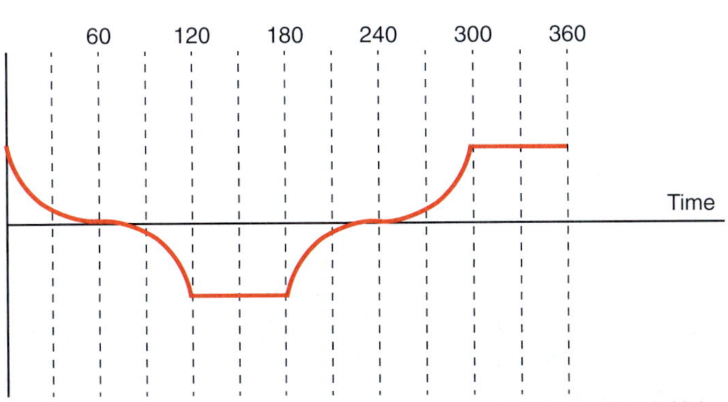

Goodheart-Willcox Publisher

Figure 30-32. The waveform created by transistors 3 and 6. This waveform is the same as the one produced by transistors 2 and 5 but is 120° out of phase with that waveform. Notice that this waveform has its positive polarity starting at 240°, which is 120° after the start of the waveform shown in **Figure 30-31**.

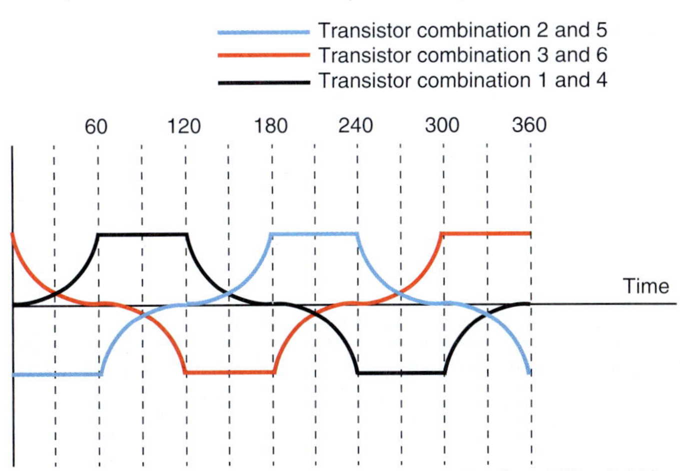

Goodheart-Willcox Publisher

Figure 30-33. Superposition of the three individual waveforms created by transistors 2 and 5, transistors 3 and 6, and transistors 1 and 4. Notice the symmetric nature of this artificial three-phase signal.

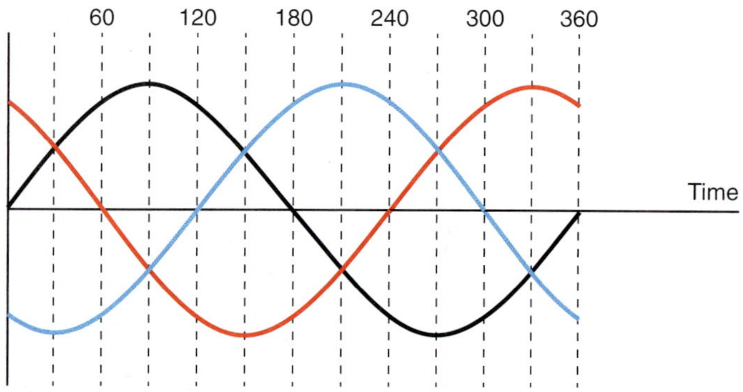

Goodheart-Willcox Publisher

Figure 30-34. The waveform illustrated in **Figure 30-32** closely resembles an ideal three-phase waveform.

single-phase ac waveforms create the three-phase signal that is ultimately supplied to the ECM, **Figure 30-33**. This waveform closely resembles an ideal three-phase waveform, **Figure 30-34**.

30.5 Frequency and Speed Adjustment

Once a three-phase signal has been supplied to the stator windings of the ECM, the motor operates in a manner very similar to a typical three-phase motor. The main difference between the ECM and a standard three-phase motor is the ability of the ECM to vary its speed based on external conditions and the operational needs of the system. The control module receives input signals from external sensors and adjusts the output signal to the motor portion of the assembly based on the input, or feedback, signals. The control module is able to make this adjustment by increasing or reducing the time gap between the control signals sent to the base terminals on the individual transistors in the inverter circuit. If the time gap is shortened, the period of the waveform is shortened, and the frequency increases. Recall that the period and the frequency of a waveform are inversely related. An increase in frequency causes the motor speed to increase. If the time gap is lengthened, the period of the waveform is increased, and the frequency decreases. A decrease in frequency causes the motor speed to decrease.

30.6 ECM Evaluation

There are many manufacturers that produce ECMs, and each has their own set of procedures to check their motors. However, there are some basic guidelines that should be followed, regardless of the motor being evaluated or who manufactured it. When troubleshooting a motor, be sure to have the manufacturer's literature on hand, as this material will provide information that is specific to the motor. In addition, some manufacturers have proprietary test instruments that can be used to quickly, effectively, and efficiently determine the cause for motor failure, if indeed the motor has malfunctioned.

Verify Motor Inputs

In order for the ECM, or any motor, to operate properly, it must be supplied with the correct voltage and control signals. Before concluding that the ECM is malfunctioning, check the following:

- Measure the voltage that is being supplied to the motor and compare this voltage to the motor's nameplate. Verify that the voltage is within the acceptable range, usually within 10% of the nameplate voltage. If the supply voltage is not correct, remedy the situation and recheck motor operation.
- Check all motor control input signals. Verify that these signals are correct by comparing the signal values to the manufacturer's literature to ensure they are within the acceptable range. If the motor control input signals are not correct, remedy the situation and recheck motor operation.

Isolate the Problem

If the main supply voltage and the motor control input signals are all within acceptable ranges, the problem lies within the motor or the motor's control module. If the ECM is a single-component motor, meaning the control module is integrated into the motor housing, the entire unit may need to be replaced. If the control module and the motor are not permanently interconnected, the problem can be remedied by replacing either the motor's control module or the motor. To determine which portion of the assembly should be replaced, disconnect the motor from the controller. Then, manually try to turn the motor's shaft to see if it turns freely and measure the winding resistances of the motor. Compare these resistance values to those provided in the manufacturer's literature.

If the motor shaft rotates freely and the winding resistances match the manufacturer's specifications, it is likely that the motor portion of the assembly is functional. The control module should be replaced.

If the motor shaft does not rotate freely or the winding resistances do not match the manufacturer's specifications, the motor portion of the assembly is defective and needs to be replaced. The control module may or may not need to be replaced. Once the motor has been replaced, motor operation should be checked. If the motor still does not operate, the control module will need to be replaced as well.

Additional general motor evaluation and troubleshooting guidelines are discussed in more detail in Chapter 33, *Motor Evaluation and Troubleshooting*.

CHAPTER 30 REVIEW

Summary

- ECMs are classified as dc three-phase brushless motors. (30.1)
- ECMs can operate as constant speed, constant torque, or constant airflow motors. (30.1)
- ECMs are soft-start motors that gradually ramp up and ramp down their speed. (30.1)
- ECMs can be configured with integrated control modules, detachable control modules, or with remote control modules that are not bolted to the motor. (30.2)
- Semiconductor materials are formed by adding impurities to silicon or germanium. Silicon is commonly used in the production of electronic circuits. (30.3)
- Diodes allow current to flow in only one direction and are made up of one layer of N-type material and one layer of P-type material. (30.3)
- Transistors are made up of three alternating layers of N-type material and P-type material. PNP transistors have two layers of P-type material on either side of the N-type material; NPN transistors have two layers of N-type material on either side of the P-type material. (30.3)
- ECMs rely on rectifiers and inverters to create the motor's power supply. (30.4)
- A dc rectifier is used to convert an ac power signal to a dc signal. (30.4)
- A signal-smoothing circuit with a bank of capacitors is used to convert a dc pulsating signal to a smoother, constant-voltage dc signal. (30.4)
- A six-step inverter circuit relies on three pairs of transistors to create each leg of a three-phase ac signal. (30.4)
- The speed of an ECM can be varied by changing the frequency of the motor's power supply. (30.5)
- The frequency of the power supply can be changed by changing the rate of control signal pulses supplied to the inverter circuit's transistors. (30.5)
- Electronically commutated motors must have the correct supply voltage and control circuit inputs in order to operate properly. (30.6)

Review Questions

Use the information in the chapter to answer the following questions.

Measuring Your Potential

1. *True or False?* An electronically commutated motor has the ability to gradually change its speed to match the load on the system. (30.1)
2. *True or False?* An electronically commutated motor uses mechanical brushes to control the polarity of a motor's power signal. (30.1)
3. Which of the following statements regarding the configuration of an ECM is *not* true? (30.2)
   A. Most ECMs consist of the motor and the control portion.
   B. ECMs with an integrated motor control have a motor and control module bolted together for installation.
   C. In an ECM, the motor is comprised of the stator windings and rotor.
   D. An ECM that uses a separate motor and control module that are not bolted together is a popular choice for installations with space restrictions.
4. A PNP transistor is made up of _____. (30.3)
   A. two layers of P-type material and one layer of N-type material
   B. two layers of N-type material and one layer of P-type material
   C. one layer of P-type material and one layer of N-type material
   D. two layers of P-type material and two layers of N-type material
5. A silicon diode will be forward biased if the polarity of the diode is _____. (30.3)
   A. incorrect and the voltage across the device is greater than 0.7 V
   B. multidirectional and the voltage across the device is less than 0.7 V
   C. correct and the voltage across the device is greater than 0.7 V
   D. correct and the voltage across the device is less than 0.7 V

6. When a diode is reverse biased, the diode acts as a(n) _____. (30.3)
   A. closed switch and allows current to flow through the device
   B. open switch and allows current to flow through the device
   C. closed switch and does not allow current to flow through the device
   D. open switch and does not allow current to flow through the device
7. A depletion region is created in a diode when the diode's _____. (30.3)
   A. cathode is connected to a battery's positive terminal
   B. cathode is connected to a battery's negative terminal
   C. collector is connected to a battery's positive terminal
   D. collector is connected to a battery's negative terminal
8. *True or False?* An inverter circuit converts ac power to dc power. (30.4)
9. *True or False?* The signal-smoothing circuit creates a ripple-shaped dc waveform. (30.4)
10. Regarding an ECM, single-phase power is found only at the _____. (30.4)
    A. inlet of the control module
    B. outlet of the control module
    C. inlet of the signal-smoothing circuit
    D. outlet of the signal-smoothing circuit
11. A full-wave bridge rectifier uses how many diodes? (30.4)
    A. 3
    B. 4
    C. 5
    D. 6
12. If the control module is found to be defective on an integrated ECM, the _____. (30.6)
    A. control module can be replaced if the motor shaft spins freely
    B. control module can be replaced if the winding resistance is correct
    C. entire motor assembly must be replaced
    D. entire motor assembly must be replaced if winding resistance is not correct

Plug-In

1. A diode's threshold voltage is best described as the _____. (30.3)
   A. maximum voltage required to create a forward-biased condition
   B. maximum voltage required to create a reverse-biased condition
   C. minimum voltage required to create a forward-biased condition
   D. minimum voltage required to create a reverse-biased condition
2. Which of the following will cause the speed of an ECM to increase? (30.5)
   A. Reducing the amount of time between base-emitter control voltage signals
   B. Increasing the amount of time between base-emitter control voltage signals
   C. Reducing the amount of time between collector-emitter control voltage signals
   D. Increasing the amount of time between collector-emitter control voltage signals
3. When troubleshooting an ECM that is not operating properly, what is the first step you should perform? (30.6)
   A. If the ECM is a single-component motor, replace the entire unit.
   B. Check to see if the motor shaft turns freely.
   C. Verify that the voltage is within the acceptable range.
   D. Measure the winding resistance of the motor.

Making the Connection

1. Explain the state of an LED if the circuit polarity is correct and the voltage across the diode is 1.2 V. (30.3)
2. Explain the similarities between a transistor and a general-purpose relay. (30.3)

SECTION

9

Motor Starting, Protection, and Troubleshooting

Goodheart-Willcox Publisher

CHAPTER

31

Single-Phase Motor Starting and Protective Devices

Suranto W/Shutterstock.com

SPARKING DISCUSSION

An electrician states that it is acceptable to replace a defective contactor with one that has higher contact amperage ratings than the failed device. When, if ever, is this acceptable? Why might this practice not be acceptable?

LEARNING OUTCOMES

After completing this chapter, you will be able to:

31.1 Explain the difference between a relay and a contactor.

31.2 Explain the terms full-load amperage (FLA), full-load current (FLC), running-load amperage (RLA), and locked-rotor amperage (LRA) as they apply to electric motors.

31.3 Explain the terms poles and throws and how they apply to control devices and switches.

31.4 Describe the difference between positive temperature coefficient (PTC) devices and negative temperature coefficient (NTC) devices.

31.5 Explain the operation of a centrifugal switch and why it is not intended for use near combustible materials.

31.6 Explain the difference between a direct-acting relay and a pilot-duty relay.

31.7 Explain the function of auxiliary contacts as they are used in conjunction with contactors.

31.8 Properly check a current magnetic relay to determine if it is operational or needs to be replaced.

31.9 Properly check a potential magnetic relay to determine if it is operational or needs to be replaced.

31.10 List three common types of motor overload devices.

31.11 Explain the differences between conventional motor protective devices and electronic motor protective devices.

31.12 Discuss the basics of properly selecting motors, motor components, and overload protection devices.

TECHNICAL TERMS

auxiliary contact
centrifugal switch
continuous coil voltage
direct-acting relay
drop-out voltage
overcurrent
pick-up voltage
pilot-duty relay
pole
thermistor
throw

Introduction

In Chapter 28, *Single-Phase AC Motors*, various types of motor configurations are discussed. In order to prolong the useful life of single-phase ac motors, it is important that they operate as effectively and efficiently as possible. As an electrician, you will need to ensure that a motor is properly installed and wired. You will also be responsible for making sure a motor starts easily and properly, without having undue mechanical or electrical stress placed upon it. While evaluating an ac motor, you will need to ensure that it runs within intended electrical parameters and is being used for applications that are defined by the manufacturer's specifications.

In this chapter, we take a closer look at motor-starting devices and expand our discussion to encompass motor protective devices that are used in conjunction with single-phase ac motors.

31.1 Relays and Contactors

The basic single-phase motor consists of the start winding and the run winding. The run winding has a lower resistance than the start winding. It is this difference in resistance that creates the magnetic field imbalance that allows single-phase motors to start. In some cases, the start winding is removed from the active electric circuit once the motor has started. In other instances, the start winding remains in the active electric circuit whenever the motor is operating. The motor's design and control strategy determine whether or not the motor's start winding remains in the active electric circuit once the motor has started and is running at, or close to, its design speed.

Single-phase motors, up to about 2 hp, can often be controlled, either manually or automatically, by snap-action switches rated for this application. Single-phase motors may also be controlled by relays or contactors. Although they perform the same basic functions and operate in a similar manner, relays and contactors have some fundamental differences. For the most part, relays are designed to control loads and circuits that draw less than 20 A, while contactors are designed to control loads and circuits that draw more than 20 A. Therefore, the motor being controlled will, for the most part, determine if a relay or a contactor will be used. In addition, relays are considered to be throwaway devices, while contactors can often be repaired or rebuilt in the field.

The size and cost of the contactor will often dictate whether or not it is economically feasible to repair. In most cases, when the contactor is small, it is less expensive to replace the device than repair it.

31.2 Motor Amperage Ratings and Values

There are four current values that are used to rate and measure electric motors: full-load amperage (FLA), full-load current (FLC), running-load amperage (RLA), and locked-rotor amerperage (LRA).

These current and amperage values are used as setpoints that aid in effective and efficient control of motors. By monitoring the various amperage values that are present at different stages of motor operation, control devices are able to open and close sets of electrical contacts to help ensure continued satisfactory motor operation. A motor's state of operation often correlates to the motor's amperage draw. For example, as the load on a motor increases, the amperage required to effectively power the motor also increases.

The motor's full-load amperage (FLA) is the current supplied to the motor when it is operating at its rated voltage, horsepower, and frequency. FLA is listed on the motor's nameplate.

A motor's full-load current (FLC) is an amperage rating listed in an *NEC* table based on the type of motor and its rated horsepower and voltage. Thus, FLC is the same for similar motors with the same horsepower rating, while FLA often varies for similar motors with the same horsepower rating.

Know the Code

The *NEC* specifies that a motor's full-load amperage (FLA), which is listed on the motor's nameplate, is used to determine the rating of the required overload protective devices for the motor. The full-load current (FLC) value, which is found in a table in the *NEC*, is used to size motor circuit conductors, short-circuit and ground-fault overcurrent protective devices, and other circuit devices.

When a motor is operating, the amount of current flowing through the motor is the running-load amperage, RLA. Thus, RLA is the actual current drawn by the motor while operating. If the motor is operating under less than full-load conditions, the RLA will be lower than the motor's FLA nameplate rating.

The locked-rotor amperage, LRA, represents the motor's current when it is initially energized, but before the motor begins to rotate. LRA is identified by a locked-rotor indicating code letter on a motor's nameplate. The code letter corresponds to a locked-rotor kVA per horsepower range.

The LRA of a single-phase motor is typically in the range from five to seven times the motor's FLA. Once the motor starts, the amperage drops from its LRA value to its normal operating current. The LRA amperage is an important current value when properly selecting motor starting and protective devices. Motors should not operate under locked-rotor conditions for extended periods of time, as doing so could damage the motor. On a properly operating motor, the LRA condition should be present for only a fraction of a second.

Shedding Light

Locked-rotor amperage (LRA) is also referred to as locked-rotor current (LRC). These two terms are used interchangeably.

31.3 Poles and Throws

Motor control or motor protective device selection must be made properly to help ensure the safe and continued operation of the motor. Control, safety, and protective devices are rated, in part, by the number of device poles and throws. The term ***pole*** refers to the number of conductive paths being simultaneously controlled by a single device, **Figure 31-1**. The term *single-pole* refers to the control of current flow through a single conductive path. The term *double-pole* refers to the control of current flow through two conductive paths at the same time. Three-pole devices have the ability to simultaneously control current flow through three conductors and are used in conjunction with three-phase circuits. Three-pole control devices are further discussed in Chapter 32, *Three-Phase Motor Starting and Protective Devices.*

The term ***throw*** refers to the number of possible current paths associated with each pole. A single-throw device offers only one possible path for electric current to take. Do not confuse the number of throws with the number of switch positions. For example, the control device in **Figure 31-2** has two positions, ON and OFF, but there is only one possible current path, namely the path between points A and B. Thus, it is a single-throw device.

A double-throw device offers two possible current paths, **Figure 31-3**. The switch position determines which of the two output conductors, B or C, is energized.

A double-pole, double-throw device is, in essence, two single-pole, double-throw switches that are controlled simultaneously, **Figure 31-4**.

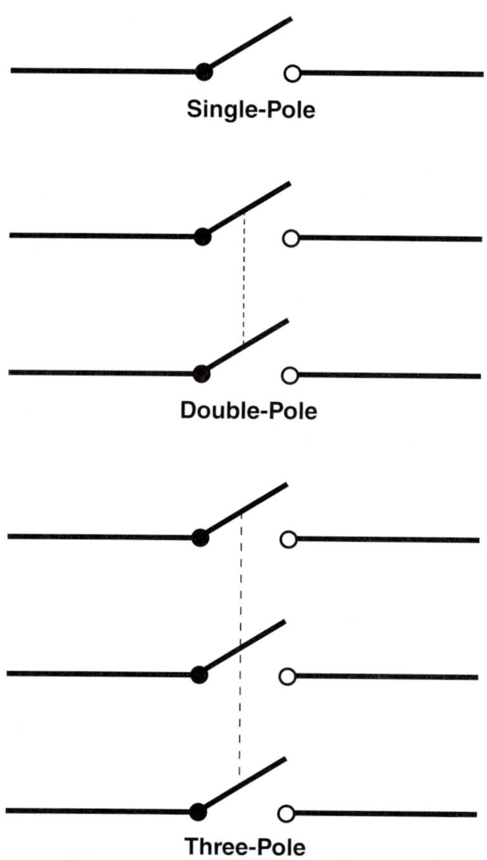

Goodheart-Willcox Publisher

Figure 31-1. A pole refers to the number of conductive paths being simultaneously controlled by a single device. The double-pole switch here, for instance, controls current through two conductive paths at the same time.

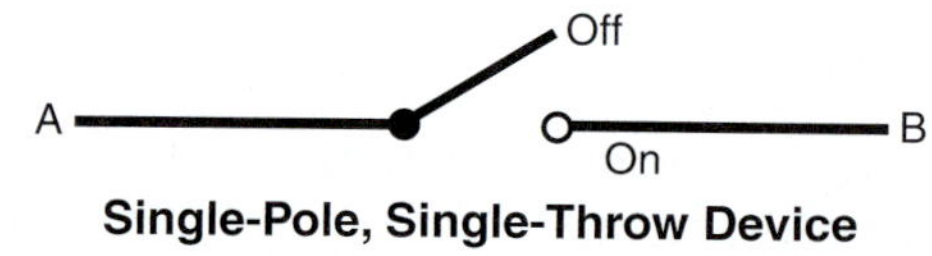

Goodheart-Willcox Publisher

Figure 31-2. A single-pole, single-throw device offers only one possible current path for electric current to follow.

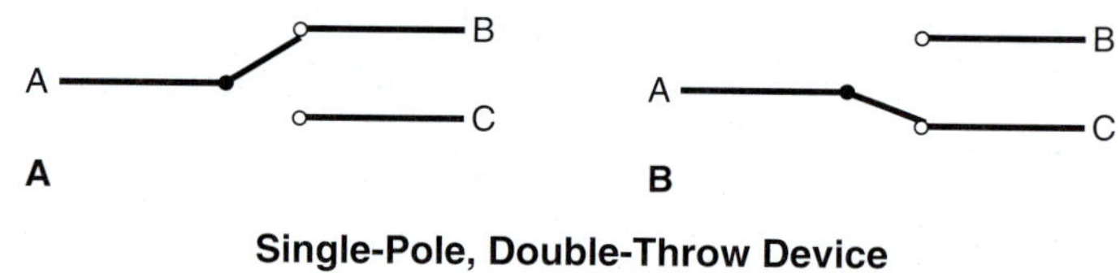

Goodheart-Willcox Publisher

Figure 31-3. A single-pole, double-throw device controls one conductor, but there are two possible current paths. A—Current is able to flow between points A and B. B—Current is now able to flow between points A and C.

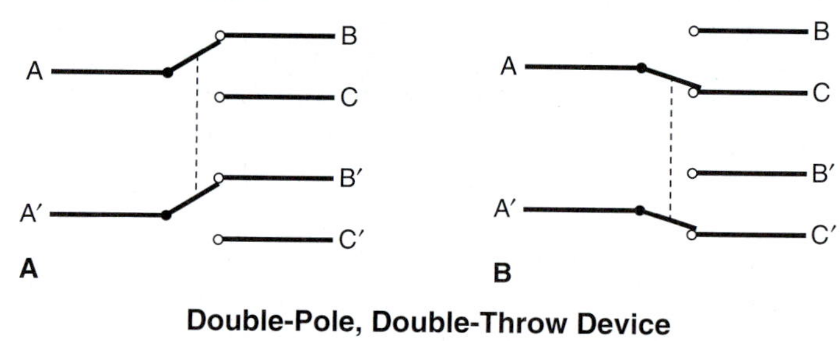

Goodheart-Willcox Publisher

Figure 31-4. A double-pole, double-throw device operates, in essence, as two single-pole, double-throw switches controlled simultaneously. A—Current is able to flow between points A and B and between A′ and B′. B—With a flip of the switch, current is able to flow between points A and C and between A′ and C′.

31.4 Temperature Coefficient Devices

For the most part, the resistance value of a conventional resistor remains relatively constant regardless of the voltage applied across it and the range of temperatures to which the device is expected to be exposed. Although the resistance of a material typically increases as the sensed temperature increases, and decreases as the sensed temperature decreases, these changes are often very small.

Special resistors, called ***thermistors***, are found in many motor control devices and are, in effect, temperature-sensitive resistors. Thermistors respond quickly to minor temperature fluctuations by making significant changes to their resistance. These changes are much more extreme than those associated with conventional resistors.

Shedding Light

The term thermistor is actually a combination of two words: *thermal*, which relates to heat or temperature, and *resistor*. So, a thermistor can be thought of as a thermally sensitive resistor.

Motor control devices that use thermistors are called temperature coefficient devices.

For some temperature coefficient devices, the relationship between the sensed temperature and its resistance is linear and can be represented by a straight line. In other instances, when the relationship is not linear, component manufacturers provide tables and charts that provide the temperature-resistance relationship. Two commonly used types of temperature coefficient devices are the positive temperature coefficient, PTC, and the negative temperature coefficient, NTC.

Positive Temperature Coefficient Devices

Positive temperature coefficient (PTC) devices use thermistors that experience an increase in resistance as the temperature acting on the PTC increases, **Figure 31-5**. The symbols commonly used to represent PTC devices in electric circuits are provided in **Figure 31-6**.

When used as a motor control device, the PTC device has a low resistance over a wide temperature range. However, when the device is heated, and a certain predetermined sensed temperature is reached, the resistance increases at a much faster rate. When the PTC device has a very high resistance, it acts like an open switch and can be used to, in effect, remove starting components from the active electric circuit.

To help gain an understanding of how the PTC device can be used to control motor operation, first examine the induction-start, induction-run (ISIR) motor. The ISIR motor uses a high-resistance start winding, a low-resistance run winding, and a starting device. A PTC device can be used to remove the start winding from the electric circuit once the motor is running, **Figure 31-7**.

When the motor is initially energized, the PTC is cool and its resistance is very low. Having a low resistance, the device acts like a closed switch and current can flow freely though the motor's start winding. As the motor starts and continues to operate, current flow through the start winding causes the PTC to heat up. As the PTC heats up, the resistance of the device increases. After a short period of time, the resistance of the PTC is high enough to act as an open switch, effectively removing the start winding from the circuit. The ISIR motor then continues to operate with only the run winding energized. Although the motor's start winding is removed from the circuit, there is still a small amount of current flowing through this winding and the PTC device. Current flow through the PTC keeps the resistance high. When there is no longer a need for motor operation, the motor is de-energized, the PTC cools down and its resistance drops in preparation for the next call for motor operation.

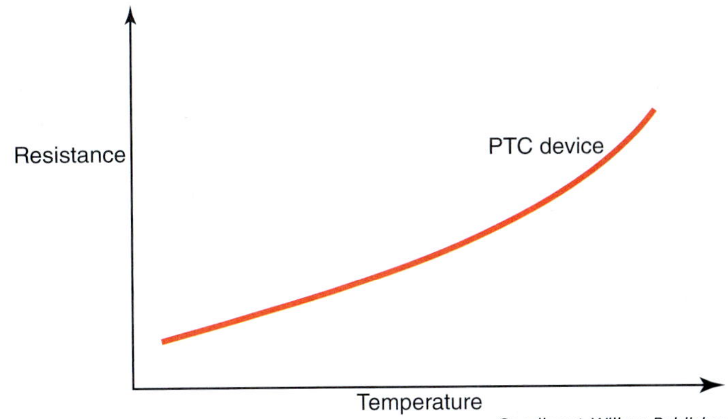

Goodheart-Willcox Publisher

Figure 31-5. For positive temperature coefficient (PTC) devices, the resistance increases as the temperature of the device increases.

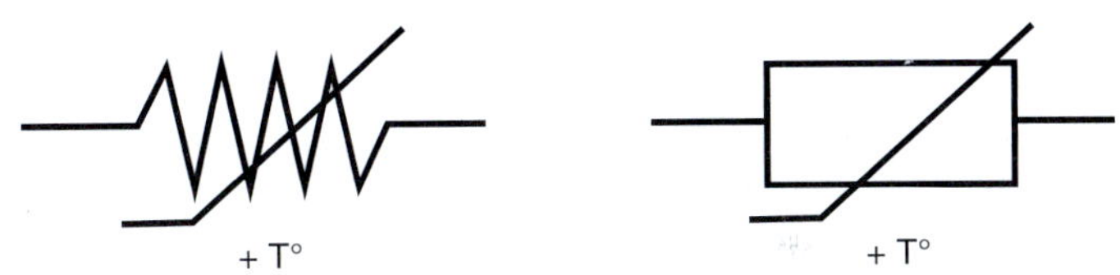

Goodheart-Willcox Publisher

Figure 31-6. Electrical symbols for PTC devices.

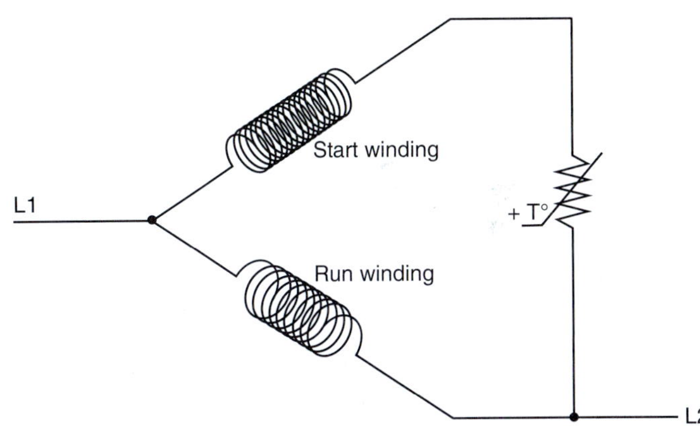

Goodheart-Willcox Publisher

Figure 31-7. Wiring diagram of an induction-start, induction-run motor using a PTC device. As the PTC device heats up, its resistance increases. At a predetermined point, the PTC has very high resistance, acts like an open switch, and removes the start winding from the circuit.

Although these devices are commonly used to control motor operation, PTC devices can also be used to sense temperatures to control electric heating elements, sense liquid levels, and perform time-delay circuit functions.

Negative Temperature Coefficient Devices

Negative temperature coefficient devices are briefly covered here, since many electronically commutated motors (ECMs) have a negative temperature coefficient (NTC) device wired in series with the motor's power supply. Negative temperature coefficient devices use thermistors to decrease their resistance as the sensed temperature increases, and increase their resistance as the temperature decreases, **Figure 31-8**. The electrical symbols for NTC devices are shown in **Figure 31-9**.

When an ECM motor is initially energized, the high resistance of the NTC reduces the amount of current flow through the motor. After a very short period of time, typically less than 0.5 seconds, the resistance of the NTC drops to 0 Ω, allowing full current to flow to the motor.

NTC devices are also commonly used as temperature sensing devices in thermostats and electronic circuits. In electronic circuits, NTCs are often used as compensating devices that help control and maintain the resistance of circuits that are expected to have exposure to extreme temperature fluctuations. This helps stabilize the performance of the circuitry. In addition, these devices can be used to protect delicate electronic circuitry from damage due to high inrush current.

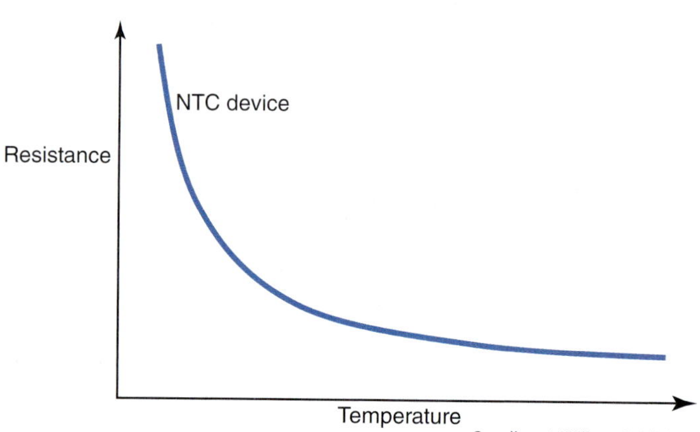

Figure 31-8. For NTC devices, resistance decreases as the temperature of the device increases.

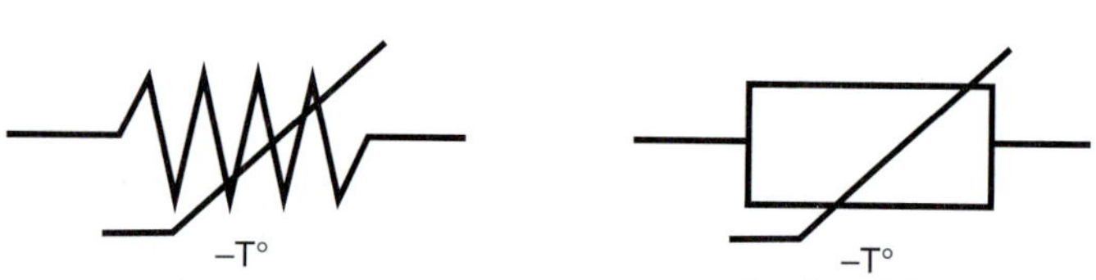

Figure 31-9. Electrical symbols for NTC devices.

31.5 Centrifugal Switches

The ***centrifugal switch*** is commonly used to remove the start winding from the motor circuit once the motor has started and has reached a speed of approximately 70% of its design operating speed. The switch's contacts are connected in series with the motor's start winding, as shown in **Figure 31-10**.

The centrifugal switch is mounted to the motor's shaft and is manufactured with weights that push outward when the motor reaches approximately 70% of its rated speed. The position of the centrifugal switch's contacts is determined by the speed of the motor. If a motor has a rated speed of 3450 rpm, the centrifugal switch contact will remain closed until the motor has reached a speed of about 2400 rpm.

When the motor is not operating, or is running at a low speed, the switch's contacts are closed and the start winding remains in the active electric circuit. Once the motor reaches an appropriately high speed, centrifugal force pushes the weights outward, causing the switch's contacts to open. This removes the start winding from the active circuit. When the motor is de-energized, the contacts on the centrifugal switch close. With the switch's contacts closed, both windings will be in the active circuit when motor operation is next called for.

Centrifugal switches are commonly found on household appliances such as washing machines. They are not desirable for use in areas where combustible materials are located or stored, as sparks are produced when the switch opens and closes its contacts. The size or intensity of the spark is determined by the amount of current flow in the circuit. Considering that the centrifugal switch is wired in the motor's power circuit, the amount of current flowing through the switch is determined by the design current draw of the motor. As a result, larger motor applications will result in larger sparks at the centrifugal switch. The operation of the centrifugal switch is easily identified as there is a loud, audible click whenever the motor starts and stops.

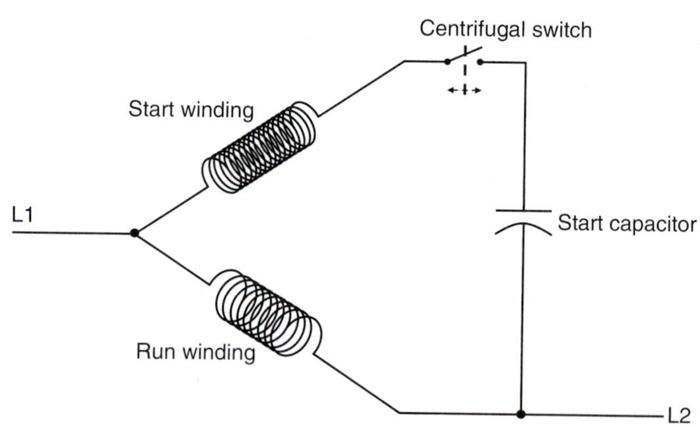

Figure 31-10. Wiring diagram of a capacitor-start, induction-run motor using a centrifugal switch. When the motor reaches approximately 70% of its design speed, the contacts of the centrifugal switch open, removing the start winding and start capacitor from the circuit.

31.6 General-Purpose Relays

General-purpose relays are devices used to control the operation of control circuits rather than main power circuit. In most cases, the control circuit is powered by a low-current, low-voltage signal, while the main circuit being controlled is a higher-current, higher-voltage circuit. General-purpose relays are made up of a magnetic coil and one or more sets of electrical contacts, **Figure 31-11**. The contacts can be either normally open or normally closed. The most basic relay has a coil and one set of normally open contacts. This relay is referred to as a single-pole, single-throw, normally open relay. The coil is often identified by terminals 1 and 3, while the normally open contacts are identified by terminals 2 and 4, **Figure 31-12**. Most relays are clearly labeled, with the wiring diagram being printed right on the device. These diagrams identify the connection terminals, the coil, and the normally open and normally closed contacts. These diagrams are useful when a system or circuit needs to be evaluated or when a relay needs to be replaced.

Shedding Light

When replacing a defective relay, you may find it helpful to draw a diagram showing all connections on the old relay before disconnecting any wires. You might also consider taking a picture of the connections or labeling the wires. If there is room to work freely, you can unmount the defective relay before removing wires from their terminals. Then, mount the new relay in place and transfer the wires from the old relay to the new device one at a time. This will greatly reduce the possibility of wiring the new control incorrectly.

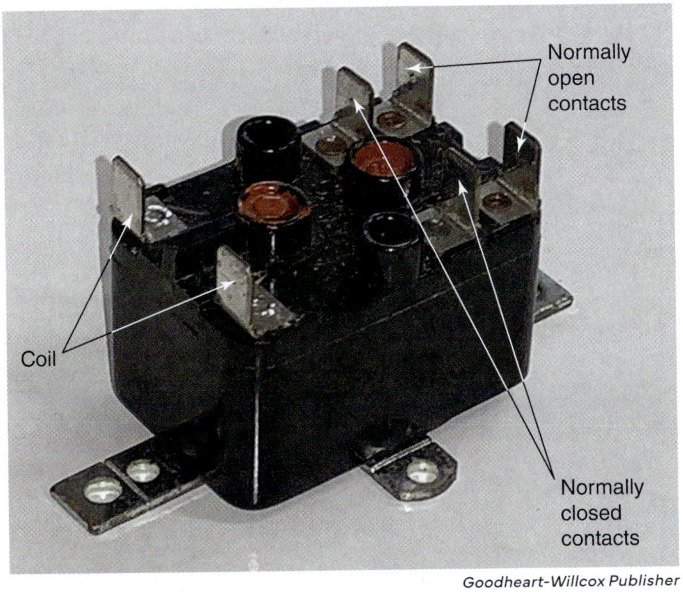

Goodheart-Willcox Publisher

Figure 31-11. A general-purpose relay with one normally open set of contacts and one normally closed set of contacts.

The internal components of the general-purpose relay are shown in **Figure 31-13**. The wiring diagram for a general-purpose relay with one set of normally closed contacts (terminals 5 and 6) and one set of normally open contacts (terminals 2 and 4) is provided in **Figure 31-14**. The relay's coil is located between terminals 1 and 3.

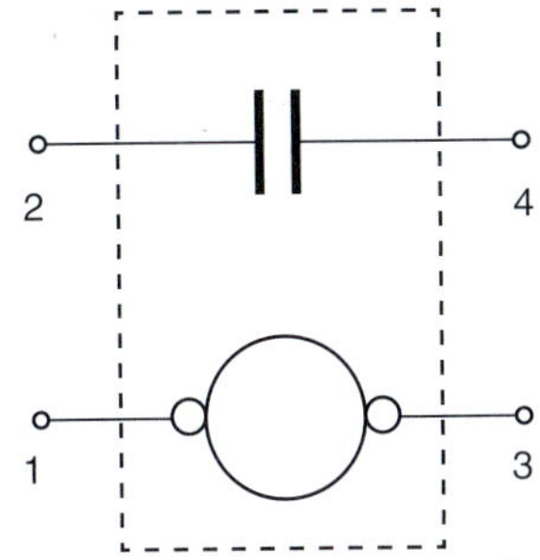

Goodheart-Willcox Publisher

Figure 31-12. Wiring diagram of a simple general-purpose relay with a coil and one set of normally open contacts. The coil terminals are 1 and 3. The normally open contact terminals are 2 and 4.

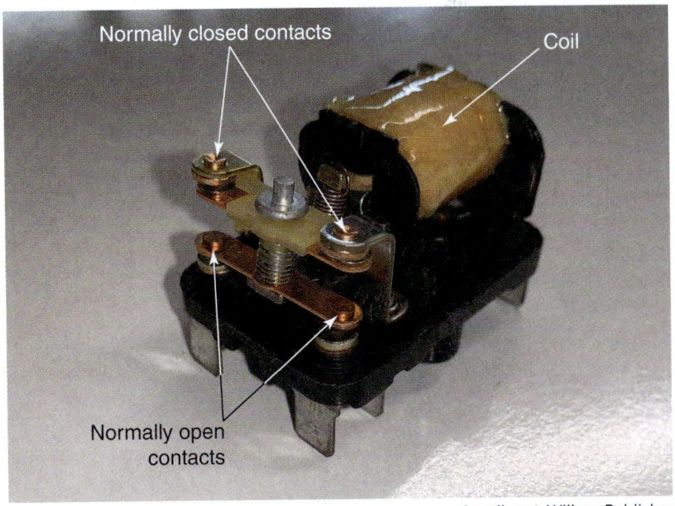

Goodheart-Willcox Publisher

Figure 31-13. The internal components of the general-purpose relay shown in **Figure 31-11**.

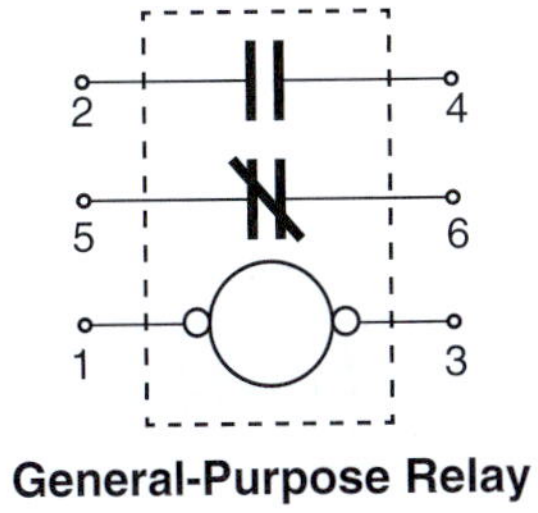

Goodheart-Willcox Publisher

Figure 31-14. Wiring diagram of a general-purpose relay with one set of normally open contacts and one set of normally closed contacts. Coil terminals are 1 and 3; the normally open contact terminals are 2 and 4; the normally closed contact terminals are 5 and 6.

Shedding Light

General-purpose relays are often sealed components. Therefore, the electrician does not have access to the relay's internally mounted coil or contacts. As a result, these relays cannot be repaired in the field and are classified as throwaway devices. If either the coil or the contacts become defective, the entire relay must be replaced. It is, however, desirable to know which portion of the device failed to be able to determine why the system malfunctioned.

Consider the wiring diagram shown in **Figure 31-15**. In this diagram, there are two circuits: a 24 V low-voltage, low-current control circuit and a 240 V line-voltage, higher-current power circuit. A low-voltage manual switch, identified in this wiring diagram as the motor operational device, is used to control the operation of the 240 V motor.

When the low-voltage switch is closed, the relay's coil is energized. This creates a magnetic field that causes the normally open contacts on the relay to close, **Figure 31-16**. The closing of the contacts completes the motor's power circuit, and the motor begins to operate. When the manual switch is opened, the relay coil is de-energized, the relay contacts open, and the motor stops operating.

Direct-Acting Relays

The relay described in the previous section is referred to as a direct-acting relay. A ***direct-acting relay*** is a control device that directly controls the operation of the circuit load. The contacts of a direct-acting relay are wired in series with the main load being controlled. The relay's contacts, therefore, must be rated at an amperage that is greater than the amperage draw of the motor. If the amperage draw of the motor exceeds the amperage rating

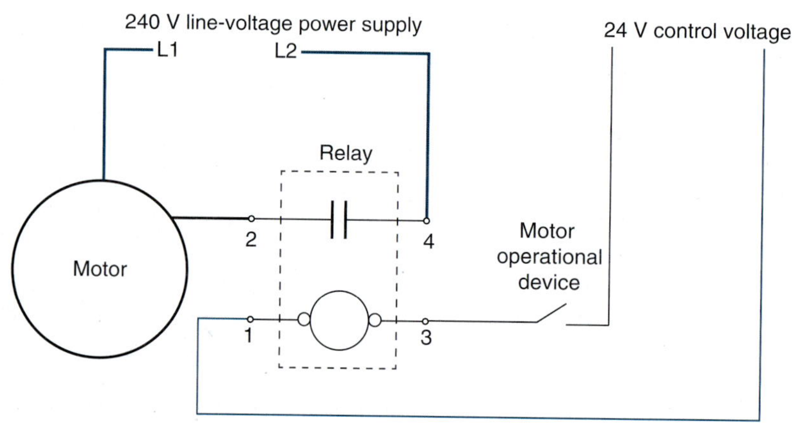

Goodheart-Willcox Publisher

Figure 31-15. A low-voltage control circuit that is used to control the operation of a single-phase motor. The relay coil is a low-voltage load, while the relay contacts are in series with the 240 V line-voltage power supply. A manual switch is used to control the operation of the motor; when the manual switch is open, the motor is off.

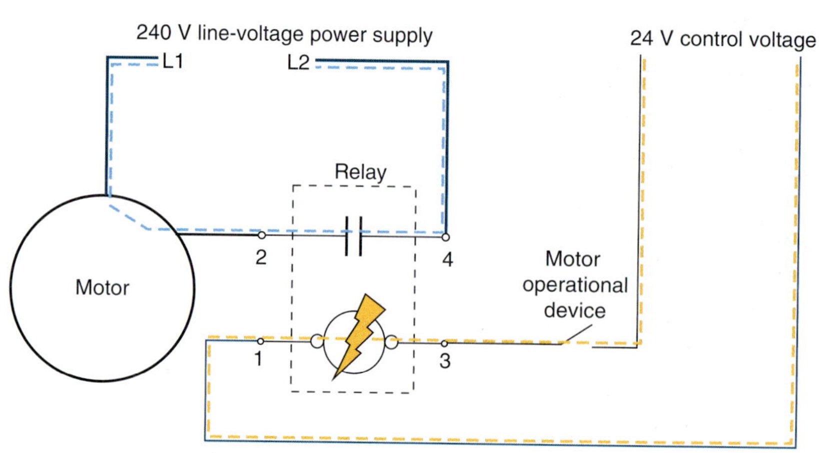

Goodheart-Willcox Publisher

Figure 31-16. Same motor and wiring as in **Figure 31-15**. Here the relay coil is energized, the normally open contacts close, and the motor circuit is energized.

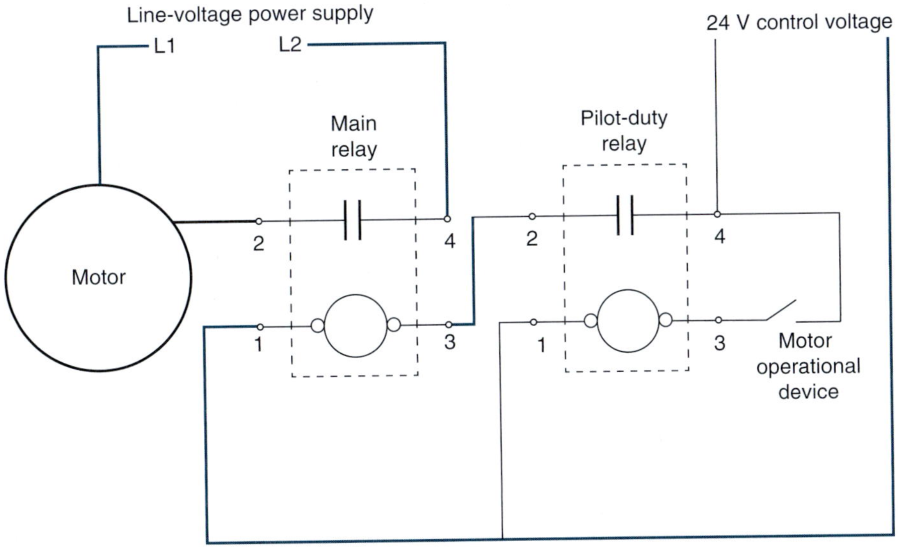

Figure 31-17. Three circuits control the operation of this motor: A pilot-duty relay is used to control the main relay, which in turn controls the line-voltage motor circuit. When the motor's operational device (manual switch) closes, the coil on the pilot-duty relay is energized. When this coil energizes, the pilot-duty relay's contacts close, and the main relay is energized, starting the motor. The contacts on the main relay are heavier duty and can carry more current than the pilot-duty relay.

of the relay's contacts, the relay will very likely fail prematurely. To ensure continued successful motor operation, the relay must be carefully selected to ensure that the contacts can safely withstand the current draw of the load being controlled.

Pilot-Duty Relays

Pilot-duty relays are low-amperage relays that are used to control motor-starting devices with higher amperage ratings. Consider the wiring diagram shown in **Figure 31-17**. This wiring diagram has three circuits that work together to control the operation of the motor. The first circuit contains the switching device that initiates motor operation by energizing the coil of the pilot-duty relay. When the contacts on the pilot-duty relay close, the coil of the main control relay is energized. This causes the contacts on the main relay to close. When the contacts on the main relay close, the motor circuit is completed and the motor starts to operate.

The contacts on the main relay are heavier duty than the contacts on the pilot-duty relay, as they are intended to carry larger amounts of current than the pilot-duty relay. Relays with high current ratings are typically physically larger than relays with lower current ratings. Never exceed the current rating of a relay's contacts.

Multispeed Fan Motor Application

General-purpose relays can be used to control a motor's operating speed. This is accomplished with a single-pole, double-throw relay as shown in **Figure 31-18** and **Figure 31-19**. This relay has one normally open set of contacts and one normally closed set of contacts. On this relay, the normally closed contacts are positioned between terminals 4 and 5, while the normally open contacts are located between terminals 2 and 4. Terminal 4, which is common to both the normally open and the normally closed contacts, is appropriately referred to as the common terminal.

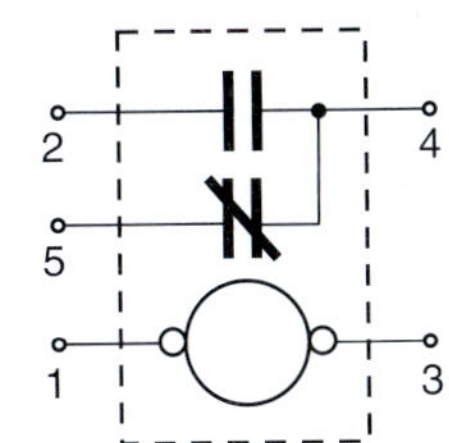

Figure 31-18. Wiring diagram of a general-purpose relay with a single-pole, double-throw set of contacts. The normally open contacts are located between terminals 2 and 4. The normally closed contacts are located between terminals 4 and 5.

The speed switch is wired in the low-voltage control circuit in series with the relay's coil. The speed switch is, in essence, a single-pole, single-throw switch, where one position causes the motor to operate at low speed, while the other position causes the motor to operate at high speed. When the switch is in the open position, the relay's coil is not energized. In this condition, the 2-4 contacts are open and the 4-5 contacts are closed. When the motor is energized, it will operate at low speed as the motor's low speed lead is energized through the terminal 4-5 connection

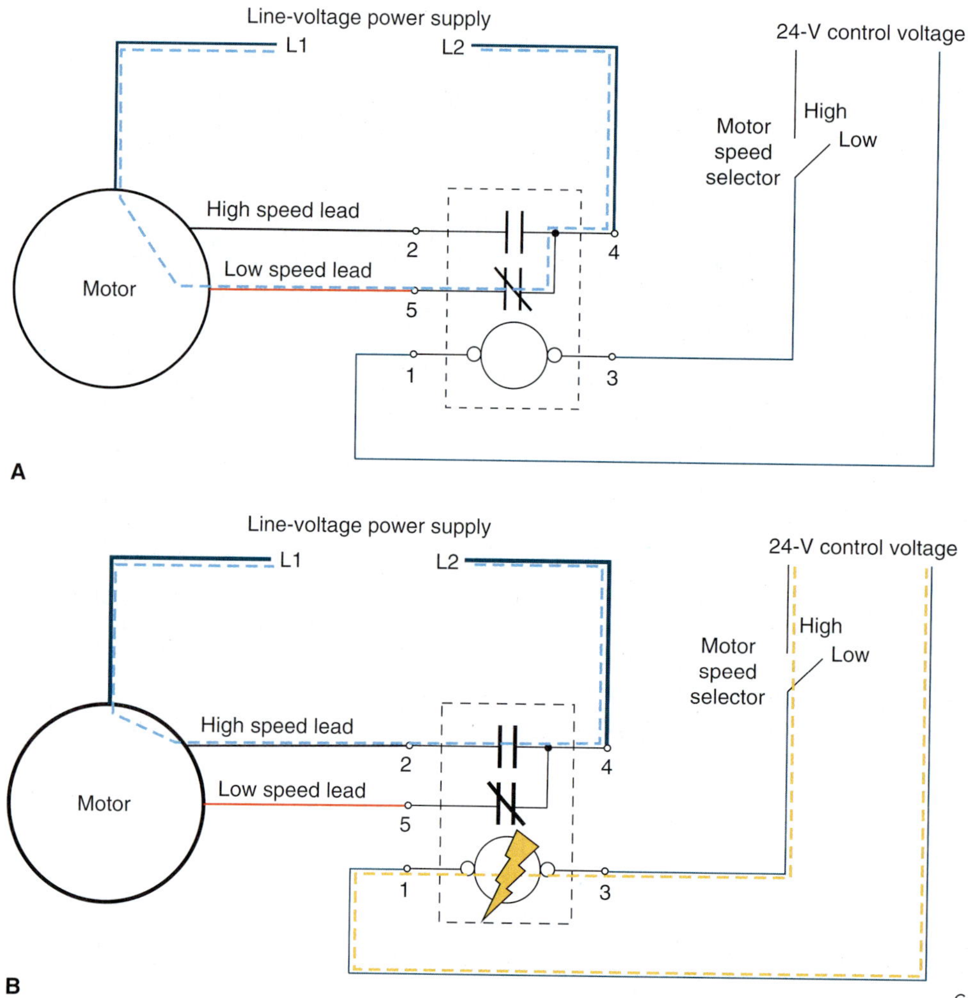

Goodheart-Willcox Publisher

Figure 31-19. Fan speed switching circuit using a relay with a single-pole, double-throw set of contacts. The fan speed motor leads are connected to the non-common terminals on the relay. A—When the switch is open, the relay coil is not energized, current passes through the normally closed contacts, and the motor will operate at low speed. B—When the switch is closed, the relay coil is energized, current passes through the normally open contacts (which close), and the motor will operate at high speed.

on the relay, **Figure 31-19A**. When the motor speed switch is in the closed position, the relay coil is energized and the positions of the relay contacts change. The 2-4 contacts close and the 4-5 contacts open. This de-energizes the motor's low speed lead and energizes the motor's high-speed lead, **Figure 31-19B**.

The application described in this section is commonly used in air-conditioning and heating appliances. In these systems, it is commonplace for the system blower to operate at low speed in the heating mode and at high speed in the cooling mode. The space thermostat controls the low voltage signal that is sent to the relay.

31.7 Contactors

The contactor is a larger version of a relay, **Figure 31-20**. Contactors have a coil and typically from one to four sets of contacts. Some contactors have up to six sets of contacts. The

Goodheart-Willcox Publisher

Figure 31-20. Contactors are, in essence, larger versions of relays. A contactor typically has from one to four sets of contacts; some have up to six sets of contacts.

main contacts on a contactor are often rated at amperages above 20 A. When selecting a contactor, make certain that the following three criteria are met:

- The contactor coil's voltage rating matches that of the voltage used to power the control circuit. Common control and coil voltages are 24 V, 120 V, 208 V, 230 V, and 240 V.
- The number of contactor poles matches the number of power lines that must be controlled. For single-phase applications, contactors typically control either one or two poles.
- The amperage rating of the contactor contacts matches or exceeds the current draw of the load being controlled. It is acceptable to select a contactor with contacts that have higher current ratings than the load but selecting a contactor with excessively high contact amperage ratings will greatly increase the cost of the control.

Since contactors are intended to control larger loads than relays, they are larger and more expensive than relays. If the contactor fails, it can often be disassembled, repaired, and reassembled. The component parts of a contactor that are easily replaced are the coil, the contacts, and the springs.

The contactor is often part of a motor starter. A motor starter is a device that includes a contactor and overload protection.

Shedding Light

If the contactor contacts become pitted, there will be a measurable voltage across them when in the closed position. A voltage reading of more than a few volts across a closed set of contacts is an indication that the integrity of the contacts might be compromised. A visual inspection of the contacts will alert you to a pitted condition. Good contacts will have a shiny, smooth appearance. Discolored or rough contact surfaces are an indication that the contacts are pitted and in need of replacement. When replacing contacts, be sure to replace the stationary as well as the moveable contacts.

Auxiliary Contactor Contacts

Some contactors have multiple sets of heavy-duty contacts that are used to start a motor, and one or more sets of lighter-duty contacts that can be used to control other circuits in the system. These lighter-duty contacts are called auxiliary contacts. ***Auxiliary contacts*** have lower amperage ratings than the main contacts and are, therefore, used to control circuits that carry less current than the main motor circuit. Auxiliary contacts can be normally open or normally closed and are often mounted to the side of the contactor, **Figure 31-21**. The physical movement of the contactor's armature is what causes the position of the auxiliary contacts to change.

Goodheart-Willcox Publisher

Figure 31-21. Auxiliary contacts mounted to the side of a contactor. Auxiliary contacts have lower amperage ratings than the main contacts and work in control circuits that carry less current than the main motor circuit.

Replacing a Contactor

When replacing an existing contactor, verify that the specifications of the replacement contactor match or, in some instances, exceed those of the device being replaced. Failure to match the new device to the old device can potentially cause premature system or equipment failure.

Consider the situation where a single-pole contactor is used to control a single-phase motor, **Figure 31-22**. When the contactor coil is energized, the contactor contacts close, and the motor is energized. When the contactor coil is de-energized, the contacts open and the motor is de-energized. Opening this single set of contacts is all that needs to be done to de-energize the motor, as there will be no current flow through the motor. Since there is only one possible current path, namely between L1 and L2, opening and closing a singular switching location is all that is necessary to start and stop a single-phase motor. A two-pole contactor can often be used instead of a single-pole contactor, **Figure 31-23**, but this is not always the case. Consider the situation where a single-pole contactor is used to control the operation of an air-conditioning compressor equipped with what is referred

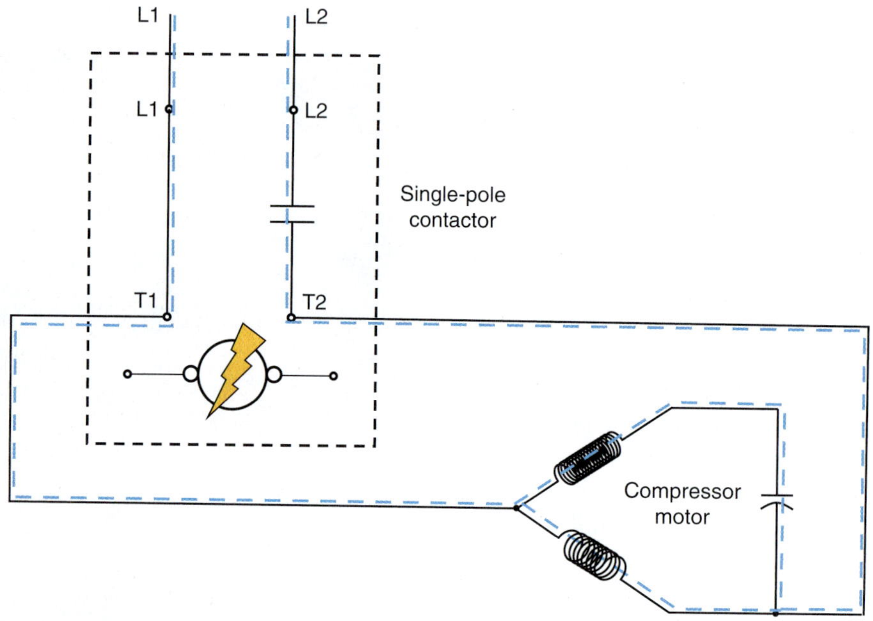

Goodheart-Willcox Publisher

Figure 31-22. Wiring diagram of a single-pole contactor being used to control the operation of a compressor motor. For single-phase applications, opening a single power lead is sufficient to de-energize a motor. When the contactor coil is energized, the motor will operate.

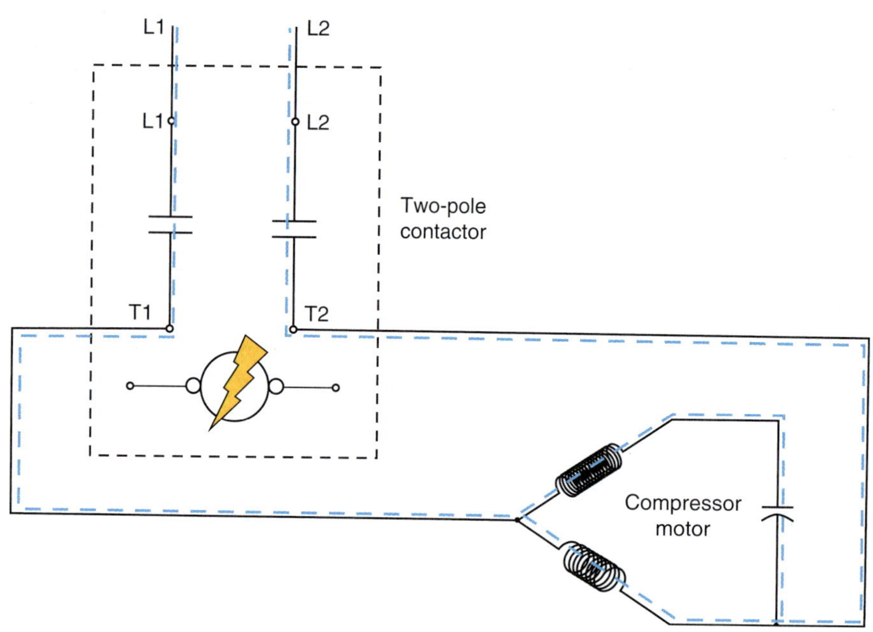

Goodheart-Willcox Publisher

Figure 31-23. Same motor as in **Figure 31-22**. In this application, a two-pole contactor can be substituted for a single-pole contactor, and no problems are likely to arise.

to as a crankcase heater, **Figure 31-24**. The heater is used to keep the compressor's motor warm when it is not operating. When the compressor cycles on, the heater cycles off. This is because the closed set of contacts allows current to bypass the heater. This wiring configuration requires the use of a single-pole contactor. When the compressor is off, the contactor contacts are in the open position, and the crankcase heater, wired in series with the motor's winding, is energized. But, because of the added resistance in series with the compressor motor, the compressor's motor cannot operate, **Figure 31-25A**. When the compressor is on, the contactor contacts are in the closed position, and the crankcase heater is bypassed by the closed contacts. The heater is now de-energized, **Figure 31-25B**, and the motor can operate. If the single-pole contactor were to fail and was replaced with a two-pole contactor, the crankcase heater would never be energized, even when the compressor is not operating, **Figure 31-26**. If the contactor contacts are closed, the heater is bypassed. If the contactor contacts are open, there is no complete current path through the heater.

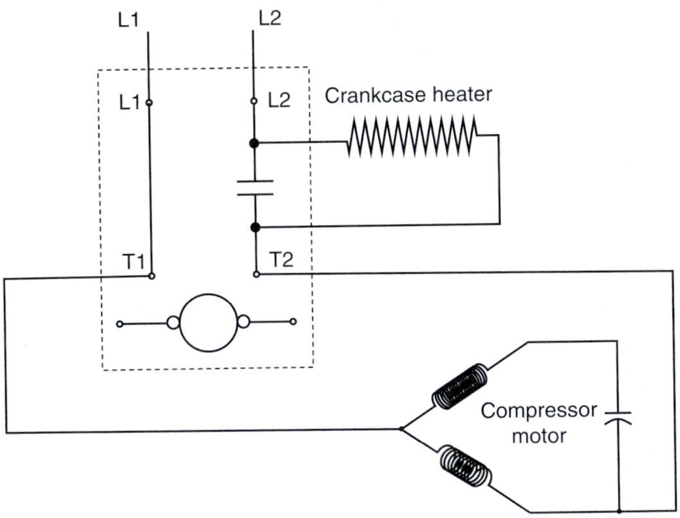

Goodheart-Willcox Publisher

Figure 31-24. Wiring diagram of a single-pole contactor being used to control the operation of a compressor motor as well as a crankcase heater. For this application, a single-pole contactor is required, as a two-pole contactor will prevent the heater from ever operating.

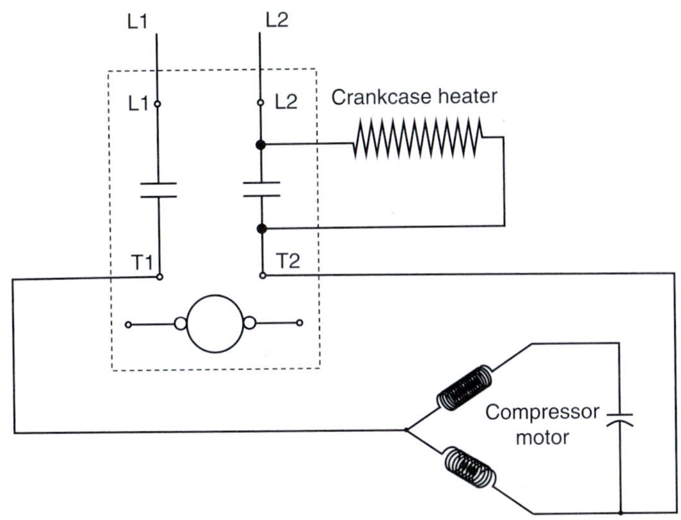

Goodheart-Willcox Publisher

Figure 31-26. Same circuit as **Figure 31-24**, but with a two-pole contactor substituted. With a two-pole contactor, the heater cannot energize. When the contacts are closed, the heater is bypassed. When the contacts are open, there is no closed current path through the heater.

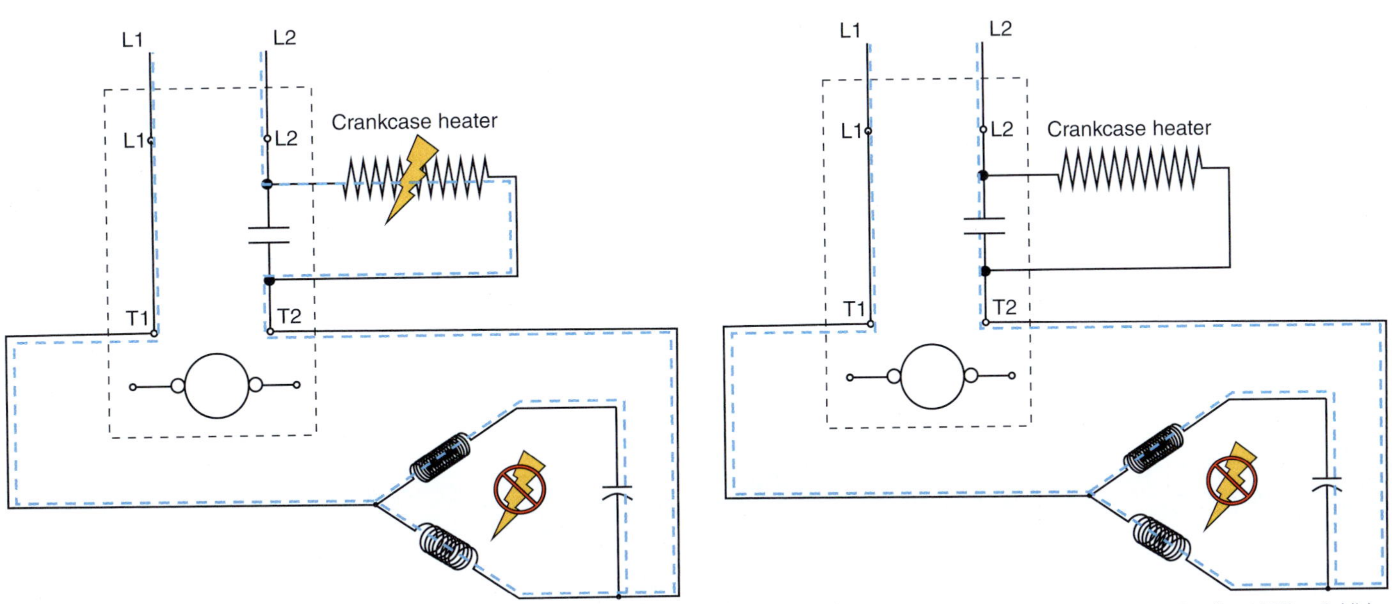

A B

Goodheart-Willcox Publisher

Figure 31-25. Same circuit as in **Figure 31-24**. When the contactor coil is not energized, the contactor contacts are open and current is directed through the heater. A—With the heater wired in series with the motor, the voltage is not sufficient to start the motor. The crankcase heater is energized. B—When the contactor coil is energized, the contactor contacts are closed and current bypasses the heater.

31.8 Current Magnetic Relays

Current magnetic relays (CMRs), **Figure 31-27**, are constructed with a very low-resistance coil and one set of normally open contacts. The resistance of the CMR's coil is typically less than 1 Ω and is wound with a thick wire. The normally open contacts on the relay are held open by gravity, so this device is position sensitive. If the relay is installed or mounted upside down, the contacts will always be closed, which can cause the motor windings to overheat, potentially leading to premature motor failure.

The CMR is wired into a motor control circuit as shown in **Figure 31-28**. The low-resistance coil is wired in series with the run winding and the normally open contacts are wired in series with the start winding for ISIR applications. For CSIR motor applications, the normally open contacts are wired in series with the start winding and the start capacitor. When a start capacitor is used, it is wired between the terminals originally connected together with the jumper wire, **Figure 31-29**.

When the motor is de-energized, the normally open contacts are open and the run winding and the coil of the CMR are in series with each other. The resistance of the series-connected run winding and CMR coil is very close to that of the run winding by itself given the low resistance of the CMR coil. When the motor is energized, only the run winding is energized, as the CMR's contacts are open and the start winding is not in the active electric circuit. With no imbalance in the magnetic field, the motor is unable to start, and the motor draws its rated locked-rotor amperage, LRA. The locked-rotor amperage flows through both the run winding and the coil of the CMR. This increased current flow through the CMR's coil produces a magnetic field that is strong enough to pull the normally open contacts closed. This adds the start winding and the start capacitor to the circuit, **Figure 31-30**. This allows the motor to

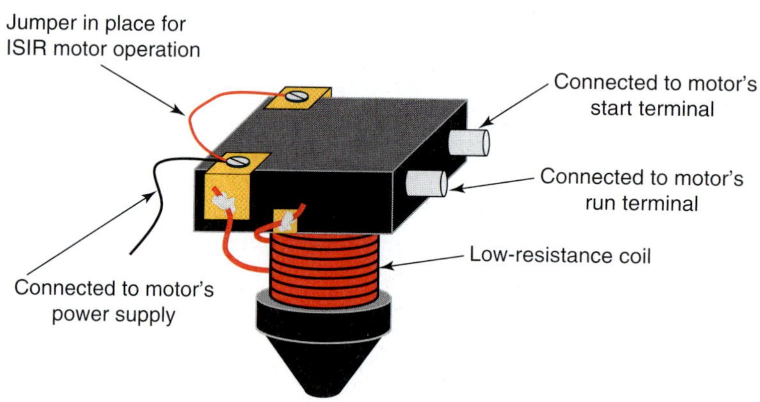

Goodheart-Willcox Publisher

Figure 31-27. A current magnetic relay is commonly used to start single-phase, capacitor-start, induction-run (CSIR) motors, or induction-start, induction-run (ISIR) motors, two fractional horsepower motors that are used for low-torque applications. Keep in mind that the CMR is position sensitive because the normally open contacts are held open by gravity.

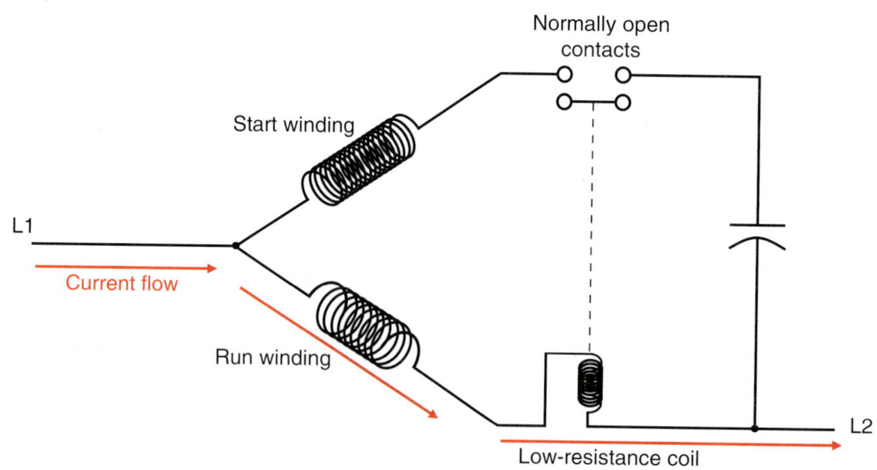

Goodheart-Willcox Publisher

Figure 31-28. In this capacitor-start, induction-run motor using a current magnetic relay, the coil of the relay is wired in series with the run winding and the normally open contacts are wired in series with the start winding, as well as the start capacitor. When the relay contacts are open and the motor is energized, only the run winding is in the active electric circuit.

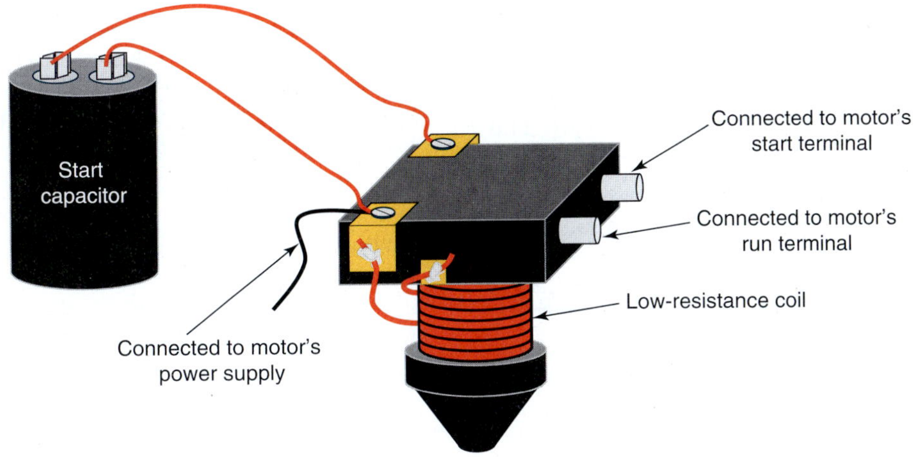

Goodheart-Willcox Publisher

Figure 31-29. A start capacitor is wired between the terminals originally connected by the jumper wire.

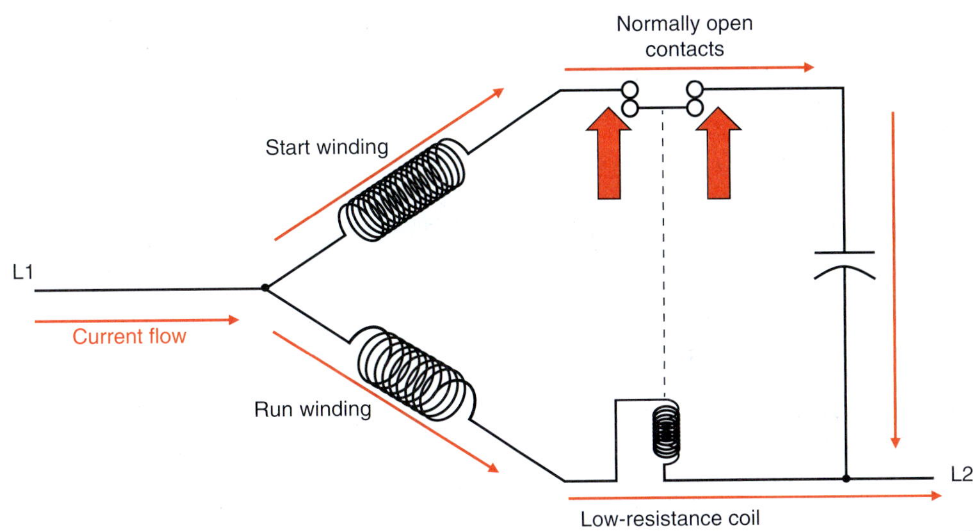

Goodheart-Willcox Publisher

Figure 31-30. Same motor as in **Figure 31-28**. The motor amperage rises to its locked-rotor amperage level; at that point, a magnetic field is generated that causes the normally open contacts on the CMR to close. The closed CMR contacts energize the start winding and the start capacitor.

start. Once started, the current draw drops to the motor's normal running-load amperage, RLA. This current is no longer sufficient to keep the contacts of the CMR closed, and gravity pulls them open. The motor continues to operate with only the run winding energized.

Selecting the Proper Current Relay

When selecting a current relay, it is important to keep the following in mind:

- **Ensure the selected relay has a cut-in amperage rating that is consistent with the locked-rotor amperage (LRA) of the motor.** If the motor's LRA is lower than the current at which the CMR contacts will close, the relay's contacts will never close. As such, the start winding and the start capacitor will never be energized and the motor will not start.
- **Check that the current rating on the relay contacts is greater than the amperage draw of the motor being controlled.** If the relay's contact amperage rating is not high enough, the relay will fail prematurely.
- **Identify how the relay will be mounted to ensure that it can be installed in the proper orientation.** Remember, the contacts on the CMR are held open by gravity. Installing the relay upside down will keep the relay contacts closed, energizing the start winding whenever the motor is powered. This can cause the motor to overheat and fail prematurely.

Evaluating a Current Relay

As with most relays, the current magnetic relay has a coil and a set of contacts. To properly check a current magnetic relay, the coil and the contacts must be mechanically and electrically tested. Before checking the relay, verify that it is electrically disconnected from the motor and any other wiring that might be connected to it. Refer to **Procedure 31-1** and **Procedure 31-2**.

Procedure 31-1

Evaluating the Coil of a Current Magnetic Relay

To check the coil of a CMR, follow these steps:

1. Turn on the digital multimeter (DMM) and set it to measure resistance.
2. Connect the meter's test leads to the meter.
3. Touch the meter's leads together and verify that the meter displays a very low reading, close to 0 Ω. This will ensure that the meter's test leads are not damaged. If the meter displays OL, check the test lead connections to the meter or replace the leads.
4. Ensure that all of the wiring from the CMR is completely disconnected. It is good practice to label the wiring connections, draw a diagram showing the connections, or take a photo of the connections before disconnecting the wires.
5. Connect the meter's test leads to the electrical terminals to which the coil is connected. The measured resistance should be very low, often below 1 Ω. If the obtained reading is high, or if the meter's display registers a reading of OL, then the relay should be replaced.

Procedure 31-2

Evaluating the Contacts of a Current Magnetic Relay

To check the contacts of a CMR, follow these steps:

1. Gently shake the CMR in an up-down motion. You should be able to hear the contacts moving inside the relay. If no sound is heard, the relay is likely damaged.
2. Turn on the digital multimeter (DMM) and set it to measure resistance.
3. Connect the meter's test leads to the meter.
4. Touch the meter's leads together and verify that the meter displays a very low reading, close to 0 Ω. This ensures that the meter's test leads are not damaged. If the meter displays OL, then check the test lead connections to the meter or replace the leads.
5. Ensure that all wiring from the CMR is completely disconnected. It is good practice to label the wiring connections, draw a diagram showing the connections, or take a photo of the connections before disconnecting the wires.
6. Hold the relay in its normal, upright position.
7. Connect the meter's test leads to the electrical terminals on the relay that connect to the motor. In most cases, the terminal will be labeled as S (start winding) and M (main or run winding).
8. Verify that the meter displays a reading of OL. This indicates that the contacts are open.
9. Ensure the test leads are still connected to the relay, then turn the relay upside down. The meter should now display a resistance value, indicating that the contacts are closed. If the meter still displays an OL reading, the relay is defective and needs to be replaced.

31.9 Potential Magnetic Relays

Potential magnetic relays (PMR) are most commonly used to start single-phase, capacitor-start, capacitor-run (CSCR) motors that are used for medium-torque and high-torque applications. Unlike the CMR discussed in the previous section, the PMR has a high-resistance coil and one set of normally closed contacts. The normally closed contacts are wired between terminals 1 and 2, and the coil is wired between terminals 2 and 5, **Figure 31-31**.

In the wiring diagram in **Figure 31-32**, the PMR's high-resistance coil is connected in parallel with the motor's start winding, while the PMR's normally closed contacts are connected in series with the motor's start capacitor. When voltage is supplied to the motor, the run winding, the start winding, the run capacitor, and the start capacitor are all in the active electric circuit. The effect of the capacitors on the motor are cumulative, as their values are added together, just as resistors connected in parallel with each other. With both the start winding and run winding energized, the motor begins to operate.

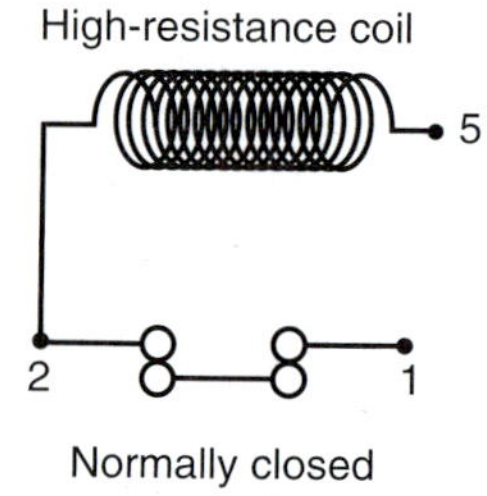

Figure 31-31. In a potential magnetic relay, the high-resistance coil is connected between terminals 2 and 5, while the normally closed contacts are located between terminals 1 and 2.

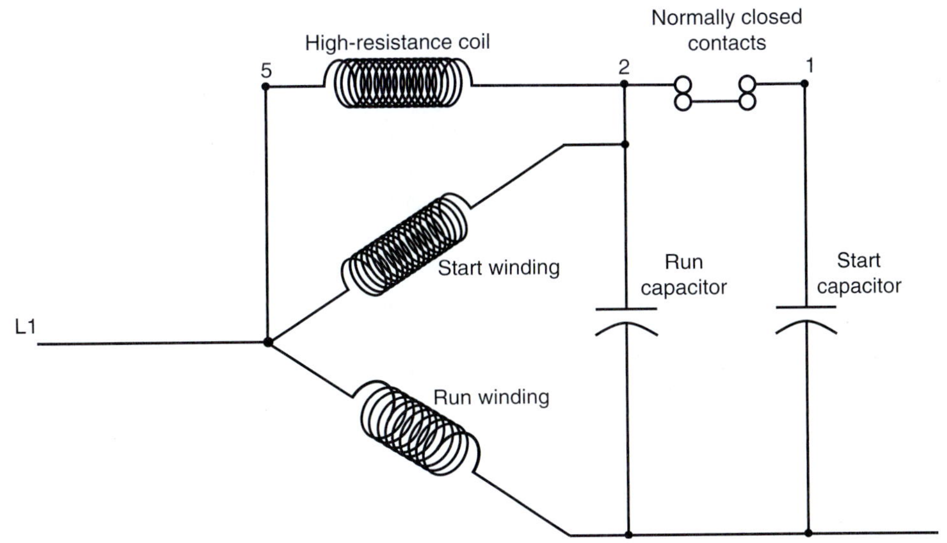

Goodheart-Willcox Publisher

Figure 31-32. The potential magnetic relay (PMR) connected to a CSCR motor. When the motor initially starts, the start winding, run winding, start capacitor, and run capacitor are all in the active electric circuit. The effects of the capacitors are cumulative. The running motor creates a counter-electromotive force, which creates an induced voltage along with the applied voltage. The PMR senses this increased total voltage, causing the normally closed contacts to open. This removes the start capacitor from the circuit.

Shedding Light

The coil of a potential magnetic relay has very high resistance. With a high resistance, the amount of current flow through the coil is very low. This is similar to a voltmeter. The resistance of a voltmeter is very high, so that it has very little effect on circuit operation or function. Consider what would happen if a voltmeter had a very low internal resistance and was used to measure the voltage supplied to a simple circuit with a single light bulb. The very act of placing the voltmeter's test leads across the load would create a short circuit and the fuse would blow or the circuit breaker would trip. The act of using a meter to test an electric circuit should not affect the operation of the circuit. The coil of the PMR is there to measure the combined voltage, applied and induced, across the motor's start winding.

The motor's rotor, which is positioned closely to the motor's windings, creates an induced voltage, or counter electromotive force (cemf) across the motor's start winding. This induced voltage, along with the applied voltage to the motor windings, results in a start winding voltage that is higher than the line voltage being supplied to the motor circuit. The coil on the PMR, which is wired in parallel with the motor's start winding and acts like a voltmeter, senses this increased voltage level. The PMR contacts react to this sensed voltage. When the motor is operating, the voltage across the start winding increases, causing the relay's contacts to open. This removes the start capacitor from the circuit and the motor continues to operate as a permanent split capacitor (PSC) motor.

Operating Voltages for the PMR

The potential magnetic relay has three voltage ratings: the pick-up voltage, the drop-out voltage, and the continuous coil voltage. When selecting a PMR, it is important that the voltage ratings are aligned with the operating voltages of the motor being controlled.

The rated ***pick-up voltage*** for a PMR is the voltage at which the relay contacts will pick-up or open. The pick-up voltage is the voltage that is measured or sensed by the relay coil wired in parallel with the start winding. This voltage is often reached when the motor reaches a speed that is about 75% of the motor's design speed. The relay's pick-up voltage is identified on the relay itself and is expressed as a range. If the sensed voltage across the start winding is never greater than the minimum pick-up voltage, the contacts between terminals 1 and 2 will never open. This will cause the start capacitor to stay in the circuit all the time, resulting in a high amperage draw. If the sensed voltage across the start winding is higher than the maximum pick-up voltage, the relay can overheat and fail prematurely.

The rated ***drop-out voltage*** for a PMR is the voltage required to keep the relay contacts open once they have been opened or picked-up. The pick-up voltage is typically higher than the drop-out voltage as it takes more voltage to initially open the contacts than it does to hold them open once they are opened. The drop-out voltage, like the pick-up voltage, is expressed as a range, with a minimum and a maximum. The drop-out voltage range is lower than the pick-up voltage range. For example, a PMR might have a pick-up voltage range from 212 Vac to 253 Vac, and a drop-out voltage range from 60 Vac to 121 Vac. This is because the motor speed drops once the relay contacts open.

The ***continuous coil voltage*** rating represents the maximum continuous voltage that can be applied to the PMR's coil without causing the relay to overheat.

Evaluating the PMR

When evaluating the potential magnetic relay, the coil and contacts must first be checked with the circuit de-energized and the wiring disconnected. Once the PMR has been disconnected and checked, a final evaluation must be completed with the relay connected and the motor circuit energized. This will ensure that the normally closed contacts between terminals 1 and 2 open as they are supposed to. If the contacts are in the closed position, the start capacitor will never be removed from the circuit. This will result in unusual motor noises and a high operating current. Refer to **Procedure 31-3**.

Procedure 31-3

Evaluating a Potential Magnetic Relay

To evaluate a potential magnetic relay, follow these steps:

1. Turn on the digital multimeter (DMM) and set it to measure resistance.
2. Connect the meter's test leads to the meter.
3. Touch the meter's leads together and verify that the meter displays a very low reading, close to 0 Ω. This ensures that the meter's test leads are not damaged. If the meter displays OL, check the test lead connections to the meter or replace the leads.
4. Verify that all wiring from the PMR is completely disconnected. It is good practice to label the wiring connections, draw a diagram showing the connections, or take a photo of the connections before disconnecting the wires.
5. Connect the meter's test leads across terminals 1 and 2 on the PMR. The measured resistance should be very low, close to 0 Ω, as the contacts between terminals 1 and 2 are supposed to be normally closed. If the obtained reading is high, or if the meter's display registers a reading of OL, the relay should be replaced.
6. Connect the meter's test leads across terminals 2 and 5 on the PMR to check the coil. The measured resistance should be very high. Depending on the relay, the reading can be many thousands of ohms. If the ohmmeter displays OL, the relay coil is open, and the relay should be replaced.
7. Connect the PMR to the motor.
8. Energize the motor and allow it to come up to speed.
9. Measure the voltage between terminals 1 and 2. If the relay is operating correctly, contacts 1 and 2 should have opened to remove the start capacitor from the circuit. If such is the case, the voltage reading should be high. If the voltage reading across the contacts is 0 V, the contacts have not opened and the relay should be replaced.

31.10 Overload or Overcurrent Protective Devices

Overcurrent refers to a current level in a circuit that is greater than the rated current or ampacity of a conductor. This overcurrent may result from a short circuit, ground fault, or an overload.

Overcurrent protective devices (OCPDs) are designed with specific trip characteristics that determine the conditions under which the overcurrent device will activate and de-energize a circuit. A trip characteristic curve combines current and time values that, together, represent the performance and behavior of the protective device, **Figure 31-33**. When the sensed current exceeds the trip current for a specified duration, the protective device operates. The greater the ground fault or short circuit, the quicker the OCPD will open to protect the motor.

Motor overload devices are intended to protect motors from overheating and overload conditions. These devices do not protect against the occurrence of ground faults and short circuits. Protection against ground faults and short circuits is provided by the motor branch-circuit overcurrent protective device.

Some overload devices protect against overheating, some protect against overload conditions, while others protect against both conditions. Motor overload devices are typically designed to monitor the current flow through a motor's windings, as well as the motor's operating temperature, and react when one or more of the sensed parameters is found to be outside the desired operating range. Overload conditions are often the result of mechanical problems within the motor or with any driven device or component that might be connected to the motor shaft. Overheating conditions are often the result of improper airflow across the motor but can also be caused by a motor's inability to start effectively and efficiently. Motor overload devices help prevent damage to the motor, help reduce system and equipment downtime, and help lengthen the useful life of the motors they protect.

When a motor operates under normal conditions, the current draw is expected to remain within a predetermined range that is established by the motor manufacturer. However, mechanical issues, such as a stuck or seized rotor, can cause the motor current to increase to a higher-than-normal level. Electrical faults, such as ground faults and short circuits, can also cause the motor current to rise to high, unsafe, or undesirable levels. When these faults occur, the motor branch-circuit overcurrent protective device (a fuse or circuit breaker), senses the overcurrent condition and reacts accordingly. A high ground fault or short circuit will cause an OCPD to open in two electrical cycles or less, which is almost instantaneous.

An overload is a condition where the current remains in its intended path at a current level that is higher than desired,

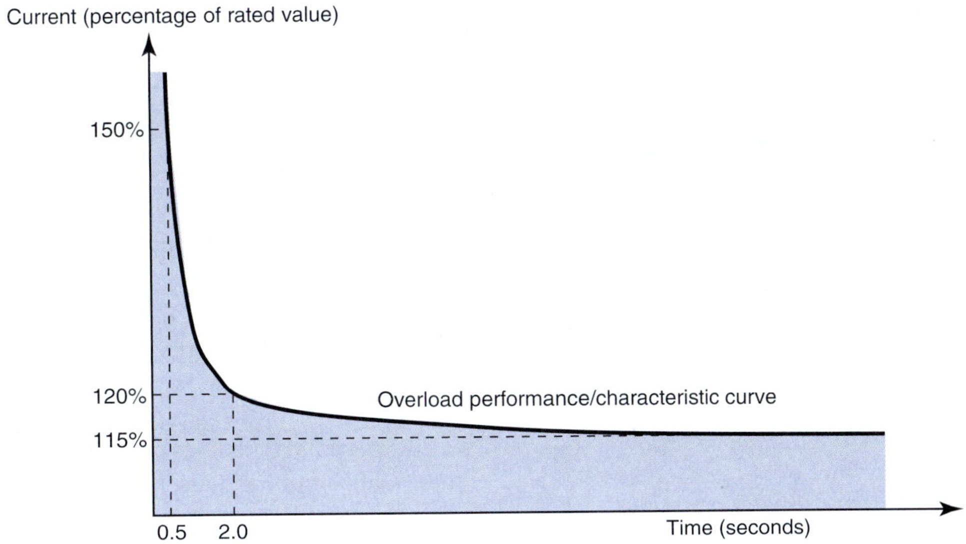

Goodheart-Willcox Publisher

Figure 31-33. A sample trip characteristic curve for an overcurrent protective device (OCPD). A motor will be able to safely tolerate lower current levels for relatively longer periods of time. As the current increases, the amount of time that a motor can withstand the exposure greatly decreases and the OCPD operation kicks in.

but lower than that of a short circuit or ground fault. Overload devices, also called running protection, react when a prolonged overload lasts for 10, 20, or 30 seconds (Class 10, Class 20, and Class 30, respectively). Many overloads are acceptable and part of the normal starting (inrush) current for a motor.

Motor overload devices are typically sized at 115% to 140% of the nameplate full-load amperage of a motor, depending on the conditions and motor service factor. Some motors may be equipped with a thermal protector that serves as the overload device. Thermal protection devices react to undesirable levels of sensed motor or motor winding temperatures. This is integral to the motor and is generally used in motors with lower horsepower ratings.

Current-Sensing Overload Devices

Current-sensing overload devices continuously monitor the current flow through a motor's windings. These devices typically use current transformers (CTs) or current sensors. Current sensors are installed in the motor control circuit and de-energize the motor if an unsafe or undesired level of motor current is detected. Some current-sensing overload devices are adjustable and can be set to a specific cut-out setting. The device settings can be adjusted based on the motor ratings and the requirements of the specific application.

The cut-out setting is often expressed as a percentage of the motor's full-load amperage and also incorporates a time delay feature. Typically, current-sensing motor overload devices are not instant-trip devices, meaning they do not de-energize a motor the instant an overcurrent condition is sensed. For example, if a motor's current draw is found to exceed the motor's full-load amperage by 20% for a period of three seconds, the overload device will de-energize the motor. The time delay feature of the motor overload differentiates between very brief current spikes during motor startup and overcurrent conditions that are present for extended periods of time. When the detected overcurrent condition persists beyond the device's preset time delay, the overload device reacts to protect the motor.

Bimetallic Overload Devices

Some motors rely on bimetallic elements for protection. Bimetallic elements consist of strips or discs of two different metals that are bonded together. The metals have different rates of expansion and contraction. When current flows through the bimetallic element, heat is generated, and the metals expand. Because the expansion rates of the metals are not the same, the element warps. Higher currents produce more heat, resulting in faster, and more extreme, warping of the element.

If the overload condition continues for an extended period of time, the element will open the circuit and de-energize the motor.

Electronic Overload Devices

Most new motors use electronic motor overload devices for protection purposes. These devices use electronic circuits to monitor and protect the motor. Just as with nonelectronic motor protective devices, electronic overload relays can open circuits to de-energize a motor if unsafe conditions are present. Electronic devices, however, are often more precise and faster. As a result, they can be used to provide additional features, such as creating an alert signal to notify the equipment user or operator that there is a problem with a motor or system.

Shedding Light

Some protective devices have a manual reset mechanism. This means that the overload will not automatically reset once the unsafe or undesirable condition is no longer present. Manual reset devices must be reset by the equipment operator. The device should only be reset once the underlying issue has been identified and resolved, and it has been determined that it is safe to restart the motor.

Safety Alert

Never "jump out" or override any safety or overcurrent devices! Doing this can allow an overcurrent condition to remain for an unsafe duration, potentially resulting in an arc fault. The high heat and arc blast in an arc fault can ignite flammable materials and cause property damage and severe personal injury. Proper use of OCPDs limits or eliminates large arc fault and arc flash situations.

31.11 Electronic Motor Protective Devices

Electronic motor protective devices perform many of the functions of mechanical and electromechanical motor protective devices, but use electronic components and circuitry instead. In addition, electronic motor protectors often have the ability to monitor a motor while it is operating and react when undesired or unsafe conditions are present. Electronic protective devices can often perform one or more of the following functions:

- Current and overload sensing
- Temperature sensing
- Electrical fault-to-ground sensing
- Short-circuit sensing

When equipped with current and overload sensing capabilities, the device is able to monitor the current flow through the motor. The sensed current is then compared to the expected current draw of the motor, based on the motor's nameplate or performance parameters. If the sensed current exceeds the motor's specifications and the maximum allowable by the *NEC*, the protective device reacts to de-energize the motor.

Temperature-sensing motor protective devices have the ability to trip the motor's control circuitry, de-energizing the motor if unsafe temperatures are sensed. Many electronic motor protective devices have a temperature-sensing feature incorporated into the control, allowing them to monitor and react to multiple motor conditions.

Short-circuit sensing devices have the ability to trip the motor's control circuitry, de-energizing the motor. The benefit of electronic short-circuit sensing devices is that they tend to react faster than nonelectronic devices.

Fault-to-ground sensing devices are used to open a motor circuit if small current leaks to ground are detected. Unlike fuses and circuit breakers, fault-to-ground detection devices measure the difference between the amount of current flowing into a circuit and the amount of current flowing out of a circuit. Electronic fault-to-ground protective devices react when current differences as low as 20 mA (0.020 A) are detected, allowing them to react quicker than nonelectronic versions of the control.

One of the most beneficial features of electronic motor protective devices is their ability to generate an alarm signal that alerts system and equipment owners and operators that a problem exists. These alerts do not always identify when a problem has occurred, but often send notifications when a motor is operating outside of its normal range. This advanced notification feature allows maintenance workers to investigate and resolve problems before major system failure results.

31.12 Selecting Motors, Motor Components, and Overload Protection

Requirements for sizing and selecting components for motor circuits is detailed in *NEC Article 430, Motors*. This text provides introductory information and an example of how to determine the required rating for two different protective devices. A thorough understanding of the requirements of *Article 430*, however, is needed to design and install motors and motor circuits.

In addition to a motor or multiple motors, a motor circuit may include several components, including the following:

- **Motor feeder.** Feeder conductors can supply multiple motor branch circuits and other loads from a single voltage source. Motor feeders have a short-circuit and ground-fault protective device sized according to the loads being supplied.
- **Motor circuit conductor.** The conductor in the motor branch circuit that supplies power to the motor.

- **Motor disconnect.** A device that disconnects the motor and motor controller.
- **Motor branch-circuit short-circuit and ground-fault protection.** An overcurrent protective device is often located with the motor disconnect.
- **Motor controller.** A device that controls the operation of the motor.
- **Motor control circuit.** A low-voltage control circuit that operates a motor controller. The motor control circuit may be tapped from the motor branch circuit conductors and requires overcurrent protection separate from that of the motor and motor branch circuit.
- **Motor overload protection.** A device that protects the motor and motor circuit conductors and devices from overload and overheating. In some smaller motors, the overload protection may be provided by a thermal overload device within the motor. In other cases, a separate overload device is required.

Part III of *Article 430* provides requirements for overload protection for motors and motor branch circuits. These requirements vary based on many factors, including motor size, motor type, supplied voltage, and device type.

Part IV of *Article 430* provides requirements for short-circuit and ground-fault protection for motor branch circuits. These devices provide protection for the motor, the motor branch circuit conductors, and all devices in the circuit. Like the overload protection requirements, these requirements vary based on many factors, including the protective device type, motor type, and circuit type.

Example 31-1 provides information on how to determine the required rating for a separate overload device and a short-circuit and ground-fault overcurrent protective device.

Know the Code

Example 31-1 provides an overly simplified sample of sizing motor protective devices. A thorough understanding of *Article 430* is needed to correctly apply the requirements for motor circuit design and installation.

Example 31-1

Determine the required rating for (1) a separate overload device and (2) a short-circuit and ground-fault overcurrent protective device for a motor circuit supplying a single-phase motor with the following specifications:

- 7.5 hp
- 230 V
- Continuous duty
- 38 A (FLA from motor nameplate)
- 1.10 service factor
- Dual-element, time-delay fuse to be used for short-circuit and ground-fault overcurrent protection

✓ Solution

Overload device. The full-load amperage listed on the motor nameplate is used to size the overload protective device. Per *Section 430.32(1)*, the overload device rating cannot be more than 115% of the FLA. Thus, the maximum rating for the overload device is 43.7 A (38 A × 1.15).

Short-circuit and ground-fault protection. *Section 430.52(C)(1)* specifies the rating should not exceed the value calculated from *Table 430.52*. If the calculated value does not match a standard device rating, *Exception No. 1* of *Section 430.52(C)(1)* allows for the next higher standard rating to be used.

From *Table 430.52*, the percentage of full-load current acceptable for a single-phase motor protected by a dual-element, time-delay fuse is 175%. The full-load current is determined from *Table 430.248* for single-phase motors. From this table, the full-load current for a 7.5 hp, 230 V motor is 40 A. Thus, the calculated value is 70 A. 70 A is a standard fuse rating listed in *Table 240.6(A)*.

CHAPTER 31 REVIEW

Summary

- Single-phase motors are commonly controlled by relays and contactors. (31.1)
- Relays are typically throwaway devices, while contactors can often be rebuilt in the field. (31.1)
- A motor operating under full load conditions draws full-load amperage (FLA). Full-load amperage is the current supplied to the motor when it is operating at its rated voltage, horsepower, and frequency. (31.2)
- When a motor initially starts, it draws locked-rotor amperage (LRA). (31.2)
- A motor's locked-rotor amperage can be as high as five to seven times greater than the motor's full-load amperage. (31.2)
- The number of poles refers to the number of conductive paths being simultaneously controlled by a single device. (31.3)
- The number of throws refers to the number of possible current paths for each pole. (31.3)
- Positive temperature coefficient (PTC) devices experience increases in resistance when their temperature increases. (31.4)
- Negative temperature coefficient (NTC) devices experience decreases in resistance when their temperature increases. (31.4)
- A general-purpose relay has a control coil and one or more sets of normally open or normally closed contacts. (31.5)
- Relay selection should include matching the relay's current rating to the motor's current rating. A direct-acting relay is a control device that directly controls the operation of the circuit load. (31.6)
- Pilot-duty relays are low-amperage relays that are used to control motor-starting devices with higher amperage ratings.
- Auxiliary contacts can be added to a contactor to control additional circuits with lower amperage ratings. (31.7)
- A current magnetic relay opens and closes its contacts in response to the current flow through the motor's run winding. (31.8)
- A potential magnetic relay opens and closes its contacts in response to the voltage measured across the motor's start winding. (31.9)
- Potential magnetic relays are rated by their pick-up voltage, their drop-out voltage, and their continuous coil voltage. (31.9)
- Common types of motor overload devices include current sensing, bimetallic devices, and electronic devices. (31.10)
- When unsafe or undesirable conditions are sensed by a motor control device, the device reacts to de-energize the motor. (31.10)
- Electronic motor protective devices have the ability to send alerts to those who maintain the equipment, potentially preventing major system or equipment failure. (31.11)
- A thorough understanding of *NEC Article 430, Motors* is necessary for sizing and selecting components for motor circuits. (31.12)

Review Questions

Use the information in the chapter to answer the following questions.

Measuring Your Potential

1. *True or False?* Because of their cost, relays are often discarded and replaced when they fail. (31.1)
2. Of the following, which represents the motor's current when it is initially energized before the motor begins to rotate? (31.2)
   A. FLA
   B. LRA
   C. RLA
   D. FLA and LRA
3. *True or False?* The number of poles on a control device or switch can never be greater than the number of device throws. (31.3)
4. A contactor with two sets of movable contacts is best described as a _____. (31.3)
   A. single-pole, single-throw device
   B. single-pole, double-throw device
   C. double-pole, double-throw device
   D. PTC device
5. A positive temperature coefficient (PTC) device is a good choice for many applications, including all of the following *except* _____. (31.4)
   A. electronic circuits
   B. ISIR motor
   C. electric heating elements
   D. time delay circuit functions

6. *True or False?* As a motor speeds up and nears its design speed, the contacts of a centrifugal switch will open. (31.5)
7. A control relay that is used to control another control relay is commonly referred to as a _____. (31.6)
   A. direct-acting relay
   B. single-pole, double-throw relay
   C. pilot-duty relay
   D. double-pole, double-throw relay
8. Of the following, which is *not* a criterion for selecting a contactor? (31.7)
   A. The contactor's coil voltage rating matches the voltage used to power the control circuit.
   B. The number of contactor poles matches the number of power lines that must be controlled.
   C. The amperage rating of the contactor contacts matches, or exceeds, the current draw of the load being controlled.
   D. The current rating is greater than the amperage draw.
9. *True or False?* A current magnetic relay opens and closes its contacts in response to the current flow through the motor's run winding. (31.8)
10. On a potential magnetic relay, the pick-up voltage is the voltage at which the _____. (31.9)
    A. PMR contacts will initially open
    B. CMR contacts will initially open
    C. PMR contacts will be held open
    D. CMR contacts will be held open
11. On a potential magnetic relay, the drop-out voltage is the voltage at which the _____. (31.9)
    A. contacts will be held closed
    B. contacts will be held open
    C. holding coil will fail
    D. holding coil will become de-energized
12. The manual reset feature on a motor _____. (31.10)
    A. helps ensure that a motor will not turn back on until an electrician has inspected the motor and verified that the unsafe condition has been fixed
    B. will reset itself once the unsafe condition has been fixed
    C. is an operational device that opens and closes its contacts during the normal operation of a motor
    D. is a safety device that will reset once power to the motor is disconnected and then restored
13. As the amount of current flow through a motor increases, the _____. (31.11)
    A. motor's winding temperature will decrease
    B. amount of time it will take for a motor protective device to react will increase
    C. motor's winding resistance will decrease
    D. amount of time it will take for a motor protective device to react will decrease

Plug-In

1. A typical wall switch that controls a light fixture is most likely a _____. (31.3)
   A. single-pole, single-throw switch
   B. single-pole, double-throw switch
   C. double-pole, single-throw switch
   D. double-pole, double-throw switch
2. While evaluating the contacts of a current magnetic relay, you noted an OL reading while touching the meter's leads together. Based on this finding, you should _____. (31.8)
   A. continue your evaluation, as the test leads are not damaged
   B. check the test lead connections to the meter or replace the leads
   C. verify that the test leads are connected to the relay
   D. replace the relay
3. When evaluating a potential magnetic relay, after you verify that the meter's test leads are not damaged, which of the following should be performed next? (31.9)
   A. Check the coils by connecting the test leads across terminals 1 and 2.
   B. Check the relay contacts by connecting the meter's test leads across terminals 1 and 2.
   C. Ensure the PMR is completely disconnected; good practice is to note the wiring connections before disconnecting.
   D. Connect the meter's test leads across terminals 2 and 5 on the PMR to check the coil.

Making the Connection

1. When replacing a relay, contactor, or other motor control device, explain why the specifications of the replacement device should match those of the old device. (31.8)
2. Explain why the time for an overcurrent protective device to de-energize a motor decreases when the amount of current flow through the motor increases. (31.12)

CHAPTER

32 Three-Phase Motor Starting and Protective Devices

Jatujak Phansuma/Shutterstock.com

SPARKING DISCUSSION

An electrician states that a soft-start motor-starting method, such as a reduced-voltage start, can be used for applications that require very high starting torque. Do you agree or disagree with this statement? Provide as much information as you can to support your response.

LEARNING OUTCOMES

After completing this chapter, you will be able to:

32.1 Explain single-phasing as it applies to three-phase motors.
32.2 Explain the basics of motor starter selection.
32.3 Discuss how heaters provide overload protection.
32.4 Explain the function and operation of an across-the-line starter.
32.5 Describe the function and operation of a primary resistor starter.
32.6 Explain the function and operation of part-winding starting.
32.7 Explain the sequence of operation for the star-delta starter technique.
32.8 Explain the function of an autotransformer starter.
32.9 Describe the function of a soft starter.
32.10 List features and benefits of an electronic protective device over a thermal protective device.

TECHNICAL TERMS

across-the-line starting
autotransformer starting
heater
part-winding starting
pawl
primary resistor starting
short-cycling
single-phasing
soft starter
star-delta starting

Introduction

Three-phase motors are used for most commercial and industrial applications and perform an integral role in many of the processes on which we all rely. Therefore, it is important that every effort be made to ensure that these large, expensive motors operate reliably, efficiently, and safely. Exceeding voltage and current limits, for example, can result in overheating, which can cause damage to the motor's winding insulation, potentially leading to premature motor failure. In addition, exceeding desirable voltage and current levels will lead to increased energy consumption and a reduction in the motor's overall operating efficiency.

There are steps an electrician can take that reduce the wear and tear on these motors as well as help make sure that they function within acceptable operating ranges. Monitoring three-phase motor operation allows the equipment operator to evaluate the motor's performance. Comparing a motor's actual operating conditions to the manufacturer's specifications can alert service and maintenance personnel of potential problems that, when caught early enough, can be acted upon to help prevent severe motor damage from occurring.

In this chapter, we examine a number of three-phase motor starting methods that are intended to help motors start in a manner that is most effective and efficient for the designated application.

32.1 Single-Phasing

Before discussing some of the most commonly used methods to control and protect three-phase motors, it is important to discuss one of the major concerns regarding the safe and controlled operation of three-phase motors. This concern is a condition known as ***single-phasing***. In a normally operating three-phase motor, current flows freely among the three line voltage motor leads and the motor's windings. Single-phasing prevents this from occurring, and happens when one of the three power lines of the motor's power supply is lost, while the other two power lines remain operational, **Figure 32-1**. This can be caused by a blown fuse, burned conductor, loose wiring connection, or any other condition that results in a loss of voltage to only one of the three power lines.

Consider the situation of a three-phase motor powered by a three-phase 240 V power supply. Voltage readings taken between any two of the three power lines will yield readings of 240 V. However, when one of the power lines is not supplying voltage, two of the three readings will be 0 V, **Figure 32-2**.

When single-phasing occurs, the motor may continue to operate, even though there is a severely imbalanced power signal being supplied to the motor. Single-phasing results in the highest degree of three-phase voltage imbalance. An imbalanced power signal also affects the motor's operating

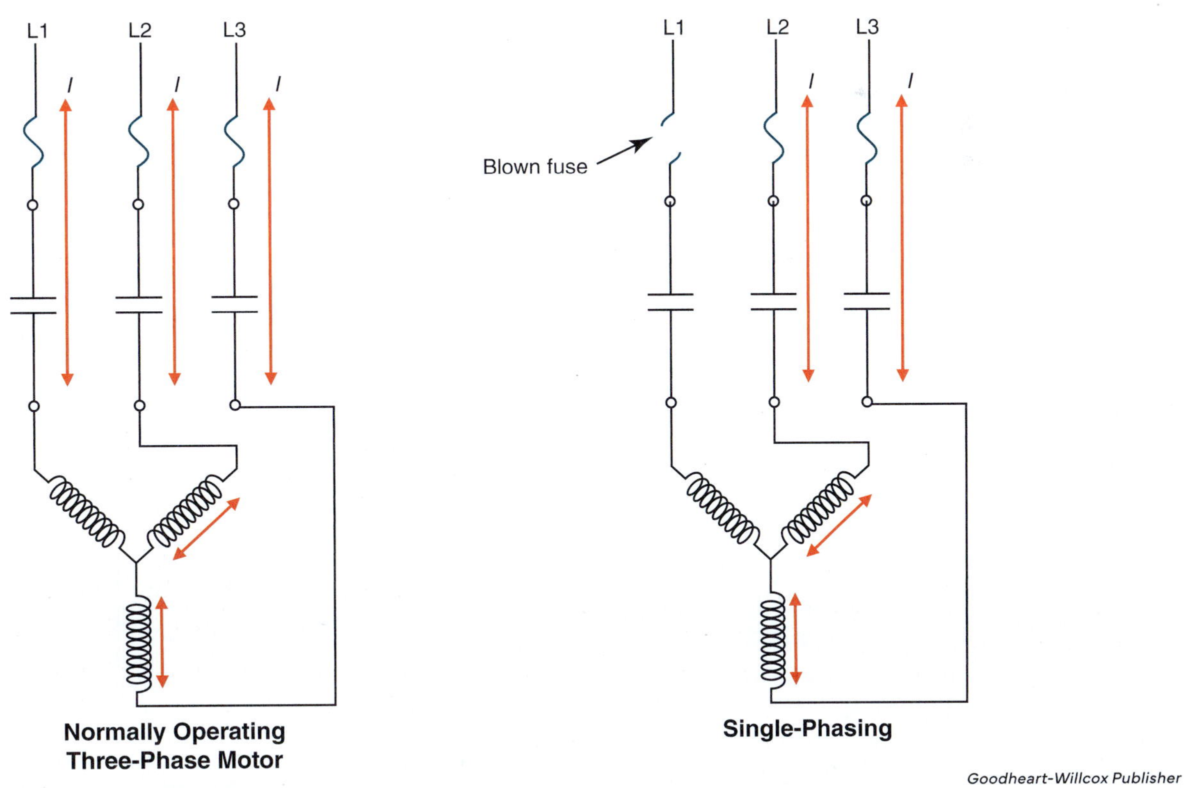

Figure 32-1. In a properly operating three-phase motor, current flows freely through all three power lines and motor windings. During single-phasing, one of the three power lines is rendered inoperable, and current flows through only two of the motor's three windings.

Sample Voltage Readings

Power Line	Normal Operation	Single-Phasing (Loss of L1)
L1 to L2	240 V	0 V
L1 to L3	240 V	0 V
L2 to L3	240 V	240 V

Goodheart-Willcox Publisher

Figure 32-2. When voltage at one power line is lost, two of the voltage readings across the three-phase power lines will drop to zero. In this chart, for example, the L1 power line has been lost and is not supplying voltage to the motor.

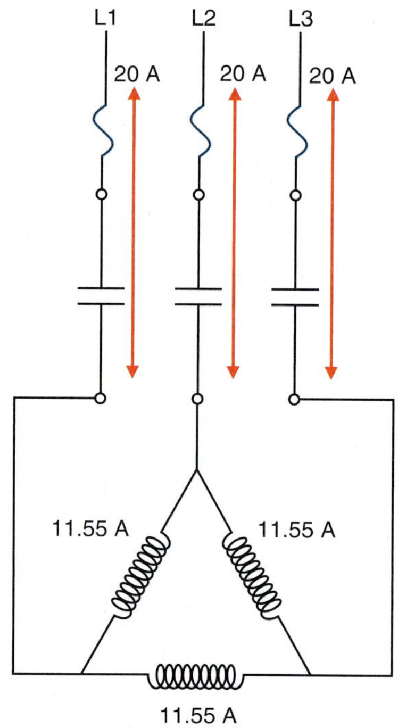

Goodheart-Willcox Publisher

Figure 32-3. In a normally operating 20-A delta configuration, the current flow through each motor winding will be 11.55 A.

efficiency, as its ability to operate at full capacity is greatly reduced. This results in increased energy consumption.

One significant symptom of single-phasing is a dramatic increase in current flow through the windings and power lines that remain in the active circuit. This is due, in part, to the motor's attempt to provide full power to the connected load. To maintain full power at a reduced voltage, the current draw of the motor increases. The current draw of a three-phase motor operating under a single-phasing condition can be as high as 173% of the motor's phase current. This higher current generally trips the overload device after a period of time.

Delta-Connected Three-Phase Motor

A normally operating delta-connected three-phase motor that operates with line currents of 20 A will operate with winding currents of 11.55 A, **Figure 32-3**. In a delta-connected motor:

$$I_{Line} = \sqrt{3} \times I_{Phase}$$

The winding currents are determined as follows:

$$I_{Line} = \sqrt{3} \times I_{Phase}$$

$$I_{Phase} = \frac{I_{Line}}{\sqrt{3}}$$

$$I_{Phase} = \frac{20\text{ A}}{1.732}$$

$$I_{Phase} = 11.55\text{ A}$$

If this same delta-connected three-phase motor experiences a single-phasing condition, the current flow through the inoperable line drops to zero, while the current flow through the remaining lines increases by a factor of $\sqrt{3}$, from 20 A to 34.64 A (20 A × $\sqrt{3}$), **Figure 32-4**. The value of 34.64 A is the mathematical sum of the two parallel-connected current flow paths that connect to each of the active power lines (23.1 A and 11.55 A). Referring to the initial circuit evaluation discussions, the total current flow into a circuit node must equal the total current flow out of the node.

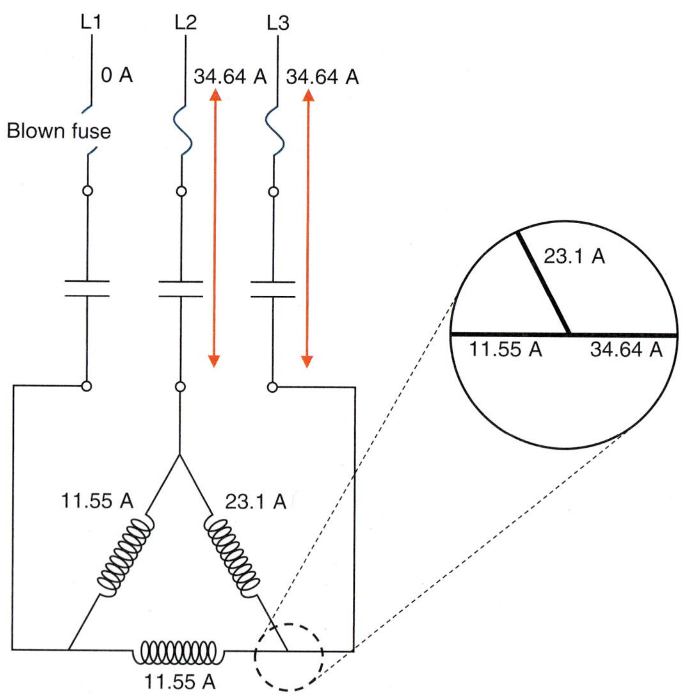

Goodheart-Willcox Publisher

Figure 32-4. When the 20-A delta configuration experiences a single-phasing condition, the line current on the operational power lines rises to 34.64 A. This happens because total current flow into a node must equal the total current flow out of the node.

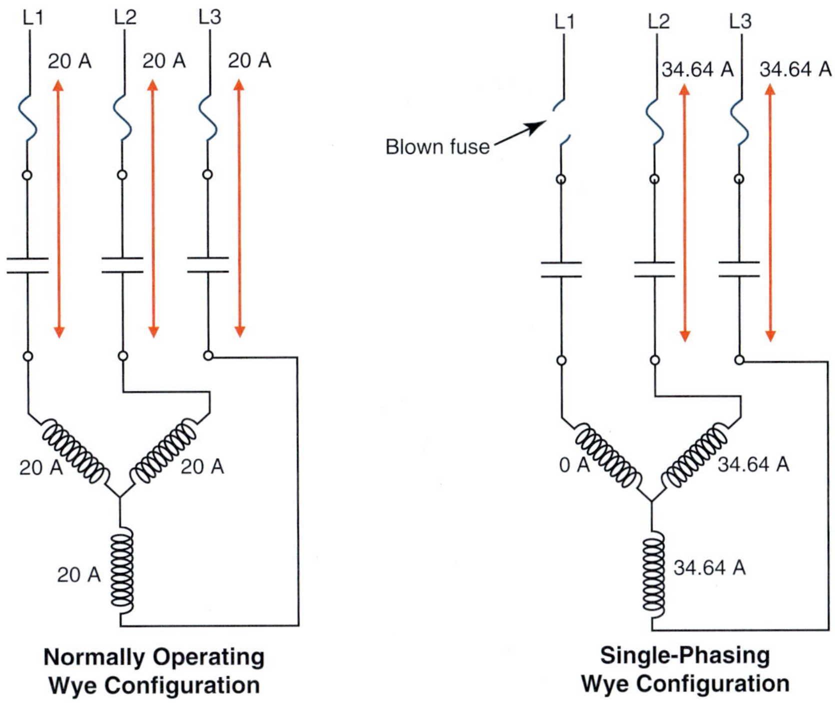

Goodheart-Willcox Publisher

Figure 32-5. On a normally operating 20-A wye configuration, the current flow through each motor winding is 20 A. The winding current is the same as the line current in a wye connection. When the 20-A wye configuration experiences a single-phasing condition, the line current through the operational power lines rises to 34.64 A. The current flow through each of the windings is also 34.64 A.

Wye-Connected Three-Phase Motor

A normally operating wye-connected three-phase motor that operates with line currents of 20 A will operate with winding, or coil, currents of 20 A. For a wye-connected motor:

$$I_{\text{Line}} = I_{\text{Phase}}$$

If this same wye-connected three-phase motor experiences a single-phasing condition, the current flow through the inoperable line drops to zero, while the current flow through the remaining lines increases by a factor of $\sqrt{3}$, from 20 A to 34.64 A, **Figure 32-5**.

If allowed to operate in a single-phasing condition, the motor will likely experience a loss in torque. Depending on the motor and the application, the motor may or may not continue to operate if the motor was running when the single-phasing condition occurred. If the motor is not operating when the loss of a power line occurs, the motor will likely not start.

If the motor continues to run, the imbalance in the power supply caused by single-phasing creates an uneven load on the motor's windings. This uneven motor load will cause the motor to overheat, possibly damaging the winding insulation, resulting in premature motor failure.

From a mechanical standpoint, single-phasing can cause excessive vibrations within the motor that can easily transmit to the driven components connected to the motor. Excessive vibrations can result in mechanical damage to the motor and its connected components, devices, and equipment.

Electronic phase loss relays are used to protect three-phase motors from single-phasing. These devices can detect single-phasing and take an appropriate action, such as opening a set of relay contacts to de-energize the motor and reduce the possibility of costly motor and equipment damage.

32.2 Motor Starters

In the simplest terms, a three-phase motor starter is a device composed of a three-pole contactor and an overload protective device, **Figure 32-6**. The starter helps ensure smooth motor operation, while helping to protect the motor and any driven devices, components, or equipment connected to it. As with contactors, motor starters are often rebuilt in the field. The coil, contacts, and springs are the most commonly replaced starter components. Starters are often represented in wiring diagrams by a diagram similar to the one provided in **Figure 32-7**. The overload protection in a three-phase motor starter is often provided by an overload relay, commonly known as a heater. Overload relays monitor the current flow through each of the motor's three phases. (Heaters are discussed further in a later section of this chapter.)

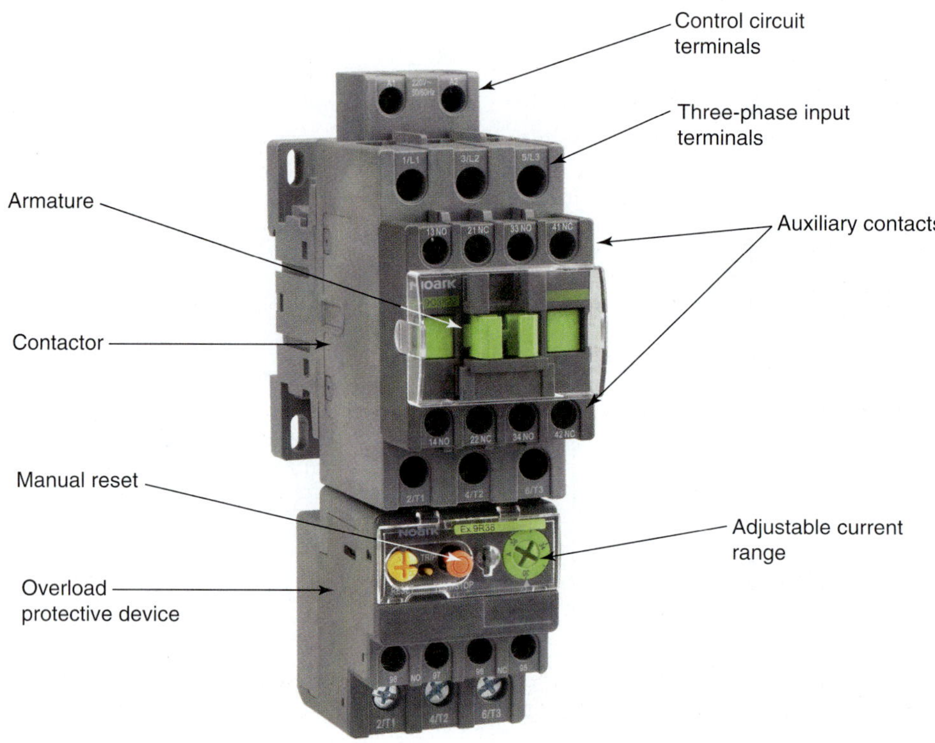

NOARK Electric North America

Figure 32-6. A three-phase motor starter. The red button at the bottom is the reset button used to reset the overload switches once the heaters have cooled.

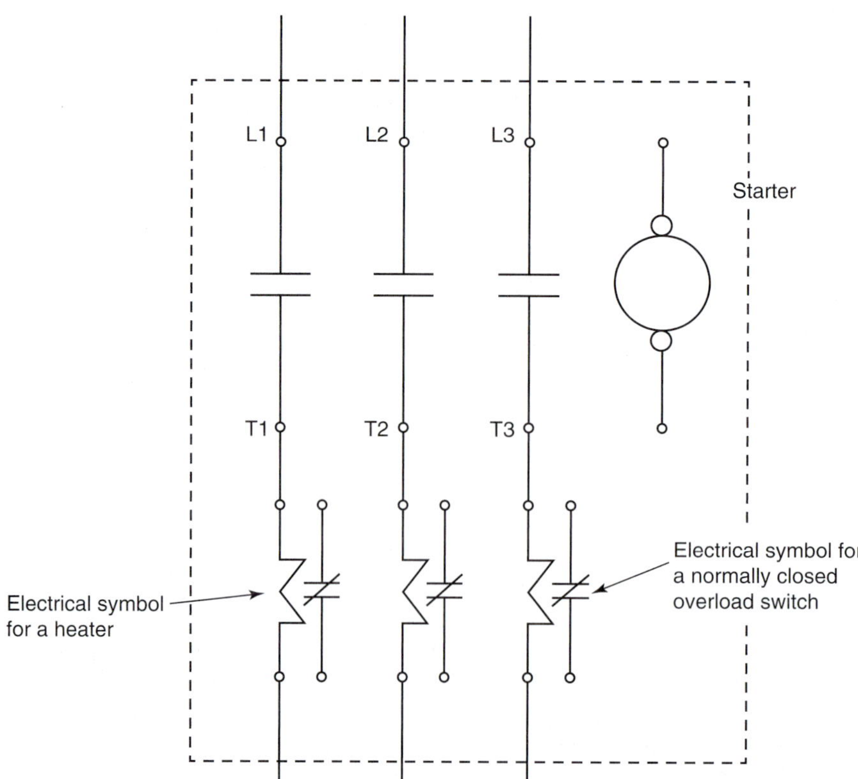

Goodheart-Willcox Publisher

Figure 32-7. Commonly used electrical symbol for a three-phase motor starter.

Motor Starter Features

Not all motor starters are created equal. Each starter is unique in that each one is designed to perform different functions. Some of the functionalities that can be incorporated into a starter or the starter's control circuitry include, but are not limited, to the following:

- **Protection from voltage fluctuations.** Some starters and control circuits monitor the voltage supplied to the motor and have the ability to disable a motor if the applied voltage is not within the specified or desired voltage range.
- **Protection from motor short-cycling.** ***Short-cycling*** is the term used to describe a condition where the motor cycles on and off at an unacceptably fast rate. Some starters and control circuits incorporate a time-delay feature that prevents the motor from attempting to restart too quickly after a shutdown.
- **Circuit isolation.** Some starters have a means to isolate the motor from the main power supply for the purposes of maintenance or repair. By isolating the motor from the main power supply, the remaining portions of the circuit can remain energized, providing at least some level of system functionality. The main circuit power should be disconnected in accordance with lockout/tagout procedures.
- **Motor reversal.** Some starters have a means to manually or automatically reverse the direction of rotation of the motor being controlled. As discussed previously, the direction of rotation of a three-phase motor can be reversed by switching the connections of any two of the three power leads on the motor's power supply. Some motor control systems will add a phase-reversal relay or this may be built into a solid-state starter or variable frequency drive (VFD). Not all motors should be run in reverse as this might cause damage to the driven load. For example, pumps and fans may be damaged if their direction of rotation is reversed. An unwanted phase reversal situation could occur if some of the wiring in the electrical distribution system is replaced and two of the three phases are changed.
- **Emergency stop feature.** Many motor starters are equipped with an emergency stop (e-stop) feature. When the button is pressed, power is quickly disconnected. This feature is used when immediate system or equipment shutdown is necessary. For example, motors powering conveyor belts, gasoline pumps, and computer room HVAC equipment may be equipped with e-stops.

Know the Code

The *NEC* classifies any device used to start and stop a motor, such as a motor starter, as a motor controller. Requirements for motor controllers are provided in *Part VI* of *Article 430*.

Motor Starter Types and Selection

The process of selecting the appropriate three-phase motor starter must be done carefully. A number of factors must be taken into account:

- The specifications (voltage, current, speed, horsepower) of the motor to be controlled
- The winding configuration of the motor being controlled
- The load that is to be controlled by the motor
- The intended functions of the starter

One cause of three-phase motor failure is improper motor starter selection. By ensuring that the voltage, frequency, and current ratings of the motor starter and its heaters match those of the motor and the power supply, the likelihood of premature motor failure is greatly reduced.

Not all motor starter types can be used with all three-phase motors; therefore, it is important to know the winding configuration of the motor being controlled, as well as the operating characteristics of the motor starter. Some three-phase motor starters, for example, are intended to energize all the motor windings upon start-up, while others energize only a portion of the motor's windings. If a particular three-phase motor has only one set of windings, using a more expensive and elaborate motor starter would not be appropriate.

Understanding the load being controlled is almost as important as knowing about the motor. In some motor starter applications, for example, the voltage initially supplied to the motor is reduced to create a soft-start condition for the motor. This results in lower motor starting torque. Before opting for a starter that creates a low starting torque condition, it is important to verify that the load being driven by the motor can safely function with low starting torque. If the driven load requires high starting torque, a motor starter that offers high torque must be selected.

There are a number of commonly used methods for starting three-phase motors:

- Across-the-line starting
- Star-delta starting
- Part-winding starting
- Autotransformer starting
- Primary resistor starting
- Solid-state starting (soft starting)
- Variable frequency drives

32.3 Heaters Used for Overload Protection

The overload protection in a three-phase motor starter is often provided by an overload relay. Overload relays monitor the current flow through each of the motor's three phases. Common types of overload relays are thermal, bimetallic, and electronic (solid-state).

Thermal overload relays are referred to as ***heaters***. As current flows through each leg of a three-phase power circuit, heat is produced. The amount of heat produced is a function of the voltage supplied to the circuit and the amount of current flow through each circuit branch. If the amount of heat generated exceeds the established setpoint for the particular application, the heater circuit's trip mechanism will activate, and the motor control circuit will de-energized.

When the control circuit is de-energized, the motor starter's coil is de-energized, and all three sets of contacts on the starter open, de-energizing the motor. Thus, overload devices play a crucial role in protecting three-phase motors.

Many overload devices must be manually reset. The manual reset feature ensures that the motor does not attempt to restart until the condition that caused the overload has been identified and resolved.

A common heater type is the melting alloy heater, also called a eutectic alloy thermal overload. This heater resembles a metallic strip with a side-mounted wheel, **Figure 32-8**. The surface of the wheel is not smooth and has fabricated notches, or teeth, machined into it. A connecting rod, referred to as a ***pawl***, **Figure 32-9**, rests within the wheel's notches. In the case of a three-phase motor starter, the pawl is responsible for keeping the normally closed control switches closed.

The pin on which the wheel is mounted extends into a cylindrical chamber on the heater. This chamber contains a eutectic alloy that melts at a specific temperature. This metal is capable of repeated cycles of melting and solidifying. Under normal operating conditions, the eutectic alloy is solid, preventing the wheel from turning, **Figure 32-10A**. When this alloy is exposed to higher-than-desired temperatures that result from an overload or single-phasing condition, the alloy melts, and the wheel is free to turn. With the resistance of the wheel removed, the pressure of the compressed spring at the base of the overload unit is now able to push the pawl upwards, causing the normally closed contacts to open, **Figure 32-10B**.

Some overload relays are constructed with bimetal strips. The two strips are fabricated from different metals and, as such, have different rates of expansion and contraction when heated or cooled. As the bimetal strip heats as a result of excessive current flow through the windings, the strip warps, causing the overload relay to open.

These overload devices are wired in series with the motor windings. As such, the current flow through the motor's windings is the same as the current flow through the overload devices.

There is an inherent time-delay feature built into the overload protection. Since it takes time for an overload device to sense the increased current, there will be a time delay between the time the overload condition presents itself and the time the overload reacts by opening the switch's normally closed contacts. The length of the time delay is affected by the magnitude of the overload condition. If the overload condition is excessively high, the amount of time required for the heaters to react will be reduced. In addition, the duration of the overload condition will also affect the overload device's response.

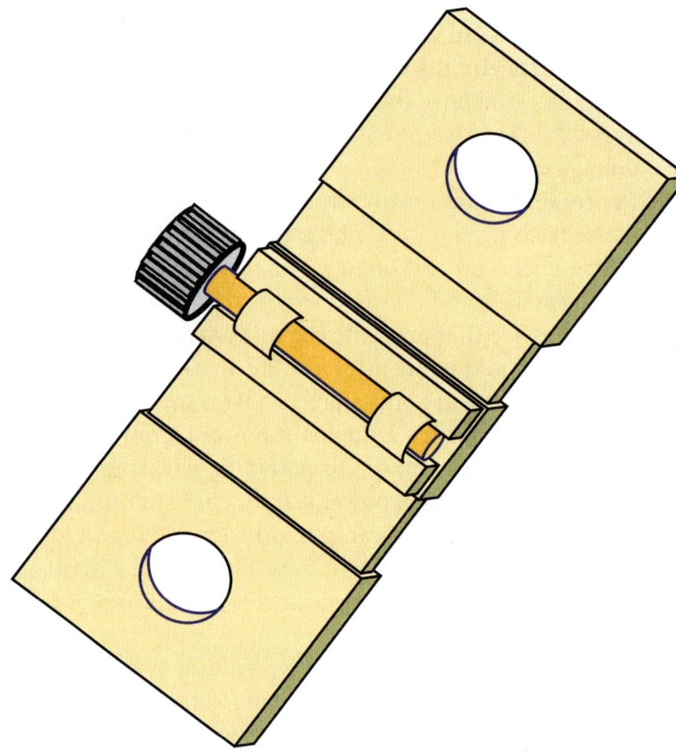

Goodheart-Willcox Publisher

Figure 32-8. A melting alloy heater. This type of heater uses a eutectic alloy that melts when an overcurrent condition is present.

Goodheart-Willcox Publisher

Figure 32-9. A pawl is the connecting rod that rests within the grooves or notches in the ratcheting wheel on the heater. In this figure, the pawl prevents the wheel from turning in the clockwise direction, while allowing it to turn in the counterclockwise direction.

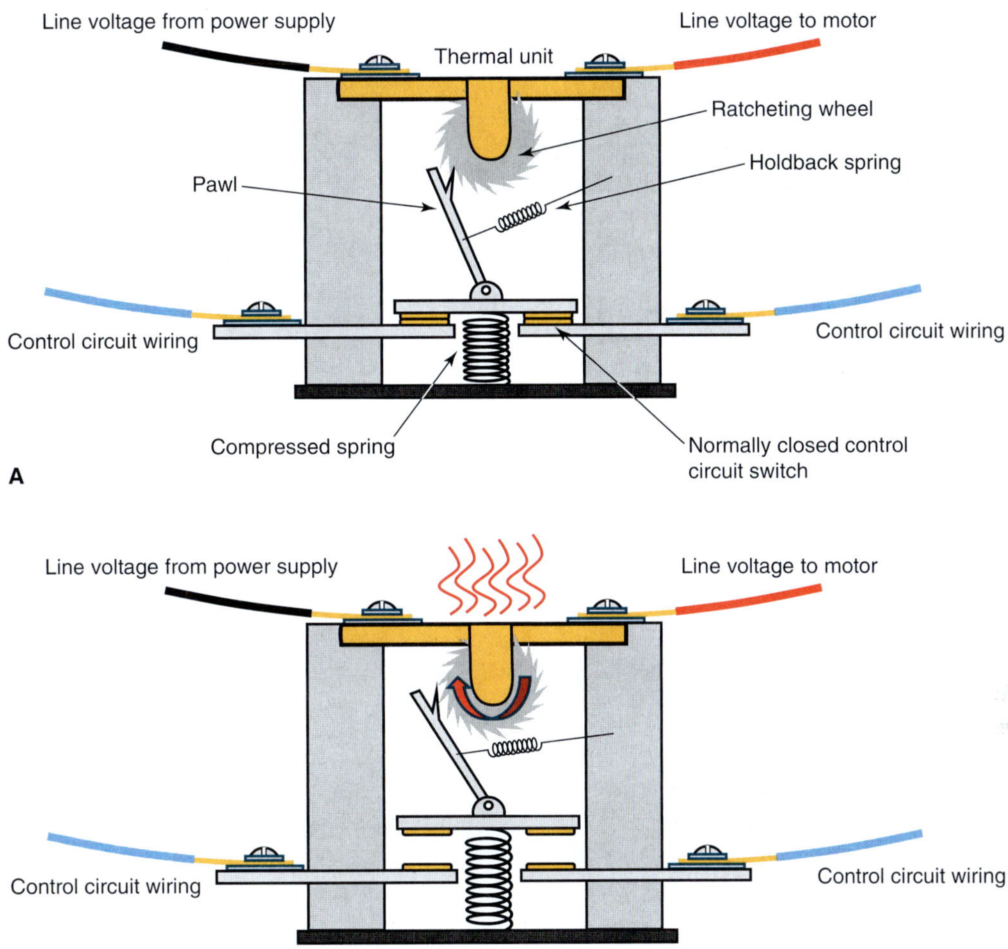

Figure 32-10. A melting alloy heater assembly. A—When the alloy in the heater is cool and solid, the ratcheting wheel cannot turn. As a result, the pawl maintains pressure on the normally closed control circuit switch, keeping it closed. B—When the alloy in the heater warms due to an overload condition, it melts, allowing the ratcheting wheel to turn. As a result, the spring pressure pushes the pawl upward and opens the normally closed control circuit switch.

Overload relays have a class rating related to the amount of time an overload condition must exist before the relay opens. A Class 20 overload relay trips within 20 seconds when a current 600% of the relay's rated current is present. A "fast trip" Class 10 relay trips within 10 seconds at the same current level. A "slow trip" Class 30 relay trips within 30 seconds at 600% of the rated current level.

Shedding Light

A motor's inrush current at start-up can be much higher than the 600% relay current rating. However, since the inrush current is present for an extremely short duration, it does not open the overload relay contacts. Overload protection is designed to allow for very high currents of a short duration and protect against relatively low overcurrent conditions that exist for a relatively long period of time.

In the event an overload condition causes the heaters to react and open the normally closed contacts, they must cool down before the starter can be reset. If the heaters have not been given enough time to cool, the starter will not reset, even if the reset button is pressed.

Overload Heater Selection

To be effective and to accurately control the operation of a three-phase motor, overload heaters must be sized for the specific motor being controlled. Motor starter selection depends on a number of factors including the motor's size, horsepower rating, voltage, rated current, type, and the unique requirements and characteristics of the load. Properly selecting overload devices helps ensure affective motor protection, while limiting nuisance trips during normal operating conditions.

As each manufacturer has their own sizing and selection charts for their heaters, it is important to refer to the literature provided by the manufacturer of the heaters being used. In addition, manufacturers provide detailed, step-by-step procedures for selecting the proper heater for the application. Since these procedures vary among manufacturers, be sure to refer to the procedures that apply to the heaters being used.

Know the Code

Requirements for sizing overload protection for motors and motor branch circuits are found in *Part III* of *Article 430* in the *NEC*. The sizing of the overloads is dependent on the motor's ability to handle heat, the motor's service factor (SF), the full-load amperes from the motor nameplate, and the starting characteristics needed to handle the load.

Overload heaters are stamped with their part numbers, **Figure 32-11**. It is important to understand that the imprinted number on the heater does not represent a current or voltage value, but simply represents the manufacturer's part number.

Shedding Light

The part number markings on an overload heater must be visible when the heater is installed on the starter. Being able to read the markings ensures that the heater is properly mounted on the motor starter. When installing heaters, make certain that all mounting screws are securely tightened and properly torqued.

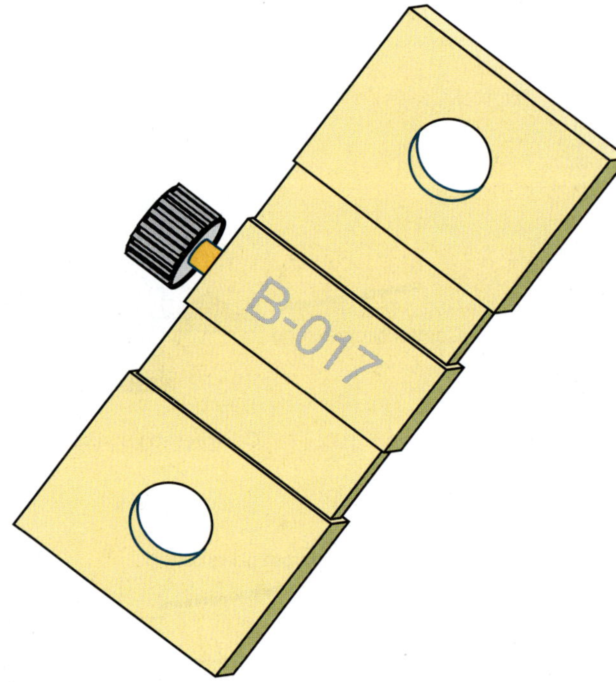

Goodheart-Willcox Publisher

Figure 32-11. The imprinted information on the heater part represents the part number, not the current, horsepower, or voltage rating of the device. When properly mounted on the heater, the imprinted information should be readable.

Overload Control Switch Connections

The three normally closed contacts on the motor starter are part of the system's control circuit. These normally closed switches are connected in series with each other. In the event any one of these three switches opens, the control circuit will open and the coil on the starter will become de-energized. It is common to have the three series-connected, normally closed contacts wired directly to the starter's control coil, **Figure 32-12**. When the power to the coil is interrupted by the overload or heater opening, the power to the motor is also interrupted, causing the motor to shut down.

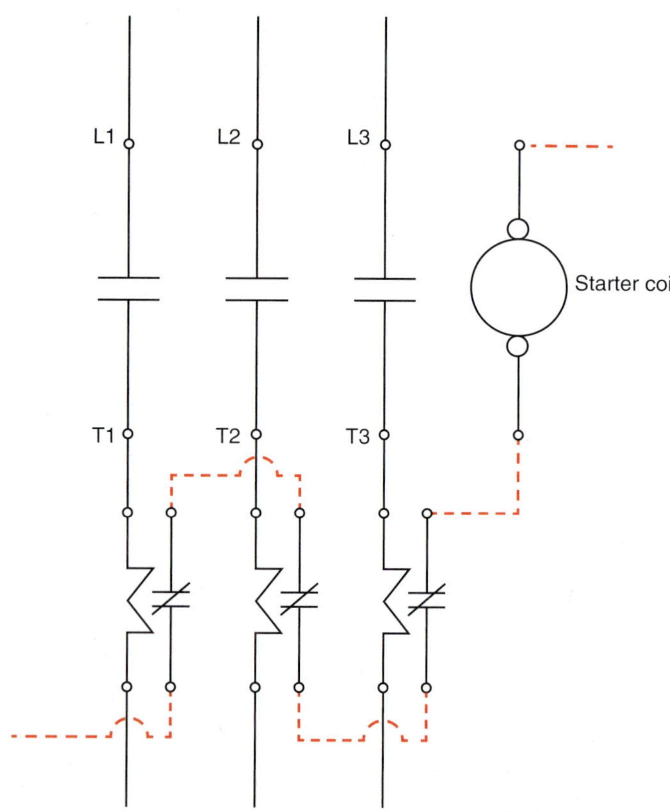

Goodheart-Willcox Publisher

Figure 32-12. The three normally closed circuit switches are wired in series with each other and the starter's control coil. In this configuration, when the power is interrupted by a heater opening, all power to the motor is interrupted.

32.4 Across-the-Line Starting

The most fundamental three-phase starter is the ***across-the-line starter***. This type of starter is called a full-voltage starter because it supplies full line voltage to all of the three-phase motor's windings during motor start-up. The line voltage power circuit is provided in **Figure 32-13**. The across-the-line starter, being among the simplest, is also, for the most part, the most economically appealing.

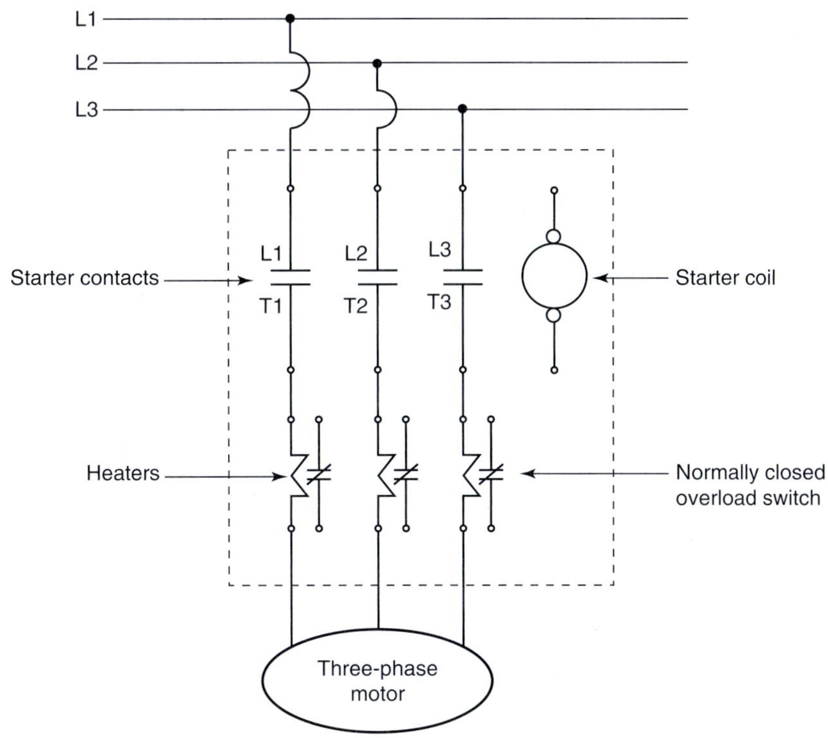

Goodheart-Willcox Publisher

Figure 32-13. Power circuit for a simple across-the-line motor starter. The heaters are wired in series with the load and are, therefore, exposed to the current flow to the motor.

The main drawback of the across-the-line starter is the high starting or inrush current that is often associated with induction motors.

Shedding Light

Across-the-line starting for large motors is, for the most part, inefficient. Higher inrush currents on large motors can cause increased energy bills. This starting method will cause voltage sags on other parts of the distribution system and voltage swells when large motors are shut off. Electrical power quality will likely be impacted in these instances.

When there is a call for motor operation, the starter contacts close, and power is passed to all of the three-phase motor's windings. The overload heaters are connected in series with the starter's contacts. As long as the motor's amperage draw does not exceed the overload device setting, the control circuit remains complete, and the motor continues to operate. Because full voltage is supplied to the motor on start-up, the motor will start rapidly with a high starting torque. The motor will continue to operate as long as there is a call for motor operation and all three normally closed overload contacts remain closed.

The operation of the across-the-line starter is straightforward and resembles that of a simple contactor or relay. The control circuit for the across-the-line starting method is provided in **Figure 32-14**. Often, the line voltage is transformed down to a safer control voltage of 24 V or 120 V.

32.5 Primary Resistor Starting

Primary resistor starting is a type of reduced-voltage starting. Reduced-voltage starting supplies reduced voltage to the motor at start-up by adding additional resistance in series with the motor. This is accomplished by using three large resistors, one in series with each of the motor's three terminals, **Figure 32-15A**. The additional resistance not only reduces the voltage being supplied to the motor, but also reduces the motor's start-up amperage. By reducing the voltage supplied to the motor as well as the amperage draw of the motor, this motor-starting configuration reduces the physical mechanical stress on the motor. This method of motor starting is commonly used on larger motors to reduce the load on the electrical system.

In addition to the starter, the contactor, and the three large resistors, the primary resistor starter configuration uses a time-delay relay (TD) wired in the control circuit, **Figure 32-15B**. Once the coil on the starter is energized, the

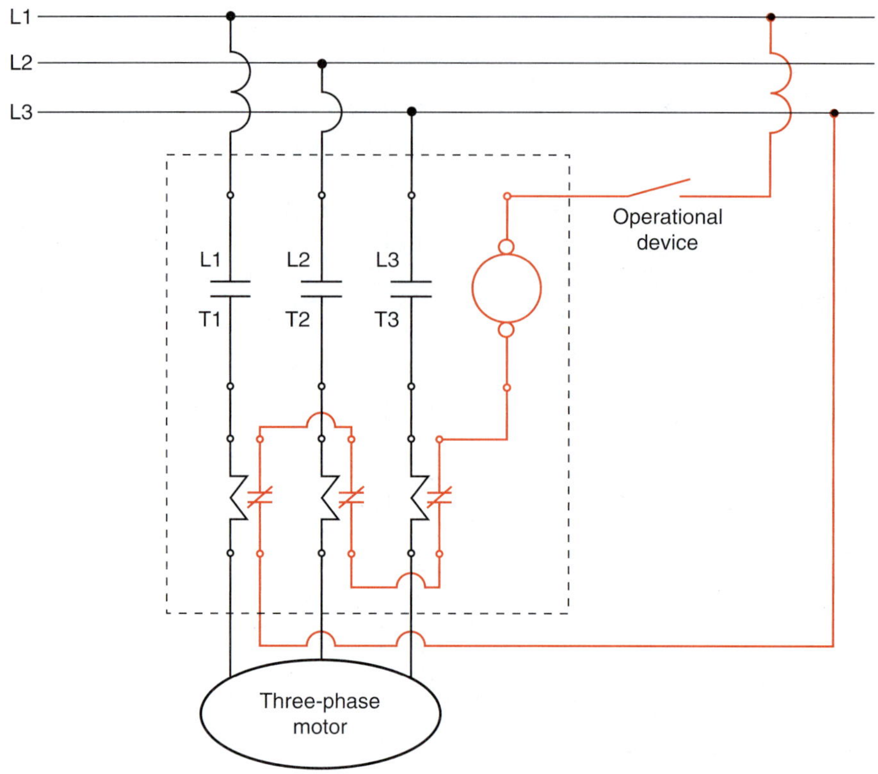

Goodheart-Willcox Publisher

Figure 32-14. Control circuit for a simple across-the-line motor starter. As long as the three normally closed control circuit switches are closed and there is power supplied to the circuit, the operational device will initiate motor operation. If the operational device, or any of the three normally closed control circuit switches open, the starter coil will become de-energized and the motor will stop operating.

three sets of starter contacts close. The closing of these contacts also energizes the coil on the time-delay relay, **Figure 32-15C**. Then, after a predetermined time delay period, the contacts on the time-delay relay close, energizing the coil on a three-pole contactor that has its contacts wired in parallel with the resistors. The closing of these three sets of parallel-connected contacts allows current to bypass the resistors, **Figure 32-15D.** This, in essence, removes the three resistors from the circuit, and full voltage is applied to the motor.

The following is the sequence of operations for the reduced-voltage start configuration:

1. The operational device closes, initiating a call for motor operation.
2. Provided all three switches on the motor starter heaters are closed, the coil on the starter is energized.
3. Energizing the starter coil causes the three starter contacts to close.
4. With the three starter contacts closed, power passes through the three large resistors (R) and on to the three-phase motor. At the same time, voltage is supplied to the time-delay relay (TD) coil.
5. The motor is energized, and the time-delay period is initiated.
6. After the time-delay period has elapsed, the contacts on the TD close.
7. The closing of the TD contacts completes the circuit through the coil on the contactor.
8. The three contactor contacts close.
9. The closing of the three contactor contacts creates bypass circuits around the three resistors.
10. By bypassing the three resistors, they are removed from the circuit, and the resistance of the motor's power circuit is reduced.
11. When the call for motor operation is no longer present, or if one of the three normally closed overload switches opens, the starter coil will no longer be energized and the three starter contacts will open, de-energizing the motor.

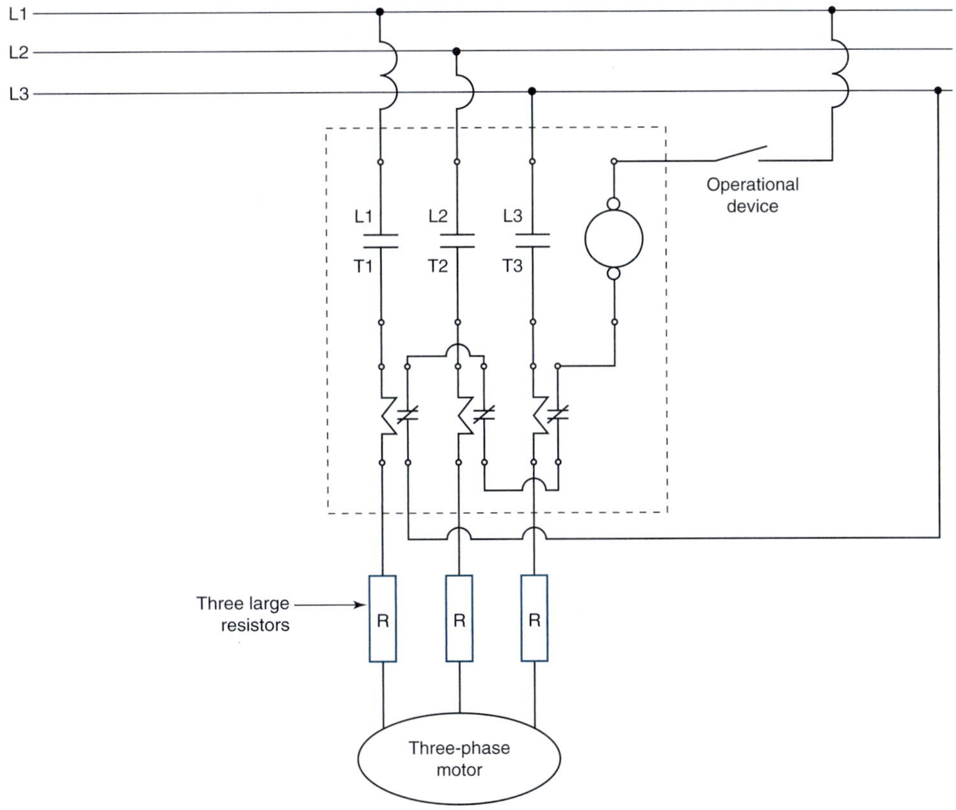

A

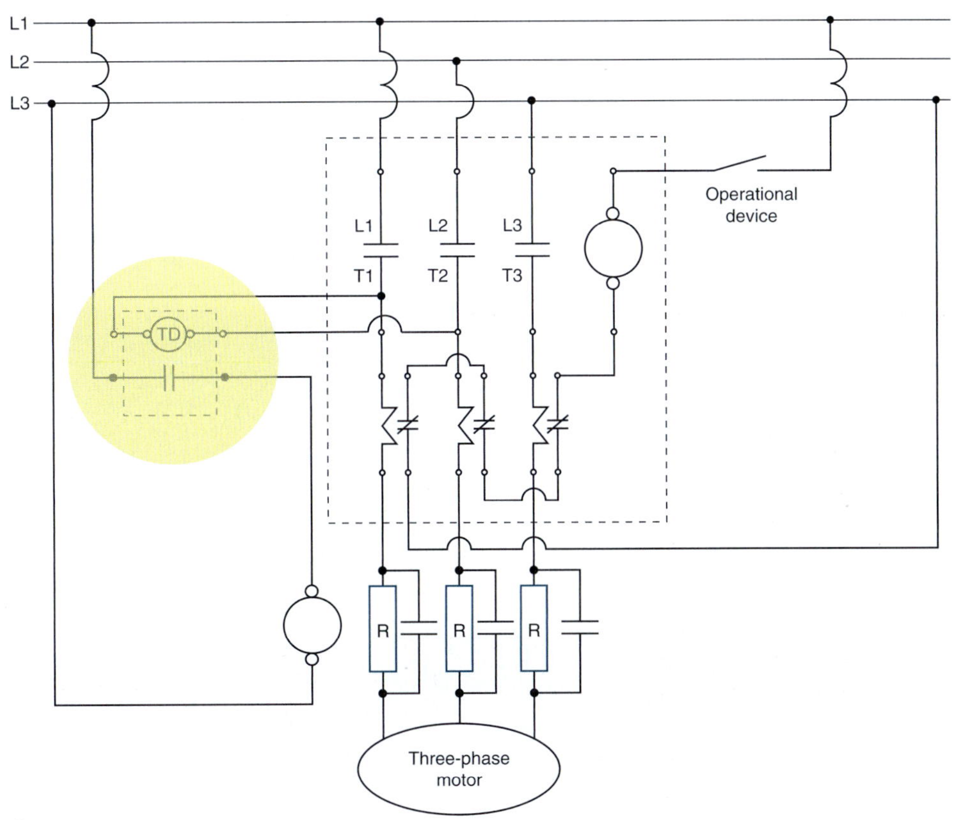

B

Goodheart-Willcox Publisher

Figure 32-15. Wiring diagram for a primary resistor start configuration. A—Three large resistors wired in series with each of the motor's three terminals. B—Time-delay relay is part of the motor's control circuit. (*Continues*)

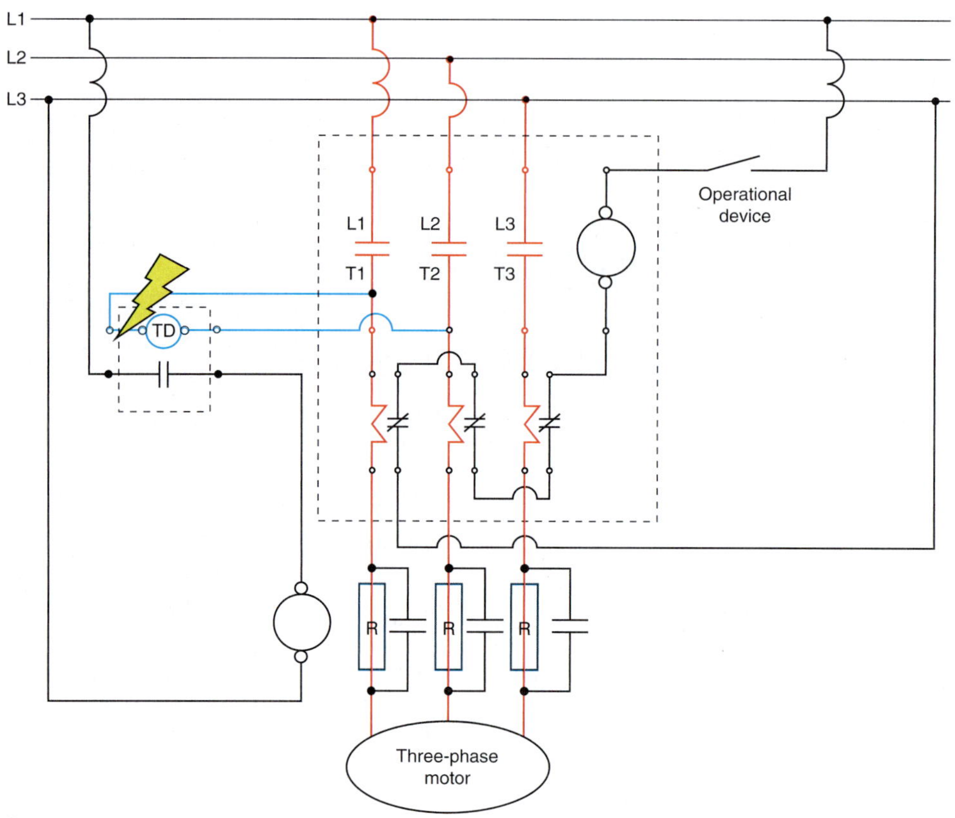

C

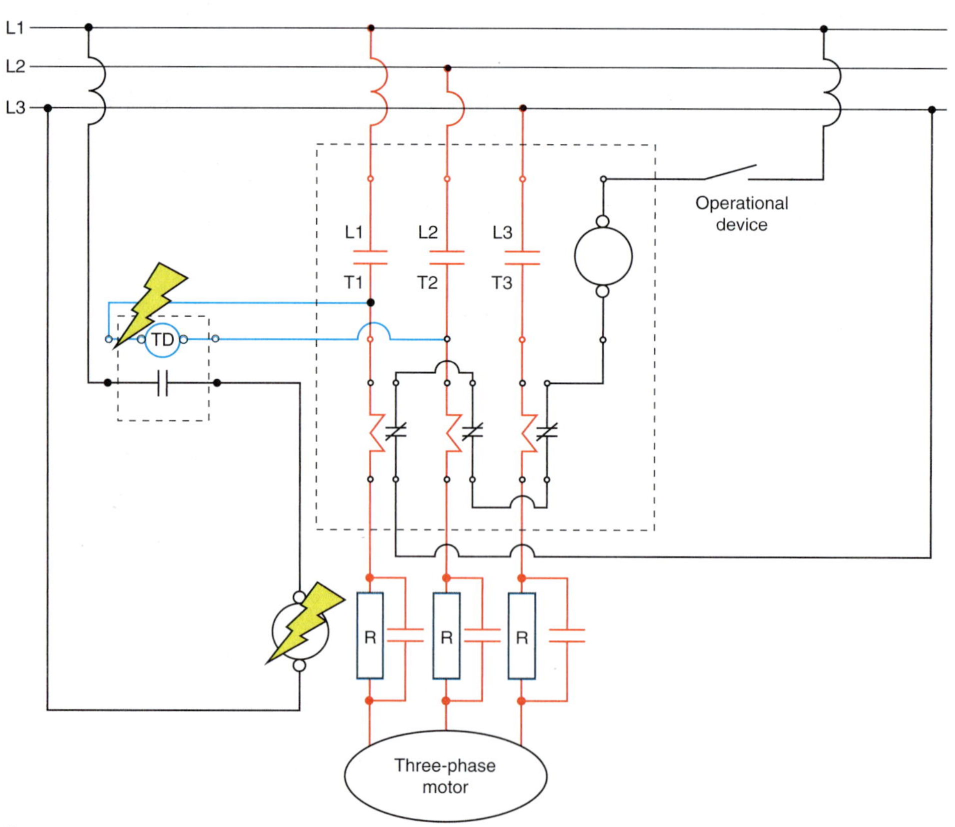

D

Figure 32-15. (*continued*) C—The time-delay relay's coil is energized when the main contacts on the starter close. D—When the three-phase contactor's coil is energized, current bypasses the three large resistors.

32.6 Part-Winding Starting

Part-winding starting, like primary resistor starting, is a reduced-voltage starting method intended to reduce the inrush current of a three-phase motor on start-up. In addition, this motor-starting method helps to reduce the physical stress on the motor and the components that are connected to it.

Unlike the primary resistor starter that provides a reduced voltage to all of the three-phase motor windings on start-up, the part-winding start configuration applies all of the voltage to only a portion of the three-phase motor windings. Once the motor has started, the remaining windings are added to the circuit as the motor reaches its design operating speed. Since only some of the windings are energized when the motor is initially powered, this method is only used for motors with multiple sets of windings, **Figure 32-16**. In the part-winding start circuit, there are two sets of windings: the primary winding set and the secondary winding set. The primary winding set is the set that is energized first, while the secondary winding set is energized a short time after the primary set. These two winding sets are electrically isolated from each other.

Power and Control Circuits

The complete wiring diagram in **Figure 32-16** can be broken down into four separate circuits: two power circuits (**Figure 32-17**) and two control circuits (**Figure 32-18**). The following provides in detail how this all-important and commonly used starting method operates:

- The first power circuit passes power through starter S1 and on to the wye-configured windings created by connecting motor leads 4, 5, and 6. This winding set is powered through motor leads 1, 2, and 3, and is often referred to as the primary winding set.
- The second power circuit passes power through starter S2 and on to the remaining wye-configured winding set. This winding set is powered through motor leads 7, 8, and 9, and is referred to as the secondary winding set.
- The first control circuit is a series circuit consisting of all six heater switches, the operation device, and the coil on S1. In order for the coil on S1 to become energized, all six heater switches as well as the operational device must be in the closed position.

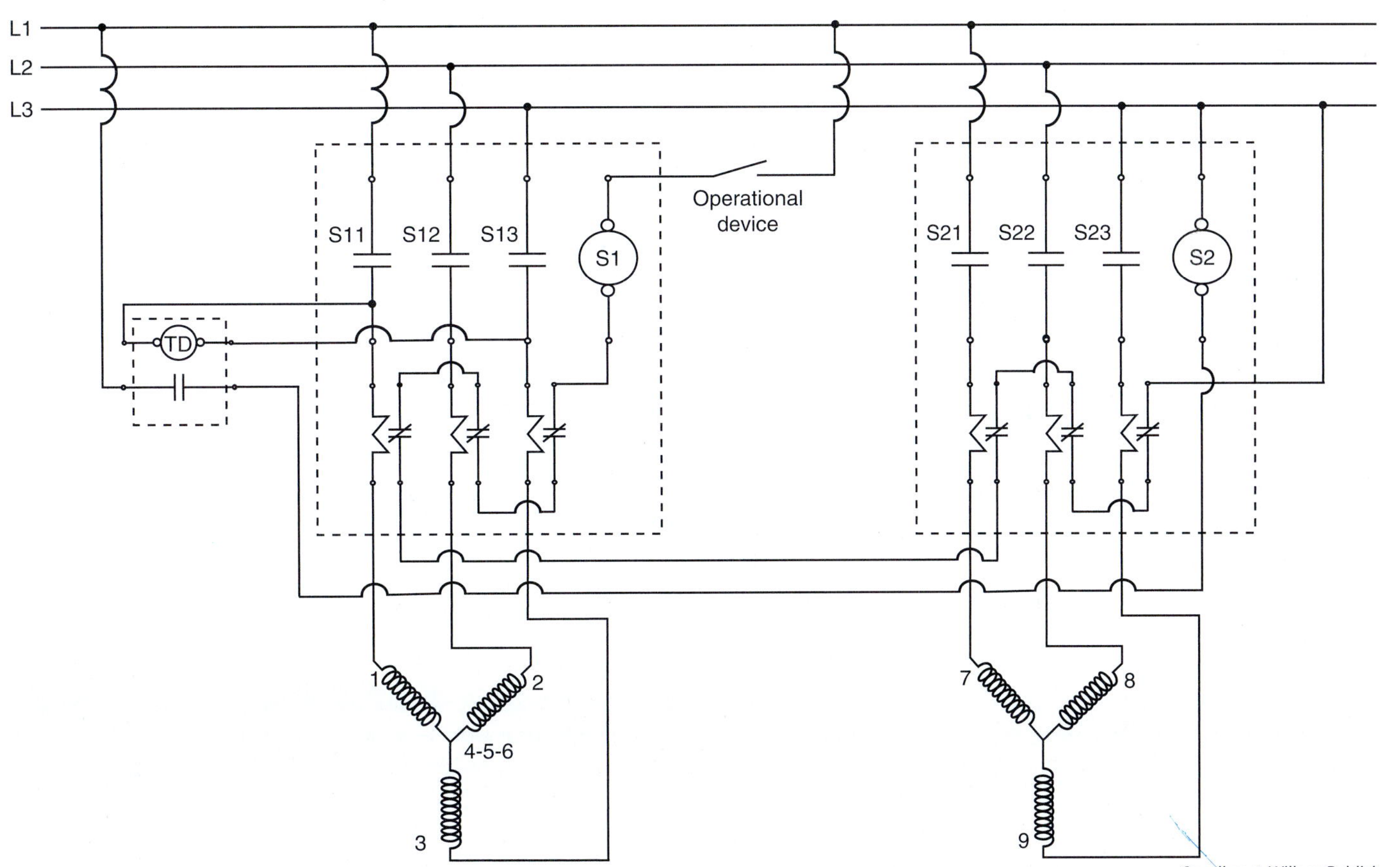

Goodheart-Willcox Publisher

Figure 32-16. Wiring diagram for a part-winding start configuration. This type of motor start strategy uses two starters, two sets of windings, and a time-delay relay. This type of starting method supplies full line voltage to only a portion of the motor's windings.

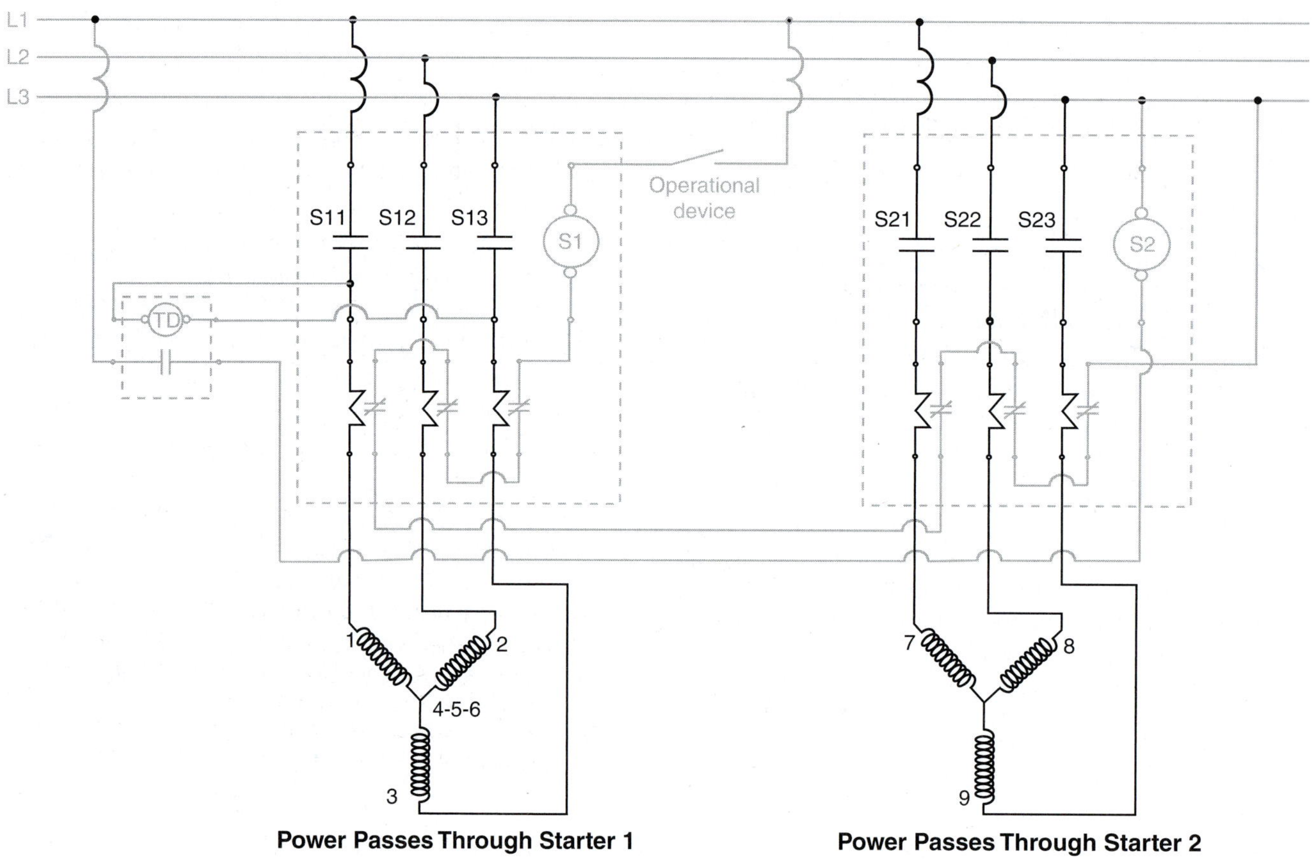

Goodheart-Willcox Publisher

Figure 32-17. The two power circuits from **Figure 32-16**. The power circuit through starter S1 delivers power to the primary winding set. This wye-configured winding set is formed when motor leads 4, 5, and 6 are connected together. The power circuit through starter S2 delivers power to the secondary winding set. Power is supplied to this wye-configured winding set at motor leads 7, 8, and 9.

- The second control circuit is a series circuit that consists of only the contacts on the time-delay relay (TD) and the coil on S2. When the contacts on the time-delay relay close, the coil on S2 will become energized.

Sequence of Operation

The following is the sequence of operation for the part-winding start configuration:

1. The operational device closes, initiating a call for motor operation.
2. Provided all six switches on the motor starter heaters are closed, the coil on S1 is energized.
3. Energizing the S1 coil causes the three power circuit contacts (S11, S12, and S13) to close.
4. With the three power circuit contacts closed, power passes to the first winding set on the three-phase motor. The first winding set on the motor is energized and the motor starts. This is sometimes called the part-winding acceleration, or starting, phase.
5. At the same time the S1 contacts close, voltage is supplied to the time-delay relay (TD) coil and the time-delay period is initiated.
6. After the time-delay period has elapsed, the contacts on the TD close.
7. The closing of the TD contacts completes the circuit through the coil on starter S2.
8. Energizing the S2 coil causes the three S2 contacts to close.
9. The closing of the three S2 contacts energizes the second winding set on the three-phase motor.

Shedding Light

In order to use a part-winding starter, the application must be evaluated to determine if it is a match for this motor-starting method. When using a part-winding starter, the motor will gradually ramp up its speed. The device or load driven by the motor must be able to function properly if gradually accelerated. If high starting torque is required for the driven load, the part-winding starter is probably not the best option.

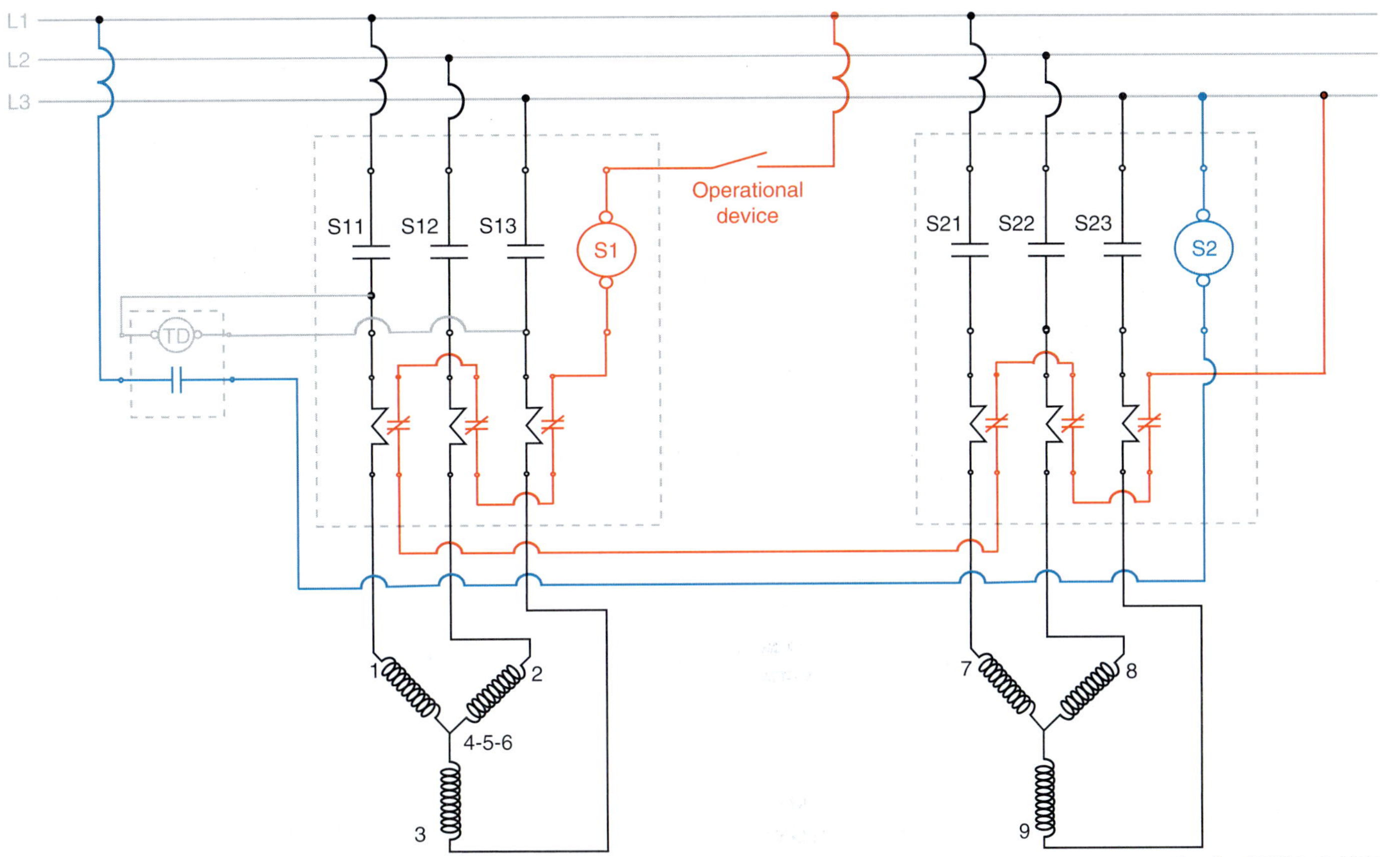

Figure 32-18. The two control circuits from **Figure 32-16**. The control circuit for starter S1 consists of all six heater switches, the starter's control coil, and the circuit's operational device. If any of the six control switches or the operational device opens, the motor will become de-energized. The control circuit for starter S2 consists of the normally open contacts on the time-delay relay and the control coil for starter S2. When the contacts on the time-delay relay close, the control coil on starter S2 will become energized.

32.7 Star-Delta Starting

Star-delta starting is another common reduced-voltage starting method used to reduce the starting current draw of larger, three-phase electric motors. This method uses both the star, or wye, winding configuration and the delta winding configuration. Starting a three-phase motor using this method will deliver 58% of the normal voltage to the motor on start-up. In addition, the amperage draw of the motor on start-up will only be about 33% of the motor's normal current. High-inertia loads, such as larger fans and heavy-duty punch presses, will work with this type of a starting method. A common characteristic of a high-inertia load is difficulty ramping up and ramping down speed. Winding configurations are discussed in Chapter 16, *Three-Phase Alternating Current Power*, Chapter 22, *Three-Phase Transformers*, and Chapter 29, *Three-Phase AC Motors*. If needed, revisiting these chapters may prove to be beneficial.

Unlike the motor-starting methods discussed earlier in this chapter, where windings or other circuit components are either added to or removed from the circuit, the star-delta motor-starting method changes the configuration of the winding connections. When the motor is initially energized, the motor windings are connected in the wye configuration. Once the motor has started, the winding configuration is changed over to the delta configuration. In this configuration, the full line voltage is supplied to each of the motor windings.

The power circuit for the star-delta motor-starting configuration is provided in **Figure 32-19**. This circuit arrangement uses a single three-pole motor starter, referred to as the master starter, and two three-pole contactors. One contactor, the delta contactor (DC), places the motor windings in a delta configuration, while the other contactor, the wye contactor (WC), places the motor windings in a wye configuration. The coil on the master starter is energized whenever the motor is operating. Only one of the coils on the two three-pole contactors is energized at any given time, depending on the desired winding configuration.

Master Starter

Figure 32-20 shows how the master starter (MS) provides power to the three-phase motor. When the three contacts (MS1, MS2, and MS3) on the master starter close, power is passed to one side of each of the three motor windings, as

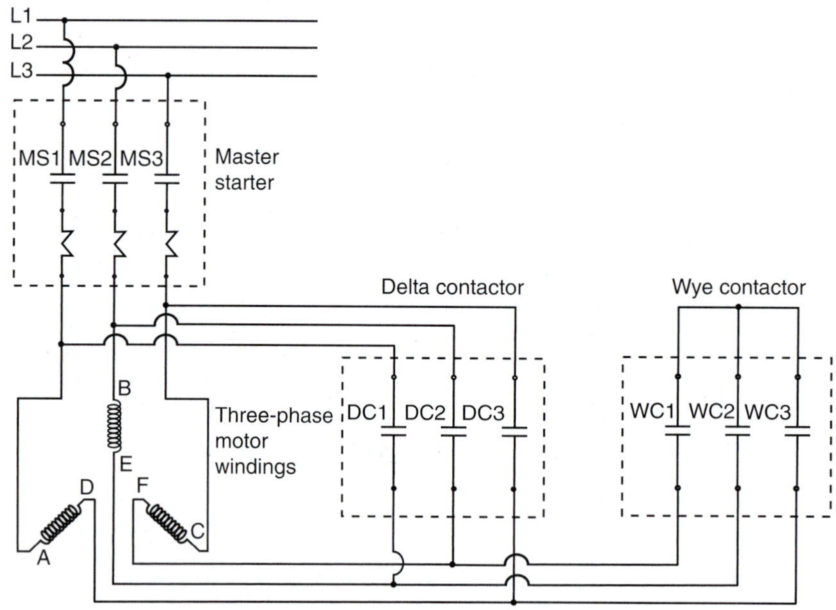

Goodheart-Willcox Publisher

Figure 32-19. Wiring diagram for the power circuit of a star-delta motor starter. This motor-starting method uses one starter and two contactors. Only one of the contactor coils can be energized at any time.

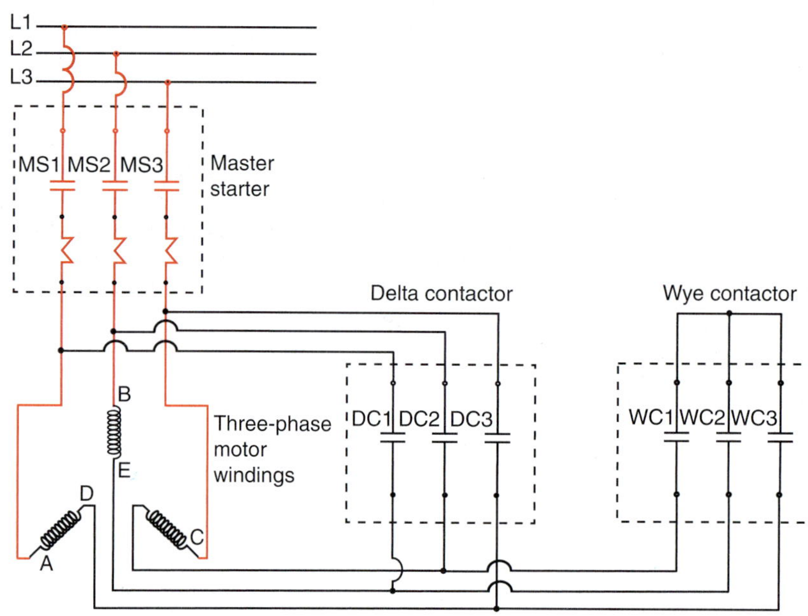

Goodheart-Willcox Publisher

Figure 32-20. Power circuit for star-delta motor starter as shown in **Figure 32-19**. When the coil of the master starter is energized, power passes to winding leads A, B, and C. Motor leads A, B, and C are powered by the master starter whenever the motor is operating.

indicated by the letters A, B, and C. The MS1, MS2, and MS3 contacts will be closed whenever the motor is operating, regardless of the motor's winding configuration.

Wye Contactor

When the motor is initially energized, the coil on the master starter will be energized, along with the coil on the wye contactor (WC). In this condition, power is supplied to the A, B, and C terminals via the MS1, MS2, and MS3 contacts, and terminals D, E, and F are connected together when contacts WC1, WC2, and WC3 close, **Figure 32-21**. As a result, the three windings are connected to each other to form a wye connection, **Figure 32-22**.

Delta Contactor

Once the motor has started, and after a brief time delay, the coil on the wye contactor will de-energize, and the coil on the delta contactor (DC) will energize. In this condition, power

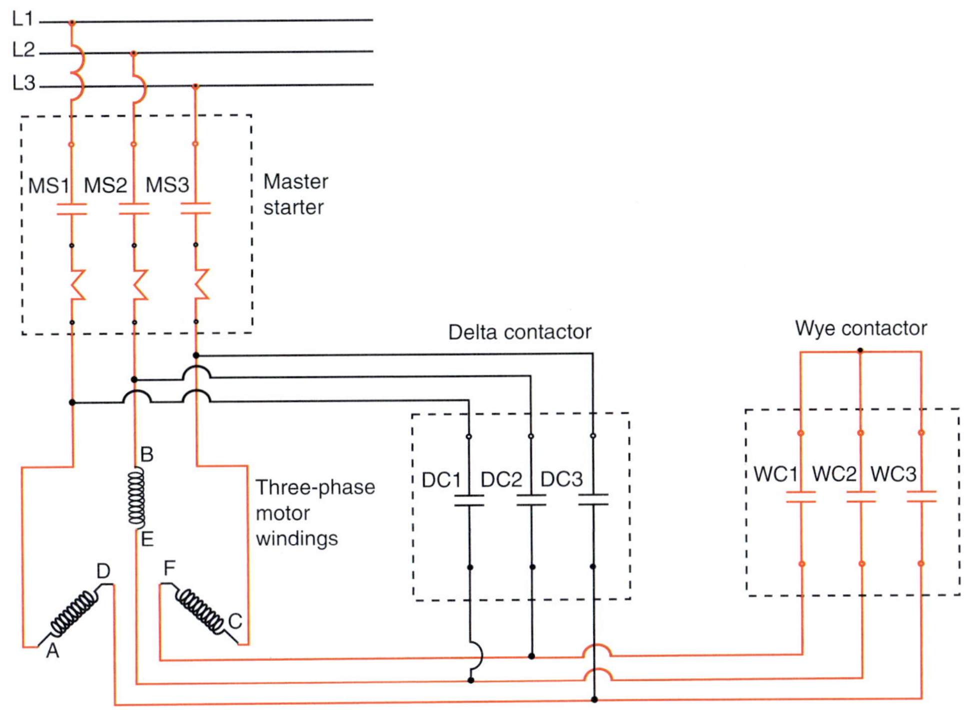

Figure 32-21. Power circuit shown in **Figure 32-19**. When the motor is initially energized, the wye contactor coil is energized and the motor windings are configured as a wye connection.

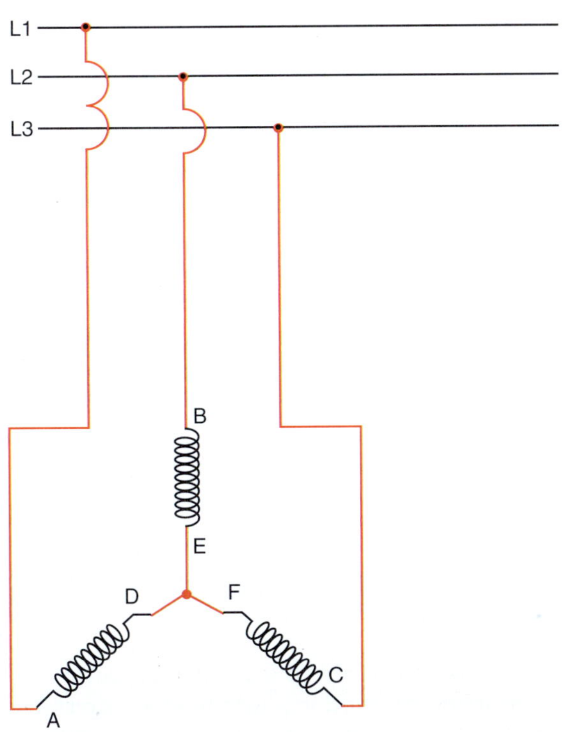

Figure 32-22. Diagram showing the wye-configuration when the motor is initially energized.

is still supplied to the A, B, and C terminals via the MS1, MS2, and MS3 contacts, but contacts DC1, DC2, and DC3 will be in the closed position, **Figure 32-23**. As a result, the three windings are connected to each other to form a delta connection, **Figure 32-24**.

Star-Delta Sequence of Operation

The following provides a detailed sequence of operation for the star-delta motor starter. Please refer to **Figure 32-25** for the ladder diagram and legend of the control sequence.

1. The operational device closes, initiating a call for motor operation.
2. Provided all three heater switches, HS, on the master starter heaters are closed, the coil on the master starter is energized. This sequence is represented by the top line in **Figure 32-25**.
3. When the coil on the master starter is energized, the normally open contacts MS1, MS2, and MS3 close, passing power to the three motor windings. (Note: This is not shown in the figure of the wire diagram.)
4. When the motor is initially energized, the windings are connected in a wye configuration, as indicated by the energized WC contactor on the second line of the wiring diagram. As such, each of the motor windings is supplied with roughly 58% of the main circuit voltage. By reducing the voltage supplied to each of the motor's windings, the current is reduced as well.
5. At the same time the motor is initially energized, the coil on the time delay relay is energized by the closing of the MS1 and MS2 contacts, as shown on line three of the wiring diagram. This starts the time-delay period.

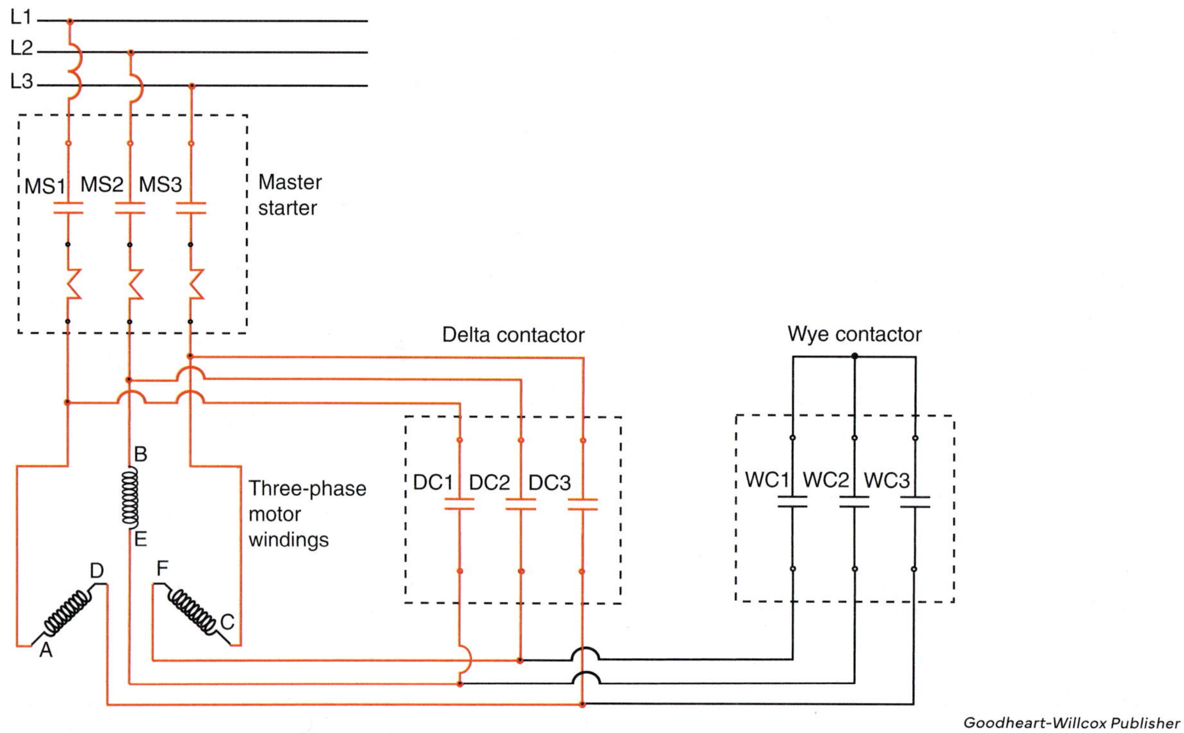

Figure 32-23. After the time-delay period has elapsed, the wye contactor coil is de-energized. At this point, the delta contactor coil is energized. This configures the windings as a delta connection.

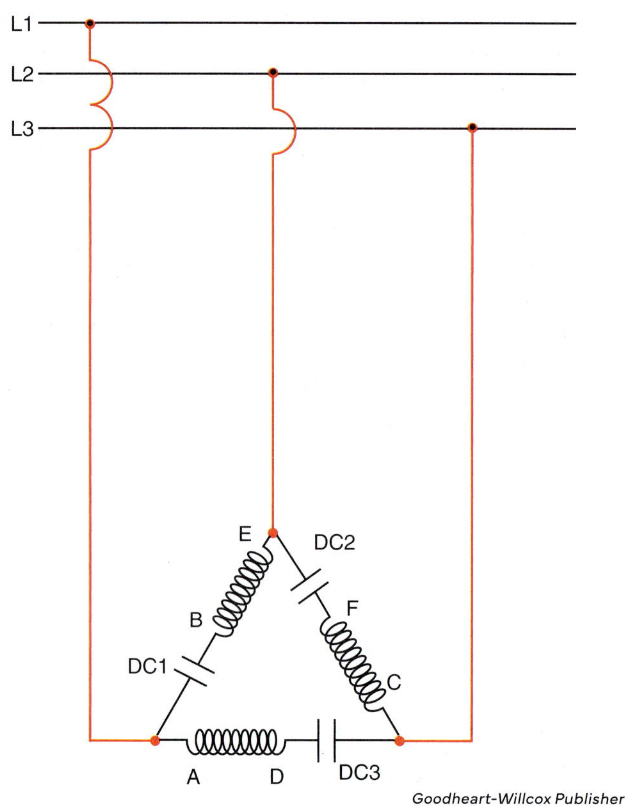

Figure 32-24. The delta configuration after the motor in **Figure 32-23** has started and switched over from the wye connection.

6. After a predetermined time-delay period has elapsed, the normally open contacts on the time-delay relay close, and the normally closed contacts on the time delay relay open. This energizes the DC coil and de-energized the WC coil, causing the wye contactor contacts to open and the delta contactor contacts to close.
7. When the time-delay relay contacts change position, the motor winding configuration changes from a wye connection to a delta configuration. In the delta configuration, each motor winding is connected directly to the corresponding phase of the power supply.
8. The motor continues to operate as a delta-connected motor until there is no longer a call for motor operation, or if one of the starter's heater switches opens.

Shedding Light

As with the part-winding starting method, when deciding whether or not to use the star-delta motor-starting method, be sure to evaluate the application to determine if it is a match for this motor-starting method. The device or load driven by the motor must be able to function properly if gradually accelerated, such as a large fan or large punch press. If high starting torque is required for the driven load, the star-delta starting method is probably not the best option.

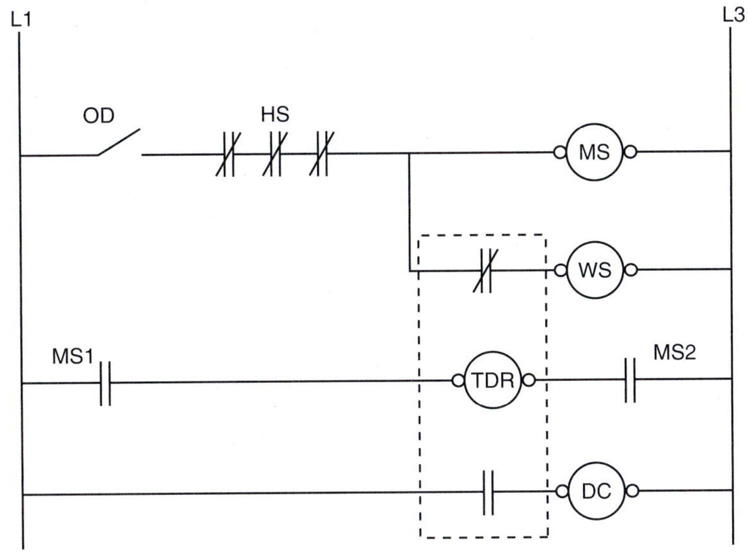

LEGEND	
DC	Delta contactor (coil)
HS	Heater switches
MS	Master starter (coil)
MS1	Master starter contacts
MS2	Master starter contacts
OD	Operational device
TDR	Time-delay relay (coil)
WC	Wye contactor (coil)

Goodheart-Willcox Publisher

Figure 32-25. Ladder diagram and legend for the control sequence of the start-delta configuration.

32.8 Autotransformer Starting

Autotransformer starting is another reduced-voltage starting method. Like other reduced-voltage starting methods, autotransformer starting is used primarily with large motors to reduce the amount of inrush current at start-up.

An autotransformer starter uses a three-phase autotransformer as a step-down transformer. It has two contactors: a main contactor and a transformer contactor. When the motor is started, the transformer contactor's contacts close. When the three transformer contacts close, power from each phase is supplied to a three-phase tapped autotransformer. The autotransformer has multiple taps, for example, 50%, 65%, and 80%. The tapped secondary connects to the phase conductors supplying the motor. At start-up, the 50% tap is connected, so the voltage supplied to the motor is one-half of the full voltage.

A relay switches between the tap points, increasing the supplied voltage as the motor's speed increases. After progressing through the autotransformer tap points, the transformer contactor is de-energized and the main contactor is energized. The contacts in the main contactor close, and the motor is supplied with full voltage.

32.9 Solid-State (Soft) Starting

Solid-state starters, or ***soft starters***, are electronic devices that use semiconductor switches to modify the voltage signal. Soft starters can reduce the line voltage and supply the motor a reduced starting voltage. A soft starter can smoothly and gradually increase the supplied voltage as the motor increases speed.

Since they are electronic, soft starters often include additional functionality such as phase monitoring, motor rotation reversal, and adjustable settings.

Shedding Light

Variable frequency drives (VFDs) are another type of motor starter and controller commonly used with three-phase motors. VFDs are covered in Chapter 25, *Motor Drives*.

32.10 Other Overcurrent Protection Devices

In addition to the starting and protection devices discussed throughout this chapter, there are other methods that can be used to protect and monitor the operation of three-phase motors. Electronic motor overload protection devices, for example, have more capabilities than thermal overload protection controls. Some of the features and benefits of electronic protection devices include the following:

- More precise control capabilities than thermal overloads
- Provide phase imbalance monitoring
- Adjustable trip settings
- Timed response to overload conditions
- Ability to communicate and send alerts to system operators
- Easily integrated into existing control systems for system monitoring
- Self-troubleshooting algorithms to help reduce system and equipment downtime.

CHAPTER 32 REVIEW

Summary

- Single-phasing is the condition where one of the three power lines on a three-phase system is lost. (32.1)
- Single-phasing can cause the motor's operating amperage to increase to levels as high as 173% of the motor's normal operating current. (32.1)
- Three-phase motor starters help ensure smooth motor operation and provide protection to the motor and any driven devices, components, or equipment connected to it. (32.2)
- Overload protection in a three-phase motor starter is provided by either electromechanical or electronic current sensors. (32.3)
- Current sensors, overloads, or running protection monitor the current flow through each of the motor's three phases. (32.3)
- The across-the-line starter supplies the entire supply voltage to all of the motor's windings. (32.4)
- Across-the-line starters subject the three-phase motor to high inrush currents. (32.4)
- The reduced-voltage starter supplies a reduced voltage to all of the motor's windings at start-up and then, after a predetermined time, provides the full voltage. (32.5)
- Primary resistor starting is a type of reduced-voltage starting. In addition to the starter, the contactor, and the three large resistors, the primary resistor starter configuration uses a time-delay relay wired in the control circuit. (32.5)
- The part-winding starter method supplies the entire supply voltage to some of the motor's windings and then, after a predetermined time, provides the full voltage to all the windings. (32.6)
- Reduced voltage starters and part-wind starters greatly reduce the inrush current experienced by three-phase motors. (32.6)
- A star-delta (wye-delta) motor starter allows a three-phase motor to start as a wye-configured motor and then switch over to a delta-configured motor. (32.7)
- Autotransformer starting is used primarily with large motors to reduce the amount of inrush current at start-up. (32.8)
- Solid-state starters, also referred to as soft starters, are electronic devices that use semiconductor switches to modify the voltage signal. (32.9)
- Other methods that can be used to protect and monitor the operation of three-phase motors include an electronic motor overload protection device. (32.10)

Review Questions

Use the information in the chapter to answer the following questions.

Measuring Your Potential

1. Which of the following is most likely to cause a single-phasing condition? (32.1)
   A. An undercurrent condition
   B. Reduced load on the motor
   C. A burned conductor
   D. An increase in motor winding resistance
2. Of the following, which is *not* a result of allowing a three-phase wye-connected motor to operate in a single-phasing condition? (32.1)
   A. A loss in torque
   B. An uneven load on the motor's winding
   C. Excessive vibrations
   D. Motor will immediately shut down
3. The forward/reverse starter on a wye-delta starter _____. (32.2)
   A. changes the voltage supplied to the motor from single-phase to three-phase
   B. changes the configuration of the motor windings from wye to delta
   C. reverses the connections between two of the three motor's power leads
   D. reverses the direction of current flow through the motor's control circuit
4. *True or False*? All three-phase starters, once they have an overload protective device that has been tripped, will automatically reset themselves. (32.3)
5. Which of the following best describes the function of an across-the-line starter? (32.4)
   A. The total supply voltage is supplied to all of the motor windings on start-up.
   B. A reduced voltage is supplied to all of the motor windings on start-up.
   C. A reduced voltage is supplied to some of the motor windings on start-up.
   D. The total supply voltage is supplied to some of the motor windings on start-up.

6. Which of the following best describes the function of a reduced-voltage starter? (32.5)
   A. The total supply voltage is supplied to all of the motor windings on start-up.
   B. A reduced voltage is supplied to all of the motor windings on start-up.
   C. A reduced voltage is supplied to some of the motor windings on start-up.
   D. The total supply voltage is supplied to some of the motor windings on start-up.
7. Which of the following best describes the function of a part-wind starter? (32.6)
   A. The total supply voltage is supplied to all of the motor windings on start-up.
   B. A reduced voltage is supplied to all of the motor windings on start-up.
   C. A reduced voltage is supplied to some of the motor windings on start-up.
   D. The total supply voltage is supplied to some of the motor windings on start-up.
8. On a part-wind start, the time-delay period is initiated when the _____. (32.6)
   A. main starter contacts close
   B. main starter contacts open
   C. secondary starter contacts close
   D. secondary starter contacts open
9. A star-delta (wye-delta) motor starter typically uses which of the following starting components? (32.7)
   A. One starter and one contactor
   B. One starter and two contactors
   C. Two starters and one contactor
   D. Two starters and two contactors
10. An autotransformer starter uses which of the following starting components? (32.8)
    A. One starter and one contactor
    B. One starter and two contactors
    C. Two starters and one contactor
    D. Two starters and two contactors
11. A(n) _____ is an electronic device that uses a semiconductor switch to modify the voltage signal. (32.9)
    A. across-the-line starter
    B. pawl
    C. primary resistor starter
    D. soft starter
12. *True or False?* Electronic protection devices provide more precise control capabilities than thermal overloads. (32.10)

Plug-In

1. If a single-phasing condition is present, with the motor de-energized at the motor disconnect, which of the following sets of voltage readings would most likely be obtained on the disconnected line terminal leads? (32.1)
   A. L1 to L2: 230 V; L1 to L3: 230 V; L2 to L3: 230 V
   B. L1 to L2: 230 V; L1 to L3: 230 V; L2 to L3: 0 V
   C. L1 to L2: 230 V; L1 to L3: 0 V; L2 to L3: 0 V
   D. L1 to L2: 230 V; L1 to L3: 120 V; L2 to L3: 120 V
2. The tripping of an alloy solder-type heater (overload) operates on the theory that as current flow _____. (32.3)
   A. increases, temperature decreases
   B. increases, temperature increases
   C. decreases, temperature decreases
   D. decreases, temperature increases
3. List the benefits of electronic protection devices. (32.10)
4. If the current draw of a three-phase motor rises to 57 A during a single-phasing condition, the motor's normal operating current is most likely in the range of _____. (32.1)
   A. 5 A to 10 A
   B. 30 A to 35 A
   C. 60 A to 70 A
   D. 95 A to 100 A

Making the Connection

1. Explain the operation of a melting alloy (eutectic) heater. (32.3)
2. Explain what would happen if the delta contactor coil and the wye contactor coil on a star-delta motor starter were to be energized at the same time. (32.7)
3. Explain the operational differences between a primary resistor starter and a part-winding starter. (32.6, 32.7)

CHAPTER

33 Motor Evaluation and Troubleshooting

Somsit/Shutterstock.com

SPARKING DISCUSSION

An electrician concludes that an electric motor is defective if it has the proper voltage supplied to it and it fails to start. Do you agree or disagree with this conclusion?

LEARNING OUTCOMES

After completing this chapter, you will be able to:

33.1 Explain troubleshooting techniques used to evaluate general motor issues.
33.2 Identify symptoms of a motor with mechanical problems.
33.3 Recognize symptoms of a motor with an open winding.
33.4 Identify symptoms of a motor with a shorted winding.
33.5 Identify symptoms of a motor with a winding shorted to ground.
33.6 Determine if a capacitor needs to be replaced.

TECHNICAL TERMS

bearing sag
megohmmeter

Introduction

Problems with electric motors can be classified as either mechanical or electrical. In some situations, mechanical problems may appear to be electrical, and electrical problems may appear to be mechanical. For example, a stuck driven assembly may prevent a motor from starting. This motor might make a humming sound and draw locked-rotor amperage, leading the electrician to conclude that the motor has a defective winding. Upon replacing the motor, the electrician observes that the new motor fails to operate as well. It is then that the electrician realizes that the seized driven device was the cause for the motor's failure to operate.

Electricians must know how to effectively, accurately, and efficiently evaluate electric motors to ensure that they are not replaced when such actions are not warranted. It is the responsibility of the electrician to prescribe only the repairs that are necessary to restore proper system and equipment operation without adding unwarranted expenses to the cost of the repair. Excessively high and unneeded repair costs lead to dissatisfied clients who will likely not become return customers.

This chapter highlights some of the most commonly encountered motor problems and provides guidance on how to quickly and accurately identify them.

33.1 General Motor Troubleshooting Techniques

If a motor fails to start, it is important to first determine if the problem is due to an external factor or if the problem lies within the motor itself. Failure to do so may result in replacing a motor that is operationally sound. The following are common external conditions that can prevent a motor from starting:

- Improper voltage supplied to the motor
- Seized or locked driven devices
- Motor brake not releasing
- Open connections on the load side

To ensure that the voltage supplied to the motor is correct, you will need to check the motor's nameplate to determine the motor's rated voltage. Then, measure the actual voltage being supplied to the motor and compare these two values. The supply voltage should be within 10% of the motor's rated voltage. If the supply voltage and the motor's rated voltage differ by more than 10%, evaluate the power supply, correct the situation, and then recheck the motor.

Another method to ensure that the proper voltage is being supplied to the motor is to verify that all power lines are energized as required. For example, if the motor is a three-phase motor, verify that the appropriate line voltage is supplied to the motor on all three lines. If one or more lines are not supplying voltage to the motor, then you will need to identify and remedy the cause for this condition. In addition, the voltage measurement between phases needs to be balanced and nearly identical. A 1% voltage unbalance can cause a significant current unbalance, often in the range from 6% to 10%.

It is also important to note something that might be relatively easy to spot, such as a motor brake that will not properly release. Motor brakes can use mechanical springs or electromagnetic fields to prevent a motor from turning. Be sure to check the motor brake mechanism to ensure that it is releasing properly before condemning the motor.

The wiring and connectors that supply power to a motor must be in good condition. Proper torque is required and essential on all bolted connections. Loose electrical connections can cause oxidation to form. Oxidation adds resistance to the motor circuit and can cause the motor's electrical connections to heat up. Once formed, oxidation will continue to get worse as additional heat causes additional oxidation to form. The additional resistance created by oxidation results in a reduction of the voltage supplied to the motor.

If possible, check the ability of the motor and its driven devices to rotate freely. To do this, verify that the power to the motor is turned off and properly locked out/tagged out before touching any rotating portion of the motor or its driven components. This will prevent the motor from starting if, upon manually turning the motor shaft, the cause of the lockup is removed. Once the motor has been turned off, attempt to rotate the motor's driven components. If the driven components are too large to turn manually, disconnect the driven components from the motor shaft or loosen the connection between the motor shaft and the driven components. With the motor shaft isolated from the driven components, the motor shaft should be able to rotate freely. If the voltage applied to the motor is correct, and the motor shaft turns freely, the motor should start when power is restored. If the motor starts, then check the driven components to determine if this is the cause of the motor's starting problem. If the motor fails to start when power to the motor is restored, the motor and its starting components must be evaluated, as the problem lies in one of these places.

33.2 Mechanical Problems with Electric Motors

Electric motors often fail due to mechanical problems that ultimately become electrical problems. In many instances, mechanical problems appear in the motor's bearings or at the point where the motor connects to the drive assembly. Bearings tend to fail when they are subjected to excessive pressure or tension, or if they are not lubricated properly. Excessive belt tension is the main cause for excessive bearing tension and premature failure. Dirt, dust, and other foreign

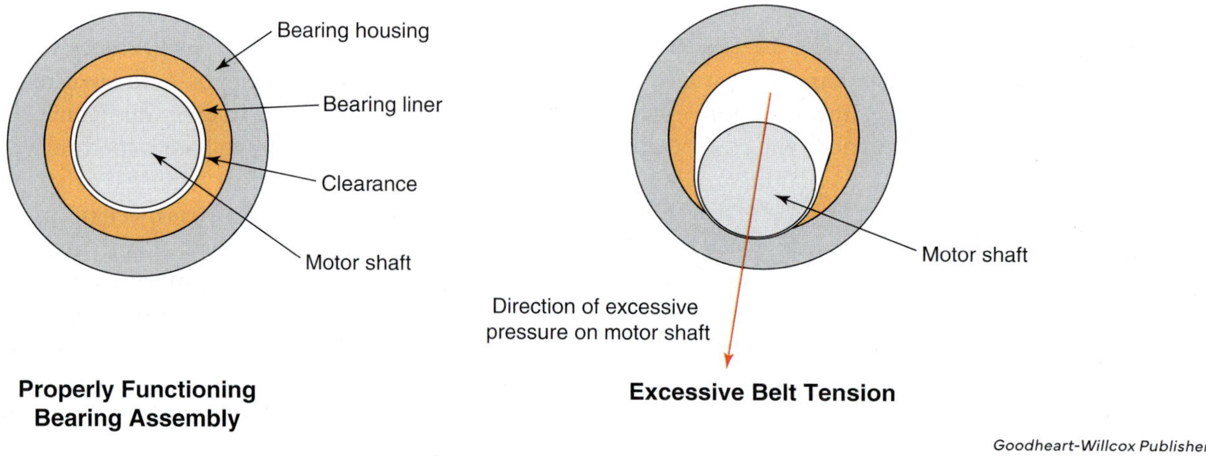

Figure 33-1. A mechanical failure can appear to be, or cause, an electrical problem. On a properly functioning bearing assembly, the bearing assembly and the motor shaft should be circular, with the motor shaft positioned in the center of the bearing assembly. When there is excessive belt tension, the bearing sags, causing damage to the bearing liner and housing. This prevents the rotor from remaining centered in the motor's stator field.

matter, if allowed to enter the bearings, can also cause motor bearings to seize, overheat, and fail prematurely.

Bearing failure can often be identified by noise coming from the bearing. When sleeve bearings fail, they either lock up and fail to turn, or they "sag." ***Bearing sag*** occurs when excessive pressure is placed on the motor shaft, pulling it against the bearing liner. The rotating shaft wears away the bearing liner and possibly the bearing's housing, **Figure 33-1**. As a result, the rotor is no longer centered within the stator field and the motor fails to start.

Shedding Light

Although motor bearings can be replaced, this is not a job typically performed by an electrician. If the motor is small, typically less than 2 hp, it is more economical to replace the motor because it would cost more to repair it than replace it. Consider the costs associated with disassembling the motor, obtaining the replacement bearings, removing the defective bearings, installing the new bearings, reassembling the motor, and providing a guarantee for the repair.

If it is determined that a motor's bearings are defective, replacing the bearings or the motor will not necessarily solve the underlying cause for the motor's failure. Bearing failure is often the effect or result of problems with the driven devices connected to the motor. The most common driven device problem is belt related. When properly installed, belts should be tight enough to prevent slippage, but loose enough to avoid excessive pressure or strain on the drive or driven bearings. Many motors fail because of overtightened belts. Chapter 26, *Motor Installation and Replacement*, explains how to properly evaluate belt tension using a belt tension gauge.

33.3 Electrical Problems with Open Motor Windings

An open motor winding is a common problem associated with electric motors. An open winding can be easily detected using the resistance function on a digital multimeter. Each winding segment on a motor should have a measurable resistance. Simply stated, taking a resistance reading across any known terminal or lead combination to which a motor winding is connected should produce a nonzero or noninfinite resistance value.

Open Motor Winding on a Single-Phase Motor

A single-phase motor that is in good working order may have the following resistance readings between the motor terminals:

- Common (C) to start (S): 6 Ω
- Common (C) to run (R): 3 Ω
- Run (R) to start (S): 9 Ω

The resistance reading between the C terminal and the R terminal provides only the resistance of the run winding. The resistance reading between the C terminal and the S terminal provides only the resistance of the start winding. The resistance reading between the S and R terminals will be the sum of the resistances of the start and run windings, which is 9 Ω (6 Ω + 3 Ω).

If the start winding were to open, the resistance readings for the same motor would be as follows:

- Common (C) to start (S): OL (infinite resistance)
- Common (C) to run (R): 3 Ω
- Run (R) to start (S): OL (infinite resistance)

An open winding on a single-phase motor will, therefore, affect two of the three resistance readings. Two of the readings will display infinite resistance, while the third reading will display a measurable, nonzero or noninfinite value.

Open Motor Winding on a Three-Wire, Delta-Connected, Three-Phase Motor

A three-wire, delta-connected three-phase motor that is in good working order may have the following resistance readings between the motor leads:

- L1 to L2: 6.67 Ω
- L1 to L3: 6.67 Ω
- L2 to L3: 6.67 Ω

On a properly operating three-wire three-phase motor, the obtained resistance readings will all be the same. The obtained readings, however, are not equal to the actual winding resistances. Consider the three-wire, delta-connected three-phase motor illustrated in **Figure 33-2**. The resistance of each of the three motor windings is 10 Ω. When a resistance measurement is being taken between motor leads L1 and L2, the meter is reading the parallel resistance connection. The total equivalent resistance is calculated as one 10 Ω load connected in parallel with a 20 Ω load, which is 6.67 Ω.

If a three-wire, delta-connected three-phase motor experiences an open winding, **Figure 33-3**, the following resistance readings between the motor leads may be observed:

- L1 to L2: 20 Ω
- L1 to L3: 10 Ω
- L2 to L3: 10 Ω

It is important to note that, because of the open winding, there are no parallel connections, and the obtained resistance reading across the open winding will be a multiple of the actual winding resistance. It is also important to note that when a three-wire, delta-connected, three-phase motor experiences an open winding, there will be no infinite resistance readings. This is because the two intact windings create an alternate path around the open winding. The resistance reading across the open winding will be twice the value of a single motor winding, because the two intact windings are in series in the alternate path.

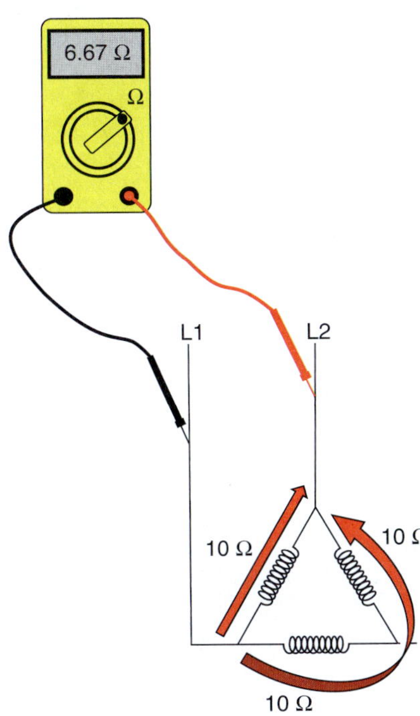

Goodheart-Willcox Publisher

Figure 33-2. A properly operating three-wire, delta-configured, three-phase motor. This motor has three 10 Ω winding segments. The resistance reading between any two motor terminals will be 6.67 Ω.

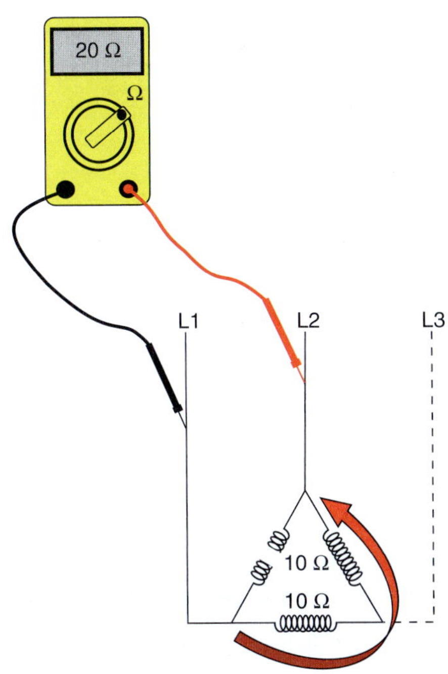

Goodheart-Willcox Publisher

Figure 33-3. A three-wire, delta-configured, three-phase motor with an open winding between motor terminals and L1 and L2. Even though there is an open winding, there is no OL or infinite resistance reading. This is due to the alternate current path created by the remaining two motor windings.

Open Motor Winding on a Three-Wire, Wye-Connected, Three-Phase Motor

A three-wire, wye-connected, three-phase motor that is in good working order may have the following resistance readings between the motor leads:

- L1 to L2: 20 Ω
- L1 to L3: 20 Ω
- L2 to L3: 20 Ω

As in the case of the three-wire, delta-connected, three-phase motor, the obtained resistance readings of the windings of a three-wire, wye-connected, three-phase motor will all be the same. Once again, the obtained readings are not equal to the actual winding resistances. Consider the three-wire, wye-connected, three-phase motor illustrated in **Figure 33-4**. The resistance of each of the three motor windings is 10 Ω. When a resistance measurement is taken between motor leads L1 and L2, the meter is reading the series resistance connection. The winding configuration is observed as two 10 Ω loads wired in series with each other, for a total of 20 Ω. The same holds true for readings taken between motor leads L1 and L3, and between motor leads L2 and L3.

If a three-wire, wye-connected, three-phase motor experiences an open winding, the following resistance readings between the motor leads may be observed:

- L1 to L2: OL (infinite resistance)
- L1 to L3: OL (infinite resistance)
- L2 to L3: 20 Ω

For a three-wire, wye-connected, three-phase motor, an open motor winding will result in two infinite resistance readings and one measurable, nonzero or noninfinite value. The two infinite resistance readings will be obtained from the two terminal lead combinations that include the motor lead that is connected to the open winding. For example, in **Figure 33-5** the winding connected to L1 is open, and both resistance readings that include L1 are infinite.

Open Motor Winding on a Six-Wire, Star-Delta-Connected, Three-Phase Motor

A six-wire, star-delta-connected three-phase motor that is in good working order may have the following resistance readings between the motor terminals:

- T1 to T4: 10 Ω
- T2 to T5: 10 Ω
- T3 to T6: 10 Ω

For this type of motor, all three resistance readings are the same and each of these readings is equal to the resistance of each winding. In **Figure 33-6**, for example, the resistance of each winding is 10 Ω and the measured resistance across each terminal pair is also 10 Ω. On a properly operating motor, there should be an infinite resistance reading between any unpaired terminal/lead combination. For example, there should be an infinite resistance reading between terminals T1 and T6. If there is a measurable, nonzero or noninfinite resistance value measured between any terminal combinations made up of three or more motor leads, the motor needs to be repaired or replaced.

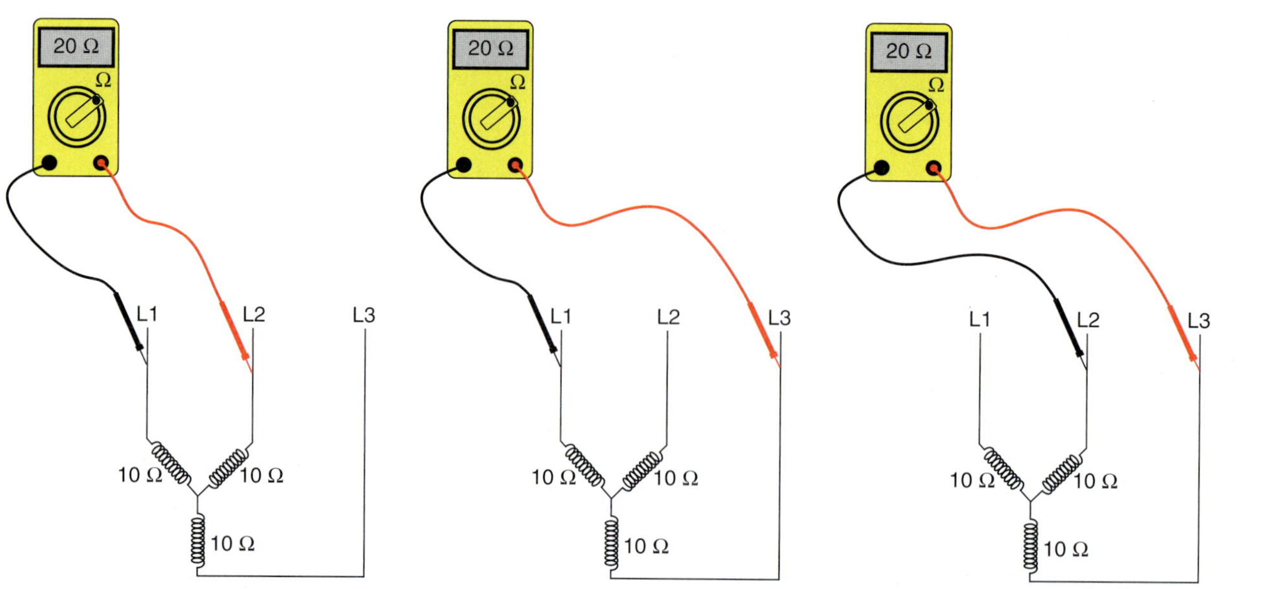

Figure 33-4. A properly operating three-wire, wye-configured, three-phase motor. This motor has three 10 Ω winding segments. The resistance reading between any two motor terminals will be 20 Ω.

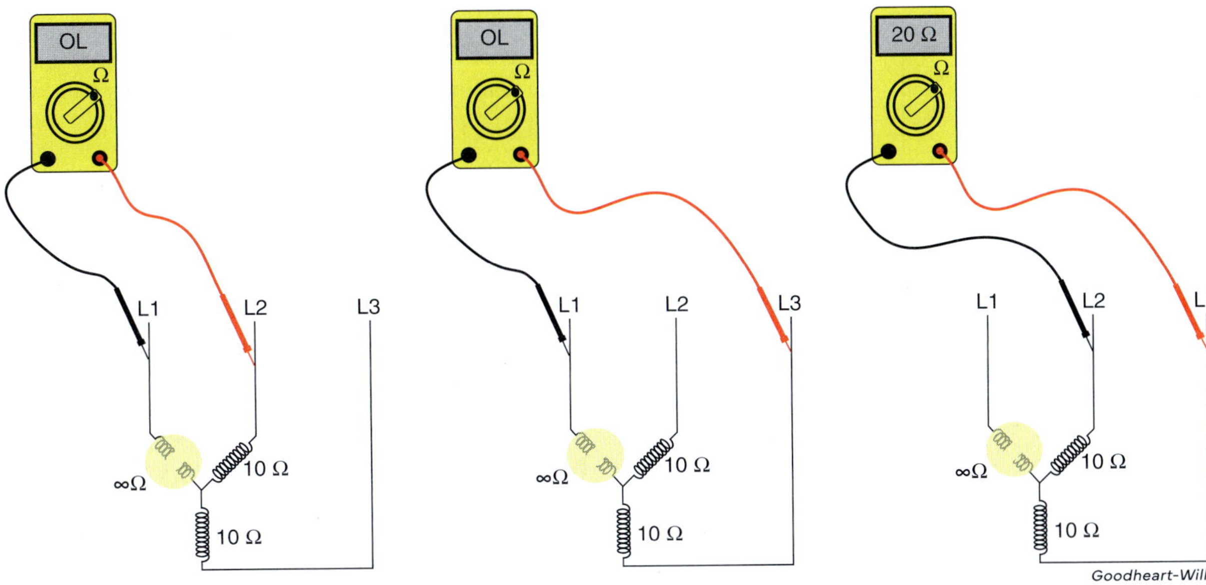

Goodheart-Willcox Publisher

Figure 33-5. A three-wire, wye-configured, three-phase motor, with an open winding connected to motor lead L1. Any resistance reading taken between L1 and any other motor lead will produce an infinite resistance reading.

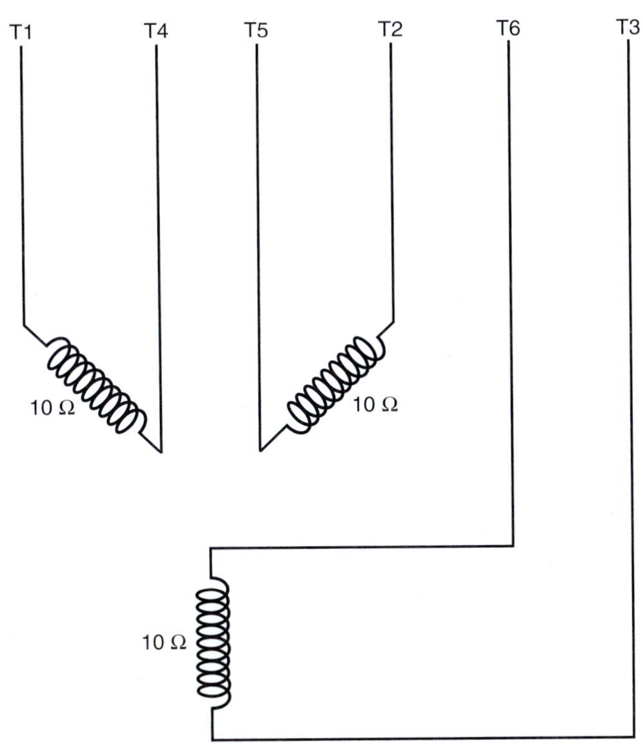

Goodheart-Willcox Publisher

Figure 33-6. A six-wire, star-delta-configured, three-phase motor in good working order with three isolated 10 Ω winding segments. The resistance reading between any two motor terminal pairs is 10 Ω.

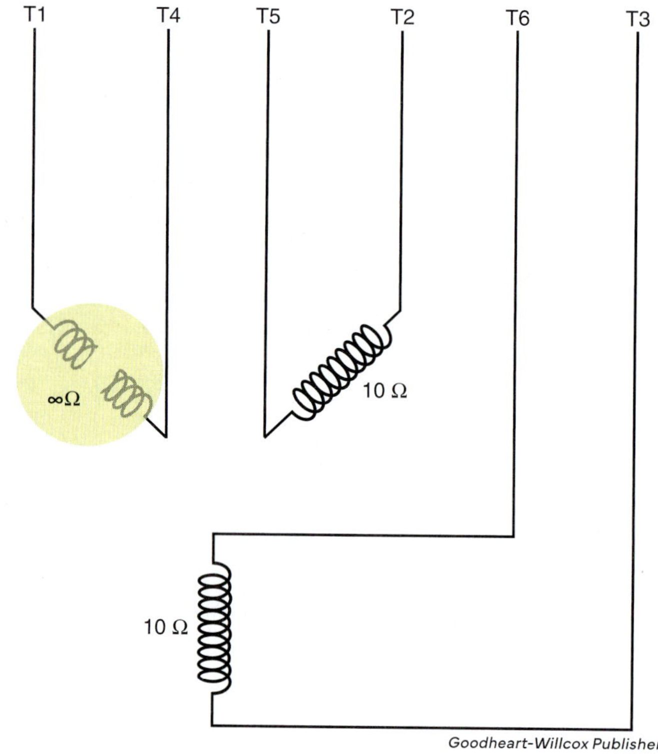

Figure 33-7. A six-wire, star-delta-configured, three-phase motor with an open winding connected to motor lead combination T1 and T4. The resistance between T1 and T4 will give an infinite resistance reading.

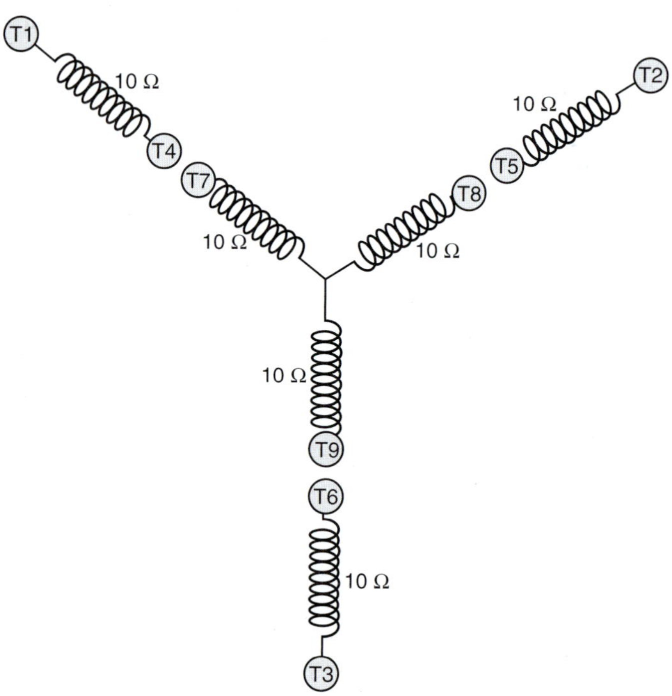

Figure 33-8. The numbering in a nine-wire, dual-voltage, three-phase motor follows a clockwise spiral that begins at the outer edge and continues inward.

If a six-wire, star-delta-connected, three-phase motor experiences an open winding, **Figure 33-7**, the following resistance readings between the motor terminals may be observed:

- T1 to T4: OL (infinite resistance)
- T2 to T5: 10 Ω
- T3 to T6: 10 Ω

In a six-wire, star-delta-connected, three-phase motor with an open motor winding, one infinite resistance reading and two measurable, nonzero or noninfinite values that are the same will be measured.

Open Motor Winding on a Nine-Wire, Dual-Voltage Three-Phase Motor

The numbering system used to identify the motor leads on a nine-wire, dual-voltage three-phase motor is highlighted in **Figure 33-8**. In a nine-wire, dual-voltage three-phase motor, there are three separate winding segments (connected to motor lead combinations T1 and T4, T2 and T5, and T3 and T6) and one wye connection (connected to motor leads T7, T8, and T9). Assuming that the motor is in good working condition and each winding segment has a resistance of 10 Ω, the following resistance readings will be observed between the motor terminals:

- T1 to T4: 10 Ω
- T2 to T5: 10 Ω
- T3 to T6: 10 Ω
- T7 to T8: 20 Ω
- T7 to T9: 20 Ω
- T8 to T9: 20 Ω

The resistance readings between motor lead combinations T1 and T4, T2 and T5, and T3 and T6 reflect the 10 Ω resistance of each of the isolated motor winding segments. The resistance readings between motor lead combinations T7 and T8, T7 and T9, and T8 and T9 reflect the measured resistances of a three-wire, wye-connected arrangement, **Figure 33-9**.

If a nine-wire, dual-voltage three-phase motor experiences an open in one of the isolated windings, the following resistance readings between the motor terminals may be observed:

- T1 to T4: OL (infinite resistance)
- T2 to T5: 10 Ω
- T3 to T6: 10 Ω
- T7 to T8: 20 Ω
- T7 to T9: 20 Ω
- T8 to T9: 20 Ω

In this scenario, the resistance reading across the motor leads T1 and T4 will change. The other resistance readings will remain the same. As illustrated in **Figure 33-10**, this combination is comprised of the two motor leads that are connected to the damaged winding segment.

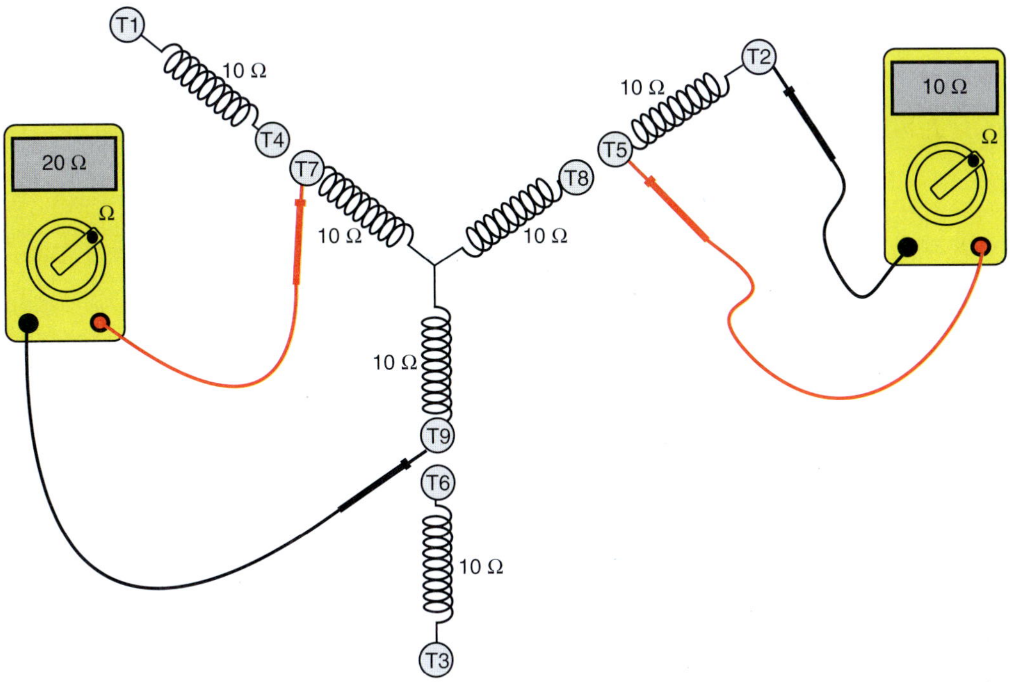

Goodheart-Willcox Publisher

Figure 33-9. Same three-phase motor as in **Figure 33-8**. There are three isolated, 10 Ω winding segments and a wye-configured winding arrangement, also with 10 Ω for each segment. The resistance reading between any two motor terminal/lead pairs on the isolated windings is 10 Ω. The resistance reading between any two motor terminal/lead pairs on the wye-connected windings is 20 Ω.

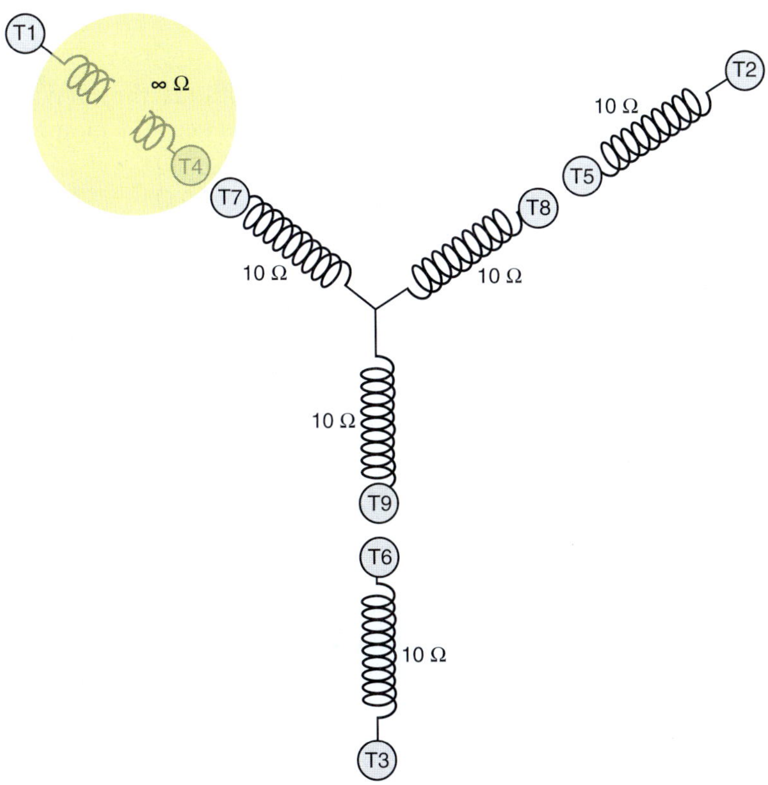

Goodheart-Willcox Publisher

Figure 33-10. In this nine-wire three-phase motor with an open winding, the resistance between motor lead pairs T1 and T4 will produce an infinite resistance reading. The resistance readings between the other motor lead pairs (T2-T5 and T3-T6) will be normal, as will the readings taken across the terminals/leads on the wye-connected winding segments.

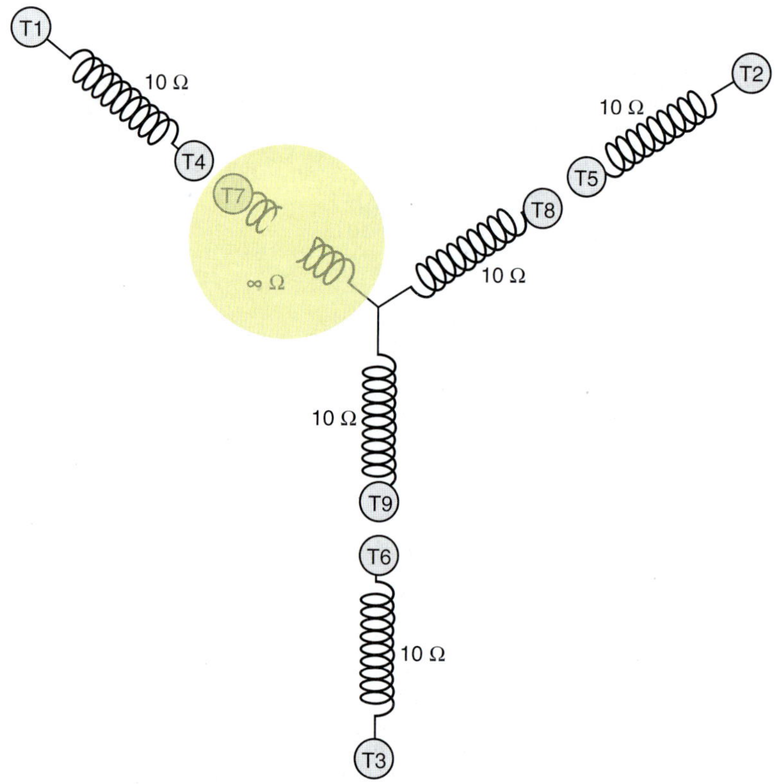

Goodheart-Willcox Publisher

Figure 33-11. In this nine-wire three-phase motor with an open winding, the resistance readings between motor leads T7 and T8, and between motor leads T7 and T9, will produce infinite resistance readings. The resistance readings between the other motor leads T8 and T9 and the resistance readings across all isolated winding segments (T1-T4, T2-T5, and T3-T6) will be normal.

If a nine-wire, dual-voltage, three-phase motor experiences an open in one of the windings that comprises the wye-connected portion of the motor, the following resistance readings between the motor terminals may be observed:

- T1 to T4: 10 Ω
- T2 to T5: 10 Ω
- T3 to T6: 10 Ω
- T7 to T8: OL (infinite resistance)
- T7 to T9: OL (infinite resistance)
- T8 to T9: 20 Ω

In this case, two resistance readings will change, **Figure 33-11**. With the open winding connected to motor lead T7, the two lead combinations that contain motor lead T7 (namely T7-T8 and T7-T9) will have infinite resistance, while all other motor terminal/lead combinations will provide measurable, nonzero or noninfinite values.

Open Motor Winding on a 12-Wire Three-Phase Motor

The numbering system used to identify the motor leads on a 12-wire, dual-voltage three-phase motor is illustrated in **Figure 33-12**. In a 12-wire three-phase motor, there are six separate winding segments. Assuming that the motor is in good working order and each winding segment has a resistance of 10 Ω, the following resistance readings will be observed between the motor leads:

- T1 to T4: 10 Ω
- T2 to T5: 10 Ω
- T3 to T6: 10 Ω
- T7 to T10: 10 Ω
- T8 to T11: 10 Ω
- T9 to T12: 10 Ω

The resistance readings of all motor lead combinations are the same.

Consider a 12-wire, dual-voltage, three-phase motor that experiences an open in one of the isolated windings, **Figure 33-13**. The following resistance readings between the motor leads may be observed:

- T1 to T4: 10 Ω
- T2 to T5: 10 Ω
- T3 to T6: 10 Ω
- T7 to T10: OL (infinite resistance)
- T8 to T11: 10 Ω
- T9 to T12: 10 Ω

In this scenario, the only resistance reading that changes is the reading across the motor leads T7 and T10, which is comprised of the two motor leads that are connected to the damaged winding segment.

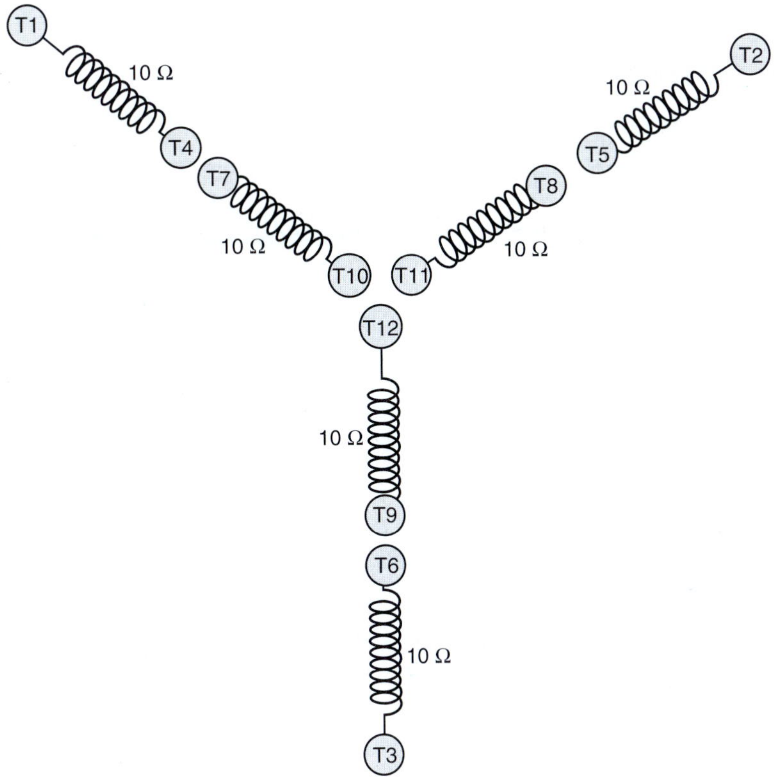

Goodheart-Willcox Publisher

Figure 33-12. A 12-wire, dual-voltage, three-phase motor. This motor's terminals/leads are numbered by following a clockwise spiral starting at the outer edge and working inward. This motor has six isolated windings and no wye-connected windings.

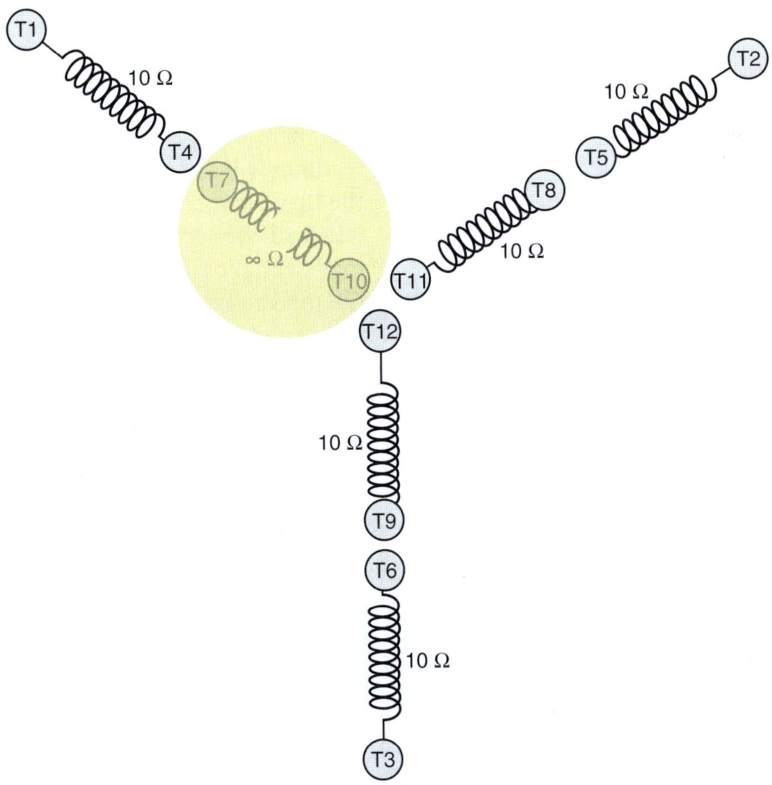

Goodheart-Willcox Publisher

Figure 33-13. In this 12-wire, dual-voltage, three-phase motor with an open winding, the resistance readings between motor leads T7 and T10 will produce infinite resistance readings. The resistance readings between the other motor leads will be unchanged.

33.4 Electrical Problems with Shorted Motor Windings

A shorted motor winding occurs when motor winding conductors come in contact with each other or the grounded metal frame, often at a point where the insulation is worn or otherwise damaged. When a shorted motor winding occurs, the resistance of the winding is reduced, which results in an increase in the amount of current flow through the shorted winding. Depending on the overall impedance of the shorted winding to ground, either the overload device will open (higher impedance) or the OCPD will open (lower impedance).

The winding can develop a short that barely affects motor operation or can develop a short that prevents the motor from operating altogether. If a motor winding develops a high amperage short circuit at a minimum level of five to seven times the full-load current of the motor, the motor will not start. The cause for the short circuit should be identified and repaired before any attempt is made to restart the motor.

If a motor with a partial short starts, it may not overheat, and may also operate without activating or tripping any motor circuit protective devices. In this case, the end user of the motor will likely not call for service, as there is no indication that a problem exists. However, when such is the case, there will be a reduction in the motor's operating effectiveness and efficiency, increasing the motor's operating costs and likely reducing the motor's useful life.

Shorted Motor Winding on a Single-Phase Motor

On a single-phase motor, there are two windings: a high-resistance start winding and a low-resistance run winding. A shorted winding can occur in either of these two windings.

If a partial short circuit occurs in the run winding, the motor could possibly continue to run; however, the motor's running amperage will increase, and its operating effectiveness and efficiency will be reduced. If this overload continues and the motor's amperage continues to increase, the motor overload devices should trip at levels varying from 115% to 170% of full-load current depending on motor type and size. By partially shorting out the run winding, the difference between the resistance of the run winding and that of the start winding will increase, resulting in a larger difference in the strength of the magnetic fields generated in the two windings. Unless the energy consumption of the motor is monitored, the motor's operator will very likely not call for service. If the winding short is more extreme, the motor's starting torque will increase, and the motor is more likely to overheat.

If the short circuit occurs in the start winding, the motor may or may not start. The reason for this is a reduction in the magnetic field imbalance that is required for the motor to start. If the start winding is partially shorted, the start winding resistance will be closer to the resistance of the run winding. This will reduce the magnetic field imbalance that is needed to start the motor.

Shedding Light

If a motor has a completely shorted winding (low impedance) where the winding resistance is 0 Ω or very close to 0 Ω, the motor will not start and will immediately trip the circuit breaker or fuse. If the motor trips the OCPD, the cause of the short circuit must be identified and resolved before attempting to restore power to the motor.

Shorted Motor Winding on a Three-Wire, Delta-Connected, Three-Phase Motor

Resistance readings in a properly operating three-wire, delta-connected, three-phase motor are described earlier in this chapter. If a three-wire, delta-connected, three-phase motor experiences a shorted winding, **Figure 33-14**, the following resistance readings between the motor terminals may be observed:

- L1 to L2: 0 Ω
- L1 to L3: 5 Ω
- L2 to L3: 5 Ω

When this motor experiences a completely shorted winding, there will be one 0 Ω reading, while the remaining two readings will be one-half of the actual winding resistance. For the L1-L2 combination, the total resistance is calculated as a 0 Ω load connected in parallel to a 20 Ω load, resulting in a total resistance of 0 Ω. For the L1-L3 and L2-L3 combinations, the total resistance for each is calculated as two 10 Ω loads connected in parallel with each other, for a total resistance of 5 Ω.

Shorted Motor Winding on a Three-Wire, Wye-Connected, Three-Phase Motor

Compare the resistance readings of a three-wire, wye-connected, three-phase motor highlighted in **Figure 33-4**, which shows a properly operating motor, to a similar three-phase motor that experiences a shorted winding, **Figure 33-15**. In this scenario, a shorted motor winding will result in two resistance readings that are one-half of the normal/expected readings, and one reading that is equal to the normal/expected winding resistance reading.

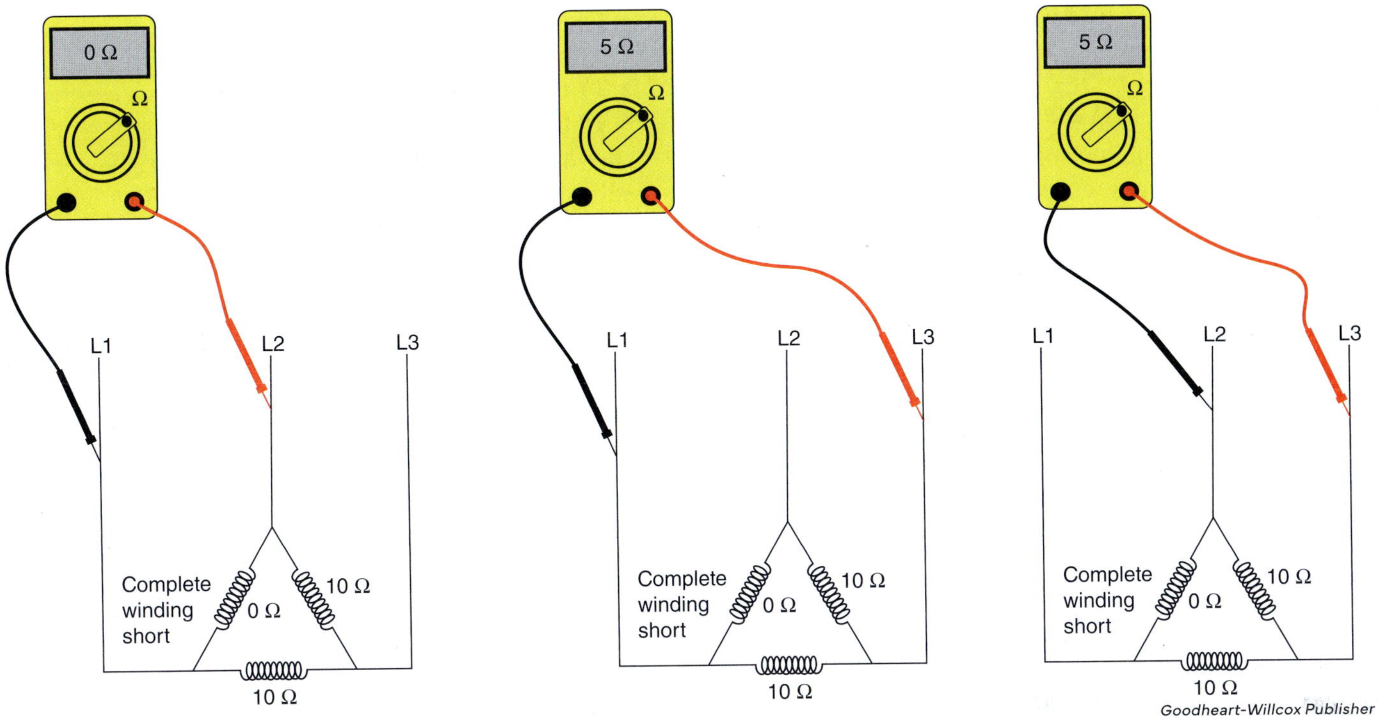

Figure 33-14. A three-wire, delta-configured, three-phase motor with a shorted winding. The resistance reading taken between motor leads L1 and L2 will be 0 Ω, while the remaining two readings will be one-half of the actual winding resistance.

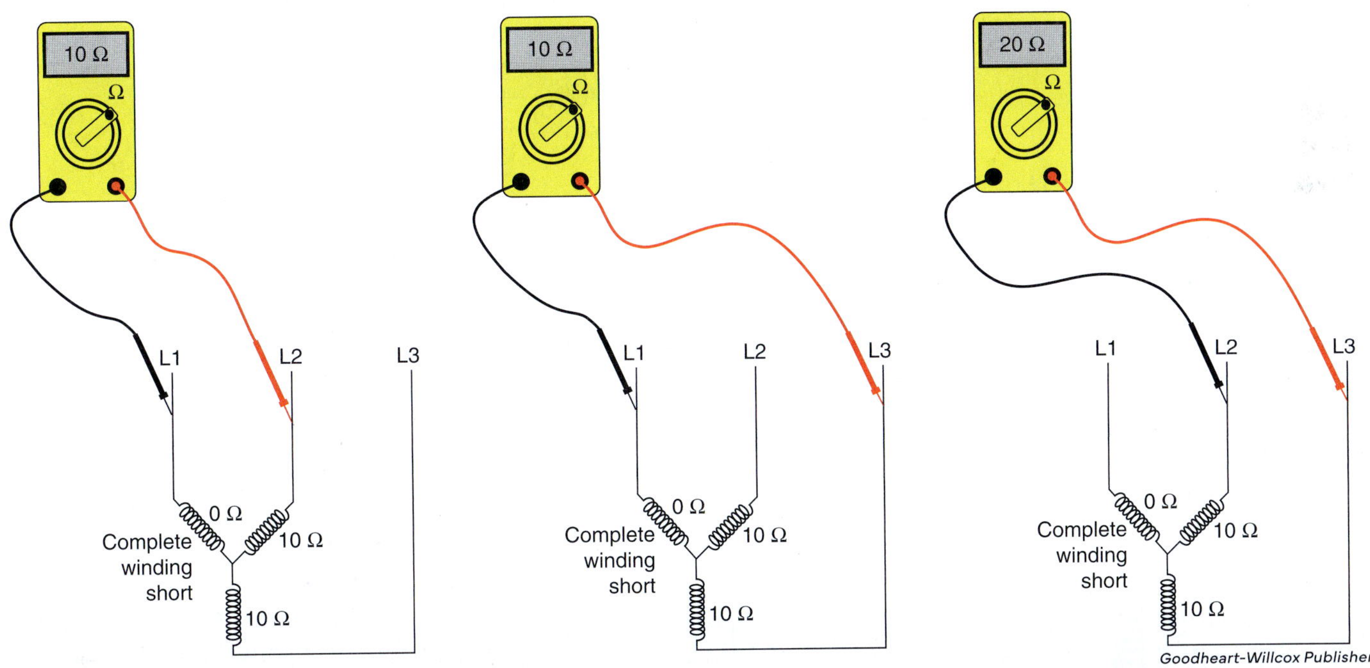

Figure 33-15. A three-wire, wye-configured, three-phase motor with a shorted winding connected to motor lead L1. The resistance readings that are taken between L1 and any other lead will produce a resistance reading of 10 Ω, while the resistance between L2 and L3 will be normal.

Shorted Motor Winding on a Six-Wire, Star-Delta-Connected, Three-Phase Motor

If a six-wire, star-delta-connected, three-phase motor experiences a shorted winding, the following resistance readings between the motor lead combinations may be observed:

- T1 to T4: 0 Ω
- T2 to T5: 10 Ω
- T3 to T6: 10 Ω

For a six-wire, star-delta-connected, three-phase motor, a shorted motor winding will result in two normal/expected resistance readings, and one 0 Ω reading, **Figure 33-16**.

Shorted Motor Winding on a Nine-Wire, Dual-Voltage, Three-Phase Motor

Recall from earlier in this chapter the readings that will be measured on a properly operating nine-wire, dual-voltage, three-phase motor. If a three-phase motor with this configuration experiences a short in one of the isolated windings, the following resistance readings between the motor terminals may be observed:

- T1 to T4: 0 Ω
- T2 to T5: 10 Ω
- T3 to T6: 10 Ω
- T7 to T8: 20 Ω
- T7 to T9: 20 Ω
- T8 to T9: 20 Ω

In this case, the only resistance reading that changes is the reading across the T1 and T4 motor lead combination, which is comprised of the two motor leads that are connected to the damaged winding segment, **Figure 33-17**.

If a nine-wire, dual-voltage, three-phase motor experiences a short in one of the windings that comprises the wye-connected portion of the motor, the following resistance readings between the motor lead combinations may be observed:

- T1 to T4: 10 Ω
- T2 to T5: 10 Ω
- T3 to T6: 10 Ω
- T7 to T8: 10 Ω
- T7 to T9: 10 Ω
- T8 to T9: 20 Ω

In this case, two resistance readings will change. With the shorted winding connected to motor lead T7, the two lead combinations that contain motor lead T7 (namely T7 and T8 and T7 and T9) will have readings of 10 Ω, while

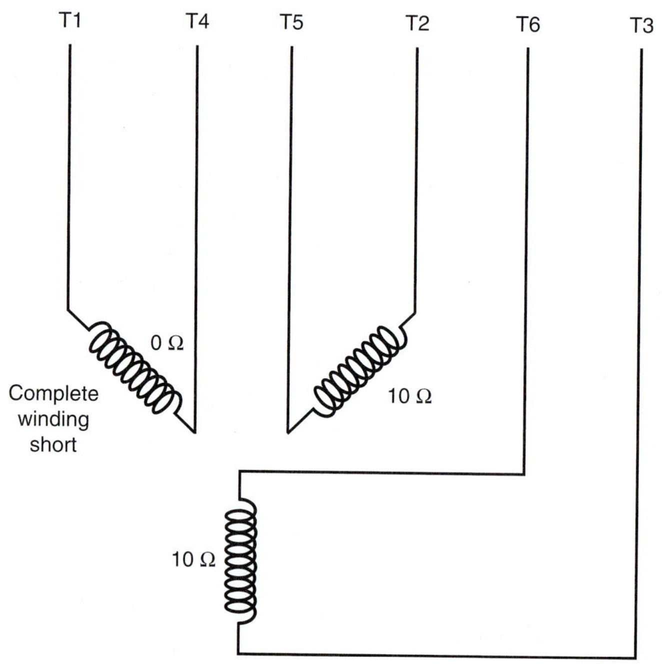

Goodheart-Willcox Publisher

Figure 33-16. A six-wire, star-delta-configured, three-phase motor with a shorted winding between motor terminals T1 and T4. The resistance across this shorted winding will be 0 Ω, while the other two readings will be normal.

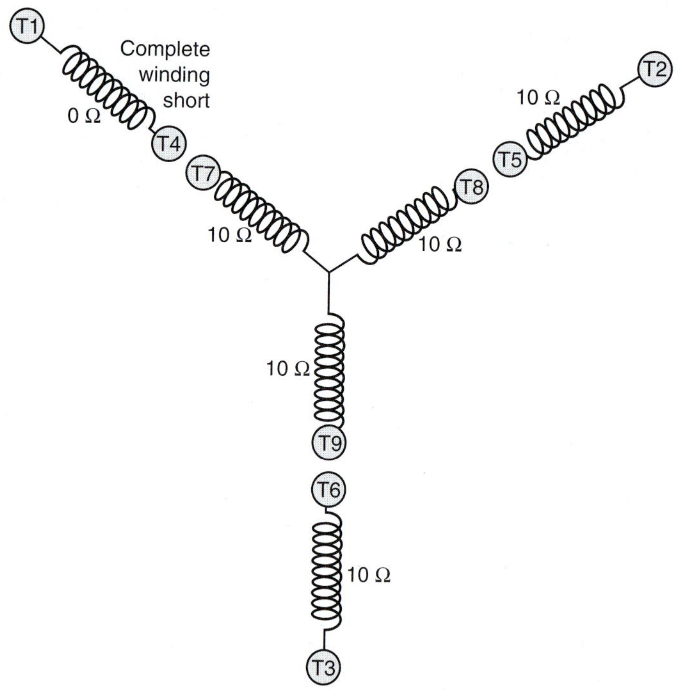

Goodheart-Willcox Publisher

Figure 33-17. A nine-wire, dual-voltage, three-phase motor, with a shorted winding between motor terminals T1 and T4. The resistance across this shorted winding will be 0 Ω, while the readings across the other isolated winding segments will be normal. The readings across the winding terminals/leads in the wye connection will also be normal.

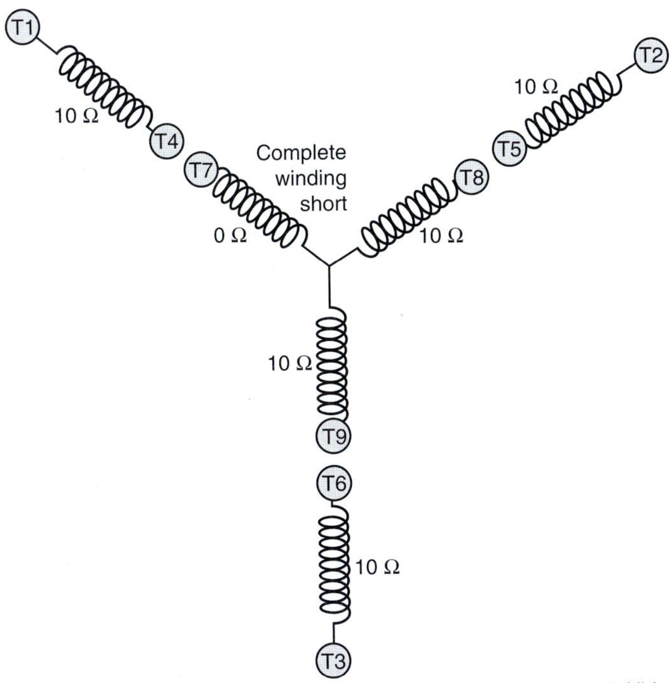

Goodheart-Willcox Publisher

Figure 33-18. A nine-wire, dual-voltage, three-phase motor with a shorted winding connected to motor terminal T7. The resistance readings across the winding terminals T7 and T8 and T7 and T9 will be 10 Ω, while the resistance reading between T8 and T9 in the wye connection will be normal. The resistance readings across all three isolated winding segments will be normal.

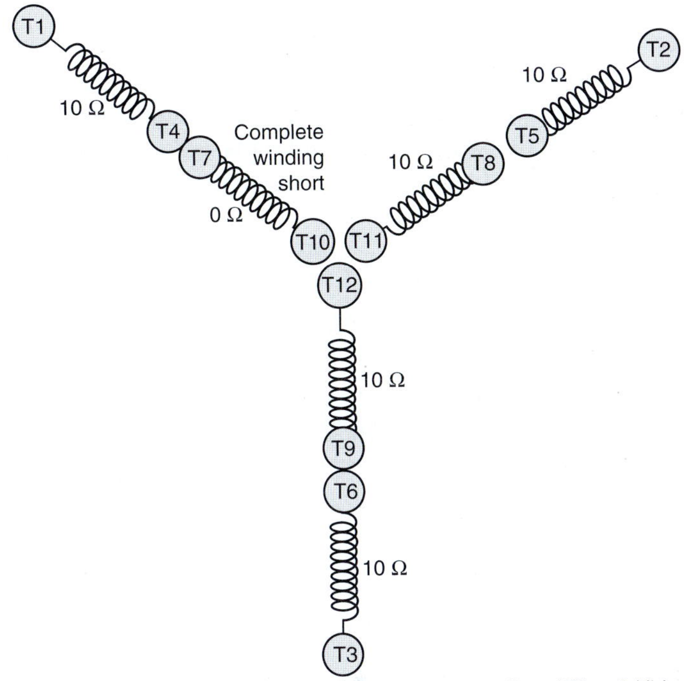

Goodheart-Willcox Publisher

Figure 33-19. A 12-wire, dual voltage, three-phase motor with a shorted winding on one of the isolated winding segments. The resistance between terminals T7 and T10 will be 0 Ω, while the resistance readings between the other motor terminal pair combinations (T1-T4, T2-T5, T3-T6, T8-T11, and T9-T12) will be normal.

all other motor lead combinations will provide measurable, nonzero or noninfinite values of the value expected, **Figure 33-18.**

Shorted Motor Winding on a 12-Wire, Dual-Voltage Three-Phase Motor

If a 12-wire, dual-voltage, three-phase motor experiences a short in one of the isolated windings, the following resistance readings between the motor leads may be observed:

- T1 to T4: 10 Ω
- T2 to T5: 10 Ω
- T3 to T6: 10 Ω
- T7 to T10: 0 Ω
- T8 to T11: 10 Ω
- T9 to T12: 10 Ω

In this scenario, the only resistance reading that changes is the reading across the motor leads T7 and T10, which is comprised of the two motor leads that are connected to the damaged winding segment, **Figure 33-19.**

33.5 Electrical Short Circuit to Ground

If an electric motor has a low impedance electrical short to ground, the fuse or circuit breaker will likely trip on a call for motor operation. However, this is not always the case. For example, a motor with a grounded winding may function normally from a mechanical standpoint if the amount of current flowing to ground is minimal. However, this shorted condition might cause an electric shock if the operator of the motor comes in physical contact with the motor. Another example to consider is a motor that might draw excessive current and overheat, resulting in further damage to the motor. Although a grounded motor winding might not be evident all the time, its presence can prove to be troublesome. It is important to check for a correctly installed equipment grounding conductor when performing preventive maintenance on equipment.

A short circuit from winding to ground or to the frame of a motor can often be detected with a high-quality digital multimeter set to read ohms. There should be no detectable current path between a motor winding and ground. Megohmmeters, called meggers, are often used for this purpose. ***Megohmmeters*** are large ohmmeters that are used to detect weak motor winding insulation and very high resistances, or small current paths, to ground on larger motors. Megohmmeters can detect resistance values in the range of tens of millions of ohms. Megohmmeters have internal, high-voltage dc batteries that enable the instrument to detect even the slightest grounding condition. The process of using a megohmmeter to check a motor's windings is commonly referred to as "megging a motor" or simply "megging".

Shedding Light

In the case of a 120 V motor with an electrical short to ground, the motor may not trip the circuit breaker if the ground is on the neutral, grounded conductor side of the circuit. If such is the case, the owner operator will likely not report a problem, as the motor will still operate. The return current will be split in proportion to the resistance of the return paths. The current on the ground return path could pose a danger to those coming in contact with the motor frame and the conduit system.

In operation, one probe of the megohmmeter is connected to one of the motor winding terminals, and the other probe is connected to ground. When a button on the megohmmeter is pushed and held, a high dc voltage is provided between its probes. This high voltage signal seeks out and measures all electrical paths to ground. If a motor winding is good, the resistance reading from the winding to ground should be at least 100,000,000 Ω (100 MΩ or 100 megohms).

Safety Note

It is important to disconnect all wires from the motor terminals before megging a motor. Also, make certain that you read the manufacturer's instructions for important information on how long to energize the megger when checking a motor's windings. It is also important that the technician does not touch the motor or any other bare metal during the test. The electrician needs to wear voltage rated rubber gloves with the appropriate leather gloves covering the rubber.

33.6 Checking Capacitors on Single-Phase Motors

If a single-phase motor fails to start, the starting components should be checked before concluding that the motor is defective and needs to be replaced. Using a digital multimeter, capacitors can be tested either under load with the motor running, or the capacitor can be removed from the circuit for testing. Regardless of the testing method used, safety should always be a top priority.

Capacitors must be discharged prior to being handled or tested. A capacitor's stored charge can damage a test instrument or cause severe injury to a technician. Capacitors should be discharged by shorting one terminal to the other using a 20,000 Ω, 5 W resistor. It is recommended to avoid using a screwdriver or other instrument to discharge a capacitor as this will cause the capacitor to discharge quickly, producing a large, potentially dangerous spark, in addition to possibly damaging the capacitor itself. Testing a capacitor that has been removed from the motor circuit and discharged can be accomplished by using the capacitor testing feature found on many digital multimeters. To use this feature, simply touch the meter's test leads to the capacitor's terminals. The meter will display the capacitance of the capacitor in microfarads. The displayed reading should be within 10% of the capacitor's nameplate rating. If a capacitor's measured capacitance differs from its nameplate rating, the capacitor should be replaced.

A fully discharged capacitor can also be tested using an ohmmeter. This test, however, will simply tell you if the capacitor is charging and discharging, not the actual capacitance value. Start by connecting the ohmmeter's test leads across the capacitor and monitoring the meter's display. The meter should display a series of changing numbers and then display OL, indicating an open lead or open circuit. If the meter initially displays OL, switch the positions of the meter's leads. If the meter displays a continuous OL regardless of the positions of the test leads, the capacitor is likely defective and in need of replacement. Similarly, if the meter's displays a consistent resistance value, regardless of the positions of the meter's test leads, the capacitor is likely defective and in need of replacement.

Summary

- Problems with electric motors can be classified as either mechanical or electrical. (33.1)
- It is necessary to ensure that the proper voltage is being supplied to the motor before concluding that the motor has failed. (33.1)
- Loose electrical connections at the motor can increase the circuit resistance and reduce the amount of voltage being supplied to the motor. (33.1)
- If the motor fails to turn, de-energize the motor and attempt to rotate the motor's shaft manually to see if the shaft is free. (33.1)
- Disconnecting the motor from its driven components will help determine if the problem lies within the motor itself, or if the problem is with the driven components. (33.1)
- Excessive belt tension can damage a motor's bearings and prevent the motor from operating. (33.2)
- Common mechanical problems include bearing failure and seized or locked driven or controlled devices. (33.2)
- An open motor winding can be easily detected by taking a resistance reading across any known terminal or lead combination to which a motor winding is connected. Resistance readings across an open winding will, in some cases, be infinite. (33.3)
- A shorted motor winding occurs when a motor winding conductor comes in contact with another conductor at a point where the insulation is worn or otherwise damaged. (33.4)
- A shorted motor may or may not start or may overheat, and the motor and circuit protection device may or may not activate. This may result in undetected motor problems. (33.4)
- A grounded motor winding might not be evident all the time. It is important to check for motor grounds when performing preventive maintenance on equipment. (33.5)
- A short circuit from winding to ground or to the grounded frame of a motor can often be detected with a megohmmeter. (33.5)
- If a single-phase motor fails to start, the starting components should be checked before concluding that the motor is defective and needs to be replaced. (33.6)
- Capacitors can be tested under load while installed on an operating motor or with a capacitor tester after the capacitor has been properly and safely discharged. (33.6)

Review Questions

Use the information in the chapter to answer the following questions.

Measuring Your Potential

1. *True or False?* A loose electrical connection on a motor will typically result in a reduction of voltage supplied to the motor. (33.1)
2. *True or False?* If the supply voltage and the motor's rated voltage differ by more than 10%, immediately replace the motor. (33.1)
3. If a sleeve bearing fails on a small, fractional horsepower motor, the most economical solution to resolve the problem is to _____. (33.2)
   A. replace the motor
   B. replace only the defective bearing
   C. replace both bearings
   D. loosen the belt tension
4. If a motor's drive belt is too tight, the _____. (33.2)
   A. motor will operate more efficiently
   B. motor bearings will likely fail prematurely
   C. motor's internal overload will likely fail
   D. drive belt will slip excessively
5. If a three-winding, three-phase motor is in good working order, which of the following statements is *true*? (33.3)
   A. All three resistance readings should be very nearly the same.
   B. The sum of two resistance readings should equal the third.
   C. One of the three resistance readings should be twice that of the other two readings.
   D. One of the three resistance readings should be half that of the other two readings.
6. *True or False?* A decrease in motor winding resistance will cause the current flow through the motor to decrease. (33.4)
7. A short circuit in a single-phase ac electric motor is characterized by a(n) _____. (33.4)
   A. infinite resistance reading through the start winding
   B. infinite resistance reading through the run winding
   C. excessively low resistance reading through one or both motor windings
   D. measurable resistance reading between any motor terminal and ground

8. *True or False?* The megohmmeter is used to measure resistances as high as millions of ohms. (33.5)
9. *True or False?* A 20,000 Ω, 5-W resistor is recommended for discharging a capacitor. (33.6)
10. Which of the following statements is *true* regarding a single-phase motor that operates with 40 µf and has a capacitance of 43.03 µf? (33.6)
    A. The actual capacitance varies by more than 10% when compared to its nameplate capacitance, so it is likely good and can remain in service.
    B. The actual capacitance varies by more than 10% when compared to its nameplate capacitance, so it should be replaced.
    C. The actual capacitance varies by less than 10% when compared to its nameplate capacitance, so it is likely good and can remain in service.
    D. The actual capacitance varies by less than 10% when compared to its nameplate capacitance, so it should be replaced.

Plug-In

1. A three-wire, wye-configured, three-phase motor has the following resistance readings:
    - L1 to L2: 12 Ω
    - L1 to L3: 12 Ω
    - L2 to L3: 24 Ω

    Which of the following is most likely correct? (33.4)
    A. The windings are likely in good condition.
    B. One winding segment has an open circuit.
    C. One winding segment has a short circuit.
    D. One winding segment is grounded.
2. A three-wire, delta-configured, three-phase motor has the following resistance readings:
    - L1 to L2: 15 Ω
    - L1 to L3: 15 Ω
    - L2 to L3: 30 Ω

    Which of the following is most likely correct? (33.3)
    A. The windings are likely in good condition.
    B. One winding segment has an open circuit.
    C. One winding segment has a short circuit.
    D. One winding segment is grounded.
3. A nine-wire, dual-voltage, three-phase motor has the following resistance readings:
    - T1 to T4: 8 Ω
    - T2 to T5: 8 Ω
    - T3 to T6: 8 Ω
    - T7 to T8: 16 Ω
    - T7 to T9: OL (infinite resistance)
    - T8 to T9: OL (infinite resistance)

    Which of the following is most likely correct? (33.3)
    A. One isolated winding segment has an open circuit.
    B. Two isolated winding segments have open circuits.
    C. One of the wye-configured winding segments has an open circuit.
    D. Two of the wye-configured winding segments have open circuits.
4. A 12-wire, dual-voltage, three-phase motor has the following resistance readings:
    - T1 to T4: 0 Ω
    - T2 to T5: 8 Ω
    - T3 to T6: 8 Ω
    - T7 to T8: 8 Ω
    - T7 to T9: 8 Ω
    - T8 to T9: 8 Ω

    Which of the following is most likely correct? (33.4)
    A. The windings are likely in good condition.
    B. One isolated winding segment has an open circuit.
    C. One isolated winding segment has a short circuit.
    D. One isolated winding segment is grounded.

Making the Connection

1. Explain why there are no infinite resistance readings on a delta-connected, three-wire, three-phase motor that has one open winding. (33.3)
2. Explain why there are two infinite resistance readings on a wye-connected, three-phase motor that has only one open winding. (33.3)

SECTION

10 Generators

Mares Lucian/Shutterstock.com

CHAPTER

34 Generator Fundamentals

Yosanon Y/Shutterstock.com

SPARKING DISCUSSION

An electrician claims that the efficiency of a generator can be greater than 100%. Do you agree or disagree with this electrician's statement? Provide as much supporting evidence as possible to support your response.

LEARNING OUTCOMES

After completing this chapter, you will be able to:

34.1 List the key components of a generator.
34.2 Describe the basic operation of a generator.
34.3 Describe the left-hand rule for generators.
34.4 Explain the difference between simplex windings and multiplex windings in armature coil configurations.
34.5 Explain how the rotating armatures affect the amount of reluctance torque.
34.6 Describe methods used to measure generator start-up torque.
34.7 Calculate the efficiency of a generator.

TECHNICAL TERMS

armature
armature stack
cylindrical armature
duplex winding
generator
multiplex winding
reluctance torque
salient-pole armature
simplex winding
stack slot
stack teeth
yoke

Introduction

In earlier chapters we discuss various types of electric motors, how they operate, and the theories on which their effective and efficient operation rely. Regardless of the type or style of motor, they all accomplish the same basic task: they convert electrical energy into mechanical energy. Electric generators are different. An electric generator, in its most basic form, performs a function that is opposite to that of an electric motor. While a motor converts electrical energy into mechanical energy, a ***generator*** converts mechanical energy into electrical energy. The electrical energy produced by a generator can be in the form of either ac or dc, depending on the construction and configuration of the generator.

The mechanical energy required for generator operation can come from any source of physical motion that is harnessed to provide the rotating torque necessary to turn the generator's rotor. For example, the required mechanical force may come from a renewable or sustainable source, such as blowing wind or flowing water.

In this chapter, we examine the construction and different types of generators, along with various winding configurations and wiring diagrams. In the two chapters that follow, we discuss dc and ac generators.

34.1 Generator Construction

The basic construction of a generator is similar to that of the electric motors discussed earlier in this text. Both motors and generators are comprised of similar components but perform opposite functions. Remember that motors convert electrical energy into mechanical energy, while generators convert mechanical energy into electrical energy.

Shaft

The shaft of a generator is designed to carry the driving torque that will be converted into electrical energy. The shaft acts as the physical link between the external mechanical force and the rotating coils of wire positioned within the generator's stationary magnetic field. As a result, the power produced by the generator is directly proportional to the speed at which the shaft rotates and the torque applied to it. As the torque applied to the generator increases or as the rotational speed of the generator's shaft increases, the output power of the generator increases as well.

Armature

The ***armature*** is the rotating part of the generator that does the work of converting the mechanical energy, in the form of torque from the shaft, into electrical energy. The armature supports the generator's windings, facilitates the connection of the windings and the generator's output, connects the generator and the external mechanical rotating force, and provides a means to rotate the electrical windings within the generator's magnetic field. The armature consists of a core, which is made up of a series of laminated iron plates, **Figure 34-1**. The core is commonly referred to as an ***armature stack***. These laminated plates are connected directly to the generator's rotating shaft. The core is sized based on the rated load of the generator. Larger generators will have larger cores, while smaller generators will have smaller cores.

The armature core is constructed with channels, called ***stack slots***, which are aligned perpendicular to the laminations, **Figure 34-2**. At the outer portion of the core between the stack

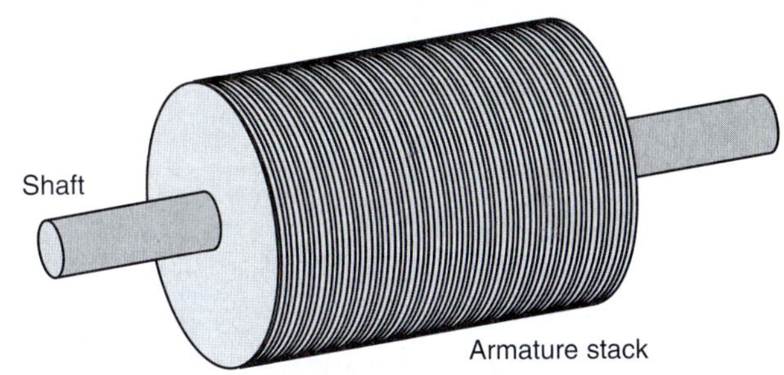

Goodheart-Willcox Publisher

Figure 34-1. The armature stack is made up of multiple laminated iron plates secured to the generator's shaft.

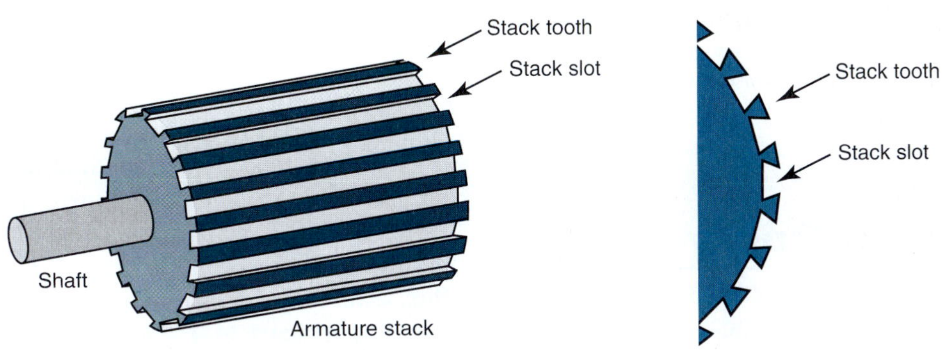

Goodheart-Willcox Publisher

Figure 34-2. The winding's coils are positioned in stack slots, which are channels cut into the armature stack, perpendicular to the laminations. Once the windings are set in place, they sit flush with the outer edge of the armature's stack teeth.

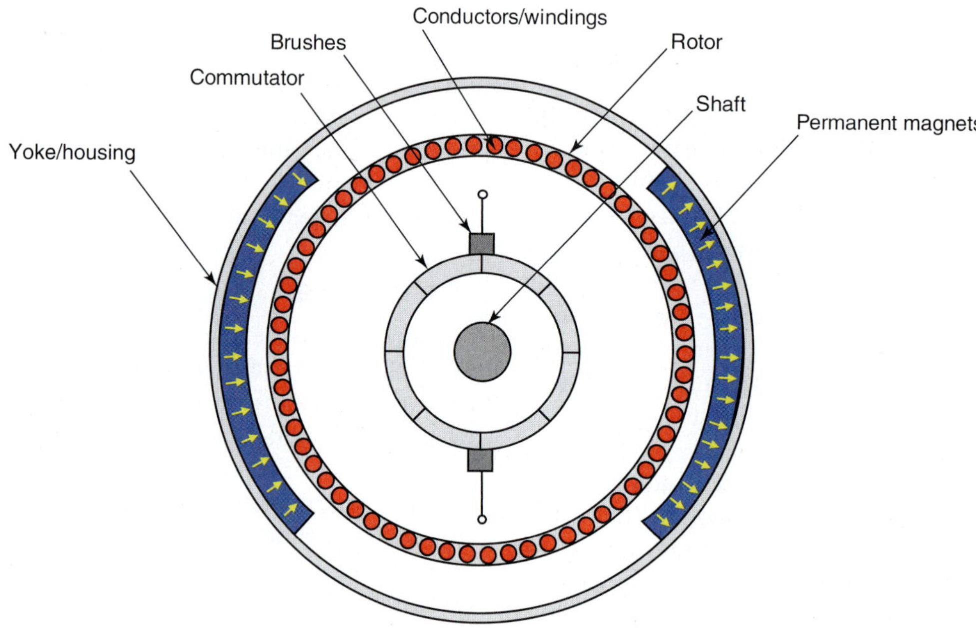

Goodheart-Willcox Publisher

Figure 34-3. Cross-sectional view of a generator. The yoke, or housing, along with the relative positions of the rotor, shaft, windings, permanent magnets, commutator, and brushes can be seen.

slots are ***stack teeth***, which serve to separate the windings. The stack slots allow for the conductors that make up the armature coils to sit flush with the surface of the core. The ends of each individual armature coil are connected to commutator segments, and coils are connected to each other to create windings. Depending on the electrical requirements of the generator, the windings can be arranged in a number of different configurations. These configurations are discussed later in the chapter.

Yoke

The ***yoke***, also known as housing, is the external structure of the generator and is normally made from cast iron or rolled steel. Cast iron is commonly found on smaller generators, while rolled steel is often used on larger machines. The yoke serves two main purposes: it provides mechanical support for the stationary field magnets or poles, and it acts to protect the windings and other internal components of the generator from physical damage. The yoke, being made of a conductive, metallic material, provides an effective means to carry the magnetic flux through the generator's poles. The magnets are arranged on the yoke to create north and south poles, **Figure 34-3**.

34.2 Generator Operation

All generators operate on the principle known as Faraday's law. Faraday's law states that when a conductor passes through a magnetic field, an electromotive force (emf) is created, which induces current flow through the conductor. The strength of the generated emf is proportional to the relative speed of the coil with respect to the magnetic field and the strength of the magnetic field. The strength of the induced emf is also proportional to the number of turns in the armature coil.

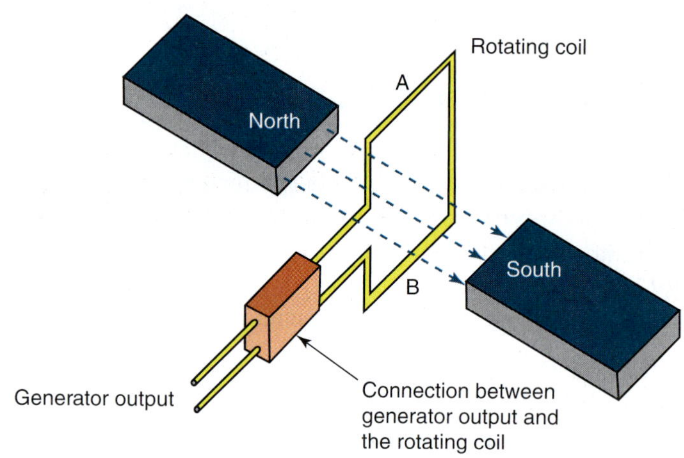

Goodheart-Willcox Publisher

Figure 34-4. A coil is positioned vertically within a magnetic field perpendicular to the direction of magnetic flux. At this point, the *A* side of the coil is positioned at the top, while the *B* side is positioned at the bottom. The rotating coil is physically connected to the generator's outlet terminals.

The operation of a simple generator is made clear by building on a basic understanding of current flow, induced emf, and flux, as discussed earlier in the text. Consider the illustration provided in **Figure 34-4**. This figure shows a single conductive loop or coil positioned vertically within a magnetic field. The coil represents the armature, which rotates within the magnetic field. The top portion of the coil is identified as the *A* side of

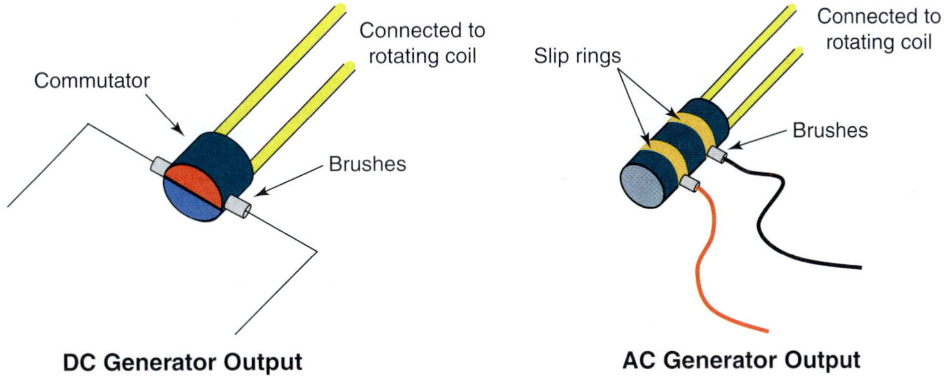

Figure 34-5. On dc generators, brushes and commutators are used to connect the rotating coil to the generator's outlet terminals. On ac generators, brushes and slip rings are used to connect the rotating coil to the generator's outlet terminals.

the coil, and the bottom portion of the coil is identified as the *B* side. In this figure, the coil is positioned perpendicular to the direction of the magnetic flux. The electrical output of the generator is measured across the two terminals that connect, in some fashion, to the rotating coil. Depending on the type of generator, the terminals at the outlet of the generator are connected to the rotating coil by either a commutator/brush assembly or by a slip ring/brush assembly, **Figure 34-5**. The commutator/brush assembly is first introduced in Chapter 27, *DC Motors*, and is revisited in Chapter 35, *DC Generators*. The slip ring/brush assembly is discussed in Chapter 36, *AC Generators*.

When the armature begins to rotate, it does so in a direction that is parallel to the direction of the magnetic flux, **Figure 34-6**. Since electromotive force is only generated when a conductor cuts or crosses lines of magnetic flux, no voltage is generated in the neutral plane. The term neutral plane describes the condition when a conductor's path and that of the lines of magnetic flux are parallel to each other.

As the armature continues to rotate, the top and bottom ends of the coil begin to cut across the lines of magnetic flux,

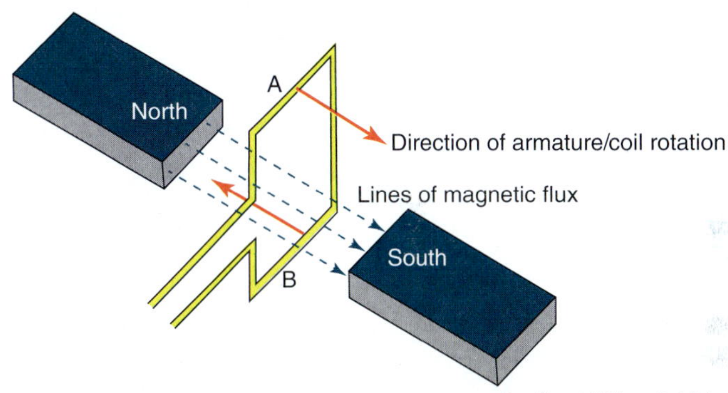

Figure 34-6. When the coil begins to rotate, the instantaneous direction of coil motion is parallel to the lines of magnetic flux. At this point, no voltage is generated, because emf is only generated when a conductor cuts or crosses lines of magnetic flux.

generating voltage. Since the angle at which the coil crosses the lines of magnetic flux is very small, the amount of voltage produced is small, **Figure 34-7**. As the coil continues to rotate, the angle at which the coil crosses the lines of magnetic flux

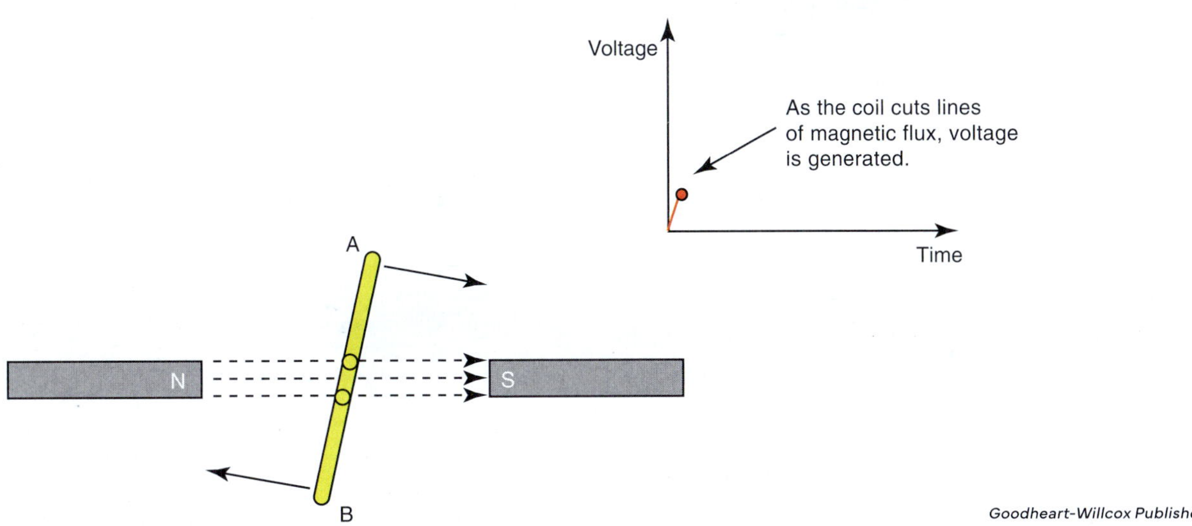

Figure 34-7. When the coil begins to rotate, the top and bottom ends of the coil begin to cut across the lines of magnetic flux, generating a small amount of voltage.

increases, and the amount of voltage generated increases. When the direction of coil motion is perpendicular to the lines of flux, the change in flux is at its highest value and, as a result, the generated voltage is at its maximum, **Figure 34-8**.

As the armature continues to rotate and moves back to a vertical position, the amount of voltage generated drops, **Figure 34-9**. This continues until the direction of coil motion is once again parallel to the direction of the lines of magnetic flux, **Figure 34-10**, and the generated voltage is once again zero. It is important to note that, even though the magnitude of the voltage produced changes, the polarity of the generated voltage remains the same.

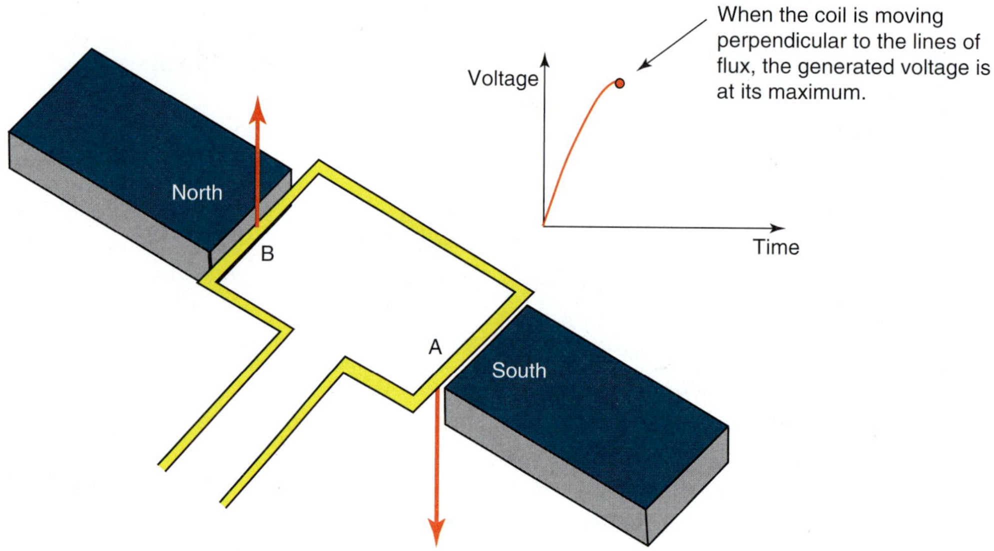

Goodheart-Willcox Publisher

Figure 34-8. When the angle at which the coil crosses the lines of magnetic flux increases, the amount of voltage generated also increases. When the coil motion is perpendicular to the lines of flux, this is when the maximum voltage is produced.

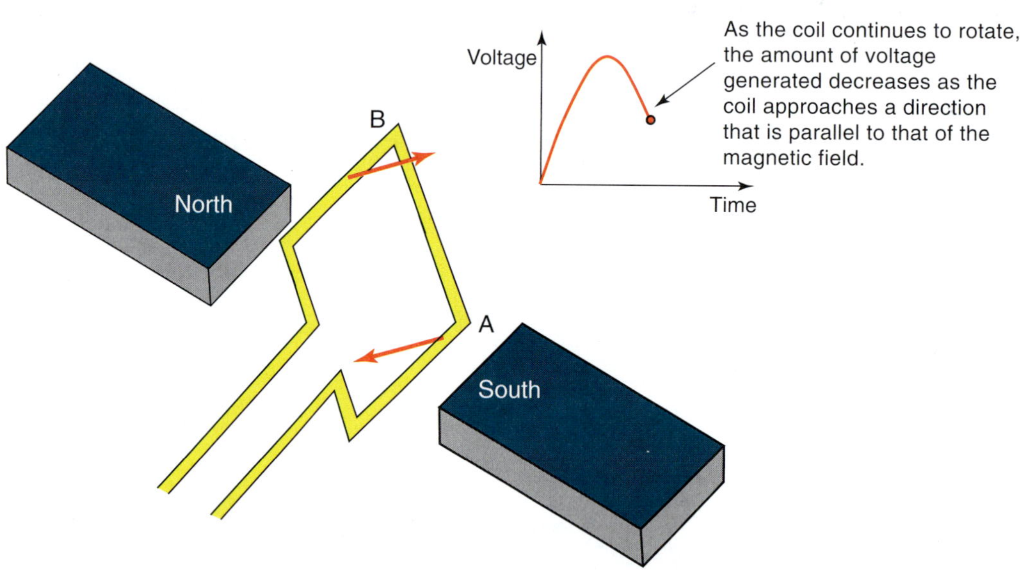

Goodheart-Willcox Publisher

Figure 34-9. As the coil continues to rotate, the amount of voltage generated decreases as the coil approaches parallel to the lines of flux.

Continued armature rotation generates additional output wave signals. For dc generators, the additional output signals will be identical to the first, **Figure 34-11**. For ac generators, the additional output signals will have alternating polarities, **Figure 34-12**. In actual generators, the armatures contain multiple coils and generate multiple output waveforms simultaneously, **Figure 34-13**.

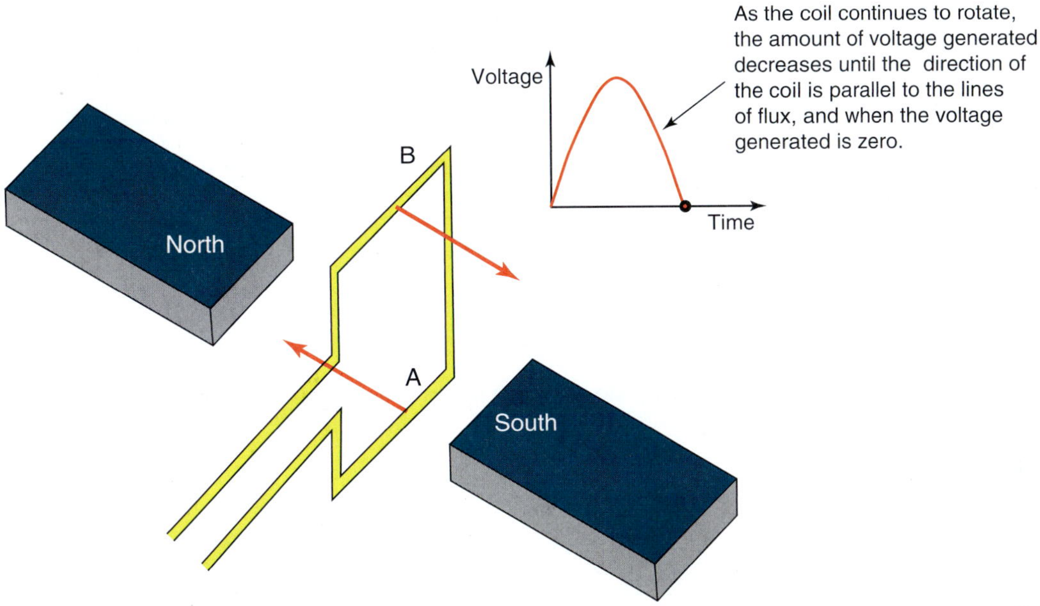

Goodheart-Willcox Publisher

Figure 34-10. When the direction of coil motion is parallel to the direction of the magnetic field, no lines of flux are crossed and the amount of voltage produced becomes zero.

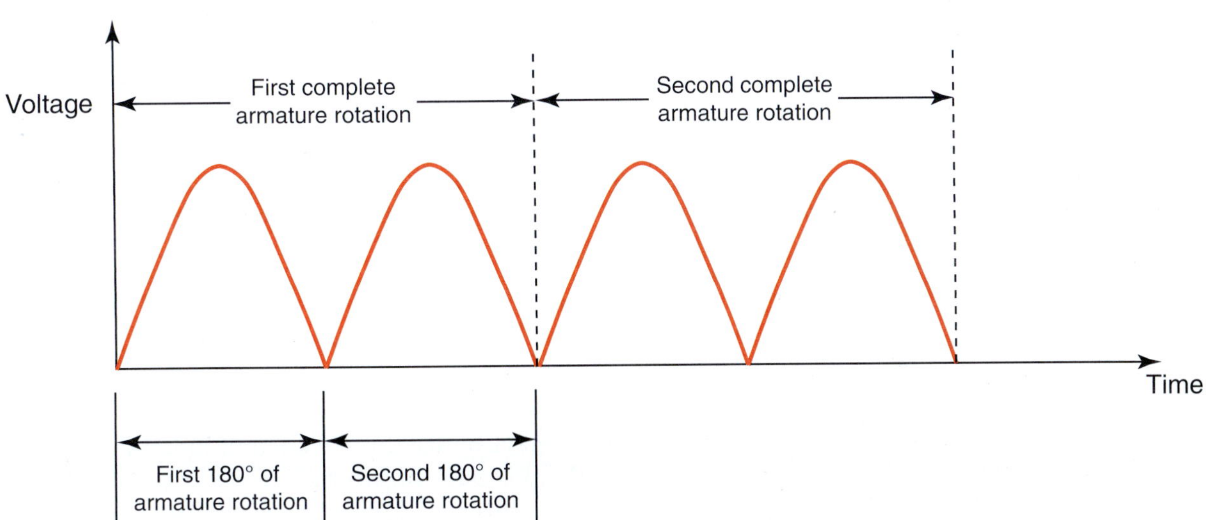

Goodheart-Willcox Publisher

Figure 34-11. As the armature continues to rotate, additional output wave signals are produced. For dc generators, the polarity of the output signal is consistent. As a result, one complete rotation of the armature creates two identical positive output waves.

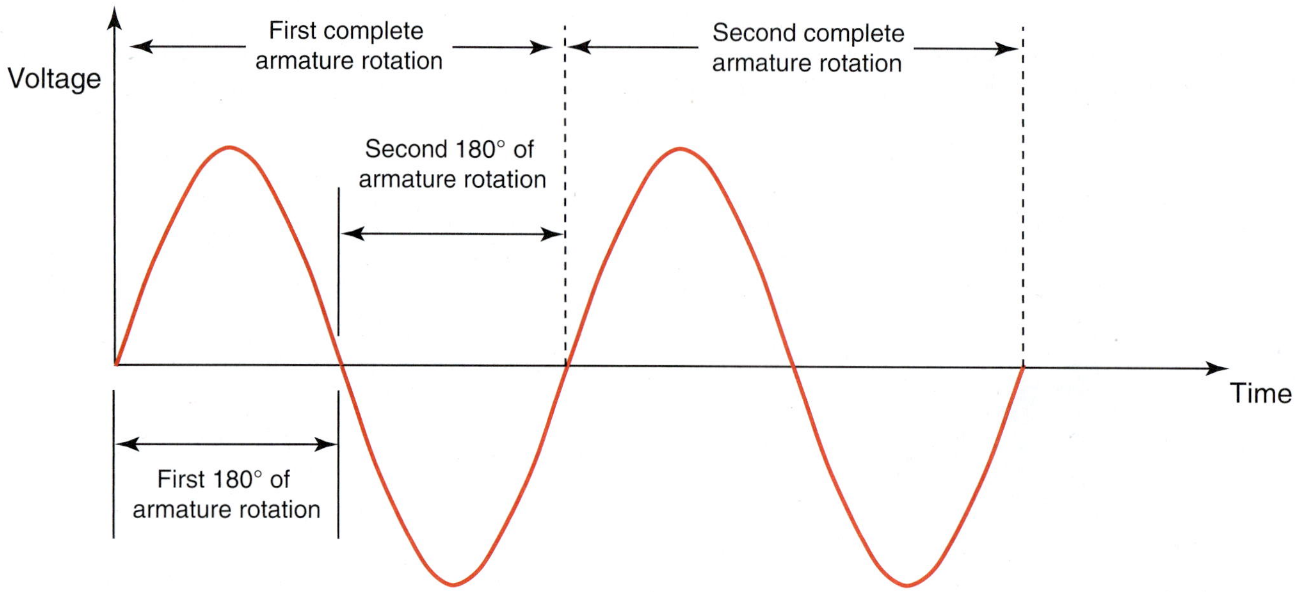

Goodheart-Willcox Publisher

Figure 34-12. For ac generators, the additional output signals have polarities changing from positive to negative repeatedly, as shown.

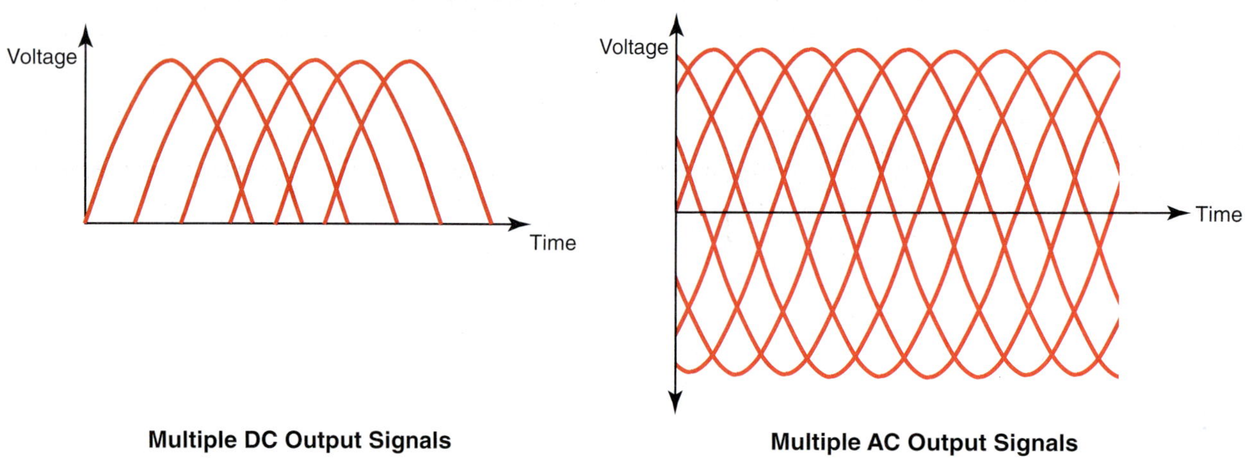

Goodheart-Willcox Publisher

Figure 34-13. In both ac and dc generators, multiple output waveforms are created simultaneously when an armature contains multiple coils.

34.3 Direction of Current Flow

The direction of the current flow produced in a dc generator is determined by the direction of the magnetic flux lines and the direction in which the conductor is moving within this field. In Chapter 27, *DC Motors*, the right-hand rule for motors is introduced, providing a visual representation of the direction of three forces: the direction of coil/winding movement, the direction of the magnetic field, and the direction of current flow. In our discussion of generators, Fleming's left-hand rule is used to help visualize the direction of induced current flow when a coil of wire is moved through a magnetic field.

To visualize the direction of induced current flow, the thumb, index finger, and middle finger of the left hand are used. By creating a 90° angle between your left thumb and index finger and extending your left middle finger to create a 90° angle with your index finger, the direction of forces can be seen. With your fingers held in this position, point your thumb in the direction of the force (the direction the coil or conductor is moving). Then, point your index finger in the direction of the magnetic field, from the north pole to the south pole. Your middle finger will then point in the direction of the induced current flow, **Figure 34-14**.

Notice how the left-hand rule for generators and the right-hand rule for motors are similar. The thumb, index finger, and middle finger represent the same quantities, but the direction of the current flow is opposite. This can be explained by Lenz's law, which states that the current induced in a coil is opposite in polarity of the magnetic flux that produced it. The right-hand rule for motors identifies the current direction when induction is creating force, as in the case of an electric motor. The left-hand rule for generators identifies current direction when force is creating current, as in the case of a generator.

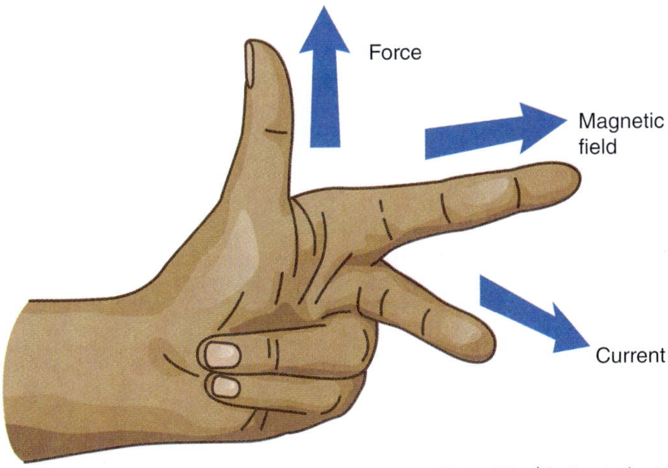

VectorMine/Shutterstock.com

Figure 34-14. In the left-hand rule for generators, the thumb points in the direction in which the conductor is moving (force), the index finger points in the direction of the magnetic field (north pole to south pole), and the middle finger points in the direction of induced current flow.

The left-hand rule for generators applies when the electron flow convention for direct current direction (negative to positive) is in use, as in this text. When the conventional flow convention (positive to negative) is used, the left-hand rule for generators is replaced by the right-hand rule for motors. When using any right-hand or left-hand rules to determine electrical behavior, be sure to apply the appropriate current direction convention.

34.4 Armature Coil Configuration

The windings that are ultimately positioned in the armature's stack slots are made up of a series of coils. These windings are commonly classified as either ***simplex windings*** or ***multiplex windings***. The term *simplex* is applied to a single-loop coil, while the term *multiplex* is applied to a coil with at least two loops. A coil with two loops is, by definition, a multiplex winding, but it is more commonly referred to as a ***duplex winding***. The term multiplex, therefore, is often applied to coils with three or more loops. A simplex winding has only one loop, which crosses each of a generator's field winding poles one time, **Figure 34-15**. A duplex winding has two loops and crosses each of a generator's poles twice.

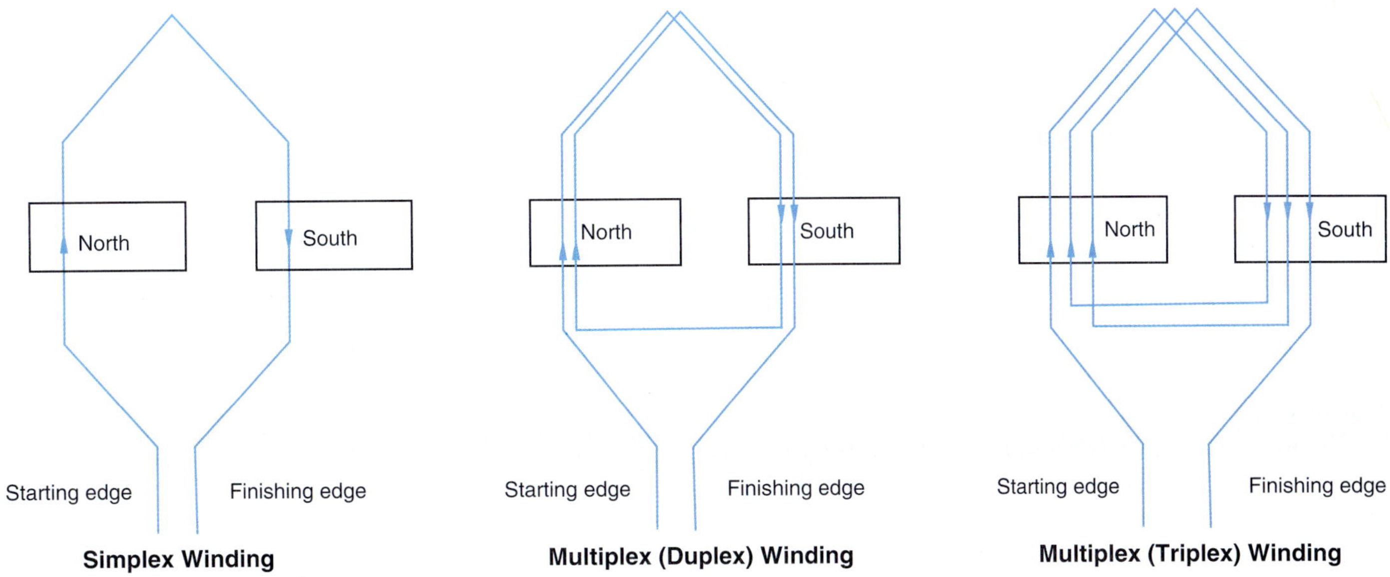

Goodheart-Willcox Publisher

Figure 34-15. In a simplex winding, the coil passes over each pole once. A duplex winding loops back onto itself to form two loops so that the coil makes two passes past each pole. A triplex winding has three loops, so each coil makes three passes past each pole.

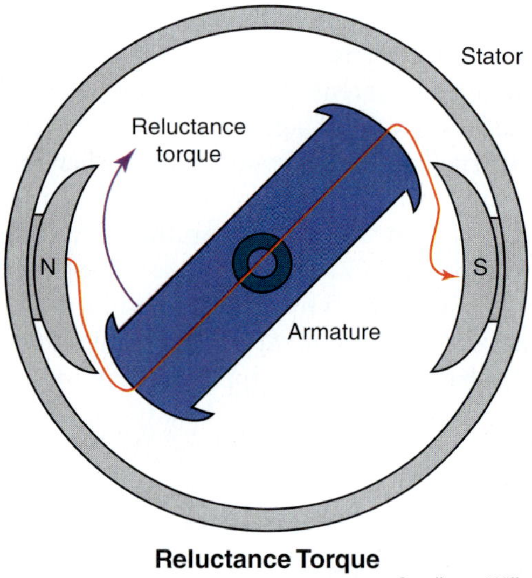

Goodheart-Willcox Publisher

Figure 34-16. Reluctance torque pushes to align a rotating object with the external magnetic field. This reluctance torque is a force that acts to resist the rotation of the armature.

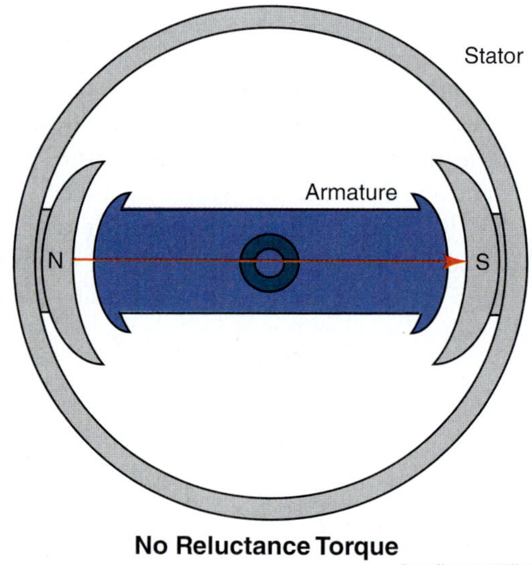

Goodheart-Willcox Publisher

Figure 34-17. When the rotating armature is aligned with the magnetic field, there is no reluctance torque.

34.5 Stator and Reluctance Torque

The stator in a generator provides the magnetic field through which the armature windings rotate to induce voltage. The magnetic field can be created with permanent magnets or electromagnets and always consists of an even number of poles. The field poles are mounted to the frame, or yoke, of the generator to reduce the ***reluctance torque*** that acts on the rotating armature in some applications. Reluctance torque, also referred to as *alignment torque*, is the equivalent of resistance in a conventional electric circuit.

Reluctance torque acts on a magnetic object, such as a generator's armature, when the magnetic object is positioned within an external magnetic field. This force can act to resist rotation of the armature, as it attempts to align the rotor with the magnetic field that is present between the stator's poles, **Figure 34-16**. The stator's external magnetic field induces an internal magnetic field in the armature, creating this opposition force or torque. When the rotating object is in alignment with the external magnetic field, the reluctance torque is zero, **Figure 34-17**.

The amount of reluctance torque depends on the position of the armature within the external magnetic field and the configuration of the rotating armature. Two common types of rotating armatures are the ***salient-pole armature*** and the ***cylindrical armature***. The salient-pole armature consists of multiple poles that extend from a central magnetic wheel, **Figure 34-18**. The generator's coils are wound around the poles. Salient-pole armatures encounter reluctance torque and are typically used in slower-speed applications where the shaft rotates at speeds up to 1500 rpm. As the name implies, cylindrical armatures or nonsalient-pole armatures are cylindrical in shape. They have parallel slots to accept windings. The armature illustrated earlier in **Figure 34-1** is a cylindrical armature. Cylindrical armatures operate without reluctance torque. Cylindrical rotors are commonly used in higher speed generators that rotate at speeds between 1500 rpm and 3000 rpm.

Shedding Light

Some motors, referred to as reluctance motors, rely on reluctance torque to operate. In the case of generators, however, reluctance torque can have a negative effect on the generator's torque and can actually result in a reduction of the generator's output power.

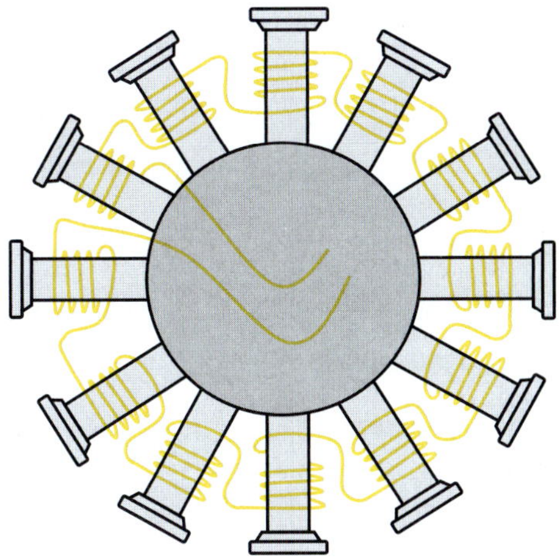

Figure 34-18. A salient-pole armature is constructed with multiple poles extending from a central magnetic wheel. This type of armature is commonly used in slower-speed applications.

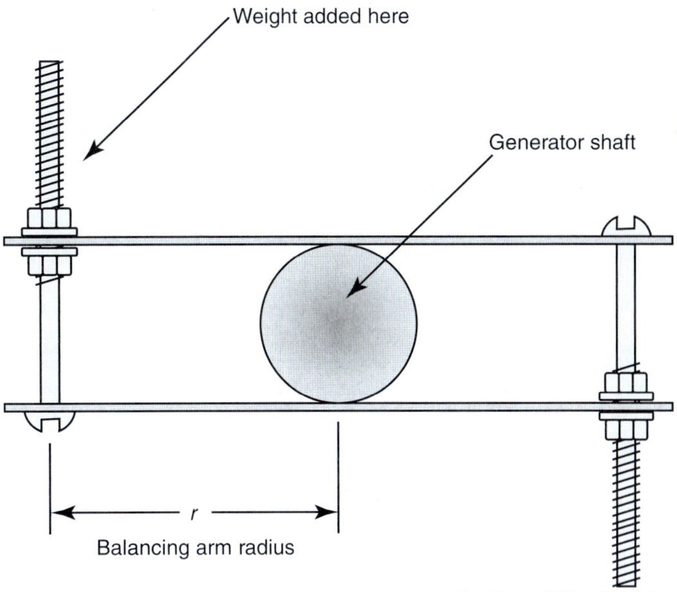

Figure 34-19. A balanced-torque measuring device used to measure a generator's starting torque. To determine starting torque, weights are applied to the arm of the device, which is connected to the generator's shaft, until the generator's shaft starts to turn. To simplify calculations, weights are typically added at a 90° angle to the direction of the shaft.

34.6 Generator Start-Up Torque

Many generator manufacturers provide start-up torque data in their specification literature. If this information is not readily available, it can easily be obtained online, measured, or calculated in the field using one of two methods. One method uses a torque wrench, while the other uses a balanced-torque measuring device.

Measuring Start-Up Torque Using a Torque Wrench

The easiest way to determine the starting torque of a generator is to use a torque wrench. With a load connected to the output of the generator, connect a torque wrench to the end of the generator's shaft. Apply force to the wrench, slowly increasing it until the generator's shaft begins to turn. At the point where the shaft begins to turn, the reading on the torque wrench will be equal to the generator's starting torque. Record this value for future reference.

Measuring Start-Up Torque Using a Balanced-Torque Measuring Device

The other method to determine a generator's start-up torque requires a balanced-torque measuring device, **Figure 34-19**, and a straightforward mathematical formula:

$$t = r \times F \times \sin\theta$$

where:

t = Torque in N·m (Newton-meters)
r = Radius of the arm, in meters, as measured from center of shaft to the end of measuring arm
F = Mass of weight, in kilograms, added to the measuring arm times the force of gravity (9.8 meters/sec^2); this result is given in Newtons (N)

The measuring arm is connected to the generator's shaft. Weights are added to the measuring arm until the generator's shaft begins to rotate. The formula is then used to calculate the generator's starting torque. Since the weights are added at a 90° angle to the direction of the shaft, the sin θ term drops out as sin 90 = 1. The torque formula then simplifies to:

$$t = r \times F \times \sin\theta$$
$$t = r \times F \times \sin 90$$
$$t = r \times F \times 1$$
$$t = r \times F$$

Refer to **Example 34-1** on how to calculate the starting torque by using this formula.

Example 34-1

Calculate a generator's starting torque if a 0.01 kg nut added to the balanced-torque measuring device shown below causes the generator's shaft to turn.

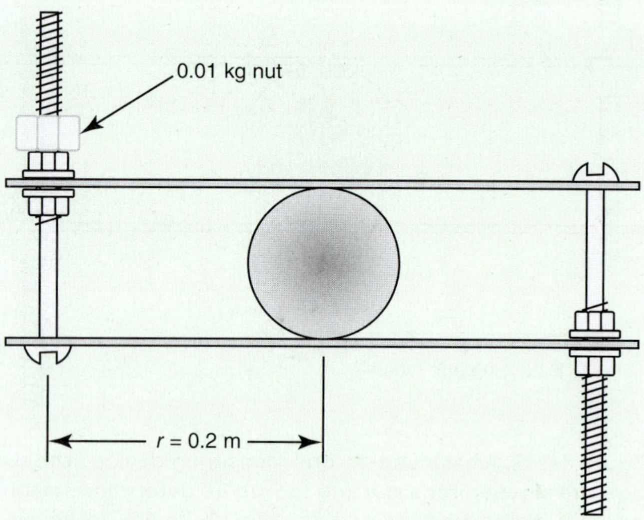

Goodheart-Willcox Publisher

✓ Solution

From the information provided, we know the following:

$$F = 0.01\text{ kg} \times 9.8\text{ meters/sec}^2 = 0.098\text{ N}$$

$$r = 0.2\text{ m}$$

$$\sin 90 = 1$$

Using the torque formula, calculate the generator's starting torque as follows:

$$t = r \times F \times \sin\theta$$

$$t = r \times F \times 1$$

$$t = 0.2\text{ m} \times 0.098\text{ N}$$

$$t = 0.0196\text{ N·m}$$

Now It's Your Turn 34-1

Calculate a generator's starting torque if the arm of a balanced-torque measuring device is 0.45 m and a 3.5 kg weight causes the shaft of a generator to turn. (Assume that the force is applied at a 90° angle to the generator's shaft.)

34.7 Generator Efficiency

Generally speaking, the efficiency of any process is determined by calculating the ratio of the process outputs to the process inputs, as shown here:

$$\text{Efficiency} = \frac{\text{Process Output}}{\text{Process Input}}$$

In terms of generators, efficiency can be calculated as the output power, or the power produced by the generator divided by the power required to produce the output:

$$\text{Generator Efficiency} = \frac{\text{Output Power}}{\text{Input Power}}$$

To express generator efficiency as a percentage, simply multiply the previous formula by 100%:

$$\text{Generator Efficiency} = \frac{\text{Output Power}}{\text{Input Power}} \times 100\%$$

Refer to **Example 34-2** on how to calculate the efficiency of a generator.

Shedding Light

In Chapter 35, *DC Generators*, more details and examples are provided for calculating the efficiency of dc generators using the actual load being powered and the generator's output power.

Example 34-2

What is the efficiency of a generator that produces 1000 W of power if it consumes 1500 W of power to produce this output?

✓ Solution

Using the formula to express generator efficiency as a percentage and substituting the known values for output power and input power, the efficiency of the generator can be calculated as follows:

$$\text{Generator Efficiency} = \frac{\text{Output Power}}{\text{Input Power}}$$

$$\text{Generator Efficiency} = \frac{1000\text{ W}}{1500\text{ W}}$$

$$\text{Generator Efficiency} = 0.667$$

$$\%\text{ Generator Efficiency} = 0.667 \times 100\%$$

$$\%\text{ Generator Efficiency} = 66.7\%$$

Now It's Your Turn 34-2

If a generator operates with an efficiency of 90%, what is the input power if the output power is to be 1800 W?

CHAPTER 34 REVIEW

Summary

- Generators convert mechanical energy into electrical energy. (34.1)
- Generators produce voltage by rotating coils of wire within a stationary magnetic field. (34.2)
- A generator has many of the same basic components as an electric motor: a shaft, a yoke (housing), an armature, and windings. (34.2)
- The left-hand rule for generators is used to help visualize the direction of conductor motion, the direction of magnetic field, and the direction of induced current flow. (34.3)
- Armature coils can be configured as simplex windings or multiplex windings. A coil with two loops is a multiplex winding but is referred to as a duplex winding. (34.4)
- The amount of reluctance torque depends on the position of the armature within the external magnetic field and the configuration of the rotating armature. Two common types of rotating armatures are the salient-pole armature and the cylindrical armature. (34.5)
- Start-up torque is the amount of force needed to rotate a generator's shaft. (34.6)
- Measuring start-up torque of a generator can be done using a torque wrench or a balanced-torque measuring device. (34.6)
- The efficiency of a dc generator can be calculated as the generator's output power divided by the power used to produce the output. (34.7)

Review Questions

Use the information in the chapter to answer the following questions.

Measuring Your Potential

1. *True or False?* A generator is a machine that transforms mechanical energy into electrical energy. (34.1)
2. The stack slots on a generator's armature are used to _____. (34.1)
   A. keep the armature balanced
   B. hold the armature windings in place
   C. prevent the stack teeth from slipping
   D. apply pressure to the commutator
3. The operation of a generator can be accurately described as _____. (34.2)
   A. rotating a coil of wire within a stationary magnetic field
   B. rotating a coil of wire within a rotating magnetic field
   C. positioning a stationary coil of wire within a rotating magnetic field
   D. positioning a stationary coil of wire within a stationary magnetic field
4. Which electrical theory states that, when a conductor passes through a magnetic field, an electromotive force, is created? (34.2)
   A. Ohm
   B. Fleming
   C. Faraday
   D. Lenz
5. *True or False?* When using the left-hand rule for generators, the thumb points in the direction of current flow. (34.3)
6. A multiplex winding will pass the generator's poles a total of how many times? (34.4)
   A. Exactly once
   B. Exactly twice
   C. Exactly three times
   D. Three or more times
7. A nonsalient pole armature is commonly referred to as a _____. (34.5)
   A. bipolar armature
   B. triangular armature
   C. rectangular armature
   D. cylindrical armature
8. Reluctance torque acts in a manner that is most similar to _____ in a conventional electric circuit. (34.5)
   A. resistance
   B. capacitance
   C. inductive reactance
   D. capacitive reactance
9. Which of the following is the correct formula to calculate torque? (34.6)
   A. $t = r \times F \times \cos\theta$
   B. $t = r \times F \times \sin\theta$
   C. $t = r \times F \times \cos^{-1}\theta$
   D. $t = r \times F \times \sin^{-1}\theta$
10. *True or False?* The efficiency of a generator is expressed as the ratio of the amount of energy used to produce energy divided by the amount of energy produced. (34.7)

Plug-In

1. A triplex-wound coil will pass the generator's poles a total of how many times? (34.4)
   A. 3
   B. 6
   C. 9
   D. 12
2. Calculate a generator's start-up torque if the arm of a balanced-torque measuring device is 0.5 m and a 1.5 kg weight causes the shaft of a generator to turn. Assume that the force is applied at a 90° angle to the generator's shaft. (34.6)
3. The efficiency of a generator that requires 2000 W of power to produce 1500 W of output is closest to _____. (34.7)
   A. 55%
   B. 65%
   C. 75%
   D. 85%
4. For a generator to have an efficiency greater than 100%, the generator's _____. (34.7)
   A. output must be greater than its input
   B. input must be greater than its output
   C. output must be exactly equal to the generator's input
   D. output must be exactly one-half the generator's input

Making the Connection

1. Explain the difference between Flemming's right-hand and left-hand rules. When is it appropriate to use each one and why? (34.3)
2. Under what circumstances would the efficiency of a generator be greater than 100%? (34.7)

CHAPTER

35 DC Generators

Alexey Seafarer/Shutterstock.com

SPARKING DISCUSSION

An electrician claims that the efficiency of a dc generator can be greater than 100%. Do you agree or disagree with this electrician's statement? Provide as much supporting information as possible to support your response.

LEARNING OUTCOMES

After completing this chapter, you will be able to:

35.1 Explain the function of commutators and brushes in dc generators.
35.2 Explain the operation and function of a dc generator.
35.3 Describe the differences among wave-wound armature windings, lap-wound armature windings, and frogleg-wound armature windings.
35.4 Explain how to calculate pole pitch, coil span, front pitch, and winding pitch of an armature.
35.5 Explain the operation of self-excited dc generators and externally-excited dc generators.
35.6 Calculate the power output of a dc generator.
35.7 Calculate the efficiency of a dc generator.

TECHNICAL TERMS

coil span
commutator pitch
compound-wound dc generator
externally-excited dc generator
fractional-pitched
frogleg-wound armature winding
front pitch
full-pitched
lap-wound armature winding
over-compounding
pole pitch
progressive wave winding
retrogressive wave winding
self-excited dc generator
series-wound dc generator
shunt-wound dc generator
under-compounding
wave-wound armature winding
winding pitch

Introduction

In Chapter 34, *Generator Fundamentals*, the basics of generators, including their construction and how they convert mechanical energy into electrical energy, are discussed. This chapter focuses specifically on the dc generator. DC generators are unique in that they produce a voltage that never changes polarity. When connected to an external, complete circuit, a dc generator produces current that flows in only one direction. This is similar to the dc current that flows through battery-powered circuits.

Although the widely accepted description of a dc generator is any machine that converts mechanical energy into dc electrical energy, a more technical definition of a generator includes the means by which the transformation from mechanical energy to electrical energy is accomplished. The dc generator's ability to transform mechanical energy to dc electrical energy is achieved by rotating a coil of wire within a stationary magnetic field, **Figure 35-1**. This is similar to the configuration of the basic dc motor.

The mechanical energy required for generator operation can come from any source of physical motion that is harnessed to provide the rotating torque necessary to turn the generator's rotor. For example, the required mechanical force may come from a renewable or sustainable source, such as blowing wind or flowing water. DC generators are becoming more popular as government regulations are having an increased effect on the use of fossil fuels. In addition, technological advancements in dc power transmission are leading to an increased demand for dc generators. Due in part to improvements in power-transmission efficiency, ocean-based wind farms, which are often located far from the power's ultimate destination, are using dc generators at an increased rate. This chapter examines the construction, types, and applications of dc generators, along with various winding configurations and wiring diagrams.

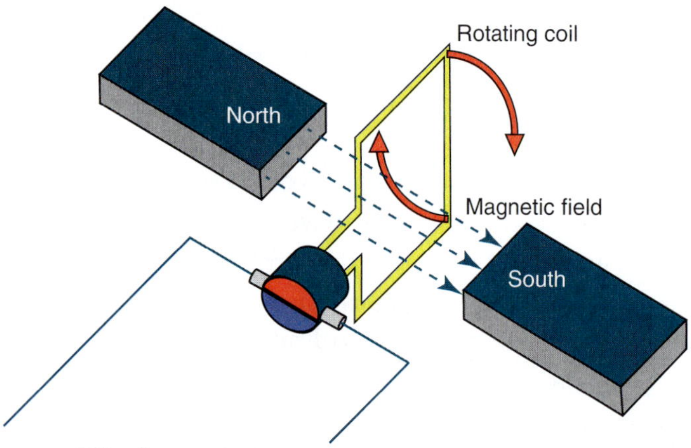

Goodheart-Willcox Publisher

Figure 35-1. A generator produces voltage by rotating coils of wire through a stationary magnetic field.

35.1 Commutator and Brushes

The construction of a dc generator is similar to that of the dc motors discussed earlier in this text. In addition to the fundamental components that are discussed in Chapter 34, *Generator Fundamentals*, dc generators use commutators and brushes to connect the rotating coils to the generator's output terminals. A generator's commutator and brushes are the components that are responsible for maintaining constant polarity of the produced dc voltage. In a dc generator, the commutator acts as a rotary switch that continually reverses the polarity of the magnetic field. The individual segments of a commutator are physically and electrically separated from each other by an insulating material to help ensure that each coil of the winding has the ability to function independently of the others.

By reversing the magnetic field's polarity as the armature rotates, the output voltage is maintained at a relatively constant dc level. The term *relatively constant* is used because the output signal from a dc generator is more of a rippled dc signal, **Figure 35-2**. The dc generator's output waveform is then smoothed out by passing the signal through a bank of large capacitors and electronic circuitry. Directly producing a pure, constant dc signal is not common in generators, as this is typically accomplished using solar panels or critically controlled chemical reactions.

Similar to dc motors, commutators are physically and permanently connected to the armature or windings of the generator, **Figure 35-3**. The number of windings in a generator's armature determines the number of segments that a commutator will have. One commutator section will be connected to each end of a winding loop. Therefore, a single-winding armature will have a two-section commutator, while an armature with two windings will have a four-section commutator. Spring-loaded connectors, called brushes, connect the generator's commutator and armature to the generator's output wiring circuitry. Brushes are constructed mainly of carbon and often have graphite and other fine metal powders added to the base material for better conduction and lubrication. Graphite, as a point of reference, is commonly used as

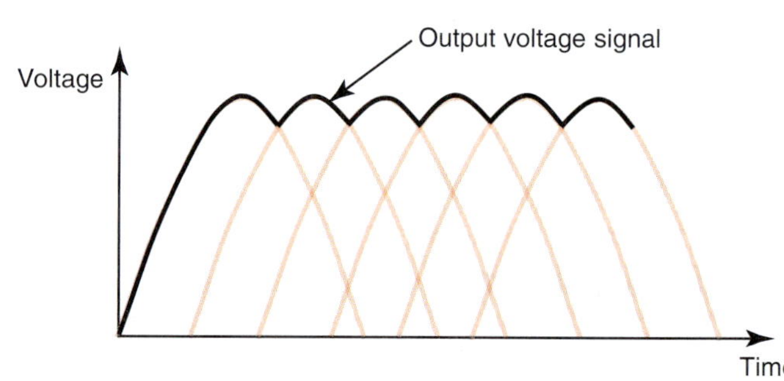

Goodheart-Willcox Publisher

Figure 35-2. A dc generator's output voltage signal is not completely horizontal. The output signal resembles more of a wave or ripple, as the magnitude of the voltage changes.

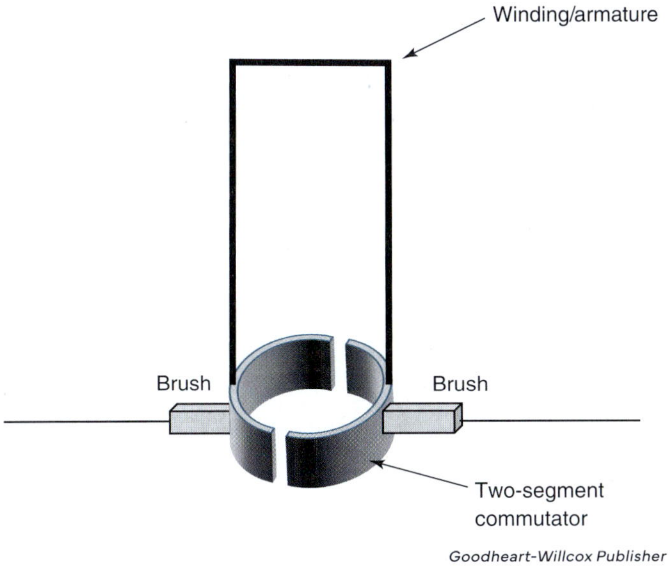

Figure 35-3. Each end of a winding is connected to a commutator segment. The commutator shown has two segments, one for each end of the generator's single winding.

a lubricant in lock cylinders. The number of brushes in a dc machine depends on the design of the generator, but always at least two and always an even number of brushes. One brush in each brush pair will be connected to the positive (+) dc side of the circuit, while the other brush will be connected to the negative (–) dc side. The number of brushes used on a generator depends on the winding strategy and configuration of the coils.

35.2 DC Generator Operation

The basic operation of a generator is discussed in Chapter 34, *Generator Fundamentals*. Discussion here builds on the content and illustrates the role that the brush/commutator assembly plays in generating a dc output signal. Consider the illustration in **Figure 35-4**. This figure shows a single, vertical conductive coil connected to the commutator. The top portion of the coil is identified as the *A* side while the bottom portion of the coil is identified as the *B* side. The output of the generator is measured across the two conductors connected to the brushes that are, in turn, in contact via the brushes with the rotating commutator.

As the armature rotates through the magnetic field in the clockwise direction, voltage is generated. The commutator and brush assembly, however, ensures that the polarity of the output signal remains the same. Consider the two coil positions shown in **Figure 35-5**. While the coil is positioned vertically in both the left and right illustrations, the coil has rotated 180°. Regardless of the coil's position, it can be seen that terminal 2 at the generator's output will always be connected to the portion of the coil that passes the magnetic field's south pole in the downward direction, while terminal 1

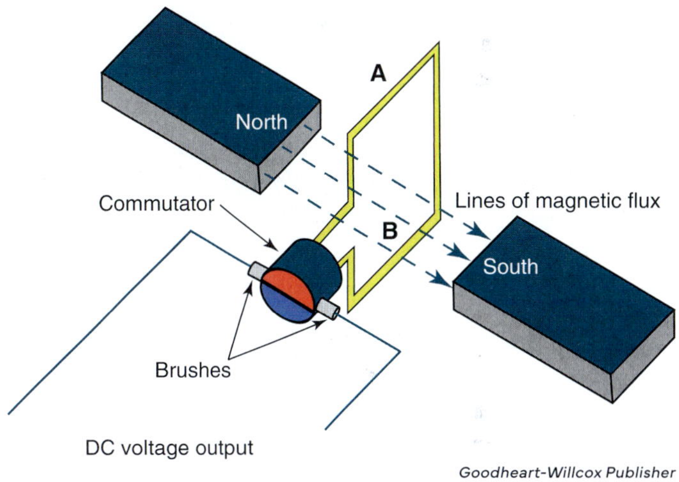

Figure 35-4. A single, vertical conductive coil connected to a commutator. Output is measured across the two conductors connected to the brushes.

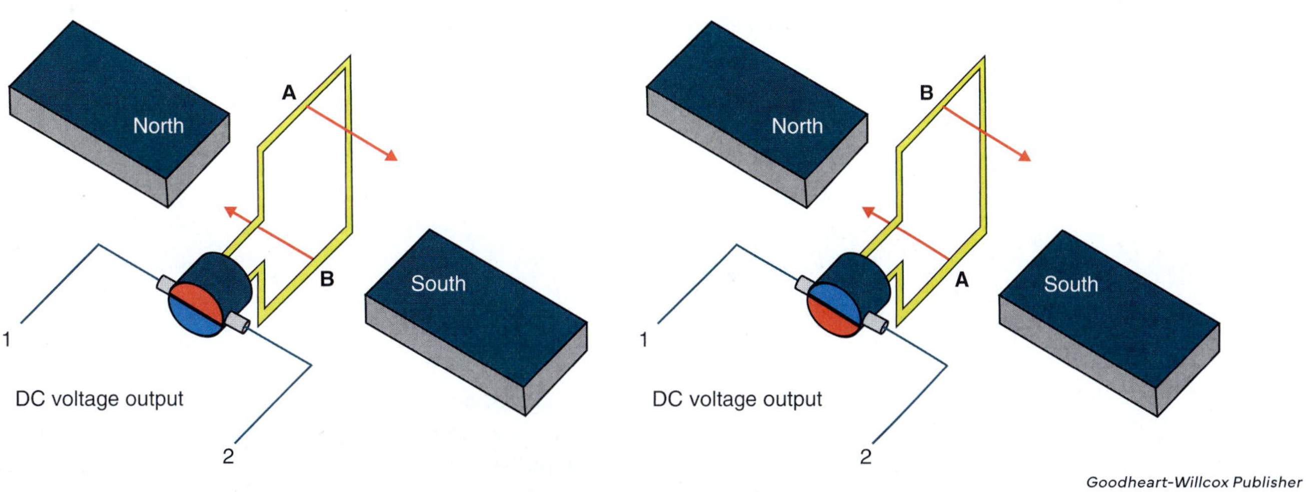

Figure 35-5. The two coil positions represented indicate 180° of rotation.

at the generator's output will always be connected to the portion of the coil that passes the magnetic field's north pole in the upward direction, **Figure 35-6**.

Continued armature rotation generates additional output wave signals as illustrated in **Figure 35-7**. In reality, the armatures contain multiple coils and generate multiple output waveforms simultaneously, **Figure 35-8**. As a result of having multiple waveforms produced simultaneously, the output voltage is higher and the waveform is smoother, though still ripple-shaped. A smoother waveform requires less manipulation to create a steadier and more constant dc signal, **Figure 35-9**. As the number of coils increases, the output waveform takes on the appearance of a more well-defined dc voltage signal, **Figure 35-10**.

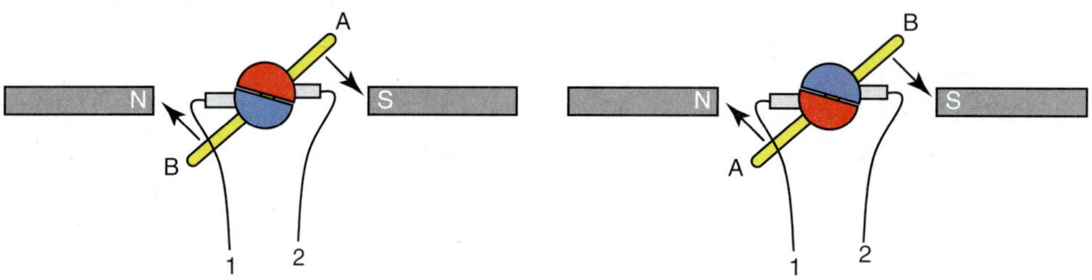

Goodheart-Willcox Publisher

Figure 35-6. By means of the commutator, the direction of coil travel through the magnetic field is always the same for each output terminal, regardless of the coil's position within the magnetic field.

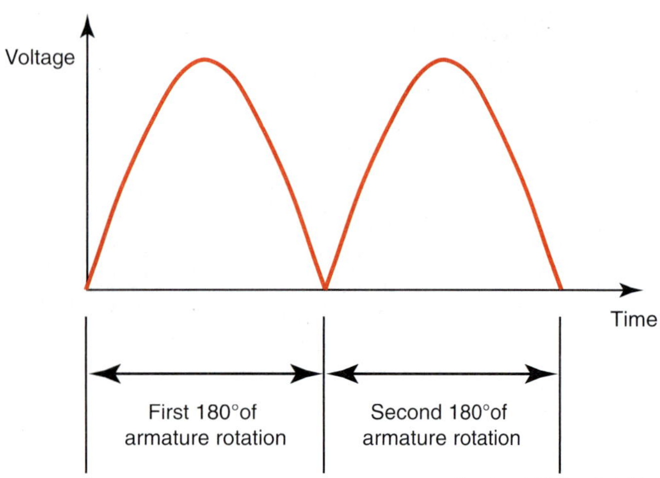

Goodheart-Willcox Publisher

Figure 35-7. As the armature continues to rotate, additional output wave signals are produced. Because the commutator is able to keep the polarity of the output signal consistent, one complete rotation of the armature creates two identical output waves.

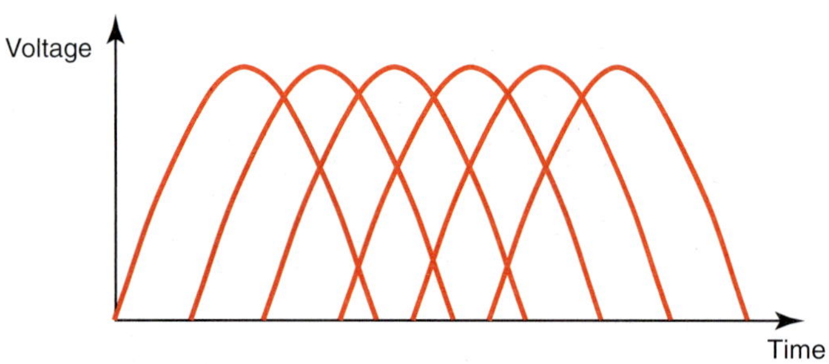

Goodheart-Willcox Publisher

Figure 35-8. Multiple output waveforms are created simultaneously when an armature contains multiple coils.

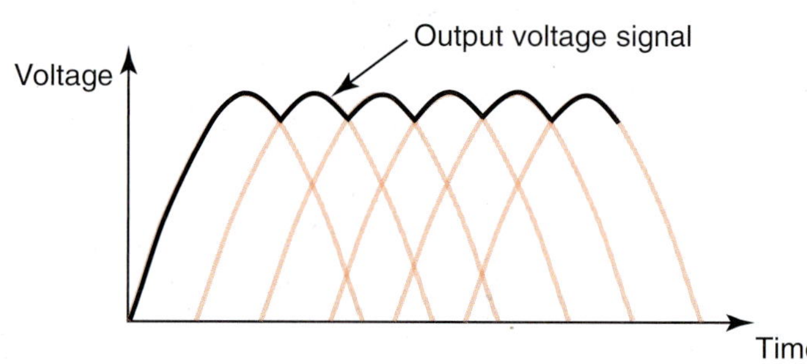

Goodheart-Willcox Publisher

Figure 35-9. The generator's output voltage waveform more closely resembles a ripple-shaped pattern than a horizontal, straight line. Although the polarity of the output voltage remains unchanged, the magnitude of the generated voltage does change.

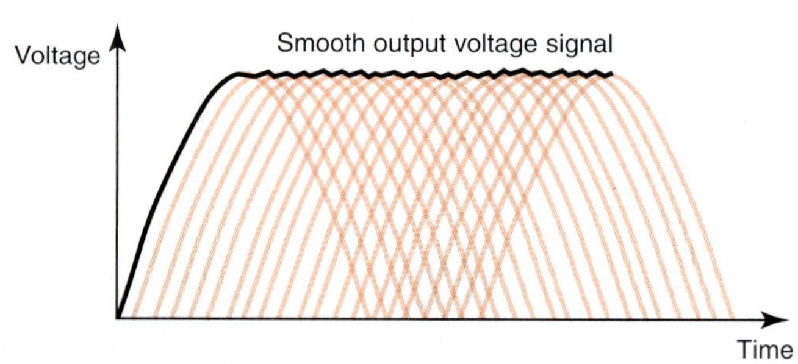

Goodheart-Willcox Publisher

Figure 35-10. As a generator's armature uses more windings, more output signals are created and the magnitude of the output voltage signal increases. Additionally, the output waveforms are slightly out-of-phase with each other, so the output voltage is much more uniform and constant.

35.3 Armature Winding Configurations

There are three commonly used armature winding configurations in dc generators: the wave-wound winding, the lap-wound winding, and the frogleg-wound winding. These configurations are determined by the manner in which individual coils of wire are connected to each other and to the generator's commutator sections.

Wave-Wound Armature Windings

Wave-wound armature windings are created when individual coils are connected end-to-end with each other, forming a large series connection. The finishing end of the first coil (Coil A) is connected to the same commutator segment as the starting end of the second coil (Coil B), **Figure 35-11**. The term *wave-wound* refers to the wave shape that is created as the coils are connected to each other. In the wave-wound configuration in **Figure 35-12**, notice how the wave shape continues to move in one direction. These end-to-end coil connections will continue until the finishing end of the last coil connects to the starting end of the first coil.

Consider the situation where a generator's armature has nine commutator segments and a coil's ends are connected to commutator segments 1 and 6, **Figure 35-13**. This coil spans five commutator segments (6 – 1). The term ***commutator pitch*** is used to describe the number of commutator segments a coil spans, so the commutator pitch for this coil is five. The commutator pitch will be the same for all coils in the winding. The next coil will have its starting end connected to commutator segment 6. Since the commutator pitch for this next coil is five as well, its finishing end will be connected to commutator segment 2, **Figure 35-14**.

The ends of the next coil will be connected to commutator segments 2 and 7, and so forth. This end-to-end connection continues until the finishing end of the last coil connects to the starting end of the first coil in the winding. Once completed, there will be exactly two coil ends connected to each

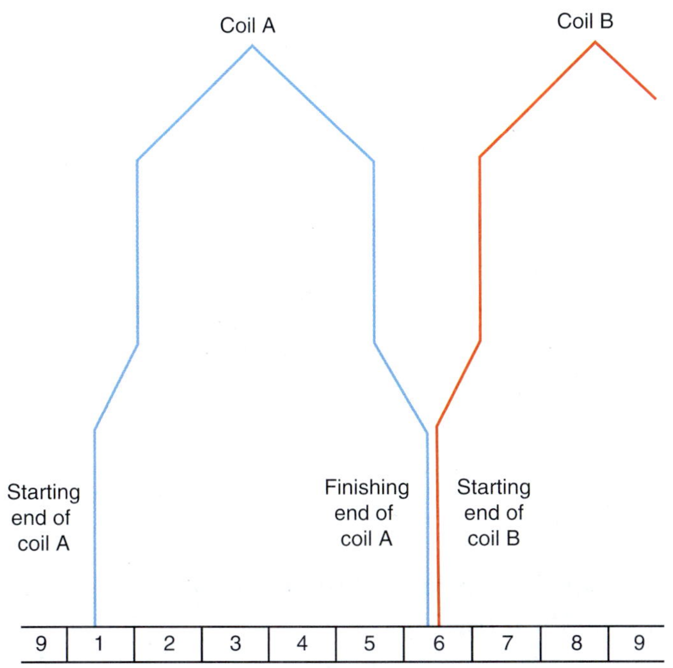

Figure 35-11. In a wave-wound configuration, the starting end of a coil is connected to the same commutator segment as the finishing end of the previous coil. In this figure, the connection between the two coils is made at commutator segment number 6.

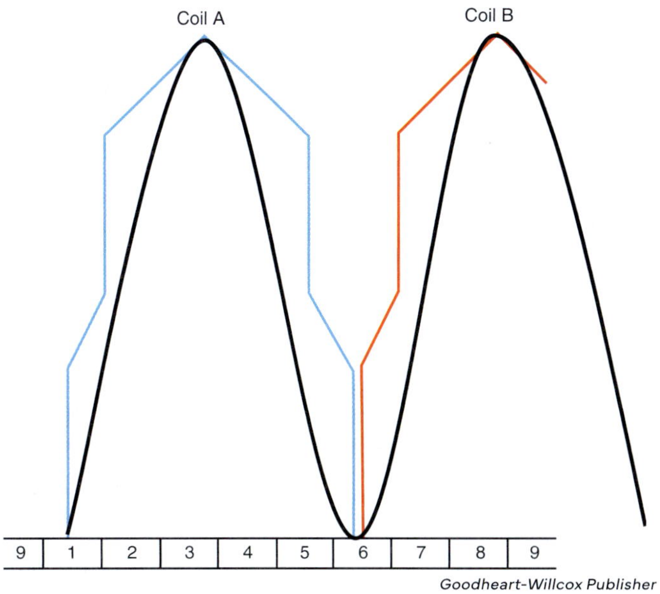

Figure 35-12. The wave-wound winding takes on the appearance of a continuous wave moving in one direction.

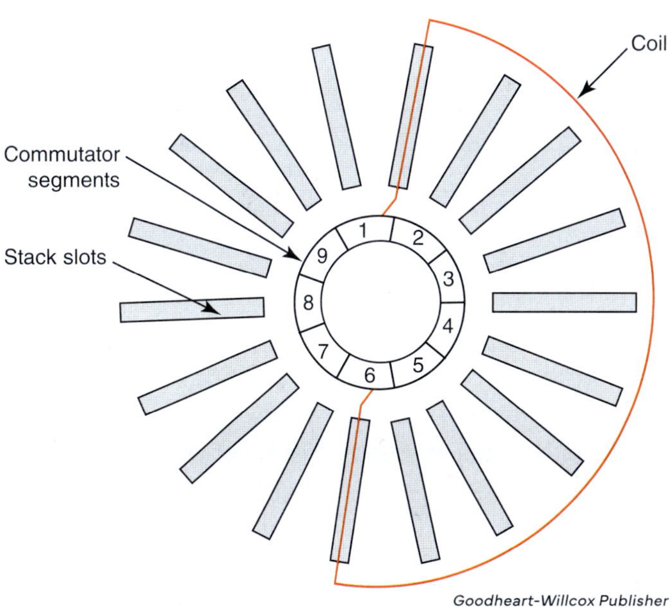

Figure 35-13. The first coil is connected to commutator segments 1 and 6. The coil has a commutator pitch of five and is positioned within the stack slots in the armature.

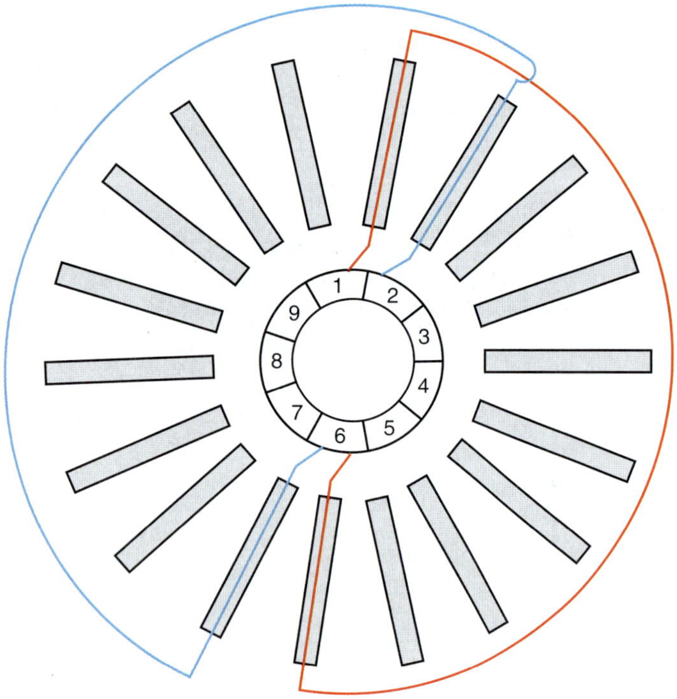

Goodheart-Willcox Publisher

Figure 35-14. The next coil has its starting end attached to commutator segment 6 to connect with the finishing end of the previous coil. With a commutator pitch of five, the coil's finishing end is connected to commutator segment 2.

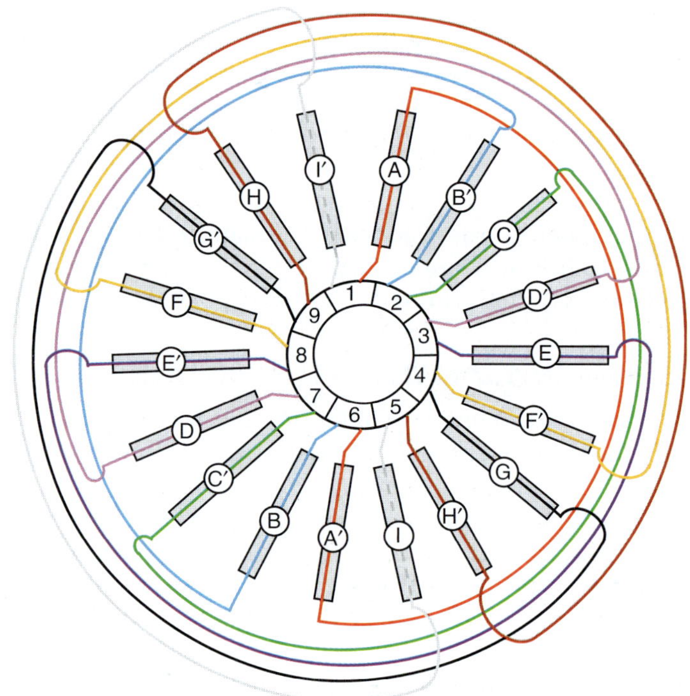

Goodheart-Willcox Publisher

Figure 35-15. A completely wired wave-wound armature. The nine windings are labeled *A* though *I* with the finishing ends identified with a prime symbol. As an example, the starting end of the *A* coil is labeled *A*, while the finishing end of the *A* coil is labeled *A′*.

Summary of Coil Connections

Coil	Starting Edge of Coil	Finishing Edge of Coil
A–A′	A, commutator section 1	A′, commutator section 6
B–B′	B, commutator section 6	B′, commutator section 2
C–C′	C, commutator section 2	C′, commutator section 7
D–D′	D, commutator section 7	D′, commutator section 3
E–E′	E, commutator section 3	E′, commutator section 8
F–F′	F, commutator section 8	F′, commutator section 4
G–G′	G, commutator section 4	G′, commutator section 9
H–H′	H, commutator section 9	H′, commutator section 5
I–I′	I, commutator section 5	I′, commutator section 1

Goodheart-Willcox Publisher

Figure 35-16. A summary of the coil connections for the armature shown in **Figure 35-15**.

commutator segment, **Figure 35-15**. For an armature with nine commutator segments, there will be a total of 18 electrical connections made to the commutator sections. Since each coil has two ends, there will be nine coils of conductive material in this armature. Simply stated, the number of coils in an armature will be equal to the armature's number of commutator segments. A summary chart of the coil connections is provided in **Figure 35-16**.

Progressive and Retrogressive Wave Windings

Notice that, in **Figure 35-14**, the coil connections do not meet up after a single path around the armature. It will take multiple paths around the armature for the finishing end of the last coil to connect with the starting end of the first coil. In the coil configuration shown in **Figure 35-15**, the finishing end of the second coil terminates after, above, or to the right of

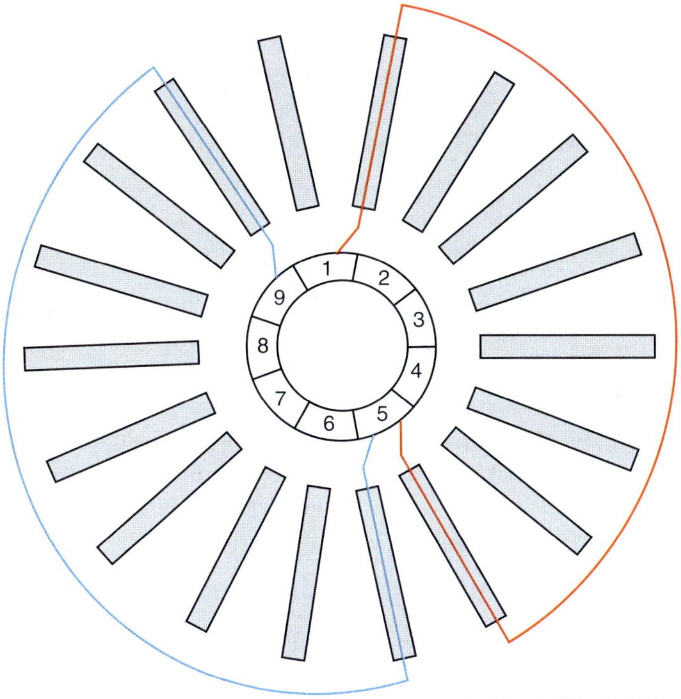

Goodheart-Willcox Publisher

Figure 35-17. In a retrogressive wave winding, the finishing end of a coil ends below, or to the left of, the starting end of the previous coil after making one path around the armature.

the starting end of the previous coil after making one rotation around the armature. In this case, the wave winding is configured as a ***progressive wave winding***. In a progressive wave winding configuration, the coil is wound to the right, and each path around the armature ends to the right of the previous path. If the finishing end of the coil terminates before, below, or to the left of the starting end of the previous coil after one rotation around the armature, **Figure 35-17**, the wave winding is configured as a ***retrogressive wave winding***. In a retrogressive wave winding configuration, the coil is wound to the right, but each path around the armature ends to the left of the previous path. Notice that, in this case, the coil has a commutator pitch of four.

Commutator Pitch

Recall that the term *commutator pitch* is used to express the distance between the ends of an individual coil. The value of the commutator pitch is a function of the number of coils in the armature and the number of poles. The formula for the commutator pitch, Y_C, for a simplex winding is:

$$Y_C = \frac{2 \times (C \pm 1)}{P}$$

where:

Y_C = Commutator pitch
C = Number of coils
P = Number of poles

The C ± 1 term can become C + 1 or C – 1, depending on whether the winding configuration is progressive or retrogressive. For a progressive simplex winding, the formula for commutator pitch is:

$$Y_C = \frac{2 \times (C + 1)}{P}$$

For a retrogressive simplex winding, the formula for commutator pitch is:

$$Y_C = \frac{2 \times (C - 1)}{P}$$

Refer to **Example 35-1** on how to calculate commutator pitch. **Example 35-2** shows a calculation for a retrogressive simplex winding.

Example 35-1

Calculate the commutator pitch for a progressive simplex winding that has nine coils and four poles.

✓ Solution

The formula for a progressive simplex winding is:

$$Y_C = \frac{2 \times (C + 1)}{P}$$

Substituting the known values of nine coils and four poles, the commutator pitch is calculated as follows:

$$Y_C = \frac{2 \times (C + 1)}{P}$$

$$Y_C = \frac{2 \times (9 + 1)}{4}$$

$$Y_C = \frac{2 \times (10)}{4}$$

$$Y_C = \frac{20}{4} = 5$$

The progressive simplex winding represented here is the same winding shown in **Figure 34-14**. Recall that this winding has nine coils and a commutator pitch of five, so it is operating with four poles.

Brushes on a Wave-Wound Armature Winding

When the wave-wound armature configuration is used, there will only be two brushes making contact with the armature's commutator. With only two brushes, the wave winding will have exactly two parallel paths between the positive and negative brushes on a simplex winding. Consider a continuous string of resistors connected to form a loop, **Figure 35-18**. If two brushes are positioned to make contact with this resistance loop, there will be exactly two parallel paths through which electric current can flow, **Figure 35-19**.

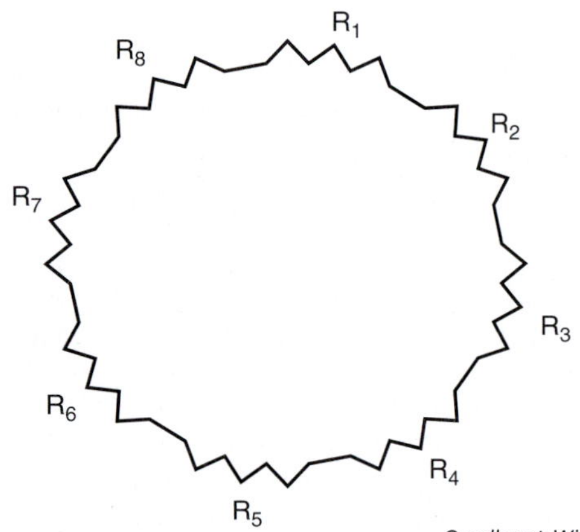

Figure 35-18. The wave-wound winding closely resembles a continuous loop of resistors that, eventually, connects back onto itself.

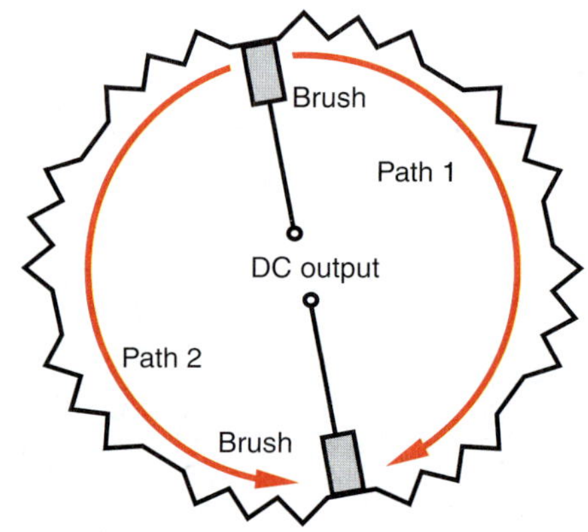

Figure 35-19. When two brushes come in contact with the continuous resistance loop, two parallel paths are created.

Example 35-2

Calculate the commutator pitch for a retrogressive simplex winding that has nine coils and four poles.

✓ Solution

The formula for a retrogressive simplex winding is:

$$Y_c = \frac{2 \times (C - 1)}{P}$$

Substituting the known values of nine coils and four poles, the commutator pitch is calculated as follows:

$$Y_c = \frac{2 \times (C - 1)}{P}$$

$$Y_c = \frac{2 \times (9 - 1)}{4}$$

$$Y_c = \frac{2 \times (8)}{4}$$

$$Y_c = \frac{16}{4} = 4$$

The retrogressive simplex winding represented here is the same winding shown in **Figure 34-17**.

Now It's Your Turn 35-2

Calculate the number of coils found on a retrogressive simplex armature winding that has four poles and is configured with a commutator pitch of seven.

Shedding Light

Even though the resistance loop depicted in **Figure 35-18** and **Figure 35-19** closes onto itself after one rotation, and the actual configuration of a wave-wound winding requires two or more rotations, the resulting number of parallel conductive paths remains constant at two.

Wave-Wound Armature Winding Properties and Applications

A wave-wound winding exhibits unique qualities and characteristics. Some of the most important are provided here:

- Used primarily for high-voltage, low-amperage applications
- Referred to as series-connected windings
- Has two parallel conductive paths for a simplex-wound armature coil
- Uses two brushes
- Has high operating efficiencies
- Generates large amounts of emf

Lap-Wound Armature Windings

Lap-wound armature windings are created in a manner similar to that of the wave-wound armature windings. The starting end of a coil is connected to the finishing end of the previous coil. However, there are significant differences between the two. These differences include the number of brushes used, the commutator pitch, the number of parallel conductive paths, and the relative positions of the coils with respect to each other. Unlike the wave-wound armature winding, which continues to move in one direction, the lap-wound armature winding creates a path that turns back onto itself, **Figure 35-20**. The lap-wound armature, therefore, resembles more of a coil than a wave, **Figure 35-21**. As in the case of the wave-wound armature, there are two electrical connections at each commutator segment on the lap-wound armature.

While the number of parallel paths in a wave-wound armature is always two, the number of parallel paths in a lap-wound armature is equal to the number of poles in the generator. The number of brushes in a lap-wound armature is also equal to the number of poles.

For a lap-wound armature with four brushes, two of the brushes will be connected to the +dc output of the generator, while the other two brushes are connected to the –dc output. Since a brush is in contact with only one commutator segment

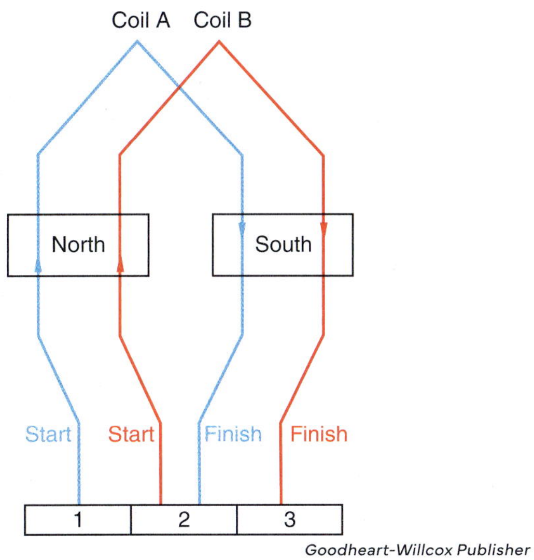

Figure 35-20. Each subsequent coil of a lap-wound armature winding laps back onto the finishing end of the previous coil.

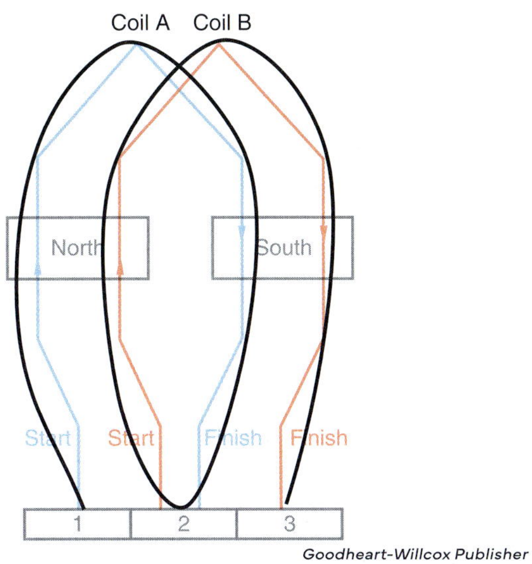

Figure 35-21. Current flow through a lap-wound armature winding follows a coil-like path.

at a time, and each commutator segment is connected to two individual coils of the winding, the two +dc brushes are connected to one end of four coils. The same holds true for the two –dc brushes, which are also connected to four coil ends. As a result, four parallel current paths are created. This is illustrated in **Figure 35-22**. It should also be noted that if you start at any point in the winding arrangement and trace out a path in either direction, you will end up where you started. This indicates that all winding segments are connected to form a continuous loop as originally shown in **Figure 35-18**.

In any motor or generator, the number of poles must be even. This is to ensure that the total magnetic flux being produced remains balanced. Since there must always be an even number of poles, the number of brushes and parallel paths must also be even. The brushes are used in pairs, where each brush pair is made up of one positive brush and one negative brush.

Lap-wound armature windings provide effective voltage and current regulation, while providing higher voltage outputs than wave-wound generators running at the same speed with the same number of armature coils.

From a manufacturing standpoint, lap-wound armature windings require more commutator segments and more complex wiring than wave-wound armatures, making them more expensive. From an operational standpoint, lap-wound armature windings are commonly found in large distribution generators due to their high current ratings and their ability to handle heavy loads and operate at lower speeds.

Lap-Wound Armature Winding Properties and Applications

Lap-wound armature windings exhibit unique qualities and characteristics. Some of the most important are provided here:

- Used primarily for low-voltage, high-amperage applications
- Referred to as parallel-connected windings
- Number of parallel conductive paths is equal to the number of generator poles and the number of brushes

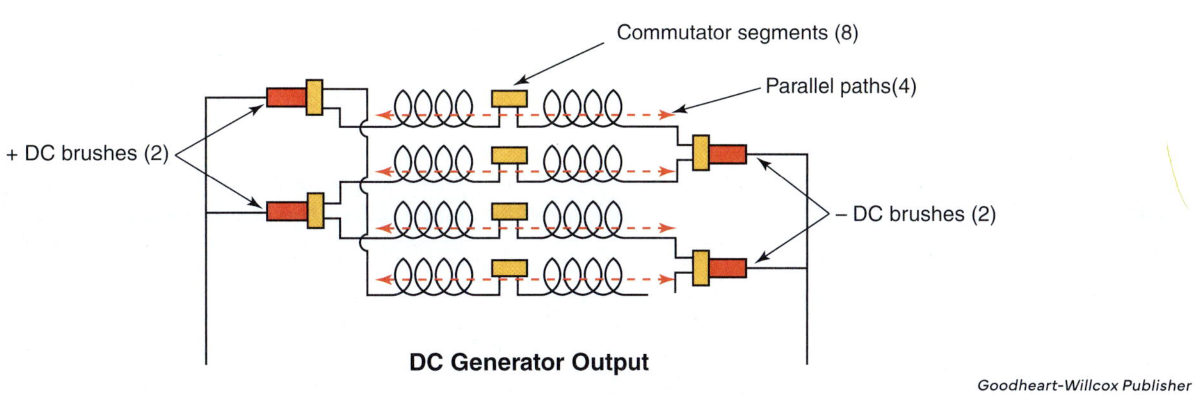

Figure 35-22. A lap-wound armature with four brushes will have four parallel paths. The number of poles is also equal to the number of parallel paths.

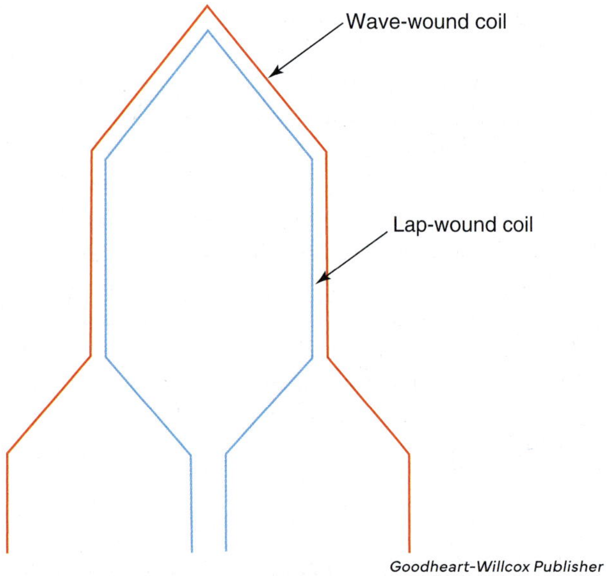

Figure 35-23. A single coil of a frogleg-wound armature made up of a wave-wound coil and a lap-wound coil.

- Commutator pitch, Y_C, for a lap-wound armature is one
- Lower operating efficiencies (as compared to wave-wound armatures)
- Generate smaller amounts of emf (as compared to wave-wound armatures)

Frogleg-Wound Armature Windings

Frogleg-wound armature windings are the most commonly used winding arrangement and are intended for applications that require moderate voltage and moderate current. The frogleg-wound configuration is a cross between the series-connected, wave-wound armature and the parallel-connected, lap-wound armature, **Figure 35-23**. The frogleg-wound winding connection is referred to as a series-parallel connection. The lap and wave coils that make up the frogleg-wound armature winding configuration are positioned in the same armature slots, even though they are insulated from each other and are wound as separate coils.

While wave-wound and lap-wound armatures have two leads for each commutator section, in the frogleg-wound connection, each commutator section has four electrical conductors or leads connected to it: two from a wave-wound winding and two from a lap-wound winding, **Figure 35-24**.

35.4 Armature Winding Terms

When discussing dc generator armatures, a number of terms are used to describe the relative positions of the individual coils with respect to each other and with respect to the stack slots in the armature itself.

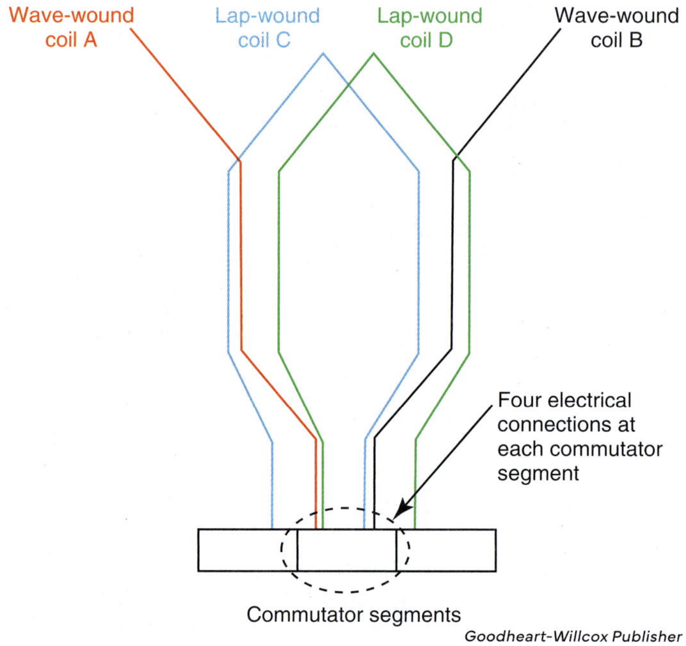

Figure 35-24. In a frogleg-wound armature, each commutator segment has four electrical connections, representing four ends of four different coils. Two of the coil ends are from the lap-wound connection, while the other two are from the wave-wound connection.

Pole Pitch, Y_P

Pole pitch, Y_P, is defined as the number of armature slots per pole, and is calculated as the total number of armature slots divided by the number of poles, P, as follows:

$$Y_P = \frac{\text{Armature slots}}{P}$$

Figure 35-25 illustrates that each pole of this generator covers six armature slots, so the pole pitch for this armature is six.

How to calculate pole pitch is explained in **Example 35-3**.

Example 35-3

What is the pole pitch of a four-pole armature if there are 48 stack slots in the armature?

✓ Solution

Using the formula to calculate pole pitch and substituting the known values for the number of armature slots and the number of poles, we get:

$$Y_P = \frac{\text{Armature slots}}{P}$$

$$Y_P = \frac{48}{4}$$

$$Y_P = 12 \text{ armature slots per pole}$$

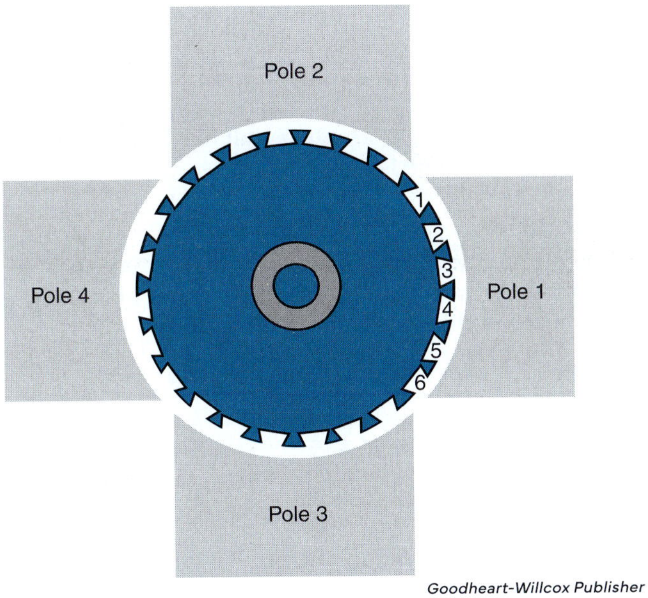

Figure 35-25. The pole pitch refers to the number of armature, or stack, slots that are covered by each pole. In this figure, the pole pitch is six because each pole covers six armature slots.

As a standalone piece of information, the pole pitch may not appear to be very useful. However, when evaluated in conjunction with other armature-related data, it becomes more important.

Coil Span, Y_S

Coil span, Y_S, is the width of an individual coil, typically expressed in terms of armature slots. Coil span is obtained at the back side of the armature, which is the side opposite the commutator, **Figure 35-26**. The commutator side of the armature is called the front side. Consider the coil shown in **Figure 35-27**. This coil has its starting end in stack slot 1 and its finishing end in stack slot 6. Therefore, the coil span is five (6 – 1).

By comparing the coil span to the pole pitch, it can be determined how the resulting emfs in each coil will interact with each other. If the coil span is equal to the pole pitch, the winding is referred to as being ***full-pitched***. This means that each of the coil's two sides will always lie under poles with opposite charges, **Figure 35-28**.

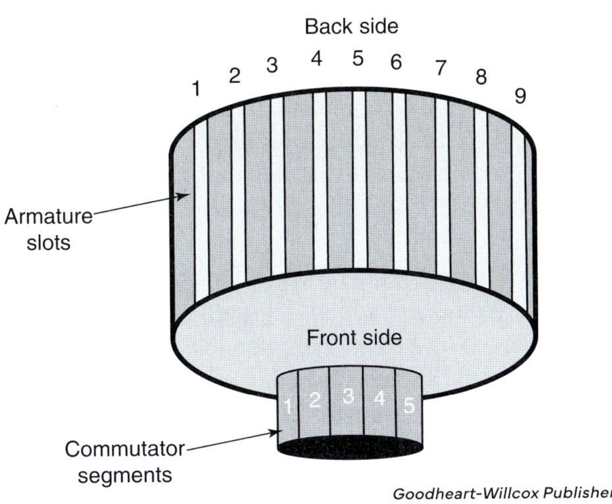

Figure 35-26. The front side of the armature is the side closest to the commutator. The armature's back side is the side opposite the commutator.

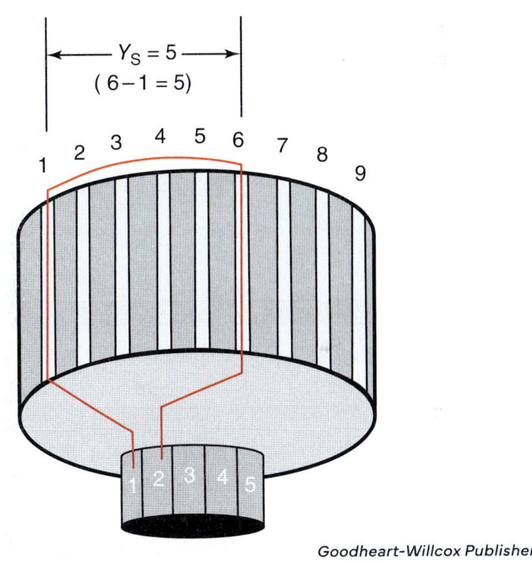

Figure 35-27. The coil span is the number of armature slots spanned by a coil. The coil span illustrated here is five.

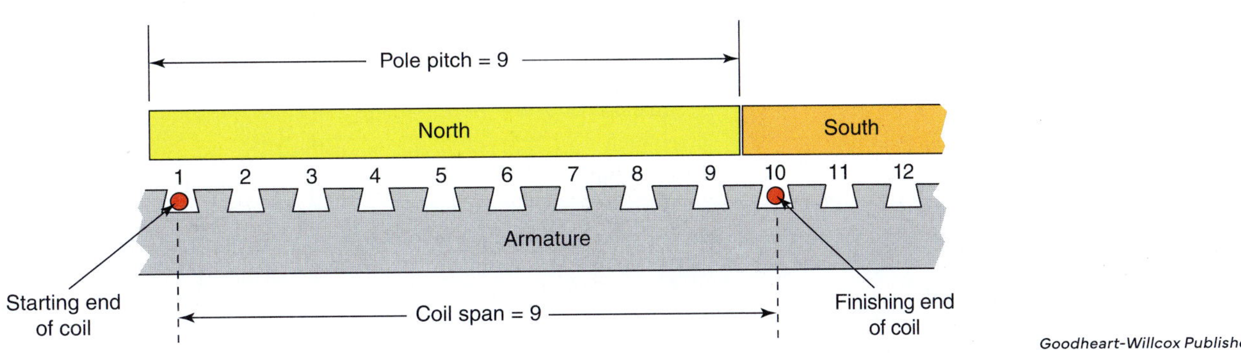

Figure 35-28. If the coil span and the pole pitch values are the same, the opposite sides of the coil will always be exposed to poles with opposite polarities. In this figure, the two red circles represent the starting and finishing sides of a coil that has a coil span of nine. As a result, both sides of the coil cannot be under the same pole at the same time, since the pole only spans nine stack slots. One side of the coil will be under a north pole, while the other side of the coil will be under a south pole.

Since the direction of current flow through each side of the coil is in the opposite direction and the polarity of the coils under which they flow are also opposites, **Figure 35-29**, the induced emfs in each conductor are completely in-phase with each other and, as a result, are additive. Therefore, the maximum emf is induced in the coil, **Figure 35-30**.

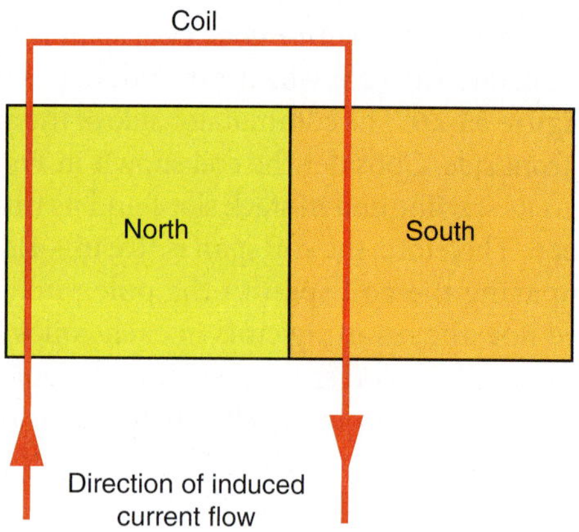

Figure 35-29. In a full-pitched winding, the direction of current flow through each side of the coil is in the opposite direction and the polarity of the coils under which they flow are also opposites. As a result, the produced emfs are in-phase with each other, and their magnitudes are additive.

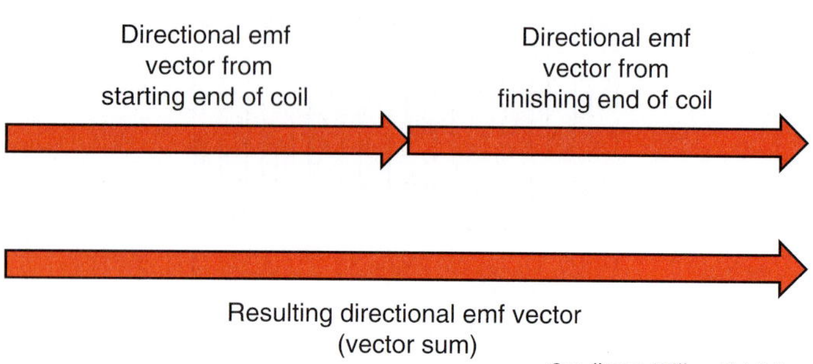

Figure 35-30. When two vectors are in-phase with each other, the direction of the resultant vector is parallel to the two initial vectors, and the magnitude of the resultant vector is simply the mathematical sum of the individual vector magnitudes.

Refer to **Example 35-4** on how to calculate coil span.

Example 35-4

Consider a four-pole armature that has 56 armature slots. If the winding is full-pitched, what is the coil span?

✓ Solution

Pole pitch is defined as the number of armature slots per pole. With 56 armature slots and four poles, the pole pitch is calculated as follows:

$$Y_P = \frac{\text{Armature slots}}{P}$$

$$Y_P = \frac{56}{4}$$

$$Y_P = 14$$

In order for a winding to be full-pitched, the pole pitch must be equal to the coil span. Therefore, the coil span is 14.

If the coil span is less than the pole pitch, the winding is classified as ***fractional-pitched***, **Figure 35-31**. In this case, both ends of the coil will be under the same pole for a portion of armature rotation, and under different poles for the remainder of the rotation, **Figure 35-32**. As a result, there will be a phase difference between the emfs produced on the two sides of the coil. This phase difference means that the total emf around the coil will be the vector sum of the emfs, **Figure 35-33**, which will be less than if the winding is full-pitched. Any time out-of-phase vectors are added together, the result will always be lower than when in-phase vectors are added, **Figure 35-34**. Although the voltage produced by a fractional-pitch winding will be lower than that produced by a full-pitched winding, the cost of producing fractional-pitch windings is lower.

Front Pitch, Y_F

The front pitch is obtained at the front end, or side, of the armature. ***Front pitch*** is defined as the distance between the finishing end of one coil and the starting end of the next coil connected to it, **Figure 35-35**.

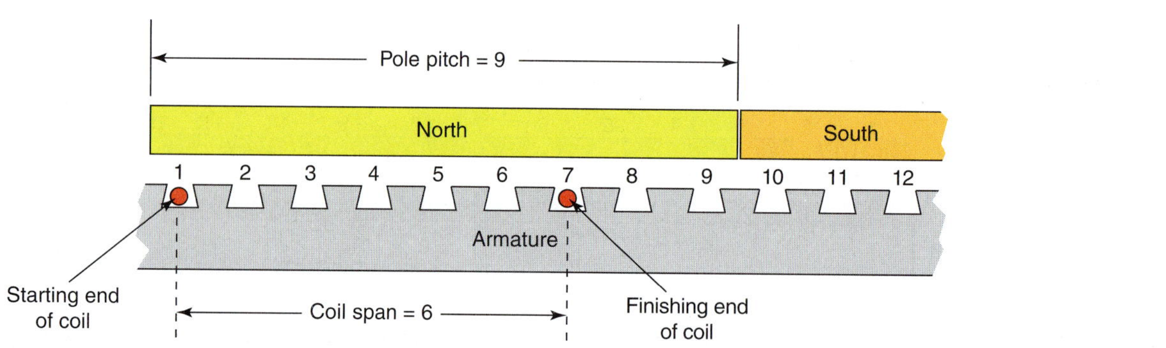

Figure 35-31. If the coil span is less than the pole pitch, the winding is said to be fractional-pitched, and both sides of the coil will be under a single pole for at least a portion of an armature's revolution.

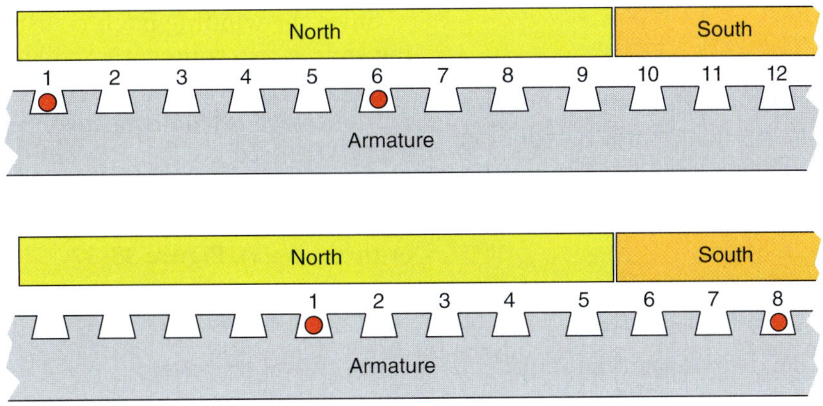

Goodheart-Willcox Publisher

Figure 35-32. In a winding classified as fractional-pitched, both coil ends will be under the same pole for a portion of armature rotation (top), and each coil end under different poles for the remainder of the rotation (bottom).

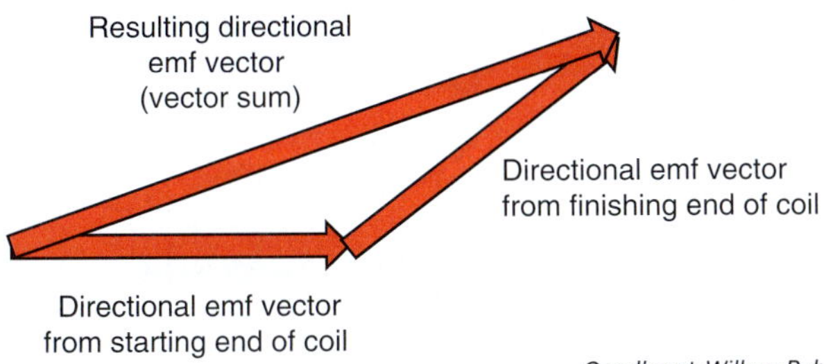

Goodheart-Willcox Publisher

Figure 35-33. When two out-of-phase vectors are added together, the tail of one vector is positioned at the head of the other. The resultant vector extends from the tail of the first vector to the head of the second.

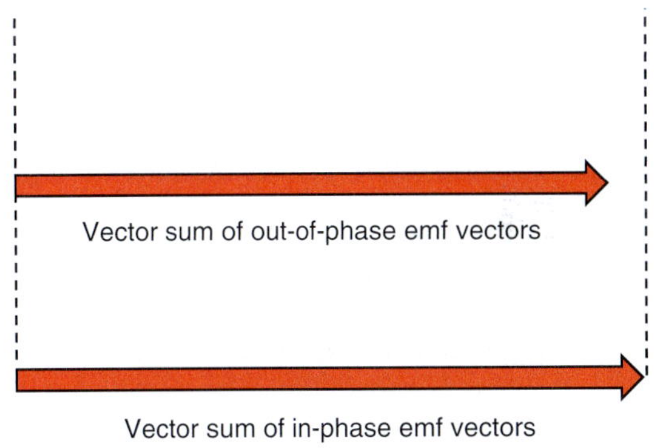

Goodheart-Willcox Publisher

Figure 35-34. The sum of two in-phase vectors will always have a magnitude that is greater than the magnitude of the sum of two out-of-phase vectors.

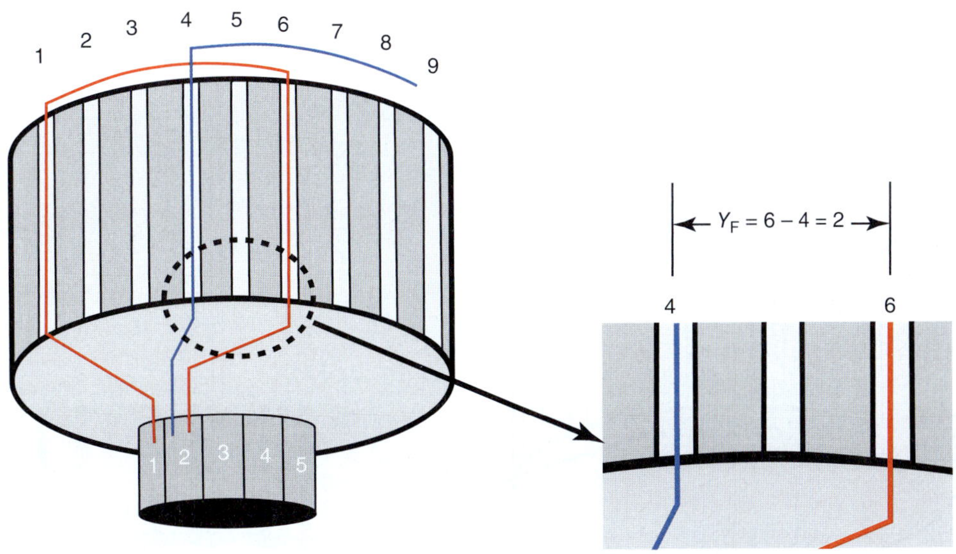

Goodheart-Willcox Publisher

Figure 35-35. The front pitch is the distance, in armature slots, between the finishing end of one coil and the starting end of the next coil. The red coil has its starting end in stack slot 1 and its finishing end in stack slot 6. The blue coil has its starting end in stack slot 4. Since the red coil finishes in stack slot 6 and the blue coil starts in stack slot 4, the front pitch for this armature is 2 (6 – 4).

Winding Pitch, *Y*

In general, ***winding pitch***, *Y*, is identified as the distance measured in number of slots between the starting ends of two consecutive coil loops. For a lap-wound winding, the winding pitch, *Y*, is calculated as the difference between the coil span, Y_S, and the front pitch, Y_F, **Figure 35-36**. This difference is expressed as follows:

$$Y\text{ (lap)} = Y_S - Y_F$$

Refer to **Example 35-5** on how to calculate winding pitch.

Example 35-5

Calculate the winding pitch for the armature represented in **Figure 35-36**.

✓ Solution

The red coil in the figure has its starting end in stack slot 1 and its finishing end in stack slot 6, so the coil span, Y_S, for this coil is five. With the red coil finishing in slot 6 and the blue coil starting end in stack slot 4, the front pitch, Y_F, is two (6 – 4). Using the formula for winding pitch and substituting the known values for the coil span and front pitch, the winding pitch is calculated as follows:

$$Y = Y_S - Y_F$$
$$Y = 5 - 2$$
$$Y = 3$$

Since the winding pitch is distance between the starting ends of two consecutive coil loops, it is often easier to simply determine the difference of the two starting ends, without first calculating the coil span and front pitch. The starting end of the blue coil is in stack slot 4 and the starting end of the red coil is in stack slot 1, so the winding pitch is the difference between the two starting ends, or three (4 – 1), **Figure 35-37**.

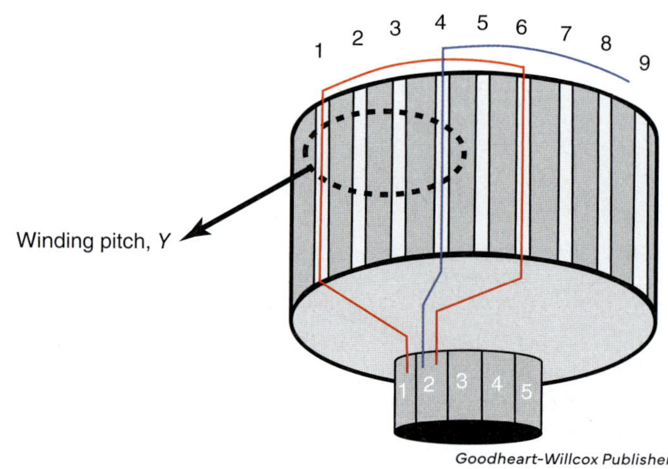

Figure 35-37. The winding pitch, *Y*, can also be calculated as the distance, in armature slots, between the starting ends of two connected coils.

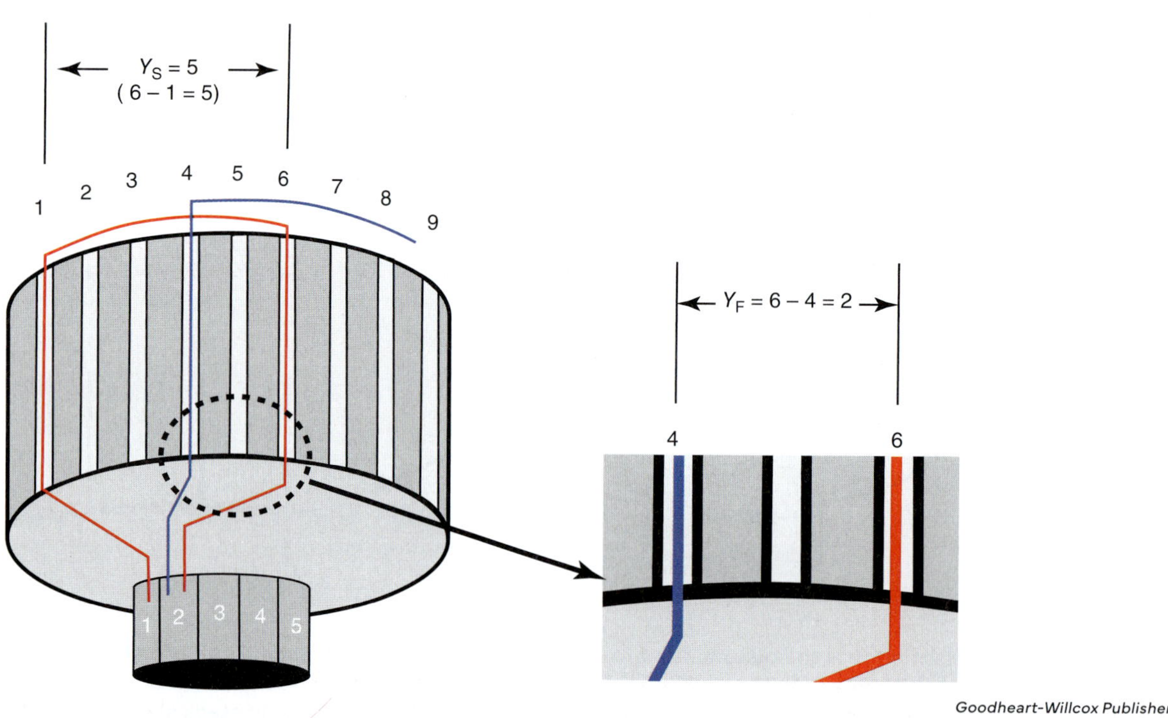

Figure 35-36. For a lap-wound armature, the winding pitch, *Y*, is calculated as the coil span, Y_S, minus the front pitch, Y_F.

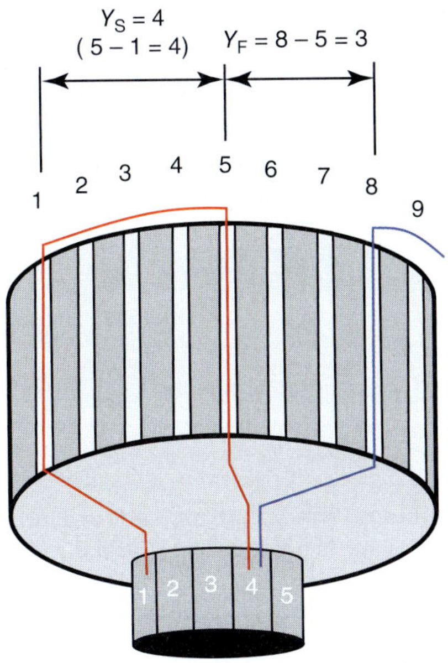

Goodheart-Willcox Publisher

Figure 35-38. For a wave-wound armature, the winding pitch, *Y*, is calculated as the coil span, Y_S, plus the front pitch, Y_F. The winding pitch can be easily determined by calculating the difference between the two starting ends. The front pitch of the blue coil is in stack slot 8 and the front pitch of the red coil is in stack slot 1, so the winding pitch is the difference between the two starting ends, or seven (8 – 1).

For a wave-wound winding, the winding pitch, *Y*, is calculated as the mathematical sum of the coil span, Y_S, and the front pitch, Y_F, **Figure 35-38**, and the formula is:

$$Y \text{ (wave)} = Y_S + Y_F$$
$$Y = Y_S + Y_F$$
$$Y = 4 + 3$$
$$Y = 7$$

Shedding Light

The various terms regarding pitch are often expressed in terms of armature conductors, but can also be expressed in terms of armature slots. If pitch information is needed for a particular project, be sure to consult the generator's literature to ensure that the proper units and selection criteria are used.

35.5 DC Generator Field Windings

As with dc motors, there are several different types of dc generators that use field windings to create the required electromagnetic fields. Two common configurations of dc generators are the self-excited dc generator and the externally-excited dc generator. If the generator supplies its own current for the field windings, the generator is classified as a self-excited machine. If the field winding's current is provided by an external source, such as another generator or a battery, the generator is classified as an externally-excited dc generator. The majority of dc generators in operation are of the self-excited variety.

Self-Excited DC Generators

Self-excited dc generators supply their own current for the field windings. There are three common configurations for self-excited dc generators:

- Series-wound generators
- Shunt-wound generators
- Compound-wound generators

The series-wound generator uses a series-connected field winding, the shunt-wound generator uses a parallel-connected field winding, and the compound-wound generator uses both series-connected and parallel-connected field windings.

In our discussion of dc motors in Chapter 27, *DC Motors*, a common labeling strategy was used to identify the connecting terminals on the armature, the series field winding, and the shunt field winding. On the armature, the terminal connections are identified as A_1 and A_2. The series field winding terminal connections are identified as S_1 and S_2, while the shunt field winding terminals are identified as F_1 and F_2, **Figure 35-39**. This same labeling strategy is used here for our discussion of dc generators.

Series-Wound DC Generators

Series-wound dc generators are constructed with their field coils in series with the armature and the load. When configured as a series connection, the current flow through the entire circuit, as measured at any given point, will be the same, **Figure 35-40**. The $E = I \times R$ form of Ohm's law can be used to verify that the voltage drops across the field coils and the connected load will be proportional to their individual resistances. The field winding is wound with few turns and thick wire to lower its resistance. By lowering the resistance of the field winding, the voltage drop across the field winding is lowered, and the generator's output power level is increased. One benefit of having the field coil connected in series with the load is that the generator will require very low torque when there is no load connected to it.

Series-wound dc generators are not the best option if precise voltage regulation is desired. Since the armature, the winding coil, and the load are all connected in series, any changes to the load will affect the current flow through the load and, as a result, the generator's output voltage. To adjust for changes in the connected load, the speed of the generator can be changed, or a variable resistor in series with the output, **Figure 35-41**, can be adjusted to maintain the output voltage at the desired level. Series-wound dc generators are commonly

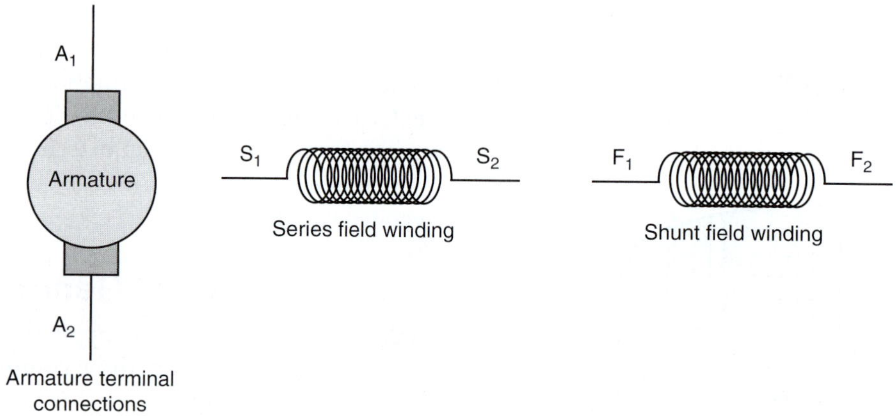

Goodheart-Willcox Publisher

Figure 35-39. The armature, series winding, and shunt winding have unique markings. The armature leads are labeled A_1 and A_2, the series field winding leads are labeled S_1 and S_2, while the shunt field winding leads are labeled F_1 and F_2.

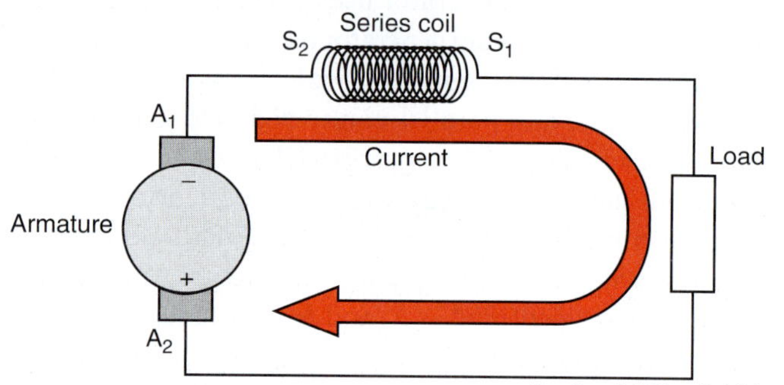

Goodheart-Willcox Publisher

Figure 35-40. In a series-wound generator, the current flow through the armature, the series field coil, and the load is the same, regardless of where an amperage reading is taken.

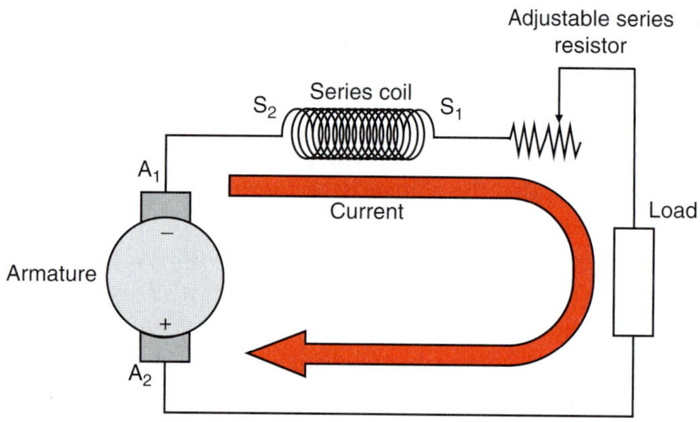

Goodheart-Willcox Publisher

Figure 35-41. A variable resistor wired in series with the load can be used to compensate for changes in the connected load to maintain the generator's output voltage at the desired level.

used in high-voltage, low-current applications such as arc welders and battery chargers, where the voltage output can be regulated or is not critical for satisfactory load operation.

How to calculate voltage drop in a series-wound dc generator is explained in **Example 35-6** and **Example 35-7**.

Example 35-6

What is the voltage drop across the 1 Ω field coil of a series-wound dc generator, similar to the one shown in **Figure 35-40**, if it is supplying 100 W at 12 V to a lamp?

✓ Solution

To solve this problem, we can use the $P = I \times E$ of the power formula to find the circuit current. Substituting the known values for power and voltage at the load, the circuit current can be determined as follows:

$$P = I \times E$$

$$I = \frac{P}{E}$$

$$I = \frac{100\text{ W}}{12\text{ V}}$$

$$I = 8.33\text{ A}$$

Since the same current flows through all components of a series circuit, the armature current, I_a, series winding current, I_s, and the load current, I_L, are all the same:

$$I_a = I_s = I_L = 8.33\text{ A}$$

The voltage drop across the field coil can be found using the $E = I \times R$ form of Ohm's law. Substituting the known values for the field coil resistance and circuit current, we get:

$$E = I \times R$$

$$E = 8.33\text{ A} \times 1\ \Omega$$

$$E = 8.33\text{ V}$$

To overcome the voltage drop of 8.33 V across the series field coil, the generator must produce 20.33 V, which is the sum of the voltage drop across the series field coil and the 12 V being supplied to the load.

Example 35-7

If the resistance of the series field coil in **Example 35-6** is changed to 2 Ω, what will the voltage drop across the field coil be if the series-wound dc generator is still intended to supply 100 W at 12 V to a lamp circuit?

✓ Solution

To solve this problem, we can use the $P = I \times E$ form of the power formula to find the circuit current. Substituting the known values for power and voltage at the load, we can solve for the circuit current as follows:

$$P = I \times E$$

$$I = \frac{P}{E}$$

$$I = \frac{100 \text{ W}}{12 \text{ V}}$$

$$I = 8.33 \text{ A}$$

Notice that the current in this example is still 8.33 A, even though the resistance of the series field coil has changed. As before, the armature current, I_a, series winding current, I_S, and the load current, I_L, are all the same:

$$I_a = I_S = I_L = 8.33\text{A}$$

The voltage drop across the 2 Ω field coil can be found using the $E = I \times R$ form of Ohm's law. Substituting the known values for the field coil resistance and circuit current, we get:

$$E = I \times R$$

$$E = 8.33 \text{ A} \times 2\ \Omega$$

$$E = 16.66 \text{ V}$$

To overcome the voltage drop of 16.66 V across the series field coil, the generator must produce 28.66 V, which is the sum of the voltage drop across the series field coil and the 12 V being supplied to the load.

When comparing the results of **Example 35-6** and **Example 35-7**, it can be seen that as the resistance of the series field coil increases, the amount of voltage required by the generator increases. As a result, it is desirable to have the resistance of the series field coil be as low as possible. As will be seen later in our discussion of dc generator efficiency, the resistance of the field coil will have a significant effect on the efficiency of the machine.

Shunt-Wound DC Generators

The ***shunt-wound dc generator*** is configured with the field coils connected in parallel to the armature and to the load, **Figure 35-42**. Because of this parallel connection, the measured

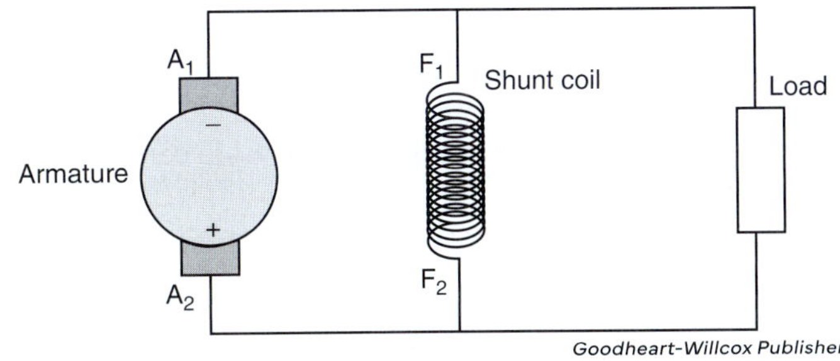

Figure 35-42. In a shunt-wound generator, the field coil is wired in parallel with the armature and the load.

voltage across the field coil will always be equal to the voltage produced by the armature and the voltage supplied to the load. This parallel arrangement provides an output voltage that is inversely proportional to the load current.

Unlike the series-wound dc generator, the shunt-wound generator requires higher running torque when no load is connected to it.

As with any parallel circuit, the branch currents are proportional to the branch resistances. The same rule holds true for the shunt-wound dc generator. As a branch resistance increases, the branch current decreases. As a branch resistance decreases, the branch current increases. Therefore, to increase the efficiency of the shunt-wound generator, the shunt field coils use thinner conductors and are wound with a large number of turns. This increases the resistance of the field coil, and maximizes the current being supplied to the connected load.

Refer to **Example 35-8**, **Example 35-9**, and **Example 35-10** on how to calculate current and power of a shunt-wound dc generator.

Example 35-8

Calculate the current through an 80 Ω field coil of a shunt-wound dc generator that is supplying 12 V at 100 W to a lamp circuit. The circuit for this arrangement is shown here:

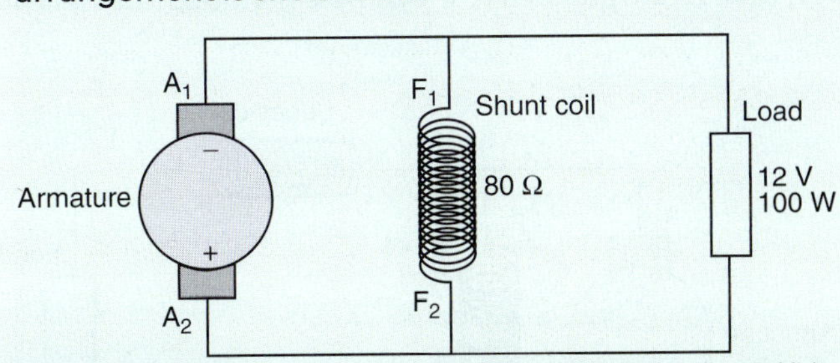

✓ Solution

Given the parallel connection of the armature, it is known that the armature voltage, V_a, the field coil voltage, V_F, and the load voltage, V_L, are all the same:

$$V_a = V_F = V_L = 12 \text{ V}$$

Continued

The current flow through the shunt coil can be found using the $I = E/R$ form of Ohm's law. Substituting the known values for the resistance of the field coil and the voltage across the field coil, the current value can be obtained as follows:

$$I = \frac{E}{R}$$

$$I = \frac{12\text{ V}}{80\ \Omega}$$

$$I = 0.15\text{ A}$$

Example 35-9

Using the results from **Example 35-8**, calculate the power that must be produced by the generator to satisfy the needs of the circuit.

✓ Solution

The generator must produce enough power to provide 100 W to the load and the power consumed by the shunt field coil. The power consumed by the shunt coil can be determined using the $P = E^2/R$ form of the power formula as follows:

$$P = \frac{E^2}{R}$$

$$P = \frac{(12\text{ V})^2}{80\ \Omega}$$

$$P = \frac{144}{80}$$

$$P = 1.8\text{ W}$$

The generator must, therefore, generate at least 101.8 W, which is the sum of the power consumed by the load and the power consumed by the shunt field coil.

Example 35-10

Using the results from **Example 35-8** and **Example 35-9**, calculate the power that must be produced by the generator to satisfy the needs of the circuit if the resistance of the shunt coil is 10 Ω instead of 80 Ω.

✓ Solution

The generator must produce enough power to provide 100 W to the load and overcome the power consumed by the shunt field coil. The power consumed by the shunt coil can be determined using the $P = E^2/R$ form of the power formula as follows:

$$P = \frac{E^2}{R}$$

$$P = \frac{(12\text{ V})^2}{10\ \Omega}$$

$$P = \frac{144}{10}$$

$$P = 14.4\text{ W}$$

The generator must, therefore, generate at least 114.4 W, which is the sum of the power consumed by the load and the power consumed by the shunt field coil.

By examining the results from **Example 35-8**, **Example 35-9**, and **Example 35-10**, it can be concluded that as the resistance value of the shunt field coil increases, the amount of power required by the generator decreases. It is, therefore, beneficial to have the shunt field coil resistance as high as possible.

Compound-Wound DC Generators

Compound-wound dc generators are self-excited generators that use both series and shunt windings. They are divided into two similar designs where the shunt field coil is wired in parallel with just the armature or in parallel with both the armature and the series field coil, **Figure 35-43**.

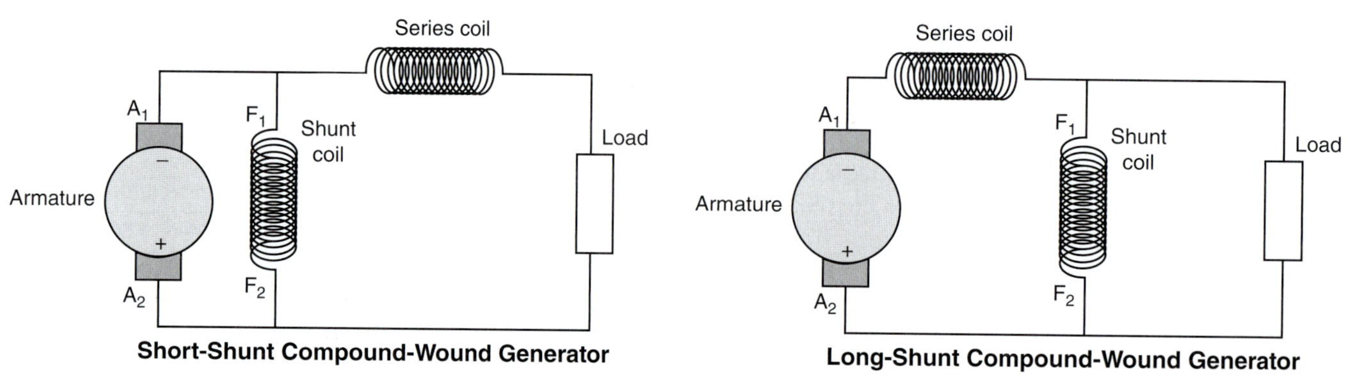

Goodheart-Willcox Publisher

Figure 35-43. Compound-wound dc generators are divided into two similar designs. In a short-shunt compound-wound generator, the shunt coil is wired in parallel with only the armature. In a long-shunt compound-wound generator, the shunt coil is wired in parallel with the armature and the series field coil.

When the shunt field coil is wired in parallel with only the armature, the generator is referred to as a short-shunt compound-wound generator. When the shunt field coil is wired in parallel with both the armature and the series field coil, the generator is referred to as a long-shunt compound-wound generator.

The following are some key points that differentiate between the long-shunt compound-wound dc generator and the short-shunt compound-wound dc generator:

- The short-shunt compound-wound dc generator increases the strength of the total field magnetic flux. This results in an increase in the generator's output voltage as load increases, helping to improve the stability of the machine.
- Short-shunt compound-wound dc generators are used in power generating and industrial environments.
- The long-shunt compound-wound configuration ensures that all of the armature current flows through the series field coil. This creates the ability to over- or under-compound the generator.

Generator ***over-compounding*** is the process of adding additional turns or loops to the series field coil. When the generator is over-compounded, additional voltage is generated when the load on the machine increases. Over-compounding is beneficial because it counters the effects of increased line losses when operating under increased load conditions. Generator ***under-compounding*** is the process of removing turns from the series field coil. When the generator is under-compounded, the strength of the shunt field is decreased as the load on the generator increases. This is beneficial as the output voltage becomes more balanced as the load on the machine increases. Long-shunt compound-wound generators with under-compounding are used when stable output voltage is required.

Shedding Light

The constant magnetic flux created by permanent magnets can sometimes cause starting problems with dc generators. With a load connected to the generator, current will start flowing as soon as the armature shaft starts to rotate. The current flow causes a magnetic field, or flux, to develop in the armature which is attracted to the magnetic field created by the permanent magnets. In order for the generator's rotating shaft to accelerate effectively and continue to rotate, this attraction force must be offset and overcome. The effect of the magnetic flux on the generator's startup torque can be minimized by switching the armature current on when the generator reaches the desired emf. This can be accomplished by using a centrifugal switch or some type of voltage regulator.

Externally-Excited DC Generators

Externally-excited dc generators use an external dc source, such as a battery or rectifier circuit, to create the current that produces the stator's magnetic field. With a separate source controlling the strength of the magnetic flux in the stator, there is greater control of voltage generated. Although the speed of the generator's rotating shaft still has a direct effect on the amount of emf generated, controlling the strength of the stator's magnetic flux with an external power source allows for more precise control of the output voltage. In addition, using an external source reduces the need to change the resistance of the load circuit to control the generator's output voltage.

Externally-excited dc generators are more expensive than self-excited dc generators. The increase in cost is attributed mainly to the external power supply and the controller needed to adjust the magnitude of the circuit current which, in turn, regulates the voltage. Due to their high cost, externally-excited dc generators are usually found in high precision applications such as medical equipment and research laboratories.

35.6 DC Generator Power

Having discussed the construction, operation, configurations, and applications for dc generators, we will now provide some examples on how to calculate the power output of dc generators.

When making power calculations, there are two important factors that affect the output power of a dc generator. These two factors are the total number of conductors in the generator and the number of parallel paths. In terms of dc generators, the term *conductor* is used to describe the portion of a wire that is positioned within a magnetic field, and has an emf produced across it. Consider the simple coil shown in **Figure 35-44**. This coil has two sections of wire that are positioned within a magnetic field and will have emfs produced

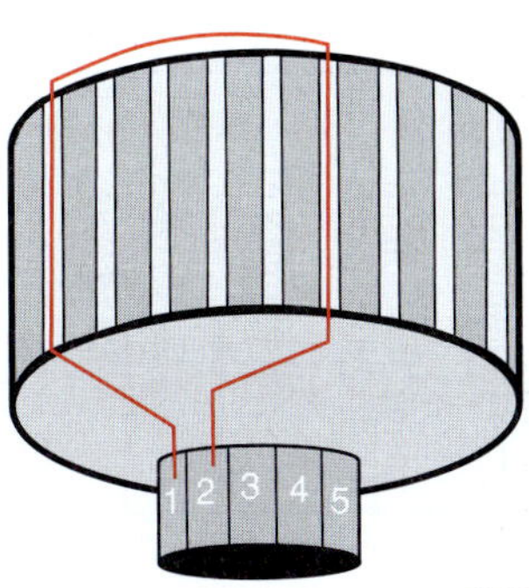

Goodheart-Willcox Publisher

Figure 35-44. This simple coil has two conductors, as it makes two passes through the armature. A simplex coil will have two conductors, while a duplex coil will have four conductors, and a triplex coil will have six conductors.

across them when rotated within a magnetic field. Therefore, this coil has two conductors. Another quantity that factors into the calculation of dc generator power is the number of conductors per path. This is calculated simply as the number of total conductors divided by the number of parallel paths. Refer to **Example 35-11**.

Example 35-11

Calculate the armature resistance, $R_{ARMATURE}$, armature voltage, $E_{GENERATOR}$, rated armature current, $I_{ARMATURE}$, and the rated output power, P_{OUTPUT}, of a 6-pole, simplex lap-wound dc generator with the following parameters:

- Number of conductors, Z_T, = 600
- The resistance per conductor, R_z, = 0.05 Ω
- The induced voltage per conductor, V_z, = 0.52 V
- The conductor current rating, I_z, = 20 A

✓ Solution

For a simplex lap-wound armature, the number of parallel paths, *a*, is given by the following:

$$a = P \times Plex$$

where the number of poles, P, is 6 and the plex is 1 for a simplex configuration. Therefore, the total number of parallel paths is calculated as:

$$a = P \times Plex$$

$$a = 6 \times 1 = 6$$

The number of conductors per path, Z_{PATH}, can be calculated as the total number of conductors, Z_T, divided by the number of parallel paths, *a*, as follows:

$$Z_{PATH} = \frac{Z_T}{a}$$

$$Z_{PATH} = \frac{600}{6}$$

$$Z_{PATH} = 100 \text{ conductors per path}$$

The configuration of this simplex lap-wound armature is shown in the following figure:

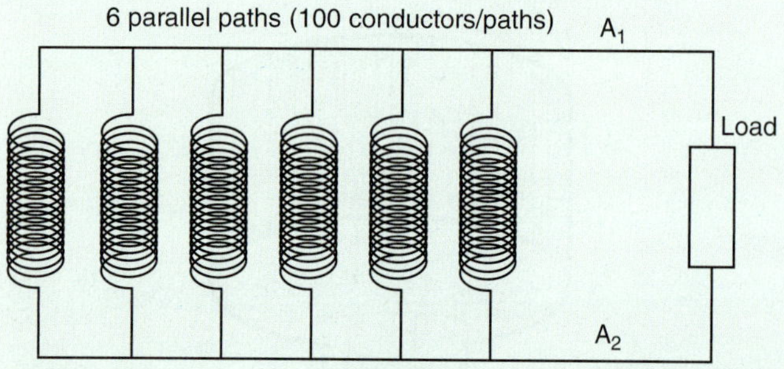

Goodheart-Willcox Publisher

The resistance of each path, R_{PATH}, can be calculated by multiplying the resistance of each conductor, R_z, and the number of conductors per path, Z_{PATH}, together as follows:

$$R_{PATH} = Z_{PATH} \times R_z$$

$$R_{PATH} = 100 \times 0.05\ \Omega$$

$$R_{PATH} = 5\ \Omega$$

With six parallel paths, each with a resistance of 5 Ω, the total equivalent resistance can be determined using the formula for multiple same-valued resistances connected in parallel, as originally introduced in Chapter 11, *Parallel Circuits*:

$$R_T = \frac{R}{\text{Number of same-valued resistances}}$$

In this case, the formula can be modified to reflect the resistance of armature, the resistance of each path, and the number of parallel paths as follows:

$$R_{ARMATURE} = \frac{R_{PATH}}{a}$$

Substituting the known values for R_{PATH} and *a*, the resistance of the armature, $R_{ARMATURE}$, is calculated as follows:

$$R_{ARMATURE} = \frac{R_{PATH}}{a}$$

$$R_{ARMATURE} = \frac{5\ \Omega}{6}$$

$$R_{ARMATURE} = 0.833\ \Omega$$

The armature voltage, $E_{GENERATOR}$, is the same as the voltage across any one path, as the voltage across any one path in a parallel circuit is equal to the supply voltage. Since each path is made up of 100 conductors, Z_{PATH}, the voltage across any one path, E_{PATH}, will be equal to the voltage across each conductor, V_z, multiplied by the number of conductors in the path:

$$E_{GENERATOR} = E_{PATH} = Z_{PATH} \times V_Z$$

$$E_{GENERATOR} = E_{PATH} = 100 \times 0.52 \text{ V}$$

$$E_{GENERATOR} = E_{PATH} = 52 \text{ V}$$

The rated armature current, $I_{ARMATURE}$, is the sum of the rated current flows through the parallel paths, as the total current flow in a parallel circuit is the sum of the branch circuit currents. Since there are six parallel paths and the conductor current rating, I_z, is

20 A, the rated armature current, I_A, can be calculated by multiplying the conductor current rating by the number of parallel paths, *a*, as follows:

$$I_{ARMATURE} = I_Z \times a$$

$$I_{ARMATURE} = 20\ A \times 6$$

$$I_{ARMATURE} = 120\ A$$

The rated output power of the armature, P_{OUTPUT}, can be calculated from the $P = I \times E$ form of the power formula introduced in Chapter 6, *Electrical Characteristics*. The rated output power of the armature is calculated as follows:

$$P_{OUTPUT} = E_{GENERATOR} \times I_{ARMATURE}$$

$$P_{OUTPUT} = 52\ V \times 120\ A$$

$$P_{OUTPUT} = 6240\ W$$

This example is summarized in the following figure.

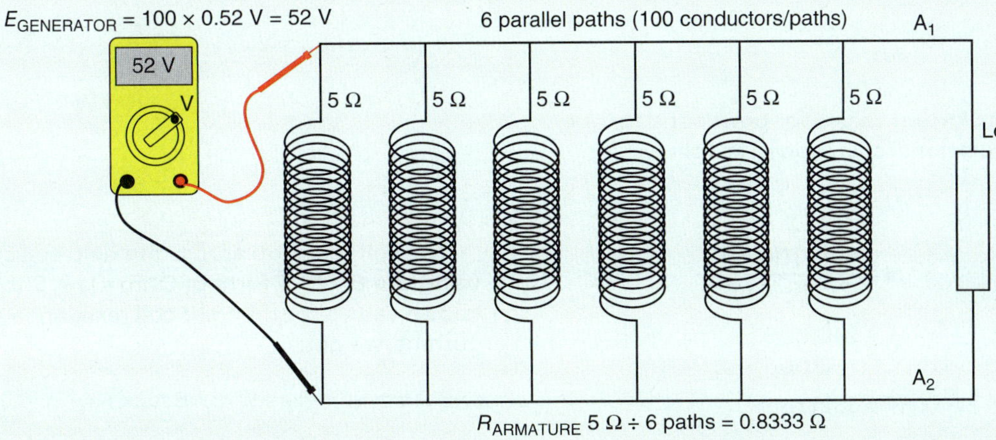

Goodheart-Willcox Publisher

The output power of a dc generator can also be calculated using the rotational speed of the shaft and the applied torque. The formula for determining the output power of a dc generator with respect to the rotational speed of the shaft and torque is:

$$P = \frac{2\pi NT}{60}$$

where:

P = Power generated in watts
N = Shaft rotational speed in revolutions per minute (rpm)
T = Torque in N·m

The $2\pi/60$ is a conversion factor that converts the shaft's rotational speed from rotations per minute (rpm) to radians/second as follows:

- 1 complete rotation = 2π radians, and
- 1 minute = 60 seconds

where:

$$1\ \text{rotation/minute (rpm)} = 2\pi\ \text{radians/minute}$$
$$= 2\pi\ \text{radians/60 seconds}$$
$$1\ \text{rpm} = (2\pi/60)\ \text{radians/second}$$

Refer to **Example 35-12** and **Example 35-13** on how to calculate the output power of a dc generator.

Example 35-12

How much power is generated by a dc generator if its shaft is turning at a speed of 1000 rpm and is operating with a torque of 3 N·m?

✓ Solution

Using the formula for generator power and substituting the known values for shaft speed and torque, we get:

$$P = \frac{2\pi NT}{60}$$

$$P = \frac{2\pi(1000\ \text{rpm})(3\ \text{N·m})}{60}$$

$$P = \frac{6000\pi}{60}$$

$$P = 100\pi$$

$$P = 314\ W$$

Example 35-13

How fast must the shaft of a dc generator turn if the generator needs to produce 4000 W of power at a torque of 10 N·m?

✓ Solution

Using the formula for generator power and solving for rotational speed, we get:

$$P = \frac{2\pi NT}{60}$$

$$60 \times P = 2\pi NT$$

$$N = \frac{60P}{2\pi T}$$

Substituting the known values for power and torque, we can solve for rotational shaft speed as follows:

$$N = \frac{60P}{2\pi T}$$

$$N = \frac{60(4000\ \text{W})}{2\pi\ (10\ \text{N·m})}$$

$$N = \frac{240{,}000}{20\pi}$$

$$N = 3821\ \text{rpm}$$

Now It's Your Turn 35-13

Calculate the torque of a dc generator if it has a rotational speed of 500 rpm and generates 3000 W.

35.7 DC Generator Efficiency

The efficiency of a dc generator can be calculated by dividing the power of the connected load by the output power of the generator.

$$\text{EFF}_{\text{GENERATOR}} = \frac{\text{Power}_{\text{LOAD}}}{\text{Power}_{\text{OUTPUT}}}$$

This result provides the efficiency of the generator in decimal form. To express the efficiency of the generator as a percentage, the result of the previous formula is multiplied by 100:

$$\%\ \text{EFF} = \frac{\text{Power}_{\text{LOAD}}}{\text{Power}_{\text{OUTPUT}}} \times 100\%$$

Refer to **Example 35-14** and **Example 35-15** on how to calculate the efficiency of a series-wound dc generator.

Example 35-14

Calculate the efficiency of a series-wound dc generator that has a 1 Ω field coil, and is supplying 12 V at 100 W to a lamp load.

✓ Solution

This problem is an extension of **Example 35-6**. In that example, the power formula was used to find the circuit current:

$$P = I \times E$$

$$I = \frac{P}{E}$$

$$I = \frac{100\ \text{W}}{12\ \text{V}}$$

$$I = 8.33\ \text{A}$$

The voltage drop across the field coil can be found using the $E = I \times R$ form of Ohm's law. Substituting the known values for the field coil resistance and circuit current, we get:

$$E = I \times R$$

$$E = 8.33\ \text{A} \times 1\ \Omega$$

$$E = 8.33\ \text{V}$$

To supply the required 12 V to the load and overcome the 8.33 V voltage drop across the field coil, the generator must produce 20.33 V (12 V + 8.33 V). The power at the generator can be calculated using the $P = I \times E$ form of the power formula and substituting the known values for voltage and current:

$$P = I \times E$$

$$P = 8.33\ \text{A} \times 20.33\ \text{V}$$

$$P = 169.35\ \text{W}$$

For this dc generator to supply 100 W of power at 12 V to the load, it must produce 169.35 W. The efficiency of the generator can be calculated as follows:

$$\text{EFF}_{\text{GENERATOR}} = \frac{\text{Power}_{\text{LOAD}}}{\text{Power}_{\text{OUTPUT}}}$$

$$\text{EFF}_{\text{GENERATOR}} = \frac{100\ \text{W}}{169.35\ \text{W}}$$

$$\text{EFF}_{\text{GENERATOR}} = 0.59 = 59\%$$

Example 35-15

Calculate the efficiency of a series-wound dc generator that has a 1 Ω field coil and is supplying 120 V at 100 W to a lamp load.

✓ Solution

This problem is, in effect, the same as **Example 35-14**, with the exception of the voltage being supplied to the load.

Using the power formula to find the circuit current, we get:

$$P = I \times E$$

$$I = \frac{P}{E}$$

$$I = \frac{100 \text{ W}}{120 \text{ V}}$$

$$I = 0.833 \text{ A}$$

The voltage drop across the 1 Ω field coil can be found using the $E = I \times R$ form of Ohm's law. Substituting the known values for the field coil resistance and circuit current, we get the following:

$$E = I \times R$$

$$E = 0.833 \text{ A} \times 1 \ \Omega$$

$$E = 0.833 \text{ V}$$

To supply the required 120 V to the load and overcome the 0.833 V voltage drop across the field coil, the generator must produce 120.833 V (120 V + 0.833 V). The power at the generator can be calculated using the $P = I \times E$ form of the power formula and substituting the known values for voltage and current:

$$P = I \times E$$

$$P = 0.833 \text{ A} \times 120.833 \text{ V}$$

$$P = 100.65 \text{ W}$$

For this dc generator to supply 100 W of power at 120 V to the load, it must produce 100.65 W. The efficiency of the generator can be calculated as follows:

$$EFF_{GENERATOR} = \frac{Power_{LOAD}}{Power_{OUTPUT}}$$

$$EFF_{GENERATOR} = \frac{100 \text{ W}}{100.65 \text{ W}}$$

$$EFF_{GENERATOR} = 0.994 = 99.4\%$$

Refer to **Example 35-16** and **Example 35-17** on how to calculate the efficiency of a shunt-wound dc generator.

Example 35-16

Calculate the efficiency of a shunt-wound dc generator that has a 90 Ω field coil and is supplying 20 V at 150 W to a lamp load as shown here.

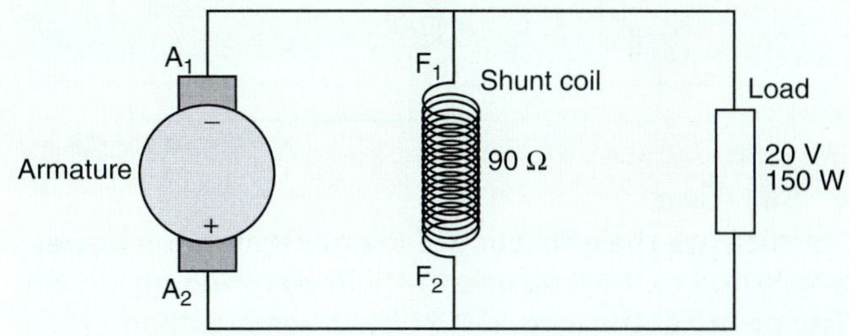

Goodheart-Willcox Publisher

✓ Solution

To calculate the efficiency of the generator, the power attributed to the load and the field coil must be known. The power at the load, 150 W, has been provided, so the power consumed by the field coil must be calculated. The power consumed by the field coil can be obtained using the $P = E^2/R$ form of the power formula.

Substituting the known values for the voltage across the field coil and the resistance of the field coil, the power is calculated as follows:

$$P = \frac{E^2}{R}$$

$$P = \frac{(20 \text{ V})^2}{90 \ \Omega}$$

$$P = \frac{400}{90}$$

$$P = 4.44 \text{ W}$$

The generator, therefore, must produce 154.44 W of power (150 W + 4.44 W). The efficiency of the generator can be calculated as follows:

$$EFF_{GENERATOR} = \frac{Power_{LOAD}}{Power_{OUTPUT}}$$

$$EFF_{GENERATOR} = \frac{150 \text{ W}}{154.44 \text{ W}}$$

$$EFF_{GENERATOR} = 0.971 = 97.1\%$$

Example 35-17

Calculate the efficiency of a shunt-wound dc generator that has a 90 Ω field coil, and is supplying 120 V at 150 W to a lamp load.

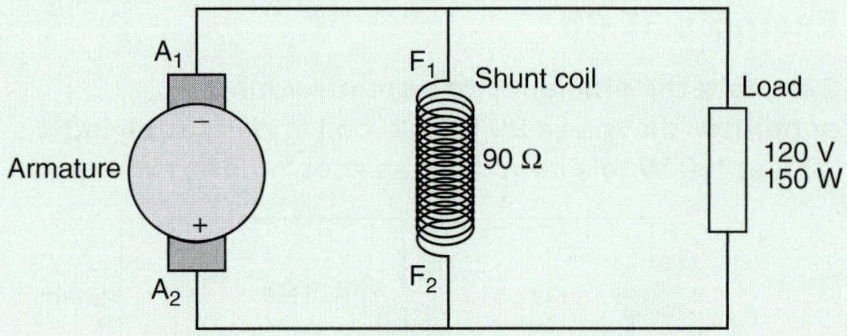

Goodheart-Willcox Publisher

✓ Solution

To calculate the efficiency of the generator, the power attributed to the load and the field coil must be known. The power at the load, 150 W, has been provided, so the power consumed by the field coil must be calculated. The power consumed by the field coil can be obtained using the $P = E^2/R$ form of the power formula. Substituting the known values for the voltage across the field coil and the resistance of the field coil, the power is calculated as follows:

$$P = \frac{E^2}{R}$$

$$P = \frac{(120\text{ V})^2}{90\ \Omega}$$

$$P = \frac{14{,}400}{90}$$

$$P = 160\text{ W}$$

The generator, therefore, must produce 310 W of power (150 W + 160 W). The efficiency of the generator can be calculated as follows:

$$EFF_{GENERATOR} = \frac{Power_{LOAD}}{Power_{OUTPUT}}$$

$$EFF_{GENERATOR} = \frac{150\text{ W}}{310\text{ W}}$$

$$EFF_{GENERATOR} = 0.484 = 48.4\%$$

By comparing the results from **Example 35-16** and **Example 35-17**, it can be seen that the shunt-wound dc generator is better suited for low-voltage applications.

CHAPTER 35 REVIEW

Summary

- The construction of a dc generator is similar to a dc motor. (35.1)
- The output of the generator is measured across the two conductors connected to the brushes that are in contact with the rotating commutator. (35.2)
- The armature windings of a dc generator are commonly configured as lap-wound, wave-wound, and frogleg-wound windings. Lap-wound windings are classified as parallel-connected windings. Wave-wound windings are classified as series-connected windings. Frogleg-wound windings are classified as series-parallel-connected windings. (35.3)
- The configuration of armature windings can be identified by the following terms: pole pitch, coil span, front pitch, and winding pitch. (35.4)
- The field windings on a dc generator can be of the self-excited or externally-excited variety. Self-excited field windings are powered internally by the generator. Externally-excited field windings rely on an external power source. Most generators are self-excited. (35.5)
- Series-wound dc generators are constructed with their field coils in series with the armature and the load. (35.5)
- The shunt-wound dc generator is configured with the field coils connected in parallel to the armature and to the load. (35.5)
- A compound dc generator is a self-excited generator that uses both series and shunt windings. (35.5)
- The rated power of a dc generator can be calculated using Ohm's law and the power formula. (35.6)
- The efficiency of a dc generator can be calculated as the either the voltage used by the load divided by the voltage produced, or as the power used divided by the power generated. (35.7)

Review Questions

Use the information in the chapter to answer the following questions.

Measuring Your Potential

1. *True or False?* The output voltage of a dc generator can change in magnitude but always has the same polarity. (35.1)
2. *True or False?* The output waveform of a dc generator can best be described as ripple-shaped. (35.1)
3. The operation of a dc generator can be accurately described as _____. (35.2)
   A. rotating a coil of wire within a stationary magnetic field
   B. rotating a coil of wire within a rotating magnetic field
   C. positioning a stationary coil of wire within a rotating magnetic field
   D. positioning a stationary coil of wire within a stationary magnetic field
4. A lap-wound armature will have how many electrical conductors at each commutator segment? (35.3)
   A. 0
   B. 1
   C. 2
   D. 3
5. *True or False?* On a typical wave-wound armature, there are two brushes in contact with the commutator. (35.3)
6. *True or False?* A lap-wound armature with four poles will have four parallel paths and four brushes. (35.3)
7. A frogleg-wound winding will have how many electrical conductors at each commutator segment? (35.3)
   A. 2
   B. 4
   C. 6
   D. 8
8. *True or False?* A lap-wound armature with six brushes will have six parallel paths. (35.3)
9. The coil span is commonly described as the _____. (35.4)
   A. width of an armature coil
   B. distance between the starting ends of two connected coils
   C. distance between the finishing end of two connected coils
   D. the distance between the starting end of one coil and the finishing end of the next coil
10. The term *pole pitch* is used to describe the number of _____. (35.4)
    A. poles a generator is operating with
    B. armature slots covered by each generator pole
    C. poles covered by each armature slot
    D. parallel paths in an armature winding

11. *True or False?* The term *fractional pitch* indicates that the coil span is less than the pole pitch. (35.4)
12. *True or False?* Winding pitch is distance between the starting ends of two consecutive coil loops. (35.4)
13. *True or False?* A self-excited dc generator uses batteries as the power source to create the required magnetic field. (35.5)
14. A(n) _____ dc generator is configured with the field coils connected in parallel to the armature and to the load. (35.5)
    A. series-wound
    B. shunt-wound
    C. compound-wound
    D. externally-excited
15. *True or False?* When calculating power of a dc generator, two factors to consider are the total number of conductors and the number of parallel paths. (35.6)
16. Of the following, which is the formula used to calculate the efficiency of a dc generator? (35.7)
    A. $\frac{P}{E}$
    B. $\frac{2\pi NT}{60}$
    C. $\frac{\text{Power}_{\text{LOAD}}}{\text{Power}_{\text{OUTPUT}}}$
    D. $\frac{2 \times (\text{C} - 1)}{\text{P}}$

Plug-In

1. Two common renewable and sustainable energy sources that can be used to directly act on a generator's rotating shaft are _____. (35.1)
    A. wind power and hydropower
    B. thermal energy and hydropower
    C. solar energy and wind power
    D. solar energy and thermal energy
2. The commutator pitch for a progressive simplex winding that has 15 coils and 8 poles is _____. (35.3)
    A. 1
    B. 2
    C. 3
    D. 4
3. A simplex wave-wound armature with nine commutator segments will have how many coils? (35.3)
    A. 3
    B. 6
    C. 9
    D. 12
4. Which of the following is the correct formula to calculate the commutator pitch of a simplex progressive wave winding? (35.4)
    A. $Y_C = \frac{2 \times (\text{C} + 1)}{\text{P}}$
    B. $Y_C = \frac{2 \times (\text{C} - 1)}{\text{P}}$
    C. $Y_C = \frac{(\text{C} + 1)}{2\text{P}}$
    D. $Y_C = \frac{(\text{C} - 1)}{2\text{P}}$
5. If a winding is classified as full-pitched, the coil span _____. (35.4)
    A. and the pole pitch are both negative
    B. is less than the pole pitch
    C. is equal to the pole pitch
    D. is greater than the pole pitch

Making the Connection

1. Explain why lap-wound armatures are more suitable for low-voltage, high-current applications. (35.3)
2. Explain how the resistance of a series-connected field coil affects the efficiency of a dc generator. (35.5)
3. Explain how the resistance of a parallel-connected field coil affects the efficiency of a dc generator. (35.5)

CHAPTER

36 AC Generators

Marina Demkina/Shutterstock.com

SPARKING DISCUSSION

An electrician claims that the commutators on a dc generator perform the same basic function as the slip rings on an ac generator. Do you agree or disagree with this electrician? Provide as much information as you can to support your opinion.

LEARNING OUTCOMES

After completing this chapter, you will be able to:

36.1 List the key components of an ac generator.
36.2 Explain how a revolving-armature and a revolving-field ac generator operate.
36.3 Describe the configuration of the windings of a single-phase ac generator.
36.4 Describe the configuration of the windings of a three-phase ac generator.
36.5 Calculate the frequency and speed of an ac generator.
36.6 List the conditions that must be met for parallel-connected ac generators.

TECHNICAL TERMS

parallel-connected generators
revolving-armature ac generator
revolving-field ac generator
slip ring

Introduction

An ac generator is a machine that converts mechanical energy into electrical energy. The main difference between the dc generator and the ac generator is the type of voltage produced by the two machines. The dc generator supplies dc voltage at its output, while the ac generator supplies ac voltage at its output. The mechanical energy that acts as the input to the ac generator can come from any source of physical motion that is harnessed to provide the rotating torque necessary to turn the generator's rotor. As in the case of any generator, the required mechanical force may come from a renewable or sustainable source, such as blowing wind or flowing water. An ac generator's rotor can also be turned by using other power sources such as steam turbines or combustion engines. In the cars we drive, the rotating force of the engine is used for multiple purposes including moving the vehicle and generating electricity. The electricity-producing component in the vehicle is a special type of generator called an alternator. The alternator, just as any other generator, relies on a rotating force at the machine's input to produce electrical energy at its output. Regardless of what they are called and how they are classified, dc generators, ac generators, and alternators convert mechanical force into electrical energy.

This chapter discusses the unique attributes of ac generators. It is recommended that you refer back to Chapter 34, *Generator Fundamentals*, for general information about generators, as much of the information contained in that chapter applies to ac generators as well.

36.1 AC Generator Construction

The construction of an ac generator is similar to that of the dc generator. The main difference is that a dc generator uses commutator sections, while ac generators do not. Instead of using commutators, ac generators use slip rings. Various components of a typical ac generator include the following:

- Field windings
- Shaft
- Armature
- Yoke, or housing
- Rotor
- Stator
- Slip rings and brushes

Field Windings

The field windings create the lines of magnetic flux that are cut by the armature windings. This cutting action occurs when the rotating armature windings cut the stationary field windings, or when the stationary armature windings are cut by the rotating field windings. The field windings are commonly made up of conductor coils that are powered by an external power source to produce the magnetic flux. In some applications, the magnetic flux can be produced by permanent magnets instead of an external voltage supply.

Shaft

The shaft of an ac generator serves as the physical connection between the external mechanical force and the coils of wire positioned within the generator's magnetic field. As with all generators and alternators, the power produced by the ac generator is directly proportional to the speed at which the shaft rotates. As the rotational speed of the generator's shaft increases, the output power of the generator increases.

Armature

The armature is the part of the generator that does the work of converting mechanical energy into electrical energy. The armature holds the generator's electrical windings in place, facilitates the electrical connections between the windings and the generator's output, and acts as the connection point between the generator and the external mechanical rotating force. Since the armature winding is connected, either directly or indirectly, to the output wiring of the generator, the armature winding conductors are designed and sized to safely carry the full-load current of the generator.

Yoke

The ac generator's yoke, or housing, is the external structure, or shell, of the generator. The yoke is typically constructed from cast iron or rolled steel and serves two main purposes in the generator. It provides mechanical support for the field magnets or poles and protects the windings and other internal components of the generator. In addition, the yoke, being made of a conductive material, effectively carries the magnetic flux through the generator's poles. Magnets are arranged within the generator's yoke to create the lines of magnetic flux that will be cut by the generator's windings.

Rotor

The rotor is the rotating component of an ac generator. In some cases, the rotor is connected to the generator's armature windings or coils. In this configuration, when the shaft rotates, the armature windings also rotate within the stationary field's lines of magnetic flux to produce electricity. In some other generator configurations, the rotor is comprised of the field winding, which rotates around the stationary armature windings to generate electricity.

Stator

The stator is the stationary portion of an ac generator. Depending on the type and construction of the generator, the stator can be used to describe the stationary field windings or the stationary armature windings. Whether the windings are found in either the stator or the rotor, it is the relative motion between the armature and field windings that results in the generation of electricity.

Slip Rings and Brushes

In a revolving-armature generator, ***slip rings*** and brushes, **Figure 36-1**, create the electrical connection points between the rotating windings and the output wiring of an ac generator. The slip rings are connected directly to the generator's shaft and completely encircle it, while the brushes push up against the slip rings. This helps ensure that good electrical contact is made between the generator's rotating windings and its output connections, **Figure 36-2**. Each end of a winding is connected to its own slip ring. Therefore, a generator that has a single winding will have two slip rings, **Figure 36-3**.

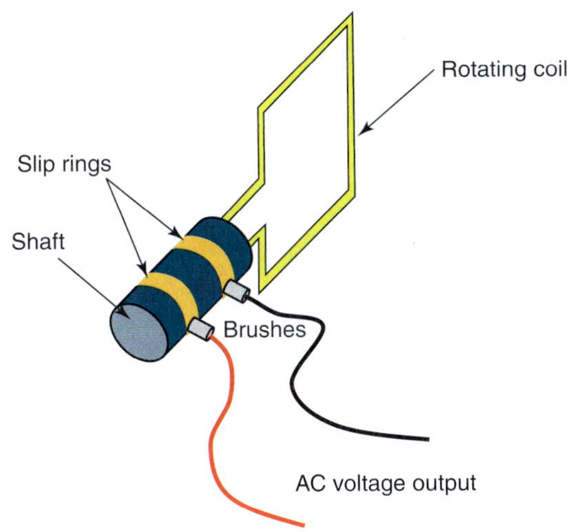

Goodheart-Willcox Publisher

Figure 36-1. In a revolving-armature ac generator, slip rings and brushes create the electrical connection points between the rotating windings and the output wiring.

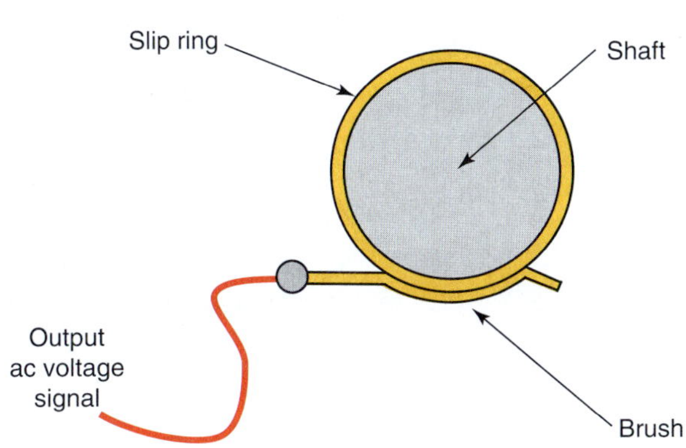

Goodheart-Willcox Publisher

Figure 36-2. The slip ring completely surrounds the generator's shaft and connects to the machine's output circuit with pressure-connected brushes.

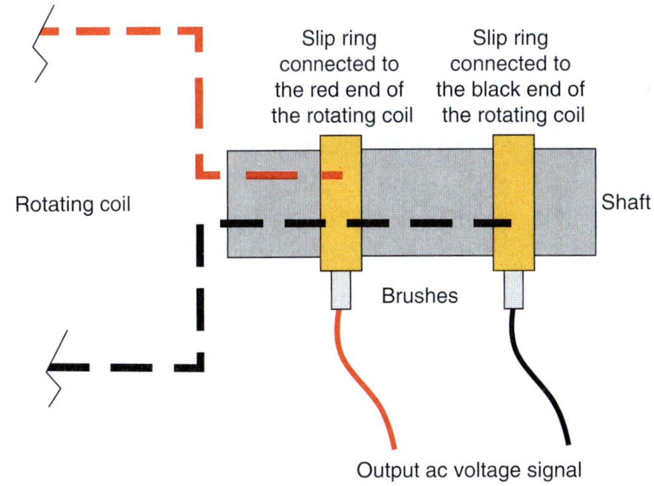

Goodheart-Willcox Publisher

Figure 36-3. The two ends of a coil are connected to different slip rings.

A three-phase ac generator, which has three windings, may have up to six slip rings.

36.2 Common Types of AC Generators

There are two common types of ac generators based on which portion of the generator rotates and which portion remains stationary. These generators are the revolving-armature ac generator and the revolving-field ac generator.

Revolving-Armature AC Generator

In a ***revolving-armature ac generator***, the armature windings, or coils, rotate within a fixed and stationary magnetic field. On this type of machine, the armature is the rotor and the stationary magnetic field is the stator. The armature of a revolving-armature ac generator closely resembles the armature of a dc generator. However, instead of connecting each end of the armature winding to a commutator segment, each end of the armature winding is connected to a slip ring. The brushes that push against the slip rings provide the electrical path to the outlet of the generator. Due to the pressure connection that exists between the slip ring and the brush, there are limitations to the output capacity of this type of generator. As a result, this type of ac generator is used for lighter-duty applications. To increase the output capacity of an ac generator, a revolving-field ac generator can be used.

Revolving-Field AC Generator

Unlike the revolving-armature ac generator, the ***revolving-field ac generator*** is constructed with a stationary armature and a rotating magnetic field. On this type of machine, the

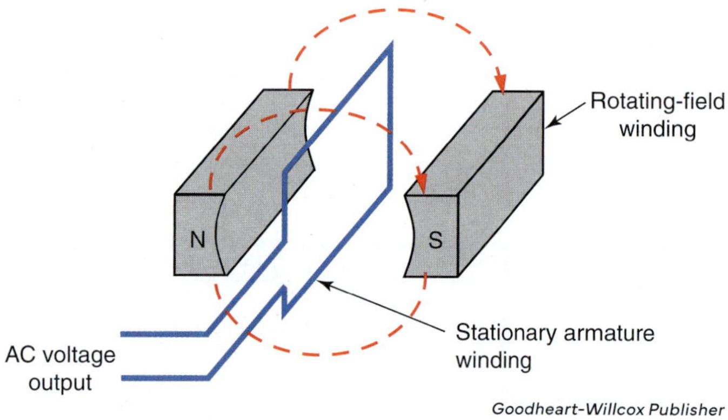

Figure 36-4. In a revolving-field ac generator, the armature is stationary while the magnetic field rotates around it. This allows the armature winding to be connected directly to the generator's output wiring.

armature is the stator and the magnetic field acts as the rotor, **Figure 36-4**. This type of ac generator does not use slip rings or brushes, as there is no relative motion or twisting between the armature windings and the generator's output wiring. The armature winding is connected directly to the generator's output wiring. With no slip rings and brushes, there are no pressure-reliant electrical connections. The revolving-field ac generator can be used for higher-duty applications and to produce higher output voltages.

> **Shedding Light**
>
> As discussed in Chapter 34, *Generator Fundamentals*, the voltage output of an ac generator is a function of the strength of the magnetic field, the speed of rotation, and the number of turns in the armature windings. Increasing any of these three factors increases the voltage output.

36.3 **Single-Phase AC Generators**

A simple ac generator is made up of one armature winding or coil positioned within a stationary magnetic field. The armature rotates within the stationary magnetic field to generate voltage. Consider the arrangement shown in **Figure 36-5**. In this figure, a single conductive loop, or coil, is positioned vertically within a stationary magnetic field. The connected coil, the shaft, and the slip rings form the armature. The ends of the armature winding are connected to the slip rings. The ac generator's output voltage is measured across the two conductors connected to the brushes that are, in turn, in contact with the slip rings.

When the armature begins to rotate in the clockwise direction, it rotates in a direction that is parallel to the direction of the magnetic flux, **Figure 36-6**. Since

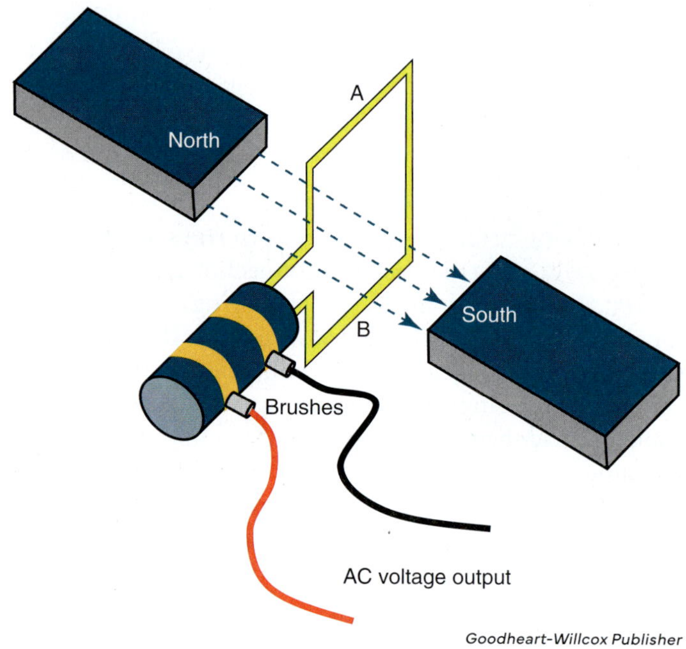

Figure 36-5. A single conductive loop positioned vertically within a stationary magnetic field. The top of the coil is identified as the *A* side of the coil, while the bottom portion of the coil is identified as the *B* side of the coil.

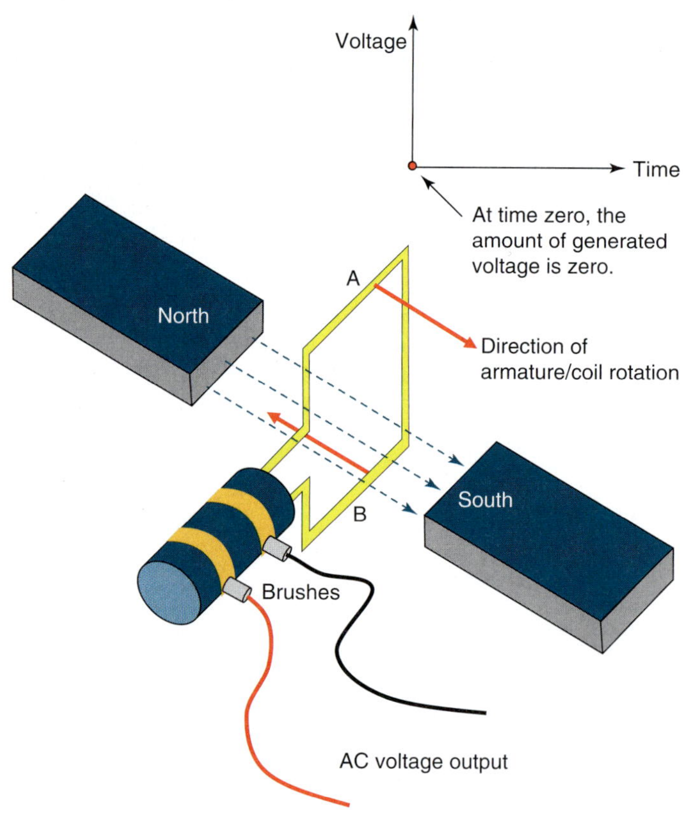

Figure 36-6. As the coil begins to rotate, the direction of the coil and that of the lines of magnetic flux are parallel to each other, and no voltage is generated. Voltage is generated only when the coil's conductors cut through the lines of magnetic flux.

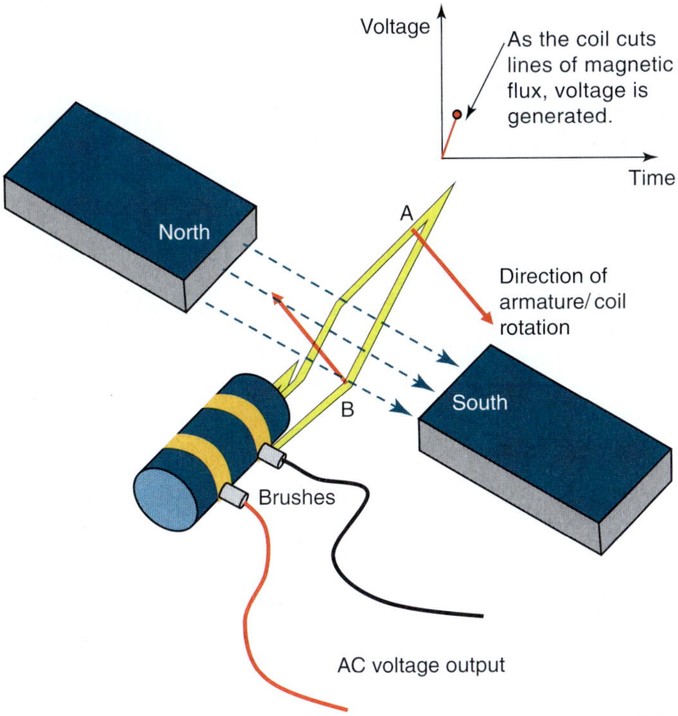

Goodheart-Willcox Publisher

Figure 36-7. As the coil begins to rotate, lines of flux are cut, and voltage is generated.

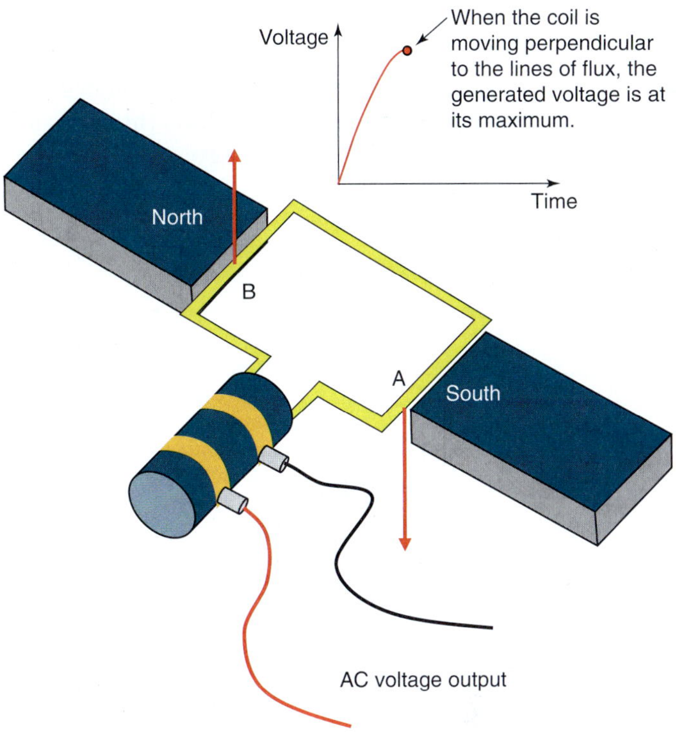

Goodheart-Willcox Publisher

Figure 36-8. The maximum lines of flux are cut when the direction of coil motion is perpendicular to the lines of magnetic flux. When the maximum number of flux lines are cut, the maximum voltage is produced.

electromotive force is only generated when a conductor cuts or crosses lines of magnetic flux, no voltage is generated, as the conductor's path and the lines of magnetic flux are parallel to each other.

As the armature continues to rotate, the top and bottom ends of the coil begin to cut across the lines of magnetic flux, generating voltage, **Figure 36-7**. The amount of voltage generated increases until the direction of coil motion is perpendicular to the lines of flux, **Figure 36-8**. It is at this point that the change in flux is at its highest value and, as a result, the generated voltage is at its maximum. At this point, the armature has turned 90° or has made one-quarter of a full rotation. As the armature continues to rotate and moves back to a vertical position, the amount of voltage generated drops. This process continues until the direction of coil motion is once again parallel to the direction of the lines of magnetic flux, and the generated voltage is once again zero, **Figure 36-9**. Up to this point, the operation of the ac generator is very similar to the operation of the dc generator. This is where the similarities between the two generator types end.

As the armature continues to rotate, lines of flux are once again being cut and voltage is being generated. However, the direction of motion of side *A* of the armature coil is now in the opposite direction, as is the direction of motion of side *B*. For the first 180° of rotation, side *A* of the coil

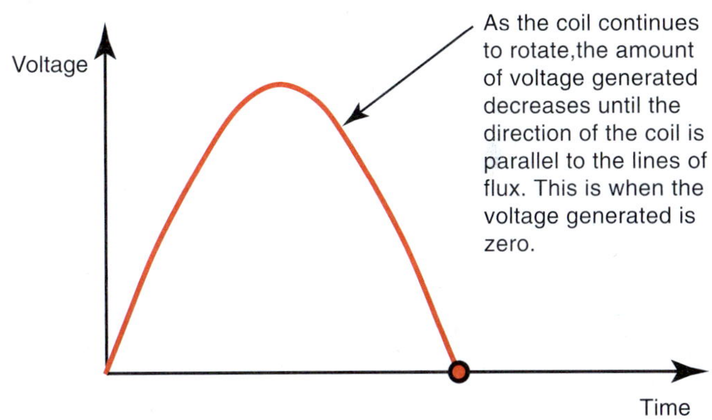

Goodheart-Willcox Publisher

Figure 36-9. As the coil continues to rotate, the amount of voltage generated decreases until the direction of the coil is once again parallel to the lines of flux. At this point, the coil has rotated 180°.

was moving in a downward direction as it cut the lines of magnetic flux, while the *B* side was moving upwards. For the second 180° of rotation, the *A* side of the coil is moving in an upward direction as it cuts the lines of magnetic flux, while the *B* side is moving downward. As a result, the polarity of the voltage being produced, as measured at the generator's outlet wiring connections, will be negative instead of positive, **Figure 36-10**.

As the armature continues to rotate, voltage with a negative polarity is produced. When the armature is parallel to the lines of magnetic flux, the maximum negative voltage is being produced, **Figure 36-11**. As the armature returns to the vertical position, the amount of negative voltage being produced drops, until it reaches zero. At this point, one complete ac sine wave has been produced, **Figure 36-12**. The production of a continuous ac signal will continue as long as the external force providing the rotation is present.

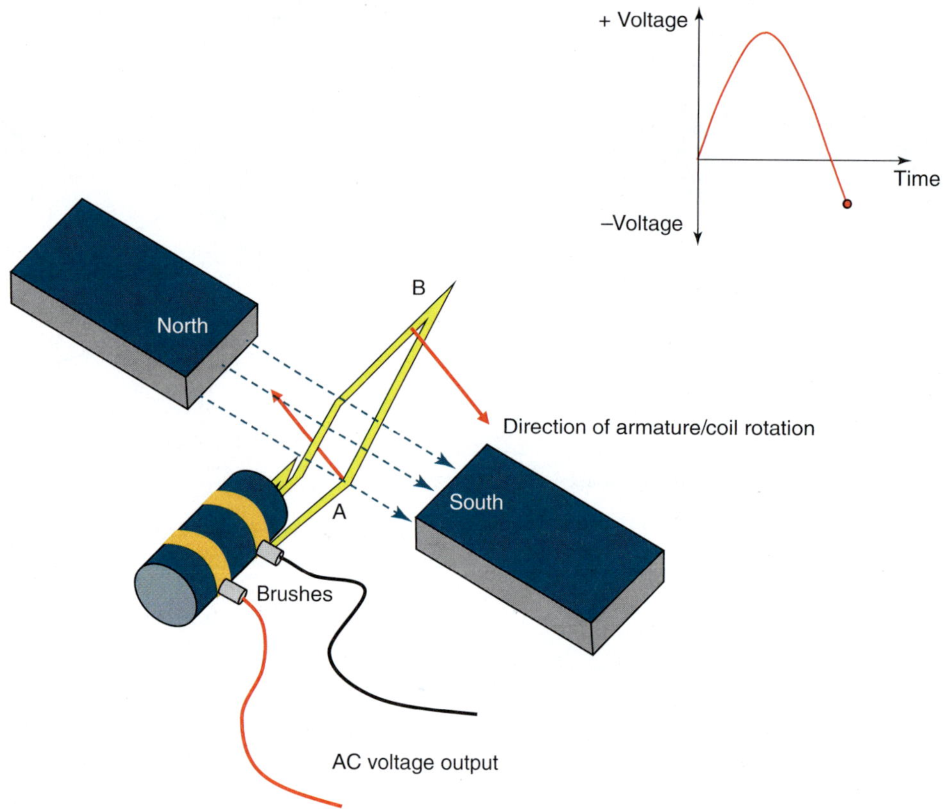

Goodheart-Willcox Publisher

Figure 36-10. As the armature continues to rotate, the polarities of the output conductors are reversed. As such, the voltage produced has an opposite polarity compared to the voltage produced during the first 180° of armature rotation.

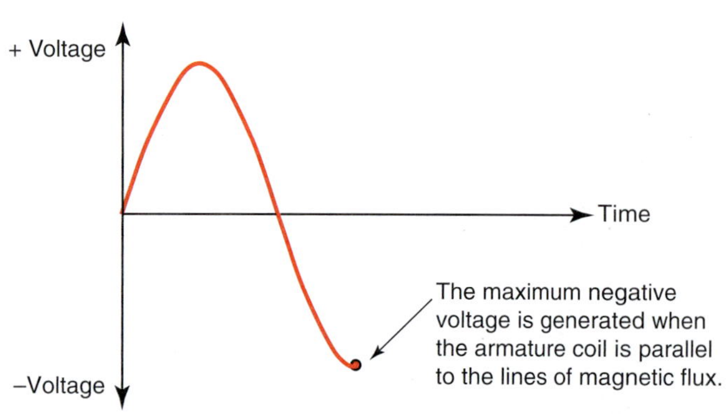

Goodheart-Willcox Publisher

Figure 36-11. When the coil is once again parallel to the lines of flux, the amount of negative voltage produced is at its maximum.

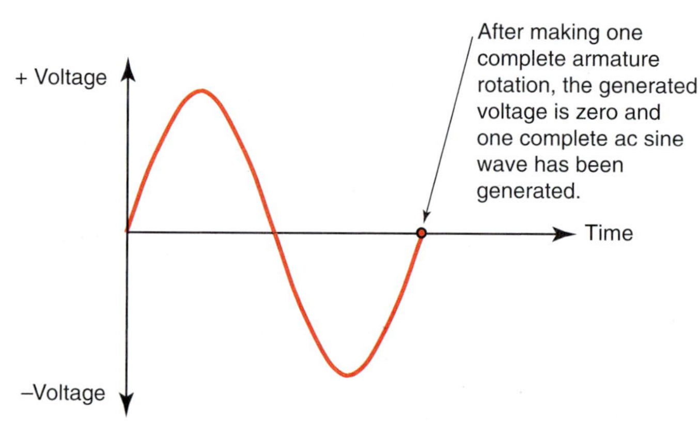

Goodheart-Willcox Publisher

Figure 36-12. After one complete armature rotation, the generated voltage is zero once again and one full ac sine wave has been generated.

36.4 Three-Phase AC Generators

A three-phase ac generator relies on three separate windings, each positioned 120° out of phase with the next, **Figure 36-13**. These windings can be configured as either a wye or delta connection. In a wye connection, **Figure 36-14**, one end of each winding is connected to the generator's output wiring, while the other ends of the winding leads are connected to each other to form the wye, **Figure 36-15**. This can be accomplished by internally connecting the three slip rings that are not directly connected to the generator's output wiring terminals. A cross-sectional view illustrating these electrical connections is provided in **Figure 36-16**.

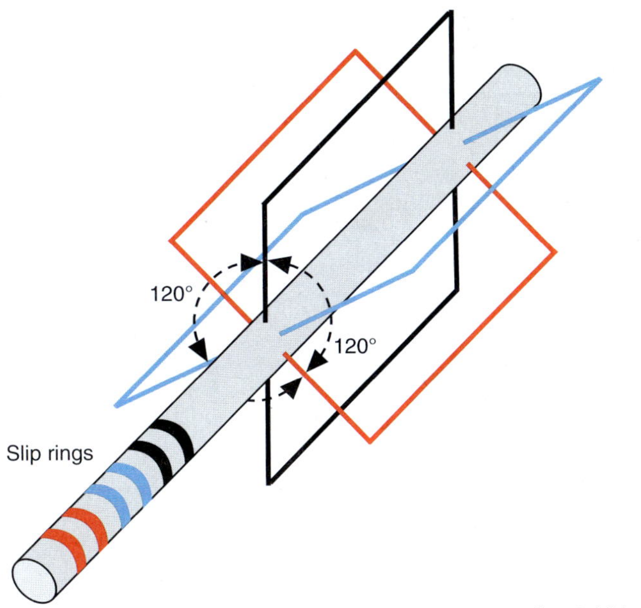

Figure 36-13. A three-phase ac generator has three armature windings, each 120° out of phase with each other. Each end of the three windings is connected to a shaft-mounted slip ring.

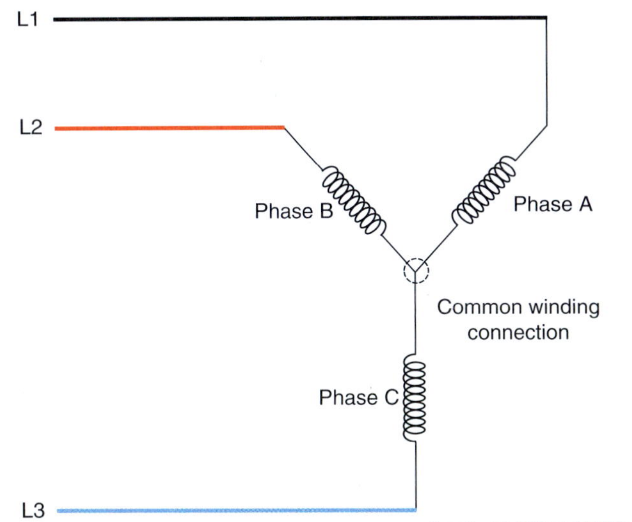

Figure 36-14. In a three-phase wye-connected generator, one end of each of the three windings is connected to the generator's output circuitry, while the other ends are joined together at a common winding connection.

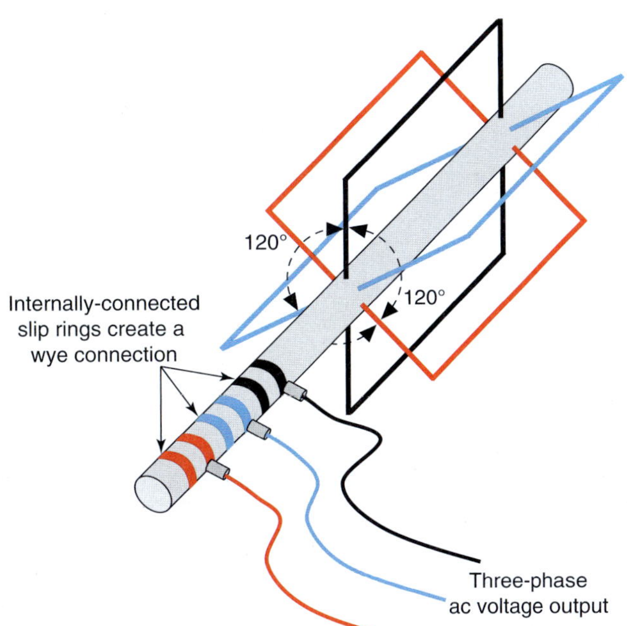

Figure 36-15. In a three-phase wye-connected generator, three slip rings (one for each phase) have a brush connection to the output circuit. The slip rings without brushes are electrically connected to each other.

In a delta connection, **Figure 36-17**, each of the generator's output wiring connections is wired to one end of two separate windings. These connections can be made by internally connecting each of the three brush-connected slip rings to one of the slip rings without a brush from a different winding, **Figure 36-18**. A cross-sectional view illustrating these electrical connections is provided in **Figure 36-19**.

Regardless of whether the generator's windings are configured as a wye or delta connection, the output power signal of a three-phase generator resembles that of a traditional three-phase power signal, which comprises three individual single-phase ac signals positioned 120° out of phase with each other, **Figure 36-20**.

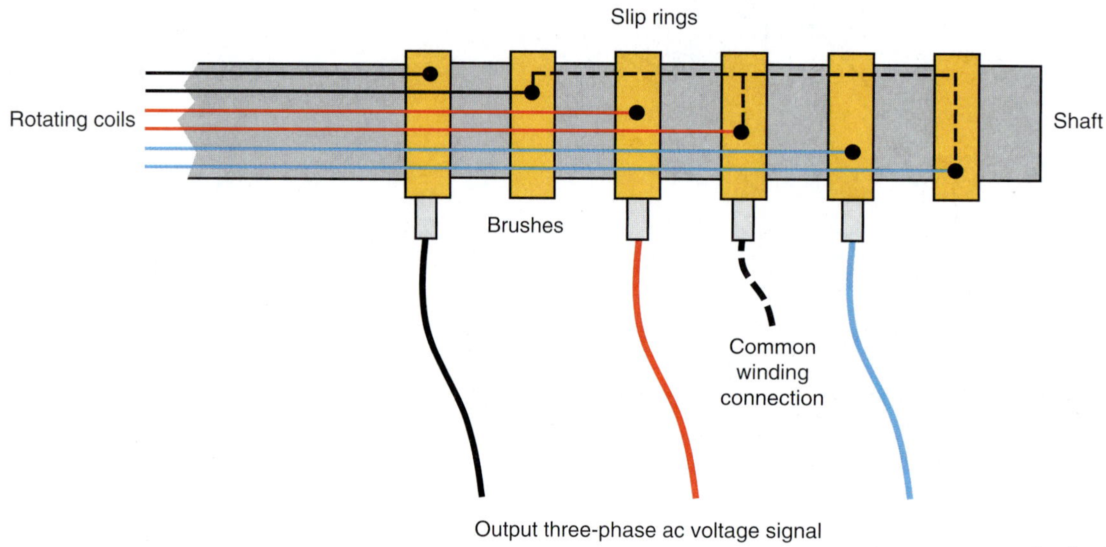

Goodheart-Willcox Publisher

Figure 36-16. Cutaway illustration of a three-phase wye-connected generator shaft and the connections between the ends of the armature coils and the slip rings.

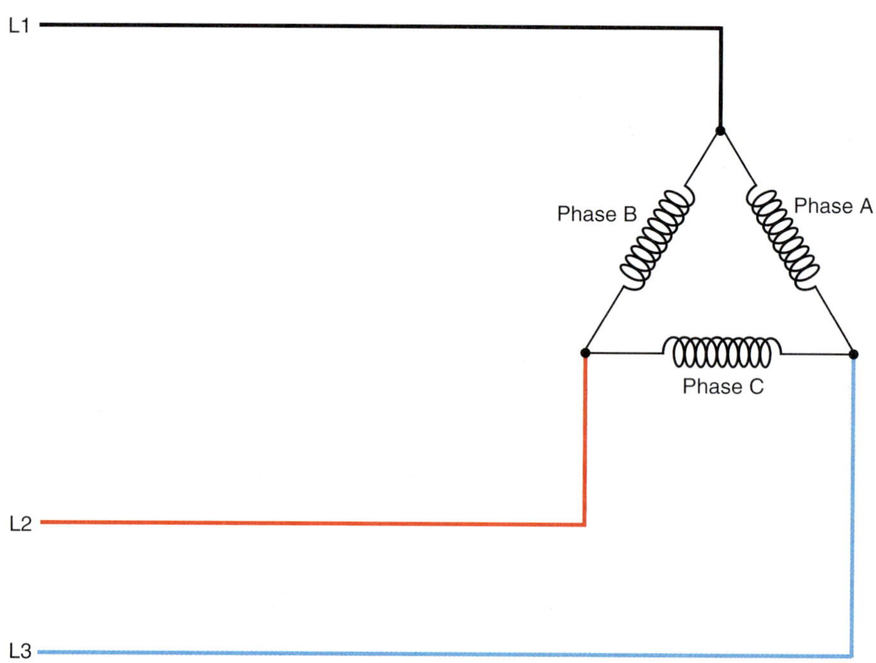

Goodheart-Willcox Publisher

Figure 36-17. In a three-phase delta-connected generator, each of the three output conductors is connected to one end of two armature windings.

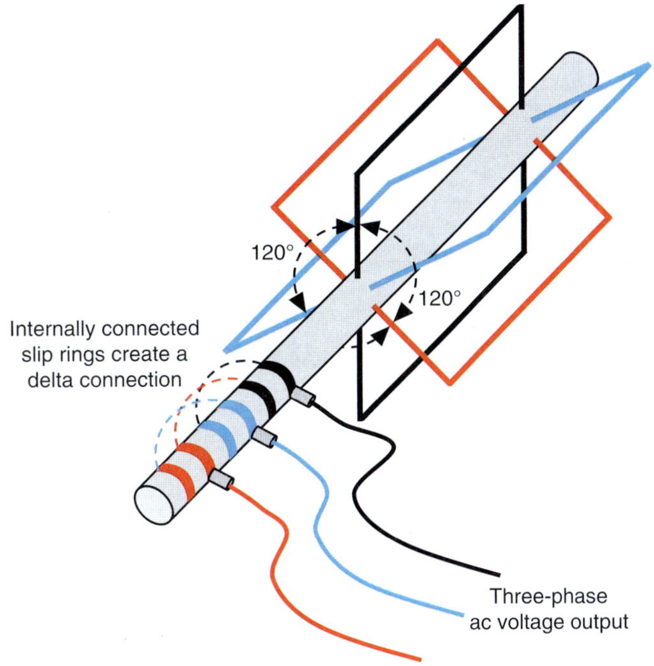

Figure 36-18. Brush connections on the three-phase delta-connected generator. The delta configuration is created by internally connecting each of the three brush-connected slip rings to one of the slip rings without a brush from a different winding.

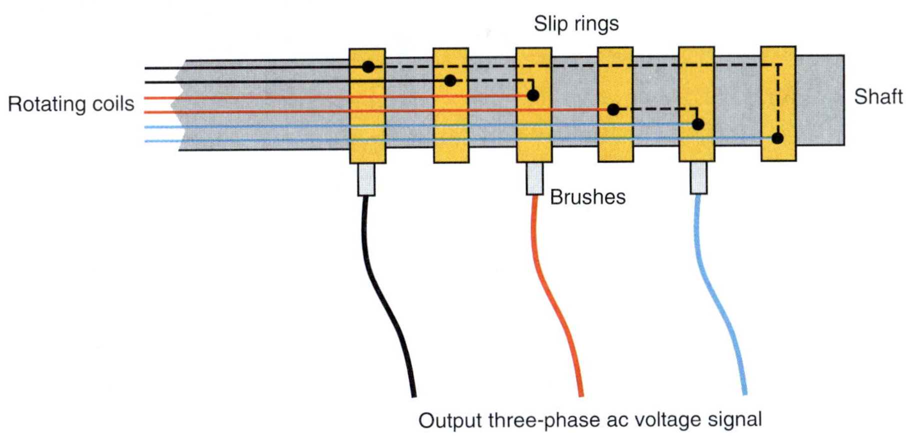

Figure 36-19. In a three-phase delta-connected generator, each of the three brush-connected slip rings is connected to a slip ring without a brush connection of a different phase.

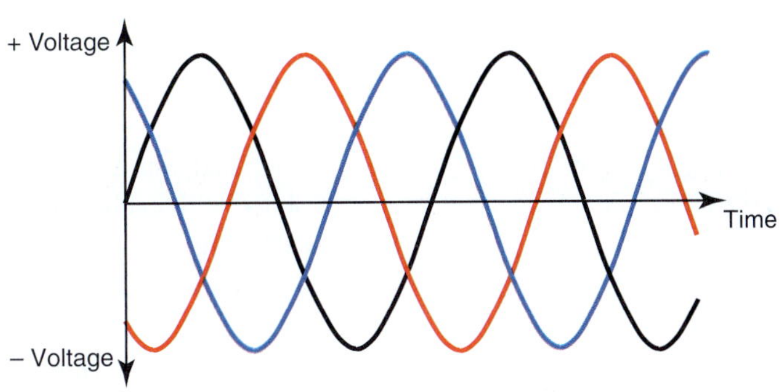

Figure 36-20. The three-phase generator's output power is configured as three single-phase ac signals that are 120° out of phase with each other.

36.5 Generator Frequency and Speed

Just as with electric motors, the speed, frequency, and number of poles associated with an ac generator are all related. The following equation describes this relationship:

$$\text{Generator speed} = \frac{120 \times f}{P}$$

where:

- Generator speed is expressed in rotations per minute, rpm
- f is frequency expressed in hertz, Hz
- P is the number of poles per phase
- 120 is a conversion constant (60 sec/min × 2 polarity changes/cycle)

In many cases, the output frequency of the generator must be controlled or regulated. In this situation, the formula can be restated as follows:

$$f = \frac{\text{Generator speed} \times P}{120}$$

Since the number of poles for a particular generator will remain constant, it can be seen from this formula that the frequency of an ac generator's output power can be controlled by adjusting the speed at which the generator's rotor spins. Refer to **Example 36-1** on how to determine the frequency in an ac generator. **Example 36-2** provides the steps to determine the speed of an ac generator.

Example 36-1

What is the frequency of a four-pole ac generator's output signal if the machine's shaft is rotating at a speed of 1000 rpm?

✓ Solution

Using the formula for frequency and substituting the known values for speed and the number of poles, we get:

$$f = \frac{\text{Generator speed} \times P}{120}$$

$$f = \frac{(1000 \times 4)}{120}$$

$$f = \frac{4000}{120}$$

$$f = 33.3 \text{ Hz}$$

Example 36-2

How fast must the rotor of a four-pole ac generator turn if the machine's output frequency is to be 50 Hz?

✓ Solution

This problem is solved by using the generator speed formula:

$$\text{Generator speed} = \frac{120 \times f}{P}$$

Next, substitute the known values for frequency and the number of poles as follows:

$$\text{Generator speed} = \frac{120 \times 50}{4}$$

$$\text{Generator speed} = \frac{6000}{4}$$

$$\text{Generator speed} = 1500 \text{ rpm}$$

Now It's Your Turn 36-2

At what speed does a six-pole generator need to turn in order to output a 60 Hz voltage signal?

36.6 Using Multiple AC Generators

In some cases, a single generator will not be able to provide the output power needed to satisfy the needs of an application. This is especially true when the requirements of the application vary. Instead of installing one large generator that will be oversized most of the time, it is often desirable to install multiple generators. During low load conditions, a single generator can be used. During high load periods, the second generator can be put into service to satisfy the increased load. When multiple generators are used, they must be installed in parallel with each other and are called ***parallel-connected generators***.

In addition to being in parallel with each other, there are a few other conditions that must be met to ensure the safe and reliable operation of the machines:

- The generators must be installed and connected so the shafts of both machines turn in the same direction and at the same speed. This will help ensure that the output voltages of the generators are the same.
- For three-phase generators, both machines must be electrically connected so that they are electrically in-phase with each other. Phase rotation was discussed earlier in this text and is vital to the proper operation of parallel-connected generators.

When opting to install multiple generators, be sure to consult with the equipment manufacturer for guidelines and recommendations specific to their machines regarding synchronizing, setup, and configuration. Generators are expensive pieces of machinery and improperly connecting them can be costly.

CHAPTER 36 REVIEW

Summary

- AC generators have some of the same basic components as dc electric motors and dc generators, namely shafts, housings, armatures, brushes, and windings. (36.1)
- Instead of the commutators found on dc machines, ac generators use slip rings and brushes to connect the rotating windings to the machine's output electrical connections. (36.1)
- AC generators can produce voltage by rotating an armature coil within a stationary magnetic field or by rotating a magnetic field around a stationary armature winding. (36.2)
- A simple ac generator is made up of one armature winding or coil positioned within a stationary magnetic field. The armature rotates within the magnetic field to generate voltage. (36.3)
- A three-phase ac generator relies on three separate windings, each positioned 120° out of phase with the next to generate voltage. (36.4)
- The output capacity of an ac generator is determined by the number of turns in the armature windings, the rotational speed of the armature, and the strength of the field winding. (36.5)
- When a single generator is not able to provide the power necessary to satisfy an application, it is often desirable to install multiple generators in parallel with each other. These generators are referred to as parallel-connected generators. (36.6)

Review Questions

Use the information in the chapter to answer the following questions.

Measuring Your Potential

1. As the rotational speed of an ac generator's shaft increases, the _____. (36.1)
   A. number of generator poles will increase
   B. frequency of the generator's output supply signal will increase
   C. number of generator poles will decrease
   D. frequency of the generator's output supply signal will decrease
2. *True or False?* On an ac generator, the slip rings completely encircle the generator's shaft. (36.1)
3. On an ac generator, which two components are pressure-connected to each other? (36.1)
   A. Slip rings and brushes
   B. Armature coils and brushes
   C. Armature coils and output wiring
   D. Armature coils and the rotating shaft
4. *True or False?* In a revolving-field ac generator, the brushes are permanently connected to the generator's output power wiring. (36.2)
5. *True or False?* All ac generators use slip rings and brushes. (36.2)
6. *True or False?* The armature winding is always attached to the rotor of an ac generator. (36.2)
7. In a revolving-field ac generator, there are _____. (36.2)
   A. the same number of brushes as there are slip rings
   B. more slip rings than brushes
   C. more brushes than slip rings
   D. no slip rings or brushes
8. When a conductor moves in a direction parallel to the lines of magnetic flux, _____. (36.3)
   A. maximum voltage is produced
   B. no voltage is produced
   C. the voltage produced is 67% of the peak voltage
   D. the voltage produced is 33% of the peak voltage
9. When a conductor moves in a direction perpendicular to the lines of magnetic flux, _____. (36.3)
   A. maximum voltage is produced
   B. no voltage is produced
   C. the voltage produced is 67% of the peak voltage
   D. the voltage produced is 33% of the peak voltage
10. Which of the following statements is true regarding the wire connections in a three-phase wye-connected generator? (36.4)
    A. Connections are made by internally connecting each brush-connected slip ring to one of the slip rings without a brush from a different wiring.
    B. Connections are made by internally connecting the slips rings that are not directly connected to the generator's output wiring terminal.
    C. Each of the generator's three output conductors is connected to one of two armature windings.
    D. One end of the three windings is connected to the generator's circuitry, while the other end is connected to a slip ring without a brush connection of a different phase.

11. *True or False?* To maintain a constant generator speed, as the number of poles on an ac generator increases, the frequency must be decreased. (36.5)
12. *True or False?* When multiple generators are connected in parallel, the shafts must turn at the same speed and in the same direction. (36.6)

Plug-In

1. At a frequency of 80 Hz, a two-pole ac generator will rotate at a speed of _____. (36.6)
   A. 2400 rpm
   B. 3600 rpm
   C. 4800 rpm
   D. 6000 rpm
2. Electrician A states that the operation of an ac generator can be accurately described as rotating a coil of wire within a stationary magnetic field. Electrician B states that the operation of an ac generator can be accurately described as rotating a magnetic field around a stationary armature winding. Which of the following statements is correct? (36.2)
   A. Only electrician A is correct.
   B. Only electrician B is correct.
   C. Both electricians are correct.
   D. Both electricians are incorrect.
3. If the rotational speed of a generator's shaft is doubled, the frequency of the output power signal _____. (36.5)
   A. will increase by a factor of four
   B. will increase by a factor of two
   C. will decrease by a factor of four
   D. will decrease by a factor of two
4. At a frequency of 40 Hz, how many poles does an ac generator have if the machine's shaft rotates at a speed of 800 rpm? (36.5)
5. At what speed must an eight-pole ac generator turn to produce an 80 Hz output signal? (36.5)

Making the Connection

1. An electrician claims that the brushes on a dc generator perform the same basic function as the brushes on an ac generator. Do you agree or disagree with this electrician? Provide as much information as you can to support your opinion. (36.1, 36.4)
2. Explain how the rotational speed of a generator's armature affects the output voltage of an ac generator. (36.5)

SECTION 11

Electricity Sources, Production, and Distribution

Piyaset/Shutterstock.com

CHAPTER

37 Electricity Production

zstock/Shutterstock.com

SPARKING DISCUSSION

Explain why solar and wind energy are classified as non-dispatchable energy sources.

LEARNING OUTCOMES

After completing this chapter, you will be able to:

37.1 Describe the law of conservation of energy.
37.2 Describe the construction, function, and operation of a fuel cell.
37.3 Explain how electricity is generated by pressure.
37.4 Describe how electricity is generated by wind.
37.5 Explain how electricity is generated by heat and friction.
37.6 Explain how electricity is generated by light.

TECHNICAL TERMS

biomass
biopower
capacity factor
catalyst
dipole
dispatchable energy source
dispatch stack
diversion
electrolyte
geothermal energy
heliostat
nacelle
net metering
oxidation
penstock
photovoltaic effect
piezoelectric effect
reduction
run-of-the-river hydroelectric system
Seebeck coefficient
Seebeck effect
true-up adjustment
turbine

Introduction

We rely on electricity in nearly all aspects of our daily lives. From powering our household appliances to lighting our cities, electricity has truly been the driving force in creating, sustaining, and advancing the world as we know it. Technological advances have led to more effective and efficient ways to produce, distribute, and use electrical energy.

Currently, approximately 60% of all electricity produced in the United States relies on the burning of fossil fuels, such as coil, oil, and natural gas, **Figure 37-1**, while nuclear power and renewable energy equally share the remaining 40%. About 48% of the electricity produced using renewable energy is produced using wind, while 18% is produced using solar, **Figure 37-2**.

The production of electricity produced from magnetism is discussed early in this text. This chapter explores some of the more diverse methods, such as friction, heat, and pressure, that can be used to produce electricity. Some of these electricity-producing methods, such as wind and solar, have already become mainstream, while others are still in their infancy and will require time to mature into viable options for widespread integration into our power grids.

As you move through this chapter, some renewable energy options, such as wind, solar, geothermal, biomass, and hydropower are introduced. As discussed throughout this text, electricity can be produced magnetically by magnetic induction. Simply stated, electricity is produced when there is relative motion between a magnetic field and a conductor. Many of the methods used to produce electricity that are discussed in this chapter rely on the relative motion between a conductor and a magnetic field to produce electricity.

37.1 Law of Conservation of Energy

In its simplest terms, the law of conservation of energy states that energy can neither be created nor destroyed. Energy can, however, be transformed from one form to another and can be transferred from one object to another. For example, the fuel in a motor vehicle's gas tank, when burned, allows the car to move. This creates exhaust as a by-product of the combustion process. In this case, chemical energy is transformed into mechanical energy, which causes the vehicle to move. It also creates thermal or heat energy, which is present in the exhaust fumes, **Figure 37-3**.

US Electricity Generation by Energy Source Type

Energy Source	2024	2050 (predicted)
Fossil fuels	60%	44%
Nuclear	19%	12%
Renewables	21%	44%

Source: US Energy Information Administration

Figure 37-1. Breakdown of electricity production in the United States by major energy source.

US Electricity Generation by Renewable Source Type

Energy Source	2024	2050 (predicted)
Solar	18%	51%
Wind	48%	31%
Hydroelectric	27%	12%
Other	7%	6%

Source: US Energy Information Administration

Figure 37-2. Breakdown of electricity production in the United States by renewable energy source.

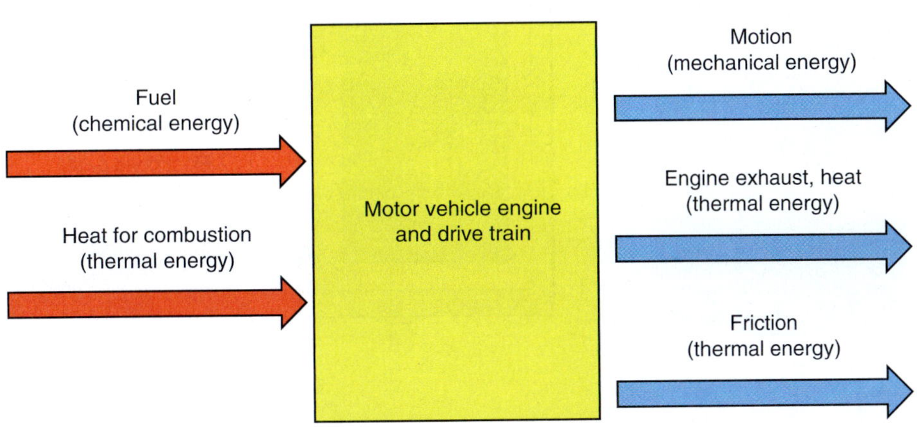

Goodheart-Willcox Publisher

Figure 37-3. Since energy cannot be created nor destroyed, the total of all energy outputs from a process must equal the total of all inputs to the process.

When energy is transferred between objects, one object gives up energy while the other receives it. For example, when ice is used to cool a glass of warm tea, the tea transfers thermal energy (heat) to the ice. In reality, the ice does not make the tea cold. Instead, the tea cools itself by transferring its heat energy to cool, and ultimately melt the ice.

Under ideal conditions, all the energy inflows will equal the energy outflows of any given process. There will be losses, but even these losses will have energy equivalents associated with them. Recall Kirchhoff's current law, which states the sum of all current flows into a node must equal the sum of all current flows out of the same node. The concepts that apply to Kirchhoff's current law are easily applied to the law of conservation of energy.

37.2 Chemically Produced Electricity

Chemically produced electricity is the result of electrochemical reactions. These electrochemical reactions are the fundamental concepts on which batteries function. Although batteries are, at the present time, the most common form of chemically produced electricity, fuel cells are becoming popular for a wide range of applications, including forklifts, buses, and semitrucks. Collectively, batteries and fuel cells are classified as electrochemical devices.

Batteries

Batteries are made up of three main components: an anode, a cathode, and an electrolyte, **Figure 37-4**. The anode and the cathode are the terminals or connection points between the battery and the external circuitry to which the battery is connected. Electrically speaking, the anode is the battery's connection point that effectively loses electrons through a process called ***oxidation***. The anode is commonly made from zinc. The cathode is the battery's connection point at which electrons are gained through a process called ***reduction***. The cathode on many batteries is made of graphite. The ***electrolyte*** provides a path that facilitates the movement of charged particles between the electrodes and allows chemical reactions within the battery to continue. The electrolyte is commonly made up of ammonium chloride. Provided that the chemical reactions continue, the battery will continue to supply power. When these reactions stop, the battery can no longer provide power, and the cell must be replaced or—in the case of rechargeable batteries—recharged.

When connected to a circuit, the anode releases electrons to the circuit. These released electrons flow through the circuit and ultimately arrive at the battery's cathode. Some batteries have an additional layer of electrolytic material, such as manganese dioxide. This layer increases the effectiveness of the cathode and, as a result, helps to increase the voltage produced by the battery while reducing the battery's physical size.

Shedding Light

A recently growing use of battery technology is battery energy storage systems (BESS). These systems are used in conjunction with wind and solar electricity production. The battery energy storage system stores and discharges electrical energy into the electrical grid to help balance supply and demand.

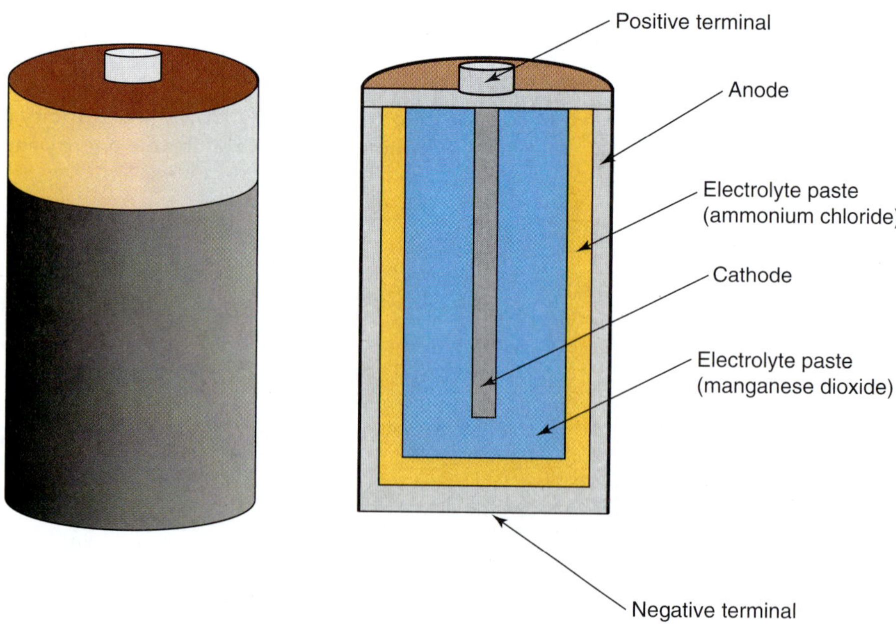

Goodheart-Willcox Publisher

Figure 37-4. Cross-sectional view of a dry-cell battery. There is a potential difference between the anode and cathode of a charged battery.

Comparison of Dry-Cell Batteries and Wet-Cell Batteries

	Dry-Cell Battery	Wet-Cell Battery
Electrolyte	Gel or paste	Liquid
Leakage Potential	Low/none	High
Physical Size	Small	Large
Portability	Yes/easy	No/more difficult
Relative Weight	Low	High
Maintenance Required	No	Yes
Power Output	Low	High
Power Density	Low	High

Goodheart-Willcox Publisher

Figure 37-5. Comparison of the properties of dry-cell and wet-cell batteries.

Batteries are typically classified as either dry cells or wet cells. Dry-cell batteries are so named because the electrolyte material in these batteries is a paste or a gel, while wet-cell batteries have liquid electrolytes. **Figure 37-5** provides a comparison between dry-cell and wet-cell batteries.

Fuel Cells

Fuel cells are similar to batteries in that they both have anodes, cathodes, and electrolytes. One of the main benefits of fuel cells over batteries is their ability to provide a continuous supply of power. There are different types of fuel cells. The most common type is the hydrogen fuel cell, which relies on the interaction between hydrogen and oxygen. Hydrogen and oxygen are the inputs to the electricity-producing process, while the only non-electricity output is water, **Figure 37-6**. Hydrogen is the fuel for the cell and is supplied to the fuel cell's anode. Oxygen is supplied to the fuel cell's cathode.

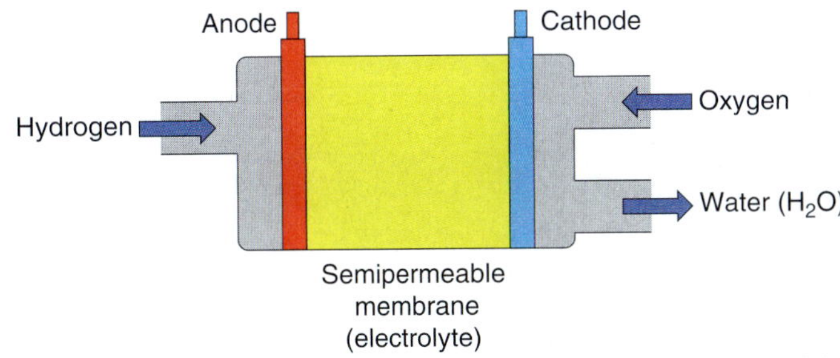

Goodheart-Willcox Publisher

Figure 37-6. The inputs of a hydrogen fuel cell are hydrogen and oxygen, while the only by-product of the process, aside from electricity, is water.

The fuel cell's anode is coated with a thin layer of platinum that is approximately one nanometer (1×10^{-9} m) thick. At the anode, electrons are separated from the hydrogen atoms, creating loose electrons (e^-) and positively charged hydrogen ions (H^+), **Figure 37-7**. The platinum-coating on the anode acts as a ***catalyst***, which is a material that helps speed up the chemical reactions without being consumed by them. This makes the process of producing electricity more efficient.

The electrons, once separated from the hydrogen atoms, flow through the connected circuit, while the positively charged hydrogen ions pass through the fuel cell's electrolyte to the cathode. The electrolyte is a substance that allows the flow of ions through it but does not allow individual electrons to pass freely. This transfer of ions allows the electrochemical

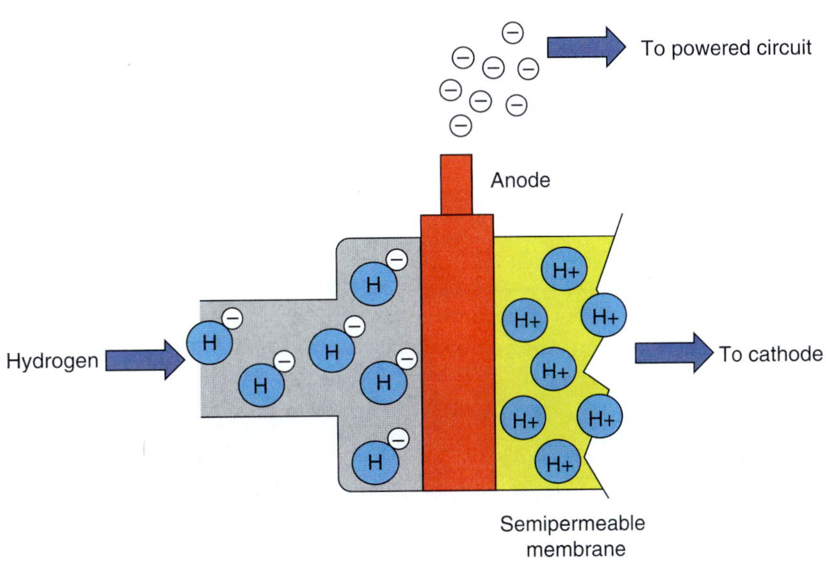

Goodheart-Willcox Publisher

Figure 37-7. At the anode of a hydrogen fuel cell, electrons are separated from hydrogen atoms. The electrons flow through the externally connected circuit, while the positive hydrogen ions flow through the semipermeable membrane to the cathode.

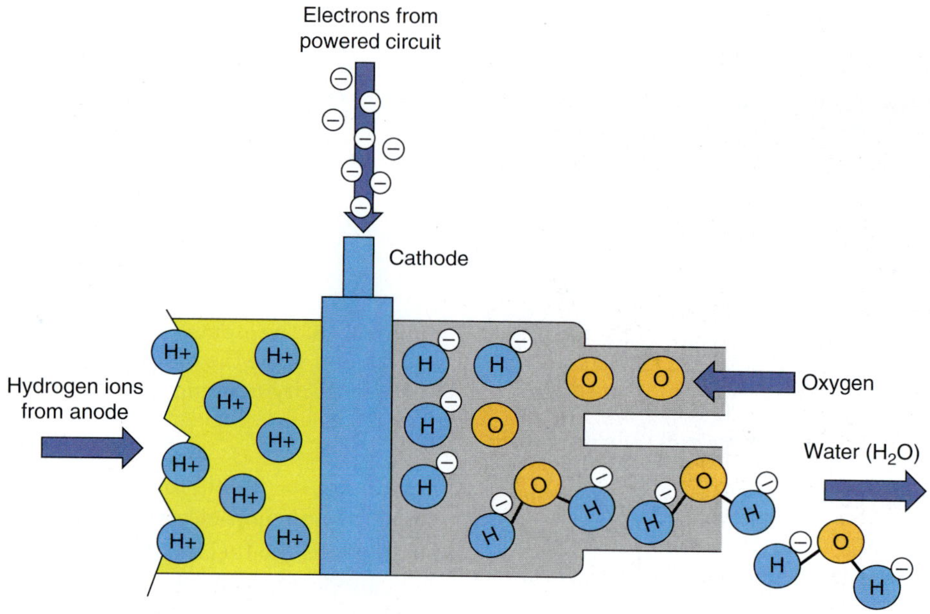

Figure 37-8. At the cathode, electrons, oxygen, and positively charged hydrogen ions react with each other to form water. The water is then discharged from the fuel cell.

processes in the fuel cell to occur. Because the electrolyte allows some particles to flow through it and not others, the electrolyte is a semipermeable membrane.

At the fuel cell's cathode, oxygen is introduced, **Figure 37-8**. The oxygen combines with the hydrogen ions and the electrons to form water and complete the reaction.

Fuel cells are typically very efficient because they convert chemical energy directly into electrical energy. Nothing is burned, so there are no combustion-related thermal losses.

The only by-products associated with hydrogen fuel cells are water and heat, making them an environmentally friendly option. The main challenge associated with hydrogen fuel cells is obtaining hydrogen. Depending on the sources of hydrogen and the processes by which it is produced and processed, fuel cells have the potential to produce large amounts of electricity with minimal environmental impact.

37.3 Pressure-Produced Electricity

A large amount of electricity is produced using pressure, most often as the result of converting mechanical energy into electrical energy. Generators, for example, rely on pressure to create the relative rotational motion between the magnetic field and the generator's windings and facilitate the production of electricity. In some electricity-generating systems, high-pressure fluids, such as air, steam, or gas are used as the rotational force that turns a ***turbine***. As the turbine rotates, the force of the flowing fluid stream is converted into mechanical energy. The turbine is mechanically joined to a generator. As the generator spins, lines of magnetic flux are cut, and electricity is generated. A turbine, therefore, is a machine that converts the energy of the flowing fluid to mechanical energy, while a generator facilitates the transformation of mechanical energy into electrical energy, **Figure 37-9**.

Hydroelectric Power

One of the most common types of pressure-produced electricity is hydroelectric power. Hydropower accounts for about 27% of all renewable-based, domestically produced electricity. Hydroelectric power relies on the kinetic energy of flowing water to turn the turbines that generate electricity. A common source of flowing water is obtained by trapping water behind a dam. In operation, water from the

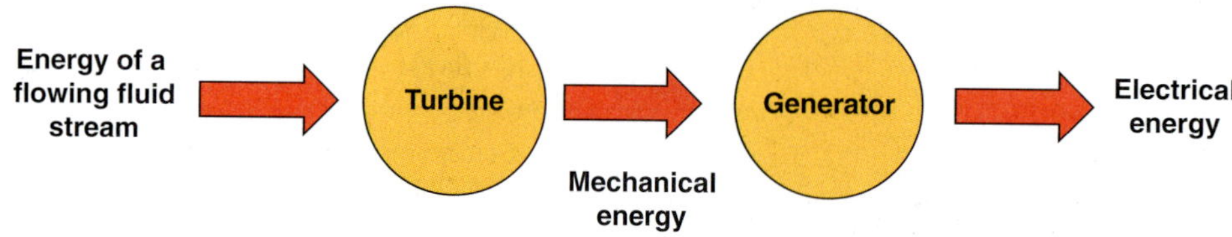

Figure 37-9. Turbines convert pressure from a flowing fluid stream to mechanical energy, while generators convert mechanical energy into electrical energy.

Goodheart-Willcox Publisher

Figure 37-10. A run-of-the-river hydroelectric power plant uses flowing water to generate electricity.

top of the dam's reservoir is directed through turbines. The extreme force of the falling water from a high level causes the turbines to spin, which causes the generator to turn and produces electricity.

Not all electricity that is generated from flowing water requires a dam. Some systems that generate electricity take advantage of the natural water flow of rivers. These systems are called ***run-of-the-river hydroelectric systems***, **Figure 37-10**. In these systems, the naturally flowing water turns turbines directly. This is accomplished by a process called ***diversion***, where a small portion of the river's flow is diverted into a pipe called a ***penstock***. From the penstock, the water is directed to a turbine, causing it to turn. The rotating turbine causes the connected generator to spin, generating electricity. After passing through the turbine, the redirected water is returned to the river at a point that is downstream of the power-generation plant. The main challenge associated with run-of-the-river hydroelectric systems is that the amount of electricity produced is not consistent. The rate of water flow through a river depends on several environmental factors, including rainfall rates.

Shedding Light

Pumped storage power plants have water reservoirs both above and below the turbine. When electricity demand and costs are low, water is pumped from the bottom reservoir to the top reservoir. When demand is high, water flows from the top reservoir down through the turbine. With the exceptions of spillover and evaporation, water is recycled and used over and over again.

Power Production in a Hydropower System

The capacity of a hydropower system, in megawatts (MW), is calculated using the following formula:

$$\text{Capacity} = \text{Water flow rate} \times \text{Net head} \times \text{Efficiency} \times \text{Conversion constant}$$

where:

Water flow rate = Water flow in ft^3/second
Net head = Vertical water flow height, expressed in feet
Efficiency = Efficiency of the system, expressed as a decimal
Conversion constant = 0.000085

The conversion constant 0.000085 is expressed in the units of MW-sec/ft^4 and is calculated as follows:

$$(62.4\ \text{lb/ft}^3) \times (33{,}000\ \text{ft-lb/min/hp}) \times (60\ \text{sec/min}) \times (746\ \text{W/hp}) \times (1\ \text{MW}/1 \times 10^6\ \text{W})$$

Refer to **Example 37-1** on how to calculate power production.

Example 37-1

How much power is generated by a hydropower system that has a flow rate of 1000 ft^3/sec, a vertical water flow height of 500′, and is operating at an efficiency of 72%?

✓ Solution

Using the formula to find the capacity of a hydropower system, substitute the known values for water flow rate, net head, efficiency, and the conversion constant as follows:

$$\text{Capacity} = \text{Water flow rate} \times \text{Net head} \times \text{Efficiency} \times \text{Conversion constant}$$

$$\text{Capacity} = 1000\ \text{ft}^3/\text{sec} \times 500\ \text{ft} \times 0.72 \times 0.000085$$

$$\text{Capacity} = 30.6\ \text{MW}$$

The Piezoelectric Effect

Another pressure-related method of producing electricity relies on what is called the ***piezoelectric effect***. The piezoelectric effect refers to the creation of an electric charge in direct response to applied pressure. Some materials, such as quartz, exhibit the piezoelectric effect when subjected to compression, force, or stress. For example, research has been done to improve the ability to harness the potential power of pressure exerted on sidewalks due to pedestrian foot traffic as well as the pressure exerted on highways due to moving vehicular traffic.

One common type of quartz used in the production of electricity is silicon dioxide (SiO_2). When configured as a crystal, silicon dioxide is shaped like a tetrahedron, **Figure 37-11**. A tetrahedron, also referred to as a triangular pyramid, has four triangular faces, six edges, and four vertices or points. The configuration of the silicon dioxide crystal is unique in that each silicon atom is bonded to four oxygen atoms, **Figure 37-12A**, while each oxygen atom is bonded to only two silicon atoms, **Figure 37-12B**. When observed from the side, the silicon dioxide crystal resembles a hexagon, **Figure 37-12C**.

One of the characteristics that make the piezoelectric effect possible in silicon dioxide is the relative size of the oxygen and silicon atoms. Oxygen atoms are physically smaller than silicon atoms and, as such, the outermost electrons of an oxygen atom are closer to the atom's nucleus when compared to the distance between a silicon atom's outermost electrons and its nucleus. This difference in distance causes the oxygen atoms to have small negative charges, and the silicon atoms to have small positive charges. These differences in charges, along

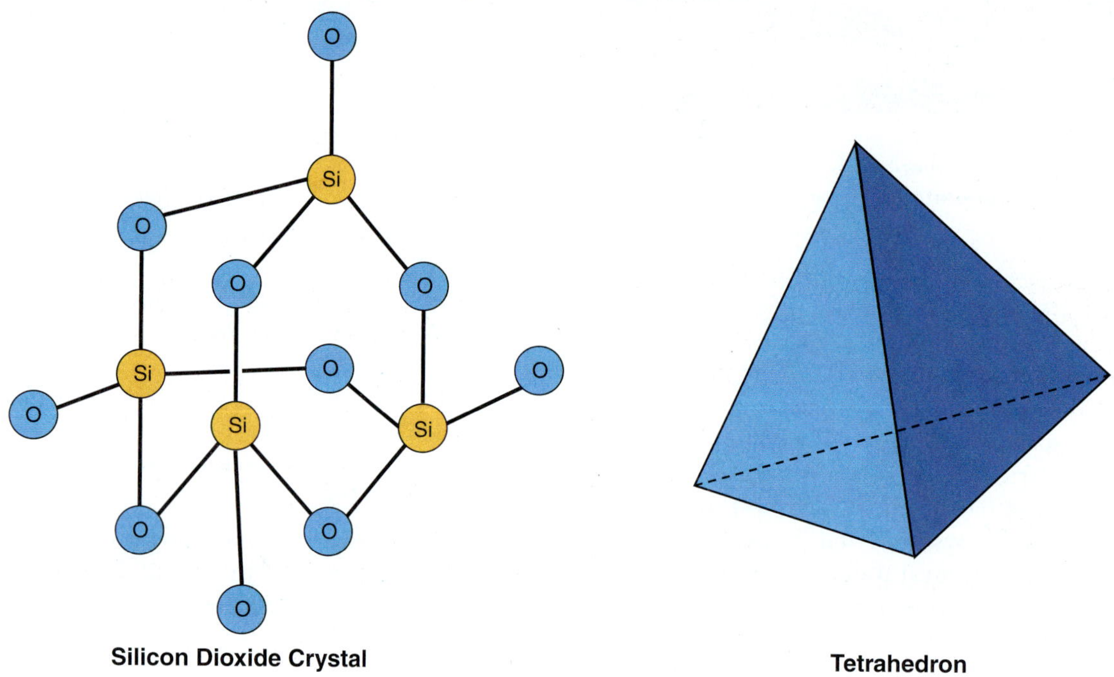

Figure 37-11. The silicon dioxide structure resembles a tetrahedron, or triangular pyramid. This structure helps make the generation of electricity by pressure possible.

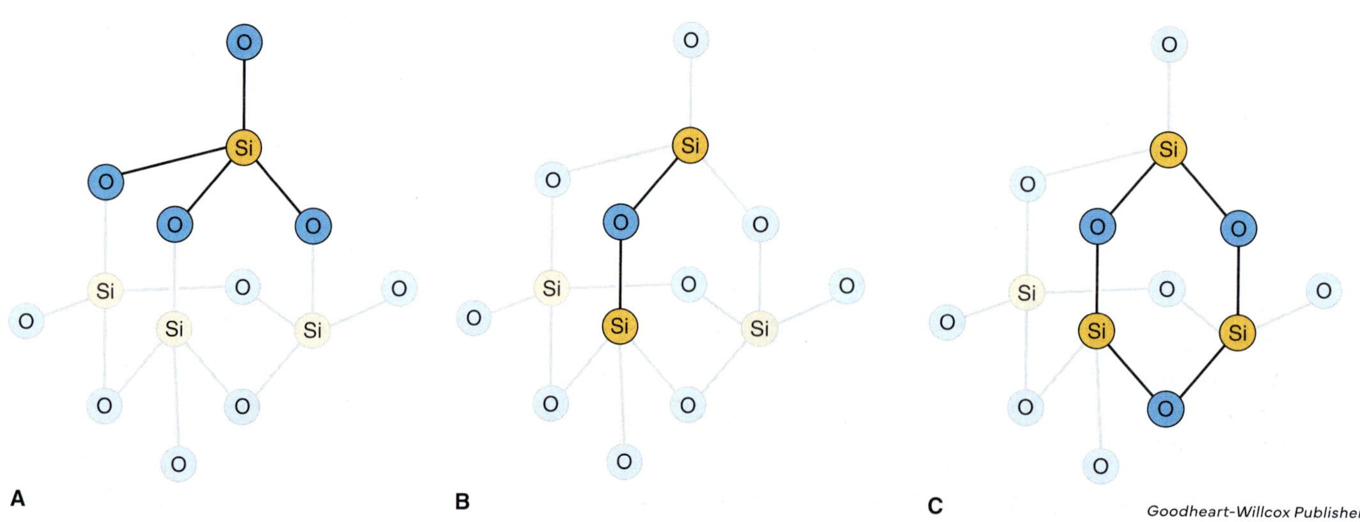

Figure 37-12. A silicon dioxide crystal. A—Each silicon atom in the silicon dioxide crystal is bonded to four oxygen atoms. B—Each oxygen atom in the silicon dioxide crystal is bonded to two silicon atoms. C—When viewed from the side, the bonding of a silicon dioxide crystal closely resembles a hexagon.

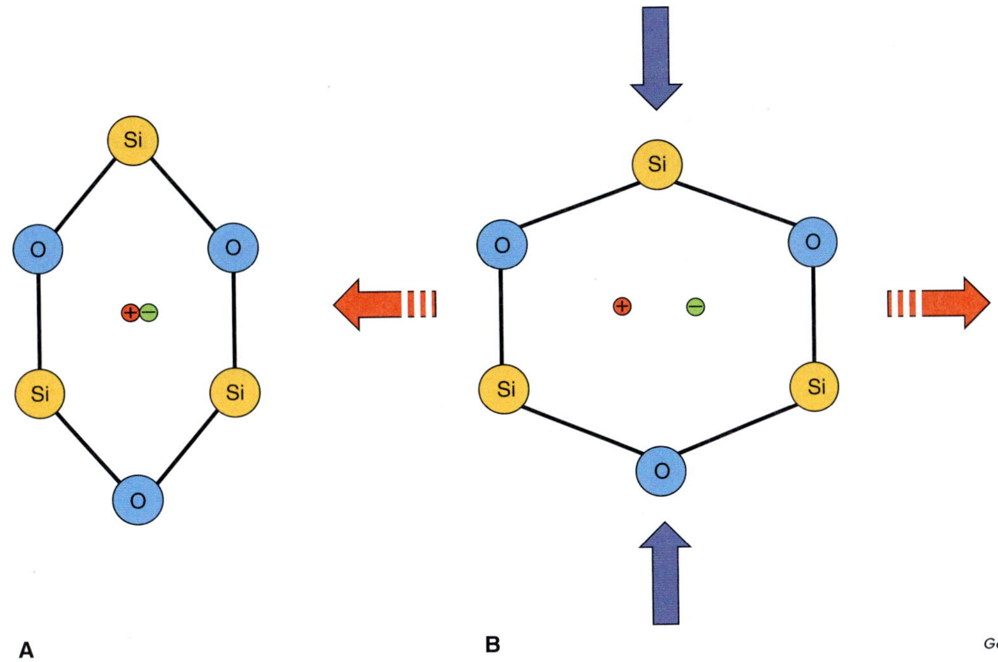

Goodheart-Willcox Publisher

Figure 37-13. Possible piezoelectric effect in silicon dioxide. A—When no pressure is exerted on a silicon dioxide crystal, the unlike charge centers are close together and no noticeable potential difference is detected between them. B—When pressure is exerted on a silicon dioxide crystal, the unlike charge centers move further apart from each other and the potential difference between them increases.

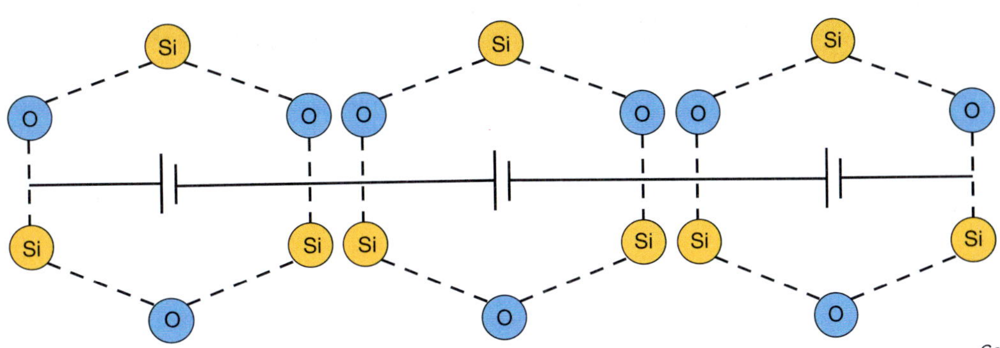

Goodheart-Willcox Publisher

Figure 37-14. This additive characteristic of the silicon dioxide crystal is similar to multiple batteries connected in series.

with the physical separation that exists between them, creates what is called a ***dipole***. A dipole is created when two (di) dissimilar charges (poles) are separated by a small distance. Dipoles are common in materials that do not have a center of symmetry.

When the crystal is at rest and not subjected to bending, stress, compression, or pressure, the effective positions of the positive and negative charges are aligned near each other, **Figure 37-13A**. When the crystal is compressed or otherwise subjected to stress, the dissimilar charges move away from each other in a process called *polarization*, **Figure 37-13B**. As a result, an electric field is created between the net negative and net positive charges, causing a small voltage to be generated between them. As the pressure exerted on the crystal increases, the amount of voltage generated increases. Since the crystal is made up of many molecules, the total voltage generated is the additive result of the individual generated voltages. This can be equated to multiple batteries connected in series with each other, **Figure 37-14**. When the pressure is reduced or eliminated completely, the amount of electricity produced is reduced as well.

37.4 Wind-Produced Electricity

Roughly 10% of the electricity used in the United States is produced by wind. This equates to about 48% of all domestically-produced renewable electricity. Just as with hydropower, electricity generated by wind is created by using pressure to turn the turbines that generate electricity.

Instead of using the kinetic energy of flowing water, however, wind-driven systems rely on the kinetic energy of wind. To maximize electricity production and reduce the costs associated with power transmission, multiple wind turbines are erected near each other to form wind farms.

Shedding Light

The location of a wind farm is selected based on data regarding the predictability and strength of the wind current, not its physical proximity to the grid. As such, wind farms are typically located far from existing power transmission networks. It would not be beneficial to locate a wind farm close to the existing power transmission network if there is no wind!

Twin Design/Shutterstock.com

Figure 37-15. Wind turbines rely on the kinetic energy of wind to generate electricity.

Wind Turbine Construction

The most common type of wind turbine is the propeller type, which is constructed with a horizontal axis, **Figure 37-15**. Most of these machines are manufactured with three blades that rotate when wind pushes on them. The overall height of a wind turbine ranges from about 200′ to 260′ (60 m to 80 m), which is the approximate height of a 20 to 26 story building. The blades on a typical land-based wind turbine can be between 100′ and 165′ (30 m and 50 m) in length, while the blades on offshore wind turbines can be over 250′ in length.

The wind turbine's rotating blades are connected to a shaft, which is connected to either a gear box or a generator. Gear boxes are used to increase or decrease the speed of the generator. Without a gear box, the generator will turn at the same speed as the rotating blades.

Once produced, power from the generator is transferred to a transformer, where the voltage is increased to the higher levels needed for effective and efficient power transmission. The shaft, gear box, and generator are contained in the ***nacelle***, which is the visible casing of the wind turbine, **Figure 37-16**. In the event of high winds, wind turbines are typically equipped with a brake to stop or slow down the turbine's rotating speed to help protect the equipment.

Inconsistency of Wind Power

One of the major concerns with wind-generated electricity is that wind is not consistent. Winds are strongest at night while the largest power demand is during the day, meaning that wind-generated power is not always needed. In addition,

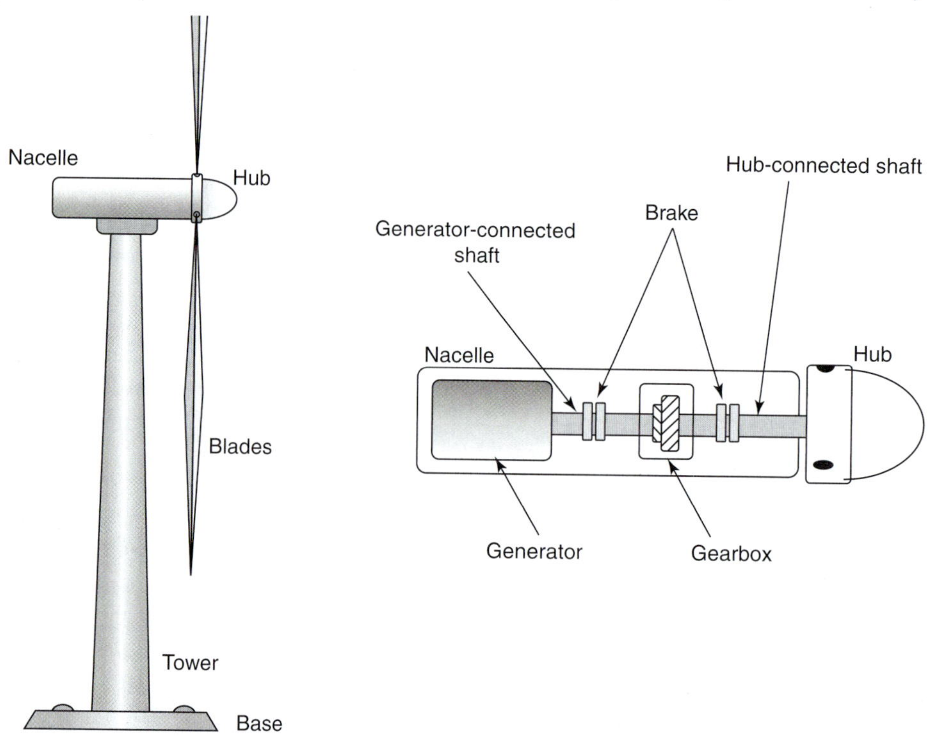

Goodheart-Willcox Publisher

Figure 37-16. Main components of a wind turbine.

the electrical energy produced is proportional to the speed of the prevailing wind. Most wind turbines require wind speeds of at least 4 mph to operate. However, excessively high wind speeds will cause the turbine to shut down completely to prevent damage to the equipment. Therefore, if wind speeds are too low or too high, the turbines cannot operate.

Due to the variability of wind power, wind-generated electricity is non-dispatchable. A ***dispatchable energy source*** is an energy source that can be accessed on-demand, based on the needs of the network.

Shedding Light

Solar power is also non-dispatchable, as sunlight is not always present, and the amount of power generated by solar systems is inconsistent. Energy sources such as hydropower and geothermal, on the other hand, are dispatchable, as supplies of running water and subsurface earth heat are constant and consistent. Simply stated, hydropower and geothermal systems can be put into operation whenever they are needed.

Wind Turbine Capacity Factor

When any power-generating project is evaluated, a number of variables are examined, with capacity factor being right at the top of the list. ***Capacity factor*** is the amount of energy produced by a process or system over a period of time divided by the amount of energy that could have been produced if the system or process operated at full capacity over the same period of time. Capacity factor is, therefore, a measurement of the efficiency of a power-producing process or facility. The higher the capacity factor, the higher the efficiency of the power-producing process. Since wind is highly variable, the capacity factor for wind turbines is very low, at approximately 27%. For example, if a 100 MW wind farm is constructed, it is expected that only 27 MW of power will be produced. **Figure 37-17** compares the capacity factors for commonly used energy sources.

Energy Source Capacity Factors

Energy Source	Capacity Factor
Biomass	85%–95%
Coal	70%
Gas	40%
Geothermal	90%
Hydropower	40%
Nuclear	90%
Solar	10%–25%
Wind	25%–30%

Goodheart-Willcox Publisher

Figure 37-17. The capacity factors of common energy sources.

37.5 Heat-Produced Electricity

Electricity can also be generated using heat. The process of producing electricity using heat is accomplished in two common ways, **Figure 37-18**. The first method is by converting thermal energy directly into electrical energy. The other method to generate electrical energy is by converting thermal energy into mechanical (kinetic) energy, which is then converted to electrical energy.

Direct Conversion of Thermal Energy to Electrical Energy

Converting heat energy directly into electrical energy relies on a phenomenon called the Seebeck effect. The ***Seebeck effect*** is the process of generating an electrical charge when a temperature difference is present across certain materials or substances. These substances, known as *thermoelectric materials*, have the ability to conduct electricity well but not thermal energy. Efficient thermoelectric materials include lead telluride (Pb_2Te_3) and silicon germanium. This property poses a challenge since good conductors of heat are also good conductors of electricity, and poor conductors of heat are also poor conductors of electricity.

Before we discuss how heat can be directly converted to electricity, it is beneficial to first understand the semiconductor materials that are used to facilitate this process.

Creation of Semiconductor Material

Silicon is commonly used to create semiconductor materials. It is important to note that, in its pure form, silicon is not a conductor of electricity. To create a material that functions as a semiconductor, silicon atoms are replaced with impurities in a process called *doping*. Doping is the process of adding

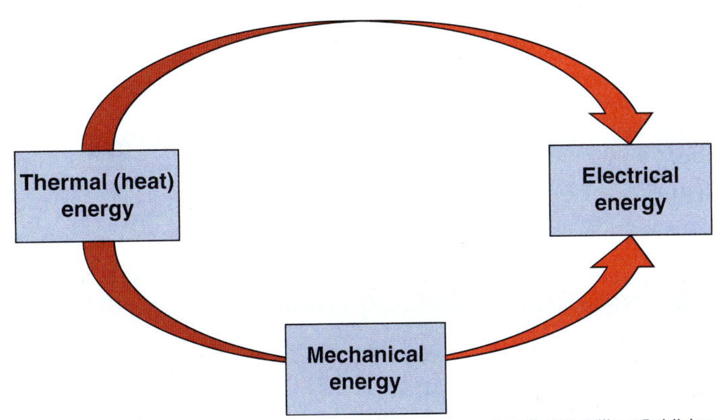

Goodheart-Willcox Publisher

Figure 37-18. Thermal energy can be converted to electrical energy either directly or indirectly. When converted indirectly, thermal energy is first converted to mechanical energy, which is then converted to electrical energy.

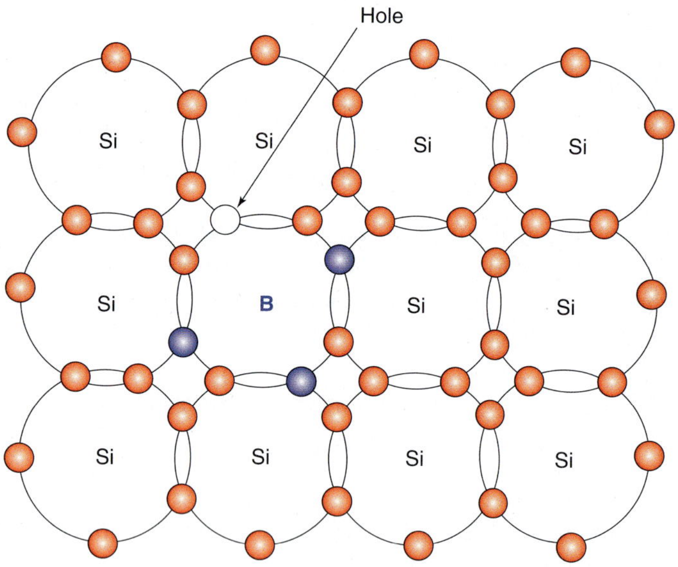

Figure 37-19. Silicon (Si) is commonly doped with boron (B) to create P-type material. P-type materials have a net positive charge.

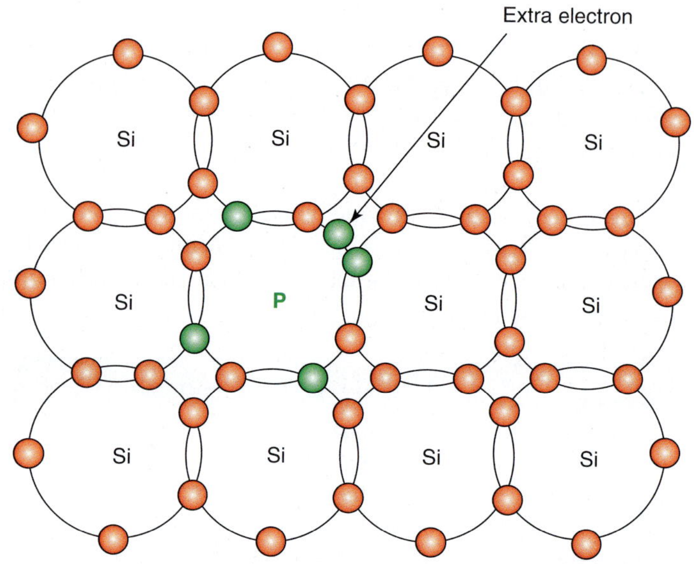

Figure 37-20. Silicon (Si) is commonly doped with phosphorus (P) to create N-type material. N-type materials have a net negative charge.

impurities to silicon to create two different materials: one that has a net positive charge (P-material) and one has a net negative charge (N-material). During the doping process, approximately 0.0001%, or one out of every million, of the silicon atoms are replaced by a boron atom to create P-type material, **Figure 37-19**. To create N-type material, silicon is commonly doped with phosphorus. In the phosphorus doping process, approximately 0.1%, or one out of every thousand, of the silicon atoms are replaced by a phosphorus atom, **Figure 37-20**.

Shedding Light

The ***Seebeck coefficient*** quantifies the amount of voltage produced across a thermoelectric material based on the temperature difference across it. The Seebeck coefficient is expressed in the units of microvolts per degree Kelvin, or μV/K. The higher the Seebeck coefficient, the greater the amount of voltage generated per unit change in temperature. The Seebeck coefficient is negative for N-type materials and positive for P-type materials.

Thermoelectric Cell Construction and Operation

Thermoelectric materials are classified as degenerate semiconductors, which have high doping levels, causing them to act more like metals than traditional conductors, such as silicon.

A typical thermoelectric cell is constructed with a section of N-type material connected in series with a section

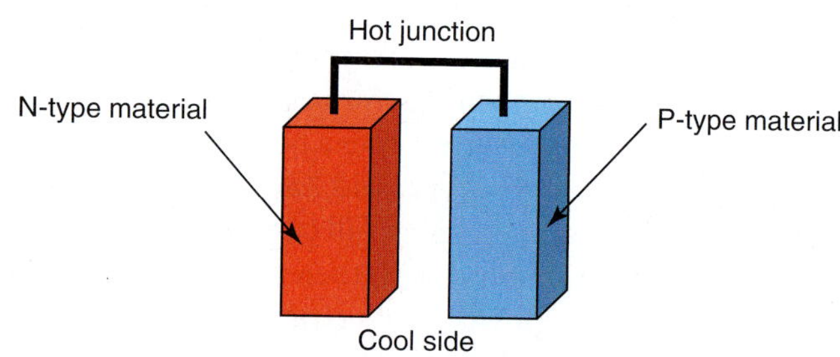

Figure 37-21. Thermoelectric cells are created by joining sections of N-type and P-type materials in series with each other. The point where the N-type and P-type materials are connected together is referred to as the hot junction.

of P-type material, **Figure 37-21**. The point at which the two materials are connected to each other is referred to as the *hot junction*, while the two remaining ends are exposed to a lower temperature. When subjected to a temperature difference, electrons flow from the cool side of the N-type material to its hot side, while electrons flow from the hot side of the P-type material to its cool side. The temperature difference, therefore, facilitates the flow of electrons through the series-connected N-type and P-type materials, **Figure 37-22**. Multiple cells can be connected in series to increase the amount of voltage generated, **Figure 37-23**. To help protect the integrity of a thermoelectric module, the sections of doped material and the conductive plates are commonly mounted to nonconductive surfaces, **Figure 37-24**. Electrical connections are made to the conductors that are fastened to the ends of the module.

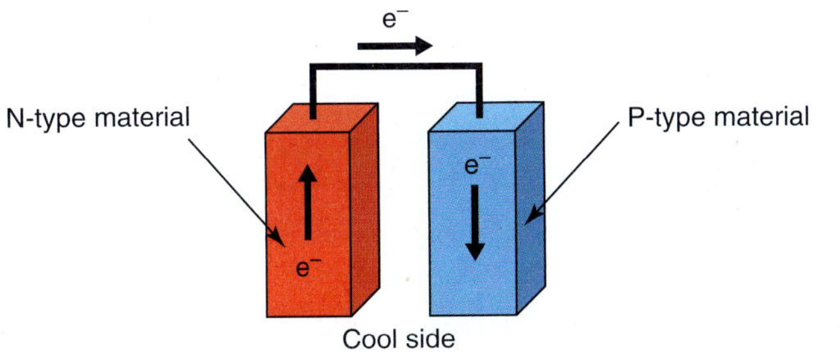

Figure 37-22. Electrons flow through the N-type material from the cool side to the hot junction, and then through the P-type material from the hot junction to the cool side.

Shedding Light

The sun is becoming an increasingly popular heat source for use in conjunction with thermoelectric materials to generate electricity. To produce electricity, sunlight is directed onto the P-N junction to create the required temperature difference. Sunlight is often referred to as a "free" heat source, although the associated equipment and installation costs of solar-based heat generation systems can be higher relative to other energy sources depending on the application and system location.

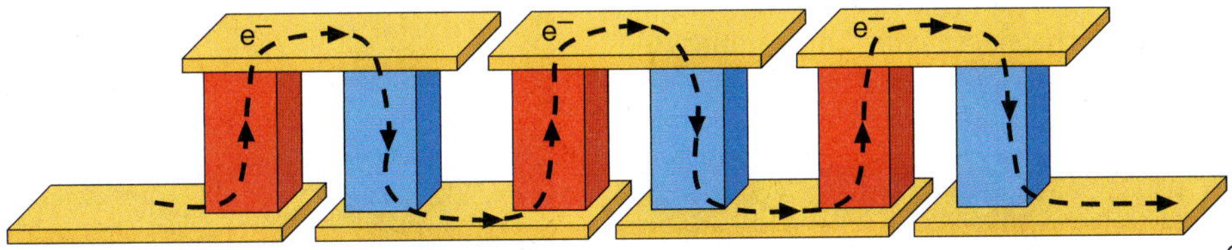

Figure 37-23. When multiple thermoelectric cells are connected in series to form a single, continuous path, the amount of generated electricity increases.

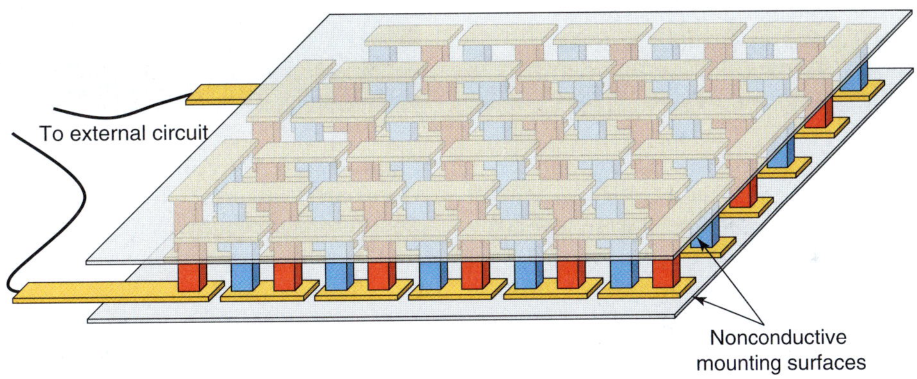

Figure 37-24. Thermoelectric cells and the interconnecting conductive plates are mounted to nonconductive plates. This helps protect the individual thermoelectric cells and provides connection points to the externally connected circuitry.

Indirect Conversion of Thermal Energy to Electrical Energy

The indirect conversion of thermal energy to electrical energy is typically a two-step process. First, heat energy is converted to mechanical energy. Second, mechanical energy is converted to electrical energy. The conversion of heat energy to mechanical energy is often facilitated by steam, which is used to turn a turbine. As discussed earlier in this chapter, the turbine's rotational force is then transferred to a generator to produce electricity. The heat required to produce steam can come from several sources, including:

- Burning coal, oil, or natural gas
- Burning biomass
- Geothermal energy
- Nuclear reaction

Traditionally, the heat required for the production of electricity came from the burning of fossil fuels such as coal, oil, and natural gas. Currently, the burning of fossil fuels accounts for approximately 60% of all electricity produced in the United States. An alternative to burning coal, oil, and natural gas is the burning of biomass.

Biopower

In its most general form, biopower relies on the burning of biomass to produce heat, which is then used to produce electricity. ***Biomass*** is organic matter that is used as the fuel source. Examples of biomass include wood, landfill gases, animal waste, sewage, and municipal solid waste. Biomass is easily converted to electrical energy by first burning the biomass to release heat, or thermal, energy. The heat energy is then used to boil water or other liquids to create steam.

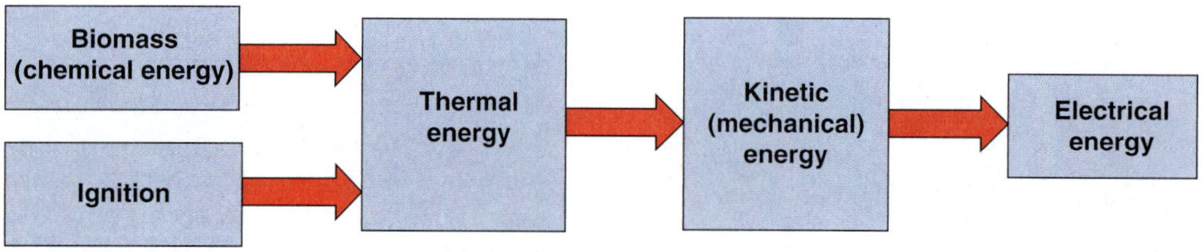

Goodheart-Willcox Publisher

Figure 37-25. Biomass is burned to release thermal energy, which is then used to create mechanical energy. Mechanical energy is then converted to electrical energy using generators.

This steam is used to turn the turbines. The rotating turbine provides the energy that is required to turn the generator, which is used to create electricity, **Figure 37-25**. The term ***biopower*** is used to describe the electricity that is generated using biomass as its fuel.

Shedding Light

The burning of municipal waste to create electricity is a viable option in some areas. The process of burning municipal waste is commonly known as waste-to-energy, or WTE. In this process, municipal solid waste, MSW, is burned to heat water and create steam, which is used to turn a generator-linked turbine. The main benefit of using WTE technologies is the reduction of the volume of waste. In some cases, the amount of waste can be reduced by as much as 90%. However, during the combustion process, a great deal of fly ash, or carbon, becomes airborne and must be captured to prevent increased air pollution levels from occurring. Presently, there are approximately 75 WTE facilities in the United States and over 500 throughout Europe.

Goodheart-Willcox Publisher

Figure 37-26. The conversion of heat extracted from the earth's subsurface to electrical power is created at a geothermal power plant.

Geothermal Energy

Geothermal energy, **Figure 37-26**, is slowly increasing in popularity as a viable heat source for use in the production of electricity. ***Geothermal energy*** refers to the conversion of heat extracted from the earth's subsurface into electrical power. A great deal of the heat that exists in the earth's subsurface is concentrated in either hot water or magma, which is molten rock. This heat is referred to as *geothermal heat*. Geothermal heat is used to create steam, which is then used to turn a generator-linked turbine.

The geothermal heat that is obtained from the earth's subsurface may not be hot enough to boil water into steam at atmospheric pressure. When this occurs, other fluids, such as isobutane or pentane, that have lower boiling temperatures can be used. The volume of these liquids, when vaporized or boiled, increases just as the volume of water increases when boiled, **Figure 37-27**. This vaporized fluid has the ability to turn turbines, just as water in the vapor state does. When liquids such as isobutane or pentane are

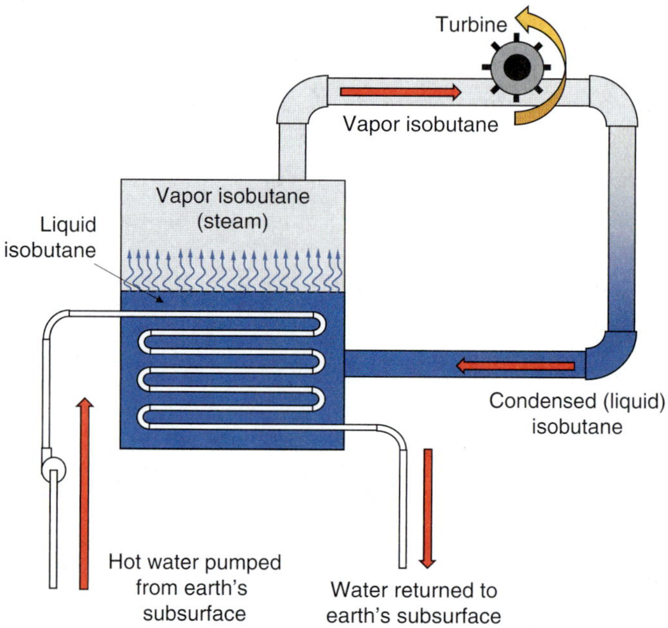

Goodheart-Willcox Publisher

Figure 37-27. When costly fluids with lower boiling temperatures are used in a geothermal system, closed piping arrangements are used for containment purposes, allowing the fluids to be reused.

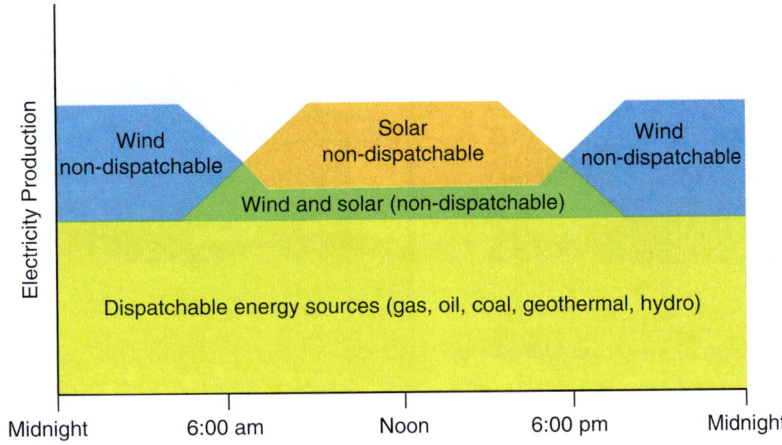

Goodheart-Willcox Publisher

Figure 37-28. Dispatchable energy sources are continuously and readily available, while energy sources that are non-dispatchable, such as wind and solar, are not. This example of a dispatch stack illustrates how wind and solar power can be used together, as they have complementary availability patterns.

used, they are contained in a closed system so they can be reused.

Shedding Light

Combined heat and power (CHP) systems have the ability to generate electricity while providing heat for a number of other purposes, such as domestic hot water, comfort heating, and a multitude of manufacturing- and materials-processing applications.

37.6 Light-Produced Electricity

With the increased popularity of solar panels, a large amount of electricity is produced by converting light energy into electrical energy. Solar panels, when positioned properly, can convert some of the sun's energy into electricity by a process known as the photovoltaic effect. The ***photovoltaic effect*** is a method commonly used to generate electricity and is the underlying principle of solar panels, or photovoltaic cells. As with wind power, sun-generated electricity is non-dispatchable, as sunlight is not always available and cannot be called upon when needed. However, solar power and wind power work hand-in-hand as solar power production peaks during the day and the majority of wind power is produced at night. As a result, it is often beneficial to combine wind and solar power generation. **Figure 37-28** illustrates a simplified ***dispatch stack***, which shows the availability of various energy sources at different times throughout the day. Dispatchable energy sources are available continuously and consistently at all times, while the availability of non-dispatchable energy sources varies greatly.

Shedding Light

The dispatch stack, in its complete form, is used by energy producers to determine the most cost-effective method to produce the required amount of energy at any given time during the day. The availability of different energy sources and their costs can vary greatly throughout the day.

Solar Panel Construction

A solar panel, or module, is made up of many interconnected individual photovoltaic cells. Each individual photovoltaic cell is made up of three layers. The base layer is a 0.2 mm thick layer of P-type semiconductor material. The middle layer is a 0.2 mm thick layer of N-type semiconductor material. The point where the N-type and P-type materials meet is called the P-N junction. The top layer, or coating, of the photovoltaic cell is a layer of antireflective coating (ARC), **Figure 37-29**. The P-N

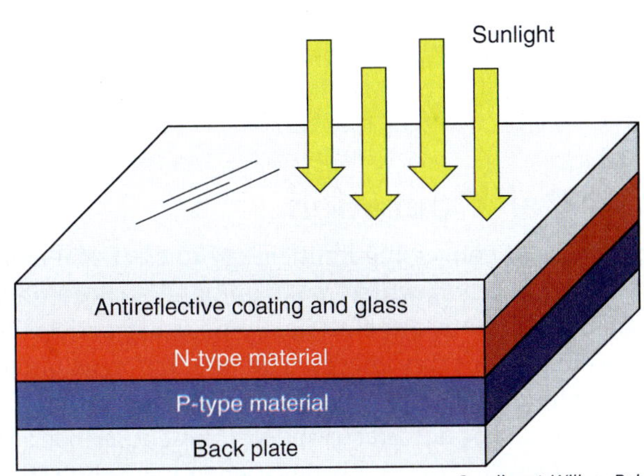

Goodheart-Willcox Publisher

Figure 37-29. A photovoltaic cell is constructed of layers of different materials.

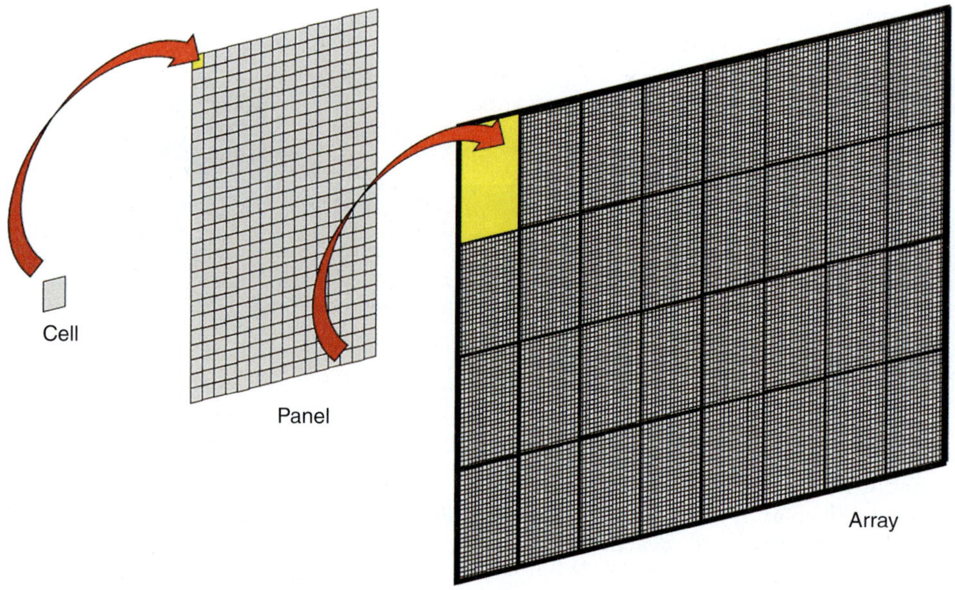

Goodheart-Willcox Publisher

Figure 37-30. Photovoltaic cells are combined to form solar panels. Multiple solar panels are connected together to form solar arrays.

junction functions as a diode. Depending on the design, a typical residential solar panel can generate between 250 W and 400 W, providing approximately 2 kWh of energy per day. Multiple solar panels can be connected to form an array, **Figure 37-30**. The power output of a solar array can be calculated as the power output of a single solar panel multiplied by the number of solar panels in the array. For example, a 20-panel solar array, with each panel producing 2 kWh per day, will have a total power output of 40 kWh per day (2 kWh × 20).

Shedding Light

Silicon is used in about 95% of the solar panels produced today. However, there are other materials that are being used to produce less expensive, but less efficient panels. These materials are cadmium telluride (CdTe), which is used in about 4% of solar panels, and copper indium gallium selenide (CIGS), which is used in about 1% of solar panels.

Solar Cell Operation

The process of converting light energy to electrical energy starts when photons, which make up light, strike a photovoltaic cell. Photons are electrically neutral particles that have no mass but carry energy. The interaction of photons with the photovoltaic cell creates what is known as the *photoelectric effect*. The photoelectric effect is best described as the release of electrons from a substance when struck by a stream of photons, or light, **Figure 37-31**. The freed electrons flow through the solar panel, which functions as diode, allowing electrons to flow through the panel in one direction, creating electric current.

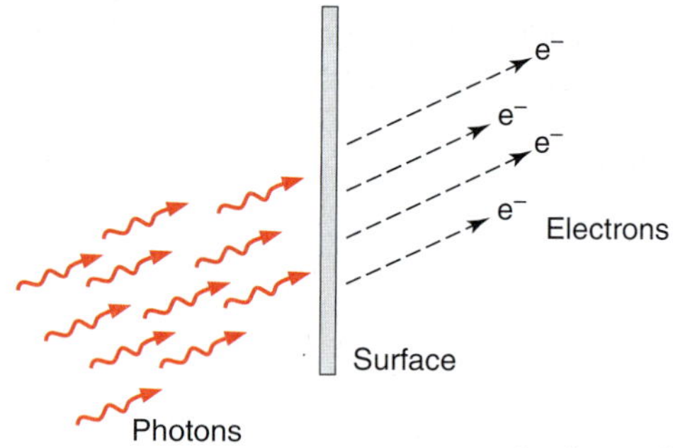

Goodheart-Willcox Publisher

Figure 37-31. When photons strike the surface of a solar cell, electrons are freed. The freed electrons create the potential difference, or voltage, across the solar panel.

Solar Cell Efficiency

The efficiency rating of solar panels is a commonly used method of comparing one solar panel to another. Mathematically speaking, solar panel efficiency is calculated as a solar panel's energy output divided by the panel's energy input received from the sun. The actual amount of electricity produced depends mainly on a cell's performance. Typically, solar panel efficiency ranges from 15% to 23%.

For the most part, sunlight that strikes a solar panel is either absorbed by the cell or reflected off the cell. The use of more effective antireflective coating materials helps to reduce reflection and increase the amount of sunlight that is absorbed by the cell.

As the efficiency of a solar panel increases, so does the cost of the panel. Therefore, it is important to carefully evaluate the costs and benefits when selecting solar panels for a particular application or project.

Example 37-2, **Example 37-3**, and **Example 37-4** explain how to evaluate the cost and benefits when determining the type of solar panels to use in a project.

Grid-Connected Photovoltaic Systems

Most photovoltaic systems that are presently installed are connected to the power grid. These systems are referred to as *grid-connected* or *grid-tied photovoltaic systems*. With a grid-connected system, end energy users can obtain their energy from the photovoltaic system or from the grid, depending on power availability. In addition, energy that is produced by the end user but not used can be directed to the power grid. Grid-connected systems do not require energy storage or batteries. This is because any energy needs that exceed the ability of the existing photovoltaic system are met by the main power grid.

Shedding Light

When a power failure occurs, residential solar panels are automatically disconnected from the grid for safety. If solar panels were not disconnected from the grid during power failures, the panels could supply voltage into the transmission lines, potentially energizing the lines and equipment being serviced by utility workers.

Net Metering

Net metering is an energy-management practice that allows energy producers to send unused energy to the power grid in exchange for credits that can be used to offset their own energy costs. In operation, net metering relies on a smart, or bidirectional, meter that measures the amount of energy being used from the power grid, as well as the amount of energy sent from the user to the grid.

If the photovoltaic system can meet the energy needs of the structure, the photovoltaic system will do so, and no power will be pulled from the grid. If the photovoltaic system is not able to meet the energy needs of the structure, energy will be pulled from the grid to satisfy the deficiency. If the amount of energy produced by the photovoltaic system exceeds the amount of energy required by the structure, the excess energy is sent into the grid, where it can be used by other energy consumers. Net metering, in its simplest form, is a means to

Example 37-2

A particular solar project requires the production of 150 kWh per day of energy. Solar panels with a 15% efficiency rating are being considered for the job. Each panel produces 1.5 kWh of energy per day. How many panels will be required for this project?

✓ Solution

If each panel can provide 1.5 kWh of energy, the total number of panels required can be determined by dividing the required 150 kWh of energy by 1.5 kWh per panel. The result indicates that 100 panels will be required (150 kWh ÷ 1.5 kWh).

Example 37-3

If 20% efficient panels with an output of 2 kWh per day were used for the project described in **Example 37-2**, how many panels will be required for the project?

✓ Solution

Since each panel has an output of 2 kWh, 75 panels (150 kWh ÷ 2 kWh) are required to produce the needed 150 kWh of energy.

Example 37-4

Using the information from **Example 37-2** and **Example 37-3**, determine the total cost of panel installation for each option if the unit costs are as follows:

- Unit cost of 15% efficiency solar panels: $200
- Unit cost of 20% efficiency solar panels: $350
- Installation cost per panel (15% or 20%): $100

✓ Solution

The cost of a 15% efficient panel installation can be calculated as follows:

100 panels @ $200 + 100 panel installations @ $100 = ?

$20,000 + $10,000 = $30,000

The cost of a 20% efficient panel installation:

75 panels @ $350 + 75 panel installations @ $100 = ?

$26,250 + $7,500 = $33,750

Based on these costs, the 15% efficient panels would be the less costly option.

track the amount of energy used and the amount of energy supplied. Based on this difference, an energy bill is generated.

The rules, guidelines, and policies that apply to net metering are not uniform across the United States. They can vary greatly from city to city, state to state, and from one power utility company to another. For example, some utility companies have limits on the amount of energy that can be sold back to the grid. In some cases, these limits are monthly limits, while in other instances, the limits are annualized. In many cases, utility companies use what is referred to a ***true-up adjustment***. On an electrical statement, the true-up adjustment reports a consumer's total energy usage over the past year against the amount of energy the consumer has sent to the power grid. Once calculated, the consumer will need to pay the difference if they used more energy than they sent to the grid or will receive a payment or credit if they consumed less energy from the grid than they sent to it. The true-up adjustment is most commonly encountered when photovoltaic systems are installed by a third-party provider that has agreed to provide power to end users for a fixed monthly fee.

Concentrated Solar Power

In concentrated solar power (CSP), sunlight is concentrated to generate high temperatures. This technology uses a large array of flat lenses or mirrors, called ***heliostats***, to reflect and concentrate sunlight onto a receiver, **Figure 37-32**. The receiver is positioned at the top of a tower. This system configuration is referred to as a *power tower*. The concentrated

Kit Leon/Shutterstock.com

Figure 37-32. The largest CSP plant in the United States is the Ivanpah Solar Electric Generating System, located in California in the Mojave Desert. This facility has the capacity to produce almost 400 MW of electricity and uses nearly 175,000 heliostats to concentrate sunlight on its three power towers.

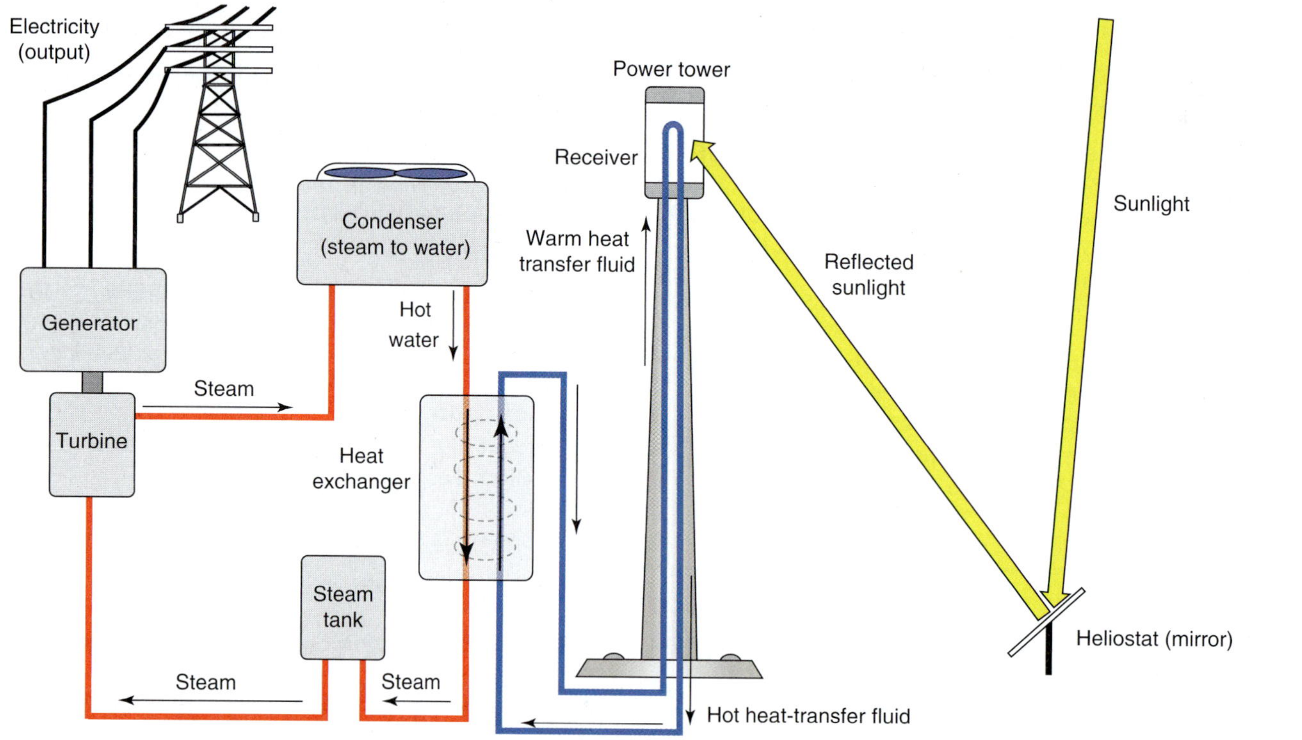

Goodheart-Willcox Publisher

Figure 37-33. Piping diagram showing how a heat transfer fluid can be used to heat the working fluid. The heat transfer fluid flows in a continuous loop between the power tower's receiver and the heat exchanger. The working fluid, in this case steam, is supplied to the turbine to generate electricity. From the turbine, the steam is condensed and then returned to the heat exchanger where it is once again vaporized for reuse in the turbine.

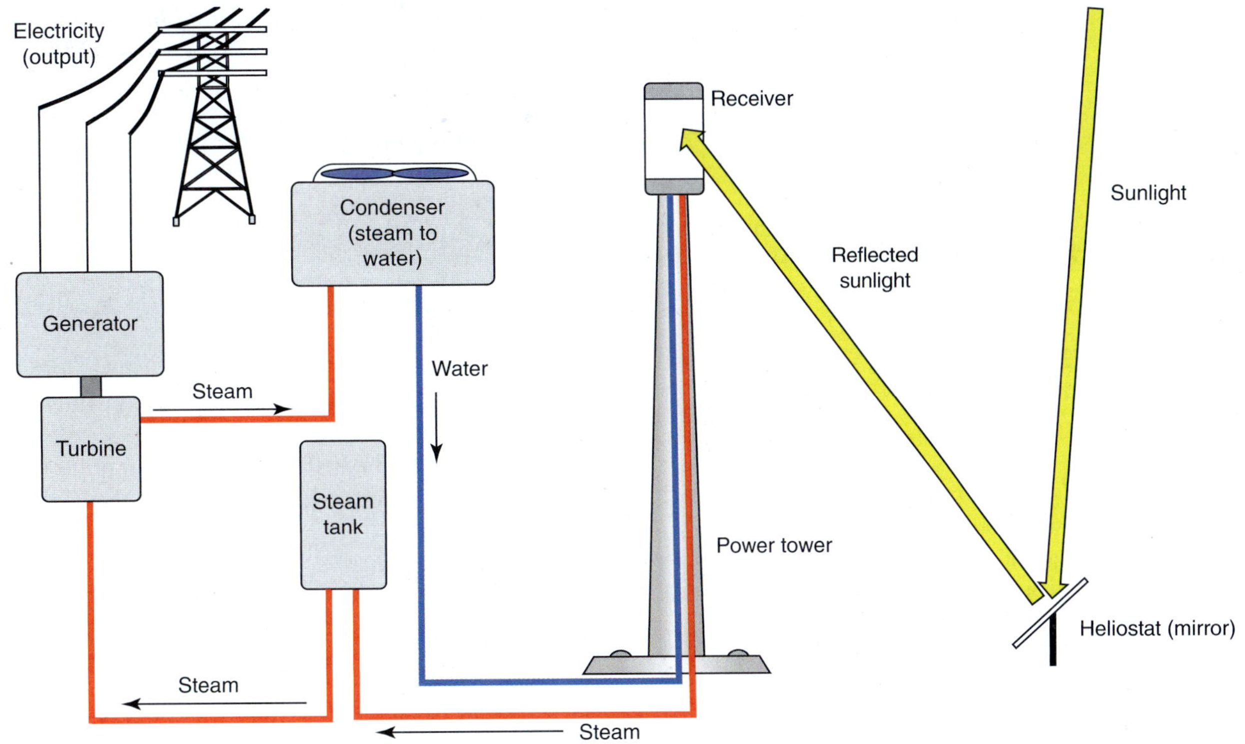

Goodheart-Willcox Publisher

Figure 37-34. Piping diagram showing how heat in the power tower's receiver is used to create steam directly. This configuration does not require the use of a heat exchanger.

sunlight heats a fluid that flows through the receiver. This heated fluid, referred to as a *heat-transfer fluid*, may be oil, brine, water, or a number of other substances. Then, by using a heat exchanger, the heat-transfer fluid heats the working fluid, which facilitates the production of electricity. In some cases, the heat-transfer fluid is used to boil water into steam, which is then used to turn turbines that generate electricity, **Figure 37-33**. In some systems, steam is produced directly by the concentrated heat in the receiver, **Figure 37-34**. In this case, no heat exchanger is required.

CHAPTER 37 REVIEW

Summary

- The law of conservation of energy states that energy cannot be destroyed nor created but can be transformed from one form to another. (37.1)
- Chemically produced electricity relies on chemical reactions that occur within devices, such as batteries and fuel cells. (37.2)
- Pressure-produced electricity relies on physical pressure to turn turbines, which ultimately generate electricity. (37.3)
- Common pressure sources for electricity production are wind and flowing water. (37.3)
- Two common sources of pressure in hydropower systems are hydroelectric dams and the natural flow of river water. (37.3)
- Two common types of hydropower systems are the hydroelectric dam and the run-of-the-river system. (37.3)
- The piezoelectric effect describes the condition where an electric charge is created in direct response to pressure. (37.3)
- Electricity generated by wind is created by using pressure to turn the turbines that generate electricity. (37.4)
- Wind and solar technologies are classified as non-dispatchable, meaning they cannot be put into operation whenever needed. (37.4)
- Capacity factor is a measurement of the efficiency of a power-producing process or facility. The higher the capacity factor, the higher the efficiency of the power-producing process. (37.4)
- Some electricity-producing technologies rely on the use of semiconductors, namely N-type and P-type materials. (37.5)
- Solar panels are made up of two primary layers: one layer of N-type material and one layer of P-type material. (37.6)
- The dispatch stack is a tool used by energy producers to determine the most cost-effective and profitable way to produce the required amount of energy at any given time. (37.6)
- Concentrated solar panel systems concentrate sunlight to heat fluid that, in turn, is used to generate electricity. (37.6)

Review Questions

Use the information in the chapter to answer the following questions.

Measuring Your Potential

1. *True or False?* Energy can be transformed from one form to another and can be transferred from one object to another. (37.1)
2. The cathode is the battery's connection point at which electrons are gained through a process known as _____. (37.2)
   A. oxidation
   B. reduction
   C. diversion
   D. impoundment
3. Which of the following is *not* true of fuel cells? (37.2)
   A. They convert chemical energy directly into electrical energy.
   B. They provide a continuous supply of power.
   C. Hydrogen used in fuel cells is easy to obtain.
   D. They have the potential to produce large amounts of electricity with minimal environmental impact.
4. *True or False?* Hydropower is desirable because the supply of flowing water is relatively consistent. (37.3)
5. *True or False?* The amount of electricity produced by wind is typically greatest during the day. (37.4)
6. The shaft, gear box, and generator of a wind turbine are contained in the _____. (37.4)
   A. hub
   B. nacelle
   C. tower
   D. base
7. The process of generating an electric charge when a temperature difference is present across a substance is best known as the _____. (37.5)
   A. Seebeck effect
   B. triboelectric effect
   C. peltier effect
   D. piezoelectric effect

8. The burning of fossil fuels accounts for what percentage of total electricity production? (37.5)
   A. 20%
   B. 40%
   C. 60%
   D. 80%
9. The construction of a solar panel most closely resembles that of a _____. (37.6)
   A. transistor
   B. diode
   C. resistor
   D. capacitor
10. Power towers use a large number of mirrors to direct sunlight to a _____. (37.6)
    A. receiver
    B. heat exchanger
    C. steam turbine
    D. condenser

Plug-In

1. Which of the following methods rely on physical pressure to generate electricity? (37.3)
   A. Photovoltaic
   B. Wind turbines
   C. Thermoelectric cells
   D. Solar panels
2. Which of the following energy sources has the highest capacity factor? (37.4)
   A. Wind
   B. Biomass
   C. Hydropower
   D. Solar
3. A thermoelectric cell is comprised of _____. (37.5)
   A. two layers of P-type material connected in series with each other
   B. two layers of N-type material connected in parallel with each other
   C. P-type material connected in series with N-type material
   D. P-type material connected in parallel with N-type material
4. To increase the amount of voltage produced by a thermoelectric module, the number of _____. (37.5)
   A. parallel-connected thermoelectric cells are reduced
   B. parallel-connected thermoelectric cells are increased
   C. series-connected thermoelectric cells are reduced
   D. series-connected thermoelectric cells are increased
5. How much power, in kWh, can be expected to be produced by a 40-panel solar array, if each panel produces 1.4 kWh of energy? (37.6)

Making the Connection

1. An electrician claims that the efficiency of an impoundment hydroelectric dam is higher than that of a run-of-the-river hydroelectric system. Do you agree or disagree with this electrician's statement? Provide as much supporting information as possible to support your response (37.3).
2. With a low-capacity factor, how would you convince a potential customer that a wind turbine is a viable option for electricity production? (37.4)

CHAPTER

38 Power Distribution: From the Generator to You

west cowboy/Shutterstock.com

SPARKING DISCUSSION

The electrical generation process and its distribution practices and philosophy are changing. This will not change the day-to-day activities of electricians. Do you agree or disagree with this statement?

LEARNING OUTCOMES

After completing this chapter, you will be able to:

38.1 Explain the importance of the power grid and the current and future challenges related to the power grid.

38.2 Describe the function of the individual components in the power grid.

38.3 Explain the components, capabilities, and implementation challenges of a smart power grid.

38.4 Describe distributed energy resources and microgrids.

38.5 Describe some of the factors that must be considered when sizing conductors.

TECHNICAL TERMS

advanced metering infrastructure (AMI)
distributed computing
distributed energy resource (DER)
distributed generation
distribution substation
distribution transformer
energy storage system
generating substation
microgrid
reactive power support
transmission substation

Introduction

If we, as consumers of electrical power, are to gain access to this valuable resource, it must first be transmitted from the point where it is produced to the point where it is ultimately used. During the transmission process, both the quality and quantity of power are somewhat reduced, so it is in the best interest of the energy producers to take actions that help minimize power degradation while providing electricity in sufficient quantities. After all, power utilities strive to provide an efficient, reliable, and profitable means to produce, transmit, and ultimately distribute power to those who rely on it.

As an electrician, it is important that you have a solid understanding of the power grid and how electricity gets from the power plant to the consumer. Our present power grid is more complicated than ever before. Conventional power distribution systems were originally designed as networks that directed energy in only one direction, from centralized power plants to consumers. The evolution of smart grids in recent years has completely changed the ways in which power is transmitted and distributed. A smart grid uses digital technologies, switching automation processes, and communication protocols to create a multidirectional flow of electrical energy and data. Today's grid is somewhat of a hybrid, incorporating the older technologies commonly associated with traditional grids and the newer technologies found in smart grids. It is those working on the front lines of the electrical trades that are responsible for bridging the gap between the legacy equipment and the cutting-edge technologies that are quickly replacing them.

This chapter discusses the fundamentals of electrical power distribution. It examines traditional power distribution networks as well as some of the capabilities of smart grids. The benefits and challenges associated with each are discussed as well.

Since our industry relies heavily on the use of conductors to transmit and distribute electricity between and among points, this chapter concludes with the topic of conductor sizing.

The advancement in electrical technology and practices over the last couple of decades has been rapid. These advancements have led to new occupations, such as wind turbine installers, solar panel technicians, smart grid operators, fiber cable installers, and energy management system integrators. The electrical skills you are learning now may be preparing you for a future occupation that does not yet exist!

38.1 The Power Grid

The power grid in the United States has delivered electricity to our homes and businesses since the 1890s. Our power grid is a complex system of interconnected transmission lines, substations, distribution lines, transformers, and a wide array of other resources and technologies that function together to provide us with the power we rely on to survive.

Our power grid affects most, if not all, aspects of our daily lives. Without our power grid, the way we work, socialize, learn, entertain ourselves, cook, clean, and get from one place to another would be greatly affected. Losing power for even a few hours has adverse effects on how we function. Imagine living without electricity for weeks, months, or even years at a time. It is not a stretch to say that our reliance on technology is, in actuality, a reliance on electricity. As the number of electric vehicles on our roads and highways increases, for example, this reliance on electricity is more profound than ever before.

Our present energy grid has an aging infrastructure, while our increased energy use and demand is outpacing the grid's ability to provide sufficient quantities of high-quality electricity. When the demand for electricity exceeds the grid's ability to produce energy, unplanned blackouts, brownouts, and other power outages occur. These outages have a negative effect on our country's ability to function, and, in some cases, can put our national safety and security at risk.

The demand for electricity has been increasing for many years and is expected to continue to increase for the foreseeable future. This increased demand is seen in all three primary sectors:

- **Residential.** Homes use more electricity today than ever before due to the large number of electrical appliances, home computers, charged devices, and electric vehicle chargers that have become commonplace. The number of houses continues to increase, and the size of houses continues to grow. Since the air-conditioning and heating systems are the largest users of electricity in a home, a larger house generally requires more electricity than a smaller house for heating and cooling.
- **Commercial.** Offices continue to implement an increasing amount of computer technology and devices. Data centers that support the computing needs of businesses and consumers require a significant amount of electricity to support both the computing devices and the cooling needs. The electrical demand by data centers is expected to continue growing over the next decade. Keeping food products refrigerated or frozen in warehouses, supermarkets, convenience stores, and restaurants requires an incredible amount of electricity. This electrical demand continues to grow as more food products are needed to feed a growing population. As more commercial electric vehicle charging stations are installed, more demand is placed on the power grid.
- **Industrial.** The increased use of automation and robotics in industry is powered by electricity. In addition, the use of electronic monitoring and control in industrial environments continues to grow.

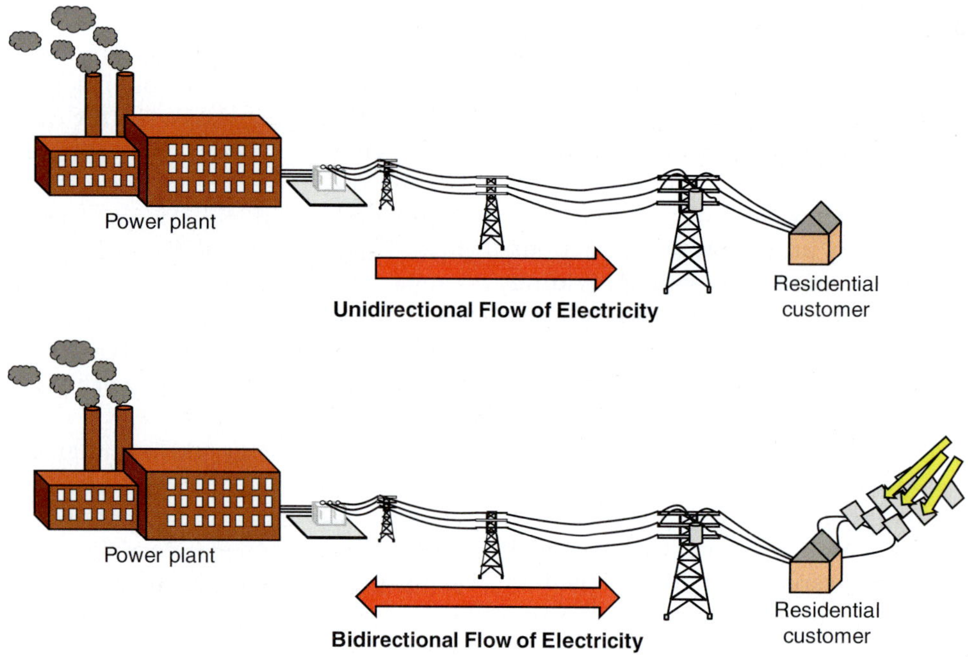

Goodheart-Willcox Publisher

Figure 38-1. Original transmission systems were unidirectional, with electricity flowing from the power plant to the end users. Today, most transmission systems are bidirectional, allowing utilities and end users to both purchase and supply electricity.

To keep up with the growing demand for electricity, the electrical industry is making major changes to the processes and procedures to generate and manage the production and distribution of power. What used to be a unidirectional flow of electricity from the power plant to the end user is fast becoming a bidirectional flow of electricity, where both the end user and the power plant are suppliers and purchasers of electrical energy, **Figure 38-1**. As there must be a way to monitor and manage this two-way flow of electricity, the need for a more advanced, smarter power grid became evident. Smart grids, which are discussed later in this chapter, are intended to better satisfy our electrical needs by increasing the reliability and stability of electricity production.

US Department of Energy

Figure 38-2. Generators such as these hydropower generators are the starting point for the transmission and distribution of electricity.

38.2 Components of an Electricity Transmission System

Just as an orchestra relies on the contributions of individual instruments to create beautiful music, an electricity transmission system relies on the effective and efficient operation of many individual components, processes, and subsystems to create a reliable energy delivery and management platform. Together, the following components transmit stable power while minimizing system losses:

- Generators
- Transformers
- Power lines
- Insulators
- Substations
- Circuit breakers and fuses
- Switches and relays
- Voltage regulators
- Communication and control systems
- Energy storage systems

Generators

System generators are responsible for generating electricity at the power plant, **Figure 38-2**. The produced energy can be the result of a number of processes and sources, including the burning of fossil fuels or harnessing the power of nuclear reactions, flowing water or wind, or capturing energy from our sun. Depending on the application and the specifications of the power plant, the voltage typically generated is between 11 kV and 33 kV.

Transformers

Transformers are responsible for increasing or decreasing the voltage levels in a power system. Step-up transformers are typically used at or near the point where energy is produced to increase the voltage in preparation for transmission. Increasing the voltage allows electricity to be transmitted at a lower current, resulting in smaller transmission conductors. These higher voltage levels also help increase the efficiency of long-distance transmissions. Step-down transformers, commonly referred to as ***distribution transformers***, are used to reduce the voltage of the transmitted power to the levels required by the end users of the energy. For example, a distribution transformer reduces voltage to single-phase 240 Vac for residential customers. Often, the voltage is reduced to a subtransmission level prior to reaching distribution transformers. Standard and typical transmission voltage are shown in **Figure 38-3**.

When a distribution transformer is used in an overhead system, it is often mounted on a power pole. Such transformers are referred to as pole-mounted transformers, **Figure 38-4**.

In areas supplied by underground distribution lines, ground-level distribution transformers called *pad-mounted transformers* are used. The underground conductors are connected to a pad-mounted transformer's primary and secondary terminals and enter the transformer from below.

Shedding Light

In addition to stepping up and stepping down voltage levels, transformers are used at substations to help enhance, boost, and maintain proper voltage levels.

Power Lines

High-voltage electricity, once produced, is transmitted from the power plant. Power lines are the conductors used to carry high-voltage electricity from the generating plants to substations and, ultimately, to the points where energy is distributed to the end users. The method used to produce electricity does not affect the nature or configuration of the distribution network. Power lines can be located either overhead (aboveground) on towers or poles, or underground in protected channels, tunnels, chases, and pipes. Some electrical cabling is designed for direct burial, such as the final wiring entering a residence.

Overhead Power Transmission and Distribution Systems

The overhead type of power distribution system is the most commonly used method to move electricity from one place to another over long distances. In the overhead configuration, electricity passes through conductors that span between ground-mounted poles or towers. Overhead power transmission and distribution systems have numerous advantages

Standard Electricity Transmission Voltages

Category	Standard Voltage Levels
High voltage (HV)	69 kV, 115 kV, 138 kV, 161 kV, 230 kV
Extra-high voltage (EHV)	345 kV, 500 kV, 765 kV
Ultra-high voltage (UHV)	1100 kV, 1500 kV
DC high voltage (dc HV)	±250 kV, ±400 kV, ±500 kV
Subtransmission voltage (typical)	34.5 kV to 161 kV
Distribution voltage (typical)	4.16 kV to 34.5 kV

Goodheart-Willcox Publisher

Figure 38-3. Substation transformers change the electrical voltage of transmitted electricity at various points in the transmission process. The higher the voltage, the more efficient the transmission.

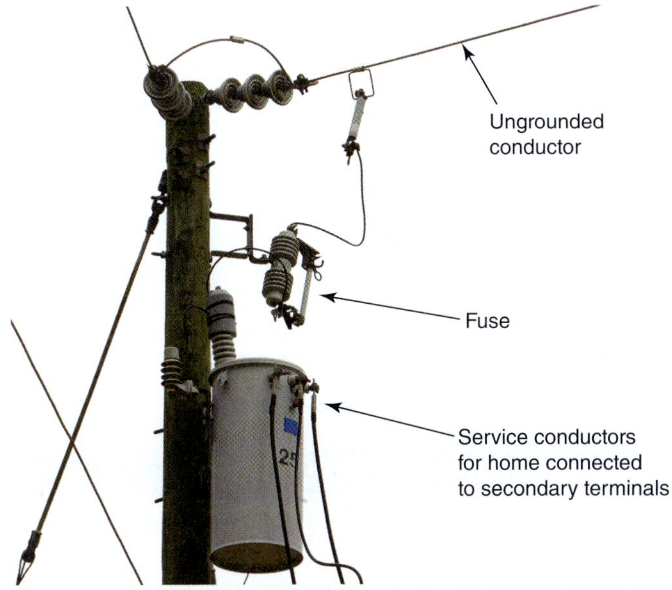

Barry Blackburn/Shutterstock.com

Figure 38-4. Pole-mounted distribution transformers are commonly used to provide single-phase power to residences.

as well as some disadvantages. The advantages of overhead systems include the following:

- Lower installation costs.
- Lines are more easily accessed.
- Lower maintenance/service costs.
- Easier to expand/upgrade the network/infrastructure.
- Bare wires with no insulation are commonly used, lowering the cost of the conductors.

Disadvantages of overhead systems include the following:

- Infrastructure is exposed to weather-related damage.
- Visibility of poles, transformers, and other equipment is not visually appealing.
- Poles are susceptible to physical damage.
- Infrastructure is more likely to experience vandalism.
- Increased potential for physical injury due to falling poles or power lines.

Underground Power Transmission and Distribution Systems

Underground power distribution systems use power cables that are buried in the ground, as opposed to the overhead cables previously discussed. Underground distribution systems are predominantly found in urban areas or in areas where extreme weather conditions, such as hurricanes, are a continuous threat.

Underground power transmission and distribution systems have numerous advantages as well as some disadvantages. The advantages of underground systems include the following:

- Infrastructure is, for the most part, protected from weather-related damage.
- System is not visually offensive as there are no poles, transformers, and other equipment in sight.
- Reduced potential for physical damage and vandalism.
- Reduced risk of physical injury due to falling equipment.

Disadvantages of underground systems include the following:

- Higher installation costs.
- Lines are more difficult to access.
- Lack of visibility creates a potential hazard when digging.
- Higher maintenance/service costs.
- More difficult to expand or upgrade the network/ infrastructure.

A relative comparison between overhead and underground power lines is provided in **Figure 38-5**.

Shedding Light

Some underground transmission and distribution systems that are contained in piping arrangements incorporate mechanical cooling or ventilation systems to help dissipate heat from the cables. The greater the current flow and the greater the number of conductors in the raceway, the more heat is generated.

Overhead and Underground Power Line Comparison

Characteristic	Overhead	Underground
Installation Cost	Lower	Higher
Maintenance Cost	Lower	Higher
Maintenance/Repair	Easier	Harder
Visual Appeal	Low (highly visible)	High (not visible)
Reliability	Lower	Higher
Useful Life Expectancy	Lower	Higher

Goodheart-Willcox Publisher

Figure 38-5. When compared, both overhead and underground power lines offer benefits and drawbacks.

Rosemarie Mosteller/Shutterstock.com

Figure 38-6. Insulators separate current-carrying power lines from support structures.

Insulators

Insulators are installed between the transmission lines and support structures such as poles, towers, and substation structures, **Figure 38-6**. They help prevent electrical leakage to ground. Insulators are typically fabricated from ceramic, glass, or polymer materials. Newer-generation polymer insulators, such as those made from silicone rubber, are being used to replace older-generation ceramic or glass insulators on overhead transmission and distribution lines. Silicone rubber insulators are lightweight and more resistant to degrading when exposed to extreme weather conditions. Because of their ability to withstand the elements better than other materials, the risks associated with electrical faults and other accidents are greatly reduced. In addition, their extended useful life expectancy helps to reduce the maintenance and repair costs of the lines.

Substations

Substations are located at various points along the power grid, **Figure 38-7**. In these substations, high-voltage electricity

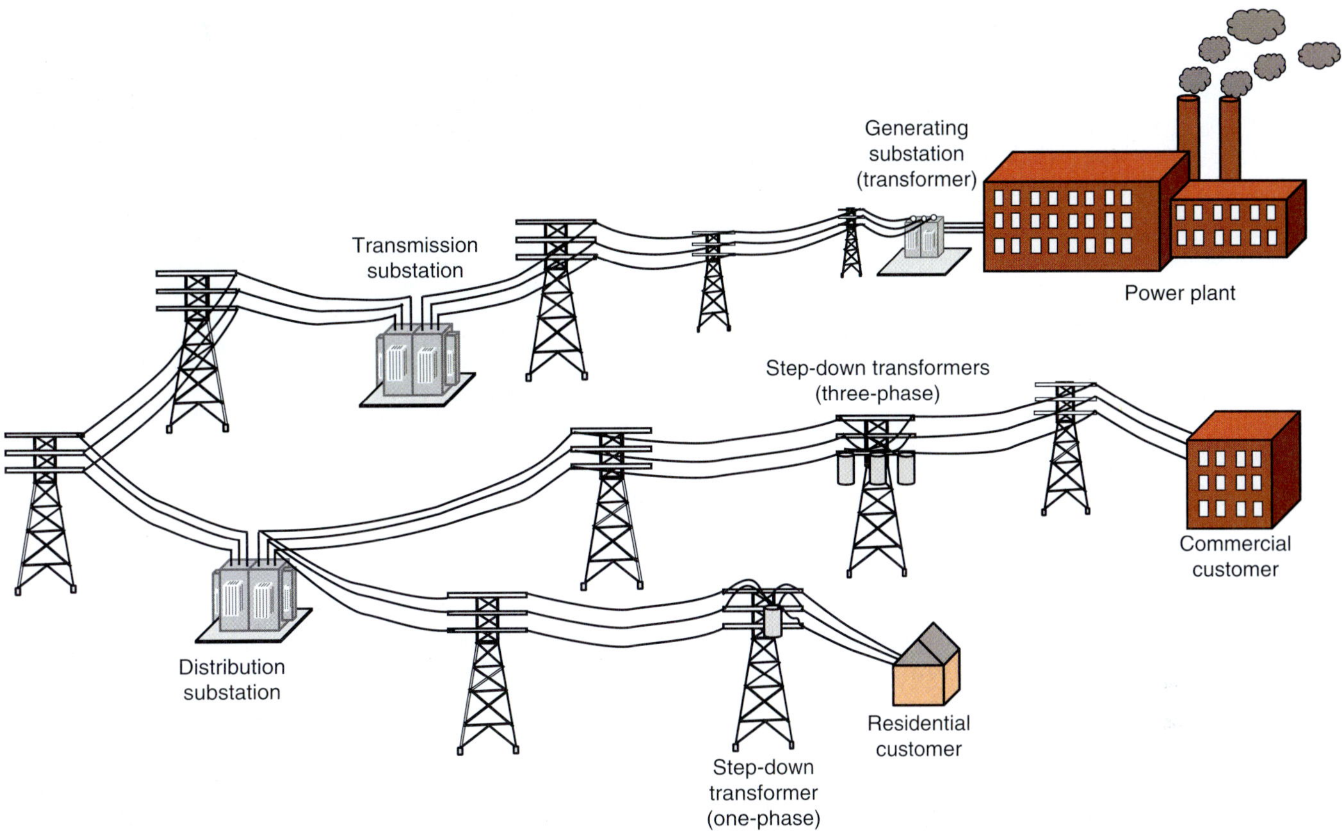

Goodheart-Willcox Publisher

Figure 38-7. Substations act as nodes in the power grid for voltage changes, switching, and other functions.

passes through transformers and other equipment where the voltage and power quality can be stepped up, stepped down, enhanced, or maintained at levels that are appropriate for distribution based on the needs of the customers being served.

In addition to quality-related factors, substations are responsible for assisting in the management and distribution of energy. It is at the substations that power can be rerouted as necessary when faults, overloads, or other undesirable conditions are present. Rerouting of power also will occur for normal maintenance.

The term *substation* is an overarching term that is used to identify a major system element with the ability to perform a number of different actions to help ensure the continued successful operation of the grid. Some of the components commonly found in substations are transformers, circuit breakers, fuses, relays, and busbars. Common types of substations are generating substations, transmission substations, and distribution substations, **Figure 38-8**.

Glen Richard/iStock/Getty Images Plus via Getty Images

Figure 38-8. Each substation has a unique design with electrical equipment tailored to the specific purpose of the substation within the power grid.

Generating Substations

Generating substations are located at the point where power leaves the power plant. It is here that the generated electricity is transformed to a higher voltage level using step-up transformers. By stepping this voltage up to a higher level, the efficiency of the transmission, especially over long distances, is increased. It is at the generating substation where generated power first enters the electricity transmission system.

Transmission Substations

Transmission substations are often characterized as the connection point for two or more sets of transmission lines.

Transmission substations can connect transmission lines that are at the same voltage level or can connect lines that have different voltages. When the entering and leaving lines carry power at the same voltage, the transmission substation often consists of flow-directing switches and controls, monitoring software, and circuit-protection devices. The flow-directing circuitry allows for the easy rerouting of power in the event that system problems arise.

When a transmission substation is to connect lines that have different voltages, additional components, controls, and equipment are needed. This additional equipment typically includes transformers, regulators, switches, monitoring software, and circuit protection devices. Data obtained from the monitoring software allows the grid operators to evaluate the operational health of the grid in real time.

Distribution Substations

A ***distribution substation*** acts as a separation point between the high-voltage transmission system and the lower-voltage distribution network. At its inlet side, the distribution substation receives high-voltage electricity from the long-distance transmission lines and reduces this voltage to the levels required for distribution. This is accomplished using step-down transformers. The lower-voltage electricity at the transformer's secondary winding is directed to the distribution lines that carry power to the customer.

Just as with transmission substations, the distribution substation may use voltage regulation equipment such as capacitors or reactors to help maintain stable voltage levels as the demand for energy fluctuates. Additionally, circuit protection devices, monitoring software, and communication systems are often found at the substation to monitor and control its operation.

A simple distribution substation has one set of incoming transmission lines and two sets of outgoing lines. More complex distribution substations may have multiple incoming transmission lines, additional transformers, and more sets of outgoing distribution lines to allow electricity to be efficiently distributed in multiple directions.

Circuit Breakers and Fuses

Just as circuit breakers and fuses protect the wiring and circuitry in a home, circuit breakers and fuses also protect the circuits in power transmission and distribution systems. If an electric fault or overload is detected, fuses and circuit breakers will interrupt the flow of electric current to protect the transmission lines, distribution lines, and other system components that are susceptible to damage from overcurrent conditions.

Fuses are single-use devices that open the circuit in an overload condition, **Figure 38-9**. The fuse must then be replaced to restore the circuit.

goldenwingsphotography/iStock/Getty Images Plus via Getty Images

Figure 38-9. Fuses provide overcurrent protection in substations.

Circuit breakers, **Figure 38-10**, are devices that trip and can be reset. Four types of circuit breakers are used in substations. They are categorized by the medium surrounding the contacts that, when the circuit breaker trips, extinguishes the arc created and absorbs the arc's energy.

- Air circuit breakers use air to extinguish the arc and are used in low-voltage and high-voltage applications.
- Vacuum circuit breakers that contain arcs in a vacuum chamber are used in medium-voltage applications.
- Oil circuit breakers rely on oil to absorb the arc's energy.
- SF6 gas circuit breakers contain arcs in a chamber filled with sulfur hexafluoride (SF6) gas, which has a high dielectric strength.

Shedding Light

An automatic circuit recloser (ACR) is an overcurrent protective device that opens during a fault condition and then attempts to close after a predetermined period of time. ACRs are helpful for one-time, self-correcting events. For example, if a tree branch falls across a power line and causes a fault but then falls away from the line on its own, an ACR will be able to reconnect the line automatically.

Switches and Relays

In a power transmission grid, a network of switches and relays controls and redirects the flow of electric current in response to signals received from the communications system. When a malfunction or system abnormality (such as an overvoltage, overcurrent, or frequency divergence) is detected, a signal is sent to the appropriate switches and relays to disable the affected portion of the system. At the same time, or within a reasonable amount of time, the holding coils on large relays are energized to redirect power from other portions of the

Oil Circuit Breakers
US Department of Energy

SF6 Gas Circuit Breakers
VG Foto/Shutterstock.com

Figure 38-10. Substation circuit breakers are designed to extinguish the arc generating when the breaker's contacts open and absorb the resulting energy.

Valmedia/Shutterstock.com

Figure 38-11. Circuit switchers control the flow of electricity through equipment and power lines in a substation. The circuit switcher shown here is currently open and is closed by swinging the disconnect blade down until it locks tightly into the jaw.

system to help meet the electrical needs of those customers in the affected areas. Storm outages and normal line maintenance will also cause the power to be redirected. Circuit switchers in substations are used to connect and disconnect lines and equipment as needed, **Figure 38-11**.

Voltage Regulators

Voltage regulators are responsible for adjusting, boosting, and stabilizing the level and quality of voltage throughout the transmission system. These regulators help ensure that consistent and uniform power is delivered to all customers along the corresponding branches of the transmission and distribution network. Voltage regulators often rely on combinations of capacitors and reactors, along with other components to effectively manage voltage levels.

Capacitors

Capacitors are used to manage the quality of the power being transmitted and delivered throughout the grid. The function of the capacitors in the system is to provide ***reactive power support***. Reactive power support refers to the process of supplying leading reactive power to the system, which offsets the effects of the lagging reactive power created by inductive loads. In a capacitive circuit, the current leads the voltage

Hawklice/iStock/Getty Images Plus via Getty Images

Figure 38-12. Capacitor banks and reactors are installed in substations to adjust and stabilize voltage levels and quality.

(ICE), and in an inductive circuit, the voltage leads the current (ELI). By providing reactive power support:

- Power factor of the system is increased.
- Line losses are reduced.
- System capacity is increased.
- Voltage stability is increased.

Reactors

Reactors, along with capacitors, are used to manage the quality of the power being transmitted and delivered throughout the grid, **Figure 38-12**. In their simplest form, reactors are inductors, or coils of wire. The function of the reactors in the system is to help control overcurrent and overvoltage conditions. To help prevent an overcurrent condition, reactors are installed in series with the power lines to increase the impedance of the circuit. This reduces the stress on the system's circuit breakers by reducing the magnitude of any fault currents that might result. To help prevent an overvoltage condition, shunt reactors are installed in parallel with the power lines to help absorb any excess reactive power. Excess reactive power might be present when there are low-load conditions in the system, causing an increase in voltage levels.

Communication and Control Systems

In order for a power grid to function as intended, a robust system must be in place that allows for high-quality, real-time communication among all of the network's individual components, devices, and subsystems. In older, conventional power grids, communication was limited, if present at all, and many of the responses to system problems were reactive, not proactive. In today's smart grids, advanced technologies such as microwave towers and fiber optic cables provide the infrastructure for effective, fast, and accurate data transfers throughout the network. Monitoring the present state of the grid and having the ability to predict future states of the grid are powerful tools that allow the grid, through its communication system, to take a proactive approach to help ensure the continued, uninterrupted supply of energy. Robust communication systems have the ability to quickly respond to changes in the demand for electricity to optimize the flow of energy within the grid.

Energy Storage Systems

In their simplest terms, ***energy storage systems*** are systems that store energy. Since energy consumption and energy production are not constant, there are many times when energy production exceeds demand. Without a means to store this excess energy, it would be lost.

By implementing some type of storage strategy, unused energy can be captured so it will be available when needed at some point in the future. For example, electricity produced by capturing energy from the sun can be stored in batteries if not needed at the time of production. This energy can then be used at a later time, such as when the sun is not shining and solar-generated electricity is not being produced. As a result, energy storage systems help ensure grid stability and reliability.

Shedding Light

New, long-duration, grid-scale energy storage technologies (including a new generation of batteries) are being developed that can meet the performance and cost requirements for more wide-scale deployment. Energy storage, when intelligently controlled, will be critical to bringing more renewable energy resources—such as solar and wind—onto the grid.

During peak demand, stored energy can be dispatched to the grid without having to ramp up the output of the power plant. There is often a significant time lapse between the time of increased demand and the time at which the plant can satisfy the increase in demand, so this process is often not very efficient.

38.3 Smart Power Grids

A smart power grid uses a wide range of technologies, sensors, monitors, and software to effectively and efficiently match the supply of electricity to the present demand of electricity based on real-time data, **Figure 38-13**. If there was only a

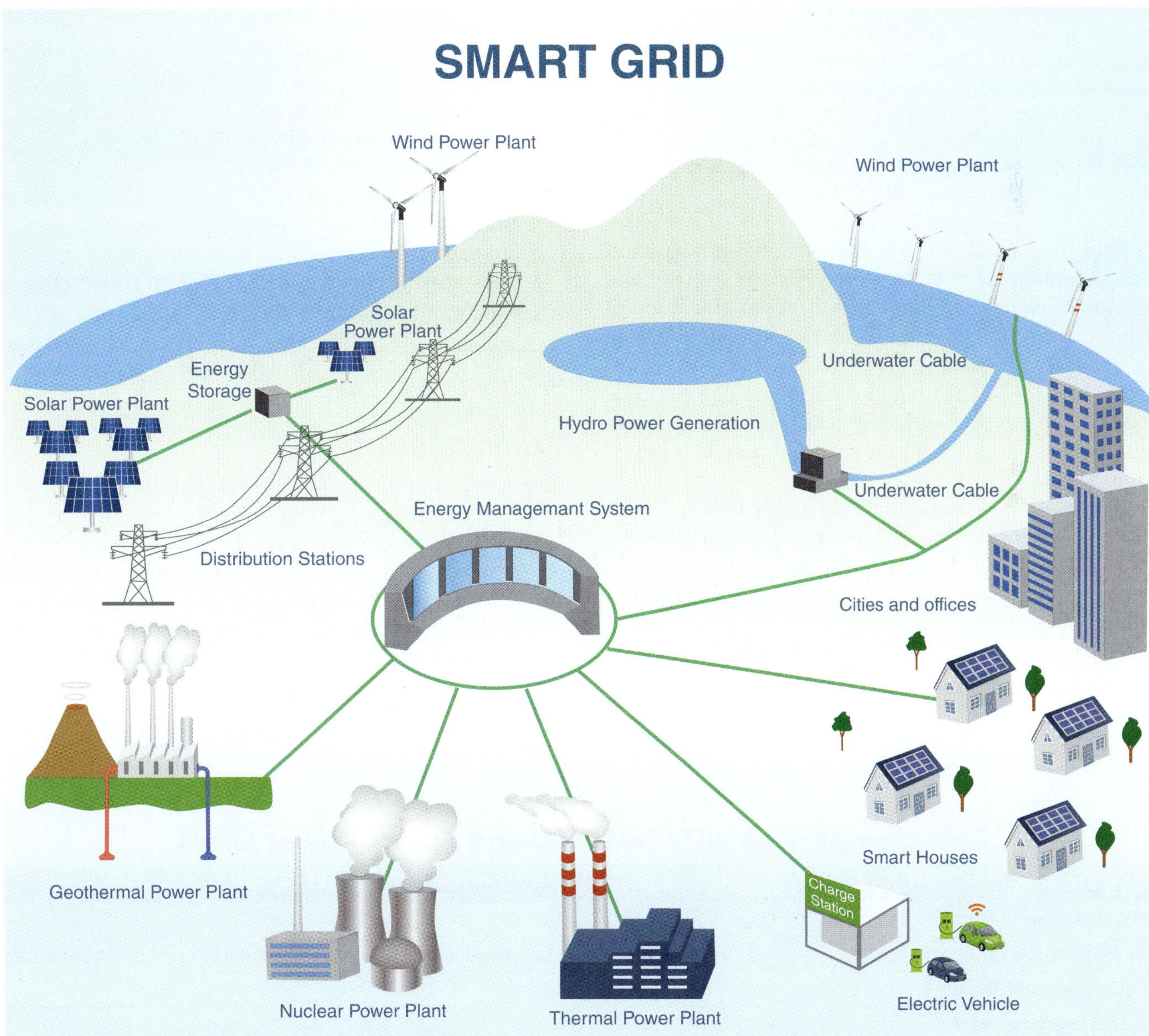

monicaodo/Shutterstock.com

Figure 38-13. In a smart power grid, the entire generation, transmission, and distribution network is automatically monitored, controlled, and adjusted by a central energy management system.

singular source for electricity, managing the generation and distribution of electricity would not pose a significant challenge. Consider, however, that there are multiple electricity generation sources, each with their own unique operational criteria, abilities, output potentials, and limitations that must be continually monitored, managed, and controlled to meet the varying demands of the end users of electricity. Simply put, a smart power grid coordinates and manages the needs, capabilities, and expectations of all energy producers, facility operators, end users, and industry stakeholders to minimize costs and environmental impacts while maintaining a reliable, resilient, flexible, and stable network.

Shedding Light

The smart power grid was first defined when the U.S. Congress passed the Energy Independence and Security Act of 2007. This piece of legislation detailed the country's desire to modernize its energy transmission and distribution system. Its goal was, in part, to strengthen the grid's infrastructure so that it would be able to adequately meet the nation's growing demand for electricity.

Within the past decade, the United States government has allocated billions of dollars to support the upgrade and expansion of its power grids. These funds are intended to enhance the flexibility of the grid and improve its ability to withstand extreme weather conditions. By doing so, the country's power sector will be strengthened, giving all Americans access to the energy they need. A general comparison of costs related to the traditional grid and smart power grid is shown in **Figure 38-14**.

Ideally, smart power grids have the ability to outperform traditional power grids due to a smart grid's capabilities:

- Allow energy consumers to voluntarily reduce electricity consumption during high-demand periods in exchange for compensation.
- Accurately predict energy demands by interpreting historical and forecasted weather data.
- Monitor consumer energy consumption patterns to predict future energy demands.
- Inform consumers how much energy costs, in real time, before they use it.
- Use advanced sensors and intelligent controls to identify, and often resolve, network irregularities before they become large-scale problems.
- Create an infrastructure that is reliable, secure, and efficient.
- Manage and control its operations with minimal human interaction and reliance.
- Configure various pieces of equipment and technology to work together seamlessly to meet electrical demands.
- Effectively balance the supply of and demand for electricity.
- Sense system overloads and reroute power as needed.
- Meet electrical demand increases without expanding production infrastructure.
- Easily incorporate and integrate additional technologies and energy sources.
- Become more resistant to natural disasters, vandalism, attacks, and extreme weather events.
- Enhance decarbonization and electrification efforts.
- Deliver sufficient quantities of high-quality energy reliably and efficiently.

A comparison of select capabilities of the traditional power grid and the smart power grid is shown in **Figure 38-15**.

Smart Meters

Smart meters are digital meters that are installed at the point of energy consumption, typically at the customer's location, **Figure 38-16**. Smart meters typically measure a customer's energy usage in real time. These meters, unlike conventional meters, can transmit data to and receive data from the utility.

An advanced metering infrastructure also uses a robust communication network to quickly transmit data among smart meters and the utility's data management system.

Cost Comparison of Traditional and Smart Power Grids

Cost Category	Traditional Grid	Smart Grid
Initial Installation Costs	Low	High
Cost of Installed Technologies	Less sophisticated and less expensive	More sophisticated and more expensive
Maintenance and Repair Costs	High	Low
System Upgrade Costs	Very high	Very low
System Operating Costs	Low	High

Goodheart-Willcox Publisher

Figure 38-14. Comparison of various costs associated with the traditional grid and the smart grid.

Comparison of Select Capabilities of Traditional and Smart Power Grids

Capability Category	Traditional Grid	Smart Grid
Energy Flow	One direction	Two directions
Communications	Poor/slow	Fast/real-time
Access to Usage Data	Slow	Real-time
Fault Detection	Very slow	Very fast
Grid Management	Reacts to problems as they arise	Predicts irregularities before they become problems
Power Restoration Time	Slow	Fast
Problem Resolution	Manual	Automated
Power Rerouting	Slow	Fast
Renewable Energy Integration	Limited ability	Inherent in system design
Overall Grid Reliability	Low	High
Cyber Attack Vulnerability	Low	High

Goodheart-Willcox Publisher

Figure 38-15. Comparison of select capabilities of the traditional grid and the smart grid.

onurdongel/iStock/Getty Images Plus via Getty Images

Figure 38-16. Smart meters allow for two-way data transmission between the meter and the utility company. In many cases, customers can monitor electrical usage on a smart device.

When smart meters are used, consumers have access to real-time usage information, allowing them to make informed decisions about their energy consumption habits and patterns. With real-time data, consumers can see the potential savings associated with shifting their energy usage from peak to off-peak periods.

In addition to the consumer-facing benefits of smart meters, these devices allow the power company to detect problems much faster, facilitating quicker reaction times and restoration of service.

Smart Grid Challenges

In order to function as desired and intended, smart grids require a combination of advanced technologies, careful planning, and ongoing innovation efforts. By developing robust communication networks, enhancing cybersecurity, improving data analytics, and investing in advanced control systems, smart grids can help ensure that future energy supplies and systems are reliable, resilient, and sustainable.

Smart grids rely heavily on digital communications and data exchange, which makes them vulnerable to cyberattacks. With millions of connected devices, each one becomes a potential entry point for hackers. ***Advanced metering infrastructure (AMI)*** is an overarching term used to describe the interconnected network of smart meters, communication systems, and data management protocols that allow for two-way communications between the power utility and its customers. This advanced metering infrastructure helps make efficient and reliable electricity distribution possible. However, attacks on this network could disrupt operations, damage equipment, or even cause large-scale blackouts. Ensuring data privacy and secure communication protocols, encrypting data, and regularly updating security measures are essential, but they are also complex and costly.

System Device Compatibility

As more devices and renewable energy sources connect to the power grid, scaling up a smart grid to accommodate increasing demand without affecting performance becomes a challenge. The grid needs to be flexible enough to adapt to new technologies. Designing and implementing a highly scalable and adaptable system requires forward-thinking architecture and substantial planning.

Since smart grids integrate a wide variety of devices from different manufacturers, ensuring that these devices can communicate effectively with each other (interoperability) is a major challenge. To help ensure effective communication throughout the entire grid, standardized protocols and interfaces are needed. Such systems have high implementation costs associated with them.

Data Management

Smart grids must process, sort, evaluate, and respond to the massive amounts of data that are generated by the system's many sensors, smart meters, and other devices in real time. Handling, storing, and retrieving data at the required near-instantaneous speed demands powerful computers and high-level algorithms. As with other elements of the infrastructure, such technologies come at a high cost.

Communication delays and failures can lead to delayed responses that can negatively affect the grid's ability to balance supply and demand, respond to faults, and prevent outages.

Grid Stability and Power Quality

Integrating renewable energy sources into the smart grid introduces variability that can have a negative effect on grid stability and power quality. The power quality characteristics that are most commonly affected are signal frequency and voltage levels. Managing these fluctuations requires advanced control systems and programs to maintain stable and high-quality power. Achieving the right balance between renewable and conventional power sources in real time is technically challenging and often requires substantial infrastructure upgrades and financial investments.

Accurately predicting the demand for energy is crucial for balancing supply and demand, but it is a complex process. Even small errors in forecasting the supply of and demand for energy can lead to significant energy overproduction or, more importantly, energy underproduction, which can greatly affect the stability of the grid and the energy prices paid by the end users.

Many existing power grid components were not initially designed to support smart grid technology or handle the complexity of integrating renewable energy sources. Upgrading or retrofitting aging infrastructure to work with new technology is challenging, as it often requires significant investments, physical space, and careful planning to avoid disruptions.

Shedding Light

Superconducting power cables reduce line losses and carry significantly more power than traditional conductors. Powered by the smart grid, advanced sensors can monitor and report line conditions in real time, enabling more power to flow over existing lines and making the grid more efficient.

Energy storage systems degrade over time, losing both capacity and efficiency. Managing this degradation and replacing these systems as needed is a technical and financial challenge. Smart grids require predictive maintenance and replacement strategies to ensure storage systems remain reliable over their lifespan.

Distributed Computing

Distributed computing refers to the process of spreading computational responsibilities over a network of connected systems. These interconnected systems work together to achieve a given task. Distributed computing benefits from the ability to harness and use the power of multiple systems to increase grid performance while expanding the scalability and functionality of the grid.

Distributed computing provides many benefits that include:

- Allowing for efficient resource use.
- Helping to increase the speed at which tasks are completed.
- Enhancing system reliability and availability.
- Providing backup and alternative resources in the event that one portion of the network fails.
- Creating a platform to easily expand systems to handle increasing demands without requiring major network alterations.

38.4 Distributed Energy Resources

Distributed energy resources (DERs) refer to small energy production plants or energy storage systems that are positioned close to the point where energy is ultimately consumed. These system elements are often operated independently, but can, in some instances, be part of the main, primary network. Distributed energy resources act to boost energy generation and storage capabilities while helping to improve reliability of the network.

When generating electricity close to the point of end use, the term ***distributed generation*** is used. Common production strategies used for distributed generation include small-scale solar arrays, wind energy systems, facility-located natural gas generators, and combined heat and power (CHP) systems. CHP systems have the ability to simultaneously produce usable electricity and heat.

Depending on the particular application, distributed energy resources can have significant benefits. By generating power close to the point where energy is used, the transmission and distribution losses can be greatly reduced. In addition, since energy production is controlled and distributed close to the point of use, dependence on the main, centralized network is reduced. If problems arise with the main, primary network, those energy consumers who receive power from distributed energy resources are less likely to be affected as much as those who rely solely on the main network.

When distributed energy resources have the ability to function alone, or as part of the main grid, the term microgrid is often used. ***Microgrids*** are smaller, independently operated grids that are installed as part of the main grid. These grids can be put into or removed from service as needed to help offset local power demands, thus reducing stress on the main grid. At times, microgrids can feed excess electricity into the main grid for other users. At other times, microgrids can disconnect from the main grid and provide downstream consumers with the electricity they need without having to rely on the main grid.

38.5 Conductor Sizing

Properly sizing the conductors used in electrical circuits and systems is an important aspect of ensuring our systems are safe, efficient, and reliable. The most important reason to correctly size electrical conductors is safety. If conductors are too small for the applied current, they can and likely will overheat, leading to melted insulation, overloads, ground faults, short circuits, and possibly fire. In addition, undersized conductors can reduce the effectiveness of circuit protection devices. If conductors are not properly sized and selected, these circuit protection devices may not de-energize the circuit as required.

Shedding Light

Conductor sizing and installation requirements for electrical transmission systems are performed in accordance with the *National Electrical Safety Code* (*NESC*). This is a highly specialized topic beyond the scope of this introductory text. The discussion in this section focuses on conductor sizing requirements in the *NEC*, which are applicable downstream from the utility termination point.

Once electrical power has been delivered from the distribution transformer by the utility company, electrical installation practices are performed in accordance with the requirements presented in the *NEC*. The *NEC* specifies the minimum conductor sizes for various applications based on a circuit's current-carrying capacity, the *NEC*-recommended voltage drop, the insulation type, the number of current-carrying conductors in a raceway, and other external conditions, such as ambient temperature that may act on the system.

Another important factor regarding proper wiring sizing is the potential for energy losses. Undersized conductors have higher resistance than properly sized conductors and, as such, have a greater voltage drop. The larger the voltage drop across a conductor, the lower the voltage and the higher the current will be at the supplied load. Improper voltage and current at the load can reduce the life expectancy of the equipment. Voltage drop with the load operating needs to be within the manufacturer's specifications, typically plus or minus 10%.

Conductors

Conductors are the wires that carry the electricity in a circuit. Most conductors are either copper or aluminum. Copper conductors are used throughout residential, commercial, and industrial electrical systems. Aluminum conductors must be larger than a copper conductor to carry similar current, but a larger aluminum conductor is lighter and less costly than the comparable copper conductor. As a result, aluminum conductors are used in applications where weight is critical, such as overhead power lines and aircraft and aerospace vehicles. Compact aluminum conductors are popular for the distribution of power in medium and large buildings.

Conductors are either bare or insulated. A bare conductor is a conductor with no covering or insulation. Bare conductors are generally allowed for grounding and bonding applications. An insulated conductor is a conductor encased in a coating. The conductor insulation provides thermal protection and serves as an insulating barrier between an energized conductor and other conductive materials with which the conductor may come into contact.

Conductor insulation is selected based on the environmental conditions to which the conductor will be exposed. Most insulation types are either thermoset, thermoplastic, or rubber materials. The different types of insulations and other coverings are covered in *Article 310* of the *NEC*.

Know the Code

The type of insulation, conductor material, and size of the conductor in AWG or circular mils determines the allowable ampacity of the conductors. *Article 310* of the *NEC* covers many variations of available wire types for residential, commercial, and industrial installations.

Wiring Methods

Wiring methods are the various materials and assemblies that are used to connect equipment and devices in an electrical system. Detailed installation requirements for all approved wiring methods are found in *Chapter 3* of the *NEC*.

Two general categories of wiring methods are cables and raceways. A cable is a premade assembly of conductors contained in a protective coating of plastic, rubber, steel, or aluminum. Some of the common cable types include nonmetallic-sheathed cable (types NM and NMC), armored cable (type AC), metal-clad cable (type MC), service-entrance cable (type SE), and underground feeder and branch-circuit cable (type UF). A raceway is an enclosed channel designed to contain conductors or cable. Common raceway types include electrical metallic tubing (EMT), rigid

Selecting *NEC* Table for Conductor Sizing

Number of Conductors	Wiring Method	Ambient Temperature	Conductor Temperature Rating	*NEC* Table to Use
No more than 3	Raceway, cable, or direct burial	86°F (30°C)	140°F (60°C) 167°F (75°C) 194°F (90°C)	*Table 310.15(B)(16)*
Single	Free air	86°F (30°C)	140°F (60°C) 167°F (75°C) 194°F (90°C)	*Table 310.15(B)(17)*
No more than 3	Raceway or cable	104°F (40°C)	302°F (150°C) 392°F (200°C) 482°F (250°C)	*Table 310.15(B)(18)*
Single	Free air	104°F (40°C)	302°F (150°C) 392°F (200°C) 482°F (250°C)	*Table 310.15(B)(19)*

Goodheart-Willcox Publisher

Figure 38-17. The *NEC* includes several conductor-sizing tables. Selecting the correct table based on the specific installation conditions is critical.

metal conduit (RMC), intermediate metal conduit (IMC), flexible metal conduit (FMC), rigid polyvinyl chloride conduit (PVC): schedule 40 and schedule 80, liquidtight flexible nonmetallic conduit (LFNC), and liquidtight flexible metallic conduit (LFMC). For other cable and raceway types, consult *Chapter 3* of the *NEC*.

Conductor Size

The American Wire Gauge (AWG) system is a standardized gauge system used to specify the required diameter of electrical conductors. The AWG is a numerical scale where the smaller the gauge number, the larger the wire's diameter. Conductors with larger diameters are used for higher current-carrying applications. For example, an 8 AWG conductor has a larger diameter and can carry more current than a 14 AWG conductor.

The largest conductor size in the AWG system is 4/0 AWG. Conductors larger than 4/0 AWG are sized based on the circular mil system. A circular mil (cmil) is a measurement of area equal to the area a circle with a 0.001″ diameter. Conductor size is identified in units of kcmil, representing one thousand circular mils. Depending on the insulation type, sizes range from 250 kcmil to 2000 kcmil.

Required Ampacity

When determining the required size of a conductor, the first step is determining the required ampacity, or current that the conductor is designed to carry. Requirements for determining the required ampacity are specified in *Article 310* of the *NEC*, and these requirements vary based on the purpose of the conductor. For example, the determination of the required ampacity for service-entrance conductors supplying a residential service panel differs from that for motor branch-circuit conductors. A thorough understanding of both the general requirements and the application-specific requirements outlined in the *NEC* is needed to determine required ampacity correctly.

Once the required ampacity is determined, the appropriate conductor size can then be determined by selecting the conductor size that provides an ampacity equal to or greater than the required ampacity for the load. Several factors must be considered when sizing conductors:

- Adjustment factor for the number of conductors in close proximity to one another in a cable or raceway
- Adjustment factor for the ambient temperature to which the conductors are exposed
- Temperature rating of the conductors with a specific type of insulation
- Type of conductor (aluminum, copper, or copper-clad aluminum)
- The distance of the run and the ampacity of the load
- The appropriate size of the OCPD per *Article 240* of the *NEC*

Based on these factors, a table in *Section 310.15* of the *NEC* is selected to find the ampacity rating for various conductor sizes. See **Figure 38-17**.

Adjustment Factors

The ampacity values found in *NEC Article 310* tables require additional adjustment if more than three current-carrying conductors are contained in the raceway or cable or if the ambient temperature differs from that specified in the tables. Refer to *Tables 310.15(B)(1)(1)*, *310.15(B)(1)(2)*, and *310.15(C)(1)* of the *NEC*.

CHAPTER 38 REVIEW

Summary

- The power grid is a complex system of interconnected transmission lines, substations, distribution lines, transformers, and other equipment that delivers electricity from the point of generation to the end user. (38.1)
- Our modern society is dependent on the readily available electricity delivered by the power grid. (38.1)
- Demand for electricity has been steadily increasing and is expected to continue to increase for the foreseeable future. (38.1)
- Power plant generators are the primary source of electricity entering the power grid. (38.2)
- In the power grid, step-up transformers increase voltage for more efficient transmission and step-down transformers reduce voltage for delivery to end users. (38.2)
- Power lines can be located overhead (aboveground) or underground. (38.2)
- Insulators are installed between power lines and support structures to prevent electrical leakage to ground. (38.2)
- Common types of circuit breakers used in substations are air, vacuum, oil, and SF6 (sulfur hexafluoride) circuit breakers. (38.2)
- Voltage regulators adjust, boost, and stabilize voltage quality in the transmission system and often incorporate capacitor banks and reactors. (38.2)
- An energy storage system stores electrical energy and then releases electricity into the power grid on demand. (38.2)
- A smart power grid uses data to automatically manage the supply of electricity to meet the instantaneous demand. (38.3)
- Advanced metering infrastructure (AMI) is an interconnected network of smart meters, communication systems, and data management protocols that allow for two-way communication between a utility company and its customers. (38.3)
- A distributed energy resource (DER) is a relatively small power plant or energy storage system located near the consumers of the produced electricity. (38.4)
- A microgrid is a small, independently operated grid connected to the main power grid. (38.4)
- Undersized conductors can overheat and reduce the effectiveness of overcurrent protective devices. (38.5)
- The *NEC* provides requirements for conductor and electrical installations downstream from the utility termination point. (38.5)
- Most conductors are made from copper or aluminum, and conductors can be bare or insulated. (38.5)
- Cables and raceways are the two categories of wiring methods. (38.5)
- The American Wire Gauge (AWG) standardized gauge system is used to express conductor diameter for smaller conductors, and the circular mils system is used for larger conductors. (38.5)
- Conductor size is determined by identifying the required ampacity using tables in the *NEC*. (38.5)
- Conductor ampacity may need to be adjusted if more than three current-carrying conductors are in close proximity or if the ambient temperature around the conductors differs from the ambient temperature for the ampacity table being used. (38.5)

Review Questions

Use the information in the chapter to answer the following questions.

Measuring Your Potential

1. Which of the following is *not* a component through which electricity passes in its transmission from generation to end user in the power grid? (38.1)
   A. Motors
   B. Substations
   C. Transformers
   D. Transmission lines
2. *True or False?* Due to increases in the efficiency of electrical equipment, demand for electricity in the United States has been declining over the past ten years. (38.1)
3. A _____ transformer provides the final reduction of voltage prior to delivery to the customer. (38.2)
   A. current
   B. distribution
   C. step-up
   D. transmission
4. A transmission voltage of 345 kV is categorized as _____ voltage. (38.2)
   A. distribution
   B. high
   C. extra-high
   D. ultra-high
5. _____ transformers are commonly used in systems with underground distribution lines. (38.2)
   A. Current
   B. Pad-mounted
   C. Pole-mounted
   D. Underground

6. *True or False?* Underground power lines are generally more expensive to install than overhead power lines. (38.2)
7. Which of the following is *not* a typical material used for power line insulators? (38.2)
   A. Aluminum
   B. Ceramic
   C. Glass
   D. Polymer
8. At what point of the electrical distribution system can power be rerouted if an outage occurs? (38.2)
   A. Automatic circuit recloser
   B. Energy storage system
   C. Pole-mounted transformer
   D. Substation
9. Step-up transformers are most likely to be found at a _____ substation. (38.2)
   A. distribution
   B. switching
   C. generating
   D. transmission
10. Substation circuit breakers are categorized by _____. (38.2)
    A. circuit breaker reset mechanism
    B. medium surrounding the contacts that extinguish the arc
    C. number of phases connected to the circuit breaker
    D. type of terminals used for connecting power lines
11. Which substation component is a type of inductor? (38.2)
    A. Capacitor
    B. Fuse
    C. Insulator
    D. Reactor
12. *True or False?* The primary purpose of an energy storage system is to supply electricity to the power grid while power plant generators are disconnected for maintenance. (38.2)
13. Which of the following is *not* an expected capability of a smart power grid? (38.3)
    A. Reduce the electrical demand of homes and businesses.
    B. Incorporate input from multiple energy sources.
    C. Monitor the transmission system and automatically reroute electricity to avoid overload conditions.
    D. Predict energy demand based on historical data and weather predictions.
14. *True or False?* Smart meters transmit data to and receive data from the utility company. (38.3)
15. Which of the following is a potential disadvantage of a smart electric meter? (38.3)
    A. Power generated by a photovoltaic system is accurately reported to the utility.
    B. Smart meters can be read remotely.
    C. Smart meters can be susceptible to cyberattack.
    D. Smart meters can guide a user to more wisely use electricity during off-peak hours.
16. _____ refers to the process of spreading computational responsibilities over a network of connected systems. (38.3)
    A. Advanced metering infrastructure
    B. Distributed computing
    C. Dual-core processing
    D. Smart grid processing
17. Small energy production plants or energy storage systems positioned near the supplied users are known as _____. (38.4)
    A. distributed energy resources
    B. distributed substations
    C. advanced metering generators
    D. home-source power plants
18. Which of the following is *not* a factor when determining a conductor's current-carrying capacity? (38.5)
    A. Ambient temperature
    B. Conductor insulation type
    C. Number of bends in the conductor between source and load
    D. Number of current-carrying conductors in a raceway
19. _____ conductors are commonly used in applications where weight is critical, such as overhead power lines and aerospace vehicles. (38.5)
    A. Aluminum
    B. Brass
    C. Copper
    D. Steel
20. Cables and raceways are two categories of _____. (38.5)
    A. conductors
    B. conduit
    C. wire fasteners
    D. wiring methods
21. *True or False?* A 12 AWG conductor is smaller than a 14 AWG conductor. (38.5)

Plug-In

1. Compare and contrast the advantages and disadvantages of overhead and underground transmission lines. (38.2)
2. Compare the traditional power grid to a smart power grid, identifying at least three similarities and three differences. (38.2)

Making the Connection

1. Describe how you would be affected by a 24-hour power outage. What additional effects would you anticipate if the power outage lasted for seven days? (38.1)
2. Discuss some challenges of smart grid implementation. What are some solutions to the challenges you identified? (38.3)

Appendix A: Commonly Used Formulas

Ohm's Law Formulas

Voltage:

$E = I \times R$

$E = \frac{P}{I}$

$E = \sqrt{P \times R}$

Current:

$I = \frac{E}{R}$

$I = \frac{P}{E}$

$I = \sqrt{\frac{P}{R}}$

Resistance:

$R = \frac{E}{I}$

$R = \frac{P}{I^2}$

$R = \frac{E^2}{P}$

Watt's Law Power Formulas

$P = I \times E$

$P = I^2 \times R$

$P = \frac{E^2}{R}$

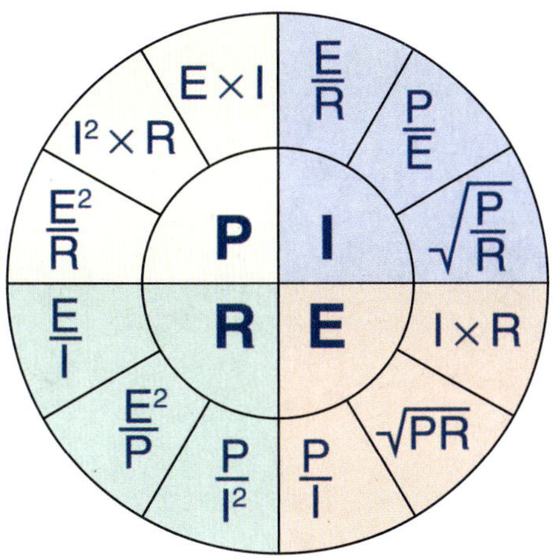

Goodheart-Willcox Publisher

Series Circuit Formulas

$R_T = R_1 + R_2 + R_3 + \ldots + R_N$

$I_T = I_1 = I_2 = I_3 = \ldots = I_N$

$E_T = E_1 + E_2 + E_3 + \ldots + E_N$

$P_T = P_1 + P_2 + P_3 + \ldots + P_N$

Parallel Circuit Formulas

$I_T = I_1 + I_2 + I_3 + \ldots + I_N$

$E_T = E_1 = E_2 = E_3 = \ldots = E_N$

$P_T = P_1 + P_2 + P_3 + \ldots + P_N$

General Resistance:

$$R_T = \frac{1}{\frac{1}{R_1} + \frac{1}{R_2} + \frac{1}{R_3} + \ldots + \frac{1}{R_N}}$$

$$\frac{1}{\mathrm{R}_T} = \frac{1}{\mathrm{R}_1} + \frac{1}{\mathrm{R}_2} + \frac{1}{\mathrm{R}_3} + \ldots + \frac{1}{\mathrm{R}_N}$$

For Two Resistors:

$$R_T = \frac{R_1 \times R_2}{R_1 + R_2}$$

For Equal-Valued Resistors:

$$R_T = \frac{R}{N}$$

Purely Inductive (L) Circuit Formulas

$$Q_L = I_L \times E_L$$

$$Q_L = I_L^2 \times X_L$$

$$Q_L = \frac{E_L^2}{X_L}$$

Series-Connected Inductance Formula

$$L_T = L_1 + L_2 + L_3 + \ldots + L_N$$

Parallel-Connected Inductance Formulas

Reciprocal Formula (works in all cases):

$$\frac{1}{L_T} = \frac{1}{L_1} + \frac{1}{L_2} + \frac{1}{L_3} + \ldots + \frac{1}{L_N}$$

Product Over Sum Formula (works with two branches):

$$L_T = \frac{L_1 \times L_2}{L_1 + L_2}$$

Equal Inductance Formula (works when all branches have the same resistance):

$$L_T = \frac{L}{N}$$

Purely Capacitive (C) Circuit Formulas

$$Q_C = I_C \times E_C$$

$$Q_C = I_C^2 \times X_C$$

$$Q_C = \frac{E_C^2}{X_C}$$

$$E_C = \frac{Q_C}{I_C}$$

$$I_C = \frac{Q_C}{E_C} = \sqrt{\frac{Q_C}{X_C}}$$

$$X_C = \frac{Q_C}{I_C^2}$$

$$X_C = \frac{E_C^2}{Q_C}$$

$$E_C = \sqrt{Q_C \times X_C}$$

$$C = \frac{1}{2\pi f X_C}$$

$$X_C = \frac{1}{2\pi f C}$$

Series-Connected Capacitance Formulas

$$E_T = E_{C1} = E_{C2} = E_{C3} = \ldots = E_{CN}$$

$$I_T = I_{C1} + I_{C2} + I_{C3} + \ldots + I_{CN}$$

General Total Capacitance Formula (works in all cases):

$$\frac{1}{C_T} = \frac{1}{C_1} + \frac{1}{C_2} + \frac{1}{C_3} + \ldots + \frac{1}{C_N}$$

Total Capacitance Formula for Two Series-Connected Capacitors:

$$C_T = \frac{C_1 \times C_2}{C_1 + C_2}$$

Total Capacitance Formula (when all capacitors have the same capacitance):

$$C_T = \frac{C}{N}$$

Parallel-Connected Capacitance Formulas

$$C_T = C_1 + C_2 + C_3 + \ldots + C_N$$

$$E_T = E_{C1} = E_{C2} = E_{C3} = \ldots = E_{CN}$$

$$I_T = I_{C1} + I_{C2} + I_{C3} + \ldots + I_{CN}$$

Resistive-Capacitive (R-C) Series Circuit Formulas

$$Z = \frac{E_T^2}{VA}$$

$$Z = \sqrt{R^2 + X_C^2}$$

$$Z = \frac{E_T}{I_T}$$

$$VA = E_T \times I_T$$

$$VA = I_T^2 \times Z$$

$$VA = \frac{E_T^2}{Z}$$

$$VA = \frac{P}{PF}$$

$$I_R = I_T = I_C$$

$$I_R = \frac{E_R}{R}$$

$$Z = \frac{VA}{I_T^2}$$

$$Z = \frac{R}{PF}$$

$$PF = \frac{R}{Z}$$

$$PF = \frac{P}{VA}$$

$$PF = \frac{E_R}{E_T}$$

$$PF = \cos\theta$$

$$R = \sqrt{Z^2 - X_C^2}$$

$$R = \frac{E_R}{I_R}$$

$$R = \frac{E_R^2}{P}$$

$$E_T = \sqrt{E_R^2 + E_C^2}$$

$$E_T = I_T \times Z$$

$$E_T = \frac{VA}{I_T}$$

$$E_T = \frac{E_R}{PF}$$

$$P = E_R \times I_R$$

$$P = \frac{E_R^2}{R}$$

$$P = I_R^2 \times R$$

$$P = VA \times PF$$

$$E_C = I_C \times X_C$$

$$E_C = \sqrt{E_T^2 - E_R^2}$$

$$I_T = I_R = I_C$$

$$I_T = \frac{E_T}{Z}$$

$$I_T = \frac{VA}{E_T}$$

$$C = \frac{1}{2\pi f X_C}$$

$$E_R = I_R \times R$$

$$E_R = \sqrt{P \times R}$$

$$E_R = \frac{P}{I_R}$$

$$E_R = \sqrt{E_T^2 - E_C^2}$$

$$E_R = E_T \times PF$$

$$I_C = I_R = I_T$$

$$I_C = \frac{E_C}{X_C}$$

$$I_R = \frac{P}{E_R}$$

$$I_R = \sqrt{\frac{P}{R}}$$

$$X_C = \sqrt{Z^2 - R^2}$$

$$X_C = \frac{E_C}{I_C}$$

$$R = \frac{P}{I_R^2}$$

$$R = Z \times PF$$

$$X_C = \frac{1}{2\pi f C}$$

Resistive-Capacitive (R-C) Parallel Circuit Formulas

$$Z = \frac{R \times X_C}{\sqrt{R^2 + X_C^2}}$$

$$Z = \frac{1}{\sqrt{\left(\frac{1}{R}\right)^2 + \left(\frac{1}{X_C}\right)^2}}$$

$$I_T = \sqrt{I_R^2 + I_C^2}$$

$$I_T = \frac{E_T}{Z}$$

$$I_T = \frac{VA}{E_T}$$

$$I_T = \frac{I_R}{PF}$$

$$I_T = \sqrt{\frac{VA}{Z}}$$

$$Z = \frac{VA}{I_T^2}$$

$$VA = E_T \times I_T$$

$$VA = I_T^2 \times Z$$

$$VA = \frac{E_T^2}{Z}z$$

$$VA = \frac{P}{PF}$$

$$E_T = E_R = E_C$$

$$E_T = I_T \times Z$$

$$E_T = \frac{VA}{I_T}$$

$$C = \frac{1}{2\pi f X_C}$$

$$Z = \frac{E_T}{I_T}$$

$$Z = \frac{E_T^2}{VA}$$

$$PF = \frac{Z}{R}$$

$$PF = \frac{1}{\left(\frac{R}{Z}\right)}$$

$$PF = \frac{P}{VA}$$

$$PF = \frac{I_R}{I_T}$$

$$PF = \cos\theta$$

$$E_T = \sqrt{VA \times Z}$$

$$Z = R \times PF$$

$$P = E_R \times I_R$$

$$P = \frac{E_R^2}{R}$$

$$P = I_R^2 \times R$$

$$P = VA \times PF$$

$$E_R = I_R \times R$$

$$E_R = \sqrt{P \times R}$$

$$E_R = \frac{P}{I_R}$$

$$E_R = E_T = E_C$$

$$I_C = \sqrt{I_T^2 - I_R^2}$$

$$I_C = \frac{E_C}{X_C}$$

$$I_R = \sqrt{I_T^2 - I_C^2}$$

$$I_R = \frac{E_R}{R}$$

$$I_R = \frac{P}{E_R}$$

$$I_R = \sqrt{\frac{P}{R}}$$

$$X_C = \frac{1}{\sqrt{\left(\frac{1}{Z}\right)^2 - \left(\frac{1}{R}\right)^2}}$$

$$X_C = \frac{E_C}{I_C}$$

$$R = \frac{1}{\sqrt{\left(\frac{1}{Z}\right)^2 - \left(\frac{1}{X_C}\right)^2}}$$

$$R = \frac{E_R}{I_R}$$

$$R = \frac{E_R^2}{P}$$

$$R = \frac{P}{I_R^2}$$

$$R = \frac{Z}{PF}$$

$$X_C = \frac{1}{2\pi fC}$$

$$E_C = I_C \times X_C$$

$$E_C = E_T = E_R$$

Resistive-Inductive (R-L) Series Circuit Formulas

$$E_T = \sqrt{E_R^2 + E_L^2}$$

$$E_T = I_T \times Z$$

$$E_T = \frac{VA}{I_T}$$

$$E_T = \frac{E_R}{PF}$$

$$PF = \frac{R}{Z}$$

$$PF = \frac{P}{VA}$$

$$PF = \frac{E_R}{E_T}$$

$$PF = \cos\theta$$

$$R = \sqrt{Z^2 - X_L^2}$$

$$R = \frac{E_R}{I_R}$$

$$R = \frac{E_R^2}{P}$$

$$R = \frac{P}{I_R^2}$$

$$R = Z \times PF$$

$$Z = \sqrt{R^2 + X_L^2}$$

$$Z = \frac{E_T}{I_T}$$

$$Z = \frac{VA}{I_T^2}$$

$$Z = \frac{R}{PF}$$

$$Z = \frac{E_T^2}{VA}$$

$$P = E_R \times I_R$$

$$P = \frac{E_R^2}{R}$$

$$P = I_R^2 \times R$$

$$P = VA \times PF$$

$$E_L = I_L \times X_L$$

$$E_L = \sqrt{E_T^2 - E_R^2}$$

$$VA = E_T \times I_T$$

$$VA = I_T^2 \times Z$$

$$VA = \frac{E_T^2}{Z}$$

$$VA = \frac{P}{PF}$$

$$E_R = I_R \times R$$

$$E_R = \sqrt{P \times R}$$

$$E_R = \frac{P}{I_R}$$

$$E_R = \sqrt{E_T^2 - E_L^2}$$

$$E_R = E_T \times PF$$

$$I_L = I_R = I_T$$

$$I_L = \frac{E_L}{X_L}$$

$$L = \frac{X_L}{2\pi f}$$

$$I_T = I_R = I_L$$

$$I_T = \frac{E_T}{Z}$$

$$I_T = \frac{VA}{E_T}$$

$$I_T = \sqrt{\frac{VA}{Z}}$$

$$I_R = I_T = I_L$$

$$I_R = \frac{E_R}{R}$$

$$I_R = \frac{P}{E_R}$$

$$I_R = \sqrt{\frac{P}{R}}$$

$$X_L = \sqrt{Z^2 - R^2}$$

$$X_L = \frac{E_L}{I_L}$$

$$X_L = 2\pi f L$$

Resistive-Inductive (R-L) Parallel Circuit Formulas

$$Z = \frac{R \times X_L}{\sqrt{R^2 \times X_L^2}}$$

$$Z = \frac{1}{\sqrt{\left(\frac{1}{R}\right)^2 + \left(\frac{1}{X_L}\right)^2}}$$

$$Z = \frac{E_T}{I_T}$$

$$Z = \frac{E_T^2}{VA}$$

$$VA = E_T \times I_T$$

$$VA = I_T^2 \times Z$$

$$VA = \frac{E_T^2}{Z}$$

$$VA = \frac{P}{PF}$$

$$X_L = \frac{E_L}{E_L}$$

$$Z = \frac{VA}{I_T^2}$$

$$Z = R \times PF$$

$$PF = \frac{Z}{R}$$

$$PF = \frac{1}{\left(\frac{R}{Z}\right)}$$

$$PF = \frac{P}{VA}$$

$$PF = \frac{I_R}{I_T}$$

$$PF = \cos \theta$$

$$X_L = \frac{1}{\sqrt{\left(\frac{1}{Z}\right)^2 + \left(\frac{1}{R}\right)^2}}$$

$$X_L = 2\pi f L$$

$$R = \frac{E_R}{I_R}$$

$$E_T = E_R = E_L$$

$$E_T = I_T \times Z$$

$$L = \frac{X_L}{2\pi f}$$

$$I_T = \sqrt{I_R^2 + I_L^2}$$

$$I_T = \frac{E_T}{Z}$$

$$E_L = I_L \times X_L$$

$$E_L = E_T = E_R$$

$$E_L = I_R \times R$$

$$E_R = \sqrt{P \times R}$$

$$E_R = \frac{R}{I_R}$$

$$E_R = E_T = E_L$$

$$R = \frac{P}{I_R^2}$$

$$E_T = \frac{VA}{I_T}$$

$$E_T = \sqrt{VA \times Z}$$

$$I_T = \frac{VA}{E_T}$$

$$I_T = \sqrt{\frac{VA}{Z}}$$

$$I_T = \frac{I_R}{PF}$$

$$I_L = \sqrt{I_T^2 - I_R^2}$$

$$I_L = \frac{E_L}{X_L}$$

$$I_R = \sqrt{I_T^2 - I_L^2}$$

$$I_R = \frac{E_R}{R}$$

$$I_R = \frac{P}{E_R}$$

$$I_R = \sqrt{\frac{P}{R}}$$

$$I_R = I_T \times PF$$

$$R = \frac{1}{\sqrt{\left(\frac{1}{Z}\right)^2 - \left(\frac{1}{X_L}\right)^2}}$$

$$P = E_R \times I_R$$

$$R = \frac{E_R^2}{P}$$

$$P = \frac{E_R^2}{R}$$

$$R = \frac{Z}{PF}$$

$$P = I_R^2 \times R$$

$$P = VA \times PF$$

Inductive-Capacitive (L-C) Series Circuit Formula

$Z = X_L - X_C$

Inductive-Capacitive (L-C) Parallel Circuit Formula

$Z = \dfrac{X_L \times X_C}{X_L - X_C}$

Resistive-Inductive-Capacitive (R-L-C) Series Circuit Formulas

$E_T = \sqrt{E_R^{\,2} + (E_L - E_C)^2}$

$E_T = I_T \times Z$

$E_T = \dfrac{\text{VA}}{I_T}$

$E_T = \dfrac{E_R}{\text{PF}}$

$Z = \sqrt{R^2 + (X_L - X_C)^2}$

$Z = \dfrac{E_T}{I_T}$

$Z = \dfrac{\text{VA}}{I_T^{\,2}}$

$Z = \dfrac{R}{\text{PF}}$

$\text{PF} = \dfrac{R}{Z}$

$\text{PF} = \dfrac{P}{\text{VA}}$

$\text{PF} = \dfrac{E_R}{E_T}$

$\text{PF} = \cos\theta$

$I_R = I_T = I_C = I_L$

$I_R = \dfrac{E_R}{R}$

$I_R = \dfrac{P}{E_R}$

$I_R = \sqrt{\dfrac{P}{R}}$

$I_C = I_R = I_T = I_L$

$I_C = \dfrac{E_C}{X_C}$

$P = E_R \times I_R$

$P = \dfrac{E_R^{\,2}}{R}$

$P = I_R^{\,2} \times R$

$R = \sqrt{Z^2 - (X_L - X_C)^2}$

$R = \dfrac{E_R}{I_R}$

$R = \dfrac{E_R^{\,2}}{P}$

$X_C = \dfrac{1}{2\pi fC}$

$X_C = \dfrac{E_C}{\text{VARs}_C}$

$X_C = \dfrac{E_C}{I_C}$

$I_T = I_R = I_L = I_C$

$I_T = \dfrac{E_T}{Z}$

$\text{VA} = E_T \times I_T$

$\text{VA} = I_T^{\,2} \times Z$

$P = \text{VA} \times \text{PF}$

$E_R = \sqrt{P \times R}$

$R = Z \times \text{PF}$

$R = \dfrac{P}{I_R^{\,2}}$

$L = \dfrac{X_L}{2\pi f}$

$X_L = \dfrac{E_L}{I_L}$

$I_T = \dfrac{\text{VA}}{E_T}$

$I_T = \sqrt{\dfrac{\text{VA}}{Z}}$

$\text{VA} = \dfrac{P}{\text{PF}}$

$\text{VA} = \dfrac{E_T^{\,2}}{Z}$

$E_R = I_R \times R$

$E_R = \dfrac{P}{I_R}$

$E_R = \sqrt{E_T^{\,2} - (E_L - E_C)^2}$

$E_R = E_T \times \text{PF}$

$E_C = I_C \times X_C$

$E_C = \sqrt{\text{VARs}_C \times X_C}$

$C = \dfrac{1}{2\pi f X_C}$

$X_L = 2\pi fL$

$E_L = I_L \times X_L$

$I_L = I_R = I_T = I_C$

$I_L = \dfrac{E_L}{X_L}$

Resistive-Inductive-Capacitive (R-L-C) Parallel Circuit Formulas

$$Z = \frac{R \times X}{\sqrt{R^2 + X^2}}$$

$$X = \frac{X_L \times X_C}{X_L - X_C}$$

$$Z = \frac{1}{\sqrt{\left(\frac{1}{R}\right)^2 + \left(\frac{1}{X_L} - \frac{1}{X_C}\right)^2}}$$

$$Z = \frac{E_T}{I_T}$$

$$Z = \frac{E_T^2}{VA}$$

$$I_L = \frac{E_L}{X_L}$$

$$VA = E_T \times I_T$$

$$VA = I_T^2 \times Z$$

$$VA = \frac{E_T^2}{Z}$$

$$Z = \frac{VA}{I_T^2}$$

$$Z = R \times PF$$

$$VA = \frac{P}{PF}$$

$$PF = \frac{P}{VA}$$

$$PF = \frac{I_R}{I_T}$$

$$PF = \frac{Z}{R}$$

$$PF = \frac{1}{\left(\frac{R}{Z}\right)}$$

$$PF = \cos\theta$$

$$X_L = \frac{E_L}{I_L}$$

$$E_T = E_R = E_L = E_C$$

$$E_T = I_T \times Z$$

$$E_T = \frac{VA}{I_T}$$

$$E_T = \sqrt{VA \times Z}$$

$$I_T = \sqrt{I_R^2 + (I_L - I_C)^2}$$

$$I_T = \frac{E_T}{Z}$$

$$E_L = I_L \times X_L$$

$$X_L = 2\pi f L$$

$$I_T = \frac{VA}{E_T}$$

$$I_T = \sqrt{\frac{VA}{Z}}$$

$$I_T = \frac{I_R}{PF}$$

$$E_L = E_T = E_R = E_C$$

$$E_L = \sqrt{VARs_L \times X_L}$$

$$L = \frac{X_L}{2\pi f}$$

$$E_R = I_R \times R$$

$$E_R = \sqrt{P \times R}$$

$$E_R = \frac{P}{I_R}$$

$$E_R = E_T = E_L = E_C$$

$$E_C = I_C \times X_C$$

$$E_C = E_T = E_R = E_L$$

$$I_R = \sqrt{I_T^2 - (I_L - I_C)^2}$$

$$I_R = \frac{E_R}{R}$$

$$I_R = \frac{P}{E_R}$$

$$I_R = \sqrt{\frac{P}{R}}$$

$$P = VA \times PF$$

$$I_C = \frac{E_C}{X_C}$$

$$I_R = I_T \times PF$$

$$R = \frac{E_R}{I_R}$$

$$R = \frac{P}{I_R^2}$$

$$R = \frac{Z}{PF}$$

$$P = E_R \times I_R$$

$$P = \frac{E_R^2}{R}$$

$$X_C = \frac{E_C}{I_C}$$

$$R = \frac{1}{\sqrt{\left(\frac{1}{Z}\right)^2 - \left(\frac{1}{X_L} - \frac{1}{X_C}\right)^2}}$$

$$R = \frac{E_R^2}{P}$$

$$P = I_R^2 \times R$$

$$X_C = \frac{1}{2\pi f C}$$

Transformer Formulas

Turns Ratio, TR:

$$TR = \frac{N_P}{N_S}$$

Voltage:

$$\frac{N_P}{N_S} = \frac{E_P}{E_S}$$

$$TR = \frac{E_P}{E_S}$$

VA Rating:

$$\text{VA rating} = E_P \times I_P = E_S \times I_S$$

Current:

$$\frac{E_P}{E_S} = \frac{I_S}{I_P}$$

$$TR = \frac{I_S}{I_P}$$

Transformer Impedance Formulas

$$\frac{Z_P}{Z_S} = \left(\frac{N_P}{N_S}\right)^2$$

$$Z_P = Z_S\left(\frac{N_P}{N_S}\right)^2$$

$$Z_S = Z_P\left(\frac{N_S}{N_P}\right)^2$$

Three-Phase Formulas

Phase Voltage and Current for a Delta Configuration:

$$E_{Phase} = E_{Line}$$

$$I_{Phase} = \frac{I_{Line}}{\sqrt{3}}$$

$$I_{Line} = I_{Phase} \times \sqrt{3}$$

Phase Voltage and Current for a Wye Configuration:

$$E_{Phase} = \frac{E_{Line}}{\sqrt{3}}$$

$$E_{Line} = E_{Phase} \times \sqrt{3}$$

$$I_{Phase} = I_{Line}$$

AC Motor Formulas

Synchronous Motor Speed:

$$N_S = 120 \times \frac{f}{p}$$

Motor Slip:

Expressed as numerical value

$$S = N_S - N$$

Expressed as a percentage

$$\%S = \frac{N_S - N}{N_S} \times 100\%$$

True, Apparent, and Reactive Power Formulas

True Power:

Power factor = Cosine (cos) of the phase angle

$$\text{Power factor} = \frac{\text{True Power}}{\text{Apparent Power}}$$

$$\text{Power factor (\%)} = \frac{\text{True Power}}{\text{Apparent Power}} \times 100\%$$

Apparent Power:

$$S = \frac{I^2}{Z}$$

$$Q = \frac{E^2}{Z}$$

Reactive Power:

$$Q_T = Q_1 + Q_2 + Q_3 + \ldots + Q_N$$

$$Q = I^2 \times X$$

$$Q = \frac{E^2}{X}$$

Appendix B: Equivalent and Conversion Charts

Time Equivalents

Unit	Equivalents
1 second	1000 milliseconds 0.01667 minute 0.000278 hour
1 minute	60 seconds 0.1667 hour 0.000694 day
1 hour	3600 seconds 60 minutes 0.04167 day
1 day	1440 minutes 24 hours 0.1429 week 0.00274 year
1 week	10,080 minutes 168 hours 7 days 0.0192 year
1 year	8760 hours 365 days 52 weeks

Goodheart-Willcox Publisher

Energy Equivalents

Energy Measurement	Equivalents
1 Btu	778 ft-lb 1054.8 joules (J)
1 horsepower (hp)	33,000 ft-lb/min 550 ft-lb/sec 746 W 2545.6 Btu/hr 42.42 Btu/min 1.014 hp (metric)
1 horsepower-hour (hp-hr)	1 hp for 1 hr 1,980,000 ft-lb 746 W/hr 0.746 kWh 2545.6 Btu
1 watt (W)	3.412 Btu/hr
1 kilowatt (kW)	1000 W 1.34 hp
1 kilowatt hour (kWh)	1 kW for 1 hr 1000 W-hr

Goodheart-Willcox Publisher

Peak, Average, and RMS Conversions

To Convert from:	Multiply by:	Or Divide by:
peak to rms	0.707	1.414
peak to average	0.637	1.57
peak to peak-to-peak	2	—
rms to peak	1.414	0.707
rms to average	0.9	1.11
average to rms	1.11	0.9
average to peak	1.57	0.637

Goodheart-Willcox Publisher

Weight Measurement Conversion

To Convert from US Customary:	Multiply by:	To Get SI:
ounces	28.349	grams
ounces	0.028349	kilograms
pounds	453.59	grams
pounds	0.4536	kilograms
To Convert from SI:	**Multiply by:**	**To Get US Customary:**
grams	0.03527	ounces
grams	0.002205	pounds
kilograms	35.274	ounces
kilograms	2.2046	pounds

Goodheart-Willcox Publisher

Linear Measurement Conversion

To Convert from US Customary:	Multiply by:	To Get SI:
inches	25.4	millimeters
inches	2.54	centimeters
inches	0.0254	meters
feet	304.8	millimeters
feet	30.48	centimeters
feet	0.3048	meters
feet	0.0003048	kilometers
To Convert from SI:	**Multiply by:**	**To Get US Customary:**
millimeters	0.03937	inches
centimeters	0.3937	inches
meters	39.37	inches
millimeters	0.003281	feet
centimeters	0.03281	feet
meters	3.281	feet
kilometers	3281	feet

Goodheart-Willcox Publisher

Common Prefixes Used in the Electrical Industry

Prefix	Symbol	Multiplication Factor
exa	E	10^{18} = 1,000,000,000,000,000,000
peta	P	10^{15} = 1,000,000,000,000,000
tera	T	10^{12} = 1,000,000,000,000
giga	G	10^{9} = 1,000,000,000
mega	M	10^{6} = 1,000,000
kilo	k	10^{3} = 1,000
hecto	h	10^{2} = 100
deca	da	10^{1} = 10
(unit)	—	10^{0} = 1
deci	d	10^{-1} = 0.1
centi	c	10^{-2} = 0.01
milli	m	10^{-3} = 0.001
micro	μ	10^{-6} = 0.000001
nano	n	10^{-9} = 0.000000001
pico	p	10^{-12} = 0.000000000001
femto	f	10^{-15} = 0.000000000000001
atto	a	10^{-18} = 0.000000000000000001

Goodheart-Willcox Publisher

Conversion Table: US Customary to SI Metric

When You Know: ⬇	Multiply By:		To Find: ⬇
	Very Accurate	Approximate	
Length			
inches	* 25.4	—	millimeters
inches	* 2.54	—	centimeters
feet	* 0.3048	—	meters
feet	* 30.48	—	centimeters
yards	* 0.9144	0.9	meters
miles	* 1.609344	1.6	kilometers
Weight			
grains	15.43236	15.4	grams
ounces	* 28.349523125	28.0	grams
ounces	* 0.028349523125	0.028	kilograms
pounds	* 0.45359237	0.45	kilograms
short ton	* 0.90718474	0.9	tonnes
Volume			
teaspoons	—	5.0	milliliters
tablespoons	—	15.0	milliliters
fluid ounces	29.57353	30.0	milliliters
cups	—	0.24	liters
pints	* 0.473176473	0.47	liters
quarts	* 0.946352946	0.95	liters
gallons	* 3.785411784	3.8	liters
cubic inches	* 0.016387064	0.02	liters
cubic feet	* 0.028316846592	0.03	cubic meters
cubic yards	* 0.764554857984	0.76	cubic meters
Area			
square inches	* 6.4516	6.5	square centimeters
square feet	* 0.09290304	0.09	square meters
square yards	* 0.83612736	0.8	square meters
square miles	* 2.58998811	2.6	square kilometers
acres	* 0.40468564224	0.4	hectares
Temperature			
Fahrenheit	*5/9 (after subtracting 32)		Celsius
Density			
pounds per cubic feet	1.602×10	16	kilograms per cubic meter
Force			
ounces (F)	2.780×10^{-1}	—	newtons
pounds (F)	4.448×10^{-3}	—	kilonewtons
kips	4.448	—	meganewtons
Stress			
pounds/square inch (psi)	6.895×10^{-3}	—	megapascals
kips/square inch (ksi)	6.895	—	megapascals
Torque			
ounce-inches	7.062×10^{3}	—	newton-meters
pound-inches	1.130×10^{-1}	—	newton-meters
pound-feet	1.356	—	newton-meters

* = Exact

Continued

Conversion Table: SI Metric to US Customary

When You Know:	Multiply By: Very Accurate	Multiply By: Approximate	To Find:
Length			
millimeters	0.0393701	0.04	inches
centimeters	0.3937008	0.4	inches
meters	3.280840	3.3	feet
meters	1.093613	1.1	yards
kilometers	0.621371	0.6	miles
Weight			
grains	0.00228571	0.0023	ounces
grams	0.03527396	0.035	ounces
kilograms	2.204623	2.2	pounds
tonnes	1.1023113	1.1	short tons
Volume			
milliliters	—	0.2	teaspoons
milliliters	0.06667	0.067	tablespoons
milliliters	0.03381402	0.03	fluid ounces
liters	61.02374	61.024	cubic inches
liters	2.113376	2.1	pints
liters	1.056688	1.06	quarts
liters	0.26417205	0.26	gallons
liters	0.03531467	0.035	cubic feet
cubic meters	61023.74	61023.7	cubic inches
cubic meters	35.31467	35.0	cubic feet
cubic meters	1.3079506	1.3	cubic yards
cubic meters	264.17205	264.0	gallons
Area			
square centimeters	0.1550003	0.16	square inches
square centimeters	0.00107639	0.001	square feet
square meters	10.76391	10.8	square feet
square meters	1.195990	1.2	square yards
square kilometers	0.38610216	0.4	square miles
hectares	2.471054	2.5	acres
Temperature			
Celsius	*9/5 (then add 32)		Fahrenheit

* = Exact

Goodheart-Willcox Publisher

Trigonometric Functions

Angle	Radian Measure	Sin	Cos	Tan
0	0.00000	0.00000	1.00000	0.00000
1	0.01745	0.01745	0.99985	0.01746
2	0.03491	0.03490	0.99939	0.03492
3	0.05236	0.05234	0.99863	0.05241
4	0.06981	0.06976	0.99756	0.06993
5	0.08727	0.08716	0.99619	0.08749
6	0.10472	0.10453	0.99452	0.10510
7	0.12217	0.12187	0.99255	0.12278
8	0.13963	0.13917	0.99027	0.14054
9	0.15708	0.15643	0.98769	0.15838
10	0.17453	0.17365	0.98481	0.17633
11	0.19199	0.19081	0.98163	0.19438
12	0.20944	0.20791	0.97815	0.21256
13	0.22689	0.22495	0.97437	0.23087
14	0.24435	0.24192	0.97030	0.24933
15	0.26180	0.25882	0.96593	0.26795
16	0.27925	0.27564	0.96126	0.28675
17	0.29671	0.29237	0.95630	0.30573
18	0.31416	0.30902	0.95106	0.32492
19	0.33161	0.32557	0.94552	0.34433
20	0.34907	0.34202	0.93969	0.36397
21	0.36652	0.35837	0.93358	0.38386
22	0.38397	0.37461	0.92718	0.40403
23	0.40143	0.39073	0.92050	0.42447
24	0.41888	0.40674	0.91355	0.44523
25	0.43633	0.42262	0.90631	0.46631
26	0.45379	0.43837	0.89879	0.48773
27	0.47124	0.45399	0.89101	0.50953
28	0.48869	0.46947	0.88295	0.53171
29	0.50615	0.48481	0.87462	0.55431
30	0.52360	0.50000	0.86603	0.57735
31	0.54105	0.51504	0.85717	0.60086
32	0.55851	0.52992	0.84805	0.62487
33	0.57596	0.54464	0.83867	0.64941
34	0.59341	0.55919	0.82904	0.67451
35	0.61087	0.57358	0.81915	0.70021
36	0.62832	0.58779	0.80902	0.72654
37	0.64577	0.60182	0.79864	0.75355
38	0.66323	0.61566	0.78801	0.78129
39	0.68068	0.62932	0.77715	0.80978
40	0.69813	0.64279	0.76604	0.83910
41	0.71558	0.65606	0.75471	0.86929
42	0.73304	0.66913	0.74314	0.90040
43	0.75049	0.68200	0.73135	0.93252
44	0.76794	0.69466	0.71934	0.96569
45	0.78540	0.70711	0.70711	1.00000

Angle	Radian Measure	Sin	Cos	Tan
46	0.80285	0.71934	0.69466	1.03553
47	0.82030	0.73135	0.68200	1.07237
48	0.83776	0.74314	0.66913	1.11061
49	0.85521	0.75471	0.65606	1.15037
50	0.87266	0.76604	0.64279	1.19175
51	0.89012	0.77715	0.62932	1.23490
52	0.90757	0.78801	0.61566	1.27994
53	0.92502	0.79864	0.60182	1.32704
54	0.94248	0.80902	0.58779	1.37638
55	0.95993	0.81915	0.57358	1.42815
56	0.97738	0.82904	0.55919	1.48256
57	0.99484	0.83867	0.54464	1.53986
58	1.01229	0.84805	0.52992	1.60033
59	1.02974	0.85717	0.51504	1.66428
60	1.04720	0.86603	0.50000	1.73205
61	1.06465	0.87462	0.48481	1.80405
62	1.08210	0.88295	0.46947	1.88073
63	1.09956	0.89101	0.45399	1.96261
64	1.11701	0.89879	0.43837	2.05030
65	1.13446	0.90631	0.42262	2.14451
66	1.15192	0.91355	0.40674	2.24604
67	1.16937	0.92050	0.39073	2.35585
68	1.18682	0.92718	0.37461	2.47509
69	1.20428	0.93358	0.35837	2.60509
70	1.22173	0.93969	0.34202	2.74748
71	1.23918	0.94552	0.32557	2.90421
72	1.25664	0.95106	0.30902	3.07768
73	1.27409	0.95630	0.29237	3.27085
74	1.29154	0.96126	0.27564	3.48741
75	1.30900	0.96593	0.25882	3.73205
76	1.32645	0.97030	0.24192	4.01078
77	1.34390	0.97437	0.22495	4.33148
78	1.36136	0.97815	0.20791	4.70463
79	1.37881	0.98163	0.19081	5.14455
80	1.39626	0.98481	0.17365	5.67128
81	1.41372	0.98769	0.15643	6.31375
82	1.43117	0.99027	0.13917	7.11537
83	1.44862	0.99255	0.12187	8.14435
84	1.46608	0.99452	0.10453	9.51436
85	1.48353	0.99619	0.08716	11.43005
86	1.50098	0.99756	0.06976	14.30067
87	1.51844	0.99863	0.05234	19.08114
88	1.53589	0.99939	0.03490	28.63625
89	1.55334	0.99985	0.01745	57.28996
90	1.57080	1.00000	0.00000	—

Goodheart-Willcox Publisher

Appendix C: NEMA Motor Dimensions, Frame Numbers, and Frame Letters

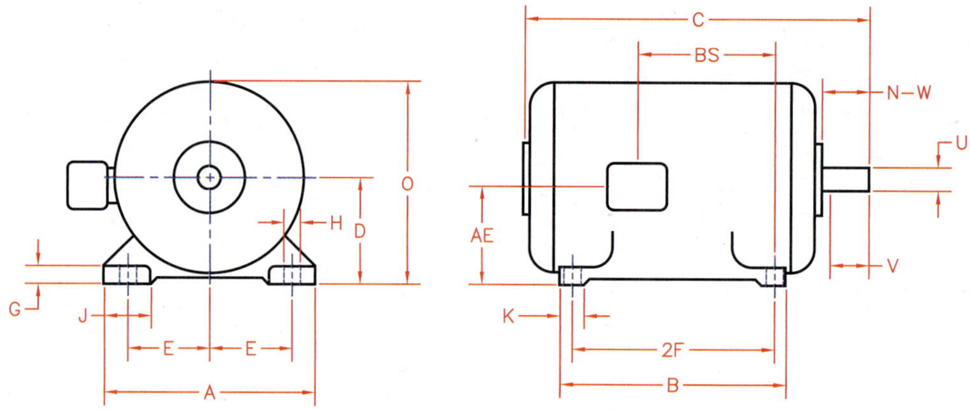

Basic NEMA Motor Dimensions

NEMA Letter	Description of Dimension
A	Outside length of feet (end view)
AE	Centerline of terminal housing to bottom of feet
B	Outside length of feet (side view)
BS	Centerline of mounting hole on shaft side to centerline of terminal housing
C	Overall length
D	Centerline of shaft to bottom of feet
E	Centerline of shaft to mounting holes in feet
2F	Distance between centerlines of mounting holes in feet (side view)
G	Thickness of feet
H	Hole diameter or slot width in feet
J	Width of feet
K	Length of feet
N-W	Length of shaft extension from shoulder
O	Top of motor to bottom of feet
U	Shaft diameter
V	Length of shaft available for coupling

Goodheart-Willcox Publisher

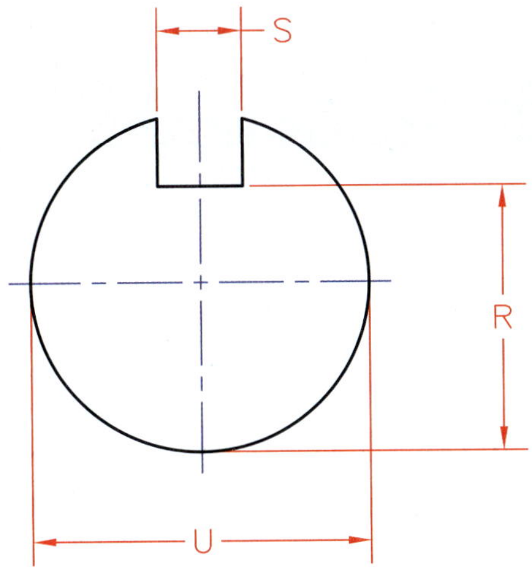

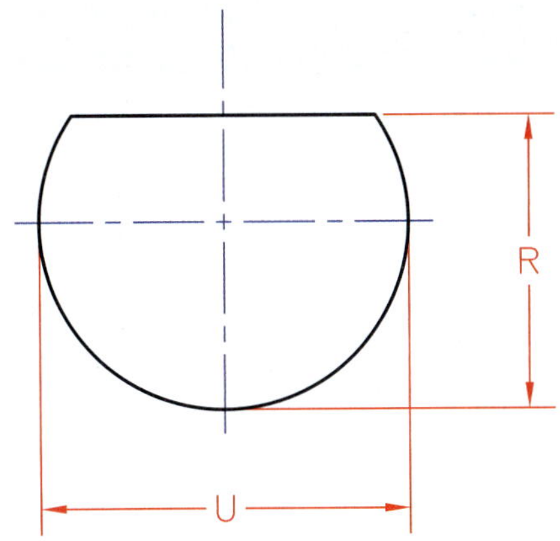

NEMA Motor Dimensions for Shaft Key Seats and Flats

U	R	S
0.375	0.328	NA
0.500	0.453	NA
0.625	0.517	0.188
0.875	0.771	0.188
1.125	0.986	0.250
1.375	1.201	0.312
1.625	1.416	0.375
1.875	1.591	0.500
2.125	1.845	0.500
2.375	2.021	0.625
2.875	2.450	0.750

Goodheart-Willcox Publisher

Common NEMA Motor Frame Numbers and Dimensions

Frame Number	A	D	E	2F	H	U	N-W
42	NA	2.62	1.75	1.69	0.28 slot	0.375	1.12
48	NA	3.00	2.12	2.75	0.34 slot	0.500	1.50
56	NA	3.50	2.44	3.00	0.34 slot	0.625	1.88
143T	7.0	3.50	2.75	4.00	0.34 hole	0.875	2.25
145T	7.0	3.50	2.75	4.00	0.34 hole	0.875	2.25
284T	14.0	7.00	5.50	9.50	0.53 hole	1.875	4.62
286T	14.0	7.0	5.50	11.00	0.53 hole	1.875	4.62
324T	16.0	8.0	6.25	10.50	0.66 hole	2.125	5.25
326T	16.0	8.0	6.25	12.00	0.66 hole	2.125	5.25
364T	18.0	9.00	7.00	11.25	0.66 hole	2.375	5.88
365T	18.0	9.00	7.00	12.25	0.66 hole	2.375	5.88
404T	20.0	10.00	8.00	12.25	0.81 hole	2.875	7.25
405T	20.0	10.00	8.00	13.75	0.81 hole	2.875	7.25

Goodheart-Willcox Publisher

Select NEMA Motor Frame Letters (Suffixes)

Frame Letter	Description
A	Industrial dc motor
C	Type C face mounting on drive end
D	Type D flange mounting on drive end
FC	Type C face mounting opposite drive end
FD	Type D flange mounting opposite drive end
G	Gasoline pump motor
J	Jet pump motor
K	Sump pump motor
M	Oil burner motor
N	Oil burner motor
T	Frame with standard dimensions
V	Vertical mount
Y	Special mounting dimensions
Z	Standard mounting dimensions except shaft extension

Goodheart-Willcox Publisher

Appendix D: Answers to Now It's Your Turn

Chapter 5

5-2: 50 At

Chapter 6

6-1: $R = 5\ \Omega$
6-2: $P = 80$ W

Chapter 7

7-1: 5
7-2: 114
7-3: 10
7-4: 32
7-5: Y = 4
7-6: 12
7-7: 21
7-8: 10°C
7-9: 10 hours
7-10: Yes
7-15: 6′2.5″, rounded up to 6′3″
7-16: 6 hp

Chapter 10

10-1: 25 A
10-2: 85 Ω
10-3: 18 V
10-4: 25 V
10-6: 54 V

Chapter 11

11-6: 1.33 Ω
11-7: 10 Ω

Chapter 12

12-4: Combination circuit
12-6: $I_T = 1.5$ A, $P_T = 45$ W
12-7: $I_T = 4$ A, $P_T = 144$ W
12-8:
$P_T = 320$ W
Series branch R_1: 288 W
Parallel branch R_2: 16 W
I through $R_2 = 2$ A
$R_2 = 4\ \Omega$
Parallel branch R_3: 10.664 W
Parallel branch R_4: 5.336 W
$P_T = 288\text{ W} + 16\text{ W} + 10.664\text{ W} + 5.336\text{ W} = 320\text{ W}$

Chapter 13

13-1: 2 A flowing out of node
13-2: $I_3 = 4$ A flowing out of node
13-4: Node F:
$I_{FG} + I_2 + I_3 = 0$
(–5 A) + 4 A + 1 A =0
Node G:
$I_T + I_{FG} + I_1 = 0$
(–8 A) + 5 A + 3 A = 0
13-5: $I_2 = 5$ A
Node K:
$I_T + I_1 + I_2 + I_3 = 0$
11 A + (–2 A) + (–5 A) + (–4A) = 0
–11 A + 11 A = 0
Node L:
$I_T + I_1 + I_2 + I_3 = 0$
(–11 A) + 2 A + 5 A + 4A = 0
–11 A + 11 A = 0
13-7: $E_S = 48$ V
13-8: $I_3 = 1.8$ A

Chapter 14

14-1: 60°
14-2: Yes
14-3: $a = 3$, $b = 5.20$
14-4: $a = 3.54$, $b = 3.54$
14-6: 39 units

Chapter 15

15-3: 50 Hz
15-4: 14 V
15-6: 28.67 V
15-7: 22.30 V
15-10: 96.6%

Chapter 16

16-1: E_{Line} = 600 V
16-2: E_{Phase} = 230.9 V
16-5: 4.39%, this is not an acceptable voltage imbalance as it is above the range of 2%.
16-6: Temperature rise (%) ≈ 12.5%
16-7: Motor temperature = 120°F
16-9: Cost increase = $1135.83
16-10: Using phase current and voltage values:
VA = 19,440 VA
Using line current and voltage values: VA = 19,440 VA
16-12: 18,285.6 W = 18.29 kW

Chapter 17

17-1: L =0.29857 H or 298.57 mH
17-2: f = 99 Hz
17-4: L = 0.2986 H
17-6: R = 8 Ω
17-8: L_2 = 0.6 H
17-10: L_T = 0.16 H
17-11: L_T = 0.2 H

Chapter 18

18-1: 0.084 C
18-2: 80 V
18-8: C_2 = 212 μF
18-10: C_T = 5 μF
18-11: C_T = 20 μF

Chapter 19

19-1: Q_L = 120 VAR
19-3: Q_C = 60.2 VAR
19-5: Q_C =1085 VAR
19-6: Total circuit voltage: X_{C1} = 132.63 Ω, X_{C2} = 88.5 Ω
Reactive power: Q_{C1} = 434 VAR, Q_{C2} = 650.8 VAR
Current flow: I_{C1} = 1.81 A, I_{C2} = 2.71 A
19-7: Q_L = 252.7 VAR
19-8: Reactive power: Q_{L1} = 176.9 VAR, Q_{L2} = 75.83 VAR
Current flow: I_{L1} = 0.88 A, I_{L2} = 0.379 A

Chapter 20

20-1: Z = 33.5 Ω
20-2: Z = 30 Ω
20-3: Z = 220 Ω
20-4: Z = 27.2 Ω
20-5: Z = 16.7 Ω
20-6: Z = 17.4 Ω
20-7: Z = 46.4 Ω
20-8: Z = 13.7 Ω (X = 33.1 Ω)
20-14: E_C = 36 V, E_L = 480 V, E_T = 444 V

Chapter 21

21-1: TR = 3:1
21-2: TR = 1:8
21-3: TR = 4:1
21-5: TR = 2:1
21-6: 50 VA
21-14: E_S = 40 V, I_S = 2 A; I_P = 0.4 A

Chapter 22

22-4: 75 kVA

Chapter 23

23-2: N_2 = 4800 rpm
23-4: S = 45 rpm, %S = 3%

Chapter 24

24-1: 5.25″

Chapter 25

25-4: 69 Hz

Chapter 26

26-3: 0.34, or 1/3 hp

Chapter 29

29-2: 8 poles
29-5: S = 3474 rpm

Chapter 34

34-1: t = 15.4 N·m
34-2: Input power = 2000 W

Chapter 35

35-2: 15
35-13: t = 57 N·m

Chapter 36

36-2: 1200 rpm

Glossary

A

across-the-line starter. A common three-phase motor starter that supplies full line voltage to a motor's windings upon start-up and through its operating cycle. (32)

acute angle. Formed when two lines intersect to create an angle that is less than 90°. (14)

additive inverse. A number that, when added to another number, will produce a result of zero. (7)

adjustable wrench. A wrench with a movable jaw that is adjusted using a thumbscrew. (8)

advanced metering infrastructure (AMI). The interconnected network of smart meters communication systems, and data management protocols that allow for two-way communication between a power utility company and its customers. (38)

affinity laws. A series of mathematical formulas that govern motor characteristics such as speed, horsepower, and torque. (25)

air-core inductor. An inductor with a nonmetallic core. (17)

air-core transformer. A transformer that has a core made of air or other insulating material. (21)

alternating current (ac). Current in an electrical circuit that reverses polarity, or direction of flow, continuously. (6)

alternator. A generator that produces ac electrical energy. (23)

ampacity. The current-carrying capacity of a conductor or circuit. (3)

ampere (A). The unit of measurement for electrical current. (6)

ampere-turn (At). Unit of magnetomotive force. (5)

amplitude. In a sine wave, the maximum positive or negative value. (15)

AND operation. A logic operation that can be represented by multiple series-connected switches, in which all switches must be closed for a load to be energized. (12)

angular misalignment. A shaft misalignment that occurs when the drive and driven shafts are not parallel to each other. (26)

anion. An ion with a net negative charge. (4)

anode. The positive terminal of a capacitor. (18)

apparent power. Power that appears to be present in a circuit when voltage and current are measured independently, expressed in volt-amperes (VA). (15)

apparent power (S). Circuit power calculated as a product of applied voltage and current, expressed in volt-amperes (VA). (19)

apprentice. An individual who works under a licensed electrician for an extended period of time while going to school to learn the trade. (2)

apprenticeship. A common path to becoming an electrician involves learning a trade under the direction and guidance of an experienced electrician. (2)

arc blast. The pressure wave created during an arc fault. (3)

arccosine (arccos). A trigonometric function used to determine an angle from its cosine value. Also called *inverse cosine*. (14)

arc fault. Occurs when electric current flows in an uncontrolled path through air, either from one conductor to another or from one conductor to ground. (3)

arc-fault circuit interrupter (AFCI). Provides protection from the effects of arc faults by de-energizing a circuit when an arc fault is detected. (3)

arc flash. An explosive and violent uncontrolled release of electrical energy that can cause severe injuries or death. (3)

arc flash boundary. The distance from an energy source within which a person is likely to receive burns in the event of an arc flash. (3)

arcsine (arcsin). A trigonometric function used to determine an angle from its sine value. Also called *inverse sine*. (14)

arctangent (arctan). A trigonometric function used to determine an angle from its tangent value. Also called *inverse tangent*. (14)

armature. The part of a relay that moves to open and close the electrical contact. (9) A generator component that holds electrical windings in place, facilitates the electrical connections between the windings and the generator's output, acts as the connection point between the generator and the external mechanical rotating force, and provides a means to rotate the electrical windings within the generator's magnetic field. (34)

armature stack. A series of laminated iron plates connected to the shaft. Also called the *armature core*. (34)

asphyxiation. When the body becomes oxygen-deprived. (3)

Note: The number in parentheses following each definition indicates the chapter in which the term can be found.

associative law. The mathematical law that states that the order in which numbers are added or multiplied can be changed without changing the value of the expression. (7)

atom. The smallest amount of an element that can exist that still maintains the properties of the element. (4)

atomic mass. Obtained from the periodic table, the number that represents the total number of protons in an atom; in an electrically neutral atom, this is also the number of electrons. (4)

atomic number. Obtained from the periodic table, the number that represents the number of protons and the number of electrons in a neutral atom. (4)

automatic reset device. A control device designed to reset automatically once the sensed condition has reached a desired or acceptable level. (9)

autotransformer. A transformer that has a single winding that serves as both the primary and secondary windings. (21)

autotransformer starting. A reduced-voltage starting method that uses a three-phase autotransformer as a step-down transformer to reduce the voltage supplied to a motor at start-up. (32)

auxiliary contact. A lighter-duty contact that has lower amperage ratings than the main contacts and has control circuits that carry less current than the main motor circuit. (31)

average dc voltage. A weighted average that takes into account the various voltages being supplied to the circuit. (6)

average voltage. The arithmetic average of multiple, individual voltage readings. (15)

axial capacitor. A capacitor with electrical connections on opposite ends of the device. (18)

axial load. A thrust force that pushes on a motor shaft in directions that are parallel to the shaft. (24)

B

back electromotive force (bemf). The opposing voltage generated when an inductor is powered. Also called *counter electromotive force (cemf)*. (17)

base. The control terminal of a transistor. (25)

bearing. Components that help reduce the friction that affects the motor shaft's ability to rotate. (24)

bearing sag. A condition in which a portion of a sleeve bearing becomes worn due to pressure from a motor shaft, resulting in improper alignment of the rotor within a stator. (33)

biofuel. Biomass fuels produced by using fast modern technologies. (1)

biomass. Organic matter that is used as the fuel source to produce electricity. (37)

biopower. Electricity that is generated using biomass as its fuel. (37)

bipolar junction transistor (BJT). Transistor type with a low saturation voltage. (25)

bleed resistor. A resistor connected across a capacitor's terminals to allow the capacitor to safely discharge. (18)

boost transformer. A transformer that is wired to increase the voltage supplied to it. (21)

box wrench. A wrench with a fixed head that completely encircles the nut or bolt head. (8)

branch. An individual current path in a parallel circuit. (11)

British thermal unit (Btu). The amount of heat energy required to raise the temperature of 1 lb of water 1°F. (7)

brush. A spring-loaded component of a dc motor that ensures a consistent electrical connection between the power source and the motor's rotating windings. (23)

buck transformer. A transformer that is wired to decrease the voltage supplied to it. (21)

bump theory. The concept that current flow occurs as electrons are repelled or "bumped" from an atom by other electrons. (4)

C

cable cutter. A tool used to cut cables or large conductors without distorting the cable or conductor. (8)

capacitance. Measure of a material's ability to hold an electric charge if a voltage is applied across it. (18)

capacitive load. In an electric circuit, a load that stores and releases electrical energy into the circuit. (9)

capacitive reactance. The resistance to current flow in a capacitive circuit created by the electrostatic charge between the capacitor's plates, expressed in ohms. (18)

capacitor. An electric circuit component that opposes a change in voltage. (18)

capacitor-start, capacitor-run (CSCR) motor. A single-phase motor that is a cross between the capacitor-start, induction-run motor and the permanent-split capacitor motor. (28)

capacitor-start, induction-run (CSIR) motor. A single-phase motor that has two distinct windings and a means to remove the start winding from the circuit. (28)

capacity factor. The amount of energy actually produced by a process or system over a period of time divided by the amount of energy that could have been produced if the system or process operated at full capacity over the same period of time. (37)

catalyst. A material or substance that helps speed up or enhance chemical reactions without being consumed by them. (37)

cathode. The negative terminal of a capacitor. (18)

cation. An ion with a net positive charge. (4)

Celsius scale. An SI scale of temperature that uses 0°C as the freezing point of water and 100°C as the boiling point. (7)

center tap. A grounded conductor connection at the center of a delta or open-delta secondary winding. (22)

centrifugal switch. A switch mounted to a motor's shaft that disconnects the start winding. (28, 31)

chalk line. A tool consisting of a string housed in a chalk-filled container, used to create a straight line when the string is snapped on a surface. (8)

choke. An inductor used to remove or filter unwanted ac ripples and fluctuations. (17)

circular saw. A type of saw that uses a round blade to make straight cuts in sheet material such as wood. (8)

clamp-on ammeter. A noncontact meter that measures electrical current by sensing the strength of the magnetic field around the conductor. (8)

coercivity. A property of magnetic materials representing the magnetic field required to demagnetize a specific material. (5)

coil span. Denoted Y_S, the width of an individual coil, typically expressed in terms of armature slots. (35)

collector-emitter junction. The conductive path between a transistor's collector and emitter. (25)

combination circuit. A circuit that contains both series-connected and parallel-connected elements. Also known as a *series-parallel circuit*. (12)

combination wrench. A two-headed wrench that combines an open-ended wrench on one side and a box wrench on the opposite end. (8)

commutative law. The mathematical law that states that the position of numbers being added or multiplied can be changed without changing the value of the expression. (7)

commutator. A dc motor component that facilitates the process of reversing the polarity of a power supply. (23)

commutator pitch. The number of commutator segments of a coil span. (35)

complementary angles. Two angles that add up to 90°. (14)

compound. A substance that is formed when different elements combine because of a chemical reaction. (4)

compound dc motor. A dc motor that has both a series winding and a shunt winding. (27)

compound-wound dc generator. A self-excited generator that uses both series and shunt windings. (35)

conductor. A material that has a small number of valence electrons and allows electrons (and thus electricity) to flow relatively easily. (4)

conduit bender. A tool used to bend smaller electrical conduit. (8)

confined space. An area with limited entry and egress points that is not suitable for human occupancy, but is large enough for a person to enter. (3)

continuous coil voltage. The maximum continuous voltage that can be applied to the coil of a potential magnetic relay without causing damage to the device. (31)

control circuit design. The process of designing a circuit to create the desired circuit output. (12)

control device. A device that opens or closes sets of electrical contacts to energize or de-energize circuits that power the main circuit loads. (9)

converter. An electronic circuit that converts an ac signal to a dc signal. (25)

core losses. The combined effects of hysteresis losses and eddy current losses. Also called *iron losses* or *armature losses*. (27)

cosine. In a right triangle, the ratio of the length of the side adjacent the angle to the length of the hypotenuse. (14)

coulomb. The unit for electric charge. (6)

counter electromotive force (cemf). The opposing voltage generated when an inductor is powered. Also called *back electromotive force (bemf)*. (17)

current. The flow of electrons through a conductor. (6)

current electricity. The continuous flow of electrons through an electric circuit. (4)

current imbalance. Measure of the current deviation from the average current in a three-phase system. (16)

current magnetic relay (CMR). A device that removes the start winding and start capacitor from the active circuit based on the current flow through the motor's run winding. (28)

current source. A device that maintains circuit current at a constant rate. (13)

cycle. The repeating portion of a wave or signal. (15)

cylindrical armature. A cylinder-shaped armature that has parallel slots to accept windings. Cylindrical armatures operate without reluctance torque and are commonly used in higher speed generators. Also called *nonsalient-pole armature*. (34)

D

dc bus. A bank of parallel-connected capacitors that filters and smooths a dc waveform. (25)

delta configuration. Wiring connection that is characterized by a continuous loop where the outlet of one winding is connected to the inlet of the next winding, creating a shape resembling the Greek letter delta, Δ. (16)

delta-delta three-phase transformer. A three-phase transformer that has both the primary and secondary windings configured as delta connections. (22)

delta-wye three-phase transformer. A three-phase transformer that has its primary windings configured as a delta and its secondary windings configured as a wye. Also called *delta-star configuration*. (22)

depletion region. The portion of a diode or transistor that does not allow electrons to pass. The depletion region is only present when the device is reverse biased. (30)

diagonal-cutting pliers. Type of pliers with an angled cutting edge, is used to cut wire, small screws, or other thin metal close to a surface. Also called *side cutters*. (8)

diamagnetic. Substances that have no magnetic field and are not attracted to magnets. (5)

dielectric. Insulating material between plates in a capacitor. (18)
dielectric constant. The unitless quantity that relates the permittivity of a substance to the permittivity of a perfect vacuum. Also called *relative permittivity.* (18)
dielectric stress. The stretching of the electron orbits of a dielectric material resulting from a potential difference across a capacitor. (18)
digital multimeter (DMM). An all-in-one device that measures volts, values of resistance in ohms, and current in amps. (8)
diode. An electronic device that allows electric current to flow in only one direction. (4)
dipole. Two dissimilar charges separated by a small distance. (37)
direct-acting relay. A relay that is used to directly control the operation of a circuit load or motor. (31)
direct current (dc). Current in an electrical circuit that flows in one direction. (6)
dispatchable energy source. An energy source that can be accessed on-demand, based on the needs of the network. (37)
dispatch stack. A chart used by energy producers to determine the most cost-effective way to produce the required amount of energy at any given time during the day. (37)
distributed computing. The process of spreading computational responsibilities over a network of connected systems. (38)
distributed energy resource (DER). A small energy production plant or energy storage system located near the point of energy consumption. (38)
distributed generation. The process of generating electricity near the point of its end use. (38)
distribution substation. A type of electrical substation that serves as a separation point between the high-voltage transmission system and lower-voltage distribution system. (38)
distribution transformer. A step-down transformer that reduces the voltage to the level required by the end user of the electricity. (38)
distributive law. The mathematical law that states when a number is multiplied by an addition (or subtraction) operation contained within a set of parentheses, the expression can be restated as the sum (or difference) of the number multiplied by each term contained within the parentheses. (7)
diversion. The process of diverting, or redirecting, a small portion of a river's flow. (37)
diversity. A global term that includes differences among people and can specifically refer to gender, sexuality, religion, race, nationality, and many other distinctions. (2)
doping. The process of introducing impurities to a pure semiconductor to create either a positively-charged or a negatively-charged material. (4, 30)
drive system. The combination of a motor and a motor drive. (25)
drop-out voltage. The voltage required across a potential magnetic relay's coil to hold the contacts open once they have already been picked-up or opened. (31)
dual-voltage single-phase motor. A single-phase motor that operates at one of two different voltages. (28)
duplex winding. A coil that loops back onto itself to form two loops, so that the coil makes two passes past each pole. (34)

E

eddy current losses. Power losses that result when currents flow around a conductor. (17)
efficiency. Expressed as a percentage, the ratio of the output produced divided by the input required to produce it. (1)
electrical charge. How a substance moves, reacts, or is acted on when exposed to a magnetic field. (4)
electricity. The property, condition, or set of conditions, that is present when there is an accumulation of an electric charge. (4)
electric shock. Occurs when an individual becomes part of an energized electric circuit. (3)
electrolyte. A substance (often liquid or gel-like) that has a high concentration of either positively or negatively charged ions. (18) The component part of a battery or fuel cell that allows chemical reactions within the device to continue by providing a path for charged particles between the device's electrodes. (37)
electrolytic capacitor. A capacitor that uses an electrolyte as its cathode. (18)
electromagnet. A temporary magnet produced by passing current through a coil of wire. (5)
electromagnetic induction. The phenomenon of creating current flow by changing magnetic flux. (5)
electromechanical control device. A control device that has a linkage or physical connection that causes sets of electrical contacts to open and close. (9)
electromotive force (EMF). The electrical pressure that pushes electrons through wires in an electrical circuit. (6)
electron. The component part of an atom that has a negative charge. (4)
electronic control device. A control device that relies on electronic circuitry and sensors to open and close a set of electrical contacts. (9)
electronically commutated motor (ECM). A motor that uses electronic components and circuitry to control the polarity of the power signal supplied to the motor windings. (30)
electronics. Classification of products that use or incorporate solid-state circuitry and components into their design. (4)

electron shell. An orbit or energy level around an atom's nucleus that contains electrons. (4)

electron spin. The rotation of electrons around their own axis while they orbit around the nucleus of an atom. (5)

electrostatic charge. A charge that is produced by an excess or deficiency of electrons in an object or substance. Also called *static charge.* (18)

element. One of 118 base substances known to exist. (4)

end bell. The part of an electric motor that supports the bearings. (24)

energy density. The amount of energy, in kilowatt-hours, that can be stored per unit weight of a battery. (1)

energy storage system. With regards to electrical power generation, a system capable of storing electrical energy and then releasing the stored energy into the power grid when needed. (38)

engine. A machine that converts various forms of energy into mechanical energy. (23)

equation. A sequence of numbers and/or variables that contains an equal sign. (7)

equilateral triangle. A triangle that has three equal sides and three equal angles. (14)

equipment grounding conductor. An intentionally grounded conductor that connects a piece of equipment to ground. The grounding conductor only carries current when a fault condition is present. (6)

explosion-proof (XP) motor. A motor designed to withstand and contain an explosion that might occur within the motor's casing. (24)

exponent. A number or symbol denoting the power to which another number is to be raised. (7)

externally-excited dc generator. A generator that relies on an external dc source to produce the stator's magnetic field. (35)

F

Fahrenheit scale. A US Customary scale of temperature that uses 32°F as the freezing point of water and 212°F as the boiling point. (7)

ferromagnetic. Substances that have a net magnetic field and are highly attracted to magnets. (5)

fish stick. A fiberglass rod that is used to fish electrical wires through walls or conduits. (8)

fish tape. A length of steel or fiberglass, usually stored in a reel, that is used to pull electrical conductors through walls and conduits. (8)

flame-proof (FLP) motor. A motor type designed to prevent flames produced inside the motor housing from traveling to its surroundings. (24)

flat. The flat portion of a motor shaft that is the resting plane for a setscrew or other device to secure the driven device to the shaft. (24)

flow-limited. The amount of energy available at any given time might be limited. (1)

form factor. The numerical value that defines the shape of a sine wave. (15)

forward bias. The condition that allows an electronic device to conduct electric current. (30)

fractional-pitched. A winding in armature that has a coil span that is less than the pole pitch. (35)

free electron. A loosely held electron that is easily released from an atom and free to travel to other atoms. (4)

frequency. The number of cycles completed in one second. Expressed in units called hertz (Hz). (15)

frogleg-wound armature winding. An armature winding that is a combination of the lap-wound and wave-wound configurations. Also called a *series-parallel winding connection.* (35)

front pitch. The distance measured in number of slots between the finishing end of one coil, and the starting end of the next coil connected to it. (35)

full-pitched. A winding in an armature that has a pole pitch and a coil span that are equal. (35)

full-wave bridge rectifier. An electronic circuit that converts an ac waveform to a pulsating dc voltage signal. (25)

G

gauss. A unit of magnetic flux of one maxwell per square centimeter, or 6.4516 lines of force per square inch. (5)

generating substation. A type of electrical substation located near the power plant or source of electrical generation at which the voltage of the generated electricity is stepped up to a higher voltage before being transmitted within the power grid. (38)

generator. A machine that converts chemical or mechanical energy into electrical energy. (23, 34)

geothermal energy. Energy produced as a result of extracting heat from the earth's subsurface. (37)

gilbert. Unit of magnetomotive force that is equal to 0.7958 ampere-turns. (5)

ground. The earth when used as a reference point for obtaining voltage readings. (6)

grounded conductor. An intentionally grounded conductor that carries current during the normal operation of a circuit. Also referred to as a *neutral* or *neutral conductor.* (6)

ground-fault circuit interrupter (GFCI). An electrical device that protects individuals from electric shock by monitoring incoming and outgoing current. (3)

H

hacksaw. A saw that is used to cut metallic materials and PVC conduit. (8)

hammer drill. A drill that uses a percussive motion to aid in penetrating brick, concrete, or other masonry material. (8)

hazardous location (HAZ) motor. A specialized motor that is classified by the type of hazard that the motor is likely to be exposed to. These factors include environmental exposure potential and maximum temperature limits. (24)

heater. A thermal overload relay. (32)

heliostat. A flat lenses or mirror used to reflect sunlight onto a receiver in a concentrated solar power (CSP) systems. (37)

henry (H). Unit of inductance that with a rate of change of one ampere per second produces a counter electromotive force (cemf) of one volt. (17)

hertz (Hz). A unit of frequency representing cycles per second. (15)

hex-key wrench. Tool used to turn bolts and screws that have hex-shaped recesses in their head. Also called *hex key, Allen wrench*, and *Allen key*. (8)

high-leg delta configuration. Modified delta wiring connection where a grounded conductor is connected to the center of one of the three phases of the system. (16)

hub. The portion of a driven device that is in contact with a motor shaft. (24)

hypotenuse. The longest side of a right triangle. This is the side of the triangle that is opposite the right angle. (14)

hysteresis losses. The losses associated with the charging and discharging of an inductor. (17)

I

impedance (*Z*). Total opposition force to current flow in a circuit. (20)

inductance. A measure of an inductor's ability to store energy in the form of a magnetic field and oppose changes in electric current. (17)

induction motor. An ac motor in which the rotor has no direct connection to a power supply and uses induced current to generate a magnetic field. (28)

induction-start, induction-run (ISIR) motor. A single-phase motor that uses a high resistance start winding, a low resistance run winding, and a starting device. (28)

inductive load. A load that uses magnetic fields to perform useful work. (9)

inductive reactance. The property of an inductor to limit current flow, expressed in ohms. (17)

inductor. An electrical component that is made up of a coil of wire. (17)

in phase. The condition when voltage and current waveforms reach their maximum levels at the same time. (15)

instantaneous voltage. A voltage at one instant in time. (15)

insulated-gate bipolar transistor (IGBT). Transistor type that has a fast-switching speed and a low saturation voltage that is commonly used on electronic motor drives. (25)

insulator. Materials that do not allow electrons to flow freely through them. (4)

interior angle. An angle formed by two adjacent sides of a triangle. (14)

internship. An arrangement made through an educational institution whereby a student works under an industry professional's direct supervision. (2)

inverse cosine. A trigonometric function used to determine an angle from its cosine value. Also called *arccosine*. (14)

inverse sine. A trigonometric function used to determine an angle from its sine value. Also called *arcsine*. (14)

inverse tangent. A trigonometric function used to determine an angle from its tangent value. Also called *arctangent*. (14)

inverter. An electronic circuit that changes a dc signal to an ac signal. Also, an ac motor drive that includes this function. (25)

ion. An atom that has either a positive or a negative charge. (4)

iron-core inductor. An inductor that has a soft magnetic core. (17)

iron-core transformer. A transformer that is manufactured with an iron core. (21)

isolation transformer. A transformer that has two separate windings that are insulated from each other and are not electrically connected to each other. (21)

isosceles right triangle. A triangle with interior angles of 90°, 45°, and 45°. (14)

isosceles triangle. A triangle that has two equal sides and two equal angles. (14)

isotope. An atom that has a different number of neutrons than the base element. (4)

J

jab saw. A saw that is used to cut holes in drywall. Also called *drywall saw*. (8)

job shadowing. The process of observing a person on the job to learn more about the job, its requirements, and what a typical workday might look like. (2)

K

key. A rectangular, prism-shaped pin that fits into a keyway to secure a driven device to a motor shaft. (24)

keyway. A channel formed by the shaft and hub grooves that accepts a key. (24)

Kirchhoff's current law. The law that states that the mathematical sum of all current flows into and out of any circuit node, connection, or point must be zero. (13)

Kirchhoff's voltage law. The law that states that the mathematical sum of all voltage drops in any complete circuit loop must be zero. (13)

L

lagging phase angle. Circuit condition when the overall circuit reactance is inductive. (20)

laminations. In a transformer, the insulated layers that constitute the transformer's iron core. (21)

lap-wound armature winding. A winding configuration that turns back onto itself to create a coil-like path. as Also called a *parallel winding connection.* (35)

law. In mathematics, an expression that quantifies relationships among electrical components. (13)

law of charges. Law that states like charges repel each other, while opposite charges attract each other. (4)

leading phase angle. Circuit condition when the overall circuit reactance is capacitive. (20)

leakage current. The trace current that flows through an electrolytic capacitor when voltage is supplied to it. (18)

left-hand rule. Used to determine the direction of the magnetic field around a conductor when holding the conductor with the left hand, the left thumb points in the direction of electron flow. (5)

Lenz's law. A law that states as the strength of a magnetic field changes, the induced current flows in a direction that is opposite to the change in magnetic field that originally created it. (17)

letter of recommendation. A letter that is written by an employer, supervisor, teacher, coach, or other adult that describes your personal characteristics and qualities that would be strengths in the workplace. (2)

lifelong learning. The process of continually updating your knowledge and skills. (2)

limited approach boundary. The farthest distance at which a worker is exposed to potential shock risk as well as significant heat-related injuries in the event of an arc flash. (3)

linear waveform. A waveform that produces voltage rises and falls at a constant rate. (15)

linesman pliers. Pliers with wider and larger jaws than conventional pliers, used to cut, bend, and twist wires together. (8)

line voltage. The voltage measured between any two ungrounded conductors connected to a three-phase system. Also called *phase-to-phase voltage.* (16)

load. In an electrical circuit, any component or device that consumes power. (6, 9)

load matching. The process of controlling a motor to provide only the required output power. (30)

locking pliers. Pliers that have the ability to lock their jaws in place, holding materials together similar to a vise. (8)

lockout/tagout (LOTO). A safety procedure that prevents equipment from being energized by mistake and alerts others that work on the equipment is being performed. (3)

lodestone. A rare form of the mineral magnetite that is a naturally occurring permanent magnet. (5)

logic theory. A set of two basic operations (AND, OR) that can be used to manipulate or evaluate binary elements. (12)

M

magnetic dipole. A microscopic magnet created by a spinning electron. (5)

magnetic domain. A physical region or space within a substance where a group of atoms all have their magnetic fields pointing in the same direction. (5)

magnetic field. An invisible field of force that is produced around a magnetic object or particle. (5)

magnetic flux. A measure of a magnetic field's strength. (5)

magnetic flux density. A measure of a magnetic field's relative strength, directly related to the concentration of magnetic field lines that pass through a given area. (5)

magnetic induction. The process of using a magnet or magnetic field to create other magnets. (5)

magnetic pole. Identified as north and south, these are the ends of a magnet and the points where the magnetic fields are strongest. (5)

magnetism. A property of matter that results in the creation of repulsive or attractive forces between substances. (5)

magnetomotive force (MMF). The force that produces a magnetic field expressed in ampere-turns. (5)

manual reset device. A control device that requires manual resetting once the position of the contact changes. (9)

mass. An object's resistance to movement when acted on by a force. (4)

matter. Any substance that has mass and occupies space. (4)

maxwell. A unit of magnetic flux equal to one line of magnetic force. (5)

mechanical losses. Motor losses resulting from factors related to the physical movement of motor components, such as friction and air resistance. (27)

megohmmeter. A large ohmmeter used to detect weak motor winding insulation and very high resistances, or small current paths, to ground on larger motors. Also called a *megger.* (33)

mentor. An individual with vast industry experience who serves as a guide in your career development. (2)

metallic cable cutter. A cable cutter used to cut across the ribbon of a metallic shielded cable to eliminate the sharp points. (8)

metal oxide semiconductor field effect transistor (MOSFET). Transistor type that has an isolated base and fast switching speed. (25)

microgrid. A relatively small, independently operated grid installed as part of a main grid. (38)

mixture. A nonchemical blend of two or more elements or compounds that can be separated without a chemical reaction. (4)

molecule. The smallest amount of a compound that can exist that still maintains the properties of the compound. (4)

motor. A machine that converts electrical energy into mechanical energy. (23)

motor and phase rotation indicator. An electronic measuring instrument that measures and indicates phase sequence and direction of rotation in a three-phase motor. Some have noncontact motor rotation capabilities. (29)

motor drive. An electronic component that controls the speed of an electric motors. (25)

multiplex winding. Term used to describe a winding that is made up of coils, each of which has two, three, or more loops. (34)

multiplicative inverse. A number that, when multiplied by another number, will produce a result of 1. Also referred to as the reciprocal. (7)

multipurpose wiring tool. An all-purpose tool that is used to cut and strip wire, in addition to crimping solderless connectors to conductors. (8)

multispeed motor. A motor that can be operated at more than one speed. (28)

multitap transformer. A transformer that is manufactured with multiple leads connected to one or both windings to allow for multiple primary and/or secondary voltages. (21)

mutual induction. The ability of the primary winding to induce a voltage in the secondary winding. (21)

N

nacelle. The outer casing or shell of a wind turbine. (37)

National Electrical Manufacturers Association (NEMA). A standards developing organization that publishes a wide range of standards to the electrical industry. (24)

needle-nose pliers. Pliers with long thin jaws that are used to reach into tight spaces that conventional pliers would not be able to reach. (8)

net metering. An energy-management practice that allows energy producers to send unused energy to the power grid in exchange for credit to offset energy costs. (37)

neutral conductor. A conductor that is connected to the neutral point of an electrical system. (6)

neutron. The component part of an atom that has a neutral charge. (4)

node. An electrical connection of two or more conductors or wires. (11)

noncontact voltage tester. An electrical device that detects the presence of voltage without having to make direct contact with the conductors. (8)

nonlinear waveform. A waveform that produces voltage rises and falls at changing rates. (15)

non-polarized capacitor. A capacitor with symmetric structure that can be connected in either orientation. (18)

nonrenewable energy. A natural resource that cannot be readily replaced in a lifetime by natural means. (1)

non-sinusoidal waveform. A waveform that does not follow a typical sinusoidal shape. (15)

nonverbal communication. Transmission of message through nonverbal ways, such as eye contact, facial expression, gestures, body language, and appearance. (2)

Norton current. The current value obtained when the voltage measured between terminals A and B is divided by the Norton resistance value. (13)

Norton resistance. The resistance measured between the circuit's output terminals if all voltage sources are replaced with short circuits and all current sources are replaced with open circuits. Equal to the Thevenin resistance. (13)

Norton's theorem. A theorem that states that any linear electric circuit can be reduced to an equivalent circuit consisting of a single current source and a single, parallel-connected resistance. (13)

N-type material. A semiconductor material that has a negative charge. (4, 30)

nucleus. The location at the center of an atom where the neutrons and protons are found. (4)

nut driver. A tool used to turn nuts, bolts, and hex-head screws. (8)

O

obtuse angle. Any angle that is greater than 90° but less than 180°. (14)

Occupational Safety and Health Administration (OSHA). Formed by Congress in 1970 as part of the Occupational Safety and Health Act to help ensure safe working conditions by setting safety-related standards for all workers and the workplace as a whole. (3)

offset misalignment. A misalignment that occurs when the drive and driven shafts are parallel each other, but the pulleys are not aligned with each other. Also called *parallel misalignment*. (26)

ohm (Ω). The unit of measurement for electrical resistance. (6)

Ohm's law. The relationship of voltage, current, and resistance in an electric circuit, expressed mathematically as $E = I \times R$. (6)

open-delta transformer. A two-transformer arrangement that produces a three-phase secondary signal. Also called *V-connected transformer*. (22)

open drip-proof (ODP) motor. A motor type that allows external air to flow directly into the motor housing, but does not allow dripping water to enter the enclosure. (24)

open-end wrench. A wrench similar to the box wrench but with an open head. (8)

operational control. A control device that opens and closes electrical contacts during the normal operation of the circuit. (9)

OR operation. A logic operation that can be represented by multiple parallel-connected switches, in which the load is energized if any one of the switches is closed. (12)

over-compounding. The process of adding additional turns or loops to a series field coil. (35)

overcurrent. Any condition in which the current draw of a circuit or device exceeds the normal or expected current level. (31)

oxidation. The process of losing electrons, typically at the anode of a battery. (37)

P

paramagnetic. Substances that have a net magnetic field and are slightly attracted to magnets. (5)

parallel circuit. An electric circuit with multiple possible current paths. (11)

parallel-connected generator. A generator that is part of a configuration where two or more generators are electrically connected to each other to increase the amount of available output power. (36)

part-winding starting. A starting method that reduces the inrush current of a three-phase motor on start-up and reduces stress on its components. (32)

pawl. The portion of a three-phase overload protective device that is connected between normally closed control circuit contacts and a ratcheting wheel. (32)

peak-to-peak voltage. The difference between the maximum and minimum voltage values. (15)

peak voltage. Maximum voltage in either the positive or negative portion of the wave. Also referred to as *amplitude*. (15)

PEMDAS. An acronym used to recall the sequence in which mathematical operations should be performed in an expression: parentheses, exponents, multiplication/division, and addition/subtraction. (7)

penstock. A section of pipe that is used to divert a portion of a river's flowing water stream for the purpose of generating electricity. (37)

period. The time required to complete one cycle. (15)

permanent magnet. A magnet that maintains its magnetic properties for a period of many years. (5)

permanent magnet dc motor. A dc motor that uses permanent magnets to create the field flux that interacts with the rotor. (27)

permanent-split capacitor (PSC) motor. A single-phase motor that has a start winding, a run winding, and a capacitor that is intended to be in the active circuit whenever the motor is energized. (28)

permeability. A property of magnetic materials representing the degree of ease with which a substance can become magnetized. (5)

permittivity (ε). Measure of a material's ability to become polarized by an electric field. (18)

personal protective equipment (PPE). The various pieces of equipment that workers wear to protect their bodies to help prevent personal injury. (3)

phase angle. The angle that exists between the true power and the apparent power. (15)

phase voltage. The voltage measured across any one phase of a three-phase system. (16)

photovoltaic effect. The creation of a potential difference (voltage) caused by exposing certain semiconductor materials to sunlight. (37)

pick-up voltage. The voltage at which the contacts of a potential magnetic relay open. (31)

piezoelectric effect. The creation of an electric charge in direct response to applied pressure. (37)

pilot-duty relay. A low-amperage relay that is used to control another higher-amperage control device, often a main relay. (31)

pinning. The ability to maintain magnetic domains in alignment after an external magnetic field is removed. (5)

plumb bob. A tool that consists of a weight with a pointed bottom supported at the top by a length of string. This tool is used to transfer points on the floor to the ceiling to ensure that an installation is vertical. (8)

point of egress. An exit or outlet from a space or area. (3)

polarized capacitor. A capacitor with asymmetric construction that must be connected in the proper orientation. (18)

pole. The number of different magnetic fields in a motor. (23, 31)

pole pitch. The number of armature slots that are covered by each pole. (35)

portfolio. A collection of documents that highlights your talents, skills, awards and abilities. (2)

positive temperature coefficient (PTC) relay. A device that is used to remove the start winding from the circuit. (28)

potential magnetic relay (PMR). A relay commonly used in capacitor-start, capacitor-run motors to remove the start capacitor from the motor circuit after start-up. (28)

powdered-iron core inductor. An inductor that has a soft magnetic core with air gaps contained within it. (17)

power. In an electric circuit, the rate at which electrical energy is transferred or converted. (6) The rate at which work is done. (7)

power analyzer. A device that calculates and displays mechanical power and efficiencies, to aid in identifying problems with the power signals. (8)

power drill. A tool that is used to create small or large holes in wood, steel, and masonry. (8)

power factor. The ratio of true power to apparent power. (15)

power supply. The source of circuit voltage. (4)

power triangle. A right triangle that illustrates the relationship among apparent power, true power, and reactive power, in which apparent power is represented by the hypotenuse of the triangle. (19)

primary resistor starting. A reduced-voltage starting method that uses resistors to reduce the voltage supplied to a motor at start-up. (32)

primary winding. The winding of a transformer that is connected to the circuit's main power source. (21)

progressive wave winding. A winding configuration where the finishing end of a coil terminates above, or to the right, of the starting end of the previous coil after making one rotation around the armature. (35)

proton. The component part of an atom that has a positive charge. (4)

P-type material. A semiconductor material that has a positive charge. (4, 30)

pulse width modulation (PWM). Digital control technique that creates an artificial ac power signal from a series of dc power signals. (25)

pump pliers. Pliers with a sliding jaw that locks in place at intervals. They are used to hold and twist nuts and conduit. (8)

Pythagorean theorem. A theorem that states that the sum of the squares of two sides of a right triangle is equal to the square of the longest side, or hypotenuse, of the triangle. (14)

Pythagorean triple. The term used to describe a set of three positive integers that make up the sides of a right triangle. The two most common Pythagorean triples are the 3-4-5 triangle and the 5-12-13 triangle. (14)

Q

quality factor (Q). A measure of the efficiency and quality of an inductor equal to its inductive reactance divided by its resistance. (17)

R

radial capacitor. A capacitor with all electrical connections on one end of the device. (18)

radial load. In motors and bearings, a force applied in a direction perpendicular to the axis of the shaft. (24)

rating. The voltage and current limits that an electrical device can be safely subjected to. (10)

RC time constant. A characteristic of a circuit with resistance and capacitance that represents the charging and discharging rate of the capacitor. (18)

reactive power (Q). The power consumed in an ac circuit that does no useful work, expressed in volt-amperes reactive (VAR). (15, 19)

reactive power support. The process of supplying leading reactive power to a system to offset the effects of lagging reactive power created by inductive loads. (38)

reactors. Large inductors that are used in power transmission applications to maintain current and stabilize voltage. (17)

reciprocal. A number that, when multiplied by another number, will produce a result of 1. Also known as a *multiplicative inverse*. (7)

reciprocating saw. A type of saw in which the blade moves in a back-and-forth motion to make a cut. (8)

rectifier. An electronic circuit that converts an ac input signal into a dc output signal. (30)

reduction. The process of gaining electrons, typically at the cathode of a battery. (37)

reflex angle. An angle that is greater than 180° but less than 360°. (14)

relative permittivity. The unitless quantity that relates the permittivity of a substance to the permittivity of a perfect vacuum. Also called *dielectric constant*. (18)

reluctance. The opposition, or resistance, to the production of a magnetic field. (5)

reluctance torque. The force that acts to align a rotating body with lines of magnetic flux. Also called *alignment torque*. (34)

renewable energy. Energy sources that cause minimal harm to the surrounding environment, are widely available, and replenish naturally. (1)

resistance. Any opposition to the flow of electric current. (6)

resistive load. A load that consumes power by converting electricity to heat and/or light. (9)

restricted approach boundary. Represents the distance from a piece of equipment that exposes a worker to a high probability of electric shock and more severe heat-related injuries in the event of an arc flash. (3)

resultant vector. The sum of two or more vectors. (14)

résumé. A brief outline of your education, work experience, and other qualifications for work. (2)

retentivity. A property of magnetic materials that describes the ability of a substance to retain a magnetic field once the initial magnetizing force is removed. (5)

retrogressive wave winding. A winding configuration where the finishing end of a coil terminates below, or to the left, of the starting end of the previous coil after making one rotation around the armature. (35)

reverse bias. The condition that prevents an electronic device from conducting electric current. (30)

revolving-armature ac generator. An ac generator configuration in which the armature windings rotate within a stationary magnetic field. (36)

revolving-field ac generator. An ac generator configuration in which the magnetic field rotates around the stationary armature winding. (36)

right angle. A 90° angle. (14)

right triangle. A triangle with one 90° angle. (14)

rms voltage. Stands for the root-mean-square voltage. The effective dc voltage of an ac power signal. (15)

rotor. The rotating portion of an electric motor. (24)

run capacitor. A motor-circuit capacitor connected in series with the start winding and in parallel with the run winding. (28)

run-of-the-river hydroelectric system. A hydroelectric system that uses a naturally flowing river to provide the force needed to turn turbines. (37)

run winding. Low-resistance winding of a split-phase motor that is connected in parallel with the start winding. (28)

S

safety control. A switching device designed to change the position of electrical contacts only when an unsafe or undesirable condition is present. (9)

safety data sheet (SDS). A document providing information about a hazardous material, including health and environmental hazards, safe exposure limits, first aid procedures, safe cleanup procedures, and proper handling and storage directions. (3)

salient-pole armature. An armature that is constructed as a central magnetic wheel with multiple poles extending from its center; typically used for lower speed generators. (34)

saturation. With respect to magnetism, the condition in which all or nearly all magnetic domains in a magnet are aligned and little or no additional magnetic strength is available. (5)

saturation voltage. The minimum transistor gate voltage that will allow the creation of a nearly resistance-free collector-emitter junction. (25)

sawtooth waveform. A waveform that is characterized as having a rise time that is significantly greater than the fall time. This waveform resembles the cutting edge of a saw. (15)

scalene triangle. A triangle with three unequal sides and three dissimilar angles. (14)

Scott transformer. A transformer with a T-connected three-phase primary and two single-phase secondary outputs. (22)

screwdriver. A tool used for the installation and removal of screws. (8)

secondary winding. The winding of a transformer in which the induced voltage is generated. (21)

Seebeck coefficient. Term used to quantify the amount of voltage produced across a thermoelectric material based on the temperature difference across it. (37)

Seebeck effect. Process of generating an electrical charge when a temperature difference is present across certain materials or substances. (37)

self-excited dc generator. A dc generator that supplies its own current for the field windings. (35)

self-motivation. An inner urge to perform well. (2)

semiconductor. A material that can either allow or prevent electron flow depending on the external conditions acting on the material. (4)

series circuit. An electric circuit that has only one possible path for electrons to follow. (10)

series dc motor. A dc motor that has a field winding wired in series with the motor's armature. (27)

shaded-pole motor. A simple single-phase induction motor that uses shading coils to facilitate motor starting. (28)

shading coil. Loops of thick copper wire or wide copper bands that are positioned around portions of a shaded-pole motor's stationary winding to facilitate motor starting. (28)

shaft. In a motor, a cylindrical rod or assembly aligned with the center of the rotor that connects to the driven device. (24)

short circuit. A condition that occurs when a current path with little or no resistance is available in a circuit, resulting in a dangerous and damaging large flow of current. (11)

short-cycling. Frequent motor starts and stops. (32)

shunt dc motor. A dc motor that has a field winding wired in parallel with the motor's armature. (27)

shunt-wound dc generator. A generator that is configured with the field coils connected in parallel to the armature and to the load. (35)

simplex winding. A winding that is made up of coils, each of which has one loop. (34)

sine. In a right triangle, the ratio of the length of the side opposite the angle to the length of the hypotenuse. (14)

sine wave. A wave that graphically represents the sine trigonometric function. (15)

single-phasing. The condition that results when a three-phase motor operates with only two of the three power lines functioning. (32)

SI system. The International System of Units; based on standard units for length, weight, and volume. Also known as the *metric system.* (7)

skinning knife. A cutting hand tool used to remove insulation from wires and cable. (8)

slip. The difference between a motor's synchronous speed and its actual speed, commonly expressed as a percentage. (23)

slip ring. An ac generator's electrical connection point between an armature's rotating winding and a brush. (36)

socket wrench. Used in conjunction with a ratchet handle and interchangeable socket heads, a wrench used to tighten and loosen bolts and nuts. Also called a *ratchet set* or *socket set.* (8)

soft-face hammer. Hammer used on a soft or finished surface. (8)

soft starter. An electronic motor starter that can vary the voltage and current supplied to a motor by modifying the ac waveform. (32)

soft-start motor. A motor that has the ability to gradually ramp up its speed, which reduces inrush current, reduces mechanical wear on bearings, reduces the amount of heat generated within the motor, and reduces the amount of noise produced. (30)

split-phase motor. An induction motor that has two distinct windings connected in parallel with each other. (28)

square waveform. A waveform that has two distinct, fixed, maximum and minimum values. Also called a *pulse train* or *pulse wave.* (15)

stack slot. The grooves in an armature core in which the winding's conductors are placed. Also called *armature slot.* (34)

stack teeth. Carved out bits resembling teeth that sit at the outer portion of the core between the stack slots. These alternating teeth and slots separate the windings and allow the conductors that make up the coils to sit flush with the surface of the core. Also called *armature teeth.* (34)

star-delta starting. A reduced-voltage starting method that supplies voltage to a motor in a wye configuration at start-up, and then switches to a delta configuration to supply full voltage. (32)

start capacitor. A motor-circuit capacitor that is wired in series with the start winding that aids in motor starting and increases starting torque. (28)

start winding. High-resistance winding of a split-phase motor that is connected in parallel with the run winding. (28)

static electricity. An imbalance between negative and positive charges in an object. (4)

stator. The stationary part of an electric motor that houses the motor's windings. (24)

step-down transformer. A transformer that changes the voltage supplied to it to a lower level; also known as a *buck transformer.* (21)

step-up transformer. A transformer that changes the voltage supplied to it to a higher level; also known as a *boost transformer.* (21)

straight angle. An angle that is equal to 180°. (14)

straight claw hammer. This hammer has a longer nose than a traditional hammer and has a smooth striking face. (8)

superposition theorem. A circuit manipulating tool that can be used to simplify and evaluate circuits with multiple voltage or current sources. (13)

supplementary angle. Two angles that add up to 180°. (14)

sustainable energy. Energy sources that have the ability to help meet the growing demand for energy today without compromising the ability of those in the future from relying on the same resource. (1)

switch. A control device in a circuit that starts and stops the flow of electrons through the circuit. (4)

synchronous speed. The calculated or theoretical speed of a motor. (23)

T

tangent. In a right triangle, the ratio of the length of the side opposite the angle to the length of the side adjacent to the angle. (14)

tape measure. A rolled metal ruler that retracts into a case or a plastic version that measures greater lengths. Also called *tape.* (8)

T-connected transformer. A three-phase transformer that is made up of a center-tapped main transformer and an appropriately sized teaser transformer. (22)

teeth per inch. The number of teeth on a saw blade per inch. The higher the number, the finer the cut. (8)

temporary magnet. A magnet that maintains its magnetic properties for only a short period of time. (5)

terminal. The connection point between circuit components and the wires or conductors that are used to connect them to each other. (10)

tesla. A unit of magnetic flux density (magnetic field strength) that is equal to 10,000 gauss. (5)

theorem. In mathematics, an expression that is proven by a natural progression of logical, previously established statements. (13)

thermistor. A resistor that experiences changes in resistance in response to changes in temperature. (31)

Thevenin equivalent circuit. An equivalent circuit consisting of a single voltage source and a single, series-connected resistance. (13)

Thevenin resistance. The resistance measured between the circuit's output terminals if all voltage sources are replaced with short circuits and all current sources are replaced with open circuits. Equal to the Norton resistance. (13)

Thevenin's theorem. A theorem that states that any linear electric circuit can be reduced to an equivalent circuit consisting of a single voltage source and a single, series-connected resistance. (13)

Thevenin voltage. Voltage measured between a Thevenin equivalent circuit's two output terminals with no external components connected to them. (13)

thread engagement. The amount of the bolt that is physically threaded into the securing nut (26).

three-phase. The simultaneous production of three evenly offset, single-phase waveforms of the same frequency. (16)

threshold voltage. The minimum voltage required to create a forward biased condition in a diode or transistor. (30)

throw. With regards to switches, the number of current paths associated with each pole. (31)

torpedo level. A small leveling tool used to check electrical conduit, equipment, appliances, and cover plates for switches and receptacles for level and plumb. (8)

torque. The twisting or rotating thrust or force produced by a motor. (27)

torque screwdriver. A tool for tightening screws, lugs, hex head, and bolt-type lugs to manufacturer's recommended torque specification; used on smaller fasteners. (8)

torque wrench. A tool that tightens hex heads, fasteners, and bolt-type lugs to the manufacturer's recommended torque specification. (8)

totally enclosed air-over (TEAO) motor. A dust-tight motor intended to be installed within an airstream, such as in an HVAC system's duct system, to provide cooling. (24)

totally enclosed blower-cooled (TEBC) motor. A motor type that has a cooling fan driven by a separate, dedicated motor. (24)

totally enclosed fan-cooled (TEFC) motor. A motor type that has a small fan attached to the shaft to move air across the exterior of the motor when operating. (24)

totally enclosed nonventilated (TENV) motor. A totally enclosed motor that has no fan and relies on the surrounding air to provide cooling. (24)

totally enclosed washdown (TEWD) motor. Motor intended for use in wet environments or areas subject to regular, thorough cleaning. (24)

totally enclosed water-cooled (TEWC) motor. A motor that uses water as a cooling medium, rather than air. (24)

trade stacking. Multiple trades working on the same jobsite at the same time. (3)

transformer bank. A three-phase transformer consisting of three individual, single-phase transformers connected together. (22)

transistor. A three-terminal semiconductor control device. (25)

transmission substation. A type of electrical substation within a power grid serving as a connection and switching point for power line that may include a wide variety of electrical equipment to ensure system safety and power quality. (38)

triangular waveform. A waveform that rises and falls at a constant rate with respect to time. The waveform resembles a triangle. (15)

true power (*P*). The actual power consumed by a circuit. Calculated as the product of voltage, current, and power factor, expressed in watts (W). Also called *active power* or *real power.* (15, 19)

true-RMS meter. A digital multimeter that measures pure sine waves and nonsinusoidal waveforms. (8)

true-up adjustment. An end-of-year reconciliation of a consumer's annual energy usage as compared to the amount of energy returned to the energy grid. (37)

turbine. A machine that converts the energy of a flowing fluid stream into mechanical energy. (37)

turns ratio. The relationship between the number of turns in a transformer's primary winding to the number of turns in the secondary winding. (21)

U

under-compounding. The process of removing turns or loops from a series field coil. (35)

ungrounded conductor. A circuit conductor that carries circuit current during normal circuit operation and has an electrical potential greater than that of ground. (6)

US Customary system. A standard system of measurement units, such as the inch, gallon, and pound. (7)

utility knife. A cutting hand tool with a folding or retractable blade that is used for general cutting. (8)

V

valence electron. An electron found in an atom's outermost shell. (4)

valence shell. The outermost shell in an atom. (4)

variable frequency drive (VFD). An electronic ac motor drive that varies a motor's speed or other operating characteristics by varying the frequency of the power signal supplied to the motor. (25)

variable transformer. A transformer that is constructed with a sliding contact that moves along the surface of the secondary winding. (21)

vector. A directional line segment that has both magnitude and direction. (14)

vector head. The ending or final point of a vector. (14)

vector tail. The starting or initial point of a vector. (14)

Venn diagram. A graphical representation of the relationships that exist among different groups, or sets, of mathematical and/or logic-related statements. (13)

verbal communication. Form of communication that involves your use of words through speaking, listening, and writing. (2)

vertex. A corner of a triangle. (14)

volt (V). The unit of measurement for voltage. (6)

voltage. The strength of an electrical charge that facilitates the flow of electrons through an electric circuit's conductive path. Also known as *electromotive force (EMF)* and *electrical potential.* (4) The difference in charge or electrical potential between two points or objects. (6)

voltage drop. The voltage measured across a circuit component. (10)

voltage imbalance. Measure of the voltage deviation from the average voltage; quantifies voltage inequality that exists among the phases of a three-phase power system. (16)

voltage source. A device, such as a battery, that has the ability to, ideally, provide continuous constant voltage to a circuit. (13)

volt-amperes (VA). Unit of measurement for apparent power. (15)

volt-amperes reactive (VAR). Unit of measurement for reactive power. (15)

volts-per-turn ratio. The relationship between the voltage at a transformer's winding to the number of turns in the same winding. (21)

W

watt (W). The unit of measurement for electrical power. (6)

Watt's law. The relationship of power, current, and voltage in an electric circuit, expressed mathematically as $P = I \times E$. (6)

wave-wound armature winding. A winding configuration that moves back and forth through the stack slots in one direction around the armature to create a continuous, wave-like path. Also called a *series winding connection.* (35)

windage losses. Mechanical loss resulting from the air surrounding a motor. (27).

winding losses. Losses due to the motor's winding resistance. Also known as *I^2R losses* or *copper power losses.* (27)

winding pitch. Denoted Y, the distance between the starting ends of two consecutive coil loops. (35)

wire stripper. A tool that is used to remove the insulation from wires without damaging the conductor itself. (8)

work. The equivalent of moving an object having a certain weight a specified distance. (7)

wye configuration. Wiring connection that is characterized by a common point that joins all three windings of a three-phase system. (16)

wye-delta three-phase transformer. A three-phase transformer that has its primary windings configured as a wye and its secondary windings configured as a delta. Also called *star-delta configuration.* (22)

wye-wye three-phase transformer. A three-phase transformer that has both primary and secondary windings configured as wye, or star, connections. Also called *star-star configuration.* (22)

Y

yoke. The exterior housing or outermost casing of a generator. (34)

Index

D

E

F

Q

R

S

U

V

W

Y